3rd International Conference on Artificial Intelligence, Automation and Control Technologies (AIACT 2019)

Journal of Physics: Conference Series
Volume 1267

Xi'an, China
25 – 27 April 2019

ISBN: 978-1-5108-9185-2
ISSN: 1742-6588

Printed from e-media with permission by:

Curran Associates, Inc.
57 Morehouse Lane
Red Hook, NY 12571

Some format issues inherent in the e-media version may also appear in this print version.

Institute of Physics
Dirac House, Temple Back
Bristol BS1 6BE UK

Phone: 44 1 17 929 7481
Fax: 44 1 17 920 0979

techtracking@iop.org

Additional copies of this publication are available from:

Curran Associates, Inc.
57 Morehouse Lane
Red Hook, NY 12571 USA
Phone: 845-758-0400
Fax: 845-758-2633
Email: curran@proceedings.com
Web: www.proceedings.com

Table of contents

Volume 1267

2019 3rd International Conference on Artificial Intelligence, Automation and Control Technologies (AIACT 2019)

25–27 April 2019, Xi'an, China

Accepted papers received: 4 June 2019
Published online: 17 July 2019

Preface

Papers

Artificial Intelligence

IOP Conf. Series: Journal of Physics: Conf. Series **1267** (2019) 011001 doi:10.1088/1742-6596/1267/1/011001

PREFACE

2019 3rd International Conference on Artificial Intelligence, Automation and Control Technologies (AIACT 2019) was held in Xi'an, China from April 25 to 27, 2019. AIACT 2019 was co-organized by is organized by Xidian University and Hong Kong Society of Mechanical Engineers, sponsored by York University and Fudan University. The conference provides a useful and wide platform both for display the latest research and for exchange of research results and thoughts in Artificial Intelligence, Automation and Control Technologies and other topics. The participants of the conference were from almost every part of the world, with background of either academia or industry, even well-known enterprise. The success and prosperity of the conference is reflected high level of the papers received.

The proceedings are a compilation of the accepted papers and represent an interesting outcome of the conference. There were 332 submissions including 293 papers and 39 abstracts. After rigorous peer review, 116 papers and 15 abstracts were accepted. This book covers 3 chapters: Artificial Intelligence; Design and Applications of Artificial Intelligence; Automatic Control.

We would like to acknowledge all of those who supported AIACT 2019. Each individual and institutional help were very important for the success of this conference. Especially we would like to thank the organizing committee for their valuable advices in the organization and helpful peer review of the papers.

We sincerely hope that AIACT 2019 will be a forum for excellent discussions that will put forward new ideas and promote collaborative researches. We are sure that the proceedings will serve as an important research source of references and the knowledge, which will lead to not only scientific and engineering progress but also other new products and processes.

Prof. Dan Zhang, York University, Canada
Prof. Xuechao Duan, Xidian University, China

IOP Conf. Series: Journal of Physics: Conf. Series **1267** (2019) 011002 doi:10.1088/1742-6596/1267/1/011002

COMMITTEES

Chairman

Dan Zhang, York University, Canada

Conference Chair

Xuechao Duan, Xidian University, China

Program Committee

Xiaochuang Xu, Fudan University, China (Chair)

Jiann-Shiou Yang, University of Minnesota, USA

Jingsong Bao, Donghua University, China

Lingfei Kong, Xi'an University of Technology, China

Songyi Dian, Sichuan University, China

Advisory Committee

Kenji Suzuki, Tokyo Institute of Technology, Japan

Toshio Fukuda, Beijing Institute of Technology, China

Wei Gao, University of New South Wales (Sydney), Australia

International Technical Committee

Chan Chun Chung Keith, The Hong Kong Polytechnic University, Hong Kong SAR

Ian R. McAndrew, Capitol Technology University, USA

Isidoros Perikos, University of Patras, Greece

Johan Potgieter, Massey University, New Zealand

Kefu Liu, Lakehead University, Canada

Manish D. Pasarkar, Sandip Foundation's , Sandip Institute of Engineering and Management, India

Li Wen, Beihang University, China

Samir N. Mekid, King Fahd University of Petroleum & Minerals, Saudi Arabia

Steven Dirven, Massey University, New Zealand

IOP Conf. Series: Journal of Physics: Conf. Series **1267** (2019) 011002 doi:10.1088/1742-6596/1267/1/011002

Ross McAree, University of Queensland, Australia

Peter Weiliang Xu, University of Auckland, New Zealand

Wei Dong, Harbin Institute of Technology, China

Zhao Tao, Sichuan University, China

Peer review statement

All papers published in this volume of *Journal of Physics: Conference Series* have been peer reviewed through processes administered by the proceedings Editors. Reviews were conducted by expert referees to the professional and scientific standards expected of a proceedings journal published by IOP Publishing.

IOP Conf. Series: Journal of Physics: Conf. Series **1267** (2019) 012001 doi:10.1088/1742-6596/1267/1/012001

A New Method for Abbreviation Prediction via CNN-BLSTM-CRF

Jianyu Zheng[1, 3], Xinge Xiao[2], Bihua Wang[1, 3], Yun Zhu[3] and Lijiao Yang[1, 3, *]

[1]Institute of Chinese Information Processing, Beijing Normal University, China
[2]Archives of Central China Normal University, Central China Normal University, China
[3]UltraPower-BNU Joint Laboratory for Artificial Intelligence, Beijing Normal University, China

[*]zheng_jianyu@126.com

Abstract. It is a crucial problem to process abbreviation in the field of natural language processing. The most commonly used way to cope with this problem is to construct the reference database by predicting the abbreviation through its fully expanded form. Previous work on abbreviation prediction mostly rely on traditional machine learning algorithms, which inevitably requires a large number of manual annotations or expert knowledge to establish a feature system. In this paper, a neural network model based on CNN-BLSTM-CRF is proposed, which can predict Chinese abbreviations better without relying too much on the feature system: Firstly, convolutional neural network extracts phrase and Chinese character information from the fully expanded form, and then BLSTM-CRF deep network is constructed to annotate the fully expanded form, so as to extract its corresponding abbreviation form. The experimental results show that the method in this paper can perform better than the state-of-art method in traditional machine learning, and the results provide a reference for abbreviation research and the construction of resource repository.

1. Introduction

Abbreviations are usually obtained by compressing and omitting longer words or phrases, such as "AI" is the abbreviation of "Artificial Intelligence" and "Bei Da" (北大) is the abbreviation of "Bei Jing Da Xue" (北京大学). Abbreviations have more concise language expression, so that they are widely used in natural language.

On the other hand, the use of abbreviations also brings some challenges to Chinese information processing. Therefore, it is helpful to the study of Chinese Word Segmentation, POS Tagging, Named Entity Recognition, Machine Translation and Information Retrieval, etc. by processing abbreviations correctly [1]. A commonly used way to cope with the challenge is to construct a large-scale fully expanded form-abbreviation reference database. The key to construct the database is to predict abbreviations through its fully complete form.

Scholars have made many efforts on Chinese abbreviations prediction. The most representative one is Xu Sun, Hou-Feng Wang, Bo Wang (2008), which used SVR to predict abbreviations: different abbreviation candidates were scored and rearranged, and the candidate with the highest score was the final choice [2]. Jiao Yan, Wang Houfeng, Zhang Longkai [3] used CRF model and Web information to predict abbreviation; Xu Sun, Wenjie Li, Fanqi Meng, Houfeng Wang [4] starting with Negative Full

Forms (NFFs), used Simple Heuristic System and Unified System to improve the accuracy of abbreviation prediction. In addition, Xu Sun, Naoaki Okazaki, Jun'ichi Tsujii [5] tried to introduce hidden variables on the basis of sequence annotation, which finally achieved good performance. Yi Zhang, Xu Sun [6] proposed a novel abbreviation prediction dataset with NFFs and evaluated it using CRF and BLSTM models.

Inspired by previous work, this paper proposes a new method based on neural network to predict abbreviations, which combines CNN, BLSTM and CRF together. Firstly we randomly generate a character embedding table and transform the fully expanded form into a character embedding matrix by looking up the table at first. Then, CNN and BLSTM are used to extract more efficient and high-dimension features and recognize characters in sequences that may exist in abbreviations respectively. Finally, CRF is used to decode the character sequence and obtain the optimal one. Through experiments, we find that this new method can improve the accuracy of abbreviation prediction, for instance, all accuracy is 0.62% higher than the state-of-are method of previous work. The main contributions of our work are:

(1) proposing a new neural network structure for abbreviation prediction;
(2) detailed discussion on the role of each module in the network;
(3) improving the accuracy of abbreviation prediction to a certain extent.

2. Problem defination

There are many ways to form Chinese abbreviations, which can be summarized as Reduction (such as "Bei Jing Da Xue-Bei Da" (北京大学-北大)), Elimination (such as "Qing Hua Da Xue-Qing Hua" ("清华大学-清华")) and Generalization (such as "Heilongjiang Sheng, Jilin Sheng, Liaoning Sheng-Dong San Sheng" ("黑龙江省、吉林省、辽宁省-东三省")). However, even if there are a limited number summed up in abbreviation, the actual situation still remains complicated. As far as the "Reduction" method is concerned, unlike English and other alphabetic writings, which form abbreviations usually by combining initial letters, the way of Reduction in Chinese is much more complicated. It is unnecessary to take the first Chinese character of each word to form a new abbreviation, for example, the "Li Shi/Yu Yan/Yan Jiu Suo" is abbreviated to "Shi Yu Suo" (历史/语言/研究所-史语所), which is related to the fact that Chinese characters have a certain meaning but alphabets do not represent special concepts. Moreover, the relationships between the fully expanded forms and its abbreviations are not always one-to-one relationship. For example, "Ren Da" (人大) can not only refer to "Ren Min Dai Biao Da Hui" (人民代表大会) but also "Ren Min Da Xue" (人民大学) [7]. In short, because the formation of abbreviations is influenced by many different factors, it is difficult for us to seek out the rules that meet all situations.

Therefore, the problem of abbreviation prediction is often transformed into a sequence labeling problem. We can make different marks on fully expanded forms according to its different characteristics. After marking the fully expanded forms, we can choose appropriate model to predict the corresponding abbreviations eventually.

Figure 1 shows a mark method commonly used in the previous research of abbreviation prediction :

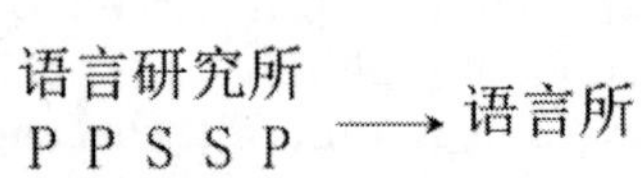

Figure 1. Chinese abbreviation generation as a sequential labeling problem.

In figure 1, each Chinese character is assigned as P (produce the current character) or S (skip the current character). The Chinese character marked P will be output finally, while the Chinese character marked S will be skipped. So the abbreviations corresponding to the fully expanded forms will be successfully generated.

Obviously, it is difficult to train an ideal model simply by using that labeling method. Xu Sun, Naoaki Okazaki, Jun'ichi Tsujii [5] took the lead by introducing Global Information labelling (GI) in the study of abbreviation prediction. Specifically, as shown in figure 2, this labeling method can better consider the global information. Experiments show that it improves the accuracy of abbreviations prediction.

	中	国	共	产	主	义	青	年	团
Orig.	S	S	**P**	S	S	S	**P**	S	**P**
GI	S0	S0	**P1**	S1	S1	S1	**P2**	S2	**P3**

Figure 2. Global Information of the fully expanded form.

According to the description of the paper, the label yi at position i attaches the information of the abbreviation length generated by its previous labels, y1, y2,···, yi−1. In this encoding, a label not only contains the produce or skip information, but also the abbreviation-length information, i.e., the label includes the number of all P labels preceding the current position [5].

In this paper, we not only use these two labeling methods, but also consider the phonetic information, tone information, location information, repetition of adjacent or separated characters and other features of each Chinese character in the fully expanded form.

3. CNN-BLSTM-CRF model

In this section, we will introduce our network model proposed in this paper. We will explain each layer of this network in top-down order.

3.1. Neural network architecture

Figure 3 shows the framework of CNN-BLSTM-CRF Model, which consists of three modules: CNN module, BLSTM module and CRF module. First, a table of character embedding is generated randomly, and the fully expanded form is transformed into the corresponding character embedding matrix by looking up the table. This matrix is convolved twice and pooled once by CNN to extract the higher dimensional features of the sequence based on the character level. A layer of BLSTM is added to recognize the characters in the sequence that may exist in the abbreviations. Finally, the output of BLSTM is decoded by CRF module, and the optimal label sequence is obtained.

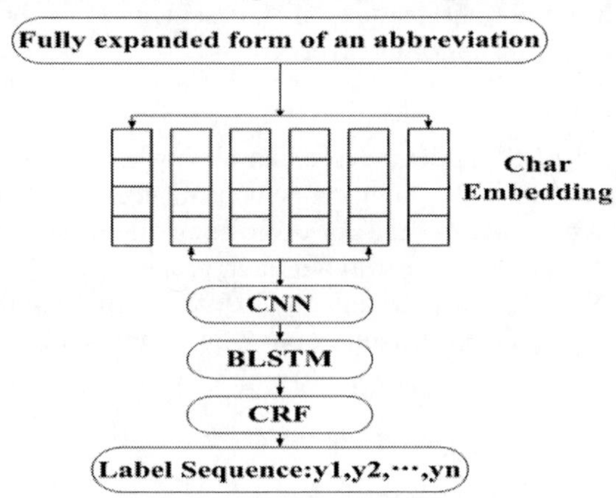

Figure 3. Model architecture diagram of CNN-BLSTM-CRF.

3.2. CNN module

The advantage of convolutional neural networks is to better extract local features of the data. Chiu and Nichols et al. [8] used CNN to extract the morphological information of characters or words and used them in the task of named entity recognition. In the CNN module, the specific network architecture is shown in figure 4:

IOP Conf. Series: Journal of Physics: Conf. Series **1267** (2019) 012001 doi:10.1088/1742-6596/1267/1/012001

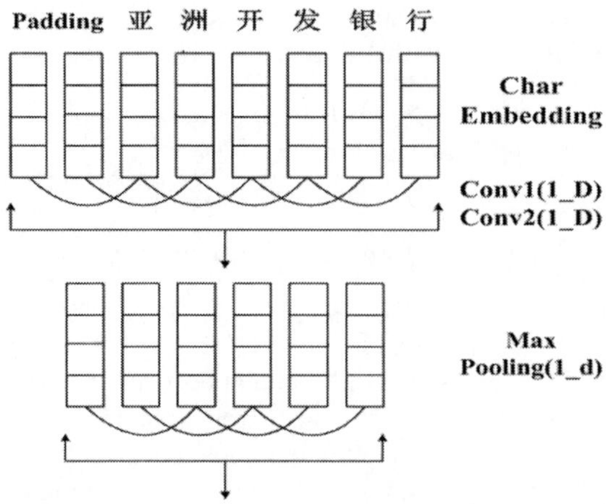

Figure 4. Network architecture of CNN module.

In figure 4, each fully expanded form is transformed into a character embedding matrix by looking up table according to the character. In order to solve the problem of different size of character embedding matrix caused by different length of fully expanded form. This paper adopts the principle of "cutting off from the long and adding to the short". That is, the full expanded form exceeding the set length will be truncated from the left, and the full expanded form that does not reach the set length will be supplemented with paddings at the left, so that all character embedding matrices are uniform in size. Finally, as shown in Char Embedding in the figure 4, the character embedding matrix is then input into the first convolution layer. Because two-character words appear more frequently in the problem of abbreviation prediction, the kernel size is set to 2. So that it can extract effective information better regardless of whether it is reduction or elimination. Next, the output of the first convolutional layer is treated as input to the next layer. The purpose of this convolution layer is to extract the corresponding character in the two-character word which is more likely to be retained in abbreviation, so the kernel size is set to 1. Finally, the important features in the sequence are better preserved and invalid information is filtered through the maxpooling layer.

3.3. BLSTM module

Recurrent neural network is a kind of depth model for simulating sequential data. Because the traditional recurrent neural network can not solve the "long-range dependence" problem of sequence data well, the phenomenon of gradient disappearance or gradient explosion will occur in practical [9].

In order to make up this disadvantage, Hochreiter et al. proposed Long short-term memory network [10]. LSTM allows the model to selectively retain contextual information through a specially designed gate structure. In general, the formalized structure of an LSTM unit at time t is as follows:

$$
\begin{aligned}
&\boldsymbol{i}_t = \sigma(\boldsymbol{W}_i \boldsymbol{h}_{t-1} + \boldsymbol{U}_i \boldsymbol{x}_t + \boldsymbol{b}_i) \\
&\boldsymbol{f}_t = \sigma(\boldsymbol{W}_f \boldsymbol{h}_{t-1} + \boldsymbol{U}_f \boldsymbol{x}_t + \boldsymbol{b}_f) \\
&\widetilde{\boldsymbol{c}}_t = \tanh(\boldsymbol{W}_c \boldsymbol{h}_{t-1} + \boldsymbol{U}_c \boldsymbol{x}_t + \boldsymbol{b}_c) \\
&\boldsymbol{ct} = \boldsymbol{f}_t \otimes \boldsymbol{c}_{t-1} + \boldsymbol{i}_t \otimes \widetilde{\boldsymbol{c}}_t \\
&\boldsymbol{o}_t = \sigma(\boldsymbol{W}_o \boldsymbol{h}_{t-1} + \boldsymbol{U}_o \boldsymbol{x}_t + \boldsymbol{b}_o) \\
&\boldsymbol{ht} = \boldsymbol{o}_t \otimes \tanh(\boldsymbol{c}_t)
\end{aligned}
\tag{1}
$$

Where, σ is the activation function sigmoid; $\otimes$ is point multiplication operation; tanh is the hyperbolic tangent function; x_t is the input vector at time t; ht is the hidden state of the LSTM unit at time t; i_t, f_t, o_t represent the input gate, the forget gate and the output gate at time t; the matrix $\boldsymbol{U}$ corresponds to the weight matrix of the input vector x_t in each gate; $\boldsymbol{W}$ and $\boldsymbol{b}$ respectively correspond to the weights and biases of the hidden states in each gate; $\widetilde{c}_t$ is an intermediate state obtained only according to the input at time t, which is used to update the current state; c_t represents the state of time t [11].

In traditional LSTM networks, the implicit state can only deal with forward transmitted sequence data. In this paper, to make better use of contextual information, we have adopted bidirectional LSTM (BLSTM) [12]. The basic idea is to obtain two different hidden states for each fully expanded form in sequence and inverse order respectively to represent the past and future information of the sequence. Finally, the two hidden states are concatenated together as the output and passed to the next module.

3.4. CRF module

In the task of sequence labeling, CRF can obtain a globally optimal label sequence by considering the relationship between adjacent labels. For example, in POS task, adjective is always followed by nouns rather than verbs. Specifically, the output of BLSTM is used as an input to the CRF layer for further solution. $X=\{x_1,x_2,\cdots,x_n\}$ is assumed to be a full expanded form of the input sequence vector, where x_n is the vector representation of the n th Chinese character in the sequence after processing by the previous module. $Y=\{y_1,y_2,\cdots,y_n\}$ represents the corresponding label sequence, y_n is the label corresponding to the n_{th} Chinese character x_n. Then, CRF probability model can be regarded as calculating the probability $p(y|x;W,b)$ by the corresponding label y under the condition of a given variable x. The formula is as follows:

$$p(y|x;W,b)=\frac{\prod_{i=1}^{n}\Psi(y_{t-1},y_t,x)}{\sum_{y'\in y(z)}\prod_{i=1}^{n}\Psi_t(y'_{t-1},y'_t,x)} \tag{2}$$

Where $\Psi_t(y',y,x)=exp(W^T_{y',y}x_t+b_{y',y})$ is a potential function, $W^T_{t',y}$ and $b_{y',y}$ represents the weight vector and bias of the tag pair (y', y), respectively. The formula for its log-likelihood function is as follows:

$$L(W,b)=p(y|x;W,b) \tag{3}$$

The optimal solution of the formula can be obtained by maximum likelihood estimation. Finally, the label sequence Y* corresponding to the optimal solution can be obtained by decoding.

In summary, the specific network architecture of the BLSTM and CRF modules is shown in figure 5.

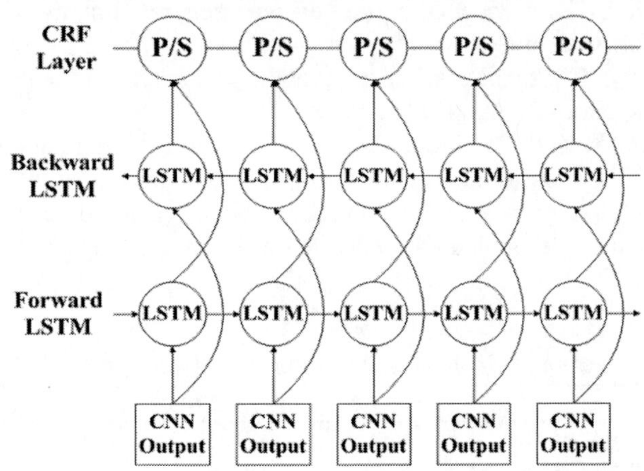

Figure 5. The specific network architecture of CNN module and CRF module.

IOP Conf. Series: Journal of Physics: Conf. Series **1267** (2019) 012001 doi:10.1088/1742-6596/1267/1/012001

4. Experiment

4.1. About the corpus

In this paper, 12338 abbreviations and their corresponding fully expanded forms are extracted from the *Chinese Abbreviations Dictionary* [13] and *Practical Abbreviations Knowledge Dictionary* [14]. Among them, 11112 abbreviations belong to reduction and elimination, which can be used in this experiment. In this paper, the experimental data are shuffled, the training set and test set are divided according to 4:1.

4.2. Experimental settings

About the optimization algorithm, this paper uses Adam as the optimizer, and the learning rate is set to 0.001. Table 1 shows the parameters of the entire network architecture. Besides, the parameters of the network can be fine-tuned according to the actual situation.

Table 1. Network parameter of CNN-BLSTM-CRF.

	Hyper-parameter	Value
Embedding	char_embedding+BMEU	200+200
	max_sequence_length	12
Conv1(1_D)	window size	2
	number of filters	250
	stride	1
	padding mode	same
Conv1(1_D)	window size	1
	number of filters	400
	stride	1
	padding mode	same
MaxPooling(1_D)	window size	2
	stride	1
	padding mode	same
BLSTM	cell size	350
	drop out	0.4

4.3. Results

To assess the effectiveness of this model, we will measure it from two indicators, all-match accuracy and character accuracy, as explained below:

- all-match accuracy: The number of correct outputs generated by the system divided by the total number of full forms in the test set [5].
- character accuracy: The number of correct labels generated by the system divided by the total number of characters in the test set [5].

We compare our model with the previous state-of-art method [5]. The operation of CRF is the same as the previous method in the experiment, considering the complete form of pinyin, tone, whether the corresponding Chinese character is a Chinese number, whether the adjacent Chinese character is the same, whether it is the same after a Chinese separated character, and the position information (BMEU feature) of each character in the word after the segment. The specific testing results are as follows in Table 2:

Table 2. Table of test results for different models.

	all acc(%)	char acc(%)
CRF	37.98	79.48
CNN-BLSTM-CRF	38.66	76.16
CRF+GI	46.4	79.64
CNN-BLSTM-CRF+GI	**47.02**	**78.45**

IOP Conf. Series: Journal of Physics: Conf. Series **1267** (2019) 012001 doi:10.1088/1742-6596/1267/1/012001

In addition, to verify the validity of our model, Table 3 shows the effects of different modules on abbreviation prediction when used separately:

Table 3. Prediction effect table of different modules of Neural Network.

	all acc(%)	char acc(%)
LSTM	31.99	75.09
BLSTM	33.71	75.89
CNN-BLSTM	35.28	76.1
BLSTM-CRF	35.61	76.15
LSTM+GI	40.96	76.66
BLSTM+GI	42.03	77.18
CNN-BLSTM+GI	43.69	77.97
BLSTM-CRF+GI	44.5	77.56

4.4. Analysis

It can be seen from Table 2 that compared with the original method, the model proposed in this paper has a certain improvement on all acc. This also shows that the end-to-end neural network model can avoid the complicated feature engineering in the traditional machine learning, and can extract some implicit high-dimensional features from the original data through the nonlinear operation of the network. It is helpful to predict the final sequence.

The effect of BLSTM in Table 3 is significantly higher than that of LSTM, which indicates that BLSTM can make full use of the context information of the sequence. CNN-BLSTM has a slightly better effect on all acc than BLSTM, and it also shows the effectiveness of CNN in extracting character-level features. In addition, BLSTM-CRF is 2% better than BLSTM in all acc, which means that linear CRF can make full use of the relationship between adjacent labels and output an optimal label sequence.

It can be seen from Tables 3 and 4 that the corresponding models are greatly improved compared with the original ones after the labels are encoded with Global Information. This is because the label incorporates non-local information when encoded, and the non-local information can not only confirm whether the current character exists in the abbreviation, but also reflect the relative position information of the character in the entire sequence [5].

It can also be seen through experiments that this model performs worse than the traditional CRF algorithm in terms of char acc indicator. This is due to the "cutting off from the long and adding to the short" operation on the fully expanded form of the input sequence: If we take the longest sequence in the whole data set as the standard to complete all other sequence, most of the sequences in the corpus will be filled with a large number of paddings, thus the prediction effect of the model will decline. However, to some extent, the truncated part of the long sequence can not be labled by the "cutting off from the long and adding to the short" method, which results in a bad char acc result.

Finally, the experimental results in Table 2 and 3 are low mainly due to uneven data. The abbreviations are collected from *Chinese Abbreviations Dictionary* and *Practical Abbreviations Knowledge Dictionary* which cover short name, abbreviations of organization names and omissions of fixed phrases, and the distribution of abbreviations is different. As a result, there are some differences in the distribution of abbreviations between the training set and the test set. In addition, the dataset also contains some long fully expanded form, among which the longest sequence contains 49 Chinese characters.

5. Conclusion

On the basis of previous researches, we propose a new abbreviation prediction model combined with CNN, BLSTM and CRF to obtain better prediction results. In experiments, we also use abbreviation corpus to evaluate some methods of abbreviation prediction proposed by predecessors and the CNN-BLSTM-CRF model with Global Information shows competitive performance. However, the amount of data in the abbreviation corpus is insufficient if we want to further get good results. To solve this

IOP Conf. Series: Journal of Physics: Conf. Series **1267** (2019) 012001 doi:10.1088/1742-6596/1267/1/012001

problem, we will try to increase the amount of data sets, improve our model and further enhance the effect of abbreviation prediction in our future work.

Acknowledgments
This work is supported by National Language Committee Research Program of China (No.ZDI135-42) and National High-Tech Research and Development Program of China (2012AA011104).

References

[1] Jiao Y, Wang H F, 2011, Abbreviation Prediction Using Machine Learning Method and Search Engine Verification, Proceedings of The Tenth China National Conference on Computational Linguistics, August 20, 64-69
[2] Sun X, Wang H F, Wang B, 2008, Predicting Chinese Abbreviations from Definitions: An Empirical Learning Approach Using Support Vector Regression, Journal of Computer Science & Technology, Vol 23, Issue 4, 602-611
[3] Jiao Y, Wang H F, Wang H F, Zhang L K, 2012, Abbreviation Prediction Using Conditional Random Field and Web Data, Journal of Chinese Information Processing, Vol 26, Issue 2. 62-69
[4] Sun X, Li W J, Meng F J, Wang H F, 2013, Generalized Abbreviation Prediction with Negative Full Forms and Its Application on Improving Chinese Web Search, International Joint Conference on Natural Language Processing, October 14-18,641-647
[5] Sun X, Okazaki N, Tsujii J,2009 Robust Approach to Abbreviating Terms: A Discriminative Latent Variable Model with Global Information, Proceedings of the Meeting of the Association for Computational Linguistics, August 2-7 905-913
[6] Zhang Y, Sun X, 2017, A Chinese Dataset with Negative Full Forms for General Abbreviation Prediction, Computer Science, 1-5
[7] Wang H F, 2011, Survey: Abbreviation Processing in Chinese Text, Journal of Chinese Information Processing, Vol 25, Issue 5. 60-68
[8] Chiu J P C, Nichols E, 2015, Named Entity Recognition with Bidirectional LSTM-CNNs, Transactions of the Association for Computational Linguistics, Vol 4, Issue 1, 1-14
[9] Pascanu R, Mikolov T, Bengio Y, 2013, On the difficulty of training recurrent neural networks, International Conference on International Conference on Machine Learning, June 16-21, 1310-1319
[10] Graves A, 1997, Long Short-Term Memory, Neural Computation, Vol 9, Issue 8,1735-1780
[11] Ma X, Hovy E, 2016, End-to-end Sequence Labeling via Bi-directional LSTM-CNNs-CRF, Proceedings of the 54th Annual Meeting of the Association for Computational Linguistics, August 7-15 ,1064-1074
[12] Dyer C, Ballesteros M, Wang L, et al, 2015, Transition-Based Dependency Parsing with Stack Long Short-Term Memory, Proceedings of the 54th Annual Meeting of the Association for Computational Linguistics, July 26-31 ,334-343
[13] Shi B Y, Xu Y W, 1990, Chinese Abbreviations Dictionary, Foreign Language Teaching and Research Press
[14] Ya K, Xiu Y, 1992, Practical Abbreviations Knowledge Dictionary, New World Press

Deep Learning for HRRP-based Satellite Recognition

W Lu[1, *], Y S Zhang[2], Y R Huo[3] and C Y Lin[4]

[1]Graduate School, Space Engineering University, No.1 Bayi Street, Huairou District, Beijing, China
[2]Space Engineering University, No.1 Bayi Street, Huairou District, Beijing, China
[3]Graduate School, Space Engineering University, No.1 Bayi Street, Huairou District, Beijing, China
[4]Space Engineering University, No.1 Bayi Street, Huairou District, Beijing, China

[*]wanglu199310@163.com

Abstract. An approach that employs deep learning technology is presented to recognize satellites based on radar high-resolution range profile (HRRP) data. We focus on extracting effective satellite recognition features in this paper. Thus, a deep learning model is constructed by gated recurrent unit (GRU) neural network and support vector machine (SVM) to extract more abstract and accurate features. Firstly, the radar HRRP data of four satellites is obtained by simulation. And data preprocessing has been done according to HRRP characteristic. Next, a GRU-SVM model is set up and some deep learning skills, such as dropout and cross validation, have been applied to improve recognition accuracy. The training results of GRU neural network show their effectiveness. In order to demonstrate the superiority of this approach, five other feature extraction methods have been used as a comparison based on clean satellite HRRP data and noisy data. The experiment results show that the presented GRU-SVM model could recognize satellites effectively and accurately, and has better recognition performance and noise robustness compared with five other methods.

1. Introduction

Space targets recognition is a primary function of space surveillance information system, and satellites recognition is of critical importance on this issue. However, few open research achievements have been reported. The difficulty of this problem is that satellites are simply too small or too far away for detailed information to be recognized, and few effective identification data can be obtained. With the development of wideband radar [1], wu could obtain lots of useful radar data, such as high-resolution range profile (HRRP), synthetic aperture radar (SAR) image, inverse synthetic aperture radar (ISAR) image, etc. HRRP is the amplitude of coherent summations of the complex time returns from target scatterers in each range cell [2], which represents the projection of the complex returned echoes from the target scattering centers onto the radar line-of-sight (LOS) [3]. It contains abundant target structure signatures, such as target size, scatterer distribution, etc. In addition, compared with SAR/ISAR image, HRRP has the advantages of easy acquisition and processing. That's why radar HRRP target recognition has received intensive attention from radar automatic target recognition (RATR) community [4-11]. So the work in this paper focuses on the satellites recognition based on the HRRP data.

Feature extraction and selection is a basic and crucial technology for radar HRRP target recognition research. It is important to adopt reasonable and effective features to improve recognition performance. For that reason, many researches have been done on feature extraction and selection methods by related scholars. In the early days, researchers often calculated the power spectrum, FFT-magnitude and various high-order spectrum of HRRP data, and used them as the features of classifier for target recognition [4-6]. Such engineered features are useful but rely on researchers' experience and skill. If we do not have sufficient prior knowledge for the applications, those features would be brittle and incomplete. In addition to features of artificial selection, machine learning algorithms have been widely utilized to represent features based on high-dimensional HRRP data by many researchers [7-11]: the principal component analysis (PCA) feature subspace is constructed to minimize reconstruction error for RATR [7]. Dictionary learning is adapted to extract the noise-robust and highly discriminative features of the HRRP [8-9]. Manifold learning is employed in target recognition of radar HRRP to reduce the feature dimensions [10-11]. These methods can work well on some occasions, but all of them are shallow architectures that cannot effectively characterize the radar HRRP. Therefore, how to automatically extract the deep abstract features, which are beneficial for target recognition, has become an important issue.

The deep learning theory [12] put forward by Hinton can effectively solve the above problem. The essence of deep learning is to construct a neural network containing multiple hidden layers to map the data in order to obtain the deep essential characteristics [13]. Some deep learning structures applied in several recent papers have been demonstrated to be useful for radar HRRP target recognition, such as autoencoder and its varieties [14-15], convolutional neural network (CNN) [16], recurrent neural network (RNN) [17-18]. Because of the unique structure of RNN, it has been widely applied to process sequential data such as action recognition [19], scene labelling [20], and language processing [21], and has achieved impressive results [22]. GRU neural network is a kind of RNNs and can learn long term reliance on information [23]. In this paper, the HRRP data of four satellites are obtained by simulation. For satellites recognition based on this sequential data, a recognition model has been designed in this paper, which contains GRU neural network as a deep features extractor and SVM as a classifier. Experiments have demonstrated that this GRU-SVM model performs well on the task of satellite recognition based on radar HRRP data and has relatively strong noise robustness.

2. Description and Preprocessing of HRRP

HRRP is the amplitude of the echo summation for target scattering centers in each range cell of wideband radar. Figure 1 shows the illustration of an HRRP sample from a satellite target. High resolution radar operates in microwave frequency band. Generally, the size of targets or their components is much larger than the wavelength of radar. For complex targets such as a satellite, the projection of an object on radar line of sight can be divided into many range cells by high resolution radar.

The radar signatures from scattering centers within the same range cell will be coherently summed into a single signature for that range cell. According to the literatures [2, 5], if the radar transmitted signal is $s(t)e^{j2\pi f_c t}$, the nth complex echo in the dth range cell ($d=1,2,\cdots D$) in the baseband can be approximated as

$$\tilde{x}_d(t,n) \approx s(t)\sum_{i=1}^{L_d}\sigma_{di}e^{-j\{(4\pi/\lambda)[R(n)+\Delta r_{di}(n)]\}} \tag{1}$$

where $s(t)$ is the complex envelop which approximates to be unvaried with the radial displacements for all scatterers in one range cell. f_c is the radar signal carrier frequency and λ denotes the wavelength of high resolution radar. L_d represents the number of target scatterers in the dth range cell. σ_{di} is the intensity of the ith scatterer in the dth range cell. $R(n)$ is the radial distance between target reference center in the nth echo and the radar. $\Delta r_{di}(n)$ is the radial displacement of the ith scatterer of the dth range cell in the nth echo. Usually, $s(t)$ is a rectangular pulse signal with unit intensity and could be omitted. After eliminating the initial phase of the nth echo $e^{-j(4\pi/\lambda)R(n)}$, the nth HRRP can be defined as

$$\mathbf{x}(n)=\left[x_1(n),x_2(n),\cdots,x_D(n)\right]$$
$$=\left[\left|\sum_{i=1}^{L_1}\sigma_{1i}e^{j\phi_{1i}(n)}\right|,\left|\sum_{i=1}^{L_2}\sigma_{2i}e^{j\phi_{2i}(n)}\right|,\cdots,\left|\sum_{i=1}^{L_D}\sigma_{1i}e^{j\phi_{Di}(n)}\right|\right] \quad (2)$$

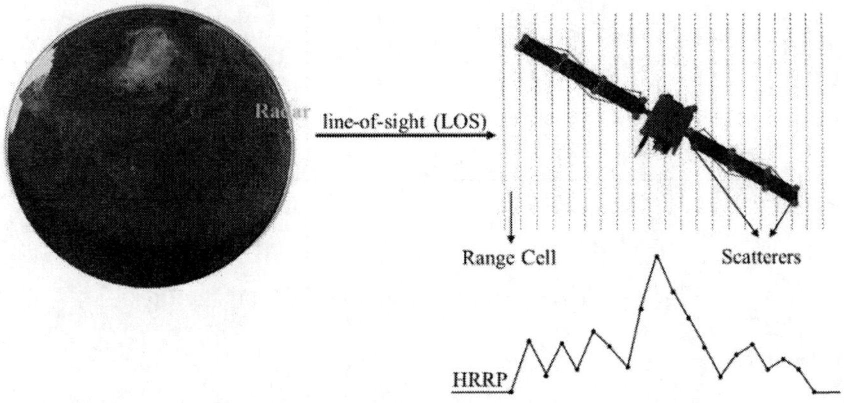

Figure 1. An HRRP sample from a satellite target.

Several issues should be considered when HRRP is applied to radar target recognition task. 1) The first one is time-shift sensitivity of HRRP. HRRP is only a part of received radar echo extracted by a range window, in which a target signal is included. So the position of target signal in HRRP may vary with the measurement. However, feature learning needs all the training samples to learn a uniform parameter model. So we adopt envelope alignment method [24] as time-shift compensation technique in this paper. 2) The second one is amplitude-scale sensitivity. It comes from the fact that the intensity of an HRRP is a function of radar transmitting power, target distance, radar antenna gain, radar receiver gain, radar system losses and so on. HRRPs measured by different radars or under different conditions will have different amplitude-scales. To deal with amplitude scale sensitivity, each HRRP is normalized by dividing maximum amplitude per frame. After the above preprocessing, the HRRP sample examples of four satellites are shown in Figure 2.

The last one is called target-attitude sensitivity. It means that the variation of target attitude will lead to different range shifts for different scattering centers on the target, even within the attitude region where the scattering center structure remains unchanged. Considering that the attitude of normal satellite is relative stable, it is expected to be solved by recognition methods.

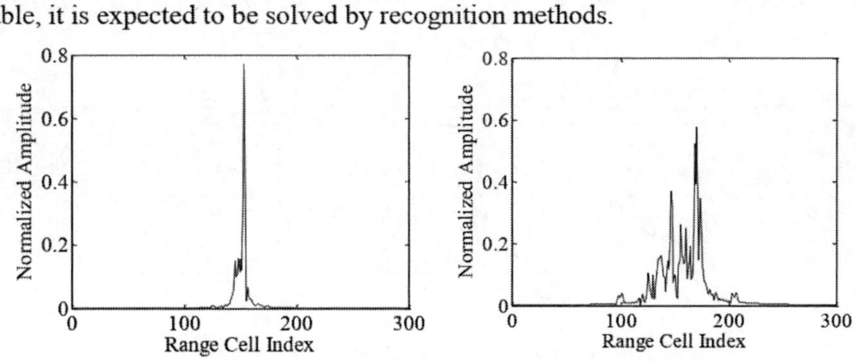

(a) Sat1 (b) Sat2

IOP Conf. Series: Journal of Physics: Conf. Series **1267** (2019) 012002 doi:10.1088/1742-6596/1267/1/012002

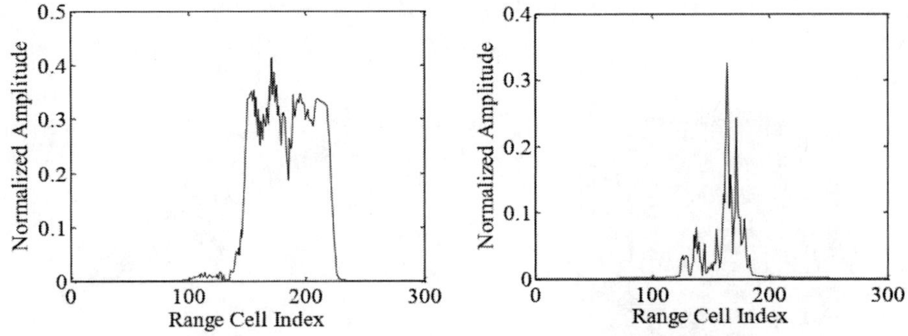

(c) Sat3 (d) Sat4

Figure 2. HRRP sample examples of four satellites.

3. GRU-SVM model

3.1. GRU

Compared with the feed-forward networks such as CNN, RNN has a recurrent connection where the last hidden state is an input to the next state [25-26]. RNN will remember the previous information and use the previous information to influence the output of following nodes, so it is better to solve problems related to time series data [27]. That is, RNN can obtain the output sequence from the input sequence at the current time step, and can also predict sequence in the next time step. However, RNN suffers from long-term dependencies problem, and gradient vanishing or exploding may occur during the training process, making it impossible to process long-term sequence data. In response to this problem, variants of RNNs such as long-short term memory (LSTM) [28-29] and GRU [30] have been proposed.

The GRU neural network is a variant of the LSTM network. There are only two gates in the GRU network unit, namely the update gate z and the reset gate r. The update gate is utilized to modulate the previous information inside the unit. The larger the value of the update gate, the more the status information of the previous moment insides. The reset door is used to control the previous state information which will be forgotten, the smaller the value of the reset gate, the more the previous state information is forgotten. Figure 3 shows a GRU model, where $\tilde{h}$ denotes candidate activation. GRU model has been proved performing better than simple RNNs in the task of processing sequential data [31].

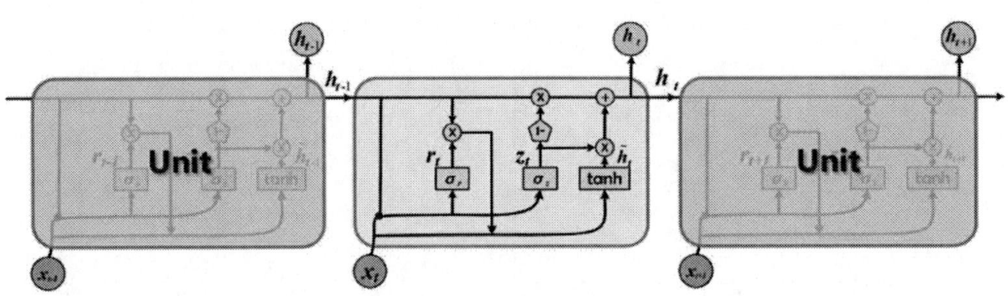

Figure 3. A Gated Recurrent Unit.

The update gate z_t is defined as

$$z_t = \sigma\left(\mathbf{W}_z \mathbf{x}_t + \mathbf{U}_z \mathbf{h}_{t-1}\right) \tag{3}$$

where $\mathbf{W}$ and $\mathbf{U}$ are weight matrix respectively. $\mathbf{x}$ denotes input data and $\mathbf{h}$ is known as hidden state. The reset gate r_t is defined as

$$r_t = \sigma\left(\mathbf{W}_r \mathbf{x}_t + \mathbf{U}_r \mathbf{h}_{t-1}\right) \tag{4}$$

The hidden state in a GRU is linearly modelled as

$$\mathbf{h}_t = (1-z)\mathbf{h}_{t-1} + z_t\tilde{\mathbf{h}}_t \tag{5}$$

The candidate hidden state is

$$\tilde{\mathbf{h}}_t = \tanh\left(\mathbf{W}_h\mathbf{x}_t + \mathbf{U}_h\left(r_t * \mathbf{h}_{t-1}\right)\right) \tag{6}$$

where $*$ is element-wise product. The $\sigma(\cdot)$ and $\tanh(\cdot)$ are two different activation functions which can be defined as

$$\sigma(x) = \frac{1}{1+e^{-x}}$$
$$\tanh(x) = \frac{1-e^{2x}}{1+e^{2x}} \tag{7}$$

In this paper, GRU is employed in the GRU-SVM model to extract effective features based on HRRP sequential data.

3.2. GRU-SVM model construction

In this paper, we build a GRU-SVM model. It consists of two parts: an encoder and a classifier. The encoder takes HRRP data of different satellites as input. Length-fixed feature vectors produced by the encoder contain sufficient information for target recognition. The classifier takes feature vectors as input for classifier and the corresponding satellites classes will be identified.

The encoder consists of an input layer, two GRU hidden layers of size 96 and a fully connected layer (activation function is ***linear*** and the number of units is 64). The output of this fully connected layer is feature vector. The classifier used here is SVM and takes feature vector as its input. The structure of GRU-SVM model is shown in Figure 4. In order to extract good features, the encoder is followed by two fully connected layers, whose activation function are ***relu*** and ***softmax*** respectively. This GRU neural network utilizes the ***Adam*** optimizer for optimization and use ***categorical crossentropy*** as a loss function to reduce the difference between the model classification value and the real value. At the same time, in order to make the training model more accurate, the bidirectional scheme as demonstrated in reference [32] is applied in GRU hidden layers.

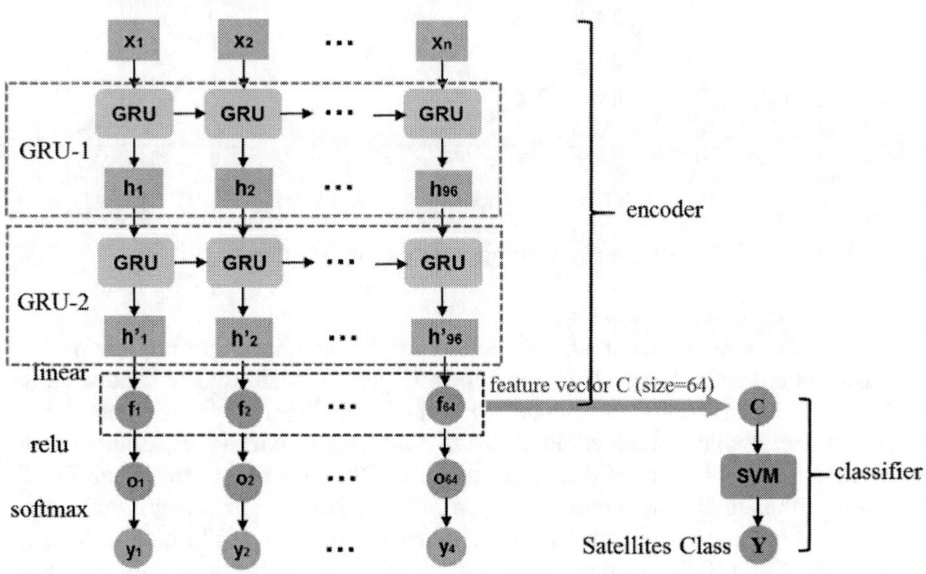

Figure 4. The structure of GRU-SVM model.

4. Experiments

In this section, testing experiments of GRU-SVM model have been carried out to obtain its recognition performance based on the HRRP data of four satellites, which all have 35000 HRRP of 300 dimensions. To better illustrate the advantages of this model, some frequently used feature extraction methods are utilized for comparison, including PCA, linear discriminant analysis (LDA), dictionary learning, autoencoder and simple RNN. In addition, noise is usually inevitable, which has great influence on the performance of radar target HRRP recognition. So recognition capability of these methods has also been tested when HRRP data with the noise of different signal-to-noise ratio (SNR) is utilized as input.

4.1. Training assessment

Most HRRP data of four satellites (about 70%~80%) will be applied as input to train the GRU neural networks and others are utilized as testing set. Meanwhile, 20% of training data is used as validation set to adjust the hyper-parameters. To avoid the problem of overfitting, ***Drop-out*** has been employed for the two GRU layers. In this paper, drop rate is set to 0.25. Activation function and loss function described in the section 3.2 are applied for GRU neural network. When the evaluation indicator is not improving, the learning rate will decrease in multiple. To accelerate the training, batch normalization inserted after each layer.

Recognition accuracy, loss and mean absolute error (MAE) are often applied to assess the training result. Recognition accuracy is the ratio of the number predicted correctly to the total testing number, and loss is defined the difference of prediction value and true value. Assuming $\hat{y}_i$ is the prediction value of *ith* sample, y_i is expected value, the MAE is defined as

$$MAE(y,\hat{y}) = \frac{1}{n_M}\sum_{i=1}^{M}|y_i - \hat{y}_i| \tag{8}$$

where M is the total number of samples. The training results of 80 times are shown in Figure 5. The recognition accuracy, loss and MAE all change as expected. The accuracy of training and validation set in the process of extracting feature reaches 99.1% and 99.3% and the loss and MAE continue to decline.

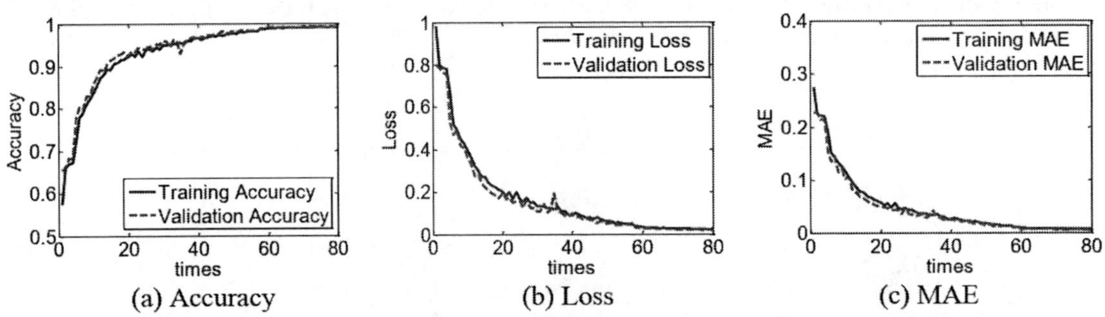

Figure 5. Training records for 80 times.

4.2. Recognition preformation comparison

After training GRU neural networks, it is important to get the recognition results of GRU-SVM model. In order to demonstrate its effect and advantages, the recognition performance for testing data has been compared when different methods are applied to identify satellites. Two deep networks (GRU and simple RNN) and four shallow models (PCA, LDA, dictionary learning and autoencoder) serve as feature vector extractor, and linear SVM is used to classify the satellites (here linear SVM only serves as a simple baseline, thus it does not employ any feature extraction). The recognition accuracy of these six methods is shown in table 1 and their confusion matrices are shown in Figure 6. Compared with the latter five methods, GRU-SVM model has good recognition performance for the four satellites. Therefore, its total recognition accuracy rate is highest among this six methods.

IOP Conf. Series: Journal of Physics: Conf. Series **1267** (2019) 012002 doi:10.1088/1742-6596/1267/1/012002

Table 1. Accuracy of six recognition methods.

recognition methods	accuracy rate (%)
GRU-SVM	99.4
simple RNN-SVM	93.4
PCA-SVM	91.9
LDA-SVM	80.6
dictionary learning-SVM	86.4
autoencoder-SVM	91.5

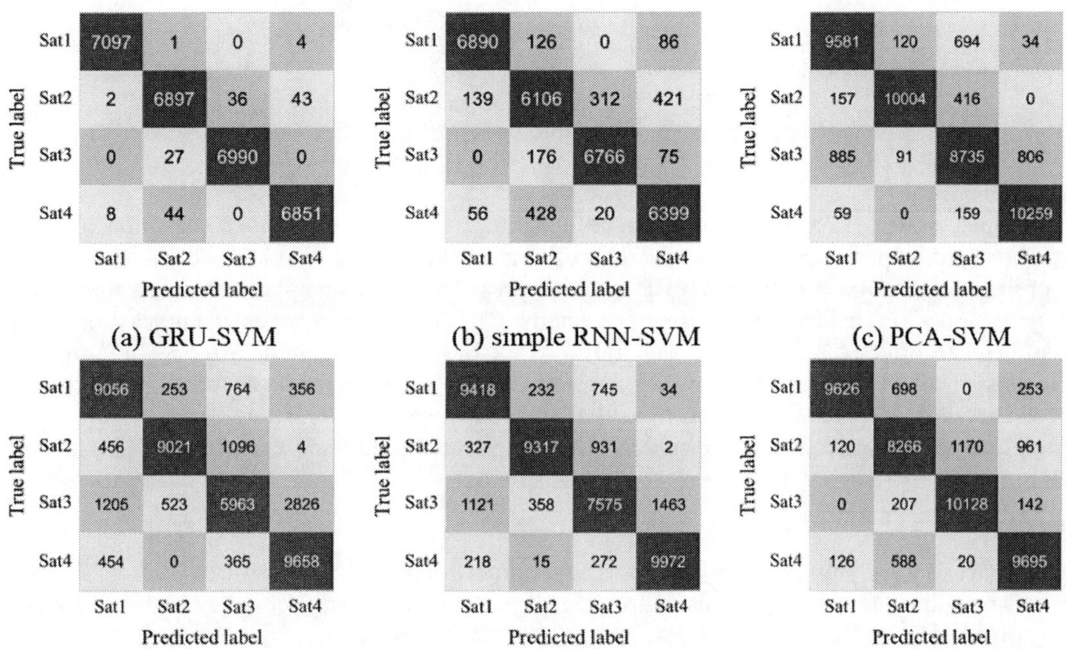

(a) GRU-SVM (b) simple RNN-SVM (c) PCA-SVM

(d) LDA-SVM (e) dictionary learning-SVM (f) autoencoder-SVM

Figure 6. Confusion matrix of six recognition methods.

To further study the noise robustness of these six recognition methods, gaussian noise with different SNR (1dB, 10dB, 20dB and 30dB) is added to clean HRRP data. The recognition accuracy rate varies with SNR and the corresponding results are illustrated in the Figure 7. It shows that the GRU-SVM model has the greater capacity of robustness than five other methods. With the increase of HRRP data SNR, its recognition accuracy gets improved. It can be seen from this figure that when the SNR of satellites HRRP data is low, for example SNR=1dB, the recognition rate of these six methods is all less than 90%. However, the recognition accuracy of GRU-SVM model is still much higher than other methods. This demonstrates that the GRU-SVM model has greater noise robustness than other methods.

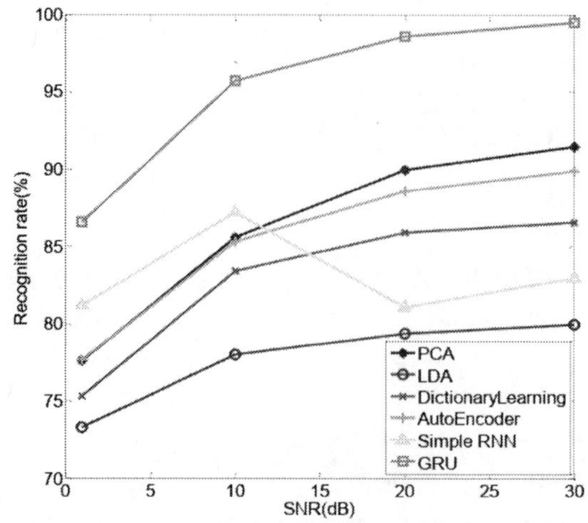

Figure 7. Recognition rate under different SNR noise.

5. Conclusion

In this paper, a GRU-SVM model has been designed for radar HRRP satellite recognition, and achieves better recognition performance than several traditional feature extraction methods in the experiments. According to the characteristic of HRRP data, we preprocess these data and establish the structure of GRU-SVM model. This model consists of two parts: GRU neural network as an encoder and SVM as a classifier. The effectiveness of GRU neural network has been validated in model training process. Performance comparison experiments are carried out to testing recognition accuracy of several shallow or deep models. The results demonstrate that the model presented in this paper achieves superior recognition performance than other models based on feature extraction, even for HRRP data with low SNR.

References

[1] Richards M A. Fundamentals of radar signal processing[M]. 2005.
[2] Du L, Liu H, Bao Z, et al. Radar automatic target recognition using complex high-resolution range profiles[J]. Iet Radar Sonar & Navigation, 2007, 1(1):18-26.
[3] Du L, Wang P, Liu H, et al. Bayesian Spatiotemporal Multitask Learning for Radar HRRP Target Recognition[J]. IEEE Transactions on Signal Processing, 2011, 59(7):3182-3196.
[4] Du L, Liu H, Bao Z, et al. Radar HRRP target recognition based on higher order spectra[J]. IEEE Transactions on Signal Processing, 2005, 53(7):2359-2368.
[5] Du L, Liu H, Wang P, et al. Noise Robust Radar HRRP Target Recognition Based on Multitask Factor Analysis With Small Training Data Size[J]. IEEE Transactions on Signal Processing, 2012, 60(7):3546-3559.
[6] Feng B, Du L, Liu H W, et al. Radar HRRP target recognition based on K-SVD algorithm[C]// Cie International Conference on Radar. 2012:642-643-644-645.
[7] Zhang X D, Shi Y, Bao Z. A new feature vector using selected bispectra for signal classification with application in radar target recognition[J]. Signal Processing IEEE Transactions on, 2001, 49(9):1875-1885.
[8] Liu H W, Feng B, Chen B, et al. Radar high-resolution range profiles target recognition based on stable dictionary learning[J]. Iet Radar Sonar & Navigation, 2016, 10(2):228-237.
[9] Li L, Liu Z. Noise-robust HRRP target recognition method via sparse-low-rank representation[J]. Electronics Letters, 2017, 53(24).

[10] Jiang Y, Han Y, Sheng W. Target recognition of radar HRRP using manifold learning with feature weighting[C]// IEEE International Workshop on Electromagnetics: Applications and Student Innovation Competition. IEEE, 2016:1-3.
[11] Yu X, Wang X, Liu B. Supervised kernel neighborhood preserving projections for radar target recognition[J]. Signal Processing, 2008, 88(9):2335-2339.
[12] Hinton, G.E.; Salakhutdinov, R.R. Reducing the dimensionality of data with neural networks. Science 2006,313, 504–507.
[13] Lecun, Y.; Bengio, Y.; Hinton, G. Deep learning. Nature 2015, 521, 436–444.
[14] Zhao F, Liu Y, Huo K, et al. Radar HRRP Target Recognition Based on Stacked Autoencoder and Extreme Learning Machine.[J]. Sensors, 2018, 18(1):173.
[15] Feng B, Chen B, Liu H. Radar HRRP target recognition with deep networks[J]. Pattern Recognition, 2017, 61:379-393.
[16] Lundén J, Koivunen V. Deep learning for HRRP-based target recognition in multistatic radar systems[C]// Radar Conference. IEEE, 2016:1-6.
[17] Xu B, Chen B, Liu H, Jin L. Attention-based Recurrent Neural Network Model for Radar High-resolution Range Profile Target Recognition[J]. Journal of Electronics and Information Technology, 2016, 38(12):2988-2995.
[18] Jithesh V, Sagayaraj M J, Srinivasa K G. LSTM recurrent neural networks for high resolution range profile based radar target classification[C]// International Conference on Computational Intelligence & Communication Technology. IEEE, 2017:1-6.
[19] Jeff Donahue, Lisa Anne Hendricks, Marcus Rohrbach, et al. Long-Term Recurrent Convolutional Networks for Visual Recognition and Description[J]. IEEE Transactions on Pattern Analysis and Machine Intelligence, 2014, 39(4):677-691.
[20] Byeon W, Breuel T M, Raue F, et al. Scene labeling with LSTM recurrent neural networks[C]// Computer Vision and Pattern Recognition. IEEE, 2015:3547-3555.
[21] Cho K, Van Merrienboer B, Gulcehre C, et al. Learning Phrase Representations using RNN Encoder-Decoder for Statistical Machine Translation[J]. Computer Science, 2014.
[22] Li S, Li W, Cook C, et al. Independently Recurrent Neural Network (IndRNN): Building A Longer and Deeper RNN[J]. 2018.
[23] Nanduri A, Sherry L. Generating Flight Operations Quality Assurance (foqa) data from the X-Plane Simulation[C]// Integrated Communications Navigation and Surveillance. IEEE, 2016:5C1-1-5C1-9.
[24] Li Y L Y , Chongyi L C L . Global correlation envelope alignment of high precision[C]// Asian & Pacific Conference on Synthetic Aperture Radar. IEEE, 2008.
[25] Li S, Li W, Cook C, et al. Independently Recurrent Neural Network (IndRNN): Building A Longer and Deeper RNN[J]. 2018.
[26] Mikolov T, Joulin A, Chopra S, et al. Learning Longer Memory in Recurrent Neural Networks[J]. Computer Science, 2014.
[27] Bengio Y, Simard P, Frasconi P. Learning long-term dependencies with gradient descent is difficult.[J]. IEEE Transactions on Neural Networks, 1994, 5(2):157-166.
[28] Jozefowicz R, Zaremba W, Sutskever I. An empirical exploration of recurrent network architectures[C]// International Conference on International Conference on Machine Learning. JMLR.org, 2015:2342-2350.
[29] Klaus Greff, Rupesh K. Srivastava, Jan Koutník, et al. LSTM: A Search Space Odyssey[J]. IEEE Transactions on Neural Networks & Learning Systems, 2015, 28(10):2222-2232.
[30] Cho K, Van Merrienboer B, Gulcehre C, et al. Learning Phrase Representations using RNN Encoder-Decoder for Statistical Machine Translation[J]. Computer Science, 2014.
[31] Chung J, Gulcehre C, Cho K H, et al. Empirical Evaluation of Gated Recurrent Neural Networks on Sequence Modeling[J]. Eprint Arxiv, 2014.
[32] Schuster M, Paliwal K K. Bidirectional recurrent neural networks[M]. IEEE Press, 1997.

IOP Conf. Series: Journal of Physics: Conf. Series **1267** (2019) 012003 doi:10.1088/1742-6596/1267/1/012003

Discrete Fireworks Algorithm for Welding Robot Path Planning

Xinyu Zhou*, Qingjie Zhao* and Dongxu Zhang

Beijing Lab of Intelligent Information Technology, School of Computer Science, Beijing Institute of Technology.

*E-mail: zhoupat1994@gmail.com, zhaoqj@bit.edu.cn.

Abstract. Welding robots are widely used in manufacturing industries, and reasonable welding path planning is an important issue in production efficiency. Welding robot path planning aims to arrange the sequence of weld joints and find the optimal welding path for the robot, which is essentially a combinatorial optimization problem. With the development of artificial intelligence, swarm intelligence algorithms shed new light on this problem. Fireworks algorithm (FWA) is a newly proposed swarm intelligence algorithm, simulating the process of fireworks explosion producing sparks to find the optimal solution. It has shown excellent performance in continuous optimization problems. In this paper, a discrete fireworks algorithm (DFWA), is proposed to solve the welding robot path planning problem. We introduce some operations to the framework of traditional FWA. In DFWA, 2-opt local search and crossover operator are applied on the explosion sparks. A new way of fireworks quality judgment is designed. Mutation operator is implemented for the generation of Gaussian sparks, and the selection strategy is improved. Simulation experiments have been made to verify our method and compare its performance with other current swarm intelligence algorithms. Experimental results show that the proposed method performs well with good convergence, stability and accuracy. It is effective for welding robot path planning.

1. Introduction

In the manufacturing industries such as automobile and shipbuilding, welding robots have been widely used to greatly increase productivity. In practical applications, the number of weld joints in a welding task is very large. Therefore, properly planning the welding sequence of these weld joints plays an important role in saving time and reducing production cost. Traditional path planning methods are mostly based on experiences of welding engineer, and cannot ensure the optimal solution. In recent years, with the rapid development of artificial intelligence (AI), more and more swarm intelligence algorithms have been born and widely used in optimization. These algorithms have provided new ways to robot path planning. Yang and Tang [1] proposed a particle swarm optimization (PSO) with quantum mechanics to realize welding path optimization. Shen [2] combined genetic algorithm (GA) and ant colony optimization (ACO) to solve the contradiction between the convergence speed and the optimization accuracy. Wang et al. [3] presented a double global optimum GA-PSO approach based on two populations to improve optimization effects of welding robot path planning.

Inspired by the natural phenomenon of fireworks explosion, Tan and Zhu [4] proposed fireworks algorithm (FWA) in 2010. FWA is a newly swarm intelligence algorithm in AI field, and performs multiple simultaneous explosion searches by simulating fireworks explosion at night. It shows high performance and efficiency in solving complex optimization problems. FWA has gradually gained attentions and follow-up researches. So far, it has been applied for solving practical optimization problems, including digital filters design [5], spam detection [6], power system reconfiguration [7], and improved in studies [8], [9], [10]. However, for the application on discrete optimization problems, e.g. path planning, the current research on FWA is in the early stage, and this direction is still a new topic. In this paper, a novel discrete FWA (DFWA) method is proposed, which combines with 2-opt local search and GA operators, in order to solve the problem of welding robot path planning efficiently.

2. Description of welding robot path planning

The welding robot path planning problem can be summarized as obtaining a reasonable sequence of weld joints for the welding robot. It is briefly described as: given N weld joints and the distance between any two joints, the manipulator starts from an initial position and returns the initial position after finishing all the assigned weld joints. The goal is to find a shortest possible route, in which each joint is welded once and only once.

The mathematical description is as follows:

Considering N weld joints $C = \{c_1, c_2, \dots, c_N\}$, the distance between each pair of joints is denoted by $d(c_i, c_j)$. Welding robot path planning aims to find a permutation $X = \{x_1, x_2, \dots, x_N\}$, $x_i \in \{1,2,\dots,N\}$ to minimize the total path length of route X.

$$f(X) = \sum_{i=1}^{N-1} d\left(c_{x_i}, c_{x_{i+1}}\right) + d\left(c_{x_N}, c_{x_1}\right) \tag{1}$$

where $f(X)$ denotes the distance when the welding sequence is X.

3. Discrete fireworks algorithm

3.1 Standard fireworks algorithm

In FWA, a firework or a spark is considered as a feasible solution in the solution space of the optimization problem. The process of fireworks explosion producing a certain amount of sparks is regarded as a neighbourhood searching process. Similar to other swarm intelligence algorithms, FWA is based on population and evolution. The optimal solution is generated through a series of iterative processes. In each iteration, the fireworks undergo explosion operations, Gaussian operations, mapping rules, and selection strategies to form the fireworks for next generation. Finally, the population converges to an optimal solution.

3.1.1. Sparks explosion. The core of the FWA is the explosion operation, which simulates the behaviour of natural fireworks exploding in the sky. The fireworks with better fitness (quality) produce more sparks in a smaller range. On the contrary, the fireworks with poorer fitness (quality) produce less sparks in a larger range. Suppose that for a D-dimensional optimization problem, n denotes the number of fireworks, $x_i = \{x_{i1}, x_{i2}, \dots, x_{id}\}$ represents the position of the ith $(i = 1,2,\dots,n)$ firework, and the explosion amplitude and sparks number are calculated by:

$$A_i = \hat{A}\frac{f(x_i) - y_{min} + \varepsilon}{\sum_{i=1}^{n}(f(x_i) - y_{min}) + \varepsilon} \tag{2}$$

$$s_i = m_s \frac{y_{max} - f(x_i) + \varepsilon}{\sum_{i=1}^{n}(y_{max} - f(x_i)) + \varepsilon} \tag{3}$$

where $f(x_i)$ represents the function value of the ith firework, A_i and s_i are the explosion amplitude and the number of sparks of the ith firework. $\hat{A}$ and m_s are constants, controlling the maximum explosion amplitude and the total number of sparks. $y_{min} = min(f(x_i))$, $y_{max} = max(f(x_i))$. ε is a very small constant, used to prevent the denominator being zero.

To avoid the overwhelming effects of outstanding fireworks, the bounds of sparks number are defined by:

$$s_i = \begin{cases} round(am_s), & if\ s_i < am_s \\ round(bm_s), & if\ s_i > bm_s,\ a < b < 1 \\ round(s_i), & otherwise \end{cases} \tag{4}$$

where a and b are constants to control the maximum and minimum of sparks.

3.1.2. Gaussian sparks explosion. Another type of sparks named Gaussian sparks are based on the Gaussian distribution, in order to increase the diversity of sparks. Suppose the position of the ith individual on the dth dimension is denoted by x_i^d, the Gaussian sparks are calculated as:

$$x_i^d = x_i^d \cdot g \tag{5}$$

Note that g is a Gaussian-distribution random number with a mean of 1 and a variance of 1, that is:

$$g \sim N(1,1) \tag{6}$$

3.1.3. Mapping rules. In the explosion operations and Gaussian operations, sparks may be outside the range of feasible domains. When spark x_i exceeds the boundary on dimension d, it will be mapped into the new location by the following mapping rules:

$$x_i^d = x_{min}^d + |x_i^d| \% (x_{max}^d - x_{min}^d) \tag{7}$$

where x_{max}^d and x_{min}^d represent the upper and lower boundaries of the dth dimension respectively, and % represents a modular operation.

3.1.4. Selection. In order to retain excellent information in the fireworks population and pass it to the next generation, after the explosion sparks and Gaussian sparks generated, a certain number of individuals (all fireworks, explosion sparks and Gaussian sparks) will be selected as the fireworks for next iteration. The individual with the best fitness is always kept for the next iteration, and the remaining $n - 1$ individuals will be selected based on their distance to other individuals. Individuals that are further away from other individuals have more chance to become the fireworks of next iteration. This selection strategy ensures the diversity of the population.

3.2 DFWA

The basic FWA is mainly applied on continuous optimization problems, however, their rules cannot be directly used to deal with discrete problems. Compared with continuous space, the objective function of discrete problems is discontinuous, and the function fitness of adjacent spaces often varies greatly. The welding robot path planning belongs to the discrete combinatorial optimization problem, so the discretization of the traditional FWA is one of the most important parts.

First, the representation of the solution of the FWA is the same as the definition of welding path, which is the sequence of weld joints. The vector $X = (x_1, x_2, \dots, x_N)$ is used to represent the sequence of weld joints in the path, and x_i is the number of the ith weld joint in the path. There is one and only

IOP Conf. Series: Journal of Physics: Conf. Series **1267** (2019) 012003 doi:10.1088/1742-6596/1267/1/012003

one chance that each weld joint is passed. Our goal is to find an arrangement X such that its path length $f(X)$ is the shortest.

Secondly, this paper redesigns the searching process of FWA while maintaining the traditional FWA framework. For the explosion operator, we introduce two new explosion methods. For the generation of Gaussian sparks, we use mutation operators which are suitable for discrete field. Moreover, the mapping rules are removed, because the solutions of discrete FWA would not exceed the definition domain of the path planning problem. Finally, the selection strategy is also adjusted. The framework of our algorithm is shown in figure 1.

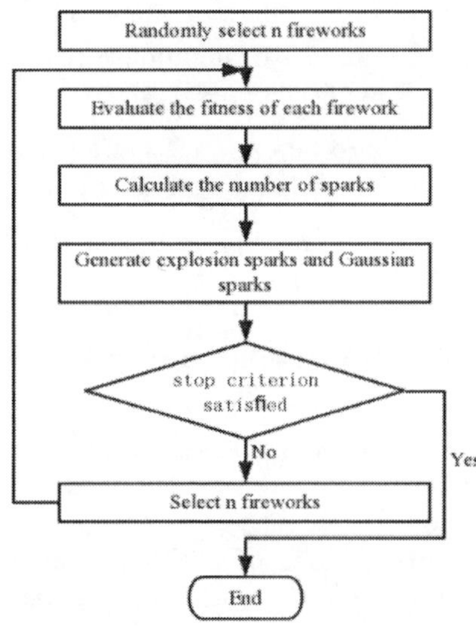

Figure 1. Framework of DFWA.

3.2.1. Modified explosion operator. The process of generating explosion sparks is an important part in FWA, which plays a role of local search and global search. Fireworks with good fitness produce more sparks in a smaller range, meaning that a local search is applied within a promising area. In the contrast, fireworks with poor fitness produce fewer sparks in a larger range, meaning that a global search is implemented.

For good fireworks, in other words, the sequence of weld joints with a short total distance, adequate exploitation is needed near their locations. Based on this idea, we use k-opt method to enhance the local search for fireworks with good fitness. As for poor fireworks, inspired from genetic algorithm (GA), we use crossover operator to deal with them. This is because the crossover damages the original gene (path), and the generated gene (path) is very different from the original. It is consistent with the idea of global search.

k-opt is a basic local search heuristic algorithm, in which core idea is to randomly select k edges and perform iterative optimization on them. When breaking k edges, there are $(K-1)!\,2^{k-1}$ ways to reconnect this path, and each new combination is a valid solution. For example, in 2-opt method, 2 edges are removed and reconnected to search better solution (Figure 2).

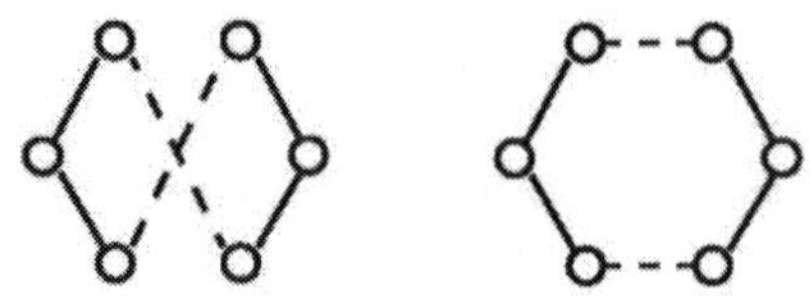

Figure 2. An example of 2-opt. The left is original path, and the right is a path possibly generated.

Increase of k gets better optimization while increases time complexity. Due to this, simple 2-opt operation is used, and it stops iterating when the maximum number of iterations is reached. $MaxItCount$ represents the maximum number of iterations set in this paper. The method using 2-opt for one robot path mainly contains following steps:

$step$ 1: Suppose the current path X_k to be optimized (e.g.$[1, 2, 3, 4, 5, 6]$) is the shortest path, the path length $f(X_k)$ is calculated according to equation (1).

$step$ 2: Randomly select two positions in the path and reverse the sub-path between the two positions, then a new path X_{new} is obtained. For example, if we randomly select the position 2 and position 4, the new path is [1, 4, 3, 2, 5, 6], as shown in figure 2.

$step$ 3: If $f(X_{new})$ is shorter than $f(X_k)$, set the new path X_{new} to be the shortest path X_k. Then return to $step$ 2.

GA treats a solution (path) as a chromosome. In GA, crossover is a main operation. In this paper, partial mapping crossover operator is used, and the current firework randomly intersects with the rest of the fireworks to create a new solution.

After redesigning the explosion operations of different fireworks, how to judge the qualities of fireworks is a special problem in discrete optimization. In accordance with the principle that the good fireworks produce a large number of sparks, we select the top 20% of fireworks as the excellent fireworks according to the calculated number of sparks, and generate their sparks through 2-opt local search. The sparks of remaining fireworks are produced by crossover operator of GA.

3.2.2. Modified Gaussian sparks generation. In the problem of combination, the Gaussian random number does not play its role. After the generation of explosion sparks, all sparks are selected to perform the mutation operator of GA, which are called mutation sparks. Corresponding to the effect of Gaussian sparks in traditional FWA, the population diversity is further enhanced.

3.2.3. Selection mechanism. After fireworks producing explosion sparks and mutation sparks, selecting a specified number of individuals to go to the next iteration is needed. In traditional FWA, the selection strategy is based on distance metrics. The individuals (fireworks and all sparks) with lower density have higher probability to be selected, in order to ensure the diversity of the population. However, this selection strategy needs to construct the Euclidean distance between any two individuals, which will lead to large time consumption and high computational cost. Therefore, in DFWA, roulette strategy is adopted, where the probability of each individual being selected is based on the fitness. It can avoid trapping in the local optimum, and has a great help to the convergence of the whole population.

3.2.4. Algorithm steps. By means of 2-opt local search and genetic algorithm (GA), a discrete fireworks algorithm (DFWA) is proposed. The specific steps of DFWA are given as follows:

$step$ 1: Initialize control parameters.
$step$ 2: Randomly select n locations for fireworks.

step 3: For each firework X_i, calculate the number of sparks of the firework s_i according to equation (3) and equation (4).

step 4: Evaluate the qualities of fireworks, and generate explosion sparks by 2-opt local search and crossover operator.

step 5: Generate mutation sparks through mutation operator.

step 6: Select the best individual and n − 1 individuals by roulette mechanism to compose fireworks for next generation.

step 7: Determine whether the algorithm stop criteria is reached. If answer is no then go to *Step* 3, otherwise the algorithm ends.

4. Simulation experiments

In the actual welding process, joints are welded by manipulator in the order. And the proposed path planning is to provide a reference welding sequence for the robot. In order to test the performance of DFWA for welding robot path planning, 26 weld joints are selected for welding robots and the joint distribution is shown in figure 3. Figure 4 shows the path planning result of the DFWA algorithm.

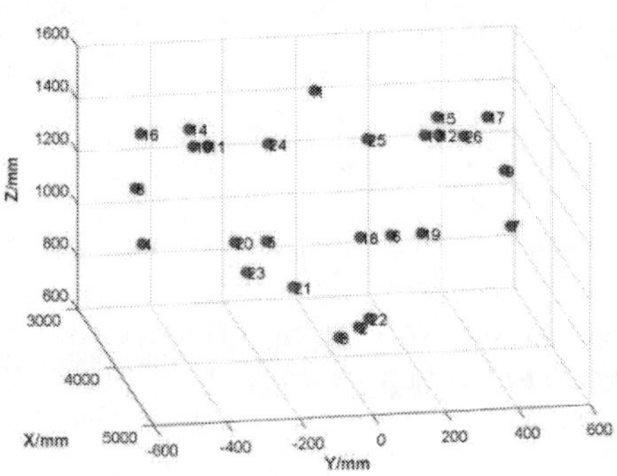

Figure 3. The distribution map of weld joints.

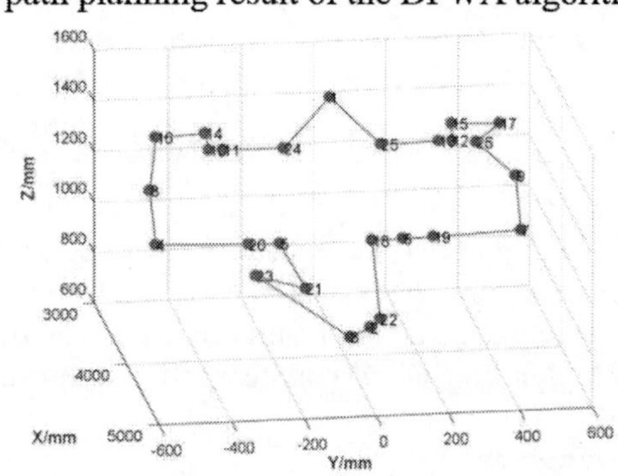

Figure 4. The optimal welding path based on DFWA.

Particle swarm optimization (PSO), genetic algorithm (GA), simulated annealing (SA) and our DFWA are applied on this example of 3-dimension path planning to show the performance of these swarm intelligence algorithms. They are respectively operated for 30 times. The parameters are given as follows. Iterations are set to 1000 for PSO and GA, inertia weight, self-learning factor and social learning factor in PSO are 0.8, 0.5 and 0.7. Crossover probability and variation probability in GA are 0.95 and 0.05. In SA, temperature $T_0 = 1000$, p = 0.9. In DFWA, n = 5, $m_s = 50$, a = 0.04, b = 0.8, and MaxItCount = 10. The experimental results are shown in table 1.

Table 1. Performance comparison of different swarm intelligence algorithms.

Algorithm	Path Length (mm)			Mean Iterations
	Optimal	*Mean*	*Maximum*	
PSO	5155.77	6472.51	7509.79	30.30
GA	4960.01	5335.90	9196.97	102.10
SA	4960.01	5352.03	6640.13	39.20
DFWA	4960.01	5050.65	5980.48	13.00

And the convergence curves of these 4 swarm intelligence algorithms are shown in figure 5.

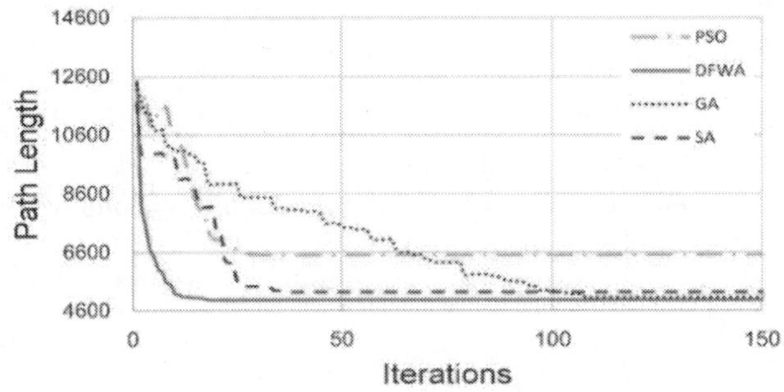

Figure 5. Comparison of convergence curves among PSO, GA, SA and DFWA.

It can be seen that DFWA performs well in convergence speed, stability and accuracy of the optimization.

Overall, the DFWA proposed in this paper is feasible and effective. It provides a new idea for the research of FWA on discrete problems. In addition, the DFWA develops a reasonable welding sequence and is significative to welding robot path planning application.

5. Conclusion

A novel discrete fireworks algorithm (DFWA) is proposed for welding robot path planning. The DFWA algorithm combines 2-opt local search and operators in GA to discretize the searching process of the FWA. As the experimental results indicate, the discrete method for FWA is very effective for welding robot path planning, and it can be applied to more discrete optimization problems.

Acknowledgments

This paper was supported by National Key R&D Program of China (2017YFB1303300).

References

[1] Y. Yang and Z. Tang, "The study of path planning of welding manipulator based on improved qpso," in *International Conference on Measuring Technology & Mechatronics Automation*, 2014, pp. 774–777.

[2] H. Shen, "A study of welding robot path planning application based on genetic ant colony hybrid algorithm," in *2016 IEEE Advanced Information Management, Communicates, Electronic and Automation Control Conference (IMCEC)*, Oct 2016, pp. 1743–1746.

[3] X. Wang, Y. Shi, D. Ding, and X. Gu, "Double global optimum genetic algorithm-particle swarm optimization-based welding robot path planning," *Engineering Optimization*, vol. 48, no. 2, pp. 299–316, 2016.

[4] Y. Tan and Y. Zhu, "Fireworks algorithm for optimization," in *International Conference on Advances in Swarm Intelligence*, 2010, pp. 355– 364.

[5] H. Gao and M. Diao, "Cultural firework algorithm and its application for digital filters design," *International Journal of Modelling Identification & Control*, vol. 14, no. 4, pp. 324–331, 2011.

[6] W. He, G. Mi, and T. Ying, "Parameter optimization of local-concentration model for spam detection by using fireworks algorithm," in *International Conference in Swarm Intelligence*, 2013.

[7] A. Mohamed Imran and M. Kowsalya, "A new power system reconfiguration scheme for power loss minimization and voltage profile enhancement using fireworks algorithm," *International Journal of Electrical Power & Energy Systems*, vol. 62, no. 62, pp. 312–322, 2014.

[8] S. Zheng, J. Li, A. Janecek, and Y. Tan, "A cooperative framework for fireworks algorithm." *IEEE/ACM Transactions on Computational Biology & Bioinformatics*, vol. 14, no. 1, pp. 27–41, 2015.
[9] J. Li, S. Zheng, and Y. Tan, "The effect of information utilization: Introducing a novel guiding spark in the fireworks algorithm," *IEEE Transactions on Evolutionary Computation*, vol. 21, no. 1, pp. 153–166, 2017.
[10] J. Li and Y. Tan, "Loser-out tournament-based fireworks algorithm for multimodal function optimization," *IEEE Transactions on Evolutionary Computation*, vol. 22, no. 5, pp. 679–691, Oct 2018.

Biologically Inspired Low Energy Clustering for Large Scale Wireless Sensor Networks

Jie Zhou[1, 2, *], Mengying Xu[1] and Yi Lu[1]

[1]College of Information Science and Technology, Shihezi University, Shihezi, 832000, China
[2]Xinjiang Tianfu Information Technology Co., Ltd.

[*]Email: jiezhou@shzu.edu.cn

Abstract. The recent technological advances in data gathering, embedded micro-devices and mobile networking has significantly advanced the applications of small-size and numerous tiny nodes. Large scale wireless sensor networks (LSWSNs) are intensively studied and used in the fields of traffic avoidance, intelligent family, medical diagnostic, environmental, multimedia surveillance, military affairs and so on. The recent success of emerging LSWSNs technology has encouraged researchers to develop new low energy clustering algorithm in this field. In LSWSNs, reducing communication energy consumption of sensor will not lead to maximize network lifetime for the total system. The low energy clustering is a typical NP-hard combinatorial optimization problem. In this paper, an immune adaptive cuckoo search algorithm (IACSA) is given to reduce total energy consumption. We first design a fitness function to evaluate energy consumption of system. The IACSA is designed to improve the energy efficiency for LSWSNs. It has the advantages of immune generator that takes into account different benefits and adaptive operator to enhance the convergence rate. Simulations are conducted to show a comparison of IACSA with the shuffled frog leaping algorithm (SFLA), particle swarm optimization (PSO) and artificial fish swarm algorithm (AFSA). Results show that the proposed IACSA has lower energy consumption compared to the SFLA, AFSA and PSO, which means that the proposed method reduces the energy consumption.

1. Introduction

The rapid development of data gathering, embedded computing and networking technologies brings a lot of advantages on wireless communication [1]-[3]. Large scale wireless sensor networks (LSWSNs) consist of a great deal of sensors, which possess limited computation, sensing and free-infrastructure capabilities. Each sensor node is made up of five components, including the information collecting component, the microprocessor component, the transceiver component, the data storage component and the energy supply component [4]-[6]. LSWSNs are growing techniques in many aspects, such as tracking, military purposes, homelands security, environment surveillance, multimedia surveillance, medical diagnostic, etc [7][8].

Low energy clustering is a strategy offering an approach that displays how to reduce communication energy consumption in LSWSNs [9]. LSWSNs have been concerned in many fields for their large size with limited communicating capabilities. Low energy clustering scheme can be an effective optimization means for achieving the desired power consumption goals, comprising communicating capability constraints. The low energy clustering problem is an NP-hard problem and still one of the most exciting challenges in the LSWSNs [10]. Nevertheless, as exhaustive search are required in every combination, the computational complexness is huge in the problem.

Heuristics which combined with swarm algorithm have shown great promise in low energy clustering. A shuffled frog leaping algorithm (SFLA) method for low energy clustering was first introduced in [11]. The proposed method is to improve the energy efficiency by using a hybrid heuristic technique. However, the rate of convergence could not reach an acceptable point. The application of artificial fish swarm algorithm (AFSA) in low energy clustering problem has been researched [12]. As of previous work of the authors, this method can reduce total energy consumption for better energy consumption optimization. The AFSA is a basic swarm strategy but this algorithm commonly yields a shorter network lifetime than evolutionary algorithms. In [13], particle swarm optimization (PSO) can improve the energy efficiency that has been investigated. The algorithm has good performance. Nevertheless, the rate of convergence could not reach an acceptable point.

In this paper, we present an immune adaptive cuckoo search algorithm (IACSA), a cuckoo search algorithm that exploits the low energy clustering scheme, and we show how to take advantage of immune theory to efficiently speed up the convergence speed. We first give a novel formulation of the objective function to maximize the energy consumption to satisfy the energy consumption optimization. We make use of the fresh immune and adaptive operators where they are employed to cuckoo search algorithm. In IACSA, an effective cuckoo search algorithm that blends the advantages of immune theory and adaptive theory is presented. So, the proposed method can raise global search capability and enhance the randomness in the process of investigation. Furthermore, with the adaptive operator strategy, IACSA can obtain higher precision solution and prevent premature convergence.
In the simulations, we compare the proposed IACSA with the SFLA, PSO and AFSA. Simulation results indicate that the IACSA provides a good choice to enhance the energy efficiency over other SFLA, AFSA and PSO schemes for LSWSNs.

2. System Model

The system model to solve the problem of low energy clustering with respect to the constraints is given in this section.

Since the LSWSNs are usually powered by a battery, and the replacement of the battery requires a relatively high cost. The energy consumption problem is an important issue in LSWSNs. If the communication energy consumption can be effectively reduced, the lifetime of LSWSNs can be prolonged, and the labour cost of replacing the battery can be reduced.

Communication energy consumption usually consists of transmitting energy consumption and receiving energy consumption. The energy consumption and reception can be calculated by (1) and (2):

$$E_{CX}(l,d) = E_{elec} \cdot l + \varepsilon_{amp} \cdot l \cdot d^n \tag{1}$$

$$E_{RX}(l) = E_{elec} \cdot l \tag{2}$$

In (1) and (2), $E_{CX}(l,d)$ indicates the energy consumed to transmit l bit to a node of distance d, and ε_{amp} is a parameter of power amplification. E_{elec} is a parameter of electronics energy. d is the distance between nodes. The value of n is determined by the specific problem. We take n=3 in this paper.

$$E_j = E_{CX} + E_{RX} \tag{3}$$

$$E_{SUM} = \sum_{j=1}^{J} E_j \tag{4}$$

In (3), E_j indicates the sum of energy consumption and reception of a node. In (4), E_{SUM} is sum of the total energy consumption of all nodes

3. An immune adaptive cuckoo search algorithm for low energy clustering in LSWSNs

The cuckoo search algorithm can solve the continuous optimization problem, but it does not solve the binary or discrete problem. For the low energy clustering problem, a discretized optimization algorithm needs to be designed. The binary IACSA is used to optimize the energy efficiency. The discrete IACSA has three advantages, simple operation, fewer parameters, and don't need to re-match parameters when deal with optimization problems.

In essence, the IACSA has three elements: optimal selection, local random flight, and global Lévy flight.

To simplify the description of the IACSA, assume the following three idealization rules.

(1) Each cuckoo has only one egg at a time, that is, there is an optimal solution, and the nest is randomly selected when it hatches;
(2) The optimal nest and optimal solution are retained to the next generation;
(3) The number of nested main birds is fixed, and the probability that cuckoo hatching eggs are found by nesting birds is also fixed as $P_a \in [0,1]$. When the cuckoos in house either throw the egg out of the nest or discard the nest to find another place and build a new nest.

3.1. Population Encoding and Initialization

Population encoding is the first step, which applies in the IACSA to solve the low energy clustering problem. According to the low energy clustering problem applied by IACSA, the corresponding encoding method, such as binary encoding, is adopted. The position of the nest is composed of binary code, that is, the variable takes 0 or 1. During the formation of the cluster, each sensor node is assigned a random number between 0 and 1. When the value is 1, the node is adopted as cluster head node. Otherwise, the node is employed as sensor node.

Supposed the population P is composed of I individuals. $P = \{P_1, P_2, \cdots\cdots, P_I\}$. The encoding population can be shown in (5).

$$P = \begin{bmatrix} p_{1,1} & p_{1,2} & \cdots & p_{1,J-1} & p_{1,J} \\ p_{2,1} & p_{2,2} & \cdots & p_{2,J-1} & p_{2,J} \\ \cdots & \cdots & p_{i,j} & \cdots & \cdots \\ p_{I-1,1} & p_{I-1,2} & \cdots & p_{I-1,J-1} & p_{I-1,J} \\ p_{I,1} & p_{I,2} & \cdots & p_{I,J-1} & p_{I,J} \end{bmatrix} \tag{5}$$

In (5), I represents the amount of cuckoos, $i \in \{1, 2, \cdots\cdots, I\}$, J indicates the quality of sensor nodes, $j \in \{1, 2, \cdots\cdots, J\}$, $p_{i,j} = 1$ indicates that the j_{th} node in the i_{th} cuckoo is the cluster head node. When $p_{i,j} = 0$, the j_{th} sensor node in the i_{th} individual cannot be applied as cluster head node.

$$\sum_{j=1}^{J} p_{i,j} = K \tag{6}$$

In (6), K is employed as the amount of cluster heads. The number of cluster heads must equal to K. In population P, each individual has energy consumption and reception. $E = \{E_1, E_2, \cdots\cdots, E_I\}$, E indicates the sum of energy consumption and reception.

3.2. Fitness Calculation

In order to optimize the communication energy consumption, we establish a fitness function based on two factors of energy consumption and reception in (7). The cuckoos will be selected to the next iteration when the energy consumption is lower.

$$f = \sum_{j=1}^{J} E_j \tag{7}$$

3.3. Lévy Flight

In discrete IACSA algorithm, a nest represents a solution, and the position of one nest is randomly initialized in the feasible solution space defined by the objective function. The most important step of the IACSA is the location update. The equations of Lévi flight of the first method are as follows:

$$Sig(step) = \frac{1}{1+\exp(-step)} \tag{8}$$

$$p_{ij}^{t+1} = \begin{cases} 1 & if\ rand \le Sig(step) \\ 0 & otherwise \end{cases} \tag{9}$$

Where $step$ is the step of the Lévi flight, $Sig()$ indicates sigmoid function. p_{ij}^{t+1} is the binary code of the j_{th} dimension and the position of the i_{th} bird's nest in the $(t+1)_{th}$ generation, and $rand$ is a completely random number in [0, 1].

The second method is as follows:

If $step \le 0$,

$$Sig(step) = 1 - \frac{2}{1+\exp(-step)} \tag{10}$$

$$p_{ij}^{t+1} = \begin{cases} 0 & if\ rand \le Sig(step) \\ p_{ij}^{t} & otherwise \end{cases} \tag{11}$$

If $step > 0$,

$$Sig(step) = \frac{2}{1+\exp(-step)} - 1 \tag{12}$$

$$p_{ij}^{t+1} = \begin{cases} 1 & if\ rand \le Sig(step) \\ p_{ij}^{t} & otherwise \end{cases} \tag{13}$$

3.4. Immune Operator

In the low energy clustering of LSWSNs, the coding sequence is long, which leads to the slow optimization process of IACSA, which easily leads to the stagnation of iterative process. The IACSA is a heuristic optimization algorithm that combines the information processing mechanism of biological immune system.

The steps of IACSA with immune operator include: encoding of antibodies, generation of initial antibodies P, calculation of antibody affinity $f(P)$, antibody selection and clonal amplification, antibody replacement and retention.

Step 1: Randomly generate the initial antibody, that is, randomly generate a candidate solution, and create a total population P, which consists of I initial antibodies.

Step 2: Calculate antibody affinity, affinity is the objective function $f(P)$.

Step 3: Select the antibody with the lowest energy consumption from the population to form a temporary antibody set.

Step 4: Cloning antibodies with low affinity: the lower the energy consumption, the larger the size of the cloned antibody.

3.5. Adaptive Operator
Taking the parameter as a fixed value is not conducive to the global search and local search. In order to make the IACSA more efficient, the IACSA with an adaptive strategy makes P_a change dynamically. In the early stage of the IACSA, a relatively small P_a is employed in the proposed method. The quantity of updated solutions is huge, so that the algorithm maintains strong global search ability, while taking into account the local search ability. In the later stage of the algorithm, a relatively large P_a is employed, so that it can be improved the convergence speed of IACSA.

3.6. Termination Condition
The most common way of termination conditions is to give a maximum iteration times in advance, or to determine whether the optimization value is continuous. In this paper, we apply the first method.

3.7. Algorithm Steps
The specific steps of the IACSA algorithm are as follows:

Step1: Randomly generate I individuals and calculate fitness. Filtering out the optimal individual, and retain to the next generation;

Step2: Lévi flight is conducted to create a new host nest location by using formula (7) to create a new population and calculate fitness.

Step3: Produce a random number $rand \in [0,1]$ obeying a uniform distribution, compared with the probability that the cuckoo is found by the bird's egg P_a=0.25, if $rand > P_a$ the bird's fitness is randomly changed, and vice versa, and then the updated bird's energy consumption, compared with the results that obtained in the previous step, the bird's nest position with lower energy consumption is taken, and the global minimum energy consumption is selected.

Step4: If the maximum number of iterations is reached, output the final solution, if not returning to step2.

4. Simulation and Results
In this section, we evaluate the efficiency of the IACSA by comparing with the AFSA, SFLA as well as PSO when deal with the problem of low energy clustering in LSWSNs. To examine the applicability for practical implementations, we evaluate the performance of the schemes on a PC with Pentium 2.40 GHz,2G RAM,WIN-7 OS, and MATLAB software.

After that, a fitness function to calculate the energy consumption is employed in this section. The coordinates of sensor nodes are then randomly chosen in the range of 0 to 400 m.

In the following experiments, the simulation parameters for the IACSA, SFLA, AFSA and PSO are given in following discussion. The results are obtained after the algorithm runs 100 times.

In IACSA, the number of nests is 50, and the probability of finding is 0.25. In PSO, the maximum velocity is definitely employed as 4. The cognitive and social values are c_1=c_2=0.5. The dynamic range of the particle is set to 0.4. The parameter values in the SFLA are given in following. The number of individuals in the population is 50, the number of subgroups is 10, and the number of individuals in the subgroup is 5. In AFSA, visual is 5. Step is 5. Try number is 10 and congestion factor is 0.6.

Figure 1 and figure 2 show the communication energy consumption of IACSA, SFLA, AFSA and PSO when the low energy clustering problem is solved with different quantity of sensor nodes. We set sensor nodes to 200 and 300, respectively, and cluster head proportion percentage is 10%. For each method, we select solution that yields the best results. The results are the average over 100 runs with different methods. To demonstrate the advantages of the IACSA, the optimization results from SFLA, AFSA and PSO approaches after 100 iterations are compared, as illustrated in the following figures.

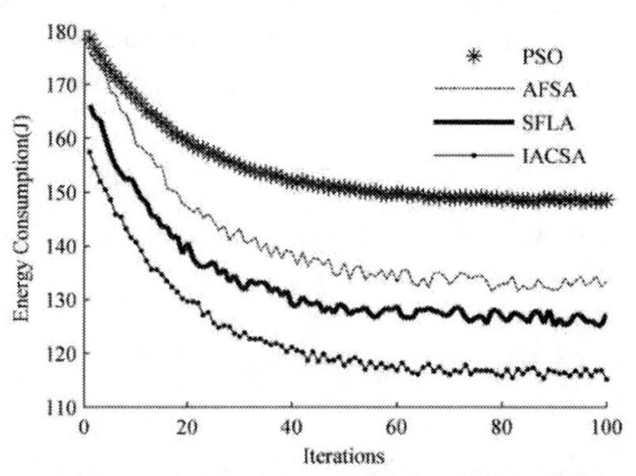

Figure 1. The number of sensor nodes is 200 and cluster head proportion is 10%

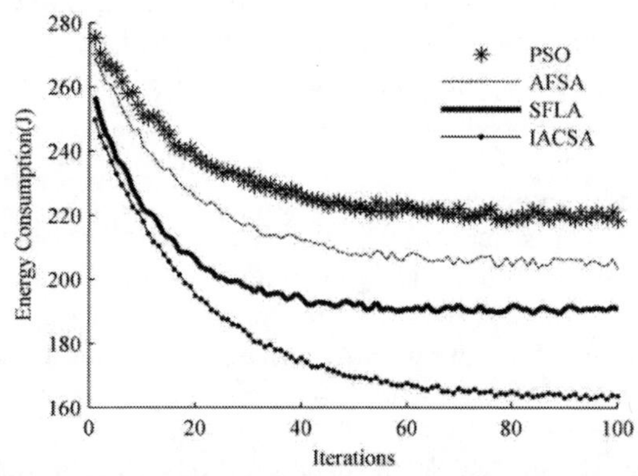

Figure 2. The number of sensor nodes is 300 and cluster head proportion is 10%

The IACSA gives the minimum energy consumption, followed by the SFLA, AFSA and PSO. The quantity of energy consumptions is close to 158 by use of IACSA in the beginning of iterations in figure 1, which is lower than that of SFLA, AFSA and PSO. It is clear that the IACSA achieves better convergence rate than the SFLA, AFSA and PSO. Meanwhile, SFLA, AFSA and PSO maintained a nearly constant result due to its premature convergence. In the end of iterations, IACSA has the best results, which is 116. In comparison with the IACSA, the proposed SFLA, AFSA and PSO provide suboptimal results, and the values are equal to 129, 134 and 150 respectively. Figure 2 has the similar performance and effect as Figure 1.

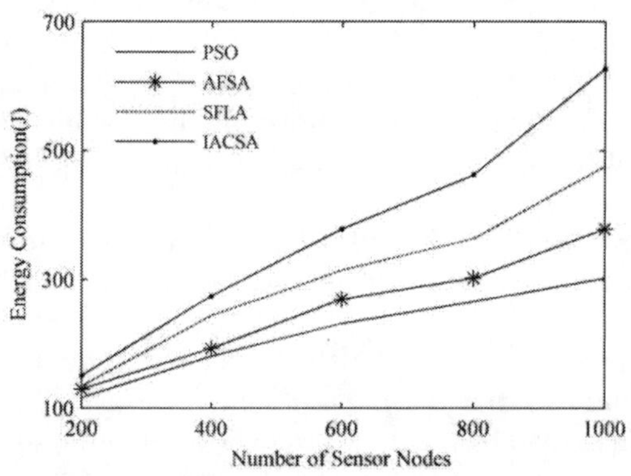

Figure 3. Cluster head proportion is 10%

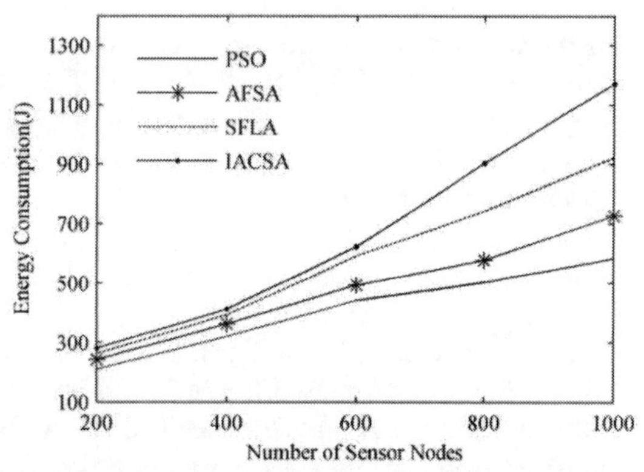

Figure 4. The cluster head proportion is 20%

Figure 3 and figure 4 demonstrate the energy consumption with different numbers of sensor nodes and the cluster head proportion is 10% and 20%, respectively. We also set sensor nodes to 200, 400, 600, 800 and 1000, respectively. Comparisons are also performed with SFLA, AFSA and PSO. After simulations, the proposed algorithm can reduce total energy consumption for better energy consumption optimization. So the proposed method has a better performance compared with SFLA, AFSA and PSO.

5. Conclusion

In this paper, an immune adaptive cuckoo search algorithm (IACSA) to reduce total energy consumption for the problem of low energy clustering is developed. In this work, we first propose a novel formulation of the objective function to improve the energy efficiency to satisfy the energy consumption optimization. The simulations are executed to display the effectiveness of the IACSA when compare with SFLA, PSO and AFSA, where indicates that the low energy consumption clustering of the proposed IACSA method is better than that of SFLA, AFSA and PSO, respectively. IACSA has the ability to compute the approximate results with adaptive operator strategy to avoid local optima that may occur in searching process.

Acknowledgments

This paper is supported by the National Nature Science Foundation of China (No. 61662063), the project of Youth and Middle-aged Scientific and Technological Innovation Leading Talents Program of the Corps (No. 2018CB006), the project of Promoting Scientific Research Cooperation and High-level Personnel Training with Americas and Oceania, the China Postdoctoral Science Foundation

References

[1] Wang C, Li J and Yang Y 2017 Combining solar energy harvesting with wireless charging for hybrid wireless sensor networks. *IEEE Transactions on Mobile Computing* **17** pp 560-76.

[2] Daskalakis S N, George G and Assimonis S D 2018 A uW backscatter-morse-leaf sensor for low-power agricultural wireless sensor networks. *IEEE Sensors Journal* **18** pp7889 - 98.

[3] Kumar N and Vidyarthi D P 2018 A green routing agorithm for IoT enabled software Ddefined wireless sensor network. *IEEE Sensors Journal* **18** pp 9449 - 60.

[4] Lai X and Wang H 2018 RNOB: Receiver negotiation opportunity broadcast protocol for trustworthy data dissemination in wireless sensor networks. *IEEE Access* **6** pp 53235 - 42.

[5] Deepak S and Bhondekar A P 2018 traffic and energy aware routing for heterogeneous wireless sensor networks. *IEEE Communications Letters* **22** pp 1608 – 11.

[6] Mekikis P V and Kartsakli E, Antonopoulos A, et al 2018 Connectivity analysis in clustered wireless sensor networks powered by solar energy. *IEEE Transactions on Wireless Communications* **17** pp 2389 - 401.

[7] Malika E, Haithem B C and Rabah A 2018 Prolonging the lifetime of large-scale wireless sensor networks using distributed cooperative transmissions. *IET Wireless Sensor Systems,* **8** pp. 229-236.

[8] Ang L M, Seng J K P and Zungeru A M 2018 optimizing energy consumption for big data collection in large-scale wireless sensor networks With Mobile Collectors. *IEEE Systems Journal,* **17** pp. 2389 - 2401.

[9] Liu X, Xiong N, Wei L, et al 2015 An optimization scheme of adaptive dynamic energy consumption based on joint network-channel coding in wireless sensor networks. *IEEE Sensors Journal,* **15** pp.5158-5168.

[10] Lee J S and Kao T Y 2017 An Improved Three-layer low energy adaptive clustering hierarchy for wireless sensor networks. *IEEE Internet of Things* Journal **3** pp. 951-958.

[11] Edla D R, Lipare A and Cheruku R 2017 An efficient load balancing of gateways using improved shuffled frog leaping algorithm and novel fitness function for WSNs. *IEEE Sensors Journal,* **17** pp. 6724 - 6733.

[12] Shen Y and Guo B 2016 Energy-efficient cluster-head selection with fuzzy logic for robotic fish swarm *Int Conf on Fuzzy Systems and Knowledge Discovery*. IEEE, pp. 513-518.

[13] Rejinaparvin J and Vasanthanayaki C 2015 Particle swarm optimization-based clustering by preventing residual nodes in wireless sensor networks. *IEEE Sensors Journal,* **15** pp. 4264-4274.

Optimization of Low Power Wireless Sensor Network Lifetime Using Improved Clone Elite Monkey Algorithm

Yi Lu[1], Jie Zhou[1,2,*] and Mengying Xu[1]

[1]College of Information Science and Technology, Shihezi University, Shihezi, 832000, China
[2]Xinjiang Tianfu Information Technology Co., Ltd.

[*]jiezhou@shzu.edu.cn

Abstract. Recent advances in efficient software algorithms, electronics and networking technologies have promoted the development of low computational complexity, low-cost, low storage, and intelligent tiny nodes. Low power wireless sensor networks (LPWSNs) are composed of some sensing units with limited communications as well as sensing capabilities. LPWSNs are approaches in a good deal of domains such as traffic control, automated assistance for the elderly monitors and so on. Recently network lifetime optimization has been receiving a lot of attention for wide applications of LPWSNs. Achieving a longer network lifetime under the restricted power source has a large calculation difficulty, which can be regard as an NP-hard problem. An improved clone elite monkey algorithm (ICEMA) to determine the network lifetime optimization in LPWSNs is given. In this paper, in order to achieve maximum network lifetime with full coverage, we first build a system model to calculate a better network lifetime. It is designed to increment the lifetime of the nodes for LPWSNs. The ICEMA has many advantages by combining the clone strategy as well as elite strategy. The simulations verify the robust and efficiency of ICEMA when compared with strategies based on evolutionary algorithm (EA), particle swarm algorithm (PSO) and artificial fish swarm algorithm (AFSA) under a LPWSNs conditions. The outcomes demonstrate that the proposed ICEMA can achieve a longer network lifetime than EA, PSO and AFSA while taking the same computational complexity.

Keywords. Wireless sensor networks, Monkey algorithm, Duty cycle design, Network lifetime

1. Introduction

The rapid development of efficient software algorithms, system-on-chip design and wireless communications has made it possible to deploy small-size, numerous, low storage, and networked wireless sensors [1]-[4]. Low power wireless sensor networks (LPWSNs) have a large number of tiny sensors with communications, sensing, processing, free-infrastructure capabilities [5]. LPWSNs support important practical applications in tracking, medical diagnostic, home automation, environment surveillance, deploying network in battlefield, commercial management, etc [6][7].

The network lifetime optimization problem attracts attention of researchers who are in industry, and military department. Sensor nodes mainly use tiny sensing units and they are limited in sensing capability [8]. They affect the effective optimization means for achieving maximum network lifetime as well as the lifetime optimization greatly.

Many heuristics techniques have been developed to improve the efficiency of the network lifetime optimization scheme through duty cycle design. In [9], a network lifetime optimization strategy that enables the performance between the computation complexity and a better lifetime is researched utilizing evolutionary algorithm (EA). In their work they increment of the number and the lifetime of the nodes without considering restricted power source constraint. However, EA suffer from low performance. A simple network lifetime optimization scheme using particle swarm algorithm (PSO) while simultaneously evaluating effective optimization means for achieving maximum network lifetime has also been attempted in [10]. Nevertheless, the computational cost is too large. In [11], an artificial fish swarm algorithm (AFSA) scheme is designed for achieving a longer network lifetime in LPWSNs. The proposed method uses a hybrid heuristic technique. However, it suffers from stagnation after only a few generations.

An improved clone elite monkey algorithm (ICEMA) is given to determine the network lifetime optimization in LPWSNs in our works. We first formulate a network optimization model as objective function. After that, in order to improve the system performance of ICEMA based network lifetime optimization issue, we put clone as well as elite strategies into the ICEMA to enhance the quality of solutions and avoid local optima.

The simulations are executed to validate the efficiency gain in terms of a better lifetime when compared with the techniques depending on EA, PSO and AFSA. Experiments indicate that the profitable optimization means for achieving maximum network lifetime of the proposed ICEMA method has a better performance than that of EA, PSO and AFSA respectively with different quantity of sensor nodes and targets.

2. System Model

In this section, the system model of the network lifetime optimization problem is built with respect to the constraints of restricted power source and the duty cycle design.

Supposed that I sensor nodes and J targets are deployed in the monitored region. The sensing relationship matrix is described as following formula.

$$E=\begin{bmatrix} e_{1,1} & e_{1,2} & \cdots & e_{1,J-1} & e_{1,J} \\ e_{2,1} & e_{2,2} & \cdots & e_{2,J-1} & e_{2,J} \\ \cdots & \cdots & e_{i,j} & \cdots & \cdots \\ e_{I-1,1} & e_{I-1,2} & \cdots & e_{I-1,J-1} & e_{I-1,J} \\ e_{I,1} & e_{I,2} & \cdots & e_{I,J-1} & e_{I,J} \end{bmatrix} \tag{1}$$

In (1), $e_{i,j}$ represents the monitoring relationship among the i_{th} sensor and the j_{th} target, $e_{i,j} \in \{0,1\}$, and $e_{i,j} = 1$ indicates that the i_{th} target located at the coverage range of the j_{th} node. $e_{i,j} = 0$ is the opposite.

Due to energy limitation, the maximum working lifetime of each sensor in the I sensor nodes is R rounds, $r \in \{1,2,\cdots\cdots,R\}$, then the maximum working lifetime of the network is IR rounds, that is, the round matrix S of the sensors can be written in (2):

$$S=\begin{bmatrix} s_{1,1} & s_{1,2} & \cdots & s_{1,I-1} & s_{1,I} \\ s_{2,1} & s_{2,2} & \cdots & s_{2,I-1} & s_{2,I} \\ \cdots & \cdots & s_{ir,i} & \cdots & \cdots \\ s_{IR-1,1} & s_{IR-1,2} & \cdots & s_{IR-1,I-1} & s_{IR-1,I} \\ s_{IR,1} & s_{IR,2} & \cdots & s_{IR,I-1} & s_{IR,I} \end{bmatrix} \tag{2}$$

Where $s_{i,j}=1$ denotes that the i_{th} node is in an active state in the r_{th} round, $s_{i,j}\in\{0,1\}$, $s_{i,j}=0$ indicates that the i_{th} sensor node is in the dormant state in the r_{th} round.

If we multiply the two matrixes, we can get the monitoring relationship among the i_{th} sensor as well as the j_{th} target in each round.

$$ES=\begin{bmatrix}\sum_{i=1}^{I}e_{1,i}s_{i,1} & \cdots & \sum_{i}^{I}e_{i,J}s_{1,i}\\ \cdots & \sum_{i=1}^{I}e_{i,j}s_{ir,i} & \cdots\\ \sum_{i=1}^{I}e_{i,1}s_{IR,i} & \cdots & \sum_{i=1}^{I}e_{i,J}s_{IR,I}\end{bmatrix} \tag{3}$$

In (3), $\sum_{i=1}^{I}e_{i,j}s_{i,r}>0$ means that the j_{th} target in the r_{th} round is monitored by at least one sensor, and $\sum_{i=1}^{I}e_{i,j}s_{i,r}=0$ means that the j_{th} target in the r_{th} round is not monitored by any one of the sensors.

In (4), the fitness function f represents the number of rows in which the first zero element in ES. The mathematical model of the lifetime optimization of the wireless sensor network can be expressed as follows:

$$f=row\_zero(ES)-1 \tag{4}$$

Subject to:

$$\sum_{ir=1}^{IR}s_{ir,r}\le R, i=1,2,\cdots\cdots I \tag{5}$$

(5) indicates that each sensor has a maximum working life, which is less than R rounds.

3. Optimization of network lifetime using improved clone elite monkey algorithm

The proposed ICEMA combines the merits of both clone operator and elite operator based on the traditional monkey algorithm. The proposed ICEMA mainly includes initialization, fitness calculation, the climbing process, the lookout process and the flipping process.

3.1. Population Encoding and Initialization

The initialization of the population has an influence on the global convergence and effect of the algorithm. Therefore, the ICEMA uses random initialization in the solution space. Discrete ICEMA is applied to solve network lifetime problem.

First define H as the size of the population P in ICEMA. A round matrix S is an individual, which consists of 0 and 1. The population P is made up of round matrixes. An individual P_h can be demonstrated by the following example in (6):

$$S_1=\begin{bmatrix}0&1&1\\0&0&1\\1&0&0\end{bmatrix}=[0\ \ 1\ \ 1\ \ 0\ \ 0\ \ 1\ \ 1\ \ 0\ \ 0]=P_1 \tag{6}$$

Where a matrix S_h (3×3) should be converted into a matrix P_h (1×9) to represent the h_{th} individual. In other words, a matrix S_h ($IR\times I$) should be converted into a matrix P_h ($1\times I^2R$) to represent the h_{th} individual. Moreover, a population which has H individuals consists of a I^2R-dimensional matrix($H \times I^2R$).

In (7), randomly generate H individuals to create an initial population, indicating the number of monkeys. The population consists of randomly generated binary codes.

$$P = \begin{bmatrix} p_{1,1} & p_{1,2} & \cdots & p_{1,I^2R-1} & p_{1,I^2R} \\ p_{2,1} & p_{2,2} & \cdots & p_{2,I^2R-1} & p_{2,I^2R} \\ \cdots & \cdots & p_{h,i^2r} & \cdots & \cdots \\ p_{H-1,1} & p_{H-1,2} & \cdots & p_{H-1,I^2R-1} & p_{H-1,I^2R} \\ p_{H,1} & p_{H,2} & \cdots & p_{H,I^2R-1} & p_{H,I^2R} \end{bmatrix} \tag{7}$$

Where $P = (P_1, P_2, \cdots, P_H)$, $h = \{1, 2, \cdots, H\}$. $P_h = (p_{h,1}, p_{h,2}, \cdots, p_{h,I^2R})$. The position can be represented as a solution to the network lifetime optimization problem.

3.2. Fitness Calculation

The objective function can be calculated in (8) to optimize the network lifetime. The monkeys will be selected to the next iteration if the network lifetime is longer.

$$f(P_h) = row\_zero(ES) - 1 \tag{8}$$

3.3. Climbing process

The climbing operator is a process in which each monkey searches for the best network lifetime of the optimization problem by iterations in the current range. Due to the binary encoding method, the climbing step is generally set to 1. For the monkey h, the climbing process is designed as following:

Step1: Randomly generate vectors $\Delta P_h^{'}$ and $\Delta P_h^{''}$. $\Delta P_h^{'} = (\Delta p_{h,1}^{'}, \Delta p_{h,2}^{'}, \cdots\cdots, \Delta p_{h,I^2R}^{'})$, $\Delta P_h^{''} = (\Delta p_{h,1}^{''}, \Delta p_{h,2}^{''}, \cdots\cdots, \Delta p_{h,I^2R}^{''})$. The climbing step w in climbing process is 1.

$$\Delta p_{h,I^2R}^{'}, \Delta p_{h,I^2R}^{''} = \begin{cases} w & probability = 0.5 \\ -w & probability = 0.5 \end{cases} \tag{9}$$

Step2: If the value of $(i^2r)_{th}$ of vectors $\Delta P_h^{'}$ or $\Delta P_h^{''}$ is 1, the value of $(i^2r)_{th}$ of P_h is unchanged; otherwise, 0 is changed to 1 or 1 is changed to 0. In this way, two new individuals $P_h^{'}$ and $P_h^{''}$ are formed according to the current individual P_h.

Step3: If $f(P_h^{'}) > f(P_h^{''})$ and $f(P_h^{'}) > f(P_h)$, set $P_h = P_h^{'}$. If $f(P_h^{''}) > f(P_h^{'})$ and $f(P_h^{''}) > f(P_h)$, set $P_h = P_h^{''}$.

Step4: Repeat step1 to step3 until the number of climbs is satisfied for a given number of executions N_c.

3.4. Lookout process

At the end of the climbing process, each monkey looks around and sees if there is a higher point than the current one. If there is, then it jumps from the current point to the higher position. The process of lookout in ICEMA is described as following.

Step1: Randomly generate a value t_{h,i^2r} that in matrix $T_h = (t_{h,1}, t_{h,2}, \cdots\cdots, t_{h,I^2R})$ is in the range of $(p_{h,i^2r} - c, p_{h,i^2r} + c)$, c represents visual of monkey that is set to 1.

Step2: Because the value of t_{h,i^2r} is within the feasible solution area [-1,2], if $t_{h,i^2r} < 0.5$, t_{h,i^2r} =0, otherwise, t_{h,i^2r} =1.

Step3: If $f(T_h) > f(P_h)$, set $P_h = T_h$; otherwise, repeat step1 until the value T_h that satisfies the condition is generated.

Step4: Use T_h as the initial position and repeat the climbing process.

3.5. Cooperation process

Assume the optimal position of the current monkeys is $P^* = (P_1^*, P_2^*, \cdots\cdots, P_H^*)$, the position of monkey h is $P_h = (p_{h,1}, p_{h,2}, \cdots\cdots, p_{h,I^2R})$. The steps of cooperation process are as follows.

Step1: Randomly generate a number α is within [0, 1].

Step2: If α <0.5, $t_{h,i^2r} = p_{h,i^2r}$; otherwise, $t_{h,i^2r} = t_{h,i^2r} = p^*_{h,i^2r}$.

Step3: If $f(T_h) > f(P_h)$, set $P_h = T_h$.

3.6. Flipping process

The main purpose of the flipping process is to ensure that the monkey can find a new search area without falling into a local search. We select the centre of gravity of all monkeys as a pivot point, and then all monkeys will flip in the direction of the pivot point.

Step1: Generate a number δ in the flipping interval $[e, f]$ of the monkey.

Step2: Randomly generate a number k, $k = 1, 2, \cdots\cdots, H$. The position of the monkey k is taken as the pivot point.

Step3: Calculate

$$t_{h,i^2r} = p_{k,I^2R} + \delta(p_{k,I^2R} - p_{h,I^2R}) \tag{10}$$

Step4: If $t_{h,i^2r} < 0.5$, t_{h,i^2r} = 0; otherwise, t_{h,i^2r} =1.

Step5: If $f(T_h) > f(P_h)$, set $P_h = T_h$; otherwise, P_h is unchanged.

In order to prevent stagnate, this chapter sets a maximum number of iterations N_f to control the algorithm to fall into local optimum.

3.7. Termination Condition

In the end of ICEMA, when the maximum number of iterations is reached, the algorithm ends and the results are output.

3.8. Clone Operator

The clone operator is inspired by the artificial immune system and has many excellent characteristics. It can improve the convergence speed of the algorithm while maintaining the diversity of the population. It can also overcome the problem of premature convergence. In this paper, the clone operator is applied in the ICEMA based on the advantages of clone strategy. The main steps of ICEMA based clone strategy is described as following:

Step1: Determine the antigen, which is the network lifetime of problem to be optimized, generating the initial antibody population P, and the initial antibody is the population of the network lifetime problem;

Step2: Calculate the affinity of the antibody, which represents the network lifetime. In ICEMA, the affinity is $f(P)$;

Step3: Select antibodies with the better network lifetime from the population P to form a collection of temporary antibodies;

Step4: Clone the individuals with better network lifetime and form a new antibody population for the next iteration calculation.

Step5: When the maximum number of iterations is reached. The algorithm terminates execution and outputs the optimal solution.

3.9. Elite Operator

In the ICEMA, the elite strategy is also introduced, which balances the relationship between global search and local search. The behaviour of monkeys in the ICEMA with elite strategy increases the diversity of the population, effectively avoids the algorithm falling into local optimum.

Every time the monkey arrives at a new location in each generation, the network lifetime of the monkey needs to be accessed again. Compare the network lifetime of the monkey in the current generation and in the previous generation, select a set of monkey with longer network lifetime, and choose the longest network lifetime from these solutions as an elite solution.

3.10. Algorithm Steps

Based on the above description of the behaviours of the monkeys, the steps that can be attributed to the ICEMA are as follows:

Step 1: Initialize the population and parameters of the ICEMA

Step 2: Calculate the network lifetime, and record the optimal network lifetime of the monkey and the optimal P_h in the population;

Step 3: Execute the climbing process, calculate the network lifetime of all monkeys according to (8), and compare the network lifetime of current iteration with last iteration.

Step 4: Execute the lookout process, search for a better network lifetime within the range of view. If a better network lifetime is found, update the network lifetime of the monkey; otherwise, execute step 3.

Step 5: Execute the cooperation process to improve the network lifetime.

Step 6: Execute the flipping process to find a new search area. If the network lifetime is improved after the maximum iterations, update the population.

Step 7: When the maximum number of iterations is reached. The algorithm terminates execution and outputs the optimal solution.

4. Simulation and Results

In this experiment, the performance of the proposed ICEMA with the EA, PSO and AFSA for network lifetime optimization in LPWSNs is accessed. To demonstrate the effectiveness of the ICEMA and the proposed objective formulation, simulations are performed. To examine the applicability for practical implementations, we evaluate the performance of the schemes on a PC with Pentium 2.40 GHz, 2G RAM,WIN-7 OS, and MATLAB software.

The ICEMA is analysed with other means over the objective function in section II. Then we generate sensor nodes and targets, and the position(*x, y*) of sensor nodes are randomly specified within the square area 500×500. Each node has their range that is 300m. In the following experiments, we set sensor nodes to 55, 75 and 100, respectively. We also set targets to 10, 20, 30 and 40, respectively. The maximum round is 10. For comparison purpose, the results of EA, PSO and the AFSA are also given out. All comparisons between ICEMA, EA, PSO and AFSA are reported using the same number of iterations and population size.

The parameters of the ICEMA, EA, PSO and AFSA are presented as following. In ICEMA, the number of monkey groups is 50, and the number of iterations is 100. The visual c is 1 and the step w is 1. The flip interval $[e, f]$ is [-1, 1]. In EA, we investigate the population size 50 that ensures a specified quality of solution. For the crossover probability and mutation probability are employed as 0.85 and 0.05. In PSO, the maximum velocity is definitely 4. The values of cognitive and social are c1=c2=0.5. In AFSA, In AFSA, visual is 5. Step is 3. Try number is 5 and congestion factor is 0.62.

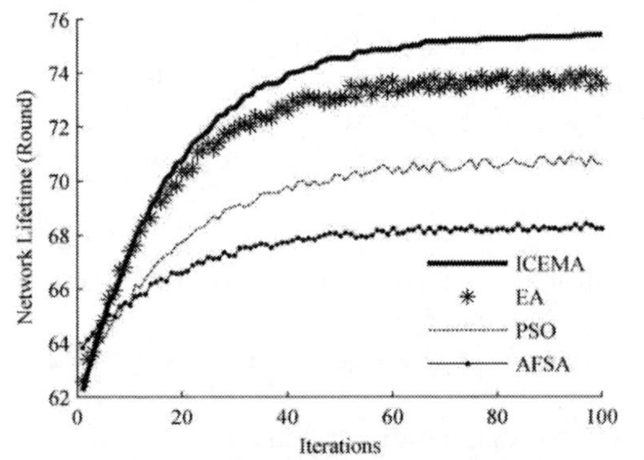

Figure1. 55 sensor nodes and 10 targets

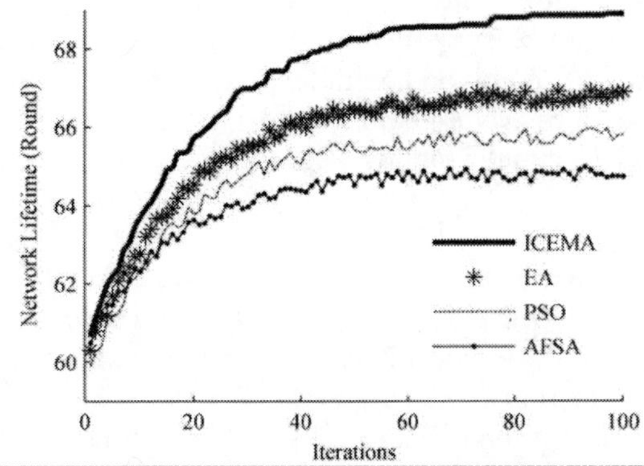

Figure2. 55 sensor nodes and 20 targets

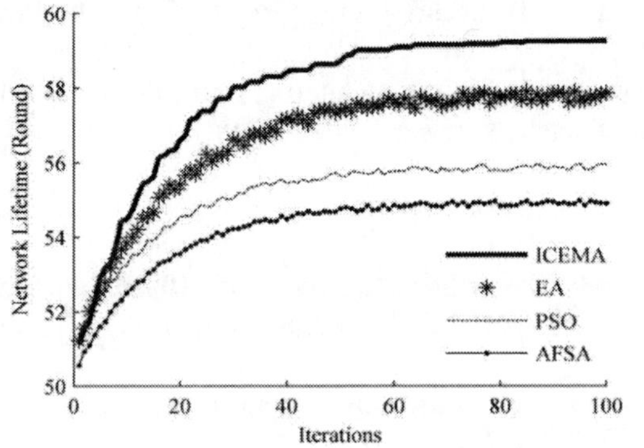

Figure3. 75 sensor nodes and 30 targets

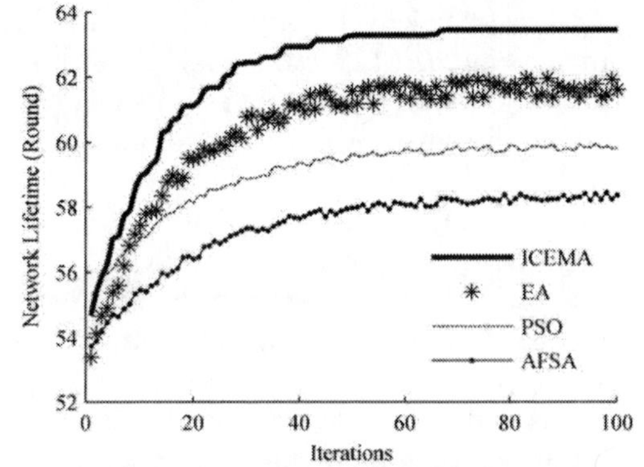

Figure4. 100 sensor nodes and 40 targets

Figure 1 shows the average number of generations of ICEMA, EA, PSO and AFSA with 10 targets and 55 sensor nodes. In the beginning, all the heuristics have shown good results when compared with a random search approach. The figure shows that the proposed ICEMA has a better convergence in the first several iterations. Meanwhile, the results of EA, PSO and AFSA are almost consistent. After the iterations, the best result of ICEMA is close to 76. Figure 2 shows the performance of network lifetime with 20 targets and 55 sensor nodes. By comparing with figure 1, when the number of nodes is unchanged and the number of targets is increased, then the round of lifetime is reduced to 69.

Figure 3 and figure 4 show the similar effective optimization for achieving maximum network lifetime of ICEMA, EA, PSO and AFSA with the 30 and 40 targets, respectively, and 75, 100 sensor nodes, respectively. After 100 iterations, the best result of ICEMA is close to 59, and 64 round, respectively. The results shows that the EAPSO has a better performance while optimize the problem of network lifetime.

5. Conclusion

Hence, an improved clone elite monkey algorithm (ICEMA) is presented and applied to network lifetime optimization problem. In this work, we first propose a novel formulation of the objective function to satisfy the lifetime optimization. Numerical simulations are conducted with ICEMA, EA, PSO and AFSA and the results are compared to verify the ICEMA. Results display that the ICEMA provides maintaining full coverage compared with previous EA, PSO and AFSA. It is more powerful and simpler than available heuristics, and can avoid local optima while searching for a better result.

Acknowledgments
This paper is supported by the National Nature Science Foundation of China (No. 61662063), the project of Youth and Middle-aged Scientific and Technological Innovation Leading Talents Program of the Corps (No. 2018CB006), the project of Promoting Scientific Research Cooperation and High-level Personnel Training with Americas and Oceania, the China Postdoctoral Science Foundation.

References

[1] Du R, Gkatzikis L and Fischione C 2017 On maximizing sensor network lifetime by energy balancing. *IEEE Transactions on Control of Network Systems* **5** pp 1206-18.
[2] Wang X and Liang Q 2015 Efficient sensor selection schemes for wireless sensor networks in microgrid. *IEEE Systems Journal,* **12** pp 539 - 47.
[3] Liu X and Zhang P 2017 Data Drainage: A novel load balancing strategy for wireless sensor networks. *IEEE Communications Letters* **22** pp 125 - 8.
[4] Osanaiye O, Alfa A S and Hancke G P 2018 Denial of service (DoS) defence for resourceavailability in wireless sensor networks. *IEEE Access* **6** pp. 6975 - 7004.
[5] Torfs T, Sterken T, Brebels S, et al 2013 Low power wireless sensor network for building monitoring. *IEEE Sensors Journal,* **13** pp 909-15.
[6] Zhao X, Sadhu V, Yang A, et al 2017 Improved circuit design of analog joint sourcechannel coding for low-power and low-complexity wireless sensors. *IEEE Sensors Journal,* **18** pp 281 – 9.
[7] Morrison D, Ablitt T and M J. Redoute 2015 Miniaturized low-power wireless sensor interface. *IEEE Sensors Journal,* **15** pp 4731-2.
[8] Frezzetti A and Manfredi S 2016 A two-layer controller scheme for efficient signal reconstruction and lifetime elongation in wirelesssensor networks. *IEEE Sensors Journal,* **16** pp 2172 – 9.
[9] Mohamed E S, Yasmine F and Hanan K 2018 Lifetime maximisation of disjoint wireless sensor networks using multiobjective genetic algorithm. *IET Wireless Sensor Systems.*
[10] Priyadharshini C and Thamarairubini K 2012 PSO based route lifetime prediction algorithm for maximizing network lifetime in MANET. *Int Conf on Recent Trends in Information Technology. IEEE,* pp 270 - 5.
[11] Song X, Wang C, et al 2010 A hierarchical routing protocol based on AFSO algorithm for WSN. *Int Conf on Computer Design and Applications.* IEEE **2** pp 635 -9.

Elite Adaptive Particle Swarm Optimization for Target Coverage Problem in High-density Wireless Sensor Networks

Mengying Xu[1], Jie Zhou[1,2,*] and Yi Lu[1]

[1]College of Information Science and Technology, Shihezi University, Shihezi, 832000, China
[2]Xinjiang Tianfu Information Technology Co., Ltd.

[*]Email: jiezhou@shzu.edu.cn

Abstract. The rapid development of intelligent sensing, micro electro-mechanical systems and communication has made it feasible to equip low computational complexity, low energy consumption, autonomous, and intelligent sensor nodes. High-density wireless sensor networks (HDWSNs) have information acquisition and communication abilities. HDWSNs are widely used in a number of areas including traffic avoidance, homelands security, target monitoring and so on. One of the major challenges in HDWSNs is to maximize the point coverage percentage. It is important since it is known that obtaining an optimal coverage target for HDWSNs is an NP-hard problem. In this paper, we use an elite adaptive particle swarm optimization (EAPSO) to solve the issue of target coverage in HDWSNs. In order to improve the effectiveness of system, a system model is provided to evaluate the monitored rate for HDWSNs. The proposed EAPSO with an efficient particle swarm optimization in discrete mode has the advantages of both adaptive as well as elite strategy. Numerical simulations are conducted with a number of nodes and targets using EAPSO, evolutionary algorithm (EA) and simulated annealing (SA). In the simulations, a better performance of EAPSO is given when it is compared with EA and SA with the same computational complexity.
Keywords. Wireless sensor networks, Particle swarm optimization, Target coverage

1. Introduction

The advance of data gathering, hardware manufacturing and networking technologies have improved the wireless sensor networks [1]-[4]. High-density wireless sensor networks (HDWSNs) have numerous sensing devices, which have many abilities, such as limited communication, computation, information acquisition, and free-infrastructure [5]. HDWSNs are a group of dense devices with limited sensing capabilities. Each sensing unit is made up of the following systems, which includes the computing system, the small power source system, the short-range radio transmitter system as well as the environmental monitoring system. HDWSNs have been widely designed in the domains of traffic control, environment surveillance, smart home, industry and many other areas [6].

Recently target coverage has been receiving a lot of attention for wide applications of HDWSNs. The target coverage scheme is usually used for HDWSNs to enhance the target coverage rate [7]. Enhancing the target coverage efficiency is an NP-hard problem [8]. Although the exhaustive search can achieve better coverage efficiency compared with other schemes, the computational difficulty is large for practical implementations [9].

IOP Conf. Series: Journal of Physics: Conf. Series **1267** (2019) 012006 doi:10.1088/1742-6596/1267/1/012006

Many heuristic algorithms have been proposed to solve the target coverage problem. In [10], a target coverage approach to maximize the target coverage rate in WSNs is investigated using evolutionary algorithm (EA). The EA has shown great performance when it comes to total system sensing range and the target coverage rate. While the EA produces some convergence troubles. In [11] a simulated annealing (SA) scheme is presented to investigate the target coverage rate in HDWSNs. This work focused on optimizing the total system sensing range with network constraints. However, in practice SA suffer from a low convergence rate.

In this paper, we propose an elite adaptive particle swarm optimization (EAPSO), a randomized swarm optimization algorithm for target coverage in HDWSNs, motivated by adaptive theory and elite theory. It is more powerful and simpler than available heuristics, and can avoid local optima while searching for a better result. The simulation results display that the EAPSO method provides a larger amount of successfully monitored targets over the current EA and SA methods.

2. System Model

The system model is given to deal with the issue of the target coverage when the monitoring capabilities and sensing radius is limited. Supposed that there are X targets and Y sensor nodes in the monitored areas. The coverage relationship could be represented by a matrix A in (1):

$$A=\begin{bmatrix} a_{1,1} & a_{1,2} & \cdots & a_{1,Y-1} & a_{1,Y} \\ a_{2,1} & a_{2,2} & \cdots & a_{2,Y-1} & a_{2,Y} \\ \cdots & \cdots & a_{x,y} & \cdots & \cdots \\ a_{X-1,1} & a_{X-1,2} & \cdots & a_{X-1,Y-1} & a_{X-1,Y} \\ a_{X,1} & a_{X,2} & \cdots & a_{X,Y-1} & a_{X,Y} \end{bmatrix} (a_{x,y} \in \{0,1\}) \quad (1)$$

In (1), the matrix represents the link among the y_{th} node as well as the x_{th} target, and $a_{x,y}=1$ suggests the x_{th} target in the sensing range of the y_{th} node. $a_{x,y}=0$ means that the target is not within the coverage of the sensor.

In equation (2), B represents the monitoring relationship matrix.

$$B=\begin{bmatrix} b_{1,1} & b_{1,2} & \cdots & b_{1,Y-1} & b_{1,Y} \\ b_{2,1} & b_{2,2} & \cdots & b_{2,Y-1} & b_{2,Y} \\ \cdots & \cdots & b_{x,y} & \cdots & \cdots \\ b_{X-1,1} & b_{X-1,2} & \cdots & b_{X-1,Y-1} & b_{X-1,Y} \\ b_{X,1} & b_{X,2} & \cdots & b_{X,Y-1} & b_{X,Y} \end{bmatrix} (b_{x,y} \in \{0,1\}) \quad (2)$$

In (2), $b_{x,y}=1$ means weather the x_{th} objective can be allocated by the y_{th} sensor. If $b_{x,y}=0$, the x_{th} objective locates in the sensing region of the y_{th} node, nevertheless, it cannot be monitored, otherwise, the x_{th} objective isn't in the sensing region of the y_{th} node.

If V targets can be monitored by each node within the coverage simultaneously, the constraint can be demonstrated in (3).

$$\sum_{x=1}^{X} b_{x,y} \le V(y=1,\cdots\cdots Y) \quad (3)$$

It is assumed that every objective must be monitored by no less than W sensors. Every sensor can detect no more than V targets at the same time. When the optimization is to increase the quantity of monitored targets, then, the fitness function in this paper to calculate the number of targets can be shown in (4).

$$\max f(b_{1,1},b_{1,2},\cdots\cdots,b_{X,Y})=\sum_{x=1}^{X}c_x \tag{4}$$

Where

$$c_x=\begin{cases}1 & \sum_{y=1}^{Y}b_{x,y}\geq W\\ 0 & \sum_{y=1}^{Y}b_{i,y}<W\end{cases} \tag{5}$$

Subjected to (3) and (6)

$$b_{x,y}\leq a_{x,y} \tag{6}$$

(3) means that every sensor can only detected a few objectives, while (6) means that only targets within the sensor coverage area can be monitored by the sensor. In (5), $c_x=1$ means that the x_{th} objective will be monitored; conversely, it will not be noticed.

3. An elite adaptive particle swarm optimization for target coverage problem in HDWSNs

Particle swarm optimization (PSO) is a swarm strategy, where each particle in this algorithm represents a solution. In this paper, the adaptive and elite strategies are employed to create a more effective binary particle swarm optimization. The main steps of EAPSO can be described as following.

3.1. Population Encoding

In EAPSO, suppose that there are X particles in the population. The EAPSO can be encoded as shown in (7), and only the matrix A with value of 1 is encoded:

$$A=\begin{bmatrix}0&\underline{1}&0&0&0\\ \underline{1}&0&\underline{1}&0&\underline{1}\\ 0&0&\underline{1}&0&\underline{1}\\ \underline{1}&0&0&0&0\\ 0&0&0&\underline{1}&\underline{1}\end{bmatrix} B=\begin{bmatrix}0&\underline{1}&0&0&0\\ \underline{0}&0&\underline{0}&0&\underline{1}\\ 0&0&\underline{1}&0&\underline{1}\\ \underline{1}&0&0&0&0\\ 0&0&0&\underline{1}&\underline{0}\end{bmatrix} \tag{7}$$

3.2. Initialization

Each row of population in the EAPSO represents a particle. The algorithm is discussed in the following description: In a Y-dimensional space, assuming there are X particles. X should be set according to the target coverage problem. The population can be represented as $B=\{b_1,b_2,\cdots\cdots,\ b_X\}$, and $b_x=\{b_{x,1},b_{x,2},\cdots\cdots,b_{x,y}\}$ represents the current target in the Y-dimensional search area.

3.3. Fitness Function

The optimization scheme to improve the coverage rate can be shown in formula (4).

3.4. Update speed and position

Suppose that S is the current speed of the particle, $S=\{s_1,s_2,\cdots\cdots,\ s_X\}$, $f=\{f_1,f_2,\cdots\cdots,f_X\}$, $x=\{1,2,\cdots\cdots,X\}$. f_x represents the position of particle, which can be also regarded as the number of successfully monitored targets of the x_{th} particle. The best position it has experienced is recorded as $l=\{l_1,l_2,\cdots\cdots,l_x\}$. f_g is the largest number of successfully monitored targets of the population,

IOP Conf. Series: Journal of Physics: Conf. Series **1267** (2019) 012006 doi:10.1088/1742-6596/1267/1/012006

which is an elite particle selected from f_x. The elite particles have advantage to improve the convergence speed. By constantly looking for f_x and f_g of particles, their speed and position is constantly updated through the following formula:

$$s_x(t+1) = ws_x(t) + z_1 d_1 (l_x(t) - f_x(t)) + z_2 d_2 (f_g(t) - f_x(t)) \tag{8}$$

$$f_x(t+1) = f_x(t) + s_x(t+1) \tag{9}$$

where $y = \{1, 2, \cdots\cdots, Y\}$. w is inertia weight coefficient. t represents iteration times; z_1 and z_2 are two constants, which mean the learning factor. This paper sets $z_1 = z_2$, so that the individuals have the same global and local search capabilities. d_1 as well as d_2 represent the random value between 0 and 1. The value of s_x is usually taken $s_x \in [-s_{\max}, s_{\max}]$, which is to ensure that particles are in the search area as much as possible during the evolution process.

In general, the monitoring relationship matrix is updated by a sigmoid mapping method, and the sigmoid function is as follows:

$$g(s_x) = \frac{1}{1 + e^{-s_x}} \tag{10}$$

$$b_{xy} = \begin{cases} 1 & if\ rand \le g(s_x) \\ 0 & otherwise \end{cases} \tag{11}$$

In (10) and (11), $rand$ is a random number between 0 and 1. $g(s_x)$ indicates the probability that b_{xy} is 1.

3.5. Termination Condition

The algorithm will be executed until the maximum iteration is reached.

3.6. Elite Operator

In EAPSO, we introduce the elite strategy, and complete the elite selection operation according to the number of successfully monitored targets. In the elite selection operation, the population is arranged in descending order, and then the optimal individuals are retained to form a new population and inherited to the next generation. In the EAPSO, the elite operation does not reduce the number of successfully monitored targets, which reduces the probability that the algorithm falls into local optimum.

3.7. Adaptive Operator

In EAPSO, the larger w is beneficial to the global search, and the smaller w is helpful to accelerate the convergence of the algorithm. The adaptive operator is used to adjust the value of w to improve the performance of the algorithm.

$$w(t+1) = \begin{cases} w(t)_{\max} \times u & f_x \ge f_{avg} \\ w(t)_{\min} \times u & f_x < f_{avg} \end{cases} \tag{12}$$

Where f_{avg} means the average number of successfully targets of the population, $w(t)_{\max}$ as well as $w(t)_{\min}$ is the best and lowest of the inertia weight coefficient. u means a constant, which takes 1.05. When f_x is greater than the average fitness, w is larger, the EAPSO prefers a wider range of search. Otherwise, the EAPSO prefers local search, which is good for balancing local search and global search.

4. Simulation and Results

In the experiment, we evaluate the number of targets of EAPSO with EA and SA for the problem of target coverage in HDWSNs. The performances of the proposed EAPSO scheme are investigated by simulation. The platform used for simulations is a Pentium IV machine with 2 GB RAM and Matlab is used as programming language.

By simulation experiments, we evaluate their performance with the objective function in section II. Then we generate sensor nodes and targets, and the position of each sensor node and each target is randomly specified within the square area, which takes 500×500. The number of targets is 200. Without considering the energy consumption by different sensing ranges, the sensing ranges from 70 to 80 is checked. The number of iterations in EAPSO is fixed at 200.

To illustrate the improvements provided by the EAPSO, the performance is compared with the EA and the SA. Its parameters of the EAPSO, EA and SA are given as following. In EAPSO, we fix the number of populations to 60 particles. The parameter values in the EAPSO are based on a parametric study. The maximum velocity is 3.8. The cognitive and social values take $z_1 = z_2 = 0.5$, w_{min} is 0.2. w_{max} is 0.9. In EA, the population have 60 individuals. Crossover rate is 0.75. Initial mutation rate is 0.08. Gap is 0.9. The temperature in the beginning of SA and the value of annealing temperature are 300 and 0.85, respectively.

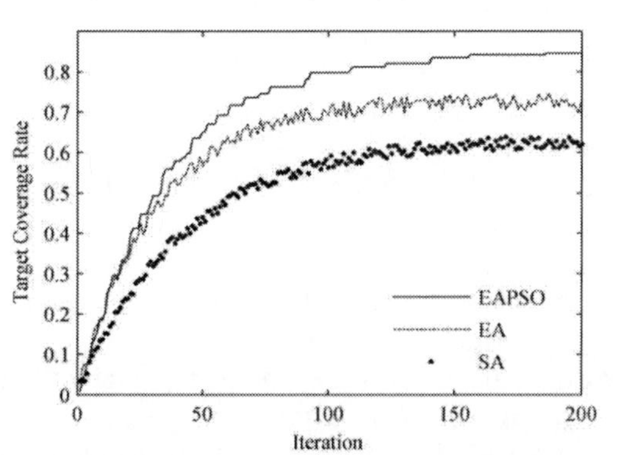

Figure 1. Sensing radius is 70m

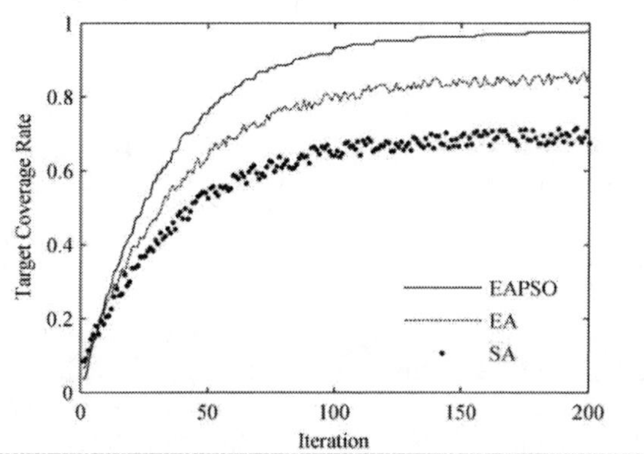

Figure 2. Sensing radius is 80m

Figure 1 shows that the target coverage rate obtained by EAPSO, EA and SA on the target coverage problem with different sensing radius when the amount of nodes is 200. For each method, the coverage rate in every generation is recorded in 200 simulation runs. After iterations, the target coverage rate in figure 1 is close to 0.85 and the target coverage rate in figure 2 is close to 0.96. Therefore, it is clear that the proposed EAPSO yield better results than the adaptive theory and elite theory in terms of the target coverage rate.

As it may be observed in figure 1, at the initial iterations, the number of the target coverage rate of all the heuristics has increased. EAPSO show the overall best behavior, while the EA and SA stagnate. Moreover, for the last 100 iterations, the performance of EAPSO becomes better than EA and SA. In particular, the fast convergence rate is shown clearly that EAPSO is prevented from premature convergence.

Figure3 shows the number of successfully monitored targets obtained by EAPSO, EA and SA for 200 iterations. It can be seen that the proposed EAPSO found its optimal value of 182 after 200 iterations. In comparison with the EAPSO, EA and SA provide suboptimal results, and the number of successfully monitored targets of EA and SA are 163 and 144, respectively. The results in figure 4 are similar as figure 3. Therefore, the performance of EAPSO is better than EA and SA.

5. Conclusion

Hence, an elite adaptive particle swarm optimization (EAPSO) for target coverage in HDWSNs is proposed. The EAPSO can improve system efficiency while maintaining total system sensing range in

IOP Conf. Series: Journal of Physics: Conf. Series **1267** (2019) 012006 doi:10.1088/1742-6596/1267/1/012006

the area. A system model is built to maximize the target coverage rate. Simulations display the performance of the EAPSO compared to EA and SA. The results demonstrate that the EAPSO method provides a larger amount of successfully monitored target over the current EA and SA methods.

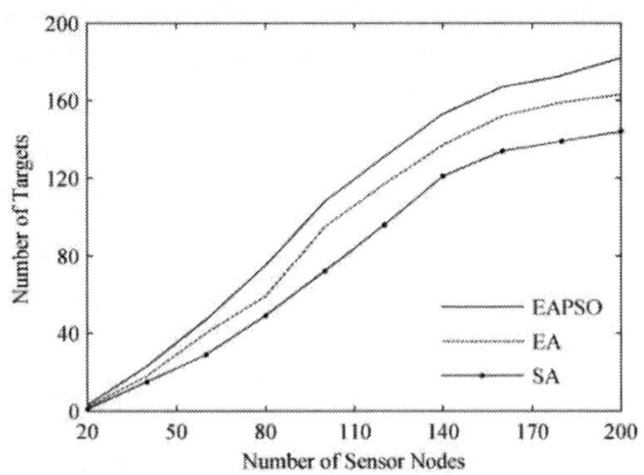

Figure 3. Sensing radius is 70m

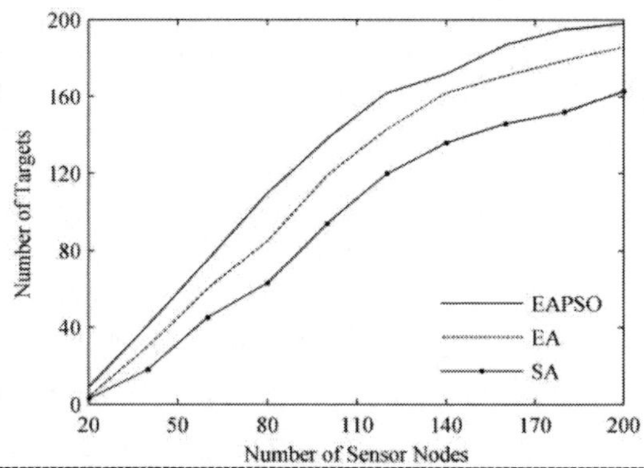

Figure 4. Sensing radius is 80m

Acknowledgments
This paper is supported by the National Nature Science Foundation of China (No. 61662063), the project of Youth and Middle-aged Scientific and Technological Innovation Leading Talents Program of the Corps (No. 2018CB006), the project of Promoting Scientific Research Cooperation and High-level Personnel Training with Americas and Oceania, the China Postdoctoral Science Foundation.

References
[1] Wang C, Li J and Yang Y 2017 Combining solar energy harvesting with wireless charging for hybrid wireless sensor networks. *IEEE Transactions on Mobile Computing* **17** pp 560-76.
[2] Tolani M, Sharma S and Singh R 2017 Two-layer optimized railway monitoring system using Wi-Fi and ZigBee interfaced wireless sensor network. *IEEE Sensors Journal* **17** pp 2241-48.
[3] Yang X, Wang L and Su J 2018 Hybrid MAC protocol design for mobile wireless sensors networks. *IEEE Sensors Letters* **2** p 1.
[4] Hasan M M and Mouftah H T 2017 Optimization of watchdog selection in wireless sensor networks. *IEEE Wireless Communications Letters* **6** pp 94-7.
[5] Kopta V, Barras D and Enz C C 2017 An approximate zero IF FM-UWB receiver for high density wireless sensor networks. *IEEE Transactions on Microwave Theory and Techniques* **65** pp 374-85.
[6] Hajjar M, Aldabbagh G and Dimitriou N 2017 Hybrid clustering scheme for relaying in multi-cell LTE high user density networks. *IEEE Access* **5** pp 4431-38.
[7] Ren X, Liang W and Xu W 2015 quality-aware target coverage in energy harvesting sensor networks. *IEEE Transactions on Emerging Topics in Computing* **3** pp 8-21.
[8] Chaturvedi P and Daniel A K 2017 Hidden markov model based node status prediction technique for target coverage in wireless sensor networks. International Conference on Intelligent Communication and Computational Techniques. *IEEE Computer Society*.
[9] Yu J, Chen Y, Ma L, et al 2016 On connected target k-coverage in heterogeneous wireless sensor networks. *Sensors* **16** p 104.
[10] Njoya A N, Abdou W, Dipanda A, et al 2015 Evolutionary-based wireless sensor deployment for target coverage. *SITIS. IEEE.*
[11] Venkatesh S, Mehata K M 2014 A competent fault tolerant system using simulated annealing approach for target tracking in wireless sensor networks. *Int Conf on Computer Communication and Informatics. IEEE.*

A Class-Incremental Learning Method Based on One Class Support Vector Machine

Chengfei Yao, Jie Zou, Yanan Luo, Tao Li and Gang Bai*

College of Computer Science, Nankai University, Tianjin, China

*E-mail: baigang@nankai.edu.cn

Abstract. Class-incremental learning refers to the problem that the number of classes increases dynamically in the training stage. In this paper, a method based on one class support vector machine (OCSVM) is proposed for class incremental learning. The basic idea of the proposed method is that one class model is used to learn the distribution of new class, and then the knowledge learned and previous models are used for class incremental learning. The proposed method does not need to store all training data and makes full use of the learned information, so it has lower memory cost and time consumption. Also, the proposed method can build model when only one class is available. Compared with OC2MC, CIL, NN, the proposed method also achieves competitive results on different datasets.

1. Introduction

Incremental learning is a hot research topic of machine learning. Incremental learning fully retains the existing model's knowledge and modifies the initial model using the knowledge learned from the new samples. The great significance of incremental learning is that when new samples are added, there is no need to retrain with all training samples, reducing memory consumption and training time cost. In terms of the classes which the new samples belong to, there are two types of incremental learning. The common type of incremental learning is that the new training samples belong to the old classes. Another is the new training samples belong to the new class, called class-incremental learning (CIL). Researchers have achieved rich results in the former [1] [2] [3]. Research on CIL is limited. CIL has a wide range of applications in the classification scene that the number of classes changes dynamically, such as text classification, sound classification. In this paper, we focus on proposing a new strategy for CIL.

A simple strategy for CIL is to use the nearest neighbour (NN) method. The flaw of NN is that all training samples are still used. [4] proposed a CIL method for SVM. One binary classifier is constructed by using the new class and all the old classes. However, this method also requires all training samples. Some researchers are devoted to solving class incremental learning by establishing multi-class classifiers through several one-class classifiers [5] [6]. [5] proposed a method for multi-classification based on OCSVM. This method trains several OCSVM models and uses the defined dissimilarity to determine the ownership of the test samples.

In this paper, we propose a class-incremental learning method based on OCSVM. The basic idea is that OCSVM models of different classes divide the feature space into different parts, which can be controversial and non-controversial parts. If only one predicted label of the sample on a part is positive, the part located by this sample is non-controversial. Otherwise it is controversial. If the enclosed surfaces of OCSVM are suitable, the hyper-plane fit well with the shape of the distribution. The support vectors held by OCSVM are used to build classifiers for handling disputes. The main contribution of our method is: 1) to make the classification process easier to explain by introducing OCSVM into class incremental learning. 2) to make support vectors as close as possible to the boundaries of the distribution by introducing an algorithm that adjusts the width of the Gaussian kernel.

The remining sections of this paper is organized as follows. Section 2 briefly introduces related work. Section 3 makes a complete presentation of the proposed method. Section 4 compares our method with other methods. Finally, section 5 summarizes this paper and looks forward to future work.

2. Related work

One-class classification builds model using only positive samples and determines whether test samples belong to the training class or not. OCSVM is proposed in 2001 to solve OCC [7]. OCSVM find a hyper-plane in high-dimensional space to separate the target sample and the origin as far as possible. The hyper-plane can be imagined as an enclosed surface of the sample, just similar to the minimum enclosing ball [8]. To separate the data from the origin, we should solve the following program:

$$\begin{aligned} &\min_{\omega\in F,\xi\in\mathbb{R}^l,\rho\in\mathbb{R}} \frac{1}{2}\|\omega\|^2 + \frac{1}{vl}\sum_i \xi_i - \rho \\ &\text{subject to } (\omega\cdot\phi(\mathbf{x}_i)) \geq \rho - \xi_i, \xi_i \geq 0 \end{aligned} \quad (1)$$

If the decision value is greater than 0, it is judged that the sample belongs to the class, otherwise it does not belong to the class.

Parameter selection, especially kernel parameter, is significant for OCSVM. There are many attempts to select fine-tuned parameters for OCSVM with RBF kernel function [9] [10] [11]. MIES algorithm is a parameter selection algorithm of Gaussian kernel for OCSVM [12]. The basic idea is that the hyper-plane of OCSVM is imagined as the enclosing surface of the samples. If the surfaces are appropriate, they should be as close as possible to the boundary point and as far as possible from the inner point.

MIES algorithm achieves the above idea by measuring the distance between the enclosing surface and the different points: internal point and the boundary point. The author adds the border-edge pattern selection method (BEPS) into MIES [13]. Equation (2) is the objective optimization function of the MIES algorithm. Because of the optimization goal of the MIES algorithm, the OCSVM with parameters determined by the MIES algorithm has a good geometric interpretation in the input space.

$$f_0(s) = \max_{\mathbf{x}_i\in\Omega_{IN}} d_N(\mathbf{x}_i) - \max_{\mathbf{x}_j\in\Omega_{ED}} d_N(\mathbf{x}_j) \quad (2)$$

3. Our method

3.1. Train OCSVM

To establish an OCSVM with Gaussian kernel, it is mainly necessary to determine two parameters s and v. The parameter s is the width of Gaussian kernel, and the parameter v mainly controls the fraction of samples against hyperplane in high dimensional feature space. In the proposed methods, OCSVM has two training objectives: one is to make SVs as close to the sample distribution boundary as possible, and another is to ensure the classification performance of the model. In order to achieve the above goal, we need to choose a suitable set of parameters (v,s) in the parameter space. Since the parameter v is only related to the proportion of the support vectors, the choice of the parameter s is crucial. We modified the optimization goal of MIES to make it more responsive to the proposed method.

As shown in figure 1, the optimization goal of the MIES algorithm is to make the enclosing surface as close as possible to the boundary points and as far as possible from the internal points. In fact, MIES does not require that the enclosing surface be far from every internal point. As a result, it is not guaranteed that the support vectors are on the boundary of the distribution. Therefore, in order to get the required support vectors, we change the optimization goal of parameter selection of Gaussian kernel to be as close as possible to the boundary point. The objective optimization function is as equation (3).

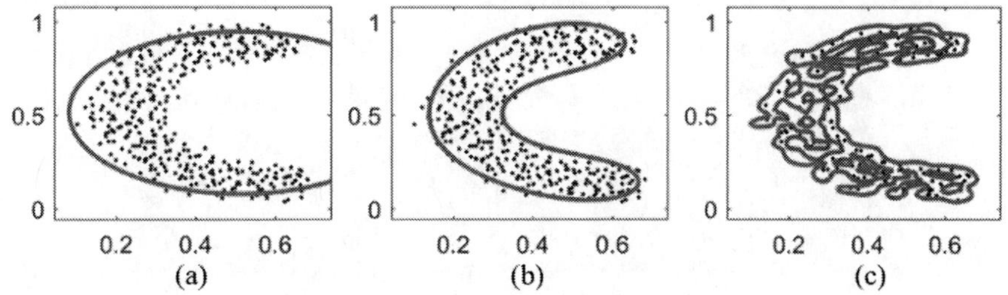

Figure 1. The comparison of enclosing surface determined by different parameter s. (a)The enclosing surface is too loose. (b)The enclosing surface is suitable. (c) The enclosing surface is too tight.

$$s = \arg\min\left(\max_{\mathbf{x}_i \in \Omega_{ED}} d_N(\mathbf{x}_i)\right) \tag{3}$$

In fact, as long as almost all of the selected boundary samples are support vectors, the optimization function can be satisfied. We limit the ratio of support vectors to prevent all samples from becoming support vectors as the value of parameter s is small. With a large number of experiments, this ratio is preferably within the range of v to 1.5v. As shown in Figure 2, the support vector obtained by the original MIES and the MIES modified through the above process (abbreviated as Modified MIES) is compared. The support vectors obtained by the modified MIES are more evenly distributed on the sample distribution boundary. The complete process of training OCSVM is shown in Algorithm 1.

Algorithm 1: Train OCSVM

Input: the training set, ν
Output: the OCSVM model
determine the width of Gaussian kernel s according to modified MIES:
    select the boundary point from the training set according to BEPS algorithm.
    exclude the candidates of s if the fraction of support vector exceeds [ν, 1.5 *ν].
    select s according to (3) from the left candidates of s.
Train OCSVM based on parameter ν and s.

3.2. Reuse SV to Train Multi-Classification Model

In practice, the distribution of different classes often has overlapping parts, so we need to establish a strategy to deal with the disputes. The strategy adopted to handle disputes is to establish SVMs by reusing these support vectors held by one-class models. SVM are generally related to boundary samples. These support vectors contain the main information needed for multi-classification. Algorithm 2 describes the detailed process of training and testing of class incremental learning.

Similar to the paper [3], we have adopted the method to establish a classifier between new and old classes to determine which test sample belongs to. The difference is that our method uses only support vectors of each class instead of retaining all training samples. Then when a new class comes, what needs to do is to train the OCSVM for this class and then train the 1VS1 SVM by using this class's support vectors and the support vectors of the old classes.

In the testing phase, first the OCSVM is used to determine whether the test sample falls into the controversial part or not. In all predicted labels of OCSVM, if only one is positive, the test sample is considered to be in the uncontroversial part. If more than one or none is positive, then it is considered to be in the disputed part. When the test sample is in the non-controversial part, the unique class label is directly assigned to the test sample. When the test sample is in the controversial part, the first step is to get the classes to which the test sample may belongs. Then support vectors of these classes are used to predict the label of test sample according to the corresponding SVM classifiers. Because the method is based on the distribution of samples, we call it SD-CIL.

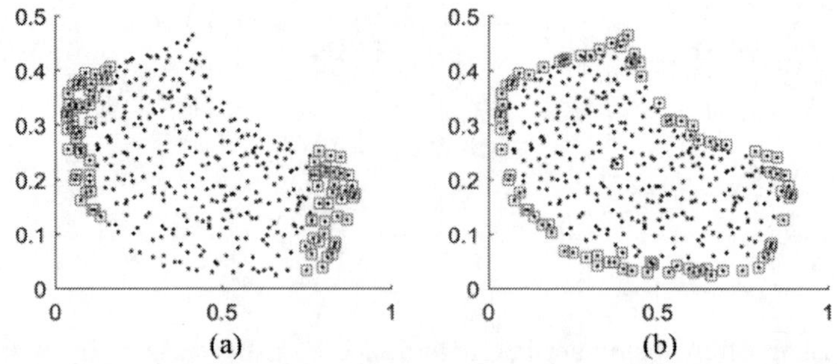

Figure 2. The comparison of support vectors determined by MIES and modified MIES. (a)MIES. (b)Modified MIES. The " • " denotes the training samples. The "□" represents support vectors.

Algorithm 2: the SD-CIL method

1: Training stage
When a new class is coming
  Train OCSVM for the class according to Algorithm 1.
  Train binary SVM for the new class and old classes by using the support vectors.

2: Test stage
Send the test sample into all OCSVM for testing.
If only one prediction is positive,
  the test sample belongs to this class.
elseif more than one is positive,
  regard a class that has a positive label as a possible class.
  get the label of the test sample using binary SVM of the possible classes.
else none is positive,
  regard all classes as possible.
  get the label of the test sample using binary SVM of the possible classes.
end if

4. Experiment

In order to test the performance of our method, experiments on different datasets are conducted. The basic information of these datasets is shown in Table 1. We compare the performance of SD-CIL with CIL, NN and the CIL method based on OCSVM in [3] (abbreviated as OC2MC).

4.1. Experiments Setup

In all of the following experiments, the parameters of our method are set as follows. When building OCSVM, the value of the parameter v is uniformly 0.3 on all datasets. The parameter s is determined by the modified MIES algorithm automatically. In the class incremental learning process, the grid search method is used for SVM. For OC2MC, the default parameters are used. For NN, the number of neighbours k is searched. For CIL, the grid search method is applied to find the optimal parameters.

4.2. Experiments on Toy Datasets and UCI Datasets

Experiments are first conducted on toy datasets. The classification map identified by SD-CIL are shown in Figure 3. Obviously, the support vectors determined by the modified MIES algorithm distribute on the boundary of the distribution. The classification map of SD-CIL is made up of two parts. The uncontroversial part is determined by the OCSVM and the controversial part is derived from the binary classifier. The results show that this composition is reasonable.

To test the performance of our method SD-CIL in practice, several UCI datasets are selected. The 70% samples of each class are used for training, and the remaining 30% are used for testing. Under the same division, we compare the accuracy of different methods by changing the number of classes from 1 to 3. We run 10 times on each dataset, and the mean accuracy are shown in Table 2. The results show that SD-CIL works well across all those datasets. SD-CIL achieves higher accuracy in the process of class incremental learning, better than other methods NN, CIL and OC2MC. Unlike NN and CIL, SD-CIL does not need to keep all the samples in the CIL process and makes full use of the previous model. In the case of only one class, only SD-CIL and OC2MC can work. In order to fully reflect the quality of the one-class model, the accuracy is calculated as$\sqrt{a^+ \cdot a^-}$, where a^+ indicates that the positive samples

Table 1. The basic information of datasets

dataset	Dimension	Training patterns	Test patterns
Toy Dataset1	2	783	336
Toy Dataset2	2	954	408
Iris	4	105	35
Seeds	7	147	63
wine	13	125	53
Waveform21	21	3500	1500
Waveform40	40	3500	1500

Table 2. Classification result on different datasets

	dataset	SD-CIL	OC2MC	CIL	NN
three classes	Toy dataset1	**0.9940**	0.9911	0.9911	0.9881
	Toy dataset2	**0.9951**	0.9926	**0.9951**	0.9877
	Iris	**1**	0.9778	**1**	0.9778
	Seeds	**0.9524**	**0.9524**	0.8413	0.9206
	Wine	**0.9811**	0.9434	**0.9811**	0.9434
	Waveform21	**0.8753**	0.8587	0.8547	0.8520
	Waveform40	**0.8560**	0.8493	0.8480	0.8313
two classes	Toy dataset1	**0.9953**	**0.9953**	**0.9953**	**0.9953**
	Toy dataset2	**1**	**1**	**1**	0.9965
	Iris	**1**	0.9667	**1**	**1**
	Seeds	**1**	0.9762	0.9286	0.9762
	Wine	**1**	0.9487	**1**	0.9231
	Waveform21	**0.9273**	0.9082	**0.9273**	0.9051
	Waveform40	**0.9243**	0.9183	0.9293	0.9213
one class	Toy dataset1	**0.8246**	0.7669	-	-
	Toy dataset2	**0.8182**	0.7467	-	-
	Iris	**0.6663**	0.5925	-	-
	Seeds	**0.7183**	0.5935	-	-
	Wine	**0.7124**	0.2387	-	-
	Waveform21	0.6961	**0.7225**	-	-
	Waveform40	**0.6752**	0.6225	-	-

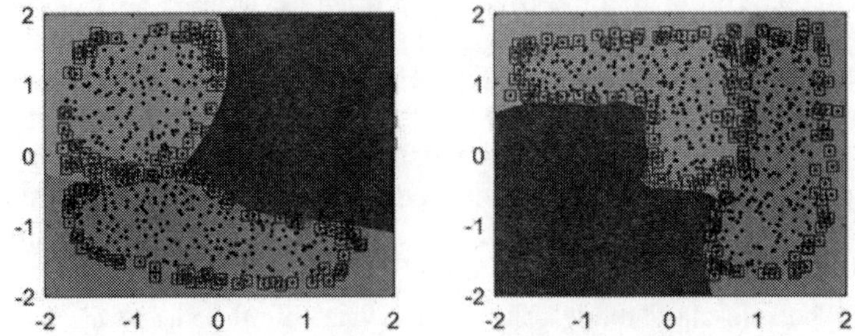

(a) (b)

Figure 3. Performance of SD-CIL on Toy Dataset. (a) Toy Dataset1. (b) Toy Dataset2. The " • " denotes the samples of different classes. The "□" represents support vectors of different classes.

are judged to be true (TP) and a^- is the negative samples to be false (FN). Because SD-CIL have taken full account of the geometric information of the distribution, it has obtained better performance.

5. Discussion

In this paper, a method SD-CIL is proposed for class-incremental learning. During the training stage, the distribution is studied by using the OCSVM. The parameter selection method modified MIES ensures that the support vectors stored by OCSVM carry sufficient information. Based on this, a class-incremental learning method from one class to multiple classes is established. These steps constitute a division of the input space. During the testing stage, using OCSVM to get the class that the test samples might belong to can also simplify the testing process for many samples.

Compared with other methods, the key advantages of SD-CIL are as follows: 1) SD-CIL do not need to store all the previous training samples. 2) SD-CIL is suitable for case where there are only one class. 3) Most test samples can get proper label by using only OCSVM, simplifying the testing process. The results on different datasets show that the proposed method is effective.

Acknowledgments

This work is partially supported by the National Natural Science Foundation (61872200), the National Key Research and Development Program of China (2016YFC0400709, 2018YFB1003405), the Natural Science Foundation of Tianjin (18YFYZCG00060).

References

[1] Syed N A, Liu H and Sung K K 1999 Handling concept drifts in incremental learning with support vector machines. *Acm Sigkdd International Conference on Knowledge Discovery & Data Mining*. ACM.

[2] Cauwenberghs G 2000 Incremental and Decremental Support Vector Machine Learning. *International Conference on Neural Information Processing Systems*. MIT Press.

[3] Diehl C P and Cauwenberghs G 2003 SVM incremental learning, adaptation and optimization. *International Joint Conference on Neural Networks*. IEEE.

[4] Zhang B F, Su J S and Xu X 2009. A Class-Incremental Learning Method for Multi-Class Support Vector Machines in Text Classification. *International Conference on Machine Learning & Cybernetics*. IEEE.

[5] Rabaoui A, Davy M, Rossignol S, Lachiri Z and Ellouze N 2007 Improved one-class SVM classifier for sounds classification. *IEEE Conference on Advanced Video & Signal Based Surveillance*.

[6] Qin Y, Karimi H R, Li D, Lun S and Zhang A 2014 A mahalanobis hyperellipsoidal learning machine class incremental learning algorithm. *Abstract and Applied Analysis,2014*, 1-5.

[7] Scholkopf B 2014 Estimating the support of a high-dimensional distribution. *Neural Computation, 13*(7), 1443-1471.

[8] Yildirim E A 2008 Two algorithms for the minimum enclosing ball problem. *Siam Journal on Optimization, 19*(3), 1368-1391.

[9] Deng H and Xu R 2007 Model Selection for Anomaly Detection in Wireless Ad Hoc Networks. *IEEE Symposium on Computational Intelligence & Data Mining*. IEEE.

[10] Khazai S, Homayouni S, Safari A and Mojaradi B 2011 Anomaly detection in hyperspectral images based on an adaptive support vector method. *IEEE Geoscience & Remote Sensing Letters, 8*(4), 646-650.

[11] Wang S, Yu J, Lapira E and Lee J 2013 A modified support vector data description based novelty detection approach for machinery components. *Applied Soft Computing,13*(2), 1193-1205.

[12] Xiao Y, Wang H and Xu W 2015 Parameter selection of gaussian kernel for one-class SVM. *IEEE Transactions on Cybernetics, 45*(5), 927-939.

[13] Li Y and Maguire L 2011 Selecting critical patterns based on local geometrical and statistical information. *IEEE Transactions on Pattern Analysis & Machine Intelligence, 33*(6), 1189-201.

IOP Conf. Series: Journal of Physics: Conf. Series **1267** (2019) 012008 doi:10.1088/1742-6596/1267/1/012008

A Novel Approach to Improve the Collaborative Representation for Face Recognition

YuJie Jia[1], JiaLu Deng[1], Xu Zhang[2], Qiong Wang[2] and Shi Yan[1,*]

[1] School of Information Science and Engineering, Lanzhou University, Lanzhou, 730000, China
[2] State Grid Gansu Information & Telecommunication Company, Lanzhou, 730000, China

E-mail: jiayj17@lzu.edu.cn; dengjl17@lzu.edu.cn; zhangxu@gs.sgcc.con.cn; wangqiong-xt@gs.sgcc.con.cn; m51@163.com*

Abstract. Collaborative representation based classification (CRC) codes a test sample as a linear combination of all the training samples. However, the recognition rate of CRC is not ideal when available training samples are insufficient, as the test sample cannot be accurately represented by limited training samples. In order to address this issue, we propose a novel idea of training samples fragmentation. First, all training samples are divided into two new training sample sets according to the similarity among them. Next, the test sample simultaneously uses the two new training sample sets to perform CRC, which ultimately uses $2M$ "nearest neighbors" from the two training sample sets to represent and classify the test sample. In addition, this method also takes advantages of a new fusion classification mechanism based on histogram similarity and Euclidean distance, which has been proven to perform better than Euclidean distance classification. The experimental results reveal that the proposed method performs better in face recognition compared with the most representation based classification methods.

1. Introduction

Face Recognition (FR) is a challenging but interesting question that attracts researchers from different fields and various techniques have been proposed to deal with different issues, such as illumination, pose, occlusion and small sample size [1]. A typical face recognition method involves two steps: feature extraction and classification. Since Wright et al. [2] proposed a face recognition method based on sparse representation based classification (SRC), SRC and its variants [3,4] have been widely used in face recognition. In the SRC, a test sample is linearly reconstructed from all training samples and classified based on the reconstruction errors associated with each class. Unlike traditional classification methods, SRC does not require feature extraction.

Recent investigations have shown that CRC [5] has achieved excellent results in face recognition by using l2 sparse constraint in the optimization problem. In the practical face recognition system, the number of available training samples is generally limited such that face recognition becomes a small sample size (SSS) problem. In order to solve this, Xu et al. [6] used symmetry of the image to generate two symmetrical virtual facial images to increase the number of training samples and then the recognition rate was greatly increased. Similarly, Tang et al. [7] used an optic flow and expression ratio image to generate 'virtual' facial expression. In addition to generating virtual samples, many researchers are committed to improving the CRC. For example, Dong et al. [8] proposed a two-stage learning algorithm, which mainly used unlabeled samples to reconstruct marker samples to obtain collaborative representation coefficients. Tian et al. [9] proposed a new fusion mechanism, which combines CRC and SRC and has the advantages of both l2 sparse constraint and l1 sparse constraint in the optimization problem.

In this paper, we propose a novel approach to improve the collaborative representation for face recognition. All training samples are firstly divided into two new training sample sets according to the similarity among them, and then the test sample simultaneously uses the two new training sample sets to perform CRC, which ultimately uses $2M$ "nearest neighbors" from the two training sample sets to represent and classify the test sample. The aim of our method is to reduce the interference of these training samples that are not same class as the test sample. Beside, since the Euclidean distance treats different attributes of samples equally, which leads to the same contribution of the residuals in different positions of two images. This sometimes fails to meet the actual requirements. Therefore, a new fusion classification mechanism based on histogram similarity and Euclidean distance is used to judge the residual. Experimental results show that the proposed method performs well in face recognition compared with the most representation-based methods.

2. The proposed method

Assume that there are c classes in all training samples, all training samples are $X=[X_1,X_2,X_3,\cdots,X_c]$, where $X_i \in R^{m\times n_i}$, n_i is the number of i class training samples.

The purpose of using the inter-class scattering is to extract the most discriminative low-dimensional features from the high-dimensional feature space, which can help different classes of samples to be separated as much as possible. The inter-class dispersion matrix of the training samples is calculated by

$$S_b=\sum_{i=1}^{c}(X_i-\overline{X})(X_i-\overline{X})^T, \quad (1)$$

where $\overline{X}$ is the average face of all training samples and X_i is the average face of the i -class training sample. Then the classes of the training samples are sorted according to the size of the discrete values. Finally, all training samples are cross-divided into two new training sample sets.

The training samples X are divided into two new training sample sets, i.e., $X=X_1\cup X_2$ ($c=c_1+c_2$, $X_1=[x_{11},x_{12},\ldots,x_{1n\times c_1}]$, and $X_2=[x_{21},x_{22},\ldots,x_{2n\times c_2}]$). Given the test sample y and two new training sets X_1 and X_2, the model of the CRC can be expressed as

$$\alpha_1=\arg\min_{\alpha_1}\left(\|y-X_1\alpha_1\|_2^2+\lambda\|\alpha_1\|_2^2\right) \quad (2)$$

and

$$\alpha_2=\arg\min_{\alpha_2}\left(\|y-X_2\alpha_2\|_2^2+\lambda\|\alpha_2\|_2^2\right), \quad (3)$$

where λ is a regularization parameter, and α_1 and α_2 are collaborative representation coefficients that linearly represent test sample y by X_1 and X_2.

The role of the regularization term is twofold: it stabilizes the solution of the least squares method and introduces some sparsity. The collaborative representation based on the regularized least squares solution can be derived as

$$\alpha_1=P_1y \quad (4)$$

and

$$\alpha_2=P_2y, \quad (5)$$

where $P_1=\left(X_1^TX_1+\lambda I\right)^{-1}X_1^T$, $P_2=\left(X_2^TX_2+\lambda I\right)^{-1}X_2^T$, $\alpha_1=[a_{11}..a_{1c_1}]$ and $\alpha_1=[a_{21}..a_{2c_2}]$.

For the residuals of the test sample and the reconstructed test samples from the two new training sample sets, a new fusion mechanism is proposed by

$$d_k = \left\| y - \sum_{i=(k-1)n+1}^{kn} a_{1i} x_{1i} \right\| + \sum_{i=(k-1)n+1}^{kn} \sqrt{pp'_{1i}} \tag{6}$$

and

$$d_q = \left\| y - \sum_{i=(q-1)n+1}^{qn} a_{2i} x_{2i} \right\| + \sum_{i=(q-1)n+1}^{qn} \sqrt{pp'_{2i}} \tag{7}$$

where p , p_1' , and p_2' represent histogram data of the test sample y , q - th training sample and k - th training sample, respectively, and the range is between [0, 1]. The closer value of the histogram similarity is to 0, the more similar the two images are.

The two residual sets are arranged in ascending order, and each of the first M residuals is taken. The class of $2M$ residuals is denoted as r_1 - th, r_2 - th,..., r_{2M} - th , and are recomposed into a new training sample set X' . The test sample y is linearly represented by the new training set X' as

$$y = f_1 x'_{r_1} + \ldots + f_{2M \times n} x'_{r_{2M} \times n} , \tag{8}$$

where f_i is the coefficient. The equation (8) is rewritten as

$$y = X'F , \tag{9}$$

where F is calculated by $F = \left(X'^T X + \gamma I\right)^{-1} X'^T$ with γ being a small positive constant and I also denotes the identity matrix. The ultimate effect on representing the test sample of the h - th class can be evaluated by

$$u_h = \left\| y - \sum_{i=(h-1)n+1}^{hn} f_i \bar{x}'_i \right\| . \tag{10}$$

If $m = \arg\min_h (u_h)$, then the test sample y is finally assigned to the m class.

3. Insight into the proposed algorithm

In this section, we give a detailed analysis to the proposed algorithm. CRC employs all training samples to linearly represent the test sample and then determines which class of training samples leads to the largest contribution of the test sample. However, a limited number of available training samples have become one bottleneck of CRC. These training samples cannot accurately reconstruct the test sample, which leads to the interference from the training samples that are not same class as the test sample. In order to avoid this kind of interference, we calculate the similarity between the training samples and put them into two new training sample sets, which lets the test sample to find its class more easily and quickly.

Take the ORL database as an example, which is a classic face database commonly used in face recognition. The second image in the ORL is used as the test sample, and the first one in each class of image is used as the training sample. **Figure 1** shows the five training samples with the closest to the test sample and test sample's coding coefficients by the two-step classification without virtual samples. **Figure 2** shows the five training samples with the closest to the test sample and test sample's coding coefficients by our method. Because the test sample is assigned to the class of minimum residual, the second image in the ORL is recognized wrongly in the two-step classification without virtual samples [8]. However, this image can be easily identified correctly by our method.

We add the idea of training samples fragmentation to the two-step classification. Compared with the two-step classification, our method reduces the interference of the training samples that are not same class as the test sample, so the recognition rate has been improved. **Figure 3** shows our method compared with the CRC [5] and the two-step classification without virtual samples in the FERET database [8].

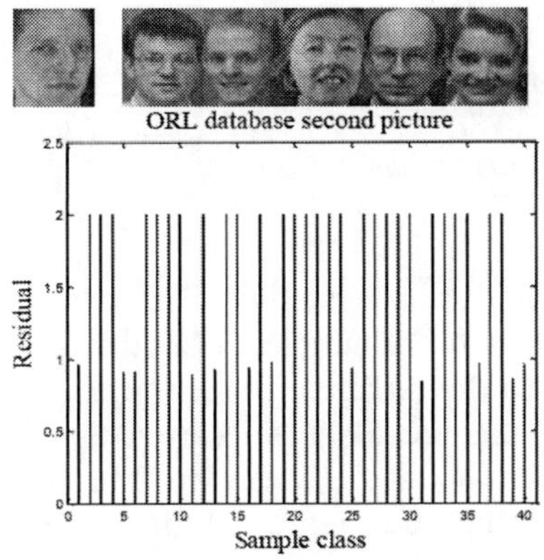

Figure 1. The closest 5 classes of face images and residuals in the second image of the ORL database by the two-step classification.

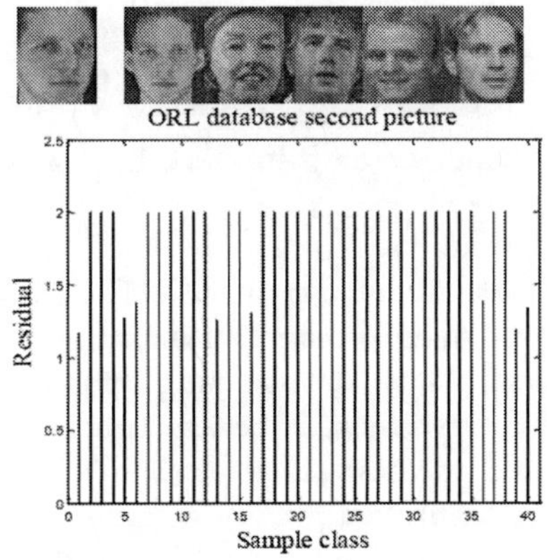

Figure 2. The closest 5 classes of face images and residuals in the second image of the ORL database by our method.

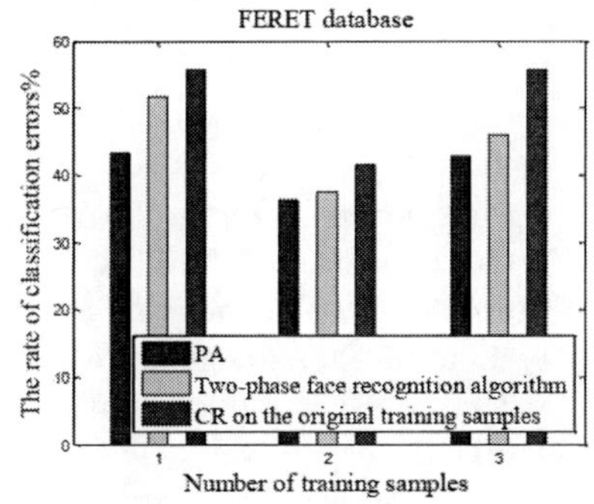

Figure 3. The rate of classification errors of different methods conducted on FERET database(%).

In addition to the inter-class scattering, histogram similarity and the new fusion method can also measure the similarity between each class of training samples. Histogram similarity obtains similarity values only by comparing the histogram information of each class of training samples. Therefore, The new fusion mechanism (6) based on histogram similarity and Euclidean distance not only contains histogram information, but also makes Euclidean distance supplement the similarity value. The recognition rates of them are better than the inter-class scattering, but their time and computational complexity are higher. Detailed experimental results are shown in Section 4.

4. Simulation Results

We conduct face recognition on three different databases including ORL, GT, and FERET database to test our method. In these training databases, our method not only effectively compensates for the defect of Euclidean distance, but also reduces the interference of these training samples that are not same class as the test sample. Besides, to show the performance of our proposed algorithm (PA), single-scale pixel processing method (MSA)[10] , linear Regression classification(LRC) [11], RBTM [12], CIRLRC [13], NFRFR [14], KRBM [15], FSSP [9], DSSR [16] are used as comparisons.

4.1. Experiments on the ORL database

The ORL database contains images from 40 people, and each person provides 10 images. **Figure 4** shows one person from the ORL face database. The first 3 samples of each class are used as training samples, and the rest are used as test samples. The experimental results are shown in **Table 1** . The training samples fragmentation reduces the interference of these training samples that are not same class as the test sample, and the new fusion mechanism compensates for the defect of Euclidean distance, which makes PA have advantage over the most representation-based methods.

Table 1. The rate of classification errors of different methods on ORL database(%)

Training sample per class	PA(inter-class scattering)	PA (histogram)	PA (new fusion method)	MSA	LRC	RBTM	CIRIRC	NFRFR	KRBM	DSRR	FSSP
1	20.00	20.28	20.00	23.06	32.25	31.94	26.39	29.44	27.22	28.06	28.62
2	8.44	7.50	6.88	12.15	20.62	20.62	13.12	12.81	11.87	13.13	14.69
3	6.43	4.64	5.71	11.43	18.57	21.79	10.36	10	10.71	10.36	10.36

Figure 4. One person from the ORL face database.

4.2. Experiments on the FERET database

We use a subset of the FERET database to test our method. The FERET database contains 700 images from 100 people, and each person provides 7 images of different poses and illuminations. Let the first 1, 2, and 3 face images of each person be used as training samples, and the remaining face images as test samples. The images of FERET database and the experimental results are shown in **Figure 5** and **Table 2** respectively. The proposed method shows strong robustness to angle variations, so the recognition accuracy rates has been improved greatly compared with the most representation-based methods.

Figure 5. One person images from the FERET database

Table 2. The rate of classification errors of different methods on FERET database(%)

Training sample per class	PA(inter-class scattering)	PA (histogram)	PA (new fusion method)	MSA	LRC	RBTM	CIRIRC	NFRFR	KRBM	DSRR	FSSP
1	37.50	38.33	37.50	47.18	54	49.83	63.08	47.17	51.33	43.67	50.83
2	30	28.80	29.88	33	35	36.60	44.50	36.20	38.60	49.50	62.30
3	40	37.05	38.25	40.25	39.95	42.75	45.87	33.75	46.25	40.50	53.50

4.3. Experiments on the GT database

The GT database includes images of 50 people. Everyone has 15 color images and the background is complex. These images show the front side with different expressions, lighting conditions and angles. **Figure 6** shows an example of the image from the GT database. The first 3 images of each person are used as training samples, and the other images are test samples. The experimental results show in **Table 3**. Because the histogram information of the images is greatly helpful to improve the recognition rate of GT database, the proposed method fuses histogram similarity and Euclidean distance to respectively compare with the most representation-based methods, so that it obtains better performance.

Figure 6. The images from the GT database

Table 3. The rate of classification errors of different methods on GT database(%)

Training sample per class	PA(inter-class scattering)	PA (histogram)	PA (new fusion method)	MSA	LRC	RBTM	CIRIRC	NFRFR	KRBM	DSRR	FSSP
1	49.86	50	49.86	64.11	64.57	66.17	60.71	64.00	66.57	64.57	63.72
2	39.85	40.62	39.38	52.62	54.15	59.85	54.77	53.38	59.69	52	50.47
3	37.00	36.67	37.83	51.33	49	60.17	50.67	51.83	51.33	47	45.67

5. Conclusions

In this paper, we have proposed the ideas of training samples fragmentation and new classification fusion mechanism. They are helpful for overcoming the drawback of limited training samples in the practical face recognition system. The training samples fragmentation has reduced the interference

IOP Conf. Series: Journal of Physics: Conf. Series **1267** (2019) 012008 doi:10.1088/1742-6596/1267/1/012008

from those training samples that are not same class as the test sample, and three methods of training samples fragmentation have measured the similarity between the training samples well. Furthermore, the new classification fusion mechanism has compensated for the weak point of Euclidean distance. These have increased the recognition rate of the CRC. The experimental results show that the proposed method is superior to the most representation-based methods.

Acknowledgments

This article is partly supported by the Natural Science Foundation of Gansu Province of China (18JR3RA286), the Fundamental Research Funds for the Central Universities (lzujbky-2018-128), CERNET IPv6 Innovation Project (NGII20170602), Key Laboratory for Research of Design and Electromagnetic Compatibility of High Speed Electronic Systems, Ministry of Education No.2017001.

References

[1] W. Zhao, R. Chellappa, P.J. Phillips, A. Rosenfeld, Face recognition: a literature survey, ACM Comput. Surv. 35 (4) (2003) 399–458.

[2] J. Wright, A.Y. Yang, A. Ganesh, S.S. Sastry, Y. Ma, Robust face recognition via sparse representation, IEEE Trans. Pattern Anal. Machine Intell. 31 (2) (2009) 210–227.

[3] M. Yang, L. Zhang, J. Yang, D. Zhang, Robust sparse coding for face recognition, in: Proc. IEEE Conf. Computer Vision and Pattern Recognition (CVPR), 2011, pp. 625–632.

[4] K. Hui, C. Li, L. Zhang, Sparse neighbor representation for classification, Pattern Recogn. Lett. 33 (5) (2012) 661–669.

[5] L. Zhang, M. Yang, X. Feng, Sparse representation or collaborative representation: which helps face recognition?, in: Proc Int. Conf. Computer Vision, 2011, pp. 471–478.

[6] Xu Y, Zhu X, Li Z, et al. Using the original and 'symmetrical face' training samples to perform representation based two-step face recognition[J]. Pattern Recognition, 2013, 46(4).

[7] B. Tang, S. Luo, H. Huang, High performance face recognition system by creating virtual sample, in: Proceedings of the International Conference on Neural Networks and Signal Processing, 2003, pp. 972–975.

[8] Dong X, Zhang H , Sun J , et al. A two-stage learning approach to face recognition[J]. Journal of Visual Communication and Image Representation, 2017, 43:21-29.

[9] Tian C, Sun G, Zhang Q, et al. Integrating Sparse and Collaborative Representation Classifications for Image Classification[J]. International Journal of Image and Graphics, 2017, 17(02):1750007.

[10] Y. Xu, B. Zhang, Z. Zhong. "Multiple representations and sparse representation for image classification", Pattern Recognition Letters, 2015, 68: 9-14.

[11] I. Naseem, R. Togneri, M. Bennamoun. "Linear regression for face recognition", Pattern Analysis and Machine Intelligence, IEEE Transactions on, 2010, 32(11): 2106-2112.

[12] Y. Xu, Q. Zhu, Z. Fan, Y. Wang, J.-S. Pan, From the idea of sparse representation to a representation-based transformation method for feature extraction, Neurocomputing. ,113,168-176,2013.

[13] Y. Xu, X.Li, J. Yang, Z.Lai, D.Zhang, Integrating conventional and inverse representation for face recognition, IEEE Transactions on Cybernetic s., 44(10), 1738-1746, 2014.

[14] Y. Xu, X. Fang, J. You, Y. Chen, H.Liu, Noise-free representation based classification and face recognition experiments. Neuro-computing, 147,307-314,2015.

[15] Y.Xu,Z.Fan,Q.Zhu,Feature space-based human face image representation and recognition, Optical Engineering. 51 (1) (2012) 017205-017205-7

[16] Xu Y, Zhong Z, Yang J , et al. A New Discriminative Sparse Representation Method for Robust Face Recognition via l2 Regularization[J]. IEEE Trans Neural Netw Learn Syst, 2017, 28(10):2233-2242.

Review Sentiment Orientation Analysis based on Deep Learning

Yixuan Liu[1] and Hailing Xiong[2,*]

[1]College of Computer & Information Science, Southwest University,400715,China
[2]Business College of Southwest University, 402460, China

[*]xionghl@swu.edu.cn

Abstract. The increase of network users has led to a large number of commentary languages on various network platforms. Traditional manual processing is time-consuming and labor-intensive. We need a mechanized way to process these commentary corpora and quickly uncover the emotional tendencies. A method of sentimental orientation analysis of comment text based on deep learning is proposed. First, we used GloVe model to train the word vector. Then, Give the different weight on word vector by using TF-IDF. Finally, the processed word vectors would be classificated by TextCNN. Experiments were carried out on the six categories of commodity review data crawled by Jingdong. This method can effectively identify the emotional tendency of the review text, which is more accurate than the traditional deep learning method.

1.Introduction

According to China Internet Network Information Center shows that the number of Internet users in China is close to 732 million. Most of them choose to surf the Internet on platform such as TaoBao , JingDong ,Weibo etc. and it will generate a lot of natural language. How to quickly and accurately capture these comments and discover the emotional tendency behind them is of great commercial value. The sentiment orientation analysis mainly focuses on the subjective commentary text and judges its emotional orientation, namely positive, negative and neutral [1]. However, natural language is an advanced human language. Unlike audio and image data, when a computer processes natural language, it needs to convert natural language text into computer-readable data. Here we use the method of word vector to represent the text vocabulary of natural language. Solving the problem of representation of natural language, we can then use the deep learning method to analyze the sentiment orientation.

2. Relate work

Sentiment orientation analysis is an important sub-task of sentiment analysis[2]. Emotional analysis was first proposed by Nasukawa T et al. [3]. They proposed that sentiment analysis is not simply to divide the whole document into positive or negative, but rather Extract specific emotions related to the subject of the document from the document. There are many classification methods for sentiment analysis. Liu B[4]systematically introduced various aspects of sentiment analysis in a book published in 2012. According to the research method of sentiment analysis, it can be divided into supervised and unsupervised methods.

A supervised approach is to manually label the training corpus and then use a machine learning model. Bermingham A et al.[5] used traditional machine learning methods to classify emotions in Twitter. This supervised method can achieve extremely high precision, and its disadvantages are also obvious. This method relies heavily on the quality and quantity of training data, but it is very difficult to obtain a large amount of high-quality corpus. Moreover, the portability of this method is poor, network new words and network buzzwords are constantly being updated, and the accuracy of supervised sentiment analysis methods will be significantly reduced. In another method, the unsupervised method generally performs emotional analysis by analyzing the data statistically and calculating the emotional distribution in the corpus by probability. Lin et al.[6] added a sentiment layer based on the LDA (Latent Dirichlet Allocation) model, and became a JST model with a 4-layer structure to obtain the emotional orientation of each topic. Taking into account the respective advantages of supervised and unsupervised methods, Maas A L et al.[7]combine these two methods to obtain emotionally significant word vectors.

However, both supervised and unsupervised methods require artificially designed features, and the design of the features requires expert domain knowledge, which means a lot of labor costs. With the continuous development of deep learning, scholars have used deep learning in the field of natural language processing. The core of deep learning is that it does not require artificially designed features but is learned through training data. Deep learning has produced very gratifying results in the tasks of natural language understanding, especially subject classification, sentiment analysis, automatic question answering and language translation[8]. Collobert R[9]proposed a unified neural network architecture and corresponding learning algorithms that can be applied to perform various natural language processing tasks. Taking into account the context information in the word Kim Y[10]tried to use CNN for sentence modeling to solve the emotional classification task. After the experiment, it also achieved good results. On this basis, Socher R et al.[11, 12]used RNN, MRNN, RNTN and other recurrent neural networks to process natural language, and considered the syntactic structure of the text into the neural network model, in sentence and The phrase level has achieved good results. There are timing problems in the processing of sentences. Tai K S et al.[13]used LSTM to solve the timing problems of sentences for sentiment analysis. The input of the above model is a word vector. The word vector is the most basic result of deep learning applied in the field of natural language processing, but the traditional word vector only contains semantic and grammatical information and can't solve the emotional problem of context. Some scholars try to add emotional information in the research process to get the corresponding emotional word vector. Tang D et al.[14] established a three-layer online learning sentiment word vector SSWE to learn from a large number of weakly supervised tweets without a lot of manual annotation. Man Lan et al.[15] proposed three models based on CNN to learn the emotional word vector(SWV), which integrates emotional information and semantic and syntactic information into word representations in three different strategies. In this paper, aiming at the word's different sentiment weight in a sentence, we creatively added weighted word vectors to get greater sentiment classification.

3. Weighted sentiment word embedding

3.1 GloVe word embedding model

With the development of deep learning, many scholars also apply deep learning, neural networks and other methods to the field of natural language processing. Since neural networks only accept numerical input types, we must use the natural language text we use. Process it and convert it to a numeric type. The word vector will appear. The emergence of word vectors is also a fundamental task for deep learning in recent years. The word vector model used in text is Stanford's GloVe (Global Vectors for Word Representation) [16]. It is very similar to our commonly used Word2vec idea, but GloVe fully considers the co-occurrence of words and is more able to distinguish the meaning of words. And it combines the advantages of LSA, CBOW, training faster, and the scalability of large-scale corpus algorithms is also very good. Performance is also good on small corpora or small vectors. The evaluation of the word vector is evaluated using the similarity of words.

IOP Conf. Series: Journal of Physics: Conf. Series **1267** (2019) 012009 doi:10.1088/1742-6596/1267/1/012009

Table 1.The 7 words most similar to "product quality"

Product quality	Similarity
Quality	0.9122
Product	0.9014
Commodity	0.8932
Good quality	0.8415
Texture	0.8387
Goods	0.7984
service	0.7481

According to Table 1, the word similarity effect of the Glove training word vector is still relatively good.

3.2. TF-IDF weighted word embedding

3.2.1 Introduce of TF-IDF TF-IDF (term frequency–inverse document frequency) is a word information weighting technique commonly used for information retrieval (data mining), as shown in the following formula 3.1. It's main idea is that the importance of a word increases proportionally with the number of times it appears in the file, but at the same time it decreases inversely with the frequency it appears in the corpus. Used to assess how important a word is to a document or a corpus.

$$\mathrm{TF-IDF}(X) = \mathrm{TF}(X) * \mathrm{IDF}(X) \quad (3.1)$$

The term TF (term frequency) in the formula represents the word frequency, that is, the number of times a given word appears in the file. IDF (inverse document frequency) represents the reverse file frequency, that is, the total number of documents in the corpus divided by the total number of articles containing the phrase plus one, and then processing by log.

After the word segmentation and de-punctuation work on the comment text corpus, the corpus of the word segmentation is trained on the one hand using the GloVe word vector model to train the word vector, and on the other hand, the TF-IDF method is used to train each word weight. Then, the obtained weights are multiplied by the trained word vectors, and finally the word vectors with weight information are obtained. The flow chart is shown in Figure 1.

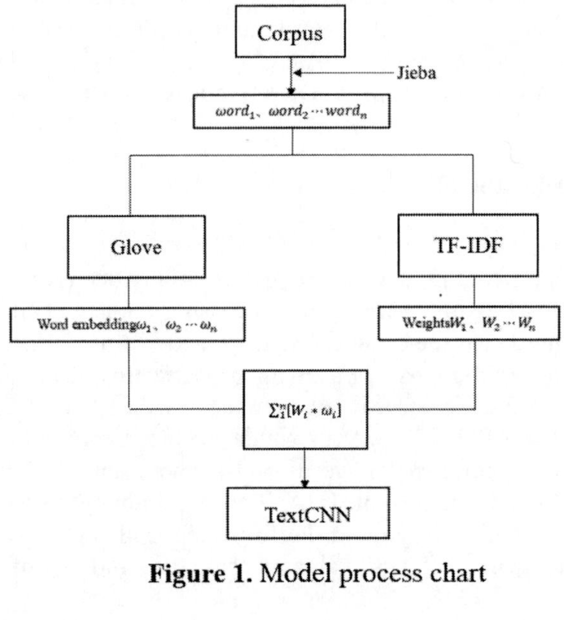

Figure 1. Model process chart

4. Sentiment Orientation classification model

Using TextCNN model to emotionally classify the weighted word vectors. and The main model process is shown in Figure 2.

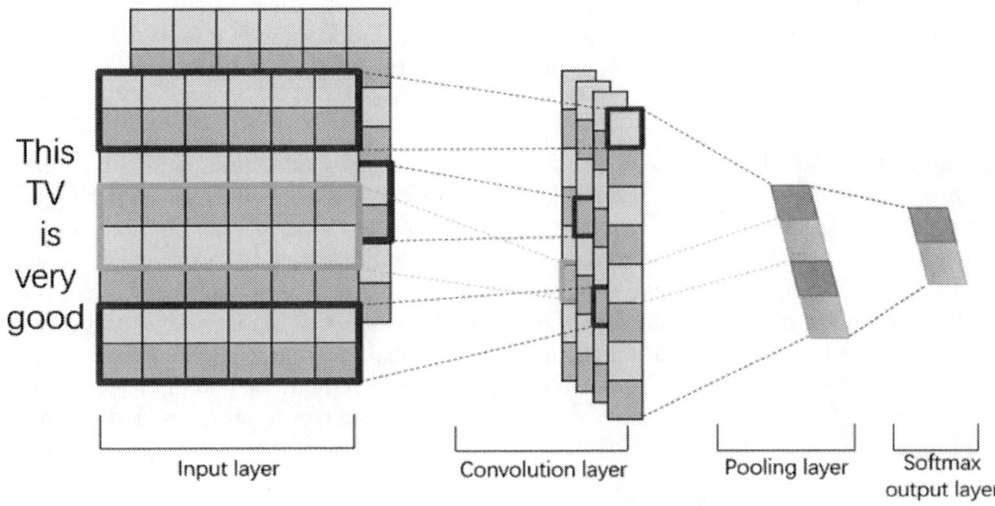

Figure 2. TextCNN model

The model consists of four layers. The first layer is the word vector input layer. Each word in the text comment corpus is mapped to the word vector space. It is assumed that there are a total of N word vectors, and each is K-dimensional . Then the word vector input layer is an N*K-dimensional image. Here, a sliding window of length H is set, and the entire word vector input matrix is swept by the sliding window to calculate a feature vector C corresponding to H, as shown in the following formula 3.2, where b is an offset.

$$c = f\left(w \bullet x_i + b\right) \tag{3.2}$$

The second layer is the convolution layer. Different filters act on the word vector input layer to generate different feature maps. The third layer is the pooling layer. The feature map generated by the convolution layer is taken. The maximum, the advantage of doing this is that it can handle comment corpus of varying lengths. The fourth layer is the softmax layer, which is the output vector of the pooling layer by Equation 3.3 below.

$$p\left(\mathbf{w} \middle| \text{Context}\left(\mathbf{w}\right)\right) = \frac{e^{y_{w,i_w}}}{\sum_{i=1}^{N} e^{y_{w,i}}} \tag{3.3}$$

Calculate the probability of derogatory and derogatory tendency. The greater the probability, the sentiment tendency of the sentence. In the TextCNN model training, we use the data set for NLPCC2014 evaluation task 2, which contains Chinese and English training corpus. For this article, only Chinese polarity comment corpus is selected as the training. In the TextCNN original text, multiple data sets were used for training. Due to the data acquisition problem, only one polar corpus training set was selected in this paper.

5. Simulation Experiment and Result Analysis

5.1 experiment data and process

The experimental data used in this paper is the commodity review corpus obtained from the Jingdong reptile. A total of six commodities (covering electrical appliances, books, hotels, daily necessities, etc.), a total of 190,533 comments. Among them, the positive evaluation is a total of 89882: The negative evaluation is 100651: The data of the direct crawling needs to be pre-processed first. 1. First, the text is de-emphasized: remove some meaningless comments and the system defaults. 2. Secondly, mechanical compression and de-emphasis: text review corpus has consecutive repeated sentences, such as: "Good, good, good, good, good, good, good, good, good, good, good." This sentence is common. The long corpus of meaning needs to be deleted. 3. Breaking sentence deletion: Delete some short text comment corpus, generally 4-8 international characters are more reasonable lower limit, combined with text comment corpus, the shortest special we set the lower limit to 4. After the pre-processing is finished, the corpus is processed for word segmentation, and when the word segmentation is completed, the punctuation marks are also removed. Here we use the jieba participle. In the word segmentation, we will remove the stop words. The stop word dictionary here is a series of commonly used stop word dictionaries summarized by organizations such as HowNet. Since there are many typos in the online commentary, we use a synonym replacement function of jieba to replace the wrong word expression with the correct word expression, which can improve the usability of our crawler corpus data. In the word segmentation, the user can also customize the dictionary. According to the corpus characteristics of the crawler, we summarize a custom dictionary to improve the accuracy of the word segmentation. After the good words are divided, the words are provided to the GloVe model for training.

5.2. Comparison of experimental results

In this experiment we will use Recall, Precision and F1-Measure as the evaluation indicators of this model.

The comparison experiments here can be compared with word2vec+TextCNN Glove+Bayesian Glove+SVM Glove+TextCNN. Ppos represents the accuracy of positive reviews, Rpos represents the recall rate of positive reviews, Fpos represents the F1 value of positive reviews, Pneg represents the accuracy of negative reviews, Rneg represents the recall rate of negative reviews, Fneg represents the F1 value of negative reviews, Farg A weighted average of the F1 values representing positive and negative comments. The performance comparison of the different methods is shown in Table 2 below.

Table 2. Performance comparison

methods	P_{pos}	R_{pos}	F_{pos}	P_{neg}	R_{neg}	F_{neg}	F_{arg}
Word2Vec+TextCNN	0.8814	0.8921	0.8916	0.9048	0.8828	0.8931	0.8880
GloVe+TF-IDF+TextCNN	**0.9024**	**0.9048**	**0.9034**	**0.9185**	**0.9037**	**0.9167**	**0.9102**
GloVe+TextCNN	0.8832	0.8974	0.8917	0.9006	0.8876	0.8954	0.8915
Word2Vec+SVM	0.8149	0.9024	0.8735	0.8969	0.8415	0.8977	0.8546
Word2Vec+Bayes	0.7941	0.8014	0.8145	0.8019	0.7965	0.7816	0.7891
GloVe+Bayes	0.7887	0.8048	0.8164	0.8047	0.7961	0.7841	0.7901
GloVe+SVM	0.8148	0.9028	0.8812	0.8971	0.8414	0.8678	0.8546

From the experimental results, we can see that the method used in this paper has a certain improvement in performance in all aspects compared with other traditional methods, which shows that the weight of the given vector can achieve a certain improvement effect of sentiment analysis accuracy. We can see that the accuracy and recall rate of negative reviews is slightly higher than positive reviews. This may be the reason why negative commentary corpus is more than positive commentary corpus.

IOP Conf. Series: Journal of Physics: Conf. Series **1267** (2019) 012009 doi:10.1088/1742-6596/1267/1/012009

6. Conclusions

This paper introduces the weight of word vector by using TF-IDF method, strengthens the weight of sentiment word vector, and makes the emotional word vector play a greater role. The effectiveness of this method is also proved by experimental verification. Here we only interpret the emotional classification of emotions. If we want to conduct further emotional analysis, we can analyze the specific emotional attributes. In this paper, the analysis of emotions only includes two kinds of meanings and derogatory meanings. If we want to consider the comprehensive emotions of the commentary, we should also add neutral emotions. The sentiment analysis is divided into neutral, derogatory and derogatory. In addition, there are many ways to express Chinese, and it is necessary to conduct in-depth research on emotional expressions containing metaphors.

Acknowledgment

This work was partially supported by "National Natural Science Foundation of China under Grant (No. 41271292)".

References

[1] K. Schouten and F. Frasincar, Survey on Aspect-Level Sentiment Analysis[M]. 2016.

[2] J. Sheng-Yi, G. Lin-Dong, W. Lian-Xi, and F. Si-Hui, "Survey on Opinion Target Extraction[J]," ACTA AUTOMATICA SINICA, vol. v.44, no. 07, pp. 15-32, 2018.

[3] T. Nasukawa and J. Yi, "Sentiment analysis: Capturing favorability using natural language processing[C]," in International Conference on Knowledge Capture, 2003.

[4] B. Liu, Sentiment Analysis and Opinion Mining[J]. 2012.

[5] A. Bermingham and A. F. Smeaton, "Classifying sentiment in microblogs: Is brevity an advantage?[J]," 2010.

[6] C. Lin and Y. He, "Joint sentiment/topic model for sentiment analysis[C]" presented at the Proceedings of the 18th ACM conference on Information and knowledge management, Hong Kong, China, 2009.

[7] A. L. Maas, R. E. Daly, P. T. Pham, D. Huang, A. Y. Ng, and C. Potts, "Learning word vectors for sentiment analysis[C]," presented at the Proceedings of the 49th Annual Meeting of the Association for Computational Linguistics: Human Language Technologies - Volume 1, Portland, Oregon, 2011.

[8] Y. Lecun, Y. Bengio, and G. J. N. Hinton, "Deep learning[J]," vol. 521, no. 7553, p. 436, 2015.

[9] R. Collobert et al., "Natural Language Processing (Almost) from Scratch %J J. Mach. Learn. Res[J]," vol. 12, pp. 2493-2537, 2011.

[10] K. Y, "Convolutional Neural Networks for Sentence Classification[J]," arXiv preprint arXiv:1408.5882, 2015.

[11] R. Socher, C. Manning, and A. Y Ng, Learning Continuous Phrase Representations and Syntactic Parsing with Recursive Neural Networks[J]. 2010.

[12] S. R, P. A, W. J. Y, and e. al, "Recursive deep models for semantic compositionality over a sentiment treebank[C]," presented at the Empirical Methods in Natural Language Processing, 2013.

[13] S. T. Kai, R. Socher, and C. D. J. C. S. Manning, "Improved Semantic Representations From Tree-Structured Long Short-Term Memory Networks[J]," vol. 5, no. 1, p.: 36., 2015.

[14] D. Tang, F. Wei, N. Yang, M. Zhou, T. Liu, and B. Qin, Learning Sentiment-Specific Word Embedding for Twitter Sentiment Classification[J]. 2014, pp. 1555-1565.

[15] L. Man, Z. Zhang, L. Yue, and W. Ju, "Three Convolutional Neural Network-based models for learning Sentiment Word Vectors towards sentiment analysis[C]," in International Joint Conference on Neural Networks, 2016.

[16] P. J, S. R, and M. C, "Glove: Global Vectors for Word Representation[C]," presented at the Empirical Methods in Natural Language Processing, 2014.

Quasi-oppositional Multi-objective Antlion Optimizer Based on Differential Evolution

Wang Yadong, Shi Quan, Song Mingchang and Song Weixing

Department of Equipment Command and Management, Army Engineering University of PLA , Shijiazhuang 050003

airesearch01@gmail.com

Abstract: In order to solve the multi-objective problem, an improved quasi-oppositional multi-objective antlion optimization algorithm based on differential evolution (DEQOMALO) is proposed. This algorithm overcomes the defect that antlion algorithm is easy to fall into local optimum. On the one hand, this algorithm uses the idea of differential evolution to make full use of the information of the ant and the elite antlion to improve the position updating method of the original algorithm. On the other hand, the population is optimized by quasi-opposite learning strategy, and the original population and its quasi-opposite individuals are mixed and then selected as the new population, which greatly increases the diversity of the population. Finally, typical benchmarks are selected to compare the algorithm with the original antlion algorithm and other MALO algorithms with traditional evolution strategies. Experimental results show that both convergence and distribution of the improved algorithm are greatly improved. The proposed DEQOMALO algorithm has good adaptability and effectiveness in solving the two-objective optimization problem.
Key worlds: multi-objective optimization,Antlion Optimizer,differential evolution,quasi-oppositional learn strategy

1. Introduction

Multi-objective optimization problem (MOP) exists widely in real world practice. In order to solve this kind of problem, we need to balance several goals which often conflict with each other. Traditional mathematical programming methods are very restrictive in solving non-convex optimization problems, so intelligent optimization methods are favored by many scholars as effective tools. At present, multi-objective optimization algorithms can be categorized in Pareto based methods, decomposition based methods and index based methods [1] The Pareto-based methods select the non-dominated solution as the optimal solution by comparing the dominating relations between different solutions. The most common used Pareto-based methods include the non-dominated sorting genetic algorithm II (NSGA-II) [2] , strength Pareto evolutionary algorithm 2 (SPEA2) [3] , and Pareto envelope-based selection algorithm II (PESA-II) [4] , and so on. The decomposition-based methods aggregate the objectives of an MOP by a scalarizing function such that a single scalar value is generated. In these algorithms, the diversity of a population is maintained by specifying a set of well-distributed reference points to guide its individuals to search simultaneously towards different optima. The decomposition based multi-objective evolutionary algorithm (MOEA/D) is one of the most common decomposition based algorithm [5]. The idea of indicator-based algorithms is to apply performance indicator to guide the search during the evolutionary process. The most common used indicators are ε indicator [6], inverted generational distance (IGD) [7], and the hyper volume (HV) indicator [8].

Current intelligent optimization methods can also be divided into swarm intelligence and evolutionary computing. The swarm intelligence is inspired by biological group behavior, while the evolutionary algorithm is inspired by biological evolutionary mechanism. Antlion algorithm, as a swarm intelligence algorithm, was put forward by S Mirjalili [9]. In order to solve the multi-objective optimization problem by using antlion algorithm, Mirjalili himself proposed a multi-objective antlion algorithm, and a large number of experiments proved that the performance of multi-objective antlion algorithm is better than that of traditional algorithms such as MOPSO, NSGA-II and so on [10].Because of the superiority of the antlion algorithm, the algorithm has been applied and popularized in lots of real world engineering problems such as parameter optimization [11], path planning problem [12],and so on.

The basic antlion algorithm is still prone to fall into local optimum and convergence slowly in the later period. Therefore, this paper tries to improve its ability of global search and local exploration in solving multi-objective problems by proposing an improved algorithm named quasi-oppositional multi-objective antlion optimization algorithm based on differential evolution.

The remaining sections of this article are organized as follows. In section 2, the basic antlion optimizer and related works are introduced. Section 3 gives a thorough description of the new proposed DEQOMALO algorithm. The performance of the proposed algorithm is tested by numerical simulation in section 4. We summarize the paper with suggestions for future improvement in section 5.

2 Related Works

2.1 Antlion optimizer

The way that antlion optimizer find the optimal solutions is simulating the process of antlions build trap and prey ants in nature. So there are two populations in the antlion algorithm, that is, the antlions and the ants. The hunting process consists of five basic steps: random walk of ants, ants fall into traps, ants slid toward antlions, antlions catch ants and antlions reconstruct traps. The basic steps of the antlion optimizer simulate the above process, and the mathematical description is as follows:

(1) The random walk of ants is formulated as follows:

$$x(t) = [0, cumsum(2r(t_1)-1), cumsum(2r(t_2)-1), ..., cumsum(2r(t_n)-1)] \quad (1)$$

where *cumsum* calculates the cumulative sum, t is the number of iterations, and n is the maximum number of iterations. $r(t) = \begin{cases} 1 & if\ rand > 0.5 \\ 0 & if\ rand \le 0.5 \end{cases}$ is a random function related to t, and $rand$ satisfies the uniform distribution of [0 ~ 1].

In order to ensure that ants move randomly in search space and prevent them from crossing the boundary, it is necessary to normalize the position of ants:

$$X_i^t = \frac{(X_i^t - a_i) \times (d_i^t - c_i^t)}{bi - ai} + c_i^t \quad (2)$$

where c_i^t and d_i^t indicates the minimum value and the maximum value of the variable i of one ant at the $t\text{-}th$ iteration respectively; a_i and b_i indicates the minimum value and the maximum value of the variable i of the ant respectively.

(2) By simulating the process of random walk around the antlion and falling into a trap, the formula is as follows:

$$c_i^t = Antlion_j^t + c^t ; d_i^t = Antlion_j^t + d^t \quad (3)$$

where c^t and d^t is the minimum and maximum values of all variables at the $t\text{-}th$ iteration; $Antlion_j^t$ is the position of antlion j at the $t\text{-}th$ iteration.

(3) By adaptively reducing the random walk range of the ants, the process of the ants to slide into the antlion is simulated as follows

$$c^t = \frac{c^t}{I}, d^t = \frac{d^t}{I} \quad (4)$$

Where $I = 1 + 10^{\omega \frac{t}{T}}$ is gradually increased as the number of iterations increases. t is the number of iterations, T is the maximum number of iterations, ω is dynamically adjusted with the number of iterations.

(4) The simulated antlions' capture process is as follows:

$$Antlion_j^t = Ant_i^t \quad if\ f(Ant_i^t) < f(Antlion_j^t) \quad (5)$$

where $Antlion_j^t$ is the position of antlion j at the t-th iteration, Ant_i^t is the position of ant i at the t-th iteration.

Select the antlions with optimal fitness as elite individual which decides the position of ants. Ants move randomly towards both a randomly selected antlion and the elite antlion, with the following formulas:

$$Ant_i^t = \frac{R_A^t + R_E^t}{2} \tag{6}$$

where R_A^t is a random movement around a antlion selected by roulette wheel selection mechanisms. R_E^t is a random movement around the elite antlion。

(5) Reconstruction the trap. Use an external archive to store the non-dominant solution set. The niche technique is used to measure the distribution of solutions in archives. The distribution of the solution is measured by the number of other solutions in the neighborhood of each solution. The more the solution in the neighborhood, the denser the individual is.

2.2 Quasi-opposition based learning strategy

Traditional meta-heuristic algorithms usually begin with a set of random initial solutions, but the randomly generated initial solutions may be far from the optimal solution. So it takes a long time for the algorithm to converge to the optimal solution. Therefore, in this paper, the quasi-opposition based learning (QOBL) strategy is used to optimize the initial population [13] . In this way, the convergence rate of the algorithm can be improved, as well as, the diversity of the population can be increased.

For a real number x with search interval $[a,b]$, its opposite number ox is defined as follows:

$$ox = a + b - x \tag{7}$$

Its quasi-opposites number is defined as follows:

$$qox = rand((\frac{a+b}{2}), ox) \tag{8}$$

Then, for the D dimensional individual $X = [X_1, X_2, ..., X_D]$, its quasi-opposites individual is $QOX = [qox_1, qox_2, ..., qox_D]$. The value of $i-th$ variable in QOX is $qox_i = rand((\frac{a_i + b_i}{2}), ox_i)$.

After the quasi-opposite population is obtained, it is mixed with the individual of the original population as the new population.

2.3 Differential evolution strategy

Differential evolutionary algorithm (DE) is proposed on the basis of genetic algorithm (GA). It constantly updates individuals and gradually finds the optimal individual population by simulating genetic variation operations, crossover operations and selection operations. Different from the traditional genetic algorithm, the mutation individual of the differential evolution algorithm is generated by the parent difference, and then the mutation individual is crossed with the other parent individuals to create a new crossover individual. Finally, the crossover individual is compared with the parent individual. The superior individual is selected as the final offspring individual.The current mutation strategies of differential evolution are as follows [13]:

(1) DE/rand/1 strategy: $v_i = x_{r1} + F \times (x_{r2} - x_{r3})$

(2) DE/rand/2 strategy: $v_i = x_{r1} + F \times (x_{r2} - x_{r3}) + F \times (x_{r4} - x_{r5})$

(3) DE/best/1 strategy: $v_i = x_{best} + F \times (x_{r1} - x_{r2})$

(4) DE/best/2 strategy: $v_i = x_{best} + F \times (x_{r1} - x_{r2}) + F \times (x_{r3} - x_{r4})$

(5) DE/current-to-best/1 strategy: $v_i = x_i + F \times (x_{best} - x_i) + F \times (x_{r1} - x_{r2})$

(6) DE/current-to-rand/1 strategy: $v_i = x_i + rand \times (x_{r1} - x_i) + F \times (x_{r2} - x_{r3})$

Where, F_i is fixed value as the control parameter, x_{ri} is a randomly selected individual from the population, x_i is mutated individuals, and x_{best} is individuals with optimal fitness function satisfied with $x_i \neq x_{best} \neq x_{ri}$.

These mutation strategies can use the information of parent population to generate new individuals.

Strategy 3, 4 , 5 also uses the information of the optimal individual in the population to control the direction of mutation of the individual. However, it can be found from the experiments (see the experiment section of this article) that, the performance of the antlion algorithm does not improve just by introducing these mutation strategies into the algorithm. Therefore, it is necessary to use the idea of differential evolution to propose a mutation strategy adapted to the antlion algorithm.

3 The proposed DEQOMALO algorithm

In order to apply the traditional differential evolution theory to the multi-objective antlion optimization algorithm, there are two difficulties need to be solved. Firstly, the mutation strategy of the traditional differential evolution is aimed at solving the single objective optimization problem, so it is necessary to transfer the strategy from single objective to multi-objective. In the single-objective optimization problem, the individual can be ranked by comparing the fitness function value. In multi-objective optimization, multiple objectives of individual should be considered at the same time, so the optimal individual can't be determined by comparing the fitness value. Therefore, the idea of non-dominated sorting is cited in this article. Firstly, the non-dominant solution is found according to the Pareto dominance relation of ants, and the non-dominant individual is placed into an external archive as the antlion population. Secondly, the elite antlion individual is selected according to the sort of antlion population in the archive by using niche technique. In this way, the elite antlion is superior to other antlion individual, and all antlion individual is superior to the ant individual.

Secondly, the ant position update formula of the original antlion algorithm only uses the information of the antlion in the external archive, and discards the information of the ant population. This may lead to premature convergence of the algorithm, which is not conducive to global exploration. The mutation strategy of traditional differential evolution emphasizes the information of ant population, and does not give full play to the information of antlion, which is not conducive to the convergence of the algorithm. So it is necessary to consider how to make full use of both the location information of ant population and the antlion at the same time. In this paper, a new ant location updating formula is proposed as follow:

$$\begin{aligned} Ant_i = \overline{Antlion_c} + F_1 \times (Antlion_{elite} - Antlion_{rand}) + \\ F_2 \times (Antlion_{elite} - Ant_i) + F_3 \times (Antlion_{rand} - Ant_i) \end{aligned} \tag{9}$$

where, $\overline{Antlion_c} = \omega_1 \times Antlion_{elite} + \omega_2 Antlion_{rand} + \omega_3 Ant_i$, $Antlion_{elite}$ and $Antlion_{rand}$ are antlions selected from the external archive, $Antlion_{elite}$ is the elite antlion selected by niche, $Antlion_{rand}$ is the random antlion selected from the archive randomly, $Antlion_{elite} \neq Antlion_{rand}$. $\omega_i \geq 0$, $\sum_{i=1}^{3} \omega_i = 1$ $i = 1,2,3$,

$\omega_i = p_i / \sum_{i=1}^{3} p_i$ $i = 1,2,3$,and $p_1 = 1$, $p_3 = rand(0.7,1)$, $p_2 = rand(0.5, p_3)$.

Formula (9) can make full use of the information of the current population and the location of the antlion to ensure the movement of the population to the elite antlion.

At the end of each iteration, the quasi-opposite population is obtained and mixed with the population of the previous generation. A new population is selected as the offspring population to enter the next iteration according to the dominance relationship.

4 Simulation experiment and result analysis

4.1 Experimental setup

For multi-objective optimization algorithm, the performance of the algorithm is mainly reflected in convergence and the distribution of the results [14].Convergence describes the degree of approximation of the result obtained by the algorithm to the true Pareto front (PF). The stronger the convergence of the algorithm, the closer the solution set is to the true optimal solution, and the more accurate the result is. The distribution describes the distribution characteristics of the obtained result in the objective space. On the one hand, the results should be distributed as much as possible on the whole PF, and on the other hand, the results should be distributed as evenly as possible. The stronger the distribution of the algorithm represents a better the global exploration ability of the algorithm.

To measure the performances of the algorithm, among the many metrics, inverted generation distance (IGD) and hyper volume (HV) were selected because they can reflect the convergence and distribution of the algorithm at the same time [15].

In order to study the properties of the improved algorithm, the test functions of ZDTs are selected to verify the algorithm [16]. The experimental parameter settings, formulations of the mutation strategy and the value of F are present at Table 1 [17].

Table.1 Variation strategy and parameter setting

mutation strategy	formulations	parameter
Classical	$Ant_i = F \times (Antlion_{elite} - Antlion_{rand})$	$F = 0.5$
DE/rand/1	$Ant_i = Ant_{r1} + F \times (Ant_{r2} - Ant_{r3})$	$F \sim N(0.5, 0.3)$
DE/best/1	$Ant_i = Antlion_{elite} + F \times (Ant_{r1} - Ant_{r2})$	$F \sim N(0.5, 0.3)$
DE/current-to-best/1	$Ant_i = Ant_i + F \times (Antlion_{elite} - Ant_i) + F \times (Ant_{r1} - Ant_{r2})$	$F \sim N(0.5, 0.3)$
Proposed method	$Ant_i = \overline{Antlion_c} + F_1 \times (Antlion_{elite} - Antlion_{rand}) + F_2 \times (Antlion_{elite} - Ant_i) + F_3 \times (Antlion_{rand} - Ant_i)$	$F_i \sim N(0.5, 0.3)$

4.2 The effect of the evolutionary strategy on the algorithm

In order to test the effect of the proposed differential evolution strategy, the proposed algorithm is compared with the classical antlion algorithm with traditional differential evolution strategy optimization algorithm. Improve the original multi-objective antlion algorithm by different mutation strategies without Quasi-Opposition strategy. The algorithms are tested on each standard test function, and the experimental results are shown in Figure 1.

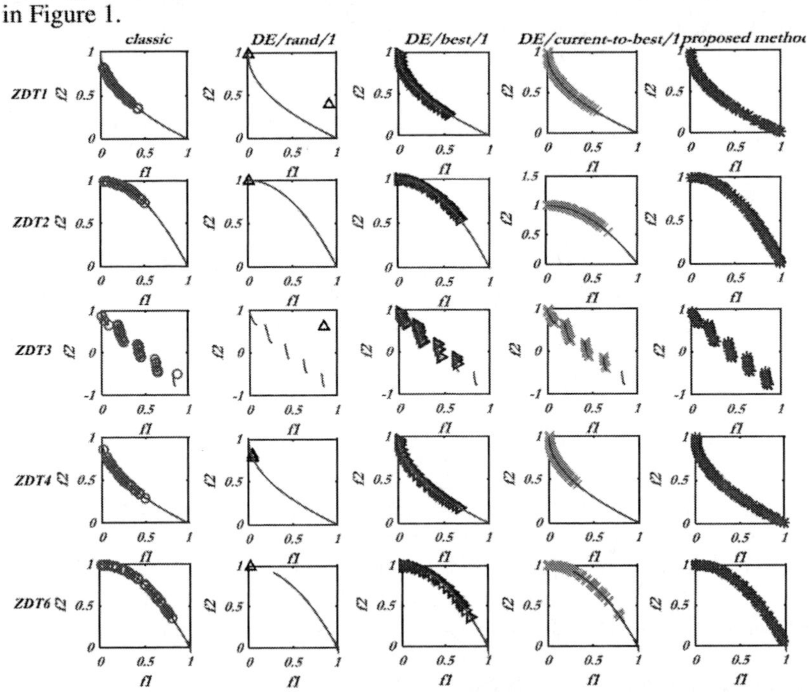

Figure 1. Optimal solution distribution of improved differential evolutionary strategies without quasi-opposites learning strategies

In Figure 1, from left to right, represent the results of the original MALO algorithm, the DE/rand/1 strategy improvement algorithm, the DE/best/1 strategy improvement algorithm, the DE/current-to-best/1 strategy improvement algorithm and the DEQOMALO algorithm proposed in this paper. Five test functions are represented from top to bottom. The distribution of the results of each algorithm on each test function can be observed intuitively from the diagram. The optimal solution set of MALO are almost fall on the true PF, but the distribution of MALO is poor. The convergence and distribution of the algorithm under DE/rand/1 strategy are very poor. This is mainly because the strategy adopts a completely random mutation strategy and completely ignores the role of elite antlions. The distribution of the algorithm under DE/best/1 strategy and DE/current-to-best/1 strategy in the objective space are also poor. The optimal solution of the proposed algorithm can be uniformly and widely distributed on the true PF in the objective space of all the test functions, and the convergence and distribution of the algorithm are both optimal.

5. Conclusion

In order to improve the precision and global search ability of antlion algorithm in solving multi-objective problem, an improved multi-objective antlion algorithm is proposed. The following improvements are made. Firstly, by introducing the idea of differential evolution, an individual updating mechanism, which can make full use of both population information and elite antlion information, is proposed to replace the original algorithm. Secondly, the quasi-opposite learning strategy is used to optimize the population. On the one hand, it can increase the diversity of populations, on the other hand, it can improve the convergence speed of the algorithm. In order to verify the performance of the proposed algorithm, the DEQOMALO is compared with the multi-objective antlion algorithm with traditional evolutionary strategy. Simulation results show that the improved algorithm has better convergence and distribution than other algorithms. The improved multi-objective antlion algorithm has higher stability, as well. Therefore, the quasi-oppositional multi-objective antlion optimization algorithm based on differential evolution proposed in this paper has good adaptability and robustness in solving multi-objective optimization problems.

Reference

[1] Li Miqing. 2015. "Evolutionary Many-objective Optimization: Pushing the Boundaries." PHD diss. Brunel University London.

[2] K Deb, A Pratap, S Agarwal, and T Meyarivan.2002. "A Fast and Elitist Multiobjective Genetic Algorithm: NSGA-II." *IEEE Transactions on Evolutionary Computation* 6 (2): 182-197.

[3] E Zitzler, L Thiele. 1999. "Multiobjective Evolutionary Algorithms: A Comparative Case Study and the Strength Pareto Approach." *IEEE Transactions on Evolutionary Computation* 3 (4): 257-271.

[4] DW Corne, NR Jerram, JD Knowles, and MJ Oates. 2001. "PESA-II: Region-based Selection in Evolutionary Multiobjective Optimization." *In Proceedings of the 3rd Annual Conference on Genetic and Evolutionary Computation (GECCO)* : 283-290.

[5] H Li, Q Zhang. 2009. "Multi-objective Optimization Problems with Complicated Pareto sets, MOEA/D and NSGA-II." *IEEE Transactions on Evolutionary Computation* 13 (2): 284-302.

[6] M Laumanns, L Thiele, K Deb, and E Zitzler. 2014. "Combining Convergence and Diversity in Evolutionary Multiobjective Optimization." *Evolutionary Computation* 10 (3): 263-282

[7] CAC Coello, MR Sierra. 2004. "A Study of the Parallelization of a Co-evolutionary Multi-objective Evolutionary Algorithm." *In Mexican International Conference on Artificial Intelligence (MICAI)*, 2972: 688-697.

[8] E Zitzler, M Laumanns, and L Thiele.2002. "SPEA2: Improving the Strength Pareto Evolutionary Algorithm for Multiobjective Optimization." *In Evolutionary Methods for Design, Optimization and Control* 95-100.

[9] S Mirjalili. 2015. "The Antlion Optimizer." Advances in Engineering Software 83 (C): 80-98.

[10] S Mirjalili, P Jangir, and S Saremi. 2016. "Multi-objective Antlion Optimizer: A Multi-objective Optimization Algorithm for Solving Engineering Problems." *Applied Intelligence* : 1-17.

[11] ZHAO Shijie, GAO Leifu, and YU Dongmei. 2016. "Antlion Optimizer with Chaotic Investigation Mechanism for Optimizing SVM Parameters" *Journal of Frontiers of Computer Science and Technology* 10 (5): 722-731

[12] P Yao, H Wang. 2016. "Dynamic Adaptive Antlion Optimizer Applied to Route Planning for Unmanned Aerial Vehicle." *Soft Computing* :1-14.

[13] Y Wang, ZZ Liu, J Li, HX Li, et.al . On the selection of solutions for mutation in differential evolution[J] . Frontiers of Computer Science , 2017 (4) :1-19

[14] R Fabregat, Y Donoso, F Solano , et.al . Multi-objective optimization scheme for multicast flows: a survey, a model and a MOEA solution [J]. International Ifip/acm Latin American Conference on Networking, 2005 :73-86.

[15] R Liu, J Li ,J Fan ,et.al . A Co-evolutionary Technique Based on Multi-swarm Particle Swarm Optimization for Dynamic Multi-objective Optimization [J] . European Journal of Operational Research , 2017 , 261 (3)

[16] S Huband, P Hingston, L Barone, et.al. A review of multi-objective test problems and a scalable test problem toolkit[J] . IEEE Transactions on Evolutionary Computation, 2006, 10 (5) :477-506.

[17] Z Meng , JS Pan , L Kong . Parameters with Adaptive Learning Mechanism (PALM) for the enhancement of Differential Evolution [J] . Knowledge-Based Systems, 2017, 141

IOP Conf. Series: Journal of Physics: Conf. Series **1267** (2019) 012011 doi:10.1088/1742-6596/1267/1/012011

Research on Duration Estimation of Rotor UAV Based on Flight Condition-Energy Consumption Identification

Wei Zhang[1,*], Qingqing Ding[1], Chao Zeng[1] and Chao Wang[1]

[1]Department of Electrical Engineering, Tsinghua University, Beijing, China

[*]E-mail: zhangw16@mails.tsinghua.edu.cn

Abstract. The quad-rotor UAV (Unmanned aerial vehicle) has a wide application market for its simple structure, easy operation and strong adaptability. During the flight, the endurance of UAV is an important parameter for flight planning, and it is of great significance to master the endurance capability of UAV. The endurance of UAV is mainly decided by the remaining capacity of the battery and the future energy consumption which changes with flight conditions. In this paper, a Flight Condition-Energy Consumption Model is established by analysing a large number of flight data with machine learning method, and the effects of different regression algorithms are compared. The validity of the model is verified by the actual flight. The endurance of the aircraft can be estimated using this model with the given remaining capacity of the battery and the future flight tasks.

1. Introduction

Nowadays, quad-rotor UAV has been widely used in civilian and military markets. In civilian market, the application of Rotor UAV in Aerial Photography, Environmental Monitoring, Agriculture, Power Line Inspection and other fields is increasing because of its simple structure, easy operation and strong adaptability [1]. Besides stable attitude control and excellent motion ability, better endurance of UAV is also required for practical application. The flight duration of UAV is an important index. To estimate the remaining flight duration of UAV accurately and to master its endurance capability is helpful for formulating UAV mission, improving effective flight time, ensuring safe return and landing of UAV, and so on [2]. For rotor UAV products on the markets, the existed functions of low-power warning and automatic return are mainly achieved by SOC estimation and minimum voltage setting of battery. For flight of rotor UAV, the endurance is not only determined by the battery capacity, but also closely related to the flight conditions, wind intensity and other factors [3]. With flight conditions considered, the remaining duration of UAV can be estimated more accurately.

Firstly, this paper analyses the flight conditions and forces of the Rotor UAV, and then excavates the relationship between the flight conditions and the energy consumption of the UAV with the method of machine learning. Afterwards, the actual endurance of UAV is estimated with the future mission, remaining battery capacity and the identified relationship. At last, the estimated endurance is revised with the wind intensity considered.

2. Analysis of Flight Conditions of Rotor UAV

Rotor UAV is composed of frame, power system, navigation system and control system. The power system includes battery, electronic speed control, motor and propeller, which provides and transfers energy to support UAV flight. Four-rotor UAV uses four motors to drive propellers to provide tension, thus overcoming gravity and air resistance and realizing flight. The vast majority of battery power is supplied to motors and a small part to other equipment consumption in navigation and control systems. The speed of four motors is controlled by electronic speed control, which makes the attitude of UAV change. With the different attitude of UAV and the speed of motor, the propellers provide tension in different directions and sizes. Ultimately, UAV can achieve different flight states shown in the Figure 1, including hover, up and down motion, forward and backward motion, left and right motion and yaw motion.

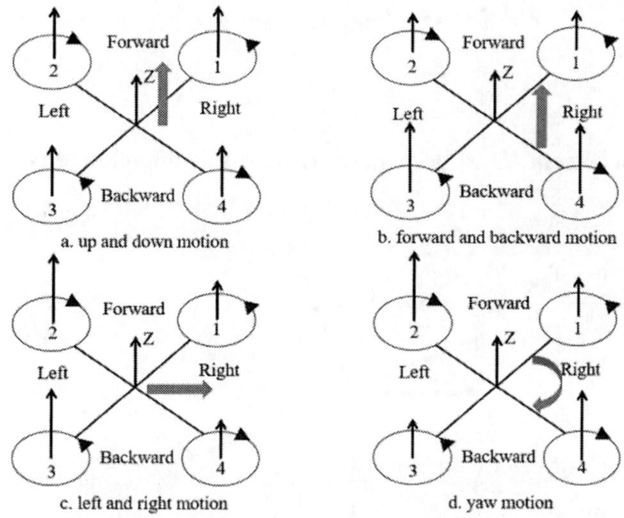

Figure 1. Different Flight Motions of UAV

By monitoring the voltage and current of the battery of UAV, the remaining power of the battery can be obtained. However, the remaining power does not represent the duration. The duration of UAV needs to be predicted according to the future flight conditions and the relationship between flight conditions and energy consumption. It is very complicated to find the relationship between flight conditions and energy consumption from the physical point of view. It is necessary to calculate the pull force provided by UAV according to the flight status of UAV. Then the power of battery, namely energy consumption, can be obtained through propeller mode, motor model and electronic speed control model [4].

The simplest case is the hovering state of the Rotor UAV. The UAV withstands gravity and tension from propeller, and they are equal. The tension provided by the propeller can be expressed as

$$T = C_T \rho \left(\frac{N}{60} \right)^2 D_p^{\,4} \tag{1}$$

Among all the parameters, C_T is the tension coefficient of propeller; ρ is air density; N is the rotate speed of propeller; D_p is the diameter of propeller.

$$N = 60 \sqrt{\frac{G}{n D_p^{\,4} C_T \rho}} \tag{2}$$

G is the gravity and n is the number of propellers.

$$M = C_M \frac{G}{n C_T} D_p \tag{3}$$

M is torque of the propeller and C_M is the torque coefficient.

$$P = M * N \tag{4}$$

P is the mechanical power of the propellers. After obtaining the torque and rotate speed, the output power of the battery can only be obtained through the motor model, the electronic speed control model

IOP Conf. Series: Journal of Physics: Conf. Series **1267** (2019) 012011 doi:10.1088/1742-6596/1267/1/012011

and the battery model [5]. The whole calculation process is complex and the efficiency of energy transfer needs to be considered, so that the final result may have a large error.

This paper uses a large number of flight data from the point of view of data and statistics, and determines the relationship between UAV flight conditions and energy consumption using the method of machine learning. In the NED coordinate system, the flight conditions of UAV can be described by the velocities and accelerations in three directions.

In the process of UAV flight, besides gravity and tension of propeller, UAV is also subject to air drag. For example, in the process of UAV flying forward at uniform speed, F_R is the air drag and its expression is as follows:

$$F_R = \frac{1}{2} C_D \rho v^2 S \tag{5}$$

C_D is the air drag coefficient; S is the windward area; v is the relative velocity between UAV and air.

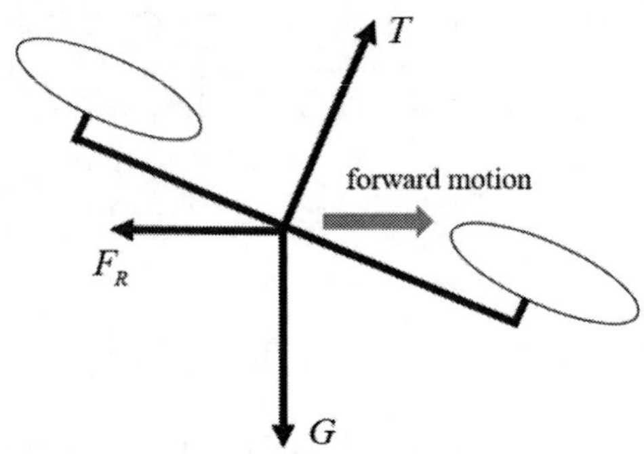

Figure 2. Force Analysis of the Rotor UAV in forward motion

The output power of battery determines the mechanical power. From the equation 4, it can be seen that the mechanical power of the propeller is non-linear with the tension. Therefore, the power of the battery is mainly related to the tension. And from the Equation 5 and Figure 2, the tension is related to the square of velocity. Therefore, from a physical point of view, the output power of battery is related to the gravity, velocity and acceleration of UAV. In the process of machine learning, it is necessary to find meaningful condition characteristics such as the square of velocity according to the physical relationship between flight conditions and energy consumption.

Next, this paper will introduce several machine learning methods for regression prediction, and an energy consumption prediction model based on condition-energy consumption identification results will be established.

3. Machine learning algorithms for regression prediction

The energy consumption of UAV is continuous numerical data. Regression algorithm is needed in machine learning algorithm for condition-energy consumption identification. At present, the basic regression algorithms of machine learning are decision tree regression, linear regression, SVM (support vector machines), KNN (k-Nearest Neighbor algorithm) and so on. The ensemble learning algorithms mainly include boosting and bagging classes and random forest, which combine weak learners into strong learners and get better results. In addition, the neural network is also an algorithm used in regression prediction. The effect and performance of ensemble learning algorithms are generally better than that of basic regression algorithms. In this paper, GBDT, XGBoost, Random Forest, LightGBM and Neural Network are used to predict the data, and their prediction results are compared.

3.1. Gradient Boosting Decision Tree

Gradient Boosting Decision Tree is the full name of GBDT, is an improved learning algorithm based on gradient boosting machine and decision tree proposed by Friedman in 1999. The decision tree used by GBDT is Classification And Regression Tree, which uses variance to find the best partition point. The predicted value is the real mean value of all samples in the leaf nodes of the decision tree. The iterative process of GBDT is as follows: On the basis of the learner obtained after the last iteration, the negative gradient of the loss function of the model is used as a sample to minimize the loss function in this iteration step, and a new weak learner regression tree $h_t(x)$ is fitted. The strong learner is updated to $f_{t-1}(x)+h_t(x)$ and the next iteration can be entered. The specific algorithm flow of GBDT is as follows:

(1) The training set sample is $T=\{(x_1,y_1),(x_2,y_2),\cdots(x_m,y_m)\}$,and the loss function is $L(y,f(x))$.

(2) Initialize the weak learner as $f_0(x)$.

(3) For each iteration $t=1,2,\cdots T$, the following calculation is performed for each sample:

$$r_{ti}=-\left[\frac{\partial L(y_i,f(x_i))}{\partial f(x_i)}\right]_{f(x)=f_{t-1}(x)} \tag{6}$$

r_{ti} is the negative gradient of the loss function of the learner updated in the last iteration, which is used as the predicted value of the sample in this iteration step.

(4) Use new sample $\{(x_1,r_{t1}),(x_2,r_{t2}),\cdots(x_m,y_{tm})\}$ fitting a new regression tree t(x) and update the strong learner.

$$f_t(x)=f_{t-1}(x)+h_t(x) \tag{7}$$

(5) The final strong learner is the sum of all regression trees.

$$f(x)=f_T(x)=f_0(x)+\sum_{t=1}^{T}h_t(x) \tag{8}$$

3.2. XGBoost

XGBoost is a scalable end-to-end machine learning system for tree boosting, and it can be considered as an improvement of tree boosting algorithm based on GBDT. The loss function of XGBoost includes not only training loss but also regularization term. The regularization term can control the complexity of the model, which helps to avoid overfitting. The first derivative of loss function with respect to f(x) is calculated as residual, which is used to generate a new decision tree in GBDT. But the model residual is obtained by calculate the first derivative and the second derivative of loss function with respect to f(x) using the method of Taylor expansion in XGBoost. In the fitting process of regression tree in each iteration, the method of finding the best segmentation point is different between XGBoost and GBDT. In addition, XGBoost has done a lot of other optimizations, such as proposing a novel sparsity aware algorithm for handling sparse data, combining cache access patterns and data compression and sharding to improve the efficiency of the algorithm and so on [6].

3.3. LightGBM

LightGBM is a framework based on gradient boosting launched by Microsoft in 2016, which also uses decision tree algorithm. The framework has the advantages of fast training efficiency, less memory and high accuracy. Conventional implementations of GBDT need to, for every feature, scan all the data instances to estimate the information gain of all the possible split points and to find the best partition point, which makes the training slow and consumes a lot of memory [7]. LightGBM improves these shortcomings accordingly:

(1) GBDT uses pre-sorting algorithm when partitioning the tree nodes, which consumes a lot of space and time. LightGBM use a method which called Exclusive Feature Bundling. Instead of using all the features to scan to get the best segmentation points, some features are bundled together to reduce the dimension of the features in this method, which reduces the consumption of searching for the best segmentation points. The bundling of exclusive features uses the histogram-based

algorithm, and it discretizes the continuous floating feature values into k integers and constructs a histogram with a width of k. Finally, the histogram is used to find the optimal segmentation points for node partition.

(2) LightGBM uses a method called "Gradient-based One-Side Sampling" to calculate the gradient of samples. Instead of using all sample points to calculate the gradient, this method sample the samples to calculate the gradient, keep those samples with large gradients, and only randomly drop those samples with small gradients.

(3) In addition, the growth strategy of decision tree, optimal segmentation of feature values and parallel learning have been optimized in LightGBM.

3.4. Random Forest

In ensemble learning, Gradient Boosting class learners are generated by serial iteration, while Random Forest learners are generated in parallel, and there is no strong dependence between individual learners in Random Forest. The training methods of Random Forest algorithm are as follows: For datasets $\{(x_1,y_1),(x_2,y_2),\cdots(x_m,y_m)\}$, new T datasets are sampled by Bootstrap method and each new dataset contains m sample $\{(x_{t1},y_{t1}),(x_{t2},y_{t2}),\cdots(x_{tm},y_{tm})\}$. A basic decision tree is trained based on each new dataset, and all the basic decision trees are combined into an ensemble learner. When new sample is input, the result of random forest output is determined by the majority voting method based on the predicted results of each decision tree. Random feature selection is introduced into the training process of decision tree. When each node of the decision tree is split, a feature subset (usually the number of features is $\log_2 k$) is randomly and uniformly extracted from all k features, and then an optimal splitting feature is selected from the subset to fit the decision tree.

After Random Forest algorithm is proposed by Leo Breiman, because of its good performance, the algorithm has been widely used in many fields such as bioinformatics, economic and financial fields, computer vision and so on.

3.5. Back-Propagation neural network

Multi-layer perceptron (MLP) neural networks are often utilized as estimation tools instead of the classical statistical methods. MLP neural network is a network structure formed by the interconnection of neurons, including input layer, hidden layer and output layer, which is shown in Figure 3. The input layer neurons receive external data input, the hidden layer and the output layer neurons deal with the data, and the final results are output by the output layer neurons. The learning process of neural network is to adjust the connection weight between neurons and the threshold of each functional neuron according to the training data.

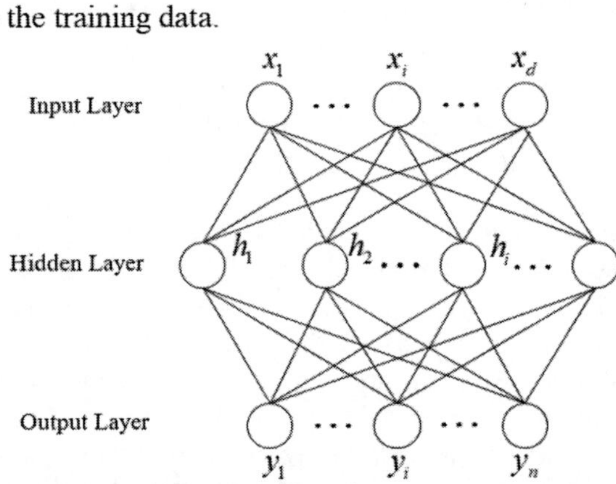

$$h_i = \sum_{i=1}^{d} \omega_i x_i$$

$$y_i = \sum_{i=1}^{m} \lambda_i h_i$$

Figure 3. The structural representation of MLP neural network

Error back propagation algorithm is the most successful learning algorithm of neural network so far. In forward propagation, the input signal is supplied to the neurons in the input layer, and then the signal is forwarded layer by layer until the result of the output layer is produced. If the output result

IOP Conf. Series: Journal of Physics: Conf. Series **1267** (2019) 012011 doi:10.1088/1742-6596/1267/1/012011

does not match the expected output, the error will be propagated backwards, and the connection weights and thresholds of neurons in each layer will be modified. Repeated training will be stopped until the output error meets the requirement.

4. The model of energy consumption prediction based on machine learning

4.1. Data preparation
Machine learning needs a lot of flight data, and the flight condition-energy consumption model is obtained by training the data. The Four-rotor flight data used in this paper is in "ulog" format, which includes position data, attitude data and battery data during flight.

4.2. Extraction of Features and Target Value
Machine learning is a process of training the relationship between features and target values. It is necessary to extract features and target values from sample data for training. In this paper, the features represent the flight condition of UAV and the target value is the energy consumption. It is necessary to use features to accurately describe the flight condition segment, and to ensure that the condition information will not be lost and distorted as far as possible. The flight conditions are described by the features in the time interval T and T is temporarily taken 0.5s. Many features related to velocity and acceleration can be selected before training begins. In the process of training, the features are adjusted according to the importance of features. The symbols and meanings of the final features after training are shown in the table 1 below.

Table1. Symbols and meanings of flight conditions features

	Feature Symbol	Feature Meaning (NED coordinate system)
Feature 1	V_{mean_N}	The mean velocity in the North direction
Feature 2	V_{mean_E}	The mean velocity in the East direction
Feature 3	V_{mean_D}	The mean velocity in the Down direction
Feature 4	V_{diff_N}	The difference between the ending speed and the beginning speed in the North direction
Feature 5	V_{diff_E}	The difference between the ending speed and the beginning speed in the East direction
Feature 6	V_{diff_D}	The difference between the ending speed and the beginning speed in the Down direction
Feature 7	V_{mean_NE}	The mean velocity on the North-East plane
Feature 8	V_{mean_NED}	The total mean velocity

The target value is the energy consumption E_T in time period T, which can be obtained according to the voltage and current data in sample data.

$$E_T = \int_{t}^{t+T} (U * I)dt \tag{9}$$

4.3. The training of sample data
After extracting feature data and target values from sample data, machine learning algorithm can be used for training. This paper uses the five machine learning algorithms introduced in Section 3 for training. There are 2217 samples in the sample set, 70% of which are trained as training set and 30% are tested as test set. The quality of the predict result is measured by three indicators, which are the determination coefficient R_2 ,the maximum deviation degree $E_{\max}$ and the mean absolute error E_{MAE} .

IOP Conf. Series: Journal of Physics: Conf. Series **1267** (2019) 012011 doi:10.1088/1742-6596/1267/1/012011

$$R_2 = 1 - \frac{\sum_{i=0}^{n-1}(y_i - \hat{y}_i)^2}{\sum_{i=0}^{n-1}(y_i - \bar{y})^2} \tag{10}$$

The determination coefficient R_2 is used to compare the predicted value error and the mean value error, and the prediction ability of the model is measured. The closer the predicted value is to the true value, the larger the R_2 is, and the maximum value is 1.

$$E_{\max} = \max(abs(y_i - \hat{y}_i)_{i\in[0,n-1]}) \tag{11}$$

Considering that the model should be applied to the energy consumption identification during the flight, there can be no large error, otherwise the final remaining duration estimation will be deviated greatly. $E_{\max}$ cannot exceed the maximum tolerance.

$$E_{MAE} = \frac{1}{n}\sum_{i=0}^{n-1}\left|y_i - \hat{y}_i\right| \tag{12}$$

The mean absolute error is the average difference between the predicted value and the real value, and also represents the accuracy of the prediction.
In the training process, the parameters of the algorithm are adjusted by controlling variables to get a better prediction model.

4.4. Experimental result

Table 2. Indicators of each algorithm model.

Algorithm model	R_2	E_{MAE} (W·s)	$E_{\max}$ (W·s)
GBDT	0.8733	1.632	4.098
XGBoost	0.8794	1.602	4.116
LightGBM	0.8691	1.657	4.138
Random Forest	0.8707	1.630	4.194
Back-Propagation neural network	0.8579	1.694	4.235

According to the training results in Table 2, it can be found that the performance gap of the five algorithms on the test set is not very large. The maximum deviation degree of the target value prediction results is about 4, the mean absolute error is about 1.6, and the determination coefficient R_2 is about 0.87. The number of hidden layers and the number of neurons in each layer of the neural network have a great influence on the prediction results, and LightGBM has the fastest training speed. Considering comprehensively, the prediction model of GBDT is selected for subsequent application.

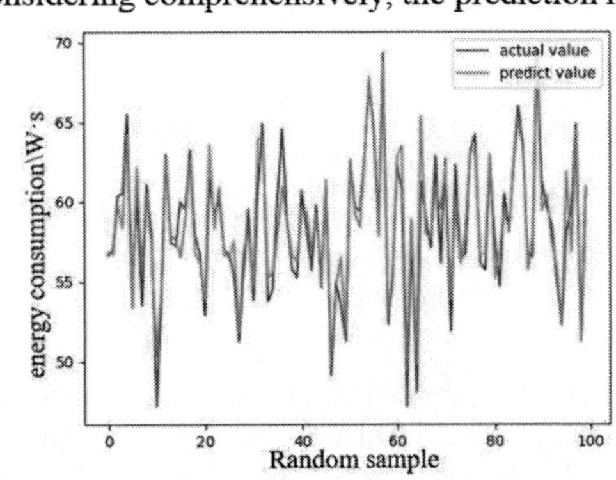

Figure 4. The predicted and actual values of GBDT in 100 samples randomly selected

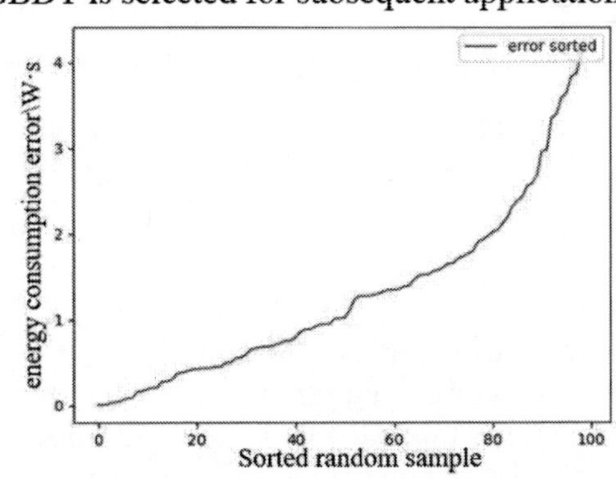

Figure 5. The absolute value of the predicted and actual errors of 100 samples

Figure 4 shows the situation of the predicted and actual values of GBDT in 100 samples randomly selected from the test set, and Figure 5 shows the absolute value of the predicted and actual errors of 100 samples, and ranks them. The results show that the prediction errors of most data samples are very small.

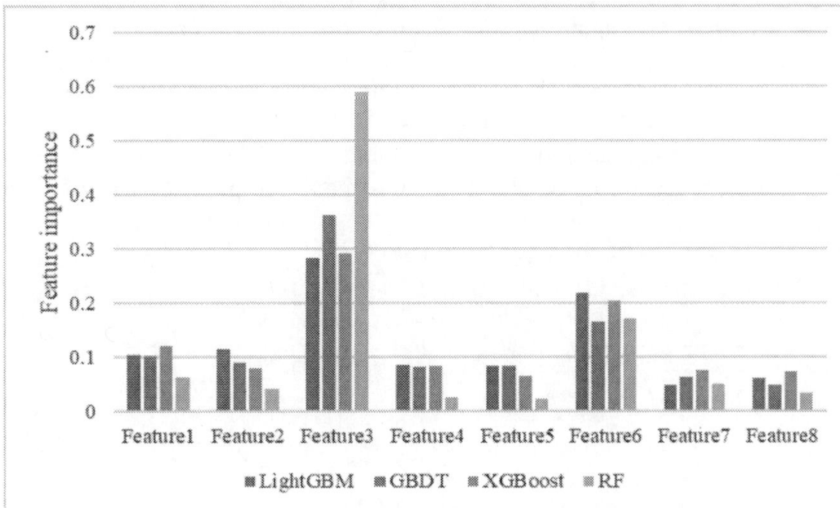

Figure 6. Feature importance of different models

Figure 6 shows the feature importance of LightGBM, GBDT, XGBoost and RF training models. The proportion of feature 3 and feature 6 is larger, because the lift motion of UAV has the greatest impact on energy consumption.

4.5. Energy consumption revision considering wind power

The air resistance caused by wind has a great impact on the flight energy consumption of UAV. It is necessary to obtain wind data during flight and revise the predicted energy consumption. The air resistance of UAV is caused by the relative velocity of UAV and air as shown in equation 5, and only horizontal direction is considered. In the case of wind, the airspeed of UAV is the vector difference between ground speed and wind speed. At this time, the air resistance is equal to F

$$F = k(\vec{v}_{ground} - \vec{v}_{wind})^2 \tag{13}$$

$$k = \frac{1}{2} C_D \rho S \tag{14}$$

The coefficient k is related to damping coefficient C_D , air density ρ and windward area of airframe. The damping coefficient and air density can be considered a constant, and the windward area can be obtained by the current attitude of UAV. So in the actual flight process, the air resistance F can be calculated according to the current wind speed and direction in the case of windy, Combining the displacement Δx , the energy consumption E obtained from flight condition prediction is revised to E_{rev} .

$$E_{rev} = E + \vec{F} \bullet \Delta\vec{x} \tag{15}$$

5. Estimation of flight duration for the UAV

Through the training of the flight condition-energy identification model, combined with the remaining capacity and cut-off voltage of the battery, the remaining endurance estimation of the UAV can be performed.

When the remaining capacity or voltage drops to a certain level, it means that the battery can no longer maintain normal flight and the UAV must complete landing before that. In this paper, we consider the estimation of the residual duration for the mission scenario which the future mission is known. When the flight mission is known, the future flight conditions of UAV, including speed, acceleration, flight duration, and so on of each flight condition, can be obtained from the flight mission [8][9]. According to the flight condition-energy consumption model, the energy consumption of each flight segment can be known, and the remaining battery capacity need be combined to estimate whether the mission can be completed. The UAV's mission over a period of time is shown below:

(1) It rises uniformly at a speed of 1m/s for 5 seconds.
(2) It hovers at fixed point and fixed altitude for 10 seconds.
(3) It flies at a constant speed of 4 m/s along the North-axis keeping the altitude unchanged for 5 seconds.
(4) It lands uniformly at a speed of 1 m/s for 5 seconds.

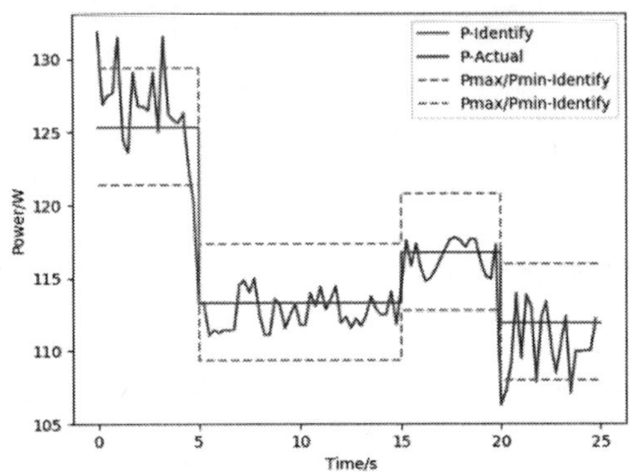

Figure 7. The predicted power and actual power in flight

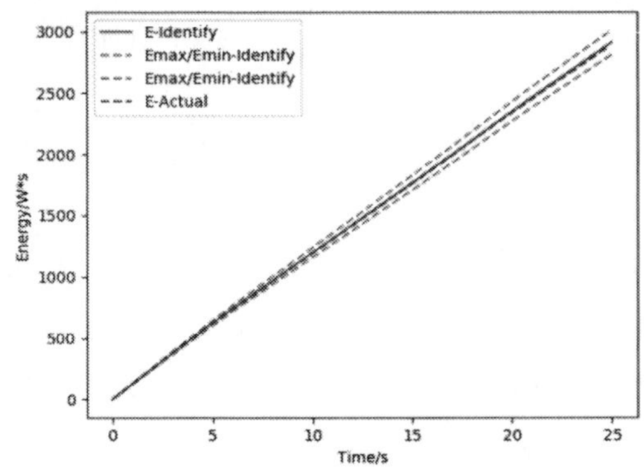

Figure 8. The predicted energy consumption and actual energy consumption in flight

The Figure 7 and Figure 8 respectively show the power and the energy consumption during flight, including the predicted value according to the flight condition, the actual value, the maximum and minimum predicted values. As can be seen from the figures, the actual value of flight is within the error range of the predicted value based on the flight condition. By predicting the energy consumption required for the mission and combining the current residual capacity of the battery, it can be estimated whether the mission can be completed in the future.

When the flight mission is unknown, the average energy consumption of UAV can be estimated by using the flight condition data of UAV in the past period of time. It replaces the energy consumption in the future flight process. In this case, the accuracy of the remaining duration cannot be guaranteed, so more safety landing time margins need to be set up to avoid UAV accidents during flight [10].

6. Conclusion

This paper analyses the flight conditions and the corresponding different energy consumption. Based on a vast amounts of flight data, this paper establishes the Flight Condition-Energy Consumption model using machine learning method, and compares the results of different regression algorithms such as GBDT, LightGBM, Random Forest and so on. According to the actual flight data, the predicted energy consumption is compared with the actual energy consumption, which proves the validity of the model. Based on the flight condition-energy consumption model, combined with the remaining capacity of battery and the future flight mission, the endurance of UAV can be estimated.

References

[1] Xiong D, Wang B, Dai L and Wang Q 2015 Prediction method of UAV inspection surplus cruise duration based on SOC *Chinese Journal of Power Sources* **39(01)** 105-7

[2] Liu F and Ma X 2010 Research for improving the endurance performance of minitype electric-powered UAV *Flight Dynamics* **28(05)** 13-5

[3] Cong S, Wang C and Jiang Y 2012 Research on Real-time Estimation Method for Range and Cruise Duration of UAV *Journal of Astronautic Metrology and Measurement* **32(2)** 62-5

[4] Liu S, Li Q, Lu N and Bin J 2014 Real-time estimation of cruise duration for electric-powered Micro-UAVs *Control & Decision Conference*. IEEE.

[5] Meng J and Zhao Z 2013 Modeling and Simulation of Microquadrotor UAV *Journal of Lanzhou Jiaotong University* **32(01)** 63-7.

[6] Chen T and Guestrin C 2016 XGBoost:A Scalable Tree Boosting System *Acm Sigkdd International Conference on Knowledge Discovery & Data Mining*.

[7] Ke G, Meng Q, Finley T, Wang T, Chen W, Ma W, Ye Q and Liu T 2017 Lightgbm: A highly efficient gradient boosting decision tree *Advances in Neural Information Processing Systems* 3146-54.

[8] Chi Q, Bole B, Hogge E, Vazquez S, Daigle M, Celaya J, Weber A and Goebel K 2013 Battery charge depletion prediction on an electric aircraft.

[9] Hogge E, Bole B, Vazquez S, Celaya J, Strom T, Hill B, Smalling K and Quach C 2015 Verification of a remaining flying time prediction system for small electric aircraft.

[10] Liu G, Ouyang M, Lu L, Han X and Gu J 2014 Driving Range Estimation for Electric Vehicles Based on Battery Energy State Estimation and Vehicle Energy Consumption Prediction *Automotive Engineering* **36(11)** 1302-9.

IOP Conf. Series: Journal of Physics: Conf. Series **1267** (2019) 012012 doi:10.1088/1742-6596/1267/1/012012

Facial-image based Age Estimation Using Imbalanced Datasets

Yasi Liu[1], Tianzi Wang[2], Limin Zheng[1,*] and Lu Yang[1]

[1]China Agricultural University, College of Information and Electrical Engineering, No. 17, Qinghua East Road, Haidian District, Beijing, China, 100083
[2]Virginia Tech, ISE department, Blacksburg, USA

[*]Limin Zheng, E-mail: zhenglimin@cau.edu.cn

Abstract. Facial image based human age estimation is of great application significance. The public-available facial image datasets used for age estimation suffer greatly from the uneven distribution of images of different age groups, which may lead to the low estimation accuracy of the under-sampled age categories and limit the usage of the age estimation in certain applications. We propose a three-stage probability adjustment based CNN algorithm to solve the imbalanced distribution problem of the dataset. In particular, we construct an ENIN neural network structure by applying the Network in Network (NIN) structure to the traditional convolution neural network (CNN) and use the probability vector adjustment to improve the classification accuracy of the under-sampled age categories. Then, we filter out the images with high possibility of being misclassified after the probability vector adjustment and reset their categories by comparing cosine similarity and retraining the ensembled ENIN classifier. We also introduce a population-age-distribution based accuracy metric *Accuracy-P* to estimate the performance of the age estimation algorithm in real-world applications. Our experimental results confirm that our algorithm can effectively improve the overall estimation accuracy by significantly improving the accuracy of the under-sampled age groups while maintaining satisfactory accuracy for the other age groups.

1. Introduction

In many age-restricted places or applications, coarse-grained facial image based human age estimation plays important roles, such as prohibiting minors from accessing the harmful information on the internet, recognizing face with a long age span and has broad application prospects in the search for lost children.

Computer vision technology leverages the common features on human faces for the facial image based age estimation. In specific, prediction models [1-3] that use the age-related features (e.g., relative distances of the main parts and the wrinkles in the specific areas of the face [4]) for categorizing the facial image into certain age groups are trained. In such model training procedures, most existing approaches use public-available datasets of facial images labeled with age.

One major problem in these approaches is that the public-available datasets they leverage are greatly imbalanced in the amount of images of different age groups, with certain age groups (e.g., the senior, the child) have fewer images than other groups (e.g., adult). Such uneven distribution in the image datasets may cause low estimation accuracy of the under-sampled age groups (i.e., the minority age groups), which will further affect the overall estimation accuracy and limit the usage of the age estimation algorithm in certain application scenarios (e.g., recognizing senior for ticket discount). Such observations are confirmed by our study, as well as the classification results obtained by the

IOP Conf. Series: Journal of Physics: Conf. Series **1267** (2019) 012012 doi:10.1088/1742-6596/1267/1/012012

commonly used classification algorithms mentioned above [4-6]. Although there are several resampling algorithms [7-17] to balance the distribution of the sample data and several cost-sensitive learning algorithms [18-21] to adjust the cost of dividing the sample into categories with different sample sizes, directly applying these algorithms won't increase the accuracy of the under-sampled groups to a satisfactory value.

To solve the problem, this paper introduces a novel accuracy metric *Accuracy-P* to evaluate how an age classification algorithm fits real-world applications, as well as a three-stage probability adjustment based CNN algorithm.

The rest of this paper is organized as follows. Section II analyzes the problems caused by the imbalanced age distribution of three popular datasets and points out the irrationality of the overall accuracy while applying it to the imbalanced datasets. Section III introduces the novel population-age-distribution based accuracy measurement *Accuracy-P*. Section IV introduces the three-stage probability adjustment based CNN algorithm. Section V shows the effectiveness of the algorithm in the image datasets. Section VII concludes this paper.

2. Datasets Analysis

2.1 Facial-Image Dataset

This article covers three datasets. The FG-NET [22] dataset ranges in age from 0 to 69. The MORPH-II [23] dataset ranges in age from 16 to 77. A general way of using the Adience [24] dataset for age classification is to divide the age into 8 categories. Some facial images in these datasets are shown in Figure 1.

The FG-NET dataset and the MORPH-II dataset are often used together as the FG-NET + MORPH dataset [4], since there are no facial images of people age 0 to 16 in the MORPH-II, and the number of facial images in the FG-NET is relatively small. The age is divided into four categories: *Children* (0-11 years old), *Teen Age* (12-21 years old), *Adult* (22-60 years old), and *Senior Adult* (> 60 years old).

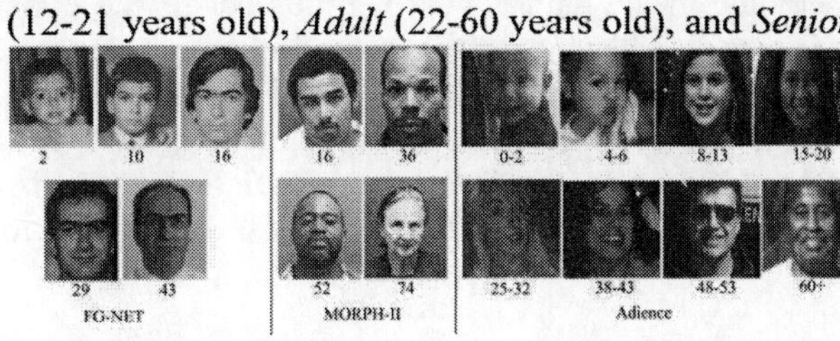

Figure 1. The facial images in the datasets used in this paper.
The number under each image represents the age.

2.2 Imbalanced Image Distribution of Age Categories

The population age distribution refers to the proportion of each age category in the total population. We use the Population Pyramids [25] to observe the population age distributions in different countries or regions in 2018, as shown in Figure 2. The population age distribution of Japan is used to represent a typical aging population distribution. The population age distribution in the world creflects the age distribution of the global population. That is, the proportion of the population in the world decreases with the increase of age. The age distributions in the FG-NET+MORPH dataset and the Adience dataset are shown in Table 1 and Table 2 , respectively. It is obvious that the population age distribution under natural conditions and the image distributions of age categories in the datasets are both imbalanced.

In the population age distributions of either Japan or the world, the *Senior Adult* (over 60) takes a much larger portion of the population (13%-34%), compared with the numbers shown in the datasets (1.6% ~ 4.7%). However, in real-world applications of general purposes, we expect the collected images will basically fit the population age distributions.

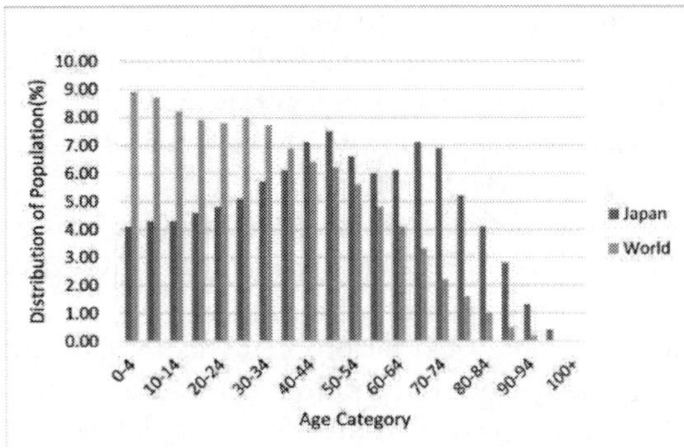

Figure 2. Population age distribution in Japan and the world in 2018.

Table 1 Age distribution in the FG-NET+MORPH dataset.

	Children	*Teen Age*	*Adult*	*Senior Adult*	Total
FG-NET	444	302	249	7	1002
MORPH-II	0	3081	11865	258	15204
FGNET+MORPH	444	3383	12114	265	16206

Table 2 Age distribution in the Adience dataset.

	0-2	*4-6*	*8-13*	*15-20*	*25-32*	*38-43*	*48-53*	*>60*	Total
Adience	1820	1586	1694	1121	3399	1598	572	583	12373

2.3 The Problem Caused by the Imbalanced Distribution

The low classification accuracies of the under-sampled age categories are ubiquitous problems of the existing age estimation methods. In the datasets used in this paper, the classification accuracy of the *Senior Adult* age category is significantly lower than that of other age categories.

In the existing image classification applications, the overall accuracy of the classification algorithm represents the probability of correctly classify an arbitrary image in the dataset. However, as the age distribution of the datasets differs greatly from the population age distribution, such accuracy measurement cannot be applied to predict the performance of the algorithm in a real-world application scenario.

3 Population-Age-Distribution based Accuracy Measurement

To evaluate the effectiveness of an age estimation algorithm in real-world applications, we introduce a population-age-distribution based accuracy measurement *Accuracy-P,* defined as:

$$Accuracy\text{-}P = \sum_{i=1}^{C} Accu_i \times Prop_i \Big/ \sum_{i=1}^{C} Prop_i \quad (1)$$

where i represents different age categories, $Accu_i$ and $Prop_i$ represent the classification accuracy and the population proportion of age category i of the natural age distribution, respectively. *Accuracy-P* estimates the probability of correctly classifying an image of an arbitrary person, whose age distribution follows the natural population age distribution.

To better understand the novelty of *Accuracy-P*, we compare it with the accuracy metrics widely used in existing approaches, the accuracy metrics can be denoted as:

$$Accuracy - D = \sum_{i=1}^{C} Accu_i \times Prop_i' \Big/ \sum_{i=1}^{C} Prop_i' \quad (2)$$

where i represents different age categories, $Accu_i$ and $Prop_i'$ represent the classification accuracy and the proportion of images of age category i in the datasets. By comparing equation (1) and equation (2), we can see that the existing accuracy metrics are affected by the imbalanced distribution of the collected image datasets. On the contrary, *Accuracy-P* can be used to reflect how accurate an algorithm is while processing facial images collected by real-world applications.

4 Three-Stage Probability Adjustment based CNN Algorithm

4.1 Algorithm Overview

In this section, we introduce our three-stage facial image based age estimation algorithm in details. The entire algorithm focuses on calculating the correct probability vector — the probabilities of an image belonging to each age category. The overall workflow of the algorithm is shown in Figure 3, which mainly contains three steps.

(1) We first calculate the probability vector $\boldsymbol{P}$ using the enhanced NIN (ENIN) network.

(2) We generate another probability vector $\boldsymbol{P'}$ considering the imbalanced image distribution in the training set.

(3) We further redefine the categorization result by filtering out the images in the under-sampled age category with a high possibility of being miscategorized using $\boldsymbol{P'}$, and reset their categorization result using a new probability vector $\boldsymbol{P''}$ constructed by applying a new cost function to $\boldsymbol{P'}$ and using a new probability vector $\boldsymbol{P_e}$ constructed by retraining an ensemble ENIN classifier.

In the rest of this section, we will introduce in details the procedure of each stage.

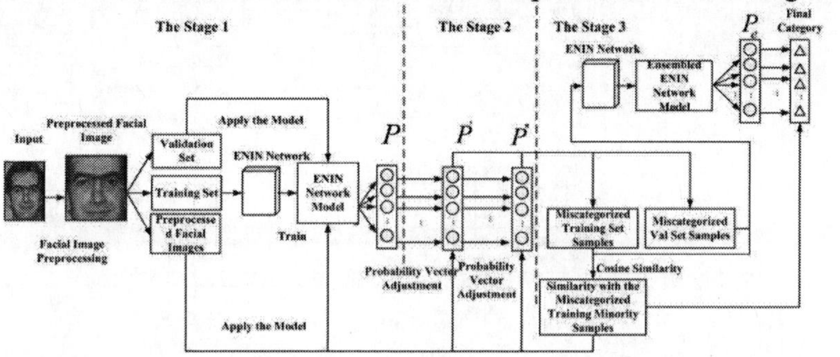

Figure 3. The workflow of the facial image based age estimation.

4.2 Stage 1: Applying the Enhanced NIN Algorithm

First, we follow the facial image preprocessing procedures proposed by Baggio [26] to effectively extract the facial features and convert the original images to the target images of 256×256 pixels. The effect of the facial image preprocessing is shown in Figure 4.

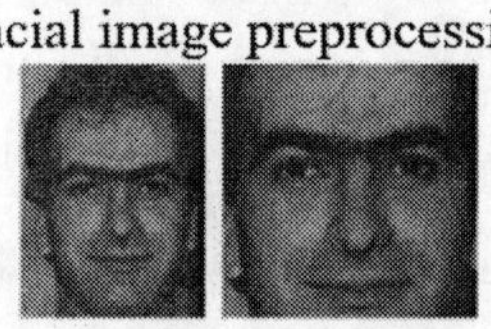

(a) (b)

Figure 4. The effect of the facial image after preprocessing. (a) The original image. (b) The preprocessed image.

We use the term ENIN to refer to the enhanced NIN network model. It enhances the effectiveness of the NIN [27] network by adding a few network structures, as shown in Figure 5. We first apply *Mlpconv* structure to the network, as the *Mlpconv* structure has strong abstraction and generalization ability and is more capable of expressing the rich features of facial images than the common convolution layer.

In particular, the structure of the slice layer and the eltwise layer is shown in Figure 6. The slice layer slices the output of the last layer into multiple output layers based on a given dimension and the slice indices. The eltwise layer uses the element-level max operation to take the maximum values of these multiple output layers to reduce the dimension of data. Our network structure is divided into two parts in the 4th Mlpconv structure. The first part contains conv4, conv7, and conv8 layers, and the second part cascades the features of the conv4 and cccp7 layers. Then, cccp9 is used to reduce the noise and select the features. The features of cccp8 and cccp9 are combined and sent into the fully-connected layer.

Each dataset in this paper is divided into three parts: the training set, the verification set, and the test set. The sample size of each part accounts for 80%, 10%, and 10% of the total sample size of the dataset. The test set is used to evaluate the accuracy of the ENIN model and our algorithm.

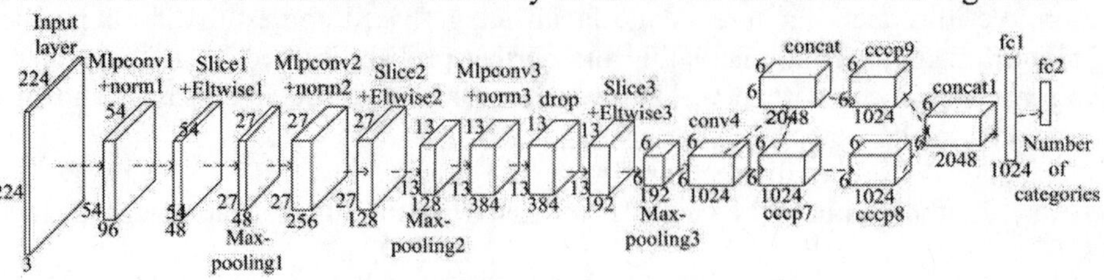

Figure 5. ENIN's network structure.

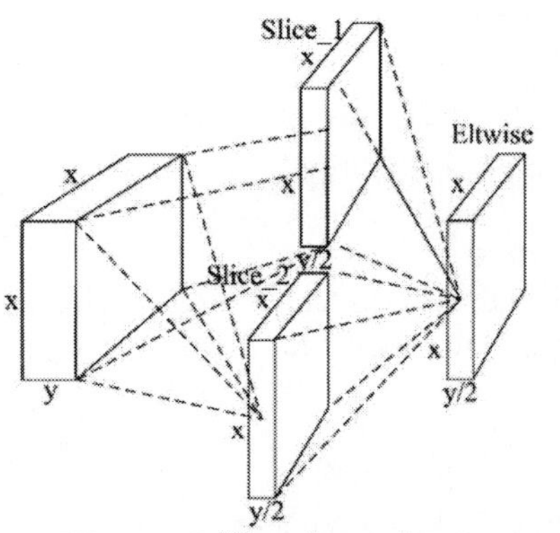

Figure 6. The *slice+eltwise* layers.

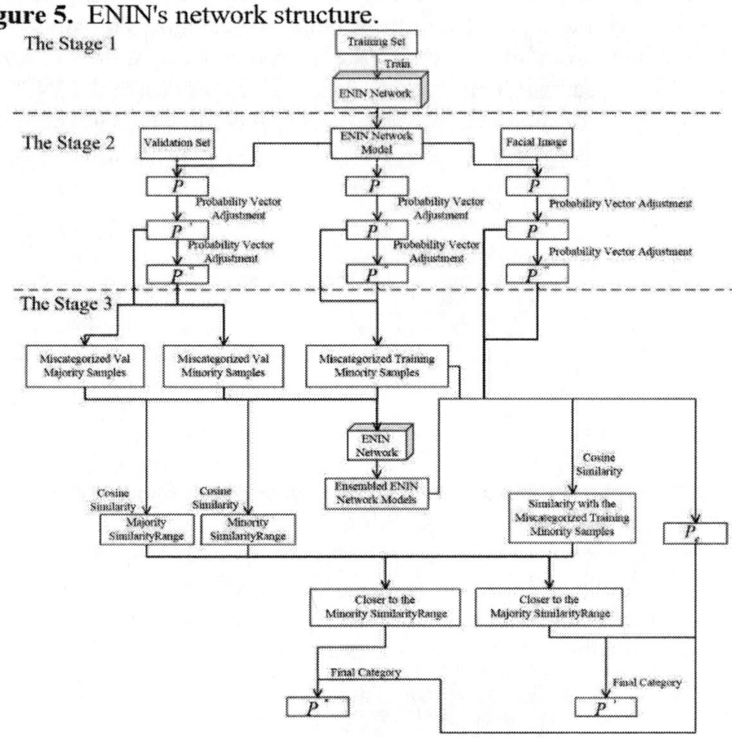

Figure 7. The workflow of our algorithm.

4.3 Stage 2: Adjusting the Probability Vector

Due to the influence of imbalanced data distribution, simply applying the ENIN may cause low age classification accuracy for the minority age categories. As a solution, we apply a cost function, which is calculated based on the image distribution of each age category, to adjust the probability vector generated by the ENIN algorithm. Figure 7 shows the detailed workflow of the second stage. The key idea behind the cost function is the threshold moving method [18].

Threshold moving uses a cost function to adjust the probability vector, as shown in equation (3).

$$P_i^* = \eta\sum_{c=1}^{C} P_i Cost[i,c] \tag{3}$$

where P_i represents the probability output of the trained CNN model, C represents the total number of categories, $i \in \{1..C\}$ represents different age categories. $\sum_{i=1}^{C} P_i = 1$, $0 \le P_i \le 1$, the predicted category value determined by the trained CNN model is $L = \arg\max_i P_i$. η is a normalization item, which guarantees $\sum_{i=1}^{C} P_i^* = 1$. *Cost*[*i,c*] represents the cost of misclassifying a sample of the category *i* into a category *c*, which affects the degree of changing the probability.

The first stage of our algorithm outputs the probability vector ***P*** that an image belongs to each age category. Then, we calculate ***P'*** according to equation (3) and equation (4).

$$Cost_1[i,c] = \begin{cases} (log(N_c/N_i+1))^2 & if \quad N_i < N_c \& c \neq i \\ log(N_c/N_i+1) & if \quad N_i > N_c \& c \neq i \\ 0 & if \quad c = i \end{cases} \tag{4}$$

where N_i and N_c represent the sample size of corresponding age categories in the training set.

From equation (4), we can observe that applying the cost function adjusts the probability vector of an image by increasing the probability of the minority age groups and decreasing the probability of the majority age groups, as the cost of misclassifying an image into a majority category is higher than that of misclassifying it into a minority category.

The experiment results indicate that the probability vector adjustment procedure improves the classification accuracies of the under-sampled age categories while maintaining the satisfactory accuracies of other categories. However, the classification error rate is still high for the under-sampled age categories.

4.4 Stage 3: Category Redefinition of the Mis-Categorized Image

To further reduce the misclassification rate of images from the under-sampled age categories, we adopt a category redefinition procedure to redefine the age categories of the images with high possibility of being misclassified after the probability vector adjustment. The procedure contains several steps.

First, we collect the misclassified images from the under-sampled age categories. We apply the ENIN model and the probability vector adjustment to the images in the training set and the validation set, and compare their classification results with their actual age categories and identify the misclassified images from the minority age categories. We name these misclassified images of different minority categories in the training set as *Samtrainmireclassi*, and name these misclassified images of different categories in the validation set as *Samvalreclassi*.

Second, for these misclassified images, we apply a second probability vector adjustment procedure to generate a new probability vector ***P''***. ***P''*** is calculated by applying a new cost function to the probability vector ***P'***, as shown by equation (5) and equation (6).

$$P'' = \eta_2 \sum_{c=1}^{C} P' Cost_2[i,c] \tag{5}$$

where *C* represents the total number of categories, $i \in \{1..C\}$ represents different age categories. ***P'*** $= \{ P_i' | \, i \in \{1..C\}, \sum_{i=1}^{C} P_i' = 1, 0 \leq P_i' \leq 1 \}$, and ***P''*** $= \{ P_i'' \mid i \in \{1..C\}, \sum_{i=1}^{C} P_i'' = 1, 0 \leq P_i'' \leq 1 \}$. η_2 is a normalization item, which guarantees $\sum_{i=1}^{C} P'' = 1$. The predicted category values of a given image determined by the twice probability vector adjustment are $L = \arg\max_i P_i'$ and $L = \arg\max_i P_i''$ respectively.

$$Cost_2[i,c] = \begin{cases} (log(N_c/N_i+1))^2 & if \quad c \neq i \\ 0 & if \quad c = i \end{cases} \tag{6}$$

where N_i and N_c represent the sample size of corresponding age categories in the training set.

We extract a vector [***P***, ***P'***, ***P''***] for each misclassified image from the minority set as features. The features of all misclassified images are compared with that of the input image, to generate similarity indexes. The cosine similarity of the vectors of a misclassified image and the input image is used as the similarity index, as shown in equation (7).

$$cos\theta = \frac{\sum_1^n (A_i \times B_i)}{(\sum_1^n A_i^2)^{1/2} \times (\sum_1^n B_i^2)^{1/2}} \tag{7}$$

where $A_i = (AP_i, AFTP_i, ASTP_i)$, AP_i represents the probability P_i that the predicted category value of sample A is i, $AFTP_i$ represents the probability P_i' that the predicted category value of sample A is i, and $ASTP_i$ represents the probability P_i'' that the predicted category value of sample A is i.

We further apply two similarity thresholds *MajoritySimilarityRange* and *MinoritySimilarityRange* to determine whether the similarity of the input image and a misclassified image is big enough that the input image should also be marked as misclassified. We calculate the similarity indexes of all images in *Samvalreclassi* and the *Samtrainmireclassi* to generate *MajoritySimilarityRange* and *MinoritySimilarityRange*. The *MajoritySimilarityRange* represents the similarity range between the correctly classified images in the validation set and the misclassified images in the training set after the probability vector adjustment. The *MinoritySimilarityRange* represents the similarity range between the misclassified images in the validation set and the training set after the probability vector adjustment.

We also train an ensembled ENIN classifier using the ENIN network structure and the misclassified images in the training set and the validation set after the probability vector adjustment. We create two datasets for each minority age category and we set that the misclassified image number of this minority category is the largest among all the categories. The misclassified images of other categories with more images are randomly selected. We train an ENIN classifier for the two datasets and use the image classification results of the ensembled ENIN classifier to assist the similarity comparison procedure to determine the final classification result of an image.

For an input image, we first apply stage 1 and stage 2 of the algorithm, to generate $\boldsymbol{P}$ and $\boldsymbol{P'}$. Then, we apply stage 3 and generate $\boldsymbol{P''}$, which improves the classification accuracies of the under-sampled age categories while decreasing those of other categories. If the classification results of $\boldsymbol{P'}$ and $\boldsymbol{P''}$ agrees with each other, we output it as the final result. Otherwise, we apply stage 3 of the algorithm and combine the classification results of similarity comparison and the classifier retraining to determine the final classification result. If any similarity index of the input image is closer to *MinoritySimilarityRange* than *MajoritySimilarityRange*, we set the classification result of the similarity comparison procedure as the classification result of $\boldsymbol{P''}$. We apply the ensembled ENIN classifier to the input image to generate the probability vector $\boldsymbol{P_e}$ and get the predicted classification results. If the classification result of the input image using ensembled ENIN classifier are consistent with the classification results of the similarity comparison and the predicted classification result belongs to the minority category, we set the final category value as the classification result of $\boldsymbol{P''}$. Otherwise, we believe that the image belongs to the majority age category, and output the classification result of $\boldsymbol{P'}$ as the final result.

5 Experiment and Results

In this section, we use the recall values and the *Accuracy-P* to display the age estimation effects.

5.1 The Classification Accuracy and the Values of the Accuracy Metrics

The age estimation effects of the various age groups at each step in the FGNET + MORPH dataset are shown in Figure 8. The Recall value indicates the correct classification proportion for a category. The Recall values of most age groups using the ENIN model are higher than those using the NIN model. Combined with the age distribution shown in Table 1 , we can see that the ENIN network model achieves better estimation accuracy in the majority age groups (e.g., the Adult age category) than the NIN.

Comparing the results using $\boldsymbol{P'}$ with the those of the ENIN, the Recall values of the minority groups (e.g., the Senior Adult age category) have improved significantly, which is consistent with the ability of the probability vector adjustment to improve the classification accuracies of the minority age groups. However, there is still a relatively high proportion of the minority samples that are misclassified.

$\boldsymbol{P''}$ further skews the age estimation results towards the minority age groups. Most of the minority samples misclassified by $\boldsymbol{P'}$ are correctly classified using $\boldsymbol{P''}$. So far the minority age groups have achieved high accuracies. However, affected by the probability vector adjustment, the Recall values of the majority age groups drop significantly.

We get the final age estimation results after applying Stage 3 of our algorithm. The Recall values of the minority age groups improve significantly than only using the ENIN model. The final age estimation effects in various age groups are relatively balanced and satisfactory.

To evaluate the overall age estimation effect of our algorithm, we calculate three accuracy metrics —*Accuracy-D*, *Accuracy-Japan,* and *Accuracy-World* at each step. The values of the accuracy metrics in the FGNET + MORPH dataset are shown in Figure 9.

Although the three accuracy metrics are dependent on the classification accuracy of each category in the dataset, they reflect the influence of different age distribution characteristics on the age estimation effect.

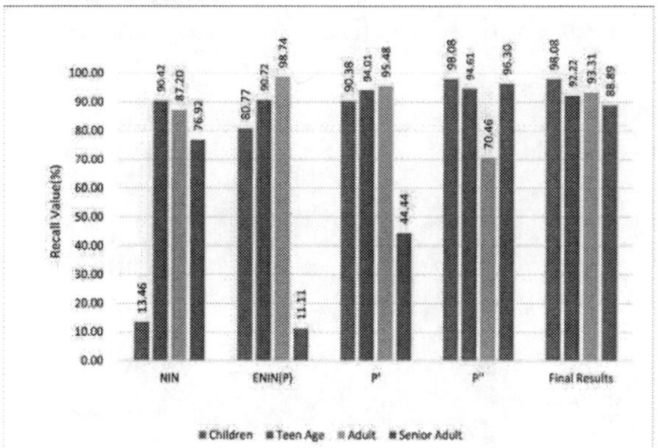

Figure 8. The results of each step of our algorithm in the FGNET + MORPH dataset.

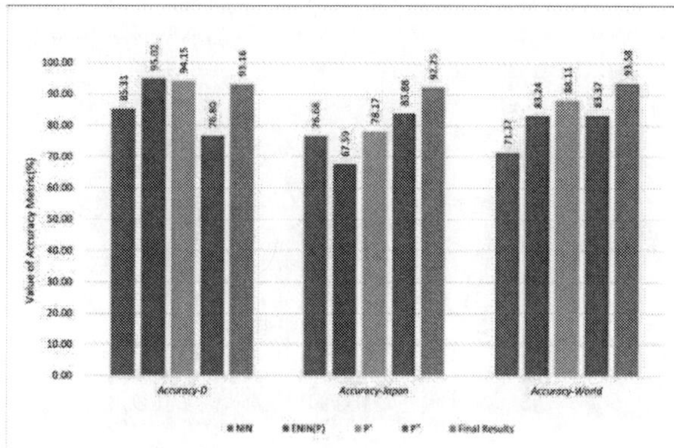

Figure 9. The values of the accuracy metrics in the FGNET + MORPH dataset.

The *Accuracy-Japan* represents the age estimation effect in real-world applications using the population age distribution of Japan. As can be seen from Figure 9, the *Accuracy-Japan* of the ENIN is lower than that of the NIN. After that, the *Accuracy-Japan* improves at each step of our algorithm. This is due to the change of the classification accuracy of each age category and the population age distribution characteristics of Japan's aging population. The population of the *Children* is small, and the total population of the *Adult* and the *Senior Adult* accounts for nearly 80% of the population of Japan. Therefore, the change of the *Accuracy-Japan* mainly follows the classification accuracy changes of the *Adult* and the *Senior Adult* categories. The *Accuracy-Japan* of the final result reaches a high value because the classification accuracies of various age groups are all satisfactory.

The *Accuracy-World* reflects the age estimation effect in the application using the age distribution of the global population. The *Accuracy-World* improves at each step of our algorithm except the step using $\boldsymbol{P''}$, which is also caused by the change of the classification accuracy of each age category and the distribution of the global age population. The population of the *Adult* age category accounts for over 50% of the total population, and the population distributions of other age groups are relatively balanced. Therefore, the change of the *Accuracy-World* mainly follows the classification accuracy changes of the *Adult* age category. The *Accuracy-World* of the final result also reaches a satisfactory level.

Similar to Figure 8 and Figure 9, Figure 10 shows the Recall values of various age groups in the Adience dataset, and Figure 11 shows the values of the accuracy metrics in the Adience dataset.

As shown in Figure 10, the changing trend of the classification accuracies at each step in the Adience dataset is similar to that in the FGNET + MORPH dataset. The ENIN model performs better than the NIN model. The probability vector adjustments skew the age estimation effects towards the minority age groups and decrease the classification accuracies of the majority age groups. The final age estimation results in various age groups after using our algorithm are relatively balanced.

IOP Conf. Series: Journal of Physics: Conf. Series **1267** (2019) 012012 doi:10.1088/1742-6596/1267/1/012012

The change of the values of the accuracy metrics in the Adience dataset is also similar to that in the FGNET + MORPH dataset, as shown in Figure 11. However, the age estimation effects in the Adience dataset are not as good as those in the FGNET + MORPH dataset. This is because the images in the Adience dataset are collected in an unconstrained environment, and the age estimation results are affected by factors such as illumination, posture, and expression, etc.

5.2 Comparison with Existing Approaches

For the FGNET+MORPH dataset, the results of our algorithm are compared with the results of the existing approaches [4-5]. As can be seen from Table 3 , our algorithm achieves the highest population-age-distribution based accuracies on the population age distribution in the world and achieves *Accuracy-Japan* that is very close to the highest *Accuracy-Japan*, which means that our algorithm performs well for any facial image in real-world applications following the natural population age distribution.

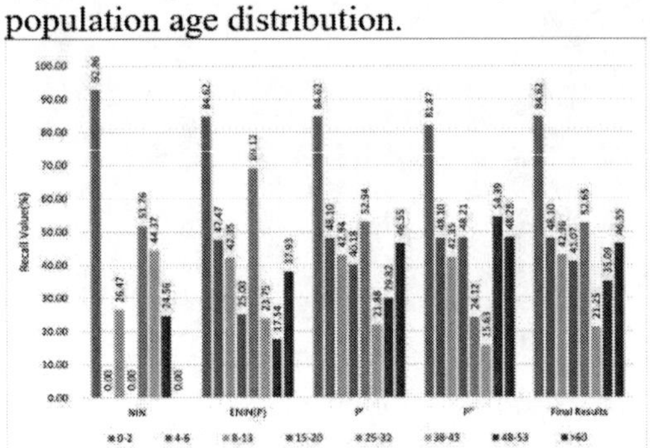

Figure 10. The results of each step of our algorithm in the Adience dataset.

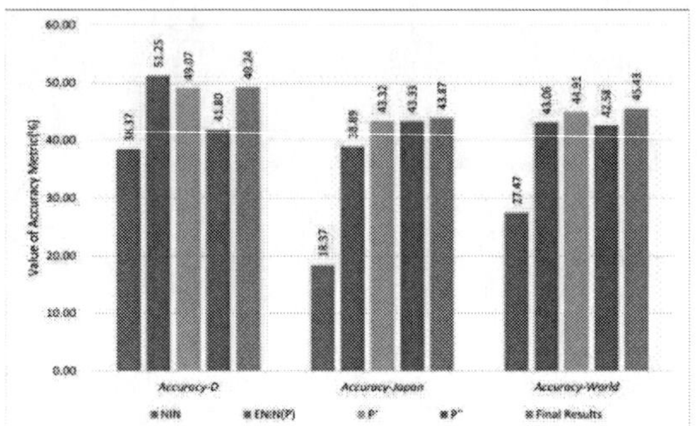

Figure 11. The values of the accuracy metrics in the Adience dataset.

Table 3 Comparison with the existing approaches in the FGNET+MORPH dataset.

Method	*Accuracy-D*(%)	*Accuracy-Japan*(%)	*Accuracy-World*(%)
E+A with L+G [6]	88.6	86.85	89.1
E+S with L+G [6]	90.2	88.6	90.7
SVM with SDM [5]	91.1	89.2	91.2
SVM with BB-FCN [5]	92.9	90.7	92.5
KNN with SDM [5]	92.3	91.1	92.53
KNN with BB-FCN [5]	92.9	**92.4**	93.21
Ours	**93.2**	92.3	**93.6**

We compare the results of our algorithm in the Adience database with the existing approaches [28-29]. As can be seen from Table 4 , our algorithm achieves the highest population-age-distribution based accuracies. Although the *Accuracy-D* of our algorithm is lower than that of the algorithm Levi & Hassncer [28] proposed, our algorithm performs better in real-world applications.

Table 4 Comparison with the existing approaches in the Adience database.

Method	*Accuracy-D*(%)	*Accuracy-Japan*(%)	*Accuracy-World*(%)
Appearance [29]	38.3	36.9	31.3
Appearance+Context [29]	42.9	39.4	36.3
Best from Ref. [28]	**54.5**	37.9	42.9
Ours	49.24	**43.87**	**45.43**

6 Conclusions and Future Works

In this paper, we propose a three-stage probability adjustment based CNN algorithm for facial-image based age estimation using imbalanced datasets. We construct an ENIN neural network and use the probability vector adjustment to improve the age estimation effect of the under-sampled age categories.

After that, the category redefinition combined with the cosine similarity comparison and the ensemble ENIN classifier retraining achieves balanced and satisfactory age estimation effects in various age groups. We also introduce a population-age-distribution based accuracy metric *Accuracy-P* to estimate the performance of the age estimation algorithm in the real-world application scenarios with the natural population age distribution. The age estimation effects of the algorithm in the two datasets FGNET+MORPH and Adience show that our algorithm achieves satisfactory age estimation accuracy and is suitable for the real-world applications.

Acknowledgments

This work was supported by the [National Key R&D Program of China] under Grant [No. 2017YFC1601800].

References

[1] Cootes, T. F, Edwards, G. J and Taylor, C. J 2001 *Pattern Analysis & Machine Intelligence IEEE Transactions on* **23** 681-85
[2] Geng, X, Zhou, Z. H and Smithmiles, K 2007 *IEEE Transactions on Pattern Analysis & Machine Intelligence* **29** 2234-40
[3] Chi, N, Quach, K. G, Luu, K, Le, H. B and Ricanek, K 2011 *Proceedings of the 36th Intl. Conf. on Acoustics, Speech and Signal Processing*
[4] Kwon, Y. H and Lobo, N. D. V 1999 *Computer Vision & Image Understanding* **74** 1-21
[5] Huang, J, Li, B, Zhu, J and Chen, J 2017 *Multimedia Tools & Applications* **76** 1-17
[6] Wang, J. G, Yau, W. Y and Wang, H. L 2009 Proceedings of the Applications of Computer Vision
[7] Buda, M, Maki, A and Mazurowski, M. A 2017 A systematic study of the class imbalance problem in convolutional neural networks
[8] Chawla, N. V, Bowyer, K. W, Hall, L. O and Kegelmeyer, W. P 2002 *Journal of Artificial Intelligence Research* **16** 321-57
[9] Mease, D, Wyner, A. J and Buja, A 2007 *Journal of Machine Learning Research* **8** 409-39
[10] Drummond, C and Holte, R. C 2003 *Proc of the Icml Workshop on Learning from Imbalanced Datasets II* 1-8
[11] He, H and Garcia, E. A 2009 *IEEE Transactions on Knowledge & Data Engineering* **21** 1263-84
[12] Han, H, Wang, W. Y and Mao, B. H 2005 *Proceedings of the International Conference on Advances in Intelligent Computing* 878-87
[13] Bunkhumpornpat, C, Sinapiromsaran, K and Lursinsap, C 2009 *Proceedings of the 13th Pacific-Asia Conference on Knowledge and Data Mining* 475-82
[14] He, H, Bai, Y, Garcia, E. A and Li, S 2008 *Proceedings of the IEEE International Joint Conference on Neural Networks* 1322-28
[15] Jo, T and Japkowicz, N 2004 *Acm Sigkdd Explorations Newsletter* **6** 40-9
[16] Liu, X. Y, Wu, J and Zhou, Z. H 2009 *IEEE Transactions on Systems Man & Cybernetics Part B* **39** 539-50
[17] Kubat, M and Matwin, S 1997 *Proceedings of the 14th International Conference on Machine* Learning 179--86
[18] Elkan, C 2001 *Proceedings of the 17th International Joint Conference on Artificial Intelligence* 973-78
[19] Zhou, Z. H and Liu, X. Y 2006 *IEEE Transactions on Knowledge & Data Engineering* **18**63-77
[20] Khan, S. H, Hayat, M, Bennamoun, M, Sohel, F. A and Togneri, R 2018 *IEEE Transactions on Neural Networks & Learning Systems* **29** 3573-78
[21] Mccarthy, K, Zabar, B and Weiss, G 2005 *Proceedings of the 1stInternational Workshop on Utility-Based Data Mining* 69-77
[22] FG-NET (Face and Gesture Recognition Network). http://www.fgnet.rsunit.com/.

[23] Jr, K. R and Tesafaye, T 2006 *Proceedings of the 7th International Conference on Automatic Face and Gesture Recognition* 341-45
[24] Eidinger, E, Enbar, R and Hassner, T 2014 *IEEE Transactions on Information Forensics & Security* **9** 2170-79
[25] Population Pyramids (https://www.populationpyramid.net).
[26] Baggio, D. L 2012 Mastering opencv with practical computer vision projects
[27] Lin, M, Chen, Q and Yan, S 2013 Network in Network
[28] Levi, G and Hassncer, T 2015 *Proceedings of the IEEE Conference on Computer Vision and Pattern Recognition (CVPR)* 34-42
[29] Gallagher, A. C and Chen, T 2009 *Proceedings of the IEEE-Computer-Society Conference on Computer Vision and Pattern Recognition (CVPR)* 256-63

Data Mining Algorithms for a Feature-Based Customer Review Process Model with Engineering Informatics Approach

A K Varudharajulu and Y Ma*

University of Alberta, Canada

*yongshen@ualberta.ca

Abstract. The data generated from online communication acts as potential gold mines for discovering knowledge for end users. Large amount of data is also generated in the form of web documents, emails, blogs, and feedback, etc. Text analytics and opinion mining are used to extract human thoughts and perceptions from unstructured texts. This paper proposes a method that focuses on analysing different classification and clustering algorithms aimed at extracting and consolidating opinions of customers from social media sites like *Facebook*, *Twitter* and through surveys, at multiple levels of granularity to monitor and measure customer satisfaction. This is an automated approach, in which algorithms aid in the process of knowledge assimilation identification and the analytics. Domain experts ratify the knowledge base and provide training data sets for the system to intuitively gather more instances for ratification. The system identifies opinion expressions as phrases containing opinion words, opinionated features and also opinion modifiers. These expressions are categorized as positive, negative or neutral. Opinion expressions are identified and categorized using localized linguistic techniques. Opinions can be congregated at any desired level of specificity i.e. feature level or product level, user level or service level, etc. It has been found that J48 classification algorithm and simple k-means clustering algorithm are most suitable for restaurant customer reviews.

1. Introduction

The World Wide Web has become the most popular way to obtain information, such as travel reviews. According to the study conducted in 2014 [1], 81% of the Internet users have searched related comments before buying a commodity at least once. 32% have provided a rating on a product, service, or person via an online rating system, and 30% (including 18% of online senior citizens) have posted an online comment or review regarding a product or service. It should be noted that these online investigations had a significant impact on the customer's decisions. In addition, Customer reviews lead to feedback for the manufacturers or service providers to benefit from the user's ideas to improve their products and services appropriate for the target group [2], which increase customer satisfaction [3]. Key objectives of the proposed feature-based process model are: (1) to enable real-time extraction of customer comments from social media and reflect uniformly the opinions of customers, and then provide a systematic basis for management review and decision making for the operation; (2) to create a quality improvement cycle with the help of a software toolkit such that the management can measure the effectiveness of process changes, and new innovations on packages and menu items; (3) to generate positive assurance on service industry quality management generically and eventually to create economic benefits in the forms of higher satisfaction rate, more customer visits, sustained business growth and reputable brand name in the sector. In the field of machine learning, there are many existing classification and clustering algorithms. This paper focuses on selecting the right algorithms for data mining of restaurant customer reviews through careful and comprehensive analysis, the analysing process of different algorithms is discussed in methodology (Section 3).

2. Literature review

Social media reviews are very noisy and full of all kinds of spelling, grammatical, and punctuation errors. Most natural language processing (NLP) tools such as parts-of-speech (POS) taggers and parsers need clean data to perform accurately. Thus, a significant amount of pre-processing is needed before any analysis [4]. To overcome the need of pre-processing during the data mining process, in this reported work, opinion mining model has been adopted [5][6]. There have been some datasets consisting of restaurant customer reviews [7][8]. Sentiment analysis is commonly performed since the management needs to qualify customers' reviews towards the restaurant's issues, activities, and features; Weka [9] is a useful tool for sentimental analysis [10][11]. Typically, customer reviews can be analysed through the steps of word parsing and tokenization (extract terms), stop-words removal (elimination of non-value adding words for sentiment analysis), lemmatization and stemming (convert all infections to their root word), term selection/feature extraction (remove the terms that have poor prediction ability). Then the cleaned customer feedback texts are classified (assign a class -positive/negative), e.g. J48 algorithm [12]. Such algorithms could result in different classification quality depending on the match to the dataset used. In comparative analysis of classification algorithms for diabetes dataset [13] and Swahili language Tweets [14], Naïve Bayes algorithm was more accurate and suitable while performance of J48 was better in an analysis done in bank dataset [15] and credit card fraud detection [16]. The next step in data mining is to design the clustering model. The dataset was categorized into three by using k-means clustering [17][18] based on the sentiment of the comments as *Improve*, *Grow* and *Continue* (i.e. keep going). This paper studies classification [19][20][21] and clustering [22] algorithms for a feature-based customer review process model.

3. Methodology

3.1. Data collection

Restaurant customer reviews data collected from the internet is used as sample data for classification and clustering. The collected data may have unstructured data for classification and clustering like emoticons, or hashtags, which were pre-processed and removed using one of the Weka tool's filters string to vector. Weka is a collection of open source Machine Learning algorithms used for pre-processing, classifiers, clustering, association rules, and visualization [23]. It is a Java-based tool used in the field of data mining, created by researchers at the University of Waikato in New Zealand. It uses flat text files to describe the data. It can work with a wide variety of data files including its own ".arff". The pre-processed training data is used to build a classifier model and compared. So, the time taken to build the model excludes the time taken for pre-processing.

3.2. Classification Algorithms.

J48 decision tree algorithm builds a decision tree after analyzing the training instances. In the decision tree, the topmost node is the root node and the leaf nodes are the class values. In general, the nodes are attributes and the branches are decisions. While classifying the test instances using the constructed decision tree, it starts from the root node and traverses to leaf node based on the condition at each node and assigns a class value. Naive Bayes Algorithm is a simple probabilistic classifier that uses Bayes theorem to calculate the probability for each instance by counting the frequency and combinations of values in a given dataset. It assumes all attributes to be independent given the value of the class variable [24].

3.2.1. Bayes Theorem [25].

$$P(C|X) = P(X|C) * P(C) / P(X) \quad (1)$$

P(C/X) is the posterior probability of class (C, target) given predictor (X, attributes). P(C) is the prior probability of class. P (X|C) is the likelihood which is the probability of predictor given class. P(X) is the prior probability of predictor. P(X) is constant for all classes.

3.2.2. Performance factors. This section includes the performance factors that have been used in the experimental analysis of classification algorithms [26]. Confusion matrix in Table 1 can be used to identify checking statuses of correlations between the predicted results and the actual judgements as shown below: True Positive (TP): When the prediction is YES and the actual value is also YES; True Negative (TN): When the prediction is NO and the actual value is also NO; False Negative (FN): When the prediction is NO but the actual value is YES; False Positive (FP): When the prediction is YES but the actual value is NO.

Table 1. Confusion matrix

	Predicted Yes	Predicted No
Actual Yes	TP	FN
Actual No	FP	TN

Correctly Classified Instances, i.e. Sum of TN and TP, measure the numbers of instances whose predicted and actual class values are equal out of the total instances. *Incorrectly Classified Instances* are measured with the number of instances whose predicted and actual class values are different i.e. Sum of FN and FP.

Precision, TP / (TP + FP), gives what fraction of the predicted positive values are actually positive. *Recall,* TP / (TP + FN), gives what fraction of actually positive values is predicted positive. F-Measure is the weighted average of Precision and Recall.

$$F = 2 *((Precision * Recall) / (Precision + Recall)) \quad (2)$$

Experimental analysis. Experiments are performed on Restaurant data by using Classification algorithms in WEKA tool. Total no of instances is 2000.

Table 2. Result based on accuracy and time

Algorithm	Correctly classified instances	Percentage of Correctly classified instances	Incorrectly classified instances	Percentage of incorrectly classified instances	Time taken to build a model (in seconds)
J48	1921	96.05	79	3.95	0.13
Naïve Bayes	1475	71.75	565	28.25	0.85

3.2.3. Results. The sample data was tested over two classification algorithms which alternatively work better in different fields. As the result shows in Table 2, though Naïve Bayes equally takes lesser time to build a model, J48 classifies most of the instances correctly (see Figure 1) and increases the accuracy. Hence the J48 algorithm was used in our model. The different performance factors for Positive and Negative customer reviews using J48 and Naïve Bayes algorithms are in Table 3. Figure 2 gives a graphical representation of the various performance factors for classification algorithms in our case.

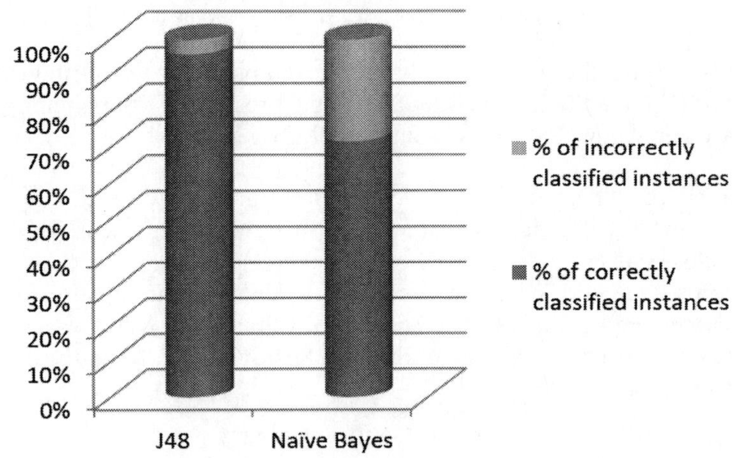

Figure 1. Graph of accuracy

3.3. Clustering Algorithms

Simple k-means is an iterative clustering algorithm in which items are moved among clusters till all instances are covered and the desired cluster is reached. Hierarchical clustering algorithm finds the similarity or dissimilarity between every pair of objects in the data set by calculating the distance between objects. Then group the objects into a binary, hierarchical cluster tree. At the next step, it determines where to cut the hierarchical tree into clusters by pruning branches off the bottom of the hierarchical tree so as to partition data. Farthest_First clustering algorithm is a variant of k-means, it places the cluster center at the point further from the present cluster. The points that are farther are clustered together first. This feature of the farthest first clustering algorithm speeds up the clustering process in many situations like less reassignment and adjustment is needed. Make_Density_Based clustering algorithm can be used when the clusters are irregular. A cluster is a dense region of points that are separated by low-density regions from the tightly dense regions. The make density-based clustering algorithm can also be used in noise and when outliers are encountered. The points with the same density and present within the same area will be connected to form clusters. The above clustering algorithms were tested with our dataset to find which is suitable for our product model by comparing various performance factors. Table 4 shows the time taken by each algorithm to cluster our dataset for various numbers of clusters.

Table 3. Performance factors for Positive and Negative using J48 and Naïve Bayes algorithms

	J48		Naïve Bayes	
	Positive	Negative	Positive	Negative
TP Rate	0.962	0.959	0.730	0.705
FP Rate	0.041	0.038	0.295	0.270
Precision	0.959	0.962	0.712	0.723
Recall	0.962	0.959	0.730	0.705
F-Measure	0.961	0.960	0.721	0.714
MCC	0.921	0.921	0.435	0.435
ROC Area	0.990	0.990	0.808	0.807

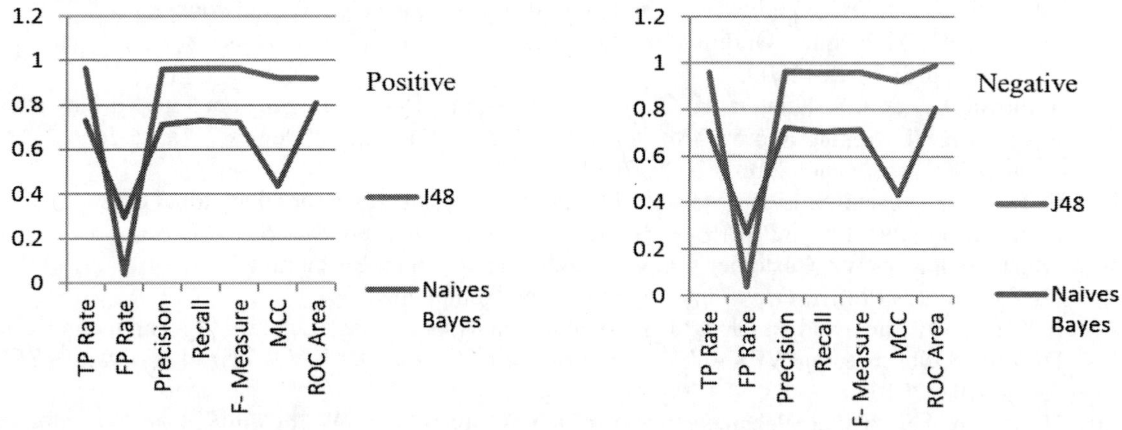

Figure 2. Performance factors for positive and negative comments (J48 vs. Naïve Bayes) algorithms

Table 4. Time taken (in seconds) to form Clusters

Cluster number	Simple k-means	Hierarchical	Farthest First	ke Density Based
10	.221	.363	.224	.230
20	.225	.356	.227	.237
50	.227	.344	.234	.244

4. Results and Conclusion

For clustering, as the cluster size increases, the time taken for clustering in *Make_Density_Based* and *Farthest_First* clustering algorithms increases. Hierarchical clustering algorithm takes more time when number of clusters is less. In all cases, simple k-means clustering algorithm remains consistent. So, *Simple k-means* is preferred for less no of clusters and *Hierarchical* clustering for large size of clusters. For classification algorithms, the two classification algorithms which alternatively work better in different fields; *Naïve Bayes* takes lesser time to build a model, *J48* classifies most of the instances correctly and increases the accuracy. Hence the J48 algorithm is recommended to be used in our model.

5. Future Work

The major future research scope is in sentiment analysis using the above classification and clustering methods. The areas identified are: (1) Sentiment Analysis on abbreviations and emoji; (2) Improving sentiment word identification algorithm; (3) Successful handling of bipolar sentiments; (4) Apply the results of sentiment analysis in marketing research; (5) Expand the technique to collective intelligent research; and (6) Generation of highly content lexicon database.

Acknowledgments

The authors would like to acknowledge the financial support: MITACS Accelerate grant (IT09867, PI: Yongsheng Ma) and NSERC Discovery grant (NSERC RGPIN 5641 Ma, PI: Yongsheng Ma). All the research works were carried out at the University of Alberta.

References

[1] B. Pang and L. Lee, "Opinion Mining and Sentiment Analysis," *Found. Trends Inf. Retr.*, vol. 1, no. 2, pp. 91–231, 2006.

[2] F. Hemmatian and M. K. Sohrabi, "A survey on classification techniques for opinion mining and sentiment analysis," *Artif. Intell. Rev.*, 2017.

[3] ISO, "ISO 9000:2015, Quality management systems — Fundamentals and vocabulary," 2015.
[4] L. Dey and S. M. Haque, "Opinion mining from noisy text data," *Int. J. Doc. Anal. Recognit.*, vol. 12, no. 3, pp. 205–226, 2009.
[5] S. Pippal, L. Batra, A. Krishna, H. Gupta, and K. Arora, "Data mining in social networking sites: A social media mining approach to generate effective business strategies," *Int. J. Innov. Adv. Comput. Sci.*, vol. 3, no. 2, pp. 22–27, 2014.
[6] D. B. Fernandez and S. Lujan-Mora, "Comparison of applications for educational data mining in Engineering Education," *2017 IEEE World Eng. Educ. Conf.*, pp. 81–85, 2017.
[7] "Restaurant Review _ Customer speaks _ Authenticate Indian Restaurant."
[8] "BBB Business Profile, DriveTime | Reviews and Complaints."
[9] R. Duriqi, V. Raca, and B. Cico, "Comparative Analysis of Classification Algorithms on Three Different Datasets using WEKA," *Embed. Comput. (MECO), 2016 5th Mediterr. Conf. on. IEEE*, pp. 96–101, 2016.
[10] V. S. Kumbhar, "Hotel Websites Facebook Data Analysis Using Weka," in *Second International Conference of Commerce and Management (ICCM)*, 2017.
[11] K. Dave, K. Dave, S. Lawrence, S. Lawrence, D. M. Pennock, and D. M. Pennock, "Mining the peanut gallery: Opinion extraction and semantic classification of product reviews," *Proc. 12th Int. Conf. World Wide Web*, pp. 519–528, 2003.
[12] S. Wakade, C. Shekar, K. J. Liszka, and C.-C. Chan, "Text Mining for Sentiment Analysis of Twitter Data," *Ike*, p. 5, 2012.
[13] S. Joshi, B. Pandey, and N. Joshi, "Comparative analysis of Naive Bayes and J48 Classification Algorithms," *Int. J. Adv. Res. Comput. Sci. Softw. Eng.*, vol. 5, no. 12, pp. 813–817, 2015.
[14] H. Seif, "Naïve Bayes and J48 Classification Algorithms on Swahili Tweets: Perfomance Evaluation," *Int. J. Comput. Sci. Inf. Secur.*, vol. 14, no. 1, p. 1, 2016.
[15] T. R. Patil, "Performance Analysis of Naive Bayes and J48 Classification Algorithm for Data Classification," *Int. J. Comput. Sci. Appl. ISSN 0974-1011*, vol. 6, no. 2, pp. 256–261, 2013.
[16] F. Alam and S. Pachauri, "Comparative Study of J48 , Naive Bayes and One-R Classification Technique for Credit Card Fraud Detection using WEKA," *Adv. Comput. Sci. Technol.*, vol. 10, no. 6, pp. 1731–1743, 2017.
[17] S. Revathi and T. Nalini, "Performance Comparison of Various Clustering Algorithm," *Int. J. Adv. Res. Comput. Sci. Softw. Eng.*, vol. 3, no. 2, pp. 2277–128, 2013.
[18] O. Abbas, "Comparisons Between Data Clustering Algorithms," *Int. Arab J. Inf. Technol.*, vol. 5, no. 3, pp. 320–325, 2008.
[19] M. N. Amin and A. Habib, "Comparison of Different Classification Techniques Using WEKA for Hematological Data," *Am. J. Eng. Res.*, no. 43, pp. 2320–847, 2015.
[20] T. C. Sharma and M. Jain, "WEKA Approach for Comparative Study of Classification Algorithm," Int. J. Adv. Res. Comput. Commun. Eng., vol. 2, no. 4, pp. 1925–1931, 2013.
[21] A. Naik and L. Samant, "Correlation Review of Classification Algorithm Using Data Mining Tool: WEKA, Rapidminer, Tanagra, Orange and Knime," Procedia Comput. Sci., vol. 85, pp. 662–668, Jan. 2016.
[22] G. Li and F. Liu, "Sentiment analysis based on clustering: A framework in improving accuracy and recognizing neutral opinions," Appl. Intell., vol. 40, no. 3, pp. 441–452, 2014.
[23] G. Kaur and A. Chhabra, "Improved J48 Classification Algorithm for the Prediction of Diabetes," Int. J. Comput. Appl., vol. 98, no. 22, pp. 13–17, 2014.
[24] A. Goyal and R. Mehta, "Performance Comparison of Naïve Bayes and J48 Classification Algorithms," Int. J. Appl. Eng. Res., vol. 7, no. 11 SUPPL., pp. 1389–1393, 2012.
[25] R. T. Bayes, P. Laplace, and S. H. Jeffreys, "Bayes ' theorem." pp. 1–12, 2018.
[26] M. Meenakshi and G. Geetika, "Survey on Classification Methods using WEKA," *Int. J. Comput. Appl.*, vol. 86, no. 18, pp. 16–19, 2014.

IOP Conf. Series: Journal of Physics: Conf. Series **1267** (2019) 012014 doi:10.1088/1742-6596/1267/1/012014

Medical Data Sharing Model Based on Blockchain

Wanghu Chen, Yuxiang Mu*, Xiaoyan Liang and Yaqiong Gao

College of Computer Science and Engineering, Northwest Normal University, Lanzhou, 730070, China

*E-mail: 1078927193@qq.com

Abstract. In the traditional medical system, individual medical data is managed by hospitals rather than individual patients. It is difficult to exchange effectively with fragmented storage, and large amounts of data are difficult to realize their potential value. With the rapid development of medical informatization, centralized storage of fragmented medical data has been unable to meet the relevant needs of the medical industry. To solve the difficulty of sharing and the complexity of confirming rights in the medical system, this paper proposes a medical data sharing model based on blockchain. The model provides reliable storage with IPFS file system, uses Proxy re-encryption to realize data sharing and ensure data proprietary rights, and uses Token economic system to measure the contribution in the sharing process, which stimulates the enthusiasm of sharing. At last, based on the existing sharing problem of medical data, the paper shows the potential solution.

1. Introduction

With the rapid development of medical informatization, the areas focusing on electronic medical records and clinical information systems have achieved comprehensive and rapid development [1] ,in the core of interconnection, information sharing and business collaboration. The extensive use of Eletronic Medical Records (EMR) has brought great convenience to the Medical field, making patients' diagnostic information (including prescriptions, laboratory lists, pathological results, MRI images, etc.) become valuable data assets [2]. If a credible inter-agency data sharing platform can be established to provide comprehensive, timely and accurate medical data for patients, doctors and relevant scientific research institutes, it can not only help patients better understand family history to make early prevention, but also help medical personnel develop optimal medical plans to achieve the best therapeutic effect, which will provide researchers accurate data sets to make more efficient and reliable disease prevention and accelerate the development of biomedicine.

However, in the existing medical system, there are still many obstacles to form a credible data sharing platform. On the one hand, patients' personal medical data is managed by different hospitals and enterprises in a decentralized manner, patients cannot control their own medical data, so data storage is discrete and fragmented, making it difficult to exchange effectively. A large amount of useful information is difficult to exert its potential value, forming an "island of informationization. On the other hand, for hospitals and enterprises stored medical data, once their data centers are damaged, it will be difficult to repair those damaged data. Moreover, companies may be motivated by their own interests to buy, sell or modify data. Therefore, the reliability of storing data in the centralized link is also a problem worthy of attention. To sum up, the current centralizing storage of fragmented medical data is no longer sufficient to meet the relevant needs of the medical industry.

Blockchain is a technical solution for collectively maintaining a reliable database in a distributed environment through decentralization and de-trusting [3]. The rise of blockchain technology provides a new idea for data sharing in the medical industry [4]. With the characters of difficulties to tamper and de-trusting, blockchain technology can ensure the reliability of stored data in the medical system. Decentralized peer-to-peer transmission can solve the problem of barriers to access and non-circulation of information between medical institutions. Cryptography-based Privacy protection avoids the disclosure of privacy of personal medical records. Therefore, this paper proposes a medical data sharing model that can ensure the authenticity and reliability of data, avoid privacy disclosure and stimulate the initiative of sharing based on blockchain.

2. Related work

At present, blockchain technology has been widely applied in finance, Internet and other fields by taking advantage of its features of decentralization, non-tampering and distributed storage. With the gradual improvement of blockchain technology, its application in the medical field has also made rapid development. Many people believe that the medical and health field is the second largest research field besides the financial field. Shrier and Chang et al. used MIT's OPAL/Enigma encryption platform combined with blockchain technology to create a secure environment for storing and analyzing medical data [5]. Based on the blockchain private chain, Kuo T et al. constructed a cross-agency medical health prediction model [6]. Literature [7,8] aims at the current problems of severe fragmentation of medical data, low sharing efficiency, insecure transmission process, lack of data integrity verification and insufficient privacy information protection, the access and sharing of medical data are realized through the Ethereum platform with smart contract. The MedRec framework mentioned in literature [9,10] combines smart contract with access control for automatic permission management, which realizes the integration and permission management of distributed medical data in different organizations. Since the MedRec framework uses the PoW consensus mechanism to maintain blockchain consistency, the computational overhead required is too large. Therefore, the MDSN[11] framework innovates the consensus mechanism using the DPoS consensus mechanism to reduce resource consumption, and the framework adopts the proxy re-encryption method to control access to medical data, which effectively improves the efficiency of data sharing while protecting privacy, but it also has the deficiency of limited data storage capacity.

Therefore, this paper designs a model for sharing medical data in blockchain mode, which proposes a storage scheme for real data transmission off-chain and operation records on-chain, and carries out data transmission off-chain with the IPFS file system. By the means of proxy re-encryption, the shared access to medical data is realized on the basis of protecting patients' privacy. Token economy system is also adopted to measure the contribution of sharing process and improve the sharing enthusiasm of participants.

3. Medical data sharing model

The medical data sharing model consists of three main modules. The first is the data storage module, which combines the InterPlanetary File System (IPFS) on the basis of the blockchain to ensure the reliability of the shared data from the source and to reduce the data users' doubts about the authenticity of the data. In the second part, Proxy re-encryption technology is used to realize the effective transmission of data among medical institutions and to achieve the purpose of data sharing based the premise of ensuring patient privacy through effective access control. Finally, in order to stimulate the enthusiasm of the sharer, Token economy system is introduced to measure the contribution of the sharer uniformly. The overall schematic of the model is shown in Figure 1.

3.1. Storage model of medical data

In the blockchain mode of operation, the lowest level of data interaction will be treated as a transaction, and exists in the blockchain environment in the form of Merkle root through the Merkle hash tree. Therefore, we regard the operation of medical data as a transaction. When data access, copy and other operations occur, we regard the operation record as a transaction and upload it to the blockchain operating environment, which realizes the running mode that the data operation record is on-chain and the real data transmitting is off-chain. Based on the characters of difficulties to tamper and de-trusting, the data operation records on-chain can accurately correspond to the corresponding data changes off-chain.

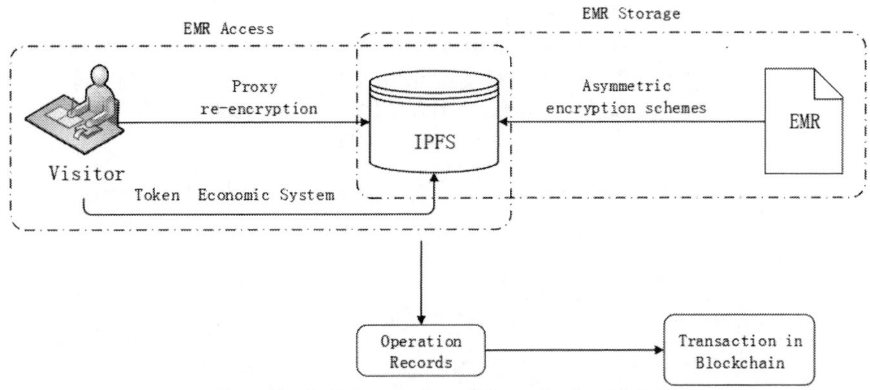

Figure 1. Schematic of the whole model

Since the operational record of the data is running on the chain, the actual data transfer takes place in the IPFS off the chain, and IPFS identifies files by generating a unique Hash value from the contents of the file, so we can divide the data operation into two types. One is the query. During the query process, the data hash value will not change, and we only need to record the data operation, including the querier and the agent (3-2 will be described in detail). The other is modification. In the process of modification, besides the operation records involved in the query, the hash of the corresponding data before and after the operation should be also added to update the data status in the IPFS file system. The corresponding data operation structure is shown in figure 2. Each data operation is accompanied by two state changes. First, an operation record is generated according to the corresponding operation type, which acts as a transaction in the blockchain. Second, after the data operation, it will update the EMR status, which retains the latest data of EMR. The operation record after each data change has a one-to-one correspondence with the data status.

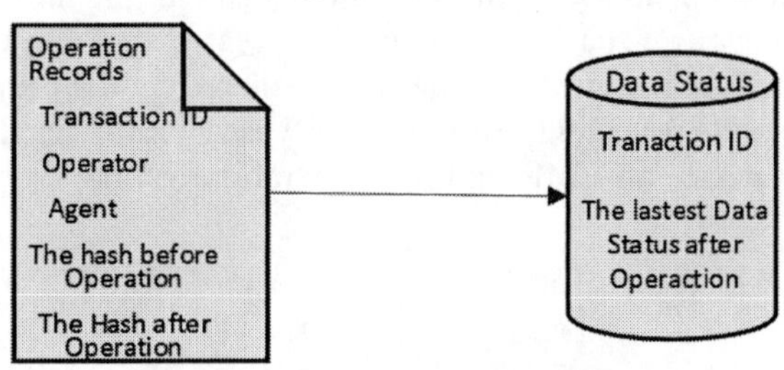

Figure 2. Data Operation Structure

The operation record and data operation structure in the blockchain can theoretically guarantee the accuracy of corresponding EMR, but the reliable storage of real EMR is still a problem. The InterPlanetary File System (IPFS) is a kind of permanent, decentralized distributed File System, which is a distributed protocol that can be used for point-to-point hypermedia through content addressing and versioning [12]. The IPFS file system has the following three characteristics: 1) Content addressability: the file is identified by generating a unique Hash value based on the file content, rather than by the file storage location. For saving storage space, the same content in the system will exist only once; 2) Versioning: trace the modification documents history; 3) Peer-to-peer hypermedia: saves various types of data through P2P. IPFS generates Hash based on content to identify files, which perfectly matches the Hash value of data in data operation structure; IPFS can trace the historical version of files, which well fits the update process of data in EMR; IPFS peer-to-peer transmission mode just fits the data transmission mode off-chain. In general, IPFS is a feasible scheme for storage data off-chain.

As shown in figure 3, the data operation record acts as a transaction, which is processed by Merkle tree and packaged into the blockchain environment in the form of Merkle root, and the corresponding data state is stored in the IPFS file system through Merkle DAG. The two are connected by a data manipulation structure. Data operation records running in the blockchain environment are guaranteed to be true and unforgeable with the guarantee of the blockchain characteristics such as non-tampering and traceability. The data operation record has a one-to-one correspondence by the data status. In this case, the real patient-related information stored in the IPFS is also guaranteed.

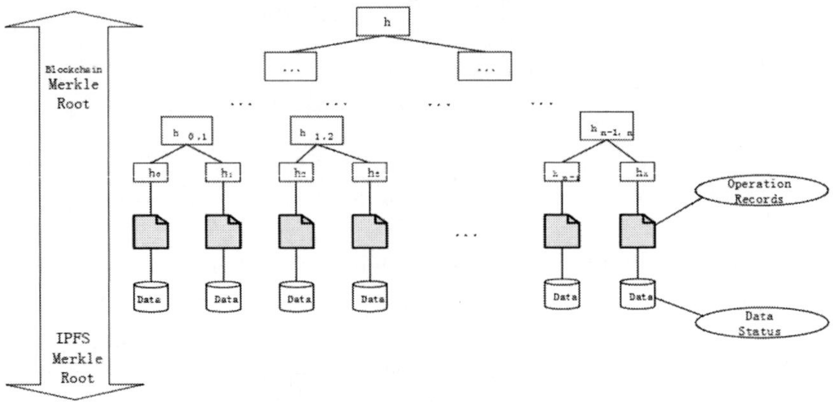

Figure 3. Schematic of Data Storage

3.2. Access control for medical data

The medical industry is very sensitive to data privacy and has a unique set of regulatory requirements for privacy protection. For example, the medical service industry in the United States must comply with the health insurance privacy and liability act (HIPAA) issued by the government of this country in 1996 [13]. Therefore, the application of blockchain in the medical industry needs to overcome the difficulties in privacy protection. In the traditional medical field, the storage of cases is managed by other institutions such as hospitals. Patients cannot have real ownership of their own cases, nor can they participate in EMR management in real time. In the shared model proposed in this paper, patients are the core, they have absolute ownership of their own cases, and other institutions such as hospitals are only third parties that operate on EMR data. All operations on the case must be authorized by the patient.

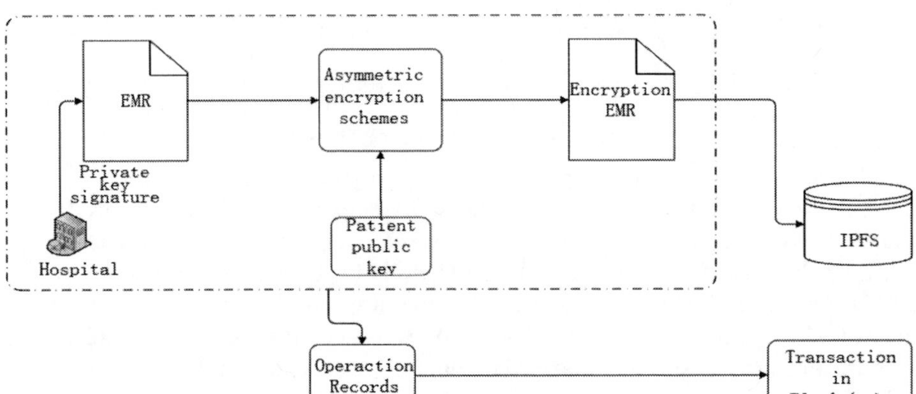

Figure 4. EMR Storage Process

As shown in Figure 4, the electronic medical record stored in the IPFS file system first needs to go through the digital signature of the hospital. The hospital signs with the private key, which avoids the possibility of fraud from the source and provides convenience for future accountability. Then, the patient public key is used to encrypt the EMR with hospital signature using asymmetric cryptography, and the final encrypted EMR is stored in the IPFS. The data operation records generated during the whole process are packaged in the blockchain operation environment in the form of blockchain transactions.

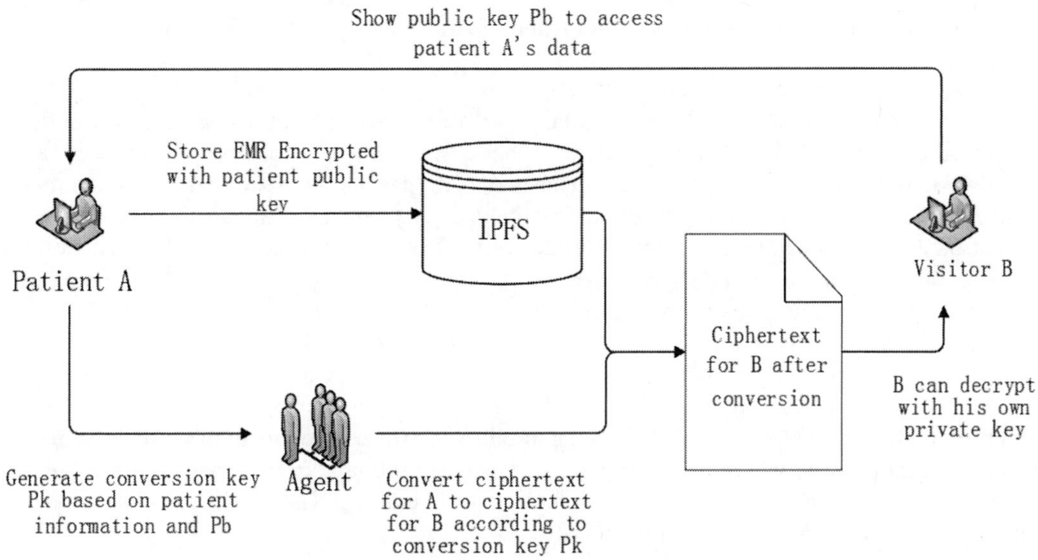

Figure 5. Proxy re-encryption Process

EMRs stored in the IPFS file system are ciphertext encrypted by the patient's public key and they can only be viewed after the patient authorizes it. In this case, the patient's ownership is guaranteed and the privacy issue is avoided. However, there are still many hidden dangers for patients to directly interact with visitors in the no-center link of the blockchain. Proxy re-encryption is a key conversion mechanism among ciphertexts. In the proxy re-encryption scheme, the semi-trusted agent can transform the ciphertext of the authorized person into the ciphertext of the receiver by using the re-encryption key granted by the authorized person. And in the process of transformation, neither external attackers nor agents can obtain the corresponding plaintext, thus it can reduce the risk of data leakage [14]. There is no trusted third party in the blockchain no-center environment and there is no unique proxy role in the traditional re-encryption scenario. Any node in the medical institution can act as an agent to complete the re-encryption operation and get the corresponding credit points as a reward. Therefore, it is feasible to use the proxy re-encryption scheme for access control. The specific case sharing acquisition process is shown in figure 5. This model uses the identity-based proxy re-encryption design protocol proposed by Jiang Mingming et al. [15] as follows:

1) Master key generation: Generate random matrix A and small norm matrix T, randomly select $l+1$ linearly independent vectors $u_0, u_1, \cdots u_l$. Then the master public key MPK=(A, $u_0, u_1, \cdots u_l$)and the master private key MSK=T.
2) User private key extraction: Enter the primary public key MPK, the primary private key MSK, and the user identity id=$(id_1, id_2 \cdots id_l)$. Calculate $u = u_0 + \sum_{i=1}^{l} id_i u_i$, use the original image sampling algorithm to generate the vector e to satisfy $Ae = u = u_0 + \sum_{i=1}^{l} id_i u_i$. Therefore, the private key of the user id is e.
3) Proxy re-encryption key generation: For the users id_1 and id_2 , use the corresponding private key $e_{id_1}$and e_{id_2} to calculate $rk_{id_1 \leftrightarrow id_2} = eid_1 - eid_2$ as a proxy re-encryption key.

4) Encryption: Enter the main public key MPK, user identity $id_1 = (id_{11}, id_{12}, \cdots id_{1l})$ and a message bit μ to calculate $u = u_0 + \sum_{i=1}^{l} id_i u_i$. Select the random vector S to calculate $y = A^T s + x, c = u^T s + x + \mu\lfloor q/2 \rfloor$. The output is ciphertext (y, c).
5) Re-encryption: Enter the re-encryption key $rk_{id_1 \leftrightarrow id_2} = eid_1 - eid_2$ and the ciphertext $(y, c_{id_1})$of user $id_1 = (id_{11}, id_{12}, \cdots id_{1l})$. Agent uses proxy re-encryption key to calculate $c_{id_2} = c_{id_1} - rk_{id_1 \leftrightarrow id_2}^T y$, Output the ciphertext$(y, c_{id_2})$of the user $id_2 = (id_{21}, id_{22}, \cdots id_{2l})$.
6) Decryption: Enter the private key e_{id_2} of user $id_2 = (id_{21}, id_{22}, \cdots id_{2l})$ to calculate $c_{id_2} - e_{id_2}^T y$. If the result is close to 0, it outputs 0, and if the result is close to $\lfloor q/2 \rfloor$, it outputs 1.

3.3. Measurement of value in the Shared pattern

In the medical sharing model, patients, hospitals, and third-party research institutes are independent from each other, and there is no direct interest to make them mutually beneficial in the EMR sharing process. As a result, the participation enthusiasm of all parties in data sharing is not high [16]. Therefore, Token economy system is adopted in this paper to make measure the contribution in the process of data sharing, which is displayed in the form of credit score. In the end, it will form a win-win whole that offers preferential treatment to patients, enhances the hospital's popularity and promotes the development of scientific research institutes.

When a third-party access agency, such as a research institution, requests to view the data, the patient is authorized to an agent, which can increase the patient's own credit score. When the patient's credit score reaches the corresponding threshold, the next diagnosis can be carried out corresponding preferential treatment. Since the EMR has hospital's signature, the hospital's credit score will be increased after the review, which has an impact on hospital rankings. In this case, patients will actively share their own cases, and hospitals can also benefit from sharing. There will form an organic whole of mutual benefit, cooperation and symbiosis.

4. Model Evaluation

This paper adopts the method of comparative analysis to evaluate the proposed medical data sharing model. On the one hand, compared with the existing sharing scheme, we analyze the advantages and disadvantages of this model, the result is shown in Table 1. On the other hand, combined with the existing problems, the solution to the analytical model is shown in table 2.

Table 1. Comparisons with existing sharing schemes

	Third parties	Private chain	Incentives	Storage	Consensus
ABE program [16]	YES	—	NO	LARGE	NO
MedRec[9][10]	NO	NO	NO	SMALL	PoW
MDSN[11]	NO	YES	NO	SMALL	DPoS
The Model	NO	NO	HAS	LARGE	Decide with actual condition

Table 1 compares the model with the existing research results from five aspects. Overall, the model has certain advantages. On the basis of the blockchain, the IPFS file system is combined to expand the storage capacity, and the incentive mechanism has been introduced to increase the enthusiasm for participation.

Table 2. Existing problems and solutions

Existing Problems	Solution
Consensus	Submit the data operation records to the public chain in the form of transactions, including Bitcoin, Ethereum and EOS
Security privacy	Use an asymmetric encryption scheme to store, use proxy re-encryption to access
Participation enthusiasm	The inclusion of Token economy makes the value in the sharing process be measured uniformly, so as to ensure that each participant gets what he needs
EMR ownership	The EMR is encrypted according to the patient's public key using the asymmetric encryption system, and the ownership always belongs to the patient
EMR fabrication and accountability	With the help of blockchain and IPFS, it can achieve traceability and accurate positioning can for hospitals that falsify data

Table 2 analyzes the solution of the model for existing problems. For the patient, it can not only ensures privacy, but also protects the ownership of their own cases. For the hospital, it not only ensures the authenticity of its own medical information, but also lays a foundation for the hospital to raise awareness. But there are still many flaws and areas that need to be optimized, such as no private medical chain, no predictive ability and so on.

5. Conclusion

In this paper, a medical data sharing model is designed in the blockchain operation mode. The paper shows an operation mode of real data transmission off-chain and operation records on-chain, which uses the IPFS to realize the efficient storage, takes asymmetric encryption system and proxy re-encryption to control data sharing access, and introduces Token economy system to measure the contribution in the sharing process to improve user engagement. However, with the continuous development of the blockchain technology, how to realize the patient's fine-grained access to the diagnosis information and the prediction of the patient's condition in the case sharing process is still an area worthy of investigation.

ACKNOWLEDGEMENT

This work is supported by the National Natural Science Foundation of China (No. 61462076).

References

[1] Huang Jian-Hua, Jiang Ya-Qin, Li Zhong-cheng, et al. Application Prospect of Blockchain in Medical Industry [J]. *Journal of Medical Intelligence*, 2018, 39(2):1-8, 13.
[2] MEI Ying. The Utilizing Blockchain-Based Method of the Secure Storage of Medical Records [J]. *Journal of Jiangxi Normal University(Natural Sciences Edition*), 2017, 41(5):484-490.
[3] HE Pu, YU Ge, ZHANG Yan-feng BAO, et al. Survey on Blockchain Technology and Its Application Prospect [J]. *Computer Science*, 2017, 44(4):1-7.
[4] YUAN Yong WANG Fei-Yue. Blockchain: The State of the Art and Future Trends [J]. *Acta Automatica Sinica*, 2016, 42(4): 481-494.
[5] Shrier A A, Chang A, Diakun-thibault N, et al. Office of the national coordinator for health information technology us department of health and human services[J]. 2016.
[6] Kuo T T, Ohno-Machado L. ModelChain: Decentralized Privacy-Preserving Healthcare Predictive Modeling Framework on Private Blockchain Networks [J]. arXiv preprint arXiv:1802.01746, 2018.

IOP Conf. Series: Journal of Physics: Conf. Series **1267** (2019) 012014 doi:10.1088/1742-6596/1267/1/012014

[7] Dagher G G, Mohler J, Milojkovic M, et al. Ancile: Privacy-preserving framework for access control and interoperability of electronic health records using blockchain technology [J]. *Sustainable Cities and Society*, 2018, 39(1): 283-297.

[8] McFarlane C, Beer M, Brown J, et al. Patientory: A Healthcare Peer-to-Peer EMR Storage Network v1 [J]. 2017.

[9] Azaria A, Ekblaw A, Vieira T, et al. MedRec: Using Blockchain for Medical Data Access and Permission Management[C]// International Conference on Open & Big Data. 2016.

[10] Ekblaw A, Azaria A, Halamka J D, et al. A Case Study for Blockchain in Healthcare: "MedRec" prototype for electronic health records and medical research data[C]//Proceedings of IEEE open & big data conference. 2016, 13: 13.

[11] Xue T F, Fu Q C, Wang C, et al. Study on Medical Data Sharing Model Based on Blockchain[J]. *Acta Automatic Sinica*, 2017, 43(9): 1555-1562.

[12] Benet J. IPFS-content addressed, versioned, P2P file system [J]. arXiv preprint arXiv:1407.3561, 2014.

[13] U U. Health Insurance Portability and Accountability Act of 1996. Public Law 104-191[J]. *United States Statutes at Large*, 1996, 110(1):1936-2123.

[14] Ge ChunPeng. Research on Several Issues of Proxy Re-encryption [D]. Nanjing University of Aeronautics and Astronautics, 2016.

[15] JIANG Ming-ming, GUO Yu-yan, YU Lei, et al. Efficient Identity-based Proxy Re-encryption on Lattice in the Standard Model [J]. Journal of Electronics & Information Technology, 2019, 41(1):61-66.

[16] Xhafa F, Li J, Zhao G, et al. Designing cloud-based electronic health record system with attribute-based encryption [J]. Multimedia Tools and Applications, 2015, 74(10):3441-3458.

IOP Conf. Series: Journal of Physics: Conf. Series **1267** (2019) 012015 doi:10.1088/1742-6596/1267/1/012015

An Approach to Edge Extraction Based on 3D Point Cloud for Robotic Chamfering

Yongzhi Wang[a], Zhijiang Du, Yongzhuo Gao, Mingyang Li and Wei Dong[b,*]

Harbin Institute of Technology

E-mail:[a]17S008047@stu.hit.edu.cn; [*, b]dongwei@hit.edu.cn

Abstract. Inconsistency of workpieces always results in low level automation of subsequent finishing process. As decreasing number of skilled workers, the demands of flexible manufacturing with robots is growing. In this paper, a kind of workpiece with complex free-form surface will be chamfered by industrial robots. Due to the intolerant differences among workpieces, traditional teach-playback method for robot is not appropriate for this application. The edge spline of each workpiece needs to be detected so that trajectories are generated to chamfer. Thus, an approach to edge extraction based on a 3D point cloud obtained by the 3D industrial camera is introduced to solve this problem. A time optimizing method is proposed to accelerate the extraction process. Finally, we decrease computing time from 55 minutes to 300 seconds, and the precision is 0.4180 mm.

1. Introduction

In many industrial area,there are many situations that the workpieces are shapes of differences and complex, and with the increasing of labor cost, flexible manufacturing is paid more and more attention to now. The workpiece with complex free-form surface such as which we will chamfer in this paper shown as figure 1 is hard to process. The traditional teach-playback is not useful for three reasons. First, they are in different shapes and sizes. Second, the first process, cost, will cause much error. At last, there is not very nice positioning reference of the workpiece with the free surface. Therefore, the visual system of industrial robot needs to be used.

Figure 1. the workpieces of different shapes and sizes

In the feature and edge identification from point cloud field, there are a lot of researches now. Their method are about contour getting straightly or the feature of surface. In the field of contour getting straightly, M J Milroy[1] et al. use an energy-minimizing active contour interactively to find the edge points and they use this method on segmenting a phone model successfully. In the field on curvature, M. Yang[2] et al, propose an automatic edge-based approach based on the computation of the surface curvatures and the principal directions, and the edge points are regarded as the curvature extremes. They also supposed an forming boundary curves algorithm based on edge-neighborhood. In the field on the feature of surface, Vanna Sam[3] et al. propose an robust algorithm for extracting a centered curve skeleton from point cloud. A lined segment sequencing algorihm proposed by Kai Wah Lee[4] is to segment the different surfaces in the point cloud of complex geometry surrounded by a developable surface.

Although there are many researches about point cloud, there is less application on the industrical robot processing. Because of the high precision in industry, it is hard to get the accurate curve to process which satisfies the requirement. In this paper, we get the workpiece from point cloud, and extract edge based on Micro-cut plane. To solve the precision problem, we propose a new iterative algorithm and get a well consequence finally.

In this paper, the program of recognizing an edge to chamfer of a workpiece from a point cloud is shown.

The rest of the paper is organized as follows: First, the related work on point cloud is introduced in Section 2. The problem is expressed in Section 3. The flow of identifying the edge from the point cloud is shown in Section 4. The result and analyzation are shown in Section 5, and Section 6 concludes the paper.

2. Problem Formulation

In our works, we will chamfer a set of workpiece with different shapes and sizes, and the precision requirement of chamfering is 0.3mmx0.3mm~0.5mmx0.5mm. For the end effector are designed to be compliant, it can compensate for errors caused by the visual system and robot. Considerating the compensating ability of the end tool, the requirement of the visual system is set as 0.3mmx0.3mm~0.5mmx0.5mm.

The industrial camera is used to obtain the point cloud of the workpiece, but there will be some other things like fixture in the point cloud. So, we must segment the workpiece from the point cloud first and then recognize the edge to be processed. Finally, a fitting curve of the edge will be got and a set of sequence points will be calculated and transferred to robot to execute.

In the paper, we set the evaluation indicators of the fitting curve as follow:

Let the fitting curve have a total of i sequence points (xs, ys, zs). One of (xs, ys, zs) nearest neighbor point in the point cloud is (x, y, z). The distance between the two points, (xs, ys, zs) and its (x, y, z), is *D*, and the precision index is that the average value of all *D*.

The problem can be described by the mathematical formula as:

$$E=\frac{\sum D}{i}=\frac{\sum\sqrt{(xs-x)^2+(ys-y)^2+(zs-z)^2}}{i} \tag{1}$$

3. Method

The flow of getting boundary is made up of four steps as follow: get workpiece from point cloud, get the boundary point, select the edge point and fit the curve of the edge. The strategy will be introduced in the following:

3.1 Extraction of Workpiece From Point Cloud

The point cloud is gotten by the industrial camera, so there is the workpiece and some other things in the point cloud which contain fixture plane mainly, and we must segment them.

In this paper, the algorithm of RANSAC[5] is used to recognize the fixture plane. RANSAC is the abbreviation of Random Sample Consensus. It can calculate the parameter of a mathematical model iteratively from a set of observation data sets containing "outliers". In this paper, the threshold t is set as 15mm, k is 6000. The point clouds after Section 4.1 is shown as Figure 2. The red presents "inliers ", and the green presents "outliers".

IOP Conf. Series: Journal of Physics: Conf. Series **1267** (2019) 012015 doi:10.1088/1742-6596/1267/1/012015

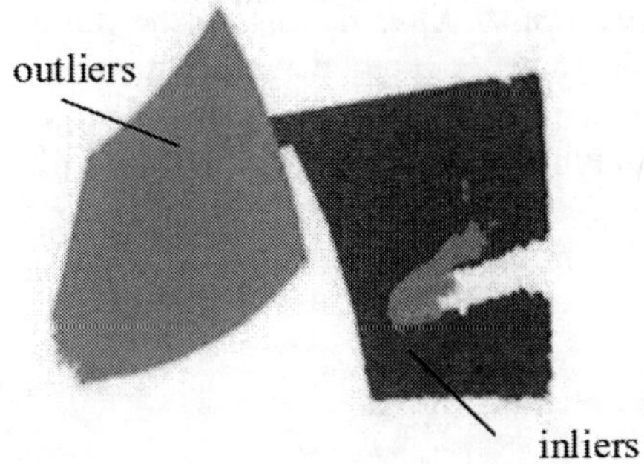

Figure 2. the point cloud after RANSAC

In the paper, we transform the pointcloud to make the fixture plane on XOY plane. And then, we cut the point whose absolute value of Z coordinate value is smaller than the threshold.

3.2 Extraction of Boundary Points

3.2.1 Denoise The Pointcloud. There will be some noise around the edge of the workpiece, for the error of the camera. In this paper, we use the Statistical Outlier Removal filter[6] to denoise the point cloud. This kind of filter has two parameters,"NumNeighbors" and "Threshold". The noise points can be caused by two reasons. One is the error of the camera and environment.These points are shown as cluster and farer from the boundary. And the other reason is that the thickness of the workspiece.These points are near the boundary. It has been seen from our practice that the larger "NumNeighbors" can delete the former, and the lower "NumNeighbors" can delete the latter.More small the value of "threshold" is, more better the result is. Therefore, we filter the point cloud twice by (10, 0.1), (50, 0.3).

3.2.2 Simplifications of the points. The number of points is in millions, and if they are all used in the next step, the time is long about 1 hour. Therefore, we must simplify the point cloud. The algorithm is described as follows, in which Workpiece_d, Bon, Boundary_range, workpiece_in, workpiece_in_d are point cloud,and Sphere_search(Co, grid) is a kind of algorithm that searches the neareast points in a sphere whose center is the point to be deal with, and radium is R=Co×Grid:

Algorithm 1:simplification to point cloud

Input:3d point cloud *Workpiece,* length of box side *grid,* coefficient *co*
Output:3d point cloud *Workpiece_simplification*
Workpiece_d←DOWNSAMPLE(*Workpiece, grid*);//as in equation (2).
Bon←BOUNDARY(*Workpiece_d*);//as in equation (3).
Calculate the neighbor points in Workspciece of each point in Bon by *Sphere_search*;
Merge all neighbor points into one point cloud *Boundary_range*;
Delete *Boundary_range* from *Workpiece*, extract point cloud *workpiece_in*;
Workpiece_in_d←DOWNSAMPLE(*Workpiece_in, grid*);
Workpiece_simplification←Merge(*Boundary_range , workpiece_in_d*);

Therein, DOWNSAMPLE is a kind of algorithm to decrease the number of points by voxelization[7], which use the center of a cube to replace the points in the cube. The width of the side of the cube is called *Grid*. We describe the algorithm as:

$$pointcloud\_new = \mathrm{DOWNSAMPLE}(pointcloud\_old,\ grid) \tag{2}$$

Grid will influence the number of points in *Workpiece_simplification*, which will eventually influence the time of Section 3.2.2~3.2.3, the mainly time in the whole flow. We change the value of *Grid*, and obtain the difference of the number and time, which will be shown in Section 4.

IOP Conf. Series: Journal of Physics: Conf. Series **1267** (2019) 012015 doi:10.1088/1742-6596/1267/1/012015

3.2.3 Extraction of The Boundary Point. After denoising the point cloud, we can get the boundary point from the workpiece. In this paper, a method named micro-cut plane projection[8] is used, as figure 3. For each point, we search its k neighbors, and make the micro-cut plane by these k neighbors. Then we project the point and its k neighbors onto the micro-cut plane, and call the projection P and N_i(i=1,…,K), spectively. We connect the P and N_i to make k vectors, and calculate the angle between adjacent two vectors. If the maximum in the set of angles is larger than the threshold t, we regard the point as the boundary point.

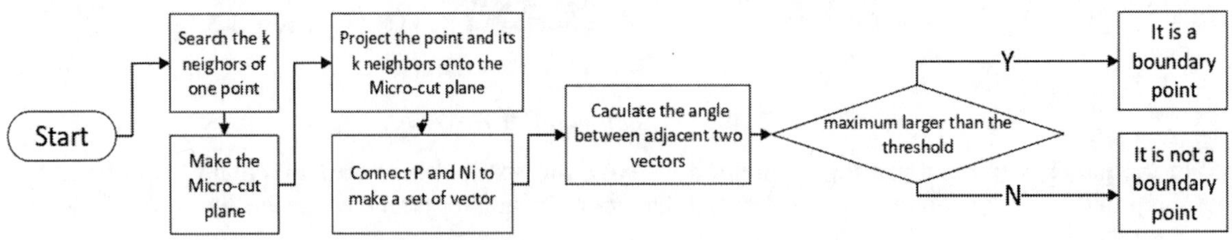

Figure 3 The program of getting the boundary point

We describe the algorithm as equation (3):

$$point\_boundary = \mathrm{BOUNDARY}(pointcloud) \quad (3)$$

3.3 Select The Edge Point

In the paper, the workpiece has 4 sides, and they will be all got from the point cloud through the above steps. But the edge to be process is not all the sides. Therefore, we must get the edge we want from the boundary points. The edge we want is a space curve and has two corners. If we get the two corners, it can be selected.

Points in our point cloud are unordered, but the work to get the corner needs a well-ordered point cloud. So the first work to do is to sort the boundary point. In this paper, we choose a point randomly as a starting point, and find its nearest point as the next point. And then, we regard the next point as the new starting point, and find the nearest point in the rest of the points as the next point. Finally, we get a set of well-ordered boundary point cloud.

We use a method named corner detection based on sliding windows[9], which judges if a point is a coner point by curvature. In our paper, the width of the window is constant and set as 17. After this step,there are some points got around the corner. Therefore, we must find the real coner point.

3.4 Fit The Curve of The Edge

After obtaining the coner of the edge to be process, we can regard the boundary points between the two corners as the point of the edge to be process. In this paper, a means named Smoothing spline is used to fit the curve.

The method, "smoothing spline"[10], is based on "least square method", but the error term is

$$\mathrm{RSS}(f,\lambda) = \sum_{i=1}^{N} \{y_i - f(x_i)\}^2 + \lambda \int \{f''(t)\}^2 dt \quad (4)$$

4. Result and evaluation

The combination of Matlab and PCL by C++ on i5-4460, 4G is used to practice the experiment. In Section 3.2.3, the *k* is set as 40 and the threshold *t* is set as 90°. The boundary point we get is shown as figure 4.

IOP Conf. Series: Journal of Physics: Conf. Series **1267** (2019) 012015 doi:10.1088/1742-6596/1267/1/012015

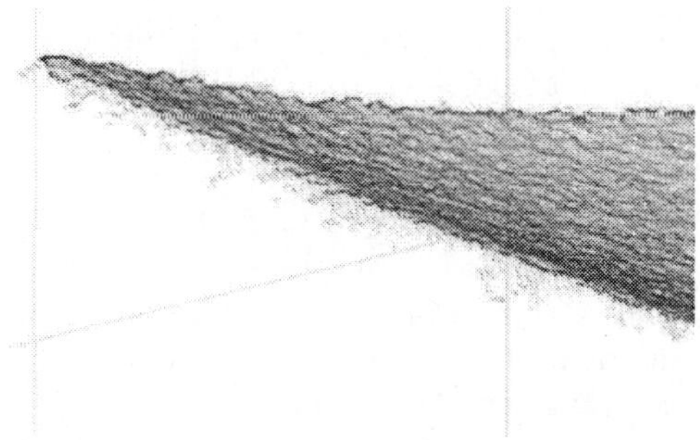

Figure 4. The boundary point and initial points

In section 3.4, we fit the curve by x, y and x, z, using equation (4) respectively. We regard x as the independent variable, and regard y, z as the dependent variable. Finally, we will get a space curve which fits the edge of the workpiece as figure 5. In this model, *E* ,as equation (1) is 0.4180 finally.

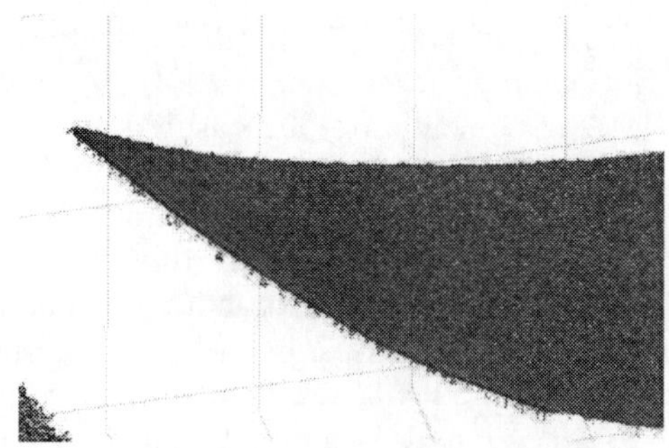

Figure 5. the fitting spline of the edge

We change the value of *Grid* in Section 3.2.3, and obtain the difference of the number and mean time which is shown in table 1.The point number of initial point cloud is 1310698, and the time is 55 min.

Table 1. The Number of points and Time with Grid from 2 to 4

Grid(Co=1.3)	2	3	4	5	6	7
Point Number	72899	68555	76350	83013	85309	86998
Time(s)	335	305	333	366	367	371

It can be seen from the table 1, that the number and time first decrease and then increase. Therefore, we set the value of *Grid* from 2 to 4, in order to find the best value.

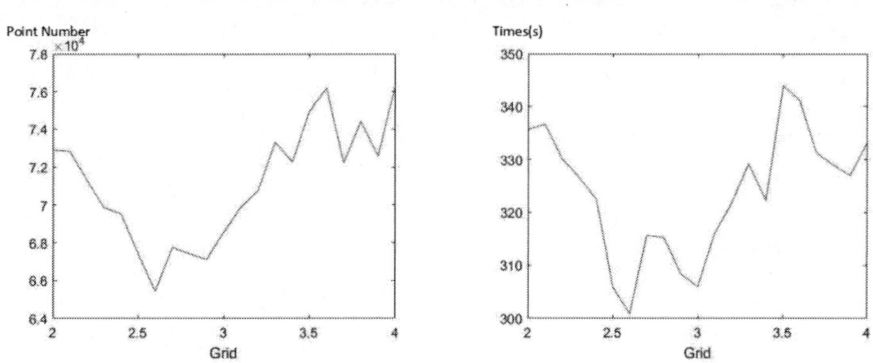

Figure 6. The Number of points and Time with *Grid* from 2 to 4

It can be seen for figure 6 that time is influenced by point number totally, and we can find that the best value of *Grid* is 2.6, whose number is 65441, time is 300s.

5. Conclusion

In this paper, an edge to be processed needs to be extracted from 3D point cloud. At first, we get workpiece from point cloud by the separation of the fixture. Secondly, the boundary point is got. In the end, the edge point we need is selected and the curve fitting of the edge is done. A time optimizing method is proposed to accelerate the extraction process. Finally, we decrease computing time from 55 minutes to 300 seconds, and the precision is 0.4180 mm.

Acknowledgement
Authors gratefully acknowledge the financial supports by National Science and Technology Major Project of China (No. 2017ZX04005001).

References

[1] Milroy, M. J. , Bradley, C. , & Vickers, G. W. . 1997 . Segmentation of a wrap-around model using an active contour[J]. *Computer Aided Design* **29(4)** 299-320
[2] Yang, M. , & Lee, E. . 1999 . Segmentation of measured point data using a parametric quadric surface approximation.[J] .*Computer Aided Design* **31(7)** 449-457
[3] Sam, V. , Kawata, H. , & Kanai, T. . 2012 . A robust and centered curve skeleton extraction from 3d point cloud[J]. *Computer-Aided Design and Applications* **9(6)** 869-879
[4] Lee, K. W. , & Bo, P. . 2015 . Feature curve extraction from point clouds via developable strip intersection[J]. *Journal of Computational Design and Engineering*, S2288430015000639
[5] Fischler, M. A. , & Bolles, R. C. . 1981. Random sample consensus: a paradigm for model fitting with applications to image analysis and automated cartography[J]. *Communications of the ACM*, 24(6), 381-395.
[6] Wolff, K. , Kim, C. , Zimmer, H. , Schroers, C. , Botsch, M. , & Sorkine-Hornung, O. , et al. 2016 . Point Cloud Noise and Outlier Removal for Image-Based 3D Reconstruction[C]. 2016 Fourth International Conference on 3D Vision (3DV). IEEE Computer Society
[7] C. Connolly. 1984. Cumulative generation of octree models from rangedata[J]. *Robotics and Automation. Proceedings*. IEEE International Conference on, vol. 1, mar 1984, pp. 25 – 32.
[8] Wu lushen, Yan Haiping, Chen Huawei, Gao Xiangqing 2014 . A Boundary Extraction Algorithm Based On Scattered Point Cloud[J]. *Computer* Applications *and Software***(11)** 264-268
[9] SUN Xiang-yang, YAO Li, WAN Yan, 2009: Adaptive Corner Detection Algorithm Based on Linear Fitting[J], *Journal of Shanghai University of Engineering Science*, **23(1)** 46-50
[10] Qiu, & Chunfu. 1995 . Nonparametric regression and generalized linear models: a roughness penalty approach[J]. *Technometrics* **37(3)** 342-343c

sEMG-based Endpoint Stiffness Estimation of Human Arm using Gene Expression Programming

Jiang Zainan*, **Yang Fan, Li Chongyang, Liu Daxiang, Wang Chenliang, Li Tianhui and Liu Hong**

State Key Laboratory of Robotics and Systems, Harbin Institute of Technology Harbin 150001 China

*E-mail:yangfan_hit@163.com

Abstract. The endpoint stiffness of the human arm has been long recognized as a key factor in the smooth contact between humans and environment. And the endpoint stiffness of the human arm is highly correlated with the surface electromyography (sEMG) produced by the contraction of the muscles. In this paper, the Gene Expression Programming (GEP) Algorithm is proposed to estimate the endpoint stiffness of human arm based on sEMG. This paper improves the traditional decoding method of GEP. Instead of generating an expression tree, it is directly decoded by looking for the effective length of the gene. And experimental results show that nonlinear models such as GEP models in this paper have higher correlation and lower RMSE (root mean square error) than regression stiffness using linear regression models. Selecting different feature of EMG signals, the correlation coefficient and the root mean square error of the model is very different. For the GEP model in this paper, WPTSVD (Wavelet Package Transform Singular Value Decomposition) and WTSVD (Wavelet Transform Singular Value Decomposition) are selected as the feature of sEMG signals have high performances and the correlation can reach $57\%\pm12.1\%$.

1 Introduction

Facing the non-deterministic and unknown environment such as post-earthquake, traditional initiative compliant control had shown insufficient performance in terms of adaptive ability and robustness [1]. But on the other hand, humans arm demonstrate versatile and stable interactions with the uncertain environment. This is achieved through the modulation of the mechanical properties of the limb, and as a consequence, task-related restoring forces are applied in response to the environmental displacements [2]. Therefore, in order to realize the transmission of the dynamic characteristics of the human arm to the robot, how to quantitatively measure the dynamic characteristics of the human arm endpoint is particularly important for Human-Robot Interaction (HRI).

Surface electromyography (sEMG) is a human body's own resource, which contains rich information about human body motion [3]. The frequency and amplitude value of the sEMG signal depend on the motion of the human arm. Currently, sEMG has become an important parameter for studying the characteristics of human arm.

One way to quantify the stability of the contact between the limb and the environment is through the estimation of the endpoint stiffness. The research of sEMG-based estimation of human arm endpoint stiffness has been discussed in many papers. The models based on sEMG to estimate the human arm endpoint stiffness are mainly divided into two types, one is linear and the other is nonlinear. R.Osu and

IOP Conf. Series: Journal of Physics: Conf. Series **1267** (2019) 012016 doi:10.1088/1742-6596/1267/1/012016

H.Gomi proposed an algorithm for estimating the human arm stiffness which relies on identifying the map between muscular activities (as coded by sEMG) and joint torques [4]. R.Osu also propose an index of contraction around the joint (IMCJ) calculated from EMG and joint torques as well as confirmed that IMCJ was correlated well with joint stiffness [4]. M.Maehara et al. propose a first-order least squares method based on the existing data to obtain the relationship between EMG signal and stiffness, and the accuracy is higher, but only on a single joint [5]. Liang et al proposed a discrete-time algorithm for stiffness extraction from sEMG and obtained the accurate value by integrating the increment [6]. Ze Cui et al estimated of upper limb muscle stiffness based on artificial neural network (ANN) but the performances was poor [7]. In addition, Catherine et al. confirmed that several factors were connected with the sEMG force relationship like muscle length, joint angle and arm posture [8]. In our experiments, subjects kept in the same posture while performing the operation, trying to avoid these influential factors.

In this paper, section 2 presents the GEP methods, including performance evaluation index of the method. Section 3 describes the experiment apparatus, the experiment procedures, data processing and the Cartesian impedance model at the end of the human arm. Section 4 presents results and discussion. Section 5 presents conclusions and the future work.

2 Methods

2.1 Gene Expression Programming Algorithm

According to the previous studies, the nonlinear model based on sEMG to estimate the human arm endpoint stiffness is more accurate and closer to the actual situation. In this paper, sEMG-based Stiffness is estimated by the model of Gene Expression Programming (GEP).

GEP is a function mining algorithm proposed by Ferreira, which combines the advantages of Genetic Algorithm (GA) and Genetic Programming (GP) [9]. Compared with GAs that solve simple problems with simple coding and GPs that solve complex problems with complex coding, GEP responds to complex problems with simple coding by using a linear string of a certain length to correspond to a flexible nonlinear tree structure of variable length. The calculation speed is 2~4 orders of magnitude faster than the traditional algorithm. This section improves the traditional GEP algorithm and establishes a mapping relationship between the sEMG signal and the stiffness in different directions of the Cartesian space.

In order to improve the efficiency of the GEP algorithm, this paper improves the traditional decoding method with reference to Xie's research. Instead of generating an expression tree, it is directly decoded by looking for the effective length of the gene.

Algorithm 1:Chromosome Decoding Algorithm

Input: Effective length of gene: len, Genotype: gene
Output: Gene expression: gene(1)

```
for i=len:-1:1
  if gene(i)== function symbol
    judging the number of parameters of the function: num
    switch num
      case 1
        perform the corresponding calculation on gene(len) and assign it to gene(i);
        len=len-1;break;
      case 2
        perform corresponding calculations on gene(len) and gene(len-1) and assign them to
gene(i);
        len=len-2;break;
    end
  end
end
```

IOP Conf. Series: Journal of Physics: Conf. Series **1267** (2019) 012016 doi:10.1088/1742-6596/1267/1/012016

In order to integrate more closely with the research object and subsequent data processing of this paper, this section designs a fitness function based on root mean square error, which is represented by (1).

$$f = M\left[\left(\frac{1}{n}\sum_{k=1}^{n}\left(e_k - m_k\right)^2\right)^{1/2}\right]^{-1} \tag{1}$$

Where M is a constant indicating that the bandwidth is selected, e_k is the value calculated by the function, m_k is the value obtained by the experiment, f is the fitness value. In addition, a suitable M value can improve the quality of evolution. The greater the fitness value, the stronger the ability of the individual to adapt to the environment and the better the performance.

Through the use of simple elite strategy, roulette selection operator, mutation, interpolation and two-point reorganization, the population is updated. While maintaining the diversity of the population, it is ensured that the optimal individual is not eliminated and the organizational structure of the new population is consistent with the original population. Table 1 lists the parameters of the GEP algorithm in this section.

Table 1. Parameter of GEP algorithm

Experimental Parameters	GEP algorithm	Experimental Parameters	GEP algorithm
Number of runs	100	Population size	20
Max number of times	100	Gene head length	4
Function set	+-*/^qcls	Gene tail length	5
Variable set	X	Number of genes	4
Constant candidate set	--	Chromosome length	36
Terminator set	X	Mutation rate	0.044
Random constant array length	10	IS displacement rate	0.1
Random constant value range	[-10,10]	IS element length	1,2
Connection function	+	Two-point recombination rate	0.3

Note: The function set '^' represents an exponential function, q denotes the square root, c denotes a trigonometric function, l denotes a logarithmic function, s denotes a square, and the variable set X is {'x1', 'x2', 'x3', 'x4', 'x5', 'x6', ' X7', 'x8'}

2.2 Performance Evaluation Criteria

After the model is established, the root mean square error (RMSE) and the correlation coefficient (CC) are used as the evaluation indicators of the model. RMSE represents the deviation between the predicted and experimental values, which can be calculated as follows:

$$\mathrm{RMSE} = \sqrt{\frac{1}{n}\sum_{k=1}^{n}\left(p_k - m_k\right)^2} \tag{2}$$

Where p_k indicates the predicted value, m_k indicates the experimental value. And the CC can be calculated by:

$$\mathrm{CC} = \frac{\mathrm{cov}(p,m)}{\sigma_p \sigma_m} \tag{3}$$

Where $\mathrm{cov}(p, m)$ represents the covariance of the predicted and measured values, σ_p indicates the expectation of the predicted value, σ_m indicates the expectation of the measured value. In general, the closer the RMSE is to 0, the closer the CC is to 1, the better the performance of the model.

3 Experiment

3.1 Experiment Apparatus

The purpose of this experimental platform is to verify the mapping between sEMG and the endpoint stiffness of the human arm. It is mainly composed of three modules: Robot System, Force Measurement System, sEMG Acquisition System. Among them, the robot system is used to provide the random disturbance displacement of the upper limb and obtain the end displacement of the human arm in real time. The force measurement system uses a six-dimensional force/torque sensor (JR3 Inc., USA) to detect the force generated at the end of the upper limb during the experiment. sEMG electrode (TrignoTM Wireless system, Delsys Inc., USA) is used to detect muscle electrical signals of human upper limbs. The platform also supporting the corresponding interface and data synchronization collection, storage and other software programs.

In the upper limb stiffness measurement experiment, 6 wireless electrodes are evenly arranged around the forearm at 1/3 of the elbow joint, 2 wireless electrodes are evenly arranged around the upper arm at 1/2 of the elbow joint, as shown in Figure 1(a). 6 wireless electrodes of the forearm mainly measure the activity of the muscles such as the radial wrist long extensor, ulnarextensor, the extensor digitorum, flexor carpi radialis muscle, flexor carpi ulnaris and the pronator teres. 2 wireless electrodes of the upper limb mainly measure the activity of the muscles such as the musculus biceps brachii and the musculus triceps brachii. The data acquisition system consists of three control computers, a KUKA manipulator, an F/T sensor and a wireless electrode. The force signal and the displacement signal are sent to the master control computer through the TCP and UDP protocols respectively, so that it can control the connection, disconnection, simultaneous acquisition, display and save of multi-modal data of each sub-computer as shown in Figure 1(b).

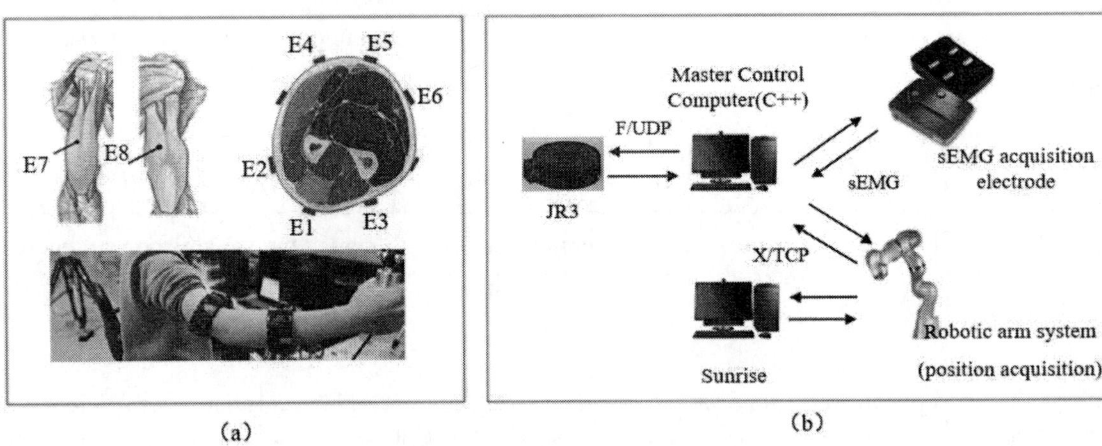

Figure 1 Electrode distribution and data acquisition system

3.2 Experiment Procedures

Five subjects (22-28 years old, 4 male and 1 female, respectively, A, B, C, D, E) were volunteered to participate in the experiment. Subjects knew the procedure before the experiment and agreed to participate in the experiment.

The robot was adjusted before the experiment to enable the subject to maintain the level of the forearm and the upper arm and the forearm at an angle of about 130-150°. We applied continuous stochastic perturbations to the subject's hand through the handle in a three-dimensional. The amplitude of the applied perturbations had the peak-to-peak value of 20mm. Data were collected from subjects at three different levels (relax, 30-50% MVC, 70-90% MVC), three experiments per muscle state, each disturbance lasting 30s. Each subject performed a total of 9 experiments (3 muscle states * 3 repetitions), and rested at any time during the experimental to avoid muscle fatigue. Including preparation time, each person takes about 40 minutes. The experimental device for predicting upper limb stiffness is shown in Figure 2.

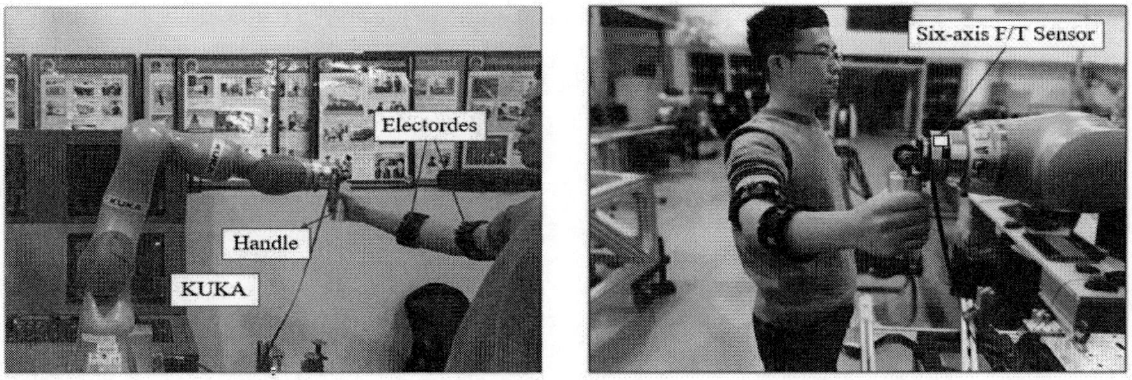

Figure 2 The schematic diagram of the experimental device

3.3 Data Processing

As described above, sEMG data were sampled at the frequency of 2000Hz and stored for offline analysis. F/T data and position data were also sampled at the frequency of 200Hz synchronously. Data preprocessing, parameter estimation, and model training were all done in MATLAB. The sEMG signal is filtered by a 4th-order Butterworth 20~500Hz bandpass filter and passed through a 50Hz notch filter to eliminate the influence of the power frequency. And the force signal and the position signal are filtered by a 4th order Butterworth 5~50Hz bandpass filter. Then according to the label of the displacement signal, find the corresponding force signal and the sEMG signal. Around this label, the mean value (MV) of force and position were extracted with the window of 250ms. The TDPSD, WPTSVD, WTSVD, WL, MDA, DASDV, RMS and MAV characteristics of sEMG were extracted with the window of 250ms. In the experiment, the horizontal grip data and the vertical grip data of the subject E are shown in Figure 3. It can be seen from Figure 3 that different levels of muscle contraction can produce different degrees of restoring force under random displacement sequences, whether horizontal or vertical.

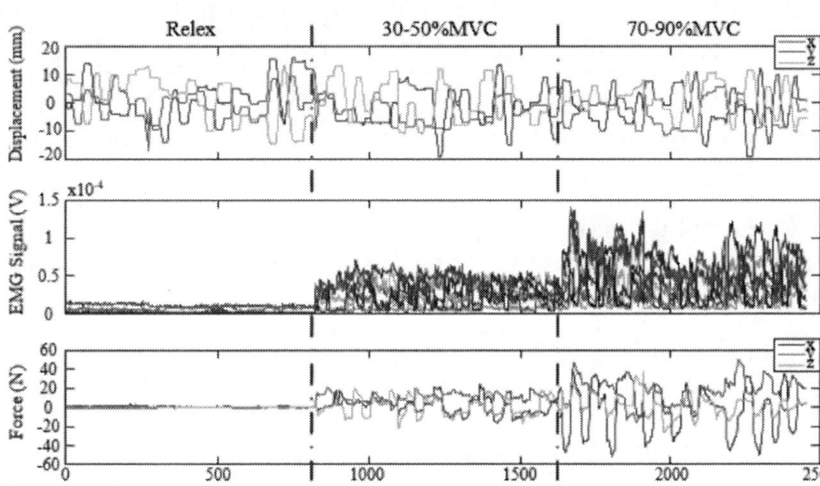

Figure 3 Partial experimental data of subject E

3.4 Modeling arm impedance

The Cartesian impedance model at the end of the human arm can be expressed by equation (4):

$$M_e\ddot{X}+B_e\dot{X}+K_e\left(X-X_u\right)=F \tag{4}$$

Where M_e B_e $K_e \in R^{3\times3}$ represented the Inertia, damping, stiffness matrices respectively. X F $\in R^{3\times3}$ are represented the end-point position and force vectors with respect to the human reference system. While X_u represented the virtual equilibrium position of the stiffness component. Equation (5) is expressed in the form of a matrix:

$$Y\pi=F \tag{5}$$

Where $\pi \in R^{27}$ was the parameter vector to be identified, given by

$$\pi=\left[M_{ex}^T \quad B_{ex}^T \quad K_{ex}^T \quad M_{ey}^T \quad B_{ey}^T \quad K_{ey}^T \quad M_{ez}^T \quad B_{ez}^T \quad K_{ez}^T\right]^T \tag{6}$$

Y was a (3×27) matrix defined by

$$Y=\begin{bmatrix} L^T & 0_{1\times9} & 0_{1\times9} \\ 0_{1\times9} & L^T & 0_{1\times9} \\ 0_{1\times9} & 0_{1\times9} & L^T \end{bmatrix} \tag{7}$$

Where $L=\left[\ddot{X}^T \quad \dot{X}^T \quad \left(X-X_u\right)^T\right]^T$.

By formula (5), we could get:

$$\pi=Y^{+}F \tag{8}$$

So according to the acquired characteristic values of the force signal and the displacement signal, the endpoint stiffness of the human arm during the experiment can be obtained. Table 2 shows the impedance characteristics of the subject E when it is gripped vertically.

Table 2 The impedance characteristics of the subject E

	M			B			K		
Relaxed	5.81	0.30	4.91	32.21	7.34	28.27	151.07	188.42	34.43
	0.13	7.14	6.20	3.17	1.89	1.33	151.40	472.57	199.86
	2.96	0.02	2.73	10.95	1.80	1.81	85.50	29.02	42.74
30-50%MVC	16.88	22.01	20.65	61.65	2.25	46.68	265.4	312.5	120.6
	18.24	32.83	16.74	90.39	41.66	8.13	316.2	353.6	238.6
	20.65	5.61	16.18	22.71	11.69	20.77	85.71	47.60	70.14
70-90%MVC	139.9	41.48	121.97	175.54	48.26	150.40	548.29	179.0	128.76
	34.33	49.01	27.46	28.96	44.67	22.67	191.47	460.86	380.01
	53.71	19.48	38.59	84.12	21.46	62.61	194.52	57.67	69.03

Taking the characteristics of the sEMG signal as input, the vector formed by the calculated diagonal value of the stiffness matrix K is output, and the regression training and testing are performed using the GEP algorithm described in the second part.

4 Results and Discussions

4.1 Results

First divide training set data and test set data, Four-fold cross-validation was used, that is, for each subject, 3/4 of the data set was randomly selected as the training set, and 1/4 of the data set was used as the test set. The GEP algorithm is used to train and test the sEMG features and the calculated stiffness data. The experimental results are shown in Figure 4.

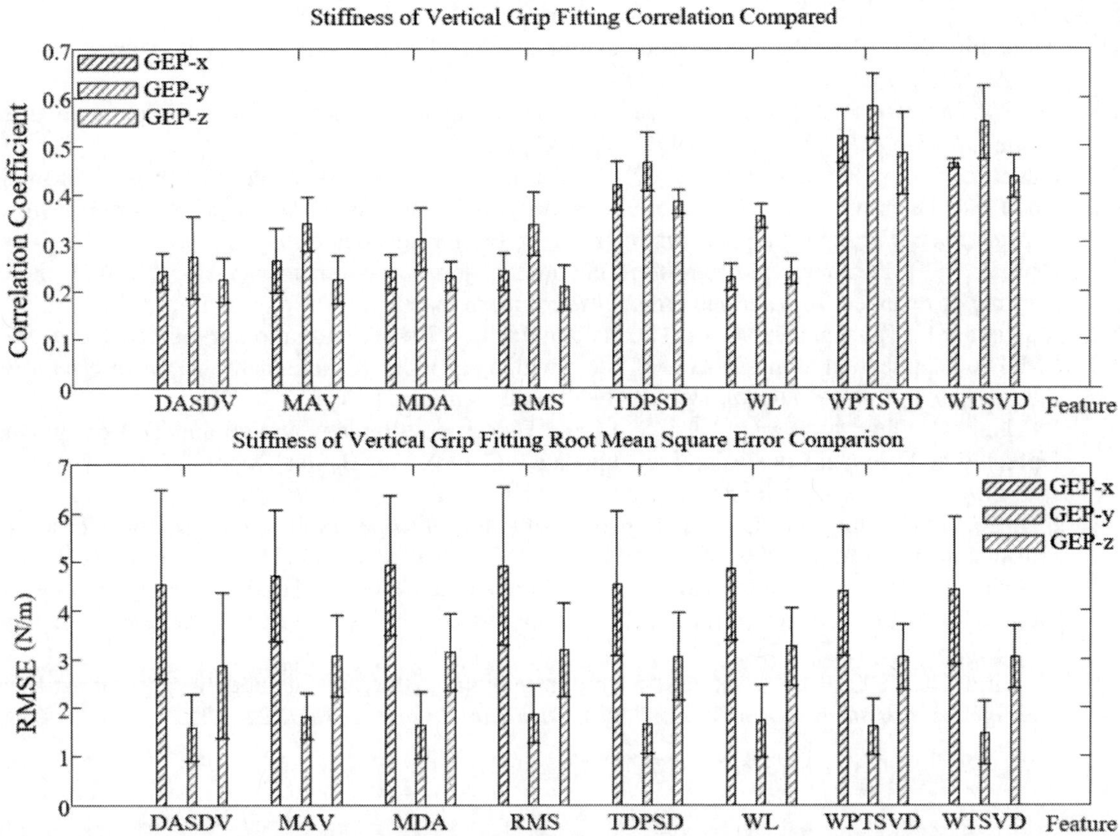

Figure 4 The endpoint Stiffness of the Human Arm Experimental Results Analysis Chart

4.2 Discussion

Among the eight selected sEMG features, the two characteristics WPTSVD and WTSVD based on singular value decomposition are the best. It can be seen that the singular value decomposition can effectively extract the differential mode information in the EMG signal to characterize the muscle stiffness. Secondly, the TDPSD feature is better because it reflects the complexity of the changes in myoelectric signals. The prediction effects of time domain features DASDV, MAV, MDA, RMS and WL are basically the same, and the performance is very stable and relatively low.

5 Conclusion

In this paper, sEMG-based stiffness is estimated by the model of Gene Expression Programming (GEP). According to the experimental results, the following conclusions can be obtained. First, nonlinear models such as GEP models have higher correlation and lower RMSE than regression stiffness using linear regression models. Secondly, selecting different feature of EMG signals, the correlation coefficient and the root mean square error of the model is very different. For the GEP model in this paper, WPTSVD and WTSVD are selected as the feature of sEMG signals, and the correlation can reach $57\% \pm 12.1\%$. In the future, we mainly focus on real-time human arm impedance estimation and more accurate estimation of human arm impedance after considering factors such as human arm configuration.

Acknowledgment

The authors gratefully acknowledge the financial supports by National Natural Science Foundation of China (No. 61803124 and No. 91848202) and Heilongjiang Provincial Postdoctoral Research Startup Fund.

References

[1] Ajoudani, A. . (2016). Transferring human impedance regulation skills to robots. *Springer Tracts in Advanced Robotics, 110.*

[2] Tsuji T, Morasso P G, Goto K, et al. Human hand impedance characteristics during maintained posture. *Biological Cybernetics*, 1995, 72(6):475-85.

[3] Kinnaird, C. R. , Ferris, D. P. . (2009). Medial gastrocnemius myoelectric control of a robotic ankle exoskeleton. *IEEE Transactions on Neural Systems & Rehabilitation Engineering A Publication of the IEEE Engineering in Medicine & Biology Society, 17*(1), 31.

[4] Osu, R. . (2002). Short- and long-term changes in joint co-contraction associated with motor learning as revealed from surface emg. *J. Neurophysiol, 88.*

[5] Maehara, M. , Tanaka, D. , Maeda, H. , Nakamura, T. . (2010). Estimation of joint stiffness using EMG and application to master-slave system with an artificial muscle manipulator. *International Symposium in Robot & Human Interactive Communication*. IEEE.

[6] P. Liang, C. Yang, N. Wang, Z. Li, R. Li and E. Burdet. "Implementation and Test of Human-Operated and Human-Like Adaptive Impedance Controls on Baxter Robot," *Taros*. 2014, pp. 109-119.

[7] Cui, Ze, Han, Wangyang, Qian, Donghai. Estimation of upper limb muscle stiffness based on artificial neural network. 2069-2074. 10.1109/ROBIO.2017.8324724.

[8] Disselhorst-Klug, C. , Schmitz-Rode, T. , & Günter Rau. (2009). Surface electromyography and muscle force: limits in semg–force relationship and new approaches for applications. *Clinical Biomechanics, 24*(3), 0-235.

[9] Ferreira, C. . (2002). Gene expression programming: mathematical modeling by an artificial intelligence. *Engineering Applications of Artificial Intelligence, 1*(3), 223–225.

IOP Conf. Series: Journal of Physics: Conf. Series **1267** (2019) 012017 doi:10.1088/1742-6596/1267/1/012017

New Research on Transfer Learning Model of Named Entity Recognition

Guoliang Guan and Min Zhu*

School of Computer Science and Software Engineering, East China Normal University, Shanghai, 200062, China

*E-mail: mzhu@cc.ecnu.edu.cn

Abstract. This paper integrates the current Google's most powerful NLP transfer learning model BERT with the traditional state-of-the-art BiLSTM-CRF model to solve the problem of named entity recognition. A bi-directional LSTM model can consider an effectively infinite amount of context on both sides of a word and eliminates the problem of limited context that applies to any feed-forward models. Google's model applied a feedforward neural network, causing its performance to weaken. We seek to solve these issues by proposing a more powerful neural network model named BT-BiLSTM. The new neural network model has obtained F1 scores on three Chinese datasets exceeds the previous BiLSTM-CRF model, especially on the value of recall. It shows the great value of the combination of large scale none-labelled data pre-trained language model with named entity recognition, which inspire new ideas on other future work.

1. Introduction

NER is also known as a proper name recognition. It generally refers to an entity with a specific meaning or strong reference in the text. Academically NER usually includes persons' names, locations or organization names. [1] The NER system extracts the above entities from the unstructured input text, and can identify more categories of entities according to needs. BERT applied a feedforward neural network, causing its performance to weaken. We seek to solve these issues by proposing a more powerful neural network model named BT-BiLSTM. A bi-directional LSTM model can consider an effectively infinite amount of context on both sides of a word and eliminates the problem of limited context that applies to any feed-forward models [2]. This article mainly focuses on chinese NER using CWS (chinese word segmentation) for sequential labelling [3]. In previous papers, named entity recognition only achieved good results in limited text types (mainly in news corpus) and entity categories (mainly names of people, places, and organizations) [4]. Compared with other information retrieval fields, entity naming evaluation is expected to be small and easy to produce over-fitting. The original BiLSTM model pays more attention to high accuracy, but the recall rate is not ideal, resulting in poor system performance for universal recognition of multiple types of named entities. In this paper, we adopted a new approach to improve the F1-score and Recall on NER.

2. Related work

2.1. Context Encoder Architectures

2.1.1. Pre-training. In the previous pre-training models (including word2vec, ELMo, etc.), word vectors are generated. From input sequence to predicted tags, a DL-based NER model consists of distributed representations for input, context encoder, and tag decoder, illustrated in figure1. But if a large number of parameters are pre-trained through a large training set such as BERT, the majority of the network structure parameters are initialized, and then the poor data volume is compared with the Fine-tuning process to adjust the parameters to make them more suitable for solving new tasks, which makes research work become simpler. In recent years, the common pre-training models include ULMFiT, GPT and BERT, etc. *The taxonomy of DL-based NER*

2.1.2. Transformer. Transformer was proposed by Vaswani et al. [5], dispenses with recurrence and convolutions entirely. Transformer utilizes stacked self-attention and point-wise, fully connected layers to build basic blocks for encoder and decoder. Experiments on various tasks [5][6][7] show Transformer to be superior in quality while requiring significantly less time to train. Based on transformer, Radford et al. proposed Generative Pre-trained Transformer (GPT) [8] for language understanding tasks. GPT has a two-stage training procedure. First, they use a language model with Transformers on unlabeled data to learn the initial parameters. Then they adapt these parameters to a target task using the supervised objective, resulting in minimal changes to the pre-trained model. Peters et al. proposed ELMo [9] representations, which are computed on top of two-layer bidirectional language models with character convolutions. This new type of deep contextualized word representation is capable of modelling both complex characteristics of word usage (e.g., semantics and syntax), and usage variations across linguistic contexts (e.g., polysemy). Unlike GPT (a left-to-right architecture), Bidirectional Encoder Representations from Transformers (BERT) [10] is proposed to pre-train deep bidirectional Transformer by jointly conditioning on both left and right context in all layers. The pre-trained BERT representations can be fine-tuned with one additional output layer for a wide range of tasks including NER and chunking.

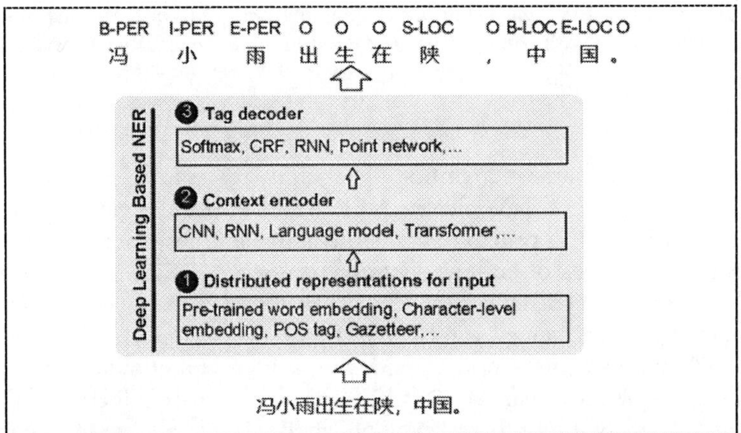

Figure 1. The taxonomy of DL-based Chinese NER.

2.2. Tag Decoder Architectures

2.2.1. Multi-Layer Perceptron. NER is in general formulated as a sequence labelling problem. With a multi-layer Perceptron +SoftMax layer as the tag decoder layer, the sequence labelling task is cast as a multi-class classification problem. Tagging for each word is independently predicted based on the context-dependent representations without considering its neighbors. A number of NER models, that has been introduced earlier use MLP + SoftMax as the tag decoder.

2.2.2. Conditional Random Fields. Based on these features, many machine learning algorithms have been applied in supervised NER, including Hidden Markov Models (HMM) [11], Decision Trees [12], Maximum Entropy Models [13], Support Vector Machines (SVM) [14], and Conditional Random Fields (CRF) [15]. A conditional random field (CRF) is a random field globally conditioned on the observation sequence. CRFs have been widely used in feature-based supervised learning approaches. Many deep learning based NER models use a CRF layer as the tag decoder, e.g., on top of a bidirectional LSTM layer [16], and on top of a CNN layer [17]. Listed in Table 1, CRF is the most common choice for tag decoder, and the state-of-the-art performance is achieved by [18] with a CRF tag decoder. The work by Huang et al. [19] is among the first to utilize a bidirectional LSTM CRF architecture to sequence tagging tasks (POS, chunking and NER).

The reasons using CRF, rather than HMM model, are given below.

- The HMM is assumed to satisfy the HMM independent assumption. CRF does not, so CRF can accommodate more contextual information.
- The CRF calculates the global optimal solution, not the local optimal value.
- CRF is the joint probability of calculating the entire marker sequence given the observed sequence. The HMM is given the current state and the next state is calculated.
- CRF compares the selection of features and the format of feature functions, and the amount of training is large.

Table 1. Summary of recent related works on neural NER

Model	Input representation			Context encoder	Tag decoder	Performance (F-score)
	Character	Word	Hybrid			
ELMo [9]	CNN-LSTM-LM	-	-	LSTM	CRF	92.2
BERT [10]	-	Word Piece	Segment, position	Transformer	MLP	92.8
BiLSTM[20]	CNN	SENNA	Capitalization, lexicons	LSTM	CRF	91.6
SS-multitask[21]	LSTM	Word2vec	-	LSTM	CRF	86.2
ID-CNNs[22]	-	SENNA	Word shape	ID-CNN	CRF	90.7
SLSM-GRSM[23]	-	SENNA	POS, gazetteers	CNN	Semi-CRF	90.9
Mc-BiLSTM[24]	LSTM	GloVe	Syntactic	LSTM	CRF	91.1
BT-BiLSTM	LSTM	Word Piece	Segment, position	Transformer	CRF	93.6

3. Model

Traditional approaches to NER are broadly classified into three main streams: rule-based, unsupervised learning, and feature-based supervised learning approaches. As to the techniques applied in NER, it briefly introduces the previous work. Deep-learning based approaches, which automatically discover representations needed for the classification or detection from raw input in an end-to-end manner. Brief descriptions are given below.

3.1. BiLSTM-CRF

In the BiLSTM-CRF model, four types of features were used for the NER task: spelling features, context features, word embeddings, and gazetteer features. Their experimental results showed that the extra features (i.e., gazetteers) boost tagging accuracy. The BiLSTM-CRF model by Chiu and Nichols [20] incorporated a bidirectional LSTM and a character-level CNN. Besides word embeddings, the model used additional word-level features (capitalization, lexicons) and character-level features (4-dimensional vector representing the type of a character: upper case, lower case, punctuation, other).

The basic recurrent LSTM functions are:

$$\begin{bmatrix} i_j^c \\ o_j^c \\ f_j^c \\ \tilde{c}_j^c \end{bmatrix} = \begin{bmatrix} \sigma \\ \sigma \\ \sigma \\ tanh \end{bmatrix} \left(W^{cT} \begin{bmatrix} x_j^c \\ h_{j-1}^c \end{bmatrix} + b^c \right) \tag{1}$$

$$c_j^c = f_j^c \odot c_{j-1}^c + i_j^c \odot \tilde{c}_j^c \tag{2}$$

$$h_j^c = o_j^c \tanh \odot (c_j^c) \tag{3}$$

Where i_j^c, f_j^c and o_j^c denote a set of input, forget and output gates, respectively. $\mathbf{W}^{cT}$ and $\mathbf{b}^c$ are model parameters. $\sigma()$ represents the sigmoid function.

$$score(l|s) = \sum_{j=1}^{m} \sum_{i=1}^{n} \lambda_j f_j(s, i, l_i, l_{i-1}) \tag{4}$$

The sentence s is the sentence we want to label the part of speech. i is used to represent the ith word in sentence s. l_i indicates the part of speech to be scored for the ith word. l_{i-1} indicating the part of speech to be scored for the $(i-1)^{th}$ word. For RNN-based models, Long Short Term Memory (LSTM) and Gated Recurrent Unit (GRU) are two typical choices of the basic units. The LSTM network is generally called LSTM[25]. At present, A new combination named bi-directional long-short term memory (BiLSTM) network is a well-recognized combination of better models.

3.2. BERT

The full name of BERT is Bidirectional Encoder Representations from Transformers, which is the encoder of the two-way Transformer, because the decoder can't get the information to be predicted. In fact, many of the design decisions in BERT were intentionally chosen to be as close to GPT as possible so that the two methods could be minimally compared. The main innovation of the model is the pre-train method, which uses Masked LM and Next Sentence Prediction to capture the word and sentence level representation respectively. In traditional pre-training, in order for a machine learning model to process words, it is first necessary to represent the words in a numerical form so that the model can be calculated, such as Word2vec word embedding. The first phase of Pre-training, similar to Word Embedding, trains a language model using existing unmarked corpus. The second phase is called fine-tuning, which uses a pre-trained language model to complete specific NLP downstream tasks. Data tag costs are always expensive because of the need for domain expertise and the process is time consuming, especially in NLP, where text understanding varies from person to person. However, there is a large amount (almost unlimited amount) of unlabeled data around us, and it can be easily extracted in order to greatly reduce the fine tune training time (the BERT is trained for 4 days on 4 TPUs). Because of this, the pre-trained network parameters, especially the underlying network parameters, are extracted from the specific task. Due to the reusability of the underlying features, the more the underlying is available, the NLP will be available regardless of the domain, and the more extracted features will be related to the task at hand. This is why we initialize the new task network parameters with the underlying pre-trained parameters. The high-level features are associated with the task. This paper uses Fine-tuning to clean up the high-level irrelevant feature extractor of BERT with the new data set.

3.3. BT-BiLSTM

This model uses the BERT model trained by Google to initialize the parameters of the model, so that the model can convergence faster in the training process. The original network layer needs to delete or add a new connection layer relationship. If all the parameters of the original network are directly changed, and then used as the pre-training model of this article, the parameter explosion phenomenon will occur, and the new training result is very poor. In this paper, the connection layer is modified and the original model is used for training. The model structure can be seen in figure2.
The method and attention are as follows:

- Delete the layer, and replace the feedforward neural network.
- The parameters of the layer that need to be left are not changed.
- Help to add new layer relationships, adjust the classifier network to have more output neurons, and then use SoftMax.
- Ensure that the data dimension relationships between the layers that are connected to each other are matched. For example, the data dimension of the original A input to the fully connected layer is 768*7*7, and the dimension input to the FC layer after modifying the network is also 768*7*7, which is consistent.

The data dimension of the output can be changed using the fully connected layer.

Previously, the BiLSTM model used word2vec. This paper mainly replaces the embedding layer of the original paper with BERT.

IOP Conf. Series: Journal of Physics: Conf. Series **1267** (2019) 012017 doi:10.1088/1742-6596/1267/1/012017

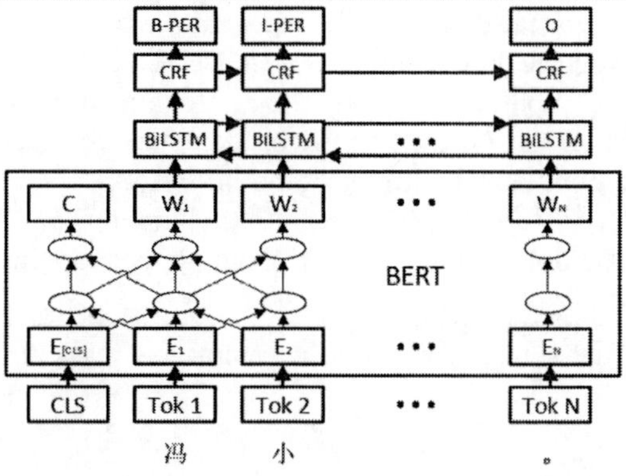

Figure 2. The architecture of Chinese BT-BiLSTM model. A bidirectional RNN produces context dependent representations, which are passed through the subsequent sequence tagging layer.

The BT-BiLSTM is used for tagging named entities. Our Model extracts a fixed length feature vector from tokens. For each word, these vectors are concatenated and fed to the BiLSTM network and then to the output layers. The output layers decode output into a score for each tag category. The extracted features of each word are fed into a forward LSTM network and a backward LSTM network. The output of each network at each time step is decoded by a linear layer and a log-SoftMax layer into log-probabilities for each tag category. These two vectors are then simply added together to produce the final output.

4. Experiment

4.1. Experiment environment and parameter

4.1.1. Batch size, learning-rate. Choosing an appropriate batch size and learning rate make the model both random and easier to converge, which makes the training the fastest convergence or the best convergence (global best) and reduces the risk of local optimization.

4.1.2. Dropout. Dropout randomly removes some of the cells of the hidden layer in order to prevent overfitting. In the training, for the neural network unit, according to a certain probability (if dropout=0.5, that is, removed with 50% probability, stop working), it is temporarily discarded from the network. In order to cover all the input information, more convolutional layers need to be added, resulting in deeper and deeper layers and more and more parameters. In order to prevent over-fitting, more regularizations such as Dropout should be taken into consideration. The hyperparameters were showed in Table 2.

Table 2. Experiment parameters

Hyperparameters	learning-rate	1E-5
	Batch size	64
	Dropout rate	0.5
	Warm up protection	0.1

4.2. Datasets

Table 3. Experiment datasets

Dataset Name	Train	dev	test
People's Daily classic data	1400K	1114K	15K
schema06	13579K	1 642K	1089K
Private data	6277K	702K	1405K

IOP Conf. Series: Journal of Physics: Conf. Series **1267** (2019) 012017 doi:10.1088/1742-6596/1267/1/012017

We use BIOES tagging scheme for Chinese NER tagging [26].The experiment datasets are showed in Table 3. All the Words in the sentence are obtained using a Chinese segment. Use BIOES data annotation mode to use People's Daily classic data. To make this compatible with Word Piece tokenization, we feed each date tokenized input word into our Word Piece tokenizer and use the hidden state corresponding to the first sub-token as input to the classifier. The popular labelling scheme is 'BIOES': 'B' for the beginning character of a word, 'I' for the internal characters, 'E' for the ending character, and 'S' for single-character word, '-' or 'O' means that the metric does not apply [27]. Example appears in Figure 3:

Text	Hayao	Tada	,	commander	of	the	Japanese	North	China	Area	Army
LOC	–	–	–	–	–	B	I	–	S	–	–
MISC	–	–	–	S	B	B	I	S	S	S	S
ORG	–	–	–	–	–	B	I	B	I	I	E
PERS	B	E	–	–	–	–	–	–	S	–	–

Figure 3. Example of how lexicon features are applied.

4.3. Experiment result

NER involves identifying both entity boundaries and entity types. With "exact-match evaluation", a named entity is considered correctly recognized only if it's both boundaries and type match ground truth. Precision, Recall, and F-score are computed on the number of true positives (TP), false positives (FP), and false negatives (FN).

- True Positive (TP): entities that are recognized by NER and match ground truth.
- False Positive (FP): entities that are recognized by NER but do not match ground truth.
- False Negative (FN): entities annotated in the ground truth that are not recognized by NER.

Precision measures the ability of a NER system to present only correct entities, and Recall measures the ability of a NER system to recognize all entities in a corpus. F1-score is the harmonic mean of precision and recall, and the balanced F1-score is most commonly used:

$$Precision = \frac{TP}{TP+FP} \qquad Recall = \frac{TP}{TP+FN} \qquad F1 = 2 \times \frac{Precision \times Recall}{Precision+Recall} \tag{5}$$

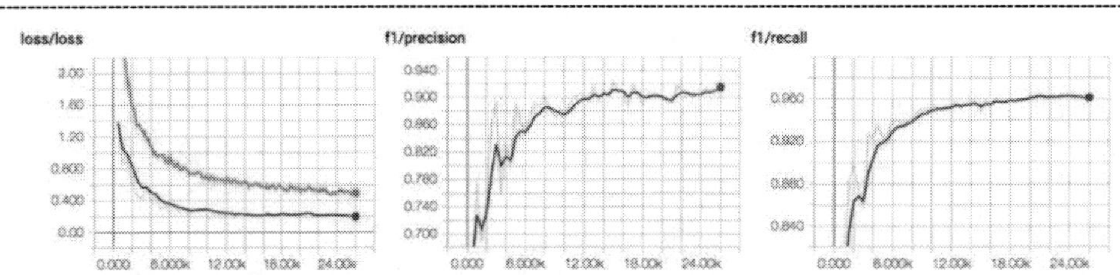

Figure 4. Loss, precision, recall within step change on dev data by BT-BiLSTM using tensor board.

Loss, precision, recall graphs can be seen in Figure 4. Test-F1 is the NER F-measure in the testing datasets. BT-BiLSTM model performed well on F1 and recall value on below three Chinese datasets. The Dev and Test scores are averaged over 5 random restarts using those hyperparameters. Results are presented in Table 4 and Table 5.

Table 4. Results on datasets.

	BiLSTM-CRF		BERT-base		BT-BiLSTM	
	daily	Schema06	daily	Schema06	daily	Schema06
Accuracy	98.7	98.5	99.0	88.7	**99.7**	**99.3**
Precision	92.1	90.6	92.4	91.5	**92.2**	**92.5**
Recall	89.8	88.7	87.5	87.1	**91.0**	**94.7**
Test F1	91.0	89.7	90.0	89.3	**91.6**	**93.6**

Table 5. NER(CWS) Results in F-measure.

		BiLSTM-CRF			BT-BiLSTM		
		Precision (%)	Recall (%)	Test-F1(%)	Precision (%)	recall (%)	Test-F1(%)
LOC[a]	Private	92.0	91.7	91.9	91.0	**95.0**	**93.0**
	Daily	93.8	90.8	92.3	**94.2**	**94.0**	**94.1**
	Schema06	93.5	88.0	90.7	**94.3**	**94.4**	**94.4**
ORG[b]	Private	89.1	83.9	86.5	81.8	**89.8**	85.7
	Daily	87.6	87.7	87.6	**88.9**	**91.8**	**90.4**
	Schema06	83.0	88.7	85.7	**84.1**	**92.8**	**88.2**
PER[c]	Private	93.8	93.6	93.7	**96.0**	**96.8**	**96.4**
	Daily	92.8	89.8	91.3	**94.1**	**90.0**	**92.1**
	Schema06	92.1	89.8	91.0	**96.1**	**96.4**	**96.3**
Average		90.9	89.3	90.1	**91.2**	**94.4**	**92.7**

[a] Location.
[b] Organization.
[c] Person.

5. Conclusion

According the experiment, our new model BT-BiLSTM has achieved nice scores on F1-score and Recall, showing a new research direction. Grammar-based systems usually get better accuracy, but at the cost of low recalls. In contrast this model improves the recall rate very well. What should be pointed is the operation speed and the time costs for the optimization. DL-based NER models achieved good performance with the cost of massive computing power. For example, ELMo representation represents each word with a 3×1024-dimensional vector, and the model was trained for 5 weeks on 32 GPUs. Google BERT representations were trained on 64 cloud TPUs. The training curve of the model shows that the convergence is very slow, and there is room for further optimization. Developing approaches to balancing model complexity and scalability will be a promising direction. On the other hand, model compression and pruning techniques are also options to reduce the space and computation time.

Acknowledgments

The authors would like to thank Kang He, Yu Liu for helpful discussions, and the anonymous reviewers for insightful comments.

References

[1] D. Nadeau and S. Sekine, "A survey of named entity recognition and classification," *Lingvisticae Investigations*, **vol. 30, no. 1**, pp.3–26, 2007.

[2] Abdel-rahman Mohamed, and Geoffrey Hinton. 2013. Speech recognition with deep recurrent neural networks. *In Proceedings of the 2013 IEEE International Conference on Acoustics, Speech and Signal Processing*, pages 6645–6649.

[3] Wenliang Chen, Yujie Zhang, and Hitoshi Isahara.2006b. Chinese named entity recognition with conditional random fields. *In Proceedings of the Fifth SIGHAN Workshop on Chinese Language Processing*. pages 118–121.

[4] Z. Huang, W. Xu, and K. Yu, "Bidirectional lstm-crf models for sequence tagging," *arXiv preprint* arXiv:1508.01991, 2015.

[5] A. Vaswani, N. Shazeer, N. Parmar, J. Uszkoreit, L. Jones, A. N.Gomez, Ł. Kaiser, and I. Polosukhin, "Attention is all you need," in *Proc. NIPS*, 2017, pp. 5998–6008.

[6] N. Kitaev and D. Klein, "Constituency parsing with a selfattentive encoder," in *Proc. ACL*, 2018, pp. 2675–2685.

[7] P. J. Liu, M. Saleh, E. Pot, B. Goodrich, R. Sepassi, L. Kaiser, and N. Shazeer, "Generating wikipedia by summarizing long sequences," *arXiv preprint* arXiv:1801.10198, 2018.

[8] A. Radford, K. Narasimhan, T. Salimans, and I. Sutskever, "Improving language understanding

by generative pre-training," Technical report, OpenAI, 2018.

[9] M. E. Peters,M. Neumann,M. Iyyer,M. Gardner, C. Clark, K. Lee, and L. Zettlemoyer, "Deep contextualized word representations," in *Proc. NAACL-HLT, 2018*, pp. 2227–2237.

[10] J. Devlin, M.-W. Chang, K. Lee, and K. Toutanova, "Bert: Pretraining of deep bidirectional transformers for language understanding", *arXiv preprint* arXiv:1810.04805, 2018.

[11] S. R. Eddy, "Hidden markov models," Current opinion in *structural biology,* ***vol. 6, no. 3***, pp. 361–365, 1996.

[12] J. R. Quinlan, "Induction of decision trees," Machine learning, **vol. 1, no. 1**, pp. 81–106, 1986.

[13] J. N. Kapur, Maximum-entropy models in science and engineering. John Wiley & Sons, 1989.

[14] M. A. Hearst, S. T. Dumais, E. Osuna, J. Platt, and B. Scholkopf, "Support vector machines," *IEEE Intelligent Systems and their applications*, **vol. 13, no. 4**, pp. 18–28, 1998.

[15] J. Lafferty, A. McCallum, and F. C. Pereira, "Conditional random fields: Probabilistic models for segmenting and labeling sequence data," 2001.

[16] Z. Huang, W. Xu, and K. Yu, "Bidirectional lstm-crf models for sequence tagging," *arXiv preprint* arXiv:1508.01991, 2015.

[17] R. Collobert, J. Weston, L. Bottou, M. Karlen, K. Kavukcuoglu,and P. Kuksa, "Natural language processing (almost) from scratch," *Journal of Machine Learning Research*, **vol. 12, no. Aug**,pp. 2493–2537, 2011.

[18] A. Akbik, D. Blythe, and R. Vollgraf, "Contextual string embeddings for sequence labeling," *in Proc. COLING, 2018*, pp. 1638–1649.

[19] V. Yadav and S. Bethard, "A survey on recent advances in named entity recognition from deep learning models," *in Proc. COLING,2018*, pp. 2145–2158.

[20] J.P. Chiu and E. Nichols, "Named entity recognition with bidirectional lstm-cnns," *Transactions of the Association for Computational Linguistics*, pp. 357–370, 2016.

[21] M. Rei, "Semi-supervised multitask learning for sequence labeling,"in *Proc. ACL*, 2017, pp. 2121–2130.

[22] E. Strubell, P. Verga, D. Belanger, and A. McCallum, "Fast and accurate entity recognition with iterated dilated convolutions,"*Proc. ACL*, 2017, pp. 2670–2680.

[23] J. Zhuo, Y. Cao, J. Zhu, B. Zhang, and Z. Nie, "Segment-level sequencemodeling using gated recursive semi-markov conditional random fields," in *Proc. ACL*, **vol. 1**, 2016, pp. 1413–1423.

[24] B. Y. Lin, F. Xu, Z. Luo, and K. Zhu, "Multi-channel bilstm-crf model for emerging named entity recognition in social media," *in Proc. the 3rd Workshop on Noisy User-generated Text, 2017*, pp.160–165.

[25] O. Kuru, O. A. Can, and D. Yuret, "Charner: Character-level named entity recognition," in *Proc. COLING, 2016*, pp. 911–921.

[26] M. E. Peters, W. Ammar, C. Bhagavatula, and R. Power, "Semi supervised sequence tagging with bidirectional language models" in *Proc. ACL, 2017*, pp. 1756–1765.

[27] Fan Yang and Paul Vozila. 2014. Semi-supervised Chinese word segmentation using partial-label learning with conditional random fields. *In Proceedings of the 2014 Conference on Empirical Methods in Natural Language Processing (EMNLP)*, pages 90–98, Doha, Qatar, October. Association for Computational Linguistics.

IOP Conf. Series: Journal of Physics: Conf. Series **1267** (2019) 012018 doi:10.1088/1742-6596/1267/1/012018

Detection of Non-Standard English Expressions by Language Sense

Li Chao and Zhang Jingxiang*

College of Information Science and Engineering, University of Jinan, Jinan 250022, Shandong Province, China

*Corresponding author's e-mail: ise_zhangjx@ujn.edu.cn

Abstract. This paper proposes an algorithm for perception English essay, in order to improve the ability of Chinese students' writing which expressing more standard. Because students are influenced by their mother tongue in second language acquisition, English expression is not accurate enough. From the lexical and syntactic aspects, this paper uses the language sense relevance analysis algorithm (N-LanSen) based on the N-model to deeply perceive the students' inappropriate expression in the essay. Different extraction methods are used to extract different features, and the corresponding feedback can be given to the places where the expression is not standard. The experimental results show that the algorithm can detect the place where English expression is not standard with high accuracy.

1. Introduction

Artificial intelligence processing is replacing the routine repetitive work handled by traditional manpower in many fields, at least it assisting manpower to carry out efficient work. English automatic scoring is no exception. At present, the number of students is much higher than that of teachers under all levels of education system in China. English teachers are unable to complete extensive corrections, while students have a strong demand in this respect, because it is difficult for students to complete a task by themselves through comparing standard answers. At present, there are three main stages in the foreign system of English essay scoring: firstly, PEG system can get scores indirectly by counting some indicators of the essay, but it can't understand the content of the composition, only simulate the scores similar to the manual marking according to the text characteristics [1]; secondly, the natural language processing technology has developed relatively, and E-rater scored by examining the organizational structure, vocabulary and syntactic structure of the essay [2]; Writing Roadmap system tends to automatically score and personalized comments [3]. Foreign automatic writing scoring system is mainly designed for native speakers of English. For Chinese students, English is a non-native language and an official language. The above foreign systems can't satisfy their assessment of English writing for the main purpose of foreign language learning.

There are also some automatic scoring systems for English essay in China. According to the principle of classification, the typical examples are the system of similarity comparison based on large English corpus to get the score of composition evaluation [4], and the system of feature extraction based on words, sentences and textual structure to get the score through Adaboost algorithm [5]. However, these algorithms have not been studied from the aspect of English language sense to test whether English expression is standard or not. With more and more frequent exchanges between countries, it may affect communication with each other which English expression is not standard.

This paper studies the problems existing in English writing based on this demand. It proposes a method of N-language sense correlation analysis, which can effectively detect the problem of essays which English expressions isn't standard. And give feedback to the existing problems in order to improve students' English expression level.

2. Language sense detection algorithms and non-standard English perception

2.1. Language sense detection algorithms

N-gram is the existing language model, which uses conditional probability to evaluate the reasonableness of sentences [6]. The N-gram model is based on the Hidden Markov Hypothesis, which assumes that the occurrence of the nth word is only related to preceding n-1 word, but not to other words. The probability of the whole sentence is the product of the probability of each word appearing in the corpus. According to the probability statistics and calculation of sentences, if the probability is less than a certain threshold, it is considered that there are grammatical errors in sentences. In the N-gram model, N isn't the larger the better, the larger N the more probability calculations. Although the current computer has been able to solve the calculation problems caused by the larger N value, but the larger N value will lead to data sparsity, the overall performance can't be improved.

The N-gram model calculates the probability of words appearing in the whole sentence, and the result is only that the sentence is judged to be grammatical error because the probability is lower than the threshold value. It is also necessary to smooth the N-tuple in the corpus so that the detected sentences can't occur at zero probability due to the absence of some phrases in the corpus.

The data in the corpus do not necessarily contain the expression of human language as a whole, and it is normal for phrases to be missing. However, the N-gram model adopts a smoothing method to deal with phrase missing, which is not necessarily reliable. For example, there are grammatical errors in sentences themselves, but the probability of sentences after smoothing is lower than the threshold of grammatical errors.

Some literatures interpret language sense as: sensitivity to following or deviating from the established usage of a language; sense of linguistic validity or appropriateness [7]. According to this idea, there should be a certain correlation between the words in the sentence. In this paper, an association rule analysis algorithm based on N-tuple, N-LanSen, is proposed for language sense measurement.

The corpus and the sentences to be tested are divided into N-tuples, and the N-tuples of the sentences to be tested are used as test sets of association rules, and the linguistic sense values of N-tuples are calculated. If the linguistic perception is less than the given threshold, the sentence expression will be problematic.

2.2. Vocabulary perception

Looking at errors in English essay, inappropriate vocabulary expression is a common problem. This is largely due to the use of English-Chinese dictionaries by students [8]. In the process of learning English, the Chinese explanations in dictionaries are regarded as the real meaning of a word, or the words are understood according to the Chinese explanations. These two approaches are unreliable. Because of the great differences between China and the West, the use of words is not the same. Using Chinese thinking to understand English vocabulary is the main reason for its mistakes. English-English dictionaries should be used as much as possible in English learning. Explaining English words in English can not only enhance the sense of English, but also realize the true meaning of English words. The perception of non-standardized vocabulary is mainly reflected in the following aspects.

2.2.1. Lexical redundant. In the process of writing, the use of words is obviously redundant, especially when they express their personal opinions. They usually use the expression “Personally speaking” or “In my opinion” with “I think”, it shows that it is redundant. Repetition is a rhetorical device in Chinese, while repetition is taboo in English [9].

Similar lexical redundancy is also reflected in the repetition of meaning caused by the use of a word and its derivatives together. The word "solve" is the verb form in English, and its derivative "solution" is a noun. "Solve a solution" is a correct expression in sentences, but "solution to solve the problem of air pollution" is redundant. When the verb "solve" is removed, it does not affect the expression of the sentence at all, and the sentence become simpler and clearer.

The use of redundant words in writing will give teachers a bad impression. First, teachers feel that students have no sense of authentic English, but they translate it directly from Chinese to English. Some teachers also feel that the redundant expressions of students may be adding the number of words, showing a phenomenon of insufficient English ability.

If the threshold of language sense is lower than the threshold of language sense, the relevant N-tuple keywords are extracted according to the rules of the same part of speech, and the cosine similarity analysis is used. If the similarity between the two N-tuples is high, the redundancy of words is considered.

Cosine similarity is the calculation of text similarity. It maps words into space by TF-IDF algorithm to obtain vector representation, and then calculates vector similarity by using cosine theorem [10]. The higher the similarity, the closer the cosine value is to 1, otherwise the cosine value is 0. The expression of cosine similarity is as follows: $\cos(\theta) = \frac{a \cdot b}{\|a\| \times \|b\|}$.

2.2.2. Improper collocation of vocabulary. The collocation of verbs and nouns in Chinese and English is different [11]. In daily life, the words "eat" can be used in meals and medicine in Chinese expressions are no more common, and also use "kick football". But in English expression, if we use "eat meals" to express eating; "eat medicine" to express taking medicine; "kick football" to express playing football, this is the so-called "apply mechanically" and translate directly according to Chinese thought. Generally, playing is used in sports such as "kick" and "play", while Chinese is more detailed. The English language is confused in eating. We can't use "eat medicine" to express "take medicine", only when eating specific food, use eat, such as eat apples.

The usage of English vocabulary is quite complex. If there is no sense of language, it is easy to make mistakes just by expressing ideas in Chinese. It will be a huge task to use rules to express English in essays which are not standard, and the correct rate is not necessarily high. Therefore, this paper adopts a statistical method to detect the improper collocation of English vocabulary.

For a certain N-tuple, the value of language sense is below the threshold, and it happens that there are verbs and nouns in the N-tuple, which is suspected of improper lexical collocation. By extracting the noun in the sentence and searching the corpus for the collocation of the noun. Since there is more than one collocation of a noun, such as "play football" mentioned above, the word "buy football" may exist in the corpus. Regardless of semantics, the verb collocation of feedback is definitely right. But it is not necessarily right to put it in the current semantics. To solve this problem, this paper adopts the strategy of editing distance. The Chinese expression of "kick football" in English is "play football"; while "buy a football" is not "buy football" in English, the article "a" must be used in the middle, otherwise it is also wrong.

Editing distance is a distance measurement method. It is the number of editing operations needed to convert one character into another character [12]. The expressions of "buy a football" and "play football" discussed in this paper need to be replaced by the verb "kick", while "buy a football" needs to add the article "a", so the editing distance is longer than "play football".

Of course, the use of editorial distance is only an idealized feedback without any other grammatical errors in the essay. If a student writes not "play football" but "buy a football", using editorial distance will not solve the problem. Idealized use of editing distance is not very reliable, but also need to extract the semantics of the sentences before and after, to obtain the correct vocabulary collocation.

2.3. Syntactic perception

IOP Conf. Series: Journal of Physics: Conf. Series **1267** (2019) 012018 doi:10.1088/1742-6596/1267/1/012018

2.3.1. Misuse of conjunctions. There are also great differences between English and Chinese in conjunction [13]. In Chinese, causality is often expressed by "because …so... ". It is easy for students with weak English sense to use "because" and "so" in the same sentence when expressing causality. This is not in accordance with grammatical rules in English expression. In a sentence, conjunctions can't exist at the same time when we use "because … So... ", similar also includes "although… but...".

For the detection of conjunctions that can't be used together, if we use N-LanSen directly, it is very likely that it can't be detected. Because "because" needs to be followed by a clause expressing the cause, if the clause is longer, according to the rule: the occurrence of the N word is only related to its preceding N-1 word. Then the N will be larger, the data sparse, or nonexistent. In order to avoid large N value and avoid sparse or non-existent data, this paper extracts the key words in sentences and uses the sense of speech analysis based on specific parts of speech.

By using clause operation in essay, we can get a complete sentence to be tested. To extract the conjunctive words in sentences, the N-gram model is used to extract the words. For example:

"Because it's raining outside, so I won't go out."

Words with conjunctive attributes extracted from part of speech are "because" and "so". The two words are put into the binary grammar model < because, so>. Since there are no such binary groups as < because, so > in the standard corpus, we can consider the use of conjunctions errors.

2.3.2. Error of passive voice. Active voice is often used in Chinese expression, but passive voice is seldom used. In Chinese, the passive voice will not be used if the sentence does not need to point out the executor of the action and there is no ambiguity between the active and passive expressions [14]. In terms of English expression, it is contrary to Chinese expression. In English expressive thinking, subjectivity is generally minimized, and the beginning of "person" is replaced by the beginning of "material name".

The conversion of active voice to passive voice will lead to great changes in sentence structure, and the use of active voice in some sentences is also in line with the habit of English expression. Therefore, this paper only detects the passive voice in the aspect of voice, and reflects whether the candidates are standardized in English writing through the aspect of passive voice.

According to various sentences with passive voice, we can find that the passive form in sentences is mainly "be + transitive verb past participle". In fact, the "be" here is meaningless. It only involves changes in person, number and tense.

When detecting the passive voice in a sentence, all verbs in the sentence are extracted according to the part of speech and combined into N-tuples. N-LanSen algorithm is used to test whether the sense of language in sentences is correct.For example: This book will be published.

The verbs in the sentence include "will, be, published". Combining them into a triple < will, be, published > and matching them in the corpus. If the sense value of the triple in the corpus exceeds the minimum threshold, the sentence writing is correct.

3. Algorithmic flows and statistical analysis of experiments

3.1. N-LanSen model

This model uses a sliding window based on part-of-speech markers to extract words, and it also needs to do clause processing before detecting the essay [15]. For example, when evaluating the sense of the sentence "He is reading English." starting with the word "He", only two words are extracted from the sentence at a time. After the extraction of the current two words "He" and "is" is completed, the extraction of the words is started iteratively. Finally, the result of binary model is < He, is > < is, reading > < reading, English >. The binary model is established as shown in Figure 1.

While extracting non-standard English syntactic features part-of-speech tags are needed to acquire N-tuples according to specific parts of speech.

For example, we use the sentences shown in Figure 1 to extract verbs. When we need to extract verbs, we ignore other parts of speech and get a binary < is, reading >. Similarly, if a noun needs to be extracted, only the binary group < He, English > composed of the words "He" and "English" is obtained. Similarly, according to feature extraction requirements, we can extract triples and so on.

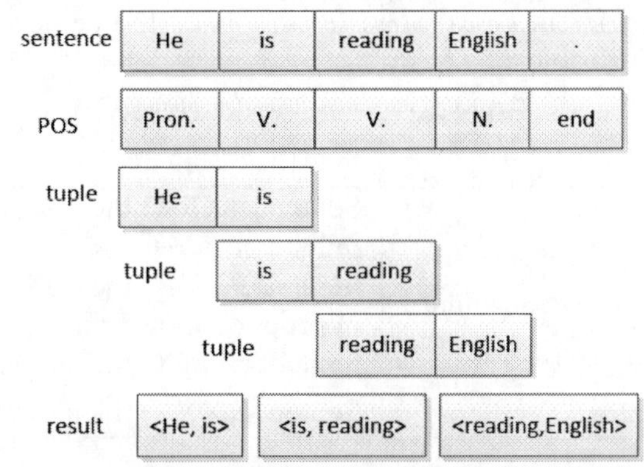

Figure 1. Tuples set-up model.

After extracting N-tuples from the sliding window, the language sense can be calculated according to the data in the corpus.

In this model, the words extracted from the N-tuple are related, which can be called language sense. For example, the tuple < He, is > extracted in Figure 1 shows that a person with strong language sense knows that the verb "be" uses "is" rather than "am" or "are" without thinking.

The extracted N-tuples are compared according to the corpus. If they exceed a certain threshold, the sense of language is correct; otherwise the sense of language is wrong. Taking the calculation of sense value of binary tuple as an example, the calculation formula is (1).

$$VLanSen(2tuples)=\frac{CorpusAppearTimesWith<a,b>}{CorpusAppearTimesWith<a,x>} \qquad (1)$$

Where the expression $<a,x>$ in the form begins with "a" and ends with any word.

3.2. Statistical analysis of experiments

This experiment use the sentences of essay which written by high school students as dataset. Some of these sentences have non-standard expressions, some are proper expression. In these non-standard sentences, there may be one or more of the above problems. Firstly, these data are labeled manually to record the problems in sentences to the database. After the system language sense detection, the correct rate of the test results is counted.

The results are shown in Table 1 by detecting N-LanSen of the sentences to be tested in the database.

Table 1. Detection result.

feature	accuracy
proper sentence	100%
lexical redundant	95.17%
improper collocation of vocabulary	97.35%
misuse of conjunctions	99.23%
misuse of passive voice	99.85%

When the system detects all the appropriate sentences, the system prompts that the sentences express correctly. But for the sentences which expressions are not so standard, the accuracy of detection isn't completely correct.

The purpose of this paper is not only to find out the shortcomings of non-standard English, but also to improve students' standardized English expression. In order to improve the standard of English expression of Chinese students, it is suggested that students should read more authentic English articles in English learning. Learning standard English can strengthen the sense of language, rather than reciting beautiful sentences or essays mechanically. Because the system adopts different feature extraction methods according to different types, it can feedback the corresponding modification suggestions according to different types when feeding back the composition results to students. The system feedback is shown in Table 2.

Table 2. System Feedback Model.

feature	feedback sentence
lexical redundant	Sentence(XXX),the word "a" and "b" is redundant
improper collocation of vocabulary	Sentence(XXX),the verb "a" and noun "b" is improper collocation
misuse of conjunctions	Sentence(XXX),the conjunction "a" is misuse
error of passive voice	Sentence(XXX),the passive voice is error

4. Conclusions

At present, N-gram is the existing grammar model, because of its large granularity, it can only get the whole sentence is correct or not, and it can't be sure that there are mistakes in some places. Starting from the smaller granularity, this paper finds out the problems existing in sentences, and uses different detection methods according to different types of errors. The results show that the improper sentence expression can be obtained by using N-LanSen algorithm, and the recognition accuracy is high. It can meet the requirement of feature extraction and feedback the inappropriate expressions in English compositions.

Reference

[1] McNamara D S, Crossley S A and Roscoe R D 2015 *J.* A hierarchical classification approach to automated essay scoring. Assessing Writing. **23** 35-59
[2] Shermis M D, Burstein J, Higgins D and Zechner K 2010 *M.* Automated Essay Scoring: Writing Assessment and Instruction. International Encyclopedia of Education (Third Edition). 20-26
[3] Xiaodong Lei 2015 *J.* Domestic Research and Application of Automatic Assessment System for English Compositions. Technology vision. **35** 43-45
[4] Xuliang He 2013 *J.* A study on the reliability and validity of Juku's English composition score. Modern Educational Technology. **23** 64-67
[5] Ting Li and Jingxiang Zhang 2018 *J.* Centralized Trend Adaptive Enhanced English Composition Scoring Algorithms. Computer Engineering and Application. **54** 151-5+256.
[6] Garner J, Crossley S, and Kyle K 2019 *J.* N-gram measures and L2 writing proficiency. System. **80** 176-187
[7] Jingxiang Zhang 2000 *J.* Talking about English Language Sense. Foreign Language Teaching. **1** 88-92
[8] Macaro E and Tian L 2015 *J.* Exploring teachers' oral explanations of new English lexical items in a Chinese university: Comparisons with dictionary information. System. **52** 78-90
[9] Jianping Zeng 2018 *J.* Provisional Translation of Chinese Semantic Repetitive Words and Sentences. Chinese Science and Technology Translation. **31** 37-40+33
[10] Al-Anzi F S and AbuZeina D 2017 *J.* Toward an enhanced Arabic text classification using cosine similarity and Latent Semantic Indexing. Journal of King Saud University - Computer and Information Sciences. **29** 189-195
[11] Gulec N and Gulec B A 2015 *J.* Lexical Collocations (Verb + Noun) Across Written Academic Genres in English. Procedia - Social and Behavioral Sciences. **182** 433-440
[12] Martin R R 2015 *J.* On the computation of edit distance functions. Discrete Mathematics. **338** 291-305
[13] Martínez A C L 2015 *J.* Use of Conjunctions in the Compositions of Secondary Education Student. Procedia - Social and Behavioral Sciences. **212** 42-46
[14] Zhi-ying LIU and Yao-hong JIN 2012 *J.* Passive sentence transformation in Chinese-English patent machine translation. The Journal of China Universities of Posts and Telecommunications. **19** 135-9
[15] Ranucci G 2006 *J.* On the significance of signal search through the "sliding window" algorithm. Nuclear Instruments and Methods in Physics Research Section A: Accelerators, Spectrometers, Detectors and Associated Equipment. **562** 433-8

IOP Conf. Series: Journal of Physics: Conf. Series **1267** (2019) 012019 doi:10.1088/1742-6596/1267/1/012019

Study on Reactive Power Imbalance and Its solution of Parallel Coupled Transmission Lines

LI Dengfeng[1, a], LIU Yuming[1,b], XIANG Bin[2,c], LI Xiaoju[1,d], SI Meng[1,e] and YANG Mincai[1,f]

[1]·State Grid Chongqing Electric Power Research Institute, Yubei District, Chongqing, 401123, China;
[2]·State Grid Chongqing Electric Power Company, Yuzhong District, Chongqing, 400015, China

E-mail: [a]597913911@qq.com; [b]148449600@qq.com; [c]xiangbin@cq.sgcc.com.cn; [d]lixiaoju@cq.sgcc.com.cn; [e]simeng@cq.sgcc.com.cn; [f]yangmincai@cq.sgcc.com.cn

Abstract. In view of the imbalance and even the abnormal phenomenon of reactive circulation of parallel high voltage transmission lines with similar measured parameters in actual operation, combined with field investigation and test, theoretical analysis and simulation demonstration, it is clear that this phenomenon is caused by the use of unconventional phase sequence and unreasonable transposition of the two-circuit lines when erecting the same tower. Based on the PSCAD simulation, the corresponding improvement measures are put forward from two aspects of optimizing the arrangement of phases and transposition. The results show that in order to solve the power imbalance problem of parallel coupled transmission lines, it is recommended to optimize the phase sequence of conductors in consideration of economy and operability.

1. Introduction

With the continuous improvement of power grid intelligence, the dependence of dispatchers on real-time on-line monitoring systems, such as supervisory control and data acquisition (SCADA) system and on-line security and stability analysis systems, is also increasing. At present, the real-time on-line monitoring of power grid is mainly based on grid model and algorithm, and the accuracy of grid parameters determines the accuracy of grid model, and then determines the accuracy and reliability of the real-time on-line monitoring system in state estimation, power flow calculation, fault analysis, stability analysis and other application functions [1] .Therefore, improving the accuracy of power grid parameters is of great significance to ensure the safe and stable operation of the power grid.

In order to alleviate the tension of transmission corridors, high-voltage transmission lines usually use double-circuit or multi-circuit parallel transmission lines on the same tower. In principle, high voltage transmission lines usually adopt transposition measures to reduce the asymmetry of current and voltage in normal operation of power system [2]. When the parallel transmission lines erected on the same tower are reasonably arranged and transposed, it is often considered that the electrical quantities of each sequence are independent and symmetrical, and the electromagnetic coupling effects among loops, phases and phases are often neglected in the actual measurement of parameters. However, in actual operation, some parallel high voltage transmission lines of the same type still have unbalanced power, even abnormal power circulation phenomena, which seriously affect the normal

IOP Conf. Series: Journal of Physics: Conf. Series **1267** (2019) 012019 doi:10.1088/1742-6596/1267/1/012019

operation of state estimation, online safety analysis and other links, and cause misjudgments to power grid operators [3]. In the first ten days of December 2017, through D5000 platform, regulator automation personnel of a power grid company found that the reactive power of two return lines (L1 and L2) of a 500 kV power plant (A power plant) was unbalanced, even the phenomenon of reactive power circulation, which seriously affected the results of state estimation. Through on-site investigation, theoretical analysis and simulation demonstration, it is clear that the reactive power imbalance of the transmission lines of the power plant is caused by the non-conventional arrangement of the two-circuit lines in the same tower but not reasonable transposition, which results in the asymmetry of the parameters, and the corresponding countermeasures are put forward.

2. Phenomenon Description

A power plant is connected to 500 KV B substation through double-circuit 500 KV lines L1 and L2. Its 500 KV system structure is shown in Figure 1.

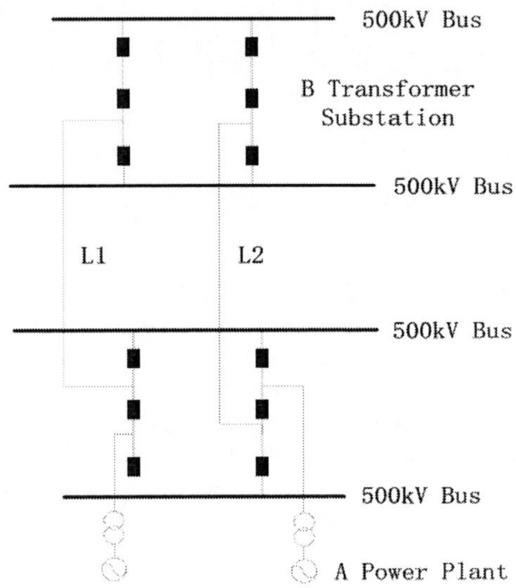

Figure 1. Primary connection diagram of outgoing lines of plant A

By inquiring D5000, the historical power records of L1 and L2 on the power plants side at several times since A power plant was connected to the grid are shown in Table 1. Historical records show that the active power of L1 and L2 is basically balanced, but the reactive power varies greatly, and even the phenomenon of different reactive power flows has existed since the grid-connected generation of A power plant.

Table 1. Historical record values of active and reactive power on power plant side for line L1 and L2

Time		L1		L2	
		P(MW)	Q(MVar)	P(MW)	Q(MVar)
2nd Feb. ,2017	21:00:00	281.98	-11.86	279.67	-35.11
3rd Feb. ,2017	11:30:00	308.99	-0.07	306.16	-25.74
29th Jun. ,2017	14:00:00	160.86	16.88	161.37	3.31
30th Jun. ,2017	11:00:00	172.09	5.87	171.22	-7.47
15th Nov. ,2017	17:00:00	636.89	6.80	632.32	-46.47
16th Nov. ,2017	12:00:00	546.94	-36.11	538.98	-80.28
20th Dec. ,2017	15:30:00	463.41	6.41	459.01	-31.31
5th Jan. ,2018	8:45:00	513.48	33.98	513.09	-7.87

3. Cause Analysis

3.1. Analysis of Accuracy of Acquisition Data

Because the above phenomenon is based on D5000 system data records, in order to check whether this phenomenon is caused by D5000 data anomalies, it is necessary to verify the accuracy of D5000 system data records.

Firstly, it is necessary to verify the acquisition accuracy of voltage transformer (PT) and current transformer (CT) in L1 and L2 power plants. The values of three-phase current, three-phase voltage and PMU (phasor measurement unit, synchronous phasor measurement unit) measured at 8:40:13 on January 5, 2018 are shown in Table 2. The results of Table 2 show that the maximum error rates of the three-phase voltage and current test values of line L1 and L2 meet the technical requirements of the relevant regulations in reference [4].

Table 2. Test results of three-phase voltage and current for line L1 and L2

	A phase voltage(kV)		B phase voltage(kV)		C phase voltage(kV)		Maximum error rate
	Test result	PMU	Test result	PMU	Test result	PMU	
L1	301.82	301.596	302.95	303.224	303.86	303.474	0.13%
L2	301.66	301.495	302.58	302.808	303.98	303.428	0.18%
	A phase current(A)		B phase current(A)		C phase current(A)		Maximum error rate
	Test result	PMU	Test result	PMU	Test result	PMU	
L1	536.66	535.896	560.12	559.089	585.22	584.724	0.18%
L2	581.02	580.452	561.09	561.531	531.63	531.013	0.12%

Next, we need to verify the correctness of D5000 system data display. The D5000 values of active power, reactive power and voltage of L1 and L2 at 8:45 and 8:50 on January 5, 2018 are randomly selected and compared with those of PMU. The results are shown in Table 3. From Table 3, it can be seen that the maximum error rate between the active, reactive and voltage values of L1 and L2 in D5000 and the corresponding values in PMU is only 2.02%. At the same time, PMU data also show that there is reactive circulation phenomenon in L1 and L2.

Based on the above analysis, it can be determined that the acquisition and display of D5000 system are normal, and the L1 and L2 reactive circulation phenomenon does exist.

Table 3. Comparison of power and voltage data from D5000 and PMU for line L1 and L2

	Active power(MW)				Active power(MW)		
	D5000	PMU	Error(%)		D5000	PMU	Error(%)
	513.48	514.12	0.12		513.09	513.27	0.04
	518.98	518.87	0.02		517.39	517.92	0.1
	Reactive power(MVar)				Reactive power(MVar)		
L1	D5000	PMU	Error(%)	L2	D5000	PMU	Error(%)
	33.98	34.04	0.17		-7.87	-8.03	-2.02
	33.39	33.2	0.58		-9.45	-9.41	-0.47
	Power plant side voltage (kV)				Power plant side voltage (kV)		
	D5000	PMU	Error(%)		D5000	PMU	Error(%)
	524	524.35	0.07		524	524.03	0.01
	524.21	524.51	0.06		524.21	524.19	0

3.2. Theoretical analysis

Under normal operation mode, L1 and L2 operate in parallel, so the two lines have the same voltage drop. Combining with power flow calculation equation of power system, the expression of reactive power on the two lines can be deduced as follows:

$$Q_1 = \frac{1}{X_1 + X_2}\left[QX_2 + P_2R_2 - P_1R_1 + \frac{U_1^2}{2}(B_2X_2 - B_1X_1)\right] \tag{1}$$

$$Q_2 = \frac{1}{X_1 + X_2}\left[QX_1 - P_2R_2 + P_1R_1 - \frac{U_1^2}{2}(B_2X_2 - B_1X_1)\right] \tag{2}$$

In the formula, P_1, P_2 , Q_1 and Q_2 are respectively the active and reactive power injected into the first end of L1 and L2 circuits (power plant side); Q is the total reactive power injected into the boost transformer outlet of A power plant; U1 is the 500 KV bus voltage of A power plant; R_1, X_1, B_1, R_2, X_2 and B_2 are the positive sequence equivalent impedance and admittance parameters of L1 and L2 circuits (including switch strings and transmission lines).

Formulas (1), (2) show that:

①When X_1=X_2, the reactive power of L1 and L2 is equalized, i.e. Q_1=Q_2, if:

$$P_2R_2 - P_1R_1 + \frac{U_1^2}{2}(B_2X_2 - B_1X_1) = 0 \tag{3}$$

②When $X_1 \neq X_2$, the reactive power of L1 and L2 is equalized, i.e. Q_1=Q_2, if:

$$Q = \frac{(P_2R_2 - P_1R_1) + U_1^2(B_2X_2 - B_1X_1)}{X_2 - X_1} \tag{4}$$

③Assuming $P_2R_2 - P_1R_1 + \frac{U_1^2}{2}(B_2X_2 - B_1X_1) > 0$, the reactive circulation occurs in line L1 and L2, i.e. the condition of Q_1=Q_2< 0 is

$$-\frac{(P_2R_2 - P_1R_1) + \frac{U_1^2}{2}(B_2X_2 - B_1X_1)}{X_2} < Q < \frac{(P_2R_2 - P_1R_1) + \frac{U_1^2}{2}(B_2X_2 - B_1X_1)}{X_1} \tag{5}$$

By inquiring the intelligent dispatching management system, it can be seen that the measured positive sequence impedance and admittance parameters of L1 and L2 circuits are very close to each other (see table 4), which basically satisfies the first case ,ignoring the influence of switching circuit. Therefore, the reactive power of the two circuits should be basically balanced, but this is not consistent with the actual situation.

Table 4. Positive sequence measured parameters for line L1 and L2

Line	*R*(Ω)	*X*(Ω)	*B*(10-4S)
L1	1.6727	16.1985	2.5202
L2	1.6553	16.2276	2.5208

By inquiring D5000 and PMU data, it can be seen that the three-phase voltages of L1 and L2 are symmetrical. According to electrical theory, under the premise of symmetrical three-phase voltages of the first and the end of parallel transmission lines, the cause of power imbalance is caused by the asymmetry of parameters between parallel lines [5] .

3.3. Asymmetric Analysis of Line L1 and L2 Parameters

Considering that line L1 and L2 are parallel lines on the same pole, with a length of 58.08km, the possible reasons for the asymmetric of their parameters are analyzed as follows:

(1) In the steady-state analysis of power system, when measuring power frequency parameters of parallel lines erected on the same tower, the traditional measurement methods are usually based on the

IOP Conf. Series: Journal of Physics: Conf. Series **1267** (2019) 012019 doi:10.1088/1742-6596/1267/1/012019

premise of symmetry of three-phase parameters of lines, usually using the lumped parameter model, ignoring the influence of mutual inductance flux between lines and between circuits [6]. Especially for short and medium-length parallel lines with the same tower (less than 100 km), affected by the specific erection methods (phase sequence arrangement and conductor transposition, etc.), there are often asymmetrical line parameters caused by the different magnitudes of mutual inductance flux between lines and between circuits [7]. In this case, the distributed parameter model is needed to calculate in order to obtain more accurate results.

(2) Ideally, the impedance of the series of switches connected to the line is usually neglected. In some special cases, the asymmetric impedance of switches, knife gates, CT and lead wires in the switching string where the two loops are located may also lead to the asymmetric impedance parameters of the two loops, such as the asymmetric arrangement of the primary wiring in the station, the different types of connected devices in the two loops such as circuit breakers or disconnecting switches, and the incomplete contact between the static and dynamic contacts of the switches due to the defects of the devices [8] -[9].

In order to find out the specific reasons for the asymmetry of L1 and L2 parameters, the operation and maintenance personnel of the power plant inspected and infrared detected the contacts between dynamic and static contacts of all circuit breakers and disconnecting switches in L1 and L2 circuits. The results showed that there was no abnormality.

By inquiring the voltage, current amplitude and phase angle curves of L1 and L2 in PMU, it is found that there are serious unbalances in the amplitude and phase angle of the three-phase current of the two lines: (1) the amplitude of line L1 increases gradually according to a-b-c phase sequence current, while the amplitude of line L2 decreases gradually according to a-b-c phase sequence current; (2) the B phase current of the two lines is basically equal; The amplitude of the A phase current of line L1 is basically equal to that of the C phase current of line L2; The C phase current of line L1 and the A phase current of line L2 are basically equal in amplitude; (3)The voltage of the two lines is three-phase symmetry. Based on the above analysis, it can be inferred that when two lines are erected, A and C phases are arranged in reverse order, which results in the asymmetry of the parameters of the two lines, while the measured parameters adopted in D5000 neglect the asymmetry caused by the influence of magnetic flux coupling between the loops and obtain approximate symmetry parameters.

4. Simulation Demonstration

In order to further demonstrate the correctness of the above inference, a detailed modeling and power flow simulation of L1 and L2 based on PSCAD software is carried out to realize the simulation reappearance of the reactive power unbalance between the two lines.

4.1. System Simulation Wiring Diagram

Figure 2 shows a simplified simulation model for the delivery system of A power plant. The equivalent parameters are set as follows:

Equivalent power supply: Rated voltage U_N=525kV. Positive sequence impedance $Z_{r1}=Z_{r2}=72.56\angle 89.9°$, zero sequence impedance $Z_{r0}=31.158\angle 89.99°$;Load: P_L=1090MW; Q_L=152Mvar.

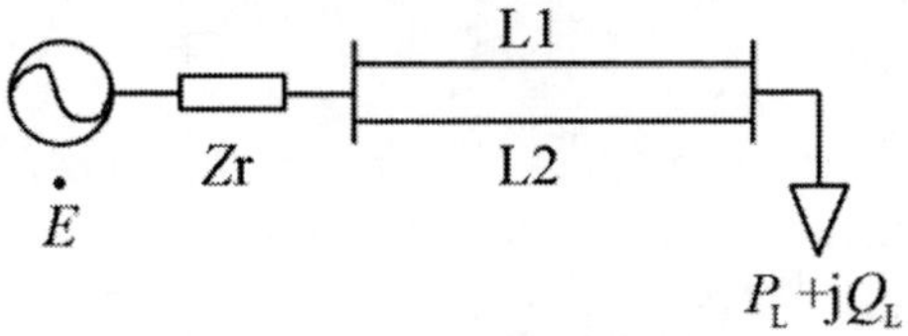

Figure 2. Simulation wiring diagram of delivery system of plant A

4.2. Transmission Line Model

L1 and L2 are double-circuit transmission lines on the same tower. In order to take into account the influence of mutual inductance between phases and loops on the degree of parameter asymmetry, it is necessary to establish a distributed parameter model of transmission lines. So Frequency Dependent (Phase) model in PSCAD software is chosen to simulate.

The design data show that the L1 and L2 double-circuit lines are erected on the same tower with a total length of 58.08 km, the conductor type is 4*JL/G1A-400/35, and the conductors are arranged in different phases (BAC/bca) without transposition. The main towers used in the line erection are double-circuit straight towers, and the relevant parameters are shown in Figure 3. The corresponding lengths of pole towers (a) and (b) account for half of the total length of the line.

Taking the power flow of the outgoing system of A power plant at 08:53 on January 10, 2018 as an example (as shown in Figure 4), the accuracy of the simulation results is verified. The specific simulation model and results are shown in Figure 5. From Figure 5, it can be seen that under the BCA/bac phase sequence arrangement of L1 and L2 conductors and the non-transposition mode of the whole line, the active power of the two lines is basically balanced, and the reactive power unbalance is relatively high. The simulation results are basically consistent with the actual results of Figure 4 (see Table 5 for details).

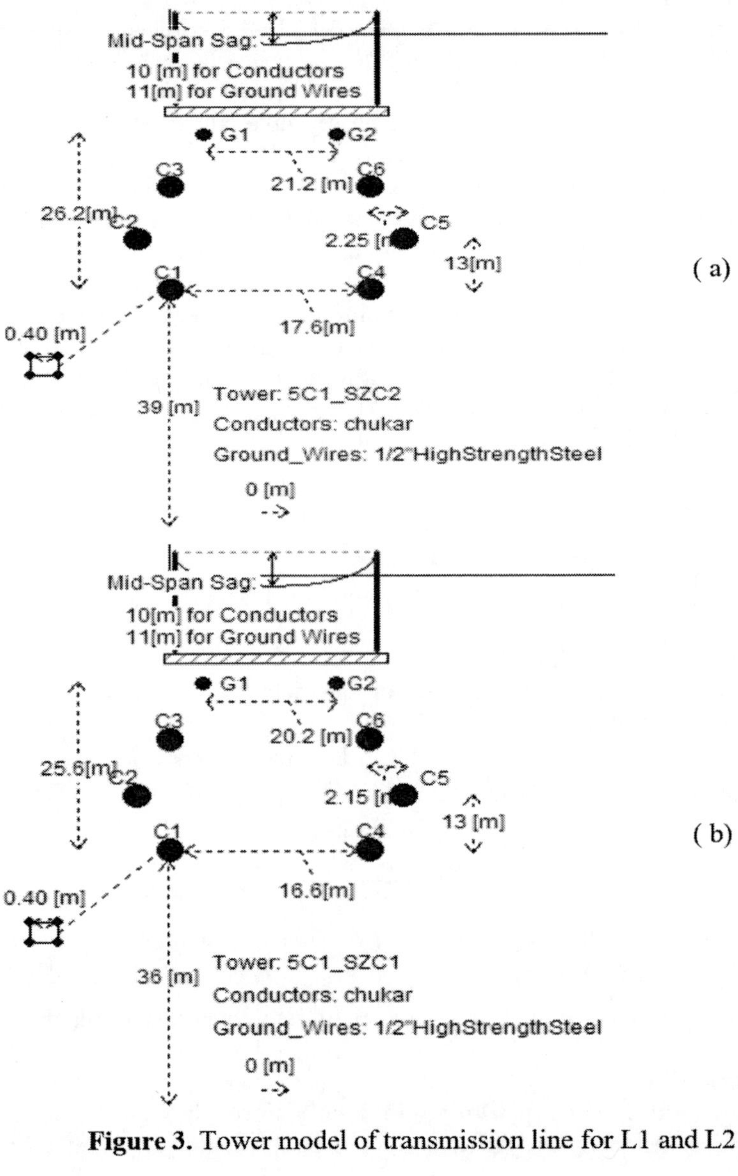

Figure 3. Tower model of transmission line for L1 and L2

IOP Conf. Series: Journal of Physics: Conf. Series **1267** (2019) 012019 doi:10.1088/1742-6596/1267/1/012019

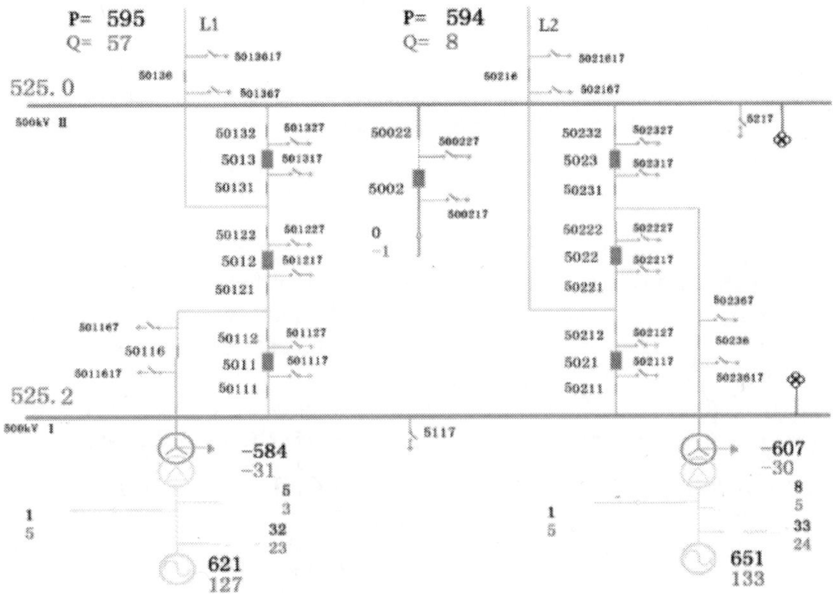

Figure 4. Power flow diagram of output system of A power plant

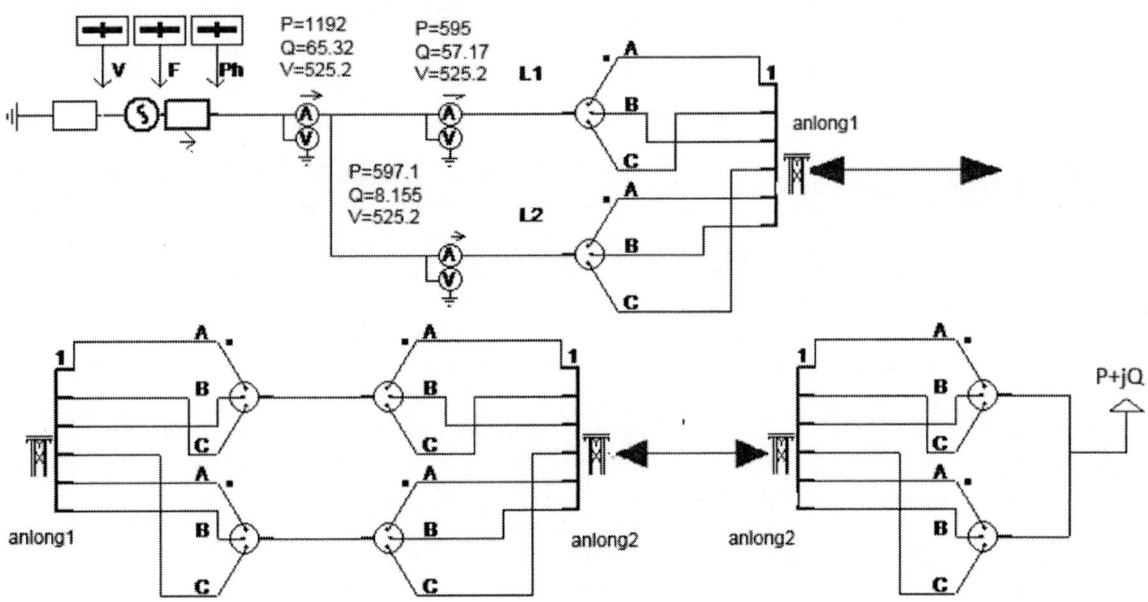

Figure 5. Reactive power simulation of Line L1 and L2

Table 5 Simulation and actual comparison for line L1 and L2

Line		P(MW)	Q(Mvar)	U(kV)
L1	Simulation value	595	57.2	525.2
	Actual value	595	57	525.2
L2	Simulation value	597.1	8.2	525.2
	Actual value	594	8	525.2

IOP Conf. Series: Journal of Physics: Conf. Series **1267** (2019) 012019 doi:10.1088/1742-6596/1267/1/012019

5. Measurement Research

Based on the analysis of the above reasons, it can be seen that in order to improve the reactive power unbalance between parallel coupled transmission lines, it is necessary to improve the asymmetry of transmission line parameters. Although there are many factors affecting the parametric asymmetry of parallel coupled transmission lines, previous studies have shown that conductor arrangement and transposition have the most significant impact on the parametric asymmetry [10].

5.1. Impact Analysis of Phase Sequence Arrangement

Because of the complex electromagnetic and electrostatic coupling effect between conductors, there exists not only a traversing unbalanced current under different phase sequence arrangement, but also a larger circulating unbalanced current under other arrangement modes besides in-phase sequence arrangement[10] - [12] .

The simulation results of L1 and L2 double-circuit power flow under seven different phase sequence arrangement without conductor transposition are shown in the table below.

Table 6 Power flow simulation of L1 and L2 double circuit lines with different phase sequence arrangements

Phase sequence	P(MW)	Q(Mvar)	Reactive power unbalance
ABC/abc (same)	596/596	34.2/34.2	0.0%
ABC/cba (inverse)	600.5/599.6	29.6/28.3	2.2%
ABC/acb (different 1)	593.5/602.4	56.4/8.8	73.0%
ABC/cab (different 2)	596.5/602.6	51.4/8.5	71.6%
ABC/bac (different 3)	595.7/598.6	7.0/58.1	78.5%
ABC/bca (different 4)	593.5/602.4	56.4/8.8	73.0%
BAC/bca(practical)	597.1/595	57.2/8.2	74.9%

Note: "/" represents L1 and L2 before and after respectively; "Reactive power unbalance" refers to the ratio of absolute value of D-value between reactive power of one loop and half of the sum of reactive power of two lines to half of the sum of reactive power of two lines, expressed in percentage.

From the results of Table 6, it can be seen that the order of reactive power unbalance of L1 and L2 double-circuit lines from small to large is: in-phase sequence < reverse sequence < different phase sequence 2=different phase sequence 1=different phase sequence 4 < practical phase sequence <different phase sequence 3. The reactive power unbalance of L1 and L2 is as high as 74.9% in the practical phase sequence. If the same or reverse phase sequence is used, the reactive power unbalance of L1 and L2 double-circuit lines will be significantly improved.

5.2. Impact Analysis of Conductor Transposition

For high-voltage parallel AC transmission lines on the same tower, if the length is long, conductor transposition is often used in engineering to reduce the unbalance between lines and circuits. It is stipulated in our country that all the lines with the length of EHV transmission lines exceeding 100 km should be transposed[13].Incomplete transposition or even non-transposition lines may aggravate the unbalanced current problem of high-voltage double-circuit transmission lines [14]. For short distance (less than 100 km) lines on the same tower, unbalanced phenomena caused by no transposition during erection are common.

In order to analyze the influence of conductor transposition on the reactive power balance of L1 and L2, based on BAC/bac phase sequence, four modes of conductor transposition are adopted as shown in Figure 6(a)~(d):

Case 1: The whole line is transposed only once, and it is still in different phase sequence after transposition.

Case 2: The whole line is transposed only once, and it is still in-phase sequence after transposition.

Case3: The whole line is transposed only once, and it is still reverse phase sequence after transposition.

Case 4: The whole line is transposed twice, and the transposition is in-phase sequence.

The simulation results of the four modes are shown in Table 7. From the results of Table 7, it can be seen that the reactive power unbalance of L1 and L2 can be significantly improved by transposition of mode 2 and mode 4.

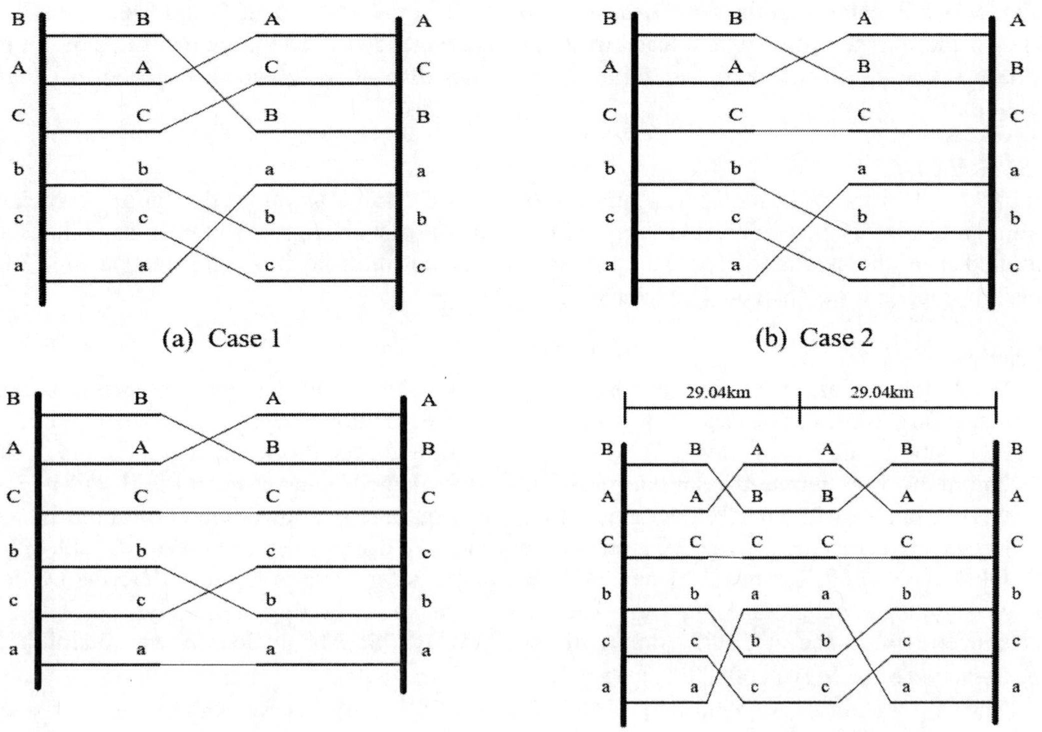

(a) Case 1 (b) Case 2

(c) Case 3 (d) Case 4

Figure 6. Transposition mode of conductors for L1 and L2

Table 7. Simulation results under four transposition modes for L1 and L2

Transposition mode	L1-*Q*(Mvar)	L2-*Q*(Mvar)	Reactive power unbalance
Non transposition	57.2	8.2	74.9%
Case 1	4.9	58.2	84.5%
Case 2	35.6	33.1	3.6%
Case 3	30.4	22.7	14.5%
Case 4	36.8	36.8	0.0%

6. Conclusions and Suggestions

6.1. CONCLUSIONS

Through on-site investigation, mechanism analysis and simulation, the following conclusions are drawn:

1) The fundamental reason for the unbalanced reactive power of L1 and L2 transmission lines in A power plant is that the two transmission lines are arranged in unconventional phase sequence while erecting on the same tower, but the parameters are not symmetrical due to the unreasonable transposition.

2) The phase sequence arrangement has a great influence on the reactive power balance of parallel coupled lines. The reactive power imbalance of L1 and L2 double-circuit lines can be reduced from 74.9% to 2.2% and 0% respectively by using in-phase or reverse-phase sequence arrangement of conductors.

3) Reasonable transposition of conductors can also effectively improve the parametric asymmetry of parallel coupling lines. If mode 3 and mode 4 in Table 6 are used to transpose L1 and L2 lines, the reactive power imbalance of the two lines can be reduced from 74.9% to 3.6% and 0%.

4) Considering the economy and feasibility, it is recommended to adopt the measures of optimizing the phase sequence of conductors to solve the problem of reactive power imbalance in L1 and L2 double-circuit

6.2. SUGGESTIONS

For new or reconstructed parallel transmission lines on the same tower in the future, especially for medium and short distance lines, it is suggested that the unbalance degree of transmission lines should be studied from the aspects of phase sequence arrangement, transposition and transposition mode at the design stage in order to avoid later transformation.

References

[1] WANG Mao-hai, BAO Jie, QI Xia, et al. Online estimation of transmission line parameters based on PMU measurements [J]. Automation of Electric Power Systems, 2010, 34(1):25-27.

[2] HU Dan-hui, CAI Han-sheng, TU Cai-qi, et al. Study on the electrical characteristic of 500 kV common-tower double-transmission line[J]. High Voltage Engineering, 2005,31(4):21-23.

[3] CHEN Jan-hua, Li Yu, ZHANG Hao, et al. Study on reactive power circulation mechanism for high voltage common-tower AC transmission lines [J]. China Electric Power,2016,49(6):83-89.

[4] DL/T448—2000, Technical Management Regulations for Electric Energy Metering Devices [S].

[5] WANG Jian-cheng, ZHANG Tao, WANG Qi-wen, et al. Testing and analysis of Three phase Current unbalance at 500kV outlet of the 2nd QIN SHAN nuclear power Plant[J]. Power System Technology , 2005, 29(5): 76—79.

[6] Zhu Ying-chun, Fan You-ping, Tang Cheng. Calculation and measurement of sequence impedance parameters of asymmetry transmission line with mutual inductance [J]. Guangxi Electric Power, 2009(6): 7-13.

[7] LI Bo-tong, LI Yong-li, CHEN Li, et al. Method for parameter decoupling and fault analysis of four-circuit transmission lines on the same tower[J]. Power System Protection and Control,2010, 38(19):1-9.

[8] WANG Xiao-ming, WANG Bin, ZHOU Wei. Analysis of three-phase current imbalance in the serial switch CT using the 3/2 connection Mode [J]. Electrical Technology, 2018(9):1-3.

[9] CHEN Hong-fu, TANG Da-yong, ZHU Feng. Analysis and treatment for three-phase current imbalance in the 4th string line of 500kV part [J]. Electricity and Energy, 2012, 33(4):337-339.

[10] LI Bin, LI Xue-bin, DING Mao-sheng, et al. Circulating unbalanced current and its influence of double-circuit UHV transmission lines [J]. Transactions of China Electro technical Society, 2012, 27(4): 202-208.

[11] ZHU Qing-gang, LIU Qian-jin, CHEN Hai-tao, et al. Study on voltage imbalance of un-transposed double-circuit lines on the same tower [J]. Power System Protection and Control,2014,42(7):73-79.

[12] Aaron Kalyuzhny, Gregory Kushnir. Analysis of current unbalance in transmission systems with short lines [J]. IEEE Transactions on Power Delivery, 2007, 22(2): 1040-1048.

[13] Power Standard of the People's Republic of China. (DL/T 5092-1999) Technical Regulations for the Design of 110-500 kV Overhead Transmission Lines [S]. Beijing: China Power Press, 1999.

[14] TAO Kai, LIU Ming-bo. Asymmetric issues caused by un-transposed transmission lines and its solution[J].Power System Protection and Control, 2010,38(16):39-43.

Cooperative Differential Game Guidance Law for Two Pursuers and One Evader

Z Q Guo[1], S L Zhou[1*] and S Yan[1]

[1]Naval Aviation University, 188 Er-ma Road, Yantai, China

*zhouslsd@sina.com

Abstract. For the nonlinear engagement between two pursuers and a single evader, in which small angle deviation assumption or other linearization conditions may not be applicable, a cooperative guidance law is analyzed and derived in the framework of a zero-sum two-person differential game. State-dependent Riccati-equation method is used to obtain the suboptimal solutions of the pursuit-evasion problem by choosing the line of sight angular rates as state variables, and closed-loop form guidance strategies that can be used online for the players are derived. Time-to-go estimations are not had to be taken into account to apply the guidance law proposed. Nonlinear two-dimensional simulations are carried out to validate the performance of the cooperative guidance law, and the comparison with a current cooperative linear quadratic differential game guidance law is made.

1. Introduction

It is well known that the interception probability will be improved and the scope of tasks will be broaden through information sharing and functions complementation, using multiple missiles to intercept targets. However, classical guidance laws such as proportional navigation (PN) guidance laws are difficult to achieve ideal results when dealing with maneuvering target, and the application of optimal guidance laws (OGL) is also limited in practice due to the dependency on the assumption about the target maneuver. The differential games guidance law (DGGL) is attracting more and more attention because of its low degree of dependency on the target information and applicability to intercept maneuvering targets. It does not need to make any assumption about the maneuvering law of the target, and optimal strategies of the players can all be obtained by differential games guidance laws [1]-[3].

Most of the current literatures on differential game guidance laws design their guidance strategies by using small angle assumption to get linear system models, however, the systems in nature are inherently nonlinear. When the scenarios are not initiated on the required collision triangle or the target has a large maneuver capability relative to its speed, one can apply state-dependent Riccati-equation (SDRE) method [3], then the complicated Hamilton-Jacobi-Bellman-Isaacs (HJBI) equation which sometimes may be unsolvable for nonlinear game problem is avoided to be considered. A nonlinear 1-on-1 pursuit-evasion engagement in which an angle constraint was imposed for the pursuer was proposed and discussed based on differential game theory and SDRE method [4]. 1-on-1 interception problems with and without angle constraint were further researched, and nonlinear pursuit-evasion strategies that were not vulnerable to unpredictability were derived in [5]-[6].

Based on the previous works about 1-on-1 engagement, a nonlinear 2-on-1 engagement is studied combined with differential games theory and SDRE method in this paper, and a guidance law that can

IOP Conf. Series: Journal of Physics: Conf. Series **1267** (2019) 012020 doi:10.1088/1742-6596/1267/1/012020

be applied in nonlinear case is proposed. The optimization problem that is usually too complicated to solve is converted into a suboptimization problem, and time-to-go estimation is not explicitly necessary in this paper.

2. Cooperative differential game guidance law

2.1. Kinematic model and guidance principle

An engagement between two skid-to-turn roll-stabilized missiles called pursuers and a single target called evader is considered. It is assumed that the motion of the players can be decoupled, and all the players have ideal control system dynamics, then the planar engagement between the two pursuers and the evader in a schematic view can be shown as figure 1, where X_I-O_I-Y_I is a Cartesian inertial reference frame. P_i (i=1,2) represents the i-th pursuer, E represents the evader. The speed, lateral acceleration, and flight path angles are denoted by V, a, γ respectively, the additional subscripts i and E correspond to pursuer P_i and the evader respectively. r_i denotes the range between pursuer P_i and the evader, φ_i denotes the line of sight angle between P_i and the evader.

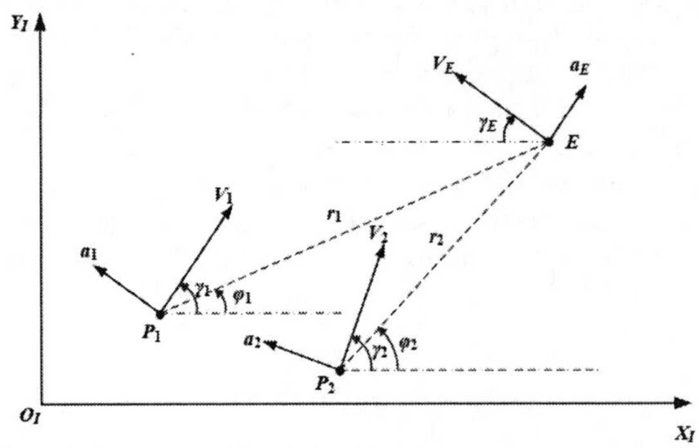

Figure 1. Planar engagement geometry of the players

The engagement kinematics can be written as

$$\dot{r}_i = -V_E \cos(\gamma_E + \varphi_i) - V_i \cos(\gamma_i - \varphi_i) \tag{1}$$

$$\dot{\varphi}_i = [V_E \sin(\gamma_E + \varphi_i) - V_i \sin(\gamma_i - \varphi_i)] / r_i \tag{2}$$

$$\dot{\gamma}_j = a_j / V_j; \qquad j = \{1, 2, E\} \tag{3}$$

The angular rate of line of sight angle can be denoted as θ_i, and its derivation is

$$\dot{\theta}_i = \ddot{\varphi}_i = -2\dot{r}_i\theta_i / r_i - u_i \cos(\gamma_i - \varphi_i)/r_i + v\cos(\gamma_E + \varphi_i)/r_i \tag{4}$$

The relationship between θ_i and zero effort miss (ZEM) [7] is

$$Z_i(t) = \frac{r_i^2\theta_i}{\sqrt{\dot{r}_i^2 + r_i^2\theta_i^2}} \tag{5}$$

From equation (5), it is found that one can control θ_i to approach zero to intercept the evader.

For the nonlinear 2-on-1 engagement, the equation of states in the form of state-dependent coefficient (SDC) can be written as

$$\dot{\boldsymbol{x}} = \boldsymbol{A}(\boldsymbol{x})\boldsymbol{x} + \boldsymbol{B}(\boldsymbol{x})\boldsymbol{u} + \boldsymbol{C}(\boldsymbol{x})\boldsymbol{v}\,, \qquad \boldsymbol{x}(t_0) = \boldsymbol{x}_0 \tag{6}$$

where $\boldsymbol{x}$ is the state vector, $\boldsymbol{A}$, $\boldsymbol{B}$, $\boldsymbol{C}$ are the state-dependent coefficient matrixes, $\boldsymbol{u}$ and $\boldsymbol{v}$ are control vectors of the pursuers and the evader respectively, and t_0 is the initial time.

The performance index of the pursuit-evasion game can be chosen as

$$J=\frac{1}{2}\int_{t_0}^{\infty}(\boldsymbol{x}^T\boldsymbol{Q}(\boldsymbol{x})\boldsymbol{x}+\boldsymbol{u}^T\boldsymbol{R}_1(\boldsymbol{x})\boldsymbol{u}-\rho^2\boldsymbol{v}^T\boldsymbol{R}_2(\boldsymbol{x})\boldsymbol{v})\mathrm{d}t \tag{7}$$

where $\boldsymbol{Q}$, $\boldsymbol{R}_1$, $\boldsymbol{R}_2$, are state-dependent weighting matrixes, and ρ is a measure of pursuer's maneuvering capability relative to that of the evader, the lower the maneuvering capability of the evader is, the higher the value of ρ is.

Choosing angular rates of line of sight angles as the system states, i.e. $\boldsymbol{x}=[\theta_1, \theta_2]^{\mathrm{T}}$, the state-dependent coefficient matrixes are

$$\boldsymbol{A}(\boldsymbol{x})=\begin{bmatrix} a_1 & 0 \\ 0 & a_2 \end{bmatrix}, \boldsymbol{B}(\boldsymbol{x})=\begin{bmatrix} b_1 & 0 \\ 0 & b_2 \end{bmatrix}, \boldsymbol{C}(\boldsymbol{x})=\begin{bmatrix} c_1 \\ c_2 \end{bmatrix} \tag{8}$$

where

$$\begin{aligned} a_1 &= -2\dot{r}_1/r_1 & a_2 &= -2\dot{r}_2/r_2 \\ b_1 &= -\cos(\gamma_1-\varphi_1)/r_1 & b_2 &= -\cos(\gamma_2-\varphi_2)/r_2 \\ c_1 &= \cos(\gamma_E+\varphi_1)/r_1 & c_2 &= \cos(\gamma_E+\varphi_2)/r_2 \end{aligned} \tag{9}$$

To obtain a controllable system, the matrix pairs $\{\boldsymbol{A},\boldsymbol{B}\}$ and $\{\boldsymbol{A},\boldsymbol{C}\}$ should satisfy the controllability requirement. The controllability matrix $[\boldsymbol{B},\boldsymbol{AB}]$ and $[\boldsymbol{C},\boldsymbol{AC}]$ have a full rank if the coefficients satisfy

$$\begin{aligned} & a_1 \neq a_2 \Rightarrow \dot{r}_1/r_1 \neq \dot{r}_2/r_2 \\ & b_1 \neq 0 \Rightarrow (\gamma_1-\varphi_1)\neq \pi/2 \\ & b_2 \neq 0 \Rightarrow (\gamma_2-\varphi_2)\neq \pi/2 \\ & c_1 \neq 0 \Rightarrow (\gamma_E+\varphi_1)\neq \pi/2 \\ & c_2 \neq 0 \Rightarrow (\gamma_E+\varphi_2)\neq \pi/2 \end{aligned} \tag{10}$$

To obtain an observable system, the matrix pair $\{\boldsymbol{Q}^{1/2},\boldsymbol{A}\}$ should satisfy the observability requirement. It is readily to find that the observability matrix $[\boldsymbol{Q}^{1/2},\boldsymbol{Q}^{1/2}\boldsymbol{A}]^{\mathrm{T}}$ has a full rank from the following equation

$$\begin{bmatrix} \boldsymbol{Q}^{1/2}(\boldsymbol{x}) \\ \boldsymbol{Q}^{1/2}(\boldsymbol{x})\boldsymbol{A}(\boldsymbol{x}) \end{bmatrix}=\begin{bmatrix} \sqrt{q_1} & 0 \\ 0 & \sqrt{q_2} \\ a_1\sqrt{q_1} & 0 \\ 0 & a_2\sqrt{q_2} \end{bmatrix} \tag{11}$$

where $\boldsymbol{Q}(\boldsymbol{x})=\begin{bmatrix} q_1 & 0 \\ 0 & q_2 \end{bmatrix}$, $q_1>0, q_2>0$.

2.2. Cooperative strategies based on SDRE

The state-dependent Riccati-equation of the differential game problem can be written as

$$\boldsymbol{A}^T(\boldsymbol{x})\boldsymbol{P}(\boldsymbol{x})+\boldsymbol{P}(\boldsymbol{x})\boldsymbol{A}(\boldsymbol{x})-\boldsymbol{P}(\boldsymbol{x})(\boldsymbol{B}(\boldsymbol{x})\boldsymbol{R}_1^{-1}(\boldsymbol{x})\boldsymbol{B}^T(\boldsymbol{x})-\rho^{-2}\boldsymbol{C}(\boldsymbol{x})\boldsymbol{R}_2^{-1}(\boldsymbol{x})\boldsymbol{C}^T(\boldsymbol{x}))\boldsymbol{P}(\boldsymbol{x})+\boldsymbol{Q}(\boldsymbol{x})=0 \tag{12}$$

Where $\boldsymbol{P}$ is the unique, symmetric, positive-definite solution of the SDRE, $\boldsymbol{R}_1$, $\boldsymbol{R}_2$ are assumed to be identity matrixes here. Then the suboptimal strategies of the players based on SDRE can be written in SDC form as

IOP Conf. Series: Journal of Physics: Conf. Series **1267** (2019) 012020 doi:10.1088/1742-6596/1267/1/012020

$$\begin{aligned} u^* &= -R_1^{-1}(x)B^T(x)P(x)x \\ v^* &= \rho^{-2}R_2^{-1}(x)C^T(x)P(x)x \end{aligned} \tag{13}$$

The Hamilton matrix of the system is

$$H(x) = \begin{bmatrix} A(x) & -F(x) \\ -Q(x) & -A^T(x) \end{bmatrix} \tag{14}$$

where $F(x) = BB^T - \rho^{-2}CC^T$ is a symmetric matrix, and its elements are

$$\begin{aligned} F(1,1) &= f_1 = b_1^2 - c_1^2/\rho^2 \\ F(1,2) &= F(2,1) = f_2 = -c_1c_2/\rho^2 \\ F(2,2) &= f_3 = b_2^2 - c_2^2/\rho^2 \end{aligned} \tag{15}$$

then the eigenvalues of H can be obtained

$$eig(H) = \left\{-\sqrt{\delta+\Delta}, \sqrt{\delta+\Delta}, \sqrt{\delta-\Delta}, -\sqrt{\delta-\Delta}\right\} \tag{16}$$

where

$$\delta = \frac{(f_1q_1 + f_3q_2 + a_1^2 + a_2^2)}{2} \tag{17}$$

$$\Delta = \frac{(a_1^4 - 2a_1^2a_2^2 + 2a_1^2f_1q_1 - 2a_1^2f_3q_2 + a_2^4 - 2a_2^2f_1q_1 + 2a_2^2f_3q_2 + f_1^2q_1^2 - 2f_1f_3q_1q_2 + 4f_2^2q_1q_2 + f_3^2q_2^2)^{1/2}}{2} \tag{18}$$

If there exists the solution of the algebraic SDRE in equation(12), H should have no eigenvalue on the imaginary axis, i.e. $\delta>\Delta>0$. According to Schur's algorithm, the matrix composed of the eigenvectors corresponding to the two negative eigenvalues of Hamilton matrix is

$$V = \begin{bmatrix} (a_2^2 + f_3q_2 - \delta - \Delta)(a_1 - \sqrt{\delta+\Delta})/(f_2q_1q_2) & (a_2^2 + f_3q_2 - \delta + \Delta)(a_1 - \sqrt{\delta-\Delta})/(f_2q_1q_2) \\ (\sqrt{\delta+\Delta} - a_2)/q_2 & (\sqrt{\delta-\Delta} - a_2)/q_2 \\ (\delta + \Delta - a_2^2 - f_3q_2)/(f_2q_2) & (\delta - \Delta - a_2^2 - f_3q_2)/(f_2q_2) \\ 1 & 1 \end{bmatrix} = \begin{bmatrix} V_1 \\ \hline V_2 \end{bmatrix} \tag{19}$$

then the solution matrix P of the SDRE can be obtained

$$P = V_2V_1^{-1} = \begin{bmatrix} p_1/\Phi & p_2/\Phi \\ p_2/\Phi & p_3/\Phi \end{bmatrix} \tag{20}$$

where

$$\begin{aligned} p_1 &= q_1[(a_2 - \sqrt{\delta+\Delta})(a_2^2 + f_3q_2 - \delta + \Delta) + (a_2 - \sqrt{\delta-\Delta})(-a_2^2 - f_3q_2 + \delta + \Delta)] \\ p_2 &= f_2q_1q_2(\sqrt{\delta+\Delta} - \sqrt{\delta-\Delta}) \\ p_3 &= q_2[(a_1 - \sqrt{\delta+\Delta})(-a_2^2 - f_3q_2 + \delta + \Delta) + (a_1 - \sqrt{\delta-\Delta})(a_2^2 + f_3q_2 - \delta + \Delta)] \end{aligned} \tag{21}$$

$$\Phi = (a_2^2 + f_3q_2 - \delta + \Delta)(a_1 - \sqrt{\delta-\Delta})(\sqrt{\delta+\Delta} - a_2) - (a_2^2 + f_3q_2 - \delta - \Delta)(a_1 - \sqrt{\delta+\Delta})(\sqrt{\delta-\Delta} - a_2) \tag{22}$$

Since P is positive-definite, the following condition is also imposed

$$(p_1 + p_3)/\Phi > 0, \qquad p_1p_3 - p_2^2 > 0 \tag{23}$$

According to equation (13), the closed-loop form suboptimal strategies of the players based on SDRE are obtained as

$$\begin{aligned}\tilde{u}_1 &= -\frac{b_1 p_1}{\Phi}\theta_1 - \frac{b_1 p_2}{\Phi}\theta_2 \\ \tilde{u}_2 &= -\frac{b_2 p_2}{\Phi}\theta_1 - \frac{b_2 p_3}{\Phi}\theta_2 \\ \tilde{v} &= \frac{c_1 p_1 + c_2 p_2}{\rho^2 \Phi}\theta_1 + \frac{c_1 p_2 + c_2 p_3}{\rho^2 \Phi}\theta_2\end{aligned} \tag{24}$$

The coordination between the two pursuers is indicated by the coupled construction of the pursuers' strategies, and it is found that the derived strategies are not explicitly related to time-to-go.

3. Simulations and performance analysis

In this section, the performance of the derived guidance law (Nonlinear Cooperative Differential Game Guidance Law Based on SDRE, SDRE-NCDG) is investigated and compared to a cooperative linear quadratic differential game (CLQDG) guidance law proposed in [1] by nonlinear simulations. It is assumed that the flight path angles, speed, and the line of sight angles can be measured or estimated, and all the players have ideal control systems. The influence of the gravity is ignored, and the simulation parameters of the engagement are shown in table 1.

Table 1 Simulation parameters

Pursuers/Evader	Weight parameters	Initial flight path angle(°)	Initial position(m)	speed(m/s)
P_1	$q_1=10^5$	30	(0,0)	600
P_2	$q_2=10^5$	−30	(0,0)	600
E	$\rho=9$	0	(5000,0)	400

3.1. Sample run

3.1.1. Nonmaneuvering target. If the evader does not maneuver, the 2-on-1 simulation results are shown in figure 2 and figure 3. Figure 2 presents the trajectories of two pursuers and one nonmaneuvering evader, and it is found that the evader is intercepted by the pursuers simultaneously using SDRE-NCDG guidance law. Figure 3 presents the acceleration profiles of the pursuers during the engagement. The acceleration of pursuers is decreasing with time until to zero at the end of the guidance.

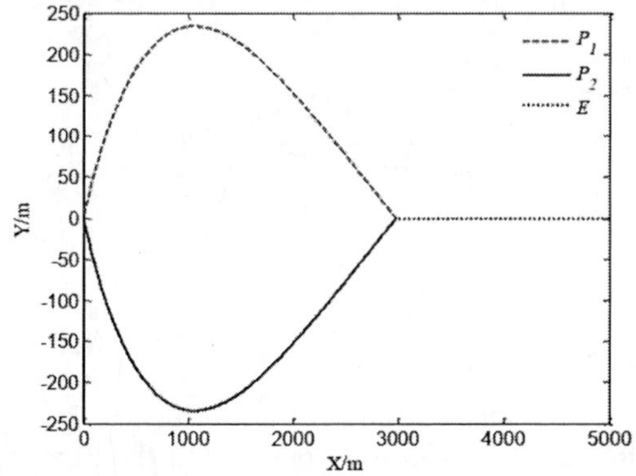

Figure 2. Trajectories of the players while the evader does not maneuver

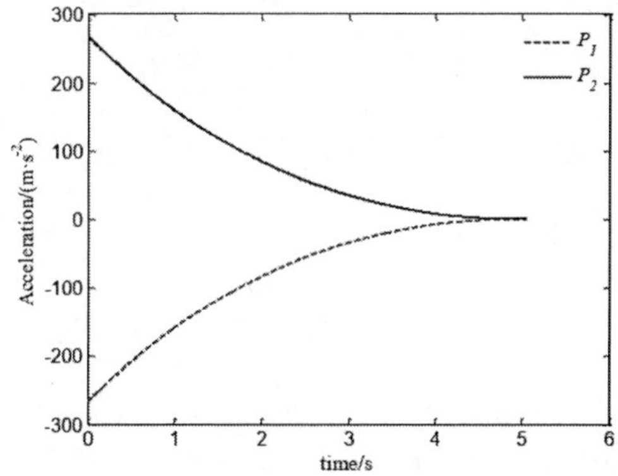

Figure 3. Acceleration profiles of the pursuers while the evader does not maneuver

3.1.2 Maneuvering target. If the evader performs a 5g evasive maneuver, the 2-on-1 simulation results are shown in figure 4 and 5. As one can see from figure 4, two pursuers don't hit the evader simultaneously (the evader is intercepted by P_1 firstly) because time constraints that are hard to impose on the differential game problem isn't taken into account in this paper either. So the distance between P_2 and the evader when P_1 hits the evader does not mean the terminal miss distance of P_2 to the evader. The acceleration profiles shown in figure 5 indicate that the acceleration of pursuers decreases with time too but they don't approach zero at the end of the guidance due to the maneuver of the evader.

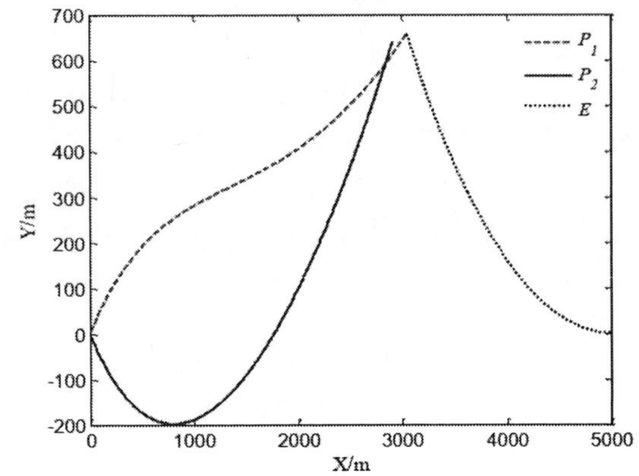

Figure 4. Trajectories of the players while the evader performs a 5g maneuver

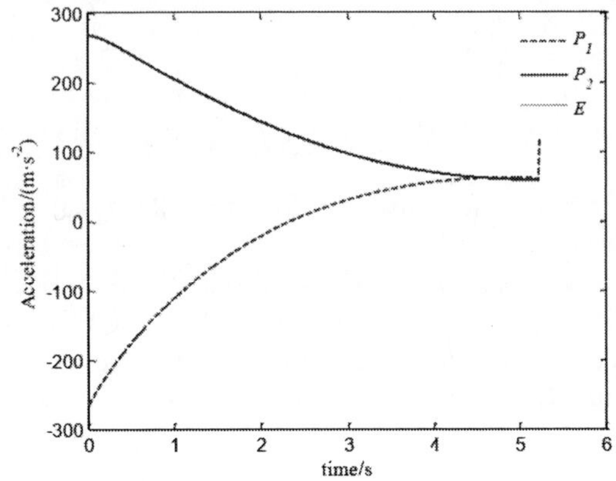

Figure 5. Acceleration profiles of the pursuers while the evader performs a 5g maneuver

3.2. Comparison to linear strategies

Figure 6 and 7 present the simulation results of a current CLQDG guidance law using the same parameters shown in table 1, the evader still performs a 5g evasive maneuver, and the curves in figure 4 and 5 are depicted again to make comparison. It is illustrated that the two kinds of guidance laws are both effective, however, the trajectories of pursuers using the nonlinear guidance law are slightly straighter than that of the CLQDG guidance law, and the maneuverability requirement on pursuers is slightly lower than that of the CLQDG guidance law. The comparison of control energy of the two different guidance laws is drawn in figure 8, in which it can be obviously seen that the energy consumed in the nonlinear case is less than the linear case (about 8% control energy reduced for P_1).

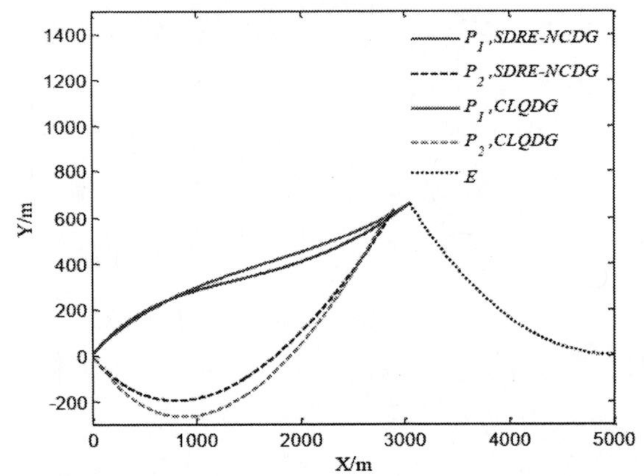

Figure 6. Trajectories of the players while the evader performs a 5g maneuver (comparison)

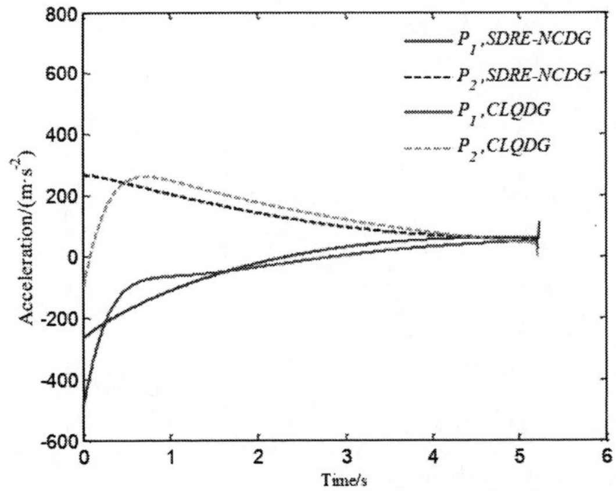

Figure 7. Acceleration profiles of the pursuers while the evader performs a 5g maneuver (comparison)

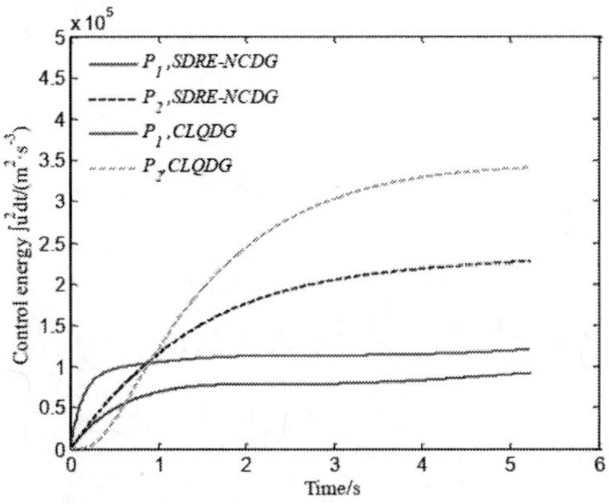

Figure 8. Control energy comparison

Acknowledgments

This paper is supported by the Project for the National Natural Science Foundation of China under Grant No. 61273058.

Reference

[1] Liu Y F, Qi N M and T Z W 2012 Linear quadratic differential game strategies with two-pursuitversus single-evader *Chinese Journal of Aeronautics* **25** 896
[2] Hua W H, Zhang Y J and Zhang J P 2016 Differential game guidance law for double missiles with cooperative intercept angle *Journal of Chinese Inertial Technology* **24** 838
[3] Shaferman V, Shima T 2017 Cooperative differential games for imposing a relative intercept angle *Journal of Guidance, Control and Dynamics* **40** 2465
[4] Bardhan R, Ghose D 2012 Intercepting maneuvering target with specified impact angle by modified SDRE technique *American Control Conference* 4613
[5] Bardhan R, Ghose D 2015 An SDRE based differential game approach for maneuvering target interception *AIAA Guidance, Navigation, and Control Conference* 1
[6] Bardhan R, Ghose D 2015 Nonlinear differential games-based impact-angle-constrained guidance law *Journal of Guidance, Control, and Dynamics* **38** 384
[7] Ratnoo A, Ghose D. 2009 State-dependent Riccati-equation based guidance law for impact-angle-constrained trajectories *Journal of Guidance, Control, and Dynamics* **32** 320

IOP Conf. Series: Journal of Physics: Conf. Series **1267** (2019) 012021 doi:10.1088/1742-6596/1267/1/012021

Research on Energy Consumption of Multi-Trains Tracking Traffic Flow Dynamics Model Under Moving Block

Yong Chen, Yao Chen and Hong He

School of Electronic and Information Engineering, Lanzhou Jiaotong University, Lanzhou, China

E-mail: edukeylab@126.com

Abstract. Based on the NaSch model, aiming at the characteristics of high-speed train tracking operation under moving block, a cellular automaton dynamics model is proposed to simulate the energy consumption of multi-train tracking operation on Railway lines. Through numerical simulation, this paper studies the influence of different line density and station stopping time on the energy consumption of high-speed railway traffic flow. The simulation results show that the energy consumption model can accurately reflect the energy consumption of multi-train tracking operation. At the same time, the phenomenon of traffic waves that sometimes go and sometimes stop is reproduced. According to the simulation results, it can be concluded that the energy consumption of multi-train tracking flow decreases gradually with the increase of line initialization density, and the average energy consumption of tracking traffic flow decreases with the increase of station stopping time. The results provide a scientific theoretical basis for efficient operation and energy-saving operation of high-speed railway. It has some guiding significance.

1. Introduction

With the continuous speed increase of railways in China, the construction of passenger dedicated lines and high-speed railways, the running speed of trains and the operating density of trains continue to increase. According to the medium and long-term development plan of China's railways, the operational mileage of China's railways will increase from 91,000 km to 120,000 km by 2020. The passenger dedicated lines and inter-city railways with speeds of 200-350 km/h will reach 18,000 km and the high-speed trains put into operation will reach 1,000 groups. Train operation energy consumption is an important economic index in railway operation, which is an important part of the railway transportation cost. Therefore, quantitative analysis of energy consumption characteristics in the operation of high-speed railway can be used to build a high-speed railway network, meet the high-density, fast and convenient, energy-saving and consumption-reducing passenger transport needs. Cellular automaton (CA) model is an effective tool to simulate nonlinear complex systems[1]. CA model can effectively simulate the microscopic movement of vehicles in traffic flow, which has become a hot spot in traffic flow research in recent years. Multi-Trains on the railway line can be regarded as a single lane traffic flow model simulation. The NaSch model is suitable for railway traffic flow simulation. In recent years, great progress has been made. Based on the NaSch model, many researchers applied the theory of cellular automata to the study of railway traffic problems, and proposed the model of railway cellular automata[2]. At present, many scholars have carried out a lot of research work on Train Energy Consumption. Kalogirou SA puts forward a train energy consumption calculation method based on neural network[3], Lucique uses theoretical model calculation and simulation method to analyze the influence of high-speed train running energy consumption[4]. At present, most of the energy consumption analysis methods are mainly for single train, and there are few studies on energy consumption analysis of multi-train tracking train flow. At present, most of the simulation studies on train energy consumption are based on the static energy consumption of single train. But in the moving block system, with the further improvement of train speed and train density, the influence between trains becomes more and more serious. The traditional single train energy consumption model has been difficult to reflect the characteristics of actual railway traffic flow energy consumption.

Aiming at the deficiency of the traditional calculation method of train energy consumption for high-speed trains, this paper presents a dynamic model of energy consumption calculation for multi-train tracking operation based on the theory of cellular automata and NaSch traffic flow model. Through computer simulation analysis, the influence of different departure time interval, line density and station stopping time on energy consumption of tracking train traffic flow is analyzed.

2. Railway energy consumption

Railway energy consumption mainly includes train operation energy consumption, station and signal infrastructure and other auxiliary energy consumption, in which railway operation energy consumption accounts for 80% of the total energy consumption. The train energy consumption operation is the energy consumption used to overcome the running resistance to acting, increase the kinetic energy of train and overcome the difference of gravity potential energy. Train traction energy consumption is mainly affected by locomotive attributes, line attributes, transport types, marshalling information and other related factors.

3. Statistical physical model of energy consumption

Under the condition of moving block, there is no fixed block partition. It improves train tracking density and running safety by advanced communication means.The control mode corresponding to traction starting, acceleration, inert speed regulation, braking deceleration and parking direction is generated. In the multi-train tracking model established in this paper, the tracking energy consumption analysis of multiple trains is composed of the sum of all single train starting energy consumption, traction acceleration energy consumption and deceleration energy consumption according to the running conditions of the train.

When the multi-train is running on the line, the calculation model of the train flow energy consumption is as follows:

$$E_t = E_q + E_r + E_o \tag{1}$$

In the formula, E_t is the total energy consumption of train flow during multi-train tracking operation, E_q is the starting energy consumption of train flow, E_r is the traction energy consumption of train flow, and E_o is the deceleration energy consumption of train flow. The calculation method of each parameter in formula (1) is as follows:

①Calculation of starting energy consumption

Starting energy consumption of single train: $e_q = k \times \frac{1}{2}(\sum M_c) \times V^2 /(3.6^3 \times 1000)$

In the formula[5], e_q for single train starting and stopping energy consumption; M_c is the train quality, the unit of which is t; k is the number of times the train tracks the starting and stopping in the energy consumption calculation section; V is the highest speed after starting operation, the unit of which is km/h; the unit conversion coefficient is $3.6^3 \times 1000$. When the multi-train tracking operation is on the line, the train flow starting energy consumption E_q is calculated as follows:

$$E_q=\sum_{t=0}^{\max t}\sum_{n=1}^{N} k\times\frac{1}{2}(\sum M_c)\times V^2/(3.6^3\times1000) \tag{2}$$

②Accelerated energy consumption of train traction

Accelerated energy consumption of single train traction $e_r=(\sum M_c w_c)\times S\times9.81\times3600$

In the formula, e_r is the energy consumption of single train traction operation; M_c is the train quality, the unit of which is t; w_c is the basic resistance of the high-speed train operation, the unit of which is N/kN, and its value is different according to different types of Electric Multiple Units; V is the train running speed; S is the traction running distance, the unit of which is km. The formula for calculating the basic resistance of commonly used high-speed trains according to different types of vehicles is as follows:

CRH1 type $w_{c(V)}=1.12+0.00542\times V+0.000146\times V^2$

CRH2 type $w_{c(V)}=0.88+0.00744\times V+0.000114\times V^2$

CRH3 type $w_{c(V)}=0.66+0.00245\times V+0.000132\times V^2$

CRH5 type $w_{c(V)}=0.69+0.0063\times V+0.00015\times V^2$

Through the statistics of the traction train acceleration energy consumption of different train types, the energy consumption of the train flow traction operation E_r on the line is calculated as follows:

$$E_r=\sum_{t=0}^{\max t}\sum_{n=1}^{N}(\sum M_c w_c)\times S\times9.81\times3600 \tag{3}$$

③Deceleration energy consumption of train

In the inertial and braking conditions of the train, the train decelerates. In order to calculate the energy consumption of trains running in this condition, the kinetic energy theorem is used to simulate the model. The kinetic energy loss caused by the above two reasons is called deceleration energy consumption. When the model is established, the kinetic energy theorem is used to calculate the energy consumption of single train and train flow in real time.

$$e_o(n,t)=\begin{cases}\frac{1}{2}(mv_n^{\,2}(t)-mv_n^{\,2}(t+1)); v_n(t+1)<v_n(t)\\ 0; v_n(t+1)=v_n(t)\end{cases} \tag{4}$$

$$E_o=\sum_{t=0}^{\max t}\sum_{n=1}^{M} e_o(n,t); \tag{5}$$

In the above formula, max$t$ represents the system energy consumption statistics time, M represents the total number of line trains, and E_o represents deceleration energy consumption of the train flow.The whole energy consumption of train traffic flow in the maximum observation period can be calculated based on Formula (5), where deceleration can be calculated from the formula (1) to (5).

4. Model establishment

For high-speed tracking train traffic flow, according to train operation condition, the following rules based on the NaSch model are used to indicate the driving of the high-speed train on the line and the station stop rules, then carry out computational simulation.

Situation 1: the front of train N is train N-1.

1) Acceleration process:

If $\Delta x_n > d_n$

$v_n(t+1) \to \min(v_n(t)+a, v_{max})$;

If $v_n(t+1) > v_n(t)$ AND $v_n(t) <> 0$; before the train accelerates, the train is run and calculates the traction energy consumption.

$E_{nr} = (\sum M_c w_c) \times S \times 9.81 \times 3600$

Else $e_{nq} = k \times \frac{1}{2}(\sum M_c) \times V^2 / (3.6^3 \times 1000)$; Before the train accelerates, the train is stationary and calculates the starting energy consumption.

Else If $\Delta x_n < d_n$

$v_n(t+1) \to \max(v_n(t)-b, 0)$;

$e_o(n,t) = \frac{1}{2}(mv_n^2(t) - mv_n^2(t+1))$; Calculate deceleration energy consumption.

Else

$v_n(t+1) \to v_n(t)$;

2) Safety protection process:

$v_n(t+1) \to \min(v_n(t+1), gap_n(t))$, Among them, $gap_n(t) = x_{n+1}(t) - x_n(t) - l_{train}$ represents the distance between adjacent trains at time t, and the rule indicates the deceleration measures taken to avoid collision with the preceding vehicle when the train is running.

3) Train movement:

$x_n(t+1) \to x_n(t) + v_n(t+1)$; The train moved forward at the adjusted speed.

Situation 2: the front of train N is the station.

Case1: The station is occupied by the adjacent train in front of N-1.

The update rule is the same as case 1.

Case2: The station was not occupied by other trains.

1) Acceleration process

If $d_s > d_m$

$v_n(t+1) \to \min(v_n(t)+a, v_{max})$

$E_{nr} = (\sum M_c w_c) \times S \times 9.81 \times 3600$; The train is far away from the station and calculates the traction energy consumption.

Else If $d_s < d_m$

$v_n(t+1) \to \max(v_n(t)-b, 0)$

$e_o(n,t) = \frac{1}{2}(mv_n^2(t) - mv_n^2(t+1))$; The train deceleration is ready to enter the station and calculate the deceleration energy consumption.

Else $v_n(t+1) \to v_n(t)$

2) Safety protection process:

$v_n(t+1) \to \min(v_n(t+1), d_s)$

3) Train movement:

$x_n(t+1) \to x_n(t) + v_n(t+1)$

Situation 3: The train N waiting for departure at the station.

Case1: When the train stops at the station less than the prescribed stopping time Td, the passenger at the stop station gets on the train and waits for departure.

$v_n(t+1) = 0$, $t<Td$, Among them, Td indicates the stopping time of train at the station.

Case2: When the train stops at the station for longer than the prescribed stopping time Td, the train departs and quickens its departure from the station platform.

1) $v_n(t+1) \rightarrow \min(v_n(t)+a, v_{max})$, $t \geq Td$.

$e_{nq} = k \times \frac{1}{2}(\sum M_c) \times V^2 / (3.6^3 \times 1000)$; When the train exceeds the stop time, the train departs,and then calculate starting energy consumption

2) Train movement: $x_n(t+1) \rightarrow x_n(t) + v_n(t+1)$, The train moved forward at the adjusted speed.

In the above formula, Δx_n denotes the distance between the train N and the train N-1, d_n is the minimum tracking interval between trains, a is the train acceleration, b is the train deceleration, $x_n(t)$ denotes the position of the train in the route at time t, d_s is the distance between the train and the station. d_m is the safe distance between train braking and driving into the station. Using formula (2) to get the starting energy consumption of track tracking train flow. The traction energy consumption of line tracking train flow is obtained by formula (3). Using formula (3) and (4) to find out the energy consumption of train deceleration on line. Finally, the overall energy consumption of train tracking traffic flow is obtained by formula (1).

5. Simulation analysis

When the model is established, the line is regarded as one-dimensional discrete lattice with the length of L. Each lattice has the same size. Each lattice is either empty or occupied by the train. The train speed is an integer between 0 and v_{max}. The length of the analog line is L cells, and the length of each cell is set to 1 m.The numerical simulation is carried out with the Harmony CRH5 Electric Multiple Units as an example. The train marshalling length is 212 m, and each train accounted for 212 cells. The system refresh interval is 1 s.This means that the actual train speed is 108 km/h when v_{max} = 30 cells/s and the actual train speed is 180 km/h when v_{max} = 50 cells/s.

In the numerical simulation, we take the length of the transmission line is L=30000 and the evolution time is T=5000. Suppose there is a station at 2/3 of the route, the length of the station is 200 m, the acceleration a and deceleration b are 1 m/s ^{2}, the safety protection distance SM = 100, and the stopping time Td = 90 s. The cellular automata model adopts the open boundary condition.In order to eliminate the influence of the randomness of the initial state, the last 5000 steps of each evolution to the steady state are recorded as the simulation time step, and the sample data are averaged after 20 iterations to obtain the space average speed of each time step, so as to ensure the maximum elimination of the effect of initial random distribution on the simulation results.

5.1. Influence of line density on energy consumption of traffic flow

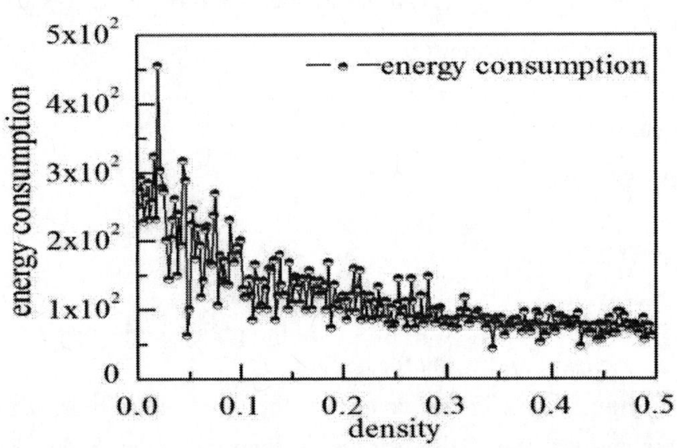

Figure 1. Relationship between the density of the same line and the energy consumption of train flow

In order to study the effect of different line density on energy consumption of tracking train traffic flow, seting the maximum train speed is 50cells/s, that is, v_{max}=180km/h. Through simulation, the energy consumption relation diagram of train traffic flow under different initial line density conditions is obtained. Figure 1 shows the relationship between the initial line density and train energy consumption under different line initialization densities. In the figure, the ordinate axis represents the average energy consumption of the tracking train, and the abscissa axis represents the initial density value of the train.

It can be seen from Figure 1 that the average energy consumption of tracking train flow on the whole line decreases gradually with the increase of initial line density. This is because as the initial traffic density continues to increase, the number of trains initially distributed on the line increases, resulting in increased interaction between CRH5 tracking trains. As new trains from upstream continue to add to the congestion, so the speed is slower and slower, more and more vehicles stop running, the energy consumption of train deceleration is becoming smaller and smaller, and the traffic condition of the whole line is gradually deteriorating. When the train flow density is increasing, but its traffic energy consumption is zero, this shows that the line has been seriously blocked, forming a traffic jam.

In order to describe the evolution process of high-speed tracking train flow more vividly, the time-space diagram of traffic flow was simulated. The simulation results are shown in Figure 2, it can be clearly seen that when the time is 300s, the trains running on the line have less interaction, the traffic flow on the line is stable in this condition, the running speed is high, and the traffic flow on the tracking train is free under moving block.

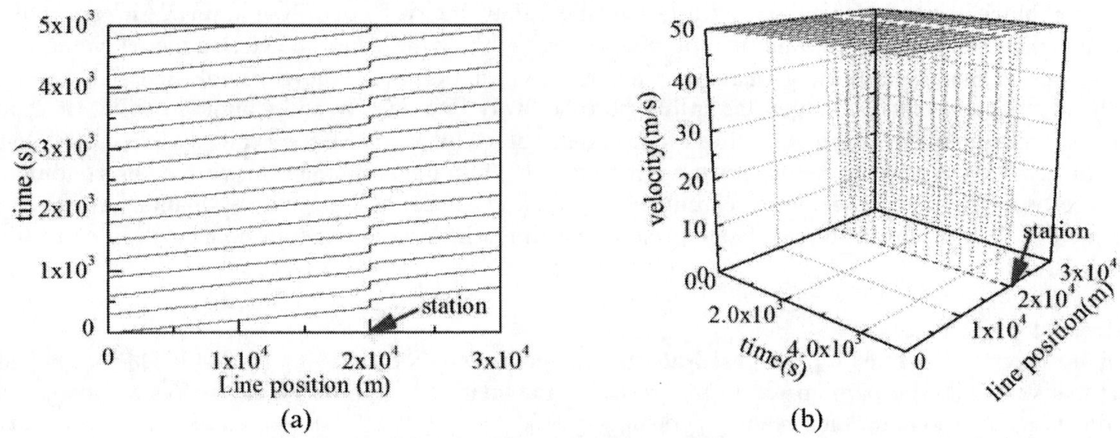

Figure 2. Spatiotemporal phase diagram of traffic flow
(a) *Tint*=300s; (b) *Tint*=300s;

5.2. Influence of station stopping time on tracking trains flow

In addition, the relationship between the different station stopping time Td and the average energy consumption of tracing train flow can be obtained by numerical simulation, as shown in Figure 3. Figure 3 is simulated for initializing line density p=0.01. As can be seen from the diagram, with the increase of station stop time, the average energy consumption of tracing train flow shows a gradual downward trend. This simulation conclusion is consistent with the actual traffic situation. That is to say, the station stop time is small at the beginning, at this time, new trains are constantly coming from the left side of the line, trains do not need to wait for more time to stop, the interaction between the trains is small, the acceleration energy consumption of train traction is large, most of the vehicles are in the free flow driving area, and the deceleration probability is small. With the increase of stopping time, the probability of train deceleration becomes larger and larger due to waiting for the train to enter the station. The number of low-speed trains increases, and the phenomenon of road congestion begins to appear. From Figure 3, it can be concluded that the longer the station stop time is, the smaller the energy consumption is at the same initialization density. This is because that the longer the station stop time is, the earlier the system enters the blocking phase; The longer the station stop time is, the larger the congestion range is. Because of the congestion, the low-speed traffic flow is formed, so the kinetic energy loss is small, and the corresponding flow and average energy consumption are also minimal.

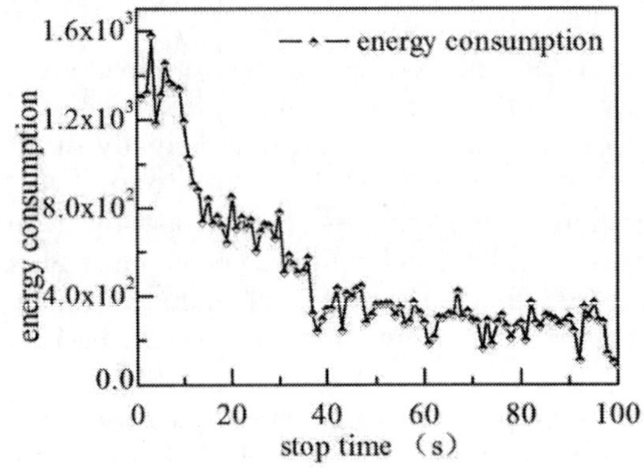

Figure 3. Energy consumption diagram of train traffic flow under different stopping time

6. Conclusion

Moving block technology is a step further in the safety interval control of trains. Aiming at the characteristics of high-speed train running under moving block, based on the NaSch model, an energy consumption model of tracking train flow under moving block condition is established. Then using cellular automata model to study the influence of railway line density and station stopping time on energy consumption of tracing train flow. The results showed that: 1) different railway line density has a significant impact on the energy consumption of multi-train tracking under moving block condition. 2) The time-space phase diagram reproduces the process of congestion and dissipation of tracking train flow. 3) with the increase of station stopping time, the average energy consumption of traffic gradually decreases.

Acknowledgments

This research was funded by National Natural Science Foundation of China (61841303). We express our thanks to all who participated in this research for their cooperation. Additionally, we thank the anonymous reviewers for their reading and suggestions.

References

[1] Nagel K and Schreckenberg M A 1992 A cellular automaton model for freeway traffic *J. Phys. I.* **2(12)** 2221 -29.
[2] Wang D T, Wang F R and Huang K 2018 Model for Tracking Operation of High-Speed Railway Train Group Based on Cellular Automa *Railw. Stand. Design* **62(04)** 87-91.
[3] Kalogirou S A and Bojic M 2000 Artificial neural networks for the prediction of the energy consumption of a passive solar building *ENERGY* **25(5)** 479-91.
[4] Lv X, Li Y and He B 2014 Analysis and Simulation of Infrastructure Influence on High-Speed Trains' Running Energy Consumption *Bridges* 10 9780784413159.327.
[5] Wang T C 2012 Research on Measure Method of High-speed Train Operation Consumption *Railw. Transp. Econ.* **34(11)** 88-92.

RGB-D Indoor Object Recognition Algorithm Based on Fusion Convolutional Neural Network

Decheng Wang[1, *], Hui Yi[1] and Feng Zhao[1, 2]

[1] Graduate School, Space Engineering University, Beijing, China
[2] 61618 Troops, Beijing, China

[*]Corresponding E-mail: 823062375@qq.com

Abstract. Aiming at the problems of insufficient network fusion and low detection efficiency in current object recognition using RGB-D images, a recognition algorithm based on the medium-level layer-by-layer fusion of dual-channel networks is proposed. First of all, the RGB and Depth networks are trained with ten labelled RGB-D indoor objects respectively, and then determine the fusion coefficients according to the identify accuracy of two types networks. Finally, two kinds of features are merged in convolutional layers step by step to obtain the final weights. By testing on the challenging NYU Depth v2 dataset, we found that the recognition accuracy of our method is 92.85%, and average detection time is 61.03ms per image. Through comparison experiments, we got the conclusion that average accuracy of the RGB-D layer-by-layer fusion network is 5.22% higher than that of the RGB network.

1. Introduction

In recent years, with the emergence of new RGB-D sensors (such as Kinect and Xtion, etc.), the application of RGB-D images in related fields has been greatly promoted. The combination of image depth information and RGB information opens a new window for solving basic problems of computer vision. Depth information provides useful additional information for scenes and target classifications of complex problems, and does not change with brightness and colour, making objects be better identified from the background. At present, the target recognition of fusing depth information has become a hot research topic in the field of computer vision. It is widely used in indoor robot visual recognition systems to realize the understanding and interaction of indoor complex scenes[1].

Socher et al.[2]first proposed a convolutional recurrent neural network (CNN-RNN) fusion model to learn RGB-D features in 2012. The original RGB-D images were directly put into CNN to obtain low-level features, and then these low-level features were put into RNN to obtain the effective high-level features, and finally the softmax method is used to obtain better classification performance. In the RGB-D target recognition algorithm based on hybrid structure proposed by [3], using the deep convolutional neural network (VGG-16) and the improved HONV descriptor Fisher Vector coding to respectively extract feature expression for the RGB and depth image of each frame, and achieved superior recognition performance. Hoffman et al.[4] proposed a medium-level fusion CNN of RGB and Depth information, and given a few types of mark depth images, realized the detection of RGB-D objects under weak supervision conditions. Xu et al.[5] proposed a new method for target recognition in RGB-D scenes. By using the shared weight strategy and the parameter-free correlation layer to jointly perform RGB-D object detection and region object recognition, through post-fusion, multimodal RGB-D object recognition effect is effectively improved. In [6], a new dual-stream convolutional network is proposed. After inputting RGB and depth images into two convolutional networks with the same structure to extract their respective features, the convolutional layer based on the optimal weights fused features of two networks, and finally the output is obtained through the fully connected layer.

In view of the current insufficient RGB-D network fusion and low detection efficiency, this paper proposes a two-channel network with layer-by-layer fusion. RGB and Depth CNN are trained respectively on the NYU Depth v2 dataset[7], and test the network after fusion. Finally we compared and evaluated quantitatively for the results.

2. Method

2.1. Convolutional neural network model structure

Convolutional neural network is excellent deep learning models in the field of image recognition and detection[8]. The paper Based on the current popular Yolo v3[9] target detection framework, in order to reduce network complexity and reduce redundant data, the original Darknet53 network is replaced by VGG16[10] to extract RGB image features. For a single Depth image, the paper designed with a convolutional network of 10 convolutional layers and 6 maximum pooling layers, so that the output feature dimension before each pooling layer corresponds to VGG16, for the two kinds of features can be effectively merged.

The training process of the network can be divided into two stages: forward propagation and backward propagation. In the forward propagation phase, information is propagated layer by layer from the input layer. The calculation formula of the output of the j neuron in the l convolution layer is:

$$a_j^l = f_c[w^l(\sum_{i\in M_j^l} a_i^{l-1} * k_{ij}^l) + b^l] \tag{1}$$

a_i^{l-1} is the output of the previous layer, $f_c()$ is ReLU activate function, k is convolution kernel, M is the set of input feature maps, w^l is weight of the l layer of the convolutional network, b_l is the bias of l layer, * represents the process of convolution. The essence of backward propagation is to iteratively adjust the weight and bias of the network according to the change of the loss function, so as to obtain the optimal network parameters. The loss function in the Yolo algorithm is defined as the sum of the coordinate positioning error, the IoU error and the classification error between the prediction data and the calibration data. The formula is as follows:

$$loss = \lambda_{coord}\sum_{i=0}^{s^2}\sum_{j=1}^{B} I_{ij}^{obj}[(x_i - \hat{x}_i)^2 + (y_i - \hat{y}_i)^2] + \lambda_{coord}\sum_{i=0}^{s^2}\sum_{j=1}^{B} I_{ij}^{obj}[(\sqrt{w_i} - \sqrt{\hat{w}_i})^2 + (\sqrt{h_i} - \sqrt{\hat{h}_i})^2]$$
$$+\sum_{i=0}^{s^2}\sum_{j=1}^{B} I_{ij}^{obj}(C_i - \hat{C}_i)^2 + \lambda_{noobj}\sum_{i=0}^{s^2}\sum_{j=1}^{B} I_{ij}^{noobj}(C_i - \hat{C}_i)^2 + \sum_{i=0}^{s^2} I_i^{obj}\sum_{c\in classes}(p_i(c) - \hat{p}_i(c))^2 \tag{2}$$

λ represents weight, $\lambda_{coord} = 5$, $\lambda_{noobj} = 0.5$, x, y, w, h, c, p are network predictive value, $\hat{x}, \hat{y}, \hat{z}, \hat{h}, \hat{c}, \hat{p}$ are calibration value, I_i^{obj} represents objects in the i grid, I_{ij}^{obj} and I_{ij}^{noobj} represents objects in and out the j bounding box of the i grid respectively.

2.2. Dual channel network fusion method

When take joint detection of RGB-D images, the information of RGB and Depth images is mainly fused by a convolutional neural network. The current simple mode for RGB-D information fusion is early and late fusion [11], but the middle fusion before the fully connected layer will achieve better recognition. Based on the [12], this paper proposes a medium-level layer-by-layer fusion method (as shown in Figure 1). The output of each convolution layer of the Depth network is fused to the feature layer of the corresponding RGB network with different weights to be the input of the next layer, and finally enter the fully connected layer of the RGB network to recognize the object category in the scene.

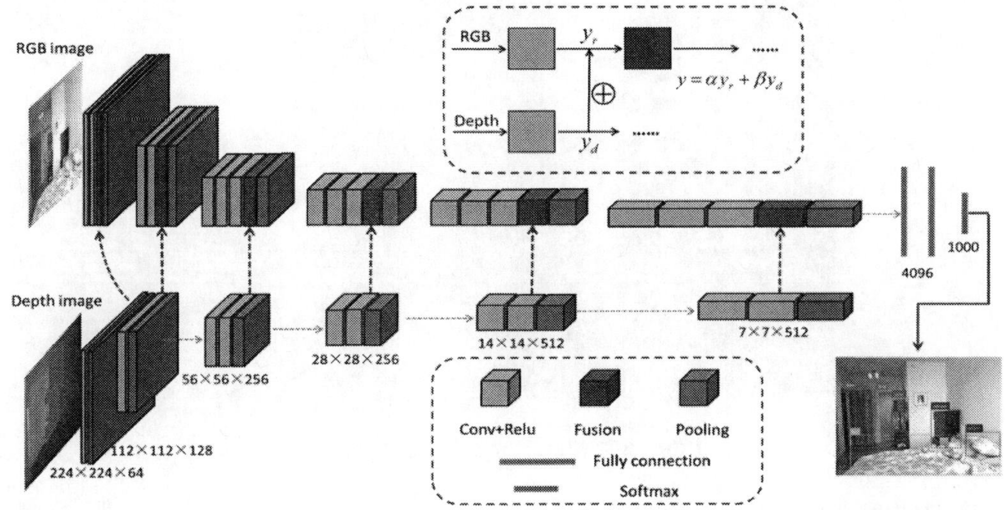

Figure 1. RGB-D medium-level layer-by-layer fusion network. y_r and y_d are output of RGB and Depth networks respectively, α and β are fusion modulus of RGB and Depth network, y is output of feature fusion.

We assume the i Neurons of the l convolution layer is a_i^l, M_r and M_d are collections of input feature maps respectively selected from the RGB network and the Depth network. According to the formula (1), the j neuron calculation formula of the fusion layer can be obtained:

$$a_j^l = f_c\{\alpha[w_r^l \cdot \sum_{i\in M_{rj}^l} (a_i^{l-1} * k_{ij}^l) + b_r^l] + \beta[w_d^m \cdot \sum_{i\in M_{dj}^m} (a_i^{m-1} * k_{ij}^m) + b_d^m]\} \tag{3}$$

In the (3), w_r^l and b_r^l are weights and bias for the RGB network l convolution layer respectively, w_d^m and b_d^m are weights and bias for the Depth network m convolution layer respectively. α and β are fusion modulus of RGB and Depth network respectively, calculation formula as follows [6]:

$$\frac{\alpha}{\beta} = \frac{R_{rgb}}{R_d} \tag{4}$$

$$\alpha + \beta = 1 \tag{5}$$

Among them, R_{rgb} and R_d are average accuracy of detecting RGB images and Depth images separately.

3. Experiment process

3.1. Experimental environment and data preprocessing

The experiment uses open source deep learning framework Darknet based on C and CUDA for model training, feature fusion and object recognition, running on Titan X graphics card and CUDA 9.0 GPU driver. The experiment used the NYU Depth v2 dataset, which contained 26 kinds of indoor scene, more than 1000 categories, and 1449 complementary depth RGB-D images.

In order to achieve the target recognition task on the NYU Depth v2 dataset, marked the original data manually with the detection box to create training samples. For reduce the time cost, our research team marked a total of 10 categories of 300 RGB-D images, and selected 80 images as test sets. Figure

2 shows that mark training samples by the Labelimg annotation tool.

Figure 2. Mark the training samples.

3.2. Training process and visualization

The experiment used batch standardization operation (BN), which trained 64 pictures per iteration, and iterated 30200 times. In the training phase, we used stochastic gradient descent with a momentum term of 0.9. The initial learning rate of the weight is 0.001, and the attenuation modulus is set to 0.0005. The RGB and Depth network parameters are separately trained. The training process visualizes the loss, IoU and Recall indicators (as shown in Figure 3) to evaluate the performance of the model. The IoU and Recall calculation formulas are as follows:

$$IoU = \frac{\text{area}(BB_{dt} \cap BB_{gt})}{\text{area}(BB_{dt} \cup BB_{gt})} \tag{6}$$

$$Recall = \frac{TP}{TP+FN} \tag{7}$$

In the (6), BB_{gt} is Ground Truth Box, BB_{dt} is Detection Truth Box, the TP (True Positive) in equation (7) represents the number that positive samples are also judged as positive samples; the FN (False Negative) indicates the number that positive samples are in fact judged as negative samples.

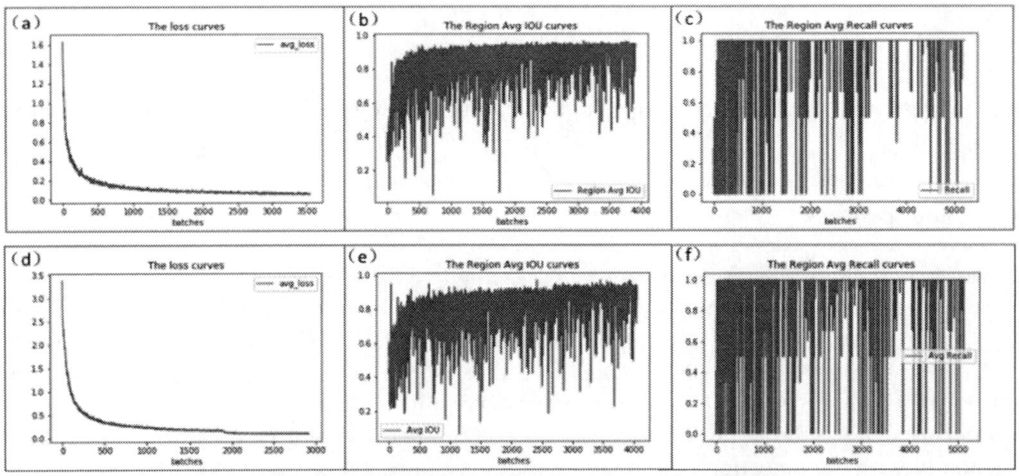

Figure 3. Visualization of the training process.

(a)(b)(c)are loss, average IoU and average Recall of RGB network respectively; (d)(e)(f) are loss, average IoU and average Recall of Depth network respectively.

In order to improve the visualization effect, the sampling rate of the loss curve in Figure 3 is 10%, and the sampling rates of IoU and Recall curves are 0.20% and 0.25% respectively. After 30200 iterations, the loss of RGB and Depth networks is finally stabilized at 0.05 and 0.10 respectively. The training indicators of the two networks are shown in Table 1. In the case of object annotation information sharing, the overall performance of the RGB network is better than Depth due to the Multidimensionality of data.

Table 1. The training index of RGB and Depth network

Network	IoU /%	Recall /%
RGB	85.60	96.48
Depth	82.96	95.80

3.3. Evaluation and comparison of experimental results

3.3.1. Qualitative evaluation. In the experiment, 80 RGB and Depth test images were tested with two trained networks, and the average accuracy was calculated. Then, the RGB-D image joint detection was performed by the medium-level layer-by-layer fusion method proposed in this paper. Finally, the recognition results are output (as shown in Figure 4).

(a) (b) (c) (d) (e) (f)

Figure 4. The recognition effects of medium-level layer-by-layer fusion method. The top part indicates the recognition result by RGB network, the middle part expresses the recognition result by Depth network, and the bottom part shows the medium-level layer-by-layer fusion network recognition effect.

Figure 4 shows the recognition effect of different scenes in the test set. The comparison results show that the recognition effect of RGB images is excellent in most cases, but the detection effect of objects such as defects and overlap in the field of view is not ideal. In the detection results of the Depth network, the contour features of the objects in the near field of view are fully extracted, and the objects such as the missing and overlapping in the near field of view can be more accurately identified. For example, the lamp in (a), the tv in (b), the Whiteboard in (c), the desk in (d)(e)(f), etc., which are not detected in the RGB network, can all be correctly identified in the Depth network. The bottom part of Figure 4 is the recognition result using the medium-level layer-by-layer fusion method. It can be seen that the method combines the recognition characteristics of the two networks on the object, which reduces the missed detection rate of the object and improves the recognition accuracy. However, there are also missed detections and false detections. For example, in (c), the projection cloth is incorrectly recognized as Whiteboard, the door in (d) is not detected and recognized. In addition, due to the limitation of the depth sensor's working distance, the Depth network only has a high recognition rate for objects in the near field of view, and objects which far away in the field are still not recognized.

3.3.2. Quantitative analysis. In order to verify the recognition effect of the method, we not only compare the results of the individual recognition of the RGB and Depth networks trained in the experiment to our way, but also compare with the methods of paper [3] and [6] are used in the recognition accuracy and detection time, the results are shown in Figure 5. It can be seen that the RGB-D fusion method proposed by us is 5.22% more accurate than the recognition result based on RGB images, and is also superior to the methods proposed in paper [3] and [6]. However, while improving the recognition accuracy, due to the complex network structure, the time cost is high, and the average time required for detection reaches 61.03 ms, which is obviously higher than the detection time of other methods.

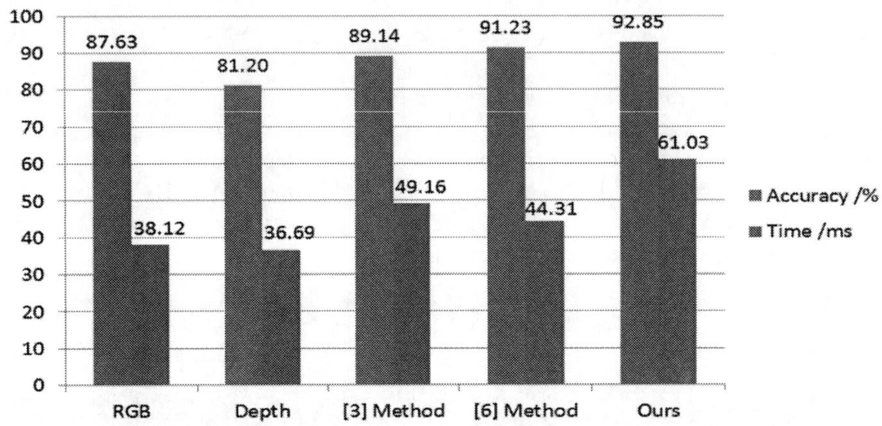

Figure 5. Comparison histogram of test results with different methods.

4. Conclusion

In the paper, we combined with the RGB and Depth images' features which are benefit to recognize objects, trained two kinds convolutional neural networks respectively, and proposed the dual-channel networks method of medium-level layer-by-layer fusion. It could fuse the features extracted by each convolutional layer with different coefficients, and obtain the final weight of the two-channel network to recognize RGB-D objects effectively.

After comparison and analysis of the results, we found that our method can realize the effective recognition of RGB-D indoor objects in the near field of view, and the recognition accuracy and recall rate are obviously improved compared with the methods used in other papers. However, there are still two problems. First, due to the limitation of the depth sensor's working distance, the Depth network has a high recognition rate only for the near-field objects of the depth image, and the distant pixel cannot be acquired. The second is complexity of the medium-level layer-by-layer fusion method is high, and the amount of its calculation is large, so that the recognition object is slower. The next work will focus on the above two aspects to research.

References

[1] Runshun L. 3D Object Recognition Algorithm Based on Deep Learning. Dalian: Dalian University of Technology, 2017: 1-4.

[2] Socher R, Huval B, Bath B, et al. Convolutional-recursive deep learning for 3d object classification. Advances in Neural Information Processing Systems, 2012: 665.

[3] Wei L. Research on RGB-D Object Recognition via Feature Learning.Wuhan: Huazhong University of Science and Technology, 2016: 25-36.

[4] Hoffman J , Gupta S , Leong J , et al. Cross-modal adaptation for RGB-D detection. IEEE International Conference on Robotics & Automation. IEEE, 2016.

[5] Xu X Y, Li Y C, Wu G S, et al. Multi-modal deep feature learning for RGB-D object detection. Pattern Recognition, 2017, 72: 300.
[6] Fan L, Pengyuan L, Junning Z, et al. Joint Detection of RGB-D Images Based on Double Flow Convolutional Neural Network. Laser & Optoelectronics Progress, 55, 021503(2018).
[7] Silberman N , Hoiem D , Kohli P , et al. Indoor Segmentation and Support Inference from RGBD Images. European Conference on Computer Vision. Springer, Berlin, Heidelberg, 2012.
[8] Lu H T, Zhang Q C. Overview of application of depth convolutional neural network in computer vision. Data Acquisition and Processing, 2016, 31(1): 1-17.
[9] Redmon J , Farhadi A . YOLOv3: An Incremental Improvement. 2018.
[10] Simonyan K, Zisserman A. Very deep convolutional networks for large-scale image recognition. arXiv preprint arXiv:1409.1556, 2014.
[11] Shuqin T, Yueju X. Yun Liang, et al. RGB-D image classification methods. Laser & Optoelectronics Progress, 2016, 53(6): 060003.
[12] Hazirbas C, Ma L , Domokos C , et al. FuseNet: Incorporating Depth into Semantic Segmentation via Fusion-Based CNN Architecture. 2016.

IOP Conf. Series: Journal of Physics: Conf. Series **1267** (2019) 012023 doi:10.1088/1742-6596/1267/1/012023

Review of Research on Task-Oriented Spoken Language Understanding

Lixian Hou*, **Yanling Li, Chengcheng Li and Min Lin**

College of computer science and technology, Inner Mongolia Normal University

*cieclcc@imnu.edu.cn

Abstract. Spoken language understanding(SLU) is an important function module of the dialogue system. Slot filling and intent detection are two key sub-tasks of task-oriented spoken language understanding. In recent years, the methods of joint recognition have become the mainstream methods of spoken language understanding to solve slot filling and intent detection. Since deep neural network has advantages such as strong generalization and autonomous learning characteristics compared with traditional methods. So far, slot filling and intent detection have been developed from traditional methods to deep neural network methods, and the performance has also been significantly improved. This paper introduces the methods of two tasks from the independent model to the joint model. It focuses on the joint modeling methods based on deep neural network, analyzes current problems and future development trend of two sub-tasks.

1. Introduction

Spoken language understanding is a key part of the dialogue system. The performance of spoken language understanding directly affects the entire dialogue system. A complete dialogue system is composed of five parts [1]: Automatic Speech Recognition (ASR), Spoken Language Understanding (SLU), Dialog Management (DM), Dialog Generation (DG) and Text to Speech (TTS). Spoken language understanding is divided into two tasks: Slot Filling (SF) (also known as Key Semantic Concept Recognition (KSCR) [2]) and Intent Detection (ID) [3]. Slot filling solves the problem of labeling of specific domain keywords and attributes [4], which belongs to sequential annotation task. Intent detection mainly analyzes user's utterance behavior in input sentences, such as booking tickets and hotels, which belongs to classification problem. The keywords and intents are expressed in a semantic framework to achieve the purpose of understanding the sentence.

In recent years, deep neural network has strong generalization and may learn the features of input text automatically. It can also capture deeper semantic information in the process of learning and training. So the deep neural network method is superior to the traditional statistical machine learning method in the field of natural language processing. At the same time, due to the interdependence of slot filling and intent detection, joint recognition has become a hot research topic in spoken language understanding.

The remainder of the paper is organized as follows. In section 2, we introduce how to solve the task of slot filling and intent detection by independent model. In section 3, it elaborates that solving the slot filling and intent detection through the joint model. Section 4 concludes the paper and analyzes the future development trend of slot filling and intent detection.

2. Independent model for slot filling and intent detection

2.1. Slot filling

The slot filling is similar to the named entity recognition. It mainly extracts the key semantic components and attributes in the expression of users and identifies them with specific symbols. This task provides useful and important information for tasks such as machine translation, topic discovery, and topic tracking. Slot filling is a richer representation of named entity recognition. It is often considered a sequential annotation task.

2.1.1. Traditional methods. There are three traditional methods to solve the slot filling, including dictionary-based method [5], rule-based method [6-8] and statistics method [9-12]. At first, the dictionary-based method is relatively simple, which mainly searches for named entities in the lexicon through string matching. But there is usually no comprehensive entity library, and it is time-consuming [5]. The rule-based method mainly adds lexical rules, syntactic rules and semantic rules in the process of entity recognition. It identifies various types of named entities by rule matching. The disadvantage of the rule-based approach is that it requires domain experts and linguists to formulate rules. The mobility is relatively weak. Once new entities appear, they will conflict with the previous rules. It is time-consuming and laborious to re-formulate rules. The statistical method takes the manually annotated corpus as training set and adjusts the parameters through multiple iterating loss functions. The disadvantage is that it needs a large number of training data labeled manually and needs to construct features manually [13].

2.1.2. Deep neural network methods. In recent years, with the rapid development of deep neural networks, researchers have begun to use deep neural networks to solve slot filling tasks, such as Recurrent Neural Network (RNN), Convolutional Neural Networks (CNN), or their combination variants. Some researchers also combine deep neural network models with traditional machine learning methods.

RNN, a deep neural network model with variable length input and long distance dependence, has shown good performance in various fields, such as speech recognition, part-of-speech tagging (POS) task [14]. Although RNN has shown good performance in various fields, it has the problem of gradient disappearance. To solve this problem, the Long-Short Term Memory (LSTM) model is proposed based on RNN. For each input sequence, its context can help to obtain the grammatical and semantic information of the current word. But the directional LSTM cannot capture the relevant grammatical and semantic information of the text. Bidirectional LSTM (BLSTM) emerges as the time requires.

2.2. Intent detection

2.2.1. Traditional methods. Intent detection extracts the intent and behavior expressed in the utterance of users, which is actually a classification task. Traditional methods include Support Vector Machine (SVM) [15] (2003) and Naive Bayesian(NB) [16] (2009), Adaboost [17] (2000) and so on.

2.2.2. Deep neural network methods. Intent detection based on deep neural networks mainly uses several kinds of neural networks to classify texts. There are two main ways to input text. One is to use the average vector of all words in the discourse as the input of the neural network. The other is to use the vector of each word in the discourse as the input of the neural network [18] (2014).

Graves et al. compared the performance of the feed-forward neural network, RNN, LSTM, and Gated Recurrent Unit (GRU) in intent detection [19] (2014). Xiao et al. proposed the combination of CNN and RNN for intent detection [20] (2016).

3. The joint model for slot filling and intent detection

Generally, slot filling and intent detection are two independent tasks. In fact, the two tasks are related.

Intent detection has a positive effect on the result of slot filling and can help identify entities more accurately. So some researchers solve two tasks simultaneously. Joint model simplifies the tasks of slot filling and intent detection by training one model. Following is an explanation of the evolution of the two tasks by joint recognition. This section mainly describes the joint recognition of two tasks on ATIS [21].

3.1. The joint model of triangular-chain CRF

Because slot filling and intent are related, and traditional parallel or cascade methods cannot capture the correlation between the two tasks. Jeong et al. used triangular-chain CRF model to solve slot filling and intent detection. It captured the internal link between them [22] (2006).

Cascade mainly refers to the sequential execution of two tasks, the latter can use the results of the former as a priori knowledge to improve performance [2]. In order to compare the joint model with the cascade model, two cascade methods are adopted.

The triangular-chain CRF of the research is shown in Figure 1: x represents the input text sequence, y represents the corresponding output sequential annotation, z represents the corresponding intent. The output of z depends on two parts: one is the input text sequence, the other is the output sequential annotation y .

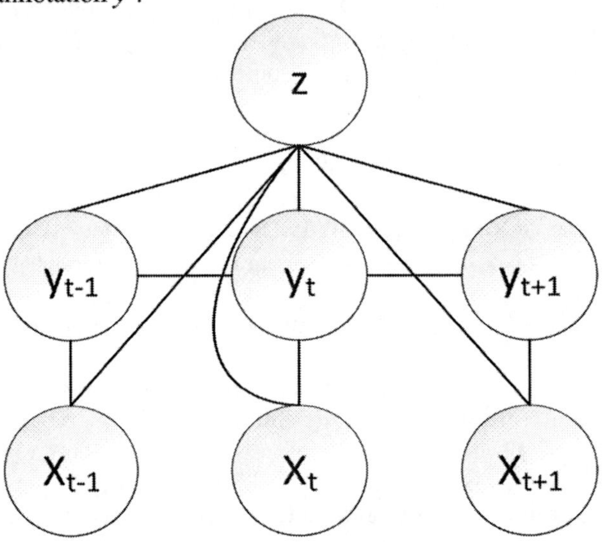

Figure 1. Triangular-chain CRF model.

Jeong's research uses a traditional statistical machine learning method CRF. Although the model contributes to joint recognition, there are still shortcomings of traditional statistical machine learning. It is time-consuming and laborious, and there must be enough training corpus.

3.2. The joint model of CNN-TriCRF

In 2013, Xu in Microsoft company used CNN-TriCRF model for solving slot filling and intent detection jointly [23]. Text features are extracted by CNN, then shared by slot filling and intent detection. TriCRF is used to normalize the whole text in slot filling task. Compared with traditional triangular-chain CRF model, this model improves the performance of intent detection and slot filling by 1% and 1.02% respectively.

Figure 2 is a joint recognition model based on CNN-TriCRF. The dotted line on the top of the graph represents the dependence between slot filling and intent detection. Feature vector h generated by CNN is shared by two tasks. The model can handle multiple sequential annotation and intent classification tasks simultaneously.

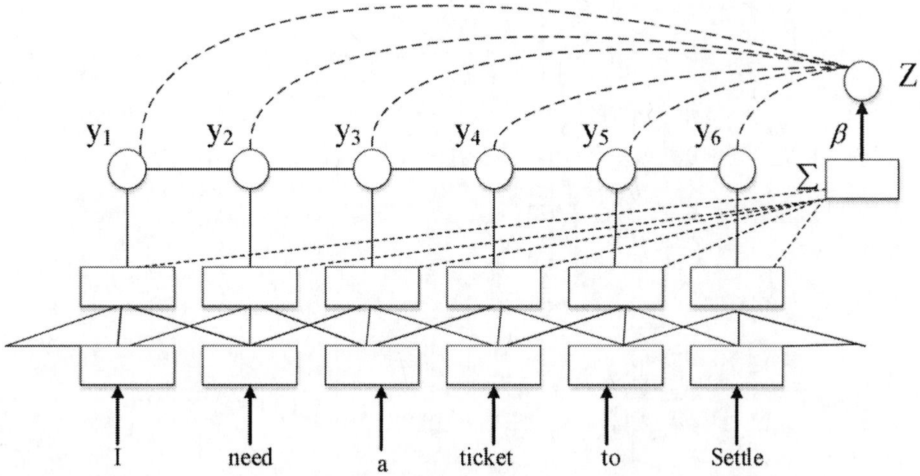

Figure 2. The joint model of CNN-TriCRF.

Compared with the traditional method of extracting text features, the CNN method does not need to define features in advance. It can learn the features of input text independently. In addition, it can also solve the variable length input sequence problem. The improved CNN is used to learn text features. Finally, the learned features are used for slot filling and intent detection tasks. Formula (1) calculates of slot filling for the improved model of this research.

$$p(Y \mid X) = \frac{e^{\sum_i (t(Y_{i-1},Y_i) + \sum_j h_{ij}(X_i,R,T)\theta_j(Y_i))}}{\sum_{Y'} e^{\sum_i (t(Y'_{i-1},Y'_i) + \sum_j h_{ij}(X_i,R,T)\theta_j(Y'_i))}} \quad (1)$$

In the above formula, $t(Y_{i-1}Y_i)$ denotes the transition score from Y_{i-1} to Y_i. h_{ij} denotes the j element extracted from the media window of X_i. θ_j denotes the feature weight corresponding to the label Y_i. The improved CNN uses the text features learned for the classification task, as shown in formula (2).

$$p(Z \mid X) = \frac{e^{\sum_i (\sum_j h_{ij}\beta_j(Z))}}{\sum_{Z'} e^{\sum_i (\sum_j h_{ij}\beta_j(Z'))}} \quad (2)$$

Xu uses CNN to learn text features, which does not depend on manual feature extraction. For slot filling, TriCRF is used to consider the global distribution of data, instead of normalizing the data locally, so as to solve the local annotation bias problem and get the global optimal solution. But compared with traditional methods, the training parameters of the model are more and the model is more complex.

3.3. The joint model of Gated Recurrent Unit

After RecNN, Xiaodong from the Institute of Computational Language of Peking University proposed to use GRU and CNN to solve the task of slot filling and intent detection jointly [24] (2016). Through GRU, each time step is learned to predict the label of slot filling. Meanwhile, the maximum pooling layer is used to capture the global features of sentences intent classification. The model is shared by two tasks. The model of GRU-CNN is shown in Figure 3.

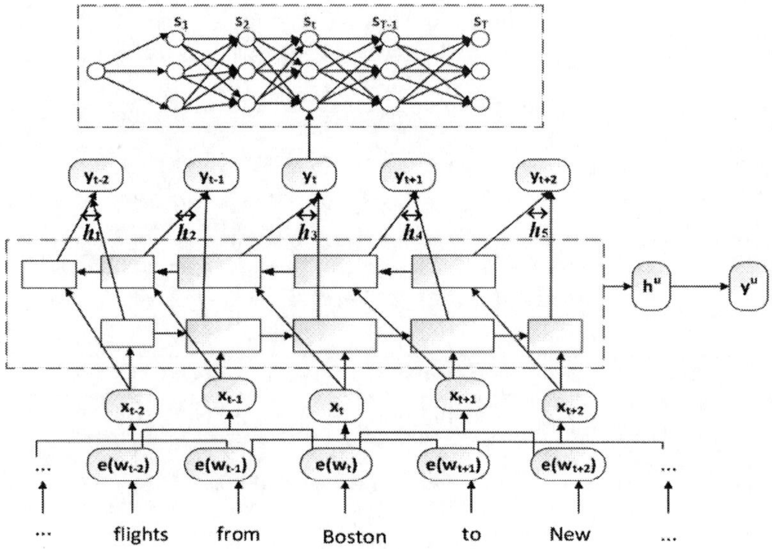

Figure 3. GRU-CNN model.

It used two data sets, ATIS, and CQUD (Chinese Question Understanding Dataset) collected from Baidu Know. The labeling process was manual annotated.

Although the performance of the model has been improved, there is still room for improvement. For intent detection task, the model uses the maximum pooling to classify intent, which will result in information loss. It should achieve intent detection by acquiring higher-level grammatical and semantic information.

3.4. The joint model of BLSTM-attention

Bing Liu used attention mechanism to solve slot filling and intent detection jointly [25] (2016). It used BLSTM model with attention mechanism. Attention parameter can capture additional dependence information which hidden state h_i cannot capture. BLSTM-attention model is shown in Figure 4. The hidden state h_i obtained by BLSTM and attention parameters c_i obtained all h_i are used as input for two tasks.

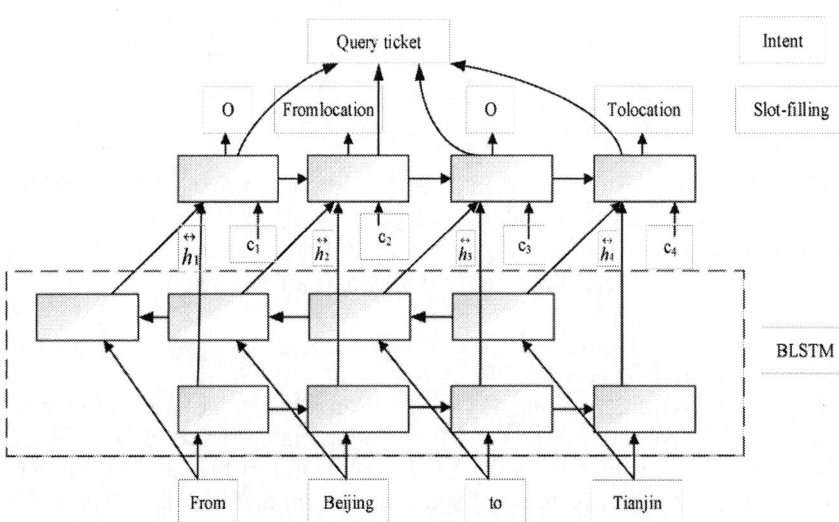

Figure 4. BLSTM-attention model.

BLSTM is used as the basic unit and attention parameters are added on the hidden layer state of BLSTM. The calculation of the attention parameter is as shown in the equation (3) - (5).

$$c_i = \sum_{j=1}^{T} a_{ij} h_j \tag{3}$$

$$a_{ij} = \frac{\exp(eij)}{\sum_{k=1}^{T} \exp(eik)} \tag{4}$$

$$e_{ik} = g(s_{i-1}, h_k) \tag{5}$$

Where c_i represents the attention vector of the i time step, which is weighted by all hidden layer states h_j. a_{ij} denotes the weight vector, which is normalized. c_i acts on the output of the decoder of the i time step. The above process realizes the transformation from the initial hidden layer to the new attention layer. The weight coefficients of the hidden layer at each time reflect the influence on the current output.

The advantage of BLSTM-attention model is that it can pay attention to the effect of different input sequences on output by adding attention mechanism. But for intent detection, information entropy will be lost if we simply take the maximum value for intent detection. For slot filling, because the global normalization of the output sequence is not taken into account, it can occur bias problem easily.

3.5. Slot-gated joint recognition model based on attention

BLSTM-attention model has achieved good performance in slot filling and intent detection tasks, but it does not clearly explain how slot and intent interact. Because slot filling and intent detection depend on each other, intent can help to predict the content of slot filling to a large extent. The slot-gated mechanism is proposed by CW Goo et al. to obtain a better semantic framework for slot filling and intent detection [26] (2018). Figure 5 is the slot-gated model based on the attention.

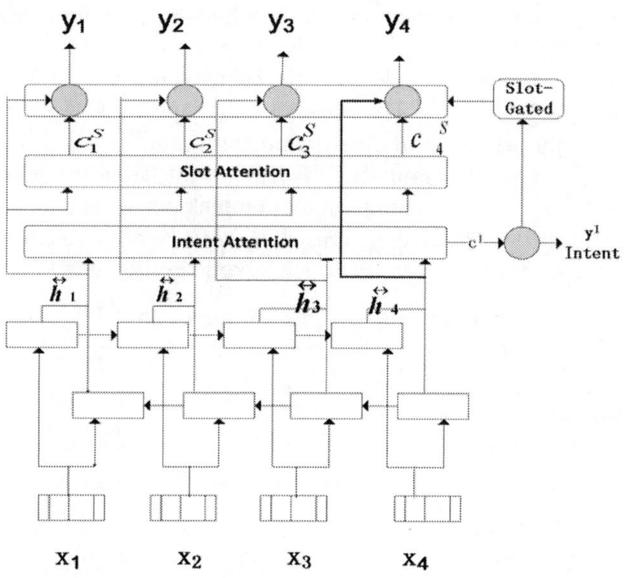

Figure 5. Attention-based slot-gated model.

Unlike the BLSTM-attention model in section 3.4, when annotating the output of the slot sequence, the intentional attention parameter of the text is added to the output of each slot filling. It contains the semantic information of the whole input text sequence. It can alleviate the problem of the local optimal solution for the slot filling. It proves that slot filling and the intent detection are interdependent. Slot-gated is shown in Figure 6. Calculation such as formula (6) - (7).

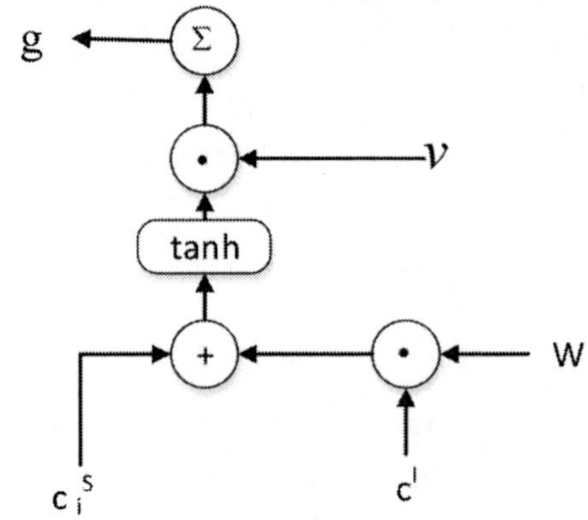

Figure 6. Slot-gated gate.

$$g = \sum v \cdot \tanh(c_i^S + W \cdot c^I) \quad (6)$$

$$y_i^S = soft\max(W_{hy}^S(h_i + c_i^S \cdot g)) \quad (7)$$

Where c_i^S represents attention context vectors of the semantic slot. c^I is the attention context vectors of the intent. c^I and c_i^S have same dimension, v and W are trainable vectors and matrices respectively. Then the sum of each time series is calculated, g is the weighted feature representation of the sum. y_i^S represents slot filling calculated by the weighted feature c_i^S and c^I.

Table 1 is the performance of the previous joint model for slot filling and intent detection tasks. It can be seen that the performance of joint recognition has been improving in recent years. From table 1 can be concluded that BLSTM-attention model is better than other models, compared with the triangular - chain CRF increased by 1.2%. Because the focus is on the basis of the original features for different input focus to study further. The model can capture a deeper semantic grammatical information, LSTM is the input of accumulative memory training. It is unable to focus on the type of different weights to different input distribution. So BLSTM-attention model has a good effect in slot filling and intent detection. GRU-CNN in intent detection task performance is superior to other models, compared with the triangular - chain CRF increased by 4.01%, the reason is that CNN can get the more advanced features, the performance has been improved significantly.

Table 1. Results of the joint model on ATIS.

Model	Slot(F1)	Intent(Accuracy)
Triangular-chain CRF	94.42	93.07
CNN-TriCRF	95.42	94.09
RecNN	93.22	95.40
RecNN+Viterbi	93.96	95.40
GRU-CNN	95.49	98.10
Attention encoder-decoder NN(with aligned inputs)	95.62	94.14
BLSTM-attention	95.78	94.40
Slot-gated	95.20	94.10

3.6. Summary of joint recognition

In addition to the above methods, there are other joint models to solve slot filling and intent detection. Dilek et al. [27] (2016) proposed using RNN-LSTM model to jointly solve slot filling and intent detection. In the process, multi-domain models were established by strengthening the learning of data in different fields, and the F1 value on ATIS datasets reached 94.7%. Zhang et al. [28] (2016) proposed using Bidirectional GRU (BGRU) and adding CRF at the same time. The model realized the joint recognition of slot filling and intent detection, but there are some shortcomings. First, for intent detection, there is information loss by choosing the maximum value of GRU output each time as the intent of user discourse. Second, it does not pay attention to which words in the input sequence have a greater impact on the output of the slot filling and intent detection. Weigelt et al. [29] (2018) proposed using context, slot filling, and intent detection tasks are solved jointly with hierarchical information.

To sum up, the joint model based on deep neural network is used to solve the tasks of slot filling and intent detection. It aims to extract higher-level semantic grammatical information from text, from the previous triangular chain conditional random field to BLSTM, from relying on manual feature extraction by continuous iterative training and learning features. The performance is constantly improved. At the same time attention is also added to focus on learning parameters on the different input on the basis of the original learning characteristics. When considering the local optimal solution of the slot filling task, a CRF or slot-gated mechanism is proposed to act on the sequential output of the slot filling task after learning the advanced features. The performance of the whole model is closely related, so it is very important to extract high-quality context grammar and semantic information.

4. Conclusion and prospect

The two key tasks have experienced the expansion from traditional statistical machine method to the deep neural network model, from independent model to joint recognition, from CRF, SVM, Adaboost to RNN, CNN, LSTM, GRU, and their variants, and the application of attention mechanism. Recognition of the performance is better BLSTM-attention model and GRU-CNN model, compared with the Triangular - chain conditional random field model on slot filling and intent recognition increased by 1% - 1.36% and 1.03% 5.03% respectively. The performance is constantly improving, and further improvement is needed.

There are following three aspects that need to improve. First, the slot filling and intent detection in some research on large-scale corpus dataset have better performance. However, in the small-scale corpus (e.g., Dutch, Spanish, and other data in the collection of text sequence annotation and the lack of language), the model has not been fully made use of an application [30]. Follow-up should consider using in this respect across language knowledge migration established between cross-language dictionary. It should combine a small amount of double language as input training a larger corpus, solving in addition to the English language and so on the large-scale corpus, other languages and annotation of corpus insufficient. Second, the joint recognition model based on the deep neural network model needs to learn better grammar and semantic information. After repeated iterative training, the training parameters of the model become more and more, and the model becomes more and more complex. How to find a simplified model equivalent to the current performance to replace the existing model needs to be further studied. Third, as can be seen from the existing joint model, based on the attention mechanism for the performance of the model under the same data set is better. But the model needs for all the hidden states under different time series based on the current moment of slot filling, which uses normalized calculating weight proportion. While it can cause the model is more complex. How to use reinforcement learning model training of text feature, so as to achieve the effect of the simplified model, further to do in-depth research.

Acknowledgments

This work is partially supported by the National Natural Science Foundation of China(Nos. 61562068, 11704229, 61640204, 61806103); Natural Science Foundation of Inner Mongolia (No. 2017MS0607);

Mongolian Informatization Special Support Sub-project of Inner Mongolia Nationalities Committee (MW-2014-MGYWXXH-01); Youth Innovation and entrepreneurship talents of Inner Mongolia "grassland talents" project; Inner Mongolia science and technology planning project (research and development operation of online interactive learning platform of Mongolian language MOOC and analysis and research of national education big data); Graduate Innovation Fund of Inner Mongolia Normal University (CXJJS18112).

References

[1] Chen H, Liu X, Yin D, et al. A Survey on Dialogue Systems[J]. ACM SIGKDD Explorations Newsletter, 2017, 19(2):25-35.

[2] LI Yanling, YAN Yonghong. Research on execution strategy about statistical spoken language understanding.Journal of Frontiers of Computer Science and Technology, 2017, 11(6): 980-987.

[3] Tur G . Spoken Language Understanding: Systems for Extracting Semantic Information from Speech[D]. Morgan & Claypool, 2011.

[4] Finkel J R, Grenager T, Manning C. Incorporating non-local information into information extraction systems by gibbs sampling[C]//Proceedings of the 43rd Annual Meeting on Association for Computational Linguistics. Association for Computational Linguistics, 2005: 363-370.

[5] wang qi. Biological named entity recognition combining dictionary and machine learning [D]. Dalian University of technology, 2009.

[6] Collins M, Singer Y. Unsupervised models for named entity classification[C]//1999 Joint SIGDAT Conference on Empirical Methods in Natural Language Processing and Very Large Corpora, 1999: 100-110.

[7] Cucerzan S, Yarowsky D. Language independent named entity recognition combining morphological and contextual evidence[C]//1999 Joint SIGDAT Conference on Empirical Methods in Natural Language Processing and Very Large Corpora, 1999: 90-99.

[8] Mikheev A, Moens M, Grover C. Named entity recognition without gazetteers[C]//Proceedings of the ninth conference on European chapter of the Association for Computational Linguistics. Association for Computational Linguistics, 1999: 1-8.

[9] Wei L, Mccallum A. Rapid development of Hindi named entity recognition using conditional random fields and feature induction[J]. Acm Transactions on Asian Language Information Processing, 2003, 2(3):290-294.

[10] Bikel D M, Miller S, Schwartz R, et al. Nymble: a High-Performance Learning Name-finder[J]. Anlp, 1998:194--201.

[11] Bikel D M, Schwartz R, Weischedel R M. An Algorithm that Learns What's in a Name[J]. Machine Learning, 1999, 34(1-3):211-231.

[12] Borthwick A, Grishman R. A maximum entropy approach to named entity recognition[D]. New York University, Graduate School of Arts and Science, 1999.

[13] Ratinov L, Roth D. Design challenges and misconceptions in named entity recognition[C]//Proceedings of the Thirteenth Conference on Computational Natural Language Learning. Association for Computational Linguistics, 2009: 147-155.

[14] Huang Z, Xu W, Yu K. Bidirectional LSTM-CRF models for sequence tagging[J]. arXiv preprint arXiv:1508.01991, 2015.

[15] Haffner P, Tur G, Wright J H. Optimizing SVMs for complex call classification [C]//IEEE International Conference on Acoustics. CiteSeer, 2003: I-632-I-635 vol.1.

[16] Chen J, Huang H, Tian S, et al. Feature selection for text classification with Naïve Bayes[J]. Expert Systems with Applications, 2009, 36(3):5432-5435.

[17] Schapire R E, Singer Y. BoosTexter: a boosting-based system for text categorization[J]. Machine Learning, 2000,39(2-3):135-168.

[18] Xu P, Sarikaya R. Contextual domain classification in spoken language understanding systems using recurrent neural network[C]//Acoustics, Speech and Signal Processing (ICASSP), 2014

IEEE International Conference on. IEEE, 2014: 136-140.
[19] Graves A, Jaitly N, Mohamed A. Hybrid speech recognition with deep bidirectional LSTM[C]//Automatic Speech Recognition and Understanding (ASRU), 2013 IEEE Workshop on. IEEE, 2013: 273-278.
[20] Xiao Y, Cho K. Efficient character-level document classification by combining convolution and recurrent layers[J]. arXiv preprint arXiv:1602.00367, 2016.
[21] Tur G, Hakkani-Tür D, Heck L. What is left to be understood in ATIS?[C]//Spoken Language Technology Workshop (SLT), 2010 IEEE. IEEE, 2010: 19-24.
[22] Jeong M, Lee G G. Jointly predicting dialog act and named entity for spoken language understanding[C]//Spoken Language Technology Workshop, 2006. IEEE. IEEE, 2006: 66-69.
[23] Xu P, Sarikaya R. Convolutional neural network based triangular crf for joint intent detection and slot filling[C]//Automatic Speech Recognition and Understanding (ASRU), 2013 IEEE Workshop on. IEEE, 2013: 78-83.
[24] Zhang X, Wang H. A Joint Model of Intent Determination and Slot Filling for Spoken Language Understanding[C]//IJCAI. 2016: 2993-2999.
[25] Liu B, Lane I. Attention-based recurrent neural network models for joint intent detection and slot filling[J]. arXiv preprint arXiv:1609.01454, 2016.
[26] Goo C W, Gao G, Hsu Y K, et al. Slot-Gated Modeling for Joint Slot Filling and Intent Prediction[C]//Proceedings of the 2018 Conference of the North American Chapter of the Association for Computational Linguistics: Human Language Technologies, Volume 2 (Short Papers). 2018, 2: 753-757.
[27] Hakkani-Tür D, Tür G, Celikyilmaz A, et al. Multi-Domain Joint Semantic Frame Parsing Using Bi-Directional RNN-LSTM[C]//Interspeech. 2016: 715-719.
[28] Zhang X, Wang H. A Joint Model of Intent Determination and Slot Filling for Spoken Language Understanding[C]//IJCAI. 2016: 2993-2999.
[29] Weigelt S, Hey T, Landhäußer M. Integrating a dialog component into a framework for spoken language understanding[C]//Proceedings of the 6th International Workshop on Realizing Artificial Intelligence Synergies in Software Engineering. ACM, 2018: 1-7.
[30] Feng X, Feng X, Qin B, et al. Improving Low Resource Named Entity Recognition using Cross-lingual Knowledge Transfer[C]//IJCAI. 2018: 4071-4077.

De-Noising Algorithm for Flight Data Recording System Based on Modified Ensemble Empirical Mode Decomposition

DAI Shaowu, CHEN Qiangqiang* and DAI Hongde

Naval Aviation University, China, Yantai, 264000.

*1195275597@qq.com

Abstract. The Flight Data Recording System (FDRS) records a lot of parameters of the aircraft during flight, which can be used for the test-flying, training mission of aircraft and so on. Effected by the working environment, information interference and its non-stability, the outliers and noise often exists in the FDRS data. These noises and outliers have a great impact on the use of FDRS. The aim of this paper is to remove outliers and de-noising of navigation data in FDRS. The causes of outliers and noise in FDRS data are analyzed firstly, with a reference suggestion proposed. Then the Letts criterion is used to remove outliers and the Modified Ensemble Empirical Mode Decomposition (MEEMD) is applied to achieve de-noising for FDRS. Results demonstrate that outliers are removed and the navigation data are de-noised effectively.

1. Introduction

FDRS were first introduced in the 1950s, which also known as black box [1]. The critical mission of an airplane FDRS is to collect, record and save flight parameters from a variety of airplane units to survive an accident [2]. Depending on the service life of an airplane, the FDRS may consists of an analog or a digital data. With the development of FDRS, which can record up to thousands of flight parameters for 25 hours, such as acceleration, air speed, time, altitude and attitude of aero engine, et al [3]. These data provide abundant resources with SINS and engine capability analysis [4].

However, FDRS is disturbed by many factors such as working environment and vibration noise when recording data [5]. For example, the electromagnetic interferences and pulse signals generated by sensor faults will cause the recorded flight data to contain abnormal signals and noise signals inevitably, which severely affect the subsequent analysis and application of flight data [6]. Therefore, it is essential to de-noising for the FDRS before data analysis.

At present, various methods have been developed to remove outliers from the obtained data. The Auto Regressive Moving Average (ARMA) model is extensively utilized in constructing a de-noising model, which can be judge whether there are outliers in the obtained data [7]. In [8], the Observer / Kalman filter identification (OKID) method is applied to remove outliers in flight data, and the estimated values are used to replace them, but the experiment data is a simulation signal, which has no practical application in FDRS. In [9], a flight data novelty detection method based on improved support vector data description (SVDD) method is proposed to realize automatic flight data interpretation. Besides, this method has a repeated process of adjustment, and the choice of parameters of SVDD has a great influence on the results.

Wavelet analysis has the capability of multi resolution decomposition and reconstruction, extracting features from the upper and lower frequencies of the data. In [10], the simulation of data de-noising to pitch angle of a type aircraft was carried out based on wavelet de-noising. In [11], wavelet neural network is applied to flight data pre-processing. However, the de-noising results of Wavelet analysis depend largely on the choice of wavelet basis and decomposition levels. In [12], combined Ensemble empirical mode decomposition (EEMD) with Hilbert Transform (HT) is used for quick access recorder data. Besides, the calculation amount grows and the completeness loses due to the white noise unneutralized completely of EEMD.

To further improve the de-noising effects, a new de-noising algorithm based on Letts criterion and MEEMD is studied in this paper. This method uses Letts criterion to eliminate the outliers of FDRS data. Finally, the MEEMD is used to eliminate or reduce the noise in FDRS data in this paper.

2. Remove outliers

Outliers are always deviation from most real values of original data. The definition of outliers is: one or more observation points in a set of observation data that violate the data variation rule are called outliers [13].

Affected by the working environment of aircraft, the FDRS data is effected by some impulse signals or vibration, which leads to the existence of outliers. Letts criterion (3σ criterion) is used for judging outliers [14]. Set the original FDRS data is S have n points. The average value of original FDRS data is $\bar{S}$. The residual series is $\tilde{S}=S-\bar{S}$ have $n-1$ points. The standard deviation of $\tilde{S}$ is:

$$\sigma = \sqrt{\sum_{i=1}^{n} \frac{\tilde{S}^2}{n-1}} \tag{1}$$

Time series follows the characteristics of Gauss distribution, the probability of the residual falling in the interval $[-3\sigma, 3\sigma]$ is about 99.7%. When the residual of a measured value is greater than 3σ, the measured value is regarded as an outlier. According to the 3σ criterion, the outliers of FDRS can be removed effectively.

After remove outliers by 3σ criterion, the location of outliers should be reconstruct. For the detected abnormal value of x_i, the first-order difference method is used to replace the abnormal information of the point [15]:

$$\hat{x}_i = 2x_{i-1} - x_{i-2} \tag{2}$$

Where $\hat{x}_i$ is the replacement value of outlier x_i. With this substitution, we can reconstruct the data without gross errors, and make preparation for data de-noising.

3. Modified Ensemble Empirical mode decomposition

3.1. Empirical mode decomposition

EMD, invented by Huang in 1998, which is a newly developed useful algorithm for non-linear and non-stationary time series analysis [16]. Based on EMD, Any complicated signal can be decomposed into several number of intrinsic mode functions (IMF) and a residual series. In physics, the IMFs represent the dynamic characteristics of the original signal, and the residual series is the trend item of the original signal. Compared with the original signal, the IMFs and the residual series have less complexity. Obviously, it can reduce the non-linear and non-stationary of original signal by EMD. The process of decomposing an IMF as follows is called "sifting" [17]:

(1) Determine all the local extrema of the original signal $x(t)$.

(2) Determine the upper envelope $e_m(t)$ with the local maxima by a cubic spline.

(3) Determine the lower envelope $e_n(t)$ with the local minima by a cubic spline.

(4) Calculate the average of the upper and lower envelope

$$m(t) = (e_m(t) + e_n(t))/2 \tag{3}$$

(5) Calculate the subtract of the original signal $x(t)$ and $m(t)$.

$$h = x(t) - m(t) \tag{4}$$

(6) Use the IMF criteria to detect h is the IMF or not. IF not, make the $x(t)=h$ and repeats the above steps; IF h is the IMF, mark the IMF $c(n)$.

$$c(n) = h \tag{5}$$

(7) Calculate the residual series r, and detect whether r meets the stopping criteria.
From the above steps, the original signal $x(t)$ can induce that:

$$x(t) = \sum_{j=1}^{n} h_n(t) + r \tag{6}$$

3.2. Modified Ensemble Empirical mode decomposition

In the EMD algorithm, a major problem in the decomposition process is the mode mixing, that is, the signals of different scales and frequencies appear in the same IMF component, or the signals of the same scale and frequency are decomposed into multiple IMF components [18]. Multiple modes are mixed in several IMF components, which affects the physical meaning of IMF components and is not conducive to the establishment of subsequent time series prediction models. Using MEEMD method for reference in dealing with mode mixing problem. The steps are as follows [19]:

(1)For the original signal s_t, white noise signal sequences $n_i(t)$ and $-n_i(t)$ with 0 mean and opposite numbers are added respectively. a_i denotes the noise amplitude and i denotes the added logarithm of white noise:

$$\begin{aligned} S_{it}^{+} &= S_t + a_i n_i(t) \\ S_{it}^{-} &= S_t - a_i n_i(t) \end{aligned} \tag{7}$$

(2)The sequence in formula (7) is decomposed by EMD, and the first order IMF components are $I_{i1}^{+}(t)$ and $I_{i1}^{-}(t)$. Integrating this pair of IMF components:

$$I_1(t) = \frac{1}{2N} \sum_{i=1}^{Ne} \left[I_{i1}^{+}(t) + I_{i1}^{-}(t) \right] \tag{8}$$

(3)The decomposed IMF components in step (2) are separated from s_t and the residual signals are obtained:

$$r(t) = S_t - \sum_{i=1}^{p-1} I_j(t) \tag{9}$$

(4)The residual signal $r(t)$ is decomposed by EMD and all IMF components are obtained.

In MEEMD, Permutation Entropy (PE) is introduced to evaluate the complexity of time series. By setting the PE value, the PE of each component obtained in the decomposition process is judged[20]. The sequence whose PE is greater than the set value is regarded as an abnormal signal, and the sequence whose permutation entropy is larger than the set value is filtered out. According to the literature, the threshold value of PE is 0.55-0.6.That is to say, if the PE value exceeds 0.6, it can be regarded as noise.

4. Results and Analysis

In this paper, we choose a set of intact flight data from FDRS. The position data output from airborne SINS were subtracted by GPS position data recorded by FDRS, Take latitude error as an example, Record latitude error as δL.

IOP Conf. Series: Journal of Physics: Conf. Series **1267** (2019) 012024 doi:10.1088/1742-6596/1267/1/012024

4.1. Remove outliers

Before data processing of flight data, outliers need to be removed first. The latitude error δL between the SINS and GPS as shown in Figure 1.

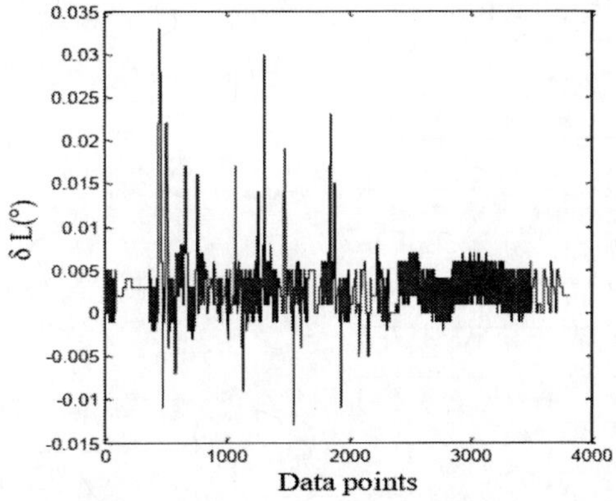

Figure 1. The original flight data

As shown in Figure 1, there are many outliers in the original flight data, which bring great errors to the subsequent flight data interpretation. In order to ensure the accuracy of flight data, it is necessary to remove the outliers.

Using the method of Chapter 2 to process the data, the results are shown in Figure 2.

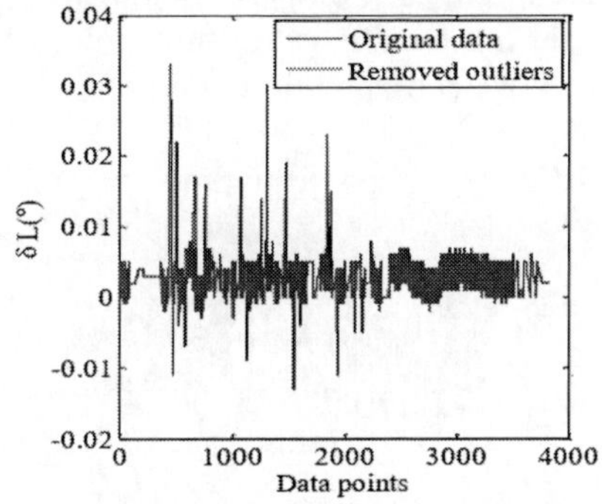

Figure 2. Flight data after removing outliers.

As shown in Figure 2, after outliers are removed, the outliers in the original flight data are effectively reduced. After eliminating outliers, We should take the second step of de-noising. Firstly, EMD is used to de-noise the original data after eliminating outliers. As shown in Figure 3.

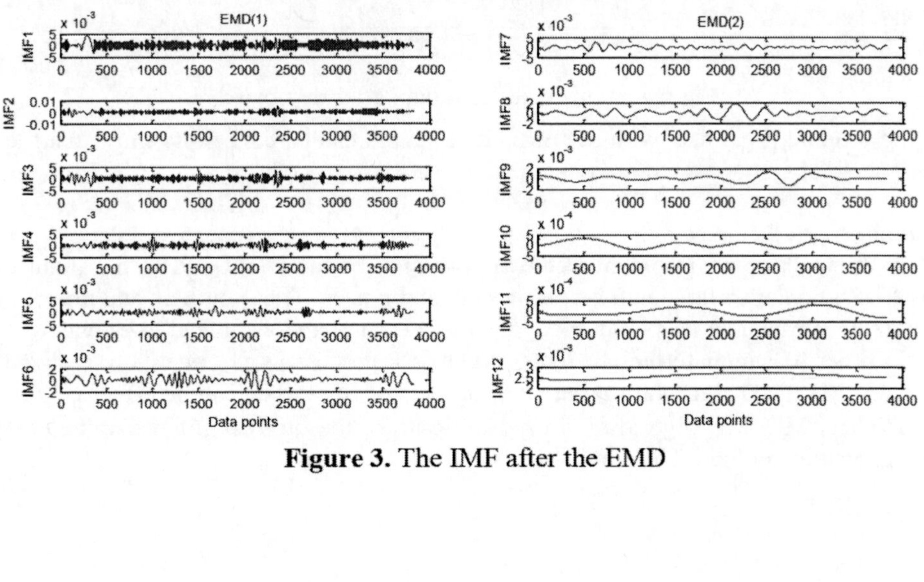

Figure 3. The IMF after the EMD

As shown in Figure 3, after EMD decomposition, 12 IMF components are obtained. There exists mode mixing in EMD decomposition, and there is no clear criterion for eliminating noise IMF components. So on this basis, the original flight parameter data after removing outliers are de-noised by MEEMD method. As shown in Figure 4.

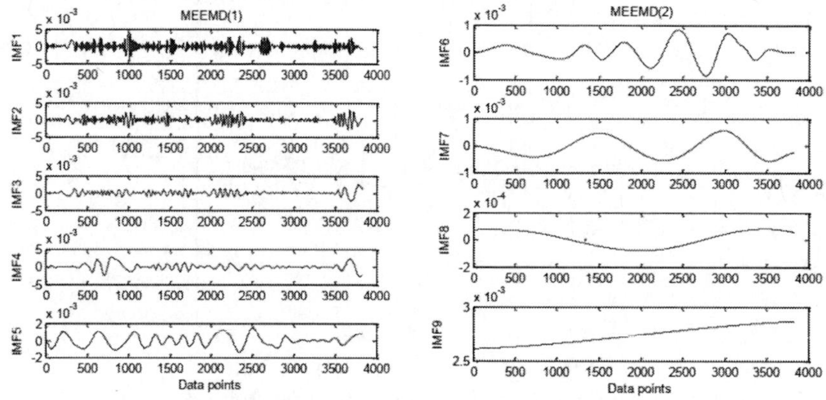

Figure 4. The IMF after the MEEMD

As shown in Figure 4, after EMD decomposition, 9 IMF components are obtained. The MEEMD method effectively alleviates the mode mixing problem in EMD decomposition, and the decomposition yields fewer IMF components. Compared with EMD, MEEMD has more advantages in mode mixing and de-noising. By setting the PE value threshold, MEEMD can effectively detect the abnormal IMF components obtained in decomposition. Therefore, the de-noising process of MEEMD is more adaptive than EMD.

The de-noising results of MEEMD are compared with the original flight data as shown in Figure 5.

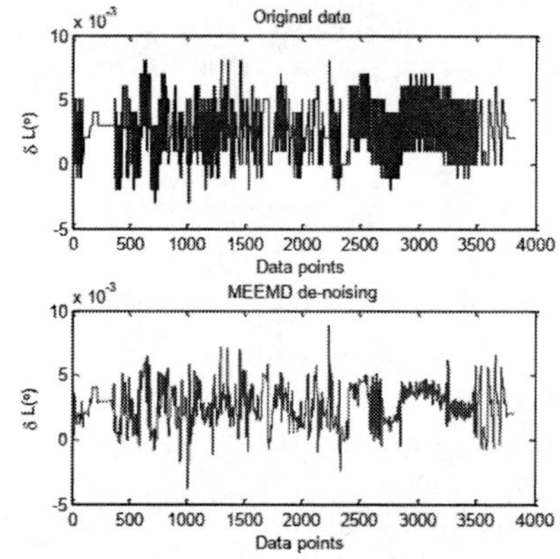

Figure 5. Original flight data after de-noised

Comparing original flight data with MEEMD de-noising results, the results show that the MEEMD method can effectively reduce the noise contained in the data.

5. Conclusions

Influenced by the working environment of the aircraft and the factors of FDRS, the flight data which recorded by FDRS contain a large number of outliers and noises. These outliers and noises bring great interference to the subsequent flight parameter data processing, and even lead to erroneous flight state estimation. In order to eliminate these effects, flight data processing is effectively realized by using 3σ criterion and MEEMD algorithm. Firstly, the outliers of flight data can be effectively removed by 3σ criterion. Then, MEEMD algorithm is used to de-noise the original flight parameter data which has been eliminated by outliers. The results show that:

(1)Through 3σ criterion, the outliers in the original flight data can be detected and removed effectively.

(2)MEEMD method can effectively solve the mode mixing problem, and has less IMF than EMD.

(3)By setting PE, abnormal IMF detection in MEEMD decomposition can be realized, and the original flight parameter data can be de-noised effectively.

References

[1] D'Angelo G, Ferretti S, Marzolla M. A Blockchain-based Flight Data Recorder for Cloud Accountability[J]. 2018.
[2] Wiseman Y. Unlimited and protected memory for flight data recorders[J]. Aircraft Engineering and Aerospace Technology. 2016, 88(6): 866-872.
[3] Timi P H, Shermin S, Rahman A. Study of flight data recorder, underwater locator beacon, data logger and flarm collision avoidance system[C]. 2017.
[4] Aborizka M, Kavi K M. An Intelligent Flight Data Recorder And Real-Time Monitoring System[J]. Journal of Public Health Management & Practice Jphmp. 2007, 9(9): 157-164.
[5] Khadilkar H, Balakrishnan H. Estimation of aircraft taxi fuel burn using flight data recorder archives[J]. Transportation Research Part D: Transport and Environment. 2012, 17(7): 532-537.
[6] Onoz B, Oguz B. Assessment of Outliers in Statistical Data Analysis[M]. 2003.
[7] Barnett V, Lewis T, Abeles F. Outliers in Statistical Data[M]. 1978.
[8] Zhu Feifei, Gao Yanhui, Xiao Qiangui, et al. Application of observer/Kalman filter in flight data processing[J]. Electronic Design Engineering. 2016(20): 91-93.
[9] Sun Wenzhu, Qu Jianling, Yuan Tao,et al. "Flight Data Novelty Detection Method Based on Improved SVDD," Chinese Journal of Scientific Instrument, 2014,35(4): 932-939.
[10] Di Yazhou, Li Furong, Yu Jianli, et al. Application of Wavelet Analysis on Flight Data De-noising[J]. Computer Simulation. 2010(10): 1-4.
[11] Tang Chongkai, Qu Jianling, Gao Feng, et al. Flight data preprocessing based on wavelet neural network[J]. Journal of Data Acquisition & Processing. 2012,27(1): 172-174.
[12] Wang Bangfeng, Lin Jianxiang, Lu Jiyun. Detection of Small Mutation Signals in Flight Data Based on EEMD-H T[J]. Journal of Vibration, Measurement & Diagnosis, 2013(03): 388-392.
[13] Peng Y, Sun H, Zu L, et al. Adaptively target tracking method based on double-Kalman filter in existence of outliers[C]. 2010.
[14] Freeman J. Outliers in Statistical Data (3rd edition)[J]. Journal of the Operational Research Society. 2017, 46(8): 1034-1035.
[15] Sun Yongshuai, Wang Shaoyun. The processing method of telemetry data based on letts criterion and wavelet transform[J]. Electronic Design Engineering. 2016, 24(20): 82-84.
[16] Huang N E, Shen Z, Long S R, et al. The empirical mode decomposition and the Hilbert spectrum for nonlinear and non-stationary time series analysis[J]. Proceedings of the Royal Society of London. Series A: Mathematical, Physical and Engineering Sciences. 1998, 454(1971): 903.
[17] Duan W, Huang L. A hybrid EMD-AR model for nonlinear and non-stationary wave forecasting[J]. Journal of Zhejiang University Science A. 2016, 17(2): 115-129.
[18] Ding Y. A novel decompose-ensemble methodology with AIC-ANN approach for crude oil forecasting[J]. Energy. 2018, 154: 328-336.
[19] ZHENG J D, CHEN J S, YANG Y. Modified EEMD algorithm and its applications[J]. Journal of Vibration and Shock 2013(21): 21-26.
[20] Bandt C. Permutation Entropy and Order Patterns in Long Time Series[M]. 2016.

IOP Conf. Series: Journal of Physics: Conf. Series **1267** (2019) 012025 doi:10.1088/1742-6596/1267/1/012025

A Review of the Research on Dialogue Management of Task-Oriented Systems

Yin Jiang Zhao, Yan Ling Li and Min Lin

College of computer science and technology, Inner Mongolia Normal University

liyanling7871397@163.com

Abstract. Scholars pay more and more attention to the spoken dialogue system after the emergence of deep learning technology. The task-based dialogue system has become one of the most important branches in the field of spoken dialogue systems. Dialogue management is the core of the task-based dialogue system, and its research theory and technology have been gained extensive attention. This paper summarizes the research progress and current situation of task-based dialogue system and dialogue management strategy. Firstly, it summarizes the status of the task-based dialogue system models and compares their advantages and disadvantages. Then it focuses on the analysis of various dialogue management research strategies from the perspective of model theory and research methods. Finally, it looks forward to the future research direction of task-oriented dialogue system combined with dialogue management.

1. Introduction

The spoken dialogue system mainly provides some kind of service for people, it is the bridge between human and intelligent machine. In recent years, the rapid development of big data and deep learning has provided great impetus for the scientific research fields such as dialogue system and computer vision. Moreover, it greatly encourages researchers to use deep learning and big data to study the spoken dialogue system from multiple perspectives. With the rise of the mobile Internet, spoken dialogue systems are tightly integrated with mobile products like Siri, Cortana, and Echo. Users can access information and services through the dialog system in a conversational manner.

Dialogue systems are generally divided into two types according to different application purposes: task-based spoken dialogue systems and non-task-based spoken dialogue systems (chattering systems). Non-task-based spoken dialogue systems are considered to be open domain chat systems, which provides gossip or entertainment to the user and does not accomplish a specific task. At present, researches on chattering systems usually include generation methods and retrieve-based methods. The purpose of a task-based dialog system is to help users complete certain tasks, such as booking train tickets, checking movie ticket, etc. It needs to collect data set for a specific domain. At present, the mainstream research methods of Task-based dialogue systems usually include pipeline method and end-to-end method. In addition, Task-oriented dialogue system has become the most prominent component in today's virtual personal assistant, which attracts more and more attention.

This paper has two objectives: (1) It summarizes the model of task-based dialog system and focuses on the latest development of dialog management; (2) It discusses possible research directions. The rest of this paper is arranged as follows: Section 2 introduces the research status of the task-based dialog system model, including pipeline method and end-to-end method; Section 3 introduces several mainstream methods of dialogue management research technology, focusing on reinforcement learning and deep reinforcement learning. It elaborates and analyzes the model theory and research progress, and points out the advantages and disadvantages of each method. In the end, it summarizes and discusses several problems to be further studied and possible future research directions.

2. The research status of a task-based dialogue system model

The task-based spoken dialogue system is an important branch of the spoken dialogue system. This section describes the pipeline method and the end-to-end method.

2.1. Pipeline-based method

The pipeline model contains the following functional modules, as shown in figure 1: Natural Language Understanding (NLU), Dialogue Management (DM) and Natural Language Generation (NLG). First, it needs to understand the semantic information expressed by humans and translate it into an internal representation. Second, it takes a series of actions based on the strategy through internal representations. Third, it translates actions into natural language and feeds them back to the user.

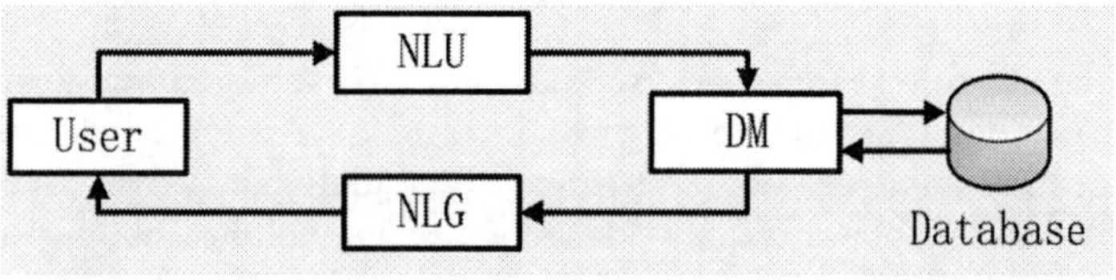

Figure 1. Traditional Pipeline for Task-oriented Systems.

The pipeline model requires that the three modules be executed sequentially and independently of each other. In this way, the pipeline model has two advantages [1]: In each module, the problem is clear and the problems to be solved in each module are detailed. Although the pipeline model has obvious advantages, it also has some problems: (1) The three modules are strongly dependent, and errors in the previous modules may accumulate to the subsequent modules; (2) The change of the output structure of the previous module will affect the design and implementation of the subsequent modules; (3) The design of models is often determined according to specific tasks, so the portability of the domain is poor.

2.2. End-to-end method

The powerful distributed representation ability of neural networks in deep learning makes many natural language processing tasks successful [2, 3]. This makes it possible to build task-oriented end-to-end dialogue systems. The idea of the end-to-end model is to design a whole derivable model, and then transfer the gradient of the output end back to the whole neural network by back propagation. Finally, model parameters are jointly optimized. The overall framework adopts the encoder-decoder structure, as shown in figure 2.

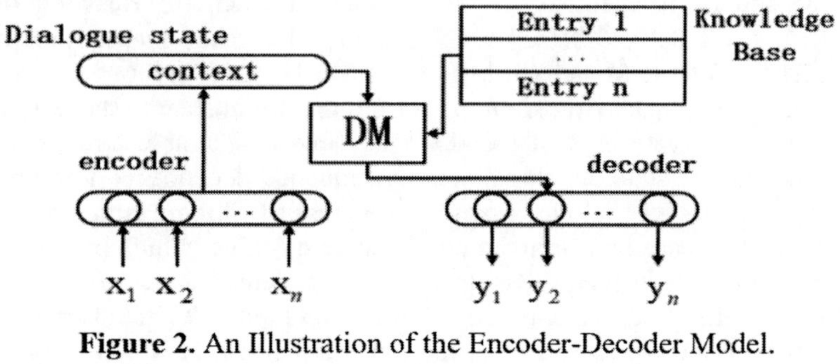

Figure 2. An Illustration of the Encoder-Decoder Model.

Task-based dialogue systems often need to query the external knowledge base. Early models retrieved entries from the knowledge base by issuing a query symbol to the knowledge base retrieve entries from the knowledge base based on attributes [3, 4, 5]. But there were two problems in the end-to-end approach: (1) It does not model uncertainty in user input; (2)The end-to-end model cannot retrieve the knowledge base, because the end-to-end model neural network does not model the query symbol, so the parser and dialogue strategy are trained separately and it is difficult to End-to-end online learning from user feedback. Scholars have done a lot of research to solve the problem of interaction with the external knowledge base. Eric et al. [6] use the replication mechanism to retrieve the knowledge base and simultaneously record entities appearing in the dialogue history. It can accurately retrieve the items in the knowledge base and achieve good results. Later Eric et al. [7] expands the existing recursive network and establishes the key-value retrieval mechanism based on attention in the knowledge base items. The knowledge base is model in the form of triples (object, attribute name, attribute value). However, not all knowledge bases are presented in the form of key-value, so the applicability is not particularly wide. Wen et al. [8] uses internal state representation based on the attention mechanism to complete the retrieval of the knowledge base by combining the advantages of the pipeline model and the end-to-end model. Finally, the replication mechanism is used to extract entities from the knowledge base to generate responses, which significantly improves the retrieval efficiency.

In addition to the above research work, scholars have also studied a task-oriented dialogue system from another perspective. They see dialogue strategies as reinforcement learning strategies. This method can significantly improve the performance of the dialogue system. Li et al. [9] firstly provides a publicly available user simulator framework. Scholars can build different user simulators according to the characteristics of the dialogue task. This greatly promotes the application of reinforcement learning in a task-based dialogue system. Dhingra et al.[10] combines the soft attention retrieval process with reinforcement learning, which can achieve a higher task success rate and higher rewards. Although user simulator can reduce learning costs, there is no uniform standard to evaluate the advantages and disadvantages of user simulator. The effectiveness of training task-based dialog agents by simulating users has always been controversial.

3. Dialogue Management

Dialogue management controls the whole dialogue process, and its design directly affects the performance of the dialogue system. Dialogue management mainly includes two tasks: (1) Dialogue state tracking (DST); (2) Dialogue policy learning (DPL).

- Dialogue state tracking: It determines the current user target based on multiple rounds of conversations between the system and the user. Dialogue state tracking provides the basis for later decisions.
- Dialogue policy selection: It selects an executable action based on the results of the dialogue state tracing. The goal is to complete the task in a short conversation as possible.

3.1. Traditional research methods of dialogue management

Scholars have proposed many methods of dialogue management in early dialogue management study[11,12,13,14,15,16,17,18,19,20,21,22].Among them, the main method includes Finite State Machine (FSM)[11,12], slot-fill Model (SFM)[13] and task-based structure method, etc. The FSM regards a dialogue process as a state transition path from the initial state to the termination state of the state machine. It can effectively control the dialogue flow and simple structure. But it has some problems such as incomplete state coverage, serious sequence dependence and poor portability[14]. The slot-filling method treats the dialog process as a process of filling a table with placeholders. This method extends the method based on finite automatic state machine, which has no strict order and the answer form is very flexible. However, SMF also has the problem of state explosion and cannot satisfy the task of multi-topic dialogue. Task-based structure method uses tree structure to express the element relations in the field, which can meet the multi-topic task [15]. But it is limited by the tree structure and difficult to extend. In addition, there are other dialogue management research methods, such as the dialogue management method based on planning [16], the Information State Update method [17,18], the Agent method [19,20] and the self-organizing dialogue management planning model based on the theme forest structure [21,22]. These methods are widely used because they have the advantages of simple structure and easy implementation. However, these methods have common shortcomings and hinder the development of dialogue system: They cannot guarantee that all cases and dialogue rules will be covered and manual rules will need to be defined. Scholars propose a method based on reinforcement learning to solve the problems of traditional dialogue management.

3.2. Reinforcement learning

The method based on reinforcement learning has always been a hotspot in research of dialogue management. More classic models including Markov Decision Process (MDP) [23, 24] and Partially observables Markov Decision Process(POMDP)[25]. The idea of reinforcement learning is that agents interact with the environment by using their own experience and environmental feedback. It is the process of learning codes of conduct through trial and error. The goal of reinforcement learning is to select the "best" behavior by maximizing the discount reward of the expected value:

$$Reward = E[\sum_{t=0}^{\infty} \gamma^t (s_t, a_t)] \tag{1}$$

Where, s_t is the state at time t, a_t is the action at time t, and γ is the discount factor. The optimal strategy can be obtained according to the maximum expected reward value:

$$\pi^* = \arg\max_{\pi} E\left[\sum_{t=0}^{a} \gamma^t r(s_t, a_t) \mid s_0 = s\right] \tag{2}$$

Reinforcement learning method can effectively solve sequence decision-making problems so that it has the following advantages: (1) It gets rid of the dependence of manually defining rules in the decision-making process and improves the generalization ability; (2) It can fully explore the state space and solve the problem of insufficient coverage. In view of these advantages, reinforcement learning has been widely used in dialogue management, and it has been proved in experiments that POMDP model is superior to MDP model [26, 27, 28, 29, 30, 31, 32, 33]. Thomson et al. [32] used a POMDP model to construct a spoken dialogue system of home care robot, and it was proved that the POMDP model performed better than the MDP model in a certain noise environment. Young and Williams et al. also did a lot of research on dialogue management based on POMDP model [34, 35, 36]. Williams et al.[34] proposed a decomposed POMDP model based on Roy's model. The auto-generated policy of this model is directly compared with the MDP model and manually formulated state machine policy. The experimental results show that the POMDP model performs better than the MDP model in the case of high speech recognition error rate. Later, Williams et al. [35] constructs a tourism navigation system based on the mixed model, which combines the decomposition model in literature [34] with slot filling method. After that, Williams et al. [36] constructed an emotional dialogue system model based on POMDP. It added user emotion items into the decomposition model in literature [34], where user emotion is divided into stressed and unstressed.

Although the reinforcement learning is superior to the traditional method, it also has the following problems: (1) It has the problem of sparse data. The MDP model and POMDP model need a lot of dialogue data to train model parameters. Gasic et al. [27] provides samples to solve this problem by using the dialogue data of user simulator and systems; (2) How to design a reasonable reward function to guide the action selection. Young et al. [31] has adopted the Inverse Reinforcement Learning (IRL) method to learn Reinforcement function from dialogue corpus, but it also needs massive dialogue data. (3) As the state space increases, the computational complexity of the model solution will increase sharply[36]. Choi et al. [33] proposed Hidden information State Model to solve this problem; (4) How to automatically improve the dialogue policy in the interaction process based on the existing dialogue policy. (5) The application of the POMDP dialog management model has limitations, and it cannot handle situations where slots or slot values are infinite.

3.3. Deep reinforcement learning techniques

3.3.1 DQN model. The traditional reinforcement learning algorithm is limited by the action space and sample space. When the state space is large or even infinite, the efficiency of learning will be greatly reduced. Later, with the development of deep learning, Mnih et al. [37] solves this problem by using deep reinforcement learning techniques. In particular, the proposal of DQN model is of ground breaking significance for the development of deep reinforcement learning techniques [38]. Before the appearance of DQN, when the neural network is used to approximate the action value function in RL, it will have the problem of unstable or non-convergence of Q value. DQN uses two techniques to solve this problem: the experience replay mechanism and the target network. The idea of the model is to approximate the value function of action with the function of a deep neural network:

$$Q(s,a) \approx f(s,a,\theta) \tag{3}$$

Where, θ is the parameter of the neural network. State s is used as input to output the Q value of each action.

Experience replay mechanism: It puts the collected samples $e_t = (s_t, a_t, r_t, s_{t+1})$ into the sample pool $D = \{e_1, e_2, ..., e_n\}$. It randomly selects small batches of samples from D each time during training. This method can reduce the correlation between samples.

Target network: It establishes two neural networks with the same structure, denoted as Q network and target value network. The Q network is used to generate the current value, denoted as $Q(s,a,\theta)$. The target value network is used to generate the target Q value, denoted as $Q(s,a,\theta^-)$. The optimization objective of the target value network is:

$$y = r(s,a,s') + \lambda \max_{a'} Q(s',a',\theta^-) \tag{4}$$

Where a' is the possible action, s' is the state at the next moment, and θ_i^- is the parameter of the target network. The loss function is expressed as follows:

$$loss(\theta) = [y - Q(s,a,\theta)^2] \tag{5}$$

In this process, the parameters θ^- in the target network are kept fixed, and then the parameters θ of Q network are updated with the stochastic gradient descent algorithm. The parameter of Q network is copied to the target value network after N rounds of iteration. The training process is shown in figure 3.

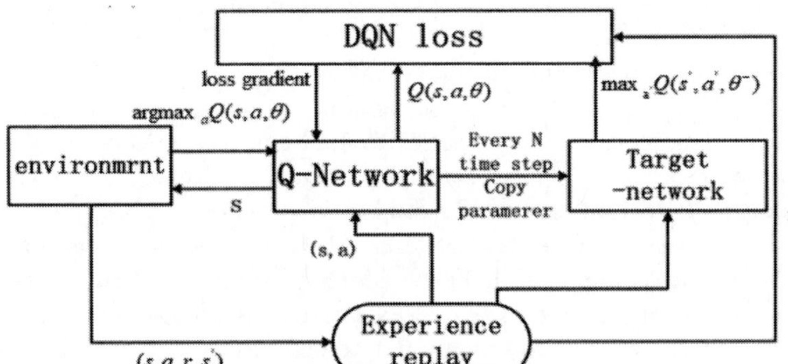

Figure 3. DQN Training Process.

The DQN model provides a new direction for the research techniques of dialogue management: (1) It provides two methods of experience replay mechanism and target network to stabilize the training process.; (2) It designed an end-to-end reinforcement learning method, which can input the state value and output the Q value of each action through the Q value network; (3) It uses the same model parameters, training algorithm and network structure in different tasks, which greatly improves the performance of the model.

Although DQN model improves the reinforcement learning algorithm and the performance of the dialogue system, it has the following disadvantages: (1) It has the problems of too long training time, low sample utilization and its reward function is hard to design; (2) It has the problem of Q network instability because Q network is sensitive to parameters; (3) It has the problem of over fitting the environment and it is difficult to adapt to the new environment.

Deep reinforcement learning technology greatly promotes the development of dialogue management research technology. For the first time, Li et al.[9] provides a publicly available framework of user simulator, which can interact with agents to learn dialogue strategies from samples. Among them, the success rate of DQN-based agents in the task of booking movie tickets is much higher than that of rule-based agents, which further illustrates the development potential of the deep reinforcement learning model in the task-based dialogue system. Volodymyr et al.[39] proposes a BBQN (Bayes-by-backdrop Q-network), which improves the DQN algorithm. It explores state space by using Thompson sampling, which greatly improves exploration efficiency.

3.3.2 Adversarial advantage actor-critic. Scholars are increasingly concerned about the use of deep reinforcement learning in task-based conversational systems [10, 27, 40, 41, 42, 43, 44, 45, 46] when the coming of deep reinforcement learning techniques. Zhao et al. [41] regards the retrieval operation of a knowledge base as a part of dialogue strategy learning through deep reinforcement learning. This method can be applied to any knowledge base and can be further optimized by supervised learning or reinforcement learning. Williams et al. [43] proposes an end-to-end task-oriented neural network model, which can improve training efficiency with only a small amount of effective data and has good mobility. However, this model requires a high-quality data set to achieve good results. Peng et al. [44] propose a hierarchical deep reinforcement learning model, which can learn dialogue management strategies in different time periods and complete multiple subtasks at the same time. Liu et al. [45] simulates the dialogue between two agents, and jointly optimizes the dialogue agent and the user simulator by using deep reinforcement learning. This approach achieves higher task success and rewards than single supervised learning and reinforcement learning.

Although these methods have achieved good results, they all have the problem of sparse reward. Scholars have come up with many different ways to solve this problem. These methods fall into two categories:(1) It learns transcendental knowledge from expert-generated dialogues and preliminary strategies from person-to-person dialogues. Peng et al. [46] and Lipton et al. [47] has proved that the pre-trained supervision strategy or the strategy based on weak rules can significantly improve the efficiency of exploration; (2) It introduces heuristic rules to guide the exploration through internal reward[48,49,50]. Jaderberg et al. [49] converts the training signals of three auxiliary tasks into intrinsic rewards, which can significantly improve the learning speed and robustness of agents. Houthooft et al. [50] added the intrinsic reward value into the reward function, which encouraged agents to explore those relatively unexplored regions. At present, the latest research method in the research field of dialogue management is the A2C model of confrontation learning proposed by Peng et al. [51]. The model combines the above two methods. The idea of the model is to use the dialogue generated by experts as the prior knowledge, but not to use the prior knowledge to construct the dialogue strategy. It trained a discriminator inspired by the antagonistic network (GAN) [52]. This discriminator can distinguish between action generated by agents and those generated by human experts. It uses the output of the discriminator as an intrinsic reward, encouraging agents to explore the state-action space like experts. This model easily leads to successful conversations and brings positive returns compared to other models. The adversarial A2C model consists of two parts, as shown in figure 4: (1) actor-critic framework; (2) adversarial learning discriminator D. Please refer to the reference [51] for the specific implementation details of the model.

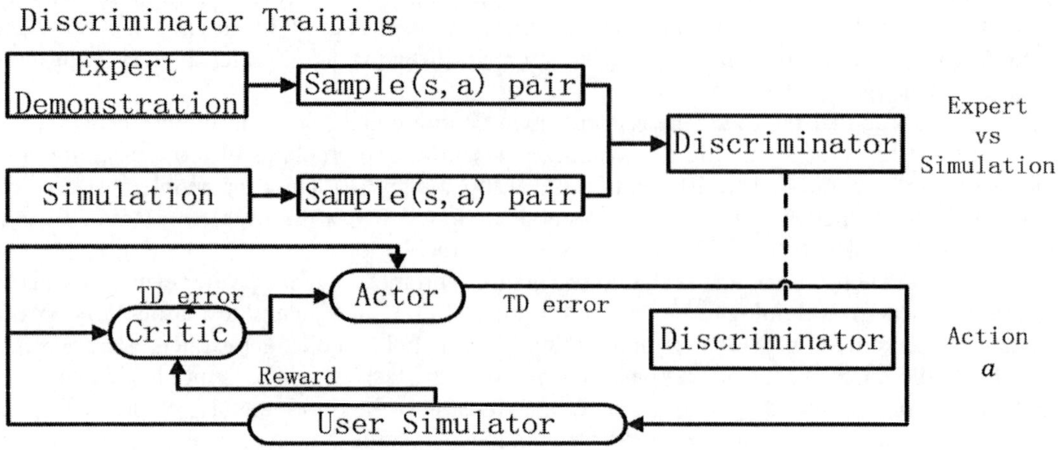

Figure 4. The adversarial advantage Actor-Critic.

4. Summary and Discussion

This paper mainly discusses the research progress of dialogue management of task-based dialogue system, and mainly introduces three kinds of dialogue management methods. At present, the end-to-end model has become a mainstream model of a task-oriented dialogue system, and the deep learning technology has become a basic technology to solve the problem of the dialogue system. In recent years, deep reinforcement learning technology has become one of the key technologies of dialogue management. Deep reinforcement learning technology can greatly improve the task success rate of end-to-end task-based dialog system. At present, there are still the following problems in the task-based dialogue system:

•The end-to-end dialogue system model does not model the external knowledge base, which affects the performance of dialogue management using the external knowledge base. How to build a knowledge base suitable for a neural network to improve the performance of dialog management is a question worth exploring.

•The dialogue management based on deep reinforcement learning requires user simulators. But there is a big gap between user simulators and real users. How to improve the user simulator based on the neural network to have more human characteristics and how to find a unified standard to evaluate the user simulator needs further attention and exploration.

•There is a problem of poor domain portability in developing dialogue systems for different domains. How to use the transfer learning to transfer the original pre-trained dialogue management model to the new dialogue system management tasks , that is also an area worth exploring.

•Data set is an important factor affecting the performance of task-based dialogue system. However, at present, the data set of the task-oriented dialogue system is scarce, which affects the development of the dialogue system. How to build and collect conversational data sets with good universality and high dialogue quality is an urgent problem to be solved.

5. Acknowledgments

This work is partially supported by the National Natural Science Foundation of China(Nos. 61562068, 11704229, 61640204, 61806103); Natural Science Foundation of Inner Mongolia (No. 2017MS0607); Mongolian Informatization Special Support Sub-project of Inner Mongolia Nationalities Committee (MW-2014-MGYWXXH-01); Youth Innovation and entrepreneurship talents of Inner Mongolia "grassland talents" project; Inner Mongolia science and technology planning project (research and development operation of online interactive learning platform of Mongolian language MOOC and analysis and research of national education big data); Graduate Innovation Fund of Inner Mongolia Normal University (CXJJS18112).

References

[1] Xu,W.Hoffmann,R.,Zhao,L.,Grishman, R.:Filling knowledge base gaps for distant supervision of relation extraction. In:Proceedings of the 51st Annual Meeting of the Association for Computational Linguistics (Volume 2: Short Papers).vol.2, pp. 665–670 (2013).
[2] Bordes, Antoine, Y.L.Boureau,andJ.Weston."Learning End-to-End Goal-Oriented dialogue" (2016)
[3] Wen TH, Vandyke D,Mrksic N,et al. A Network-based End-to-End Trainable Task-Oriented dialogue System[J]. 2016.
[4] Li X, Chen Y N, Li L, et al. End-to-End Task-Completion Neural dialogue Systems[J]. 2017.
[5] J. D. Williams and G. Zweig.End-to-end lstm-based dialogue control optimized with supervised and reinforcement learning. arXiv preprint arXiv:1606. 01269.
[6] Eric M, Manning C D. A Copy-Augmented Sequence-to-Sequence Architecture Gives Good Performance on Task-Oriented dialogue[J]. 2017.
[7] Eric M, Manning C D. Key-Value Retrieval Networks for Task-Oriented dialogue[J]. 2017.
[8] Wen H, Liu Y, CheW, et al. Sequence-to-Sequence Learning for Task-oriented dialogue with dialogue State Representation[J]. 2018.
[9] Xiujun Li, Zachary C. Lipton, Bhuwan Dhingra, Lihong Li, Jianfeng Gao, Yun-Nung Chen. 2016. A User Simulator for Task-Completion dialogues.
[10] Dhingra B, Li L, Li X,et al. Towards End-to-End Reinforcement Learning of dialogue Agents for Information Access[J]. 2016.
[11] Branting K,Lester J, Mott B.dialogue Management for Conversational Case-Based Reasoning [C]//Advances in Case-based Reasoning, European Conference, Eccbr, Madrid, Spain, August 30 September. DBLP, 2004.
[12] McTear M F. Modeling spoken dialogues with state transition diagrams: experiences with the CSLU tookit[C]//Pro 5th International Conference on Spoken Language Processing. 1998:5-7
[13] Yuan Yan, tian Huaifeng, du bo, et al. Research and implementation of frame-based dialogue management model [J]. Computer engineering, 2005, 31(13):212-214.
[14] Abe K, Kurokawa K, Taketa K, etal.A new method for dialogue management in an intelligent system for information retrieval[C]//Proc of the 6th international conference on spoken language processing(ICSLP). Beijing, China, 2000:118-121
[15] Rosenblum DS ,Wolf AL. A design framework for Internet-scale event observation and notification[J]. ACM SIGSOFT Software Engineering Notes, 1997, 22(6):344-360.
[16] Bohus D, Rudnicky A I. The RavenClaw dialogue management framework: Architecture and systems[J]. Computer Speech &Language , 2009 , 23 (3):332 -361
[17] Bos J, Klein E, Lemon O, et al. DIPPER: Description and formalisation of an information-state update dialogue system architecture [C]//Proc of 4th SIGdial Workshop on Discourse and dialogue. 2003: 115 - 124
[18] Larsson S, Traum D R. Information state and dialogue management in the TRINDI dialogue move engine toolkit [J]. Natural language engineering, 2000, 6:323-340
[19] Nguyen A. An Agent-Based Approach to dialogue Management in Personal Assistants[C]// International Conference on Intelligent User Interfaces. DBLP, 2005.
[20] Xu kaihua, zhang degan, yao Lin. Design and implementation of agent-based human-machine dialogue system [J]. Computer engineering, 2007, 33(16).
[21] Wu X, Zheng F, Xu M. Topic Forest: a plan-based dialogue management structure[C]//IEEE International Conference on Acoustics. 2001.
[22] Wu xiaojun, zheng fang, xu mingxing. A dialogue management model based on thematic forest structure [J]. Acta automatica sinica, 2003, 29(2):275-283.

[23] Levin E,Pieraccini R, Eckert W. Using Markov decision process for learning dialogue strategies. [C]// IEEE International Conference on Acoustics. IEEE, 2002.
[24] Howard, R. A. Dynamic programming and markov processes[J]. 1960.
[25] Cassandra A R. Exact and approximate algorithms for partially observable Markov decision processes[M]. Brown University, 1998.
[26] HENDERSON M S. Discriminative methods for statistical spoken dialogue systems[D]. [S.I]: University of Cambridge, 2015.
[27] Young S, Gasic M, Thomson B, et al. POMDP-Based Statistical Spoken dialogue Systems:A Review[J]. Proceedings of the IEEE, 2013, 101(5):1160-1179.
[28] WILLIAMS J D. A case study of applying decision theory in the real world :POMDPS and spoken dialogue systems[J].Decision Theory Models for Applications in Artificial Intelligence: Concepts and Solutions. 2010:315-342
[29] YASAVURU. Statistical dialogue management for health interventions[D]. [S. I.]:FLORIDA INTERNATIONAL UNVERSITY, 2014.
[30] Roy N. Spoken dialogue Management Using Probabilistic Reasoning[J]. Acl, 2000.
[31] Choi J,Kim K E. Inverse Reinforcement Learning in Partially Observabl Environments[C] //International Jont Conference on Artifical Intelligence. Morgan Kaufmann Publishers Inc. 2009.
[32] Thomson B, Young S. Bayesian update of dialogue state:A POMDP framework for spokken dialogue systems[J]. Computer Speech and Language, 2010, 24(4):562-588.
[33] Young S, Ga?I? M, Keizer S, et al. The Hidden Information State model: A practical framework for POMDP-based spoken dialogue management[J]. Computer Speech and Language, 2010, 24(2):150-174.
[34] Williams J D,Young S. Partially observable Markov decision processes for spoken dialogue systems[J]. Computer Speech and Language, 2007, 21(2):393-422.
[35] Williams J D, Young S. Scaling POMDPs for Spoken dialogue Management[J]. IEEE Transactions on Audio, Speech and Language Processing, 2007, 15(7):2116-2129.
[36] Williams, Jason D. Challenges and Opportunities for State Tracking in Statistical Spoken dialogue Systems:Results From Two Public Deployments[J]. IEEE Journal of Selected Topics in Signal Processing, 2012, 6(8):959-970.
[37] Mnih V, Kavukcuoglu K, Silver D, et al. Playing Atari with Deep Reinforcement Learning [J]. Computer Science, 2013.
[38] Human-level control through deep reinforcement learning[J].Nature,2015, 518(7540):529-533.
[39] Lipton Z C, Li X, Gao J, et al. BBQ-Networks:Efficient Exploration in Deep Reinforcement Learning for Task-Oriented dialogue Systems[J]. 2016.
[40] Fatemi M, Asri L E, Schulz H, et al. Policy Networks with Two-Stage Training for dialogue Systems[J]. 2016.
[41] Zhao T,Eskenazi M.Towards End-to-End Learning for dialogue State Tracking and Management using Deep Reinforcement Learning[J]. 2016.
[42] Su P H, Gasic M, Mrksic N, et al. Continuously Learning Neural dialogue Management[J]. 2016.
[43] Williams J D, Asadi K, Zweig G. Hybrid Code Networks:practical and efficient end-to-end dialogue control with supervised and reinforcement learning[J]. 2017.
[44] Peng B,Li X,Li L ,et al. Composite Task-Completion dialogue Policy Learning via Hierarchical Deep Reinforcement Learning[J]. 2017.
[45] Liu B, Lane I. Iterative Policy Learning in End-to-End Trainable Task-Oriented Neural dialogue Models[J]. 2017.
[46] Baolin Peng, Xiujun Li, Jianfeng Gao, Jingjing Liu, and KamFai Wong, "Integrating planning for task-completion dialogue policy learning, "arXiv preprint arXiv:1801. 06176, 2018. .
[47] Lipton Z C,Li X,Gao J,et al.BBQ-Networks:Efficient Exploration in Deep Reinforcement Learning for Task-Oriented Dialogue Systems[J]. 2016.

[48] Singh S, Lewis R L, Barto A G, et al. Intrinsically Motivated Reinforcement Learning: An Evolutionary Perspective[J].IEEE Transactions on Autonomous Mental Development, 2010, 2(2):70-82.
[49] Jaderberg M,Mnih V,Czarnecki W M ,et al. Reinforcemnt Learning with Unsupervised Auxiliary Tasks[J]. 2016.
[50] Houthooft R, Xi C, an D, et al. Variational Information Maximizing Exploration[J]. 2016.
[51] Peng B,Li X,Gao J,etal. Adversarial Advantage Actor-Critic Model for Task-Completion dialogue Policy Learning[J].2018.
[52] Goodfellow IJ, Pouget-Abadie J, Mirza M, et al. Generative adversarial nets[C]// International Conference on Neural Information Processing Systems. 2014.

IOP Conf. Series: Journal of Physics: Conf. Series **1267** (2019) 012026 doi:10.1088/1742-6596/1267/1/012026

A Method for Selecting Decision Center of Heterogeneous Operational Networks Based on Operational Capability Analysis

Kebin Chen[1], Yunjun Lu[1*], Yiqiao Jin[1] and Mengyao Han[1,2]

[1]College of Information and Communication, National University of Defense Technology, Wuhan 430019, China

[2]Department of Defense Economics, Army Logistical University of PLA, Chongqing 400030, China

*lu_yunjun@hotmail.com

Abstract. In order to improve the operational capability of heterogeneous combat networks (HCN), a decision center (DC) selecting method is proposed in this paper. According to the characteristics of modern information warfare, the entities in HCN are divided into influencer entities, decider entities, sensor entities and DC entity. The optimal distributed set of information flow links is obtained based on genetic algorithm. In light of the degree of each entity in this set, the entity with highest degree is selected as DC. Since DC has higher decision capability, highest degree means that the operational capability of majority operational chain is enhanced. And thus, the overall HCN operational capability is improved.

1. Introduction

With the development of information technology, the operational entities of different regions can participate in the battle at the same time, which makes the combat system-of-systems more and more complex [1]. Complex networks are important tools for portraying complex systems, which can help us understand the system deeply and then guide the construction and improvement of the system [2]. Different from the general complex network, the entities in the combat network have different functions and task division, so the combat network is a heterogeneous network, and the nodes are heterogeneous [3]. Researchers focus on heterogeneous combat network (HCN) in recent years, different researcher has proposed different entity classification methods, but decider entities always appeared in different methods [4-6]. However, they did not further divide the decider nodes. In modern warfare, there must be a decision center (DC) entity among decider nodes. DC is a must to ensure the unified coordination of the joint forces. Moreover, as the amount of information in the war increases, only the DC can effectively process massive data. Therefore, according to HCN, the decider entities are divided into DC and ordinary decider entities in this work, which helps to further reveal the rules of modern warfare.

One of the research purposes of HCN is to improve the operational capabilities of the network. The measures of combat capabilities include the network connectivity measures, spectrum measures, and operational capability index (OCI) measures. The OCI reflects the heterogeneity of nodes and links and is suitable for HCN [7]. OCI is based on the operational chain which is originated from the OODA theory [8]. In this paper, to improve HCN operational capability, genetic algorithm is used to select DC among the decider entities.

2. HCN combat capability analysis

HCN can be expressed as $G=(V,E)$, where V is a node set, E is a link set. The node set $V=(v_1,\cdots,v_i,\cdots,v_n)$ represents a set of entities in the HCN. Link set $E=(e_1,\cdots,e_i,\cdots,e_m)$ represents an information flow between two entities in HCN. Li [7] divided V for three categories: sensor nodes S, decider nodes D, and influencer nodes I, as shown in Figure 1.

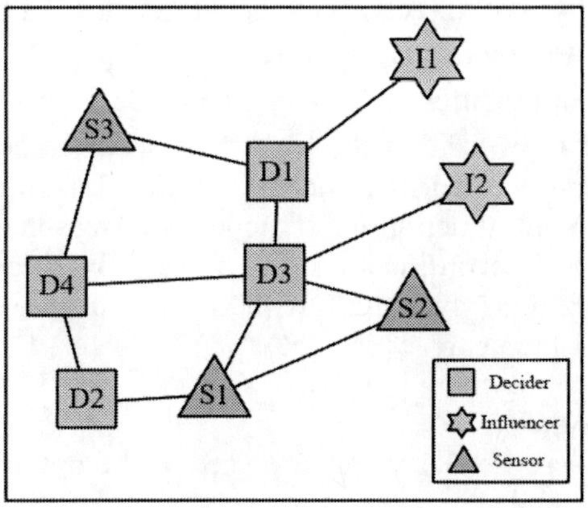

Figure 1. Diagrammatic sketch of HCN

On the basis of Figure 1, the entities are divided into four categories in this work, namely, the sensor entities S, the DC D_c, the decider entities D and the influencer entities I. Because of the obvious differences in the physical forms of sensor, decider and influencer, the distinction of S, D and I is apparent. However, to figure out D_c from D does not depend entirely on the entity itself, but also on the location of the entity, the battlefield environment, and other entities around it. The main work of this paper is to make reasonable calculations on the decider entities based on Figure 1, and further divide the decider nodes into D_c and D. The OCI measures can reflect the characteristics of HCN effectively. The method to calculates the operational capabilities of one operational chain, and then sum multiple OCs to obtain the operational capabilities of the entire network. A single operational chain operational capability $C(l_j)$ is defined as [7]:

$$C(l_j)=\frac{1}{|l_j|}\times\max Cs(s_j)\times\max Cd(d_j)\times\max Ci(i_j) \tag{1}$$

Where $|l_j|$ is the length of OC, $Cs(s_j)$ is the sensor capability, $Cd(d_j)$ is the decider capability, and $Ci(i_j)$ is the influencer capability. Then the operational capability of HCN can be expressed as:

$$H=\sum_{lj\in LG} C(l_j) \tag{2}$$

In above definition, the decider capability of the decieder entities is an important part of the overall network's combat capability. The greater the decider capability in the operational chain is, the stronger the operational capability of the entire network. Obviously, the DC has stronger d decider capability than other common decider entities. Therefore, the DC must be the convergence node of the information link. And then more operational pass through the DC. That means $\max Cd(d_j)$ in (1) is easy to obtain the maximum value, and the whole network has the best operational capability.

3. HCN decision center selection algorithm

In order to improve the operational capability of the HCN, it is necessary to ensure that the location of the DC is the convergence node of the information flow. This section describes the method to obtain the location.

3.1. Information flow aggregation analysis

The information flow in HCN is not arbitrary. The link with suitable condition is suitable for more information to flow. In addition, in HCN, according to the OODA theory, the information flow in decider entities, sensor entities and influencer entities are not homogeneous. for example, for the wisdom of the group, the information flow between decider entities will be frequent; The influencer entities are usually the end node of the information flow and is only connected to the decider entities; the sensor entities are usually not directly connected to the influencer entities, and they can communicate the information to other sensor entities or multiple decider entities. Therefore, for S, D, I, D usually have more information links, and the number of connected information links for S is medium and the number of links for I is the least. Moreover, in certain warfare, different pair of entities have different attraction coefficients, and information links are more likely to be generated between the two nodes with high attraction coefficients. The reasonable distribution of information links in HCN could be obtained according to above factor. By calculating the degree of different decider entities in this distribution, the entity with highest degree indicate that it has the most information flow, and is more suitable to act as a DC.

3.2. Objective function construction

Considering three factors discussed in 3.1 which are used as the optimization targets of the HCN, the objective function can be expressed as:

$$F=\sum_{j}^{E} e_j \cdot r_j \cdot w_j \cdot \sqrt{k_j^{start} \cdot k_j^{end}} \tag{3}$$

where e_j is the information link in HCN and its value is 0 or 1. 1 indicates that the link exists in the network, 0 indicates that the link does not exist. r_j is the stability factor of the link. w_j is the attraction coefficient between the two entities of the link. k_j^{start} and k_j^{end} are information convergence factor for two endpoints of the link e_j respectively. When the endpoint is the decider entity, the k_j value is the largest. And when the node is influencer entity, the value k_j is the smallest. Therefore, if the number of information flow links is constant, when the value of (3) is the largest, the information flow is most reasonable in the HCN. The entity with the largest degree is more suitable to serve as the DC. Since the operational capability of OC generally increased, the operational capability of HCN has also been improved.

3.3. Algorithm implementation

As discussed in 3.2, the key step in this paper is to use (3) as the objective function to find the distribution of information flow links for each entity. This work is similar to the network topology optimization problem. And genetic algorithm (GA) is suitable for solving those problems. GA is a non-deterministic intelligent search algorithm that simulates the natural selection of biological evolution. Different organisms have different genes, so their ability to adapt to the outside world is also different. Individuals with strong adaptability are usually able to survive and reproduce through natural selection. The gene may be mutated during its transmission to the offspring and can enhance or reduce the genetic adaptability. According to the above theory, the GA encodes the solution of the problem space. Each solution represents a chromosome. Through the selection and crossover of chromosome genes, fitness function values are used as targets for natural selection.

In HCN, the chromosome of the GA is a possible distribution of information flow links, which is encoded by common binary coding. The selection operator uses the tournament operator, which is more sensitive to the difference of the fitness values of the children. In a group of individuals, the optimal individual can always be selected, which is suitable for the optimization problem under multi-constraint conditions. The crossover operator uses the single-point interchange cross method. The method steps are: selecting two parent chromosomes, randomly selecting the position with a value of 1 of one parent, and judging whether the same position gene in the other parent is 0, and if so, exchange two genes. Otherwise, re-select the position. The mutation operator uses the single point variation method, that is, selecting one 1 and one 0 in a certain chromosome, and changing the position of these two genes.

3.4. Calculation process

The process of determining the DC of the HCN is shown in Figure 2. First, according to the characteristic of HCN, especially heterogeneous entities, determine the stability factor r_j of each information link, the attraction coefficient w_j between entities and the information convergence factor k_j of entities. Second, based on genetic algorithm, using (3) as the objective function, the optimal distribution set of the information flow links is calculated. Third, calculating the degree of each decider entity according to the optimal distribution set of the information flow link, and the decider entities are ranked by according to the degree. Fourth, the decider entity with highest degree indicates that more information flows through it, and it is more suitable to act as a DC.

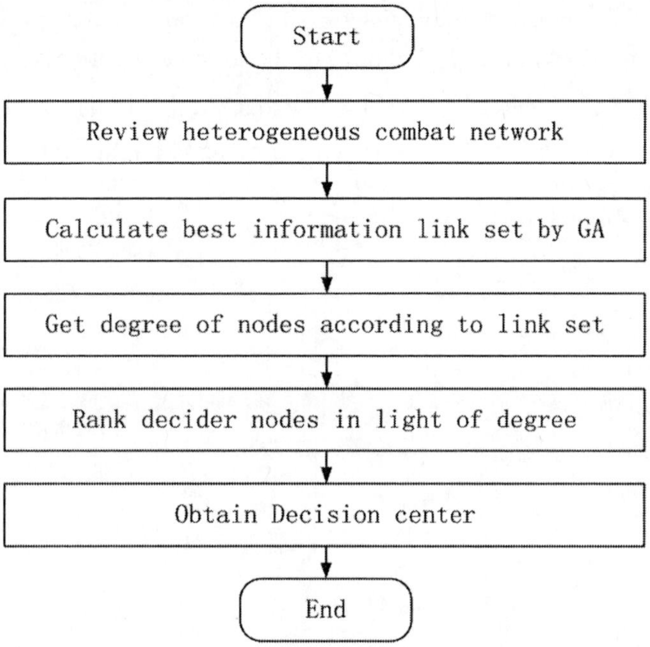

Figure 2. DC selection process

4. Numerical examples and simulation

Taking the HCN in Figure 1 as an example, one decision center node is determined from four decider entities. The position of the node in Figure 1 reflects the distance relationship of different nodes. According to the general rule of communication technology, the closer the communication link of the node is, the more stable the information flow link stability factor r_j is. In addition, the information aggregation factors k_j of three types of nodes S, D, I are 5, 10, and 1, respectively. The majority of the attraction coefficients between the nodes in Figure 1 are 1. However, the attraction coefficient between D4 and S3, D1 and I1 is 5.

The convergence of the GA used in our method are proved in Figure 3. The GA runs in the situation of 10, 15, 20, and 25 information flow links respectively, and the fitness change of the objective function of (3) is obtained. It can be seen from Figure 3 that the GA used in this work has good convergence under different link numbers, indicating that the obtained information flow link distribution set is the optimal one.

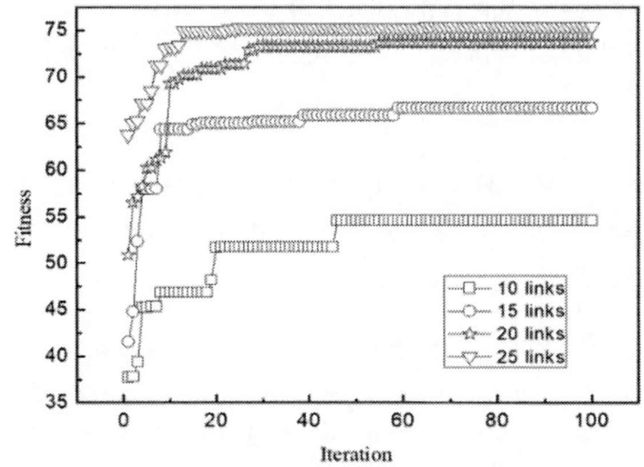

Figure 3. Fitness curve under different information flow links

Figure 4 shows the optimal distributed set of information flow links of Figure 1 under different number of information flow links. Since the k_j of the influencer entity is small, no matter how many the information flow link increases, only a few links pass through the two entities. The k_j of decider entity is the largest, so the information flow link is more likely to connect to the decider entity. In addition, because the decider entity is located at the geographic center, there are more links connected to it. The links of D1 and I1, D4 and S3 appear in all four figures due to their large attraction coefficient.

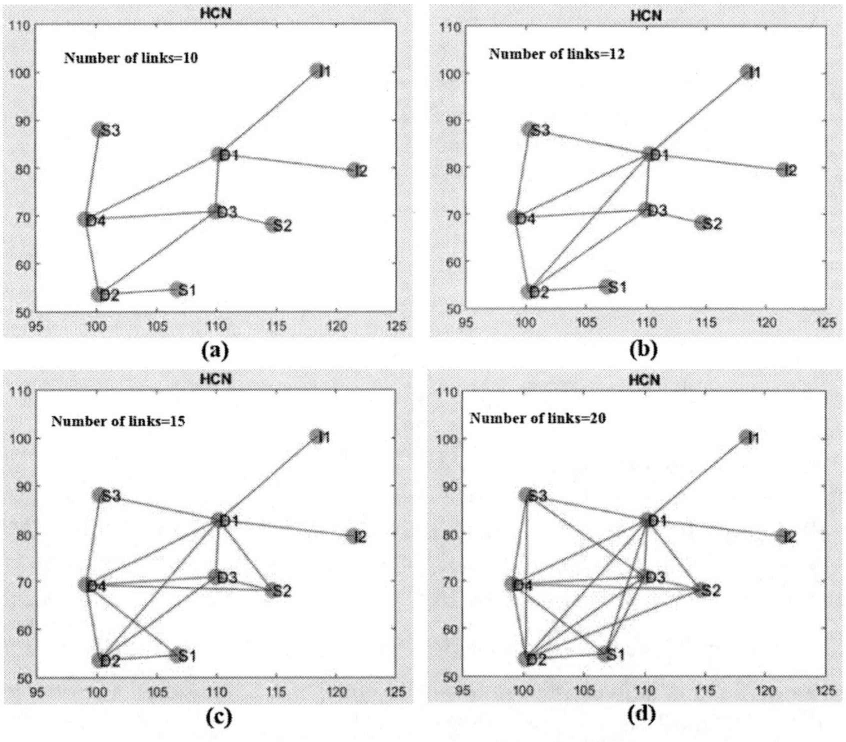

Figure 4. The optimal distributed set of information flow links

Table 1 records the degree of each entity in Figure 4. It can be seen that the degree of the four decider entities is significantly larger than others, and the degree of the D1 is the largest, indicating that the D1 has more information flowing, which is most suitable to be the DC. Figure 5 is a diagrammatic sketch of HCN after the D1 is selected as a DC entity.

Table 1. Degree of each node in a heterogeneous operational network.

Number of links	S1	S2	S3	D1	D2	D3	D4	I1	I2
10	1	1	1	4	3	4	4	1	1
12	1	1	2	6	4	4	4	1	1
15	2	3	2	7	4	4	6	1	1
20	4	4	4	8	6	6	6	1	1

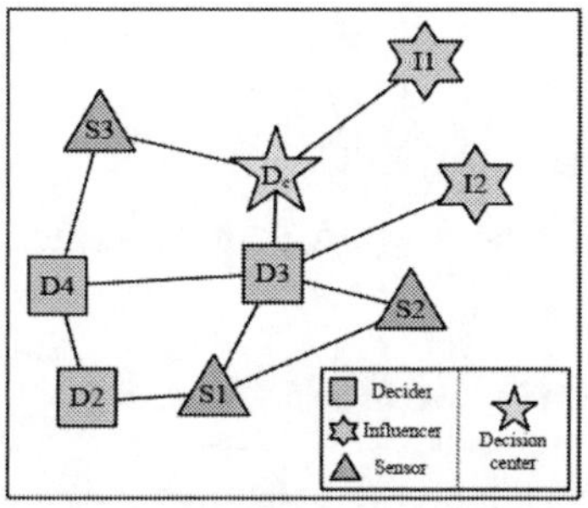

Figure 5. Diagrammatic sketch of improved HCN

According to the topology of Figure 5, the people and devices with strong decision-making capability are concentrated in the DC entity, so that the decision-making capability of the DC is higher than that of other decider entities. Since more OCs pass through DC, according to (1), the operational capability of overall HCN is improved finally.

5. Conclusions

In order to improve the operational capability, this paper presents a new method for partitioning HCN. Based on the influencer entities, decider entities and sensor entities, the DC is added. By using the genetic algorithm, the information flow of each entity is obtained by calculating the distribution of the optimal information flow links. The decider entity with the largest information flow is selected as DC. The selection method of the DC ensures that more operational chain pass through the DC, thereby improving the operational capability of overall HCN.

References

[1] Z. R. Bogdanowicz and K. Patel, Quick collateral damage estimation based on weapons assigned to targets[J]. IEEE Trans. Syst., Man, Cybern., Syst., vol. 45, no. 5, 2015: 762–769.
[2] M. Q. Pasta, F. Zaidi. Topology of Complex Networks and Performance Limitations of Community Detection Algorithms[J]. IEEE Access, 2017: 10901-10914.
[3] Q. Zhao, S. Li, Y. Dou, X. Wang, An approach for weapon system-of-systems scheme generation based on a supernetwork granular analysis[J]. IEEE Syst. J. (2015) 1–12. published online.
[4] A. H. Dekker, Measuring the agility of networked military forces. Journal of Battlefeild Technology[J]. J. Battlefield Technol., vol. 9, no. 1, 2006:19–24.
[5] J. Li, B. Ge, K. Yang, et al. Meta-path based heterogeneous combat network link prediction[J]. Physica A-statistical Mechanics and Its Applications, 2017: 507-523.
[6] J. Li, J. Jiang, K. Yang, et al. Research on Functional Robustness of Heterogeneous Combat Networks[J]. IEEE Systems Journal, 2018: 1-9.
[7] J. Li, D. Zhao, B. Ge, J. Jiang and K. Yang. Disintegration of Operational Capability of Heterogeneous Combat Networks Under Incomplete Information[J]. in IEEE Transactions on Systems, Man, and Cybernetics: Systems. to be published. doi: 10.1109/TSMC.2018.2867532
[8] M. Zhang, L. Ma, and M. Ji. Networked system-of-systems combat OODA command circulation time measuring modeling and experiment[J]. J. Command Control, vol. 1, no. 1, 2015: 50–55.

An Improved K-Means Algorithm Based on Kurtosis Test

Tingxuan Wang* **and Junyao Gao**

Beijing Institute of Technology, 5 South Zhongguancun Street, Haidian District, Beijing, China.

*460336022@qq.com

Abstract. Clustering is a process of classifying data into different classes and has become an important tool in data mining. Among many clustering algorithms, the K-means clustering algorithm is widely used because of its simplicity and high efficiency. However, the traditional K-means algorithm can only find spherical clusters, and is also susceptible to noise points and isolated points, which makes the clustering results affected. To solve these problems, this paper proposes an improved K-means algorithm based on kurtosis test. The improved algorithm can improve the adaptability of clustering algorithm to complex shape datasets while reducing the impact of outlier data on clustering results, so that the algorithm results can be more accurate. The method used in our study is known as kurtosis test and Monte Carlo method. We validate our theoretical results in experiments on a variety of datasets. The experimental results show that the proposed algorithm has larger external indicators of clustering performance metrics, which means that the accuracy of clustering results is significantly improved.

1. Introduction

Data mining is a hot research direction nowadays. It is the process of mining valuable information and knowledge from a large number of irregular data [1]. Now, clustering is an important exploratory analysis method in data mining and a statistical analysis method for classification problems. It is a method that divides some physical or abstract objects into several clusters, which satisfies the conditions that the similarity of objects in the same cluster is high, while the similarity of objects in the different clusters is low.

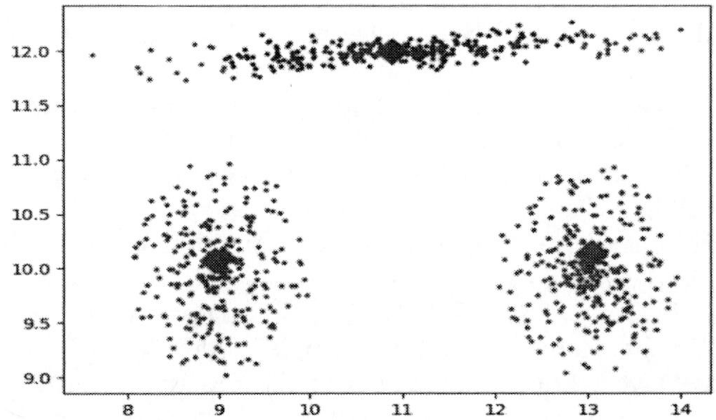

Figure 1. Limitations of the traditional K-means algorithm.

The K-means algorithm is one of the most classical clustering algorithms. This algorithm places the data to be clustered in the European space and considers that each cluster is composed of the data which is close to each other. The algorithm is widely used because of its simplicity, ease to be understood, high efficiency, and its applicability to large datasets [2]. However, the K-means algorithm also has some defects: for example, it is susceptible to noise points and isolated points. At the same time, the Euclidean distance characteristic determines that the K-means algorithm can only be applied to spherical datasets with similar size and density. Figure 1 shows the limitations of the traditional K-means algorithm.

Aiming at the limitation of the traditional K-means algorithm, researchers have made some attempts [3]. In the literature [4], the author uses the I-divergence measure and the Max Entropy measure as similarity measure index, which makes the algorithm suitable for datasets of various complex and irregular shapes. In the literature [5], the author proposes an improved outlier detection method in the iteration process of updating cluster centers, which improves the accuracy of clustering. In the literature [6], based on the original K-means, the author gives a weight to each dimension of the eigenvector. In this way, for the dimension of noise, its weight should be very small, so that the overall cost function value will be lowest. To sum up, in view of the limitations of the traditional K-means algorithm, domestic and foreign scholars have made a lot of improvement studies and achieved some remarkable results. However, the most obvious disadvantage of the current research is that most of the improved algorithms only focus on the improvement of one aspect of the traditional K-means algorithm, which results in the improvement effect of K-means accuracy rate is not obvious. In the process of reading relevant literature, we learn that data kurtosis is a measure of outlier degree of outlier data [7]. The larger the outlier degree of outlier data is, the larger the kurtosis value of the whole data is. We decided to apply this conclusion to the improvement of the K-means algorithm.

In this paper, we propose an improved K-means algorithm based on kurtosis test. This method performs kurtosis test on the traditional K-means running results, and finds the "abnormal data" that are mistakenly divided into each cluster. Then, the kurtosis calculation is used to find suitable clusters for these data. For those data which can't find the appropriate cluster, we consider them to be outlier data in the dataset. In theory, the method can take into account both the shape of the dataset and the influence of the outlier data on the clustering results, so that the accuracy of the algorithm can be improved. The comparison experiments show that compared with the results in other literatures, the external indicators of clustering performance metrics of the new algorithm are improved. The proposed algorithm is advanced and innovative. The results of this paper can be applied to improve the clustering effect and classification accuracy.

The rest of this paper is organized as follows: Second 2 introduces the principle and method of normality test based on kurtosis, Monte Carlo simulation experiment and the Kurtosis-Kmeans algorithm flow. In Second 3, the experimental results are presented, which show the superiority of the new algorithm. Second 4 is the summary and conclusion.

2. Methods

According to the literature [7], data kurtosis is a measure of outlier degree of outlier data. We assume that when the number of samples is large enough, the samples in each cluster obtained by clustering algorithm are approximately normal distribution, which is also in line with our objective cognition. In this way, we can apply statistical normality test based on kurtosis to K-means clustering algorithm [8]. From Section I, we know that for complex and irregular shape datasets, the traditional K-means algorithm does not work well, as shown by dividing many data originally belonging to another cluster into the wrong cluster according to the Euclidean distance. These error-divided data are abnormal for the cluster, like the outlier data, which will cause the sudden change of the kurtosis value of the cluster. According to this character, this paper proposes an improved K-means algorithm, which can detect these "abnormal data" by using kurtosis value, and then classify these data correctly and find out the true outlier data among them, until we think that all the clusters obey the multi-dimensional normal distribution.

IOP Conf. Series: Journal of Physics: Conf. Series **1267** (2019) 012027 doi:10.1088/1742-6596/1267/1/012027

2.1. Multi-dimensional normality test based on M-type multi-dimensional kurtosis

The multi-dimensional normality test of the sample is an important basis for determining whether the algorithm still needs to be iterated. However, there is no uniform definition of multi-dimensional kurtosis. At present, many statisticians have given the generalization of one-dimensional kurtosis in multi-dimensional cases. This paper intends to adopt the definition of multi-dimensional kurtosis proposed by statistician Mardia. In 1970, Mardia first gave a generalization of one-dimensional kurtosis in multi-dimensional cases, called M-type multi-dimensional kurtosis, and constructed multi-dimensional normality test statistics and their asymptotic distribution.

Let **X** denote a set of n points in R^p. Its mean $E(X)=\mu$, and its covariance matrix $\Sigma=E\left[(X-\mu)(X-\mu)^T\right]$ is non-singular matrix. We define equation (1) and equation (2) as the sample mean and sample covariance matrix respectively.

$$\bar{\chi}=\frac{1}{n}\sum_{i=1}^{n}\chi_i \tag{1}$$

$$S=\frac{1}{n}\sum_{i=1}^{n}\left(\chi_i-\bar{\chi}\right)\left(\chi_i-\bar{\chi}\right)^T \tag{2}$$

After deducing the formula, Mardia obtains the measure of M-type multidimensional kurtosis β_ρ, as shown in equation (3).

$$\beta_\rho=E\left[(X-\mu)^T\Sigma^{-1}(X-\mu)\right]^2 \tag{3}$$

The corresponding M-type multi-dimensional sample kurtosis is shown in equation (4).

$$b_\rho=\frac{1}{n}\sum_{i=1}^{n}\left[\left(\chi_i-\bar{\chi}\right)^T S^{-1}\left(\chi_i-\bar{\chi}\right)\right]^2 \tag{4}$$

Mardia not only gives the definition of multidimensional kurtosis, but also obtains the asymptotic distribution of related statistics in multi-dimensional normal case, as shown in equation (5).

$$M_2=\frac{b_p-\rho(\rho+2)}{\left[8\rho(\rho+2)n^{-1}\right]^{1/2}}\xrightarrow{D}N(0,1)(n\to\infty) \tag{5}$$

According to the multidimensional kurtosis definition and related statistics proposed by Mardia, we can get the method of multidimensional normality test: take the significance level α, compare the statistics M_2 and the upper α quantile of the corresponding distribution, when the former is larger than the latter, reject the hypothesis: $H_0: X\sim N_\rho(\mu,\Sigma)$. In addition, we can also test the hypothesis: $H_0:\beta_\rho=\rho(\rho+2)$ directly. The decision to accept or reject the hypothesis is made by calculating b_ρ and comparing it with the significance level quantile of the corresponding empirical distribution.

This paper intends to use the method of using the significance level quantile of the empirical distribution as the test method. The advantage of this method is that it can be used in the case of large samples and small samples, while the method of using the asymptotic distribution of statistics can only be used in the case of large samples. Obviously, the former is more suitable for the application of clustering algorithm.

2.2. Monte Carlo method

The Monte Carlo method is a very important numerical method. It is based on a probabilistic model, and the results of the simulation experiments are used as approximate solutions to the problem according to the process described by the model [9]. In this paper, for the purpose of obtaining the upper α quantile of the b_ρ empirical distribution, the Monte Carlo simulation is used.

Firstly, 10,000 sets of samples from the normal distribution $N_\rho(0, I)$ with sample size n are generated by using MATLAB. Then, each set of samples is substituted into the expression of b_ρ, and the value of b_ρ corresponding to each set of samples is obtained. The 10,000 sets of samples correspond to the values of 10000 b_ρ. What we need is the upper α quantile of the b_ρ empirical distribution. It is known from its probability meaning that there should be 10,000α samples in the

IOP Conf. Series: Journal of Physics: Conf. Series **1267** (2019) 012027 doi:10.1088/1742-6596/1267/1/012027

10,000 sets of samples larger than this value. Finally, the values of the 10000 b_ρ are arranged in descending order, and the value of the sequence number 10000α can be approximated as the upper α quantile of the b_ρ empirical distribution. Table 1 shows the upper 5% quantile of the b_ρ empirical distribution.

Table 1. The upper 5% quantile of the b_ρ empirical distribution.

ρ	n	The upper 5% of the b_ρ	ρ	n	The upper 5% of the b_ρ
	50	9.46		50	25.88
	100	9.17		100	25.74
2	150	8.99	4	150	25.58
	180	8.93		180	25.42
	300	8.73		300	25.15
	1000	8.41		1000	24.69
ρ	n	The upper 5% of the b_ρ	ρ	n	The upper 5% of the b_ρ
	50	16.70		50	193.73
	100	16.49		100	196.75
3	150	16.31	13	150	197.32
	180	16.18		180	197.36
	300	15.98		300	197.29
	1000	15.55		1000	196.69

2.3. The proposed algorithm flow

With the method of multi-dimensional normality test and the upper α quantile of b_ρ empirical distribution obtained by Monte Carlo method, we propose an improved K-means algorithm based on kurtosis test. The proposed algorithm flow is as follows.

Algorithm 1. An improved K-means algorithm based on kurtosis test.

Input: sample set $D=\{\chi_1,\chi_2,\ldots,\chi_m\}$, cluster number k.
Output: k clusters, outlier dataset C'.

```
use the traditional K-means algorithm to get the initial k clusters C={C_1,C_2,...,C_k};
initialize the outlier dataset C'=∅;
repeat
    for i=1; i≤k do
        calculate the kurtosis value of C_i——Kurtosis(C_i);
        find the upper α quantile of the bρ empirical distribution according to table 1;
        if Kurtosis(C_i)>bρ then
            for j=1; j≤|C_i| do
                set threshold ε_1, and find out the data C_{i_j} that will make the kurtosis
              of the cluster C_i suddenly change in the direction of decreasing;
                C_i = C_i − {C_{i_j}}, C' = C' ∪ {C_{i_j}};
            end for
        end if
    end for
```

```
    for p=1; p ≤ |C'| do
        for q=1; q ≤ k do
            set threshold ε_2;
            if C'_p doesn't make the kurtosis of cluster C_q suddenly change then
                C' = C' − {C'_p}, C_q = C_q ∪ {C'_p};
            end if
        end for
    end for
until the kurtosis values of all clusters are less than the upper α quantile of the b_ρ empirical
      distribution
```

3. Results

3.1. Experimental environment and datasets

The experimental hardware environment is Intel(R) Core(TM) i5-3320M CPU @ 2.60GHz, 8G RAM, and the software environment is python3.7. The datasets used in the experiment are simulation dataset containing elliptical data, simulation dataset containing outlier data and Iris and Wine datasets in UCI database. The simulation dataset two contains eight outlier data. Table 2 shows the information of the four datasets mentioned above.

Table 2. Experimental datasets.

No.	Dataset	Number of instances	Number of attributes	Size of classes
1	Simulation dataset one	900	2	300,300,300
2	Simulation dataset two	908	2	300,300,300,8
3	Iris	150	4	50,50,50
4	Wine	178	13	59,71,48

3.2. Simulation experiment results and analysis

In order to verify that the Kurtosis-Kmeans algorithm can break through the defect that the traditional K-means algorithm is only applicable to spherical datasets with similar size and density, it is applied to the simulation dataset one. Figure 2 shows the running process of Kurtosis-Kmeans algorithm and the results of each step. Red, green and blue point sets represent clustering results, and black point set represents outlier dataset.

It can be seen from figure 2 (e) that after the Kurtosis-Kmeans algorithm is iterated twice, compared with figure 2 (a), most of the data that are incorrectly divided due to the limitations of Euclidean distance are divided into the correct clusters. The experimental results show that the Kurtosis-Kmeans algorithm is more suitable for complex and irregular shape datasets, which improves the clustering accuracy.

In order to verify that the Kurtosis-Kmeans algorithm can break through the defect that the traditional K-means algorithm is susceptible to noise points and isolated points, it is applied to the simulation dataset two. Figure 3 shows the running process of Kurtosis-Kmeans algorithm and the results of each step. Red, green and blue point sets represent clustering results, and black point set represents outlier dataset.

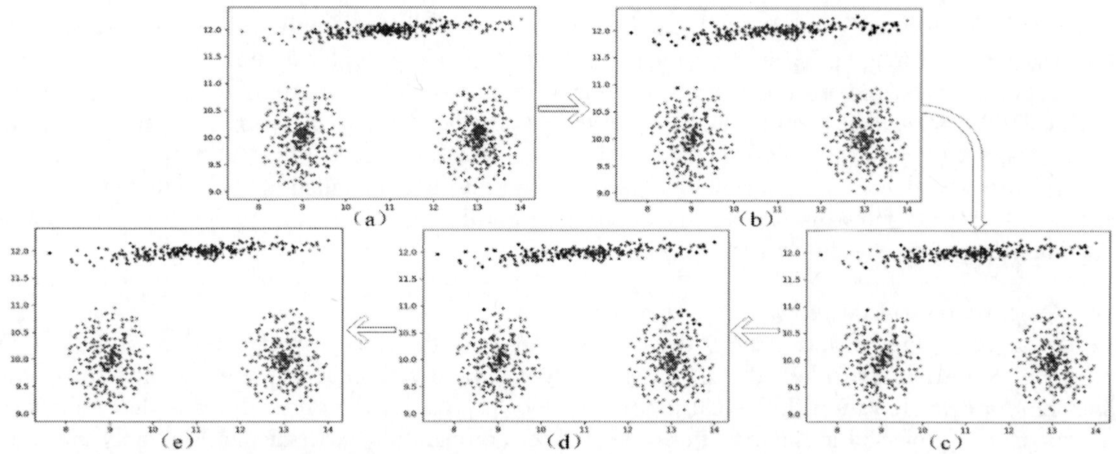

Figure 2. The running process of Kurtosis-Kmeans algorithm on simulation dataset one.

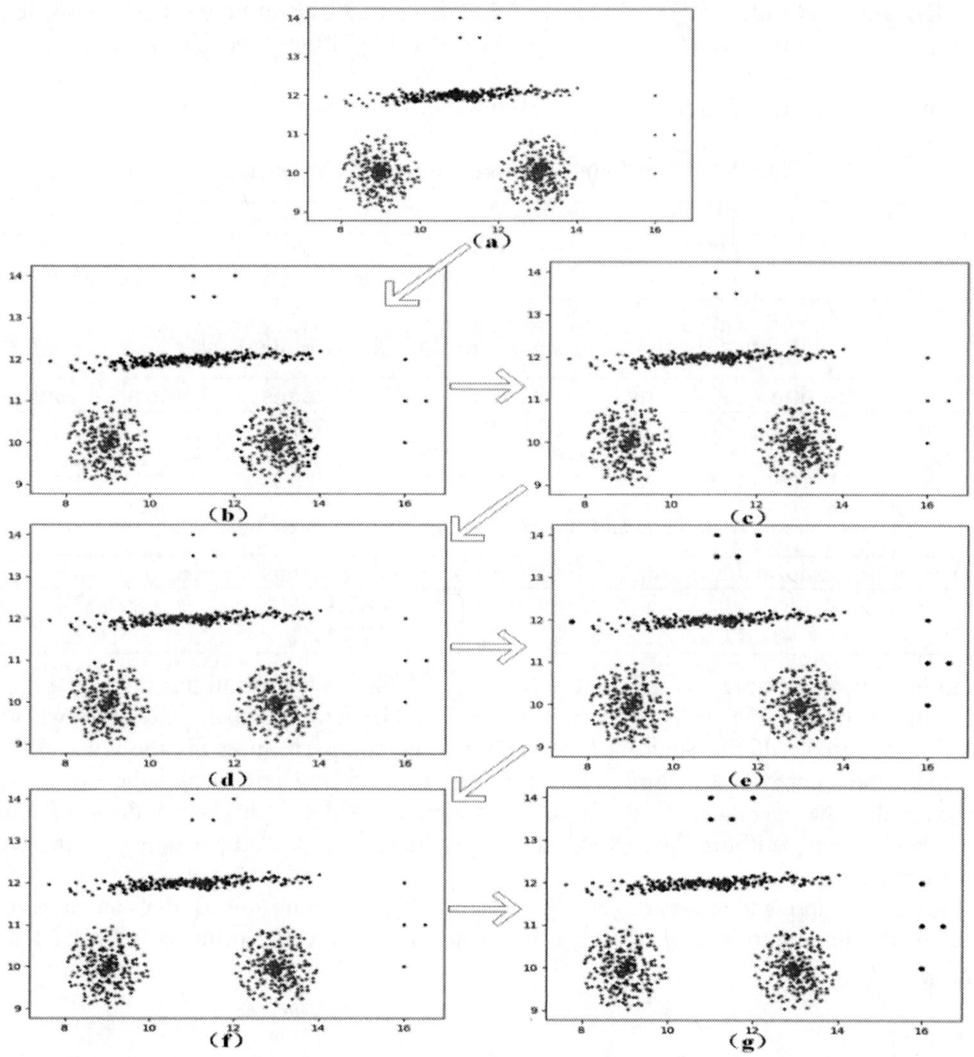

Figure 3. The running process of Kurtosis-Kmeans algorithm on simulation dataset two.

IOP Conf. Series: Journal of Physics: Conf. Series **1267** (2019) 012027 doi:10.1088/1742-6596/1267/1/012027

As can be seen from figure 3, due to the influence of the addition of outlier data, the kurtosis value of the cluster containing outlier data becomes sensitive. When looking for outlier data for the first time, many data are considered as outlier data, as shown in figure 3 (b). Therefore, in the following iteration, the threshold ε should be appropriately increased. After the Kurtosis-Kmeans algorithm is iterated three times, the figure 3 (g) is compared with the initial situation figure 3 (a), not only the majority of data are correctly divided as expected, but also the eight outlier data added are correctly detected. The experimental results show that the Kurtosis-Kmeans algorithm can detect the outlier data in the dataset, which improves the clustering effect.

3.3. *Comparisons with results in other researches*

In order to verify the advancement of the new algorithm, in the same experimental environment, the traditional K-means algorithm, the ID-Kmeans algorithm in reference [4], the improved K-means algorithm for discrete values (DV-Kmeans) in reference [5], and the Kurtosis-Kmeans algorithm based on kurtosis test proposed in this paper are selected for comparative experiments. We apply the above four algorithms to the Iris and Wine datasets of the UCI database. Then, the external indicators of clustering performance metrics of the four algorithms Jaccard Coefficient (JC), Fowlkes and Mallows Index (FMI), and Rand Index (RI) are calculated and compared respectively [10]. The result values of the above performance metrics are all in the [0, 1] interval, and the larger the value is, the better the performance of the algorithm is.

The results of the external indicators of clustering performance metrics are shown in table 3 - table 5.

Table 3. JC of four algorithms in Iris and Wine datasets.

Dataset	Traditional K-means	ID-Kmeans	DV-Kmeans	Kurtosis—Kmeans
Iris	68.23%	71.00%	64.59%	73.85%
Wine	41.20%	41.44%	41.74%	45.94%

Table 4. FMI of four algorithms in Iris and Wine datasets.

Dataset	Traditional K-means	ID-Kmeans	DV-Kmeans	Kurtosis—Kmeans
Iris	81.12%	83.06%	78.57%	84.98%
Wine	58.35%	58.59%	58.89%	62.96%

Table 5. RI of four algorithms in Iris and Wine datasets.

Dataset	Traditional K-means	ID-Kmeans	DV-Kmeans	Kurtosis—Kmeans
Iris	87.37%	88.59%	85.15%	89.88%
Wine	71.87%	72.04%	72.21%	75.11%

As can be seen from table 3 - table 5, the Kurtosis—Kmeans algorithm has improved the external indicators of clustering performance metrics Jaccard Coefficient, Fowlkes and Mallows Index and Rand Index compared with the other three algorithms. The comparison results show that the Kurtosis-Kmeans algorithm increases the similarity between the clustering results and the original samples, which means that the accuracy of the Kurtosis-Kmeans algorithm is higher, and the effect is better than the other three algorithms. The experimental results reflects the advancement of the Kurtosis—Kmeans algorithm.

Through simulation and comparison experiments, it has been proved that the new algorithm proposed by us can improve the clustering effect and eliminate the influence of outlier data. It is advanced and innovative.

4. Conclusion

As a classical unsupervised learning algorithm, the improvement of K-means algorithm is of great significance. In this paper, we use the statistical multi-dimensional normality test method to improve the clustering effect, and propose an improved K-means algorithm based on kurtosis test. This algorithm can improve the adaptability of clustering results to complex shape datasets while reducing the impact of outlier data on clustering results. The experimental results show that the Kurtosis-Kmeans algorithm has larger external indicators of clustering performance metrics than other algorithms, and has achieved good clustering results. The effectiveness and advancement of the Kurtosis-Kmeans algorithm are also demonstrated by the experimental results. In future work, more attention will be paid to the threshold setting in the process of the algorithm. In addition, using the kurtosis value directly as a dynamic indicator in the algorithm iteration process is also a positive research direction and is worth researching.

References

[1] Huang Z. Extensions to the k-Means Algorithm for Clustering Large Data Sets with Categorical Values[J]. *Data Mining and Knowledge Discovery*, 1998, **2**(3): 283-304.
[2] Jain A K. Data clustering: 50 years beyond K-means[J]. *Pattern recognition letters*, 2010, **31**(8): 651-666.
[3] Fahim A M, Salem A M, Torkey F A, et al. An efficient enhanced k-means clustering algorithm[J]. *Journal of Zhejiang University Science A*, 2006, **7**(10): 1626-33.
[4] Li H. Improved K-means clustering method and its application[D]. Northeast Agricultural University, 2014.
[5] Zhu J. Research and Application of K-means algorithm[D]. Dalian University of Technology, 2013.
[6] Huang J Z, Ng M K, Rong H, et al. Automated variable weighting in k-means type clustering[J]. *IEEE Transactions on Pattern Analysis and Machine Intelligence*, 2005, **27**(5): 657-668.
[7] Xiao-Ran Z. Discussions of kurtosis' statistical meanings[J]. *Journal of Yanshan University*, 2006, **30**: 57-60.
[8] Tian Y. Tests for normality based on Skewness and Kurtosis[D]. Shanghai Jiao Tong University, 2012.
[9] Pagès G. The Monte Carlo Method and Applications to Option Pricing[M]//Numerical Probability. Springer, Cham, 2018: 27-47.
[10] Zhou Y, Ji C, Zhang C. A Clustering Algorithm based on Internal Constrained Multi-view K-means[C]//Proceedings of the 12th Chinese Conference on Computer Supported Cooperative Work and Social Computing. ACM, 2017: 137-144.

Positioning and Tilt Correction of Milk Production Date Based on Mathematical Morphology

Fei Lei[1, a], Kang Sun[1, b] and Xueli Wang[1]

[1]Beijing University of Technology, Beijing, China

[a]myislj@126.com; [b]skang0502@163.com

Abstract: Milk is one of the most common drinks in people's lives. Its quality is related to people's health and safety. Therefore, food regulatory authorities and milk manufacturers are increasingly strict with milk. The production date on the packaging is an important part of milk testing. One. Aiming at the problems of long time and high cost of manual detection, this paper proposes a method based on mathematical morphology for milk production date location and tilt correction based on computer vision related technology. The method firstly preprocesses the collected milk image, then uses the mathematical morphology method to obtain the candidate region of the milk production date, then analyzes the candidate region, extracts the accurate milk production date region, and finally uses the least squares principle. The straight line is fitted to calculate the tilt angle of the milk production date, and the milk image is tilt corrected using the rotation formula. In order to verify the comparison with the Hough transform method, the experiment proves that the proposed method can accurately predict the milk production date and tilt correction.

1. Introduction

With the development of social economy, the dairy industry has become an important part of China's modern agriculture and food industry, and the variety of dairy products represented by milk has become an indispensable food in the daily life of Chinese people.

Therefore, the requirements of the national food regulatory authorities and manufacturers for the detection of milk have become more and more strict. The visual identification and analysis of the milk packaging production date is one of the important means of milk detection. As a kind of computer vision, visual inspection is used in milk production date character recognition [1], which can effectively improve production efficiency and reduce defective rate. It is of great significance in terms of efficiency and reliability of milk production date detection. Therefore, this paper combines advanced computer pattern recognition and digital image processing technology and other related theoretical knowledge, combined with mathematical morphology edge detection to have the ability to suppress noise and facilitate regional extraction. Based on the analysis of connected regions, a mathematics-based approach is proposed. Morphology and regional analysis of milk production date localization method, and the disadvantages of computational complexity and poor stability of Hough transform. This paper proposes to use the least squares principle to fit the straight line to calculate the milk production date tilt angle, and finally use bilinear interpolation. The milk image is rotated for correction.

2. Positioning of milk production date

IOP Conf. Series: Journal of Physics: Conf. Series **1267** (2019) 012028 doi:10.1088/1742-6596/1267/1/012028

2.1. Pretreatment of milk image

In the process of collecting milk, the image quality is degraded due to illumination changes, printer vibration and other factors, which is not conducive to accurate positioning of production date. In order to improve the image quality and improve the accuracy of milk production date positioning, it is necessary to pre-process the milk image.

2.1.1. Grayscale. The process of converting a color image into a grayscale image is the grayscale of the image. A color image refers to an image in which each pixel is composed of red, green, and blue, where R, G, and B are described by different gray levels. A grayscale image refers to an image in which each pixel is described by one quantized gradation.

There are many methods for graying out color images. In this paper, the milk image is grayed out according to formula 1.

$$G = 0.299R + 0.587G + 0.114B \quad (1)$$

The result of the grayscale processing is shown in Figure 1.

Figure 1. Milk image before and after grayscale processing

2.1.2. Binarization. In order to improve the accuracy of milk production date positioning and recognition, the target and background should be separated, so that the entire image shows a clear black and white visual effect. The process of transforming a grayscale image into a binary image is the binarization process of the image. In this paper, the Otsu algorithm is used to binarize the image [2].

2.1.3. Edge detection. Edge detection is a method of highlighting the edges of an image, weakening the image area outside the edge, and highlighting the outline of the image. It can greatly reduce the amount of processed data while preserving the useful structural information about the boundary of the object, thus simplifying the analysis process of the image [3].

Commonly used edge detection operators are: the Roberts edge operator [4], Prewitt edge operator [4] and Sobel edge operator [4].

The Roberts edge operator is an operator that uses the local difference operator to find the edge. The edge of the processed image is not very smooth. The Prewitt edge operator uses the gray level difference between the pixel points up and down and the left and right adjacent points. The extreme value detection edge is reached at the edge; the Sobel operator is improved based on the Prewitt operator, and the 4-neighbor domain is weighted to calculate the difference. There are two Sobel operators that detect horizontal and vertical edges, respectively. This operator introduces a similar local average operation, which can suppress the noise well and is better than the Prewitt operator and the Roberts operator.

Therefore, this paper uses the Sobel operator to convolve each pixel in the image. The Sobel operator uses a directional template for convolution to add edge features in one direction. The Sobel operators in different directions are shown in Figure 2.

$$G_x = \begin{pmatrix} -1 & 0 & 1 \\ -2 & 0 & 2 \\ -1 & 0 & 1 \end{pmatrix} \qquad G_y = \begin{pmatrix} 1 & 2 & 1 \\ 0 & 0 & 0 \\ -1 & -2 & -1 \end{pmatrix}$$

Figure 2. Horizontal and vertical templates for Sobel edge operators

The effect after the Sobel operator detection is shown in Figure 3. It can be seen from the figure that the contour information of the milk production date area can be well preserved after the Sobel operator edge detection.

Figure 3. Results of Sobel edge operator processing

2.2. Morphological processing

Mathematical morphology is currently a commonly used subject of image processing and analysis [5]. Its principle is to use a certain structural element to measure and extract the corresponding shape in the image to satisfy the analysis and recognition of the image. Mathematical morphology algorithms can improve the speed of image analysis and processing. Its basic operations are four, namely: expansion, corrosion, open and closed operations.

Dilation: The structural element B is translated to the x to obtain B_x. If B_x and A are not empty, then the x is recorded. All the sets of x satisfying the above conditions are called the result of the dilation of A by B. Expressed as:

$$D(A) = \{x | (B_x \cap A) \neq \varnothing\} = A \oplus B \tag{2}$$

Erosion: can be regarded as an expanded dual operation, that is, the structural element B is translated to the x to obtain B_x. If B_x is included in A, the x is recorded, and all the sets of x satisfying the above conditions are called the result of A being erosion by B. Expressed as:

$$E(A) = \{x | B_x \subseteq A\} = A \Theta B \tag{3}$$

Opening: A is first eroded by B and then dilated to be called A is opened by B. Expressed as:

$$O(A) = D(E(A)) = (A \Theta B) \oplus B \tag{4}$$

Closing: A is first dilated by B and then eroded is called A is closed by B. Expressed as:

$$C(A) = E(D(A)) = (A \oplus B) \Theta B \tag{5}$$

In this paper, the milk image after edge detection is first closed, that is, the milk image is first expanded, the production date is filled into a connected area, and then the etching operation filters out some smaller areas. Then the milk image is opened. Through the previous closing operation, although the image basically conforms to the production date area, the image leaves a part of the thin line, first performs the etching operation, filters it out, and then expands to restore the size of the connected area. Smooth the image. After the above treatment, most of the interference area is removed, and the rest may be the milk production date area. The processed image is shown in Figure 4.

Figure 4. Morphological processing results

2.3. Analysis of connected areas

After the previous step, multiple candidate areas for milk production date are obtained, which need to be analyzed and the parameters of different areas are compared to locate the milk production date area. Specifically, it is based on the method of positioning such as license plate [6-8]. Firstly, the height and width of the minimum circumscribed rectangle of each candidate area are obtained, and then the height, width and aspect ratio of the production date area are within a certain range. Geometric features to get the real milk production date area.

The algorithm for extracting the milk production date region based on the region analysis is as follows:

- Marking each candidate area in the image;
- Successively scanning the marked candidate domains to obtain the height, width and coordinate values of the minimum circumscribed rectangle of each marked area;
- Using the nature of the milk production date is a fixed value, combined with the height of the milk production date, extracts the milk production date area.

The height and width of the same milk production date area in China are fixed. For example, the milk production date used in this experiment is 4 mm in height, 19 mm in width and about 4.75 in aspect ratio. Because of the possibility of tilting, etc., after experiment, the aspect ratio threshold interval of the milk production date area is [4, 5], and the candidate areas that do not satisfy this condition are eliminated during the analysis. At the same time, considering the single condition judgment error, this paper selects the milk production date height as the second judgment condition. After two screenings, the area that satisfies the aspect ratio and height conditions is the real milk production date area. The extracted milk production date area is shown in Figure 5.

Figure 5. Milk production date positioning results

3. Tilt correction

The image after the milk production date is positioned may be skewed, and the character segmentation may be directly performed, which may result in inaccurate segmentation results and affect subsequent recognition operations. Therefore, it is necessary to perform a tilt correction process on the milk production date.

3.1. Calculate the tilt angle

Let the coordinates of the vertical edge points of the milk production date area be $(x_i, y_i)(i=1,2\cdots,n)$, where n is the number of vertical edge points, and the principle of least squares fits the set of data with a smooth curve $y=f(x)$ and makes the curve has the smallest sum of squared deviations from all vertical edges, and the curve reflects the tilt direction of the production date [8].

Let the linear exponential equation be:

$$y=ax+b \tag{6}$$

In the above formula, a represents the slope and b represents the intercept. For each x, there are two corresponding y, one is the measured value y_i and the other is the value y_i' obtained by fitting the equation. The difference between the measured value y_i and the fitted value y_i' is recorded as a deviation:

$$\delta_i = y_i - y_i' \tag{7}$$

The principle of the least squares method described above is to find the minimum value of the sum of the squares of the deviations, to find the value of $\min(\delta_i^2)$, and the definition is:

$$\phi=\sum_{i=1}\delta_i^2 = \sum_{i=1}(y_i - ax_i - b)^2 \tag{8}$$

It can be seen from Equation 8 that the value of a and b when the demand ϕ takes the minimum value. First, let the first-order partial derivative of ϕ for a and b be equal to 0; then find the second-order partial derivative of ϕ for a and b. When the second-order partial derivative is greater than 0, the obtained values of a and b can make the square of the error. The smallest, get:

$$a=\frac{n\sum_{i=1}^{n}x_iy_i-(\sum_{i=1}^{n}x_i)(\sum_{i=1}^{n}y_i)}{n\sum_{i=1}^{n}x_i^2-(\sum_{i=1}^{n}x_i)^2}\qquad b=\frac{(\sum_{i=1}^{n}x_i^2)(\sum_{i=1}^{n}y_i)-(\sum_{i=1}^{n}x_iy_i)(\sum_{i=1}^{n}x_i)}{n\sum_{i=1}^{n}x_i^2-(\sum_{i=1}^{n}x_i)^2} \tag{9}$$

Therefore, the inclination α of the characteristic line is:

$$\alpha=\arctan(a) \tag{10}$$

In this way, to calculate the inclination, the inclination can be calculated simply by multiplying the coordinates of the vertical edge points by simple multiplication and summation.

After obtaining the α, it is only necessary to rotate the picture according to the rotation formula. The tilt correction matrix is of the form:

$$[X' \quad Y' \quad 1]=[X \quad Y \quad 1]\begin{bmatrix}\cos\alpha & -\sin\alpha & 0\\ \sin\alpha & \cos\alpha & 0\\ 0 & 0 & 1\end{bmatrix} \tag{11}$$

Where X and Y represent the x and y coordinate matrices of the oblique image, respectively, representing the x and y coordinate matrices of the corrected image, respectively.

3.2. Milk production date correction

The tilt milk production date correction is divided into horizontal direction correction and vertical direction correction. According to the method of the above section, the inclination angle α in the horizontal direction of the milk production date and the inclination angle β in the vertical direction are respectively calculated.

The bidirectional interpolation method is used to correct the horizontal tilt. If $\alpha>0$, the clockwise rotation is performed. Otherwise, the transposition is performed to obtain the corrected image.

For vertical tilt correction, it is necessary to shift the pixel offset between the same line to determine the coordinate point of the corrected pixel. The specific formula is as follows:

$$y_{new}=y+(x+1)\tan\beta, \beta>0 \tag{12}$$

$$y_{new}=y+(h-x)\tan\beta, \beta<0 \tag{13}$$

After the corrected pixel point coordinates are calculated, the grey value is calculated by bilinear interpolation.

4. Experimental results and analysis

In this experiment, we not only used this method for the tilt correction of milk production date image, but also realized the Hough transform method for milk production date image tilt correction. The tilt correction results of the two methods are shown in Figure 6.

(a) Tilt image (b) Hough transform corrected image (c) Least squares corrected image

Figure 6. Tilt corrected contrast image

The milk image containing the date of manufacture produced by the random manufacturer of the pixel was tested, and the accuracy and running time of the least square method and Hough transform in calculating the inclination angle were analysed in matlab7.11. The specific analysis results are shown in Table 1.

Table 1. Algorithm comparison

	Correct rate (%)	Time (s)
Hough	83.3	0.2934
LSM	96.7	0.0643

The experimental results show that the accuracy and speed of the calculation of the milk production date tilt angle by the least squares method are better than the Hough transform.

5. Conclusion

In this paper, a method based on mathematical morphology for milk production date location and tilt correction is designed. Firstly, the milk production date candidate area is obtained by edge detection and mathematical morphology processing. Then, the milk production date aspect ratio analysis is used to remove the interference. The connected domain is tilt corrected for the milk production date by fitting the line based on the least squares method. The experimental results show that the positioning and tilt correction method is simple, fast and accurate, which can lay a good foundation for the next step of milk production date segmentation and recognition.

References

[1] Ni C X. A review of computer vision research [J]. Electronic world, 2018(01):91+93.
[2] Chen S. Binary Image of License Plate Image Based on Otsu Algorithm and Its Matlab Realization [J]. Journal of Changchun Teachers College, 2012, 31(03):33-35.
[3] Wei W B, Rui X T. Research on Image Edge Detection Method [J]. Computer Engineering and Applications, 2006(30):88-91.
[4] Guan H R, Ding H. A Survey of Edge Detection in Image Processing [J]. Journal of Capital Normal University (Natural Science Edition), 2009, 30(S1):66-69.
[5] He D J, Geng N and Zhang Y K. Digital image processing [M]. Xi'an University of Electronic Science and Technology Press, Xi'an. 2008.
[6] Chen L, Chen W. Method for license plate location and tilt correction based on mathematical morphology [J]. Science & Technology Vision, 2015(23):10+50.
[7] Wang Y. A Mathematical Morphology Based on Algorithm for Elevator Registration Number Locating Process [J]. China elevator, 2018, 29(08):42-46.
[8] Xue D. Research the algorithms of license plate location and slant correction under the complicated conditions [D]. Zhengzhou University, 2013.

IOP Conf. Series: Journal of Physics: Conf. Series **1267** (2019) 012029 doi:10.1088/1742-6596/1267/1/012029

A Target Detection Method Based on Statistical Mean Value Difference Model

Fei Lei[1, a], Kai Long[1, b] and Xueli Wang[1]

[1]Beijing University of Technology, Beijing, China

[a]myislj@126.com; [b]longkai225@163.com

Abstract: In the field of oil pipeline monitoring applications, the guarantee of high accuracy target detection is required. Aiming at the problem of moving target detection under static camera, a method of mean background model based on probability and statistics is proposed. The background model is quickly constructed. The background threshold is adaptively used to eliminate the influence of illumination changes. The target is detected quickly and accurately. Finally, the proposed target is proposed. The background update method updates the background model in real time. The simulation results show that the better detection results can be obtained under the influence of light changes. The background model has strong stability and adaptability, and can be applied to the real-time monitoring system of the field oil pipeline.

1. Introduction

Due to its safety, economy and speed, the oil pipeline has received attention and rapid development in China in recent years. While the oil pipeline is developing rapidly, some lawless elements have made huge profits by stealing national crude oil, which has hindered the development of the oil pipeline transportation industry and the loss of state property [1]. Therefore, it is necessary to monitor multiple parts of the oil pipeline by means of computer vision image processing technology. Common moving target detection methods are optical flow method [2], adjacent frame difference method [3], and background difference method [4]. Although the optical flow method can detect moving objects independently, it is very time consuming and difficult to meet real-time. The adjacent frame difference method can adapt to the dynamic environment quickly, but the segmented moving objects are incomplete. The background difference method can make up for the first two. The shortcomings of the methods are the best in terms of speed and effect.

In this paper, an improved algorithm for mean background modeling is proposed. The mean model background is constructed by mathematical modeling method of probability and statistics. The threshold is adaptively selected according to the illumination change of the image. The image to be detected is compared with the background image, and then the foreground target is extracted. In the process of constructing the background model, the background image is updated in real time, effectively eliminating the interference of environmental factors, and thus obtaining an ideal background image.

2. Background model

2.1. Mean background model

The mean background model is the averaging of multiple images based on time, that is, the background image is obtained by obtaining the average of the sum of the continuous video images over a period of time.

Initialize the image sequence, use the nearest neighbor interpolation method to adjust the image size to $H \times W$, and the mean background model formula is as follows:

$$B_k(i,j) = \frac{1}{N}\sum_{k=1}^{N} A_k(i,j) \tag{1}$$

In formula (1): $B_k(i,j)$ is the pixel value of the background image frame at time k at (i,j) position; $A_k(i,j)$ is the pixel value of the input image frame at time (i,j) at time k ; N is the total number of frames of continuous images.

2.2. Improved mean background model

The mean background model is sensitive to ambient light changes and branching, and sometimes "smear". In order to reduce the influence of these factors, a probabilistic method is used, and each of the pixel positions may be generated by using a statistical histogram. The number of occurrences of the pixel value is counted, and several gray values with a large number of occurrences are added and averaged as the pixel value of the background.

Assuming a sequence of N frames of images, the pixel value that may occur at each pixel location is $c = \{c_1, c_2, ..., c_l\}$,The number of occurrences of each pixel value is $m = \{m_{c1}, m_{c2}, ..., m_{cl}\}$, The total number of times is equal to N, namely:

$$m_{c1} + m_{c2} + ... + m_{cl} = N \tag{2}$$

Judging each pixel value traversal, if the number of occurrences meets the requirements, accumulate the pixel value, the formula is as follows:

$$\begin{cases} m' = m' + m_c, \dfrac{m_c}{N} > \varepsilon \\ B(i,j) = B(i,j) + c * m_c \end{cases} \quad \text{(Where } c = \{c_1, c_2, ..., c_l\}, m' \text{ initial value is 0)} \tag{3}$$

In equation (3), ε is the empirical value, generally about 0.3~0.4. The improved mean background model is as follows:

$$B_k(i,j) = \frac{B(i,j)}{m'} \tag{4}$$

The image to be detected is subtracted by the background model of the structure, and the pixel value after subtraction is binarized. It is important to set the segmentation threshold. If the threshold is too large, some foreground pixels will be judged as the background; if the threshold is taken The value is too small, and some background pixels will be judged as foreground. In this paper, the adaptive threshold selection method is adopted to make the threshold T with the largest variance between classes the most suitable threshold. The pixels smaller than T are set to 0, and the pixels larger than T are set to 255 to obtain the foreground image. The formula is as follows:

$$D(i,j) = \begin{cases} 255, |I_k(i,j) - B_k(i,j)| > T \\ 0, |I_k(i,j) - B_k(i,j)| \le T \end{cases} \tag{5}$$

3. Adaptive background differential target detection

3.1. Process of target detection algorithm

The basic steps of target detection: image frame analysis of the input video, the background model is established by the mean method, the foreground area is obtained by adaptive threshold background difference, and the post-processing is finally performed and the detection target is identified. The flowchart is shown in Figure 1.

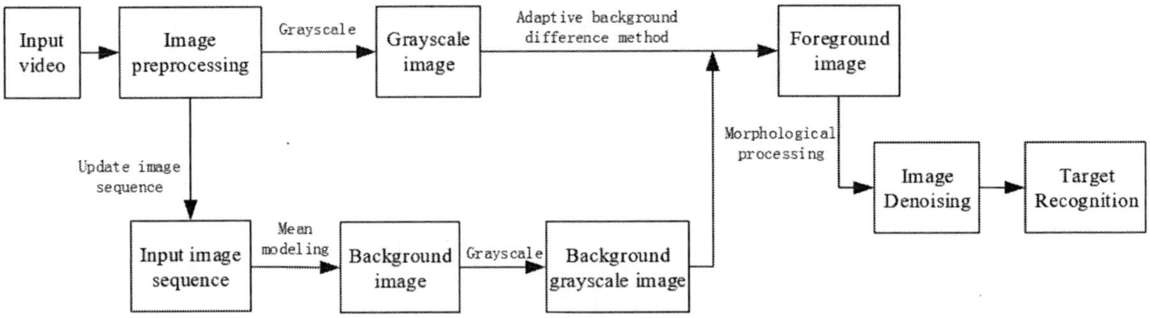

Figure 1. Algorithm flowchart

3.2. Specific method of detection

- Obtain an input image as a video sequence image $\{A_k(i,j)\}_{k=1}^{N}$ at each of the first N days of the current time,Where k is the sequence of frames and N is the total number of sequence images.
- The statistical histogram is used to count the number of pixel values that may occur at each pixel position, and several pixel values with a large number of times are summed and averaged as the background pixel of the pixel position, that is, the background model $B_k(i,j)=\frac{B(i,j)}{m'}$ is obtained.
- Each pixel of the image is traversed, and each pixel is binarized. A pixel larger than the threshold T is set as a foreground; a pixel smaller than the threshold T is set as a background, that is, a binary foreground image is obtained.
- Real-time update of the built background model, background update rate is $\alpha\in(0,1)$,The background update model is $B_{k+1}(i,j)=\alpha B_k(i,j)+(1-\alpha)I(i,j)$,The next frame background image is related to the current frame image and the previous image.
- Identify the detection target of the foreground image and determine the type of the target.

3.3. Selection of threshold T

The value of the image binarization segmentation threshold is very important, and the use of a fixed threshold cannot adapt to the effects of ambient illumination changes. If the split value is too small, the background pixel will be mistaken for the foreground pixel; if the value is too large, some foreground pixels will be mistaken for the background pixel.

The largest inter-class variance method was first proposed by Otsu, also known as the Otsu threshold algorithm [5]. The idea of the algorithm is to use the principle of least squares to determine an optimal threshold, so that the image is binarized, and the variance between the foreground and background pixels is guaranteed to be the largest.

Assuming that the total number of pixels of the image to be processed whose gray level is L is N, and the total number of pixels whose gray level is i is n_i , the frequency at which each gray value appears is expressed as:

$$P_i=\frac{n_i}{N}, i=1,2,...,L-1 \tag{6}$$

IOP Conf. Series: Journal of Physics: Conf. Series **1267** (2019) 012029 doi:10.1088/1742-6596/1267/1/012029

It is assumed that the image is divided into foreground and background regions with grayscale K as the threshold, pixels with gradation of $0 \sim k$ and pixels with gradation of $k+1 \sim L-1$ constitute region B (background) and F (foreground)), the frequencies are:

$$P_B = \sum_{i=0}^{k} P_i \text{ and } P_F = \sum_{i=k+1}^{L-1} P_i \tag{7}$$

Then the average gray levels of area B and area F are:

$$\mu_B = \frac{\sum_{i=0}^{k} i * P_i}{P_B} \text{ and } \mu_F = \frac{\sum_{i=k+1}^{L-1} i * P_i}{P_F} \tag{8}$$

Then the inter-class variance of the image can be defined as:

$$\mu_k = P_B(\mu_B - k)^2 + P_F(\mu_F - k)^2 \tag{9}$$

The traversal method is used to obtain a threshold T that maximizes the variance between classes, which is the most appropriate threshold.

3.4. Background update

The background model is constructed based on a sequence of images over time. Over time, it may be disturbed by factors such as changes in illumination, branching, and sudden motion of stationary scenes. In order to make the modified background model have certain adaptability to external changes, it is necessary to update the background model to overcome the interference of these factors.

The basic method of background update is to traverse each pixel of the image and determine that the pixel is the foreground pixel, and keep the original background pixel value without updating; if it is judged as the background pixel, it is updated, and the background update formula [6] is:

$$B_{k+1}(i,j) = \alpha B_k(i,j) + (1-\alpha)I(i,j) \tag{10}$$

In equation (10), α is the background update rate, and its range is $(0,1)$, where $I(i,j)$ is the current frame image, $B_k(i,j)$ is the gray value of the current background frame image, and $B_{k+1}(i,j)$ is the gray value of the background frame image of the next time. It can be seen from equation (10) that the background model is related to the current frame image and the previous frame image, effectively avoiding sudden phenomena, and will remain relatively stable over time, and can adapt to the effects of illumination changes, branch swings and the like.

4. Experimental results and analysis

4.1. evaluation standard

The evaluation criteria of image quality are mainly divided into subjective and objective evaluation. The subjective evaluation is mainly based on the human eye to observe the image, and it is evaluated by the feeling, which is easily affected by personal factors, so the objective evaluation standard is generally used.

The objective evaluation standard is mainly to establish a mathematical model for the image, and measure the information index through the specific formula. This paper studies the information entropy [7] and the edge ratio of the whole image.

Information entropy is the probability that the random variable appears. The size of the information is measured by the magnitude of the probability. Generally, the amount of information can only be represented by the change. Definition of information entropy: Assume that the gray level of the image is N, and the value of the random variable is $p_{xi} = p\{X = x_i\}, i = 1, 2, ..., N$, x_i represents the gray value that appears, and the information entropy is expressed as:

$$H[S] = -\sum_{x=0}^{N-1} p_x \log p_x \tag{11}$$

The edge ratio represents the edge information of the image. When the number of targets is large, the edge ratio of the image is larger, and the more edge information is retained. The edge ratio is expressed as:

$$R = P_{edge} / (W * H) \tag{12}$$

In equation (12), P_{edge} represents the number of edge information with image size $W * H$. In this paper, the sobel edge detection algorithm [8] is used to detect the number of edge pixels in the image.

4.2. Experimental results and analysis

In order to verify the effectiveness of the proposed algorithm, simulation experiments were carried out on the experimental data collected. The experimental platform runs on the operating system Windows7 Professional, the video image resolution is 320×280, with the suffix name in avi format, and the programming environment is vs2010 and the opencv2.4.10 machine vision library.

The improved mean background modeling effect is shown in the following figure:

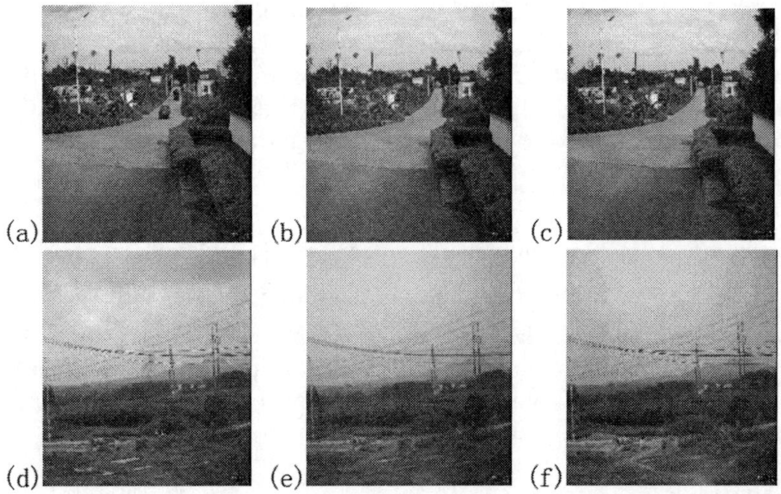

Figure 2. Background modeling method comparison effectdiagram,(a) Scene 1 image of the 200th frame,(b) Mean background image,(c) Statistical mean background image,(d) Scene 2 image of the 256th frame,(e) Mean background image,and (f) Statistical mean background image.

As shown in the above figure, Figure 2(b) and (e) are background models constructed by accumulating the mean values of a sequence of pixels. It can be seen that the branches are swaying, the wires are swaying, and the motion of the moving objects will appear "smear", which will reduce the background image. quality. Figure 2(c) and (f) show an improved mean background image, which significantly suppresses the effects of these factors and achieves better results.

As can be seen from Figure 3 below, the effect of using the adaptive threshold is better than selecting a fixed threshold for binary processing, and the adaptive threshold can be selected to select the most appropriate threshold for segmentation. In Figure 3(c), some interfering pixels appear in the image, and some background pixels are mistaken for foreground, while graph Figure 3(d) achieves better experimental results. It can be seen from Table 1 that the improved algorithm adopted in this paper is slower in processing time, but the richness of the internal state of the constructed background image and the edge information achieve better results.

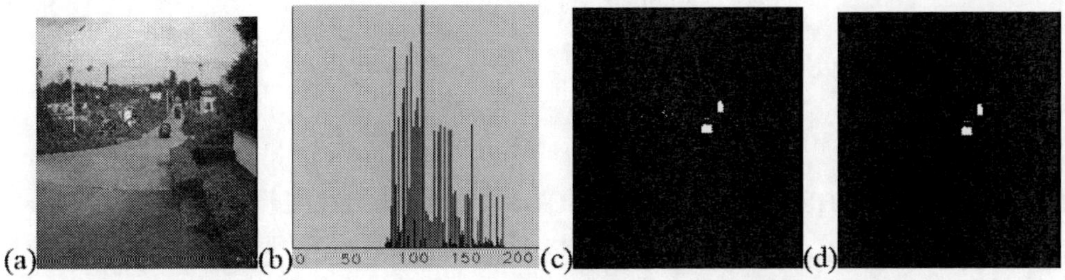

Figure 3. Target Detection.(a) Original image,(b) Grayscale frequency histogram,(c) Fixed threshold detection,and (d) OTSU algorithm detection.

Table 1. Experimental comparison of background modeling algorithms

Detection image	Background modeling method	Algorithm processing time(ms/f)		Information entropy	Edge ratio
		average	maximum		
Figure 2. Scene one the 200th frame image	Mean background modeling	1.120	1.341	7.0140	0.0691
	Background modeling of statistical mean	1.323	1.516	7. 5776	0.0733
Figure 2. Scene two the 256th frame image	Mean background modeling	0.974	1.271	6.2420	0.0715
	Background modeling of statistical mean	1.212	1.456	7. 6661	0.0761

5. Conclusion

The background model based on statistical mean is improved on the mean background model method. The background model is simpler to implement, the edge information of the image is better preserved, and the negative effects caused by sudden changes in background can be better eliminated. The model is more stable and improves the accuracy of target detection.

References

[1] Xufeng Yu.Research on Pattern Recognition Technology of Burglary Police in Anti-theft Detection of Oil Pipeline [D]. Nanjing Aerospace University, 2004.1.1.

[2] Hiraiwa A,Fuse K,Komatsu N,et al. Accurate Estimation of Optical Flow for Fully Automated Tracking of Moving-Objects within Video Streams[J].IEEE International Symposium on Circuits and Systems,1999,4:515-519.

[3] Lipton A,Fujiyoshi H,Patil R.Moving Target Classification and Tracking from Real-time Video[C]//Applications of Computer Vision,1998.WACV'98.Proceedings,Fourth IEEE Workshop on.Princeton,NJ,USA:IEEE,1998:8-14.

[4] HAN Hong-zhe.Adaptive Background Modeling with Shadow Suppression[J].The IEEE 6th International Conference On Intelligent Transportation Systems,2003,1:720-724.

[5] Qian Wang,Zhong Peng,Li Liu. An Image Segmentation Algorithm Based on Adaptive Threshold[R], Transactions of Beijing Institute of Technology,2003,23(4):521-524.

[6] Chong Wang,Zhihong Xi,Chunli Xiao.Moving Target Detection Method Based on Background Difference[J].Applied Science and Technology,2009,36(10):16-19.

[7] Junbo Wang,Yinglei Song,Bing Zhang,Taoying Sun.Target Detection for Background Model Selection[J].Modern Electronics Technique,2018,41(13):34-37.

[8] Meijing Zhang. Research on sub pixel edge detection technology [D]. Shenyang: Shenyang Ligong University, 2013.

IOP Conf. Series: Journal of Physics: Conf. Series **1267** (2019) 012030 doi:10.1088/1742-6596/1267/1/012030

Multi-agent Consensus Problem Based on Observer and Considered Lur'e Nonlinearity in Directed Topology

Shaolei Zhou, Xueyuan Zhao*, Shi Yan and Shuailei Wang

Naval Aviation University, Yantai 264001, Shandong, PR China

*1607527622@qq.com

Abstract. This paper studies the multi-agent systems consensus problem considering Lur'e-type nonlinearity under directed network. Compared with the conclusions of the existing undirected communication topology, the conclusions obtained have greatly reduced the communication topology requirements and are more general. This paper designs a control protocol based on the observer, which solves the problem that the system state information is unknown. By decomposition of a specific form of the Laplacin matrix, consensus problem is converted into stability problem for low-dimensional systems. In this paper, the method of solving the control protocol is given and proved by the piecewise Lyapunov function. The simulation results show that the designed control protocol can solve the consistency problem.

1. Introduction

In recent years, the collaborative control of multi-agent systems has received extensive attention and research, especially in the field of group behavior such as swarming, assembly and formation control [1-4]. As a fundamental problem in the control of multi-agent systems, consistency-related research has achieved a lot of results. Literature [5] studies the consistency control problem of multi-agent systems with first-order dynamics model. It is concluded that the communication topology map needs to contain a spanning tree if the system is agreed upon. In the literature [6], the controller of the second-order time-delay multi-agent system in the directional topology is designed. By using a multi-leader framework, the condition of the formation controller of the system to achieve a fixed formation is obtained. Literature [7] studies multi-agent system with linear time-invariant dynamics. The literature [8] proposes a unified framework for dealing with the consistency of nonlinear agent networks that satisfy certain sector conditions. In literature [9], a class of nonlinear agent systems is standardized by feedback linearization method, then a consistency protocol is designed, and the performance index of the protocol is analyzed according to the properties of the lower triangular matrix.

Actual control systems often need to be modeled as nonlinear systems. Lur'e type nonlinearity is a typical nonlinear link and has a wide range of applications in many fields such as aircraft control, hydraulic servo control, and chaotic synchronization. The literature [10] studies the control problem of Lur'e type nonlinear link system under undirected topology and gives the global synchronization region. The literature [11-12] studies the robust consistency problem with leaders in an undirected topology.

In many practical applications, full state information cannot be used for controller design. Similarly, in a multi-agent system, a single agent often cannot effectively obtain relative state information of neighbor nodes, and even cannot obtain its own measurement output. In response to this situation, the introduction of an observer is generally used to solve such problems [13-15]. In literature [16], a finite-time consistency control algorithm based on sliding mode observer is proposed for the multi-agent system with undetectable speed of the agent. The literature [17] designs a robust consistency control protocol for the nonlinear multi-agent system with unknown interference using the state estimation information of the observer.

This paper studies the multi-agent consistency control based on observer-based Lur'e-type nonlinearity under directional topology. The existing literature has relaxed the topological conditions of the undirected topology, making the application conditions of the research more general. This paper proposes a consistency controller based on the observer type .Different from literature [18-20] need to solve the algebraic Riccati equation (ARE) to solve the controller parameters. In this paper, the consistency control problem is transformed into several sets of linear matrix inequality (LMI) solving problems, which makes the problem solving more convenient and reduces the conservativeness of the results.

2. Problem Formulation

The symbol $\mathbb{R}^{n\times n}$ and $\mathbb{C}^{n\times n}$ respectively represent $n\times n$ dimensional real matrix and complex matrix. $\mathrm{Re}(\mu)$ represent the real of $\mu\in\mathbb{C}$. I_n $n\times n$ dimensional represent unit matrix. $A\otimes B$ represent the Kronecker of A and B . $A>B$ and $A\geq B$ respectively represent $A-B$ is positive or semi-positive.

$\mathcal{G}=(\mathbb{V},\mathbb{E},\mathbb{A})$ represent the directed topology during the agents, where $\mathbb{V}=\{v_1,v_2,\dots,v_N\}$ is node set of topology, $\mathbb{E}\subseteq\mathcal{V}\times\mathcal{V}$ is edge set of topology, $\mathbb{A}=\left[a_{ij}\right]_{N\times N}$ is adjacency matrix of topology and $a_{ij}\geq 0$. a_{ij} is the connection weight of note v_i and v_j ,and it represent the note v_i can receive information from v_j ,otherwise $a_{ij}=0$ .If there is a note v_i , from this point, along the directed edge, it can reach any other point in the graph, we call the graph $\mathcal{G}$ contain a directed spanning tree, and v_i is root node. When connection weight $a_{ij}=1$, $\mathcal{N}_i:=\left\{j\in\mathbb{V}:a_{ij}=1\right\}$ represent the neighbor of v_i .The Laplacian of $\mathcal{G}$ is $\mathcal{L}=[\mathcal{L}_{ij}]_{N\times N}$,where $\mathcal{L}_{ii}=\sum_{j\neq i}a_{ij}$ $\mathcal{L}_{ij}=-a_{ij},i\neq j$

Lemma 1 [1] The Laplacian of $\mathcal{G}$ a least have one zero eigenvalue, other non-zero eigenvalues have positive real parts. If the directed topology $\mathcal{G}$ have a directed spanning tree, then 0 is simple eigenvalue of $\mathcal{L}$, and 1_N is the feature vector of $\mathcal{L}$

Lemma 2 [21] For any given $x,y\in\mathbb{R}^n$ and the matrix with compatible dimensions $P>0, D$ and S .We can get that

$$2x^T DSy\leq x^T DPD^T x+y^T S^T P^{-1} Sy$$

Lemma 3 [22] Shur complement lemma of matrix

Assume $S(P)$ is $n\times n$ dimensional, its block is expressed as

$$S(P)=\begin{bmatrix} S_{11}(P) & S_{12}(P) \\ S_{21}(P) & S_{22}(P)\end{bmatrix} \tag{1}$$

Where, $S_{11}(P)$ is $r\times r$ dimensional, if $S_{11}(P)$ is non-singular, then $S_{22}(P)-S_{21}(P)S_{11}^{-1}(P)S_{12}(P)$ is the Shur complement of $S_{11}(P)$ in $S(P)$. Then the following three conclusions are equivalent

1) $S(P)<0$;

2) $S_{11}(P)<0$, $S_{22}(P)-S_{21}(P)S_{11}^{-1}(P)S_{12}(P)<0$;

3) $S_{22}(P)<0$, $S_{11}(P)-S_{12}(P)S_{22}^{-1}(P)S_{21}(P)<0$.

Lemma 4 [23] For the Laplacian $\mathcal{L}\in\mathbb{R}^{N\times N}$, there is a matrix $M\in\mathbb{R}^{N\times N-1}$, which let $\mathcal{L}=ME$,where $E\in\mathbb{R}^{(N-1)\times N}$,

IOP Conf. Series: Journal of Physics: Conf. Series **1267** (2019) 012030 doi:10.1088/1742-6596/1267/1/012030

$$E=\begin{bmatrix}1 & -1 & 0 & \cdots & 0\\ 0 & 1 & -1 & \cdots & 0\\ \vdots & \vdots & \ddots & \vdots & \vdots\\ 0 & 0 & 0 & 1 & -1\end{bmatrix} \tag{2}$$

Further, if the graph contains a directed spanning tree, then M is full rank and the eigenvalues of EM are the eigenvalues of $\mathcal{L}$ .So, $\mathrm{Re}(\lambda(EM))>0$

Lemma 5 [24] If the graph contains a directed spanning tree, then there is a symmetric positive definite matrix Q and a positive scalar parameter α

Let $(EM)^T Q+QEM>\alpha Q$, where $0<\alpha<2\min\{\mathrm{Re}(\lambda(EM))\}$

Consider the following n multi-agent systems with nonlinear terms, the model of the i agent is

$$\begin{cases}\dot{x}_i\ t = Ax_i\ t + Bu_i\ t + Df\ y_i\\ y_i\ t = Cx_i\ t\end{cases} \tag{3}$$

where $x_i\ t$ is the status of agent i ,the control input of agent i is $u_i\ t\in\mathbb{R}^m$,the output of agent i is $y_i\ t$. A, B, C, D are system matrix, control matrix, and output matrix. Nonlinear function $f(y_i(t))=(f(y_{i1}),\ldots,f(y_{iq}))^T$ need slope condition $[0,\Delta]$, which is

$$0\le(f_i(b)-f_i(a))\le\delta_i(b-a), f_i(0)=0, \forall b\ne a, i=1,\ldots,\mathrm{q}.$$

For general slope conditions $[\Delta_1,\Delta_2]$,we can converted to a slope condition $[0,\Delta]$ by a reasonable loop transformation, so we only need consider slope condition $[0,\Delta]$.

In some practical systems, considering the cost, often do not directly get the full state of the agent, this paper designs the following observer for each agent to observe the system status.

$$\begin{aligned}\dot{z}_i(t)&=Az_i(t)+Bu_i(t)+G(y_i(t)-\xi_i(t))\\ \xi_i(t)&=Cz_i(t)\end{aligned} \tag{4}$$

Where $z_i(t)$ and $\xi_i(t)$ respectively represent the status and output of the observer i . G is feedback matrix to be sought.

Assumption 1 Linear multi-agent system , (A,B) is controllable, and (A,C) is considerable.

So, for the agent i , based on the above observer relative output information, the following control protocol is designed.

$u_i=K\sum_{j\in N_i}a_{ij}(\xi_j(t)-\xi_i(t))$, where K is feedback matrix to be sought.

Agent state and observer observation error is $h_i=x_i-z_i$ So

$$\begin{aligned}\dot{h}_i(t)&=Ah_i(t)-G(y_i(t)-\xi_i(t))+Df(y_i)=(A-GC)h_i(t)+Df(y_i)\\ \dot{x}_i(t)&=Ax_i(t)+BKC\sum_{j\in N_i}a_{ij}(h_i(t)-h_j(t))-BKC\sum_{j\in N_i}a_{ij}(x_i(t)-x_j(t))+Df(y_i)\end{aligned} \tag{5}$$

Then the system can be described as

$$\begin{aligned}\dot{x}(t)&=(I\otimes A-L\otimes BKC)x(t)+(L\otimes BKC)h(t)+(I\otimes D)f(y)\\ \dot{h}(t)&=(I\otimes(A-GC))h(t)+(I\otimes D)f(y)\end{aligned} \tag{6}$$

Definition 1 The system can get consensus, when for arbitrary finite initial state $x_i(0)$, there is control protocol $u_i(t)$ let $\lim_{t\to\infty}\|x_i(t)-x_j(t)\|=0, \forall i,j=1,2,\ldots\mathrm{N}$

Consider system consistency issues, let $\hat{x}_i=x_i-x_{i+1}\ i=1,2,\ldots N-1$

So we can get $\hat{x}(t)=(E\otimes I)x=[\hat{x}_1,\hat{x}_2,\ldots,\hat{x}_{N-1}]^T$

The same reason we can get $\hat{h}(t)=(E\otimes I)h=\left[\hat{h}_1,\hat{h}_2,\ldots,\hat{h}_{N-1}\right]^T$,where $\hat{h}_i=h_i-h_{i+1}$, $i=1,2,\ldots N-1$.
$\phi\ y\ =\ E\otimes I\ f\ y\ =\left[\phi_1^T\ y\ ,\phi_2^T\ y\ ,\ldots,\phi_{N-1}^T\ y\ \right]^T$,where $\phi_i(y)=f(y_i)-f(y_{i+1})\leq\delta_i\ y_i-y_{i+1}$, $i=1,2,\ldots N-1$.

So we can get

$$\begin{aligned}\dot{\hat{x}}&=\left(I_{N-1}\otimes A-EM\otimes BKC\right)\hat{x}+\left(EM\otimes BKC\right)\hat{h}+\left(I_{N-1}\otimes D\right)\phi(y)\\ \dot{\hat{h}}&=\left(I_{N-1}\otimes\left(A-GC\right)\right)\hat{h}+\left(I_{N-1}\otimes D\right)\phi(y)\end{aligned}\tag{7}$$

And $\lim_{t\to\infty}\left\|\hat{x}_i(t)\right\|=0,\forall i=1,2,\ldots\mathrm{N}-1\Leftrightarrow\lim_{t\to\infty}\left\|x_i(t)-x_j(t)\right\|=0,\forall i,j=1,2,\ldots\mathrm{N}$

3. Theoretical results

The control protocol design process is as follows

a. Select scalar parameter $0<\alpha<2\min\left\{\mathrm{Re}\left(\lambda\left(EM\right)\right)\right\}$, use lemma 5,we can get Q

b. Select the appropriate nonlinear slope parameter σ_0 , solve linear matrix inequality (6),we can get P_1

$$\begin{bmatrix}A^TP_1+P_1A-\dfrac{\alpha}{2}P_1BB^TP_1 & P_1D & C^T\\ D^TP_1 & -I & 0\\ C & 0 & -\dfrac{1}{\sigma_0^2}I\end{bmatrix}<0\tag{8}$$

c. solve linear matrix inequality (7),we can get P_2

$$\begin{bmatrix}A^TP_2+P_2A+\dfrac{\alpha}{2}P_1BB^TP_1 & P_2D & C^T\\ D^TP_2 & -I & 0\\ C & 0 & -\dfrac{1}{\sigma_0^2-2}I\end{bmatrix}<0\tag{9}$$

Prove: Consider the following piecewise continuous quadratic Lyapunov function

$$\begin{aligned}V(t)&=V_1(t)+V_2(t)\\ V_1(t)&=\hat{x}(t)^T\left(Q\otimes P_1\right)\hat{x}(t)\\ V_2(t)&=\hat{h}(t)^T\left(Q\otimes P_1\right)\hat{h}(t)\end{aligned}\tag{10}$$

let $K=B^TP_1C^T\left(CC^T\right)^{-1}$, use lemma 5,we can get

$$\begin{aligned}&\dot{V}_1(t)=\hat{x}(t)^T\left(Q\otimes\left(A^TP_1+P_1A\right)-\left(\left(EM\right)^TQ+QEM\right)\otimes P_1BB^TP_1\right)\hat{x}(t)\\ &+2\hat{x}(t)^T\left(QEM\otimes P_1BB^TP_1\right)\hat{h}(t)+2\hat{x}(t)^T\left(Q\otimes P_1D\right)\phi(y)\\ &\leq\hat{x}(t)^T\left(Q\otimes\left(A^TP_1+P_1A-\alpha P_1BB^TP_1\right)\right)\hat{x}(t)+\hat{x}(t)^T\left(Q\otimes\alpha P_1BB^TP_1\right)\hat{h}(t)+2\hat{x}(t)^T\left(Q\otimes P_1D\right)\phi(y)\end{aligned}\tag{11}$$

Let $\tilde{x}(t)=\left(U^T\otimes I\right)\hat{x}(t)$, $\tilde{h}(t)=\left(U^T\otimes I\right)\hat{h}(t)$ We can get

$$\begin{aligned}&\dot{V}_1(t)\leq\tilde{x}(t)^T\left(\Omega\otimes\left(A^TP_1+P_1A-\alpha P_1BB^TP_1\right)\right)\tilde{x}(t)+\tilde{x}(t)^T\left(\Omega\otimes\alpha P_1BB^TP_1\right)\tilde{h}(t)\\ &+2\tilde{x}(t)^T\left(\Omega\otimes P_1D\right)\left(U^T\otimes I\right)\phi(y)\end{aligned}\tag{12}$$

Use lemma 2, we can get

$$
\begin{aligned}
&2\tilde{x}(t)^T(\Omega\otimes P_1D)(U^T\otimes I)\phi(y)\\
&\le \tilde{x}(t)^T(\Omega R\Omega\otimes P_1DD^TP_1)\tilde{x}(t)+\left[(U^T\otimes I)\phi(y)\right]^T(R^{-1}\otimes I)\left[(U^T\otimes I)\phi(y)\right]\\
&\le \tilde{x}(t)^T(\Omega R\Omega\otimes P_1DD^TP_1)\tilde{x}(t)+\sigma_0^{\,2}\tilde{x}(t)^T(R^{-1}\otimes C^TC)\tilde{x}(t)
\end{aligned}
\tag{13}
$$

Let $R=\Omega^{-1}$, so

$$
\begin{aligned}
&\dot{V}_1(t)\le \tilde{x}(t)^T\left(\Omega\otimes\left(A^TP_1+P_1A-\alpha P_1BB^TP_1+P_1DD^TP_1+\sigma_0^{\,2}C^TC\right)\right)\tilde{x}(t)\\
&+\tilde{x}(t)^T(\Omega\otimes\alpha P_1BB^TP_1)\tilde{h}(t)
\end{aligned}
\tag{14}
$$

The same reason, we can get

$$
\dot{V}_2(t)=\hat{h}(t)^T\left(Q\otimes\left(P_2(A-GC)+(A-GC)^TP_2\right)\right)\hat{h}(t)+2\hat{h}(t)^T(Q\otimes P_2D)\phi(y) \tag{15}
$$

Let $G=P_2^{-1}C^T$, so

$$
\dot{V}_2(t)=\tilde{h}(t)^T\left(\Omega\otimes\left(A^TP_2+P_2A-2C^TC\right)\right)\tilde{h}(t)+2\tilde{h}(t)^T(\Omega\otimes P_2D)(U^T\otimes I)\phi(y) \tag{16}
$$

Use lemma 2 ,we can get

$$
2\tilde{h}(t)^T(\Omega\otimes P_2D)(U^T\otimes I)\phi(y)\le\tilde{h}(t)^T(\Omega R\Omega\otimes P_2DD^TP_2)\tilde{h}(t)+\sigma_0^{\,2}\tilde{h}(t)^T(R^{-1}\otimes C^TC)\tilde{h}(t) \tag{17}
$$

So,

$$
\dot{V}_2(t)\le\tilde{h}(t)^T\left(\Omega\otimes\left(A^TP_2+P_2A-2C^TC+\sigma_0^{\,2}C^TC+P_2DD^TP_2\right)\right)\tilde{h}(t) \tag{18}
$$

We can get

$$
\begin{aligned}
&V(t)=V_1(t)+V_2(t)\le\tilde{x}(t)^T\left(\Omega\otimes\left(A^TP_1+P_1A-\alpha P_1BB^TP_1+P_1DD^TP_1+\sigma_0^{\,2}C^TC\right)\right)\tilde{x}(t)\\
&+\tilde{x}(t)^T(\Omega\otimes\alpha P_1BB^TP_1)\tilde{h}(t)+\tilde{h}(t)^T\left(\Omega\otimes\left(A^TP_2+P_2A-2C^TC+\sigma_0^{\,2}C^TC+P_2DD^TP_2\right)\right)\tilde{h}(t)
\end{aligned}
\tag{19}
$$

Let

$$
\zeta(t)=\left[\tilde{x}(t)\quad\tilde{h}(t)\right]^T,\mathrm{H}=\begin{bmatrix}\mathrm{H}_1 & \mathrm{H}_3\\ \mathrm{H}_3 & \mathrm{H}_2\end{bmatrix},V(t)\le\zeta(t)^T\mathrm{H}\zeta(t) \tag{20}
$$

Where

$$
\begin{aligned}
&\mathrm{H}_1=A^TP_1+P_1A-\alpha P_1BB^TP_1+P_1DD^TP_1+\sigma_0^{\,2}C^TC\\
&\mathrm{H}_2=A^TP_2+P_2A-2C^TC+\sigma_0^{\,2}C^TC+P_2DD^TP_2\\
&\mathrm{H}_3=\frac{\alpha}{2}P_1BB^TP_1
\end{aligned}
\tag{21}
$$

Solve linear matrix inequality ,we can get $\mathrm{H}_1+\mathrm{H}_3<0$ and $\mathrm{H}_1<0$

Solve linear matrix inequality ,we can get $\mathrm{H}_2+\mathrm{H}_3<0$ and $\mathrm{H}_2<0$.

Further we can get $\mathrm{H}_2-\mathrm{H}_3^{\,T}\mathrm{H}_1^{\,-1}\mathrm{H}_3\le\mathrm{H}_2+\mathrm{H}_3^{\,T}\mathrm{H}_3^{\,-1}\mathrm{H}_3=\mathrm{H}_2+\mathrm{H}_3<0$

Use lemma 3,we can get $\mathrm{H}<0$, which is $V(t)<0$ So, the system reach a consensus.

Note 1 Compare with literature [25], the method used in this paper is applicable to more common communication topologies, releasing the conditions of the undirected graph to the directed graph. By transforming, the Laplacian matrix of the system is similar to the approximate array, and it is impossible to perform coupling analysis on the multi-agent system with external disturbances. This paper uses a special variable substitution to successfully transform the consistency problem of multi-agent systems into the stability problem of dimensionality reduction systems. The method used in this paper has more intuitive physical meaning and a more concise derivation process.

Note 2 Compare with literature [11-12], the consensus problem of the leader without the leader in this paper. Compared with the non-navigate consistency problem, the consistency control problem of the leader can often define the tracking error of the follower and the leader. It's easier to turn it into a stability problem with an equivalent system. We can deal with mature M matrix theory.

IOP Conf. Series: Journal of Physics: Conf. Series **1267** (2019) 012030 doi:10.1088/1742-6596/1267/1/012030

Note 3 Inspired by the processing method of matrix M in the derivation process of the literature [26], it is necessary to introduce linear unitary transformation. In reference [24], the linear unitary transformation is not used in the derivation process. Although the correct conclusion can be obtained, the derivation process in this paper is more rigorous and reasonable.

4. Simulation results

This section demonstrates the validity of the conclusions obtained through a numerical experiment. Considering a multi-agent system consisting of a set of Chua's circuits, the Chua's circuit can be written in the form of a Lur'e type nonlinear system.

$$x_i = \begin{bmatrix} x_{i1} \\ x_{i2} \\ x_{i3} \end{bmatrix}, A = \begin{bmatrix} -a(m_1+1) & a & 0 \\ 1 & -1 & 1 \\ 0 & -b_0 & 0 \end{bmatrix}, B = \begin{bmatrix} 0 \\ 0 \\ d \end{bmatrix}, C = \begin{bmatrix} 1 & 0 & 0 \end{bmatrix}, \; , D = \begin{bmatrix} -a(m_2 - m_1) \\ 0 \\ 0 \end{bmatrix} \tag{22}$$

Nonlinear link $f(y_i) = \frac{1}{2}(|x_{i1}+1| - |x_{i1}-1|)$, slope condition is [0, 1]. Select parameter

$a = 9, b_0 = 14, m_1 = -\frac{3}{4}, m_2 = -\frac{4}{3}, d = 10$

The directed topology is shown in Figure 1. Easy to verify, the topology contains a directed spanning tree. Let α=1.9 Solve Q with the LMI toolbox. Further, the feedback matrix can be calculated.

$$K = \begin{bmatrix} -1.438 & -1.4750 & -0.1051 \end{bmatrix}$$

$$G = \begin{bmatrix} 1.4570 & 3.3999 & 26.1296 \end{bmatrix}^T$$

$$Q = \begin{bmatrix} 32.2857 & -0.2398 & 0.4793 \\ -0.2398 & 3.4716 & 3.2651 \\ 0.4793 & 3.2651 & 3.3234 \end{bmatrix}$$

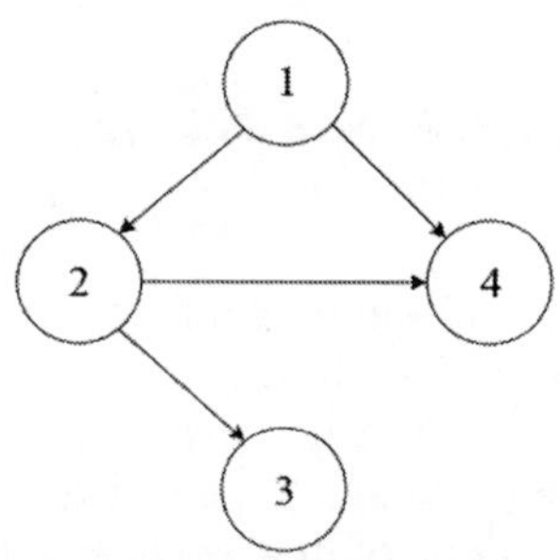

Figure 1. Topology

The states of each agent in a multi-agent system are shown in Figure 2, Figure 3 and Figure 4 respectively. It can be seen from the graph that all three states of an agent can achieve consensus

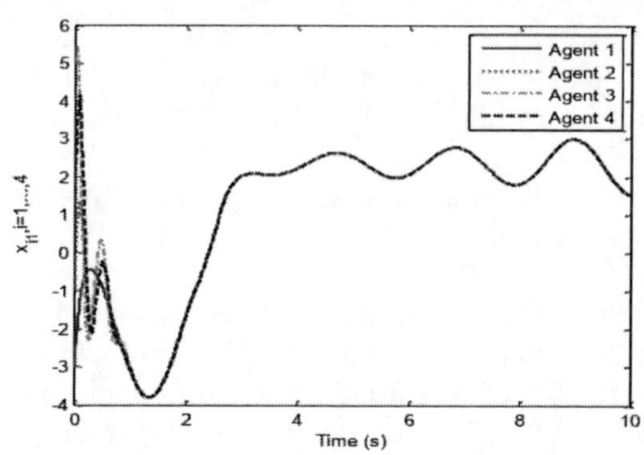

Figure 2. Agent status $x_{i1}, i = 1,2,3,4$

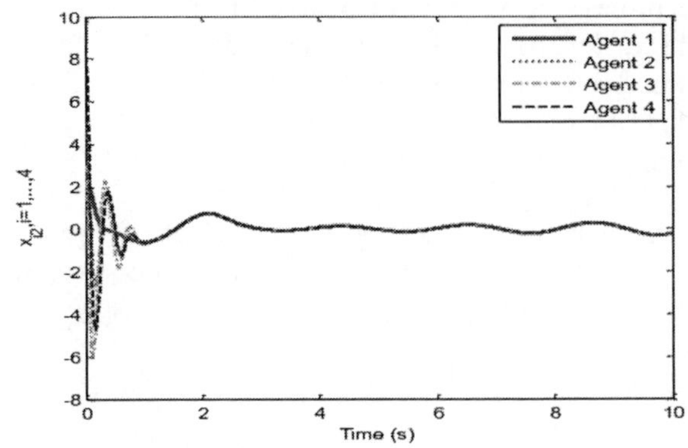

Figure 3. Agent status $x_{i2}, i = 1,2,3,4$

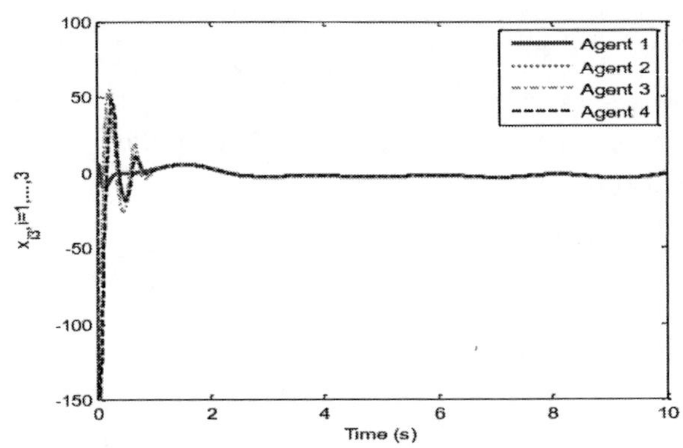

Figure 4. Agent status $x_{i3}, i = 1,2,3,4$

5. Conclusions

In this paper, the consensus problem of multi-agent systems with Lur'e-type non-linear links is transformed into the stability problem of low-dimensional systems by decomposing and substituting the Laplacian matrix in a specific form under general communication topology. By constructing Laypunov function, this paper gives and proves the consistency control protocol of the system. The simulation results verify the effectiveness of the proposed method. Compared with the existing conclusions, this paper reduces the communication topology conditions of multi-agent systems, considering that a single agent can not effectively obtain the relative state information of neighbor nodes. It makes the conclusion of this paper more practical.

References

[1] Dong Y, Huang J. Flocking with connectivity preservation of multiple double integrator systems subject to external disturbances by a distributed control law[J]. Automatica,2015, 55: 197-203.

[2] Dong Y, Huang J. leader-following connectivity preservation rendezvous of multiple double integrator systems based on position measurement only[J]. IEEE Trans on Automatic Control, 2014, 59(9): 2598-2603.

[3] Wu Z, Guan Z, Wu X, et al. Consensus based formation control and trajectory tracing of multi-agent robot systems[J]. J of Intelligent and Robotic Systems, 2007,48(3): 397-410.

[4] Ma Z, Liu Z, Chen Z. Distributed formation control for a multi-agent system with dynamic role assignment[C]. Proc of the 34th Chinese Control Conf. Hangzhou: IEEE Press,2015: 7118-7123.

[5] Ren W, Beard R W. Consensus seeking in multiagent systems under dynamically changing interaction topologies[J]. IEEE Trans on Automatic Control,2005,50(5):655-661.

[6] OH K,AHN H. Formation control for second-order multi-agent system with time-varying delays under directed topology[J].Automatica,2011,47(10):2306-2312.
[7] S.Tuna. Synchronizing linear systems via partial-state coupling[J]. Automatica,2008,44(8):2179-2184.
[8] M.Arcak. Passivity as a design tool for group coordination[J].IEEE Transactions on Automatic Control,2005,52(8):1380-1390.
[9] Z Qu, J Wang, R Hull. Cooperative control of dynamical systems with application to autonomous vehicles[J].IEEE Transactions on Automatic Control,2008,53(4)894-911.
[10] Guo L, Nian X, Zhao Y, et al. Cluster synchronisation of Lur'e dynamical[J]. IET Control Theory and Applications. 2012, 6(16): 2499-2508.
[11] Zhao Y, Duan Z, Wen G, et al. Robust consensus tracking of multi-agent systems with uncertain lur'e-type non-linear dynamics[J]. IET Control THEORY And Applications. 2013, 7(9): 1249-1260.
[12] Gao L, Xu B, Li J, et al. Distributed reduced-order observer-based approach to consensus problems for linear multi-agent systems[J]. IET Control Theory & Applications. 2015, 9(5): 784-792.
[13] Chu H, Yuan J, Zhang W. Observer-based adaptive consensus tracking for linear multi-agent systems with input saturation[J]. IET Control Theory & Applications,2015, 9(14):2124-2131.
[14] Mesbahi A, Velni J M. Distributed observer-based cooperative control for output regulation in multi-agent linear parameter-varying systems[J]. IET Control Theory & Applications, 2017,11(9):1394-1403.
[15] Zhang H, Feng G, Yan H, Chen Q. Observer-based output feedback event-triggered control for consensus of multi-agent systems[J]. IEEE Transactions on Industrial Electronics, 2014,61(9):4885-4894.
[16] Jin Z Q, Niu Y G, Zou Y Y. Consensus for multi-agent systems with sliding-mode observer[J]. Control Theory & Applications,2017,34(2):251-259.
[17] Gao L, Xu B, Li J, et al. Distributed reduced-order observer-based approach to consensus problems for linear multi-agent systems[J]. IET Control Theory & Applications. 2015, 9(5): 784-792.
[18] Li Z, Liu X, Lin P, et al. Consensus of linear multi-agent systems with reduced-order observer-based protocols[J]. Systems & Control Letters. 2011, 60(7): 510-516.
[19] Zhang H, Lewis F L, Das A. Optimal Design for Synchronization of Cooperative Systems: State Feedback, Observer and Output Feedback[J]. IEEE Transactions on Automatic Control. 2011, 56(8): 1948-1952.
[20] Gao L, Xu B, Li J, et al. Distributed reduced-order observer-based approach to consensus problems for linear multi-agent systems[J]. IET Control Theory & Applications. 2015, 9(5): 784-792.
[21] Zhao Y, Li Z, Duan Z. Distributed consensus tracking of multi-agent systems with nonlinear dynamics under a reference leader[J]. INTERNATIONAL JOURNAL OF CONTROL. 2013, 86(10): 1859-1869.
[22] Wu Min, He Yong. Stability Analysis and Robust Control of Time-Delay Systems [M]. Science Press, 2008
[23] Zhou S, Liu W, Wu Q, et al. Leaderless consensus of liner multi-agent systems: matrix decomoposition approach[C]. Proc of the 7th Int Conf on Intelligent Human-Machine Systems and Cybernetics. Hangzhou: IEEE Press,2015:327-331.
[24] Liu W, Zhou S, Qi Y, et al. Leaderless consensus of multi-agent systems with Lipschitz nonlinear dynamics and switching topologies[J]. Neurocomputing. 2015.
[25] Wang J, Duan Z, Zhao Y, et al. H_∞ and H_2 control of multi-agent systems with transient performance improvement[J]. Int J of Control,2013,86(12):2131-2145.
[26] Horn R A, Johnson C R. Matrix Analysis[M]. People Post Press, 2005: 1-17.

Analysis and Research on Influencing Factors of Haze Weather

Hua Lan* **and Yun Pan**

School of Computer Science Community University of China, Beijing, China

*1195252827@qq.com

Abstract. In recent years,the causes, prevention and control methods and prediction system of continuous haze weather have become the focus of social research.Taking the meteorological observation data of Hebei province as an example, this paper studies the influencing factors of continuous haze weather.Firstly, the improved TOPSIS algorithm is used to preprocess the data. Then, it combines the data mining methods such as hierarchical analysis and grey correlation analysis to carry out modeling analysis , and it is concluded that the particulate matter in the pollution source has the greatest impact on the haze weather. Moreover, as the aerodynamics factor affects the diffusion and aggregation of the haze pollution source, it is found that it has a linear influence on the formation of haze weather.Finally, through the current meteorological observation data, BP neural network is used to predict haze weather changes.A large number of experimental results show that, on the premise of allowing a certain error rate, the prediction effect of the BP neural network model is relatively accurate,and it also indicates that the formation of haze weather is closely related to the air quality index factor and meteorological index factor.

1. Introduction

In recent years, with the rapid development of heavy industry in our country, haze weather in our country has been frequent, and the total number of days of haze weather has also increased year by year, and the hazards caused by haze weather have become more and more serious.Especially since 2013, the haze weather has led to great economic losses in China and affected the health of residents[1].In the winter of 2016, several provinces in north China, including Beijing, Shanxi, Hebei and Tianjin, suffered from severe haze weather[2].For example, due to the haze weather, the air is cloudy and the visibility drops sharply, so the flights in China are forced to stop sailing, even to the point of being forced to land on the spot.Besides, the haze weather has led to cars being unable to drive normally. Vehicles have been forced to stop their journey, and even some places have closed down the expressway.In addition, it is also worth noting that as the regions affected by haze weather are expanding year by year, the harm degree of haze is also increasing. More and more people are affected by haze weather, especially the elderly and children with low immunity[3].During the periods of frequent haze weather in spring and winter, they would fall ill to various degrees due to the influence of haze weather.In view of this, the whole society's attention to haze weather is constantly increasing, and therefore, the research on the influencing factors of haze weather becomes a hot topic.

Based on the mechanism of the formation of haze weather and prompted haze weather form the influence of the factors can be roughly divided into two groups[4],one is the source of pollution in the composition of haze weather (i.e., sulfur dioxide, carbon monoxide, nitrogen dioxide, ozone and

PM10 and PM2.5), The other(i.e., humidity, temperature, wind velocity, visibility, and atmospheric pressure) is through the diffusion of the composition of the haze weather, which in turn affects the weather conditions formed by the haze weather.Based on this, this article takes Hebei province meteorological monitoring system to collect the relevant data as the research object.Firstly, the AHP (i.e.,the analytic hierarchy process) method and the improved TOPSIS algorithm are used to preprocess the abnormal data in the original data.Then,data mining methods such as grey correlation analysis and correlation analysis are used to explore the influencing factors of haze weather from the aspects of pollution sources and meteorological indexes respectively,Finally,by using the historical data of meteorological monitoring system, the BP neural network prediction model is constructed to analyze the future haze weather.

2. *The improved TOPSIS algorithm*

TOPSIS (Technique for Order Preference by Similarity to an Ideal Solution) algorithm was first proposed by two foreign scholars in 1981, namely c.l.wang and K.Yoon.TOPSIS algorithm is to calculate the positive ideal solution and the negative ideal solution of each indicator by calculating the distance from each solution to the positive and negative ideal solution.Therefore, it is sometimes called the ideal solution [5].

Due to the wide variety and complexity of the influencing factors of haze weather, some abnormal data, such as repeated data, negative data, inverted data (PM2.5>PM10) and zero value data, often appear in the data acquired by the actual meteorological monitoring system.How to carry out data cleaning for these abnormal data is the primary task to study the correlation of haze weather factors.In this paper, the algorithm idea of TOPSIS is used to build an improved TOPSIS model, and the ideal solution of each abnormal data was obtained, and the ideal value was used to replace the abnormal data, so as to achieve the goal of removing the abnormal data.

The steps to build an improved TOPSIS model are shown below[6].

(1) The vector normalization method is used to obtain the normative decision matrix,Set the decision matrix of multi-attribute decision problem as $\mathrm{A}=(a_{ij})_{m\times n}$, Set the normalized decision matrix as

(2) , and

$$\mathrm{b}_{ij}=a_{ij}\,/\sqrt{\sum_{n}^{m}a_{ij}^{2}} \tag{1}$$

Where $\mathrm{i}=\{i=1,2,...,m\}$; $\mathrm{j}=\{1,2,...,\mathrm{n}\}$.

(2)Form the weighted canonical matrix $\mathrm{C}=(\mathrm{c}_{ij})_{m\times n}$. Set the weight vector of each attribute given by the decision maker $\mathrm{w}=[w_1,w_2,...,w_n]^{\mathrm{T}}$, and

$$\mathrm{c}_{ij}=w_j\bullet b_{ij} \tag{2}$$

Where $\mathrm{i}=\{i=1,2,...,m\}$; $\mathrm{j}=\{1,2,...,\mathrm{n}\}$.

The positive ideal solution and negative ideal solution are determined, respectively, C^* and C^0 .Set the JTH attribute value of the positive ideal solution C^* to be $\mathrm{c}_{\mathrm{j}}^*$,Set the JTH attribute value of the negative ideal solution C^0 to be $\mathrm{c}_{\mathrm{j}}^0$,Then the positive and negative of ideal solution are $\mathrm{c}_j^*=\begin{cases}\max\limits_i c_{ij}\\ \min\limits_i c_{ij}\end{cases}$ and $\mathrm{c}_j^0=\begin{cases}\min\limits_i c_{ij}\\ \max\limits_i c_{ij}\end{cases}$

Where the j of $\min\limits_i c_{ij}$ is the attribute of benefit type,the j of $\max\limits_{\mathrm{i}} c_{ij}$ is the attribute of cost type,

$j = \{1,2....,n\}$.

(3)Using the vector projection method, the best ideal solution is selected to replace the abnormal data.The normalized original data vector and ideal solution vector are $\partial = (a_1, a_2, ...a_n)^T$ and $\beta = (b_1, b_2, ...b_n)^T$ respectively,and vector projection is

$$\gamma_\beta^\partial = \|\partial\| \times \frac{\partial^T \beta}{\|\partial\|\|\beta\|} = \frac{\sum_{i=1}^{n} a_i b_i}{\sqrt{\sum_{i=1}^{n} b_i^2}} \tag{3}$$

(4)Sort γ_β^∂ and select the ideal solution as the replacement value of the abnormal data.

3. Experimentation

At present, China's ministry of environmental protection regulates the use of AQI (air quality index) to evaluate air quality.In this paper, the correlation degree between sulfur dioxide, carbon monoxide, nitrogen dioxide, ozone, PM10 and AQI is studied by using the grey relational analysis model, so as to analyze the influence of air quality influencing factors on the formation of haze weather and predict the haze weather.

This paper takes the air quality monitoring data monitored by century park monitoring station in a city of Hebei province from September 1 to 30, 2014 as the experimental data, and designed the experimental process according to the research objectives as shown in figure 1 below

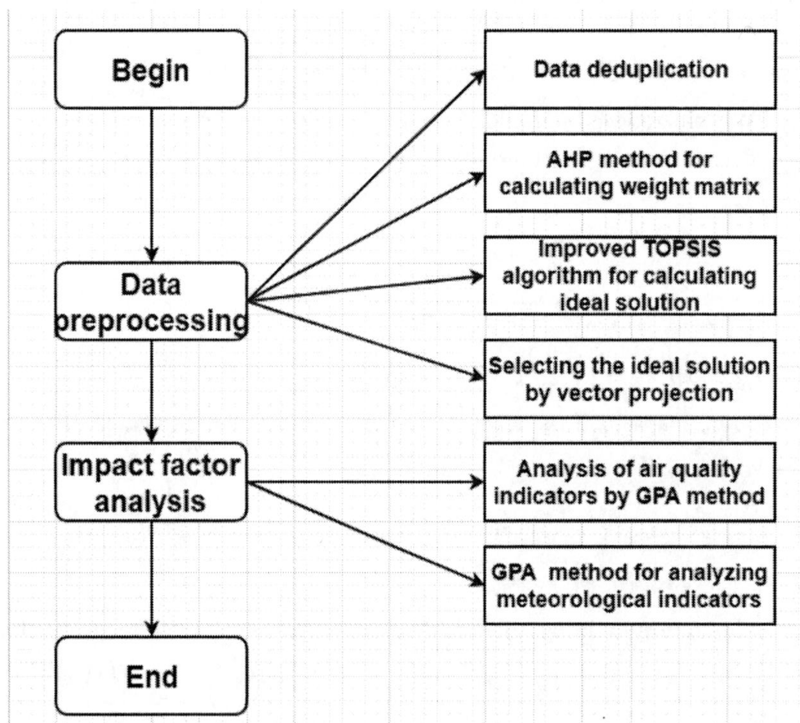

Figure 1. Experimental flow graph.

4. Experimental results and analysis

This experimental scheme is implemented on the platform of Matlab 2016b, the 64-bit operating system of Windows 10.According to the experimental steps, the data preprocessing is carried out first, and then the grey correlation analysis model is constructed to analyze the air quality index factors. Table 1 shows the correlation ranking of air quality indexes on haze weather.

Table 1. Ranking of the influence degree of air quality index on formation of haze weather.

Air quality index	Correlation degree	Order of influnence
PM2.5	0.6734	1
SO_2	0.5261	2
PM10	0.5127	3
O_3	0.5072	4
NO_2	0.4999	5
CO	0.4974	6

According to the data in table 1,it can be concluded that PM2.5 has the greatest impact on haze formation, followed by sulfur dioxide, while the influence of carbon monoxide in air quality index is minimized,carbon monoxide to form fog weather correlation is 0.4974 > 0.3, that is, between the two is strongly related. Therefore sulfur dioxide, carbon monoxide, nitrogen dioxide, ozone and PM10 and PM2.5 has greater impact on forming haze weather.

Figure 2 is the histogram drawn by the correlation coefficient matrix of sulfur dioxide, carbon monoxide, nitrogen dioxide, ozone, PM10 and PM2.5 and AQI in a ten-day period in September 2014.

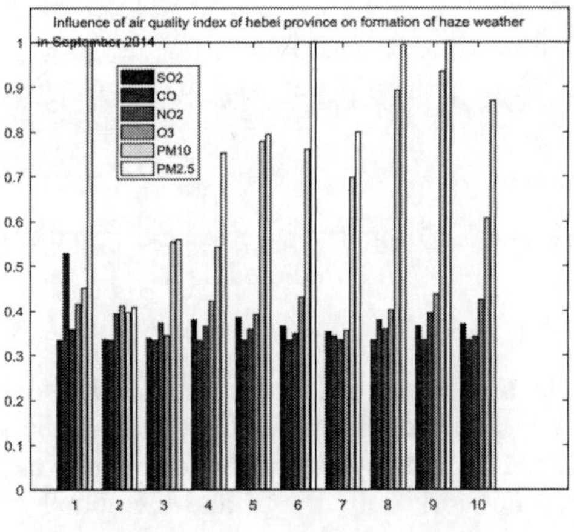

Figure 2 . Correlation degree of air quality index to haze weather.

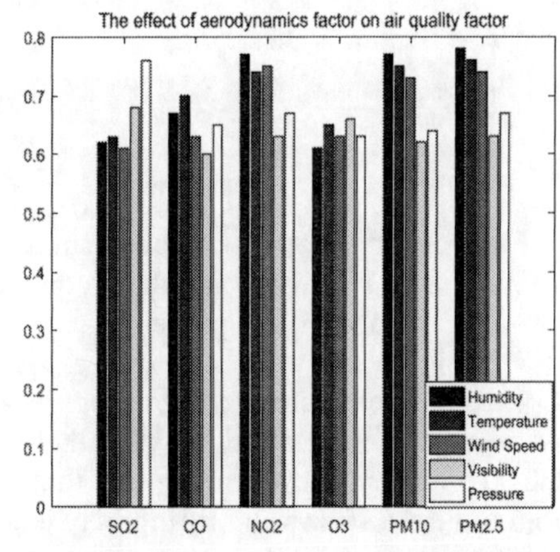

Figure 3. Influence of meteorological factors on air quality index factors.

From the above figure 2, it can be seen that compared with ozone, nitrogen dioxide and carbon monoxide, PM10, PM2.5 and sulfur dioxide are highly correlated with AQI, and then it can be inferred that PM10, PM2.5 and sulfur dioxide have more significant effects on the formation of haze weather.According to the relevant knowledge of aerodynamics, humidity, temperature, wind speed, visibility and air pressure in meteorological indicators affect their concentration by influencing the speed of aggregation and diffusion of air quality indicators such as sulfur dioxide, carbon monoxide, nitrogen dioxide, ozone, PM10 and PM2.5.Therefore, it is necessary to analyze the influence of meteorological factors on air quality index factors at the same time. Figure 3 is a histogram of the correlation degree between meteorological factors and air quality index factors.

From the above figure 3,compared with other meteorological factors, humidity, temperature and wind speed have the most significant influence on air quality index factors, that is, the change of humidity, temperature and wind speed has the most obvious influence on air quality index factor concentration.

Finally, BP neural network model is built to predict the autumn haze weather, the input layer as sulfur dioxide, carbon monoxide, nitrogen dioxide, ozone and PM10 five air quality index factors and the humidity, temperature, wind velocity, visibility and air pressure five meteorological index factor, the output layer of BP model is AQI and PM2.5, number of hidden layer nodes is 10 in the middle.In addition, the input sample is the value of each indicator in September 2014, the test sample is the value of AQI and PM2.5 in September 2014, and the prediction sample is the value of AQI and PM2.5 in September 2015.After several trainings until the ideal effect is achieved (that is, the error between the network monitoring value and the actual monitoring value is up to the requirement), figure 4 shows the correlation between the actual measurement value and the predicted value of the model.

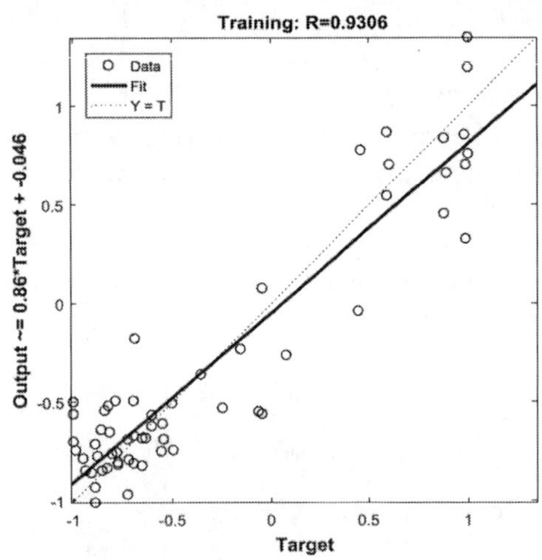

Figure 4. The correlation between the actual measured value and the predicted value of the model.

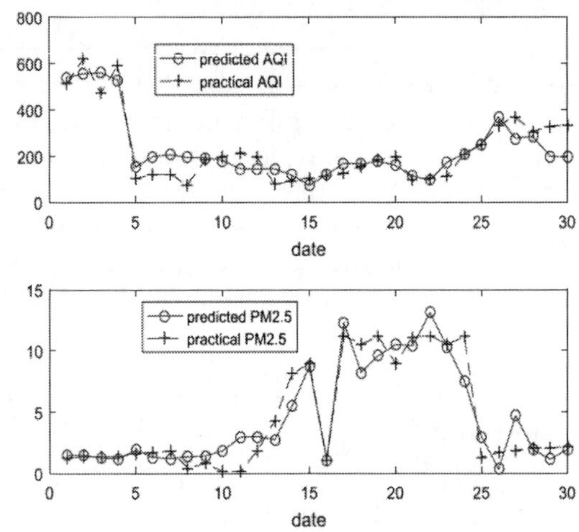

Figure 5. Comparison between original data and predicted data.

As can be seen from the above figure 4, after the BP neural network model has been trained for many times, the correlation between the actual measured value and the predicted value of the model reaches 0.93.Therefore, it can be said that the trained model can be used to predict the haze weather in autumn.Figure 5 shows the fitting between the predicted network output AQI and the actual AQI by the model.

It can be seen from figure 5 that the established BP neural network model has an excellent prediction effect on the autumn haze weather. Within a certain error range, the predicted value is almost consistent with the actual monitoring value. Therefore, this model can be used to predict the future autumn haze weather in Hebei province.

5. Conclusion

In this paper, based on the correlation study of haze weather influence factors in Hebei Province, Fistly, TOPSIS model is constructed, and the abnormal data are replaced by ideal solution. Then, after the collection of data preprocessing, and the influence degree of each influence factor on haze weather is quantitatively analyzed by grey correlation method.Finally, according to the autumn weather data collected, The BP neural network model is established for training and learning, predicting the autumn haze weather conditions.In the research on the correlation analysis of haze weather influencing factors, there are still many areas that need to be continued in the future, such as the following aspects:

IOP Conf. Series: Journal of Physics: Conf. Series **1267** (2019) 012031 doi:10.1088/1742-6596/1267/1/012031

(1) from the perspective of haze weather impact factors, in addition to the study of the pollutants that affect haze formation provided by this region, other pollutants provided by adjacent regions should also be considered.

(2) from the perspective of establishing the forecast model of haze weather, it is necessary to build a forecast model capable of predicting haze weather in spring, summer and winter in the future.

References

[1] Gao Ge.Climatic Characteristics and Variation Analysis of Haze Days in China from 1961 to 2005.*Journal of Geography*, 2008,63(7):761-748.

[2] Guo lijun. A study on the types, structural characteristics and mechanism of persistent smog-haze weather in north China. Nanjing, *nanjing university of information engineering*, 2015.

[3] Tao J,Chan C Y,Lin M,et al.Regression analyses between recent air quality and visibility changes in megacities at four haze regions in China. *Aerosol and Quality Research*, 2012, 12(6):1049-1061.

[4] Bi xueyan, deng xuejiao, li fei, wu wei, et al. Study on the visibility reduction caused by atmospheric haze in the pearl river delta,2015.

[5] Si Shoukui, Sun Xi Jing. Data modeling algorithm and application. Beijing,*National Defense University press*, 2014.

[6] Zhang Kunlong, Yang Wenguang. Computational intelligence. Beijing, *Coal Industry Press*, 2015.

Research of OpenMV Intelligent Monitoring and Disease Identification System

WANG Xiu-qing[1,a], CHEN Qi[1,b], YAO Shi-jia[1,c] and JIA Xiao-yun[1,d]

[1]College of Electronic Information and Automation, Tianjin University of Science & Technology, Tianjin 300222, China

E-mail: [a]lwxqly@163.com; [b]18322031620@163.com; [c]yaoshijia6@163.com; [d]1094622439@qq.com

Abstract. Aiming at the requirement of video surveillance and disease identification in modern greenhouse, this paper designed a system consisted by greenhouse intelligent monitoring and disease identification system. At first, this paper using OpenMV camera designed a disease monitoring system which could recognize and collect disease images automatically by multi-color thresholds tracking. Then, this paper designed a cuckoo search and BP neural network collaborative search (CSBP-CS) algorithm, the algorithm combined the global search capability of Cuckoo Search(CS) and back-propagation algorithm of BP algorithm to optimize weights and thresholds collaboratively. this paper took three tomato diseases and normal leaves as research objects, firstly step was to separate the disease spots from disease images, then was to extract 56 classification features and select 47 excellent classification features by relief F to construct CSBP-CS RF classifier. Finally, this paper compared the classification accuracy of CSBP-CS with CSBP-CS RF network and analyzed the effectiveness of relief F. The simulation results showed that the average correct recognition rate of the CSBP-CS RF was approximately equal to CSBP-CS under the same conditions, but CSBP-CS RF is more simple, so Relief F can help to speed up the efficiency of CSBP-CS.

1. Introduction

Tomato diseases can lead to a decline in crop yield and quality seriously. Therefore, when diseases occurred, agricultural producers need to respond quickly to employ pesticide rationally. so it is indispensable to monitoring greenhouse diseases and research about disease classification models .

In recent years, OpenMV intelligent monitoring system is mainly used in car tracking[1,2]. In the field of video surveillance in Greenhouse, Liu Wenhui et al[3] combined cloud computing technology with video surveillance technology to realize the storage and calculation of massive video files, but plant diseases do not occur all the time, it resulting in the waste of memory resources.

On the other hand, the research on crop disease identification model has attracted more and more scholars' attention[4-5]. Sun Chen[6] put forward the method of CS-BP to forecast the stock market. MA Wei[7] combined cuckoo with SVM, and proposed CS-SVM algorithm for human body recognition.Therefore, the Cuckoo Search(CS) has less optimization parameters and strong global search ability, but easy to fall into local optimization and too slow to convergence in the later stage. So this paper constructed the cooperative search algorithm of CS and BP network (CSBP-CS) and used it for tomato disease identification models.

Firstly, this paper separated the disease spots and extracted 56 classification features as a feature description. Then, Relief F was used to select 47 excellent classification features to construct CSBP-CS classifier. Besides, this paper designs a disease monitoring system based on OpenMV intelligent monitoring system, which can automatically recognize and collect disease image. The collected disease image is recognized by classifier. Finally, this paper compared the recognition accuracy rate of CSBP-CS with CSBP-CS RF network and analyzed the effect of relief F.

2. Disease Image Monitoring

In this paper, OpenMV intelligent monitoring system was applied to the field of greenhouse disease monitoring to realize the function of automatically tracking and collecting disease images with multi-color thresholds tracking.

Because of the disease leaves generally consist of normal disease-free green and brown or green and white or green and yellow spots, this paper combined this characteristic, and used the camera to collect diseased leaves by double color recognition, when tracing the coexistence of green and yellow, or green and brown, or green and white in the surveillance video, the LED warning light turned bright, and the camera automatically took photos and saves pictures. At the same time, the platform was controlled by the steering gear to rotate continuously for 180° from top to bottom and from right to left. Finally, the experiment achieved the effect of the steering gear to rotate continuously and multi-color recognition to take pictures and save pictures. The image copy is saved to the file system in the path by the automatic accumulation method of the image name taken each time and the collected disease images will be recognized by the classifier the sort of disease.

3. Image Segmentation and Feature Extraction

3.1 Disease Image Acquisition

This paper collected 260 images including 65 powdery mildew disease images, 65 gray mould disease images, 65 late blight disease images and 65 normal leaves from the network as training set of classifier, and the disease category in training set was verified by experts. Then the collected disease images by OpenMV intelligent monitoring module will be the test set for testing the effect of classifier, the test set includes 35 powdery mildew disease images, 35 gray mould disease images, 35 late blight disease images and 35 normal leaves, and they are also verified by experts. Most of the selected images are natural backgrounds, so the clipping of complex natural backgrounds when segmenting diseased parts is more likely to cause errors, but it is more suitable for practical applications.

3.2 Disease Image Segmentation

This paper separated the disease spots by the difference of chroma information between disease spots and green chroma based on the RGB-HSI transformation algorithm. HSI model reflects the way human visual system observes color, and separate the color image to chrominance (H), saturation (S), and intensity (I), so it has great advantages in the field of image segmentation. The conversion formula of RGB coordinates to HSI coordinate chrominance H is as follows:

$$H=\begin{cases}\theta, & G\ge B\\ 2\pi-\theta, & G<B\end{cases} \tag{1}$$

in formula (1):

$$\theta=\cos^{-1}\left(\frac{(R-G)+(R-B)}{2\sqrt{(R-G)^2+(R-B)(G-B)}}\right) \tag{2}$$

Then the chrominance component is normalized to a chromaticity adjustment factor h. By adjusting the range of h, the paper separated the disease spot. This paper selected $h\in[1/6,0.5]$ as the basic value, and fine-tune between [1/6,0.6] when the effect is poor. The segmentation result is shown in Figure 1.

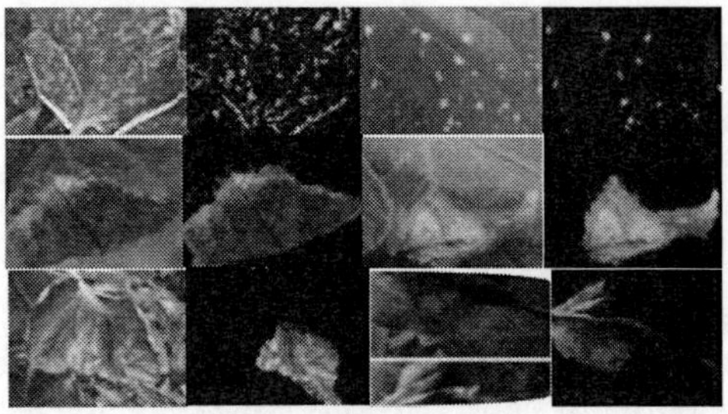

Figure.1 Segmentation of disease spots

It can be seen from the figure that HSI model can quickly and excellently segment the three kinds of disease images in this paper. By experiment, the algorithm only takes 2.036 seconds to divide 100 disease images in MATLAB, it can be known that the adaptability of HSI model is wide and it is suitable for quickly extracting a large number of diseased leaf images.

3.3 Feature Extraction

In this paper, 400 images were segmented for extracting features set. Firstly, 9 color features was extracted including the R, G, and B color components of the color histogram RGB, the hue, saturation, and brightness of the color space HSV, and the chroma, saturation, and intensity of the color model HSI. Then, the gray-gradient concurrence matrix method[8] was used to extract Gray Level Average, Inverse Gap, Inertia, Mixed Entropy, Gradient Entropy, Gray Level Entropy, Relevance, etc, respectively in RGB histogram, the H, S and V three channels of HSV color space, and H, S, I three channels of HSI color model, a total of 45 texture features. Finally, the tamura method[9] was used to extract the roughness and contrast of the disease plot in RGB coordinates. In total, the 56 features of every samples were extracted as the feature set of the dimension space [400,56], then relief F algorithm was used to select excellent features.

3.4 Excellent Feature Selection by Relief F

The relief proposed at the earliest time is only can deal with two labels of data. So Kononeill extends it and gets the relief F algorithm which can deal with multi-class data. When dealing with multi-class problems, each time a sample R is randomly extracted from the training sample set, then K near-Hits of R are found from the same sample set as R, and K near-Hits are found from the different sample sets of each R. Then the weight of each feature is updated by formula, which is as follows:

$$W(A)=W(A)-\frac{\sum_{j=1}^{k}diff(A,R,H_j)}{mk}+\frac{\sum_{C\in class(R)}[\frac{p(C)}{1-p(class(R))}\sum_{j=1}^{k}diff(A,R,M_j(C))]}{mk} \quad (3)$$

In the formula above, $M_j(C)$ denotes the jth nearest neighbor sample in class C.

This paper choose k = 20 and executes 40 times to get the average weight of each weight, then used 0.05 as the threshold value. Then, 9 redundant features were removed. They are H component, gray level heterogeneity, heterogeneity of gradient distribution in HSV color space; gray level heterogeneity, heterogeneity of gradient distribution, Inertia in RGB color space; gray level heterogeneity, heterogeneity of gradient distribution, inverse difference moment in HSI color space. Finally, the excellent feature set of dimension space [400,47] is obtained.

4. The Detailed Design of CSBP-CS

4.1 Cuckoo Search

The standard cuckoo search algorithm is a group intelligent search algorithm inspired by the behavior of brood parasitism and designed by Xin-She Yang and SuashDeb. Each nest represents a candidate solution, which is based on random walk mode of levy flight (Equation 4). The random walk mode searches for a new solution instead of the old solution, and finally searches for the optimal solution in the nest group.

$$levy = \alpha * s * (x_n - x_b) \tag{4}$$

In (4), x_n is old nest, x_b is current optimal solution, levy flight step is (5)

$$s = u \Big/ |v|^{1/\beta} \tag{5}$$

The variable is given by $\beta \in [0.3, 1.99]$, u and v subject to normal distribution:

$$\{u \sim N(0, \partial_u^2)\}, \{v \sim N(0, \partial_v^2)\}$$

4.2 BP Neural Network

The BP neural network is a Multi Layer Perceptron (MLP) consisting of multiple neurons, which is trained by the BackPropagation algorithm. It consists of input layer, hidden layer and output layer. The neural network model constructed in this paper is shown in Figure 2:

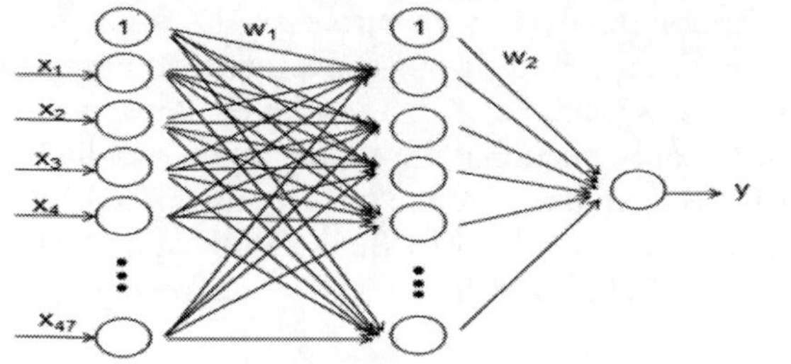

Figure.2 Model of BP

In Fig.2, In output layer L, for each training set output label value y^t ,the error is $\delta^L = a^L - y^t$, Layer l error is calculated by back propagation algorithm as follows:

$$\delta^l = ((\theta^l)^T \delta^{l+1}).*a^l.*(1-a^l) \tag{6}$$

At this time, the weight updating formula for each neuron is as follows:

$$W_{ij}^l = W_{ij}^l + \lambda a_j^l \delta_i^{l+1} + \mu C_{ij} \tag{7}$$

Where W_{ij}^l denotes weights between i, j of l layer neurons. In order to speed up the learning efficiency, the correction matrix mechanism is introduced as follows:

$$C_{ij} = a_j^l \delta_i^{l+1} \tag{8}$$

4.3 The Design of CSBP-CS

Because of convergence speed of CS algorithm is slower when in later stage and easy to fall into local minimization, so this paper combined the back propagation ability of BP neural network with global search capability of CS to propose a cuckoo search and BP cooperative search algorithm (CSBP-CS). CSBP-CS can help CS to jump out of the local minimum when in at the later stage of optimization and decrease the disadvantage of easily falling into the local minimum in CS.

In the collaborative network, this paper combined the thresholds and weights {w1}{b2}{w2}{b3}(Fig.4) of three-layer BP network into a nest vector x^{bp} by formula (9), where {b2}{b3} are the thresholds of hidden layer and output layer respectively,

$$x_{bp} = [w^1, b^2, w^2, b^3] \tag{9}$$

Then compared the best fitness value with x^{cs}'s that is searched by CS algorithm and judged what was better nest according to formula (10):

$$x^b = \begin{cases} x^{bp} & \text{if } f(x^{bp}) < f(x_b{}^{cs}) \\ x_b{}^{cs} & \text{if } f(x_b{}^{cs}) < f(x^{bp}) \end{cases} \tag{10}$$

So the better solution vector x^b was obtained. Then iterative updating of x^b from formula (4) (7) and continuous optimization until the end condition, finally, the optimal solution vector in the CSBP-CS network is searched.

IOP Conf. Series: Journal of Physics: Conf. Series **1267** (2019) 012032 doi:10.1088/1742-6596/1267/1/012032

5. The Result and Analysis of Classification

5.1 Simulation Experiment

In order to analyze the Relief F's effectiveness in feature selection, the simulation experiment is implemented in MATLAB 2014. The training set is features of 260 images, The test set is 140 images that is collected by OpenMV. The optimal bird's nest x^b obtained by cooperative search is used as the weight threshold of the CSBP-CS recognition model. Under the same conditions, Then the CSBP-CS network and CSBP-CS RF network are compared to analyze the effectiveness of Relief F algorithm. The number of iterations is all 50, the search nest of CSBP-CS is 30, but the search nest of CSBP-CS RF is only 25.

5.2 Result and Analysis

The two algorithm classification models were successively implemented for many consecutive independent experiments, this paper mainly from convergence time and tomato disease recognition accuracy rate to analysis the performance of RF. The model fitness value changes with the iteration times as shown in Figure 3. It can be seen that the fitness value of CSBP-CS is reduced to the lowest after 27 iterations, but CSBP-CS RF reduced to the lowest only after 19 iterations, which is faster then the speed of CSBP-CS, and the fitness value of CSBP-CS RF and CSBP-CS has little difference, so reducing feature dimension by Relief F has no effect on recognition efficiency, but accelerate convergence rate.

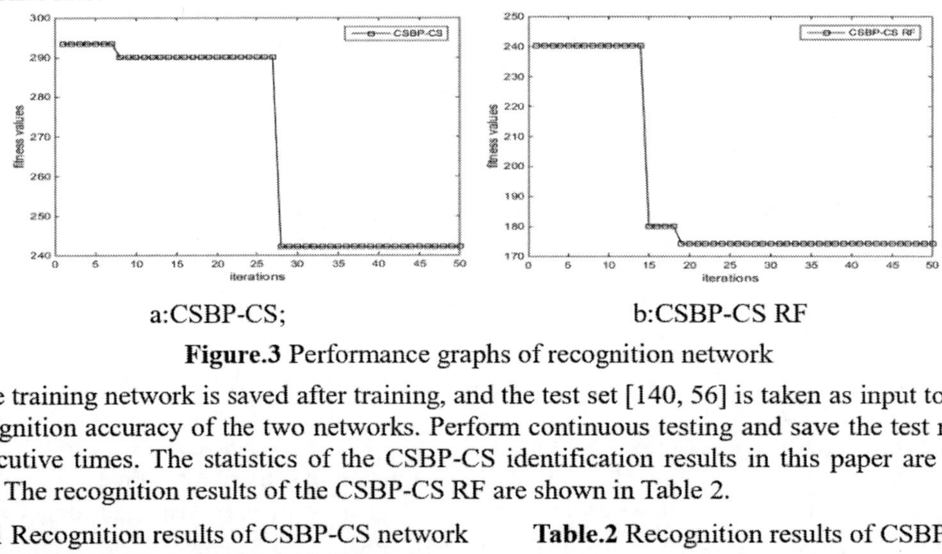

a:CSBP-CS; b:CSBP-CS RF

Figure.3 Performance graphs of recognition network

The training network is saved after training, and the test set [140, 56] is taken as input to calculate the recognition accuracy of the two networks. Perform continuous testing and save the test results for 5 consecutive times. The statistics of the CSBP-CS identification results in this paper are shown in Table 1. The recognition results of the CSBP-CS RF are shown in Table 2.

Table.1 Recognition results of CSBP-CS network

classifier	powdery mildew	gray mold	late blight	normal leaf	recognition accuracy	average accuracy	average iteration time
	35	35	35	35			
CSBP-CS	32	31	31	34	91.4%	90.8%	81.5s
	30	30	33	35	91.4%		
	31	28	32	35	90%		
	32	30	33	34	92.1%		
	30	28	32	35	89.3%		

Table.2 Recognition results of CSBP-CS RF

classifier	powdery mildew	gray mold	late blight	normal leaf	recognition accuracy	average accuracy	average iteration time
	35	35	35	35			
CSBP-CS RF	31	28	34	34	90.7%	90.6%	67.9s
	30	32	33	35	92.9%		
	31	31	30	35	90.7%		
	33	28	31	34	90.0%		
	31	29	30	34	88.6%		

It can be seen from the Table.1 and Table.2 that the average accuracy of CSBP-CS is same as CSBP-CS RF, but the number of nest needed by CSBP-CS RF to research optimal is five fewer than CSBP-CS's because the input of CSBP-CS is 56, but CS BP-CS RF is only 47, so the average iteration time of CSBP-CS RF is less than CSBP-CS. So we can know that reducing feature dimension by Relief F makes the network run faster and won't reduce the recognition accuracy. In conclusion, RF algorithm can be used in this network to accelerate the speed of optimization without affecting the recognition accuracy, and RF can simplify the model.

IOP Conf. Series: Journal of Physics: Conf. Series **1267** (2019) 012032 doi:10.1088/1742-6596/1267/1/012032

6. Summary

In this paper, CSBP-CS RF was designed to be a disease classifier, 260 disease images were used for training set and OpenMV intelligent monitoring system was used to collect 140 disease images for testing set. The RGB-HSI conversion algorithm was selected to realize the segmentation of the disease spot, and 56 features such as color features and texture features were extracted from the disease spot, the relief F was used to select excellent features with large weights and the excellent feature set was used to construct the disease classification model. Than CSBP-CS RF network was proposed to be disease classifier. Simulation experiments showed that the CSBP-CS RF network performance was better than CSBP-CS network, the use of relief F is effective.

References

[1] ZHUANG Qiongyun. Design and Implementation of the Auto-Tracking Smart Car Based on OpenMV, J. Journal of Liming Vocational UniversityActa Agriculturae Zhejiangensis, 2018(4).

[2] MEI Yanpin, FU Rong. Research on the 3D Identification System Based on OpenMV, J. New Technology & New Process, 2018.

[3] Liu Wenhui, Wang Zhe. Design and implementation of green house video monitoring system based on cloud computing, J. Application of Electronic Technique, 2015(6):139-142.

[4] WANG Lin-ming, GAO Xiao-yang, LI Hong-ling,el. Barley diseases discrimination based on neural network, J. Journal of Gansu Agricultural University, 2015,50(02):173-176.

[5] Soni P , Chahar R . A segmentation improved robust PNN model for disease identification in different leaf images[C]// IEEE International Conference on Power Electronics. IEEE, 2017.

[6] Sun Chen, Li Yang, Li Xiaoge,Yu Jiaoyan. Stock forecasting model based on optimising BP neural network with cuckoo search, J.Computer Applications and Software,2016,33(02),276-279.

[7] MA Wei. Human Behavior Recognition Based on Cuckoo Search Algorithm Optimizing Features and Classifier Parameters[J].Microelectronics & Computer, 2016, 33(5):102-105.

[8] XIAO Da, WANG Runmin, ZOU Xiao,el. A Noninvasive Temperature Measurement Based on Gabor Transform and Gray Level Gradient Co-Occurrence Matrix Using Ultrasound, J. Chinese Journal of Sensors and Actuators, 2017, 30(11):1684-1688.

[9] Xiao-Qi L , Jin-Ge G , Yu-Hong Z . Research and realization of Tamura texture feature extraction method based on image segmentation, J. Chinese Journal of Tissue Engineering Research, 2012.

IOP Conf. Series: Journal of Physics: Conf. Series **1267** (2019) 012033 doi:10.1088/1742-6596/1267/1/012033

Exploring Multiple Embedded Features on Event Extraction

Shi-Xiang Yi* **and Chun-Yan Li**

Key Lab of the Ministry of Education, School of Automation, Chongqing University, Chongqing, 400044, China

*Corresponding authors: yishixiang@gmail.com

Abstract. In recent years, the neural network method can automatically learn effectively features. Unlike traditional discrete features, neural network features are mostly continuous features and can be automatically combined to build higher-level features. The efficiency of the features has been proven in numerous tasks in natural language processing and has led to breakthroughs. In this paper, we propose a event extraction system based on combination of multiple embedded features. Our work is mainly based on the three aspects: (1) traditional pipeline systems have serious error propagation problems; (2) there are several different event descriptions in the text; (3) representation learning can provide rich semantic and syntactic representation. As a result, we achieve competitive performance, specifically, F1-measure of 60.25 in event extraction. Meanwhile, evaluation results point out some shortcomings that need to be addressed in future work.

1. Introduction

In such an era of information overload, how to obtain valuable information that people want from massively fragmented text information is a huge challenge. Moreover, a lot of inflammatory titles make readers more suspicious towards news content. Information extraction technology can effectively solve this problem. The general goal of information extraction is to extract specific structured information for text. Therefore, this method can automatically mine key information from massive news data and help to further classify, extract, fuse and verify information. Research in this field is based primarily on Automatic Content Extraction (ACE). The ACE defines the information to be extracted for the event extraction task:

- Event Mention: A phrase or sentence describe the event, containing a trigger and any number of arguments.
- Event Trigger: The word in the event description that best represents the occurrence of the event, and it is generally a verb or a noun.
- Event Argument: Important information of an event consists of fine-grained units that express complete semantics, such as entities and attribute values.
- Argument Role: The role of the event element in the event, the semantic relationship between the event element and the event.

A number of event extraction frameworks are feature-based approaches to apply to sentence-level or document-level. Feature-based methods mainly rely on a large number of feature engineering for machine learning models, from local features to high-level feature representations. Gupta [1] chose to find the cross-document anchoring features to seek the event extraction model applicable to different domains. Liao [2] tagged the easier cases, and use such knowledge to help tag the harder cases.

In recently, state-of-the-art event extraction system uses neural network models to improve event extraction. Neural networks have recently been introduced into event extraction with the goal of overcoming the two basic limitations of traditional feature-based methods: (1) feature engineering for rich feature sets; (2) error propagation caused by feature extraction phase. All of these models use word embedding, a generic word representation generated through unsupervised training on deep learning models. Jagannatha and Yu [3] use bi-directional recurrent neural network to extract instances of events in health records. Feng [4] developed a hybrid neural network (CNN and RNN) that captures sequence information and block information from sentences, enabling training of an event detector that does not rely on any manual features.

However, to our knowledge, there is no comprehensive research on concatenating embedded features in a pipeline model. The remainder of this paper is organized as follows: We present the motivation for building models by evaluating existing models (section 2). Subsequently, we explain the model proposed in this paper in detail (section 3), including the trigger classification model and the argument role classification model. Finally, we conduct experiments based on the model proposed in this paper (section 4) and conclude (section 5).

2. Motivations

In this section we analyze the event extraction method mentioned above and present our solutions based on error analysis for the baseline event extraction.

2.1. Error Propagation

The traditional event extraction method divides the whole process into event recognition task and argument role classification task. And there are multiple subtasks in each task, such as trigger recognition, event classification, etc. If the previous task is biased, subsequent work will no longer be feasible and effective. Björne's research shows that more than 60% of the errors in event extraction are derived from trigger recognition tasks. Therefore, the completion of the trigger word recognition with high quality is a guarantee for subsequent public safety event extraction.

In this work, we use the representation learning method of embedded features to solve this problem. Embedding essentially casts features into a higher-dimensional space based on the tasks. This allows the classifier to learn representation better and more comprehensively. Embedding means converting data into feature representations, some of which can be represented by the distance of the vector distance.

2.2. Feature Engineering

Feature-based event extraction systems require natural language processing toolkits to extract various types of local features and global features. We summarize the commonly used features below.

(1) Local features of argument: context of the argument mention, type of argument, type of event, common ancestor node between arguments in the parse tree, relative position of current argument and trigger and whether there are candidate arguments of the same type. (2) Local features of trigger: type of trigger, distance between current argument and trigger, syntactic dependency type with trigger words, the closest entity type with the trigger and the path from the leaf node of the trigger to the root in the parse tree. (3) Cross-domain features: entity consistency, trigger consistency, co-occurrence with trigger in clause and entity-subtypes of arguments.

The above features rely on traditional external NLP tools such as dependency analysis, syntax analysis, and part-of-speech tagging in the feature extraction process. Therefore, it would cause error accumulation. In addition, some languages and fields do not have such processing tools, and the features need to be manually set. A well-designed manual feature can bring significant improvements, but in contrast, it can make the system slow and difficult to handle in large-scale applications.

In order to solve this problem, we choose Bi-LSTM (Bi-directional Long Short Term Memory) and CNN (Convolutional Neural Network) to adaptively extract text features. Deep model can perform well on these types of tasks without any complicated and time consuming feature engineering. Most of the time, these features require domain knowledge, creativity, and a lot of trial and error. A more detailed model structure is described in Section 3.

2.3. Multiple events Per Sentence

Event extraction methods mainly include the extraction method based on trigger words and the extraction method based on event instances. The detection method based on event instances is to use sentences instead of words as identification instances. Then, it is transformed into a sentence clustering problem by clustering method, and event sentences are obtained by clustering. So this method could only be used to extract text with an event and cannot solve the event extraction problem of long text in practice. Consider the following sentence,

In Baghdad, a cameraman *died* when an American tank *fired* on the Palestine Hotel.

There are two events in the above sentence, include Life.Die and Conflict.Attack. Since a sentence may contain multiple events, the event extraction method based on trigger words is more accurate and comprehensive for event extraction. In this paper, we regard semantic role labeling as a classification task in which candidate arguments forms a phrase with the trigger word. Then, we concatenate the embedding feature from the proposed trigger classification model with other embedding features. Our model is different from the traditional pipeline model, and it can fully utilize the event type information represented by the trigger word embedding.

3. System Approach Overview

3.1. Event trigger classification

The Bi-LSTM based event trigger classification method proposed in this study is shown in figure 1. The system can be divided into three parts: (1) The input part completes the splicing of the word embedding vector and the remaining word features; (2) The high-level semantic representation acquisition part is completed by the Bi-LSTM network model; (3) the forward neural network model to complete the final trigger word classification task, and its input combine the word embedding and the semantic features of the text. Then we will briefly describe the various parts of the model.

For each word $w_i \in W$ in the text, we concatenate two features. $E_w \in R^{n_w * d_w}$ represents pre-trained word embedding, and $E_o \in R^{n_o * d_o}$ represents other features of a word (entity type, part of speech type, and syntactic type). n_w, n_o respectively represent the length of the vocabulary and the number of other feature types. d_w, d_o respectively represent the dimension of the word embedding and the dimension of the embedded features. Each word in the text is mapped to a dense vector representation by two word-feature embedded vector lookup tables.

3.2. Argument classification

We models the argument prediction task as a relation extraction problem, pairs the trigger and the argument, and classifies the relationship. We use convolutional neural networks to classify arguments by concatenating multiple embedded features (word embedding, position embedding, trigger word embedding) that is shown in figure 2. Given a piece of text T with a length of l, each candidate argument $w_{Argu} \in T$ can construct a contextual word pair $\left(w_{Argu}, w_{Tri}\right)$ with the trigger word w_{Tri} in text. The main parts of the model are explained below.

(1)Word embedding table: construct word embedding with pre-trained word embedding $word_i$, and the length is m_i. (2)Relative position embedding table: the text can be represented as $T=\left[w_1, w_2, w_{Argu}, \ldots, w_{Ti}, \ldots, w_l\right]$. The relative position mainly measures the distance between the word and the candidate argument and the distance between the word and the trigger word. Then the relative distance ranges from $-l+1$ to $l-1$, and the length of position embedding p_i is m_p .(3)Trigger word embedding table: trigger word embedding Tri_i is determinated by the input of forward neural network in the previous event trigger classification model.

The feature vectors $x_i = \left[\text{word}_i, p_i^{(\text{argu})}, p_i^{(tri)}, Tri_i\right]$ of each word in the text are obtained from the three types of features mentioned above. Thus, the whole text could be represented by X with size of $(m_w + 2m_p + m_t) \times l$. The matrix X is used as an input to perform a convolution operation with the convolution kernel. After that, we choose max-pooling layer to retain the strongest feature of the feature vectors obtained after each convolution, and concatenate these features to get the final high-level feature representation. In the end, softmax function predicts the type of argument. The selection of the max-pooling layer is based on the following considerations: The operation of max-pooling ensures the invariance of position and rotation, condenses variable-length feature map into a representation of fixed length, and effectively reduces the number of hyper-parameter of the model.

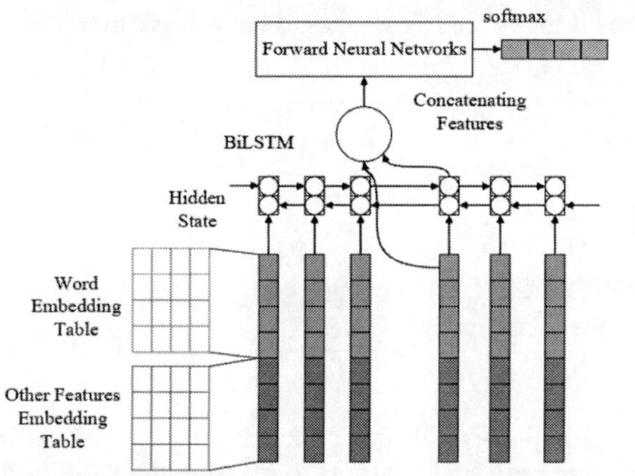

Figure 1. Trigger word classification model

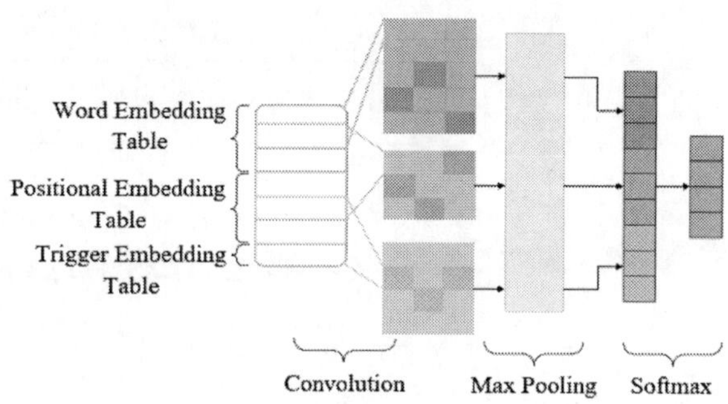

Figure 2. Argument classification model

4. Experimental Results

The experimental environment of this paper is as follows: the operating system is Ubuntu-16.04, the CPU is Intel Core i7-8700K, the GPU is GeForce GTX-1070Ti, and the development environment is Tensorflow-1.10. We adopt the CEC (Chinese emergency corpus) built by Shanghai University. It contains five social news reports of earthquakes, fires, traffic accidents, terrorist attacks and food poisoning. Similar to the dataset provided by ACE2005, CEC marks the six most important event extraction information in the corpus: Event, Denoter, Time, Location, Participant, and Object.

4.1. Event trigger classification

We use training set and development set to select the hyper-parameter of the model. The forward neural network has two hidden layers and the number of neurons in the hidden layers is 130 and 110 respectively. Cross entropy is selected as the loss function and the probability of dropout is set to 0.3. Furthermore, the number of LSTM layers is 2 and the number of cells in the hidden layer is 200. We choose the pre-training model of Bert [5] and integrate it into the task model.

In table 1, the model proposed in this paper has a certain improvement under the F1-score compared with the previous benchmark model. And the performance of the support vector machine model is obviously inferior to our model. It proves that the heavy dependence on the NLP tool would continuously introduce errors, and the features obtained by the empirically determined feature engineering cannot effectively represent the text information. CNN-based trigger classification method significantly reduces recall rate due to fixed input dimension in CNN model. It means that the convoluted object must have local correlation, and CNN-based model cannot learn long dependencies.

Table 1. Comparison of trigger classification model

Model	Precision	Recall	F1-score
Our model	78.68	77.93	78.30
CNN [6]	79.21	76.53	77.85
SVM [7]	76.45	71.04	73.65
SVM+Embedding [8]	77.85	73.62	75.68

In table 2, we also focus on the impact of different features on model performance. E_w, E_o respectively represent word embedding and other embedding features (lexical, syntactic, entity). It can be seen that the syntactic features are the most helpful for the event trigger word classification task and the part of speech information is usually used as an auxiliary feature of tasks such as semantic role labeling and entity recognition. Furthermore, the entity type can represent the components of the event to a certain extent, such as time, place and object, thus it helps the trigger classification task more.

Table 2. Comparison of features

No.	Features	F1-score
1	E_w, g	76.61
2	$E_w + E_o, g$(part of speech tag)	77.59
3	$E_w + E_o, g$(entity type)	78.81
4	$E_w + E_o, g$(dependence relationship)	78.47

4.2. Argument classification

In this section, we compare the model proposed in this paper with other models of classical work. What's more, we compare the model effects after not combine such features to measure the validity of the event trigger embedding and the position distance embedding. The size of the convolution kernels in the convolution layer are $\{3,4,5\}$, and there are 400 convolution kernels of each size. The Position embedding set to 50 dimensions and word embedding set to 100 dimensions. The training optimization method uses the Adam algorithm with a learning rate of $1e^{-3}$.

The results are shown in table 3. Accuracy rate of across documents model and feature-based event extraction model are quite competitive owing to combine multiple features (Syntactic and semantic). But in comparison the model proposed in this paper only relies on automatic feature extraction, which has low dependence on various natural language processing tools. Simultaneously, our model has the highest recall rate, because it not only utilizes the semantic information of the sentence, but also uses the event type information introduced by the trigger word embedding feature, which can effectively guide the classification of argument characters. It should be noted that information about event types is also used in the model proposed in [9]. However, since its feature is discrete, the representation of event type information is still not as rich as that of dense embedding.

Table 3. Comparison of arguments classification model

Model	Precision	Recall	F1-score
Our model	63.14	57.62	60.25
Cross-event [2]	45.17	44.12	44.64
Cross-entity [9]	51.65	45.58	48.43
Joint learn with local and global features [10]	62.72	48.55	54.73
Dynamic multi pooling CNN [11]	61.28	45.86	52.46

In order to analyze the importance of embedded features, we set the base-line Model, in which only word embedding and position embedding of candidate argument are used. Then we add the trigger embedding feature and the trigger word distance feature in turn and compare the performance of the model. The result is shown in table 4. Trigger embedding has the biggest effect on the model. This is because the trigger embedding can effectively represent the type information of the event, which is of great help to the classification of argument characters.

Table 4. Comparison of the importance of features

Model	Precision	Recall	F1-score
Our model	63.14	57.62	60.25
+ Trigger embedding	59.38	51.45	55.13
+ Trigger position embedding	34.22	21.61	26.49
Base-line model	28.74	21.53	24.62

5. Conclusion

In this work, we classify the argument role as a classification task that forms a phrase with the trigger word. In order to get rid of the dependence on feature engineering, the model concatenates word embedding, trigger embedding and distance embedding to construct the representation of text. As a result, we achieve competitive performance, specifically, F1-measure of 60.25 in event extraction. The experimental results also show that the existing trigger classification model does not reach the ideal state, so the accuracy of the event extraction model cannot be further improved. Our future work will take advantage of the nonlinear combination of neural networks and try to build a joint model.

Acknowledgments
This work is supported by the National Natural Science Foundation of China (Grant Nos. 61773080) and the Basic Science and Research Technology Major Project of Chongqing (Grant Nos. cstc2015jcyjB0569).

References

[1] Gupta R and Sarawagi S 2009 Domain adaptation of information extraction models *ACM SIGMOD Record* **37(4)**: 35-40
[2] Liao S and Grishman R 2010 Using document level cross-event inference to improve event extraction *Proceedings of the 48th ACL* 789-797
[3] Jagannatha A N and Yu H 2016 Bidirectional RNN for medical event detection in electronic health records *Proceedings of the conference. Association for NAACL* Chapter 473
[4] Feng X, Qin B, and Liu T 2018 A language-independent neural network for event detection *Science China Information Sciences* **61(9)**: 092106
[5] Devlin J, Chang M W, Lee K, et al. 2018 Bert: Pre-training of deep bidirectional transformers for language understanding (*Preprint* arXiv:1810.04805)
[6] Wang J, Li H, An Y, et al. 2016 Biomedical event trigger detection based on convolutional neural network *International Journal of Data Mining and Bioinformatics* **15(3)**: 195-213
[7] Pyysalo S, Ohta T, Miwa M, et al. 2012 Event extraction across multiple levels of biological organization *Bioinformatics* **28(18)**: i575-i581
[8] Zhou D, Zhong D, He Y 2014 Event trigger identification for biomedical events extraction using domain knowledge *Bioinformatics* **30(11)**: 1587-1594
[9] Hong Y, Zhang J, Ma B, et al. 2011 Using cross-entity inference to improve event extraction *Proceedings of the 49th ACL: Human Language Technologies* **1**: 1127-1136
[10] Li Q, Ji H, and Huang L 2013 Joint event extraction via structured prediction with global features *Proceedings of the 51st ACL* **1**: 73-82
[11] Chen Y, Xu L, Liu K, et al. 2015 Event extraction via dynamic multi-pooling convolutional neural networks *Proceedings of the 53rd ACL* **1**: 167-176

Power System Stochastic Unit Commitment Model for Large-scare Wind Power Integration Considering Demand Side Resources

Wenxia Liu[1], Suhua Lou[2], Shuyun Wang[2,*], Xianggang He[1] and Xinfu Song[3]

[1] Grid Planning & Research Center, Guizhou Power Grid Co., LTD., Guiyang 550003, Guizhou Province, China
[2]State Key Laboratory of Advanced Electromagnetic Engineering and Technology (College of Electrical and Electronic Engineering, Huazhong University of Science and Technology), Wuhan 430074, Hubei Province, China
[3]Economy and Technology Research Institute of State Grid Xinjiang Electric Power Co., LTD., Urumqi 830002, Xinjiang, China

[*]E-mail: 1282333872@qq.com

Abstract. As the uncertainty output characteristics of wind power, the large-scale wind power integrated to the grid, has brought new challenges to power system operation. Based on the uncertainty characteristics of wind power, this paper establishes the wind power output model, then focuses on the positive effects of price-based demand response (PDR) and incentive-based demand response (IDR) ,which are allowed to participate in the power balance, on the unit commitment (UC) with high wind power penetration, and establishes the PDR and IDR models. Based on the chance constrained programming, this paper proposes a stochastic UC model with large-scale wind power integration considering demand side resources(DSR). Stochastic simulation combined with particle swarm optimization (PSO) algorithm was presented to solve this model. At last, the model is proved feasibly and effectively by testing the IEEE 10 machines system.

1. Introduction

In recent years, wind power has developed rapidly worldwide as a clean energy source. However, the uncertainty of wind power output has brought great challenges to the power system. Therefore, it is of great significance to excavate the flexibility resources on demand side and coordinately optimize resources on the supply side and the demand side. Reference [1] studied the influence of the day-ahead PDR and the in-day IDR on the wind power consumption in wind power integrated system. Based on the risk constraint theory, reference [2] studies the peaking benefits of demand response(DR) such as excited-load and interruptible load. However, the above references are based on the traditional deterministic scheduling model, and the DR they consider is not comprehensive. Based on the chance constrained method, reference [3] studied the system scheduling model that uses the demand side response to improve wind power consumption, but the model does not optimize resources on the supply side and the demand side jointly.

Under the above background, this paper adopts stochastic chance-constrained programming to establish a UC model of large-scale wind power integration system with consideration of coordinated optimization of resources on the supply side and the demand side, and studies the effect of time-of-use pricing(TOU) and IDR on improving the flexibility of power system and promoting the consumption of wind power. Stochastic simulation combined with PSO algorithm was presented to solve this model.

2. Wind Power Output Forecasting Model

Based on the prediction of wind speed and wind power output in reference [4], this paper predicts the wind power output curve by point forecast, and obtains the interval prediction of wind power output by considering the wind power prediction error of ±20% up and down deviation. Figure 1 is a diagram of the prediction interval of the total output of a wind farm with 80% confidence probability.

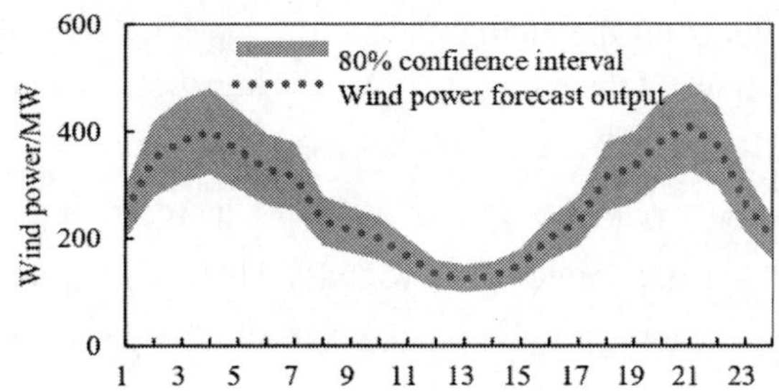

Figure 1. Diagram of 80% confidence probability for certain wind farm total power output prediction.

3. Response Mode of DSR Participating in Scheduling Optimization

3.1. Response Mode of PDR

PDR refers to a load response method in which the user responds to changes of the retail price and adjusts the demand for electricity accordingly[5]. TOU pricing is an electricity price response method that is divided into several different price levels according to the time characteristics of the load curve. The price is utilized to guide the user to actively adjust the power demand to achieve load transfer and peak load shifting. The price elasticity matrix of electricity demand is used to establish a DR model which is based on TOU pricing, and the self-elasticity coefficient $\varepsilon_{t,t}$ and cross-elastic coefficient $\varepsilon_{t,h}$ are used to measure the impact of TOU pricing on electricity demand.

$$\varepsilon_{t,t} = \frac{\partial d_t / d_t}{\partial \rho_t / \rho_t} \tag{1}$$

$$\varepsilon_{t,h} = \frac{\partial d_t / d_t}{\partial \rho_h / \rho_h} \tag{2}$$

where d_t denotes the electric power demand at time t, and the electricity prices and h are ρ_t and ρ_h, respectively. d_t can be formulated as follows:

$$d_t = d_{t0} \times \left[\varepsilon_{t,t} \times \frac{\rho_t - \rho_{t0}}{\rho_{t0}} + \sum_{\substack{h=1 \\ h \neq t}}^{T} \varepsilon_{t,h} \times \frac{\rho_h - \rho_{h0}}{\rho_{h0}} \right] \tag{3}$$

where d_{t0} denotes the initial electric power demand before the response. ρ_{t0} and ρ_{h0} respectively represent the reference electricity price and h before the electricity price changes.

3.2. Response Mode of IDR

IDR refers to a DR method in which the DR implementing agency motivates users to immediately reduce or increase the load demand when the electric power supply is tight or excessive by formulating policies. The IDR studied in this paper mainly includes direct load control(DLC), interruptible load(IL), electric vehicles(EV) with two-way adjustment capability and energy storage, etc. The scheduling cost of IDR is divided into two parts: capacity compensation cost and power compensation cost. And the power compensation is in the form of stepped compensation.

The invoking cost of IDR is as follows:

$$F_{IDR}=\sum_{\substack{d=1\\ d\in IDR}}^{N_d}(C_d\cdot D_d^{IDR})+\sum_{t=1}^{T}\left[\sum_{\substack{d=1\\ d\in IDR}}^{N_d}(C_{E,dm}^{+}\cdot d_{dm,t}^{+})+\sum_{\substack{d=1\\ d\in IDR}}^{N_d}(C_{E,dm}^{-}\cdot d_{dm,t}^{-})\right] \tag{4}$$

where the first and second items on the right side are the capacity cost and the power cost of IDR respectively. F_{IDR} is the total scheduling cost of IDR during the scheduling period. C_d is the unit capacity response cost of IDR ($/MW) . D_d^{IDR} is the response capacity of the user d participating in IDR. $C_{E,dm}^{+}$ and $C_{E,dm}^{-}$ are respectively the cost of per unit increase and decrease generated energy corresponding to segment m on IDR price curve ($/(MW•h)), and $d_{dm,t}^{+}$ and $d_{dm,t}^{-}$ are respectively the increase and decrease generated energy. N_d is the number of users participating in IDR.

4. A Chance-constraint-programming-based UC Model Considering Large-Scale Wind Power and DSR

4.1. Stochastic Chance Constrained Programming with Uncertain Theory

Power system scheduling optimization with large-scale wind power is a nonlinear integer programming problem with random variables. This paper uses chance constrained programming to describe this randomness and its impact. Its mathematical description can be expressed as follows:

$$\begin{aligned}&\min \bar{f}(x)\\ &s.t.\begin{cases}P_r\{f(x,\xi)\le \bar{f}(x)\}\ge\beta\\ P_r\{g_i(x,\xi)\ge 0, i=1,2,\cdots,p\}\ge\alpha\end{cases}\end{aligned} \tag{5}$$

where x is an n-dimensional decision vector. ξ is a random vector with a known probability density function. $f(x,\xi)$ is an objective function. $g_i(x,\xi)$ is constraint function. $P_r\{\cdot\}$ is the probability that an event is established in $\{\cdot\}$. α and β are respectively the confidence level of the constraint condition and the objective function which is pre-set by the decision maker. $\bar{f}(x)$ is the minimum value of the objective function when the probability level is β .

4.2. Objective Function

The optimization goal of the model established in this paper is to minimize the total operating cost.

$$F=\min E\,[F_G+F_{IDR}+F_{pena}] \tag{6}$$

where F denotes the total operating cost. F_G denotes the operating cost of units, including fuel cost and on-off cost. F_{IDR} denotes the invoking cost of IDR. F_{pena} denotes the cost of the wind abandonment and load shedding when the system has insufficient peak capacity.

F_G can be formulated as follows:

$$F_G=\sum_{t=1}^{T}\sum_{i=1}^{N_g}\left[f_{i,t}(P_{Gi,t})+u_{i,t}(1-u_{i,t-1})S_{i,t}\right] \tag{7}$$

IOP Conf. Series: Journal of Physics: Conf. Series **1267** (2019) 012034 doi:10.1088/1742-6596/1267/1/012034

where N_G is the number of units and $f_{i,t}(\bullet)$ is the operating cost of the unit i during the period t. $P_{Gi,t}$ denotes the active output of unit i. $u_{i,t}$ is a 0-1 variable. The power-on is 1 and the power-off is 0. $S_{i,t}$ denotes the on-off cost of unit i.

F_{pena} can be formulated as follows:

$$F_{pena} = \sum_{t=1}^{T} C_w^{pena} * q_t^{wind} + \sum_{t=1}^{T} C_i^{pena} * q_t^{load} \tag{8}$$

where C_w^{pena} and C_i^{pena} are the unit penalty cost for the wind abandonment and load shedding respectively. q_t^{wind} and q_t^{load} are the abandoned wind electricity the load shedding electricity respectively.

4.3. Constraint Condition

1) System Electricity Power Balance Constraint

$$\sum_{i=1}^{N_g} P_{Gi,t} + (P_{w,t} - q_t^{wind}) = P_{Ld,t} + d_t^{TOU} + (\sum_{\substack{d=1 \\ d\in IDR}}^{N_d} d_{dm,t}^{+} - \sum_{\substack{d=1 \\ d\in IDR}}^{N_d} d_{dm,t}^{-}) - q_t^{load} \tag{9}$$

Where $P_{w,t}$ is the total output of the wind farm. $P_{Ld,t}$ is the load of the system. d_t^{TOU} is the amount of change of the response power to TOU pricing.

2) System Rotation Reserve Constraint

$$P_r\left\{\sum_{i=1}^{N_g} u_{i,t}(P_{Gi}^{\max} - P_{Gi,t}) + \sum_{\substack{d=1 \\ d\in IDR}}^{N_d} D_d^{-} \ge R_{dup,t} + R_{wup,t}\right\} \ge \beta_1 \tag{10}$$

$$P_r\left\{\sum_{i=1}^{N_g} u_{i,t}(P_{Gi,t} - P_{Gi}^{\min}) + \sum_{\substack{d=1 \\ d\in IDR}}^{N_d} D_d^{+} \ge R_{ddn,t} + R_{wdn,t}\right\} \ge \beta_2 \tag{11}$$

where $P_{Gi}^{\max}$ and $P_{Gi}^{\min}$ are respectively the upper and lower limit of the electricity power provided by unit i. D_d^{-} and D_d^{+} are respectively the upper limit of the load reduction and increment that can be provided by the user d in the form of IDR. $R_{dup,t}$ and $R_{wup,t}$ are respectively the standby requirements of positive rotation for load and wind power. $R_{ddn,t}$ and $R_{wdn,t}$ are respectively the standby requirements of negative rotation for load and wind power. $P_r\{\bullet\}$ is the probability that the event in $\{\bullet\}$ is established. β_1 and β_2 respectively denote the confidence level of positive and negative rotation reserve of given system.

3) Electricity Power Constraint in Each Time Period

$$-D_d^{TOU} \le d_t^{TOU} \le D_d^{TOU} \quad \forall t \in T, d \in D_{TOU} \tag{12}$$

where D_d^{TOU} is the limit value of response power for the user participating in TOU pricing.

4) Constraint of IDR

$$\begin{cases} 0 \le d_{d,t}^{+} \le D_d^{+} \\ 0 \le d_{d,t}^{-} \le D_d^{-} \end{cases} \tag{13}$$

where D_d^{+} and D_d^{-} respectively denote the upper and lower limits of the load that the IDR user d is allowed to increase and decrease.

IOP Conf. Series: Journal of Physics: Conf. Series **1267** (2019) 012034 doi:10.1088/1742-6596/1267/1/012034

5. PSO Algorithm Based on Stochastic Simulation

In this paper, the stochastic simulation technique is combined with the quantum-inspired binary particle swarm optimization (QBPSO) algorithm to solve the chance constrained programming model.

The specific solution process is as follows:

1) Set system parameters, random simulation parameters and particle swarm parameters.

2) Optimize the load change of the PDR and obtain the optimized load curve.

3) Using random simulation technology, randomly generate N_s strips 24-hour wind power output curve according to the wind power output model.

4) Using PSO to randomly generate N_p feasible original conventional unit on-off population matrices.

5) Verify the particle swarm matrix using stochastic simulation technology, and check whether the particle on/off state at each moment satisfies the confidence level β_1 and β_2 for the N_s samples. If not, correct the on/off combination of particles at that moment. Check again until the condition is met.

6) Calculate the fitness of each particle, update the individual extremum of the particle and the global extremum of the particle swarm.

7) update the position and speed of particles, and repeat steps 5) and 6) until the termination condition is met.

8) obtain the optimal day-ahead UC results meeting random programming constraints.

9) Based on the results obtained in step 8), the real-time economic dispatching of the system is carried out. According to the wind power output probability distribution, the Monte Carlo simulation method is used to obtain N strips wind power random output curves as wind power output scenarios for real-time economic dispatch. And obtain a new net load curve based on day-ahead PDR optimization;

10) According to the day-ahead UC strategy, the real-time economic dispatching is carried out to verify whether the on-off strategy under the confidence level can meet the real-time peaking demand within the day. If not, invoke the IDR first. Abandon wind or cut load, or increase the confidence level of system if still not satisfied.

11) Output system economic dispatch results under N wind power output scenarios in step 9), and calculate the expectation value of system scheduling under the confidence level.

6. Simulation Results

In this section, the IEEE 10 machine system example is used. The number of random simulations is set to 5000, the particle population size is set to 40, and the number of algorithm iterations is set to 100. The self-elastic coefficient $\varepsilon_{t,t}$ is -0.2, and the cross-elastic coefficient $\varepsilon_{t,h}$ is 0.033.

In order to explain the impact of PDR and IDR on the operation cost of large-scale wind power system, this section compares the operation costs of the system under different operation cases. The results are shown in Table 1.

Case 1: The IDR cost and TOU cost are not included. Case 2: The TOU cost is included. Case 3: The IDR cost is included. Case 4: The IDR cost and TOU costs are included.

Table 1. UC costs under different cases (Confidence level 0.85, Unit: $)

case	total	fuel	On-off	Wind abandonment	Load shedding	IDR
1	493 020	435 970	5 550	6 000	45 500	-
2	435 072	427 442	5 380	2 250	0	-
3	442 960	433 354	6 450	2 900	0	255.6
4	437 500	431 378	5 900	0	0	221

It can be seen from Table 1 that under the power supply structure of this paper and the same confidence level, when considering DSR into the UC, the total operating cost and fuel cost are reduced. And wind abandonment and load shedding no longer occur.In order to analyse the reasons for the above results, this section further analyses the impact of TOU and IDR on operational scheduling.

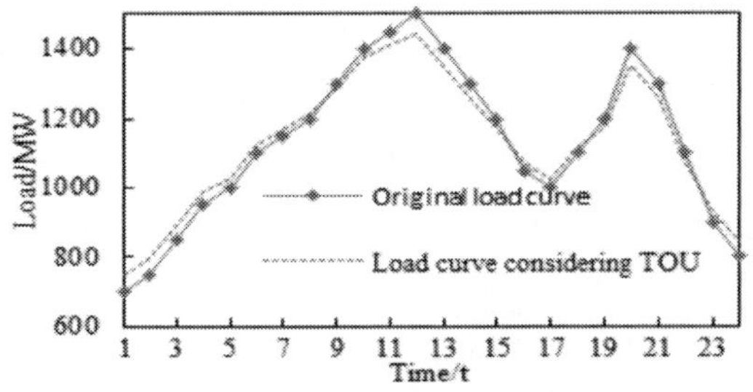

Figure 2. Load curves with or without TOU

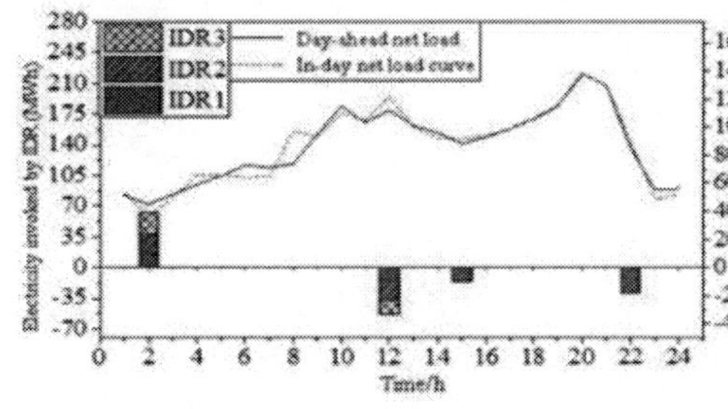

Figure 3. Diagram of system operation with IDR participated

Figure 2 shows the load curves with or without TOU. It can be seen that the system load curve has a corresponding change after the TOU is considered into UC. This is because TOU is an interactive way between the power generation side and the user side. By increasing the electricity price during the peak load period and lowering the electricity price during the load period, the power consumption mode can be affected, and the load curve is optimized.

Figure 3 shows the effect of IDR. The day-ahead net load curve does not completely coincide with the in-day net load curve, which is due to the forecast error of wind power output. During original low load period(e.g. 2h), The actual output of wind power is greater than the predicted output, and the negative rotation is insufficient. Therefore, the system increases the load by invoking IDR to avoid abandoning the wind power. During original peak load period(e.g. 12h), The actual output of the wind power is less than the predicted output, and the active rotation is insufficient. Therefore, the system interrupts the load by invoking IDR to avoid random load shedding.

7. Conclusion

Based on comprehensively optimizing DSR and power generation side resources, this paper establishes a stochastic UC model based on opportunity constrained programming with large-scale wind power integration, and uses stochastic simulated PSO algorithm to solve the model. Through the analysis of the example, the following conclusions are drawn: 1) Deal with the uncertainty of wind power output by using opportunity constrained programming, and optimize the total operating cost of the system by setting a reasonable confidence level. 2) Improve the wind power consumption space, reduce the system load shedding, and reduce the total cost of system operation total cost by rational allocation and appropriate invoking of DSR.

Acknowledgement

This paper is supported by China Southern Grid Corporation Science and Technology Project (067600KK52170001).

References

[1] Liu X C, Wang B B, Li Y, et al.Stochastic unit commitment model for high wind power integration considering demand side resources [J].Proceedings of the CSEE, 2015, 35(14): 3714-3723.
[2] Yang N, Wang B, Liu D C, et al.An integrated supply-demand stochastic optimization method considering large-scale wind power and flexible load [J].Proceedings of the CSEE, 2013, 33(16): 63-69.
[3] Ai X, Liu X.Chance constrained model for wind power usage based on demand response[J].Journal of North China Electric Power University: Natural Science Edition, 2011, 38(3): 17-22.
[4] R. Doherty, M. O'Malley.A new approach to quantify reserve demand in systems with significant installed wind capacity[J].IEEE TRANSACTIONS ON POWER SYSTEMS, 2005, 20(2): 587-595.
[5] Zhang Qin, Wang Xifan, Wang Jianxue, et al. Survey of demand response research in deregulated

electricity markets[J]. Automation of Electric Power Systems, 2008, 32(3): 97-106(in Chinese).

A Method for Determining Access Node of Microwave Relay Network Based on Genetic Algorithm

Ji Qi* **and Kebin Chen**

College of Information and Communication, National University of Defense Technology, Wuhan 430019, China

*qiji6812078@qq.com

Abstract. In order to improve the deployment efficiency of the microwave relay network, a method for determining access node of microwave relay network based on genetic algorithm is proposed in this paper. System analysis of network structure characteristics, node attributes and deployment process, the access node deployment issues is transformed into continuous area location issues. After discretization the area, the set of candidate access nodes is obtained considering boundary points and boundary chords. Comparing the coverage set of each access node, the candidate set is de-redundancy. Then taking coverage ratio maximum and the deployment cost minimum as the optimization goal, the optimal set of access nodes is obtained from candidate set by using genetic algorithm. A case with 50 user terminals is simulated by our method. Results show that after de-redundancy and genetic algorithm calculation, the number of access nodes is optimized from 240 to 17, and only 9 user terminals are repeated covered by different access node.

1. Introduction

Modern warfare is increasingly demanding the real-time, accuracy and flexibility of equipment organization. How to use engineering methods to quickly integrate limited resources and accurately deploy operational elements is an important reason for determining the victory of war. The microwave relay network nodes are mainly divided into a backbone node (BN) and an access node (AN). The user terminal accesses the network through the access nodes [1], and the access nodes are interconnected through the backbone nodes [2]. In theory, the access node's ability to access the user terminal is not limited, and there is no directionality. The distribution of user terminals can be regarded as known. In actual use, the bottom-up approach is used to determine the access nodes and the backbone nodes. That is, firstly, according to the location of the user terminals, the location of the access node is determined, and then the location of the backbone node is determined by the location of the access nodes. Due to the large number of user terminals, the location of the access node is difficult to determine. It is necessary to ensure the full coverage of the user terminals and to pursue the minimum number of access nodes. This is a combination optimization problem, and genetic algorithm has advantage to solve those problems [3-5]. Based on the concept of boundary points and boundary chords in Ref. [6], the set of candidate access nodes is obtained by discretization and de-redundancy, and the concept of coverage ratio is introduced. And then the optimal set of access nodes is obtained from candidate set by using genetic algorithm.

2. Feasibility analysis and assumptions

2.1. Feasibility analysis

In this paper, the geometric coverage model is established, and the effective coverage ratio is used as the objective function. The candidate nodes are formed in the candidate region of the two-dimensional plane discretization processing, and the user terminal position is considered to be relatively fixed. Regarding whether the initial conditions and model selection are consistent with the actual combat, the following three explanations are made:

(1) Effective coverage issues. If the geographic area of the combat area is small, the deployment cost of the global coverage is not high, and the cellular grid (virtual cell) method can be used to ensure the minimum overlapping coverage area, which is not the research focus of this paper. The main concern of this paper is how to achieve effective coverage better, and all user terminals access the upper-level network under low deployment cost conditions in a relatively wide operational area with the number of access nodes cannot meet the global coverage requirements.

(2) The influence of terrain. In practical applications, the candidate node position determined by the modelling solution is a reference position and is not fixed. In actual using, it can be flexibly adjusted according to the local terrain and in the vicinity of the reference position to avoid problems such as poor visibility and unsuitable opening.

(3) User terminals status. In practice, the user terminal is often in motion, but this does not mean that the access node and the backbone node are also in motion. On the one hand, it is a limitation of the technical system, for example, the microwave link cannot be connected in a moving manner; on the other hand, the movement of the backbone and the access node will cause the communication quality to decrease.

2.2. Model hypothesis

Hypothesis 1: The coverage of the access nodes is a circular area with the radius of r, and the quality of the communication signal is not distinguished in the area;

Hypothesis 2: The access capability of the access node is not limited, and the connectivity problem between the access nodes is not considered;

Hypothesis 3: The deployment cost of all access nodes is the same, and the total cost of deployment can be represented by the number of access nodes;

Hypothesis 4: The priority of the user terminal is not differentiated according to the level.

3. Access node determination method

The problem of the deployment of the microwave relay network access node is a continuous area location problem. Due to the large size of the alternative location, the efficiency of the genetic algorithm is reduced. In order to narrow the selection range of the alternative solution, firstly, the deployable area of the access nodes should be compressed and discretized, and then the redundant candidate nodes should be removed to improve the accuracy and effectiveness of the subsequent calculation. The optimal access node set should cover all user terminals with a minimum number of access nodes.

3.1. Candidate area and candidate node set

The candidate area is the initial compression of the access node deployable area in the combat area. As shown in Figure 1, the distance of adjacent two user terminals v_1, v_2 is $\text{dis}(v_1, v_2)$ and the access node coverage radius is r. When $\text{dis}(v_1, v_2) \le 2r$, with v_1, v_2 as the center and r as the radius, there must be an intersection area (a deep shadow part in Figure 1). If an access node is deployed in the area, two user terminals can be covered at the same time; if an access node is deployed in a lightly shadowed part, and only one user terminal can be covered; If an access node is deployed in a blank area, none of user terminal can be covered. The dark shadow portion of the intersection is the candidate deployment area for the two nodes. Similarly, it can be extended to other user terminals, and a candidate area can be generated between any two user terminals whose distance is smaller than $2r$. So, the range of the

IOP Conf. Series: Journal of Physics: Conf. Series **1267** (2019) 012035 doi:10.1088/1742-6596/1267/1/012035

suitable area to deploy access node is greatly compressed when only consider the candidate area.

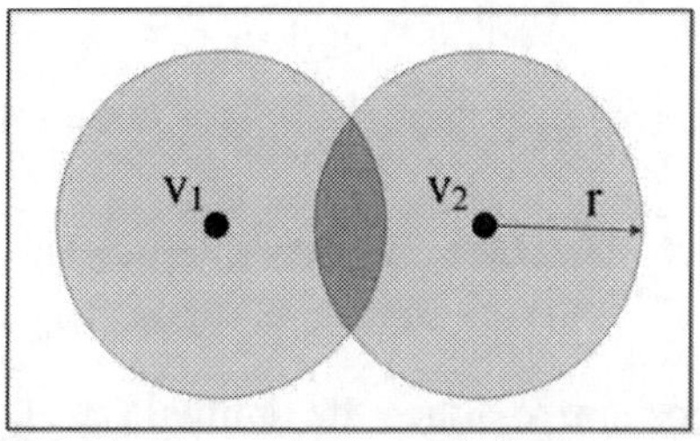

Figure 1. Two-node candidate area

However, the above method narrows the scope of the alternative area, but still belongs to the continuous area. It is necessary to select some representative points in the candidate area, and construct a set of discrete points to facilitate modeling. Based on this idea, reference [6] proposes the concept of boundary points and boundary chords according to geometric principles. As shown in Figure 2, the user terminals v_1 and v_2 of $\mathrm{dis}(v_1,v_2)\le 2r$ form a set of boundary points. The boundary points are connected as boundary strings. The boundary string is symmetric. The boundary points k_1, k_2 which have distance r away from v_1, v_2 become the candidate nodes. For example, at k_1, k_2, the access node can simultaneously cover two v_1, v_2 user terminals and convert the continuous space into discrete points.

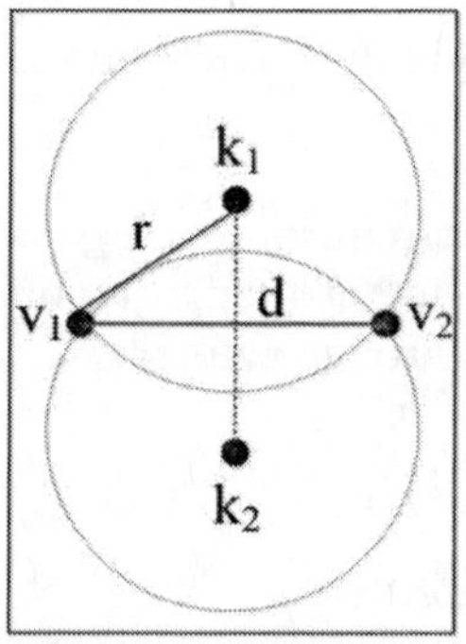

Figure 2. Alternative node formation method

If the distance between a user terminal and all other user terminals is greater than r, it is an isolated node. In the actual opening, the access node should be deployed separately at the node to ensure that all user terminals can be covered by the access node.

3.2. Redundancy removal

The size of the candidate node set depends on the number and location relationship of the user terminals. When the number of user terminals is large, the user terminals covered by different access nodes may be included in each other, resulting in redundancy of the candidate nodes. When the above redundancy occurs, deleting the candidate nodes that cover the user terminal is relatively small, and the size of the candidate node set can be further compressed to reduce the amount of calculation.

3.3. Access node selection process

The access node selection process flow is shown in Figure 3. The specific process is implemented as follows:

Step1. According to the method of Figure 2 in 3.1 section, initially determine the set of candidate access nodes $A_{initial}$ in light of the location of the user terminals;

Step2. According section 3.2, compares the user terminals of each candidate access node covers. If the user terminal set of access node AN_i is a subset of the user terminal set of AN_j, the AN_i is a redundant node and should be deleted. And then, the access node set A_s after the redundancy removal is obtained.

Step3. The candidate node in A_s can still be further filtered. The reason is that some user terminals covered by a candidate access node AN_i are repeatedly overwritten twice. This selecting step can be implemented by using genetic algorithm and then the optimal set A_f of access nodes can be obtained.

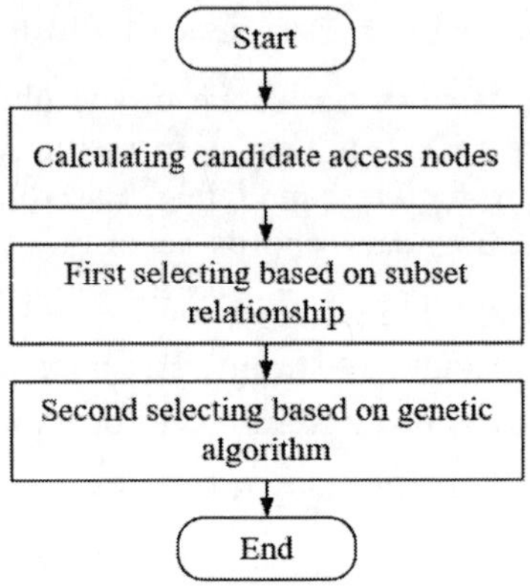

Figure 3. Access node selection process

4. Genetic algorithm implementation

Genetic Algorithm (GA) is a non-deterministic intelligent search algorithm that simulates the natural selection of biological evolution. Different living organisms have different genes, so the ability to adapt to the outside is also different. Through natural selection, individuals with strong adaptability are usually able to survive and reproduce.

4.1. Fitness function

The fitness function of the genetic algorithm is

$$\min \sum_{a=1}^{m}\sum_{t=1}^{n}\theta_{at} \tag{1}$$

$$\text{s.t.} \quad \sum_{a=1}^{m}\theta_{at} \geq 1, \forall t \in C \tag{2}$$

Where θ_{at} is the coverage ratio, it means the times of each user terminal covered by different candidate access nodes, m is the number of candidate access nodes, n is the number of user terminals, and C is the set of user terminals. The value of θ_{at} is obtained as follows, when the user terminal t is covered by the access node a, $\theta_{at}=1$, otherwise $\theta_{at}=0$. The goal of genetic algorithm optimization is that equation (1) is the smallest. And the ideal situation is that all user terminals were covered by access node only once (namely $\forall\sum_{a=1}^{m}\theta_{at}=1$).

4.2. Encoding

Binary encoding is used in this paper. A possible subset of candidate node set A_s is used as the chromosome $P=[p_1, p_2,...p_i..., p_m]$. The length of the chromosome is the length of the node set A_s, when the access node i is selected, $p_i=1$, otherwise $p_i=0$.

4.3. Operator selection

The tournament operator is used as selection operator. This selection operator is more sensitive to the difference of the adaptive values of the children. Single point crossover method is used as crossover operator. Mutation operator uses basic position variation.

5. Analysis of case and simulation results

The algorithm is implemented by MATLAB. The parameters are set as follows: initial population size NP=20, number of iterations GEN=40, crossover rate PC=0.6, and mutation rate PM=0.8. The number of user terminals is 50, distributed in a square area of 10 square kilometres, and the distribution of user terminals (purple circle) is shown in Figure 4(a). First, according to the access node generation method in Section 3.1, all the candidate nodes (open triangles) set $A_{initial}$ are obtained as shown in Figure 4(b). At this time, there are a total of 240 alternative access nodes, and the number of redundancy nodes is large. Subsequently, according to Section 3.2, the access node is filtered, and 206 nodes are filtered out, leaving only 34 nodes (Figure 4 (c), blue square); Finally, using the genetic algorithm, the remaining 34 access nodes are selected. Under the requirement that each user terminal has more than one coverage, the coverage ratio is optimized, and the optimal access node set (Figure 4(d), red five-pointed star) is obtained. It can be seen that the access node is further reduced, leaving only 17 nodes.

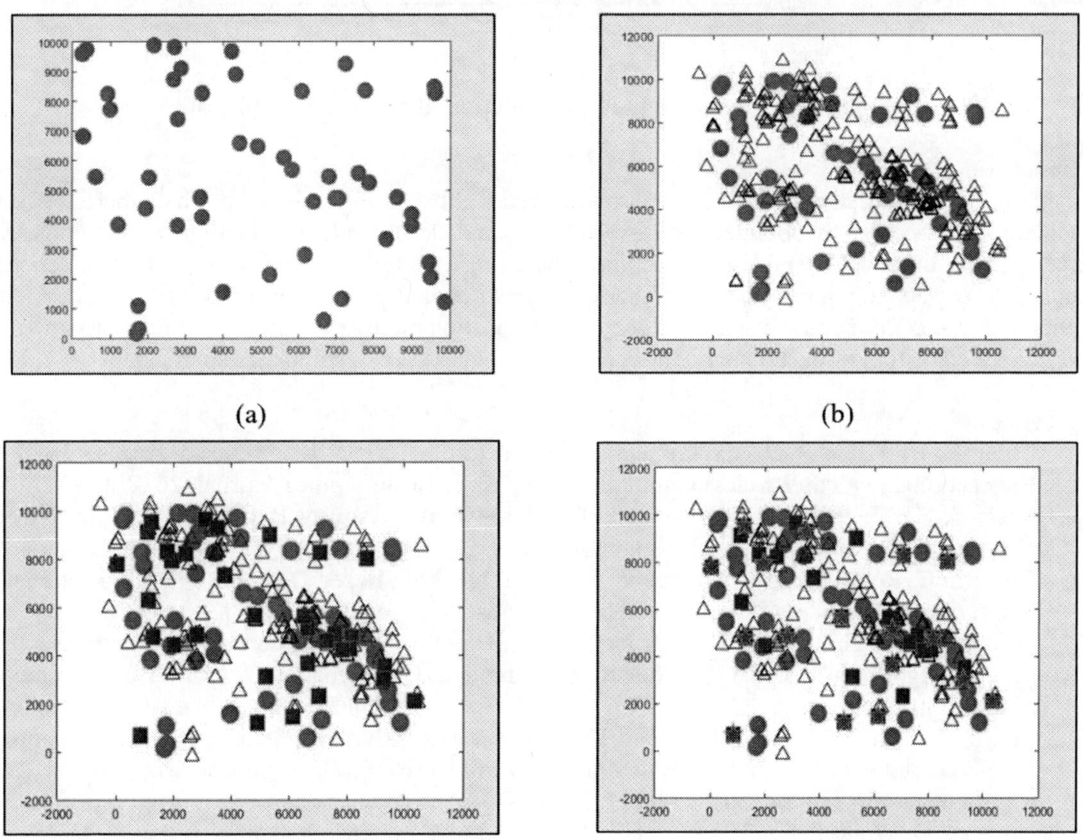

(a) (b)

(c) (d)

Figure 4. Generation and selecting of access nodes ((a) distribution of user terminals (b) initialization of access nodes (c) redundancy of access nodes (d) further filter of access nodes using genetic algorithms)

The curve of fitness (total coverage ratio θ) with the number of iterations is shown in Figure 5. After the number of iterations reaches 18, it gradually converges to the global optimal solution, and the coverage is closer to 60 from 130. The optimal total coverage ratio value is 59 (ideally, the coverage is 50, that is, all user terminals are only covered once). The repeated coverage of all user terminals are only 9 times, which achieves the purpose of effective coverage.

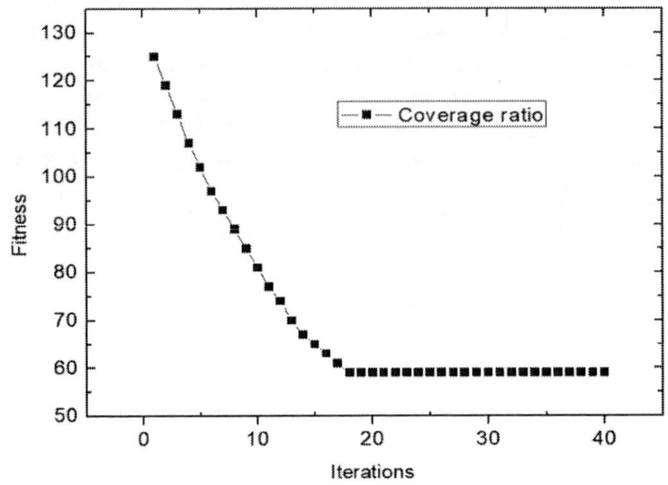

Figure 5. Coverage ratio as a function of the number of iterations

6. Conclusion

A method of selecting access node for microwave relay network is proposed in this paper. The set of candidate access nodes is obtained by discretization and de-redundancy. In light of coverage ratio concept, the optimal access node set is acquired by using genetic algorithm from candidate set. After calculating by genetic algorithm, the number of access nodes is reduced from 240 to 17. The method of this paper can be further used in other communication networks. In the future work, we will determine the position of backbone node according the access nodes calculated in this paper.

References

[1] Chou H H, Hsiao J H. Experimental Demonstration of an LCoS-based Access Node for Bidirectional Optical Wireless Communications[J]. IEEE Photonics Journal, 2018, PP(99):1-1.
[2] Smys S. STAB-WIN: Self Organized, Topology Control Ability Backbone Node in Wireless Networks[J]. Wireless Personal Communications, 2012, 63(3):529-548.
[3] Bharathi C, Rekha D, Vijayakumar V. Genetic Algorithm Based Demand Side Management for Smart Grid[J]. Wireless Personal Communications, 2017, 93(2):481-502.
[4] Yoshimura M, Shimoyama K, Misaka T, et al. Topology optimization of fluid problems using genetic algorithm assisted by the Kriging model[J]. International Journal for Numerical Methods in Engineering, 2017, 109(4).
[5] Chang Y, Yuan X, Li B, et al. A Joint Unsupervised Learning and Genetic Algorithm Approach for Topology Control in Energy-Efficient Ultra-Dense Wireless Sensor Networks[J]. IEEE Communications Letters, 2018, 22(11):2370-2373.
[6] Huang S Q, Wang G C, Zhang Z, et al. A Method of Geometric K-center Gateway Deployment of Wireless Mesh Networks [J]. Chinese Journal of Computers, 2013, 36(7).

Common Sensors in Industrial Robots: A Review

Peng Li[1] and Xiangpeng Liu[1,*]

[1] School of Mechanical Engineering, Shanghai Jiao Tong University, 800 Dongchuan Road, Shanghai 200240, China

[*]Corresponding E-mail: x.liu@sjtu.edu.cn

Abstract. The application of industrial robots has greatly promoted the development of industry in the past decades. Now with the proposal and prevalence of Industry 4.0, industrial robots are required to be more independent and intelligent to accomplish more complex and flexible tasks. The advancement of industrial robots relies on the development and progress of multiple technologies, among which sensors are the indispensable part. They can acquire abundant information to help industrial robots implement their functions. This paper reviews the recent literatures and gives a summary and introduction of the commonly used sensors in industrial robots. Additionally, the applications of these sensors in diverse functions of industrial robots are also presented. Finally, the developing direction and challenges of industrial robots in the future are discussed in the last part of this article.

1. Introduction

Nowadays many countries focus on the development of Industry 4.0 to promote the transition and upgrade of the manufacturing sector. Industry 4.0, which is considered to be the fourth industrial revolution, is proposed by Germany at the industrial fair in Hannover in 2013. Its objective is to realize the intelligent, flexible and personalized process of manufacture with the help of Internet of Things (IOT), digital information technology, and computer technology [1]. In Industry 4.0, industrial robots are the indispensable component.

Industrial robots have a history of more than 60 years [2]. The development and utilization of them have vastly improved the productivity and promoted the progress of industry, as well as relieved workers of onerous and iterative tasks. What's more, with the advancement of sensors technology, automation and computer information technology in recent years, industrial robots are developed to be more intelligent. Now they can implement multiple functions and are extensively employed in many fields of manufacture, i.e., welding, assembly, materials transportation, mechanical processing, etc [2].

Sensors are the essential parts for industrial robots to accomplish their tasks. The information acquired by sensors can be used to judge the state of industrial robots and the circumstances of external environment, thus helping control and regulate robots to execute appointed functions. This paper reviews the recent literatures and summarizes the frequently-used sensors and their applications in industrial robots.

The structure of this article is organized as follows: The introduction of background in Section 1 is followed by the presentation and analysis of the common sensors employed in industrial robots in Section 2. Section 3 introduces the concrete applications of these sensors in industrial robots. And the last section concludes this article and makes a discussion about the future works.

IOP Conf. Series: Journal of Physics: Conf. Series **1267** (2019) 012036 doi:10.1088/1742-6596/1267/1/012036

2. Common Sensors in Industrial Robots

Sensors are of great importance to industrial robots. They are equivalent to the sensorium of industrial robots to help recognize the motion status of the robot and the circumstance of environment. With the help of these information, the controller can issue corresponding instructions to make the robot complete the required operations. After a survey on literatures in recent years, the common sensors deployed in industrial robots are presented as follows.

2.1. Tactile Sensors

Just as the significance of tactile perception for people to perceive multiple properties of external objects and interact with the environment, industrial robots also need the sense of touch to cognize the outer world. Therefore, tactile sensors, which endue industrial robots with the ability of tactile sensation, are vital components to make them more intelligent.

Tactile sensors have been developed for several decades, and they are gradually used in various robots now. The types of tactile sensors mainly include capacitive, piezoelectric, piezo-resistive and optical [3]. The capacitive tactile sensor utilizes the change of capacitance to measure the contact force. It has great spatial resolution and low power consumption, but its robustness to interference is poor [4]. The piezoelectric tactile sensor depends on the theory of piezoelectric effect which means that the electrical charge will be emerged on the surface of piezoelectric materials when exterior load is exerted. The frequency response of it is good and its measuring range is large, but its resolution is not very satisfying [3]. The piezo-resistive tactile sensor bases on the principle of piezo-resistive effect which refers to the variance of resistance when outer force is implemented. It has large measuring range and good robustness, however, it is impressionable to hysteresis [4]. The optical tactile sensor can transduce the external contact information into the variance of parameters of light. It is resistant to the external interference and possesses great spatial resolution [5]. Furthermore, some researchers have developed artificial skins to imitate human's skin, which may be the development trend of the future. Multiple sensors are deployed on each cell of the skin to provide abundant data.

Although tactile sensors are attracting more and more attention and interest, their performance, like versatility and adaptability, is not very satisfying at present. The development of them depends on the advancement of various domains of technology, such as materials, electronics, relevant algorithms, and so on [5]. To achieve the level of human's tactile perception, there is still a long way to go.

2.2. Visual Sensors

Visual sensing technology has developed rapidly over the past few years, and nowadays it is extensively utilized in many fields, such as face recognition, three-dimensional reconstruction and multiple robots, etc. Images captured by visual sensors are processed by the processor to extract useful information for specific tasks.

Visual sensors mainly include various kinds of cameras, like RGB cameras, multispectral cameras, and depth cameras [6]. The photosensitive elements equipped in multiple cameras are usually CCD or CMOS, which can transduce the light information into electrical signals based on the principle of photoelectric effect. The flexibility and pictures' quality of CCD cameras are better than that of CMOS cameras, but CMOS has its superiority in the aspects of cost and power consumption [7].

Different types of cameras can provide diverse information. RGB cameras, the mostly used one in people's daily life, capture chromatic images on the basis of the theory that each kind of visible color can be obtained by three colors: red, green, blue and their combination. Multispectral cameras are able to acquire images in diverse bands of spectrum, involving visible and invisible wavelength, consequently, they can get the information that cannot be provided by RGB cameras [8]. Depth cameras add the distance information to two-dimensional images, thus realizing stereo imaging. They can be classified as RGB binocular, structured light and TOF according to their operational principles.

Visual sensors enjoy great popularity for their advantages of low cost, luxuriant information supplied and ease of use [6]. Nonetheless, the processing of data from visual sensors are complex and time-consuming. Although many researchers have put forward several algorithms, the applicability and flexibility are still not very satisfying.

2.3. Laser Sensors

Laser is invented in twentieth century and has outstanding monochromaticity, directivity and brightness [9], thus being widely employed for diverse applications. Laser sensors are the kind of sensor that utilizes laser technology to accomplish measuring tasks. They are usually composed of laser emitter, detector and measuring circuit. The working substance of the emitter mainly contains four categories: solid, liquid, gas and semiconductor [10].

Laser sensors are mostly utilized to measure physical parameters, like distance, velocity and vibration. The common types are laser range finder, laser displacement sensor, laser scanner, laser tracker, etc. The fundamentals of laser range measurement mainly include three kinds: time of flight (TOF), triangulation method and optical interference [11]. TOF refers to the time from projecting laser to receiving the catoptric light. It is often used in laser range finder for long distance measuring, and the accuracy depends on the measuring precision of the flying time because of the high speed of light. The triangulation method utilizes the homothetic triangle theory and trigonometric function to calculate the distance of objects. Laser displacement sensor is based on this method to implement short range measuring. Optical interference is a phenomenon that the superposition of two light beams with different phase will generate bright and dark fringes. It is employed in the laser tracker to gauge the moved distance of the target fitted with a reflector.

Laser sensors can realize contactless telemetering, and the measuring speed and accuracy are satisfying [6]. However, the wavelength of laser can be influenced by temperature, atmospheric pressure and air humidity, so the compensation will be required when the above parameters change.

2.4. Encoder

The encoder is a kind of sensor that can transduce the angular displacement or velocity into electrical impulse or digital quantity. It can be divided into four categories: photoelectric, magnetic, inductive and capacitive according to the detection principles [12].

The frequently-used type is the photoelectric encoder, which utilizes the theory of photoelectric effect to complete signal conversion. It is usually composed of an optical coded disc and the photoelectric detecting appliance. In terms of the calibration mode of the coded disc, photoelectric encoders can be classified into incremental encoders and absolute encoders. The output of incremental photoelectric encoders is a series of square wave pluses. The rotation angle can be calculated on the basis of the amount of the pluses, and a zero reference position is required to determine the absolute position of the rotating shaft. Absolute photoelectric encoders output the binary digital quantity that corresponds to each location of the axis, therefore, they can get the absolute position directly.

Encoders have been widely used for many years on account of their advantages of compactness, long service life, ease of use, and mature technology [13]. The resolution of encoders relies on the amount of scribed lines on the coded disc in one circle. Smaller angle can be discriminated with more lines, thereby generating higher resolution, in the meanwhile, the cost will be higher too.

2.5. Other Sensors

Apart from the above four kinds of sensors, some other sensors are also deployed in industrial robots to realize several functions, such as proximity sensors, inertial sensors, torque sensors, acoustic sensors, magnetic sensors, ultrasonic sensors, etc.

The proximity sensor is a kind of non-contact device that can detect the approach of objects and output homologous switching signals. It can be categorized into capacitive, inductive and photoelectric on the basis of the operational fundamentals [14]. Capacitive proximity sensors utilize the alteration of the circuit status caused by the capacity variance of the detecting electrode to perceive the approaching objects. Inductive proximity sensors are based on the electromagnetic induction to work. The sensing elements of them are the detecting coils, whose inductance will change when a metallic conductor comes near. Photoelectric proximity sensors are usually composed of the light emitting diode and the photoelectric device. The light emitted from the diode will be reflected to the photoelectric device when objects approach, then the detecting circuit will generate corresponding output signals.

Inertial sensors usually refer to the accelerometer, gyroscope and magnetometer. They are widely employed to measure the kinematical parameters of moving objects, including the acceleration, angular speed and azimuthal angle. Specially, the triaxial combination of them is known as the inertial measuring unit (IMU). The measurement principle of inertial sensors is dead reckoning (DR) that utilizes the method of integration to calculate the amount of objects' movement. The precision of inertial sensors is satisfying in a short time, but the error will be larger and larger as time goes by [15].

Table 1. A brief summary of the presented sensors.

Sensor type	Principle	Information obtained	Applications in industrial robots
Tactile sensors	capacitive, piezoelectric, piezo-resistive, optical	contact force, area, position	human-robot collaboration (HRC), objects grasping, quality monitoring
Visual sensors	CCD or CMOS imaging	images	human-robot collaboration (HRC), navigation, manipulator control, assembly, robot programming
Laser sensors	time of flight (TOF), triangulation, optical interference	distance, displacement	human-robot collaboration (HRC), navigation, manipulator control
Encoders	photoelectric, magnetic, inductive, capacitive	angular displacement	navigation, manipulator control
Proximity sensors	capacitive, inductive, photoelectric	approaching of objects	human-robot collaboration (HRC), objects grasping
Inertial Sensors	dead reckoning (DR)	acceleration, angular speed, azimuthal angle	navigation, manipulator control
Torque sensors	inductive, resistance strain	torque	human-robot collaboration (HRC), objects grasping, robot programming
Acoustic sensors	capacitive	sound signals	human-robot collaboration (HRC), welding
Magnetic sensors	Hall Effect	magnetic field intensity	navigation
Ultrasonic Sensors	time of flight (TOF)	distance	obstacles avoidance

Torque sensors are mostly used to measure the torque exerted to mechanical axis. The common types are the inductive torque sensor and the resistance strain torque sensor. The structure of them usually consists of a torsion bar and detecting elements like coils or resistance strain gage. The input shaft and the output shaft are connected by the torsion bar, and the torsional deformation of the torsion bar caused by a torque can be transduced into electric signals via the variance of parameters of detecting elements, thus realizing the torque measurement.

Acoustic sensors are able to transduce the sound waves into electric signals. A capacitive electret microphone is installed in them, and sound waves can cause the vibration of the electret membrane in the microphone, thus resulting in the variance of capacity and generating weak voltage. Then the voltage is transformed for subsequent processing.

Magnetic sensors are mainly deployed to detect the magnetic field intensity. The principle of them is Hall Effect which refers to the phenomenon that when the electricity flows through a conductor, an

IOP Conf. Series: Journal of Physics: Conf. Series **1267** (2019) 012036 doi:10.1088/1742-6596/1267/1/012036

electric field which is perpendicular to the direction of the magnetic field and the electricity will be generated, thus causing a potential difference on the surface of the conductor.

Ultrasonic sensors are often used to detect obstacles. They estimate the range of objects according to the time from projecting the ultrasonic waves to detecting the echo. The size and weight of them are small, and their cost and power consumption are low [16].

Table 1 provides a brief summary of the above presented sensors.

3. Concrete Applications

Sensors are frequently deployed to assist the controller to control industrial robots to implement appointed functions, which involve various aspects, such as human-robot collaboration (HRC), autonomous navigation, manipulator control, etc. In the following part, this article gives a presentation and analysis of the above sensors' concrete applications in industrial robots.

3.1. Human-Robot Collaboration

Recently the concept of human-robot collaboration (HRC) has been brought out to realize the collaborative operation of workers and robots. This mode of production can combine the cognitive ability and strain capacity of human beings with the accuracy and tirelessness of robots to improve the flexibility and adaptability of manufacture systems [17].

The primary problem needed to be considered in HRC is safety [18]. Robots should be able to detect and recognize objects to avoid collisions or stop moving immediately when collisions occur. The common sensors applied to implement this function are vision sensors, laser sensors, proximity sensors, torque sensors and tactile sensors. For example, in [19, 20], tactile sensors are deployed to perceive the physical contact and locate the position of collisions to protect workers cooperating with the robot. And the internal joint torque sensors are utilized to detect and classify collisions by estimating the external forces in [21]. However, tactile sensors and torque sensors have a shortcoming that they can only detect but cannot avoid collisions. The other three kinds of sensors can fill the gap. In [22, 23], the capacitive or photoelectric proximity sensors are deployed to detect the approaching humans or objects so as to help the industrial robots react timely before collisions happen. Depth cameras are used to recognize humans in the operating scope of industrial robots to ensure safety in [24]. And in [25], laser scanners are equipped to monitor the working circumstances of robots for collisions avoidance. Furthermore, multiple sensors fusion is applied to guarantee the collaborative safety in HRC. For instance, in [26], a RGBD camera, proximity sensors and a laser range finder are utilized to monitor the working area of industrial robots and detect the approaching human beings. In [27], RGBD cameras and laser scanners are employed to recognize and track the workers in the working range of robots.

The interaction of humans and robots is also an important part in HRC. Workers can effectively control computer programming. For example, speech recognition assisted by acoustical sensors is exploited to help humans interact with robots in [28]. But the performance of this method may not be satisfying when the environment is noisy. Targeting this problem, hand gestures recognition with the help of vision sensors is proposed to control the robots in [29].

3.2. AGV Navigation

Nowadays AGV (Automated Guided Vehicle) is widely deployed in industrial environment for materials transportation, flexible production lines and some other applications [30]. Autonomous navigation is the primary function of it, and the current technology can be classified into two main types: physical paths and virtual paths [31].

Physical paths mainly refer to magnetic tapes or optical lines. In this method, magnetic sensors [32] or vision sensors [33] are commonly used to detect the path to estimate the position deviation of AGV, then the controller can make homologous adjustments to ensure AGV travels along the predetermined trajectory. The physical path has advantages of low cost, high reliability and maturity, but its flexibility is poor [31]. In contrast, the mode of virtual paths is more convenient to alter the route of AGV, and the sensors frequently utilized in it are vision sensors, laser sensors, encoders and inertial sensors. For instance, in [34], a TOF camera is deployed to perceive the environment and locate and guide a forklift AGV. And in [35], laser sensors are employed to calculate the position of AGV robots based on the triangular measurement method. Encoders and inertial sensors are also used to estimate the shift amount of AGV to execute navigation [36]. Additionally, the mode of multiple sensors fusion is utilized to improve the performance of navigation. In [37], magnetic sensors are appended to calibrate the location of AGV and modify the cumulative deviation of encoders and inertial sensors. And TOF cameras and laser range finders are used together to complete autonomous navigation in [38].

In addition to the above navigation technology, obstacles detection and avoidance is also an essential component in automated traveling of AGV. Ultrasonic sensors [39] or laser sensors [40] are usually deployed to detect obstacles to help the controller re-plan the route of AGV for collision avoidance.

3.3. Manipulator Control

Manipulator control is of great importance to the accuracy of various tasks implemented by industrial robots, such as mechanical processing, assembly, welding, etc. It involves the position and motion trajectory control of the end-effector. In order to get high precision of industrial robots, errors caused by several factors like compliance of joints, transmission of gears and temperature variation are needed to be gauged and modified [41].

The commonly used sensors in this respect are laser sensors, vision sensors, inertial sensors and encoders. For example, in [42], laser trackers are deployed to measure the current location of the end-effector to get the absolute position deviation, then the controller can make a real time calibration to improve the accuracy of industrial robots. And in [43, 44], the method of visual servoing is utilized to compensate errors of the robot actuator timely and realize the accurate path tracking. IMU is also used to measure the joint angles to estimate the position of the end-effector [45], but its precision is not very satisfying because of the accumulated errors. In addition, the accuracy of encoders equipped on servo motors of the robot joints is also not good because they cannot detect the deviation generated by the mechanical deformation or transmission of gears. Targeting this problem, the secondary encoders are deployed on each output axis to obtain high position accuracy of the actuator and improve the performance of mechanical processing [46], however, their price is a little expensive. Furthermore, the combination of inertial sensors and encoders is also employed to get better location estimation of the robot end-effector in [47].

3.4. Objects Grasping and Assembly

Objects grasping is a common operation in industrial manufacture. To complete this task, robots need to recognize the target object and determine its position coordinates, then the actuator should pick it in a proper force in case of damage or slippage. The frequently applied sensors in this process are usually vision sensors, proximity sensors, torque sensors and tactile sensors. In [48], depth cameras are deployed to identify the location and posture of the objects to help the controller plan the motion path of the robot end-effector. And in [49], proximity sensors are also configured to help modify the approaching deviation and improve the accuracy of grasping. On the other hand, grasping force of the end-effector are ordinarily estimated and controlled by torque sensors [50]. Additionally, in order to detect and avoid the slippage of the grasped object, tactile sensors are often employed to help adjust the picking force [51, 52].

Assembly is an important component of the industrial production lines. Apart from the above presented operation of parts grasping, the recognition of assembly position is also essential to autonomous assembling. In [53, 54], vision sensors are utilized to detect and localize the installation site to guide the end-effector to implement assembly. Moreover, torque sensors [55] or laser sensors [56] are also configured to help compensate the location error of visual sensors to improve the accuracy and efficiency of assembling.

3.5. Other Applications

In addition to the above introduced applications, sensors presented in section 2 are also applied in some other aspects of industrial robots, such as welding, quality monitoring, robot programming, etc.

For instance, in [57, 58], visual sensors are deployed to capture the real time welding images which will be processed to obtain the position of welding seam, then the controller can control the end-effector to execute welding accurately. Furthermore, acoustic sensors are also used to realize welding seam tracking by analyzing the characteristics of arc sound signals in [59]. And in [60], vision sensors and acoustic sensors are integrated to monitor the welding penetration status to control the quality of welding.

In the aspect of quality monitoring on production line, in [61], stereo vision sensors are utilized to inspect the quality of products via the processing and features extraction of the acquired images. And tactile sensors are also deployed on the robot end-effector to implement quality control by perceiving the surface information of products in [62].

Currently some intuitive methods are employed with the help of several sensors to simplify the process and improve the efficiency of robot programming. For example, in [63, 64], torque sensors are utilized to measure forces and torques to estimate the motion parameters of the robot manipulator during the procedure of manual guidance programming. And in [65], cameras and a luminous marker are used to track the movement of the human wrist to realize the demonstration programming of industrial robots. Additionally, in [66], tactile sensors are also deployed to recognize the tactile gestures to implement robot programming.

4. Conclusion

This article reviews the common sensors and their concrete applications in industrial robots in recent years. From the literatures, it can be seen that industrial robots are developing towards the direction of more independent and intelligent at present, which will help realize more flexible, accurate and personalized industrial manufacture. However, one challenge in the development of industrial robots is to develop sensors with better properties to improve the robots' perception of the external environment and their own status, and this relies on the development of multidisciplinary technologies. Additionally, the algorithms for handling the information acquired from sensors are also needed to be optimized to improve the processing speed and accuracy to meet the higher performance requirements of industrial robots.

References

[1] Keliang Zhou, Taigang Liu, Lifeng Zhou. Industry 4.0: Towards Future Industrial Opportunities and Challenges. *12th International Conference on Fuzzy Systems and Knowledge Discovery (FSKD)*, Zhangjiajie, China, August 15-17, 2015

[2] Tianmiao Wang, Yong Tao, 2014. Research status and industrialization development strategy of Chinese industrial robot. *Journal of Mechanical Engineering*, **50(9)**, 1-13.

[3] Liang Zou, Chang Ge, Z. Jane Wang, Edmond Cretu, Xiaoou Li, 2017. Novel tactile sensor technology and smart tactile sensing systems: a review. *Sensors*, **17**, 2653.

[4] Uriel Martinez-Hernandez, 2016. *Tactile sensors* (Paris: Atlantis Press).

[5] Mohsin I. Tiwana, Stephen J. Redmond, Nigel H. Lovell, 2012. A review of tactile sensing technologies with applications in biomedical engineering. *Sensors and Actuators A: Physical*, **179**, 17-31.

[6] Francisco Yandun Narvaez, Giulio Reina, Miguel Torres-Torriti, George Kantor, Fernando Auat Cheein, 2017. A Survey of Ranging and Imaging Techniques for Precision Agriculture Phenotyping. *IEEE/ASME Transactions on Mechatronics*, **22(6)**, 2428-2439.

[7] Dave Litwiller, 2001. CCD vs. CMOS: Facts and Fiction. *Photonics Spectra*, Laurin Publishing Co. Inc.

[8] Francisco Yandun Narvaez, Giulio Reina, Miguel Torres-Torriti, George Kantor, Fernando Auat Cheein, 2017. A survey of ranging and imaging techniques for precision agriculture phenotyping. *IEEE/ASME Transactions on Mectronics*, **22(6)**, 2428-2439.

[9] Avanish KumarDubey, VinodYadava, 2008. Laser beam machining—A review. *International Journal of Machine Tools and Manufacture*, **48(6)**, 609-628.

[10] J.D. Majumdar, I. Manna, 2003. Laser processing of materials. *Sadhana*, **28(3-4)**, 495-562.
[11] Markus-Christian Amann, Thierry M. Bosch, Marc Lescure, Risto A. Myllylae, Marc Rioux, 2001. Laser ranging: a critical review of unusual techniques for distance measurement. *Optical Engineering*, **40(1)**, https://doi.org/10.1117/1.1330700.
[12] Josef Janisch, 2006. Summary of Untouched Circumgyrate Coder. *Global Electronics China*, **4**, 53-54,56.
[13] Shan Chen, 2018. The review of patented encoder technology. *China Science and Technology Information*, **22**, 18-21.
[14] Dick Johnson, 2007. Proximity sensors. *Computer Engineering & Software*, **11**, 34-38.
[15] Hossein Mousazadeh, 2013. A technical review on navigation systems of agricultural autonomous off-road vehicles. *Journal of Terramechanics*, **50(3)**, 211-232.
[16] He Zhao, Zheyao Wang, 2012. Motion Measurement Using Inertial Sensors, Ultrasonic Sensors, and Magnetometers With Extended Kalman Filter for Data Fusion. *IEEE Sensors Journal*, **12(5)**, 943-953.
[17] GeorgeMichalos, et al, 2014. ROBO-PARTNER: Seamless Human-Robot Cooperation for Intelligent, Flexible and Safe Operations in the Assembly Factories of the Future. *Procedia CIRP*, **23**, 71-76.
[18] Ales Vysocky, Petr Novak, 2016. Human-Robot Collaboration in Industry. *MM Science Journal*, **2**, 903-906.
[19] John O'Neill, Jason Lu, Rodney Dockter, Timothy Kowalewski. Practical, Stretchable Smart Skin Sensors for Contact-Aware Robots in Safe and Collaborative Interactions. *IEEE International Conference on Robotics and Automation (ICRA)*, Seattle, WA, USA, May 26-30, 2015.
[20] Markus Fritzsche, Jose Saenz, Felix Penzlin. A Large Scale Tactile Sensor for Safe Mobile Robot Manipulation. *11th ACM/IEEE International Conference on Human-Robot Interaction (HRI)*, Christchurch, New Zealand, March 7-10, 2016.
[21] Dmitry Popov, Alexandr Klimchik, Nikolaos Mavridis. Collision Detection, Localization & Classification for Industrial Robots with Joint Torque Sensors. *26th IEEE International Symposium on Robot and Human Interactive Communication (RO-MAN)*, Lisbon, Portugal, Aug 28-Sept 1, 2017.
[22] Alwin Hoffmann, Alexander Poeppel, Andreas Schierl, Wolfgang Reif. Environment-aware Proximity Detection with Capacitive Sensors for Human-Robot-Interaction. *IEEE/RSJ International Conference on Intelligent Robots and Systems (IROS)*, Daejeon, Korea, October 9-14, 2016.
[23] Nicola Maria Ceriani, Andrea Maria Zanchettin, Paolo Rocco, Andreas Stolt, Anders Robertsson, 2015. Reactive Task Adaptation Based on Hierarchical Constraints Classification for Safe Industrial Robots. *IEEE/ASME Transactions on Mechatronics*, **20(6)**, 2935-2949.
[24] Abdullah Mohammed, Bernard Schmidt, Lihui Wang, 2017. Active collision avoidance for human–robot collaboration driven by vision sensors. *International Journal of Computer Integrated Manufacturing*, **30(9)**, 970-980.
[25] Dario Antonelli, Giulia Bruno, 2017. Human-Robot Collaboration Using Industrial Robots. *Advances in Engineering Research*, **86**, 99-102.
[26] Stephan Kallweit, Robert Walenta, Michael Gottschalk, 2016. ROS Based Safety Concept for Collaborative Robots in Industrial Applications. *Advances in Robot Design and Intelligent Control*, **371**, 27-35.
[27] Joséde Gea Fernández, et al, 2017. Multimodal sensor-based whole-body control for human–robot collaboration in industrial settings. *Robotics and Autonomous Systems*, **94**, 102-119.
[28] Dr. Jayashri Vajpai, Avnish Bora, 2016. Industrial Applications of Automatic Speech Recognition Systems. *Journal of Engineering Research and Applications*, **6(3)**, 88-95.
[29] Gilbert Tang, Seemal Asif, Phil Webb, 2015. The integration of contactless static pose recognition and dynamic hand motion tracking control system for industrial human and robot collaboration. *Industrial Robot: An International Journal*, **42(5)**, 416-428.

[30] B.Y. Qi, Q.L. Yang, Y.Y. Zhou. Application of AGV in intelligent logistics system. *5th Asia International Symposium on Mechatronics (AISM)*, Guilin, China, October 7-10, 2015.

[31] Alberto Vale, Rodrigo Ventura, Pedro Lopes, Isabel Ribeiro, 2017. Assessment of navigation technologies for automated guided vehicle in nuclear fusion facilities. *Robotics and Autonomous Systems*, **97**, 153-170.

[32] BoYang Xu, Dongqing Wang. Magnetic Locating AGV Navigation Based on Kalman Filter and PID Control. *Chinese Automation Congress (CAC)*, Xi'an, China, November 30- December 2, 2018.

[33] Chunfu Wu, Xiaolong Wang, Qingxie Chen, Xiaowei Cai, Guodong Li. Research on Visual Navigation Algorithm of AGV used in the Small Agile Warehouse. *Chinese Automation Congress (CAC)*, Jinan, China, October 20-22, 2017.

[34] Ulrich Behrje, Marian Himstedt, and Erik Maehle. An Autonomous Forklift with 3D Time-of-Flight Camera-Based Localization and Navigation. *15th International Conference on Control, Automation, Robotics and Vision (ICARCV)*, Singapore, Singapore, November 18-21, 2018.

[35] Hang Li, Andrey V. Savkin, 2018. Robotics and Computer Integrated Manufacturing. *Robotics and Computer Integrated Manufacturing*, **54**, 65-82.

[36] Hyunhak Cho, Eun Kyeong Kim, Eunseok Jang, Sungshin Kim, 2017. Improved Positioning Method for Magnetic Encoder Type AGV Using Extended Kalman Filter and Encoder Compensation Method. *International Journal of Control, Automation and Systems*, **15(4)**, 1844-1856.

[37] Pingbao Yin, Wenfeng Li, Ying Duan. Combinatorial Inertial Guidance System for an Automated Guided Vehicle. *IEEE 15th International Conference on Networking, Sensing and Control (ICNSC)*, Zhuhai, China, March 27-29, 2018.

[38] Andreas Do¨mel, et al, 2017. Toward fully autonomous mobile manipulation for industrial environments. *International Journal of Advanced Robotic Systems*, 1-19.

[39] J. Sankari ; R. Imtiaz. Automated guided vehicle(AGV) for industrial sector. *10th International Conference on Intelligent Systems and Control (ISCO)*, Coimbatore, India, January 7-8, 2016.

[40] Elena Cardarelli, Valerio Digani, Lorenzo Sabattini, Cristian Secchi, Cesare Fantuzzi, 2017. Cooperative cloud robotics architecture for the coordination of multi-AGV systems in industrial warehouses. *Mechatronics*, **45**, 1-13.

[41] Christian Moeller, et al, 2017. Real Time Pose Control of an Industrial Robotic System for Machining of Large Scale Components in Aerospace Industry Using Laser Tracker System. *SAE Int. J. Aerosp*, **10(2)**, 100-108.

[42] J. R. Diaz Posada, et al. High Accurate Robotic Drilling with External Sensor and Compliance Model-Based Compensation. *IEEE International Conference on Robotics and Automation (ICRA)*, Stockholm, Sweden, May 16-21, 2016.

[43] Tingting Shu, Sepehr Gharaaty, WenFang Xie, Ahmed Joubair, Ilian A. Bonev, 2018. Dynamic Path Tracking of Industrial Robots with High Accuracy Using Photogrammetry Sensor. *IEEE/ASME Transactions on Mechatronics*, **23(3)**, 1159-1170.

[44] Mohammad Keshmiri, Wenfang Xie, 2017. Image-Based Visual Servoing Using an Optimized Trajectory Planning Technique. *IEEE/ASME Transactions on Mechatronics*, **22(1)**, 359-370.

[45] Luciano Cantelli, Giovanni Muscato, Marco Nunnari, Davide Spina, 2015. A Joint-Angle Estimation Method for Industrial Manipulators Using Inertial Sensors. *IEEE/ASME Transactions on Mechatronics*, **20(5)**, 2486-2495.

[46] A. Klimchik, A. Pashkevich, 2018. Robotic manipulators with double encoders: accuracy improvement based on advanced stiffness modeling and intelligent control. *IFAC PapersOnLine*, **51(11)**, 740–745.

[47] Benigno Munoz-Barron, Jesus R. Rivera-Guillen, Roque A. Osornio-Rios, Rene J. Romero-Troncoso, 2015. Sensor Fusion for Joint Kinematic Estimation in Serial Robots Using Encoder, Accelerometer and Gyroscope. *Journal of Intelligent & Robotic Systems*, **78(3-4)**, 529-540.

[48] Ryo KABUTAN, et al. Development of Robotic Intelligent Space Using Multiple RGB-D

Cameras for Industrial Robots. *ICT-ROBOT*, Busan, Korea, September, 2016.

[49] Om Prakash Sahu, Bibhuti Bhusan Biswal, Saptarshi Mukharjee, Panchanand Jha, 2015. Development of Robotic End-effector Using Sensors for Part Recognition and Grasping. *International Journal of Materials Science and Engineering*, **3(1)**, 39-43.

[50] Eduardo Moreira, Luís F. Rocha, Andry M. Pinto, A. Paulo Moreira, Germano Veiga, 2016. Assessment of Robotic Picking Operations Using a 6 Axis Force/Torque Sensor. *IEEE Robotics and Automation Letters*, **1(2)**, 768-775.

[51] Rocco A. Romeo, Calogero M. Oddo, Maria Chiara Carrozza, Eugenio Guglielmelli, Loredana Zollo, 2017. Slippage Detection with Piezoresistive Tactile Sensors. *Sensors*, **17**, 1844.

[52] Giuseppe De Maria, Pietro Falco, Ciro Natale, Salvatore Pirozzi. Integrated Force/Tactile Sensing: The Enabling Technology for Slipping Detection and Avoidance. *IEEE International Conference on Robotics and Automation (ICRA)*, Seattle, Washington, USA, May 26-30, 2015.

[53] Chyi-Yeu Lin, Le Thai Son, Yu-Lun Chang, Ya-Shiun Shiue, 2017. Image-Sensor-Based Fast Industrial-Robot Positioning System for Assembly Implementation. *Sensors and Materials*, **29(7)**, 935-945.

[54] Rafiq Ahmad, Peter Plapper, 2016. Safe and Automated Assembly Process using Vision assisted Robot Manipulator. *Procedia CIRP*, **41**, 771-776.

[55] Mustafa W. Abdullah, Hubert Roth, Michael Weyrich, Jürgen Wahrburg, 2015. An Approach for Peg-in-Hole Assembling using Intuitive Search Algorithm based on Human Behavior and Carried by Sensors Guided Industrial Robot. *IFAC-PapersOnLine*, **48(3)**, 1476–1481.

[56] Zhengke Qin, Peng Wang, Jia Sun, Jinyan Lu, Hong Qiao, 2016. Precise Robotic Assembly for Large-Scale Objects Based on Automatic Guidance and Alignment. *IEEE Transactions on Instrumentation and Measurement*, **65(6)**, 1398-1411.

[57] Yanling Xu, Gu Fang, Na Lv, Shanben Chen, Jujia Zou, 2015. Computer vision technology for seam tracking in robotic GTAW and GMAW. *Robotics and Computer-Integrated Manufacturing*, **32**, 25-36.

[58] Yanling Xu, et al, 2017. Welding seam tracking in robotic gas metal arc welding. *Journal of Materials Processing Tech.*, **248**, 18-30.

[59] Na Lv, et al, 2017. Real-time monitoring of welding path in pulse metal-inert gas robotic welding using a dual-microphone array. *The International Journal of Advanced Manufacturing Technology*, **90(9-12)**, 2955–2968.

[60] Tao Zhu, Yonghua Shi, Shuwan Cui, Yanxin Cui, 2019. Recognition of Weld Penetration During K-TIG Welding Based on Acoustic and Visual Sensing. *Sensing and Imaging*, **20**, 3, https://doi.org/10.1007/s11220-018-0224-9.

[61] Daniel Frank, Jimmy Chhor, Robert Schmitt. Stereo-vision for autonomous industrial inspection robots. *IEEE International Conference on Robotics and Biomimetics, Macau SAR*, China, December 5-8, 2017.

[62] Nathan F. Lepora, Benjamin Ward-Cherrier, 2016. Tactile Quality Control with Biomimetic Active Touch. *IEEE Robotics and Automation Letters*, **1(2)**, 646-652.

[63] Rebecca Hollmann, Arne Rost, Martin Hägele, Alexander Verl. A HMM-based Approach to Learning Probability Models of Programming Strategies for Industrial Robots. *IEEE International Conference on Robotics and Automation*, Anchorage, Alaska, USA, May 3-8, 2010.

[64] Daniele Massa, Massimo Callegari, Cristina Cristalli, 2015. Manual guidance for industrial robot programming. *Industrial Robot: An International Journal*, **42(5)**, 457-465.

[65] Marcos Ferreira, Paulo Costa, Luís Rocha, A. Paulo Moreira, 2016. Stereo-based real-time 6-DoF work tool tracking for robot programing by demonstration. *The International Journal of Advanced Manufacturing Technology*, **85(1-4)**, 57-69.

[66] Daniel Kubus, Arne Muxfeldt, Konrad Kissener, Jan Haus, Jochen Steil. Robust Recognition of Tactile Gestures for Intuitive Robot Programming and Control. *IEEE/RSJ International Conference on Intelligent Robots and Systems (IROS)*, Vancouver, BC, Canada, September 24–28, 2017.

IOP Conf. Series: Journal of Physics: Conf. Series **1267** (2019) 012037 doi:10.1088/1742-6596/1267/1/012037

An Improved Negatively Correlated Search Inspired by Particle Swarm Optimization

Niu Qun, Liu Bing, Jiang Kecheng and Wang Han

School of Mechanical Engineering Automation,Shanghai Key Laboratory of Power Station Automation Technology, Shanghai University, Shanghai, China

E-mail: nq@shu.edu.cn; liubing7891@hotmail.com; jiangkecheng1994@hotmail.com; dawnbreakwh@gmail.com

Abstract. Negatively correlated search (NCS) is a latest intelligence algorithm which maintains multiple individual search processes in parallel and models the behavior of individual search processes using probability distributions. In this paper, an improved NCS inspired by particle swarm optimization (PSO) is proposed, namely PSO-NCS. This algorithm introduces the global best solution (gbest) and personal best solution (pbest) from PSO to calculate the Bhattacharyya Distance (BD) of individuals in the population, which greatly reduces the computational time. Experimental results on six 100 dimensional functions and 10 CEC2005 benchmark functions show that the proposed PSO-NCS has outstanding capability in solving the optimization problems and outperforms the conventional NCS and other methods on most of the test functions. The computational time has been greatly shortened, which saves about 65% in average compared to NCS.

1. Introduction

With the advent of the big data era, the computational models used in real world engineering fields are becoming more and more complicated. How to choose the optimal parameters to obtain higher accurate results in the field of complex system modeling can be treated as optimization problems which are often challenging due to such features as discontinuity, non-differentiability, multiple local minimums, and flat areas, etc. Conventional nonlinear optimization methods, such as the gradient descent method [1] and lagrangian relaxation [2] suffer from sever limitations in solving such challenging problems. For example, the conventional methods have difficulties in solving non-linear discontinuous objective functions or objective functions with no mathematical expressions. Further, the performance of conventional methods depend heavily on the selection of initial values. If the initial values are not well selected, they are easily trapped in the local extremums, and there is no generic solutions to guide the selection of initial values. In the last decades, a number of intelligent optimization algorithms such as the Genetic Algorithm (GA) [3], Particle Swarm Optimization (PSO) [4], the Differential Evolution (DE) [5] and Artificial Bee Colony algorithm (ABC) [6] have been proposed to solve such challenging optimization problems. These intelligent algorithms have the features of simple principle, strong robustness and no special requirements for the objective function and constraints, and they have become indispensable tools in solving the complex optimization problems.

The Negatively Correlated Search (NCS) [7] was a recent intelligent optimization tool that was originally proposed by Tang etc., and it has attracted a lot of attentions since 2016. The NCS

IOP Conf. Series: Journal of Physics: Conf. Series **1267** (2019) 012037 doi:10.1088/1742-6596/1267/1/012037

is characterized by its information sharing and cooperation scheme, which explicitly promotes negative correlation search behavior for more effective exploration in the search space. Empirical results shows that NCS has a stronger global searching capability and the higher robustness in comparison to other existing algorithms on complex benchmark problems. NCS has been applied in deep learning to decide the sizes of the deep belief network (DBN) and learning rates during the training processes, and it exhibits excellent performance [8].

Although NCS has a number of merits, it also has some shortcomings. NCS employs the Bhattacharyya distance (BD) to calculate the difference and the correlation, where each individual calculates the BD with all other individuals. Therefore, the algorithm spends a lot of time in calculation, and inefficiency therefore becomes a shortfall of the algorithm. In order to improve the efficiency of NCS, this paper proposes an improved NCS method based on the idea of PSO, namely PSO-NCS. Each individual does not need to calculate the Bhattacharyya distances with all other individuals and only needs to calculate the distance between the mean of global best solution (gbest) and personal best solution (pbest). For example, there are ten individuals in the population, the NCS needs to calculate the distance by 10*10 times, but the PSO-NCS only needs to calculate 10*1 times, which greatly reduces the computation time. The improved PSO-NCS not only significantly shortens the computation time, thus improving its efficiency, but also improves the performance. In order to verify the effectiveness, PSO-NCS is applied to test six 100 dimensional benchmark functions and 10 CEC2005 functions in comparison with other algorithms.

The reminder of this paper is organized as follows: In section 2, it introduces the conventional NCS and details the PSO-NCS algorithm. Section 3 presents the simulation results. Finally, Section 4 concludes the paper.

2. PSO-NCS

2.1. NCS(Negatively Correlated Search)

In 2016, Tang etc. proposed the NCS method motivated by the interpretation of human behavior of cooperation and avoiding duplication of same work. It adopts the Bhattacharyya Distance (BD) [9] to estimate the difference and the correlation between two continuous probability distributions. BD is defined in the Eq. (1).

$$D_B(p_i, p_j) = -\ln\left(\int \sqrt{p_i(x)p_j(x)}dx\right) \tag{1}$$

where p_i and p_j are the probability density functions of two distributions.

In addition, the distance between each individual and the associated group $Corr(p_i)$ can be defined as (2) :

$$Corr(p_i) = \min_j \{D_B(p_i, p_j) | j \neq i\} \tag{2}$$

A new solution $x_i{}'$ is generated by Guassian mutation on a current solution as (3) :

$$x_{id}{}' = x_{id} + \mathrm{N}(0, \sigma_i) \tag{3}$$

where the $N(0, \sigma_i)$ stands for a Guassian random variable of zero mean and standard deviation σ_i and $x_{id}{}'$ is the dth element of x_i.

The step size σ_i follows the 1/5 successful rule [10] for every iteration as (4) :

$$\sigma_i = \begin{cases} \frac{\sigma_i}{r} & if \quad \frac{c}{epoch} > 0.2 \\ \sigma_i * r & if \quad \frac{c}{epoch} < 0.2 \\ \sigma_i & if \quad \frac{c}{epoch} = 0.2 \end{cases} \tag{4}$$

where r is a parameter less than 1, and c is the times that x_i' is discarded in the past iterations. This indicates that the larger c, the smaller search step-size, and the solution is closer to the global optimum. On the contrary, a large step size will more likely make the search being trapped in a local optimum.

In NCS, a smaller objective function value $f(x_i')$ with a larger $Corr(p_i')$ is preferred. To obtain a better population, strategy considering the $f(x_i')$ and $Corr(p_i')$ is used in NCS, which is defined as Eq. (5) :

$$x_{new,i} = \begin{cases} x_i & if \frac{f(x_i')}{Corr(p_i')} < \lambda \\ x_i' & otherwise \end{cases} \tag{5}$$

where λ is a time-varying parameter which can influence the search process and the capability of NCS as Eq. (6) :

$$\lambda_{iter} = N(1, 0.1 - 0.1 * \frac{t}{T_{\max}}) \tag{6}$$

where the $T_{\max}$ is the total number of iterations in executing the NCS, and t is the current iteration.

2.2. PSO-NCS

In the conventional NCS, each individual in the population needs to calculate its BD with all other individuals in the population, which is computationally intensive. In order to enhance the efficiency of NCS, this paper proposes an improved NCS, namely PSO-NCS. PSO is initialized with a group of random particles which are then updated using several operators to produce new generation. At every iteration, each particle is updated using two "best" values. The first one is the personal best solution (fitness) it has achieved so far. The fitness value is also stored. This value is called "pbest". Another "best" value that is tracked by the particle swarm optimizer is the global best value, obtained so far by any particle in the population. This best value is a global best and called "gbest". In PSO-NCS, each individual calculates the sum of distances between itself to the pbest and to the gbest. Therefore, the BD in PSO-NCS can be defined as (7) [11] :

$$D_B(p_i, p_j) = \frac{1}{8}(x_i - x_q)^T(x_i - x_q) + \frac{1}{2}\ln\left(\frac{\det \sum}{\sqrt{\det \sum_i \sum_q}}\right) \tag{7}$$

where the x_q is the mean of gbest and pbest in the iteration, $\sum = \frac{\sum_i + \sum_q}{2}$, $\sum_i = \sigma_i^2 I$, and I is an n-dimensional identity matrix.

In Eq. (6), it can be found that for each individual, it calculates only two distances rather than all distances in NCS. The calculation time is therefore greatly reduced. Meanwhile, the introduction of the pbest and gbest can improve the solution quality.

3. Simulations And Results

In this section, two case studies are presented to validate the effectiveness and efficiency of PSO-NCS. In case 1, the proposed PSO-NCS is tested on six typical shifted benchmark functions with 100 dimensions which listed in Table 1 to verify its performance compared to conventional NCS and other well-known algorithms such as the PSO-CF [12], HS [13], and weight-PSO [14]. In case 2, the PSO-NCS is tested on the CEC2005 test functions in comparison to the conventional NCS. All the cases are run for 25 independent trials under the same number of function evaluations(FES) (case 1 is set to 1000000 and case 2 is set to 300000) for a fair comparison. The control parameters used in NCS follow the references in the literature while the parameters of the PSO-NCS are the same as the NCS. All the experiments are carried out using the MATLAB 2016a on a PC of 3.70GHz with 32GB RAM.

Table 1. Formulations of Benchmark functions

Function	Formulation	Boundary	$f_{\min}$
Shifted Sphere	$f(x) = \sum_{i=1}^{Dim} (x_i - o_i)^2$	[-100,100]	0
Shifted Rosenbrock	$f(x) = \sum_{i=1}^{Dim-1} (100((x_{i+1} - o_{i+1}) - x_i^2) + (x_{i+1} - o_{i+1} - 1)^2)$	[-30,30]	0
Shifted Rastrigin	$f(x) = \sum_{i=1}^{Dim-1} ((x_i - o_i)^2 - 10\cos(2\pi(x_i - o_i) + 10)$	[-5.12,5.12]	0
Shifted Griewank	$f(x) = \sum_{i=1}^{Dim} \frac{(x_i-o_i)^2}{4000} - \prod_{i=1}^{Dim} \cos(\frac{x_i-o_i}{\sqrt{i}}) + 10$	[-600,600]	0
Shifted Axis	$f(x) = \sum_{i=1}^{Dim} i(x_i - o_i)^2$	[-5.12,5.12]	0
Shifted De Jong	$f(x) = \sum_{i=1}^{Dim} i(x_i - o_i)^4$	[-1.28,1.28]	0

Table 2. Test results in case1

		PSO-CF	W-PSO	HS	NCS	PSO-NCS
Shifted Sphere	mean	7.42E02	9.08E01	4.87E03	4.76E03	**3.11E-27**
	std	3.92E02	5.66E01	5.35E02	8.21E02	**7.79E-28**
Shifted Rosenbrock	mean	7.75E04	1.44E04	1.56E06	9.90E01	**8.89E01**
	std	6.86E04	2.14E04	2.94E05	1.51E01	**3.50E00**
Shifted Rastrigin	mean	7.14E02	4.27E02	**1.46E02**	7.42E02	6.30E02
	std	6.53E01	7.20E01	**1.30E01**	5.95E02	6.03E01
Shifted Griewank	mean	1.55E01	3.91E00	4.60E01	4.24E01	**3.15E-03**
	std	5.78E00	1.87E00	5.78E00	7.51E00	**5.53E-03**
Shifted Axis	mean	4.42E00	**9.38E-01**	2.51E00	3.58E00	3.89E00
	std	8.84E-01	2.79E-01	**1.36E-01**	4.75E-01	5.62E-01
Shifted De Jong	mean	9.60E00	2.47E00	4.47E00	1.10E00	**5.63E-01**
	std	1.56E00	5.88E-01	4.22E-01	5.30E-01	**3.78E-01**

3.1. Case1:PSO-NCS in benchmark function test

Table 2 lists the means and the standard deviations of the 25 runs of the five different algorithms on the six benchmark functions with 100 dimensions. Two different types of benchmark functions (a) shifted sphere(unimodal) (b) shifted griewank (multimodal) are selected to demonstrate the convergence characteristics of five algorithms as illustrated in Figure. 1. It can be seen that the PSO-NCS is easier to jump out of local optimum and to avoid premature convergence than other algorithms regardless of the types of the benchmark functions.

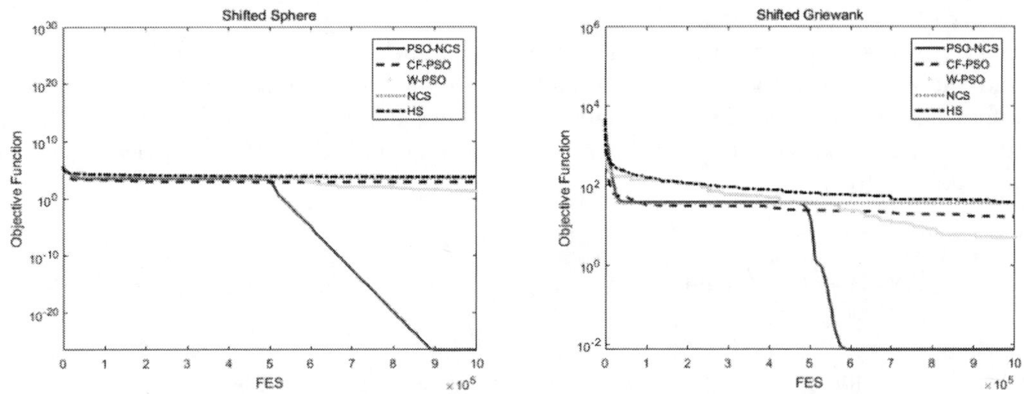

(a) sphere (b) griewank

Figure 1. Convergence curves for five algorithms.

3.2. Case2:PSO-NCS in CEC2005 test

Multi-modal optimization problems are more attractive since they have motivated the invention of a number of population-based search methods. 10 multi-modal continuous functions proposed in the CEC2005 competition in [15] are used in the study. Table 3 shows that the means, the standard deviations and the executing time of the 25 runs of the PSO-NCS and the NCS in the Ref.[7] on the 10 CEC2005 functions with 30 dimensions. It can be found in the Table 3 that PSO-NCS outperforms the conventional NCS on most of the test functions. It should be pointed out that the computational time of PSO-NCS has been greatly decreased on the all test functions, achieving about 65% reduction in average.

Table 3. Test results in case 2

		f6	f7	f8	f9	f10
NCS	mean	2.08E01	1.69E-02	2.00E01	9.36E01	9.03E01
	std	3.61E00	**1.38E-02**	1.22E-02	**1.38E01**	1.79E01
	time	35.2	35.92	37.28	37.88	37.16
PSO-NCS	mean	**1.98E01**	**1.47E-02**	**2.00E01**	**8.23E01**	**8.76E01**
	std	**2.23E00**	1.51E-02	**9.44E-03**	1.78E01	**1.74E01**
	time	**9.12**	**9.56**	**9.52**	**9.56**	**9.72**
		f11	f12	f13	f14	f15
NCS	mean	1.37E01	**1.57E03**	4.54E00	**1.24E01**	**3.15E02**
	std	**1.27E00**	1.52E03	**8.04E-01**	3.31E-01	**5.68E01**
	time	55.32	44.08	38.04	38.44	158.68
PSO-NCS	mean	**1.32E01**	1.71E03	**4.18E00**	1.25E01	3.27E02
	std	2.05E00	**1.26E03**	8.43E-01	**2.69E-01**	6.59E01
	time	**28.68**	**10.32**	**10.6**	**11.68**	**130.52**

4. Conclusion and future work

This paper proposes an improved NCS algorithm, namely PSO-NCS, which introduces the pbest and the gbest from the PSO to calculate the BD for each individuals in the population. The simulation results on six 100-dimenional benchmark functions and ten CEC2005 functions demonstrate that PSO-NCS has a stronger global search capability and a higher computational effectiveness than NCS.

Acknowledgment

This work is supported by the National Natural Science Foundation of China (61773252).

References

[1] Beck A and Teboulle M 2009 Fast gradient-based algorithms for constrained total variatio image denoisingand deblurring Problems *IEEE Transactions on Image Processing* **18** 2419-2434.

[2] C,Jognson E L,Nemhauser G L, Savelsbergh M W P and Vance P H 1998 Column generation for solving huge integer programs *J. Operations Research* **43** 316-329.

[3] Jones G, Willett P, Glen R C, Leach A R and Taylor R. 1997 Development and validation of a genetic algorithm for flexible docking *J. Joural of molecular biology.* **267** 727-748.

[4] Kennedy J and Eberhart R 1995 Particle swarm optimization *Proceedings of ICNN'95-International Conference on Neural Networks* **4** 1942-1948.

[5] Storn P and Price K 1997 Differential evolution-a simple and efficient heuristic for global optimization over continuous spaces **11** 341-359.

[6] Karaboga D, Akay B 2009 A comparative study of artificial bee colony algorithm *J. Applied Mathematics and Computation* **214** 108-132.

[7] Tang K, Yang P and Yao X 2016 Negatively correlated search *J. IEEE Journal on Selected Areas in Communications* **34** 542-550.

[8] Lin Y, Liu H, Xie G and Zhang Y 2018 Time series forecasting by evolving deep belief network with negative correlation Search *2018 Chinese Automation Congress (CAC)* 3839-3843.

[9] Kailath T 1967 The divergence and bhattacharyya distance measures in signal selection *J. IEEE Transactions on Communication Technology* **15** 52-60.

[10] Beyer H G and Schwefel H P 2002 Evolution strategies, a comprehensive introduction *J. Nat. Comput.* **1** 3-52.

[11] Fukunaga K 1900 Introduction to statistical pattern recognition *2nd. ed. New York, NY, USA:Academic.*

[12] Ganguly S and Sahoo N C 2011 Mono- and multi-objective planning of electrical distribution networks using particle swarm optimization *J. Applied Soft Computing* **11** 2391-2405.

[13] Geeem Z W, Kim J H and Loganathan G V 2001 A new heuristic optimization algorithm:Harmony search *J. Simulation* **76** 60-68.

[14] Kadhiravan D, Venkatalakshmi K and Praisie R S 2015 Comparison of k-Means, variant of k-means, genetic algorithm, particle swarm optimization, weighted particle swarm optimization and firefly algorithms for image classification *J. International Journal of Applied Engineering Research* **10** 8803-8813.

[15] Suganthan P. N. et al. 2005 Problem definitions and evaluation criteria for the CEC 2005 special session on real-parameter optimization *Tech. Rep in Nanyang Technological University (NTU).*

Research on a Fast Continuous Vialog Algorithm

Zhang Jinning[1], Zhou Yishu[1*], He Huayang[1] and Ren Lishuo[1]

[1]Research Institute of Highway Ministry of Transport No. 8 Xitucheng Road, Haidian District, Beijing

[*]Corresponding Author E-mail: ys.zhou@rioh.cn

Abstract. Surface evenness affects vehicle safety, service life, and passenger comfort. The road surface with poor flatness will produce resistance to the vehicle driving, making the vehicle vibrate and bump, hindering the safety of the vehicle driving, and will also cause local water accumulation, speed up the road damage and so on. In order to make the vehicle safe, passengers feel comfortable, road life growth, road flatness detection is particularly important.In order to make the vehicle safe to travel, passengers feel comfortable, the road service life growth, and the detection of road surface evenness is particularly important. In order to overcome the shortcomings of the existing algorithms for pavement evenness, this paper has put forward a new fast algorithm that can be applied to the continuous vialog, and according to the elevation value obtains the evenness standard deviation σ. The results show that the algorithm is high reliability, which provides a reference for the upgrade of instruments and equipment, and makes up for the shortcomings that the continuous vialog cannot be tested quickly.

1. Introduction

In recent years, with the rapid development of the national economy, the transportation industry has also developed rapidly, especially the highway transportation mode has played an important role in the vigorous development of the national economy. At the same time, there are problems such as highway detection and maintenance have emerged, especially in the highway sections build and opened to traffic before, there have been obvious damage cracks, uneven pavement and other phenomena, which directly affect the service life of the highway and the safety of the vehicles. Therefore, it is necessary to establish perfect highway pavement maintenance management for scientific and efficient understanding of the condition of the paved road surface[1]. It also has great significance for the scientific evaluation of highway construction grade and road surface operation status to detect, such as the detection, evaluation and analysis of the evenness, pavement crack and rutting size of the road surface that has been completed[2]. Evenness affects vehicle safety, service life and passenger comfort. The road surface with poor evenness will produce resistance to the vehicle driving, make the vehicle to generate vibration bumps, hinder the driving safety of the vehicle, and cause localized water accumulation and accelerated road damage speed. In order to make the vehicle safe to drive, passengers feel comfortable, the road service life growth, the detection of road surface evenness is particularly important.

The developed countries in the world, such as the United States and Japan, have taken the pavement evenness as the main index to measure the service performance and maintenance quality of the road surface, and conducted regular testing it as an important basis for formulating road pavement maintenance management strategies. The pavement evenness can be used not only to measure the quality of pavement construction and the quality of pavement maintenance, but more importantly, the index of pavement evenness can be used to reflect the degree of human comfort in the process of driving. Therefore, this paper focuses on the research of the evenness algorithm.

2. Research Status of Evenness Algorithm

There are many different indexes in the aspect of road evenness evaluation indicators. For example, ARS can adjust the average slope, which is measured by the reaction equipment at a fixed speed; BI turbulence accumulation index, measured by a road gauge developed by the British TRRL; CP evenness index is proposed by the Belgian Road Research Institute (CRR), is calculated after moving average filtering of different widths for the measured value of APL72 measuring instrument; QI is 1/4 vehicle index, the predecessor of international evenness index (IRI), which is based on the elevation of the longitudinal section of the road; NAASRA index, proposed by the Australian Road Research Institute; International evenness index based on road section data and calculated by fixed parameter 1/4 vehicle model[3].

In order to study the correlation between the indicators, foreign researchers[4] simulated the longitudinal section of the road based on the road power spectrum density irregularity, and calculated the international evenness index (IRI) by Newmark continuous integration method and mathematical statistics method. According to the simulation and regression analysis, there is a very good linear correlation between the IRI and the standard deviation of the relative speed of the axle and suspension system. If the power spectral density of PSD evenness can be expressed by a polynomial function of the wave number, IRI can be calculated by the root mean square of the sum of the weighted coefficients in the PSD evenness regression analysis, which enables the highway management department to obtain the road IRI data based on the know road PSD evenness data,the inspection data of the highway management department and the automobile manufacturer can be compared with the same standard. Benneit[5] pointed out through research that the reaction evenness meter needs to carry out calibration for the international evenness index IRI to ensure the accuracy of its measurements. Old and Peter[6] pointed out through research that if there is irregular evenness power spectral density on the road, it needs to be analyzed and described by different independent parameters. In this case, IRI can't distinguish the significant difference in the longitudinal section of the road, and pointed out that the standard deviation of vertical elevation RMSH is better than the singular index IRI, mainly because RMSH reflects the vehicles in the short, medium and long range vibration situation of road evenness. Wei Xiaodan used IRI as the evaluation index of pavement flatness. In the study, the IRI and the continuous flatness meter σ value were calculated by using the level gauge and the hand-inferred surface meter, and the correlation test was carried out. The test results show that the IRI and σ indicators can be converted by the formula σ=0.6IRI. However, Xu Yongxin and Zhou Yan analyzed the use method and working principle of the evenness test in the "Research on the method of pavement evenness detection". In this paper, they have established the correspondence between the flatness and the test data and given some usual conversion methods: IRI =1.0319 sigma +0.2711. Zhang Yi and Ma Ronggui[7] analyzed the road evenness detection method based on inertial reference, and on the basis of a quarter car model in the research of "laser pavement evenness detection based on inertial reference system research" , In this paper, laser range finder is presented and the acceleration sensor output signal for road surface profile measurement and calculation formula of the relative elevation to achieve the multi-channel data acquisition and joint processing, introduces the data processing method of the sensor output signal, through the contrast test with precision level measurement method, results show that the detection method based on inertial reference has feasibility in the road surface evenness detection.

When most of the domestic laser evenness meters find the standard deviation σ, the usual algorithm is σ=0.6IRI, and the calculation is not performed by the original elevation. In this paper, a new fast algorithm that can be applied to continuous vialog is proposed, which provides a reference for equipment upgrade and makes up for the shortcoming that continuous flatness meter cannot be tested quickly.

3. Demonstration of laser flatness analyzer algorithm

The laser evenness meter is the most commonly used detection equipment in highway inspection. It has many advantages, such as being unaffected by the detection speed and not affected by the type of the vehicle body, so it plays an important role in the evenness detection market. Generally, the laser flatness meter is add an inertial reference unit to a laser sensor module. The working process is as follows: firstly, the vertical elevation of the equipment to the road surface is measured by a laser sensor. Then, the absolute displacement is generated in the direction, and the error is corrected by the accelerometer measuring device to obtain the real road pavement longitudinal elevation information. Finally, the international evenness standard deviation σ is calculated based on IRI.

According to the research literature, most of the algorithms for laser flatness meters at this stage is to calculate the IRI first, and then calculate the σ by σ=0.6IRI. This algorithm is derived from Wei Xiaodan's research on the international flatness index IRI as the evaluation index of pavement evenness. IRI and σ were calculated by using a level gauge and a hand-inferred face meter, and then the correlation test was performed on the two indicators. The test results show that the IRI and σ indicators can be converted by the formula σ=0.6IRI.

But this algorithm is calculated by statistics. There is no mathematical correlation between the two indicators. In different literatures, the conversion methods of this two indicators are different. Among them, Xu Yongxin and Zhou Yan in "research on the detection method of road surface evenness" have analyzed the operation method and working principle of the evenness test vehicle, and established the corresponding relationship between the flatness and the test data, and gave different conversion methods: IRI = 1.0319 σ+0.2711.

In this paper, it is considered that the accuracy of the laser flatness meter and the accuracy of the continuous flatness meter cannot be verified before the establishment of the laser flatness equipment standard in the literature. As a result, the measured values and correlation of such equipment cannot be used as the algorithm for fast continuous evenness.

Due to the lack of theoretical correlation, this paper uses the level gauge to collect the road surface actual data, take some actual data and the original measured value of the continuous vialogvalue, and calculateσ based on the elevation value of the road surface. Then correlation between σ and IRI was calculated. The correlation curve is shown in Figure 1. We can get the relationship between σ and IRI: σ=0.844IRI-0.234, R=0.98.

Since the actual road segment data are relatively close and the values are relatively concentrated, the computer is used to simulate the road surface with different values in accordance with the actual segment . And obtains the original measured value of the continuous flatness value according to the simulated road surface, and then obtains the relationship between σ and IRI by correlation. As shown in Figure 2, it was found that σ = 0.741 IRI - 0.045, and R > 0.99.

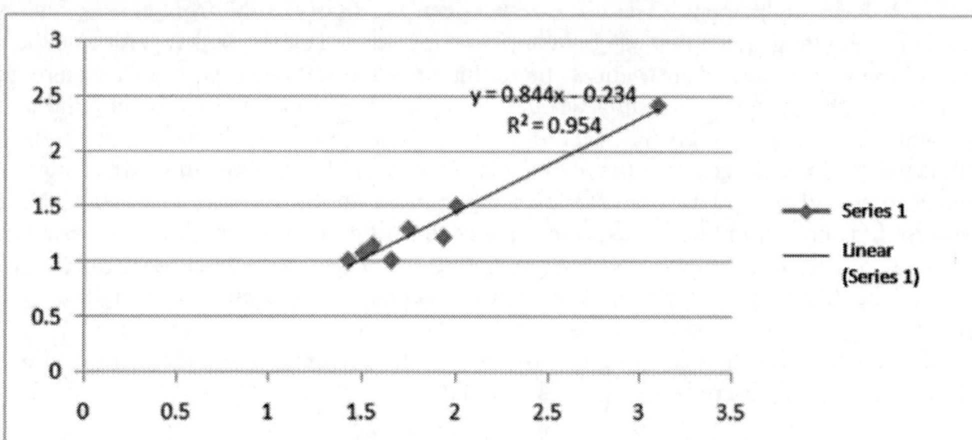

Figure 1. Correlation curve between σ and IRI

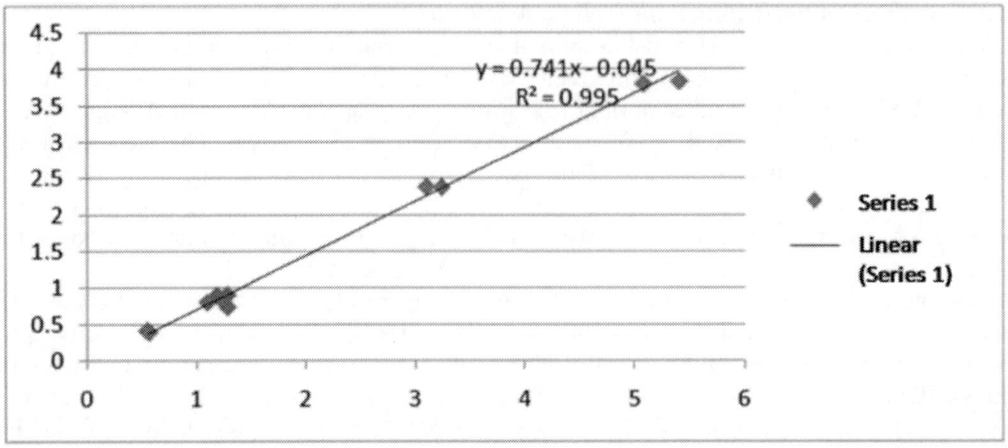

Figure 2. Computer simulation of the relationship between sigma and IRI on the pavement

Based on this, it is found that different devices use IRI value conversion σ, there will be different correlations, simple correlation conversion, the value can not guarantee the time, and the algorithm for obtaining σ according to IRI does not have universality. Therefore, this paper proposes a fast continuous vialog algorithm for calculating σ by elevation values.

4. Research on algorithm of fast continuous vialog

According to the research, the cross-section value of the existing laser evenness meter does not meet the requirements of continuous evenness. Therefore, this paper chooses to improve the algorithm design of the continuous vialog.

According to the characteristics of the continuous vialog, the evenness of E60 is calculated as a calculation interval per 100m, and one detection section (usually 1~3km) has several calculation intervals. After the evenness of each calculation interval is obtained from the measured value. The evenness of the test section is assessed by the control standard and the acceptance quality standard during the construction process. Based on this, we have put forward the method of three-meter vertical distance measurement to improve the fast continuous evenness eight-wheel meter. The equipment is used to measure the road surface elevation and then calculate σ through the value of the road surface elevation.

The standard length of the fast continuous vialogeight-wheel meter is 3m, and there are four rows of wheels in front and rear. The two pairs are arranged in pairs to form a balanced type, which has the functions of balance and support. The distance between the shafts of the front and rear wheels is 3m. The middle is a 3m long frame that can be telescopic or folded. The frame is regarded as the follow-up reference of the eight-wheeled instrument. Its midpoint is the reference point of measurement and is located at the center of eight supporting wheels. The rack is equipped with a battery and a detachable test box, which can be used to display, record, print or draw the measurement results. In the middle of the rack, there is a measuring wheel that can rise and fall with the ground, and a detecting device such as a displacement sensor and a distance sensor is mounted thereon. When measuring, the measuring wheel is kept in contact with the road surface under the action of the spring, and the variation between the reference and the road surface is measured, and displacement data is automatically acquired. The front end of the frame is equipped with a traction mechanism, which can be pulled by human or automobile.

According to the measurement principle of the fast continuous vialog, the following algorithm is applied to the directly measured elevation value and σ:

$$d_i = \left(d_{0(i-15)} + d_{0(i+15)}\right)/2 - d_{0i} \quad (1)$$

$$\sigma_i=\sqrt{\frac{\sum_{i=1}^{N}(d_i-\bar{d})^2}{N-1}} \tag{2}$$

Among them: σ_i——the calculation value of the evenness of each calculation interval (mm); d_{0i}——the original value (mm) of the road surface concave-convex deviation displacement acquired at a certain distance with 100m; d_i——100m A calculation interval, the correction value of the road surface concave-convex deviation displacement (mm) collected at a certain distance; $\bar{d}$——the average value (mm) of the N road surface concave-convex deviation displacement values collected in a calculation interval; N— a calculation interval The number of test data used to calculate the standard deviation.

5.Verification of fast continuous vialog analyzer algorithm

In order to verify the scheme of fast continuous vialog, the algorithm was verified by the model established in this paper. The standard elevation value was measured by the level gauge and the measured elevation value is measured by the fast continuous vialog, then σ was calculated by formula 1 and 2. The results are shown in Table 1.

Table 1. Calculation results of continuous flatness meter.

Standard elevation value	Measured elevation value	Error	Standard elevation value	Measured elevation value	Error
2.420903	2.383224	-0.03768	1.088615	1.070034	-0.01858
0.793996	0.779551	-0.01444	1.228306	1.24026	0.011954
1.304706	1.28118	-0.02353	1.513746	1.483591	-0.03015
1.034822	1.013604	-0.02122	1.0246	1.005239	-0.01936

It can be seen from Table 1 that the error between the standard value and the measured value does not exceed 0.04, and the error range satisfies the design requirements of the algorithm. It shows that the algorithm has higher reliability, and the algorithm obtains σ directly from the IRI by the elevation value. It is proved that this algorithm has higher reliability and better versatility.

6. Conclusion

In this paper, we first demonstrated that the algorithm relationship between the standard deviation of the σ and IRI of the laser evenness meter has no universality. Different devices correspond to different algorithms. Therefore, it is not recommended to use IRI for the conversion of the σ. But when it is used a reference value, can use σ= 0.6 IRI for simple conversion.

In this paper, a new algorithm is proposed by studying the fast continuous flatness algorithm. The algorithm calculates the continuous flatness standard deviation σ by the elevation value. The algorithm formula is as follows: $d_i=\left(d_{0(i-15)}+d_{0(i+15)}\right)/2-d_{0i}$、 $\sigma_i=\sqrt{\frac{\sum_{i=1}^{N}(d_i-\bar{d})^2}{N-1}}$. The algorithm is verified by the established model, and the results show that the algorithm is more accurate. This algorithm can be applied to the fast continuous vialog, which has a good reference for the upgrade of the instrument and equipment, and also makes up for the shortcomings that the continuous flatness meter cannot be tested quickly. The proposal of the fast continuous vialog algorithm can provide valuable experience for the research of other evenness calculations in the transportation industry and play a positive role in promoting the development of detection work in the transportation industry.

IOP Conf. Series: Journal of Physics: Conf. Series **1267** (2019) 012038 doi:10.1088/1742-6596/1267/1/012038

References

[1] Xu Yan.2019. Science & Technology Ecnony Market. 415.1:14-16.
[2] Wang Gang,Yu Shanshan.2017. Transpoworld. 36.032:72-73
[3] Iván Moyano.2016. Mathematics of Control, Signals, and Systems. Vol.28 (4)
[4] Lu Sun. 2003.Mathematics,Computers in Simulation. 2(61):77-88.
[5] Christopher R.Bennett.1996. Journal of the Transportation Research Board. 1536(1): 73-81.
[6] Oldřich Kropáč, Peter Múčka.2005. Journal of Sound and Vibration. 287: 989-1003.
[7] Zhang Yi , Ma Ronggui, Ding Hui.2007. Transportation and computer. 25(5):13-16.

IOP Conf. Series: Journal of Physics: Conf. Series **1267** (2019) 012039 doi:10.1088/1742-6596/1267/1/012039

Research and Optimization of Real-time Simultaneous Localization and Mapping of Indoor Robot Based on Binocular Vision

Qiwei Zhang[1], Guanglu Zhou[1], Pengpeng Wang[1,2] and Fengguang Wu[2]

[1] School of Computer Science and Technology, Harbin Institute of Technology, 2 Wenhuaxi Road, 264209 Weihai, China

[2] Luoyang Electronic Equipment Test Center of China, 104 Tianjin Road, 471003 Luoyang, China

1498588668@qq.com

Abstract. For the problem of inaccuracy and cumulative error of visual odometer, The research and optimization of real-time Simultaneous Localization and Mapping of indoor robot based on binocular vision are studied. Based on ORB-SLAM2, key-frame map is created. First, the ORB feature is extracted from each frame of the input image and matched by fast approximation nearest neighbour(FLANN). Then, perform the preliminary pose estimation using EPnP, and optimize it with bundle adjustment and key-frame maps. When the tracking fails, apply key-frame maps and bag of words model to relocate. Finally, for the input binocular image, the SGBM is used to solve the parallax and then the depth, which will be converted to radar format data to create a map. In the research and optimization of real-time Simultaneous Localization and Mapping of indoor robot based on binocular vision, propose a method of assisted positioning with key frame map, and a method of feature matching optimization and relocation, which combines various pose optimization to achieve the accuracy of the robot indoors positioning and map construction.

1. Introduction

With the deepening of intelligence, unmanned driving, virtual reality, augmented reality, and three-dimensional reconstruction have received extensive attention in both academic and general usage. However, the products in these areas are not yet mature. One of the key issues is the accuracy and efficiency of three-dimensional positioning. Also, the Simultaneous Localization and Mapping (SLAM) of visual robots is another core issue in three-dimensional positioning. The visual robot SLAM is divided into visual odometer and mapping. The development of current odometers mainly includes the method of the multi-sensor fusion, the method of deep learning, the method of optical flowing (OF) and the method of feature point. In the method of the multi-sensor fusion, the estimation the accuracy of the pose is directly related to the sensor used. In general, sensors with higher accuracy are also more expensive. Therefore, cost is the main reason that restricts the development of this method. The method of deep learning is a new method proposed in recent years. The main problem with this method is that in known environments, the estimation has high accuracy, while in unknown ones, the estimation accuracy of the pose is poor. The OF method, which is more severely affected by illumination, can only be used in specific environments. With the increase in computing speed and the maturity of visual algorithms, the feature point method has become the most widely used method. This method refers to feature extraction of an image taken by a camera to estimate the pose of the camera. It has the advantages of rich information, low cost and good real-time performance. At present, the mainstream visual odometer system based on feature point method is also relatively mature. Representative methods such as SVO[1], PTAM[2], ORB-SLAM, LSD-SLAM[3,4], RTAB-MAP[5] have achieved good results in certain scenarios. The reconstruction mainly refers to three-dimensional reconstruction and two-dimensional reconstruction. Accurate maps ensure the accuracy of robot location navigation. This study is based on the ORB-SLAM2[6] framework and GMapping framework to create the key frame map and optimize the relocation method to solve the pose. Using EPnP[7] and BA to solve the robot pose, the precise positioning and map construction of the robot in the room is realized.

2. Basic knowledge

2.1. Binocular camera

A binocular camera is a device that interacts directly with the external environment and is a visually positioned eye. The physical model of the binocular camera is shown in Figure 1(a), and the geometric model is shown in Figure 1(b). The binocular camera is divided into a left eye camera and a right eye camera. In Figure 1(b), the optical centers of the left eye camera and the right eye camera are respectively recorded as o_L and u_R, and are connected by a baseline b. The pixel points of the point P in the left and right imaging planes are denoted as P_L and P_R, respectively, u_L and u_R are the abscissas of the point P on the left and right imaging planes. Because the ΔPP_LP_R is similar to the ΔPO_LO_R , (2.1) is obtained. In (2.1), d is recorded as the parallax of the spatial point in the left and right imaging planes. The depth of the spatial point can be obtained after the parallax is obtained.

$$z = \frac{fb}{d}, d = u_L - u_R \tag{2.1}$$

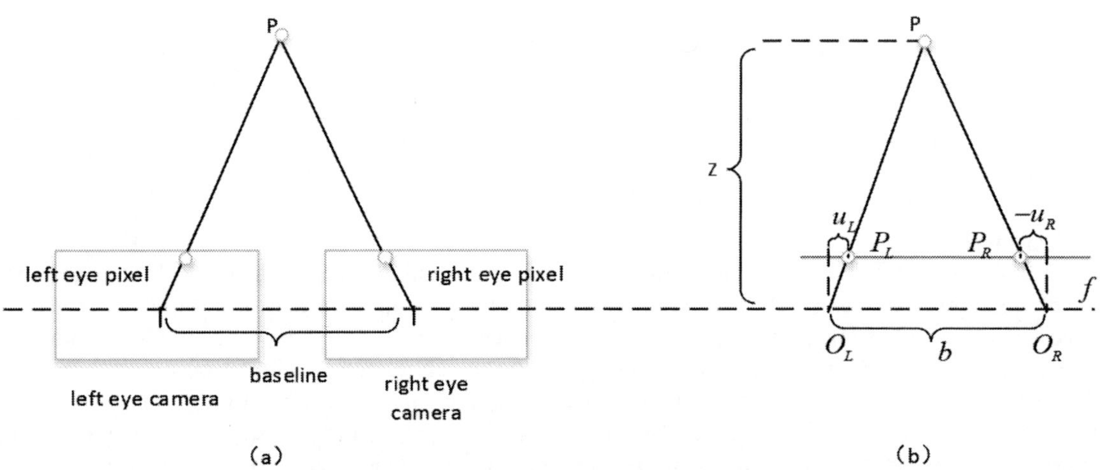

Figure 1. Binocular vision camera model.

2.2. Use the bag of word to match images

The bag of words, as dictionary data in relocation, generates a database of feature vectors for the images. First, collect get enough images, perform ORB[8] feature extraction on the images, and describe the feature with descriptors. Then, in [9], the dictionary is expressed by the K-tree in Figure 2. Finally, an unsupervised clustering algorithm is applied to form a word that represents a feature (or a collection of similar features)

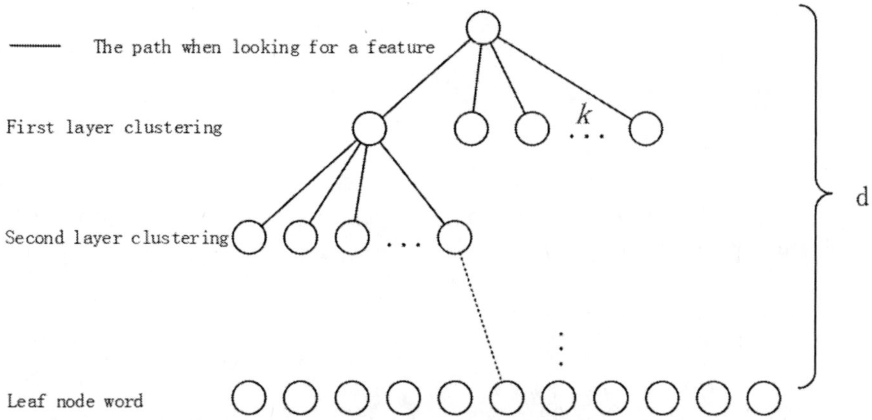

Figure 2. Schematic diagram of the K-tree dictionary.

3. Research and design of simultaneous localization and mapping of indoor robot based on binocular vision

The research and optimization of Simultaneous Localization and Mapping of indoor robot based on binocular vision is divided into three aspects: creating key frame map, visual odometer and mapping. First, save key-frames, landmarks, and related information acquired through the ORB-SLAM2 framework as key-frame maps. Secondly, load the key frame map in the visual odometer and extract the ORB feature for each frame of the input images and matched using fast approximate nearest neighbour (FLANN) [10]. Perform preliminary pose estimation using EPnP and optimize using bundle adjustment(BA) [11] and key-frame maps. When the tracking fails, an accurate estimation of the camera position is made using the key-frame map. Finally, a grid map is created based on the estimated camera pose and depth information of the image.

3.1. Create a key-frame map

Use a binocular camera to capture images from the environment, enter the obtained images into the ORB-SLAM2 frame to create a key-frame map and save it. When the first frame is input, the current frame is initialized. The camera coordinate system of the first frame is set as the origin coordinate of the world coordinate system. EPnP is applied to solve the pose of the camera for each frame of image input later. Determining key-frames and inserting them into the key-frame queue for optimization, finally, use bundle adjustment and closed loop detection to optimize all key-frames and save key-frame maps.

3.2. Computing visual odometer

Load the key-frame map and perform ORB feature extraction on the input image. Initial pose is estimated by EPnP and optimized with bundle adjustment. EPnP calculates the camera pose to represent the 3D coordinates of the world coordinate system as a weighted sum of a set of virtual control points. In general, the number of control points in the EPnP algorithm is four, and the four control points cannot be coplanar. Because the external parameters of the camera are unknown, the coordinates of the four control points in the camera reference coordinate system are unknown. The coordinates of the four control points in the camera reference coordinate system are solved, and the pose of the camera can be calculated. EPnP can be used to refer to the reference [7].

In the bundle adjustment optimization process, the coordinates of the spatial point are $P_i=[X_i \quad Y_i \quad Z_i]^T$, and the coordinates of the pixel point are $u_i=[x_i \quad y_i]^T$, and ξ indicates that the initial value is Lie algebra form of camera pose, which is solved by EPnP. s_i is the scaling factor. The relationship between pixel coordinates and spatial coordinates is shown in equation (3.1).The text should be set to single line spacing.

$$s_i\begin{bmatrix}u_i\\v_i\\1\end{bmatrix}=K\exp(\xi^\wedge)\begin{bmatrix}X_i\\Y_i\\Z_i\\1\end{bmatrix} \tag{3.1}$$

Write (3.1) as a matrix form as $s_i u_i = K\exp(\xi^\wedge)P_i$. To sum the errors, construct a least squares problem, and its reprojection error is shown in Figure 3. A preliminary pose estimation of P1 and P3 is performed on the matching points using EPnP.

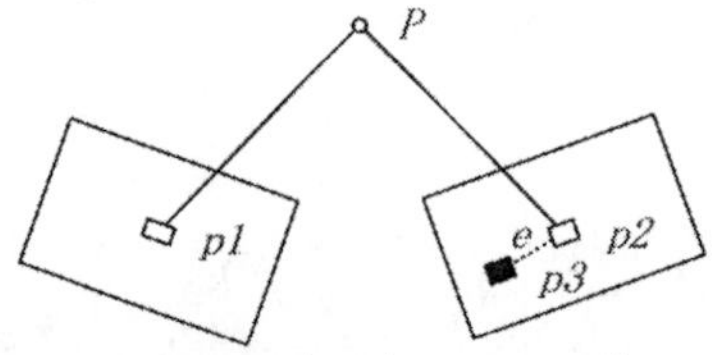

Figure 3. Reprojection error diagram.

In the next frame image, the distance between P2 to P3 minimized by adjusting the camera pose to optimize the camera pose to optimize the camera pose shown in Equation (3.2).

$$\xi^*=\arg\min_{\xi}\frac{1}{2}\sum_{i=1}^{n}\| u_i-\frac{1}{s_i}K\exp(\xi^\wedge)P_i\|_2^2 \tag{3.2}$$

Calculate the derivative of each error term with respect to the optimization variable, where $e(x+\Delta x)\approx e(x)+J\Delta x$ is a Taylor expansion of the error term. e is the pixel coordinate error. x is the camera pose. J is a matrix of two rows and six columns. P' is the coordinates of the space point in the camera coordination system. The formula $s_i u_i = KP'$ can be obtained. For the $\xi^\wedge$ left multiplication disturbance $\delta\xi$, Calculate the derivative of the change amount of e with respect to the disturbance amount, and see equation (3.3) using law of the chain rule. (3.4) and (3.5) are the results of (3.3)

$$\frac{\partial e}{\partial\delta\xi}=\lim_{\delta\xi\to 0}\frac{e(\delta\xi\oplus\xi)}{\delta\xi}=\frac{\partial e}{\partial P'}\frac{\partial P'}{\partial\delta\xi} \tag{3.3}$$

$$\frac{\partial e}{\partial P'}=-\begin{bmatrix}\frac{\partial u}{\partial X'} & \frac{\partial u}{\partial Y'} & \frac{\partial u}{\partial Z'}\\ \frac{\partial v}{\partial X'} & \frac{\partial v}{\partial Y'} & \frac{\partial v}{\partial Z'}\end{bmatrix}=-\begin{bmatrix}\frac{f_x}{Z'} & 0 & -\frac{f_xX'}{Z'^2}\\ 0 & \frac{f_y}{Z'} & -\frac{f_yY'}{Z'^2}\end{bmatrix} \tag{3.4}$$

$$\frac{\partial e}{\partial\delta\xi}=-\begin{bmatrix}\frac{f_x}{Z'} & 0 & -\frac{f_xX'}{Z'^2} & -\frac{f_xX'Y'}{Z'^2} & f_x+\frac{f_xX^2}{Z'^2} & -\frac{f_xY'}{Z'}\\ 0 & \frac{f_y}{Z'} & -\frac{f_yY'}{Z'^2} & -f_y-\frac{f_yY'^2}{Z'^2} & \frac{f_xX'Y'}{Z'^2} & \frac{f_xX'}{Z'}\end{bmatrix} \tag{3.5}$$

This Jacobian matrix $\frac{\partial e}{\partial\delta\xi}$ describes the first derivative of the reprojection error with respect to the Lie algebra of camera pose. In addition to optimizing the pose, the research and optimization of SLAM of indoor robots based on binocular vision also optimizes the spatial position of feature points. Therefore, it is necessary to calculate the derivative of e with respect to the spatial point P. The calculation results are shown in formula (3.6) (3.7) (3.8).

$$\frac{\partial e}{\partial\delta\xi}=\lim_{\delta\xi\to 0}\frac{e(\delta\xi\oplus\xi)}{\delta\xi}=\frac{\partial e}{\partial P'}\frac{\partial P'}{\partial\delta\xi} \tag{3.6}$$

$$P'=\exp(\xi^\wedge)P=RP+t \tag{3.7}$$

$$\frac{\partial e}{\partial P}=-\begin{bmatrix} \frac{f_x}{Z'} & 0 & -\frac{f_x X'}{Z'^2} \\ 0 & \frac{f_y}{Z'} & -\frac{f_y Y'}{Z'^2} \end{bmatrix} \tag{3.8}$$

So far, a derivative matrix of pixel coordinate errors with respect to camera pose and feature point position is obtained. SLAM's workers recognize the sparsity of the H matrix by the cumulative pixel coordinate error with respect to camera pose and feature point pose, which can be represented by graph optimization, which allows the use of this matrix in combination with Gauss-Newton method to real time optimize camera position and feature point positions are possible.

After obtaining the current frame pose by the Gauss-Newton method, the rotation matrix R and the translation vector t are obtained from the pose matrix. Use Rodriguez formulas (3.9) and (3.10). Since the camera rotates around the z axis, taking $n=\begin{bmatrix}0 & 0 & 1\end{bmatrix}$. At the same time, the rotation matrix is converted into Euler angles and the value of θ is determined by combining the correspondence between sin θ and the rotation matrix R. The θ is summed to obtain the rotation angle θ_{sum} of the current camera, and the calculation is given by Equation (3.11).

$$R=\cos\theta I+(1-\cos\theta)nn^T+\sin\theta n^{\wedge} \tag{3.9}$$

$$\theta=\arccos(\frac{tr(R)-1}{2}) \tag{3.10}$$

$$\theta_{sum}=\sum_{i=0}^{i=n}\theta \tag{3.11}$$

When tracking the previous frame failure, use the created bag of word to determine the word bag vector of the current frame, and compare the similarity between the word bag vector of the current frame and the word bag vector of the key frame by using formula (3.12), and select similarity. The key frame, which is with the highest degree and is greater than the set threshold is used as the candidate key frame of the current frame. After determining the key frame of the current frame, feature extraction is performed on the current frame and the candidate key frame respectively, and the transformation matrix of the pose of the current frame relative to the pose of the key frame is calculated by matching the feature points, and the current frame posture is determined by transforming the matrix.

$$s(a,b)=1-\frac{1}{w}\|a-b\|_1 \tag{3.12}$$

a is the word bag vector of the current frame, b is the word bag vector of the candidate key frame, and w is the proportion of each word in the word bag dictionary, and different weights are assigned according to the influence of different words on the result.

3.3. Create radar format data and build grid map

Converting deep data to radar format data is better compatible with current mainstream mapping frameworks and navigation frameworks. At the same time, since the general navigation positioning uses a two-dimensional grid map, the conversion of depth data is also an important part of the research and optimization of indoor robot SLAM based on binocular vision. According to different application scenarios, the binocular image is first converted into a depth map and projected onto a two-dimensional plane to generate a two-dimensional image, and then the two-dimensional image is converted into radar format data. Finally, radar format data is released to create grid maps and navigation against obstacles.

4. Research result

The research and optimization of real-time Simultaneous Localization and Mapping of indoor robot based on binocular vision are divided into three aspects: creating key frame map, locating with key frame map and creating grid map.

IOP Conf. Series: Journal of Physics: Conf. Series **1267** (2019) 012039 doi:10.1088/1742-6596/1267/1/012039

4.1. Create a key-frame map

Create a key-frame map with the ORB-SLAM2 framework. Taking an indoor scene as the background, the binocular camera captures the image of the room, and the image is input into the ORB-SLAM2 frame to process and save the key frame map. Key-frame maps using key-frames, landmarks, and related information are basically consistent with indoor scenes. The results of creating a key-frame map are shown in Figure 4(b). A key-frame map is created. The blue box in Figure 4(a) is the extracted feature point

(a) (b)

Figure 4. Create a key-frame map.

4.2. Location use key-frame maps

When the current frame is tracked successfully, the pose of the current frame is obtained by the research and optimization of real-time Simultaneous Localization and Mapping of indoor robot based on binocular vision; Figure 5(a) indicates that the feature points are correctly tracked, and the pose of the current frame is displayed in Figure 5(b). When a frame fails on frame tracking, the key frame map is used for relocation. Figure 5(c) indicates that the tracking failed, but after a short time, this success.

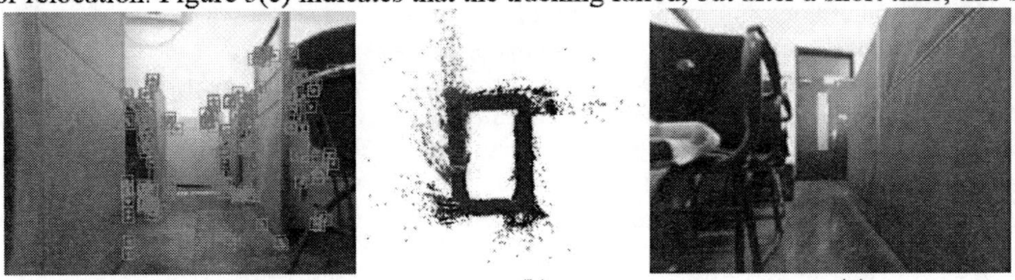

(a) (b) (c)

Figure 5. Relocation when tracking fails.

4.3. Create grid map

Firstly, the camera pose is obtained by using the method of solving the odometer in the research and optimization of real-time Simultaneous Localization and Mapping of indoor robot based on binocular vision. Then, the binocular camera is used to solve the depth map and the 3D depth map is projected into the 2D plane to create the grid map. The final study results for the visual odometer are shown in Figure 6(a) visual odometer. The grid created map is shown in Figure 6(b)

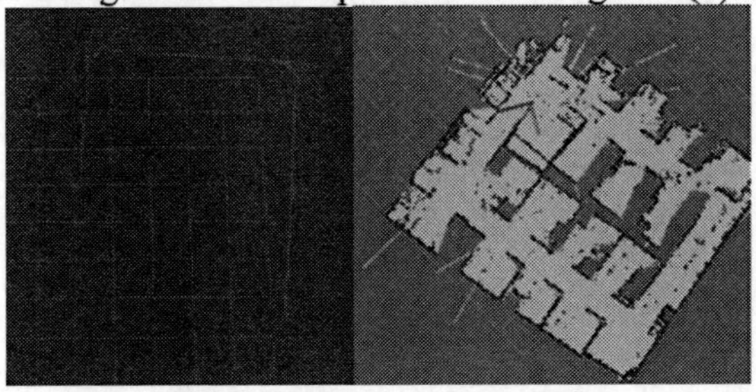

(a) (b)

Figure 6. Creating a map.

5. Conclusion

Research and optimization of real-time Simultaneous Localization and Mapping of indoor robot based on binocular vision with an indoor scene as background, using a binocular camera sensor, a key-frame map is created based on the ORB-SLAM2 framework. In the visual odometer section, EPnP and BA are used to solve the camera pose. Use key-frame maps to assist in positioning at the same time. When the tracking fails, the real-time and accuracy of relocation is ensured by adding key frame maps and word bag dictionaries. Finally, the acquired binocular image is converted into a depth image and then converted into radar format data to construct a grid map. Research and optimization of real-time Simultaneous Localization and Mapping of indoor robot based on binocular vision integrates mainstream framework and stability algorithm of visual robot, In the computational visual odometer section, a key-frame map is used to assist in solving the odometer, optimize the relocation pose. It can effectively solve the problem of location and mapping of mobile robots.

References

[1] C. Forster, M. Pizzoli, and D. Scaramuzza, "Svo: Fast semi-direct monocular visual odometry," in Robotics and Automation (ICRA), 2014 IEEE International Conference on (rs, ed.), pp. 15–22, IEEE, 2014.

[2] G. Klein and D. Murray, "Parallel tracking and mapping for small ar workspaces," in Mixed and Augmented Reality, 2007. ISMAR 2007. 6th IEEE and ACM International Symposium on, pp. 225–234, IEEE, 2007.

[3] J. Engel, T. Schöps, and D. Cremers, "Lsd-slam: Large-scale direct monocular slam," in Computer Vision–ECCV 2014, pp. 834–849, Springer, 2014.

[4] J. Engel, J. Sturm, and D. Cremers, "Semi-dense visual odometry for a monocular camera," in Proceedings of the IEEE International Conference on Computer Vision, pp. 1449–1456, 2013..

[5] M. Labbé and F. Michaud, "Online global loop closure detection for large-scale multi-session graph-based slam," in 2014 IEEE/RSJ International Conference on Intelligent Robots and Systems, pp. 2661–2666, IEEE, 2014.

[6] Mur-Artal R, Tardos J D. ORB-SLAM2: An Open-Source SLAM System for Monocular, Stereo, and RGB-D Cameras [J]. IEEE Transactions on Robotics, 2017:1-8.

[7] Lepetit V, Moreno-Noguer F, Fua P. EPnP: An AccurateO(n) Solution to the PnP Problem[J]. International Journal of Computer Vision, 2009, 81(2):155-166.

[8] E. Rublee, V. Rabaud, K. Konolige, and G. Bradski, "Orb: an efficient alternative to sift or surf," in 2011 IEEE International Conference on Computer Vision (ICCV), pp. 2564–2571, IEEE, 2011. Pham C C, Jeon J W. Domain Transformation-Based Efficient Cost Aggregation for Local Stereo Matching [J]. IEEE Transactions on Circuits & Systems for Video Technology, 2013, 23(7):1119-1130.

[9] D. Galvez-Lopez and J. D. Tardos, "Bags of binary words for fast place recognition in image sequences," IEEE Transactions On Robotics, vol. 28, no. 5, pp. 1188–1197, 2012.

[10] M. Muja and D. G. Lowe, "Fast approximate nearest neighbors with automatic algorithm configuration," in VISAPP (1), pp. 331–340, 2009.

[11] B. Triggs, P. F. McLauchlan, R. I. Hartley, and A. W. Fitzgibbon, "Bundle adjustment: a modern synthesis," in Vision algorithms: theory and practice, pp. 298–372, Springer, 2000.

Analysis of Competitive Situation in Basic Research of Global Intelligent Robots

Li Rong, Liu Ru*, Luo Zichao and Liu Jing

Beijing Institute of Science and Technology Information, No.140, Xizhimenwai Street, Xicheng District, Beijing,100044, P R CHINA

*Corresponding author e-mail: liur@bjstinfo.com.cn

Abstract. Practically grasp the global development trend of the basic research on intelligent robots and provide support for the scientific development of intelligent robots in China. Relying on bibliometrics, this paper has proposed a four-dimensional competitive situation analysis system for basic research based on activeness and talent distribution, etc., which can effectively analyze the global situation of basic research on intelligent robots, able to reveal and describe the competition status quo and situation of the basic scientific research in specific fields. There is a big gap between China and the United States in basic research on intelligent robots, mainly including such aspects as follows: uneven distribution of high-level basic research talents; insufficient international cooperation in high-level scientific research; lower research activeness; layout of key research and development.

1. Introduction

Basic research is the source of the whole scientific system and the cornerstone of building a powerful country in science and technology. The total output of scientific papers in China has ranked the first in the world and the number of citations of international papers has also risen to global second, but it must be noted that the key technologies in many fields are constrained by others and in a deeper sense, there are various shortcomings highlighted in the field of basic scientific research, such as lack of major original achievements, lack of top talents and teams, inadequate investment and unreasonable structure [1]. The new round of scientific and technological revolution and industrial transformation are ceaselessly deepening, the global competition for scientific and technological innovation is unprecedentedly fierce and the basic research is becoming the commanding height of global competition. China is speeding up the implementation of innovation-driven development strategy; moreover, Shanghai and Beijing are striving to build science and technology innovation centers with global influence. This requires policy makers to have an international vision for competition. On the one hand, they should be able to accurately grasp the trend of scientific evolution and the global competitive situation of scientific and technological innovation; on the other hand, they should also clarify their advantages and gaps between their regions and leading countries and regions in basic research, so as to optimize the layout of regional basic scientific research.

2. Research Approach

The competitive situation is a relatively broad concept and the competitive situation of basic research mentioned in this paper is not only different from the capability evaluation at the level of scientific and technological competitiveness, but also different from the latest progress at the level of micro-knowledge unit. Actually, it is to grasp the overall development status of the basic research in specific scientific and technological fields from the perspective of competition.

As the main achievement form of basic research, the econometric analysis based on scientific papers has been widely used in the relevant research topics of scientific and technological innovation decision-making, such as research situation analysis, research hotspot finding, innovation effect evaluation and talent selection. Moreover, the econometric analysis method based on massive academic data not only makes it possible to grasp the scientific structure and evolutionary model from a micro perspective, but also becomes a tool for effective management of the forward-looking planning in scientific research [2]. At present, there have been numerous analysis results and studies made on the competitive situation of the basic research in specific fields by using scientific econometric methods. Liu Ya has carried out the analysis and research on the international cooperation situation of basic research based on the distribution of scientific papers at the levels of disciplines, countries and institutions, etc. [3]; Ma Tingchan, et al. has put forward the comprehensive competitiveness index of basic research based on the data of the National Natural Science Foundation of China to make a comparative analysis of the competitiveness and its changes of the provincial basic research in China [4]; based on such three indexes as "activeness index", "attraction index" and "efficiency index", Chen Kaihua, et al. has constructed a "comprehensive research capability index" and used such indexes to compare and describe the relative position and competition pattern of different countries in the field of science and technology in terms of the basic research [5]; Zhang Zhiqiang, et al. has chosen such two perspectives as global overall R&D situation and representative research units and consequently proposed a set of competitive situation analysis indexes for basic research from four aspects, such as strategic positioning, paper output, talent team structure and scientific research mode[6]; Li Zexia, et al. has classified the field situation analysis into four aspects: strategic layout, scientific research, technological development and industrial development, focusing on a set of indexes for analyzing technological development situation [7]; Bian Wenyue, et al. has put forward an information analysis framework of the development situation in the frontier of science and technology by combining the micro-analysis of knowledge units with the development situation analysis at the macro level [8].

To sum up, there are two main trends in the development of the original econometric research heavily relying on the external metrological characteristics of literature: Firstly, from a single analysis perspective based on the external characteristics of literature to a multi-comprehensive perspective such as discipline structure, cooperation network and knowledge units; secondly, the dimensions of competitive situation analysis in the field of basic research are diversified and index-oriented. However, the author believes that the existing research perspectives continue to expand and the analysis indexes are more systematic and comprehensive, but still there are two aspects of shortcomings: Firstly, the comparative analysis mostly stays at the horizontal comparison among different countries at the macro level, but pays insufficient attention to the regional meso-level; on the other hand, the existing analysis framework mostly targets the evaluation of the basic research capacity, but less research has been made to propose a bibliometric analysis framework for answering the basic research competitive situation and pattern in a specific area at the global and regional levels.

Accordingly, the current research, focusing on the optimization of regional basic scientific research layout in the perspective of global competition, has proposed to follow such four dimensions as research activeness, distribution of high-level talents, research cooperation and key R&D layout for grasping the competitive situation of the global basic research from the perspective of urban (regional) strategies; moreover, it has made an attempt to proposed a new analytical framework on the basis of expanding and optimizing the existing relevant analysis indexes and select the technological field of intelligent robots for an empirical research, thus to verify the scientificity and effectiveness of this analytical framework.

3. Design of Competitive Situation Analysis Framework for Field Basic Research

IOP Conf. Series: Journal of Physics: Conf. Series **1267** (2019) 012040 doi:10.1088/1742-6596/1267/1/012040

The basic research, as one of the most complex intellectual activities in human society, is subject to many factors affecting its level development. Generally speaking, there are four aspects: Firstly, primitive accumulation, including academic output, basic research talents and capacity basis, etc.; secondly, investment in basic research and government guidance; thirdly, distribution of high-level research groups, mainly referring to research institutions and researchers with certain academic influence; fourthly, high-level research cooperation network [9, 10]. Based on the above knowledge, this paper has put forward four dimensions for competitive situation analysis (namely, activeness level of basic research, distribution of high-level research talents, network for international cooperation research and layout of key R&D projects), focusing on horizontal comparison among domestic and foreign cities (regions) to refine relevant analysis indexes and strive to describe the macro-competitive situation for international competition at the urban or regional level (as shown in Figure-1). Compared with the existing research, the analytical framework proposed in this paper has the following three characteristics: Bibliographic representations are used to define the high-level talents for basic research and simultaneously the distribution of such talents is taken as an important research index for competitive situation analysis; the correlation analysis is carried out at three levels: nations, cities and institutions, with the regional basic research level compared in the perspective of global competitive environment; the dimensions have been expanded for analyzing the competitive situation in the field of basic research, with such key elements as R&D layout, network for international research cooperation and distribution of high-level talents brought into the same analysis index system.

3.1. Situation analysis of research activeness

From such two dimensions as national geospace and time, it can reflect the activeness level of the basic research in specific fields. The national geospatial analysis can clarify the overall competition pattern of global basic research in the field and find out the competitive potential of the research object; the time dimension analysis can help us to realize the trend relationship between the basic scientific research in the field and the evolution of the applied technology.

3.2. Situation analysis of high-level talents distribution

The number of high-level research talents determines the level of basic research, so the distribution of high-level talents has become the key factor to judge the competitive situation of the basic research. Based on the bibliometric characteristics of high-level talents for basic research, this paper has comprehensively measured the rationality of the distribution of high-level talents in target areas under their own strategic positioning from three levels, such as countries, cities and institutions.

3.3. Situation analysis of international cooperation research

With the continuous strengthening of the integration between different disciplines and fields, international cooperation has become one of the prominent features of contemporary basic research [11]. The research also proposes to judge the dependence of the target area on international cooperation and whether the selection of partners is conducive to the improvement on the basic research level in this field through statistical analysis of the cooperation frequency from three levels: countries, cities and institutions.

3.4. Situation analysis of key R&D layout

The improvement of the original innovation level needs not only to follow the laws of the science itself, but also to be supported and guided by the government. The support of the scientific fund can reflect the layout and competitiveness of a country or region in terms of basic research. In this paper, the types and proportion of scientific funds corresponding to SCI papers are included in the proposed analytical framework. At the same time, the competition analysis of regional basic research supported by scientific funds has been extended to the main hotspots in the established fields.

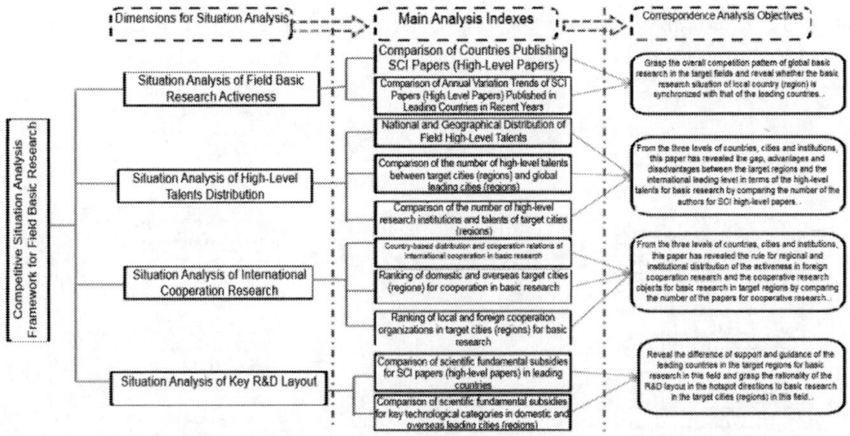

Figure 1. Framework created in this paper for competitive situation analysis of basic research.

4. Analysis of Global Competitive Situation for Basic Research in the Field of Intelligent Robots

Intelligent service robots are a kind of high-tech integrated intelligent equipment which provides essential services for human beings in an unstructured environment. As a new interdisciplinary application technology combining artificial intelligence and automation technology, it has become one of the focus areas of global response to the new round of scientific revolution and industrial transformation. At present, our country is lagging behind in several key technology fields of intelligent robots, such as bodywork and drive, perception and information fusion, intelligent control and human-machine interaction. The core components of high-precision reducers, servo motors and so on have been long relying on imports, and the technological breakthroughs in such fields must be driven by basic sciences [12]. For this reason, this paper has chosen the intelligent robot technology as the field for empirical verification of the analysis framework and index system proposed by the research institute.

4.1. Data Sources and Processing Methods

Intelligent robots belong to a comprehensive technology field obvious in cross-disciplinary characteristics. Therefore, the accurate retrieval of the scientific research papers in this field must be achieved by constructing a key technology system. As shown in Fig. 2, based on the development status and trend of global intelligent robot technology as well as the suggestions of experts in the field, the current research has preliminarily put forward the key technology system of intelligent robots consisting of 8 key technical categories and 43 sub-technologies, such as new materials, interaction and sensing.

The research has chosen the WOS Core Collection (SCIE) as the source of literature data and taken secondary sub-technologies as the key words for retrieval to carry out literature retrieval for eight key technical categories respectively. Moreover, the research targets the latest research situation and simultaneously the attention of the basic science circles to each technical category is chronologically different, so the time span varies from 5 to 8 years for retrieving the data of papers in different technical categories. With the help of Derwent Date Analyzer (DDA), we have cleaned the data and finally obtained a total of 15094 literatures containing effective original data.

In addition, according to the research needs, this paper has defined the basic high-level talents under the bibliometric characteristics. Referring to the ESI definition of highly cited papers, the research has taken the papers featuring concentrated research data and ranking top 10% in terms of key technologies, years of retrieval and citation frequency in the same year as the high-level papers in this research, and then identified the first authors and corresponding authors of such papers as high-level research talents. To have this strategy more scientific, the research has used relevant literature retrieval tools to automatically identify the papers from SCI journals not ranked among 10% of the original data, and to supplement such papers to the data set for high-level papers research.

Figure 2. Key Technologies System of Intelligent Robots.

4.2. Situation Analysis of Basic Research Activeness

4.2.1. Comparison of Countries Leading in SCI Papers. The research has chosen SCI and its high-level papers as two analysis indicators to analyze the output of papers in order to grasp the global macro-competition pattern of basic research in the field of intelligent robot technology. As can be seen from Figure-3, the United States and China have formed the first competitive echelon, which has opened a certain gap with other countries. In this field, the output level of SCI papers in China has approached that of the United States. In terms of high-level SCI papers, China has become the second largest country in the world, but still there is a big gap to the United States.

4.2.2. Comparison of the trend for SCI papers in leading countries. In addition, from the statistical analysis of the annual publications of SCI and its high-level papers in the past eight years by robotic powers such as the United States, Japan and Germany as shown in Figure-4 (1), the competitive situation is clearer between the two camps in the world. Compared with the trend of relatively stable rising in the United States, the output level of basic scientific research in the field of intelligent robotics in China has shown a rapid upward trend since 2012 and surpassed the United States in 2017. In terms of high-level achievements, as shown in Figure-4 (2), China emerged from the second camp in 2015, and the number of high-level papers showed a linear growth trend in the following two years, surpassing the United States in 2017. In summary, we can see that the output scale of basic scientific research in the field of intelligent robot technology in China is close to that in the United States, but still there is a gap to the United States in the level of academic influence.

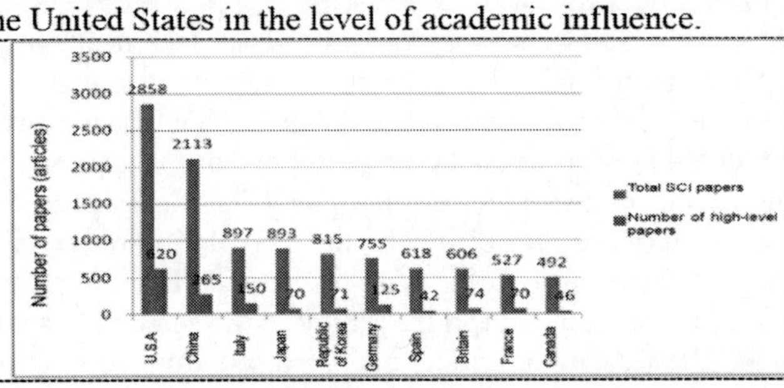

Figure 3. Distribution of Nations with SCI and High-level Papers.

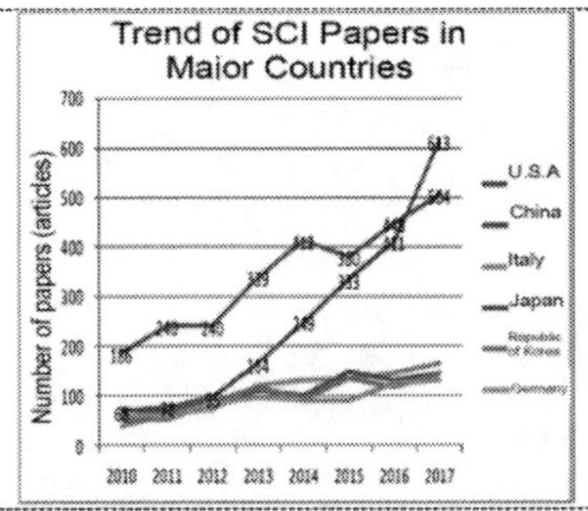

Figure 4(1). Trend of SCI Paper Publications in Leading Countries.

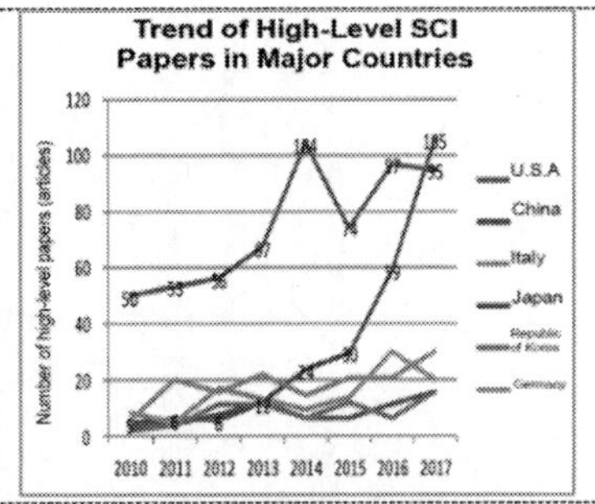

Figure 4(2). Trend of High-level SCI Paper Publications in Leading Countries.

4.3. Analysis of Distribution Situation of High-Level Talents

4.3.1. National and Regional Distribution of Field High-Level Talents. The empirical case study in this paper has made a statistical analysis of the country, city and institution-based distribution of the first and corresponding authors of the high-level papers. At the national level, as shown in Figure-5, corresponding to the number of high-level papers, the number of high-level talents for basic research in the field of intelligent robot technology in China is the second highest in the world, only second to the United States, but still there is a certain gap in the number with the United States. Other eight countries, such as Italy and Germany, are basically at the same level.

4.3.2. City and Region-based Distribution of Field High-Level Talents. At the urban level, as shown in Table-1, Beijing ranks first in the world in the number of high-level talents, which surpasses Cambridge in the United States, but it also accounts for only one quarter of the total number of the talents in this field in China; moreover, Beijing and several other domestic cities in the global TOP20 account for only 55% of the total number of domestic high-level talents in this field. This group of data reflects that the geographical distribution is relatively scattered for high-level talents in basic research in the field of intelligent robots in China. Under the strategic background of building a national science and technology innovation center, Beijing's advantage in gathering high-level talents in this field has not yet appeared.

4.3.3. Distribution of high-level research institutions in the field. The number of high-level talents directly represents the competitiveness of scientific research institutions. At the institutional level, we have ranked domestic and overseas research institutions according to the number of high-level talents and compared the top 10 institutions respectively. As can be seen from Table-2, several famous universities in the United States have obvious advantages in high-level talents for basic research in the field of intelligent robots; however, there is a huge gap between domestic and international leading institutions in the scale of talents; the advantages are not obvious for high-level talents of scientific research institutions in Beijing area, and Harbin Institute of Technology is more prominent at the level of domestic institutions.

IOP Conf. Series: Journal of Physics: Conf. Series **1267** (2019) 012040 doi:10.1088/1742-6596/1267/1/012040

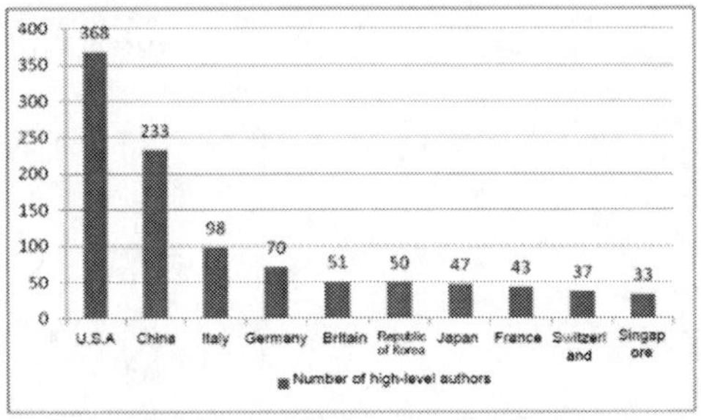

Figure 5. Nation-based distribution of high-level paper authors.

Table 1. Ranking of global cities leading in high-level paper authors.

S/N	Leading City	Country	Number of High-Level Talents (Persons)	Proportion of Local High-Level Talents in the Total
1	Beijing	China	61	26%
2	Cambridge	U.S.A	56	15%
3	Singapore	Singapore	33	100%
4	Seoul	Republic of Korea	28	56%
5	Shanghai	China	23	10%
6	Pisa	Italy	21	21%
7	Lausanne	Switzerland	20	54%
8	Pittsburgh	U.S.A	20	5%
9	Harbin	China	19	8%
10	Genoa	Italy	18	18%
11	Teheran	Iran	18	75%
12	Tubingen	Germany	16	23%
13	Stanford	U.S.A	15	4%
14	Nanjing	China	15	6%
15	Tokyo	Japan	14	30%
16	London	Britain	13	25%
17	Paris	France	13	30%
18	Zurich	Switzerland	13	35%
19	Los Angeles	U.S.A	12	3%
20	Guangzhou	China	12	5%

Table 2. Distribution of high-level talents in domestic and overseas leading institutions.

Distribution of High-Level Talents in Overseas Leading Research Institutions					Distribution of High-Level Talents in Domestic Leading Research Institutions				
S/N	Name of Institution	Number of High-Level Talents (Persons)	Ranking of Global Research Institutions	Local City of Institution	S/N	Name of Institution	Number of High-Level Talents (Persons)	Ranking of Global Research Institutions	Local City of Institution
1	Harvard University	34	1	Cambridge (USA)	1	Harbin Institute of Technology	16	2	Harbin
2	MIT	21	2	Cambridge (USA)	2	Tsinghua University	15	4	Beijing
3	Stanford University	15	3	Palo Oddo (USA)	3	Shanghai Jiao Tong University	9	12	Shanghai
4	Nanyang University of Technology	13	6	Singapore	4	Beijing Institute of Nanoenergy and Nanosystems, Chinese Academy of Sciences	8	18	Beijing
5	Carnegie Mellon University	12	7	Pittsburgh (USA)	5	University of Science & Technology Beijing, USTB	6	26	Beijing
6	National University of Singapore	12	8	Singapore	6	South China University of Technology	6	28	Guangzhou
7	University of Michigan	12	8	Ann Arbor (USA)	7	Sichuan University	6	33	Chengdu
8	University of Tuebingen	10	10	Tubingen (Germany)	8	Nanjing University	6	34	Nanjing
9	SEOUL National University	9	11	Seoul (Korea)	9	Zhejiang University	5	35	Hangzhou
10	Cornell University	9	14	Ithaca (USA)	10	Huazhong University of Science and Technology	5	36	Wuhan

4.4. Situation Analysis of Basic Research Cooperation in the Field

The frequency of international cooperation in publishing SCI papers was counted, and a network of cooperation was formed (see figure-6). The United States is the most active country in the field of basic research and international cooperation in the field of intelligent robotics. The volume of cooperation with other countries is generally high, especially with China. However, China's international cooperation is ranked fifth in the world with fewer publications than Italy, Germany and Italy. In addition to more cooperative research with the United States, China also has a certain degree of research cooperation with the United Kingdom and Japan.

IOP Conf. Series: Journal of Physics: Conf. Series **1267** (2019) 012040 doi:10.1088/1742-6596/1267/1/012040

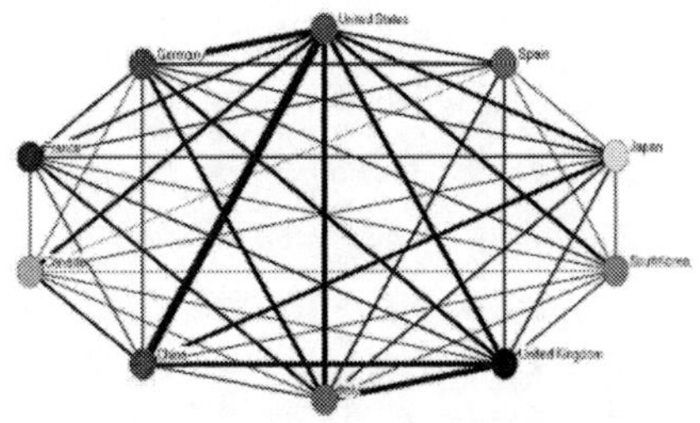

Figure 6. Network for Cooperation of SCI Co-authored Papers among Major Countries.

4.5. Analysis of Key R&D Layout Situation

4.5.1. Comparison of Scientific Fund Subsidies for SCI Papers in Leading Countries. In order to reveal the global layout of basic scientific research and development in the field of intelligent robot technology, the current research has quantitatively analyzed the scientific fund subsidies for SCI and its high-level papers published by major leading countries. As shown in Figure-7 (1), more than 80% of SCI papers in the field of intelligent robot technology in China are subsidized by relevant scientific funds, before the United States, but only after Korea; besides, more than 50% of subsidies are funded by national funds. Compared with Figure-7 (2), the proportion of scientific fund subsidies for high-level SCI papers in various countries has generally increased, especially in China and Korea. From the perspective of fund types, the National Science Foundation (NSF) and the National Institutes of Health (NIH) are the main funders in the United States; besides relying on their own local research funding institutions, European countries also have important access to funding from various EU R&D programs; many large-scale scientific research funds in Korea have provided funding for robot scientific research; the robot scientific research in China has mainly be subsidized by the "Natural Science Foundation of China" and the "National Key R&D Program of China".

4.5.2. Comparison of Scientific Fund Subsidies for Key Technology Categories between Domestic and Foreign Leading Cities. Furthermore, based on the eight key technology categories proposed in the current research in the field of intelligent robot technology, it has counted the number of papers classified in different technology categories and supported by scientific funds in the leading cities of SCI papers, compared with Cambridge, the United States, which ranks first in the world, and achieved the results as shown in Figure-8. By virtue of the geographical advantages in scientific and technological resources, Beijing's scientific research institutes have gained much more support from scientific funds than other domestic cities in basic research of all key technology categories. Relevant research shows that sensing, driving and control, interaction and recognition are three major important categories of intelligent robot technology for fostering strategic emerging industries and developing high-tech industries in the future [13]. Compared with U.S. Cambridge, Beijing has been paying more attention to the original innovation of the sensing technology, but the basic research layout in such two directions as driving and interactive technology needs to be further strengthened.

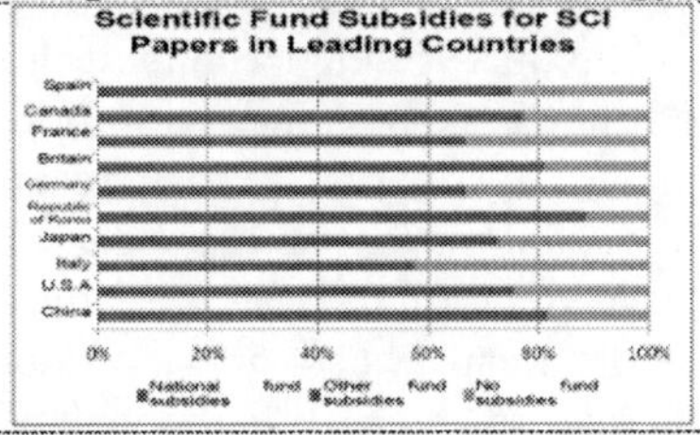

Figure 7(1). Scientific Fund Subsidies for SCI Papers in Leading Countries.

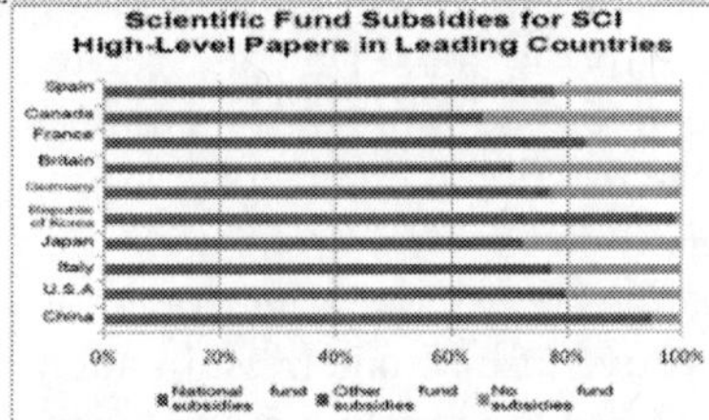

Figure 7(2). Scientific Fund Subsidies for SCI High-Level Papers in Leading Countries.

IOP Conf. Series: Journal of Physics: Conf. Series **1267** (2019) 012040 doi:10.1088/1742-6596/1267/1/012040

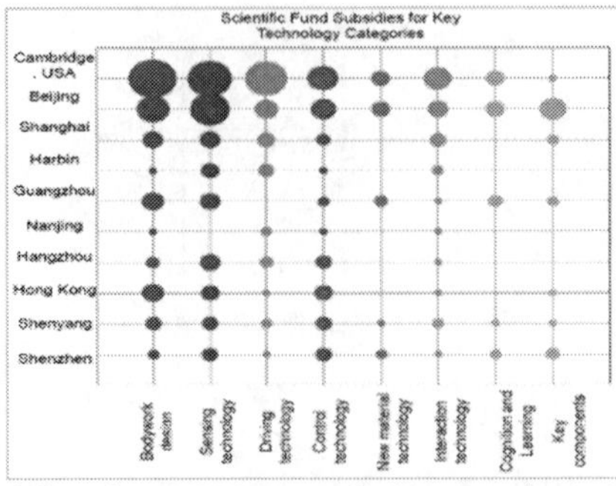

Figure 8. Scientific Fund Subsidies for SCI Papers in Domestic and Overseas Leading Cities in Key Technological Fields

4.6. Conclusion

As can be seen from the analysis results, China is at the same level as the United States in the output of basic research achievements in the field of intelligent robot technology, and the research achievements show a rapid upward trend in both "quality" and "quantity", but still there is a certain gap to the United States in the level of influence. In China, the geographical distribution of high-level talents for basic research in this field is relatively scattered. Beijing is leading in domestic cities, but the advantages for gathering high-level research talents have not yet emerged under the dual strategic positioning of building a national science and technology innovation center and forging a high-grade, precision and advanced economic structure. The systematic support of various scientific funds is an important guarantee for the output of high-level basic research achievements. The scientific research institutes in Beijing have obvious advantages over other cities in obtaining scientific funds, but they need to strengthen the research layout of original innovations in such two key technologies as driving and interaction.

5. Summary and Prospect

Orienting at the application of supporting the strategic planning for regional scientific research and targeting comprehensively grasping the competitive situation of regional basic research in a global perspective, this paper has carried out multi-dimensional correlation analysis from such three levels as countries, regions and institutions based on bibliometrics, incorporated such key competitive elements as R&D layout, international cooperation network and distribution of high-level talents into the same analysis framework in order to reveal and describe the competition status quo and situation of the basic scientific research in specific fields, and simultaneously has verified the practicability and effectiveness of the proposed analysis framework, taking the intelligent robot technology as an example.

The analytical framework proposed in the current research has helpfully expanded and improved the existing research in terms of analytical perspectives and dimensions, but, as mentioned in this paper, many factors may affect and determine the development of basic research and its competitive situation, especially in answering the related strategic layout of science and technology from the regional perspective. The four analysis dimensions and related analysis indexes proposed in this paper have improve the existing methods of grasping the competitive situation of basic research by means of bibliometric analysis, but no systematic comparison or measurement has been made, and some scientific limitations may exist in grasping the competitive situation of basic research in a certain field from the perspective of bibliometrics. In addition, some of the analysis indexes proposed in this paper have only considered too single factors, for example, the part of international cooperative research has only revealed the geographical distribution of cooperative partners by the number of cooperative publications, but not considered the influence of co-authored papers; simply identified the first and corresponding authors of highly cited papers as high-level authors, but not incorporated authors' H index or average number of citations in the analysis indexes together.

With the integration of basic research, applied research and industrialization, the weakness and one-sidedness starts to appear for bibliometric methods applied in basic research capacity evaluation or competitive situation analysis [14]. In the future, the innovation and application of relevant analysis methods should organically integrate the national objectives for scientific development, have bibliometrics mutually complemented and corrected with such indexes as scientific research project funds, scientific awards, patents and achievement transformation, so as to further improve the effectiveness and scientificity in analyzing the competitive situation of the basic research.

Acknowledgments

This work was supported by Beijing Municipal Science and Technology Project Grant Z181100001018035. The authors thank the WOS Core Collection (SCIE) for sharing the data.

References

[1] Sina Focus. Ministry of Science and Technology: China's basic research lacks major original achievements, top talents and teams [EB/OL]. [2018-02-11]. http://k.sina.com.cn/article_1645705403_621778bb020003q3v.html
[2] Santo Fortunato, Carl T. Bergstrom, et al. The whys and wherefores of SciSci [J].science,2018,359(02): 1004-1006
[3] Liu Ya. Analysis of International Cooperation Situation in the Field of China's Basic Research Based on Bibliometrics [J]. Forum on Science and Technology in China, 2010 (03): 149-155
[4] Ma Tingcan, Wang Guifang and Wanyong. Comparative Analysis of Medical Basic Research in China Based on National Natural Science Foundation of China [J]. Science and Technology Management Research, 2015 (17): 71-76
[5] Chen Kaihua, Zhang Yi, Mu Rongping. International Comparison of Basic Research Capacity in Science and Technology – Evidence from the Field of Energy Storage [J]. Studies in Science of Science, 2017 (01): 34-44
[6] Zhang Zhiqiang, Chen Yunwei, Tao Cheng, et al. Bibliometric Analysis on International Competitive Situation of Quantum Information Research [J].World Sci-Tech R&D, 2018 (01): 37-49
[7] Li Zexia, Liu Xiaoping, Huang Longguang, et al. Technology Analysis of Research and Development Based on Trend Analysis of Disciplines :A Case of the Field of Nuclear Materials [J]. Library and Information Service, 2013 (24): 90-94
[8] Bian Wenyue, Li Zexia, Leng Fuhai. Construct a Framework for Information Analysis on the Frontier of Science and Technology Including Knowledge Element Analysis – A Case for Studying Direct Ethylene Production from Methane [J]. Library and Information Service, 2016 (10): 87-94
[9] Chen Yalan, Han Longshi, Wang Jinxiang, et al. Research on Influencing Factors and Evolving Mechanism of Original Innovation [J]. Studies in Science of Science, 2003 (04): 433-437
[10] Tang Xianming. Affecting Factors and Spatial Distribution Analysis of Provincial Basic Research Activities Based on Funded Projects from Natural Science Foundation of China [D]. Chongqing: Chongqing University: 2010
[11] Liu Yun, Changqing. Scientometric Measurement and Evaluation of International Cooperation in Basic Research in China [J]. Journal of Management Sciences in China, 2001 (01): 64-73
[12] Wang Tianmiao, Liu Jinchang. Study on the Development Strategy of Robot Technology Subject [J]. Machine Design and Manufacturing Engineering, 2003 (01): 31-36
[13] Yuan Like, Yin Zhixin. Importance Evaluation and Cluster Analysis of Key Technologies for Intelligent Robots [J]. Scientific Research Management, 2017 (04): 502-509
[14] Huang Baosheng. Analysis of the Application of Literature Metrology to Basic Research Evaluation [J]. R&D Management, 2008 (12): 108-111

Intelligent Recognition of Production Date Based on Machine Vision

Xiaona Sun, Guohua Gao, Boxing Tian, Bowen Li, Sai Zhang and Juan Huang

College of Mechanical Engineering and Applied Electronics Technology, Beijing University of Technology, Beijing 100124, China

E-mails: 1436435550@qq.com, ggh6768@126.com, tianboxing@126.com, 164783613@qq.com, 1184862964@qq.com and 942290908@qq.com

Abstract. Aiming at the problems of large work intensity, low intelligence level and low efficiency, the production date intelligent recognition algorithm was designed. The algorithm optimizes the image preprocessing algorithm, combines the grid statistical method and the projection density method to extract the feature vector, and uses the support vector machine algorithm to identify the production date. The experimental results show that the algorithm can accurately identify the production date and achieve the expected recognition effect. Has practical significance.

1. Introduction

Machine vision [1-2] is one of the important directions in the development of artificial intelligence. Using the machine instead of the human eye to complete the judgment and detection has a broad application prospect, converting the acquired target into an image, and transmitting the signal to the image processing software, using digital image processing technology [3] to obtain various information of the target, according to the identification The result is to control the operation of the device.

In recent years, with the increase of residents' income and health concerns, people are paying more and more attention to the safety of food. The date of production of food is an important factor in determining food safety, which directly determines whether the food has reached the edible condition. Accurate identification of production date information is particularly important.

This design proposes a code identification algorithm applied to the outer packaging production date. Through the acquired image preprocessing, ROI region extraction, segmentation correction, recognition and other technologies, the production date information is obtained, and the production date is recognized. The research on the quality inspection of dairy products provides a reference.

2. Overall overview of production date intelligent identification

2.1. Intelligent identification system overall framework

The intelligent identification of production date mainly includes two parts, hardware acquisition system and software processing system. The hardware acquisition system mainly includes camera, light source, photoelectric sensor, sealing cover, industrial computer, etc. The distribution of hardware system is shown in Figure 1 below. Hardware acquisition system The collected image is transmitted to the software processing system, and processed by the image preprocessing algorithm and the image recognition algorithm to identify the production date. The overall frame of the identification system is shown in Figure 2.

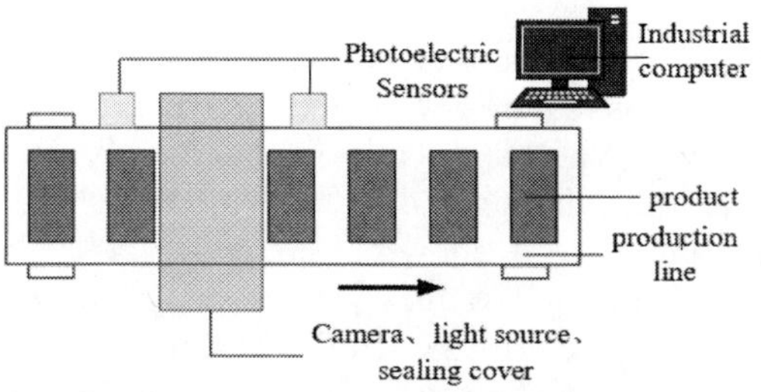

Figure 1. Hardware acquisition system map

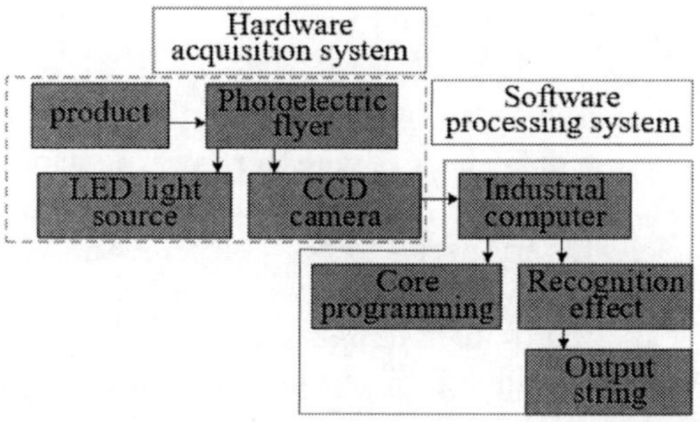

Figure 2. Intelligent identification overall framework

2.2. Intelligent identification system algorithm flow

The intelligent recognition algorithm based on production date mainly includes two kinds of algorithms, image preprocessing and character recognition algorithm. The former mainly lays the foundation for the latter's implementation, and directly affects the recognition effect, including image denoising and removal due to hardware facilities. Noise interference caused by the environment, improve the contrast of the image; image positioning, used to locate the region of interest, extract the production date area; character correction, because the segmented character will have a certain direction of rotation, correct it to the horizontal direction It is convenient for character segmentation; character segmentation divides the production date characters into single characters to facilitate the extraction of character features. This paper is based on SVM [4-6] neural network algorithm to identify characters, including character normalization, feature extraction, and character training. The algorithm flow is shown in Figure 3.

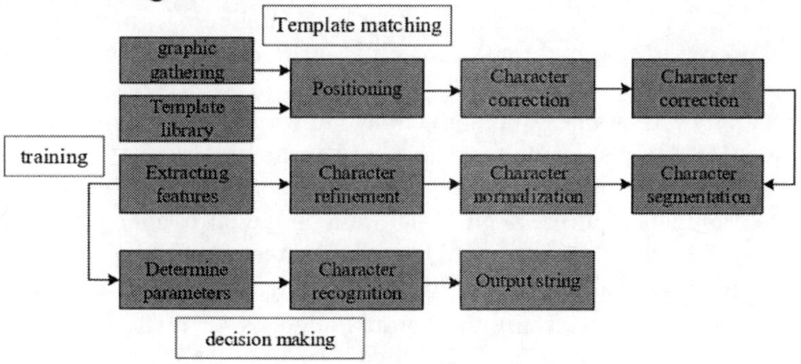

Figure 3. Test flow chart

3. Preprocessing algorithm implementation

3.1. Location Algorithm

In this paper, the adaptive anti-rotation template matching method is used for positioning. Firstly, the image is filtered to reduce the noise interference during matching. Then, the multi-angle template library is built by rotating the image to be recognized to solve the false contour caused by the template rotation. Finally, the normalized product correlation algorithm optimized by FFT and integral graph principle is used to measure the similarity. Combined with the image pyramid search strategy and the indented search step method, the computational complexity is reduced and the robustness is guaranteed.

3.1.1. Filter processing As the first step of production date identification, positioning detection filter processing is needed to reduce the interference of noise on subsequent images. In this paper, traditional mean filtering, Gaussian filtering, median filtering and block filtering are used to denoise the source image. After comparison of effects, this paper uses median filtering to suppress noise interference.

3.1.2. Create a regional template First, you need to create a standard template image. Since the production date changes with the date, the template feature has feature invariance. herefore, the year in the production date is selected as the area template library. When the year is updated, only the area template library needs to be replaced. The year date is ok, reducing the amount of work based on template matching. The regional template library used in this paper is shown in Figure 4.

If the multi-angle template is created in the rotating area template library, a false contour will be generated, resulting in a matching failure. Since the region of interest generally does not exist at the edge of the image to be detected, the false contour does not interfere with the matching process, so Rotate the image to be recognized to eliminate interference from false contours.

(a) Source image (b) false contours

Figure 4. Regional template

3.1.3. Positioning matching process In order to improve the matching precision and reduce the amount of calculation, this paper proposes a method of indenting the angle step to optimize the matching process. On the highest pyramid *Y*, the image to be detected is rotated by (0.25×2^{Y}) degrees to create an angle library. The template image is matched, and the optimal angle θ_Y is returned according to the NCC algorithm and the coordinate information of the center position is recorded.

On the *Y*-1 pyramid, take the angular subgraph in the range of ($\theta_Y\pm0.25\times2^{Y-1}$), perform rotation matching every ($0.25\times2^{Y-1}$) degrees, return the optimal angle θ^{Y-1} and record Center position coordinates. In turn, on the lowest pyramid, a match is made every 0.5° to return the final matching angle θ_1.

Taking the 5-layer pyramid as an example, the image to be detected needs to be rotated every 8° on the fifth layer pyramid to establish a multi-angle template library, as shown in Fig. 5, returning the optimal angle θ5=8°, on the fourth layer pyramid, Rotate the template library every 4° from an angle range of 4° to 12°, returning the optimal angle θ_4=4°, and on the third layer pyramid, take the image to be detected within 2°~6° angle every 2° Rotate to build the template library, return the optimal angle θ_3=3°, and build the angle template library on the second layer pyramid by rotating the image to be detected within 2°~4° every 1° , returning the optimal angle θ_2=2° . The image is rotated by 1.5° to 2.5° per 0.5° on the lowest pyramid, and the final matching angle θ_1=2.5° is returned, thereby achieving template matching under rotational interference.

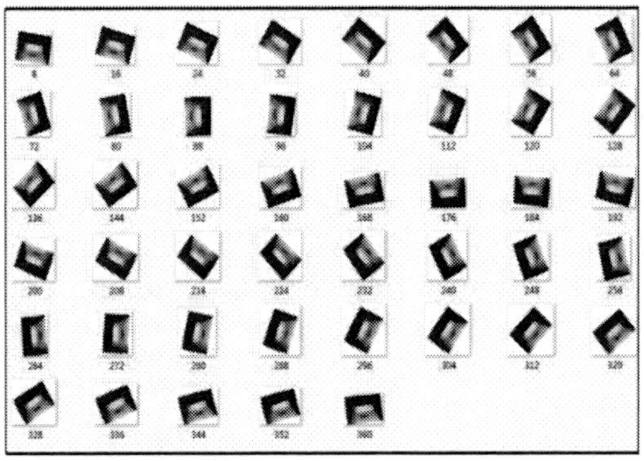

Figure 5. Multi-angle template library for 5-layer pyramid images

IOP Conf. Series: Journal of Physics: Conf. Series **1267** (2019) 012041 doi:10.1088/1742-6596/1267/1/012041

3.2. Character correction

In the production date defect detection system, the production date character will have a certain inclination, which will bring difficulties for subsequent character segmentation. It must be tilt corrected before segmentation, by calculating the tilt angle of the production date, and then the production date character. The image is rotated to achieve the purpose of proof of production date tilt correction. The calibration results are shown in Figure 6.

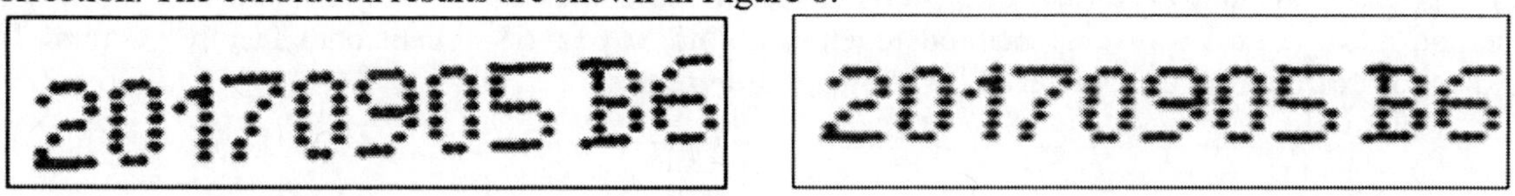

(a) Before correction (b) After correction

Figure 6. Character correction

3.3. Character segmentation

In this paper, the vertical projection method and the horizontal projection method are used to divide the binary image of the production date. Before the segmentation, the character needs to be expanded to become a single connected character. The segmentation process and segmentation effect of the character are shown in Fig. 7.

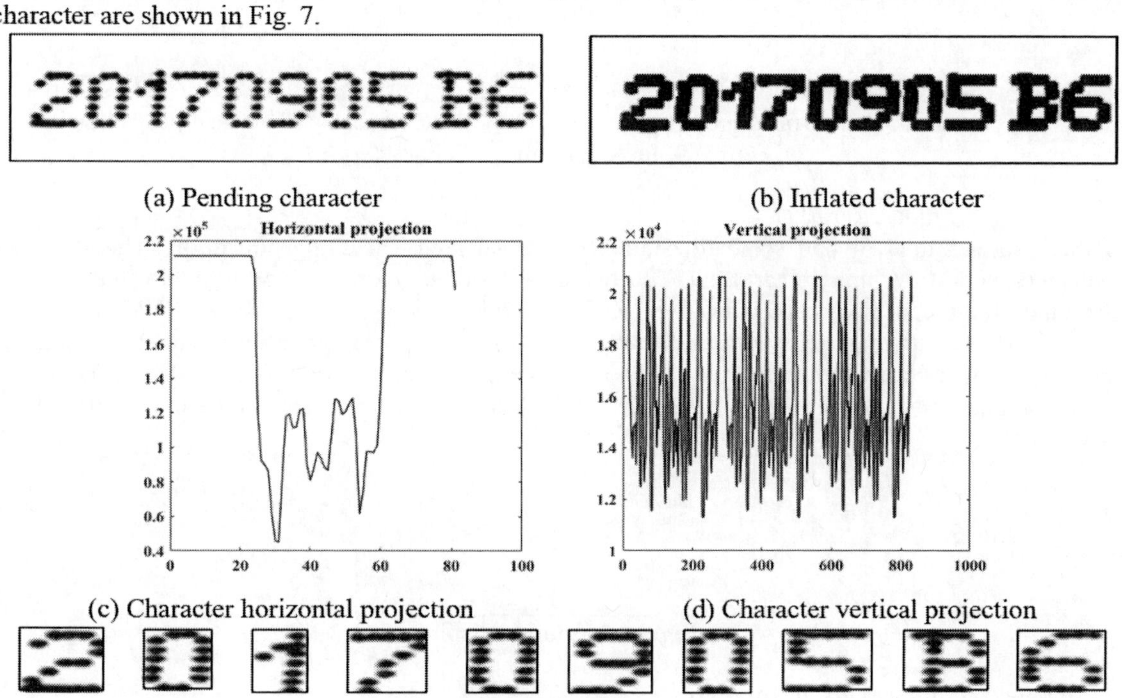

(a) Pending character (b) Inflated character

(c) Character horizontal projection (d) Character vertical projection

(e) Split effect

Figure 7. Character segmentation

4. Recognition algorithm implementation

4.1. Character normalization

Since the size of the divided characters is different, in order to facilitate the recognition of characters, the characters need to be normalized to the same size. This paper uses the centroid algorithm to normalize the characters. In general, the ratio of the height to the width of a character is 2:1. By calculating the centroid and divergence of the character, the character is normalized into a character block of 12 pixels wide and 24 pixels high.

4.2. Feature extraction

The purpose of feature extraction is to extract the classification features of characters as the basis for recognition. In this paper, the combination of character grid statistical method and feature projection density method is used as the classification feature. The projection density method obtains 12+24=36 dimensional feature vectors in total. The characters are divided into 32 grids, and the pixel values of each grid are counted for a total of 32-dimensional feature vectors. Therefore, the projection density method and the grid statistical method together obtain 36+32=68-dimensional feature vectors. The projection feature of the character "1" is shown in Figure 8.

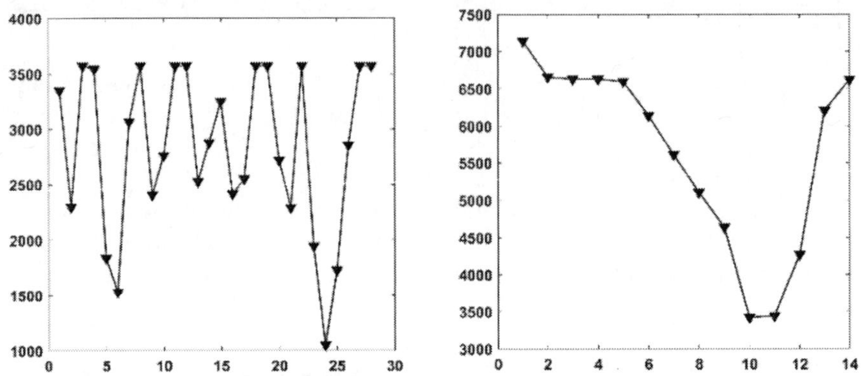

(a) Character "1" horizontal projection (b) Character "1" vertical projection

Figure 8. Projection map of characters

4.3. SVM recognition algorithm

In this paper, a total of 650 standard characters are selected, including 500 digital characters, 50 characters, and 100 Chinese characters. The feature vectors are extracted and trained using SVM. In this paper, the radial basis kernel function (RBF) is used as the kernel function of SVM. For a two-classifier combination of one, it is necessary to form three SVM classifiers, namely a character classifier, a letter classifier and a Chinese character classifier. The feature vectors of the characters are tested and matched to the three SVM classifiers by a decision function. Part of the sample set is shown in Figure 9.

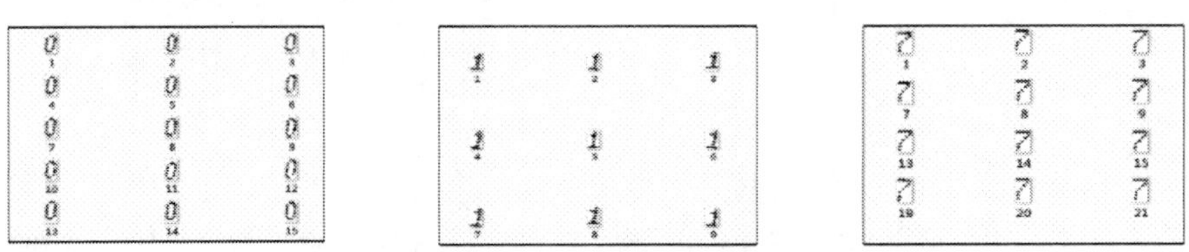

Figure 9. Sample display

4.4. Implementation results and analysis

In this paper, 100 milk box photos are selected for testing. The total number of characters is 1200, and the correct number of characters is 1183, and the correct rate is 98.58%. The recognition effect is shown in Figure 10.

Figure 10 Recognition result

5. Conclusions
Correct identification of production date has a crucial impact on consumers' health and corporate image. Traditional manual identification methods have certain limitations, high cost and low efficiency. This paper uses support vector machine method to make production date. The algorithm for image preprocessing is optimized and improved. The results show that this method is robust to the identification of production date and provides technical and physical support for the smooth implementation of the food and drug recall system.

Acknowledgments
This research work is supported by the Ministry of Science and Technology's innovative methods, project number 2016IM030400, and the title of the project is the innovative method research and application demonstration of the whole chain quality control of dairy safety.
This research work is supported by the Ministry of Science and Technology's innovative methods, project number SQ2016IM03640007, and the title of the project is Systematic Application and Demonstration of Intelligent Manufacturing Innovation Method for Agricultural Machinery Equipment

References
[1] Luna Wang,"The application of using vision sensor to check medicines and chemical reagents production date", Journal of CAM, Jinan: Shandong University [D] 2008: 84-84.
[2] Wenjun Wu, Research on the character detection and recognition system of beer bottles based on machine vision [D]. Wuhan: Hubei University of Technology, 2005:1-3.
[3] Steger C, Ulrich M, Wiedemann C. Machine Vision Algorithms and Applications [M]. 2007.
[4] Jin-Jun W U,Shu-Xin D U. Application of SVM to character recognition of vehicle license plate [J]. Journal of Circuits & Systems, 2008.
[5] Wei Wang, Yongqiang, Qiang Peng, Application of SVM multi-classifier on license plate character recognition. [J]. Computer engineering and design, 2011, 32(9):3166-3169..
[6] Laide Guo, Yan Zhang, Yang Cao, Application of SVM in motor vehicle license plate recognition. [J]. Computer and digital engineering, 2012, 40(4):105-106.

IOP Conf. Series: Journal of Physics: Conf. Series **1267** (2019) 012042 doi:10.1088/1742-6596/1267/1/012042

The Spread Behavior Analysis of a SIQR Epidemic Model under the Small World Network Environment

Yuan Liu and Yan-Yong Zhao*

School of Statistics and Mathematics, Nanjing Audit University, Nanjing, 211815, China

*zhaoyanyong1987@163.com

Abstract. In this paper, we focus on the behavior of virus spreading on complex networks. A SIQR epidemic model on small-world network is established. We find the spread of infectious diseases threshold, which can determine the dynamic behavior of the system in a positive invariant set. If $R_0>1$, there are two equilibriums in the system, the disease-free equilibrium and the endemic equilibrium. If $R_0<1$, There is only the disease-free equilibrium in the system. It is globally stable in the positive invariant set. Furthermore, we analyze the equilibrium and its stability. Finally, some numerical simulation studies are provided to illustrate the effectiveness of the results.

1. Introduction

Infectious diseases have always been an enemy to human health，history and reality tell us：human beings are facing the long-term and serious threat of infectious diseases. The research on the pathogenesis of infectious diseases, regularity of infectious diseases, prevention and control strategy is becoming important. It is a significant problem that needs to be urgently addressed in today's world.

The reemergence of tuberculosis (TB) from the 1980s to the early 1990s instigated extensive researches on the mechanisms behind the transmission dynamics of TB epidemics. The earliest mathematical models describing the TB dynamics appeared in the 1960s and focused on the prediction and control strategies using simulation approaches. Reference [1] provides a detailed review of the work on the dynamics and control of TB. In the late 20th century, complex network theory developed rapidly, and became a new trend to study infectious diseases in complex networks. Scientists focused on SIS epidemic model and SIR epidemic model. And it is generally assume that that total number of individual is constant. Reference [2] discusses the SIS epidemic model in local area networks. However, the birth and death of some diseases seriously affect the total number of individual, so it is necessary to consider the birth and death of individuals in such circumstances [3]. Jiang et al. investigated the self-affirmation effect on formation of public opinion in a directed small-world social network [4]. Christensen et al. developed a framework for simulating a realistic, evolving social network (a city) into which a disease is introduced. They compared the results to prevaccine era measles data for England and Wales, and finded that they captured the quantitative and qualitative features of epidemics in populations spanning two orders of magnitude [5]. Discrete epidemic models with periodic epidemiological parameters in [6] are formulated and studied to take into account seasonal variations of infectious diseases. In the process of transmission of infectious diseases，pre-exposure prophylaxis (re-exposure prophylaxis (PrEP) consists in the use of an antiretroviral medication to prevent the acquisition infection by uninfected individuals. It can significantly reduce the spread of infectious diseases. In [7], Xiao and Ruan study an epidemic model with nonmonotonic incidence rate, which describes the psychological effect of certain serious diseases on the community when the number of infective is getting larger.

Small world networks have small perturbation degree, approximation for the average degree <k> [8]. Assume that the total number of network nodes remains unchanged. And the state nodes are evenly distributed. It can simulate the spread of infectious diseases well. However, the above studies ignored the spread environment, just thinking about the epidemic model alone, or in building an infectious disease model without considering isolation and immunity. This is deviate from the spread of infectious diseases. Therefore, in the small world networks, the method of isolating the source of infection and enhancing immunity should be adopted. This makes the conclusion true and reliable. In this paper we analysis and research on the dynamics of SIQR infectious disease models in small world networks, we find the threshold $R_0=(\delta+\gamma)/\beta<k>$, which determines the dynamic behavior of the system in a positive invariant set D. If $R_0>1$, there are two equilibriums in the system, the disease-free equilibrium E_0 and the endemic equilibrium E_*. E_0 in the set D is unstable. E_* in the set D is locally asymptotically stable. If the disease is present at the beginning, then the disease will be stable in the population, and it forms endemic disease. If $R_0<1$, there is only a disease-free equilibrium E_0 in the system. It is globally stable in the set D. The disease will eventually die out. Finally numerical simulations are provided to illustrate the effectiveness of the results.

2. Establishment of SIQR Infectious disease model

There are four types of nodes in the network: susceptible S, infectious I, quarantine Q, recovery R. Small world networks have small perturbation degree, approximation for the average degree <k>. Assume that the total number of network nodes remains unchanged. And the state nodes are evenly distributed. S(t), I(t), Q(t), R(t) represent the proportion of each state node at time t. At the same time, β represents the infection rate of the infected state node, At time t, the probability of the node in the susceptible state infected with the infected node is $<k>\beta IS$, The transformation relationship between states is shown in Figure 1, in which δ represents the probability that the infected node is isolated, ε indicates the probability that the isolation node is cured, γ stands for the probability that the infected node is cured, and α is the probability that the immune node loses immunity and becomes susceptible.

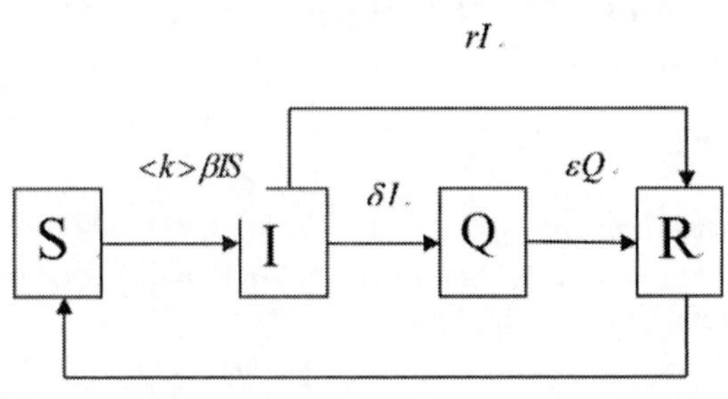

Figure 1. The transformation relationship between the states of SIQR models in a uniform network.

In the model, if a node in a susceptible state is connected to a node in the state of infection, it is infected with the probability of $<k>\beta$. So at time t, the susceptible node with the degree of k is infected with a disease node with the probability of $k\beta$, since the proportion of the susceptible nodes is S(t). So the susceptible node with the degree of k is infected with a disease node at a rate of $<k>\beta S(t)$. At the same time, the removal of the immune node has been redeveloped into a node of the susceptible state with the probability of α. Therefore, the probability of the susceptible node in the total node is $-<k>\beta S(t)+\alpha R(t)$ at the time t. Similarly, the probability of infected nodes, the quarantine nodes and the recovery nodes are respectively $<k>\beta S(t)-\gamma I(t)-\delta I(t)$, $\delta I(t)-\varepsilon Q(t)$, $\varepsilon Q(t)+\gamma I(t)-\alpha R(t)$. According to the mean-field theory, the propagation dynamics equation can be obtained as follows:

IOP Conf. Series: Journal of Physics: Conf. Series **1267** (2019) 012042 doi:10.1088/1742-6596/1267/1/012042

$$\begin{cases} \dfrac{dS(t)}{dt} = -<k> \beta I(t)S(t) + aR(t), \\ \dfrac{dI(t)}{dt} = <k> \beta I(t)S(t) - \gamma I(t) - \delta I(t), \\ \dfrac{dQ(t)}{dt} = \delta I(t) - \varepsilon Q(t), \\ \dfrac{dR(t)}{dt} = \varepsilon Q(t) + \gamma I(t) - aR(t). \end{cases} \tag{1}$$

3. The equilibriums and stability analysis of SIQR epidemic Model

3.1. The existence of the equilibriums

Use the normalized condition $S(t)+I(t)+Q(t)+R(t)=1$, we can simplify (1) to the following model:

$$\begin{cases} \dfrac{dI(t)}{dt} = <k> \beta I(t)\left(1-I(t)-Q(t)-R(t)\right) - \gamma I(t) - \delta I(t), \\ \dfrac{dQ(t)}{dt} = \delta I(t) - \varepsilon Q(t), \\ \dfrac{dR(t)}{dt} = \varepsilon Q(t) + \gamma I(t) - aR(t), \end{cases} \tag{2}$$

$D=\{I(t), Q(t), R(t) \mid I(t), Q(t), R(t)>0, I(t)+Q(t)+R(t)\leq 1\}$ is the positive invariant set of system (2). Let the right-hand side of (2) be equal to zero, obviously the zero point is the equilibrium point. Then, the system (1) has disease-free equilibrium $E(1,0,0,0)$. Let the right-hand side of (2) be equal to zero, other variables are indicated by I, $S(t)= (\delta+\gamma)/\beta<k>$, $Q(t)= \delta I(t)/\varepsilon$, $R(t)= (\delta+\gamma)I(t)/\alpha$, then we have $(\delta+\gamma)/\beta<k>+I(t)+\delta I(t)/\varepsilon+(\delta+\gamma)I(t)/\alpha=1$ and $I(t)=(1-R_0)/[1+\delta/\varepsilon+(\delta+\gamma)/\alpha]$, where $R_0=(\delta+\gamma)/\beta<k>$. If $R<1$, there is a unique positive equilibrium $E_*=(S_*, I_*, Q_*, R_*)$ of the system (1) in the set of D, where $S_*(t)= (\delta+\gamma)/\beta<k>$, $I_*(t)= (1-R_0)/[1+\delta/\varepsilon+(\delta+\gamma)/\alpha]$, $Q_*(t)= \delta I_*(t)/\varepsilon$ and $R_*(t)= (\delta+\gamma)I_*(t)/\alpha$.

Theorem 3.1 *The model (1) always has a disease-free equilibrium E, while $R_0<1$, the model still has an endemic equilibrium E_*.*

Proof. The proof of this theorem can be found in [9].

3.2. The stability analysis of equilibrium

Theorem 3.2 *If $R_0\geq 1$, E is local asymptotic stability. If $R_0<1$, E is unstable.*

Proof. The Jacobian matrix of model (3.1) at *E* is:

$$J_0 = \begin{pmatrix} <k> \beta - \gamma - \delta & 0 & 0 \\ \delta & -\varepsilon & 0 \\ \gamma & \varepsilon & -a \end{pmatrix}$$

The characteristic equation of J_0 is $(\lambda-<k>\beta+\gamma+\delta)(\lambda+\varepsilon)(\lambda+\alpha)=0$, and eigenvalues are $\lambda_1=<k>\beta-\gamma-\delta$, $\lambda_2=-\varepsilon$, $\lambda_3=-\alpha$.

If $R_0\geq 1$, we have $\lambda_1\leq 0$, $\lambda_2<0$, $\lambda_3<0$, then E is local asymptotic stability. If $R_0<1$, we have $\lambda_1>0$, then E is unstable in the set of D. In other words, if $R_0\geq 1$, the system (1) converges to the disease-free equilibrium $E(1,0,0,0)$.

Theorem 3.3 *If $R_0\geq 1$, E_* is local asymptotic stability.*

Proof. The Jacobian matrix of model (2) at E_* is

$$J_0=\begin{pmatrix} <k>\beta S_*(t)-<k>\beta I_*(t)-\gamma-\delta & -<k>\beta I_*(t) & -<k>\beta I_*(t) \\ \delta & -\varepsilon & 0 \\ \gamma & \varepsilon & -a \end{pmatrix}.$$

The characteristic equation of J_0 is

$$\lambda^3+(<k>\beta I_*(t)+\varepsilon+a)\lambda^2+(<k>\beta I_*(t)(\varepsilon+a+\delta+\gamma)+a\varepsilon)\lambda+<k>\beta I_*(t)(\varepsilon a+a\delta+\varepsilon\delta+\gamma\varepsilon)=0.$$

Let $\rho_1=<k>\beta I_*(t)+\varepsilon+\alpha$, $\rho_2=<k>\beta I_*(t)(\varepsilon+\alpha+\delta+\gamma)+\alpha\varepsilon$, $\rho_3=<k>\beta I_*(t)(\varepsilon\alpha+\alpha\delta+\varepsilon\delta+\gamma\varepsilon)$.

According to *Hurwitz*

$H_1=\rho_1=<k>\beta I_*(t)+\varepsilon+\alpha=<k>\beta(1-R_0)/[1+\delta/\varepsilon+(\delta+\gamma)/\alpha]+\varepsilon+\alpha>0$,

$H_2=\rho_1\rho_2-\rho_3=[<k>\beta I_*(t)+\varepsilon+\alpha][<k>\beta I_*(t)(\varepsilon+\alpha+\delta+\gamma)+\alpha\varepsilon]-<k>\beta I_*(t)(\varepsilon\alpha+\alpha\delta+\varepsilon\delta+\gamma\varepsilon)$,

where $<k>\beta I_*(t)(\varepsilon\alpha+\alpha\delta+\varepsilon\delta+\gamma\varepsilon)>0$, we have

$H_3=\rho_1\rho_2\rho_3=[<k>\beta I_*(t)+\varepsilon+\alpha][<k>\beta I_*(t)(\varepsilon+\alpha+\delta+\gamma)+\alpha\varepsilon][<k>\beta I_*(t)(\varepsilon\alpha+\alpha\delta+\varepsilon\delta+\gamma\varepsilon)]>0$.

Hence, all the characteristic roots of J_0 have a negative real part. If $R_0\geq1$, E_* is local asymptotic stability.

Theorem 3.4 *If $R_0\geq1$, E is globally stable.*

Proof. Since S(t)+ I(t)+Q(t)+R(t)=1, then S(t)≤1, we have

$$\begin{aligned}\frac{dI(t)}{dt}&=<k>\beta I(t)S(t)-\gamma I(t)-\delta I(t)\\ &\leq<k>\beta I(t)-\gamma I(t)-\delta I(t)\\ &=I(t)[<k>\beta-(\gamma+\delta)]\\ &=I(t)<k>\beta(1-R_0).\end{aligned}$$

If $R_0\geq1$, we have $\frac{dI}{dt}\leq0$, $\lim_{t\to+\infty}I(t)=0$. If $t\to+\infty$, the extreme equations of model (1) is

$$\begin{cases}\dfrac{dS(t)}{dt}=aR(t),\\ \dfrac{dQ(t)}{dt}=-\varepsilon Q(t),\\ \dfrac{dR(t)}{dt}=\varepsilon Q(t)-aR(t).\end{cases}\qquad(3)$$

Then, we have $Q(t)=C_1e^{-\varepsilon t}$, $\lim_{t\to+\infty}Q(t)=0$. It can be calculated that $\lim_{t\to+\infty}R(t)=0$, $\lim_{t\to+\infty}S(t)=1-\lim_{t\to+\infty}I(t)-\lim_{t\to+\infty}Q(t)-\lim_{t\to+\infty}R(t)=1$.

Therefore, the equilibrium of the extreme equations of model (1) is globally stable. By the extreme equation theory [10], we can obtain that any solution E of the model (1) satisfy $\lim_{t\to+\infty}E=(1,0,0,0)$ when $\lim_{t\to+\infty}I(t)=0$. So, if $R_0\geq1$, the disease-free equilibrium of the model (1) is globally stable.

4. Numerical simulations

In the numerical simulations, the global stability of disease-free equilibrium and the local asymptotic stability of local equilibrium is verified. And better understand the development trend of diseases. Let parameter N=5000, m=3,<k>=6 to construct a small world network in the system (1). In addition, let parameter $\varepsilon=\delta=0.8$, $\gamma=0.2$, $\alpha=0.4$, $\beta=0.05$, and calculate that $R_0<1$, the model (1) has a disease-free equilibrium E, E is globally stable. Let the initial values are S(0)=0.2, I(0)=0.2, Q(0)=0.3, R(0)=0.1; S(0)=0.1, I(0)=0.2, Q(0)=0.3, R(0)=0.3; S(0)=0.3, I(0)=0.2, Q(0)=0.1, R(0)=0.2. Numerical simulation of system (1) by *Matlab*, we can get the following Figure 2.

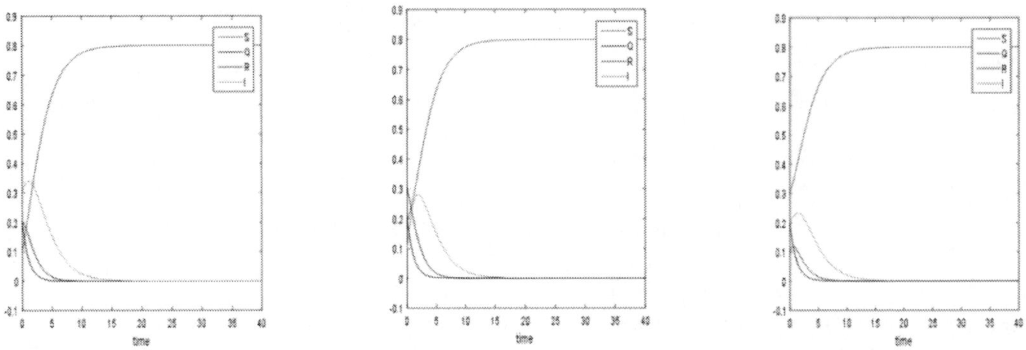

Figure 2. Global stability of the disease-free equilibrium when R_0>1and three different initial values.

To construct the same small world network in the system (1). Let parameter $\varepsilon=\delta=0.8$, $\gamma=0.2$, $\alpha=0.4$, $\beta=0.5$. Calculating $R_0<1$, the model (1) has a endemic equilibrium E_*, and E_* is local asymptotic stability. Let the initial values are S(0)=0.2, I(0)=0.2, Q(0)=0.3, R(0)=0.1; S(0)=0.1, I(0)=0.2, Q(0)=0.2, R(0)=0.3; S(0)=0.3, I(0)=0.2, Q(0)=0.1, R(0)=0.2. Numerical simulation of system (2.1) by *Matlab*, we can get Figure 3.

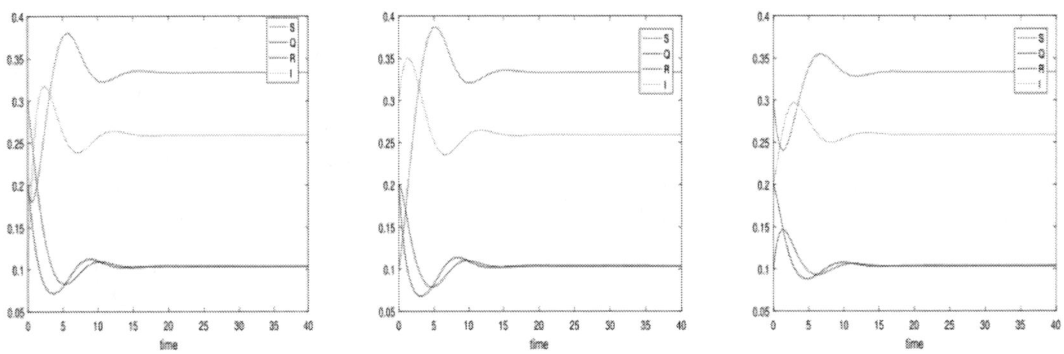

Figure 3. Local asymptotic stability of the endemic equilibrium when $R_0<1$ and different initial values.

5. Conclusion

We study the dynamics of SIQR infectious disease models in small world networks, and find the threshold $R_0=(\delta+\gamma)/\beta\langle k\rangle$, which can determine the dynamic behavior of the system in a positive invariant set D. If $R_0>1$, there are two equilibriums in the system, the disease-free equilibrium E_0 and the endemic equilibrium E_*. E_0 in the set D is unstable, E_* in the set D is locally asymptotically stable. If the disease is present at the beginning, then the disease will be stable in the population, and it forms endemic disease. If $R_0<1$, There is only E_0 in the system. It is globally stable in the set D. The disease will eventually die out.

Reference

[1] Castillo-Chavez C, Song B. Dynamical models of tuberculosis and their applications. Mathematical Biosciences & Engineering. 2004, 1(2): 361-404.
[2] Xia C Y, Liu Z X, Chen Z Q et al. Epidemic spreading behavior in local-world evolving networks. Progress in Natural Science, 2008, 18: 763-768.
[3] Mena-Lorca J, Hetheote H W. Dynamic models of infectious diseases as regulators of population biology. J.Math.Biol.1992, 30: 693-716.
[4] Jiang L L , Hua D Y, Zhu J F, Wang B H, Zhou T. Opinion dynamics on directed small-world

networks. The European Physical Journal B, 2008, 65(2): 251-255.
[5] Christensen C, Albert I, Grenfell B, Albert R. Disease dynamics in a dynamic social network. Physica A, 2010, 389: 2663-2674.
[6] Cao H, Zhou Y. The basic reproduction number of discrete SIR and SEIS models with periodic parameters. Discrete & Continuous Dynamical Systems-B. 2013, 18(1): 37-56.
[7] Xiao D M, Ruan S G. Global analysis of an epidemic model with nonmonotone incidence rate. Mathematical Biosciences, 2007, 208(2): 419-429.
[8] Watts D J, Strogatz S H. Collective dynamics of small-world networks. Nature, 1998, 393(6684): 440-442.
[9] Ma Z, Zhou Y. Stability and stability of ordinary differential equations. Beijing: Science Press, 2001.
[10] Zhang H, Xin C. Spreading of epidemics on scale free networks with nonlinear infectivity. Nonlinear Analysis Theory Methods & Applications, 2009, 70 (9): 3273-327.

IOP Conf. Series: Journal of Physics: Conf. Series **1267** (2019) 012043 doi:10.1088/1742-6596/1267/1/012043

Neural Detection of Foreign Objects for Transmission Lines in Power Systems

Pengfei Xia[1], Jin Yin[1], Jingsong He[1*], Li Gu[1] and Kejun Yang[2]

[1]University of Science and Technology of China, Hefei, China
[2]Anhui Nari Jiyuan Electric Power System Tech CO., LTD, Hefei, China

[*]Email: hjss@ustc.edu.cn

Abstract. Ensuring the normal operation of the transmission lines, which provides a path for directing the transmission of energy from one place to another, is a prerequisite for delivering power to cities and enterprises. A major threat comes from foreign objects, which may cause interruption of power transmission. Compared with traditional manual method, which not only consumes a lot of manpower, but more importantly, affects the safety and efficiency of power network, in this paper, we apply a neural detection of foreign objects for transmission lines. Transfer learning and data augmentation are used to mitigate data shortages. Experimental results show that even with small training data, the neural detection with transfer learning and data augmentation is an effective method for this task without loss of real-time property.

1. Introduction

Transmission lines are one of the most important part of power systems, which provides a path for directing the transmission of energy from one place to another, and is an essential guarantee for national life and enterprise production [1]. Suspended foreign objects is the main threat to the normal operation of transmission lines and is the primary task of power inspection. Nowadays, manual inspection is still the most common method in power inspection. However, except for requiring a lot of labor and is susceptible to environment and climate [2], more importantly, with high-intensity work, low detection accuracy affect the safety and efficiency of power network. Automatic detection of foreign objects for transmission lines in power systems is a vital part of the smart power line, and is significant for ensuring operational safety and efficiency.

Neural networks, which is to mimic the nerves of animals, have achieved remarkable results in traditional visual perception tasks, such as image classification [3,4], objects detection [5,6] and semantic segmentation [7,8]. Foreign object detection can be regarded as a visual perception task. For foreign objects detection in transmission lines, using the aircraft to photograph and then processing the images with neural based methods is a promising way, which saves lots of resources and is less affected by the environment.

Our investigation centers on the foreign objects detection on acquired images from reality. There are some pivotal issues worth concerning when considering the foreign objects detection in transmission lines:

Transmission lines are located in diverse environments, such as countryside, city, hill and so on. As a result, the background of the images is complicated;

- The images are taken by the aircraft, so the views are uncertain and variable;
- Less training data available than other tasks;

- Real-time property needs to be considered.

In this paper, we apply neural network to detect foreign objects for transmission lines in power systems and the results is good. In view of the issues listed above, YOLOv3 [9], a unified, real-time framework is adopted. Meanwhile, when considering the shortage of training data, which is a realistic difficult, transfer learning and data augmentation are employed to combat with it. The experimental results show that the neural foreign objects detection we adopted in this paper is a effective method for transmission lines in power systems.

2. Neural Foreign objects Detection

2.1. Task Description

Figure 1 shows three examples of foreign objects detection for transmission lines. Simply put, the task is to determine whether there are any foreign objects in the image and mark the location if exist.

(a) (b) (c)

Figure 1. Three examples of foreign objects detection for transmission lines.

2.2. Architecture

YOLOv3, which is an update of YOLO [10] and YOLO9000 [11], performs the state-of-the-art performance on objects detection, both in accuracy and inference time. Different from two-stage methods, such as R-CNN [12], Fast R-CNN [5] or Faster R-CNN [6], YOLOv3 is an one-stage method and can be trained end to end. The main idea is to regard objects detection as a single regression problem, straight from image pixels to bounding box coordinates and class probabilities [10].

The architecture can be roughly divided into two parts: the image feature extraction and the objects detection, as shown in Figure 2(a) and Figure 2(b) respectively. The image feature extraction is used to represent image into some fixed size tensors. Darknet-53 is adopted in YOLOv3 and three different size image features F_a, F_b and F_c are extracted for detecting objects. To detect objects with various sizes, multi-scale detection is used in this method, as shown in Figure 2(b), where S_1, S_2 and S_3 are three scales predictions.

We take S_1 as an example to explain the whole process detailedly. Assuming that the input image is resized to 416×416, and after inference we acquire S_1, which is 13×13×18. 13×13 means that the input image is divided into 13×13 grid. Each grid cell predicts six value: x, y, w, h, confidence and class probabilities. The (x, y) coordinates represent the center of the box relative to the bounds of the grid cell. The w and h are width and height of bounding box. Thus, the size of S_1 should be 13×13×6. But for more accurate, there are three anchors for each grid cell, so the size is 13×13×18.

IOP Conf. Series: Journal of Physics: Conf. Series **1267** (2019) 012043 doi:10.1088/1742-6596/1267/1/012043

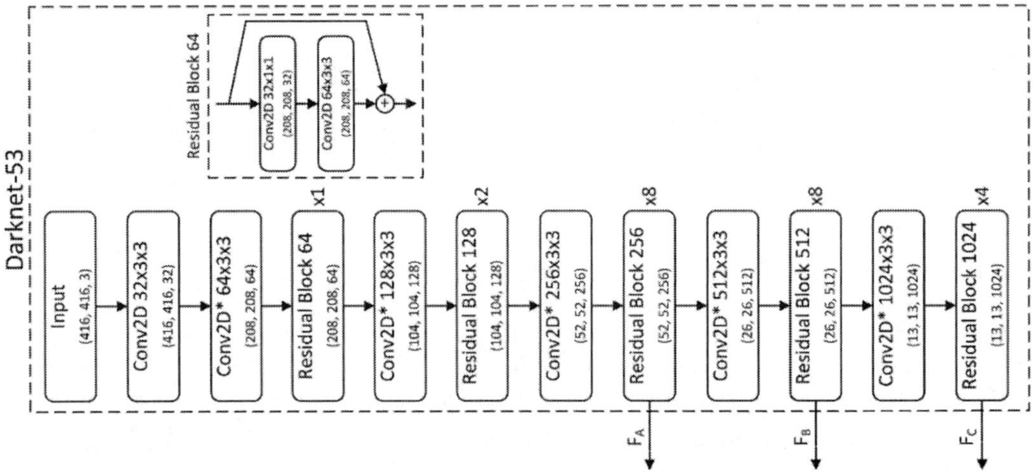

(a)

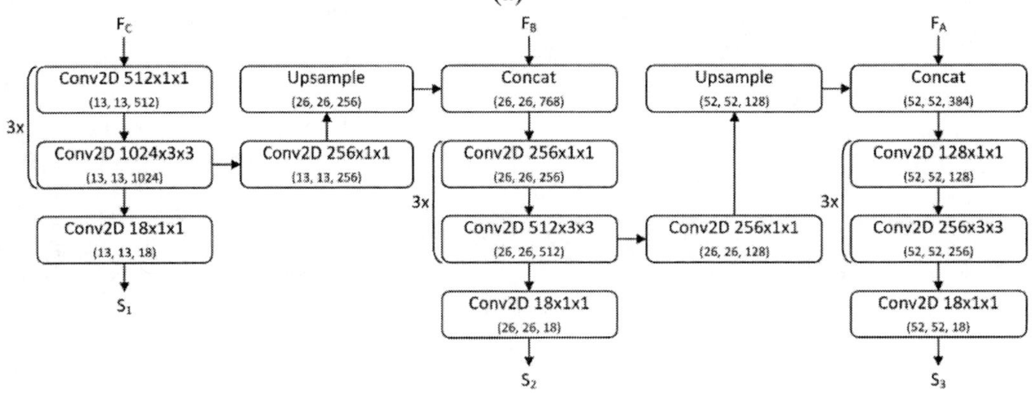

(b)

Figure 2. Two part of YOLOv3. (a) is the image feature extraction and (b) is the objects detection.

2.3. Transfer Learning and Data Augmentation

Transfer learning and data augmentation are used for mitigating data shortage.

Transfer learning. We first train Darknet-53 in imagenet [13], which consists of more than 14 million images and is one of the biggest image dataset. Then we move the fully connected layers and keep the convolutional layers for foreign objects detection.

Data augmentation. Horizontal flip, scale and random crop are applied to data, as shown in Figure 3. It is noteworthy that the coordinate position of the box will also change after horizontal flip and random crop operation.

(a) (b) (c)

Figure 3. An example of data augmentation.

3. Experiments

3.1. Training Details

We implement all code of our model based on PyTorch [14], which is one of widely adopted deep learning frameworks. 387 images are divided randomly into two parts: 300 for training and 87 for testing. Adam optimizer [15] is used to optimize the parameters of the neural network. The initial learning rate is 0.0001 and the batch size is set to 24. Weight decay [16] with 0.00005 is used to prevent overfitting. The max epoch is set to 300. All training is finished with a NVIDIA GTX 1080Ti GPU.

3.2. Results

YOLOv3 (Yv3), YOLOv3 with transfer learning (Yv3+TL), Yv3 with data augmentation (Yv3+DA), YOLOv3 with transfer learning and data augmentation (Yv3+TL+DA) are included in experiments. The results of precision and recall-precision curves are shown in Table 1 and Figure 4. The results show effective performance of detection in the test data, and transfer learning and data augmentation improve the generalize of the model in small training data. The average inference time is 46 ms at 416×416 resolution.

Table 1. Test precision of Yv3, Yv3+TL, Yv3+DA and Yv3+TL+DA.

	Yv3	Yv3+TL	Yv3+DA	Yv3+TL+DA
Precision	0.558	0.638	0.674	0.724

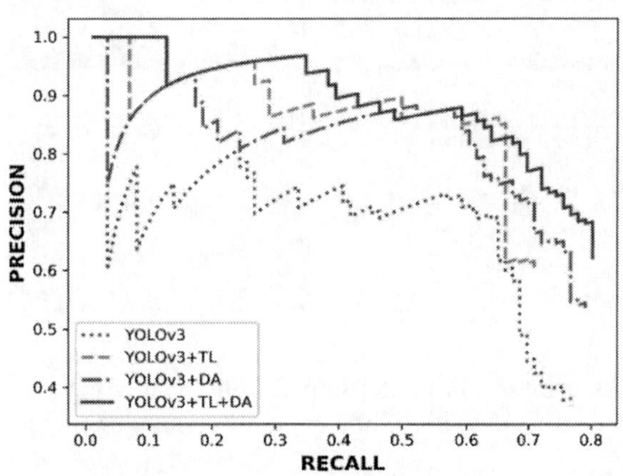

Figure 4. Precision-recall curves.

In order to better show the performance of detection, three types of results, i.e., correct detection, excessive detection and false detection are shown in Figure 5(a), Figure 5(b) and Figure 5(c) respectively. The results indicate generalization, such as the objects in row 1 and column 4, row 3 and column 3 are not present in the training set. However, there are some mistakes, which mainly reflect in the case of metal brackets and background interference. This situation can be alleviated with more training data.

IOP Conf. Series: Journal of Physics: Conf. Series **1267** (2019) 012043 doi:10.1088/1742-6596/1267/1/012043

(a)

(b)

(c)

Figure 5. Qualitative results from test data. (a) shows some results with correct detection, (b) shows some results with excessive detection, and (c) shows some results with false detection.

4. Conclusion

Foreign objects detection is an important protection measure to ensure that the normal work of the transmission lines in power systems. Compared with conventional methods, neural detection, which is a data-driven approach, can be able to deal with complex environments and save a lot of labor resources. In this paper, we apply a neural detection of foreign objects for transmission lines in power systems. Transfer learning and data augmentation are used to mitigate data shortages. Experimental results indicate that the neural detection is effective in detecting foreign objects with keeping the real-time property.

Acknowledgments

This study was funded and supported by the National Natural Science Foundation of China through Grant No.61273315.

IOP Conf. Series: Journal of Physics: Conf. Series **1267** (2019) 012043 doi:10.1088/1742-6596/1267/1/012043

References

[1] Wadell B C 1991 *Transmission line design handbook* (Artech House)
[2] Koshelev V and Kozlov D 2015 Wire recognition in image within aerial inspection application *Embedded Computing (MECO), 2015 4th Mediterranean Conference on* (IEEE) pp 159–162
[3] Krizhevsky A, Sutskever I and Hinton G E 2012 Imagenet classification with deep convolutional neural networks *Advances in neural information processing systems* pp 1097–1105
[4] He K, Zhang X, Ren S and Sun J 2016 Deep residual learning for image recognition *Proceedings of the IEEE conference on computer vision and pattern recognition* pp 770–778
[5] Girshick R 2015 Fast r-cnn *Proceedings of the IEEE international conference on computer vision* pp 1440–1448
[6] Ren S, He K, Girshick R and Sun J 2015 Faster r-cnn: Towards real-time object detection with region proposal networks *Advances in neural information processing systems* pp 91–99
[7] Ronneberger O, Fischer P and Brox T 2015 U-net: Convolutional networks for biomedical image segmentation *International Conference on Medical image computing and computer-assisted intervention* (Springer) pp234–241
[8] Badrinarayanan V, Kendall A and Cipolla R 2015 *arXiv preprint arXiv:1511.00561*
[9] Redmon J and Farhadi A 2018 *arXiv preprint arXiv:1804.02767*
[10] Redmon J, Divvala S, Girshick R and Farhadi A 2016 You only look once: Unified, real-time object detection *Proceedings of the IEEE conference on computer vision and pattern recognition* pp 779–788
[11] Redmon J and Farhadi A 2017 *arXiv preprint*
[12] Girshick R, Donahue J, Darrell T and Malik J 2014 Rich feature hierarchies for accurate object detection and semantic segmentation *Proceedings of the IEEE conference on computer vision and pattern recognition* pp 580–587
[13] Deng J, Dong W, Socher R, Li L J, Li K and Fei-Fei L 2009 Imagenet: A large-scale hierarchical image database *Computer Vision and Pattern Recognition, 2009. CVPR 2009. IEEE Conference on* (Ieee) pp 248–255
[14] Paszke A, Gross S, Chintala S, Chanan G, Yang E, DeVito Z, Lin Z, Desmaison A, Antiga L and Lerer A 2017 Automatic differentiation in pytorch *NIPS-W*
[15] Kingma D P and Ba J 2014 *arXiv preprint arXiv:1412.6980*
[16] Krogh A and Hertz J A 1992 A simple weight decay can improve generalization *Advances in neural information processing systems* pp 950–957

IOP Conf. Series: Journal of Physics: Conf. Series **1267** (2019) 012044 doi:10.1088/1742-6596/1267/1/012044

CNN-LSTM Combined Network for IoT Enabled Fall Detection Applications

Jun Xu[1, a], Zunwen He[1, b] and Yan Zhang[1, c]

[1]School of Information and Electronics, Beijing Institute of Technology, Beijing 100081, China

[a]2220170443@bit.edu.cn; [b]hezunwen@bit.edu.cn; [c]zhangy@bit.edu.cn

Abstract. An accidental fall could do a great damage to the health of elderly. Failure to provide timely assistance after a fall may cause injury or even death. In this paper, a fall detection algorithm based on Convolutional Neural Network (CNN)-Long Short Term Memory (LSTM) combined network is proposed, which makes full use of the powerful feature extraction ability of CNN and the excellent time series processing ability of LSTM. Data required by the algorithm is only the resultant acceleration from a low cost three-axis acceleration sensor. The experimental results show that compared with the algorithms based on Support Vector Machine (SVM) and CNN, the proposed algorithm has higher detection accuracy with a small data volume, which is very suitable for Internet of Things (IoT) enabled fall detection applications.

1. Introduction

As the global aging population continues to increase, more and more attention has been paid to the damage of falls on the health of elderly. According to the data from World Health Organization, about 28~35% of people over the age of 65 fall every year. For people over the age of 70, this number has increased to around 32~42%. Therefore, an efficient and accurate algorithm of fall detection is of great social significance.

At present, a large number of algorithms have been applied to fall detection, these methods can be divided into two categories according to the deployment methods, i.e., visual-based detection method and wearable device-based detection method. The visual-based method [1]-[2] acquires scene information through the camera, without the need for the elderly to wear additional equipment. Thanks to the rich information acquired by the camera, this kind of approaches can achieve a very high precision. However, video images are greatly affected by environmental factors such as light, and the scope is more limited, so visual-based methods are more suitable for special occasions such as hospital wards. Compared with the visual-based methods, wearable device-based solutions are less affected by the environment and have a wider range of application scenarios. Threshold based detection algorithm [3]-[4] has been widely used in wearable devices, where one or more thresholds are used to determine various motion states of the human body, according to some specific combinations of actions to determine whether a fall event occurred. These methods put high requirements on the installation position and mounting angle of the sensor, thus have a relatively poor robustness. With the development of machine learning technology, SVM [5]-[6], random forest and other machine learning algorithms are used for fall detection, and get better detection accuracy, but this kind of algorithms still need to manually extract features for training, which requires a lot of prior knowledge, and the process of feature extraction is cumbersome, meanwhile the computation is large.

In recent years, deep learning has also been increasingly applied to fall detection. Fakhrulddin et al. (2017) [7] proposed an algorithm that directly converts acceleration sensor data into RGB images and classifies them by CNN network, simplifying the process of data preprocessing, with a final detection accuracy of 92.3%. Deep learning based methods need to transmit data to the server through network for processing because it's difficult to be deployed directly on embedded devices. Although this solution may bring some delays, in such scenarios as nursing home, it facilitates the unified monitoring of the elderly's movement status in the park to ensure timely rescue in case of falls. However, the method in [7] still needs a large amount of data to detect a fall, which will consume a lot of network bandwidth in an IoT system.

In order to solve the problems above, we propose a fall detection algorithm based on CNN-LSTM combined network, which uses neural network to automatically extract features and classify them, avoiding the cumbersome preprocessing process. The raw data from three-axis accelerometer is firstly calculated as the resultant acceleration value to reduce data usage, and then sent to the CNN-LSTM network for classification. It only needs a small amount of data to realize the end-to-end fall detection application, which makes it very suitable for IoT applications. The accuracy evaluated on the MobiAct dataset [8] is 98.98%.

The rest of the paper is organized as follows. Section 2 describes the scheme and network structure used in this study. Section 3 mainly introduces the settings of dataset and experimental parameters. Section 4 shows the results of this study and compares them with other works. The conclusions and future works are drawn in Section 5.

2. CNN-LSTM network architecture

In this study, a CNN-LSTM network is proposed. The network architecture is shown in Figure 1.

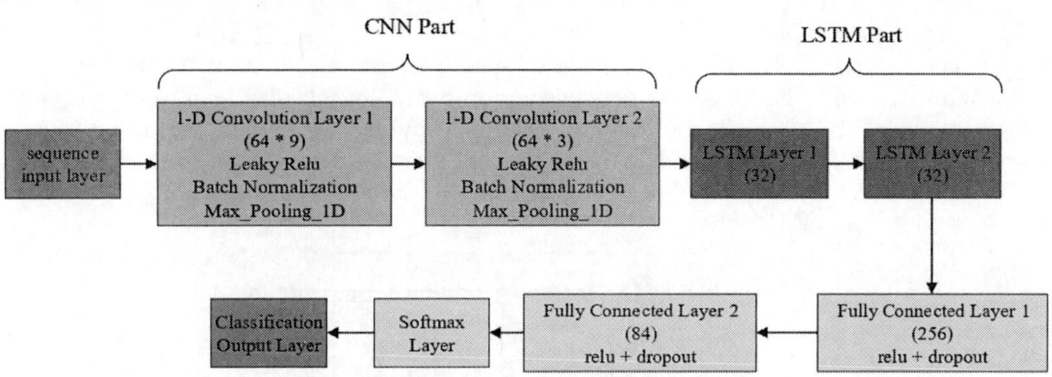

Figure 1. Proposed network architecture.

The network input layer receives the sequence of the resultant acceleration values from the sensors. Suppose a_x, a_y and a_z are the value obtained from 3-axis accelerometer, then the resultant acceleration a_r can be calculated by equation:

$$a_r = \sqrt{a_x^2 + a_y^2 + a_z^2} \tag{1}$$

In the CNN part, each convolution layer in the network consists of a one-dimensional convolution, a linear activation layer, and a maximum pooling layer. Moreover, batch normalization layer [9] is also applied to the network to optimize the convergence speed and model accuracy. The essence of batch normalization is to use the optimization to change the size of the variance and the mean position, so that the new distribution is more in line with the true distribution of the data, and the nonlinear expression ability of the model is guaranteed.

Two LSTM layers are used to ensure that the model can learn enough features. Highly abstract features are generated by the CNN and LSTM layers, and two fully connected layers with a softmax layer are used to map the learned features to the sample marker space.

The loss function in this network is given as follow.

$$\text{Cost} = -\frac{1}{N}\sum_{i=0}^{N-1}[y_i ln a_i + (1-y_i)ln(1-a_i)] + \lambda\omega_2 \tag{2}$$

where N is the number of categories, y_i is the expected output and a_i is the actual output of neurons. A L2 regularization term is used to improve generalization ability of the network, λ is the proportional parameter of the regularization term, and ω represents the weight of the network.

2.1. 1-D CNN description

CNN has been widely used in many areas for its powerful feature extraction capabilities. Two-dimensional and three-dimensional CNNs play an important role in the fields of image and video processing. Meanwhile, one-dimensional CNNs are best suited for learning the feature in 1-D time series.

In this study, two 1-D convolutional layers are used to extract multiple features in acceleration sequence data. These convolutional layers use the Leaky ReLu activation function to avoid the problem of gradient disappearance. The Leaky ReLu is given as:

$$f(x) = \begin{cases} x, & x > 0 \\ \lambda x, & x \leq 0 \end{cases} \tag{3}$$

The Leaky ReLu activation function was first proposed in the acoustic model, which has a better performance for 1-D time series. Each layer of convolution is followed by a maximum pooling layer of length 2, which reduces the size of the feature vector, so that the subsequent network obtains higher level abstract feature information.

2.2. LSTM Network description

RNN plays an important role in the processing of speech signals and time series. However, it has a poor ability to deal with the long-term dependency problems, which may lead to problems such as gradient exploding. LSTM solves the long-term dependency problems of the sequence by adding a memory gate to the RNN, resulting in better results. The structure of LSTM is shown in Figure 2.

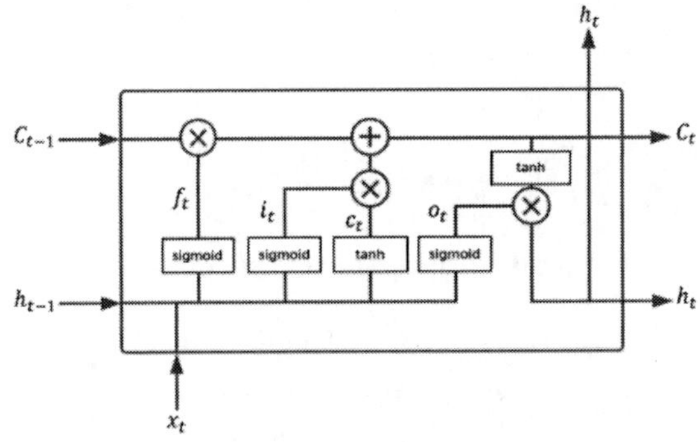

Figure 2. Structure of LSTM.

The output response h_t of the nerve cell at time t is determined by the input x_t at that moment and the output response h_{t-1} at the previous moment:

$$h_t = \theta(W_{xi}x_t + W_{hh}h_{t-1} + b_h) \tag{4}$$

where θ(·) represents the activation function, b_h represents the offset vector, W_{xh} represents the weight matrix between the input and the neuron, and W_{hh} represents the cyclic weight matrix of the inner self-loop of the neuron. The operations of the various parts of the neuron at time t are as follows.

$$i_t = \sigma(W_{xi}x_t + W_{hi}h_{t-1} + W_{ci}c_{t-1} + b_i) \tag{5}$$

IOP Conf. Series: Journal of Physics: Conf. Series **1267** (2019) 012044 doi:10.1088/1742-6596/1267/1/012044

$$f_t = \sigma(W_{xf}x_t + W_{hf}h_{t-1} + W_{cf}c_{t-1} + b_f) \quad (6)$$
$$c_t = f_t \otimes c_{t-1} + i_t \otimes tanh(W_{xc}x_i + W_{hc}h_{t-1} + b_c) \quad (7)$$
$$o_t = \sigma(W_{xo}x_t + W_{ho}h_{t-1} + W_{co}c_{t-1} + b_o) \quad (8)$$
$$h_t = o_t \otimes tanh(c_t) \quad (9)$$

where $\sigma(\cdot)$ denotes the sigmoid function: $\sigma(x) = (1 + e^{-x})^{-1}$, W_{**} denotes a weight matrix between any two units, and b_* denotes the offset of a certain unit.

3. Experiment setup

3.1. Dataset

In this section, the feasibility and performance of the proposed approach are evaluated on the public MobiAct dataset. The MobiAct dataset uses the built-in LSM330DLC inertial sensor module from Samsung Galaxy S3 smartphone to collect data from three-axis accelerometer, gyroscope and direction meter at a frequency of 200 Hz.

In this experiment, the data set is down sampled to obtain the 50 Hz sampling rate signal, three-axis acceleration data is used to calculate the resultant acceleration. Other data from the gyroscope and direction meter is ignored to make sure that the model uses the least amount of data. Fall behaviour and other six kinds of Activities of Daily Livings (ADL)s are selected to train the model. Figure 3 shows the acceleration curve of part of the behaviours.

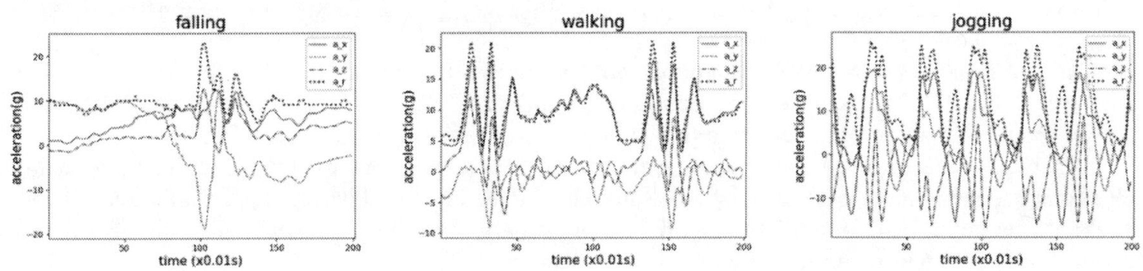

(a) (b) (c)

Figure 3. Acceleration curves of different ADLs: (a) the acceleration curve of falling, (b) the acceleration curve of walking, and (c) the acceleration curve of jogging.

Since the fall process lasts about 2 to 4 seconds, we intercept 4 seconds from the moment the fall occurs, i.e., 200 sampling points. The experimental data set is divided into train set (80%) and test set (20%), a total of 5,891 train samples and 1473 test samples are obtained. During the training process, 15% of the train set is used as validation set to verify the generalization ability of the model.

3.2. Parameter setup

In this experiment, we use the Keras deep learning framework to build the CNN-LSTM network model. The network parameters are shown in table 1. We also train a pure LSTM and a 1-D CNN with the same parameters to make comparisons.

Table 1. Parameter setup for training.

Parameter	Value
Epoch	300
Optimizer	Adam
Batch size	32
Learning rate	0.001

IOP Conf. Series: Journal of Physics: Conf. Series **1267** (2019) 012044 doi:10.1088/1742-6596/1267/1/012044

4. Results and Discussion

We have tested several network structures on MobiAct dataset. The accuracy curve of the training process is shown in Figure 4. The experimental results show that the accuracy of CNN-LSTM network is higher than that of 1D CNN and LSTM. In addition, as shown in the results, the convergence speed of the CNN-based network is much faster than that of the LSTM network.

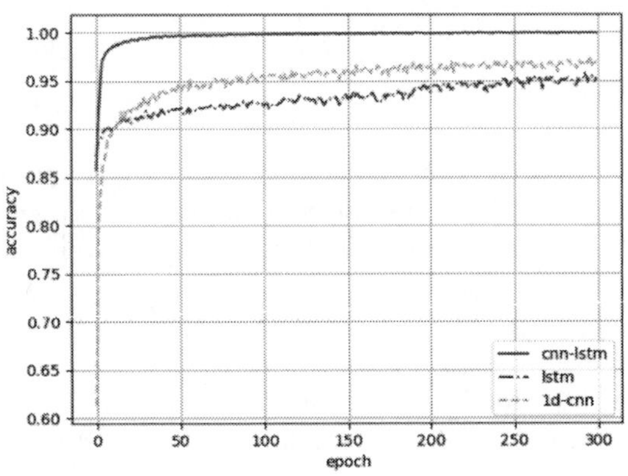

Figure 4. Accuracy curve of the training process.

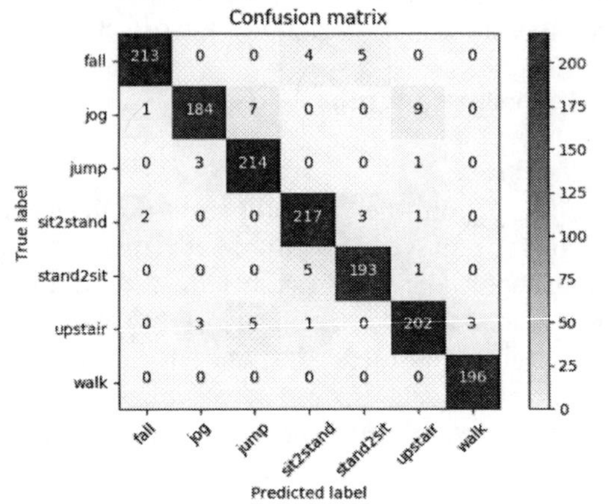

Figure 5. Confusion matrix.

The test results on CNN-LSTM network are shown as a confusion matrix in figure 5. It is shown that the CNN-LSTM achieves a classification accuracy of 98.98% in seven types of daily behaviours. As for the fall behaviour, the detection precision and specificity are 98.61% and 99.76%, respectively.

Meanwhile, we compare the results with several other algorithms evaluated on MobiAct dataset. We compare the number of categories, the sampling frequency and the accuracy of ADL classification. It should be noted that some of the algorithms are binary classification method, so the classification accuracy is consistent with the precision of fall detection.

In particular, to evaluate the complexity of the corresponding solutions, the number of sampling points used per seconds and the required sensors are also included in the comparison range to assess whether the model is suitable for IoT applications. The results of each method are shown in table 2.

Table 2. The comparison of fall detection methods.

Study	categories	sampling rate	Mothed	Classification accuracy	Sensors	Points per seconds
Bayat et al. 2014 [5]	4	100 Hz	SVM	91.15%	Accelerometer	-
Ivan et al. 2018 [10]	5	100 Hz	DNN	89.51%	accelerometer magnetometer gyroscope	-
Xiang et al. 2017 [7]	2	50 Hz	CNN	92.30%	Accelerometer	150
Ours	**7**	50 Hz	CNN-LSTM	**98.98%**	Accelerometer	**50**

It can be seen from the result that the proposed algorithm achieves a higher precision than most other algorithms. Among the listed deep learning based algorithms, CNN-LSTM requires a least amount of data and sensors, which makes it possible for IoT applications.

5. Conclusion

In this paper, we proposed a fall detection algorithm based on CNN-LSTM network. The average classification accuracy of 7 types of ADLs including fall on MobiAct reaches 98.98%. And the detection precision and specificity for fall detection are 98.61% and 99.76%. The method does not need to perform artificial feature extraction, the detection results can be directly obtained from the raw data from the low cost three-axis accelerometer, thereby realizing end-to-end application.

It is worth noting that the algorithm requires only the resultant acceleration rather than 3 axis acceleration data, which means a very low data volume. In the following works, the CNN-LSTM network will be deployed on a gateway of an IoT system to achieve an AI-enabled IoT (AIoT) application, which will not only bring more convenience to places such as nursing home but also collect large amounts of measured data for academic research.

Acknowledgement
This work was supported in part by National Natural Science Foundation of China (No. 61871035, 61201192), National High Technology Research and Development Program of China (No. 2015AA01A706).

References

[1] Anishchenko L 2018 Machine learning in video surveillance for fall detection *Ural Symposium on Biomedical Engineering, Radioelectronics and Information Technology (USBEREIT)* pp 99-102

[2] Ge C, Gu I Y and Yang J 2017 Human fall detection using segment-level cnn features and sparse dictionary learning *IEEE 27th Int. Workshop on Machine Learning for Signal Processing (MLSP)* pp 1-6

[3] Razum D, Seketa G, Vugrin J and Lackovic I 2018 Optimal threshold selection for threshold-based fall detection algorithms with multiple features *41st Int. Convention on Information and Communication Technology, Electronics and Microelectronics (MIPRO)* pp 1513-6

[4] Guo H W, Hsieh Y T, Huang Y S, Chien J C, Haraikawa K and Shieh J S 2015 A threshold-based algorithm of fall detection using a wearable device with tri-axial accelerometer and gyroscope *Int. Conf. on Intelligent Informatics and Biomedical Sciences (ICIIBMS)* pp 54-7

[5] Bayat A, Pomplun M and Tran D A 2014 A Study on Human Activity Recognition Using Accelerometer Data from Smartphones *Procedia Computer Science* **34** 450-7

[6] Wang R, Zhang Y, Dong L, Lu J, Zhang Z and He X 2015 Fall detection algorithm for the elderly based on human characteristic matrix and SVM *15th Int. Conf. on Control, Automation and Systems (ICCAS)* pp 1190-5

[7] Fakhrulddin A H, Fei X and Li H 2017 Convolutional neural networks (CNN) based human fall detection on Body Sensor Networks (BSN) sensor data *4th Int. Conf. on Systems and Informatics (ICSAI)* pp 1461-5

[8] Vavoulas G, Chatzaki C, Malliotakis T, Pediaditis M and Tsiknakis M 2016 The MobiAct Dataset: Recognition of Activities of Daily Living using Smartphones *Proc. of the 2nd Int. Conf. on Information and Communication Technologies for Ageing Well and e-Health (ICT4AWE)* pp 143-51

[9] Ioffe S and Szegedy C 2015 Batch normalization: accelerating deep network training by reducing internal covariate shift *Proc. of the 32nd Int. Conf. on Machine Learning (ICML)* vol 37 (Lille, France: JMLR.org) pp 448-56

[10] Pires I M, Garcia N M, Pombo N, Flórez-Revuelta F, Spinsante S and Teixeira M C 2018 Identification of activities of daily living through data fusion on motion and magnetic sensors embedded on mobile devices *Pervasive and Mobile Computing* **47** 78-93

IOP Conf. Series: Journal of Physics: Conf. Series **1267** (2019) 012045 doi:10.1088/1742-6596/1267/1/012045

Parametric Design of Grasshopper Based on Moulding Characteristics of Longitudinal Profile of Shoe Last

Yujing Tian[1], Yuxin Miao[1], Ying Yu[1,*] and Ziran Zhang[2]

[1] Donghua University, Fashion and Design College, 200051 Shanghai, P.R. China
[2] Shanghai University of Engineering Science, School of Art and Design, 201620 Shanghai, P.R. China

*yuyin513@163.com

Abstract. In order to achieve low-cost, high-efficiency customized footwear design and service to meet the individual needs of customers, quickly producing various types of shoe last based on different needs has become the key to customization. According to the moulding characteristics of the longitudinal profile of the shoe last, the key characteristic parameters and morphogenesis logic are found. Combined with the parameterized nonlinear thinking feature, the parameter logic is constructed. Rhino's parametric design plug-in Grasshopper is used to develop the shoe last design process. The automatic moulding process of the sectional view, combined with the parametric design of the shoe last and the cross-section of the shoe last, can form a quick custom shoe last method. A professional and rapid design method is formed to enhance the design and production efficiency of customized shoe lasts.

1. Introduction

Under the environment of market globalization, economic regionalization, information network and individualized demand of shoe industry, enterprises urgently need to explore an efficient and flexible design and processing method as a new competitive means to meet the needs of consumers[1]. Shoe last, as the matrix of a shoe, is also the soul of a shoe and the moulding of a shoe. It not only determines the shape and style of the shoe, but also determines the comfort and other special functions of the shoe. In the shoe last design process, quickly producing various types of shoe last based on different needs has become the key to customization.

As a nonlinear design method, parametric design can form a variety of constraint relations in the model through constraint equations, and generate different design forms quickly by using the change of parameters, so as to improve the speed of model generation and modification[2-3]. Although some scholars have proposed the use of CAD, Catia, UG and PRO-E to parametrically design shoe last by interpreting the distribution of plantar pressure and establishing the constraint of nonlinear sizes or other methods, designers lack direct visual experience in the design process. However, the parametric design of the Grasshopper plug-in can visualize the design process and record the entire modeling process at the same time. Designers can easily change the form of the model by adjusting the parameters[4-6].

According to the moulding characteristics of the shoe last profile, key feature parameters and morphological generation logic, and parameterized nonlinear thinking features, Rhino software's parametric design plug-in Grasshopper can build an automatic prototyping program for the

IOP Conf. Series: Journal of Physics: Conf. Series **1267** (2019) 012045 doi:10.1088/1742-6596/1267/1/012045

longitudinal profile of the shoe last as a feasible design method to enhance the design and production efficiency of customized shoe lasts.

2. Traditional shoe last longitudinal profile moulding process

Based on the foot shape, the shape of shoe last must follow the foot shape, so the design of the shoe last is closely related to the parameters of the foot[7]. This article gives an example of a standard female shoe last with the length of 240mm, the last toe allowance of 20mm and the heel-height of 40mm to briefly introduce the drawing process of shoe last longitudinal section, as follows:

2.1. Take the auxiliary points of the profile of the shoe last for moulding

Draw a horizontal line OA of 240mm length and divide it into 12 equal parts, 20mm each square; determine the point B of 135mm from the rear end point O as the tread point; heel height OC, 40mm; Toe spring AD, 9mm; Draw an arc with a radius of 135mm at tread point B, find the intersection E of the horizontal line of the heel height and the arc, as the back point of the last. Connect BE and divide it into 5 parts equally; connect BD and DE; find 2/5 of BE to make a perpendicular to DE, and find the midpoint F; over point E draw a vertical line of DE drawn up 60mm, and divide it into 3 segments, and get G, H and I; I 'is 4mm ahead of point I, G' is 4.5mm behind point G, and GG 'is the back tolerance. Draw an arc with a radius of 90mm over I '; Draw an arc with a radius of 85mm over F, and find the intersection point J of two arcs; Move the rear straight line EI forward to 2/3 of the DE in parallel; move DE up to I and find the intersection K with the parallel line just drawn; the toe depth LD, 15mm; the toe width LM, 20mm; connect MI', intersect with the vertical line of the 3/12 bottom line at N, as shown in Figure 1.

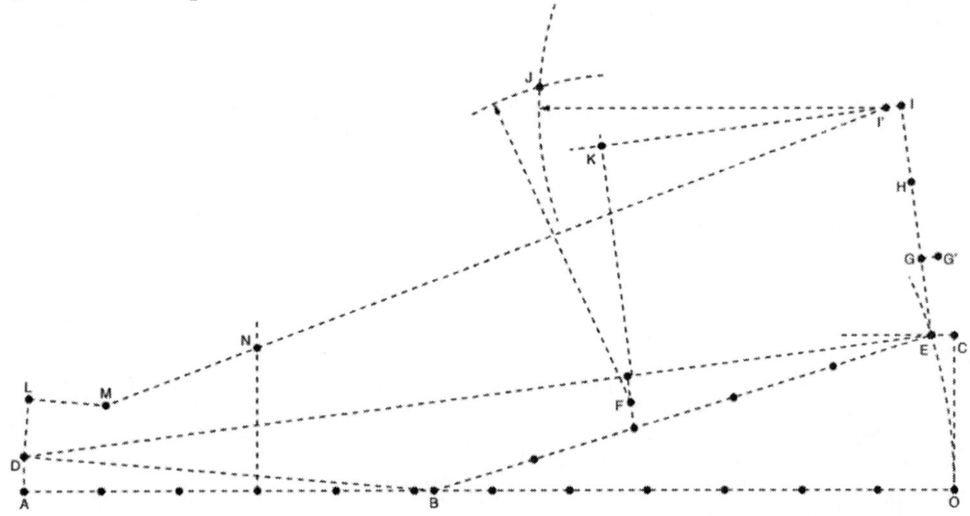

Figure 1. Take the auxiliary point of the profile of the shoe last for moulding.

2.2. Generate profile auxiliary line of shoe last profile

As shown in Figure 2, E, G', H, I', K, J, N, M, L, D, B, F, and E are connected with a polyline according to the direction of the arrows to generate an auxiliary polyline of the longitudinal profile of the shoe last. That is the basic frame of the longitudinal profile of the shoe last.

IOP Conf. Series: Journal of Physics: Conf. Series **1267** (2019) 012045 doi:10.1088/1742-6596/1267/1/012045

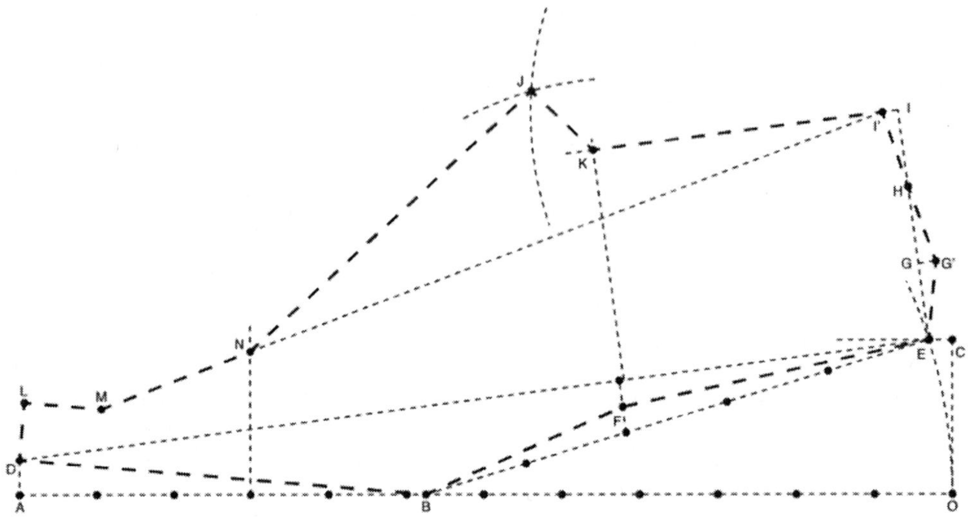

Figure 2. Generate profile auxiliary polyline of shoe last profile.

2.3. Moulding shoe last longitudinal section

As shown in Figure 3, according to the curvature reference directions indicated by the arrows, the auxiliary polyline of the longitudinal profile of the shoe last is converted into a curve, and the profile of the standard female shoe with the length of 240 mm, the last toe allowance of 20 mm, and the heel height of 40 mm is shown.

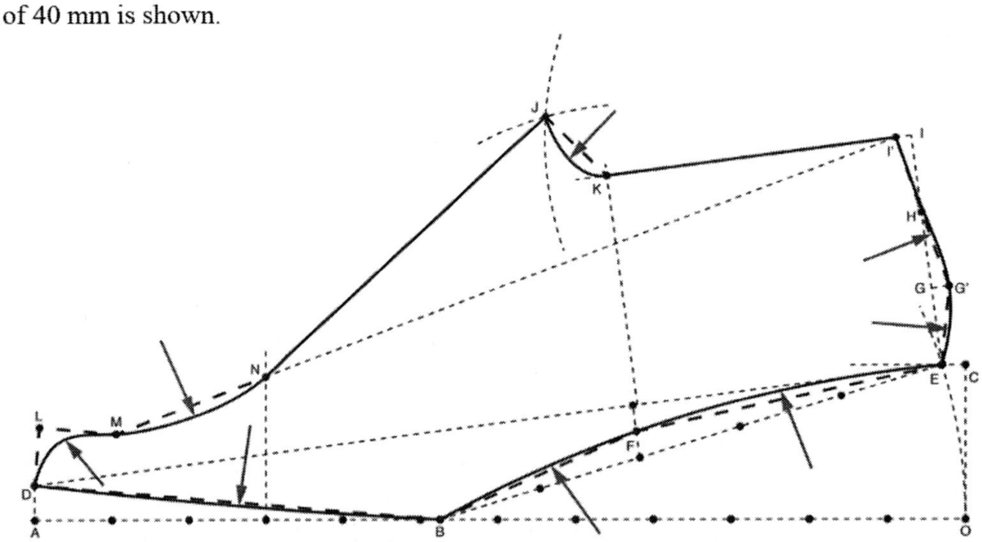

Figure 3. Moulding shoe last longitudinal section.

3. Extracted key characteristic parameters of profile of shoe last

By analysing the drawing process of the traditional shoe last longitudinal section drawing, 10 key characteristic parameters and characteristic points of the longitudinal profile of shoe last are extracted, which are shoe last length OA, heel height OC, back height EI, last toe allowance LM, toe depth DL, toe spring AD, back tolerance GG', back point of topline I', tread point B, and length of topline I'J, where the characteristic parameters 1-6 can be adjusted according to the measurement data of the foot type during customization. In this study, the given reference value is used as the known data. The characteristic parameters 7-9 are based on the proportional algorithm of the measured data of the foot type. The characteristic parameter 10 is the reference value of the known national standard data, which are label values used by last factories, as shown in Table 1.

Table 1. Summary table of moulding characteristic parameters and characteristic points of longitudinal profile of shoe last.

No.	Parameter name	Calculation method	Length (unit: mm)	Marked in the figure
1	Shoe last length	Known data, can be adjusted according to the user's design requirements and the measurement data of the foot type during customization	240	OA
2	Heel height		40	OC
3	Back height		60	EI
4	Last toe allowance		20	LM
5	Toe depth		15	DL
6	Toe spring		9	AD
7	Back tolerance	Foot length *2%	4.5	GG'
8	Back point of topline	Foot length *2%	4.5	II'
9	Tread point	Foot length *59%	135	OB
10	Length of topline	Known national standard data reference value	90	I'J

4. Parametric logic construction based on Grasshopper

4.1. Hierarchical logic combing of characteristic parameters

As shown in Table 2, according to the importance level of the shape control the parameters are divided into a primary parameter, secondary parameters, and tertiary parameters. The first-level parameter refers to the length of the foot, which is the basic parameter that should be determined firstly to draw the longitudinal section of the shoe last; the secondary parameters are the key characteristic parameters extracted through the drawing process of the standard shoe last section, but can also be modified by the target style through the process of customizing the shoe last; the third-level parameters are based on the secondary parameters, for the accuracy of the curve shape of the longitudinal profile, and the shape control points are added for the customization process in order to modify the shape details. The corresponding step is to connect the key feature points with the last curve in the shoe last drawing. The shape of the connecting curve is often determined by the personal experience of the drafter, so there is some uncertainty. The third-level parameters have no determined parameter values, and it is still necessary to adjust the location of each point through a certain experience to adjust the curves of each segment. The logical relationship is shown in Figure 4.

Table 2. Parametric variables required for longitudinal profile drawing of shoe last using Grasshopper.

Parameter level	Parameter name	Effect
Primary parameter	Foot length	Control overall foot length for different users
Secondary parameter	Heel height (back point of the last)	Key characteristic parameters of longitudinal profile of shoe last
	Back height	
	Back tolerance (maximum point of heel curve)	
	Back point of topline	
	Length of topline(front point of topline)	
	Last toe allowance	
	Toe depth	
	Toe spring(front point of the bottom)	
	Tread point	

IOP Conf. Series: Journal of Physics: Conf. Series **1267** (2019) 012045 doi:10.1088/1742-6596/1267/1/012045

Tertiary parameter	Morphological control point of back	Controls the shape of the back curve between the maximum point of heel curve and the back point of topline
	Morphological control point of front of topline	Control the shape of the curve behind the front point of topline
	Morphological control point of last surface	Control the curve shape of the shoe last surface at the metatarsal-phalange
	Morphological control point of toe	Control the shape of the toe curve
	Morphological control point of shoe last bottom	Control the curve of shoe last ahead of tread point

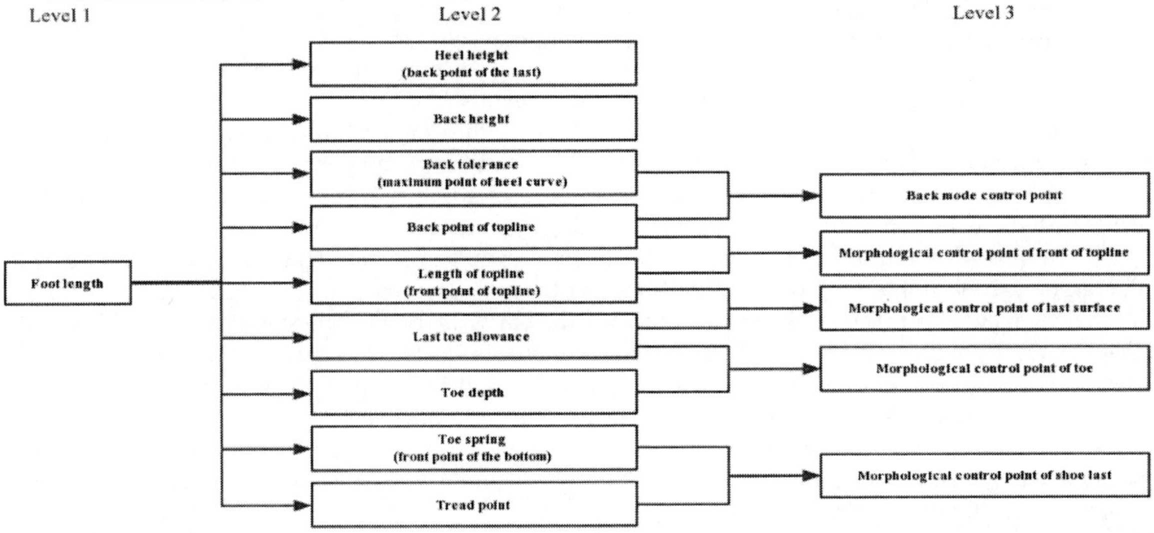

Figure 4. Logical relationship diagram of primary, secondary and tertiary parameters.

4.2. Edit input logic relationship through Grasshopper

The above characteristic parameters are logically connected in combination with the conventional moulding process of the longitudinal profile of shoe last. It includes the length calculation method and shape connection mode of each feature part. The longitudinal profile of the shoe last is obtained according to the personal data of the customized measurement and individual requirements, and the basic operation method is used for logic construction in the parametric plug-in Grasshopper of the three-dimensional modeling software Rhino. The result can be seen instantly on the Rhino interface during the construction process. The final program interface of the logical relationship and the longitudinal profile of the shoe last is shown in Figure 5.

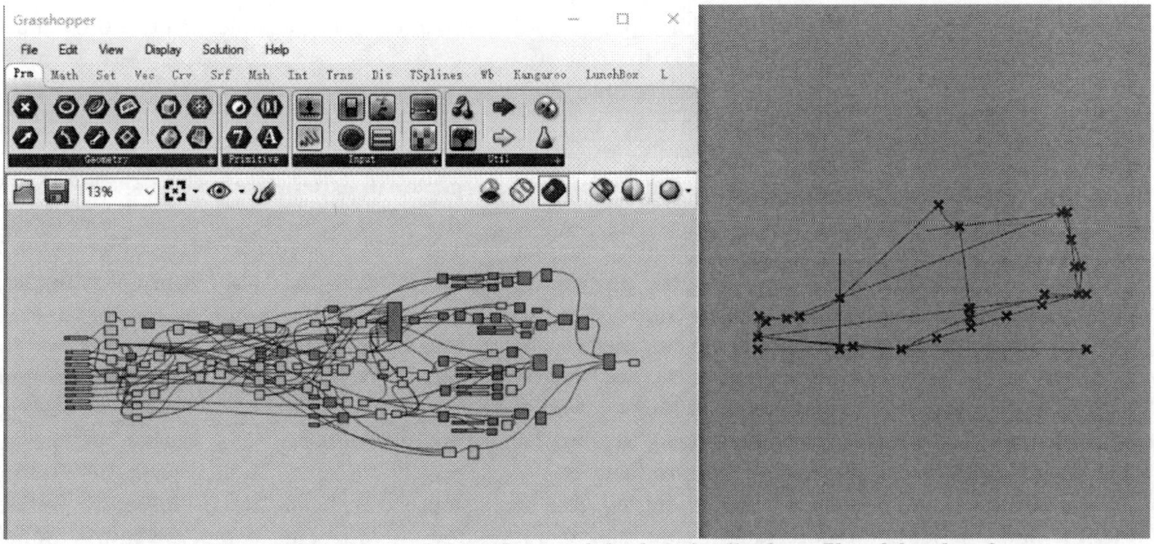

Figure 5. Grasshopper work interface and the longitudinal profile of the shoe last.

4.3. Operation and inspection of drawing effect

After modifying the shape of the curve by three levels of parameters to make the last profile accurate, the final last curve will be baked separately to the Rhino interface, as shown in Figure 6. In the Rhino interface, you can open the edit point to verify again. In order to facilitate the next 3D design and production of the last, the corresponding formats can be derived based on different moulding software.

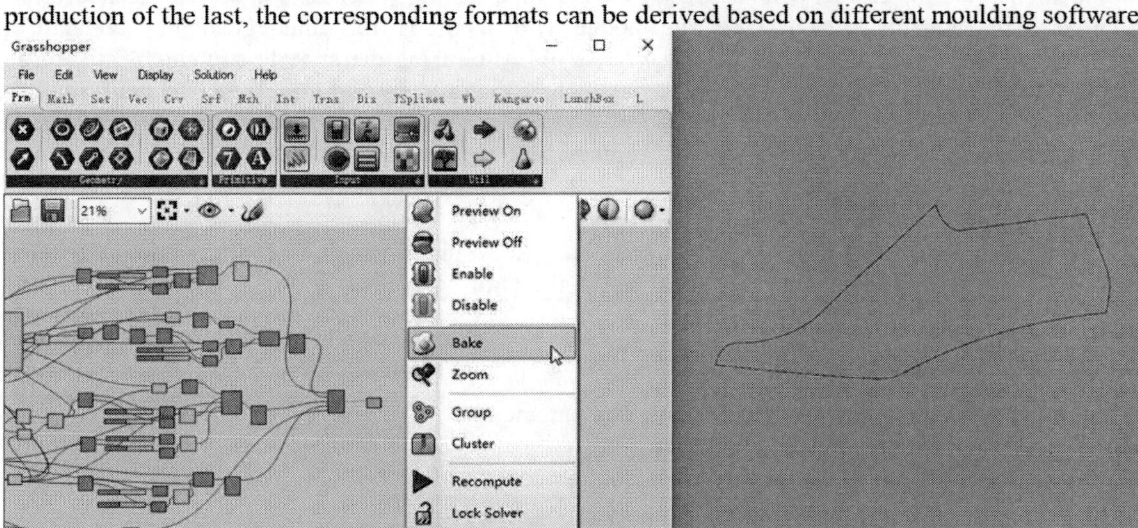

Figure 6. Bake the last profile curve in Grasshopper to Rhino.

5. Comparative analyses of the traditional longitudinal section drawing method and the parametric drawing method

5.1. Step comparison

According to the traditional custom process of personal shoe last, the steps of drawing the traditional shoe last profile are as follows:

(a) Measure the target user's foot length, and according to the individual and the style needed determine the length of the foot, heel height (back point of the last), back height, the maximum point

IOP Conf. Series: Journal of Physics: Conf. Series **1267** (2019) 012045 doi:10.1088/1742-6596/1267/1/012045

of the heel curve, back point of the last, back point of topline, front point of topline, last toe allowance, toe depth, toe spring (front point of the last), tread point.
(b) Measure and obtain the position of each feature point.
(c) Connect each feature point with a polyline.
(d) Connect each feature point with a smooth curve.
(e) Adjust the curve shape, curvature, etc. according to personal experience to make the shape smoother and more accurate.
(f) Scan the draft and import into the computer, and then trace the curve for further 3D production.

Using the parameterized shoe last profile procedure that has been already written in Grasshopper, the steps to customize the personal last profile are as follows:
(a) Measure and determine characteristic parameters, as in the traditional method.
(b) Input the measurement data directly into the primary and secondary parameters in the program, and the initial shoe last longitudinal profile curve is automatically generated.
(c) Adjust the third-level parameters to make the curve shape more accurate.
(d) Export curves directly for further 3D production.

After comparison, the parametric drawing process has less process of measure each feature points, polyline and smooth curve connection, computer scanning and curve tracing compared to traditional drawing.

5.2. Efficiency comparison

The parametric drawing process is only two steps less than the traditional drawing process, but the efficiency is completely different. In order to avoid the influence of the painter's experience, this article invites a shoe learner who understands the shoe lasting process, but the experience is not familiar. The primary students use the above two methods for time-consuming comparison experiments, and finally found that in the process of customizing the shoe last moulding measuring and determining the various characteristic parameters required by the target user to generating an electronic version of the longitudinal profile of the shoe last that can be further used, the cumulative time of the conventional method is about 37 minutes; and the parametric method takes two fewer steps and takes about 9 minutes and 30 seconds, greatly reducing the time it takes to customize a brand new shoe last profile， as shown in Table 3.

Table 3. Time-consuming comparison of traditional and parametric methods for drawing longitudinal profile of shoe last.

	Time consuming	Traditional drawing method	Parametric drawing method
Measure and determine the various characteristic parameters required by the target user	5 mins	√	√
Measure and obtain the position of each feature point	10 mins	√	×
Connect each feature point with a polyline	2 mins	√	×
Input the measurement data directly into the primary parameter and the secondary parameters in the program, and the preliminary shoe last longitudinal profile curve is automatically generated	2 mins	×	√
Connect various feature points with smooth curves	5 mins	√	×
Adjust the curve shape, curvature, etc. according to personal experience to make the shape smoother and more accurate	5 mins	√	×
Adjust the third-level parameters to make the curve shape more accurate	2 mins	×	√
Scan the draft and import into the computer, and then trace the curve for further 3D production	10 mins	√	×

Export curves directly for further 3D production	0.5 mins	×	√
Cumulative time		37 mins	9 mins 30 secs

5.3. Result comparison
In order to compare the differences between results realized by two methods, this article compares the results of shoe last longitudinal profile drawn by primary trainees with two different methods mentioned above in 5.2. It is found that the effect of parametric drawing is smoother, while the effect of drawing by traditional method still has much sketch mark. It can be seen that it is much easier for a normal primary person to achieve more precise and more beautiful shoe last longitudinal profile as shown in Figure 7.

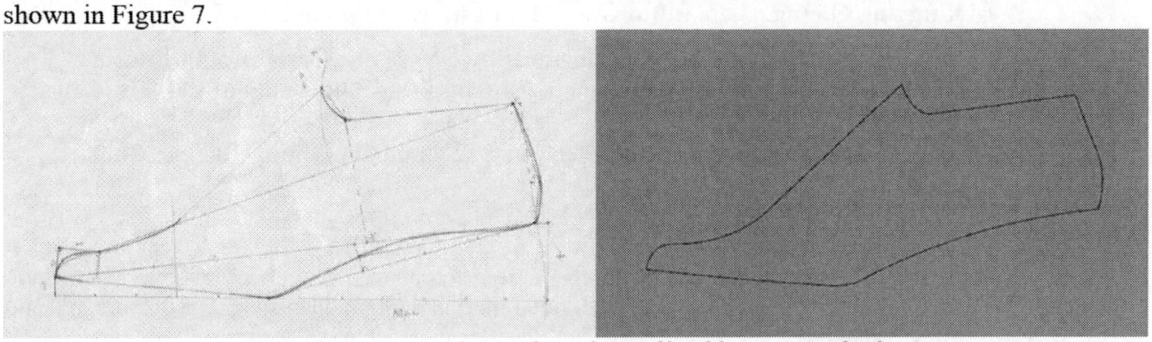

Figure 7. Comparison of results realized by two methods.

6. Summary
The aesthetic and comfort of the customized shoes depend on the shoe last. In order to ensure the accuracy of the shape, it is a useful method to follow the moulding characteristics of the longitudinal section of the shoe last and combine with the parametric design plug-in Grasshopper to construct the automatic moulding program of the last profile. The efficiency of the shoe last design is greatly improved, and it is also a technological innovation with commercial value, which will bring more possibilities for the personalized design of shoes in the future.

Acknowledgments
Project Support by the Fundamental Research Funds for the Central Universities, Shanghai Style Fashion Design &Value Creation Collaborative Innovation Center & the Fundamental Research Funds for the Central Universities & Shanghai Summit Discipline in Design (DB18212).

References
[1] Pandremenos J, Georgoulias K, Chryssolouris G, Jufer N and Bathelt J 2016 a shoe design support module towards mass customization *Technology Management Conf.* (Trondheim: IEEE)
[2] Dai X 2016 the application and exploration of parametric design used in product design based on grasshopper *Des.* 122-3
[3] Chen J 2015 *Art J.* 111-5
[4] Sun C 2018 *Research on Parametric and Rapid Manufacturing of 07B Shoe Last Model* Hebei University of Science & Technology
[5] He S and Liu B 2017 parametric design of shoe last based on feature size *J. Mech. Eng.* **53** 118-27
[6] Hu X, Zhao B, Zhang Q and Dong L 2014 *China Mech. Eng.* **25** 3044-9.
[7] Mahmoodi-Rad A, Molla-Alizadeh-Zavardehi S, Dehghan R, Sanei M and Niroomand S 2015 *Int. J. Adv. Manuf. Technol.* **85** 455-67

A Novel Line Feature Description and Matching Method for Visual-Aided Inertial Navigation System

Xingxing Guang [1,2,*], Yanbin Gao [1], Pan Liu [3] and Guangchun Li [1]

[1]College of Automation, Harbin Engineering University, Harbin 150001, China;
[2]Department of Electrical and Computer Engineering, University of Calgary, Calgary, AB T2N1N4, Canada;
[3]Beijing Institute of Control and Electronic Technology, Beijing 100032, China;

[*]Correspondence: guangxingxing@hrbeu.edu.cn

Abstract. With the rapid development of machine vision technology, more and more attention has been paid to the visual-aided inertial navigation system. It is important that to extract and track the line features at the dynamic situation in the visual-aided inertial navigation system which is based on visual line feature information to compensate attitude errors. A novel line feature description is proposed that use the SURF points to mark the LSD lines. Then, through coarse matching and fine matching, the function of continuously tracking the one line features in different images was realized. These line feature description and tracking method are applied in the visual-aided inertial navigation system, and its effectiveness is verified by the vehicle experiment.

1. Introduction

It is known the reason why GPS cannot offer accurate position information at the center of city is that denseness buildings will obstruct some, or all, of the satellite signals. Do not worry about it, there are other methods used to position and navigate, like inertial navigation system (INS), and other methods that fusion inertial and other sensors into the integrated navigation system. INS is to obtain the motion state and position of the object by integrating the measurement information of gyroscopes and accelerometers. However, the integral calculation method will bring a serious error that increases with time. The most widely used method is the INS and other sensors integrated navigation system when GPS is unavailable.

One of the most widely method is inertial and doppler radar integrated navigation system [1][2][3]. This method used the radar information to compensate the velocity error of Strap-down Inertial Navigation System (SINS). Celestial and inertial integrated navigation system is a good method to improve the accuracy of INS [4][5][6]. However, complex environment and severe weather have great influence on the Celestial sensor. INS/Wifi hybrid localization system is used to smart-phone and robotic localization indoors [7][8][9]. This method will be affected by Wifi signal quality and vehicle speed when it used outdoors. Another important method is inertial/visual integrated navigation system [10][11]. With the development of computer vision, visual technology will be more applied in the navigation system.

In reference [11], we proposed a method combining inertial with visual to solve the positioning and navigation problem when GPS is unavailable. The visual information to compensate the attitude error of SINS. However, we just discussed the static condition. At motion condition, how to recognize line feature is not only to get the line feature information, it is also to track one line feature during it appears to disappears. In this paper, we proposed a new method to describe line features, to ensure the visual-aided inertial navigation system works at motion condition.

2. How to describe the line feature

First of all, to extract the LSD features [12] and SURF point features , respectively. These two features are used to define the new description line features in dynamic condition. Through Coarse matching and fine matching, the function of continuously tracking the same line features in dynamic images was realized.

2.1. The line feature description method

Hough transform is used to extract the line feature for visual-aided inertial navigation system when vehicle stops [11]. In this study, we use the faster and more accurate line segment detector, LSD [12], to replace Hough transform. L is the line feature symbol, it includes two parameters of line feature, θ and ρ [13].

Images in vehicle data recorder is changing with the vehicle moves. So that the line features in different time are not entirely same. Even if a line feature exists in different images at the different time, its feature angle in different images may be different. To describe the same line feature in different images, we introduce SURF [14] points with rotational invariant to label line feature. Set P_{SURF} includes all of the SURF points in one image.

In one image, for each line feature L, the neighbourhood S_L is the set includes the pixels located in the rectangle H_L. H_L has the same midpoint and slope with L, and its width and length are $2d$ and $L+2d$, where d defined as the neighborhood width parameter. So S_L can be described as below:

$$S_L : \{Pixel | Pixel \subset I, Pixel \subset H_L\} \tag{1}$$

Therefor, we define the set S_{LP}, includes all pixels located sets both P_{SURF} and S_L.

$$S_{LP} : \{p_i | p_i \in S_L, p_i \in P_{SURF}, i = 1,2,\dots,n\} \tag{2}$$

Then, we use both L and S_{LP} to describe the line feature at dynamic condition. The description is named as line marked by SURF points. For convenience, we still use L to name line features.

2.2. Line feature matching

Line feature matching includes two steps: Coarse matching and Fine matching.

2.2.1. Coarse matching rule. Through SURF feature matching between the neighbouring two images' S_{LP}. Lines corresponded to the matched SURF feature points were classified as the same line feature.

After coarse matching, we can get the appearance time and disappearance time of the line feature L_j.Define the existence time interval of L_j as Δl_j.

The result of coarse matching is expressed as $L_j : \{l_j^1, l_j^2, \dots, l_j^{\Delta l_j}, L_m^x\}$, where L_m^x is the lines set which is not the same line with L_j, $m \neq j$. The lines L_m^x mixed into the set of L_j is cased by the matched SURF points in different images marked with different lines. In order to solve this problem, we will continue to fine matching.

2.2.2. Fine matching rule. For coarse matching results error that put the different line feature into one set, we give a precise matching rule. For SURF point p_i

$$f = \alpha D(p_i, L_j) + (1-\alpha)\upsilon(p_i, P_{SURF}) \tag{3}$$

IOP Conf. Series: Journal of Physics: Conf. Series **1267** (2019) 012046 doi:10.1088/1742-6596/1267/1/012046

Where f is the trust value of fine matching. Line feature $L_j : u\sin\theta_j + v\cos\theta_j - \rho_j = 0$, SURF point $p_i(u_i, v_i)$, $D(p_i, L_j)$ is the distance between p_i and L_j, $D(p_i, L_j) = |u_i \sin\theta_j + v_i \cos\theta_j - \rho_j|$. We introduce $\upsilon(P_i, M_P)$ to enhance the robustness of the matching process [15].

$$\upsilon(P_i, M_P) = -\sum_{j=1}^{N_{\Delta l_j}} N_{Pi}^{\Delta l_i} \log N_{Pi}^{\Delta l_i} \tag{4}$$

Where, during Δl_j, the number of line features is $N_{\Delta l_j}$, the times of SURF point P_i is $N_{Pi}^{\Delta l_i}$.

The description of the SURF feature is a 16×4 vector, in which four kinds of intensity variation: $\sum dx$, $\sum dy$, $\sum|dx|$ and $\sum|dy|$, are calculated over sixteen regions. We use P_i, a 64×1 vector to store the 1-64 element in the description of SURF, suppose there are N SURF features during Δl_j, put them in matrix M_P,

$$M_P = [P_1, P_2, \ldots, P_N] \tag{5}$$

Equation (4) is built in the spirit of Fisher information which is a classic method to evaluate the compactness of a probability distribution. In general, a low score of υ implies that elements in SURF point P_i tends to be identical, indicating that the feature is less distinctiveness. In the process of matching identity lines, the matching results with the higher f value is preferred.

3. Visual-aided inertial navigation system

3.1. The visual-aided inertial navigation system

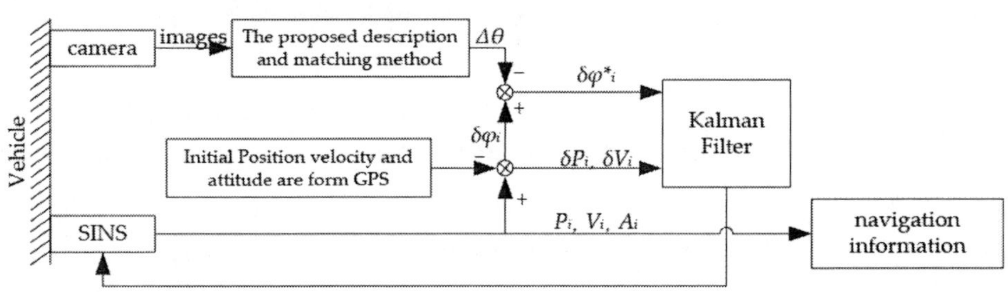

Figure 1. The schematic diagram of inertial-visual integrated navigation system [11].

The SINS and camera are fixed on the vehicle. $\Delta\theta$ is the attitude error information from camera. It is uesd to compensate the SINS attitude errors. The Kalman Filter (KF) is used to fusion the visual attitude information and SINS attitude information together. The schematic diagram of inertial-visual integrated navigation system is shown in Figure 1.

3.2. The attitude information from camera

As the proposed method, first of all, to extract the LSD lines and SURF points for all images. Then, using the SURF points to mark LSD lines. The marked LSD lines are the line features which can be traced. Next, to calculate the $\Delta\theta$ of each line feature. In the overlap part of lines, the $\Delta\theta$ of is replaced with the mean of overlap $\Delta\theta$. Working like these, the visual attitude error is obtained.

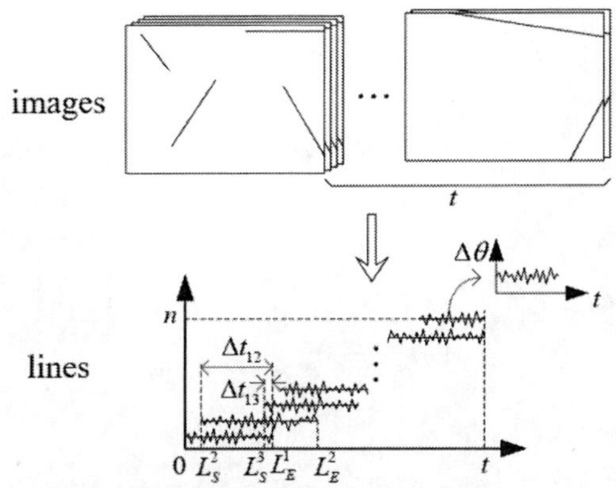

Figure 2. Processing of line feature information at dynamic conditions.

In Figure 2, L_S^2 and L_E^2 are the start time and end time of line L_2, n is the total number of line features. Δt_{12} is the overlap time of lines L_1 and L_2. During the overlap time, $\Delta\theta$ is the mean of the overlap lines'.

4. Experiment

4.1. Visual-aided inertial navigation system

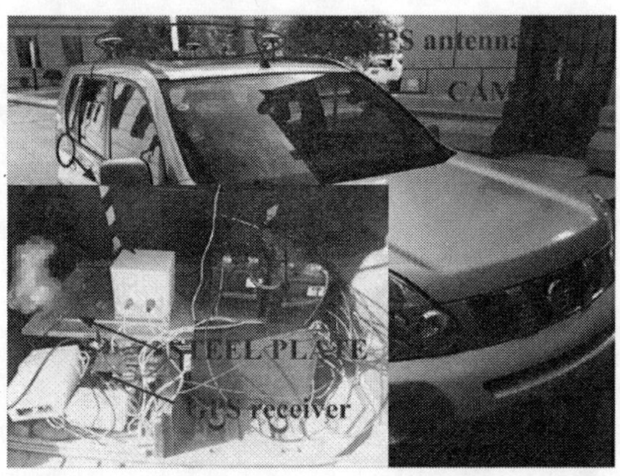

Figure 3. The equipment used in the experiment.

During the experiment, the vehicle is started and stop. In this experiment, the GPS is used to provide a coarse alignment of position system and to provide a position reference. In the experiment, the frequency of the IMU is 2Hz, the frequency of the video is 2Hz, and the run time length is 50s.

4.2. LSD line marked by SURF points

For the extraction of LSD line features in the experiment, in order to reduce the amount of calculation and increase the availability of line features, we limited number n_l of the LSD, to a certain extent to reduce the useless line features, $n_l \geq 2$.

The LSD lines marked by SURF points in one image are shown in Figure 3.

Figure 4. LSD line marked by SURF points.

The neighborhood width d affects the the marketability of line features. During the experiment, it was found that not all LSD lines could be marked by SURF points. As shown in figure 4, there are six line features but only two marked by SURF points. In this case, the neighborhood width parameter d could be modified to select an appropriate range and mark enough line features.

5. Results

5.1. The attitude error

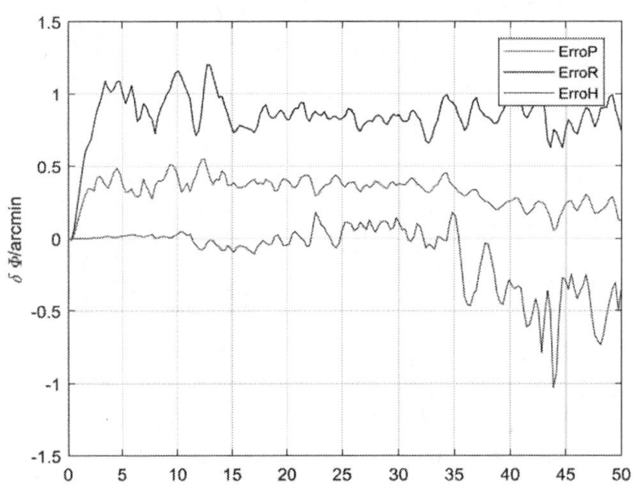

Figure 5. the attitude error of only INS.

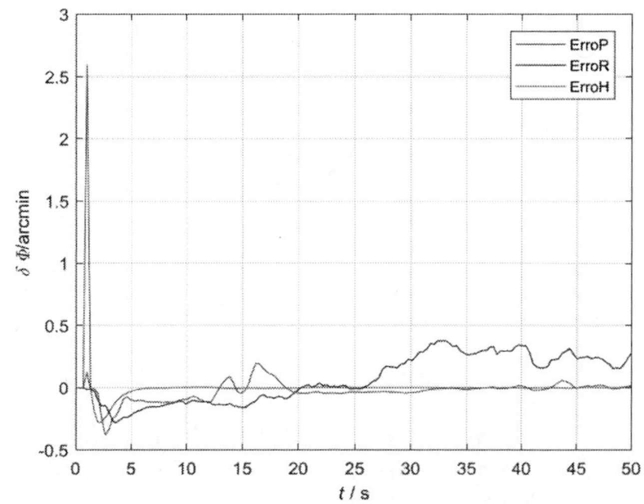

Figure 6. the attitude error of visual-aided inertial navigation system.

As shown in Figure 5 and 6, in the case of only INS, the attitude error given by the positioning system has obvious zero deviation. After 35 seconds, the rolling error and heading error are diverging. The visual-aided inertial method can restrain the deviation of zero and the divergence of rolling error and heading error, although the accuracy of pitch is sacrificed.

5.2. Position information

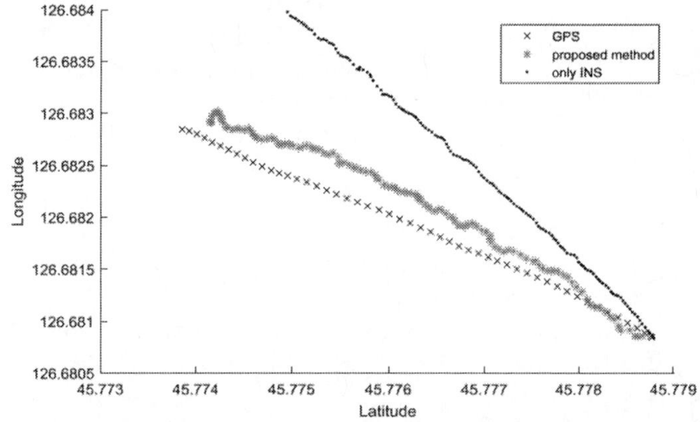

Figure 7. the position information of only INS, visual-aided inertial navigation system and GPS.

The GPS position is the refer information. It can be clearly seen from the figure 7 that in the case of only INS, the position has a significant divergence, that is, with the time goes, the position error of only INS is increase. However, the position divergence of visual-aided inertial system is obviously smaller than only INS.

6. Conclusion

In this paper, it is proposed that a novel line feature description for visual-aided inertial navigation system. LSD and SURF are used to describe the line feature. Through Coarse matching and fine matching, the function of continuously tracking the same line features in dynamic images were realized. This feature tracking method is applied to the visual-aided inertial navigation system, and its effectiveness is verified by experiments. The experiment results shown that the proposed line description works well for visual-aided inertial navigation system. The visual-aided inertial navigation system can work to offer position information when the GPS is unavailable.

Acknowledgments
Xingxing Guang is sponsored by the China Scholarship Council (CSC) grant number [201706680019] for her joint Ph.D. research training program at the University of Calgary, Calgary, AB, Canada. This research was funded by [Supported by the National Natural Science Foundation of China (NSFC)] grant number [61803118], [Supported by the Science and Technology Research Program of Chongqing Municipal Education Commission] grant number [KJZD-K201804701], [Supported by the Post Doc. Foundation of Heilongjiang Province] grant number [LBH-Z17053], and [the Fundamental Research Funds for the Central Universities] grant number [HEUCFJ180402].

References

[1] Fu Q and Liu Y Liu 2018 High-Accuracy SINS/LDV Integration for Long-Distance Land Navigation[J] *IEEE/ASME Transactions on Mechatronics* 23(6) pp 2952-2962
[2] Bo Y and Yue-gang 2018 Accurate integrated position and orientation method for vehicles based on strapdown inertial navigation system/Doppler radar[J] *Measurement and Control* vol 51(9-10) pp 431-442
[3] Emami M and Taban M R 2018 A Low Complexity Integrated Navigation System for Underwater Vehicles[J] *The Journal of Navigation* pp 1-17
[4] Ning X and Zhang J 2018 A Fast Calibration Method of the Star Sensor Installation Error Based on Observability Analysis for the Tightly Coupled SINS/CNS-Integrated Navigation System[J] *IEEE Sensors Journal* vol 18(16) pp 6794-6803
[5] Zhu M and Yu F 2018 Research on GFSINS/Star-sensor Integrated Attitude Estimation Algorithm Based on UKF[J] *Engineering Letters* vol 26(4)
[6] Zhao F and Chen C 2018 A filtering approach based on MMAE for a SINS/CNS integrated navigation system[J] *IEEE/CAA Journal of Automatica Sinica* vol 5(6) pp 1113-1120
[7] Chen J and Ou G 2018 An INS/WiFi Indoor Localization System Based on the Weighted Least Squares[J] *Sensors (Basel, Switzerland)* vol 18(5) p1458
[8] Zhu N and Zhao H 2015 A novel particle filter approach for indoor positioning by fusing WiFi and inertial sensors[J] *Chinese Journal of Aeronautics* vol 28(6) pp 1725-1734
[9] Li Y and Zhuang Y 2017 An improved inertial/wifi/magnetic fusion structure for indoor navigation[J] *Information Fusion* vol 34 pp 101-119
[10] Sun Q and Zhang Y 2017 An improved FAST feature extraction based on RANSAC method of vision/SINS integrated navigation system in GNSS-denied environments[J] *Advances in Space Research* vol 60(12) pp 2660-2671
[11] Guang X and Gao Y 2018 An autonomous vehicle navigation system based on inertial and visual Sensors[J] *Sensors* vol 18(9) p 2952
[12] Von Gioi R G and Jakubowicz J 2010 LSD: A fast line segment detector with a false detection control[J] *IEEE transactions on pattern analysis and machine intelligence* vol 32(4) pp 722-732
[13] Corke P 2017 Robotics, Vision and Control: Fundamental Algorithms In MATLAB®[M] Second, (*Completely Revised* Vol 118) Springer chapter 13 pp 438-443
[14] Bay H and Ess A 2008 Speeded-up robust features (SURF)[J] *Computer vision and image understanding* vol 110(3) pp 346-359
[15] Yu M and Cui H 2017 Database construction for vision aided navigation in planetary landing[J] *Acta Astronautica* vol 140 pp 235-246

IOP Conf. Series: Journal of Physics: Conf. Series **1267** (2019) 012047 doi:10.1088/1742-6596/1267/1/012047

Multi-layer Feature Fusion Network with Atrous Convolution for Pedestrian Detection

You Li[1,*], Qingxuan Zhang[1] and Yulei Zhang[1]

Beijing Lab of Intelligent Information Technology, School of Computer Science, Beijing Institute of Technology.

[*] ly541911136@163.com

Abstract. In this paper, we present a simple but effective framework K-AFPN that incorporates feature pyramid method for small-size pedestrian detection, fully utilizing the lower-layer detail features and higher-layer semantic features. The method not only enhances the robustness of the features, but also improves the discrimination of the feature maps, achieving competitive accuracy. In addition, atrous convolution is used to optimize the network for high-resolution feature maps, avoiding information loss caused by frequent down or up sampling. On top of the backbone network, K-means algorithm is used to obtain optimal initial anchor base sizes, which reduces computational costs and improves location accuracy. Hence, our method pays more concentration on pedestrians, especially those of relatively small size. Comprehensive experimental results on two classic pedestrian benchmarks illustrate the effectiveness of the proposed approach.

1. Introduction

Pedestrian detection is one of the significant research tasks in computer vision, involving multiple technologies, such as computer graphics, artificial intelligence, image processing and deep learning. It plays a critical role in several real-world applications including robotics, automatic driving, human behaviour analysis, video surveillance and intelligent transportation [1][2].

Different from general object detection, pedestrian detection suffers from the interference caused by the complex backgrounds, various lighting conditions, different camera shooting angles and other factors. While many CNN based approaches have been proposed, pedestrian detection task remains challenging problems because that the appearance of pedestrians is susceptible to diversified appearances, walking postures, low resolution, occlusion issues and so on. So, it is a valuable and challenging subject in the field of computer vision.

We propose an effective pedestrian detection approach on the Faster R-CNN framework in this paper, which can overcome the limitations of current pedestrian detection methods, especially for the small-size pedestrian detection, and get rid of traditional hand-crafted features. At the same time, by optimizing the anchor scales, the detection miss rate is reduced and the generalization ability of the model is improved. We performed experiments on the Caltech and INRIA datasets respectively, and the results demonstrate that our method achieves competitive performance. The performance of small-scale objects is improved by 6% on the Caltech dataset, which is significantly better than the traditional methods and other Faster R-CNN based methods.

2. Related Works

Many efforts have been made to improve detection performance in recent years. Existing methods fall into two main categories: hand-crafted features based [3][4][5][6] and deep learning features based. With the establishment of various large-scale image datasets and the improvement of hardware computing capability, the application of convolutional neural network (CNN) makes computer vision tasks obtain state-of-the-art results in terms of accuracy [7][8][9][10]. The advantage of CNN is that it can autonomously learning deep semantic features from raw pixels. Thus, it is more robust for detecting objects from complex backgrounds. While deep learning features have improved object detection performance [11], there are still several shortcomings for pedestrian detection. In practical applications, such as autonomous driving and intelligent surveillance, pedestrians are of small sizes (e.g., 28*70 for Caltech). Small objects are turned to low-resolution feature maps after layer-by-layer down-sampling, and these features are not discriminative for detection [12].

Faster R-CNN [13][14][15] is a particularly successful approach based on CNN, which has leading accuracy on general object detection benchmarks. Therefore, we propose a multi-scale deep feature learning detector for pedestrian detection, which adopting Faster R-CNN as the basic framework, aiming at getting rid of hand-crafted features. Specifically, the detector is focus on pedestrian regions for its scale-awareness. We use the feature pyramid method to fuse multi-layers features in the stage of feature extraction, so that the lower-layer detail features and higher-layer semantic features can be fully utilized. The feature pyramid method enhances the robustness of the features, and improves the detection accuracy. Meanwhile, we adopt atrous convolution (i.e., dilated convolution) [16][17][18], avoiding continuous subsampling and increasing higher-layer feature map size. In addition, we also use K-means algorithm to obtain the optimal initial scales of anchors for fine-tuning the network parameters to improve the location accuracy and reduce the computational cost.

3. Approach

Our goal in introducing the K-AFPN method is to provide a better way to solve the small-scale objects detection problem in practical application. An overview of our approach is depicted in Figure 1. Given an input image, the network generates multiple convolutional feature layers with high resolution. The backbone of K-AFPN is a Resnet-101 network, which is fast and accurate for object recognition [19].

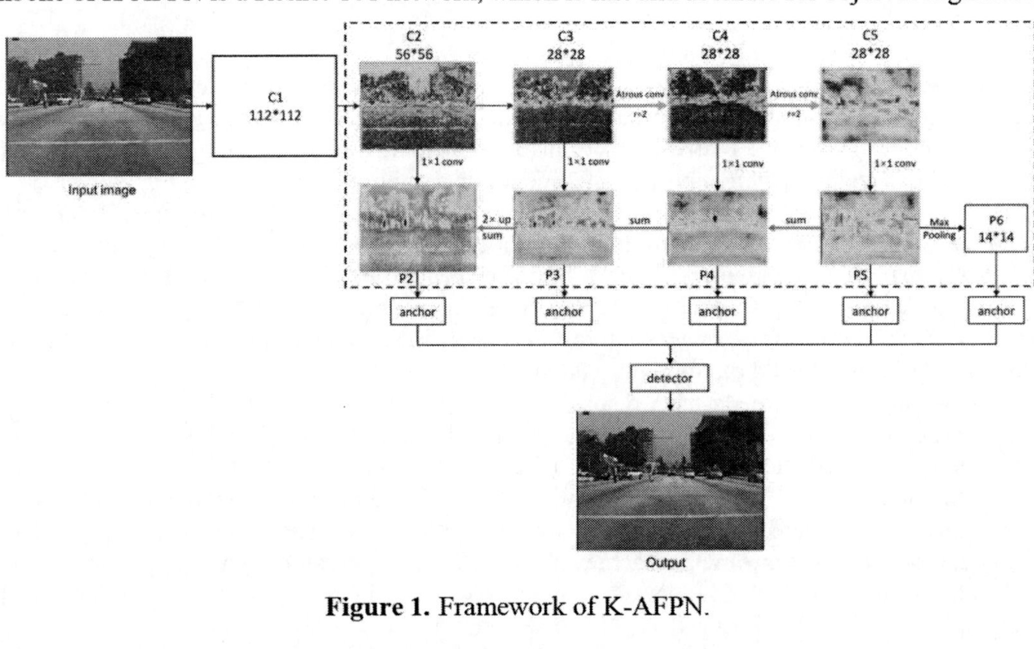

Figure 1. Framework of K-AFPN.

3.1 Feature Pyramid Module with Atrous Convolution

A multitude of previous object detection methods used only top-layer features for prediction, because the deeper features tend to encode more global and semantic information of objects that is robust

IOP Conf. Series: Journal of Physics: Conf. Series **1267** (2019) 012047 doi:10.1088/1742-6596/1267/1/012047

against appearance variations. Nevertheless, although the outputs of lower layers contain less semantic information, the features provide more precise localization. We use deep residual network (ResNet) as the backbone, which is an extremely deep convolutional neural network model [19]. It can avoid problems such as gradient disappearance and precision degradation caused by simple stacking of convolutional layers. The model is easier to be optimized and has excellent performance.

Generally speaking, the deeper the network is, the stronger the translational and rotational invariance it has, which is of positive significance for ensuring the robustness of the model. For detection problems, the object locating task requires models to have a good perception of position information, but excessive translational and rotational invariance will weaken this performance. For deeper full convolutional neural networks, an obvious drawback of the Faster-RCNN detection framework [15] is that the sensitivity of the detector to the object position information and the detection accuracy decreases.

Region proposal network (RPN), the bounding box proposal generation module of Faster R-CNN, is a sliding-window object detector. Two paratactic 1*1 convolutions for classification and regression are designed to be a head, which follows a 3*3 convolutional layer on top of the final feature map. Object classification and bounding box regression are defined towards the anchors of multiple hand-crafted scales and aspect ratios. The anchors are designed to cover targets of different sizes.

In general, the most intuitive solution to the drawback of Faster R-CNN is to move the position of the RPN to shallower layers. However, this method will significantly increase the amount of computation of the down-stream classifier Fast-RCNN, making the detection speed slower and the feature robustness lower. This problem is alleviated in our method, as it is not constrained to use features of a certain layer or several shallow layers like previous method. We adopt Feature Pyramid Network (FPN) [20] to improve RPN for extracting region proposals, and attach the head to each level for prediction on every convolutional block independently. And we obtain 3 areas of the anchors through K-means clustering on {P2, P3, P4, P5, P6} respectively. So, it is unnecessary to have multiple aspect ratio on each level. Instead, we set the aspect ratio to be {1:1}. Totally, there are still 15 anchors over the framework.

We fix the number of feature dimension in all the feature maps to be 256-channel according to [20], and distribute the training labels as follows: An anchor is considered as a positive label, which has the highest IoU ratio for a given ground-truth bounding box or an IoU over 0.7 with any ground-truth bounding box. While if it has IoU lower than 0.3 for all ground-truth boxes, it is considered as a negative label.

It is a critical issue in pedestrian detection that detecting small-size objects is inclined to bring about feeble contrast and motion blur, while FPN can effectively handle small targets. The reason is that the structure of FPN can be described as two parts: the bottom-up part is the forward propagation process of neural networks as usual, and the feature map becomes smaller and smaller through convolution calculations; the top-down process is to up-sample the deeper features that are more abstract and more semantic. The horizontal connection is used to change the number of channels of the shallower-layer feature maps. Then the processed deep features and the processed shallow features are merged, and independent prediction is performed at each layer. Therefore, features from shallower but higher-resolution layers can provide more accurate details, meanwhile the deep-layer features provide stronger representation. The feature map of each prediction layer combines features of different resolutions and semantic intensities, so that the features are enhanced to be more discriminative, ensuring that objects of different sizes can be detected, especially for small targets. This method does not change the size of the acquired features by image scaling like image pyramid, which greatly improves the detection performance without increasing the calculation amount of the original model too much.

However, information loss occurs during multiple subsampling and up-sampling operations, so the location information is still not completely refined. Hence, how to minimize the information loss in the process of acquiring a larger perceptive field is the key to making detection more accurate.

Standard convolution has several problems in certain scenes such as image semantic segmentation. Reducing calculation amount and increasing the perceptive field through pooling operation, and expanding the image to the original size by deconvolution (up-sampling) lead to information loss, especially spatial structure information. And another problem caused by these operations is that small object information cannot be reconstructed. Dilated convolution [16][17][18] can enlarge the perceptive field without pooling, avoiding the compression of the image resolution, and the data structure inside the image is retained, which can obtain a better performance than the standard convolution.

There is an additional dilation rate in the dilated convolution to control the size of the kernel size expansion. The larger the dilation rate parameter is, the larger the perceptive field of same convolution kernel size dilates. We use dilated convolution to increase higher-layer feature map size.

Assuming that the input signal is two-dimensional, for each position i on an input y and the filter ω, the calculation formula for applying the atrous convolution on the input feature map x is:

$$y[i] = \Sigma_k x[i + r \cdot k]\omega[k] \quad (1)$$

where the hole rate r is the sampling stride, that is, to insert r-1 zero points between two neighbouring convolution kernel values in each spatial dimension of the filter. Standard convolution is a special form of r = 1 for atrous convolution. The atrous convolution can modify the perceptive field of the filter by changing the value of the hole rate r.

Atrous convolution can increase the perceptive field, but because the convolution kernel values are discontinuous, continuity of information may be lost, which has an impact on small objects detection. So, we only apply the atrous convolution to Conv4_x layer and conv5_x layer of ResNet101. The final feature map for ResNet101 is 32 times smaller than the size of the input image, so the output stride is 32. We set the step size of the last reduced resolution pooling layer or convolutional layer to 1 to avoid signal extraction, and the subsequent convolutional layers are replaced by atrous convolutional layers of r=2, which makes the final output stride remaining 8 and no longer increasing to 32, that can increase resolution and reduce stride. Thus, the network is able to extract denser features without having to learn any additional parameters. In addition, the deep feature map eliminates the up-sampling operation when merging, and only needs to change the dimension by 1*1 conv to merge with the upper feature map, which reduces information loss. In addition to the improvement of FPN with atrous convolution, we adopt the adaptations in the same fashion as in [20].

3.2 Anchor Clustering

The original RPN with FPN has anchors of hand-crafted aspect ratios and numbers. The parameters are adjusted to approach the ground truth with the iteration during training. Anchors of inappropriate aspect ratios are associated with few examples, so are noisy and harmful for detection accuracy and speed. We address this problem by adopting K-means algorithm.

Therefore, we conduct dimension clustering of ground truth boxes with K-means method. Taking K as the number of anchors, the sizes of the K clustering center boxes are set to be the dimensions of the anchors. The optimal number and dimensions of anchors can improve positioning accuracy and reduce unnecessary calculation. The traditional K-means uses the Euclidean distance function, which means that larger boxes will generate more errors than smaller ones. Hence, we use IOU so that the results are irrelevant to the size of the candidate boxes.

The distance function is set according to the method in [21]:

$$d(box, truth) = 1 - IOU(box, truth) \quad (2)$$

The objective function of the clustering operation is:

$$S = \min \Sigma_{i=0}^{i=k}(1 - IOU(box - truth)) \quad (3)$$

where box is the candidate bounding box, truth is the ground truth box, and K is the number of anchors.

IOP Conf. Series: Journal of Physics: Conf. Series **1267** (2019) 012047 doi:10.1088/1742-6596/1267/1/012047

We clustered the Caltech dataset and the optimal dimensions are [[[9,22],[13,25],[12,35]], [[18,32],[17,42],[21,53]], [[34,37],[26,56],[23,70]], [[31,74],[36,97],[60,52]], [[53,129],[156,82],[100, 244]]], distributed on the five prediction layers according to the results of K-means algorithm.

4. Experiments

4.1 Datasets

We evaluated the proposed approach on two benchmarks: Caltech [1][22] and INRIA [4]. Here we give a brief description of these datasets. The log-average miss rate is used to evaluate the detection performance and is calculated by averaging miss rate on false positive per-image (FPPI) in the range of $[10^{-2},10^{0}]$, denoted MR.

Acting as a representative benchmark for pedestrian detection, Caltech pedestrian is currently the most challenging and large-scale dataset. Over 70% of the annotated pedestrian instances have a height smaller than 100 pixels, including extremely tiny instances under 50 pixels. Pedestrians in the dataset are grouped into 3 subsets by their height in pixels: near (80 or more pixels), medium (between 30-80 pixels), and far (30 pixels or less). Assuming an urban speed of 55km/h, an 80-pixel person is just 1.5s away, while a 30-pixel person is 4s away. Thus, detection in the medium scales is essential for automotive applications. In the experiment, we extract 42782 images for training by sampling every 3rd frame from training sets, and evaluate the framework on a standard test set consisting of 4024 frames. And the performance is measured under the Medium subset [22].

The INRIA dataset is a classic pedestrian dataset. It has various backgrounds and high-quality annotations of pedestrians. The training set contains 2416 pedestrians, distributed in 614 positive training images. The test set contains 1126 pedestrians, distributed in 288 positive sample images [4].

4.2 Evaluation Metrics

To evaluate the detection performance, we plot miss rate against false positives per image (FPPI) (using log-log plots) by varying the threshold on detection confidence, according to the evaluation metrics defined by the Caltech Pedestrian method. We set the threshold to be 50 percent. When the IoU of a detected Bounding Box and a ground truth bounding box exceeds the threshold, then we consider that they are matched. The lower the MR-FPPI curve is, the better the performance of the pedestrian detection method will be [22].

4.3 Results and Analysis

For our experiments, the model was implemented using the TensorFlow machine learning library, and it was trained on a NVIDIA GeForce GTX 1080Ti GPU. For the feature extraction network, we used the ImageNet pre-trained ResNet-101 as initial parameters.

To train the network, the standard stochastic gradient descent with momentum was employed for optimization, where the initial learning rate, momentum, and weight decay were set to 0.001, 0.9, and 0.0005. By default, an IoU threshold of 0.5 was used for determining "true positives". The NMS was adopted both region proposal stage and classification stage to avoid redundant detections.

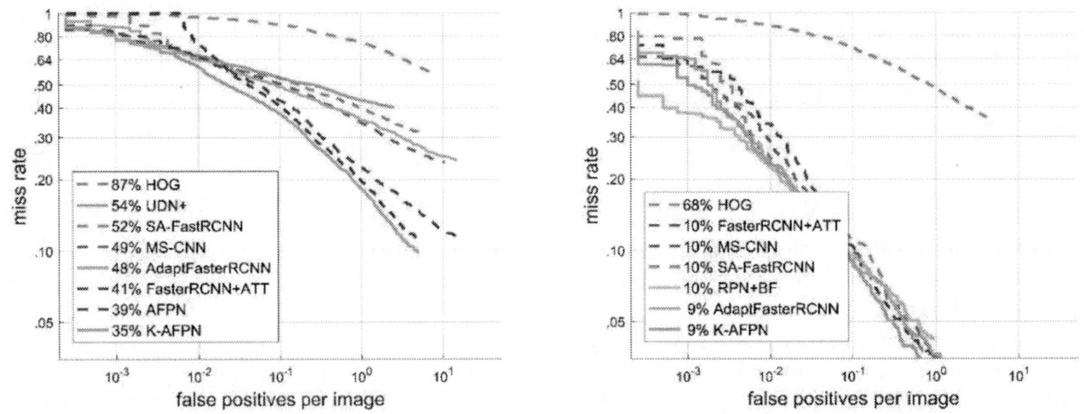

(a) Medium (b) Reasonable

Figure 2. Comparison with state-of-the-art methods on the Caltech dataset under Medium and Reasonable evaluation protocols (lower is better).

Figure 2 indicates that the ROC plot of miss rate against FPPI for the available top performing approaches based on Faster RCNN and several other outstanding methods reported on Caltech medium and reasonable subsets. In medium scale case, as shown in Figure 2 (a), our K-AFPN achieves 35% miss rate, which significantly outperforms the other detectors. Using K-means to initialize the anchor scale instead of hand-crafted settings improve the performance by 4%. As depicted in Figure 2 (b), our detector is slightly better than the previous best methods on Caltech reasonable subset. Although the performance improvement is not very remarkable on the reasonable subset, since our approach is mainly aim at small targets, but it still acquires the state-of-the-art performance with 9% miss rate.

For INRIA, we trained our model with 614 positive images by excluding negative images and evaluated on the test set. Figure 3 shows the comparison with previous methods on INRIA test dataset in terms of miss rate, which illustrates that our method gets a compelling result with 9% miss rate, better than the competitive approaches by 2%. The performance reveals that our approach, taking full advantage of multi-level features, has strong generalization ability, which yields state-of-the-art results even if the training set is very limited.

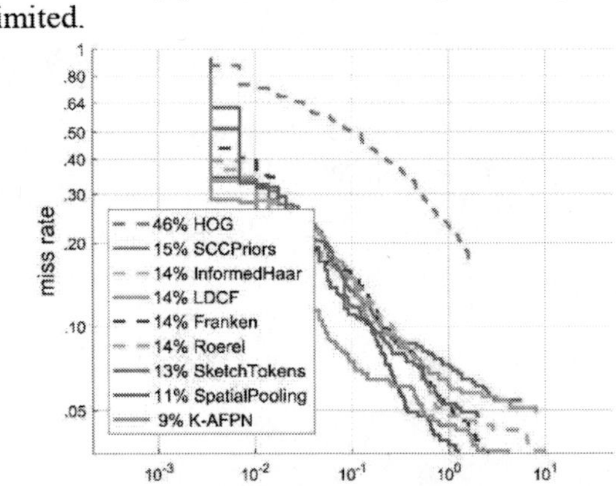

Figure 3. Comparison results of K-AFPN with state-of-the-art methods on INRIA dataset (lower is better).

5. Conclusion

A K-AFPN method for pedestrian detection is proposed in this paper, which merges detail information on shallower layers into higher-layer features to gain more discriminative features for detection, while most existing pedestrian detection methods only consider low resolution features extracted from the higher-layer feature maps. Atrous convolution is used to optimize the network for high-resolution feature maps. In addition, we use k-means to obtain appropriate anchor number and dimensions for pedestrian. Hence, our method pays more concentration on pedestrians, especially those of relatively small size. Experimental results on the two widely used pedestrian benchmarks demonstrate the advantages on detection robustness and efficiency of the proposed method. As for the future work, we will focus on the overall performance, and occlusion handling will be involved.

References

[1] Dollár P, Wojek C, Schiele B and Perona P. Pedestrian detection: a benchmark. In *Computer Vision and Pattern Recognition (CVPR)*, 2009, pp. 304–311.
[2] Benenson R, Omran M, Hosang J, et al. Ten years of pedestrian detection, what have we learned? In *European Conference on Computer Vision (ECCV)*, 2014, pp. 613-627.
[3] Dollár P, Tu Z W, et al. Integral channel features. In *British Machine Vision Conference*, 2009.
[4] Dalal N, Triggs B. Histograms of oriented gradients for human detection. In *Computer Vision and Pattern Recognition (CVPR)*, 2005, pp. 886-893.
[5] Zhang S, Bauckhage C, Cremers A B. Informed haar-like features improve pedestrian detection. In *Computer Vision and Pattern Recognition (CVPR)*, 2014, pp. 947–954.
[6] Felzenszwalb P, McAllester D, Ramanan D. A discriminatively trained, multiscale, deformable part model. In *Computer Vision and Pattern Recognition (CVPR)*, 2008, pp. 1-8.
[7] Tian Y L, Ping L, Wang X G, et al. Pedestrian detection aided by deep learning semantic tasks. In *Computer Vision and Pattern Recognition (CVPR)*, 2015, pp. 5079-5087.
[8] Zhang S, Benenson R, Omran M, et al. How far are we from solving pedestrian detection? In *Computer Vision and Pattern Recognition (CVPR)*, 2016, pp. 2-10.
[9] Li C, Wang X, Liu W. Neural features for pedestrian detection. In *Neurocomputing*, vol. 238, no. 2017, pp. 420-432, 2017.
[10] Girshick R, Donahue J, Darrell T, et al. Region-based convolutional networks for accurate object detection and segmentation. In *IEEE Trans on Pattern Analysis & Machine Intelligence (TPAMI)*, vol. 38, no. 1, pp. 142, 2016.
[11] Krizhevsky A, Sutskever I. Imagenet classification with deep convolutional neural networks. In the Advances in *Neural Information Processing Systems (NIPS)*, 2012, pp. 1097-1105.
[12] Zhang L, Lin L, Liang X, He K. Is faster r-cnn doing well for pedestrian detection? Proceedings of the *European Conference on Computer Vision (ECCV)*, 2016, pp. 443-457.
[13] Girshick R, Donahue J, Darrell T, et al. Rich feature hierarchies for accurate object detection and semantic segmentation. In *Computer Vision and Pattern Recognition (CVPR)*, 2014, pp. 580-587.
[14] Girshick R. Fast R-CNN. In *International Conference on Computer Vision(ICCV)*, 2015, pp. 1440-1448.
[15] Ren S Q, He K M, Girshick R, et al. Faster R-CNN: towards real-time object detection with region proposal networks. In *IEEE Trans on Pattern Analysis and Machine Intelligence (TPAMI)*, vol. 39, no. 6, pp. 1137-1149, 2017.
[16] 才 Yu F, Koltun V. Multi-scale context aggregation by dilated convolution. In *International Conference on Learning Representations (ICLR)*, 2015, arXiv preprint arXiv: 1511.07122.
[17] Chen L C, Papandreou G, et al. Rethinking atrous convolution for semantic image segmentation. In *Computer Vision and Pattern Recognition (CVPR)*, 2017, arXiv preprint arXiv: 1706.05587.
[18] Li Y H, Zhang X F, Chen D M. CSRNet: dilated convolutional neural networks for understanding the highly congested scenes. In *Computer Vision and Pattern Recognition*, 2018, pp. 1091-1100.
[19] He K, Zhang X, Ren S, et al. Deep residual learning for image recognition. In *Computer Vision and Pattern Recognition (CVPR)*, 2016, pp. 770-778.
[20] Lin T Y, Dollár P, Girshick R, et al. Feature pyramid networks for object detection. In *Computer Vision and Pattern Recognition (CVPR)*, 2017, pp. 936-944.
[21] Redmon J, Farhadi A. YOLO9000: better, faster, stronger. In *IEEE Conference on Computer Vision & Pattern Recognition (CVPR)*, 2017, pp. 6517-6525.
[22] Dollár P, Wojek C, Schiele B, et al. Pedestrian detection: an evaluation of the state of the art. In *IEEE Trans. Pattern Anal. Mach. Intell (TPAMI)*, vol. 34, no. 4, pp. 743, 2012.

IOP Conf. Series: Journal of Physics: Conf. Series **1267** (2019) 012048 doi:10.1088/1742-6596/1267/1/012048

An Improved Method for Calibration Between a 2D Lidar and a Camera Based on Point-Line Correspondences

Xuepeng Chu, Jun Zhou, Li Chen and Xiaobin Xu

School of Mechanical and Electrical Engineering, HoHai University, Changzhou, Jiangsu, China

footms@hhu.edu.cn

Abstract. The paper depicts an improved method for calibration parameters, which is a 2D homography, using point-line correspondences. The correspondences are the edge lines of the board provided by camera and the intersection points obtained by lidar. Particularly, a data preprocessing method is put forward and exploited to dramatically reduce the lidar's measurement error. Afterwards, we use a triangular board to establish constraints. At least 8 correspondences are employed to compute the initial value, and the final result is optimized by L-M algorithm. Massive experimental results show that proposed method has about 1 pixel improvement, compared with the previous methods.

1. Introduction

Light detection and ranging sensors (Lidars) and visual sensors are widely used in robotic perceptron, particularly simultaneous localization and mapping (SLAM). A 2D lidar is capable of providing high-accuracy point clouds, whilst the visual sensors are able to acquire the rich environmental information, such as the colour, shape and texture information of objects. Through fusing above two kinds of sensors, a 2D map in SLAM system, containing richer information, can be achieved [1]. The primary problem of data fusion is the united calibration. In fact, united calibration is a process that estimates mapping relations between point clouds and images. Generally, this problem can be divided into two parts, camera intrinsic calibration and extrinsic calibration of both sensors. In the term of intrinsic calibration, Zhang developed a method which is the most popular approach now to obtain the camera's intrinsic parameters [2]. In the aspect of extrinsic calibration, there are many calibration methods based on distinct geometric features. In this paper, an approach based on Kiho's and Daniel's method [3] is enhanced by two improvements. One improvement is a data preprocessing method which is proposed to decrease the ranging error according to distribution property of the lidar measurement [4]. Another improvement is that we estimate 2D homography directly without calculating the intrinsic and extrinsic parameters in advance.

2. Problem Formulation

As figure 1 shows, $O_C - X_C Y_C Z_C$ is the camera coordinate system and $O - UV$ is the pixel coordinate system [5]. In addition, $O_L - X_L Y_L Z_L$ is the lidar coordinate system. The extrinsic parameters represent the rotation and translation from $O_L - X_L Y_L Z_L$ to $O_C - X_C Y_C Z_C$. The intrinsic parameters denote the transformation from $O_C - X_C Y_C Z_C$ to $O - UV$. According to pinhole camera model, the projection from lidar coordinate $P_L = (x_L, y_L, z_L)^T$ to pixel coordinate $P_{uv} = (u, v)^T$ satisfies:

$$s\begin{pmatrix}u\\v\\1\end{pmatrix}=K\begin{pmatrix}R&t\\0&1\end{pmatrix}\begin{pmatrix}x_L\\y_L\\z_L\\1\end{pmatrix} \quad (1)$$

Where K is a 3×3 matrix that represents camera intrinsic parameters, R is a 3×3 matrix that represents rotation, t is a 3×1 matrix that represents translation and s is a scale factor.

While the pose of lidar and camera are fixed, the extrinsic parameters$(R,\ t)$ will be a definite constant matrix. Moreover, 2D lidar can only provide two dimensions information, so we assume $z_L = 0$. The Eq. (1) is rewritten as follows:

$$\begin{pmatrix}u\\v\\1\end{pmatrix}=H\begin{pmatrix}x_L\\y_L\\1\end{pmatrix}=\begin{pmatrix}h_1&h_2&h_3\\h_4&h_5&h_6\\h_7&h_8&h_9\end{pmatrix}\begin{pmatrix}x_L\\y_L\\1\end{pmatrix} \quad (2)$$

Where H is a 2D homography that represents the transformation from lidar scanning plane to image pixel plane. It is a 3×3 matrix and has 8 DOF.

Obviously, the united calibration parameter is a 2D homography. In order to estimate the 8-DOF homography, we need at least 8 pairs of the correspondences and each triangular board pose can provide two correspondences. Thus, at least 4 distinct triangular board poses are required to compute H.

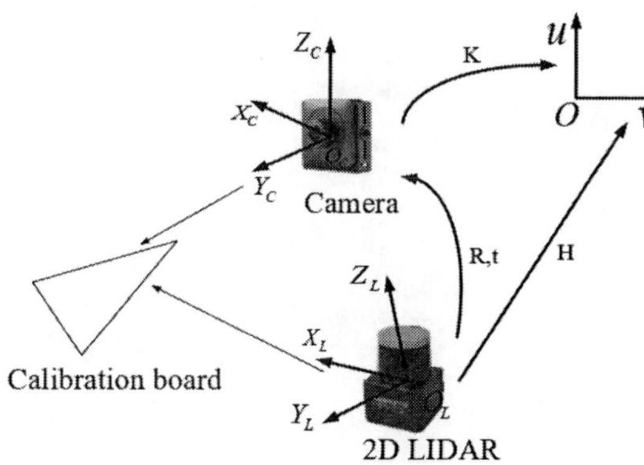

Figure 1. Lidar-Camera calibration system

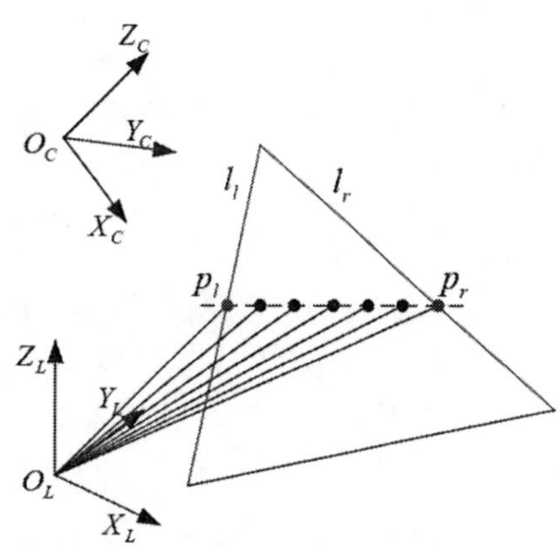

Figure 2. Triangular calibration board used for united calibration: The red lines, left(l_l) and right(l_r) represent the line features that are extracted in the images and the red points, left(p_l) and right(p_r) describe the intersection points that are acquired from laser scanning

3. Method description

3.1. Data Collection and Preprocessing

The correspondences adapted in proposed method are the point-line features. As is shown in figure 2, the intersection points are provided by 2D lidar, depicted as their homogeneous form $p_l = (x_l, y_l, 1)^T$ and $p_r = (x_r, y_r, 1)^T$. The edge lines are exacted from images manually and formulated as $l_l = (a_l, b_l, c_l)$ and $l_r = (a_r, b_r, c_r)$.

The calibration system is composed of a HOKUYO UST-10LX single line lidar and a HIKVISION DS-2CS55D0B-NA colour camera with the resolution 640×480. Above all, the edge lines in images are exacted manually. By selecting 5 points on each line, the line parameters are fitted by least square method. What is more, the intersection points are exacted by using lidar scanning data. With above edge lines and intersection points, the corresponds are explored to estimate 2D homography. However, the ranging error of the lidar is within 40mm. This error along the direction of the laser beams can be not ignored and it will lead to an imprecise calibration consequence. As an example, figure 3(a) presents 5 group of random scanning points, lying on calibration board, are collected at the same pose and the ranging error of scanning sequences reaches$\pm$20mm. Intuitively, the accuracy of intersection points, directly using only one group of scanning data, is not adequate. Thus, a preprocessing method is exploited to reduce the ranging error, and it is depicted as follows:

1. Repeat sampling the scanning points lying on calibration board for n times. Let m be the quantity of the scanning sequences lying on board and the sequences are denoted as $\{q_1,\cdots,q_i,\cdots,q_m\}$. Each sequence q_i has n ranging values, which is described as $\{q_i^1,\cdots,q_i^k,\cdots q_i^n\}$.
2. For each sequence, find the maximum value q_i^{max} and minimum value q_i^{min} among n measurements. From q_i^{min} to q_i^{max}, take the step of 1mm and count the times of every value. Then, a scatter plot is drafted with above information. The horizontal axis of the plot is the values from q_i^{min} to q_i^{max} and the vertical axis is the statistic times of every value. For instance, after 1000 measurements of one sequence, the plot in figure 3(b) exhibits the characteristic of a Gaussian curve. So we regard the mean of the Gaussian distribution as this sequence's true distance value.
3. For every sequence, repeat step 1 and 2. Take the means of the Gaussian distribution as their true values. With this method using in the same scenario of figure 3(a), the true point clouds are fitted as the figure 3(c):

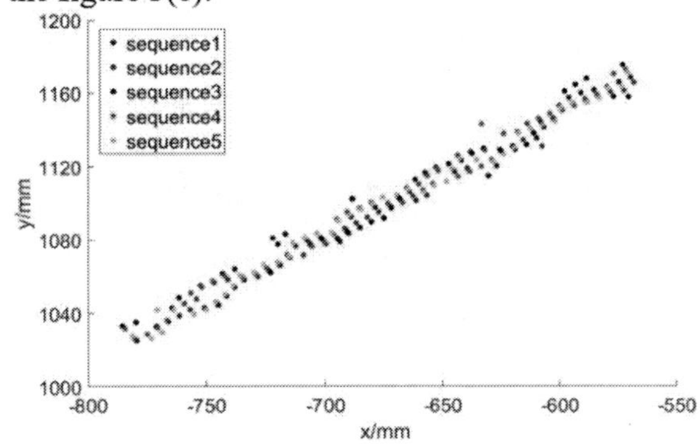

(a)

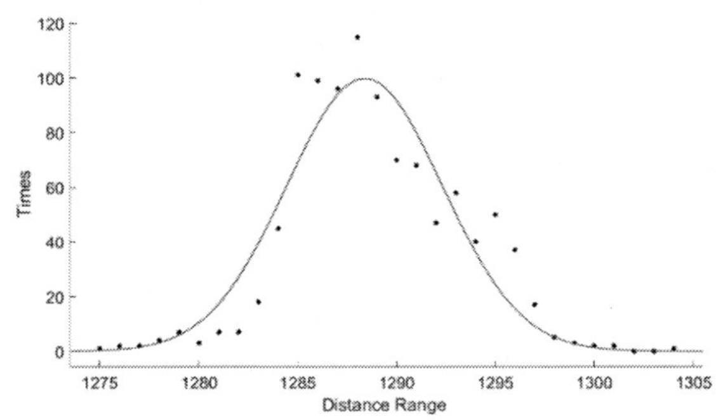

(b)

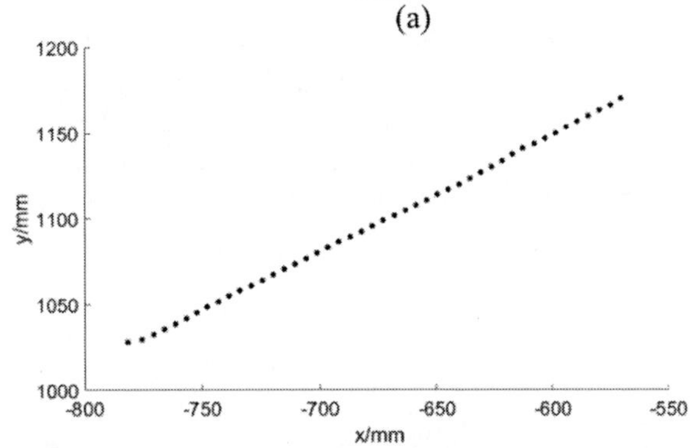

(c)

Figure 3. The preprocessing used to reduce ranging error; (a)the scanning points lying on calibration board with noise; (b) the distribution of a laser sequence's ranging values(n=1000); (c) the points lying on calibration board after data preprocessing

Some experiments are conducted to validate the effect of the data preprocessing. As figure 4 shows, we mount the lidar-camera system and a rectangular board onto an optical platform. The ground truth of the distance between lidar and rectangular board is acquired by more precise measurement, which is a high-accuracy range finder. Compared with the ground truth, the measurement error after data preprocessing decreases to $\pm$5mm.

Figure 4. The experimental platform designed to validate the accuracy of the ranging data

3.2. Establish Constraints

In this section, we describe the intersection points and their correspondent pixel coordinate as $p_i = (x_i, y_i, 1)^T$ and $q_i = (u_i, v_i, 1)^T$. The edge lines are represented as $l_i = (a_i, b_i, c_i)^T$. Obviously, the points lie on its correspondent edge lines.

$$l_i^T q_i = 0 \tag{3}$$

And the $q_i = (u_i, v_i, 1)^T$ can be denoted by $p_i = (x_i, y_i, 1)^T$ and H via Eq. (2), which is $u_i = \frac{h_1x_i+h_2y_i+h_3}{h_7x_i+h_8y_i+h_9}$, $v_i = \frac{h_4x_i+h_5y_i+h_6}{h_7x_i+h_8y_i+h_9}$.

So Eq. (3) can be denoted as follows:

$$A_i h^T = 0 \tag{4}$$

Where $h = (h_1 \quad h_2 \quad h_3 \quad h_4 \quad h_5 \quad h_6 \quad h_7 \quad h_8 \quad h_9)^T$,
$A_i = (a_ix_i \quad a_iy_i \quad a_i \quad b_ix_i \quad b_iy_i \quad b_i \quad c_ix_i \quad c_iy_i \quad c_i)$.

Since the homography H has 8 DOF, at least 8 correspondences are needed to solve H. Namely, we need to find $i \geq 8$ point-line correspondences. The initial value of H with closed form can be computed by SVD method.

3.3. Optimization

The homography H is estimated optimally by minimizing the distance between the intersection points and corresponding edge lines projected onto image plane. Given i pairs different corresponding features, the ith distance formula $f_i(l, q)$ satisfies:

$$f_i(l, q) = \frac{|a_iu_i+b_iv_i+c_i|}{\sqrt{a_i^2+b_i^2}} \tag{5}$$

Projected scanning points p_i onto image plane q_i, Eq. (5) is rewritten as:

$$f_i(l, p) = \frac{|A_ih^T|}{\sqrt{a_i^2+b_i^2}(x_ih_7+y_ih_8+h_9)} \tag{6}$$

The error function, defined as the square sum of these distance over n pairs of features, is

$$E(h) = \sum_{i=1}^{n} f_i^2(l, p) \tag{7}$$

The optimal h is available by minimizing the error function $E(h)$, which is:

$$\min_h E(h) = \min_h \sum_{i=1}^{n} f_i^2(l, p) \tag{8}$$

Obviously, Eq. (8) is a non-linear least square problem without the closed-form solution. It can be minimized by Levenberg-Marquardt algorithm, which is a generally iterative optimization algorithm. The L-M algorithm needs an initial estimation, which can be calculated by SVD method in section 3.2.

IOP Conf. Series: Journal of Physics: Conf. Series **1267** (2019) 012048 doi:10.1088/1742-6596/1267/1/012048

4. Experiments and results

In this section, we use a set of 26 scan/image pairs to optimize H. 4 pairs are used to calculate initial value and H is optimized by another 22 pairs. Since we are lack of the ground truth, we compare the proposed method with Zhou's method [6] which lacks data preprocessing. Let H_1 represent the homography obtained by proposed method and H_2 represent the homography obtained by Zhou's method. Both H_1 and H_2 are computed by the same 26 scan/image pairs. To conduct these experiments, we collect another 30 scan/image pairs according to the distance range and the objects' shapes. The distance between lidar and object ranges from 0.5m to 3m, and the objects are the triangular or rectangle boards. Then, we project the scanning points onto image using H_1 and H_2, respectively. The results show H_1 and H_2 are almost similar. Nevertheless, for the sake of brevity, two samples in figure 5 presents that proposed method has about 1 pixel improvement at the boundaries.

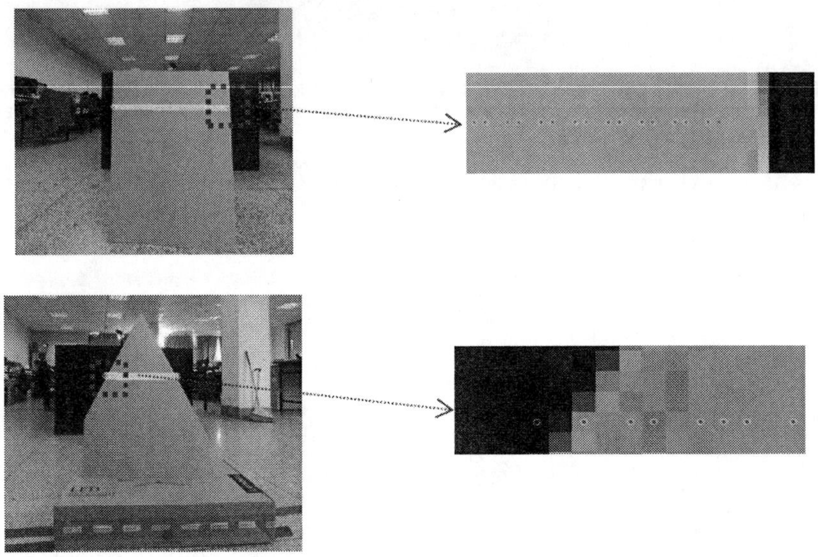

Figure 5. Two samples of the experimental results: laser scanning points projected onto image by Zhou's method (yellow circles) and proposed improved method (red dots)

5. Conclusion

In this paper, an improved method for united calibration between a 2D lidar and a colour camera is addressed. In this method, the point-line correspondences are exploited to establish constraints. In contrast to the previous work, we put forward a data preprocessing method to enhance the accuracy of laser scanning points. The experimental results show this method can improve about 1 pixel in calibration system. Since we estimate the homography directly without obtaining the intrinsic parameters in advance, not only triangular board, but V-shape, rectangular board and so on, all of them can be used as calibration tools in proposed method, which makes the method more flexible.

Acknowledgements

This work was supported by the Postgraduate Research & Practice Innovation Program of Jiangsu Province no. KYCX18_0540 under grant agreement no.2018B733X14.

References

[1] Ganhua Li, Yunhui Liu, Li Dong, Xuanping Cai and Dongxiang Zhou 2007 An algorithm for extrinsic parameters calibration of a camera and a laser range finder using line features. *IEEE/RSJ International Conference on intelligent Robots and Systems*. pp.3854–3859.

IOP Conf. Series: Journal of Physics: Conf. Series **1267** (2019) 012048 doi:10.1088/1742-6596/1267/1/012048

[2] Z. Zhang 2000 A flexible new technique for camera calibration. *IEEE Trans. Pattern Analysis and Machine Intelligence*. pp. 1330-1344.

[3] Kiho Kwak, Daniel F. Huber, Hernan Badino, and Takeo Kanade 2011 Extrinsic Calibration of a Single Line Scanning Lidar and a Camera. *IEEE/RSJ International Conference on intelligent Robots and Systems*. pp. 3283–3289.

[4] Xiaobin Xu, Minzhou Luo, Ronghao Pei, Zhiying Tan 2018 A novel laser ranging method based on adaptive kalman filter technology. *Microwave and Optical Technology Letters*. pp. 2881-2887.

[5] Heng Yang, Xiaolin Liu, Ioannis Patras 2012 A simple and effective extrinsic calibration method of a camera and a single line scanning lidar. *21st International Conference on Pattern Recognition*. pp. 1439-1442.

[6] Lipu Zhou and Zhidong Deng 2014 A new algorithm for the establishing data association between a camera and a 2-D LIDAR. *TSINGHUA SCIENCE AND TECHNOLOGY*. pp. 314-322.

Text Modeling Based on the Analysis of Categorical Words

Xiaoli Wang

No.113 XueFu Road, KunMing City, YunNan Province, China

Email: 2931678760@qq.com

Abstract. Recently the most essential research in machine learning is to discover correct topics from large scale of electronic documents, and many studies about text modeling have been made in natural language process with the environment of big data. This paper discusses a fundamental way for people to form words while talking and describing ideas, the document is considered as words convergence and a new method for text modeling based on the relational analysis about verbs, nouns, and modifiers is proposed. Additionally the paper introduces a new algorithm for estimating parameters in the generative probabilities of the model, which can alleviate the process about statistics-based text analysis from computational complexity. Experimental results show that the method improves the quality for learning a document.

1. Introduction

With the popularity of Internet and computer technology, large amount of electronic documents have been accumulated and the underlying information from these texts has become a treasure of inestimable value. If data from these documents can be organized and transformed into an efficient text model, the machine learning to categorize, research, and extract topical keywords is available. So it's great potential for people to explore important information or to get abstracts from the large scale of corpus.

Considerable efforts have been made in text modeling, Vector Space Model(VSM) [1] was proposed by Salton in 1969, its basic idea is to transform the texts into vectors of identifiers, then the information from texts can be computed by values in the vectors. The definition of values is usually the TF-IDF [2] weighting, so it neglects the semantic message from texts. On the other hand, VSM usually get a high dimensionality by the number of distinct words occurring in a document, and the values in the vector space are always sparse.

Topic model is a statistical model for discovering hidden "topics" in documents[3], and convergences of particular words in documents would be considered to represent a certain topic. Nowadays, topic model is widely used for abstracts from journals or newspapers, and also used to analyze changes in topics over time.

There are two ways for topic modeling, Probabilistic Latent Semantic Analysis (PLSA) [4] considers a document as "bag of words", then for each document, a latent class which is viewed as representations of "topic" is chosen conditionally to the document, and the word is generated from a single topic, thereby the document can be reduced to a probability distribution on different topics. Formally, the latent topics in the model can be learned by EM algorithm.

Latent Dirichlet Allocation (LDA) [5] is proposed to discover topics in large scale of corpus, each document can be a mixture of various topics, and each topic is characterized by a distribution over words. The words is exchangeable in the document, then the text model can be represented as a three-level hierarchical Bayesian model, in which each class is modeled as a finite mixture over a set of topics, and each topic is modeled as an infinite mixture over a set of topic probabilities.

Recently, the studies on text modeling are mostly focused on how to recognize the latent topics in documents efficiently with less cost of computational complexity and storage space even in the big data phenomenon. This paper presents a new method for text modeling based on the lexical semantics. In section 2, the relationship among verbs, nouns, modifiers is discussed and a probability of sentence can be generated. In section 3, we describe in detail the process of text model. In section 4, we report the experimental results by the measurement of Perplexity on a public dataset. Conclusions and ideas for future work are given in section 5.

2. Basic Concepts

2.1. Semantics and Information

This paper considers a document as language convergence for people to express some topics. In general, there are always many topics in one document, in order to express ideas more clearly, we are accustomed to describing them by sentences, and every sentence can express part of idea which is relevant to the topics.

From the perspective of Computational Linguistics[6], a sentence can be viewed as a particular logical expression. Then every word in a sentence is a computational symbol, and a word is regarded as the basic unit in a document. When people attempt to write down a topic, they have to organize the words, so words about the topic will appear in a document more or less frequently[7]. Studying the convergence of these particular words may capture some abstract notions, therefore, these words are semantic code for the topic.

Since the essence of language is information coding, it can express the existence characteristics about the objects, and also can express the motion characteristics about the objects. However, a single word can only represent either existence or motion. If we need to describe a topic completely, there are at least two words with different nature in a sentence[8].

Noun+Verb or Noun/ Adjective(Adverb) +Verb (1)

With the change of the times, the development of language has become more and more mature, and the sentence structure has been more diverse than shown in formula (1), types and quantities of vocabulary have also been greatly expanded. However, the fundamental logical mode of people thinking about the world is still in the form of "Ontology-Attribute", consequently the language we use is basically in accordance with "Reference-Attribute", that is, specifying the describing object, then describing the characteristics of the object. Therefore, we can use a notional word as the center word, and then state its attribute characteristics with other words, in this way, a message can be expressed. If a sentence is regarded as a complete unit of description or information expression, all the words with actual meaning in a sentence is a vocabulary-centered word aggregation, the order of the words is not important.

2.2. Generative Probability of Sentence

There are many words in a sentence, we can divide these words into three categories: verbs used to describe dynamic changes, nouns used to describe things regarding the changes, and adjectives used to modify the nouns or the verbs.

Then the process of forming a sentence is considered as: firstly, we write out the dynamic changes of things, then we mark the objects involved in the actions, lastly we modify the attributes of the actions or objects.

Assuming that there is a document D that can be segmented, and parts of speech of the words in the document can be specified to form three sets, the set of nouns denoted by $n = \{n_1, n_2, \dots, n_s\}$, the set of verbs denoted by $v = \{v_1, v_2, \dots, v_r\}$, the set of adjectives and adverbs denoted by $adj = \{adj_1, adj_2, \dots, adj_t\}$.

IOP Conf. Series: Journal of Physics: Conf. Series **1267** (2019) 012049 doi:10.1088/1742-6596/1267/1/012049

Then every sentence S in the document is a triple made up of the different kinds of vocabulary in the document:

$$S=\{ v_{1-r} \in D, n_{1-s} \in D, adj_{1-t} \in D\}$$

On the basis of formula (1), we can deduce the generative probability of statement S as:

$$P\ (S) = P_S\ (n, v, adj) \tag{2}$$

And according to the chain rule[9] of conditional probability, the formula can be:

$$P_S\ (n, v, adj) = P_S\ (adj|v, n) P_S\ (n|v) P_S\ (v) \tag{3}$$

Formula (3) is important not only for it simplifies the generative probability with three parameters, but also for it illustrates the interpretation of text model: The probability of generating a sentence is divided into three steps: the verb is chosen as a center word from the vocabulary set, then the nouns which involved in the dynamic change, also known as the objects of the action, are selected, finally, the modifiers are chosen conditionally on the dynamic change or the objects.

3. Text Modeling

3.1. Probability Resolution

In terms of formula (3), the semantic model established in this paper is referred to as a hierarchical probability model with three-levels. When estimating the model, we can assume that the vocabulary within a "word bag model" where words are relatively independent for each other. This property will facilitate the parameter estimation in the following work. Therefore, when we describe the reference of "verb-noun", we can simplify the formula (3) as follows:

$$P_S\ (n|v) = \prod\nolimits_{v\in s, n\in s} P(n|v) \tag{4}$$

As for the semantic relationship between modifiers and verbs/nouns, when the words are relatively independent, we can attain:

$$P_S\ (adj|v, n) = P_S\ (adj|v) P_S\ (adj|n) = \prod\nolimits_{v\in s, adj\in s} P(adj|v) \prod\nolimits_{n\in s, adj\in s} P(adj|n) \tag{5}$$

Taking the products of the probabilities of words, we obtain the generative probability of a sentence:

$$P_S\ (n, v, adj) = \prod\nolimits_{v\in s, n\in s, adj\in s} P(adj|v) P(adj|n) P(n|v) P(v) \tag{6}$$

As a result, the generation probability of sentence can be realized by the conditional probability of just two words in vocabulary set.

There are many sentences in a document, and a complete meaning for a human to express is just based on the grouping of verbs, nouns and modifiers in a sentence, then we can get the generative probability of a document D with triple words:

$$P_D\ (n, v, adj) = \sum\nolimits_{S\in D} P_S\ (n, v, adj) \tag{7}$$

Due to the chain rule, the $P_D\ (n, v, adj)$ can also be decomposed, then based on formula (6) and (7), we can obtain the following calculations:

The probability of a verb is just as selecting one item from all:

$$P_D\ (v) = \frac{1}{count(verbs)} \tag{8}$$

Where $count(verbs)$ means the number of verbs in the document.

Given a verb, the probability of a noun relevant to the verb in a sentence can be:

$$P_D\ (n|v) = \sum\nolimits_{S\in D} \frac{P_S\ (n, v)}{P_S\ (v)} = \frac{count(v \in \text{s and } n \in \text{s})/count(s)}{count(v \in \text{s})/count(s)}$$

Then we get:

$$P_D\ (n|v) = \frac{count(v\in \text{s and } n\in \text{s})}{count(v\in \text{s})} \tag{9}$$

In the same way, the calculation formulas can be attained:

$$P_D\,(adj|v) = \frac{count(v\in s \text{ and } adj\in s)}{count(v\in s)} \tag{10}$$

$$P_D\,(adj|n) = \frac{count(n\in s \text{ and } adj\in s)}{count(n\in s)} \tag{11}$$

3.2. Parameter Estimation

In order to calculate the conditional probabilities in the above calculations, it is necessary to count the numbers of words appearing in a sentence at the same time. A common method is to build a matrix, since the probability involves three kinds of words, the word probabilities are parameterized by a three-dimensional matrix of $n \times v \times adj$, and the values in the matrix is represented by the times of words appearing in the same sentence.

However, it's complicated to estimate parameters from such a three-dimensional matrix because the large amount of vocabulary in a document may lead to computationally intractable problems. Thus we consider an algorithm which can reduce this three-dimensional matrix by decomposition with one kind of word so that the computation can be available from a normal two-dimensional matrix.

Algorithm De_RC

(1) Choose the least of the three kinds of words.(In this paper, we assume that modifier is the least word among vocabulary in a document)

(2) The row of the matrix consists of both one kind of words and the least words chosen by previous step, such as verbs and modifiers. Hence the row can be defined as follows: $row_j \in \{n_1 \ldots n_s\}$, $row_k \in \{adj_1 \ldots adj_t\}$

(3) The column of the matrix consists of another kind of words and the least words chosen by step 1, such as nouns and modifiers. Then the column can be defined as follows: $col_i \in \{v_1 \ldots v_r\}$, $col_k \in \{adj_1 \ldots adj_t\}$

(4) The values in the matrix can be denoted as $mat_{rwo,col}$, then the matrix is shown in figure 1.

$$\begin{bmatrix} & v_1 & \cdots & v_r & adj_1 & \cdots & adj_t \\ n_1 & & & & & & \\ \cdots & & & & & & \\ n_s & & & mat_{col,row} & & & \\ adj_1 & & & & & & \\ \cdots & & & & & & \\ adj_t & & & & & & \end{bmatrix}$$

Figure 1. A diagram of the matrix

(5) From the matrix, we can get:

$$P_D\,(n,v,adj) = \frac{mat_{row_k,col_i}}{\sum_1^r count(v_i)} \frac{mat_{row_j,col_k}}{\sum_1^s count(n_j)} \frac{mat_{row_j,col_i}}{\sum_1^r count(v_i)} \frac{1}{count(verbs)} \tag{12}$$

In this way, the parameters in formula (7) can be computed generally, we can achieve the generative probability for a document at a cost of additional items in both rows and columns, however vastly simplify the computational complexity from a three dimensional matrix.

4. Evaluating the Language Model

A Language Model is always expected to predict the k+1 word while given the first k words of a sentence. In a sense, we hope the model can give the probability distribution of the possible occurrence of the K+1 word.

IOP Conf. Series: Journal of Physics: Conf. Series **1267** (2019) 012049 doi:10.1088/1742-6596/1267/1/012049

Perplexity[10] is such a criterion that can be used to measure the quality of Language Models in the field of Natural Language Processing, it can estimate the probability of a sentence based on each word.

$$Perplexity(sentence) = P(w_1 w_2 \dots w_N)^{-\frac{1}{N}} \quad (13)$$

The perplexity is algebraiclly equivalent to the inverse of the geometric mean per-word likelihood. A lower perplexity score indicates better generalization performance. Especially, when the language model is a unigram model, the perplexity can be:

$$Perplexity(sentence) = 2^{-\frac{1}{N}\sum \log(P(w_i))} \quad (14)$$

In this paper, $P(v_i) = \frac{count(v_i) \in sentences}{count(vocabulary)}$.

The nouns and modifiers are generated as the verbs' relevant objects or attribution, so:

$$P(w_i) = P(v_i) \prod_{j=1}^{s} P(n_j|v_i) \prod_{k=1}^{t} P(adj_k|v_i) \quad (15)$$

From formula(14) and formula(15), the perplexity score for text model which is based on linguistic analysis with different parts of speech can be obtained. Then we trained all the variables in models using the 20newsgroups[11] to compare the generalization performance with unigram, bigram, and trigram.

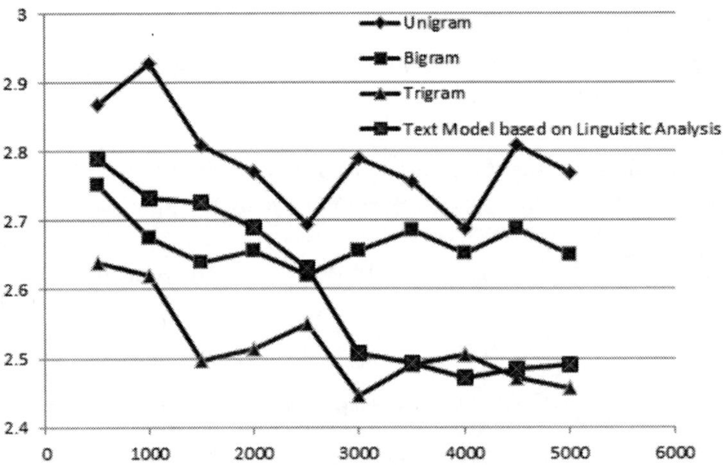

Figure 2. Perplexity from text models

The experiment begins at segmentation of parts of speech, the words except verbs, nouns, and modifiers are removed. Figure 2 indicates that the generative probability of a sentence proposed in this paper is approximated to the probability in a bigram when the number of words is small. However, with the increase of vocabulary, it plays a better and more accurate performance than unigram and bigram. Additionally, our text model is approximately to trigram with less relevant words involved and the changeability of the words makes the computation simpler.

Furthermore, we conduct experiments on different types of documents in the 20 newsgroup with their sizes varying from 200 to 5000 words.

From table 1, the experiment results present higher scores when the news document contains too many proper nouns and parameters, especially in the type of Motorcycles. These professional words occurred frequently in the document often represent a name for the objects' in corresponding area, which causes very small probabilities of $P(n_j|v_i)$ and the Perplexity grows.

Table 1. Perplexity from various documents

	N=200	N=500	N=1000	N=2000	N=5000
Graphics	2(1.59197)	2(1.55205)	2(1.21148)	2(1.35249)	2(1.28547)
Pc.hardware	2(1.42305)	2(1.6931)	2(1.16733)	2(1.25861)	2(1.26879)
Motorcycles	2(1.75798)	2(1.60706)	2(1.75008)	2(2.0178)	2(1.81253)
Electronics	2(1.55432)	2(1.86683)	2(1.5993)	2(1.51268)	2(1.47268)
Space	2(1.58746)	2(1.57852)	2(1.46556)	2(1.36985)	2(1.25478)

5. Conclusion

In this paper, we assume that corresponding words in a sentence contribute to describing certain meaning of the topic, and all kinds of description by human is fundamentally based on verbs, nouns, and modifiers. Consequently the text can be modeled according to the relationship among these three kinds of words. The generative probability of the model is given and an algorithm for dimensionality reduction is proposed to overcome the common problems of matrix built in text modeling.

In the future work, we plan to extend the text modeling to the application for document classification and feature extraction. On the other hand, the improvement for the sparse matrix in this paper is essential. More researches need to be done in the analysis and selection of semantic words.

References

[1] Salton G , Wong A. and Yang C. S. 1975 A vector space model for automatic indexing *Communications of the ACM* pp. 613-620

[2] Chowdhury G. 2003 Natural Language Processing *Annual Review of Information Science & Technology* pp.51-89

[3] Blei, David M. 2012 Probabilistic Topic Models *Communications of the ACM.* pp.77–84

[4] Thomas H. 2000 Learning the Similarity of Documents : an information-geometric approach to document retrieval and categorization *Advances in Neural Information Processing Systems* MIT Press pp.914-920

[5] Blei, David M. and Andrew Y. 2003 Latent Dirichlet Allocation, *Journal of Machine Learning Research.* pp.993–1022

[6] Stanford University, Computational Linguistics, *Stanford Metaphysics Research Lab, Encyclopedia of Philosophy*, Apr, 2017

[7] Goldberg Y. 2016 A Primer on Neural Network Models for Natural Language *Processing. Journal of Artificial Intelligence Research* pp.345–420

[8] Zheng J. 2017 , *Principle and Practice of NLP Chinese Natural Language Processing*, Publishing House of Electronics Industry pp 68-71

[9] Russell, Stuart J and Norvig 2003 *Artificial Intelligence: A Modern Approach,* Upper Saddle River, New Jersey: Prentice Hall, 2nd edition

[10 Peter E, Vincent J 1992 An Estimate of an Upper Bound for the Entropy of English *Computational Linguistics* pp.31-40

[11 Xiaona W., Jia Z. and Jianfeng Y. 2014 Finding Better Topics: Features,Priors and Constraints *Pacific-Asia Conference on KnowledgeDiscovery and Data Mining*. pp. 296-310, New York, Springer

IOP Conf. Series: Journal of Physics: Conf. Series **1267** (2019) 012050 doi:10.1088/1742-6596/1267/1/012050

Semantic Segmentation of Remote Sensing Images via Stepwise-Refined Large-Kernel Deconvolutional Networks

Zhao Xinwei[1], Li Haichang[2], Wang Rui[2], Zheng Changwen[2] and Shi Song[3]

[1] Institute of Software Chinese Academy of Sciences, University of Chinese Academy of Sciences, Beijing, China
[2] Science and Technology on Integrated Information System Laboratory, Institute of Software Chinese Academy of Sciences, Beijing, China
[3] Teleware Info & Tech (Fujian) Co.,LTD, Fujian Province, China

*E-mail: rayarrow@qq.com

Abstract. Deep CNN based semantic segmentation has been developed for several years and many models are proposed. However, most of them are designed for natural scene images such as PASCAL VOC, and cannot perform very well on remote sensing images, in which objects are much smaller and more densely distributed than those in natural scene images. In this paper, we demonstrate the importance of high-resolution feature maps and the problem of large dilated convolutional kernels in semantic segmentation of remote sensing images. Furthermore, we propose a Stepwise-Refined Large-Kernel Deconvolutional Network with a focus on small and densely-distributed objects such as houses and buildings, or long and narrow ones such as roads and rivers. Experiments on a public available ISPRS Vaihingen Challenge Dataset and our self-compiled Fujian Dataset show that our model outperforms the state-of-the-art models in semantic segmentation of remote sensing images.

1. Introduction

Semantic segmentation is an important topic in computer vision, which can be seen as a per-pixel classification task. Specifically, given an image, we need to classify each pixel within a fixed set of classes, and produce a segmentation mask at last. See figure 1. Semantic segmentation has been widely applied in many fields such as autonomous vehicles, medical image analysis and land cover monitoring.

For image classification task, only semantic information is needed. But in semantic segmentation, spatial information is also required, in ordered to locate objects and recover their boundaries. Although deep learning based methods for semantic segmentation have been developed for many years, localization still remains a tough task especially for boundaries of objects.

Most proposed networks for semantic segmentation employs an encoder-decoder structure, in which encoders (such VGG16 [1] and ResNet [2]) are used to extract spatial and semantic information, while decoders are used to gather them to recover the results. FCN [3] (Fully Convolutional Neural network) was the first model trained end-to-end to perform the dense pixel-wise prediction tasks. Although FCN largely speeded up the training process and improved the accuracy, it cannot avoid the loss of spatial information caused by pooling layers, and this is the major reason to restrict the performance.

To tackle this problem, [4] proposed dilated convolution to replace pooling and standard non-dilated convolutional layers, which in turn expands the receptive fields in addition to preserving the resolution of feature maps without increasing the number of parameters. U-Net [5] suggests concatenating low-level features to high-level ones to compensate for the loss of spatial information. Deeplab v3 [6] and PSPNet [7] use pyramid pooling to capture multi-scale contextual information. Based on Deeplab v3, Deeplab v3+ [8] merged low-level feature maps to the pyramid pooling module like U-Net.

Among these models, Deeplab v3+ achieves state-of-the-art performance on PASCAL VOC dataset. It captures multi-scale contextual information at the highest layer of the encoder using ASPP (Atrous Spatial Pyramid Pooling), which employs 6-rated, 12-rated and 18-rated dilated kernels. However, objects in remote sensing images are small and densely-distributed, so regions covered by the kernels of ASPP often share similar patterns and not much discriminative semantic information can be captured. Worse still, noise are introduced occasionally. An example will be given to further illustrate this problem.

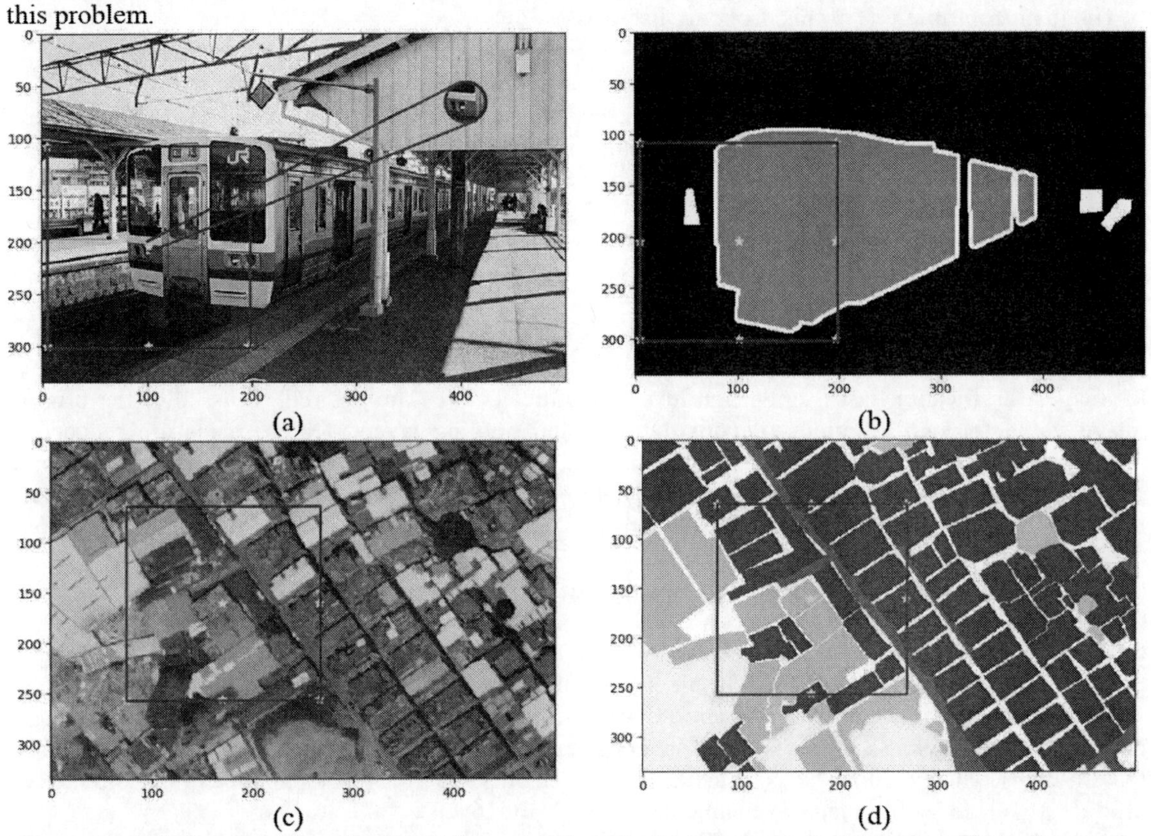

Figure 1. Effects of 6-dilated convolution on 16-stride feature maps ignoing the effect of receptive field.. Axes are kept on purpose to measure the sizes of objects. (b) and (d) are the corresponding segmentation mask of (a) and (c). (a) and (b) come from PASCAL VOC while (c) and (d) come from Fujian Dataset. Red rectangles are approximate visualization of the 6-rate dilated convolutional kernel of ASPP on 16-stride feature maps and yellow stars are non-zero weights in the kernel. The 6-dilated convolutional kernels are up sampled 16 times instead of down sampling original images by 16 times, as 16-time down sampled images are too blur and objects are hard to be identified.

In figure 1(a), it's not easy to tell if the central 3x3 region covered by the 6-rate dilated kernel's center weight belongs to a bus front or a train front when the surrounding areas are obscured (as a 3x3 non-dilated kernel perceives). The dilated kernel captures rich contextual information that the vehicle runs on railways and platforms are on both sides. With this information, this vehicle is much more likely to be classified as a train rather than a bus. In contrast, the dilated kernel in figure 1(c) is too aggressive and it reaches too far to capture helpful information. Areas 50m away has little to do with the class of the target building concerned. 12-rate and 18-rate dilated kernels are much more

IOP Conf. Series: Journal of Physics: Conf. Series **1267** (2019) 012050 doi:10.1088/1742-6596/1267/1/012050

aggressive and likely to capture irrelevant noise. Rather, dilated kernels with small rates that capture the boundary and adjacent buildings or roads may help. Thus, this structure loses its advantages in remote sensing images.

Although U-Net utilizes feature maps of all sizes and recovers the result in a stepwise fashion, it extracts surplus semantic information for our task and is hard to train. Besides, some feature maps to be concatenated in the decoder from earlier layers may be harmful. The goal of this paper is to demonstrate the importance of spatial information for the overall accuracy of our task and propose a method to deal with small and densely-distributed objects. Our model picks only useful feature maps from the encoder. In addition, large kernels make it shallow and easy to train without decreasing the receptive fields.

The main contributions of this paper are listed as follows:

- We modify popular state-of-the-art models in recent years to demonstrate the importance of spatial information in semantic segmentation of remote sensing images.
- A Stepwise-Refined Large-Kernel Deconvolutional Network is proposed to deal with small and densely distributed objects in semantic segmentation of remote sensing images.

2. Proposed Models

In this section, we present our model from encoder and decoder aspects respectively. Here we denote *stride* of feature maps as the ratio of the size of the original image to the size of these feature maps. For example, 4-stride feature maps indicate their size are 1/4 of the original input image.

2.1. Encoder

In image classification tasks, only high-level semantic features are desired. Thus, after the original image going through a serious of convolutional and pooling layers, feature maps are reduced to vectors and all spatial information are discarded. But in semantic segmentation, spatial information is vital to recover the segmentation mask Therefore, the encoder should preserve more spatial information at the expense of losing some semantic information. A balance has to be struck between the two kinds of information.

Reduced VGG16: FCN (with VGG16 as the encoder) reduces the size of feature maps to 1/32 of the original image's size and produces a coarse segmentation mask, in which boundaries of objects are overly smoothed [3]. However, it becomes a fiasco on remote sensing images. Consider a house in figure 1(d), which covers less than 30 pixels. It will reduce to covering less than 1 pixel in the 32-stride feature maps. Furthermore, semantic richness in remote sensing images is much less than that in natural scene images. So we remove all layers after the 4th pooling layer and reduce the number all feature maps' channels to 64. As a result, the model preserves only 3% parameters of VGG16 and produce 16-stride feature maps. We name it reduced VGG16.

Dilated Resnet: Dilated convolution is quite powerful in preserving spatial information and increasing receptive fields without increasing parameters. Deeplab v3 and PSPNet, take dilated convolution as an essential part. We use the same dilated Resnet as that in Deeplab v3.

2.2. Decoder

Stepwise-Refined Large-Kernel Deconvolutional Network: Compared to semantic segmentation of natural scene images, the effect of semantic information decreases a lot when spatial information is scarce in remote sensing images. So we design a Stepwise-Refined Large-Kernel Deconvolutional Network to focus on spatial information. This structure is composed of 4 blocks, with each block made up by a 2x upsample layer, followed by 64 7x7 convolutional kernels, followed by a batchnorm layer and a ReLU activation function. An overview of our model is shown in figure 2.

We recover the segmentation mask stepwise instead of in one step as FCN and Deeplab v3 do, considering that objects are too small and their boundaries are hard to be sharply recovered by a single upsample layer. In addition, large 7x7 convolutional layers is employed instead of their lightweight equivalent – 3 consecutive 3x3 convolution layers, as the latter deepen the network and take up more memory to store feature maps, which in turn causes most GPUs of 12G memory running out of memory, including ours, despite the fact that the latter has less parameters. At last, 4-stride and 8-stride feature maps from the encoder are concatenated in the decoder to compensate for lost spatial information. 2-stride feature maps are ignored due to the large gap between them and feature maps in the decoder. Including 2-stride feature maps does harm to our model.

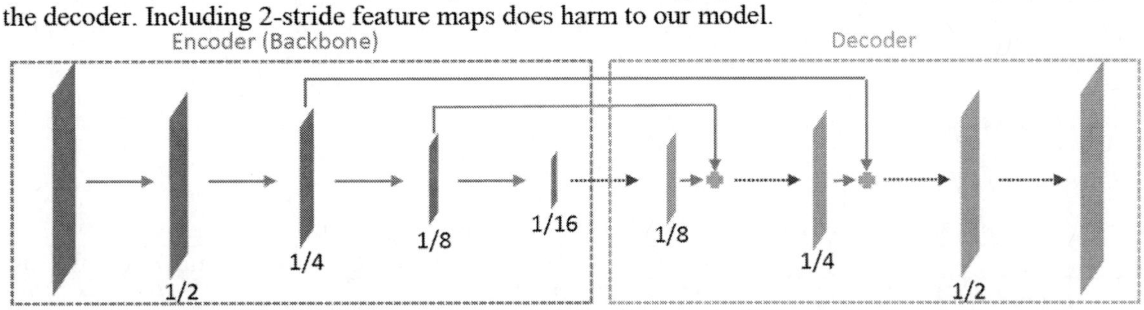

Figure 2. Overview of our Stepwise-Refined Large-Kernel Deconvolutional Network.

3. Experiments

3.1. Dataset

The datasets in our experiments are ISPRS Vaihingen Challenge Dataset [9] and Fujian Dataset.

Vaihingen Dataset consists of 3-band IRRG (Infrared, Red and Green) images. These images vary in size and average around 2500×2000. We choose 5 images (11, 15, 28, 30, 34) as test set and the other 11 images as training set, like most online submitter did.

Fujian Dataset contains 50 images captured by a satellite at 0.5m/pixel spatial resolution over a city in Fujian Province, China. The image sizes are 1000×1000. 10 classes are annotated and the colors for each class are specified in Table 1. White color are undefined regions and ignored when training. We conduct a 5-fold cross validation to evaluate our model on this dataset.

Table 1. Annotation details. Each cell in 1st and 3rd rows are RGB values followed by a color block. Each cell in 2nd and 4th rows indicates the class that the color above represents.

140,255,90	155,255,100	160,255,110	255,255,20	200,50,255
Plantation	Woodland	Lawn	Soil	Road
92,92,92	200,200,200	229,229,229	180,180,180	100,200,250
Shanty town	Simple buildings	Roof	Facade	water

3.2. Configuration

Training policy, and parameter settings are detailed in the following paragraphs.

Fine-Tune: VGG16 and Resnet-101, the encoders in our model, are pre-trained on ImageNet. All decoders and the reduced VGG16 are trained from scratch.

Data augmentation: Due to the insufficiency of data, data augmentation is necessary to train a decent model. We augment our data in the following three ways. First, images are left-right flipped randomly with a probability of 0.5. Second, images are rescaled randomly from 0.5 to 2.0 times the original size. Third, crop a 513×513 patch from the image produced in the second phase. Before the third phase, an image might be smaller than 513×513 if shrunk too much in the second phase. In this case, some pixels should be padded to the right and bottom of the image, in order to guarantee it's not smaller than 513×513. These pixels will be ignored when the loss is calculated as they are useless for training.

Optimization: Our models are implemented using Tensorflow and trained on a single NVIDIA TITAN Xp. We use SGD with batch size of 8, momentum of 0.9 and weight decay of 0.0005 to train our models. Each model is trained for 9000 epochs (Data augmentation substantially reduces the risk of overfitting if over-trained). Initial learning rates for VGG16 and Resnet-101 are set to 0.001 and

0.007 respectively. For each model, the learning rate is controlled by the polynomial learning rate policy:

$$\alpha_k = \alpha * \left(1 - \frac{k}{m}\right)^p$$

where α_k is the learning rate at iteration k, α is the initial learning rate, k is the current iteration, m is the number of iterations to finish the training process and p is power mentioned above.

In our experiments, we report a common metrics from semantic segmentation and machine learning: mean Intersection over Union (mean IoU).

3.3. Results and Discussion

Table 2. Results of models on the dataset. "+" separates the encoder and decoder.

Model	mIoU V[a]	mIoU F	Model	mIoU V	mIoU F
VGG16+UP[b] (FCN)[c]	46.46%	47.15%	Res+UP	55.11%	56.16%
Reduced VGG16+UP	46.88%	47.37%	Res+SRLKD[f]	55.21%	56.47%
Res[d]+ASPP (Deeplab v3)	54.68%	56.10%	**Res+SRLKD-F8-F4**	**59.18%**	**58.86%**
Res+ASPP-F4[e] (Deeplab v3+)	58.69%	57.71%	Res+SRLKD-F8-F4-F2	58.60%	57.95%

[a] "mIoU V" and "mIoU F" stand for mean IoU on Vaihingen Dataset and Fujian Dataset respectively.
[b] "UP" stands for a single upsample layer.
[c] If a combination has been proposed before, we put its name inside the parentheses.
[d] "Res" stands for Resnet101 which employs dilated convolution, producing 16-stride feature maps.
[e] F8, F4 and F2 means 8-, 4- and 2-stride feature maps are concatenated in the decoder.
[f] SRLKD stands for our proposed Stepwise-Refined Large-Kernel Deconvolutional Network.

Table 2 list the performance of previous state-of-the-art models and our proposed models.

The reduced VGG16, with only 3% parameters of FCN and trained from scratch, outperforms the original FCN by a small margin. It indicates that most semantic information carried by VGG16 is redundant, and it doesn't benefit recovering the boundaries of objects in remote sensing images. In contrast, doubling the size of the last feature maps and exchanging more spatial information with semantic information helps improving the accuracy.

It is also worth noting that Res+UP outperforms Deeplab v3 by a little margin. Note that ASPP in Deeplab v3 is a superset of UP, but performs worse than that. This implies that the contextual information captured by ASPP is unnecessary or even harmful. Deeplab v3+ saves the situation purely by merging 4-stride feature maps based on Deeplab v3, and improves the performance obviously.

These aforementioned experiment results all demonstrate the insignificance of surplus semantic information and the importance of spatial information, which in turn encourages us to further focus on the spatial information. This is where our model comes in.

The best results are obtained by Res+SRLKD-F8-F4. SRLKD is designed to be quite shallow as it has no need to further extract semantic features. It recovers the segmentation mask in a stepwise fashion and boundaries of objects are much more refined. See figure 3(d). Besides, the main cause of the mixing of adjacent houses in the shanty town is the undefined white areas, as seen in figure 3(b). Further annotation are required to solve this problem. It's noteworthy that with additional 2-stride features concatenated, the performance of Res+SRLKD-F8-F4-F2 drops by a small amount. The gap between 2-stride feature maps and high-level feature maps largely accounts for this. 2-stride feature maps only go through a single 7x7 convolutional layer and they differ little from the original images, which contain little semantic information. In this case, these low-level feature maps cannot be concatenated to high-level feature maps seamlessly. Thus, they are more likely to introduce noise and do harm to our model rather than compensate for spatial information.

4. Conclusion

In this paper, we demonstrate that the spatial information in remote sensing images is more important than that in natural scene images, as objects in remote sensing images are small and densely distributed. Based on this, we propose a Stepwise-Refined Large-Kernel Deconvolutional Network to focus on spatial information. During the deconvolution process, we employ a large-kernel but shallow deconvolutional structure to recover the segmentation mask stepwise instead of in one step like FCN and Deeplab v3 do. Our model recovers the objects' boundaries more sharply and outperforms the previous ones in remote sensing images. In the future, we plan to simplify our network to produce 8-stride feature maps and further preserve the spatial information given the constraints of GPU memory.

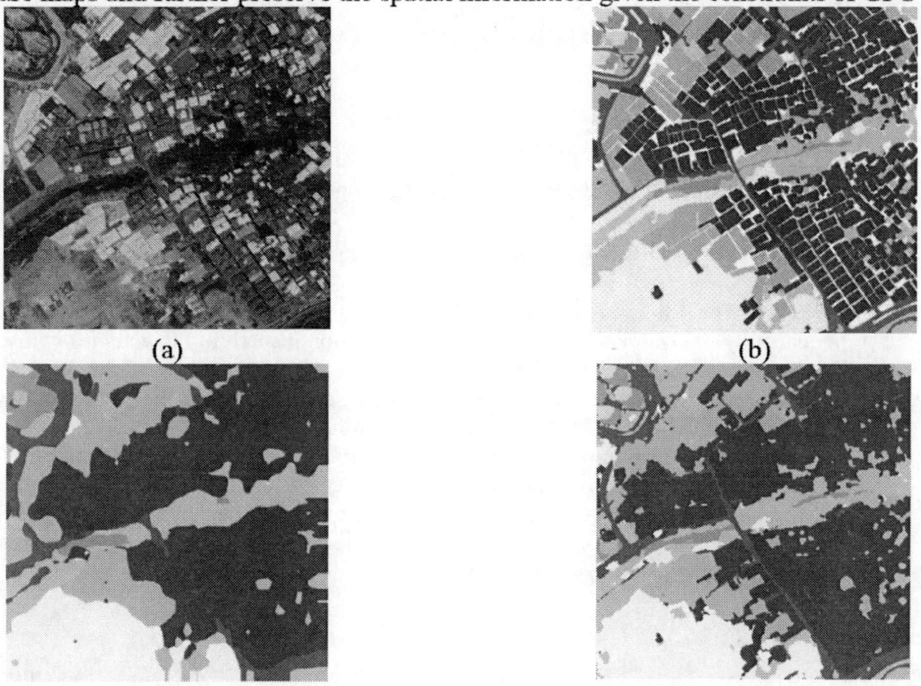

(a) (b) (c) (d)

Figure 3. Visual results on the remote sensing dataset. (a) is the original image. (b) is the ground truth label. (c) and (d) are predictions of FCN and Res+SRLKD-F8-F4 respectively.

Reference

[1] Simonyan K, Zisserman A. Very deep convolutional networks for large-scale image recognition. arXiv preprint arXiv:1409.1556. 2014 Sep 4.
[2] He K, Zhang X, Ren S, Sun J. Deep residual learning for image recognition. InProceedings of the IEEE conference on computer vision and pattern recognition 2016 (pp. 770-778).
[3] Long J, Shelhamer E, Darrell T. Fully convolutional networks for semantic segmentation. InProceedings of the IEEE conference on computer vision and pattern recognition 2015 (pp. 3431-3440).
[4] Yu F, Koltun V. Multi-scale context aggregation by dilated convolutions. arXiv preprint arXiv:1511.07122. 2015 Nov 23.
[5] Ronneberger O, Fischer P, Brox T. U-net: Convolutional networks for biomedical image segmentation. InInternational Conference on Medical image computing and computer-assisted intervention 2015 Oct 5 (pp. 234-241). Springer, Cham.
[6] Chen LC, Papandreou G, Schroff F, Adam H. Rethinking atrous convolution for semantic image segmentation. arXiv preprint arXiv:1706.05587. 2017 Jun 17.
[7] Zhao H, Shi J, Qi X, Wang X, Jia J. Pyramid scene parsing network. InProceedings of the IEEE conference on computer vision and pattern recognition 2017 (pp. 2881-2890).
[8] Chen LC, Zhu Y, Papandreou G, Schroff F, Adam H. Encoder-decoder with atrous separable convolution for semantic image segmentation. InProceedings of the European Conference on Computer Vision (ECCV) 2018 (pp. 801-818).
[9] Cramer M, The DGPF-test on digital airborne camera evaluation–overview and test design. Photogrammetrie – Fernerkundung – Geoinformation 2010 (pp. 73-82).

IOP Conf. Series: Journal of Physics: Conf. Series **1267** (2019) 012051 doi:10.1088/1742-6596/1267/1/012051

Design of a Synchronous Control System for Mobile Lift Based on Wireless Network

Lu Zhuang and Haijin Chen

School of Information Science and Technology, Nantong University, Nantong, China

Email: chen.hj@ntu.edu.cn, 1535611336@qq.com

Abstract. The lift used in automobile maintenance industry is usually installed in a place permanently. It occupies the place, and also has poor adaptability to vehicles with different size. Furthermore, there exists physical connection between lifting columns, which affects maintenance operation. To solve these problems, a mobile hydraulic lift control system based on wireless network has been designed. It uses the STM32F103 microcontroller and nRF905 wireless communication module to form a control network for several mobile lifts. The height of each lift is measured with laser displacement sensors, and synchronously controlled through the wireless control network. The control system was verified by lifting the Jinlong bus with wirelessly connected four lifts. The lifting process is stable, which demonstrates a bright prospect for practical usages.

1. Introduction

Automobile lift is a vehicle lift equipment used in the process of vehicle maintenance. When the vehicle is driven to the lifting machine, it can be lifted to a certain height by manual operation, which is convenient for vehicle maintenance. It is a necessary equipment for garage. In recent years, car ownership in China has been on the rise, and the corresponding demand for car maintenance has also risen sharply, resulting in a very strong market demand[1]. The traditional lift has column type, shear type, trench type and other structures. The technology is relatively mature, but there are also obvious shortcomings. Take the double-column lift as an example, two vertical columns are fixed and installed in the maintenance site, and there is a fixed pipeline connection between the vertical columns to achieve synchronous lifting. Its disadvantages are as follows:

(1) The fixed installed column and the pipeline connecting the column occupy the valuable space of the maintenance site. When there is no maintenance task, the station cannot be used for other purposes;

(2) Once the column is installed, the maximum size of the vehicle it can lift will be limited, and the flexibility is poor;

(3) The connected pipeline between the columns is easy to trap the maintenance personnel, which is a potential safety hazard.

The mobile lift can overcome the above disadvantages of the traditional lift. A mobile lift corresponds to a freely movable lift column. The power is provided by battery, and through wireless communication, a number of lifting columns can be netted for operation on site. Through precise state detection, communication and control, multiple lifting columns can be simultaneously lifted (for example, 2 lifting units can lift family cars, 4 lifting units can lift common buses and trucks, and 6 lifting units can lift heavy-duty trucks). This is convenient to meet a variety of vehicle maintenance operations.

The related studies reported in the literature mainly involve traditional lifts, such as mechanical design, hydraulic design, PLC control and other technologies [2-6]. There are few reports on mobile lifts. Literature [7] reported the design of a mobile lift, and analyzed the mechanical model and hydraulic system. However, the control system only mentioned the lifting function and safety protection control, and did not further explain the implementation issue. In this paper, a wireless network-based synchronous control system for mobile lift is designed. The network control of four mobile lifts is realized, and a bus has been used as the load to successfully carry out the experimental verification.

2. System scheme

The lift studied in this paper uses hydraulic transmission with a 24V battery for power supply. The normal lifting speed is 3.5cm/s. The design requires that the maximum height deviation of each lift during lifting is less than 2.5cm, so as to maintain the stability of the lifted vehicle. The synchronous lifting working mode of the mobile lift for networking operation is shown in figure 1. The wireless network is composed of several lifts (4 lifts in the figure). One of them is set as the host lift, which is used to collect the information of each lift in real time, make decisions, and transmit control instructions to the slave lift. In order to meet the requirement of maximum height deviation, a high-precision displacement sensor is installed on each lift. The height is detected synchronously according to the command of the host lift and sent to the host lift in turn. The host lift makes judgment according to the height data collected. If the maximum height deviation reaches the speed regulation threshold, it will start the speed regulation program. If it is less than the stop speed threshold, it will stop the speed regulation program.

The schematic diagram of the hydraulic speed regulation system of a single lift is shown in figure 2. In the figure, YA1 branch is the main branch with a large flow. YA2 branch is speed regulation branch with small flow. During normal lift operation, both YA1 and YA2 are turned on. When speed regulation is required, YA2 is closed to reduce the speed. If the motor is started, and YA3 is turned off, the lift moves up. If the motor is stopped, and YA3 is turned off, the lift stops. If the motor is stopped, and YA3 is turned on, the lift moves down.

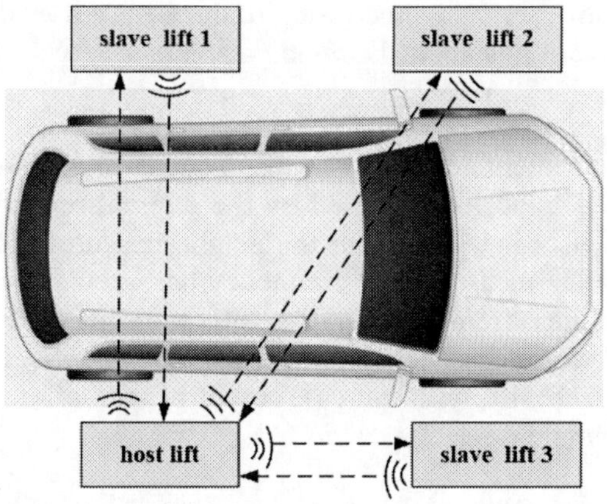

Figure 1. Schematic diagram of wirelessly grouped lifts

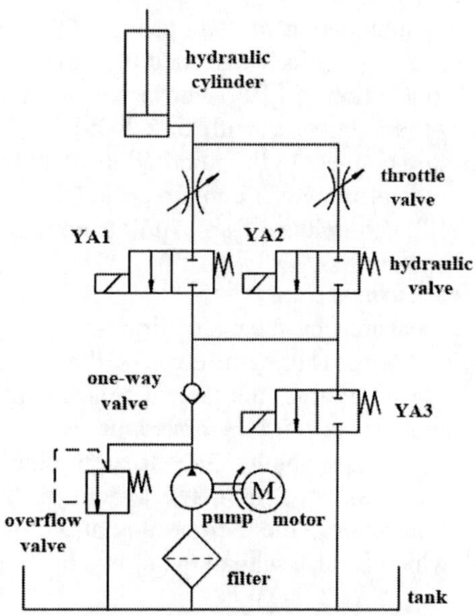

Figure 2. Schematic diagram of hydraulic speed regulating system of a single lift

Each lift is equipped with a controller, which can be configured by the user into stand-alone mode or networking mode on the job site as required. The power supply of the controller comes from 24V battery, and all the control is realized by microcontroller programming. Considering the real-time requirements of motion control for data processing, the 32-bit microcontroller STM32F103 is selected as the processing core with the highest clock frequency of 72MHz. Wireless communication uses nRF905 wireless transceiver module working in ISM(Industrial Scientific Medical) frequency band. Height measurement is made using laser displacement sensor. The lifting control and speed regulation signals are sent to YA1, YA2 and YA3 through relays. The touch screen is used to display the necessary status information and receive the parameter input during networking and other operation steps. The mechanical key inputs are set corresponding to the up, down, and stop command, respectively. The overall structure of the controller is shown in figure 3.

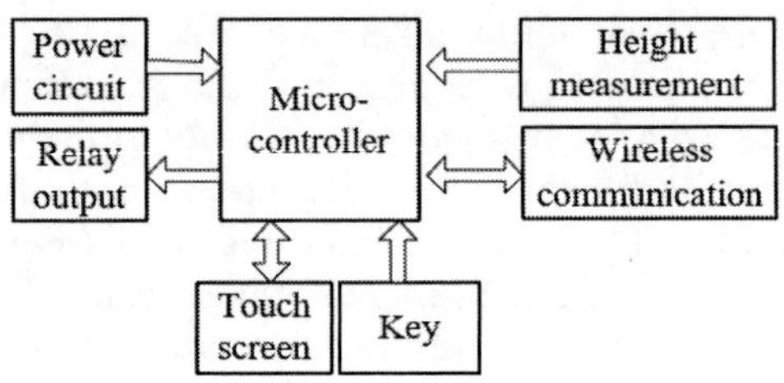

Figure 3. Block diagram of the controller

3. Controller circuit design

3.1. Power circuit

The operating voltage of each module of the controller is not completely consistent, which requires 5V and 3.3v. In this paper, the switching power supply circuit is used to convert the 24V DC input voltage into 5V, and then the linear voltage regulator is used to convert the 5V voltage into 3.3v. The circuit uses conventional technology and is not repeated here.

3.2. Wireless communication

The wireless communication module uses nRF905 wireless transceiver chip of Nordic company, and only a few peripheral circuits are needed to complete the wireless communication task. This is suitable for industrial applications [8][9], whose carrier frequency is 433MHz/868MHz/915MHz. The band can be used without license, with 512 bands and a wireless communication rate of 50kbps. SPI interface is used to connect to the microcontroller. The operating voltage is 1.9v-3.6v, and 3.3v is used in this paper. Its data frame can contain up to 32 bytes of user data. According to the size of user data, the time required to complete a frame of data communication is about 1.3ms-6.4ms.

3.3. Height measurement

Height can be measured by means of draw-wire displacement sensor, laser displacement sensor, etc. Considering that the measurement error of draw-wire displacement caused by the internal rewinding mechanism is apt to be large, the laser displacement sensor is selected for the height measurement. A laser displacement sensor with ±1mm accuracy and 30Hz measurement frequency has been selected. In other words, it takes about 33ms to complete a measurement, and the lifting distance of the corresponding lift is 3.5cm/s x33ms=0.115cm. In the worst case, if a lift is in the stop state and the other lifts rise normally, the height deviation of each lift will increase by 0.115cm after one measurement, which is far less than the allowable maximum deviation value.

3.4. Input and output module

The input and output module includes key, touch screen and relay output part. Keys are used to receive input signals such as up, down, stop and networking. Common key circuit structure is adopted.

IOP Conf. Series: Journal of Physics: Conf. Series **1267** (2019) 012051 doi:10.1088/1742-6596/1267/1/012051

The General Purpose Input Output (GPIO) port of MCU is directly connected to the key input signals. The touch screen is used to display working status information and provide necessary input interface in networking and other operation steps. The relay output is used to control the hydraulic valve and the input is from the MCU GPIO pin. The rated current of relay coil is about 15mA. In order to match the driving ability, a driving circuit is designed. The driving circuit of the corresponding relay of YA1 is shown in figure 4. The control signal from MCU controls the on-off state of transistor Q1, and then drives the relay U1 to the corresponding state. In the figure, diode D1 is used to provide a path for the current in relay coil to flow during the transition of Q1 being turned off.

3.5. Main control chip
The main control chip uses the 32-bit microcontroller STM32F103, which is made by the ST Microelectronics Company. It uses advanced RISC architecture, with built-in rich peripherals and large capacity FLASH storage space. The highest working frequency is 72MHz, which can meet the requirement of high-speed data processing, so as to make a quick response to all kinds of emergencies during the lifting process. The designed controller is shown in figure 5.

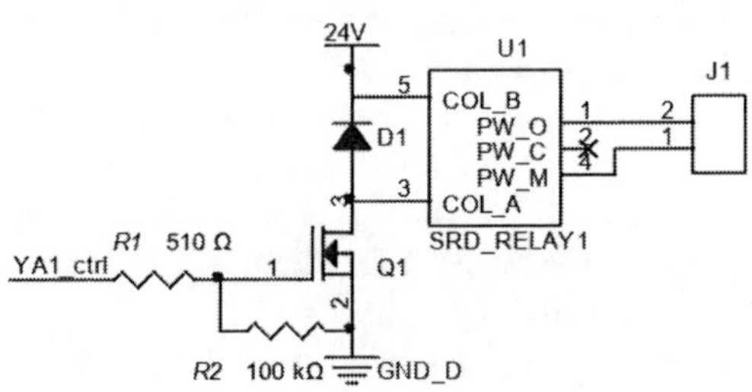

Figure 4. Relay interface circuit

Figure 5. The designed controller

4. Control method and software design

4.1. Wireless communication and networking control
The data frame of wireless module nRF905 includes four fields, as is shown in figure 6[10]. The preamble is automatically added by the hardware during transmission, and the address is used for the receiving end to match the transmitting end. The payload is the actual user data to be transmitted, and the CRC is used to verify the data. Time On Air (TOA) of a frame can be estimated as

$$TOA = t_{startup} + t_{preamble} + \frac{N_{addr}+N_{payload}+N_{CRC}}{BR} \quad (1)$$

Where, $t_{startup}$ is the RF startup time, and its value is 550 μs for the switching of receiving and transmitting modes; $t_{preamble}$ is the preamble time, which is 200 μs; N_{addr}, $N_{payload}$ and N_{CRC} are the binary bits of address, payload and CRC check code, respectively; BR is baud rate, which is fixed at 50kbps. In this paper, N_{addr} takes 4 bytes. The data to be transmitted each time mainly includes command encoding and height data, including reserved fields. The total length is 10 bytes. N_{CRC} is set to 1 byte. According to figure 6 and formula (1), the value of TOA is 3150μs.

preamble	address	payload	CRC

Figure 6. nRF905 data frame format

The first step is to set the network for the lift unit starts to work. The method is to set the grouped lift unit with the same address, which is uniformly assigned by the controller manufacturer to ensure

IOP Conf. Series: Journal of Physics: Conf. Series **1267** (2019) 012051 doi:10.1088/1742-6596/1267/1/012051

that the network address will not overlap. In the practical operations, it may happen that other irrelevant devices using the nRF905 module with the same address will disturb the function of the grouped lifts. To prevent it, additional information used for authentication is added in the user data field. During the rising or falling process, the host lift sends the broadcast command firstly, and each lift receives the command at the same time to measure the height. Then the host lift communicates with each slave lift in turn, and commands each slave lift to upload height data in turn. Finally, the host lift makes decision whether to start the speed adjustment process based on the collected height value. The time t_{cycle} required for the grouped lifts to complete a round of communication is

$$t_{cycle} = t_{broadcast} + (n-1) \cdot (t_{down} + t_{up}) \quad (2)$$

In the formula, $t_{broadcast}$, t_{down} and t_{up} are the time of broadcasting command, downloading command and uploading data, respectively; n is the number of grouped lifts. In a worst-case scenario, $t_{broadcast}$、t_{down} and t_{up} are all set as 3150μs of the maximum TOA value. If n = 4, then t_{cycle}=22050μs, or 22.05ms. It is less than the time the height sensor takes to measure for one time.

4.2. Speed regulation

Two thresholds, th_{start} and th_{stop}, are used for speed regulation. When the maximum height deviation is greater than th_{start}, the speed regulation program is started. For the lift in the speed regulation state, when its height deviation is less than th_{stop}, speed regulation is stopped.

There is a delay t_d from issuing the command to the speed actually being regulated. The delay consists of air transmission time TOA, relay response time, solenoid valve and hydraulic response time, program processing time and other parts. The response time of the relay in the design is about 10ms. Solenoid valve and hydraulic response time cannot be accurately measured. It is generally in tens to hundreds of milliseconds. In the study, a harsh case of 500ms is used for estimation. The program executing time is at the microsecond level, which is relatively small and negligible. The resulting td value is 513.15ms. During the td period, each lift moves at the original speed, and the maximum deviation is required to be less than the allowable value th_{max}, then

$$\Delta v \cdot t_d < th_{max} - th_{start} \quad (3)$$

In the formula, Δv is the speed deviation between the lifts, which is due to the influence of manufacturing process deviation, load difference and other factors. It is difficult to obtain Δv accurately. In this study, 30% of the rated speed of 3.5cm /s is taken, that is 1.05cm /s. The th_{max} is 2.5cm according to the design requirements. According to formula (3), th_{start} < 1.96cm can be obtained.

The determination principle of speed regulation threshold th_{stop} is similar to that of th_{start}, with the purpose of avoiding excessive regulation and forming new deviation. The calculation of th_{stop} can be expressed as

$$\Delta v \cdot t_d < th_{stop} \quad (4)$$

The calculated result is th_{stop} > 0.54cm. Under the premise that the values of th_{start} and th_{stop} meet the above calculation requirements, the closer to the conditional boundary, the easier it is to meet the requirements of non-deviation. However, the solenoid valve switches more frequently if the two thresholds are close to the condition boundary.

4.3. Program design

The working mode of the controller is divided into the host and the slave. The host mode needs to manage the lift in the whole network, and the software flow is shown in figure 7. Firstly, the hardware is initialized. Before the first synchronous lifting, all the lifts in the wireless network are adjusted to the same level and the height value is saved. Then the program enters key query state. when the "up" or "down" key is pressed, the host controller will send broadcast command to all slaves to query the height. At the same time, the host lift height is detected, and the status information of each slave height is queried successively. Then the host controller determines whether to start the speed regulation program. When the "up" or "down" key is released or the "stop" key is pressed, the host controller will send a broadcast to all the slaves, and command them to stop. The host lift is also stopped at the same time.

IOP Conf. Series: Journal of Physics: Conf. Series **1267** (2019) 012051 doi:10.1088/1742-6596/1267/1/012051

In slave mode, the controller performs corresponding actions according to the received host commands, and its software flow is shown in figure 8. It firstly initializes the hardware, then enters receive mode and waits for the wireless command from the host. After receiving the command, the command is parsed and the corresponding operation is completed. The slave remains in receive mode after executing the command to wait for the wireless command from the host.

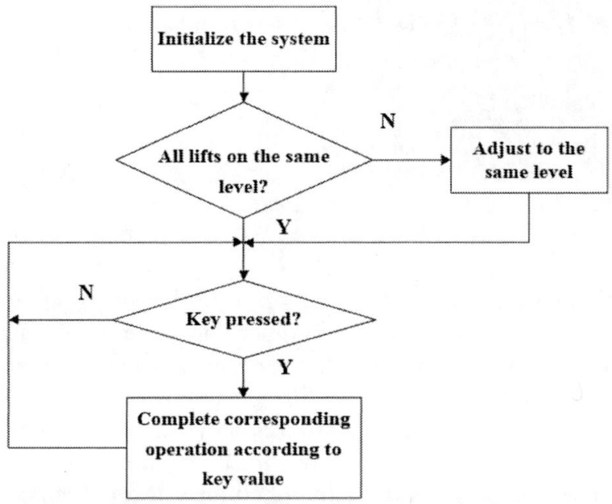

Figure 7. Flow chart of the host controller

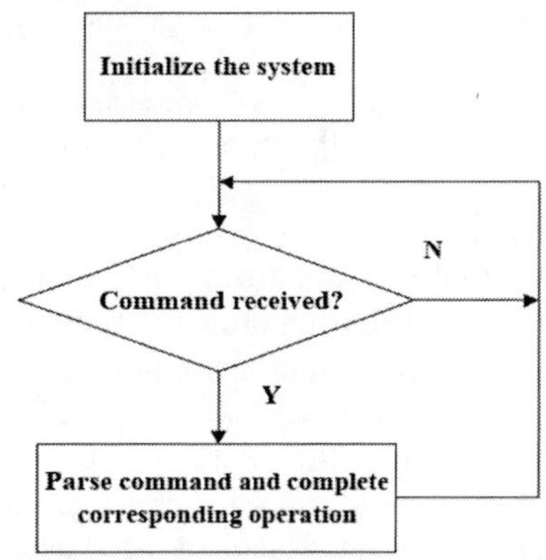

Figure 8. Flow chart of the slave controller

5. Experimental results

After the completion of the controller design, it is mounted into the control cabinet of the lift, and connected with the electrical and hydraulic systems. The following experimental tests have been conducted.

(1) Wireless communication test in the working environment: Two controllers are placed across a bus and continuously send and receive 1000 frames of data through the wireless module to test the communication stability. The experimental results show that the communication is smooth and the frame loss rate is zero.

(2) Repeatability test: Four lift units are grouped to lift 20 times with even load of 4 tons, and the total number of repeated experiments is 200 times after battery charging. During the experiment, the control system behaves normally.

(3) Partial load test: Two lifts are grouped, one lift with a load of 3.3 tons and the other with a load of 1.1 tons. In the process of lifting, the lift operates smoothly with the height deviation being within 2.5cm.

(4) Actual working condition test: four mobile lifts are used to carry out the lifting test on the bus outdoors, as is shown in figure 9. The experimental result shows that the control system works correctly. As can be seen from the figure, the mobile lift has low requirements for the working site and is flexible and convenient to use.

IOP Conf. Series: Journal of Physics: Conf. Series **1267** (2019) 012051 doi:10.1088/1742-6596/1267/1/012051

Figure 9. Test of the mobile lifts using a bus as the load

6. Conclusion

This paper analyzes and designs the wireless networking, communication, speed regulation and synchronous control of mobile lifts by using ISM band wireless communication technology. The corresponding circuit and program are designed to realize the information interconnection and synchronous control of several mobile lifts. And it has been applied for lifting the bus outdoors. The experimental results show that it is feasible to realize the network operation and synchronous control of mobile lift based on wireless communication and microcontrollers. In this paper, only the networking control of 4 sets of lift has been tested experimentally. In the future, the networking operation of more sets of lift can be studied to meet the needs of a wider range of applications, such as lifting ultra-long vehicles.

7. References

[1] Guo W, Huang ZX, Wu SS 2011 Analysis of the development trend of automobile lifts *Mechanical Design and Manufacturing Engineering* **40** pp 75-79.

[2] Tian H Y and Zhang ZY 2011 Design and simulation based on Pro/E for a hydraulic lift platform in scissors type *Procedia Engineering* **16** pp 772-781.

[3] Zhou YW, Zhang Q and He WD 2017 Simulation of hydraulic continuous lifting system of offshore platform based on fuzzy PID control *International Symposium on Computational Intelligence & Design* IEEE pp 91-94.

[4] Dou HB, Wang SP 2011 Synchronization control of hydraulic horizontal regulation system *Industrial Electronics & Applications* IEEE pp 1922-1927.

[5] Nuño F, Prieto, Díaz J, Diaz, Sáiz P J V, Carracedo J and Río C 2014 Electronic control of a novel system for lifting large structures *Transactions on Industry Applications* IEEE **50** pp 4260-4271.

[6] Ding HG, Zhao JY, Zhu QM 2018 A new hydraulic speed regulation scheme: valve-pump parallel variable mode control *Journals & Magazines* IEEE **6** pp 55257-55263

[7] Guo W 2011 Research on combined mobile hydraulic lift for heavy vehicles

[8] Guan W, Wang C, Cai YQ and Zhang HZ 2016 Design and implementation of wireless monitoring network for temperature-humidity measurement *Journal of Ambient Intelligence and Humanized Computing* **7** pp 131-138.

[9] Li HY and Tian Y 2013 The development of the wireless vehicle detector data receiving and processing host system *Emerging Technologies for Information Systems, Computing, and Management* (Springer New York) chapter 64 pp 571–577.

[10] Nordic Semiconductor ASA 2008 nRF905 product specification v1.5

A Phoneme Sequence Driven Lightweight End-To-End Speech Synthesis Approach

Zite Jiang, Feiwei Qin and Liaoying Zhao

College of Computer Science, HangZhou Dianzi University,Hangzhou 310018,China

zhaoly@hdu.edu.cn

Abstract. This paper develops an end-to-end neural network model for text-to-speech (TTS) system based on phoneme sequence. Inspired by the Tacotron-2, the proposed model adopts an encoder-decoder model with attention mechanism and applies mel-spectrogram to measure the intermediate acoustic feature. Phoneme sequence is used to replace the character sequence in order to overcome the shortage of the character feature used in Tacotron-2. Unlike the conventional concatenate methodology based TTS system, our model can generate waveform directly from phoneme sequence. In addition, analogue from text analysis, a new analysis methodology is proposed for phoneme analysis. Experiment result on LJ Speech dataset shows that, compared with char-based model, our model can get a comparative or better performance.

1. Introduction

Speech synthesis system has great potential in developing intelligent system like smart home devices, smart guided tours and human-like robots in the future. However, it is still a challenging work to synthesize an *intelligibility* and *naturalness* voice from text. Generally, a text-to-speech (TTS) system could be divided into two pieces: *acoustic feature extraction* and *waveform generation* (also called Vocoder).

Conventional TTS uses a concatenated methodology like **unit selection synthesis [1] and** statistical parametric based **subband synthesize [2]**. However, a concatenated speech synthesis requires complex model design and large database setting which is hard to be deployed in build-in systems. A seq-to-seq neural network model is proposed for machine translation and succeeds in **speech synthesis [3]**.

How to efficiently extract expressive feature from the raw text is still a challenging issue in data-driven speech synthesis. **Tacotron [4]** and **Tacotron-2 [5]**, two end-to-end models, are conditioned on character feature. A **deep voice 3 [6]** considers both character and phoneme feature.

In addition, acoustic feature extraction has a great influence on Vocoder process. **STRAIGHT [7]** is conditioned on fundamental frequency (F0), spectral envelope and aperiodicity features. **WaveNet [8],** a fully autoregressive model, is a robust generative model. However, it requires long training stage and is in low synthesis speed. A fast **Griffin-Lim [9]**, an estimation of origin waveform, is based on spectrogram.

Inspired by the sequence-to-sequence char-based Tacotron-2, we develop a phoneme sequence mode that has more expressive feature than the character sequence in speech synthesis.

2. Related Work

IOP Conf. Series: Journal of Physics: Conf. Series **1267** (2019) 012052 doi:10.1088/1742-6596/1267/1/012052

There is no universal accepted linguistic feature as input and acoustic feature as output in neural network model currently. **Deep Voice 1 [10]** uses phonemes, looking up from the phoneme dictionary, to predict the phoneme duration and F0. Tacotron proposes a character sequence as input to predict linear spectrogram. Deep Voice 3 considers both phoneme and character feature as input. In paper **[11]**, a phonetic posteriorgrams is used in speech synthesis stage. In our model, we consider phoneme which have the same representability to text, as a sort of text.

Conventional acoustic features extractor is based on **HMM [12]**. A **Bi-LSTM [13]** and **CNN [14]** model performs well in sequence feature extraction in text processing. Attention mechanism performs well in text-based **machine translation [15]**, **speech recognition [16]** and **speech synthesis [17]**. Inspired by text processing methodology, we adopt a model like text processing model and hope it work well in phoneme processing.

Generally, researchers conduct the process of short-time Fourier transformation (STFT) on the raw waveform to obtain the linear spectrogram. However, STFT is an irreversible operation, which causes an open issue to design a better **signal estimation** approach. A neural network model like WaveNet could generate smooth voice. However, it is tough to train the model. Griffin-Lim uses an iterative algorithm to generate voice. Hence in this paper, we adopt Griffin-Lim as Vocoder.

3. Overview of Approach

This section gives a brief introduction of the proposed model. As shown in Fig.1, the whole procedure can be divided into two stages: acoustic feature extractor and Vocoder. The phoneme sequence is converted from text (also called grapheme-to-phoneme). We will discuss how to obtain phoneme sequence from text in Section 3.1 and the details of acoustic features extractor in Section 3.2.

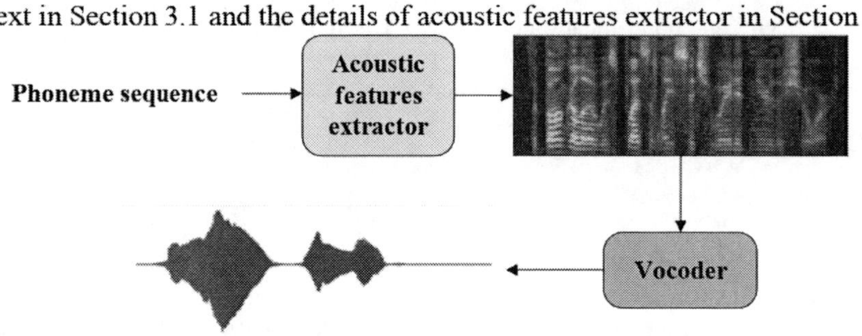

Figure 1. Speech synthesis architecture. The Model uses phoneme sequence as input to predict spectrogram and then estimate the signal.

Grapheme-to-phoneme (G2P) transcription [18] converts word to phoneme. There are two kinds of approaches respectively based on dictionary look-up and model. Since a model-based approach could not achieve hundred percentages to predict the phoneme, to reduce the unnecessary error causes in system, a dictionary look-up is used to convert grapheme to phoneme. **Fig. 2** shows an example of the dictionary look-up.

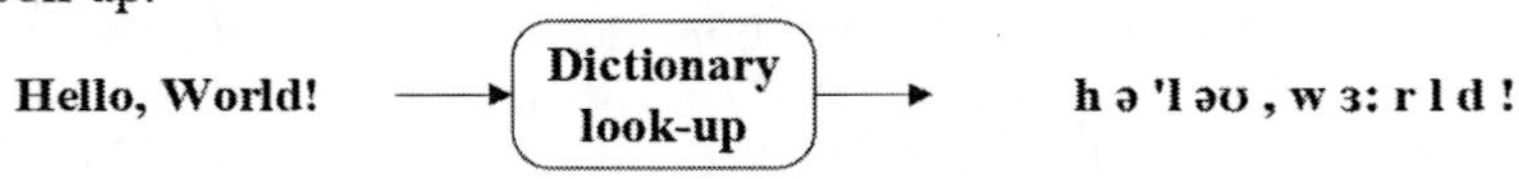

Figure 2. A dictionary look-up example.

Since a word could have multiple phonemes in different context and sound-link is an inevitable phenomenon in real human spoken, many sequence-to-sequence models are proposed to extract contextual information to synthesize more natural voice. Since an encoder-decoder model works well in generative processing, we consider the similar process. The acoustic feature extractor could be divided into three parts: encoder, decoder and postprocess net, as shown in **Fig. 3**.

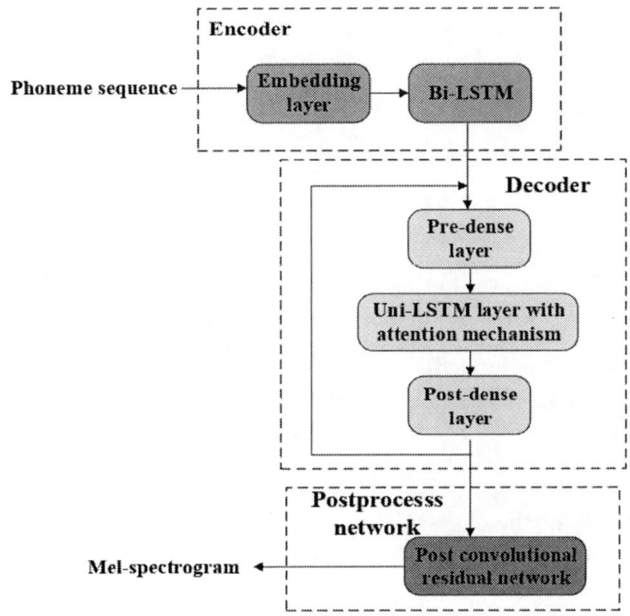

Figure 3. Architecture of mel-spectrogram estimation from phoneme sequence.

3.1. Encoder and decoder

In the encoding stage, the model should be able to extract enough features that present the target spectrogram. The phoneme is embedded into a trainable vector, and then we pass the embedded phoneme into a bi-directional LSTM layer. Finally, encoded vector is generated after encoder.

In the decoding stage, we use the same architecture descripted in Tacotron-2. Tacotron-2, a local-sensitive attention based autoregressive neural network, performs well in dynamically decoding the spectrogram information. The input of decoder each time step is concatenated by attention context vector and previous time step's output. The attention context vector is calculated from the previous step using the same mechanism as the **Tacotron-2 [5]**. After pre-dense layer (fully connected neural network), the result is passed into uni-directional LSTM layer. Finally, the raw frame features are created.

3.2. postprocess net and audio generation

In the postprocess net, the output of the decoder, presenting the time features, are passed to a post **convolutional residual network [19]** to predict mel-spectrogram. The linear spectrogram, transformed from the mel-spectrogram, then convert to the waveform by Griffin-Lim algorithm.

4. Experimental Results

4.1. Experiment setup

The **LJ Speech** dataset was used in the experiment. There is totally about 23 hours human speech length. In the experiment, only half of the data (about 10 hours' length) is used, the sample rate is 22,050Hz.

The training data occupies 95% in the whole data. This dataset has mappings between the text and waveform, but lacks the phoneme information. **Carnegie Mellon university pronouncing dictionary (CMUdict)** is used in grapheme-to-phoneme transcription. The dictionary has more than 130, 000 words. Since the dictionary doesn't include all vocabularies in LJ Speech dataset, these sentences are removed in our training and test stages.

In acoustic feature extractor, the parameter settings of our model are shown in **Table. 1**. In the post dense layer, a linear transform is used to fix the target length to 80-dimension, which is the same size as mel-frequency cepstrum coefficients (**MFCC**) **[20]** dimension. During the training, we also predict the stop token in the dynamic LSTM to synthesize the reasonable length of the voice.

Table 1. Parameter settings in acoustic feature extractor.

Module	Unit size	*channels*	depth	*Kernel size*
Embedding layer	512	-	-	-
Encode bi-LSTM	256	-	-	-
Pre-dense layer	[256, 256]	-	2	-
uni-LSTM	1024	2	-	-
Post convolutional residual network	-	512	5	5

At the signal estimation stage, Griffin-Lim algorithm is used to synthesize the final waveform presentation. In addition, both character sequence and phoneme sequence are used as input in the experiment for comparison.

4.2. Experiment Analysis

In English word, the length of the character is much longer than phoneme's. When this phenomenon reflects in a sequence-to-sequence analysis, a phoneme-only based model will benefit from it since the length is shortened. We conduct a small evaluation on the length of the two different features in LJ Speech, the results are shown in **Table 2**. From which we can see that the length of the phonemes is shorter than characters.

Table 2. A brief comparison of two sequence features.

Index	Value
Number of Phonemes	730172
Number of Characters	867121
Ratio	0.842
Total words	191628

Character feature requires stronger contextual information to map the pronunciation. Here is an example, the 'o' in 'hello' and 'world' have different pronounce. The phoneme feature could eliminate the unnecessary contextual-dependence and will reduce the prediction error.

4.3. Experiment Results

In our experiment, we didn't train our model for enough time to converge it. But it still takes about 70 hours training in Tesla K80 GPU.

The mel cepstral distortion (MCD) value and spectrum distortion value were used to evaluate the results objectively. Mel-scaled coefficient distortion can evaluate the human auditory-level perception distortion and spectrum distortion can evaluate the signal-level distortion. The formula of MCD is shown as (1), where T means the frames of the sentence, $c_d^{target}(t)$ means the d-th coefficient of the targets, $c_d^{predicted}(t)$ means the d-th coefficient of the predicted results, D means the number of coefficients. The formula of SD is shown as (2), where $s_f^{target}(t)$ means the f-th frequency of the target results, $s_f^{predicted}(t)$ means the f-th frequency of the predicted results.

$$\mathrm{MCD} = \alpha * \mathrm{T}^{-1} * \sum_{t=0}^{T-1}\left(2 * \sum_{d=0}^{D-1}\left(c_d^{target}(t) - c_d^{predicted}(t)\right)^2\right)^{\frac{1}{2}}, \alpha = \frac{10\sqrt{2}}{ln10} \approx 6.14 \quad (1)$$

$$\mathrm{SD} = \mathrm{T}^{-1} * D^{-1} * \sum_{t=0}^{T-1}(\sum_{f=0}^{F-1}(s_f^{target}(t) - s_f^{predicted}(t))^2)^{1/2} \quad (2)$$

In the experiment, we also consider using the **WaveNet [12]** as our Vocoder. However, we trained the WaveNet more than 30 hours, the generated voice is still noisy. Instead, the Griffin-Lim algorithm can get a good performance without a long-time training. **Fig.4** shows the evaluate result.

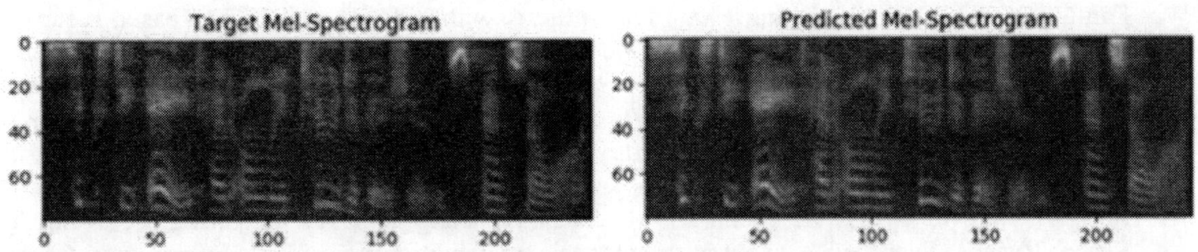

Figure 4. Result of mel spectrogram prediction.

For comparison, we adopted the same architecture of the phoneme-based sequence model, while the only difference is the character feature to phoneme feature. Models with character sequence inputs take a bit more time than phoneme sequence in training under the same training conditions, and same thing happens in synthesis stage, however, it didn't get a better performance. In addition, we also train the Tacotron-2 model before the WaveNet, and test in the same condition. The result is summarized in **Table. 3**. The phoneme-based model have slightly overperform than the char-based model. The Griffin-Lim distortion is calculated to estimate the Griffin-Lim algorithm's error.

Table 3. Comparison of char-based and phoneme-based model

System	MCD(dB)	SD(dB)
Char-based (our model)	11.4083294573	0.45788899644
Phoneme-based (our model)	11.1987140227	0.45884787579
Tacotron-2 (char-based)	11.2795017916	0.457523534425
Tacotron-2 (phoneme)	11.8089875978	0.459087886076
Griffin-Lim distortion	-	0.454450131282

5. Conclusion and Future Work

A new model using a phoneme sequence feature is proposed in this paper. The new model can achieve comparative or better performance to the character sequence-based model. Compared with the character-based model Tacotron-2, our model is faster in the training and get a lower MCD value. The result shows that the phoneme can be analysed as the text in speech signal processing and this feature is better than character in the speech synthesis.

By eliminating the error caused by the text prediction from phoneme, using phoneme as intermedia feature will make a progress in voice conversion system.

Acknowledgments

This work was supported in part by the Key Research & Development Program of Zhejiang Province (No. 2019C03127). The authors would like to thank the reviewers in advance for their comments and suggestions.

References

[1] Merritt T, Clark RA, Wu Z, Yamagishi J, King S. Deep neural network-guided unit selection synthesis. In *2016 IEEE International Conference on Acoustics, Speech and Signal Processing (ICASSP2016)*, 2016 Mar 20 (pp. 5145-5149).

[2] Okamoto T, Tachibana K, Toda T, Shiga Y, Kawai H. An investigation of subband WaveNet vocoder covering entire audible frequency range with limited acoustic features. In *2018 IEEE International Conference on Acoustics, Speech and Signal Processing (ICASSP2018)*, 2018 Apr 15 (pp. 5654-5658).

IOP Conf. Series: Journal of Physics: Conf. Series **1267** (2019) 012052 doi:10.1088/1742-6596/1267/1/012052

[3] Fan Y, Qian Y, Xie FL, Soong FK. TTS synthesis with bidirectional LSTM based recurrent neural networks. In *Fifteenth Annual Conference of the International Speech Communication Association 2014*.

[4] Wang Y, Skerry-Ryan RJ, Stanton D, Wu Y, Weiss RJ, Jaitly N, Yang Z, Xiao Y, Chen Z, Bengio S, Le Q. Tacotron: Towards end-to-end speech synthesis. *arXiv preprint arXiv*:1703.10135. 2017 Mar 29.

[5] Shen J, Pang R, Weiss RJ, Schuster M, Jaitly N, Yang Z, Chen Z, Zhang Y, Wang Y, Skerrv-Ryan R, Saurous RA. Natural tts synthesis by conditioning wavenet on mel spectrogram predictions. In *2018 IEEE International Conference on Acoustics, Speech and Signal Processing (ICASSP2018)*, 2018 Apr 15 (pp. 4779-4783).

[6] Ping W, Peng K, Gibiansky A, Arik SO, Kannan A, Narang S, Raiman J, Miller J. Deep voice 3: Scaling text-to-speech with convolutional sequence learning. *arXiv preprint arXiv*:1710.07654. 2017 Oct 20.

[7] Kawahara H. STRAIGHT, exploitation of the other aspect of VOCODER: Perceptually isomorphic decomposition of speech sounds. *Acoustical science and technology*. 2006;27(6):349-53.

[8] Oord AV, Dieleman S, Zen H, Simonyan K, Vinyals O, Graves A, Kalchbrenner N, Senior A, Kavukcuoglu K. Wavenet: A generative model for raw audio. *arXiv preprint arXiv*:1609.03499. 2016 Sep 12.

[9] Griffin D, Lim J. Signal estimation from modified short-time Fourier transform. *IEEE Transactions on Acoustics, Speech, and Signal Processing*. 1984 Apr;32(2):236-43.

[10] Arik SO, Chrzanowski M, Coates A, Diamos G, Gibiansky A, Kang Y, Li X, Miller J, Ng A, Raiman J, Sengupta S. Deep voice: Real-time neural text-to-speech. *arXiv preprint arXiv*:1702.07825. 2017 Feb 25.

[11] Sun L, Li K, Wang H, Kang S, Meng H. Phonetic posteriorgrams for many-to-one voice conversion without parallel data training. In *2016 IEEE International Conference on Multimedia and Expo (ICME2016)*, 2016 Jul 11 (pp. 1-6).

[12] Wen Z, Tao J, Pan S, Wang Y. Pitch-scaled spectrum based excitation model for HMM-based speech synthesis. *Journal of Signal Processing Systems*. 2014 Mar 1;74(3):423-35.

[13] Zhou C, Sun C, Liu Z, Lau F. A C-LSTM neural network for text classification. *arXiv preprint arXiv*:1511.08630. 2015 Nov 27.

[14] Abdel-Hamid O, Mohamed AR, Jiang H, Penn G. Applying convolutional neural networks concepts to hybrid NN-HMM model for speech recognition. In *2012 IEEE International Conference on Acoustics, Speech and Signal Processing (ICASSP2012)*, 2012 Mar 25 (pp. 4277-4280).

[15] Luong MT, Pham H, Manning CD. Effective approaches to attention-based neural machine translation. *arXiv preprint arXiv*:1508.04025. 2015 Aug 17.

[16] Chorowski JK, Bahdanau D, Serdyuk D, Cho K, Bengio Y. Attention-based models for speech recognition. In *Advances in Neural Information Processing systems 2015* (pp. 577-585).

[17] Bulyko I, Ostendorf M. Joint prosody prediction and unit selection for concatenative speech synthesis. In *ICASSP 2001* May 7 (pp. 781-784).

[18] Bisani M, Ney H. Joint-sequence models for grapheme-to-phoneme conversion. *Speech communication. 2008* May 1;50(5):434-51.

[19] Laina I, Rupprecht C, Belagiannis V, Tombari F, Navab N. Deeper depth prediction with fully convolutional residual networks. In *2016 Fourth International Conference on 3D Vision (3DV)*, 2016 Oct 25 (pp. 239-248). IEEE.

[20] Zheng F, Zhang G, Song Z. Comparison of different implementations of MFCC. *Journal of Computer science and Technology*. 2001 Nov 1;16(6):582-9.

IOP Conf. Series: Journal of Physics: Conf. Series **1267** (2019) 012053 doi:10.1088/1742-6596/1267/1/012053

An Improved Vision-Based Lane Departure Warning System under High Speed Driving Condition

X T Liu[1, a] Y Zou[1,b] and H W Guo[1, c]

[1]School of Mechanical Engineering, National Engineering Laboratory for Electric Vehicles, Beijing Institute of Technology, NO.5 Zhongguancun south Street, Haidian, Beijing, China.

[a]bitlxiantao@163.com; [b]zouyuanbit@vip.163.com; [c]guohongwei@bit.edu.cn

Abstract. Recently, Lane departure warning system has attracted great attention as it contributes to vehicle active safety. In this paper, a vision-based lane departure warning system under high-speed driving is proposed. The system consists of two functional parts: lane markings detection and vehicle departure identification. The Hough Transform is applied to detect lane boundaries, which is a most effective detection method with high reliability. Based on the road line, a method using several Euclidean-distance-related parameters to calculate vehicle's position and its deviating status is proposed, which addresses the problem of efficient detection of lane departure under high-speed driving condition. The algorithm is tested and verified in various real driving conditions and proved a reliable and steady performance.

1. Introduction

In this paper, a technique for the identification of vehicle's unintentional lane departure behaviour is proposed based on vision-based systems. Nowadays, traffic accidents cause hundreds of deaths, thousands of injuries and billions of dollars in loss every year. Among them, unintentional lane departure behaviour as a result of driver's distraction, drowsiness and fatigue takes quite a considerable proportion [1]. Therefore, Advanced Driver Assist System (ADAS) such as Lane Departure Warning (LDW) system is considered as an effective method to help driver with real-time reminder and assistance. The vision-based LDW system uses cameras installed on vehicle to collect surrounding information and send alarms to driver. For its convenience and its high reliability, it's in great demand of automotive industry and widely studied in recent years [2].

To generate warning signals for divers, lane departure warning system mainly consists of vision capture, lane detection and departure status identification. There have been many studies in this area over the past decades. Effort have been widely done related to image-based video pre-processing, Hough-based image processing, feature-based and model-based lane detection, vehicle departure identification, etc. [3]. Many other researchers concentrate on any other aspects, such as lighting conditions [4], varying road conditions [5] and driving environments [6]. In the meanwhile, the system should provide precise and timely alerts responding to potential driving dangerous, which is a challenging task. So robust and efficiency need to be recognize as a fatal evaluation of LDW system.

The main contribution of the paper is to provide an LDW system with high accuracy and low complexity under high-speed condition. The figure 1 shows the overall procedures of our LDW system. Since images captured, pre-processing including image conversion, smoothing and edge detection is applied. Then, Region of Interest (ROI) extraction and Hough Transform (HT) are used to extract lane

IOP Conf. Series: Journal of Physics: Conf. Series **1267** (2019) 012053 doi:10.1088/1742-6596/1267/1/012053

markings, which is described in Section 2. Based on the lane information, the technique using Euclidean distance parameters to determine vehicle departure identification status is applied, which is presented in Section 3. Finally, Experimental results are shown in Section 4 and conclusion is presented in Section 5.

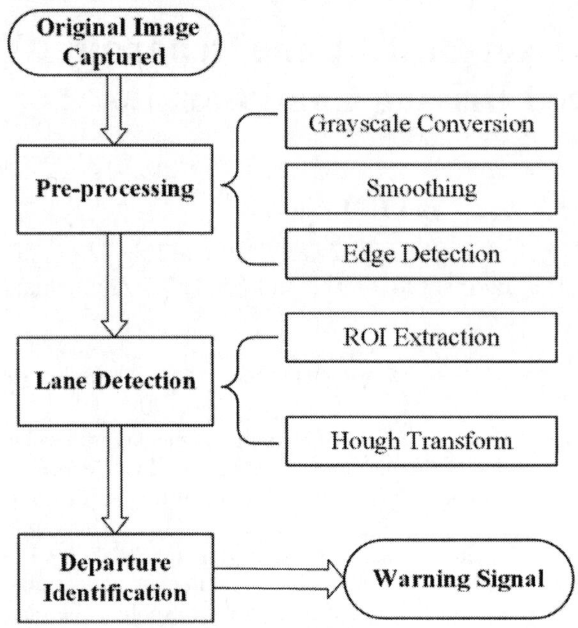

Figure 1. Block diagram of proposed LDW system

2. Lane Detection

In the LDW system, accurate identification of lanes plays a key role to the overall system. Only when the identification and parameter calculation of the lane markings have been settled, the vehicle position and motion parameters can be accurately determined.

2.1. Pre-processing

Pre-processing is a vital step after the original image is captured from monocular camera, as it enhance the frame by reducing the noise and strengthen the contrast. Additionally, image smoothing and region of interest extraction help to remove the unnecessary part with interference information, which can improve the efficiency of calculation.

Generally, the input is firstly converted to grayscale image using a weighting equation to increase the pixel intensity value as follows:

$$I(x,y)=\omega_R R(x,y)+\omega_G G(x,y)+\omega_B B(x,y) \tag{1}$$

Where $R(x, y)$, $G(x, y)$ and $B(x, y)$ are the intensities of the red, green, and blue colours at (x, y), respectively. And ω_R, ω_G and ω_B are the weights for $R(x, y)$, $G(x, y)$, and $B(x, y)$, respectively.

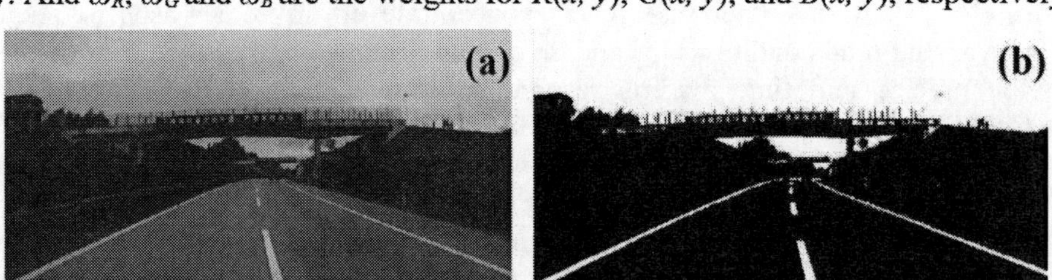

Figure 2. (a) Original image captured on Highway (b) converted binary image

By setting a threshold for grayscale, the part with grayscale intensity $I(x, y)$ larger than the limit is defined as 1 and the rest is set to 0. Therefore, the grayscale image can be converted into a binary image automatically, as shown in figure 2. In order to filter out noise effect, there are many filters used by researchers, such as Median filter, Gaussian filter, dilation and erosion filters, etc. To simplify calculation, a 1D-Gaussian filter is applied to the grayscale image, which smoothing the image considered in X direction and Y direction separately.

The introduction of edge detection helps to enhance the contrast and highlight the edge information so that improve the quality of pre-processing. Commonly, edge operators such as Sobel operator [7], Canny operator [8] and steering filter [9] are widely applied to detect edge lines. Since other operators may cause high frequency noise, which affects edge feature extraction and leads to false lane detection, the Sobel operator is applied to edge detection for its strong noise suppression and smooth results.

2.2. Region of Interest extraction

In order to reduce redundant part of the entire input image that contains of useless information, so it is important to extract the Region of Interest to reduce the computational complexity [10]. Faced with the standard lane markings on highway, especially detecting road lanes under high-speed driving condition, we directly complete the ROI extraction by intercepting the partial images in the horizontal and vertical directions, respectively. Although this approach may reduce the accuracy of recognition, it can greatly improve the computational efficiency under high-speed condition.

At the same time, the remaining part of the image is divided into two sub-areas and line feature detection will be completed in each sub-areas independently, as shown in figure 3. The Hough origin H_0 is located at the midpoint of the lateral width of the ROI. Therefore, the calculation in each sub-areas makes the distance parameters of lanes acquired easily.

Figure 3. Select of ROI and sub-area segmentation

2.3. Hough Transform

When vehicle is travelling at high speed on the structural road, road lanes can usually be characterized as a straight line. Therefore, HT can be used to extract road information in such case. The HT adopts the form of spatial mapping, which maps the lines in the image space to the points in the parameter space separately [11]. The feature lines can be effectively detected by searching for the point-to-line correspondence, making the system has a great fault tolerance and robustness.

In the pixel space, the point on the edge line can be represented by a polar coordinate equation, as follows:

$$\rho = x\cos\theta + y\sin\theta \tag{2}$$

Where (x, y) is the coordinate value of a pixel, θ is the angle between the x-axis and the normal line, and ρ is the distance between the origin and the fitted line as shown in figure 4.

The line to be detected in pixel space corresponds to a certain array (ρ, θ) in Hough space. After traversing all the points on the edge line, the local maximum accumulator cell $A(\rho, \theta)$ can be found. Thus, the expression of feature line in the pixel space is obtained.

3. Identification of Lane Departure

3.1. Evaluation measurements of lane departure

Lane identification is just the premise for deviating detection, the identification of the vehicle's driving status counts more in LDW system. Several evaluation parameters are defined based on the results of former image processing steps. And the Hough origin H_0 is the basic reference for all other parameters. The road lines in Hough-parameter form is obtained, which is describe in an array (ρ, θ). Here, ρ is the exact vertical distance from the midpoint Hough origin H_0 to the road lane edge. As shown in figure 4, P_l and P_r represents the crossing points of perpendicular lines connecting the Hough origin to the edge line respectively as shown in figure 4.

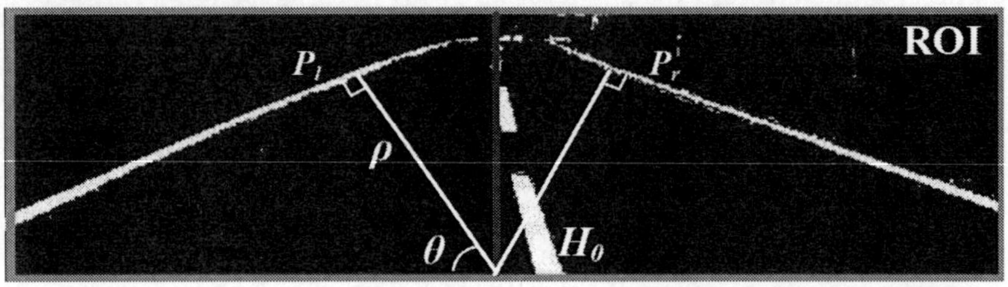

Figure 4. Line detection based on Hough Transform in two sub-regions

Thus, positon-related parameter $\mathbf{D_l}$ is defined as the Euclidean distance between the left crossing point P_l to the Hough origin H_0, which is a vector with a direction and its value is represented as follows:

$$\boldsymbol{D_l} = \|\boldsymbol{P_l} - \boldsymbol{H_o}\| = (\|\boldsymbol{P_l}\|^2 + \|\boldsymbol{H_o}\|^2 - 2\|\boldsymbol{P_l} \cdot \boldsymbol{H_o}\|)^{1/2} \quad (3)$$

Similarly, another position-related parameter $\mathbf{D_r}$ is defined as the Euclidean distance between the right crossing point P_r to the Hough origin H_0, which is a vector with a direction and its value is represented as follows:

$$\boldsymbol{D_r} = \|\boldsymbol{P_r} - \boldsymbol{H_o}\| = (\|\boldsymbol{P_r}\|^2 + \|\boldsymbol{H_o}\|^2 - 2\|\boldsymbol{P_r} \cdot \boldsymbol{H_o}\|)^{1/2} \quad (4)$$

According to the aforementioned two parameters, the distance-related parameter $o(\boldsymbol{D_l}\text{-}\boldsymbol{D_r})$ that determine whether the vehicle has been deviated from lanes is estimated. The parameter is calculated as follows:

$$o(\boldsymbol{D_l} - \boldsymbol{D_r}) = o\|\boldsymbol{D_l} - \boldsymbol{D_r}\| = (\|\boldsymbol{D_l}\|^2 + \|\boldsymbol{D_r}\|^2 - 2\|\boldsymbol{D_l} \cdot \boldsymbol{D_r}\|)^{1/2} \quad (5)$$

If a certain threshold ζ_0 of the lane departure evaluation $o(\boldsymbol{D_l}\text{-}\boldsymbol{D_r})$ is exceeded in several consecutive frames, it can be assumed that the vehicle has been deviated from lane at the current position.

In the meanwhile, the lane departure may happen in a short time when travelling fast, even though it does not deviate at the current moment. Therefore, a status-related parameter is introduced to describe the driving state of the vehicle in the near future as follows:

$$\psi = \frac{do(\boldsymbol{D_l} - \boldsymbol{D_r})}{dt} = \frac{do\|\boldsymbol{D_l} - \boldsymbol{D_r}\|}{dt} \quad (6)$$

Where Ψ is expressed as changing rate of the above parameters in successive multi-frame images. Only in consecutive several frames can Ψ be calculated. With the sliding window detection method, it reflects the tendency of the vehicle to deviate from lanes for some seconds to come. If a certain threshold ε_0 of the lane departure evaluation Ψ is exceeded, the risk of deviation of the vehicle in the coming seconds can be determined.

3.2. Identification of departure status

Lane departure has been defined as occurring when one of the front wheels of the vehicle crosses a lane marking. The following two parameters are used to identify the lane departure states, which are $o(\boldsymbol{D_l}\text{-}\boldsymbol{D_r})$ and Ψ. The two parameters need to be verified synchronously to estimate the lane departure state.

In the proposed LDW system, if $o(\boldsymbol{D_l}\text{-}\boldsymbol{D_r})$ is less than the threshold ζ_0, the vehicle is considered to remain within the lane at this moment. Simultaneously, if Ψ is greater than the threshold ε_0, it is considered that the vehicle may be unconsciously and quickly exiting the current lane due to driver's fatigue or the like, which may leads to traffic accidents. If $o(\boldsymbol{D_l}\text{-}\boldsymbol{D_r})$ is greater than the threshold ζ_0 for a certain period of time, it is considered that the vehicle has deviated from the current driving lane at this state, at which the driver should be warned in time. When the vehicle is moving at high speed, the state of the vehicle is changing all the time. Therefore, that is not enough only when the position of the current time meets the safety condition: $o(\boldsymbol{D_l}\text{-}\boldsymbol{D_r})$ is less than the threshold ζ_0 .It's necessary to analyse $o(\boldsymbol{D_l}\text{-}\boldsymbol{D_r})$ and Ψ together to determine whether the vehicle is involuntarily deviated in the next period of time.

4. Experimental Results

Being able to evaluate the accuracy and real-time performance of the proposed system, sample lanes on Highway was selected. The test video sequences were captured from a camera mounted on vehicle, whose speed was between 80 and 110km/h. The proposed system was implemented on a computer equipped with a CPU Core i5-6400 2.70GHz in the running environment of MATLAB. The video sequences were pre-processed before lane detection and departure identification. To evaluate the proposed algorithm, different driver's behaviours with intentional and unintentional deviations was tested.

Table 1. Performance of proposed LDW system.

Scenarios	Total Frame	Line Detected Frame	Line detection Rate	Ground truth Departure	Detected Departure	True Warning Rate
Left-shifted driving	1059	1046	98.8%	106	99	93.4%
Right-shifted driving	939	924	98.4%	93	88	94.6%
Straight driving	1659	1640	98.9%	0	1 [a]	—
Total	3657	3610	98.7%	199	187	94.0%

[a] There is a false alarm in straight driving condition.

As illustrated in table 1, the proposed LDW system successfully detects lane boundaries under Highway condition. The proposed algorithm is applied to detect the deviation state at different driving behaviours. For details, lane markings are identified accurately among left-shifted, right-shifted and straight driving conditions with a total true detection rate of 98.7%. In addition, the performance of vehicle departure identification tested in the same three experimental scenarios. The result shows that the system works on well in different deviating conditions with a total true warning rate of 94.0%. Compared with other state-of-art LDW systems, algorithm based on the two parameter criterion performs a better true warning rate than [10] does, which fits more under high speed condition. Therefore, the proposed algorithm is proved to achieve a fast processing with high accuracy.

5. Conclusion

In this paper, a novel vision-based lane departure warning system with low computational complexity and high accuracy has been presented. The system is approved to detect road lanes and identify unintentional lane deviation behaviours effectively, especially under high-speed condition. When verified, the frame sequences captured from real-time driving condition are processed through image conversion, ROI segmentation and Hough Transform. For vehicle position and motion status identification, an algorithm with several Euclidean-distance-related parameters criterion is proposed. The system performs well with a 94.0% true warning rate, which demonstrates that the proposed method

IOP Conf. Series: Journal of Physics: Conf. Series **1267** (2019) 012053 doi:10.1088/1742-6596/1267/1/012053

for lane departure system is effective and accurate. Since the approach presented only has a minimal requirement of information to characterize driving status, this reduce the amount of calculation and makes it more feasible in a real-time embedded system application.

Acknowledgments
The authors would like to thank the colleagues for their selfless support and helpful suggestions for the submission of paper.

References
[1] J Navarro, J Deniel, E Yousfi, C Jallais, M Bueno and A Fort 2017 Influence of lane departure warnings onset and reliability on car drivers' behaviors *Appl. Ergon.* **59** 123
[2] J C McCall and M M Trivedi 2006 Video-based lane estimation and tracking for driver assistance: survey, system, and evaluation IEEE *Trans. Intell. Transp. Syst.* **7** 20
[3] H Zhu, K V Yuen, L Mihaylova and H Leung 2017 Overview of environment perception for intelligent vehicles *IEEE Trans. Intell. Transp. Syst.* **18** 2584
[4] J Son, H Yoo and S Kim 2015 Real-time illumination invariant lane detection for lane departure warning system *Expert Syst. Appl.* **42** 1816
[5] J G Wang, C Lin and S Chen 2010 Applying fuzzy method to vision-based lane detection and departure warning system *Expert Syst. Appl.* **37** 113
[6] Z Kim 2008 Robust lane detection and tracking in challenging scenarios *IEEE Trans. Intell. Transp. Syst.* **9** 16
[7] C Mu and X Ma 2014 Lane detection based on object segmentation and piecewise fitting *Electr. Eng.* **12** 3491
[8] P C Wu, C Chang and C H Lin 2014 Lane mark extraction for automobiles under complex conditions *Pattern Recognit.* **47** 2756
[9] X An, E Shang, J Song Li and H He 2013 Real-time lane departure warning system based on a single fpga *Image Video Process.* **38** 1
[10] V Gaikwad and S Lokhande 2015 Lane departure identification for advanced driver assistance *IEEE Trans. Intell. Transp. Syst.* **16** 910
[11] Y Otsuka, S Muramatsu, H Takenaga, Y Kobayashi and T K Monji 2002 Multitype lane markers recognition using local edge direction *IEEE Intell. Veh. Symp.* **2** 604

IOP Conf. Series: Journal of Physics: Conf. Series **1267** (2019) 012054 doi:10.1088/1742-6596/1267/1/012054

Anomaly Detection of ICS based on EB-OCSVM

R B Zhang[1,a], L H Xia[1,b] and Y Lu[1,c]

[1]Hefei University of Technology, Hefei, Anhui

[a]zhangrb@hfut.edu.cn; [b]1428023665@qq.com; [c]luyang.hf@126.com

Abstract. Industrial process anomaly detection mechanisms have been proposed to protect industrial control system to minimize the risk of damage or loss of resources. In this paper, an one-class Support Vector Machine based extended boundary (EB-OSCVM) is used to detect anomalies in industrial multivariate time series data from a simulated Tennessee Eastman Process (TEP) with many cyber attacks. In detail, determine the change points of each process variable and capture the causality relationship between the variables based on the location and time delay of the change points. Then, by monitoring the leaf nodes in the causality graph, we can know whether the system is abnormal, it can effectively reduce the dimension of process data. The EB-OSCVM extend classification boundary of OCSVM in order to reduce the error of noise, if data is outside the boundary of EB-OCSVM, there is an anomaly. Finally, tracing the anomaly source according to causal direction. An experiment is used to verify the effectiveness of the proposed approach, the results demonstrate that the approach presents a high-accuracy solution and traces the source of anomaly correctly.

1 Introduction

The rapid growth of information and communication technologies has motivated the traditional industrial control system (ICS) to seek tighter integration between physical process and cyberspace. However, integrating the cyber and physical domain significantly reduces isolation of the physical system from the outside world, which increases their vulnerabilities and triggers a number of security problem [1,2].

Unlike information technology (IT) system, the main target of cyber-attacks on industrial control system is to cause a catastrophe by disrupting the physical process. An arbitrarily attack on the physical process would potentially cause the whole system to shut down. If only consider cyber information, the attack might be neglected when the activities are hidden or the evidence of the attack is insufficient to be identified as an anomaly in the cyber domain. Therefore, the information of relevant physical process for anomaly detection must be taken into consideration.

Different approaches have been proposed to detect anomalies in industrial physical process data, the goal of anomaly detection is find the unusual behaviors, i.e. behaviors that are not exhibited under normal operation.

Constructing a mathematical model of ICS for both physics and control dynamics can detect anomaly [3-7]. Literature [3] describes a process-oriented technology for detecting network attacks against programmable logic controllers (PLCs), using autoregressive methods to model specific process variables. Based on process attribute invariants, Urbina et al. [4-7] used Linear Dynamical State Space (LDS) to model physical processes and detect anomalies caused by network attacks. Unfortunately, creating a precise model of complex physical processes is a very challenging task, arising from the tight integration of algorithmic control and complex physical processes. It requires an in depth understanding of the system and its implementation, which is a time consuming and cannot scale up to large and complex systems.

Most existing machine learning approaches focus on detecting anomalies in feature space, i.e., looking at data points with large deviation from normal space. These require little system knowledge and can detect a large range of attacks.

The RNN is one of the machine learning approaches used for anomaly detection in the SWaT system. However, due to the expensive training time, they only consider the first out of the six stages of the system [8]. As a follow-up work, the Deep Neural Network (DNN) and the one-class Support Vector Machine (SVM) models have been applied for anomaly detection in physical process data. All stages and attack scenarios are considered in this work. But, the DNN and one-class SVM can only detect fewer attacks [9]. Kravchik et al.[10] used a variety of deep neural networks architectures including different variants of convolutional networks to detect anomaly. The method can effectively detect the physical process anomalies of ICS and significantly superior to method in [9]. An RNN-based forecasting approach detected early anomalies in industrial multivariate time series data from a simulated Tennessee Eastman Process (TEP) with many cyber-attacks [11].

However, A significant shortcoming of the currently applied machine learning methods is that they provide little insight into the system and no explanation of detection results. All of the above articles can only detect anomalies and cannot Trace the cause of the anomaly.

In [12], a novel graphical model-based approach is proposed to learn the local behavior of a complex water treatment plant, which is used for anomaly detection. Timed automa are learned as a model of regular behaviors in sensors signal and Bayesian networks are learned to discover dependencies between sensors and actuators. It can detect cyber-attacks with high precision, and trace the cause of anomaly according to correlation between sensors and actuators. However, the method depends on the periodicity and linearity of industrial process data, and the practical industrial data usually do not satisfy.

In this work, we present a machine learning method of one-class SVM [13] based on extend boundary (EB-OCSVM) to detect attacks of Tennessee Eastman Process [14] (TEP), and trace the anomaly source. First, we use change points (CPs) to analysis causality between TEP variables and construct graph of causality. This method is suitable for both linear and non-linear system, and does not need to construct an accurate mathematical model to describe the correlation between variables. Then, obtaining normal classification boundary by training OCSVM, the boundary can be extended as anomaly detection control limit in order to reduce the error caused by noise. If process data exceeds control limit, anomaly is detected. Finally, if we capture anomaly in process data, trace the root cause of anomaly according to graph of causality.

2 Methodology

2.1 A Causal Discovery based Change Points

With the increase in scale and complexity of process operations in large industrial plants, anomaly may occur on any of the thousands of components, easily propagate along information and material flow pathways and affect other parts of the system. To determine the root cause of certain abnormality, it is important to capture the process connectivity.

In a complex industrial process, elements are not only connected to each other, they are also mutually dependent. The concept of causality has been introduced to describe the cause-effect relationships between variables or events. There are two characteristics of causality: (1) the cause occurs before the effect; (2) the cause contains information about the effect, we focus on the feature of causality that the cause occurs before the effect and the cause inevitably impel the effect to change.

2.1.1 Determination of change points. In this section, we propose a change-point detect method, determining unknown number of change points. The goal of change-point detection is to detect abrupt changes in the distribution of samples. Industrial process variables have characteristics of high delay and high noise, even the same variable have different level of noise, so it is difficult to capture causality.

The most critical step of the proposed method is to determine change points of a given data sequence according to the change of tendency among process variables. It can be seen that a point in a data sequence with the maximum distance to a line connecting the starting and ending point of the sequence has the greatest impact on the tendency. Thus, we need to detect some special points, referred to as change points, representing the change of tendency among process variables.

First, in order to prevent that some process variables with large amplitudes erroneously play dominate roles, the proposed method preprocesses the raw data by normalizing the observations of each process variable in order to make data between 0 and 1. Given the raw data of a multivariate time series $\widetilde{\mathbf{T}}=\{\widetilde{X}_i(n)\}$ for i=1, ..., M and n=1, ..., N, the normalized time series **T** is obtained as

$$\mathbf{T}=\{\mathbf{X}_i(n)\}=\{(\widetilde{\mathbf{X}}_i(n)-\min(\widetilde{\mathbf{X}}_i(n)))/(\max(\widetilde{\mathbf{X}}_i(n))-\min(\widetilde{\mathbf{X}}_i(n)))\} \tag{1}$$

Where min and max denote the sample maximum and minimum value, respectively. The orthogonal distance from a point X_i to the line AB between the point X_A and X_B is defined by equation (2), namely, the distance between P and its projection on AB.

$$D_i=[(\left\|\overrightarrow{Ai}\right\|_2\left\|\overrightarrow{AB}\right\|_2)^2-(\overrightarrow{AB}\cdot\overrightarrow{Ai})^2]^{1/2}\Big/\left\|\overrightarrow{AB}\right\|_2 \tag{2}$$

A potential change point P on a segment [s, e] is given by:

$$D_P=\underset{s\le i\le e}{\arg\max}\,\overset{P}{\mathbf{D}_i} \tag{3}$$

It can be shown that p is the maximum distance of the change-point location in interval [s, e]. The value D_p is tested against a threshold Wt in order to decide whether the null hypothesis of no change-point is rejected or not, recursively applying the above method on [s, p] and [p+1, e]. The algorithm stops in each current interval when no further change point are detected, that the obtained distance fall below threshold Wt.

2.1.2 Determination of threshold Wt. The Wt is the threshold to terminate recursive process, if Wt is to small, there are many false positive, and if Wt is to high, we will miss many change points. That is to say, a rational threshold has great influence on the determination of change point. Meanwhile, process variables have different level of noise, a fixed threshold used in different process variables is unpractical, We propose a method that can adaptively determine the threshold Wt.

Given a sequence T={X(i)} for s<=i<=e, s and e are start point and end point of sequence, respectively, two special indices s=1 and e=N.

(1)we calculating orthogonal distance from a point to a line connecting the starting and ending points of the sequence, obtain a change point P that makes orthogonal distance maximum. Then calculating first-order differential cumulative sum, which can represent the change of trend and the direction of change, in two segments [s, p] and [p+1, e], respectively, denoted as *sum_l* and *sum_r*. Last, we calculate the absolute value of the difference between *sum_l* and *sum_r*, denoted as *p_th*.

(2) calculating the maximum orthogonal distance of subsequence [s, p] , the process is as step (1), we can get the absolute value of the difference as *pl_th*;

(3)calculating the maximum orthogonal distance of subsequence [p+1, e], the process is as step (1), we can get the absolute value of the difference as *pr_th*;

(4) If *p_th* > *pl_th* and *p_th* > *pr_th*, then stop, otherwise, recursively applying the above method on [s, p] and [p+1, e].

Thresholds vary in the iteration process, and suitable for different levels of noise, so we can get change points without fixed threshold.

IOP Conf. Series: Journal of Physics: Conf. Series **1267** (2019) 012054 doi:10.1088/1742-6596/1267/1/012054

2.1.3 Causal analysis based change points. To describe the causal relationships between all the variables, a graph can be constructed with nodes denoting variables and arcs denoting their causal direction. Causality analysis provides an effective way to localize root cause of plant-wide abnormalities since a causal graph can represent the direction of anomaly propagation.

We construct a causal graph based on the number and location of change points, which can be used to reflect the causal relationship between process variables, even be used for anomaly traceability. Given variable X and Y, if X is the cause of Y, then Y must change when X changes. Since the industrial process variable has time delay and noise, the change points of Y will occur after X. If each change point of X can cause Y changes, and in the each change interval of X, the number of change points of Y remains constant, we can accept the hypothesis that X cause Y. Moreover, if Y change after X change, and X change after Y change, we can get the direction of correlation: X cause Y and Y cause X, that is to say, X and Y are mutually causal.

The change points method, a bivariate analysis, can capture a significant causality or an indirect causality. In order to distinguish direct or indirect causality between two variables, we use time delay as a rule. Given an underlying model is X->Y->Z, if the delay of X cause Z is less than the delay of Y cause Z, then we can say there is a direct causality between X and Z, otherwise, X indirectly caused Z conditioned on Y.

2.2 Anomaly Detection Based on EB-OCSVM

Recently, unsupervised machine learning was shown to be effective for detecting ICS attacks. In this paper, we apply EB-OCSVM algorithm to detect anomaly in industrial control system. The EB-OCSVM builds a model from training on normal data and then classifies test data as either normal or attack based on its classification boundary.

Physical process data of ICS has high noise, we used OCSVM to learn classification boundary and obtain label after training, then stripped noise data where label is -1 as normal data, we can get upper control limit (UCL) and lower control limit (LCL), where UCL, LCL is maximum and minimum value of normal data, respectively. Lastly, we extended control limit as the abnormal boundary of EB-OCSVM in order to reduce false positive because of strong noise, denoted by

$$UCL = UCL + \alpha \times \delta(x) \tag{4}$$

$$LCL = LCL - \alpha \times \delta(x) \tag{5}$$

Where x is normal data after noise elimination,δ is standard deviation of x, α is parameter that determine the allowable error range. So, we obtain boundary control to determine that an anomaly has occurred, i.e. if data is greater than UCL or lower than LCL, we determine that an anomaly is detected. In this work, we define α as 3.

By monitoring the leaf nodes, we can monitor the whole variables of causality graph whether is abnormal. It is reduce dimension of process data. If a leaf node is abnormal, the parent node of the leaf node should be analyzed, until the parent node of the current node is normal, so, the current node is the source of the anomaly.

3 Experiments

3.1 TEP Description

In this experiment, we used the TEP dataset [15]. TEP is a benchmark simulation model to test process control and monitoring approaches, represented in Figure 1. It was simulated at different normal modes and under cyber-attacks. The TEP has five major units: a reactor, a compressor, a separator and a stripper, the dataset contains 11 manipulated variables, 22 process measurements and 19 analyzer measurements. Our experiment focuses on reactor unit, including 4 measurements: reactor level, reactor pressure, reactor temperature, reactor cooling water temperature and 1 manipulated variable: reactor cooling water flow.

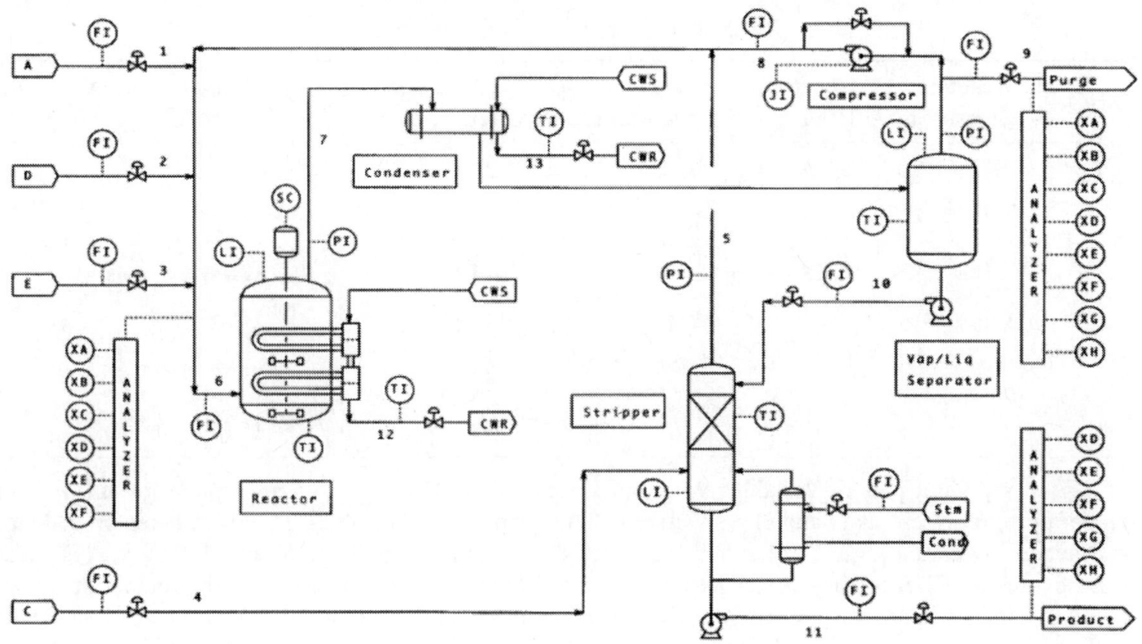

Figure 1. Tennessee Eastman Process

3.2 Causality graph based on change-points

We use change points to capture causality between reactor variables. According to experiment, we can get change points as circle points in Figure 2, as can be seen from the figure, the location of the change points are accurate.

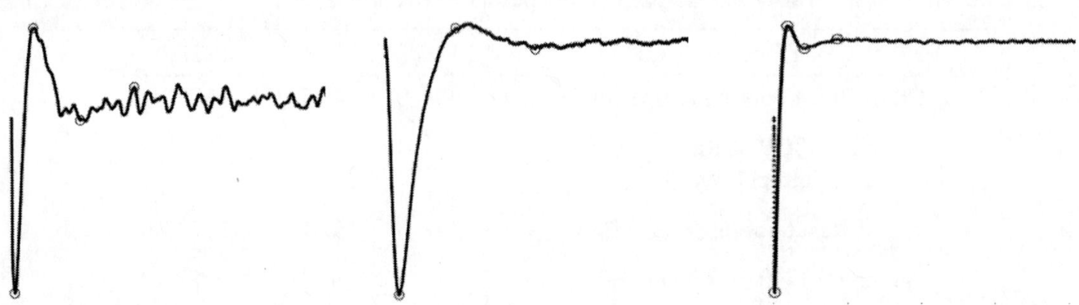

(a) reactor pressure (b) reactor level (c) reactor temperature

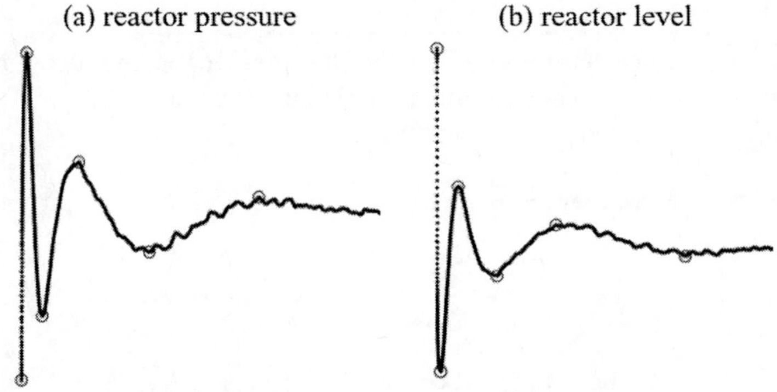

Figure 2. Change points of reactor variables

(d) reactor cooling temperature (e) reactor cooling water flow

Then, analyse causality according to the number and occurrence time of change points. The occurrence time and number of change points shown in Table 1.

IOP Conf. Series: Journal of Physics: Conf. Series **1267** (2019) 012054 doi:10.1088/1742-6596/1267/1/012054

Table 1. Information of change points

NO	Reactor pressure	Reactor level	Reactor temperature	React cooling temperature	Reactor cooling water flow
1	292	1146	58	23	4
2	1728	5752	867	345	248
3	5548	12283	2002	1394	1348
4	9859		4245	3828	3812
5				8483	7528
6				15755	15739

First, we get a pair of variables with same number of change point: react pressure and reactor temperature, represent as A and B respectively. The change points occurs time of A takes precedence of B according chronological order, We can conclude that the change in A affects B, that is, A is cause of B denoted A->B. Similarly, the cause of reactor cooling water flow is react cooling temperature.

Next, we compare variables with different number of change points, like reactor level and react cooling temperature. The first change point of reactor level occurs at 1146, we search for the change point of react cooling temperature that is closest and greater than 1146 at 1394. By same ways, we show the results in Table 2. Calculating the number of change points of reactor cooling temperature between the previous change point and the current change point of reactor level, because of different numbers of change point, reactor level is not affect react cooling temperature and react cooling water flow. The causal relationship analysis process between other variables is as described above.

Table 2. Analyze change points of reactor level

CP of Reactor level	1146	5752	12283
CP of React cooling temperature	1394	8483	15755
Number of CP of React cooling temperature		2	1
CP of React cooling water flow	1348	7528	15739
Number of CP of React cooling water flow		2	1

Finally, for direct causality and indirect causality, we consider that the smaller the time delay is, the more direct causation is. As mentioned above, the causal graph is shown in Figure 3.

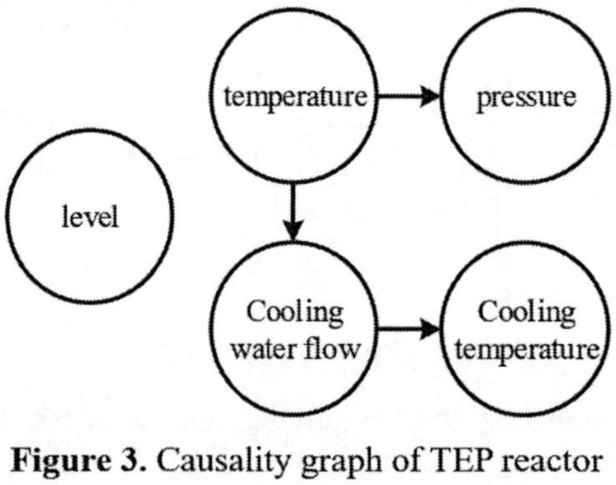

Figure 3. Causality graph of TEP reactor

3.3 Anomaly detection based on EB-OCSVM

In this paper, we use EB-OCSVM algorithm to detect anomaly, seek boundary of normal data. In order to thoroughly evaluate anomaly detection techniques, four evaluation indexes are adopt: accuracy (A), false positive rate (FPR) , false negative rate (FNR) and recall rate (R), where T represents a normal state and F represents a abnormal state. Their definitions are described as follows:

$$A=(TP+TN)/(TP+TN+FP+FN) \tag{6}$$

$$FPR=FP/(FP+TN) \tag{7}$$

$$FNR=FN/(FN+TP) \tag{8}$$

$$R=TP/(TP+FN) \tag{9}$$

There is no anomaly of reactor level, Table 3 shows the experimental results of TEP variable except reactor level.

Table 3. Effect of anomaly detection

	Reactor pressure	Reactor Temperature	React cooling Temperature	Reactor cooling water flow
A	99.93%	99.87%	98.59%	99.72%
FPR	0.06%	0.13%	1.38%	0.16%
FNR	1.7%	0%	4.77%	12%
R	98.3%	100%	95.23%	88%

It can be seen from Table 3 that EB-OCSVM-based anomaly detection method has higher accuracy and recall rate. Although the partial variable has a high false negative rate, EB-OCSVM can effectively detect process data anomalies.

3.4 Analyze the source of anomaly

In section, we obtained causality graph of reactor unit, combining with anomaly detection results, we can capture the source of the anomaly. If EB-OCSVM detect an abnormality in the reactor pressure, it needs to analyze the cause of the reactor pressure in the causality graph. If the node has an abnormality, continue to analyze the parent node of the cause node until the parent of the current variable is normal. So anomaly is generated by the current variable. Because of the cause of reactor pressure is abnormal and the cause variable does not have a parent node, the anomaly is caused by the reactor temperature.

4 Conclusion

In this paper, EB-OCSVM was proposed to detect anomaly of industrial control system process. We used change points capture causality between process variables, which change points were determined by maximum distance to a line connecting the starting and ending point of the sequence. It can obtain an unknown number of change points and be used for periodic and non-periodic systems. By analyzing the number and time delay of change points, the correlation between the variables and the causal direction are obtained. The process variable , which is the leaf node in the causal graph, were classified into normal or abnormal ones by EB-OCSVM, if there is an anomaly detected, trace the causality graph to determine the source of anomaly. The results illustrated the effectiveness of the proposed method.

REFERENCES

[1] RAVAL S, BlackEnergy a threat to Industrial Control Systems network security [J], *International Journal of Advance Research in Engineering, Science & Technology (IJA- REST),* 2015, 2(12).

IOP Conf. Series: Journal of Physics: Conf. Series **1267** (2019) 012054 doi:10.1088/1742-6596/1267/1/012054

[2] LIU N, YU X H, ZHANG J H, Coordinated Cyber-attack: Inference and Thinking of In- cident on Ukrainian Power Grid[J], *Automation of Electric Power Systems*, 2016, 40(6):144-147
[3] Adepu S, Mathur A, Detecting Multi-Point Attacks in a Water Treatment System Using Intermittent Control Actions[C]// Inaugural Singapore Cyber Security R&d Conference, 2016.
[4] Adepu, S., Mathur, A., Distributed detection of single-stage multipoint cyber attacks in a water treatment plant, In: Proceedings of ACM Asia Conference on Computer and Communications Security (AsiaCCS 2016), pp. 449–460, ACM (2016).
[5] D. I. Urbina, J. A. Giraldo, A. A. Cardenas, et al, Limiting the impact of stealthy attacks on industrial control systems. In Proceedings of the ACM SIGSAC Conference on Computer and Communications Security (CCS), pp: 1092–1105. ACM, 2016.
[6] D. I. Urbina, J. Giraldo, A. A. Cardenas, et al, Survey and new directions for physicsbased attack detection in control systems. US Department of Commerce, National Institute of Standards and Technology, 2016.
[7] H.R.Ghaeini, et al, State-Aware Anomaly Detection for Industrial Control Systems, SAC 2018: Symposium on Applied Computing, 2018.
[8] Goh J, Adepu S, Tan M, et al, Anomaly Detection in Cyber Physical Systems Using Recurrent Neural Networks[C]// IEEE, *International Symposium on High Assurance Systems Engineering. IEEE*, 2017:140-145.
[9] J. Inoue, Y. Yamagata, Y. Chen, C. M. Poskitt, and J. Sun, Anomaly detection for a water treatment system using unsupervised machine learning, in Proc. IEEE International Conference on Data Mining Workshops (ICDMW 2017): Data Mining for Cyberphysical and Industrial Systems (DMCIS 2017). IEEE, 2017, pp.1058–1065.
[10] Moshe Kravchik, Asaf Shabtai, Detecting Cyber-attacks in Industrial Control Systems Using Convolutional Neural Networks, 2018
[11] Filonov P, Kitashov F, Lavrentyev A. RNN-based Early Cyber-Attack Detection for the Tennessee Eastman Process, ICML 2017 Time Series Workshop, 2017.
[12] Qin Lin, Sridha Adepu, Sicco Verwer, and Aditya Mathur. 2018. TABOR: A Graphical Model-based Approach for Anomaly Detection in Industrial Control Systems. (2018).
[13] Maryamsadat Hejazi , Yashwant Prasad Singh, ONE-CLASS SUPPORT VECTOR MACHINES APPROACH TO ANOMALY DETECTION, *Applied Artificial Intelligence*, 27:5, 351-366, 2013.
[14] Downs, J and Vogel, E. A plant-wide industrial process control problem, *Computers & chemical engineering,* 17(3):245–255, 1993.
[15] Tennessee Eastman Process with cyber-attacks dataset, URL: https://kas.pr/ics-research/dataset_tep_59, 2017.

A Dynamic Hand Gesture Recognition Algorithm Based on CSI and YOLOv3

Qiang Zhang*, **Yong Zhang and Zhiguo Liu**

Hefei University of Technology, Hefei 230001, China

*1378082472@qq.com

Abstract. Object detection algorithms based on convolutional neural networks are generally suitable for static gesture recognition. For actual hand gesture scenes, dynamic gestures are also widely used. A dynamic hand gesture recognition algorithm based on Channel State Information (CSI) and You Only Look Once: Version 3 (YOLOv3) is proposed for continuous dynamic hand gesture recognition. The data acquisition adopts a CSI-based radio frequency method. The adaptive weighted fusion, Kalman filtering, threshold segmentation and data conversion are used to generate gray value images. Finally, the YOLOv3 object detection algorithm is used to train and identify the grayscale image which include the information of continuous dynamic hand gestures. The effectiveness of the proposed method is verified by the recognition confusion matrix. And the proposed method has an average recognition accuracy of 94% for four custom dynamic hand gestures.

1. Introduction

With the continuous development and advancement of computer technology, the way of human-computer interaction (HCI) has also been diversified, and dynamic hand gestures have become an important way of HCI. In recent years, with the continuous development and advancement of WiFi technology, the way of dynamic hand gesture data acquisition has also undergone fundamental changes, and non-contact HCI is no longer limited to depth sensors, infrared cameras, etc. [1-2] . The channel state information (CSI) can represent the amplitude and phase information of each subcarrier corresponding to the space in the frequency domain, and fully utilizes the idea of Orthogonal Frequency Division Multiplex (OFDM) to make the wireless signal enables efficient communication between the transmitter and the receiver. Even if the object only performs a relatively small movement, CSI can be sensitively perceived and displayed on each subcarrier[3]. Therefore, some researchers use CSI to do human behavior recognition [4] and indoor intrusion detection [5].

How to classify the collected dynamic hand gesture data is also a research hotspot of dynamic hand gesture recognition technology. At present, the object detection algorithm based on CNN can effectively identify static hand gestures. The YOLOv3, a kind of object detection algorithm proposed by Redmon et al. [6] is one of the best object detection algorithm at present. Currently, YOLOv3 has been successfully applied to pedestrians detection [7], vehicle detection [8] and other fields[9-10]. However, the input of the YOLOv3 algorithm receives a static image, and the original CSI data cannot be used as an input.

In order to solve this problem, this paper adopts adaptive weighted fusion to extract a set of data from multiple links, adopts Kalman filter to filter the noise, and adopts threshold segmentation to extract the CSI valid data of the hand gesture action area. The process of preprocesses the raw CSI

IOP Conf. Series: Journal of Physics: Conf. Series **1267** (2019) 012055 doi:10.1088/1742-6596/1267/1/012055

data and data conversion to convert the CSI data into gray value images, and finally adopts YOLOv3 for training and recognition.

2. Proposed method

The algorithm of this paper mainly includes CSI data acquisition, CSI data preprocessing, data conversion, YOLOv3 training and recognition. The algorithm flowchart is shown in Figure 1.

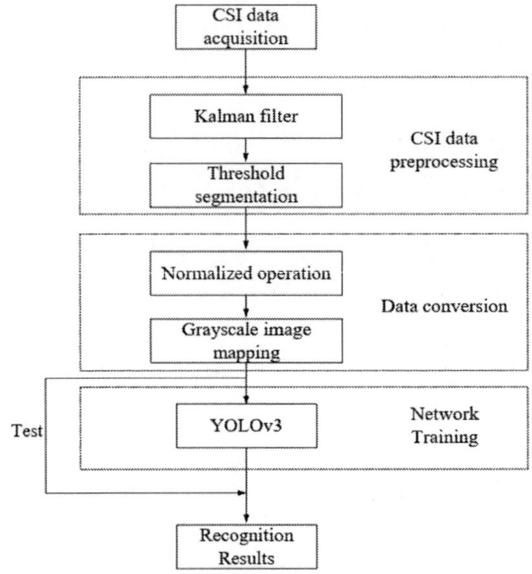

Figure 1. Flowchart of the proposed method.

2.1. CSI data acquisition

The Intel 5300 NIC is used as the receiving AP, and the TL-WR845N dual-antenna router is used as the transmitting DP. Multi-input Multi-output (MIMO) model is adopted to save bandwidth when actually collecting CSI data.

In order to collect more hand gesture CSI data, the AP uses three antennas to transmit signals, and the DP uses two antennas to receive signals to form a MIMO array. The data collectors made four custom hand gestures (waving to the left, waving to the right, waving forward, and waving backwards) between the DP and the AP. The experimental scene was located at the B1908 Laboratory of Science and Technology Building, Hefei University of Technology. The data collection scenario is shown in Figure 2.

Figure 2. Schematic diagram of collecting dynamic hand gesture

Each original CSI data packet is divided into 6 data streams, or 6 links, and each link has 30 subcarriers. Therefore, the obtained CSI data form can be expressed as:

$$CSI = \begin{bmatrix} H_{1,1} & H_{1,2} & \cdots & H_{1,30} \\ H_{21} & H_{22} & \cdots & H_{2,30} \\ \vdots & \vdots & \vdots & \vdots \\ H_{6,1} & H_{6,2} & \cdots & H_{6,30} \end{bmatrix} \quad (1)$$

IOP Conf. Series: Journal of Physics: Conf. Series **1267** (2019) 012055 doi:10.1088/1742-6596/1267/1/012055

Where $H_{i,j}$ represents the CSI value of the signal state information of the i th link and the j th subcarrier.

2.2. CSI data preprocessing
The original CSI data obtained in this way is co-dimensional data, and the too high dimension means that the complexity is too large, and the data needs to be dimensionally reduced. First, the signals of 30 subcarriers on each link are averaged to represent the data values of the link, and then adaptive weighted fusion [11] is applied to the six links to obtain a value. Kalman filtering is performed on the merged CSI data to remove the noise. Finally, the threshold segmentation is performed to obtain CSI data of the hand gesture action region.

2.2.1. Adaptive weighted fusion. Taking the hand gesture waving to left for example, the data of the 30 subcarrier signals on the six links after taking the tie value is shown in Figure 3. The abscissa represents time and the ordinate represents amplitude (in dB).

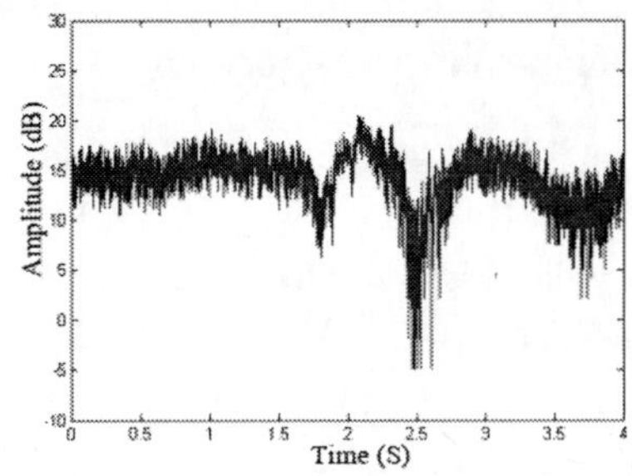

(a) CSI data waveform diagram of Link 1

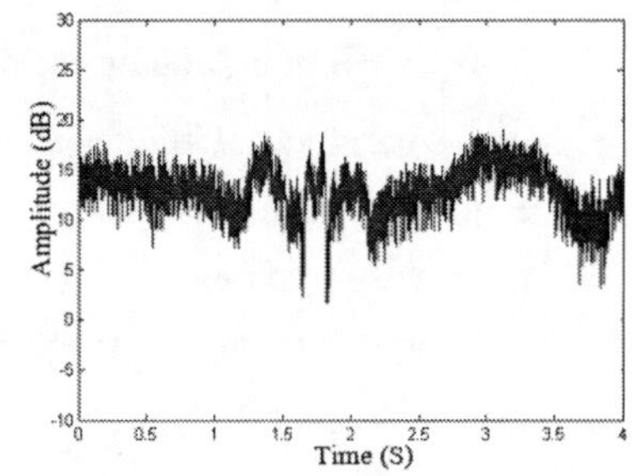

(b) CSI data waveform diagram of Link 2

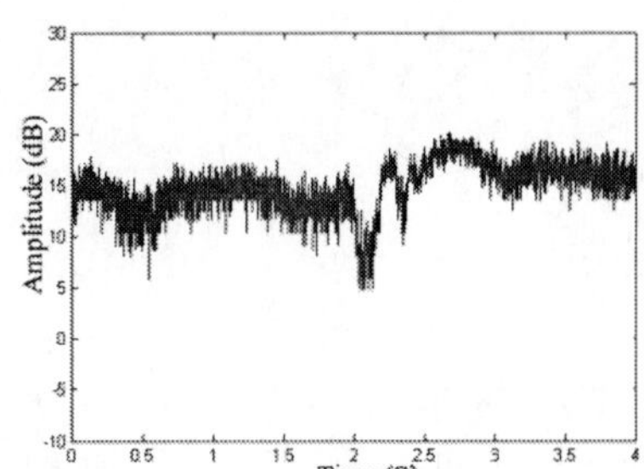

(c) CSI data waveform diagram of Link 3

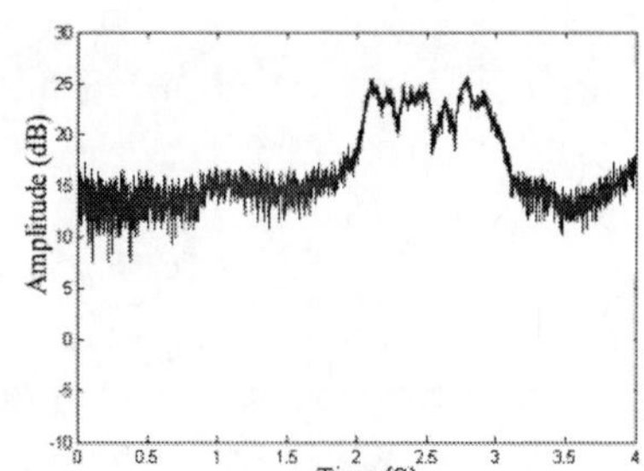

(d) CSI data waveform diagram of Link 4

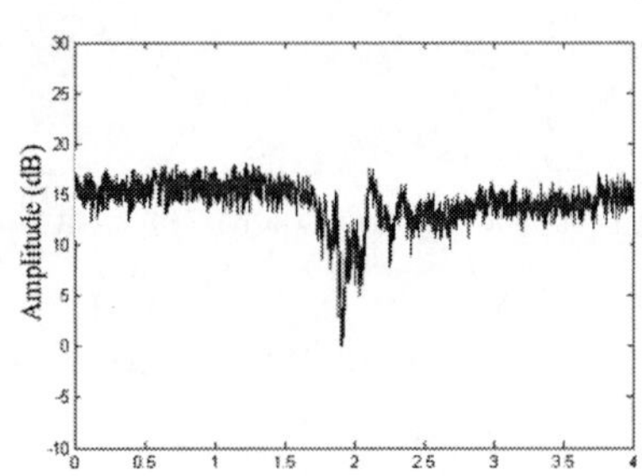

(e) CSI data waveform diagram of Link 3

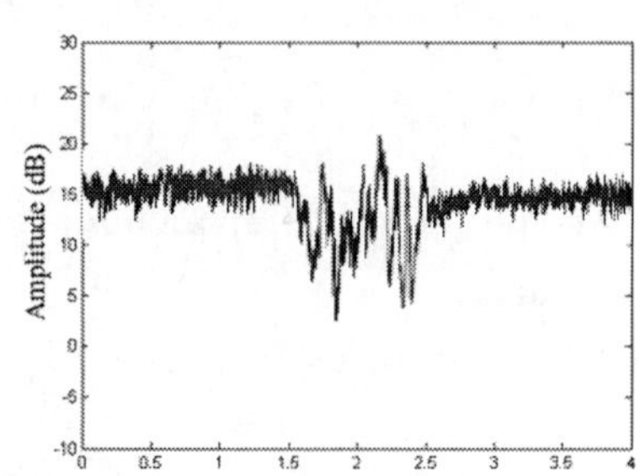

(f) CSI data waveform diagram of Link 4

Figure 3. CSI data waveform diagram of different links

Next, adaptive weighted fusion of the CSI data obtained by the six links is performed to eliminate the possible redundancy between multiple data links, and reduce complexity and uncertainty.

Assuming the number of links is n, the corresponding n groups of CSI data is $X_1, X_2, \cdots, X_n$ respectively, adaptive weighted fusion is to use an adaptive method to obtain an optimal weighting factor matching the set of data, so that the total variance of the set of data is the smallest. $\hat{X}$ represents the value after adaptive weighted fusion. The adaptive weighted fusion algorithm model is shown in Figure 4.

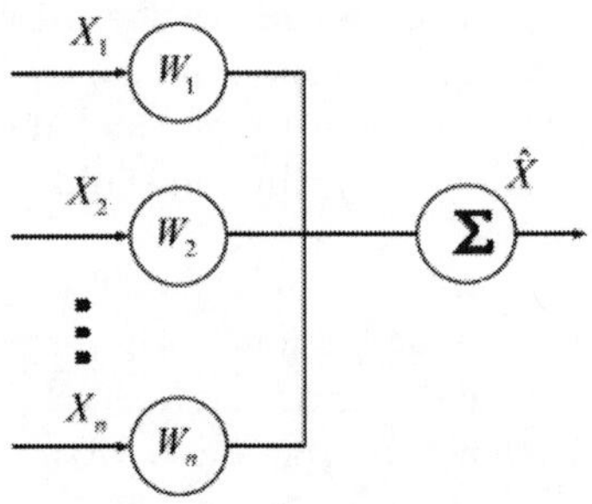

Figure 4. Diagram of adaptive weighted fusion algorithm

Assume that the variance of the n groups of CIS data is $\sigma_1^2, \sigma_2^2, \cdots, \sigma_n^2$ respectively. X is the actual value to be estimated. The weighting factors for the n groups of CIS data are $W_1, W_2, \cdots, W_n$ respectively and the fused values $\hat{X}$ can be expressed by the following formula:

$$\begin{cases} \hat{X} = \sum_{i=1}^{n} W_i X_i \\ \sum_{i=1}^{n} W_i = 1 \end{cases} \tag{2}$$

The total variance is:

$$\sigma^2 = E[(X - \hat{X})^2] = E[(\sum_{i=1}^{n} W_i X - \sum_{i=1}^{n} W_i X_i)^2] \tag{3}$$

After finishing:

$$\sigma^2 = E[\sum_{i=1}^{n} W_i^2 (X - X_i)^2] = \sum_{i=1}^{n} W_i^2 \sigma_i^2 \tag{4}$$

Where σ_i^2 represents the variance of the i th group CSI data. The minimum value of the σ^2 is $\sigma_{\min}^2$:

$$\begin{cases} \sigma_{\min}^2 = \min\left(\sum_{i=1}^{n} W_i^2 \sigma_i^2\right) \\ \sum_{i=1}^{n} W_i = 1 \end{cases} \tag{5}$$

where $\min(\bullet)$ represents the minimum operation. The optimal weighting factor is obtained when the variance is minimum:

$$W_i^* = \frac{1}{\left(\sigma_i^2 \sum_{i=1}^{n} \frac{1}{\sigma_i^2}\right)} \quad (i = 1, 2, \cdots, n) \tag{6}$$

The corresponding total variance is the minimum at this time:

$$\sigma_{\min}^2 = \frac{1}{\sum_{i=1}^{n} \frac{1}{\sigma_i^2}} \tag{7}$$

The fused value of 6 links of CSI data obtained by adaptive weighted fusion is shown in Figure 5.

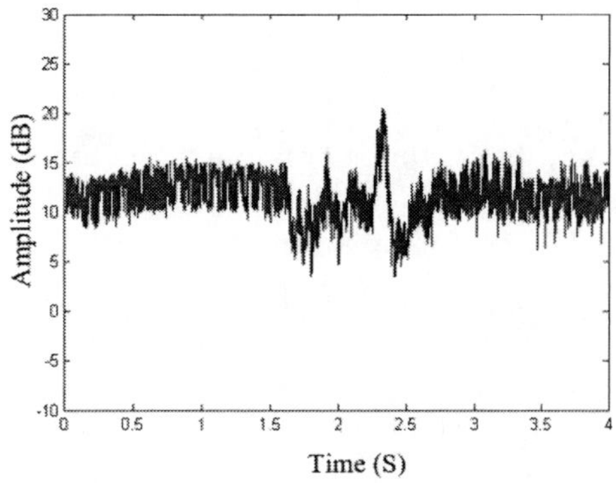

Figure 5. Waveform diagram of adaptive weighted fusion CSI data

2.2.2. Kalman filter. Although the general low-pass filter can remove the noise signal, it also filters out some useful signals, and the filtering effect is not good when the gesture is slow. Kalman Filtering is a state equation of a linear system. The basic idea is to use the input and output of the system to observe the data, and then to estimate the state of the entire linear system [12]. The calculation principle is as follows:

$$Z_t = F_t Z_{t-1} + B_t U_t + Q_t \tag{8}$$

Where Z_t denotes the state vector at the time of t, F_t denotes the state transition matrix, B_t denotes the control matrix, U_t denotes the control vector, Q_t denotes the noise matrix of the n_1 dimension. S_t is adopted to represents the observation vector of system at the time of t and the equation can be expressed as:

$$S_t = P_t Z_t + V_t \tag{9}$$

where P_t is the $n_1 \times n_2$ dimensional observation matrix, V_t is the n_2 dimensional observation noise matrix. Both V_t and Q_t are Gaussian white noise, which have have the following properties:

$$E[Q_t] = 0, E[V_t] = 0 \tag{10}$$

$$E[Q_t Q_t^T] = C, E[V_t V_t^T] = D \tag{11}$$

where C represents the prediction noise covariance matrix, and D represents the measurement noise covariance matrix.

Kalman filter is used to estimate the state quantity of the time system. The state prediction value $Z_t^{'}$ at the system time of t can be expressed as:

$$Z_t^{'} = F_t Z_{t-1} + B_t U_t \tag{12}$$

The covariance matrix prediction value $G_t^{'}$ at the system time of t can be expressed as:

$$G_t^{'} = F_t G_{t-1} F_t^T + C_t \tag{13}$$

Where G_{t-1} represents the covariance matrix value calculated at the system time of t-1. The Kalman gain matrix K_t can be expressed as:

$$K_t = \frac{G_t^{'} P_t^T}{P_t G_t^{'} P_t^T + D_t} \tag{14}$$

The optimal estimate of the state quantity of the system time of t is:

$$\hat{Z}_t = Z_t^{'} + K_t (S_t - P_t Z_t^{'}) \tag{15}$$

Update the covariance matrix:

$$G_t = (I - K_t P_t) G_t^{'} \tag{16}$$

where I represents the unit matrix.

The result of performing the above Kalman filtering on the CSI data obtained by adaptive weighting is shown in Figure 6.

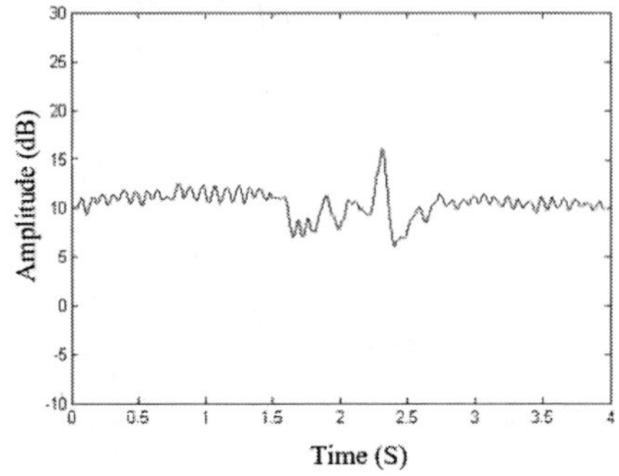

Figure 6. Kalman filtered CSI data waveform

2.2.3. Threshold segmentation. The CSI data obtained by the Kalman filter obtained in the previous section is a hand gesture of moving to left motion of 0 to 4 seconds. Although the noise is filtered, there are many meaningless data. Threshold segmentation separates the CSI data of the hand gesture action occurrence area from the whole data. In the process of CSI data analysis, it is found that the state change of the phase difference variance between CSI antennas is closely related to the hand gesture motion detection. Therefore, the threshold difference-based sliding window method is used to detect the phase difference of the CSI signal [13]. Specific steps are as follows:

(1) Collect CSI data in a quiet environment without any action, and calculate the average value μ and normalized standard deviation σ of the signals in the sliding window.

(2) Determine the threshold δ based on the μ and σ calculated in the previous step, and the relationship between the three is satisfied:

$$\mu + 6\sigma \le \delta \tag{17}$$

(3) Acquire the signal flow in the sliding window, compare the average value of the signal in the sliding window with the threshold, and if the average value is greater, determine that the signal is a signal in the hand gesture action area. In the actual experiment process, since the dynamic hand gesture is a continuous process, the signal in the hand gesture action area should be continuous. There is a start point and an end point, and after determining the start point and the end point of the hand gesture action, the CSI data is segmented.

After threshold segmentation, the CSI data of the hand gesture action area obtained is as shown in Figure 7.

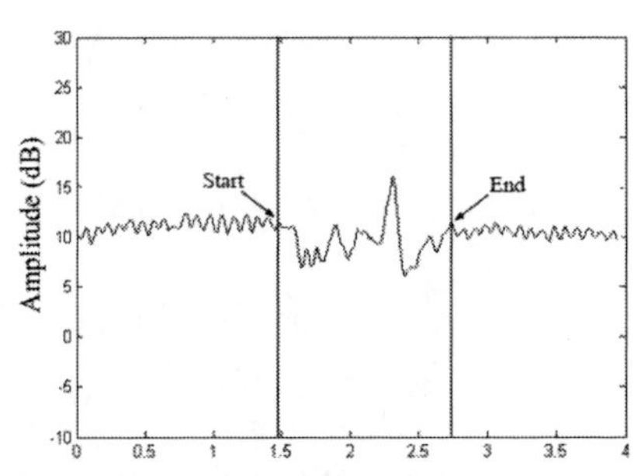

Figure 7. Schematic diagram of CSI data after threshold segmentation

IOP Conf. Series: Journal of Physics: Conf. Series **1267** (2019) 012055 doi:10.1088/1742-6596/1267/1/012055

2.3. Data conversion

After the threshold segmentation, the CSI data of the hand gesture action area is obtained. In order to facilitate the YOLOv3 training and recognition, the CSI data is converted into an image as an input of neural network for training and recognition. The CSI data is sampled to obtain 100 sets of data, and then the 100 sets of data are normalized and converted into gray value data, and finally mapped to the grayscale image.

2.3.1. Normalized operation. The CSI data of the hand gesture action area obtained in previous section is equally spaced to obtain 100 sets of CSI data $A=\{a_1,a_2,\cdots,a_{100}\}$. Find the maximum and minimum values, and then normalize the 100 sets of data. The new data is $L=\{l_1,l_2,\cdots,l_{100}\}$:

$$l_i=\frac{a_i-\min}{\max-\min}\quad(i=1,2,\cdots,100) \tag{18}$$

Then the new data obtained is converted into gray value data $O=\{o_1,o_2,\cdots,o_{100}\}$:

$$o_i=l_i\times 255\quad(i=1,2,\cdots,100) \tag{19}$$

The final result is 100 sets of gray value data between 0 and 255. The 100 sets of data contain all the information of the continuous dynamic hand gesture of waving to left in the sample.

2.3.2 Grayscale image mapping. The 100 sets of gray value data obtained in the previous section are arranged in a 10×10 matrix:

$$\begin{bmatrix} o_1 & o_2 & \cdots & o_{10} \\ o_{11} & o_{12} & \cdots & o_{20} \\ \vdots & \vdots & \vdots & \vdots \\ o_{91} & o_{92} & \cdots & o_{100} \end{bmatrix} \tag{20}$$

The 10×10 matrix data is mapped into the image of the 10×10 grids so that it corresponds to the gray value in the image grids. Figure 8 shows the grayscale value image obtained after the normalized operation of the CSI data of the hand gesture action area and the mapping of the grayscale image.

Figure 8. CSI data mapping grayscale image

2.4. YOLOv3 training and recognition

Through the above operations, the CSI data of the original continuous dynamic hand gesture action is converted into image data, and the images are assigned a training set and a test set in a ratio of 3:1. Next, the training set image is trained using the object detection algorithm YOLOv3 based on CNN, and the test set image is tested and identified.

The basic feature extractor of the YOLOv3 network uses the Darknet-53 model, the training set and the test set adopt the CSI hand gesture data to map the gray value image. The specific process is as follows:

IOP Conf. Series: Journal of Physics: Conf. Series **1267** (2019) 012055 doi:10.1088/1742-6596/1267/1/012055

(1) Copy each gray value image in the training set and adjust their resolutions to 300 × 225, 400 × 300, 500 × 375, 600 × 450, respectively. In this way, the number of images are expanded four times.

(2) Each of the gray value images obtained in the previous step is manually labeled in accordance with the format of the VOC data set, and the category information and the position information of the hand gestures are annotated to generate a hand gesture tag file.

(3) Combine all the gray value images for training obtained in the previous step and the corresponding hand gesture tag files, and train the YOLOv3 network as a training set (VOC format).

(4) Identify and detect the gray value image in the test set with the trained YOLOv3 model.

3. Experimental results and analysis

During the experiment, 1000 sets of CSI data were generated by waving to left, waving to right, waving forward, and waving backwards. After CSI data preprocessing and data conversion process, 4000 gray value images were obtained. For each kind of hand gesture motion, 750 images were extracted as a training set, and 250 images were used as a test set.

In this paper, the recognition confusion matrix is used to evaluate the experimental results. Each row of the matrix represents the probability that a real category is identified as each category, and each column represents the probability that each real category is identified as a particular category.

Assuming $P_{i,j}$ represents the value of the cell of row i and column j, then the value indicates the probability that the true category of the row i is identified as the predicted category of the column j. The recognition confusion matrix of the four categories of CSI hand gesture motion gray value images is shown in Figure 9.

	Moving to left	Moving to right	Moving forward	Moving backward
Moving to left	0. 95	0. 04	0. 01	0. 00
Moving to right	0. 05	0. 93	0. 01	0. 01
Moving forward	0. 02	0. 01	0. 94	0. 03
Moving backward	0. 01	0. 00	0. 06	0. 93

Figure 9. Recognition Confusion matrix

It can be seen from the recognition results that the recognition accuracy rates of the four dynamic hand gestures of waving to left, waving to right, waving forward, and waving backward are 95%, 93%, 94%, 93%, respectively, and the average recognition accuracy is about 94%. Although there are only a few types of gestures, in the case of only 1000 sets of data for each type of action, recognition accuracy of 94% can be obtained.

4. Conclusion

This paper proposed a dynamic hand gesture recognition algorithm based on CSI and YOLOv3. The CSI data of the hand gesture action is collected by the network card and the router. The 30 subcarrier signals are averaged, and adaptive weighted fusion is performed, so that a set of data is combined from the CSI data of the six links. Then Kalman filter is used to perform the CSI data. By Filter denoising and threshold segmentation, the CSI data of the hand gesture action area is obtained. Next, the CSI data is converted into gray value image, and the YOLOV3 network is used for training and recognition. Recognition accuracy of 94% is obtained, which shows the feasibility and effectiveness of the proposed method.

IOP Conf. Series: Journal of Physics: Conf. Series **1267** (2019) 012055 doi:10.1088/1742-6596/1267/1/012055

References

[1] Wu X,Chu Z, Yang P, Xiang C, Zheng X, Huang W. Tw-see: Human activity recognition through the wall with commodity wi-fi devices[J]. IEEE Transactions on Vehicular Technology, 2019, 68 (1): 306–319.
[2] Man D, Wang W, Wang X, Lv J, Du X, Yu M. Pwig: A phase-based wireless gesture recognition system[C]. 2018 International Conference on Computing, Networking and Communications (ICNC), 2018: 837–842
[3] Liu X, Cao J, Tang S, Wen J. Wi-sleep: Contactless sleep monitoring via wifi signals, 2014 IEEE Real-Time Systems Symposium, 2014: 346–355.
[4] Mei H F. Research and implementation of human behavior recognition based on CSI [D]. Wuhan University of Technology, 2017.
[5] Zhou J. Research on indoor intrusion detection and behavior recognition based on CSI in wireless sensing network [D]. Nanjing University of Posts and Telecommunications, 2018.
[6] Redmon J, Farhadi A. Yolov3: An incremental improvement [Z/OL]. Computer Vision and Patern Recognition (CVPR), 2018, pp.126-134.
[7] Wang D W, He Y H, Li D, Liu Y, Xu Z J, Wang, J. Improved pedestrian detection algorithm for YOLOv3 infrared video images[J]. Journal of Xi'an University of Posts and Telecommunications, 2018, 23(04): 48-52+67.
[8] Zhang F K, Yang F, Li C. Fast vehicle detection method based on improved YOLOv3[J]. Computer Engineering and Applications, 2019, 55(02): 12-20.
[9] Shuo Y, Zhang J X, Bo C J, Wang M, Chen L J. Fast vehicle logo detection in complex scenes[J]. Optics and Laser Technology,2018.
[10] Liu B, Wang S Z, Zhao J S, Li M F. Ship tracking recognition based on Darknet network and YOLOv3 algorithm [J/OL]. Computer application, pp.1-7
[11] Hua S Z. Research and application of multi-sensor data fusion algorithm [D]. Northeastern University, 2017.
[12] Liu J, Kong Y C. Fusion Positioning of WiFi Inertial Navigation Using Linear Kalman Filter[J].Electronic Measurement Technology, 2017,40(4):1-4.
[13] Liu J L. Research on human motion detection in complex scenes based on CSI [D]. Dalian University of Technology, 2017.

IOP Conf. Series: Journal of Physics: Conf. Series **1267** (2019) 012056 doi:10.1088/1742-6596/1267/1/012056

A Human-Robot Interaction for a Mecanum Wheeled Mobile Robot with Real-Time 3D Two-Hand Gesture Recognition

Xueling Luo[1], Andrea Amighetti[1,2] and Dan Zhang[1,*]

[1]Department of Mechanical Engineering, York University, Toronto, Ontario, Canada
[2]Department of Industrial Mechanical Engineering, University of Brescia, Brescia, Italy

* Email: dzhang99@yorku.ca

Abstract. Human interaction with mobile robot becomes a popular research area and its applications are widely used in industrial, commercial and military fields. A two-hand gesture recognition method with depth camera is presented for real-time controlling the mecanum wheeled mobile robot. Seven different gestures could be recognized from one hand for mobile robot navigation and three gestures could be recognized from the other hand for controlling the gripper installed on the robot. Under the proposed control scheme, the mobile robot system can be navigated and can be operated at the same time for achieving missions by two different groups of hand gestures. The accuracy of the gesture recognition is about 94%. During mobile robot control experiment, the system works timely, accurately and stably for certain tasks such as directional movement, grasping and cleaning obstacles.

1. Introduction

With the development of human computer interaction (HCI), hand gesture recognition based HCI becomes one of the most popular implementation since it is the most similar way as interaction between human and human daily operation activity [1]. In the past decades, there are two main type of sensors for hand gesture recognition: contact sensors such as 'Cyberglove' [2], body-worn inertial sensors [3] and non-contact sensors such as biplanar videoradiography system [4], camera systems [5]- [7]. Those contact sensors are measuring gestures into electrical signals for HCI which could bring accurate detected result and it already been applied into military, industry and commercial fields. However, wearable devices are not always comfortable or welcomed by users and this kind of devices are hard to be manufactured and maintained since they normally consist by several parts and requires complex calibration [5].

While contact sensors for gestures recognition are still widely applied, non-contact sensors got more attention in recent years by its features such as natural interaction, measuring without physical contact and funny user experience. Non-contact sensors show more advantages comparing with traditional sensors in certain applications such as it will not take any risk to damage the object and it has high accuracy. Many application systems with gestures recognition become available in the market for smart surveillance, teleconferencing, HCI, etc. [8]. Besides, gesture recognition also be applied in a visual environment for very large scale biomolecular modelling [9] and works as integrated controller for the virtual environment Battle Field [10]. The most common method for such applications is color spaces recognition which could effectively reach the desired result and could also solve problems like partially occluded [11], [12]. Since vision system was designed by inspiration from human eyes which could be applied for 3-dimension measurement. A few 3D sensors have been published in recent years such as Kinect, Wii, Intel RealSense 3D camera. Among of numerous application, one of the most interesting application is using gesture for mobile robot control.

Zhao [12] presented a gesture-based control method for mobile robot. To apply pattern recognition method such as mean filtering, HSV color space and to use morphological image processing for static gesture recognition. Five different gestures are encoded with their feature information. A library is created by this process and it is also used for later recognition with contour correspondence method. During the control for the robot, a gesture-based state-feedback control strategy is applied. This method only provides five gestures for mobile robot control which is hard to satisfy a complex operation for real application. Besides, the experiment for hand gesture recognition method is in an experimental environment, the detection background is pure white which is easy for the detection. Kundu [13] proposed a hand gesture-based classifier algorithm to control an omnidirectional wheel-chair system. Seven motions are corresponded to seven different gestures to control maneuverability of the wheelchair. Wearable IMU sensor and two EMG sensors are used to extract wrist movement and two forearm muscle activities to encode gestures and for later gestures recognition. The system accuracy is depended on the wearable sensor which need special training to make sure the sensor is placed right for normal user. And the sensor need to be handled carefully to keeping it function. Besides, the training method for the gesture will take a lot of time. Gao's [13] robotic wheelchair control system based on hand gesture control for the disabled is using Kinect sensor for mobile robot operation. The method used Fuzzy control to create a corresponding from gestures to speed and direction for the wheelchair. This method provides a recognition method for gesture movement trend to command the robot's moving trend. During the experiment, there must have some vibration of the hand since users have to hold their hand in the air through the whole time which could affects the control strategy. Besides, during the high-speed model the control method is not working as it desired [13].

We developed a two-hand gesture recognition method with a depth camera from the Kincet sensor for controlling the mecanum wheeled mobile robot. A gripper was designed and equipped for the mobile robot. A model based on 3D geometry is designed for gesture recognition. Seven different gestures could be recognized from right hand for mobile robot movement operation and three gestures could be recognized from left hand to control the gripper. In the experiment, certain tasks like moving in a designed map, gripping an item in the path and put the object in certain place. The system works timely, accurately and stably.

2. Proposed method

We proposed a mobile robot operational control method with gesture recognition which can be described as figure 1. The system consisted of two main processes: image processing and human-robot interaction. During image processing, there are two different groups of gestures defined respectively for robot navigating and robot operational control. In this paper, we presented a hand-finger poses estimation method based on pose-geometry and depth information.

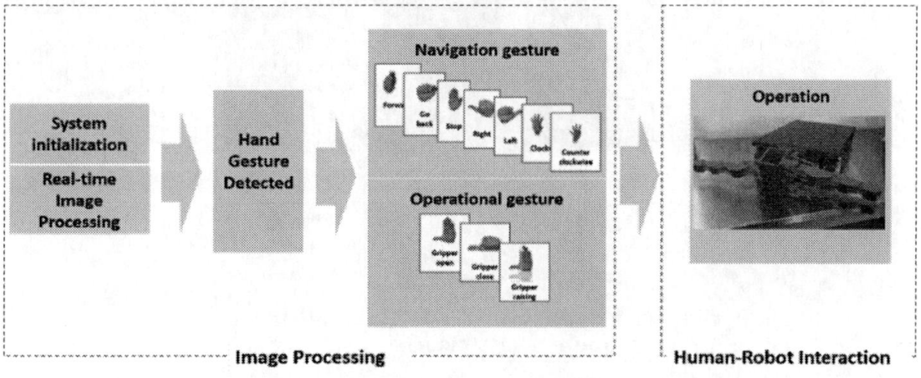

Figure 1. The system working flowchart

IOP Conf. Series: Journal of Physics: Conf. Series **1267** (2019) 012056 doi:10.1088/1742-6596/1267/1/012056

In order to discriminate 7 different gestures, a coordinate system of a hand was defined which is shown in figure 2. In figure 2(a), h, w and d are the length of the mass centre to the middle finger in vertical direction, the length of the mass centre to the thumb in horizontal direction and the length of the mass centre to the thumb in depth direction.

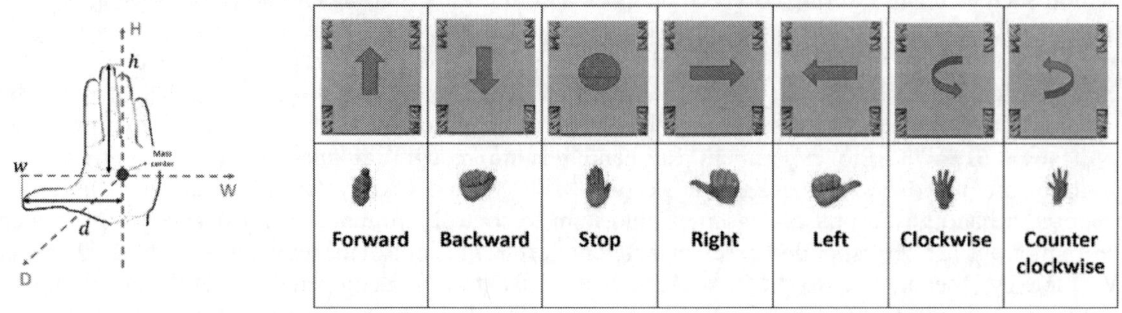

(a) **(b)**

Figure 2. (a) Hand gesture coordinate, (b) Hand gesture for mobile robot operation

There are 7 different hand gesture defined for mobile robot operation which are shown in figure 2(b). Based on gesture geometry, a fast and accurate method could be presented as function (1)

$$G(w,h,d) = \begin{cases} w(T_{WN}) \ \&\& \ h(T_{HN}) & \ldots Backward \\ w(T_{WX}) \ \&\& \ h(T_{HN}) & \ldots Right \\ w(-T_{WX}) \ \&\& \ h(T_{HN}) & \ldots Left \\ w(T_{WN}) \ \&\& \ h(T_{HX}) & \ldots Stop \end{cases} \quad (1)$$

where, $G(w,h,d)$ is the current gesture geometry. When there is no rotation trend detected, the gesture would be classified in logic equation (1). There are two gestures is required for initialization which are the gesture "Left" with thumb open and the other four fingers close and the gesture "Stop" with thumb close and the other four fingers upright. This process is to determine ranges of thresholds that thumb close - T_{WN}, other four fingers close - T_{HN}, thumb open - T_{WX} and the other four fingers open - T_{HX}, respectively. When both $T_W < T_{WN}$ and $T_H < T_{HN}$ are detected, the state of the detection is "Forward". During the operation, if the depth of the mass centre is changed from the initial value, all thresholds will change with the same ratio.

When the rotation trend appearing in the system, the rotation direction of the system can determinate by checking the movement of thumb in depth direction. For discriminating the rotational signal, we use the famous method "convexity defects" [14] [15] for detection. When T_{WN} is detected, the command signal for navigation is clockwise rotation. When $-T_{WN}$ is detected, the command is counter clockwise rotation.

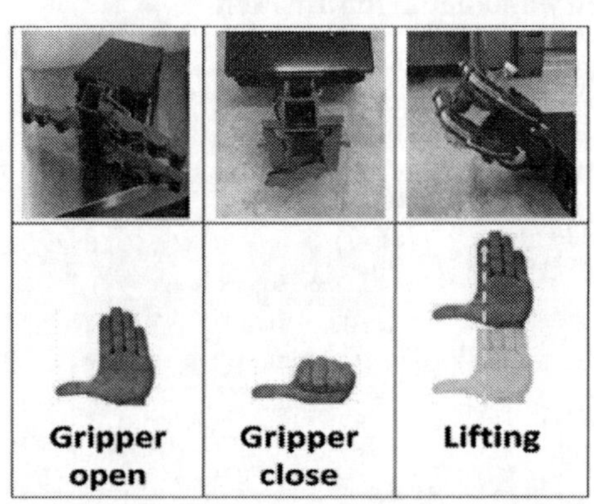

Figure 3. Hand gesture for mobile robot operation

IOP Conf. Series: Journal of Physics: Conf. Series **1267** (2019) 012056 doi:10.1088/1742-6596/1267/1/012056

Operation for gripper control is less complex than movement control. There are only three gestures for the gripper's control. In the figure 3, the map between those three gestures and the gripper commands is shown. The detection logic of the operational gestures is similar as the navigation gestures detection. When "Gripper open" and "Gripper close" are detected, if additional vertical direction movement of the gesture centre is also detected, the system can respond by sending gripper "lifting" or "putting-down" signal.

3. Mobile robot kinematic

Mecanum wheels have been designed with passive rollers around the wheel circumference at $45^{\circ}$of angle. In the figure 4, the configuration of the robot is shown.

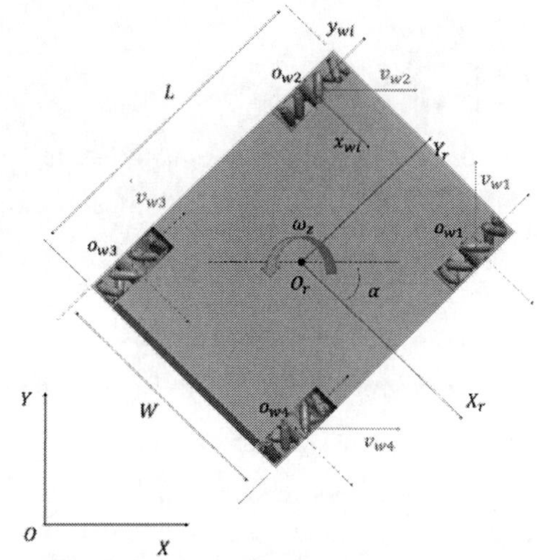

Figure 4. The configuration of the robot

Then, the inverse kinematics of the robot could be expressed as:

$$\begin{bmatrix} \dot{x}_r \\ \dot{y}_r \\ \dot{\alpha}_r \end{bmatrix} = \frac{R}{4} \begin{bmatrix} -1 & 1 & -1 & 1 \\ 1 & 1 & 1 & 1 \\ \frac{1}{\frac{W+L}{2}} & -\frac{1}{\frac{W+L}{2}} & -\frac{1}{\frac{W+L}{2}} & \frac{1}{\frac{W+L}{2}} \end{bmatrix} \begin{bmatrix} \dot{\theta}_1 \\ \dot{\theta}_2 \\ \dot{\theta}_3 \\ \dot{\theta}_4 \end{bmatrix} \quad (2)$$

where, the coordinates OXY is the fixed coordinate and the $O_rX_rY_r$ is the moving coordinates for the robot. The position of the robot is $S_r = [x_r, \ y_r, \ \alpha_r]^T$, R is the radius of the wheel, W and L are wide and length of the robot, $[\dot{\theta}_1, \dot{\theta}_2, \dot{\theta}_3, \dot{\theta}_4]$T is the angular velocity of each wheel. In the control of the robot, the signal could be sent in as the coordinate$[\dot{x}_r, \dot{y}_r, \dot{\alpha}_r]$.

4. Experiment

For hand gesture recognition, a laptop with windows 10 system, Inter I7-6700 CPU is working as the main processor. In the figure 5(a), the mobile robot with mecanum wheels is shown and a gripper is installed on it. Inside of the robot there is an Arduino chip working as the main processor for the robot. The gripper is an underactuated gripper with two motors with 2-DOF: open - close and lifting - putting down. The communication between the laptop and the robot is based on a pair of Bluetooth chips. The set up for the gesture recognition system is shown in figure 5(b). And a Kinect is used as the visual sensor of the system. The figure 5(c) is the experiment area for the mobile robot to verify the proposed method could order the robot to act as it has been commanded. The experimental field is label as a map with straight line and curve which is for verifying flexibility of the proposed control method for the mobile robot system. During the experiment, there are several obstacles placed on the map for robot to pick up and remove.

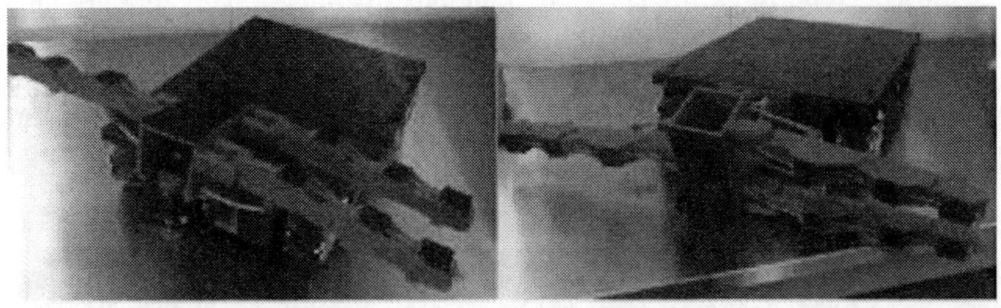

(a) The mecanum wheeled robot with a gripper system

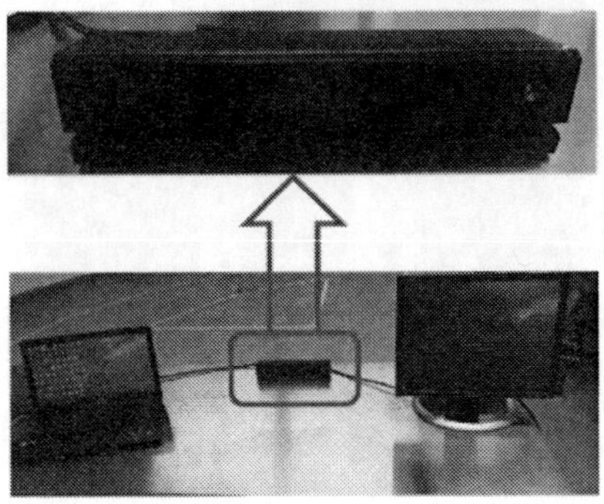

(b) The visual system for gesture recognition **(c)** The experimental field

Figure 5. The experiment setup

In the table 1, the accuracy of the gestures recognition for both hands are show. It is a continue experiment with a series of gesture changing in turn. The number of detected gestures is shown in the table and the number of correctly detected gestures is also shown in the table. The accuracy of the navigation gestures is 94.39%. And the accuracy of the operational gestures is 96.34% since the gesture group is less complex than the navigation gesture group.

Table 1. Accuracy of hand recognition algorithm

	Detected (Number)	Correctly detected (Number)	Accuracy (%)
Forward	217	206	94.93
Backward	194	185	95.36
Right	226	213	94.25
Left	207	196	94.69
Stop	191	183	95.81
Clockwise rotation	213	197	92.49
CCW rotation	192	179	93.23
Gripper close	230	221	96.09
Gripper open	217	209	96.31
Gripper lifting	208	201	96.63

For the initial value setup, we detected the 'stop' and the 'left' gesture from the right-hand. And the 'gripper close' gesture was detected for operational gesture initial value. This process must be done manually to make sure those gestures could be detected clearly. In the figure 6, a few photos took from the right monitor of the experiment setup in figure 5(a). Figure 6 are different gestures corresponding to operation forward, backward, right and counter clockwise rotation, respectively. The robot could respond to the gesture command accurately and timely.

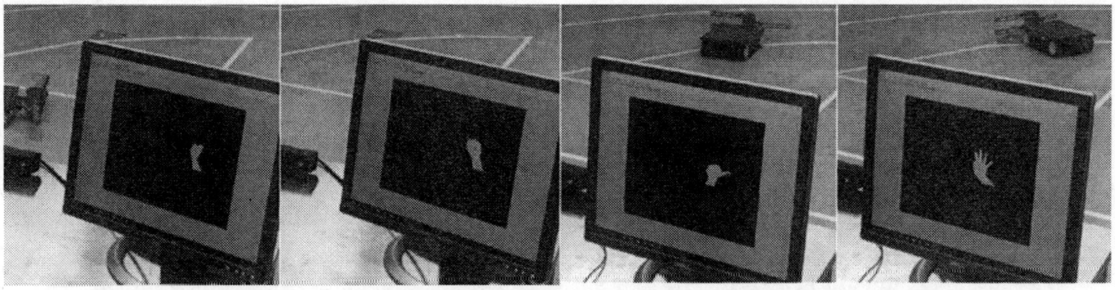

Figure 6. The result of mobile robot navigation with hand gesture

To verify the system could work for certain tasks, we put a few objects in the map and use two-hand to operate the robot for directional movement, picking up the object and putting the object into a certain place. In figure 7, photos of navigation and operational processes are shown. In figure 7(a), the robot is responded to command lifting gripper and counter clockwise rotation in the same time. In current moment, the robot tries to move the obstacle away from its path. In figure 7(b), the robot is responded to command closing gripper and stop moving in the same time. Under this stage, the robot met an obstacle and it had to stop and grip the obstacle in front of it.

(a) To rotate the robot CCW and to lift the gripper **(b)** To stop the robot and to close the gripper

Figure 7. The result of mobile robot navigation and operation with hand gesture

In the experiment, for different users we must initialize the system for each of them to make sure the high accuracy of this method. The user should try to keep each gesture as stable as possible. However, small rotation angle of each gesture could be adjusted by the threshold ranges and the slow change in depth direction could also be accepted by the algorithm.

5. Conclusion

The hand gesture method we proposed could recognize seven commands from right-hand and three commands from left-hand for the mecanum wheeled mobile robot control. The system works timely, accurately and stably for the desired tasks such as directional movement, grasping the item and carrying it to the certain place. The accuracy of the gesture recognition is 94.4% for this application. The experiment result proved that the designed system could be control remotely and it has well performance on those certain tasks. This kind of system could be useful in industry workshop or for certain risky task such as carry hot, sharp, chemical or radioactive items. The camera system is easy to install or maintain. The training of this method only needs three well placed gesture which would easy for users to learn. Besides, the algorithm is using the depth camera which would not affect by the light condition. This feature of the proposed method will allow it to be used in special working space where the light condition is strictly controlled.

However, this work still could be improved. The hand recognition method is designed for a relatively clean background which is not allowed much interference. If the human body parts such as main body or human face is too closed to the hand or their position are too close to the camera than the hand. The algorithm would not detect the gesture accurately.

Acknowledgments
The authors would like to thank the financial support from the Natural Sciences and Engineering Research Council of Canada (NSERC) and gratefully acknowledge the financial support from the York Research Chairs (YRC) program.

References

[1] Ren Z and Yuan J. "Depth Camera Based Hand Gesture Recognition and its Applications in Human-Computer-Interaction", *IEEE 8th Int. Conf. on Information, Communications and Signal Processing*, Singapore, 2011.

[2] Lee C and Xu Y, "Online, interactive learning of gestures for human/robot interfaces", *IEEE Int. Conf. on Robotics and Automation*, Minneapolis, 1996.

[3] Junker H, Amft O, Lukowicz P and et. al. "Gesture spotting with body-worn inertial sensors to detect user activities", *Pattern Recognition*, 41: pp. 2010-2024, 2008

[4] Miranda D.L., Rainbow M.J., Crisco J.J. and et. al. "Kinematic differences between optical motion capture and biplanar videoradiography during a jump–cut maneuver", *Journal of Biomechanics*, 43: pp. 567-573, 2013

[5] Ren Z, Yuan J, Meng J and et. al. "Robust part-based hand gesture recognition using Kinect sensor", *IEEE Trans. on Multimedia*, 15(5): pp. 1110-1120, 2013.

[6] Shimada N, Shirai Y, Kuno Y, and et. al. "Hand gesture estimation and model refinement using monocular camera-ambiguity limitation by inequality constraints", *In Processing of Third IEEE Int. Conf. on Face and Gesture Recognition*, Nara, 1998.

[7] Chua C, Guan H, and Ho Y. "Model-based 3d hand posture estimation from a single 2d image", *Image and Vision Computing*, 20: pp.191 – 202, 2002.

[8] Hsieh C, Liou D and Lee D. "A real time hand gesture recognition system using motion history image", *The 2nd Int. Conf. on Signal Processing Systems*, pp.394-439, Dalian, 2010.

[9] Zeller M, Phillips J.C., Dalke A and et. al. "A visual computing environment for very large scale biomolecular modeling", *IEEE 24th Int. Conf. on Application-Specific Systems, Architectures and Processors*, pp. 3-12, Zurich, 1997.

[10] Wu C, Chen W and Lin C. "Depth-based hand gesture recognition", *Multimedia Tools and Applications*, 75(12), pp. 7065 - 7086, 2016.

[11] Soleimanizadeh S and Mohamad D, Saba T, Rehman A. "Recognition of partially occluded objects based on the three different color spaces (RGB, YCbCr, HSV)", *3D Research*, 6(3): pp.1–10, 2015.

[12] Zhao H, Hu J and et. al. "Hand Gesture Based Control Strategy for Mobile Robots", *2017 29th Chinese Control and Decision Conf.*, pp: 5868-5872, Chongqing, 2017.

[13] Kundu S, Mazumder O and et. al. "Hand Gesture Recognition Based Omnidirectional Wheelchair Control Using IMU and EMG Sensors", *Journal of Intelligent & Robotic Systems*, pp. 1-13, 2017.

[14] Manresa C, Varona J, Mas R and et. al. "Perales Hand tracking and gesture recognition for humancomputer interaction", *Electron Letters Computer Vision and Image Analysis*, 5(3), pp. 96–104, 2005

[15] Yeo H, Lee B and Lim H. "Hand Tracking and Gesture Recognition System for Human-computer Interaction using Low-cost Hardware", *Multimedia Tools and Application*, 74, pp. 2687-2715, 2015.

Research and Application of Target Tracking Algorithm in Ship Safety

Bailin Lin[a] and Zheng'an Xiao[b]

School of Physics and Mechanical&Electrical Engineering, Hubei University of Education, Wuhan 430205, China

[a]home12345@163.com; [b]285106243@qq.com

Abstract. Target tracking is often used in tasks such as video surveillance. Its purpose is to obtain the target of interest in the sequence image accurately, robustly and in real time, and to establish the connection between the moving objects in each image. In recent years, with the continuous development of the global shipping economy, ships have gradually developed into large-scale and high-speed, and how to ensure the safety of ships sailing in these sports has become an important issue. Based on the principle of particle filter and mean shift algorithm, this paper takes the moving ship as the research object and uses the polynomial fitting method to analyse the performance of the two algorithms. A large number of experimental results show that the mean shift algorithm has higher accuracy, and the particle filter method has better real-time performance. And as the number of particles increases, the accuracy of particle filtering will gradually increase.

1. Introduction

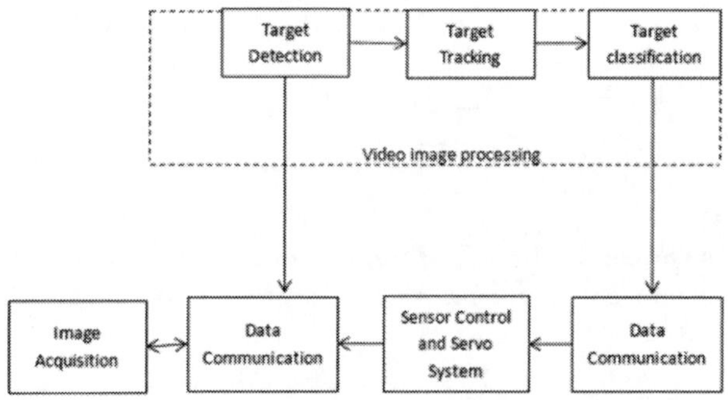

Figure 1. Block Diagram of Ship Automatic Tracking System.

The purpose of video target tracking [1] is to analyse the video sequence captured by the sensor and correlate the same moving target in different frames of the image sequence to obtain the complete motion track of each moving target. The essence of target tracking is to analyse the video sequence captured by the image sensor and calculate the position, size and movement speed of the target in each frame of the image. As a main branch of the field of computer vision, the research of video target tracking method is increasingly applied to security, intelligent video surveillance, intelligent traffic management, military robot vision and other fields. This article takes sports ship target tracking as an example to carry out relevant algorithms and experimental research. The system that is formed is shown in Figure 1. The shipping monitoring system collects images or video and transmits it to image processing equipment and then detects the target according to ship template matching and Track your goals.

In this paper, the sports ship is taken as the tracking target. Each frame of image in the video is taken as the object of processing and the image is enhanced. Then, the target ship is tracked by Kalman and particle filter respectively.

2. Algorithm principle analysis

2.1. Mean Shift algorithm

Mean Shift [2] is a typical algorithm for the offset vector. In the target tracking, the probability density function is used to iteratively search for the target colour feature to track the target. Now commonly referred to as Mean Shift algorithm belongs to the search and match category in the tracking. According to the maximum similarity of the tracking template, the mean value of the current point and the mean value of the shifted point are automatically found in the neighbouring area. This is a relatively simple density estimation model. Constant search until the calculated offset point is the last pixel to scan the entire image [3]. The specific process of the algorithm is as follows.

1) The target model of the initial frame

In the target search window space, the interval is divided into K equal intervals, and the central pixel coordinate of the search window in the initial frame is set as the coordinate of the i-th pixel in the image space area. In addition, $k\left(\|x\|^2\right)$ is used to represent the kernel function selected during processing and the window width is h. Then the probability of the u-th eigenvalue can be expressed by formula (1).

$$\hat{q} = C\sum_{i=1}^{n} k\left(\left\|\frac{x_0 - x_i}{h}\right\|\right)\delta[b(x_i - u)] \tag{1}$$

In (1), the membership of the color corresponding to the characteristic value u in the coordinates can be judged by b. In the color space, C is the constant coefficient to make the probability normalized.

$$\sum_{i=1}^{n} \hat{q} = 1 \tag{2}$$

2) Current frame model

According to the establishment of the initial frame model, the probability of the u-th feature value of the search window in the current frame can also be expressed as:

$$\hat{p}(y_0) = C_h \sum_{i=1}^{n} k\left(\left\|\frac{y_0 - x_i}{h}\right\|\right)\delta[b(x_i - u)] \tag{3}$$

In equation (3), the central pixel coordinates corresponding to the current frame search window inthe initial frame model correspond to C in equation (1).

3) Similarity function

The similarity between the target model and the current frame model in the initial frame can be described by a similarity function, which is defined as (4):

$$\rho(\hat{p}(y),\hat{q}) = \frac{1}{2}\sum_{u=1}^{m}\sqrt{\hat{p}_u(y)\hat{q}_u} + \frac{C_h}{2}\sum_{i=1}^{n_h} w_i k\left(\left\|\frac{y - x_i}{h}\right\|\right)^2 \tag{4}$$

Where $w_i = \sum_{i=1}^{m} \sqrt{\frac{\hat{q}_u}{\hat{p}_n(y_0)}} \delta[b(x_i - u)]$.

In the feature interval, the desired Mean Shift vector can be obtained when the similarity function ρ(y) is at the maximum value [4]:

$$m_{h,G}(y) = y_1 - y_0 = \left[\sum_{i=1}^{n_h} x_i w_i g\left(\left\| \frac{\hat{y} - x_i}{h} \right\|^2 \right) \Big/ \sum_{i=1}^{n_h} w_i g\left(\left\| \frac{\hat{y}_0 - xi}{h} \right\|^2 \right) \right] - y_0 \quad (5)$$

2.2. Particle Filter Fundamentals

The basic idea of particle filtering is to firstly generate a set of random samples in the state space based on the empirical conditional distribution of the system state vectors. These samples are called particles. Then the weights and positions of the particles are continuously adjusted according to the measurements, and the particles are adjusted. The information corrects the initial distribution of experience conditions. The essence of this approach is to approximate the associated probability distributions using discrete random measures consisting of particles and their weights and recursively update the discrete random densities according to the algorithm. When the sample size is large, this Monte Carlo description approximates the true nonlinear stochastic system of the state variable. The precision can be approximated to the optimal estimate. It is a very effective nonlinear filtering technique [4].

Particle filtering, also known as sequential Monte Carlo method (SMC), refers to finding a series of random samples that can approximately express the probability density function $p(x_k \mid z_k)$ in the state space and replacing the integral operation with an arithmetic mean value to obtain the state minimum variance of the process of estimation. These random samples are called "particles".

Combining the SIS with the particle resampling method, a complete particle filter algorithm can be obtained. Specific steps are as follows:

1) Initialization

Based on the distribution of, random sample in the vicinity of to obtain a series of particles.

2) SIS

Usually the importance distribution $q(x_k \mid x_{0:k-1}, z_{1:k})$ can be simplified to $p(x_k \mid x_{k-1})$ and N particles $x_k^i \propto p(x_k \mid x_{k-1})$ can be sampled.

3) Calculate weights

Calculate the weight of each particle $w_k^{(i)}$ and normalize it $w_k^{(i)} = w_k^{(i)} / \sum_{i=1}^{N} w_k^{(i)}$.

4) Posterior probability estimation

Output a set of weighted particles $\{(x_k^i, w_k^i), i = 1,2,...,N\}$. According to the weighted average or maximum posterior probability of the particle, the posterior probability estimate of the current time k can be obtained.

5) Particle resampling

According to the particle weights, the sample set is re-sampled. The large-weighted particles are divided into multiple particles while the small-weighted particles are deleted and the new N particles are obtained.

6) State transfer

When k+1 arrives, record the observations and repeat from 2 to 6.

IOP Conf. Series: Journal of Physics: Conf. Series **1267** (2019) 012057 doi:10.1088/1742-6596/1267/1/012057

3. Sports ship tracking experiment

3.1. Mean Shift tracking algorithm

The traditional Mean Shift algorithm was used for target tracking in the experiment. The first step is the initialization of the target tracking, which can be used to obtain the circumscribed rectangle of the initial target that needs to be tracked. However, the main research focus of this paper is not on the target detection module but to track the algorithm, so the manual selection method by the mouse is selected. Then the histogram distribution of the search window weighted by the kernel function is calculated and with the same method the histogram distribution of the corresponding window of the Nth frame is calculated. Based on the principle of maximum similarity between the two target template distributions, the search window is moved in the direction in which the density increases the most, and the true position of the target is obtained. The tracking steps are as follows.

1) Calculate the probability density $\{q_u\}_{u=1..m}$ of the target template, the target estimated position y_0 and the nuclear window width h.

2) Using the target position y_0 of the initial frame to calculate the candidate target template $\{p_u(y_0)\}_{u=1..m}$.

3) Calculate the weight value of each point in the current window.

4) Calculate the new position of the target:

$$y_1 = \frac{\sum_{i=1}^{m} x_i w_i g\left(\left\|\frac{y_0 - x_i}{h}\right\|^2\right)}{\sum_{i=1}^{m} w_i g\left(\left\|\frac{y_0 - x_i}{h}\right\|^2\right)} \tag{6}$$

The Mean Shift algorithm continuously iterates and finally finds the optimal position of the target in the image sequence.

In the experiment, the tracking window is first determined and the tracking target model of the initial frame is set. Select the target to be tracked in the first frame, and then frame a rectangle containing the ship's target, that is, the ship feature search window, also called the tracked target area. Cut the area and the result is shown in Figure 2. Observe the record tracking results after the end of the trace, as shown in Figure 3 to 6.

It can be seen from the experimental results that the two ships are staggered at 50 frames. The tracking frame is only slightly offset at a video around the 100th frame in the middle, but then the tracking frame is restored again until the final 273th frame. Observing the tracking trajectory basically conforms to the motion situation, but the upper and lower hopping occurs. On the one hand, there is tracking error, but the main reason is that the camera does not use a tripod during video capture and the image is shaken, so it is in an acceptable error range.

Figure 2. Image Selection.

Figure 3. Frame 20.

Figure 4. Frame 50.

Figure 5. Frame 70.

Figure 6. Frame 100.

3.2. Ship tracking based on particle filter

Particles [5] used in target tracking usually refer to target status information, including target position, size, color, and motion information [6]. The specific implementation steps are as follows.

1) Initialization

The initial position of the target X_0 is selected and N particles are randomly distributed near X_0.

2) Get particle weight

Assuming the system is in a state such as $X_1, X_2, ...X_n$, where X_i, i = 1, 2, ..., n, which we call a particle, represents a particle-centric image block. For each particle, the position, color, edge, and other information together form the current state of the moment. Z is the observed value of the system at this time, and $p(z_{1:k}|x_k)$ is obtained based on historical information. To estimate the exact position of the target X, it is necessary to operate on the particles. Each particle X_i corresponds to a rectangular frame, and the color in the frame is counted to obtain a 16-dimensional color histogram vector H_i. Each color histogram has 16 bins containing 16 grayscale values (0~255/16). After the statistics, the H_i is normalized so that the sum of the squares of each bin is 1, which is a 16-dimensional unit vector.

The weight W_i of each H_i can be understood as the inner product of the color histograms H and H_i of the target template, that is, the quantities of each bin are multiplied and added. This represents the similarity between the image region represented by the current particle and the target template. The higher the weight, the more similar the particle is to the target. The weights $W_1, W_2, ...W_n$ of $X_1, X_2, ...X_n$ are calculated from $W_i = \exp(10 * < H, H_i >)$, where $< H, H_i >$ represents the inner product operation.

3) Predict the current position

From the weighted average of the obtained particles $X_1, X_2, ...X_n$ and the corresponding weight

$W_1, W_2, ...W_n$, the system state can be estimated: $X = X_1 * W_1 + X_2 * W_2 + ... + X_n * W_n$, where X_i represents the location information of the center of the area. The resulting X is approximately equal to the posterior probability estimate and can be used to predict the current position of the target.

4) Particle resampling

As the number of video dumps increases, the weight will be more and more concentrated on a small number of particles, and most other particles have very small weights. Before entering the next iteration, the particle weights are redistributed to make them all the same. Therefore, the particles need to be re-sampled, i.e., N particles $X_1', X_2', ...X_n'$ are regenerated. The regenerated particles all come from the original particle set $X_1, X_2, ...X_n$, in which particles with a large weight generate more new particles, and particles with smaller weights have fewer new particles.

5) State transfer

At the next moment, the particles are updated from X_i' to X_i''. Since the probability that the system state is transferred transition from X_i' to X_i'' according to the state matrix is $p(x_k|x_{k-1})$, the probability of changing from X_i' to X_i'' is equal to $p(x_k|x_{k-1})$.

6) Repeat 2 to 5

In the experiment, the number of particles was selected as 200. The particle filter algorithm of this paper also adopts the method of manually selecting the tracking target. First display the initial frame and use the marquee to select a rectangle containing the ship's target, which is the ship feature search window. The feature search box is cut out as far as possible from the edge of the ship and cut out with the right mouse button. The cut is performed using the [temp, rect]=imcrop(I) format, and the image matrix of the ship model is stored in the temp variable. The histogram is then calculated for this variable. The initial position coordinates of the ship are stored in the variable rect and for calculating the ship center position coordinates.

Figure 7. Frame 20.

Figure 8. Frame 53.

Figure 9. Frame 70.

Figure 10. Frame 80.

After the end of the tracking, the tracking results are observed. The results are shown in Figure 7 to Figure 10.

4. Conclusion

In this paper, the tracking accuracy and real-time performance of the two algorithms are compared. It is found that the accuracy of the particle filter tracking algorithm is lower than that of the mean shift algorithm, but the real-time performance is better. Then, the influence of particle number on particle filter is studied. It is found that as the number of particles increases, the accuracy of tracking increases, but the running time becomes longer and the real-time performance deteriorates.

Acknowledgments

This work was supported by Hubei Natural Science Foundation 2016CFB681.

References

[1] Bourque A E, Bedwani S, Carrier J F, et al. *Particle Filter–Based Target Tracking Algorithm for Magnetic Resonance–Guided Respiratory Compensation: Robustness and Accuracy Assessment[J]*. International Journal of Radiation Oncology* Biology* Physics, 2018, 100(2): 325-334..

[2] Comaniciu D, Meer P. *Mean shift: A robust approach toward feature space analysis*[J]. IEEE Transactions on pattern analysis and machine intelligence, 2002, 24(5): 603-619.

[3] Cheng Y. *Mean shift, mode seeking, and clustering[J]*. IEEE transactions on pattern analysis and machine intelligence, 1995, 17(8): 790-799.

[4] Collins R T. *Mean-shift blob tracking through scale space[C]*. Computer Vision and Pattern Recognition, 2003. Proceedings. 2003 IEEE Computer Society Conference on. IEEE, 2003, 2: II-234.

[5] Hans R. Künsch. *Particle filters[J]*.Statistics,2013

[6] Pitt M K, Shephard N. *Filtering via simulation: Auxiliary particle filters*[J]. Journal of the American statistical association, 1999, 94(446): 590-599.

IOP Conf. Series: Journal of Physics: Conf. Series **1267** (2019) 012058 doi:10.1088/1742-6596/1267/1/012058

Design and Performance Analysis of a Lateral Thrust Measuring Device

Zongjin Ren[1,*], Mei Hua[1], Jun Zhang[1] and Yang Li[1]

School of mechanical engineering, DaLian University of Technology, China

[*]E-mail address: renzongjin@dlut.edu.cn

Abstract: This paper takes multi-dimensional force measurement of lateral force power unit as research objective, and a piezoelectric multi-dimensional force measuring device with piezoelectric quartz crystal as force sensitive component is designed. First, the mathematical model between sensors' output and measured lateral force is established by theoretical derivation. The relationship between the force applied to measured object and measuring device's output is obtained through the finite element analysis, which verifies the measurement performance of measuring device. On the other hand, through ANSYS model analysis, a conclusion that the first natural frequency of measuring device is 1467.8Hz can be obtained, and the dynamic performance of measuring device is good, which enables accurate measurement of real-time varying lateral force.

1. Introduction

To achieve accurate control of aircraft under high altitude and low pressure, direct lateral force control technology is introduced into traditional trajectory and attitude control system. The lateral thrust generated by attitude control engine generates handle torque. Lateral force control technology is less affected by flight altitude and flight speed, greatly improving the manoeuvring performance of aircraft[1]. As a key performance parameter of lateral force control technology, lateral thrust is of great significance for ensuring control precision and improving target hit rate[2]. Therefore, accurate measurement of lateral thrust is particularly significant.

For force measurement technology, there are the related research. Princeton University developed an inverted pendulum thrust test device, which measured the thrust generated by engine based on the displacement signal measured by displacement sensor[3]. R.John Stephen and K.Rajanna et al. developed a film strain gauge ion engine thrust test system to measure the thrust of engine through strain gauge sensors attached to elastic strut and a corresponding test system[4]. Ann M Wright et al. studied a six-component thrust measurement, and the load of each component was measured by six force measuring arms fixed by test stand[5]. Dwarakanath and Kang et al. developed a six-dimensional force sensor based on the Stewart parallel mechanism[6]. From the published literature, it can be obtained strain measuring devices are mostly used for force measurement. That is, the force is measured according to the deformation amount of elastic element, and the sensor of different structure is placed between measured object and fixing device. In order to satisfy the requirements of sensitivity, force-sensitive elastic links with low stiffness are generally used in strain force measuring devices. However, the low-stiffness will result in complicated structure, low natural frequency, more test links, adjustment difficulties and so on.

Aiming at the spatial six-dimensional thrust vector measurement problem of lateral force power unit, a thrust vector measuring device is designed. And the problems of traditional strain measuring device such as complex structure, low natural frequency and more test links are solved. In addition, the performance of measuring device is verified through theoretical analysis and ANSYS simulation.

2. Designing of measuring device

2.1. Researching on measurement principle

Since the lateral force control technology requires multiple power units to work together to achieve precise trajectory and attitude control, and the thrust performance of each power plant and multi-dimensional thrust performance of composite ignition are research focus, which resulting in more measurement factors and more complicated of measured object. Therefore, the tri-direction force sensor is chosen as the basic unit of measuring device here. However, the measurement of spatial six-dimensional force is not possible with only one tri-direction force sensor. Multi-point spatial arrangement method based on the tri-direction force sensor is an effective way to solve the problem of multi-dimensional force measurement. To ensure measurement results are not affected, the sensitivity of each force sensitive component is required to be the same. Meanwhile, the force acting at any point on platform should be assigned to force sensitive component in accordance with "leverage principle". Therefore, ensuring the stiffness of each force sensitive component being the same in the same direction is necessary, which ensures the structural symmetry and the performance consistency of each fulcrum. So square arrangement is selected here. In addition to ensuring the symmetry of structure, square arrangement can avoid affecting measurement result and reducing cross interference of sensors where the position of external force action point changes greatly. The arrangement of sensors is shown in figure1.

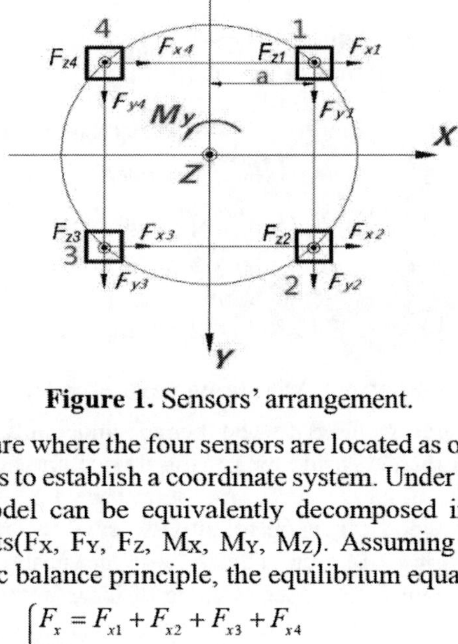

Figure 1. Sensors' arrangement.

Taking the centre of the square where the four sensors are located as origin O, the XY plane coincides with the lower surface of sensors to establish a coordinate system. Under the reference coordinate system, the lateral force acting on model can be equivalently decomposed into three unidirectional forces components and three moments(F_X, F_Y, F_Z, M_X, M_Y, M_Z). Assuming the distance between adjacent sensors is 2a, according to static balance principle, the equilibrium equation can be obtained.

$$\begin{cases} F_x = F_{x1} + F_{x2} + F_{x3} + F_{x4} \\ F_y = F_{y1} + F_{y2} + F_{y3} + F_{y4} \\ F_z = F_{z1} + F_{z2} + F_{z3} + F_{z4} \\ M_x = a(F_{z_1} - F_{z_2} - F_{z3} + F_{z4}) \\ M_y = a(F_{z1} + F_{z2} - F_{z3} - F_{z4}) \\ M_z = a(-F_{x1} + F_{x2} + F_{x3} - F_{x4} - F_{y1} - F_{y2} + F_{y3} + F_{y4}) \end{cases} \tag{1}$$

IOP Conf. Series: Journal of Physics: Conf. Series **1267** (2019) 012058 doi:10.1088/1742-6596/1267/1/012058

2.2. Force sensitive component

Force sensitive component is mechanically sensitive components of measurement system[7]. Real-time variation of the force generated by lateral force power unit requires force sensitive component to have sufficient sensitivity and natural frequency in addition to meeting the requirements of force measurement. Meanwhile, the force sensitive component must ensure measuring device has enough installation space for measured device. The physical properties of quartz crystal is relatively stable, and its sensitivity varies little with time. Moreover, quartz crystal has advantages of high stiffness, high natural frequency, fast transient response and long-time stability[8]. Meanwhile, quartz crystal is anisotropic material, so quartz crystal can be cut in different directions to obtain different geometric cutting-type, and different geometric cutting-type can measure different types of force. Normal load can be measured by X0 cutting-type, and tangential load can be measured by Y0 cutting-type. Tri-direction force sensor is the combination of one X0 crystal group and two Y0 crystal group (figure 2)[9]. Besides, quartz crystal has high Curie points and good temperature stability, so it's ideal for environment with large temperature changes. Therefore, piezoelectric quartz crystal is selected as force sensitive component of measuring device.

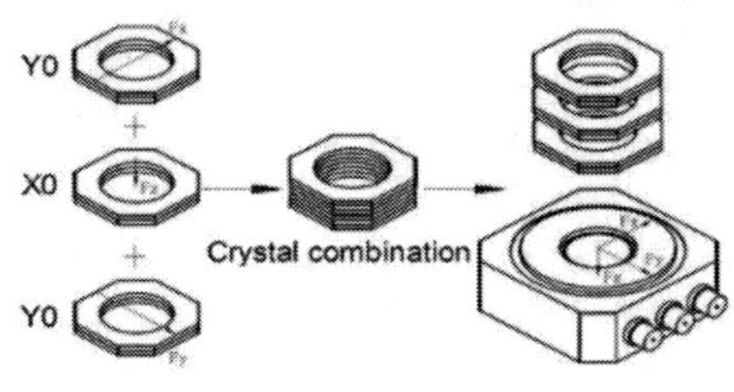

Figure.2 Combined crystal group of tri-direction force sensor.

2.3. Mechanical structure of measuring device

The structure of measuring device is shown in figure 3.

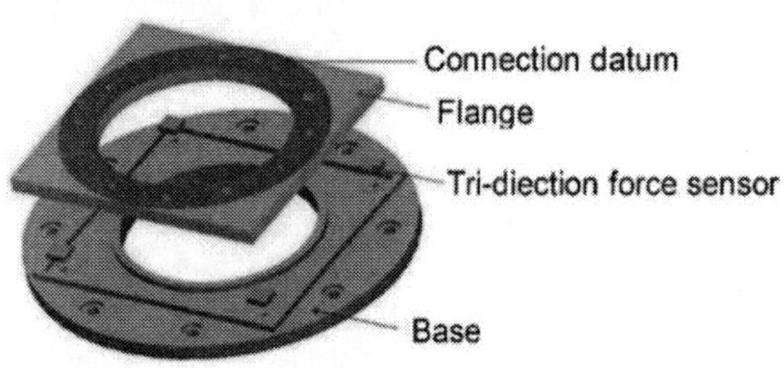

Figure 3. Measuring device model.

The base of measuring device is fixed to test bench, and connection datum is connected with measured object. Sensors are arranged inside measuring device and are integral with flange and base. Compared with traditional strain gauge force measuring device, this measuring device has the advantages of simple structure, good precision retentivity, easy to dismount, fully sealed package of sensors, high degree of protection and so on. The base of measuring device is fixed with measuring object, and the rest of space is completely open. Centre of measurement is the interface of measuring device and measured object. Datum is positioned, so the positional consistency is good during measuring. Therefore, piezoelectric force measuring device is more suitable for the measurement of spatial six-dimensional thrust vector.

3. Measuring principle

3.1. Mathematical model

Measuring simulation is shown in figure 4. Lateral force generating device is circumferentially distributed. When lateral force generating device works, the measured object will generate two forces and three moments, which are lift(F_s), lateral force(F_c), rolling moment(M_g), yawing moment(M_p), pitching moment(M_f).

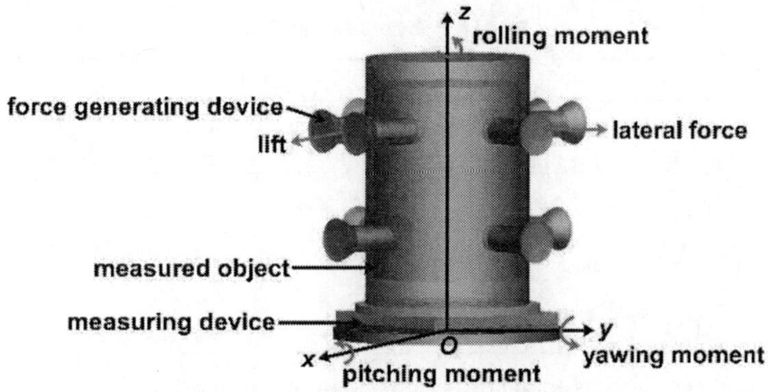

Figure 4. Analog measurement model.

Assuming lateral force generating device generates a lateral force F and the coordinate if acting point is (m, n, l), the expressions of force and moment generated by device under measurement are as follows:

$$\begin{bmatrix} F_c \\ F_s \\ M_p \\ M_f \\ M_g \end{bmatrix} = \begin{bmatrix} 1 & 0 \\ 0 & 1 \\ 0 & -l \\ l & 0 \\ -n & m \end{bmatrix} \cdot \begin{bmatrix} F_{xc} \\ F_{yc} \end{bmatrix} \tag{2}$$

Since lateral force generating device is circumferentially distributed, so $m^2 + n^2 = r^2$. Where F_{XC} and F_{YC} are component forces of lateral force(F) on x axis and y axis, and r is the distance from the point of action to z axis. When the output of measuring device is known, the magnitude, direction and point of action of lateral force can be derived by expression(2).

3.2. Simulation analysis

Simulating measuring device with ANSYS Workbench. The 3D model built by Pro/E is imported into ANSYS Workbench static analysis. A lateral force(F) where $F_X = F_Y = 1000N$ is applied to measured object. The lateral force is perpendicular to z-axis and the coordinate of action point is (50,256,375.5). The sensors' output is obtained by ANSYS simulation analysis. The simulation analysis diagram is shown in Figure 5.

Figure 5. ANSYS simulation analysis.

It is known that $2a = 242mm$. Bring the outputs of tri-direction force sensors into formula(1), the three-axis forces and three-axis moments can be calculated. The output of measuring device and calculation results are shown in table 1.

Table 1. Sensors' output

$F_X=F_Y=1000N, F_Z=0$					
X1	X2	X3	X4	Resultant force(N)	Moment(Nm)
654.75	-583.28	1112.3	-183.74	F_x=1000.03	M_x=-374.315
Y1	Y2	Y3	Y4		
1099.6	-581.05	663.85	-182.44	F_y=999.96	M_y=373.877
Z1	Z2	Z3	Z4		
1544.7	2.0122	-1547	0.20174	F_z=0.08606	M_z=-204.665

Bringing the three-axis forces and three-axis moments measured by measuring device into equation(2), the magnitude, direction and action point of lateral force can be reserved. Calculation results are shown in table 2.

Table 2. Measured value of lateral force

F_X(N)	F_Y(N)	m	n	l
1000.03	999.96	51.12	255.78	374.0

It can be concluded from table 2 that the maximum difference between force values simulated by ANSYS and applied to measuring device is 0.04N. Simultaneously, the maximum difference between measured force point coordinate value and practical coordinate value is in millimeter level. Therefore, the measuring device can be used for lateral force measurement.

4. Model analysis

Since lateral force generating device has the characteristics of starting, shutting down frequently, short working time, and the force generated by lateral force generating device changes in real time, so measuring device should have good dynamic performance in addition to satisfying measuring accuracy[10]. As a modern method to study the dynamic characteristics of structure, model analysis can be applied to evaluate dynamic characteristics of existing structural system[11]. Modal analysis is actually the solution to the natural frequency of system. At present, natural frequency is mostly used to evaluate the dynamic performance of measuring device. Generally considered that the higher natural frequency of measuring device, the better dynamic performance of measuring device. Therefore, the natural frequency is required to be as large as possible. The first six-order modal analysis of measuring device is performed by ANSYS Workbench. The first six natural frequencies of measuring device are shown in table 3. Figures 6~11 show the first six-stage arrays and deformations of measuring device.

Table 3. First six natural frequencies

Mode	Frequency[Hz]
1	1467.8
2	1507
3	1617.7
4	1620.4
5	1779.7
6	2551.7

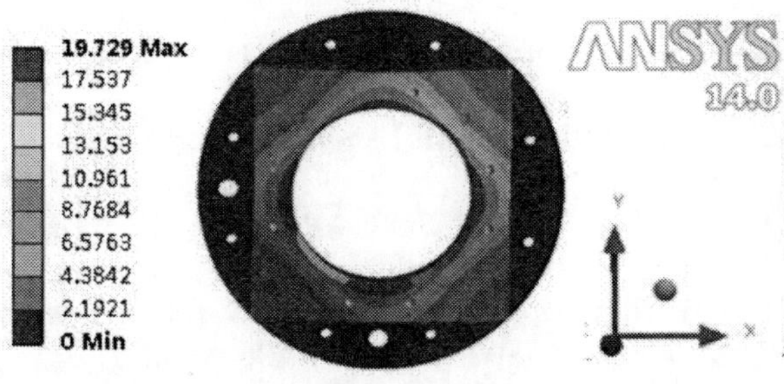

Figure 6. First-order mode

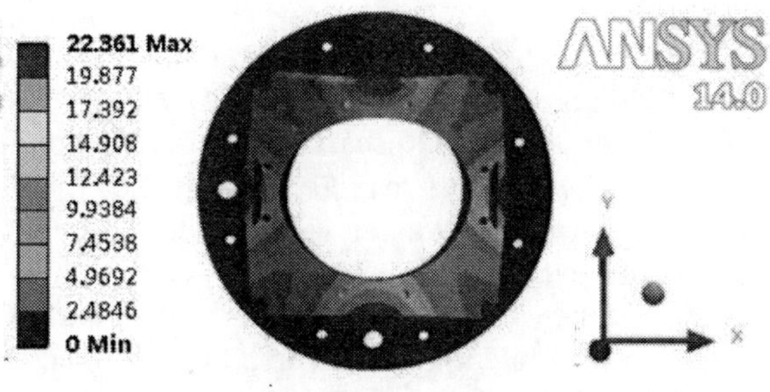

Figure 7. Second-order mode

Figure 8. Third-order mode

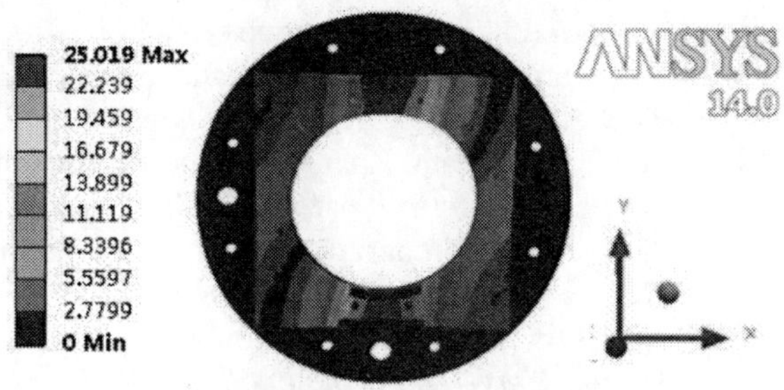

Figure 9. Forth-order mode

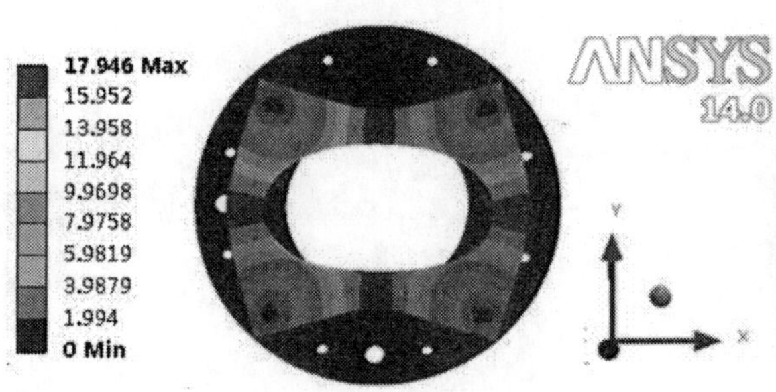

Figure 10. Fifth-order mode

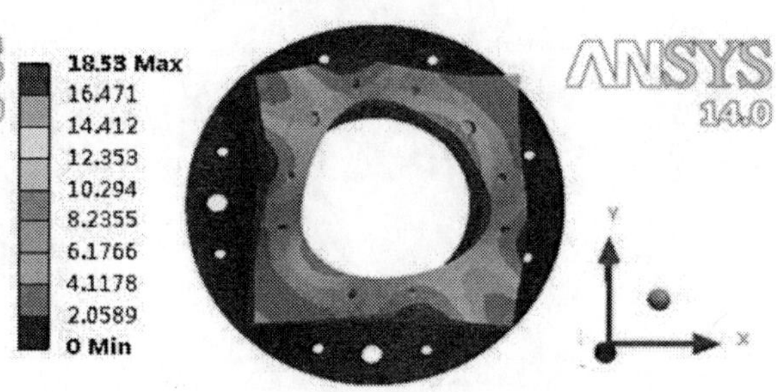

Figure 11. Sixth-order mode

It can be concluded from model analysis that the first natural frequency of measuring device is 1467.8Hz, which can satisfy measurement requirement and avoid resonance during measure.

5. Conclusion

This paper takes multi-dimensional force measurement of lateral force power unit as research objective. Taking piezoelectric quartz crystal as force sensitive material, and the principle and method of lateral force measurement were described. A lateral force measuring device was designed and theoretical, finite element analysis of measuring device was carried out.

(1) The magnitude, direction and application point of the force calculated by theoretical derivation and ANSYS simulation are basically equivalent to the magnitude, direction and action point of the force applied to the measured object. Therefore, the measuring device can satisfy the precision requirement of lateral force measurement.

(2) Through modal analysis, it can be concluded that the natural frequency of measuring device is large and the dynamic performance is good. So the measuring device can accurately measure the lateral force that changes in real time.

IOP Conf. Series: Journal of Physics: Conf. Series **1267** (2019) 012058 doi:10.1088/1742-6596/1267/1/012058

Reference

[1] Wu Shaoying. Research on Attitude Control Method of Direct Lateral Force and Aerodynamic Composite Control Missile [D]. *Harbin Institute of Technology*, 2012.

[2] Chen Xiuping. Research on Thrust Vector Test System of Lunar Soft Landing Rocket Engine [D]. *Dalian University of Technology*, 2013.

[3] Gillard J H, Walker M L R and Pencil E J. THRUST STAND FOR APPLIED FIELD MPD THRUSTERS [J]. 2005.

[4] Stephen R J, Rajanna K and Dhar V. Thin film strain gauge sensors for ion thrust measurement[C]// Sensors, 2002. *Proceedings of IEEE. IEEE*, 2012:1702-1706 vol.2.

[5] Wright A M, Wright A B and Born T. A six degree-of-freedom thrust sensor for a lab scale hybrid rocket [J]. *Measurement Science & Technology*, 2013, 24(24):5104.

[6] Dwarakanath T A, Dasgupta B and Mruthyunjaya T S. Design and development of a Stewart platform based force–torque sensor [J]. *Mechatronics*, 2001, 11(7), pp 793-809.

[7] Ren Zongjin, Sun Baoyuan and Zhang Jun. Thrust Vector Dynamic Test of Rail/Situ Control Rocket Engine [J]. *Acta Aeronautica Sinica*, 2009, 30(11).

[8] Sun Baoyuan and Zhang Yugong. Piezoelectric Quartz Force Sensor and Dynamic Cutting Force Tester [M]. *Metrology Press*, 1985.

[9] Zhao Liqi. Comparative study on performance of LGS and quartz piezoelectric sensors [D]. *Dalian University of Technology*, 2015.

[10] Xing Qin. Development of Pulsating Thrust Test System for Attitude Control Rocket Engine [D]. *Dalian University of Technology*, 2008.

[11] Liang Jun and Zhao Dengfeng. Overview of Modal Analysis Methods [J]. *Modern Manufacturing Engineering*, 2006(8), pp 139-141.

Review of Intent Detection Methods in the Human-Machine Dialogue System

Jiao Liu, Yanling Li* and Min Lin

College of computer science and technology, Inner Mongolia Normal University

*liyanling7871397@163.com

Abstract. Spoken language understanding is an important part of the human-machine dialogue system, intent detection is a sub-task of spoken language understanding, and it is very important. The accuracy of intent detection is directly related to the performance of semantic slot filling, and it is helpful to the following research of the dialogue system. Considering the difficulty of intent detection in human-machine dialogue system, the traditional machine learning method cannot understand the deep semantic information of user's discourse. This paper mainly analyzes, compares and summarizes the deep learning methods applied in the research of intent detection in recent years, and further considers how to apply deep learning model to multi-intent detection task, so as to promote the research of multi-intent detection methods based on deep neural network.

1. Introduction

With the advent of the era of artificial intelligence, more and more intelligent products have been widely used in our daily life, such as emotional care robot, personal phone assistant Siri, voice assistants Google Now and intelligent chat robot Xiao Bing from Microsoft Research Asia, etc. These intelligent dialogue systems bring a lot of convenience to users' lives. The dialogue system is mainly composed of five parts: Automatic Speech Recognition (ASR), Spoken Language Understanding (SLU), Dialogue Management (DM), Dialogue Generation (DG) and Text to Speech (TTS) [1], as shown in figure 1. In order to better understand the user's expression, and then feedback the correct information for the user, spoken language understanding plays an extremely important role. Intent Detection (ID), as a sub-module of spoken language understanding, is also the key to human-machine dialogue system. Traditional spoken language understanding is mainly divided into two sub-tasks -- intent detection and semantic slot filling. Because early research was constrained by application scenarios, data, and computing power, most spoken language understanding was limited to the specific domain.However, with technological innovation and the emergence of the multi-domain dialogue system, the current spoken language understanding is often divided into three tasks -- domain recognition, intent detection and semantic slot filling [2].

In a dialogue system, intent detection is crucial. The intent is the will of the user, that is, what the user wants to do. Intents are sometimes referred to as "Dialog Act" [3], which is the action of information that users share in the dialogue and are constantly updated. The intent is generally named "verb + noun", such as query weather, book a hotel, etc. Intent detection, also known as intent classification, classifies user utterances into previously defined intent categories according to the domains and intents involved in user utterances [4].

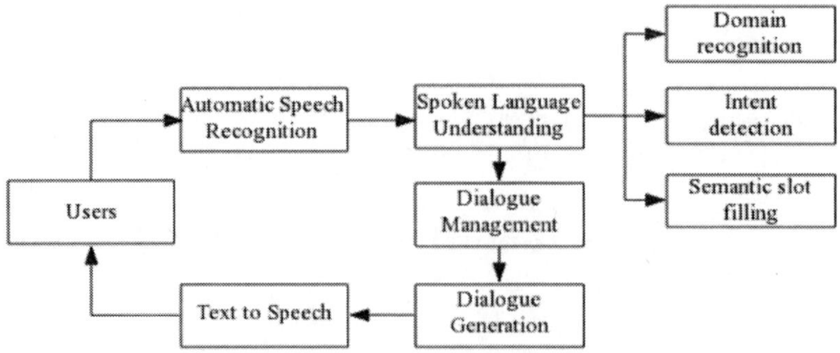

Figure 1. Schematic diagram of the dialogue system.

Nowadays, in the application process of the human-machine dialogue system, users may have multiple intents in different occasions, which will trigger multiple domains in the human-machine dialogue system, including task-oriented vertical domains and chats, etc. In the dialogue system, only when the user's topic domain is clearly defined, can the specific needs of the user be correctly analyzed, otherwise it will lead to wrong intent detection behind. Figure 2 is an instance diagram of the application of three tasks in spoken language understanding. When a user enters a query, it first needs to clear the user's input text belonging to a topic domain to "train" or "flight", due to the intent category is finer-grained than the topic domain, so we need to determine the user's intent, which is booking the ticket or refunding the ticket or querying time, according to the user's specific semantic information. And the semantic slot filling can also help the user intent judgment.

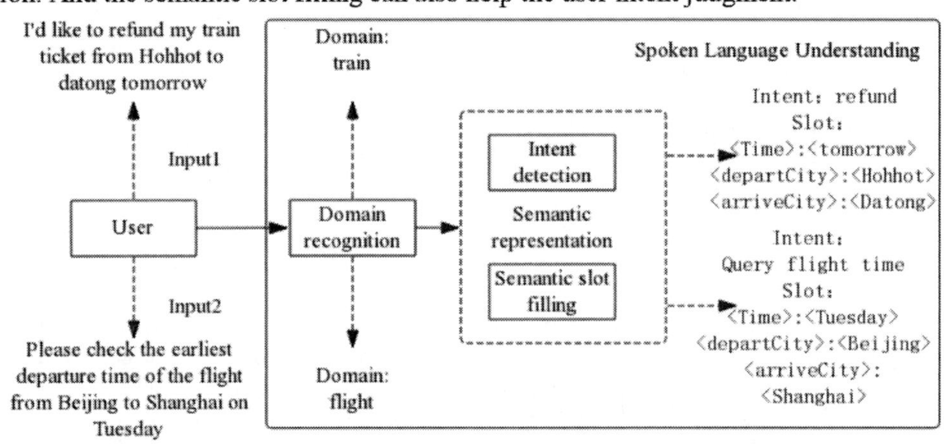

Figure 2. An instance diagram of intent detection.

2. The difficulties ofintent detection

2.1. Lack of data sources

With the development of artificial intelligence technology, large Internet companies have launched chat robots. Due to the less user experience, it is difficult for most researchers to obtain the chat text between users and robots, which leads to the limited amount of dialogue text to be studied, which has become a major problem faced by intent detection tasks [5]. In the actual process of intent detection, there are very few intent texts with annotations and they are very difficult to obtain, which also brings challenges to the research and development of intent detection [6].

2.2. The irregularity of user expression

In the chat system, the user's intent text is generally characterized by colloquial expression, short sentences and broad content, which makes it difficult to identify the user's intent. For example, "I want to look for a dinner place", the corresponding intent of this colloquial daily expression is "looking for a restaurant", so the colloquial intent text makes the domain topic is not clear, which is not conducive to the identification of user intent. For "Hanting", this semantically poor expression of the intent text, although "Hanting" often appears with "Hotel", it is very difficult for the machine to identify the user's topic domain as "Hotel". For "I want to book a ticket", which may be to book air tickets, train tickets, bus tickets and so on. Due to the user's expression is too broad, the machine cannot give feedback to the user in time.

2.3. Implicit intent detection

With the continuous expansion of the application scope of the human-machine dialogue system, there are more and more ways to express intent. According to the types of expression, intents can be divided into explicit intents and implicit intents [7]. Explicit intents refer to that the user clearly points out his or her intent requirements in text content, including topic domain, intent category and so on. Implicit intent refers to the fact that the user does not have clear intent requirements and it is necessary to infer the user's real intent by analyzing the user's potential intent [8]. Such as explicit intent text "Book a hotel near the People's Park for one night" and implicit intention text "I'm going to Shenzhen for two days next week". Although their intent is booking a hotel, the latter needs to judge the user's potential intent and speculate on the user's true intent. Therefore, implicit intent detection without explicit topic domain and category information is very difficult in the intent detection task.

2.4. Multiple intents detection

Multi-intent detection is similar to multi-label classification, but it is different from multi-label classification. Multi-label classification usually deals with long text, while multi-intent detection mainly deals with short text. How to detect multiple intents of users in short text is another difficulty of intent detection. In the process of multi-intent detection, we need to pay attention to three problems. Firstly, how to find that the user's intent text contains multiple intents; Secondly, how to determine the number of intents contained in the user's intent text; Finally, it is worth thinking about how to accurately identify users' various intents.

3. Main methods of intent detection

3.1. Traditional intent detection methods

In recent years, most scholars regard intent detection as a Semantic Utterance Classification (SUC) problem [9]. Traditional intent detection mainly includes rule-based template semantic recognition method [10] (1993) and classification algorithm based on statistical features [11,12] (2002-2014). Although the rule-based template matching method does not require a lot of training data, it can guarantee the accuracy of detection, but it cannot solve the problem of high cost of template reconstruction caused by changing the intent category. However, the method based on statistical feature classification needs to extract the key features of corpus text. However, the method of manually extracting features is not only costly, but also the accuracy of features cannot be guaranteed, which also leads to data sparse problems. Common methods include Naive Bayes [13] (1998), Adaboost [14] (2000), Support Vector Machine(SVM) [15] (2003) and logistic regression [16] (2007). Considering the non-standard and ambiguous information of user text, the traditional intent detection method cannot accurately understand the deep semantics of user text. How to accurately identify the user's real intent, this is still a very challenging task.

3.2. The current mainstream methods

With the development of deep learning, more and more scholars apply word embedding, Convolution Neural Networks (CNN), Recurrent Neural Networks (RNN), Long Short-Term Memory (LSTM) Network, Gated Recurrent Unit (GRU), Attention Mechanism and Capsule Networks to intent detection task. Compared with the traditional machine learning methods, the deep learning model has a great improvement in detection performance.

3.2.1. Intent detection based on word embedding. In recent years, word embedding is gradually used in semantic analysis tasks, due to the use of original lexical features will lead to data sparse problems, and continuous representation learning can solve data sparse problems [17] (2003). Kim et al. [18] (2015) took word embedding as lexical features for intent classification. Compared with the traditional word bag model, the intent classification method based on word embedding has better representational ability and domain extensibility for different classification contents. Considering the insufficiency of semantic information of word embedding, Kim et al. [19] (2016) used semantic vocabulary dictionary information to rich word embedding and improved the semantic representation of intent text. This model achieved good results, and show that rich word embedding will be helpful to improve the performance of intent detection.

3.2.2. Intent detection based on convolution neural network. CNN was originally used for image processing [20]. With the emergence of word embedding technology, CNN has been widely used in the field of natural language processing and achieved good research results. Kim et al. [21] (2014) attempted to use CNN in text classification tasks and achieved very ideal results. Hashemi et al. [22] (2016) used CNN to extract text vector representation as query classification feature to identify the user's intents earch query. Compared with traditional artificial feature extraction method, this method not only reduces a lot of feature engineering, but also obtains deeper feature representation. However, CNN has representational limitations.

3.2.3. Intent detection based on recurrent neural networks and their variants. RNN is different from CNN, it represents a word sequence and can learn semantic information of word order according to the context. Bhargava A[23] (2013) reduced the error rate of intent detection by incorporating context information into intent detection tasks, indicating that context information is beneficial to intent detection. A simple RNN has problems such as gradient explosion or gradient disappearance, which cannot well simulate long-term dependence.

LSTM [24] (1997) can solve this problem by introducing a memory unit into the RNN structure, which can control the information to be retained and forgotten. This model is often also used to solve the problem of intent detection. Ravuri et al. [25] (2015) proposed using RNN and LSTM to solve the problem of intent classification. Experiments on Air Travel Information System (ATIS) dataset show that the error rate of intent detection of LSTM is 1.48% lower than that of RNN. The main reason is that LSTM has a good ability to modeling the temporal relationship of text, and has a good memory function for the input of long text.

GRU is an improvement of the LSTM model [26], which has the ability to retain information on long sequences and can learn contextual semantic information. For intent detection task, Ravuri et al. [27] (2016) used GRU and LSTM to comprehensively compare the performance of intent detection on ATIS and Cortana datasets. Experiments show that GRU and LSTM have almost the same performance in the intent detection task, but GRU has fewer parameters and the model is simpler.

3.2.4. Intent detection based on the combination of deep learning models. Considering the advantages and disadvantages of various deep learning models, most researchers combine the deep learning models with different advantages to classify users' intents. Qian et al. [8] (2017) proposed a travel consumption intent detection model based on Convolutional-LSTM, which took advantage of CNN can extract intent text features at a deeper level and LSTM can build the temporal relationship of text, and achieved good performance. Yu et al. [28] (2018) aimed at the problem of data sparse caused by the short text, and proposed a multi-turn dialogue intent detection model based on Biterm Topic Model (BTM) and Bidirectional Gated Recurrent Unit (BGRU). This combined model has achieved good results in the users' medical intent detection, and is superior to the performance of literature [8]. Huang [29] (2018) proposed the Character-CNN-BGRU deep learning combined model. The combined model not only makes use of the character-based method to make the list of words smaller, but also can solve the problem of unknown words, coupled with CNN can extract local features of the intent text and BGRU can guarantee the temporal relationship of text, highlighting the advantages of the combined model in the intent detection task. However, the structure of the combined model is complex and the training time is long, how to simplify the combined model is a problem worth considering.

3.2.5. Intent detection based on Bidirectional Long short-term Memory (BLSTM) self-attention model. With the development of deep learning models, the expression of various sentence level vectors has emerged, such as: using the maximum pooling or average pooling of CNN to obtain sentence vectors, using the hidden state or the final hidden state of RNN to create sentence representation, etc. Lin et al. [30] (2017) proposed an improved method by introducing a self-attention mechanism to extract sentence representation. Sentence vectors were represented by two-dimensional matrices and different semantic information of sentences was represented by multiple vectors. The model is implemented on BLSTM, and obtains the sentence vector representation by weighted summation of the hidden state of LSTM, and realizes intent classification. This model can obtain various semantic information of sentence through the self-attention mechanism, which is helpful to the research of multi-intent detection.

Figure3 shows the self-attention model, as shown in figure3(a), assume that the input sequence is $S=(w_1,w_2,...,w_n)\in R^{n\times d}$, input S into BLSTM, and the hidden state of the front and rear terms of the t word is calculated as follows:

$$\overrightarrow{h_t}=\overrightarrow{LSTM}(w_t,\overrightarrow{h_{t-1}}) \tag{1}$$

$$\overleftarrow{h_t}=\overleftarrow{LSTM}(w_t,\overleftarrow{h_{t+1}}) \tag{2}$$

If the number of LSTM hidden cells is k , and the hidden state of the preceding and trailing items is connected to get h_t , and $h_t\in R^{2k}$, $H\in R^{n\times 2k}$ represents the set of all hidden states h_t , and $H=(h_1,h_2,...,h_n)$.

The self-attention weight matrix is expressed as:

$$a=soft\max(W_{s2}\tanh(W_{s1}H^T)) \tag{3}$$

As shown in figure3(b), $W_{s1}\in R^{m\times 2k}$ and $W_{s2}\in R^{1\times m}$ are the self-attention weight matrix. m is the number of hidden cells in self-attention model, which is a super parameter and can be set arbitrarily. $H^T\in R^{2k\times n}$ is the transpose matrix of H . $a\in R^{1\times n}$ can be obtained by calculation, and finally normalization by softmax function. Each dimension in a represents the attention of the corresponding words in the sentence. So the intent text vector can be expressed as: $d=a\cdot H$, then $d\in R^{1\times 2k}$.

If we want to extract r semantic features from the intent text, we need r self-attention headers to extract semantic features. Then $W_{s2}\in R^{r\times m}$, gets self-attention weight matrix $A\in R^{r\times n}$, and the final intent text vector can be expressed as : $D=A\cdot H$, then $D\in R^{r\times 2k}$.

3.2.6. Intent detection based on the capsule network model. The concept of "capsule" was first proposed by Hinton et al. [31] (2011) to solve the representation limitations of CNN. A capsule contains the vector representation of a group of neurons, the direction of the vector represents the entity attributes, and the length of the vector represents the probability of the existence of the entity. Sabour et al. [32] (2017) proposed the capsule network, replacing CNN scalar output feature detector with vector output capsule, and replacing max-pooling with protocol routing. Compared with the original CNN, the capsule network will maintain the accurate location information of the entity in the region. Therefore, Zhao et al. [33] (2018) used capsule network for text classification tasks for the first time, and proposed three dynamic routing strategies to improve the performance of dynamic routing process, and reduced the interference of noise (stop words and words unrelated to categories) capsules. Experiments on six standard datasets show that the capsule network performs well in text classification tasks, and the capsule network also has good performance in multi-label text classification tasks.

In the intent detection task, Xia et al. [34] (2018) proposed an intent capsule model based on capsule network, which utilizes the advantages of capsule model in the text modeling to process text hierarchically. As shown in figure 4, first of all, using the method of literature [30] in the intent text extracted semantic features with the self-attention mechanism, named semantic vector. Since different users express the same intent in different ways, but they contribute more to one intent than others, the appropriate contribution of each semantic is dynamically allocated by using dynamic routing mechanism to aggregate them into a higher level prediction vector, that is, the semantic expression of intent, and classify intents. The model achieves good results in the intent detection task.

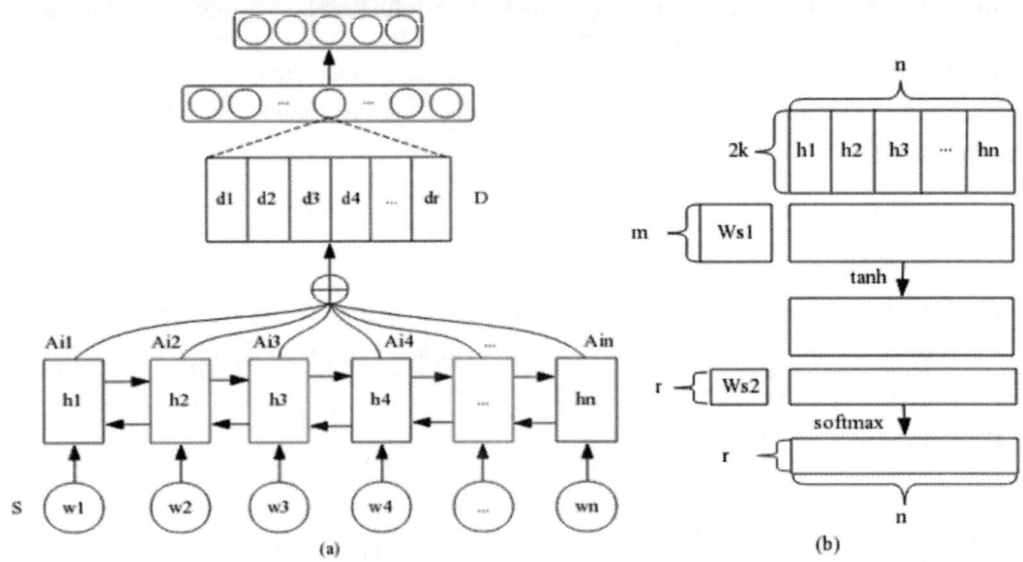

Figure 3. Self-attention model diagram [30].

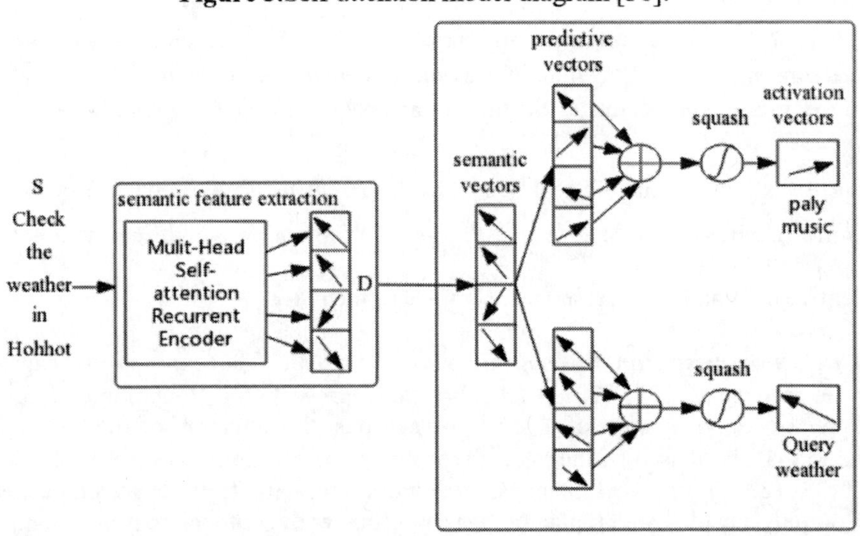

Figure 4. Intent detection process diagram based on intent capsule.

3.2.7. Intent detection based on the method of joint recognition. With the continuous research and improvement of intent detection methods, considering that single-task research is prone to error propagation due to its independent model, some scholars have proposed a joint model of semantic slot filling and intent detection. Li et al. [35] (2017) conducted a joint model of intent detection and semantic slot filling through triangular chain conditional random fields. Compared with the cascade model that takes semantic slot filling result as intent detection feature, the joint model performs well in the intent detection task, highlighting the relevance between them. Liu et al. [36] (2016) captured important semantic components of sentences by adding attention mechanism on the hidden layer of bidirectional recurrent neural network (BRNN), and improved the accuracy of intent detection. In ATIS dataset, the error rate of intent detection of BRNN based on attention mechanism is 2.35%. Through the joint experiment of semantic slot filling and intent detection, the error rate of intent detection in ATIS dataset is reduced to 1.79%. It can be seen that semantic slot filling is helpful for intent detection.

3.2.8. Multiple intent detection methods. For multi-intent detection tasks, Yang et al. [37](2018) relied on syntax analysis to determine whether the user's intent text contains multiple intents, then determined the number of intent by using the word frequency-inverse document frequency (TF-IDF) and trained word embedding to calculate the matrix distance. Finally, classified the intents by combining syntactic features and CNN, and determined the user's multiple intents. However, there are few studies on multi-intent detection based on deep learning model, so this is a direction worth studying.

4. The evaluation methods of intent detection

4.1. The evaluation method of single intent detection

At present, intent detection is generally regarded as a semantic discourse classification problem, so the performance of intent detection method is evaluated by the one which used in text classifier [28]. That is, accuracy, recall rate, F1-score and error rate, classification speed, etc.

(1) Accuracy: $$Accuracy = \frac{TP+TN}{TP+TN+FP+FN} \tag{4}$$

(2) Precision: $$Precision = \frac{TP}{TP+FP} \tag{5}$$

(3) Recall: $$Recall = \frac{TP}{TP+FN} \tag{6}$$

(4) F1-score: $$F1-score = \frac{2 \times Precision \times Recall}{Precision + Recall} \tag{7}$$

Among them, TP indicates True Positive, that is, class A samples were correctly classified into class A. TN indicates True Negative, which do not belong to the class A samples were classified correctly into classes other than class A. FP indicates False positives, which does not belong to the class A samples were classified erroneously into class A. FN indicates False negatives, which belongs to the class A samples were erroneously into the classes other than class A.

4.2. The evaluation method of multiple intent detection

The evaluation method of multi-intent detection is different from that of single intent detection, it needs to judge the types of intent prediction in sentences. This paper adopts evaluation index of literature [37]. Suppose that there are N multiple intent samples recorded as (x_i, A_i), $0 \le i \le N$. A_i denotes the set of correct intent results of the sample set. $B_i = S_{(x_i)}$ denotes the set of predicted intent results of the sample set to be tested. The specific calculation formulas are as follows:

(1) Multi-Intent Accuracy(MIA): $MIA = \frac{1}{N}\sum_{i=1}^{N}\frac{|A_i \cap B_i|}{|A_i \cup B_i|}$ (8)

(2) Multi-Intent Precision(MIP): $MIP = \frac{1}{N}\sum_{i=1}^{N}\frac{|A_i \cap B_i|}{|B_i|}$ (9)

(3) Multi-Intent Recall(MIR): $MIR = \frac{1}{N}\sum_{i=1}^{N}\frac{|A_i \cap B_i|}{|A_i|}$ (10)

5. Comparison of experimental performance results

Literature [34] has experimented with various deep learning models mentioned above on SNIPS English dataset and Commercial Voice Assistant (CVA) Chinese dataset, and the performance results of intent detection were shown in table 1:

Table 1. Performance comparison results of various deep learning models on intent detection tasks.

Model	SNIPS (Five kinds of intents)		CVA (80 kinds of intents)	
	Accuracy	F1	Accuracy	F1
CNN	0.9595	0.9595	0.8223	0.8210
RNN	0.9516	0.9518	0.8286	0.8275
GRU	0.9535	0.9534	0.8239	0.8216
LSTM	0.9569	0.9569	0.8319	0.8306
BLSTM	0.9501	0.9502	0.8428	0.8419
Self-attention BLSTM	0.9524	0.9522	0.8521	0.8513
Intent Capsnet	0.9621	0.9620	0.9088	0.9023

According to the results in the table, BLSTM can make full use of context to represent sentence features, which is better than the local feature representation of CNN, and the self-attention model can more fully capture the deep semantic information of sentences. Due to intent capsule can not only capture the deep semantics of intent text, but also ensure the exact location of semantic information of intent text, its intent detection performance is better than other deep neural network models.

6. Summary and prospect

This paper mainly introduces the difficulty and method of intent detection in the human-machine dialogue system. It summarizes and compares the intent detection methods of the deep learning model. Traditional intent detection methods can't understand user's intent in depth, while the deep learning model shows its advantages. The capsule network model achieves good performance in the intent detection task, and also has a good effect on multi-label classification. The self-attention model can extract various semantic features of sentences in the process of intent detection, thus contributing to the research of multi-intent detection.

At present, intent detection is not only applied in various fields such as e-commerce, travel consumption, medical treatment and chat, but also applied to network intrusion, network fraud and air target combat fields, to provide the guarantee for network security problems. Traditional dialogue systems are mainly oriented at single intent detection in specific fields. With the increasingly frequent interaction between humans and machines, the users' discourse expression is not limited to only one intent. How to accurately identify multiple intents of users will be our next research work.

Acknowledgments

This work is partially supported by the National Natural Science Foundation of China (Nos. 61562068,11704229,61640204,61806103); Natural Science Foundation of Inner Mongolia (No.

2017MS0607); Subproject of Mongolian Language Informatization of Inner Mongolia People's Committee(MW-2014-MGYWXXH-01); Youth Innovation and entrepreneurship talents of Inner Mongolia "grassland talents" project; Inner Mongolia science and technology planning project (research and development operation of online interactive learning platform of Mongolian language MOOC and analysis and research of national education big data); Graduate Innovation Fund of Inner Mongolia Normal University (CXJJS18112).

References

[1] Hongshen Chen, Xiaorui Liu, Dawei Yin, et al. A survey on dialogue systems: recent advances and new frontiers[J]. *Acm Sigkdd Explorations Newsletter*, 2017, 19(2):25-35.

[2] Tur G. Spoken Language Understanding: Systems for extracting semantic information from speech[D]. NewYork, NY: John Wiley and Sons, 2011.

[3] Austin J A. *How to do things with words*[M]// Harvard University Press. Cambridge: 1962.

[4] Celikyilmaz A, Hakkani-Tur D, Tur G, et al. Exploiting distance based similarity in topic models for user intent detection[C]// *Automatic Speech Recognition & Understanding*. IEEE, 2011:425-430.

[5] Bingfeng Luo, Yansong Feng, Zheng Wang, et al. 2018.Marrying up regular expressions with neural networks:A case study for spoken language understanding. arXiv preprint arXiv:1805.05588.

[6] Yanling Li, Yonghong Yan. Weakly-supervised training method about Chinese spoken language understanging[J]. *Application of Computers*, 2015, 35(7):1965-1968.

[7] Zhiyuan Chen, Bing Liu, Hsu Meichun, et al. Identifying intent posts in discussion forums[C]// *Proc. of Conference of the North American Chapter of the Association for ComputationalLinguistics-HumanLanguageTechnologies*.Atlanta,2013:1041–1050.

[8] Yue Qian. Research on the identification method of users' travel consumption intent in chat robot[D]. Harbin: Harbin Institute of Technology, 2017.

[9] Dauphin Y N, Tur G, Hakkani-Tur D, et al. "Zero-shot learning for semantic utterance classification," arXiv preprint arXiv:1401.0509, 2013.

[10] Appelt D, Bear J, Cherny L, et al. GEMINI: A natural language system for spoken-language understanding[C]// *Proc. Meeting of the Association for Computational Linguistics*. 1993:54-61.

[11] Yan Pengju. Research on natural language understanding in conversational systems[D]. Beijing: Tsinghua University, 2002.

[12] Ahmad A S, Hassan M Y, Abdullah M P, et al. A review on applications of ANN and SVM for building electrical energy consumption forecasting[J]. *Renewable & Sustainable Energy Reviews,* 2014, 33(2):102 -109.

[13] Andrew McCallum, Kamal Nigam, et al.A comparison of event models for naive bayes text classification[C]// *In AAAI-98 workshop on learning for text categorization*. 1998:41–48.

[14] Schapire R E, Singer Y. BoosTexter: a boosting-based system for text categorization[J]. *Machine Learning*, 2000, 39(2-3):135-168.

[15] Haffner P, Tur G, Wright J H. Optimizing SVMs for complex call classification[C]// *IEEE International Conference on Acoustics*. IEEE, 2003:632-635.

[16] Genkin A, Lewis D D, Madigan D. Large-Scale Bayesian Logistic Regression for Text Categorization[J]. *Technometrics*, 2007, 49(3):291-304.

[17] Bengio Y, Ducharme R, Vincent P, et al. A neural probabilistic language model[J].*Journal of Machine Learning Research*, 2003, (3): 1137-1155.

[18] KimD, LeeY, ZhangJ, et al. Lexical feature embedding for classifying dialogue acts on Korean conversations[C]//*Proc. of 42th Winter Conference on Korean Institute of Information Scientists and Engineers,* 2015: 575–577.

[19] Kim JK, Tur G, Celikyilmaz A, et al. Intent detection using semantically enriched word embeddings[C]// *Spoken Language Technology Workshop*. IEEE, 2016:414-419.

[20] Lecun Y L, Bottou L, Bengio Y, et al. Gradient-Based Learning Applied to Document Recognition[J]. *Proceedings of the IEEE*, 1998, 86(11):2278-2324.
[21] KimY.Convolutional Neural Networks for Sentence Classification[C]// *Proc. of the 2014 Conference on Empirical Methods in Natural Language Processing (EMNLP)*, 2014:1746–1751.
[22] HashemiHB., AsiaeeA, KraftR, Query intent detection using convolutional neural networks[C]// *International Conference on Web Search and Data Mining, Workshop on Query Understanding* , 2016.
[23] Bhargava A, Celikyilmaz A, Hakkanitur D, et al. Easy Contextual Intent Prediction and Slot Detection[C]// *IEEE International Conference on Acoustics*. IEEE, 2013: 8337-8341.
[24] Hochreiter S, Schmidhuber J. Long Short-Term Memory[J]. *Neural Computation*, 1997, 9(8):1735-1780.
[25] Ravuri S V, Stolcke A. Recurrent neural network and LSTM models for lexical utterance classification[C]// *16th Annual Conference of the International Speech Communication Association*. 2015:135-139.
[26] Dey R, Salemt F M. Gate-Variants of Gated Recurrent Unit (GRU) Neural Networks[C]//*IEEE 60th International Midwest Symposium on Circuits and Systems*.IEEE, 2017:1597-1600.
[27] Ravuri S, Stolcke A. A comparative study of recurrent neural network models for lexical domain classification[C]// *Proc. of the 41th IEEE International Conference on Acoustics, Speech, and Signal Processing*. IEEE, 2016: 6075-6079.
[28] Hui Yu, Xupeng Feng, Lijun Liu, et al. Identification method of user's medical intent in chatting robot[J].*Journal of Computer Applications*, 2018, 38(8): 2170-2174.
[29] Jiawei Huang. Research on the classification method of user intent in the human-machine dialogue system[D]. Wuhan: Central China Normal University, 2018.
[30] Zhouhan Lin, Minwei Feng, Santos CN D, et al. A structured self-attentive sentence embedding. arXiv preprint arXiv:1703.03130, 2017.
[31] Hinton G E, Krizhevsky A, Wang S D. Transforming Auto-Encoders[C]//*In International Conference on Artificial Neural Networks*, 2011:44–51.
[32] Sabour S, Frosst N, Hinton G E. Dynamic routing between capsules[J]. *In Advances in Neural Information Processing Systems*, 2017:3859–3869.
[33] Wei Zhao, Jianbo Ye, Min Yang, et al. Investigating capsule networks with dynamic routing for text classification.Computing Research Repository, arXiv:1804.00857. Version 1, 2018.
[34] Congying Xia, Chenwei Zhang, Chenwei Yan, et al. Zero-shot User Intent Detection via Capsule Neural Networks[C]//*Proc. of the 2018 Conference on Empirical Methods in Natural Language Processing*. 2018:3090-3099.
[35] Yanling Li, Yonghong Yan. Research on execution strategy about statistical spoken language understanding[J]. *Computer Science and Technology*, 2017, 11(6):980-987.
[36] Liu B, Lane I. Attention-Based Recurrent Neural Network Models for Joint Intent Detection and Slot Filling[C]//*17th Annual Conference of the International Speech Communication Association*. 2016:685-689.
[37] Chunni Yang, Chaosheng Feng. Multi-intent recognition model with combination of syntactic feature and convolution neural network[J]. *Journal of Computer Applications*, 2018, 38(7): 1839-1845.

IOP Conf. Series: Journal of Physics: Conf. Series **1267** (2019) 012060 doi:10.1088/1742-6596/1267/1/012060

Impact Analysis of Track Slip in Sediment on Submarine Tracked Vehicle Steering Performance

Zhang Hao[1,*,a], Duan Yubing[1,b], Shi Xin[2,c], Hu Xiaoli[1,d], Su Jianjun[1,e], Yao Jinxia[1,f] and Liu Hui[1,g]

[1]State Grid Shandong Electric Power Research Institute, Jinan, China
[2]State Grid Shandong Electric Power Company, Jinan, China

[*,a]sdqzzh@163.com; [b]dyb83@126.com; [c]shixin@geidco.org;[d]13964020109@163.com; [e]suj2000@163.com; [f]yaojinxia@geidco.org;[g]hwi.te@163.com

Abstract. In combination with the requirements of the Tracked Vehicle-Cable Salvage project, which is aiming at the characteristics of the submarine crawler slip, the impact of the slip rate on the tangential drive to steering radius and steering angular velocity of the submarine crawler chassis. it is analyzed from the theoretical and experimental levels. Simplify the force model of the working submersible during steady-state steering, and it can obtain the steady-state steering body dynamics equation; in order to solve the arbitrary shear rate and shear displacement of the grounding surface in the steering process. According to the steering geometry relationship, it can combined with the empirical experience of seafloor sediments. The model derives the longitudinal driving. lateral resistance and drag torque of the left and right side of the track when it is during steady-state steering; when we analyzes the instantaneous lateral shift of the track shear rate during steady-state steering; it derives the instantaneous rate of the instantaneous steering rate, which based on the geometric relationship of the steering motion The relationship between the offset and the steering radius of the body, the steering angular velocity, and the sliding rate as a function of the steering radius and steering angular velocity of the body; when the steering model test is carried out to verify the influence of the slip rate on the steering radius and the steering angular velocity; The slip rate has an effect on both the steering radius and the steering angular velocity. It was measured steering radius becomes larger, the steering angular velocity becomes smaller, and the theoretical prediction curve is basically consistent with the measured data.

1. Introduction

With the development of offshore oil and gas, offshore wind power can be development, coastal and island construction, it can submarine cable pipelines have increasingly become an important channel for the transportation of materials, which is the information and energy in the sea area. The construction speed of submarine cable pipelines is accelerating, and the demand for operation and maintenance of submarine cable pipelines has increased significantly[1-3], submarine cable operation and maintenance equipment was born. When the Sub Marine Tracked Vehicle-Cable Salvage is mainly used to complete the task related to faulty cable salvage repair[4,5]; the equipment is mainly composed of submarine crawler running mechanism, which we called excavation clearing system, operation hooking system, hydraulic power system, electronic control and sensing monitoring and so on[6]. System and other components, which became the performance of the submarine crawler walking mechanism directly determines the success or failure of the equipment operation[7-8].

In this paper, the characteristics of the submarine crawler slip are analyzed from both theoretical and experimental aspects. The influence of the slip rate on the tangential drive, which is steering radius and steering angular velocity during steady-state steering, when is calculated and analyzed. The validity of the theoretical analysis is verified by the scale model test; it provides a reference for the late design, we call it the performance prediction and offshore application of the submersible.

2. Tracked Vehicle - Cable Salvage Steady Steering Mechanics Modeling

Sub Marine Tracked Vehicle-Cable Salvage (SMT-CS) is powered by a submarine self-propelled chassis as a carrier. The left/starboard track drive is adjusted to overcome the steering resistance torque, which is shown in Figure 1.

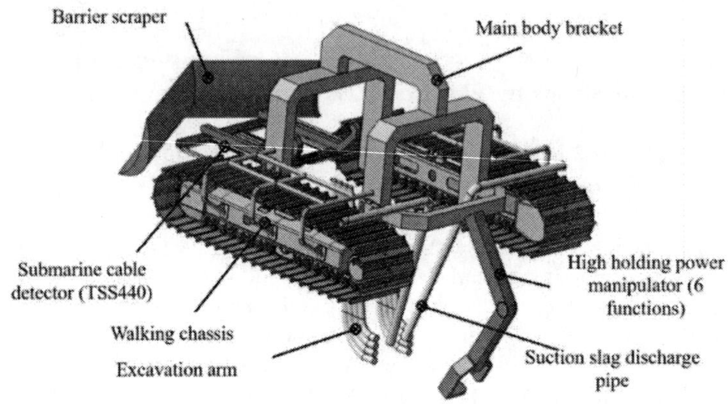

Figure 1 Schematic diagram of the working principle and composition of the cable salvage equipment

The lateral resistance R_{CL} and R_{CS} expressions for the left and right-side tracks are:

$$\begin{cases} R_{CL}=\int_{-(L_H+D_{IC})}^{L_H-D_{IC}} f_{TL}dx_L=\int_{-(L_H+D_{IC})}^{L_H-D_{IC}}\int_{-\frac{1}{2}B_S}^{\frac{1}{2}B_S}\tau\left(s_{yL}\right)dy_L dx_L \\ R_{CS}=\int_{-(L_H+D_{IC})}^{L_H-D_{IC}} f_{TS}dx_S=\int_{-(L_H+D_{IC})}^{L_H-D_{IC}}\int_{-\frac{1}{2}B_S}^{\frac{1}{2}B_S}\tau\left(s_{yS}\right)dy_S dx_S \end{cases} \tag{1}$$

The left and right-side track resistance torques M_{RL} and M_{RS} are expressed as:

$$\begin{cases} M_{RL}=\int_{-(L_H+D_{IC})}^{L_H-D_{IC}} f_{TL}x_L dx_L=\int_{-(L_H+D_{IC})}^{L_H-D_{IC}}\int_{-\frac{1}{2}B_S}^{\frac{1}{2}B_S}\tau\left(s_{yL}\right)x_L dy_L dx_L \\ M_{RS}=\int_{-(L_H+D_{IC})}^{L_H-D_{IC}} f_{TS}x_S dx_S=\int_{-(L_H+D_{IC})}^{L_H-D_{IC}}\int_{-\frac{1}{2}B_S}^{\frac{1}{2}B_S}\tau\left(s_{yS}\right)x_S dy_S dx_S \end{cases} \tag{2}$$

The shear displacement component is:

$$\begin{cases} s_{xL}=\left[\left(L_H-D_{IC}\right)-y_{AL}\right]\cdot\left[1-\frac{\omega'_{TZ}\left(R'-D_{TG}+y_{AL}\right)}{R_{SP}\omega_{DL}}\right] \\ s_{yL}=\left[\left(L_H-D_{IC}\right)-y_{AL}\right]\cdot\frac{\left(-\omega'_{TZ}x_{AL}\right)}{R_{SP}\omega_{DL}} \end{cases} \tag{3}$$

3. Analysis of the effect of slip rate on tangential drive and steering radius

3.1 The effect of slip rate on tangential drive

Take the Sub Marine Tracked Vehicle-Cable Salvage (SMT) as an example for calculation and analysis. The general characteristics of SMT are shown in Table 1.

IOP Conf. Series: Journal of Physics: Conf. Series **1267** (2019) 012060 doi:10.1088/1742-6596/1267/1/012060

Table 1. General characteristics of SMT

Name	Performance parameter description	Name	Performance parameter description
Dimensions ($L_O \times W_O \times H_O$)	5.8m×5m×2.5m	Water weight	20 ton (buoyancy adjustment)
		Chassis half gauge	1.85m
Strip length and width	1.3m	Track pitch	0.19m
Grounding half length	1.9m	Bearing wheel spacing	0.38m

During the steady-state steering of the walking chassis, the tangential traction force $F_T = F_{TL} + F_{TS}$ and the tangential walking resistance $R_C = R_{CL} + R_{CS}$ are both related to the track slip ratio i. Therefore, the tangential effective driving force $F_D = F_T - R_C$ of the traveling chassis is also necessarily a function of the slip ratio. Because the actual working conditions are more complicated and the calculation requirements are high, the special working conditions are taken here for analysis. It is assumed that the self-propelled submersibles have a steady-state differential left turn, and the left and right-side track slip ratios satisfy $i_S = -i_L = i$. The $F_T - i$ curve, the $F_D - i$ curve, the $R_{CL} - i$ curve, and the $R_{CS} - i$ curve are obtained, as shown in Figure 2 (half logarithmic display).

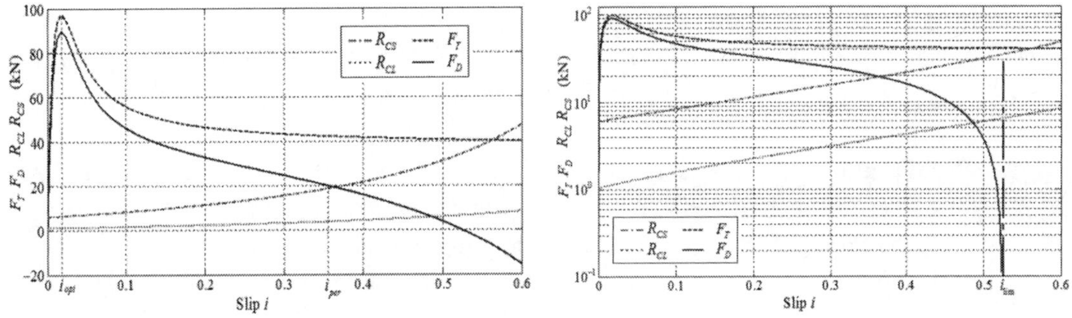

Figure 2 Slip curves and semi logarithmic curve of F_T, F_D, R_{CL} and R_{CS} plotted against i

From the slip curve of Figure 2, it can be seen that:

(1)Tangential traction force F_T, effective towing force F_D The slip curve has a hump-shaped distribution; in the initial stage, the slip rate increases, the tangential traction force and the effective drag force increase to the peak value. At this time, the optimal slip ratio $i_{opt} = 0.018$, the corresponding peak value Tangential traction force $F_{T\max} = 96.85$ kN, at this time peak effective drag force $F_{D\max} = 89.36$ kN;

(2)As the slip rate continues to increase, the tangential traction force F_T decreases significantly and finally stabilizes at 39.92 kN, while the effective drag force F_D decreases with increasing resistance (R_{CL}, R_{CS}). At $i = i_{\lim} = 0.523$, the effective drag force F_D decays. Zero, at this time, although the strip continues to shear the deposit, there is no actual propulsion effect;

(3)The design of self-propelled submersible crawler is the running mechanism the maximum tangential working load during the steering process, it cannot be lower than 15kN, and the design margin of 1.3 is taken. At this time, corresponding effective drag force $F_D = 19.5$ kN, it is rounding takes 20kN, corresponding slip The lower limit of the transfer rate is $i = 0.002$ and the upper limit is $i = 0.354$; therefore, the design allowable slip rate interval is $i_{per} \in (0.002, 0.354)$; when the track trencher is in normal operation, the load change slip rate changes within the allowable interval; the allowable slip rate interval In addition, the tangential effective drag force during steady-state steering cannot meet the design requirements of the workload.

3.2 Influence of slip rate on steering radius and steering angular velocity

The slip rate can be calculated by measuring the rotational speed of the drive wheel and the absolute displacement of the track relative to the ground; the relationship between the slip ratio and the drive ratio and the steering radius can be used to analyze the steering-related performance. The relationship between R'/R and K_{Dr} i is shown in Figure 3.

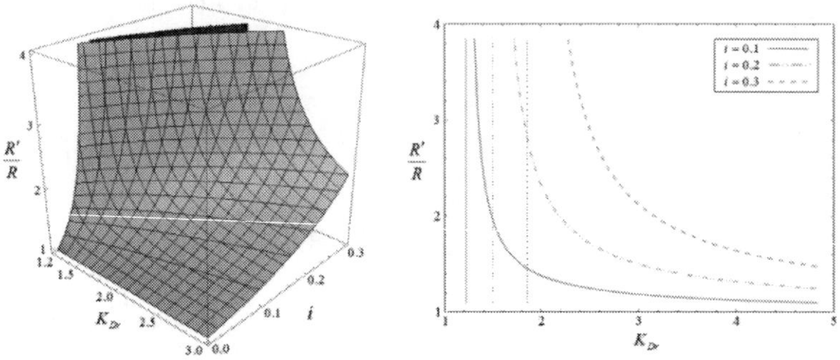

Figure 3 Relation curve of K_{Dr}, i and R'/R

The left figure of Figure 3 is the change relation of R'/R within the interval of $K_{Dr} \in [1.2,3]$ and $i \in [0,0.3]$, the right figure is $i = 0.1, 0.2, 0.3$, and R'/R varies with K_{Dr}. It can be seen from Figure 3 that:

(1)Due to the influence of the slip rate i of the left and right side tracks, the ratio R'/R is always greater than 1, which means that when the driving ratio K_{Dr} is constant, the steering radius increases under the action of the sliding drive and the sliding resistance;

(2)When the driving ratio K_{Dr} increases, the effect of the slip rate on the steering radius becomes smaller;

(3)There is a steady-state steering minimum driving ratio $K_{Dr\_min} = (1-i_L)/(1-i_S)$, and when $K_{Dr} \le K_{Dr\_min}$, due to the existence of the slip rate $R'/R \to \infty$, the body steering effect at this time not obvious.

4. Sub Marine Tracked Vehicle - Cable Salvage Model Steering Test

The Sub Marine Tracked Vehicle-Cable Salvage test was carried out is using a scale model.

The model steering test solution is shown in Figure 4. The entire system consists of a lifting platform, a circular mud pool and a submersible scale model.

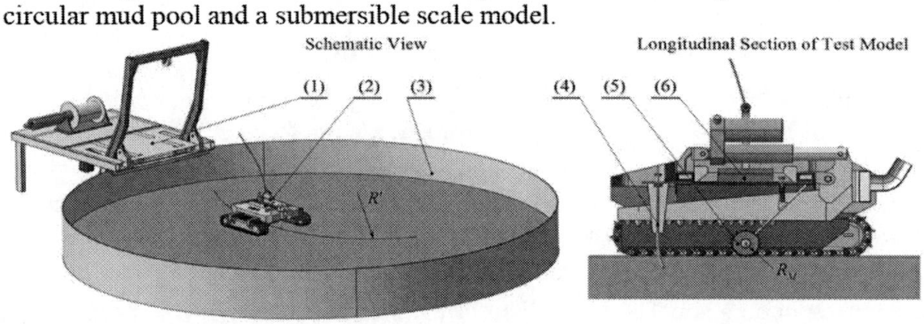

(1) Lifting platform; (2) Self-propelled submersible test model; (3) Circular mud pool; (4) Load hook; (5) Meter wheel; (6) Instrument box (digital compass).

Figure 4 Steering test solution for SMT scaled model

The geometric relationship between the measured quantities of the steady-state steering process is shown in Figure 5.

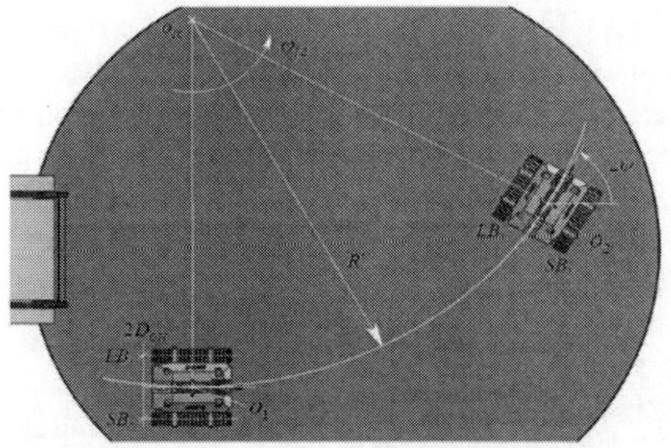

Figure 5 Geometric relationship of motion parameters during a steady-state turn for SMT scaled model

During the steady-state steering process, the traction force and the load are in equilibrium. Therefore, the depth of the load can be inserted into the soil during the test to obtain different slip rates.

According to the heavy-duty slow-moving, medium-load constant speed, when light-load fast port-side steering design test variables, it can be found the scene photo in Figure 6.

(1) Scale model test site (2) Port side track slip resistance rut

Figure 6 Steady-state tearing tests of SMT scaled model

According to the previous theoretical derivation, the results are shown in Table 2.

Table 2 Data Processing results of the steady-state steering test

Test number	Track slip rate		R (m)	ω_{TZ} (rad/s)	R' (m)		ω'_{TZ} (rad/s)		R'/R	ω'_{TZ}/ω_{TZ}
	i_L	i_S			Theory	Measured	Theory	Measured	Theory	Measured
1	-0.098	0.013	3.45	0.013	4.524	4.939	0.010	0.008	1.431	0.615
2	-0.102	0.049	3.45	0.013	5.189	5.352	0.009	0.009	1.551	0.692
3	-0.188	0.093	2.07	0.053	3.190	3.344	0.034	0.030	1.615	0.566
4	-0.145	0.143	2.07	0.079	3.287	3.332	0.048	0.044	1.609	0.557
5	-0.207	0.223	1.38	0.264	2.058	2.256	0.158	0.146	1.634	0.553

In Table 2, the ratio R'/R of the measured data to the ideal state is greater than 1, which indicates that the slip rate has an influence on the steering radius, and its existence makes the actual turning radius larger; the analysis results are consistent with the previous theoretical predictions; the measured data and the ideal The ratio ω'_{TZ}/ω_{TZ} of the state is less than 1, which indicates that the slip rate has an influence on the steering angular velocity, and its existence will slow the steering speed of the body. The previous theoretical predictions are consistent with the experimental results.

According to the test analysis results, the slip rate-steering radius curve and the slip rate-steering angle speed curve are shown in Figure 7 and Figure 8.

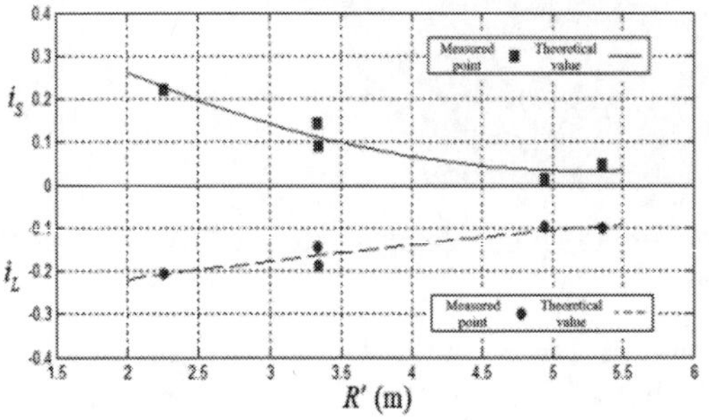

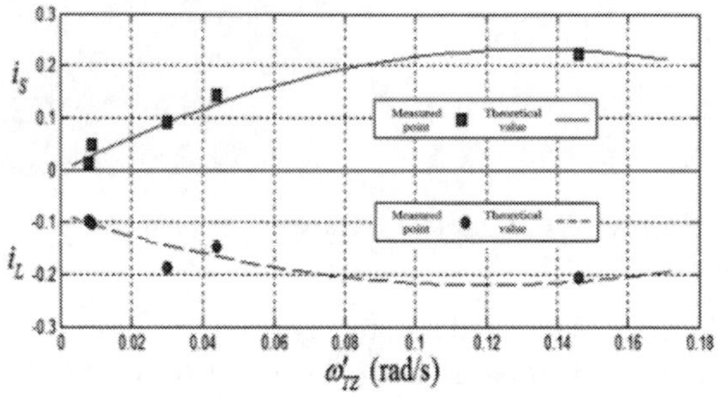

Figure 7 Relation curve of i_L, i_S - R'

Figure 8 Relation curve of i_L, i_S - ω'_{TZ}

It can be seen from Figure 7 and Figure 8 that the theoretical calculation curves of the relationship between the left and right side track slip ratios (i_L, i_S) and the body turning radius R' and the body steering angular velocity ω'_{TZ} are consistent with the trend of the measured results, so when the foregoing relates is the steering radius and the steering angular velocity. It can be seen the theoretical analysis is effective.

5. Conclusion

(1)The dynamic model is established by the force analysis of Tracked Vehicle-Cable Salvage during steady-state steering; the calculation shows that the track slip rate i has an effect on the steering drive performance of the walking chassis during steady-state steering; tangential traction F_T, The tangential effective drag force F_D has a hump-shaped distribution with the change of the slip rate i; there is a limit slip rate. When $i \geq 0.523$, the strip plate continues to shear the sediment, but there is no actual propulsion effect;

(2)According to the geometric relationship analysis of the amount of walking chassis movement during the steady-state steering process, which can be shown the relationship between the steady-state steering special operating condition slip rate i, the driving ratio K_{Dr} and the actual turning radius R' is derived; the simulation calculation shows that the track Under the action of slipping, the steering radius R' becomes larger and the steering angular velocity ω'_{TZ} becomes smaller. The analysis results show that the effective allowable slip ratio of the Tracked Vehicle-Cable Salvage is $i_{per} \in (0.002, 0.354)$, beyond which the chassis drive cannot meet the workload demand;

(3)Laboratory model test results show that the slip rate has an effect on the steering radius and steering angular velocity. The measured data is consistent with the trend of the theoretical prediction curve, which can provide reference for the later offshore engineering application.

Acknowledgements
This work was supported by Research on Operation Technology of Cross-sea High Voltage DC Submarine Cable and Key Technologies of Detection Robot, which was funded by the Technology Projects of State Grid Corporation of China. This support is very gratefully acknowledged.

References

[1] Le A T, Rye D C, Durrant-Whyte H F. Estimation of track-soil interactions for autonomous tracked vehicles. Proceedings of IEEE International Conference on Robotics and Automation. Vol. 2 (1997) No.28, p. 1388-1393.

[2] Song Z B, Zweiri Y H, Seneviratne L D, et al. Non-linear observer for slip estimation of tracked vehicles. Proceedings of the Institution of Mechanical Engineers - Part D. Vol. 2 (2008) No.22, p. 515-533.

[3] Iagnemma K, Kang S, Shibly H, et al. Online terrain parameter estimation for wheeled mobile robots with application to planetary rovers. Robotics. Vol. 5 (2004) No.20, p. 921-927.

[4] Tchamna R, Youn I. Yaw rate and side-slip control considering vehicle longitudinal dynamics. International Journal of Automotive Technology. Vol. 1 (2013) No.14, p. 53-60.

[5] Liu Wei, Wu Hongyun, Jiang Min, et al. Research on differential steering performance of submarine self-propelled four-track vehicle. Mining research and development, Vol. 11 (2017) No.18, p. 81-85.

[6] Wong J Y, Preston-Thomas J. On the characterization of the shear stress-displacement relationship of terrain. Journal of Terramechanics. Vol. 4 (1983) No.19, p. 225-234.

[7] Choi J, Hong S, Kim H W. An Experimental Study on Steering Performance of Tracked Vehicle on Cohesive Soft Soil by DOE (Design of Experiment) Using Orthogonal Arrays. Vol. 4 (2004) No.28, p. 103-105.

[8] Edwin P, Shankar K, Kannan K. Soft soil track interaction modeling in single rigid body tracked vehicle models. Journal of Terramechanics, Vol. 7 (2018) No.7, p. 1-14.

IOP Conf. Series: Journal of Physics: Conf. Series **1267** (2019) 012061 doi:10.1088/1742-6596/1267/1/012061

Mechanics Modeling and Kinetic Analysis of Sub Marine Tracked Vehicle-Cable Salvage during Steady-state Steering

Zhang Hao[1,*a], Su Jianjun[1,b], Hu Xiaoli[1,c], Duan Yubing[1,d], Shi Xin[2,e], Yao Jinxia[1,e] and Gu Chao[1, f]

[1]State Grid Shandong Electric Power Research Institute, Jinan, China
[2]State Grid Shandong Electric Power Company, Jinan, China

[*a]sdqzzh@163.com, [b]suj2000@163.com, [c]13964020109@163.com, [d]dyb83@126.com, [e]shixin@geidco.org, [f]yajinxia@geidco.org, [g]13969231958@163.com

Abstract. Based on the demand for deep-sea submarine cable salvage in Northeast Asia, a mathematical model of Sub Marine Tracked Vehicle-Cable Salvage (SMT-CS) was designed in this study. Considering the slip characteristics of the submarine track, the mechanical model of the SMT-CS during steady-state steering was simplified to obtain the kinetic equation of the vehicle body during steady-state steering. According to the steering geometrical relationship, the arbitrary shear rate and shear displacement of the ground plane during the steering process were solved. Combined with the empirical shear model of bottom sediments, the longitudinal driving force, lateral resistance and resistance moment of the larboard and starboard were derived during steady-state steering. The functional relationship between the lateral shift of instantaneous center of the conventional steering rate and the turning radius and angular velocity of the vehicle body as well as that between the slipping rate and the turning radius and angular velocity of the vehicle body was derived according to the geometrical relationship of steering motion. These are of great reference value to the design, performance prediction and maritime application of SMT-CS in the future.

1. Introduction

Northeast Asia is one of the most developed regions in Asia. The power grid interconnection project in Northeast Asia is going to realize the cross-border power grid interconnection between China, South Korea and Japan, which will increase the percentage of cleaner alternative energy sources, adjust the energy structure and optimize energy resource allocation in surrounding areas. Planning to enter the Yellow Sea from Weihai city, the China-South Korea submarine cable project will be laid on the seabed eastward and landed in Incheon, South Korea, with an average water depth of more than 100 meters. Since the cross section of the large-capacity submarine cable conductor and the dead weight of the submarine cable are quite large, whereas the submarine cable salvage technology in China is rather limited, it is of great significance to study the submarine cable salvage technology.

In this study, a steady-state steering mechanics model was designed based on the operational requirements on repairing the faulted submarine cable salvage in Northeast Asia. This model theoretically analyzed the impact of slipping rate on the tangential driving force, turning radius and angular velocity of Sub Marine Tracked Vehicle-Cable Salvage (SMT-CS), simplified the force analysis of SMT-CS during steady-state steering, and conducted a kinetic analysis of the mechanical model.

2. Mechanics Modeling of SMT-CS during Steady-state Steering

2.1 Structure and Assumptions of SMT-CS

The vehicle body of a SMT-CS(Sub Marine Tracked Vehicle-Cable Salvage)primarily consists of track chassis and functional modules, which is shown in Figure1. The track chassis is mainly rigidly connected by the track and the bracket, and a turn is realized by overcoming the steering resistance through adjusting the driving forces of the larboard and starboard tracks.

Figure 1 Track chassis of a SMT-CS

For the convenience of calculation, the following assumptions were made during the steering operation of the SMT-CS [1-3]:

- The steering of the track chassis is limited to a two-dimensional horizontal plane. When the force is analyzed, the working resistance and resistance moment are projected in the turning plane.
- The SMT-CS has a relatively large turning radius during its operation and maintenance along the submarine cable, with the steering mode mainly dominated by differential steering.
- The submarine cable salvage equipment is symmetric longitudinally and laterally relative to the intermediate plane, and the projection on the horizontal plane of the center of gravity CG coincides with the geometrical center o of the vehicle body.
- The ground pressure of the submarine cable salvage equipment is evenly distributed on the surface of the horizontal seabed.

2.2 Mechanics Modeling of SMT-CS during Steady-state Steering

The mass of the SMT-CS is assumed to be m_T, point o_{IC} denotes the instant turning center, D_{IC} is the eccentric distance between the turning center and the geometrical center o, ω_{TZ} and R' are the turning angular velocity and radius, respectively.

The operating velocity of the SMT-CS is set as 300m/hr, the turn is steadily made at a low velocity, and the turning radius is large. Therefore, the force borne by the vehicle is considered balanced ($m_T\ddot{x}=0, m_T\ddot{y}=0$), the centripetal force can be neglected ($m_T V_T^2 / R' = 0$), and the moment of force relative to point o_{IC} is also balanced (the accelerated velocity of the turning angle $\dot{\omega}_{TZ}=0$); a force balance equation was established for the SMT-CS, $\sum F_x = 0$, $\sum F_y = 0$, and a moment balance equation was set up relative to point o_{IC}.

$$\begin{cases} F_{DL} + F_{DS} = \left(F_{TL} - R_{TL}\right) + \left(F_{TS} - R_{TS}\right) = R_{Wx} \\ \int_{-(L_H+D_{IC})}^{L_H-D_{IC}} f_{TL} \cdot dl + \int_{-(L_H+D_{IC})}^{L_H-D_{IC}} f_{TS} \cdot dl = R_{Wy} \\ F_{DL}\left(R' - D_{TG}\right) + F_{DS}\left(R' + D_{TG}\right) = \left(\int_{-(L_H+D_{IC})}^{L_H-D_{IC}} f_{TL} l \cdot dl + \int_{-(L_H+D_{IC})}^{L_H-D_{IC}} f_{TS} l \cdot dl\right) + R_{Wy}\left(D_{WR} + D_{IC}\right) \end{cases} \quad (1)$$

2.3 Calculation of the Driving Force, Resistance and Resistance Moment during Steady-state Steering

o_{IC} is the instant turning center of the track chassis of SMT-CS; R' is the actual turning radius; ω_{TZ} is the turning angular velocity; o_{VC} is the projection of o_{IC} on the longitudinal axis ox of the vehicle body; D_{IC} is longitudinal eccentric distance of the projection; o_L and o_S are the intersection points between the line connecting o_{IC} and o_{VC}, the trajectories of the larboard and starboard, respectively; $o_L x_L y_L$ and $o_S x_S y_S$ are the body-fitted coordinates established on o_L and o_S, respectively; $o_{IC} o_L o_S$ is the lateral axis of the turning center; o'_L and o'_S are the instantaneous velocity centers of turning shear of the larboard and starboard tracks, respectively, both of which are on the lateral axis of the turning center.

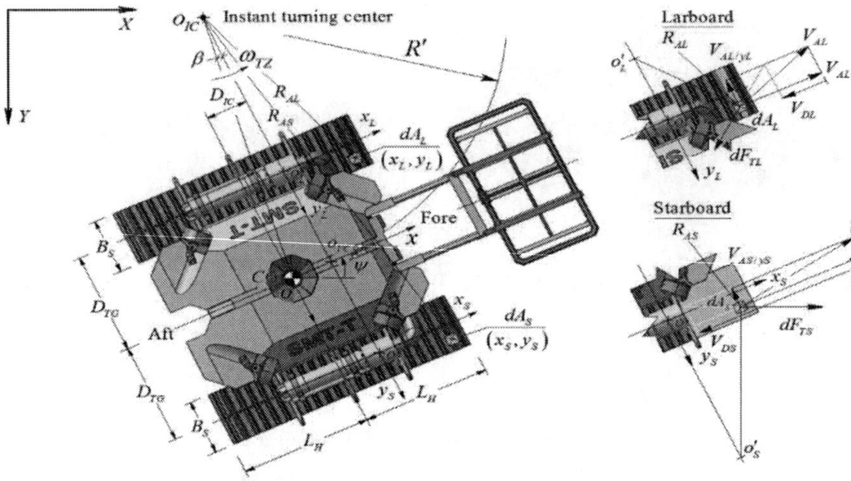

Figure 2 Geometrical relationship of parameters of SMT-CS during steady-state steering

As shown in Figure 2, F_{TS} and F_{TL} are the driving forces obtained by the larboard and starboard tracks' shearing of the soil respectively; R_{TS} and R_{TL} are the resistances caused by the deformation of the soil longitudinally squeezed by the larboard and starboard tracks, respectively; f_{TS} and f_{TL} are the distribution of the lateral earth pressure of the larboard and starboard tracks along the grounding direction respectively; wage resistance caused by clearing, additional drag resistance caused by umbilical cable, resistance generated by the mechanical arm and the force of underflow on the vehicle body are combined into the work disturbance resistance R_W whose projections along the body-fitted coordinates are $\vec{R}_{Wx}$ and $\vec{R}_{Wy}$; D_{WR} is the eccentric distance at the action point of the work disturbance resistance.

$dA_L(x_{AL}, y_{AL})$ is a random infinitesimal element on the ground plane of the larboard track, the corresponding shear displacement of which is assumed to be s_L; $dA_S(x_{AS}, y_{AS})$ is a random infinitesimal element on the ground plane of the starboard track, the corresponding shear displacement of which is assumed to be s_S. The shear force of the selected infinitesimal elements can be obtained based on the shear stress-displacement empirical equation. In this study, the classic shear model[4] proposed by Wong JY was used to solve the interaction force between the tracks and the sediments. Relevant researches have verified that this model is effective for bottom sediments[5,6].

2.3.1 Longitudinal Driving Forces of the Larboard and Starboard Tracks. As shown in Figure 2, s_{x*} is the shear displacement of random infinitesimal element along the x axis of the body-fitted coordinates, with that for the larboard track and starboard track denoted as s_{xL} and s_{xS}, respectively. dF_T is the longitudinal driving force of the infinitesimal element, and F_T, the driving force generated by the tracks' shearing of bottom sediments, can be obtained with a longitudinal integration along the ground plane.

$$F_{T*} = \iint_{A_*} \tau\left(s_{x*}\right) dA_* = \int_{-\frac{1}{2}B_S}^{\frac{1}{2}B_S} \int_{-\left(L_H+D_{IC}\right)}^{L_H-D_{IC}} \tau\left(s_{x*}\right) dx_* dy_* \tag{2}$$

Where $\tau(s)$ is the shear model proposed by Wong JY.

2.3.2 Lateral Resistance and Resistance Moment of the Larboard and Starboard Tracks. s_{y*} is assumed to be the shear displacement of random infinitesimal element along the y axis of the body-fitted coordinates.

M_{R*}, resistance moment of the tracks, can be obtained with a longitudinal integration along the ground plane.

$$M_{R*} = \int_{-\left(L_H+D_{IC}\right)}^{L_H-D_{IC}} f_{T*} x_* dx_* = \int_{-\left(L_H+D_{IC}\right)}^{L_H-D_{IC}} \int_{-\frac{1}{2}B_S}^{\frac{1}{2}B_S} \tau\left(s_{y*}\right) x_* dy_* dx_* \tag{3}$$

3. Kinematics Analysis of SMT-CS during Steady-state Steering

3.1 Lateral Shift of Instantaneous Velocity Center of Shear Rate of Tracks

Due to the slip drive and slip resistance, the instantaneous velocity centers of the larboard and starboard tracks experienced a lateral shift. As shown in Figure 3, V_{o*} is the traction velocity of the tracks, with the traction velocity of the larboard and starboard tracks denoted as V_{oL} and V_{oS}, respectively, V_{D*} is the velocity of the drive wire, V_{s*} represents the shear rate.

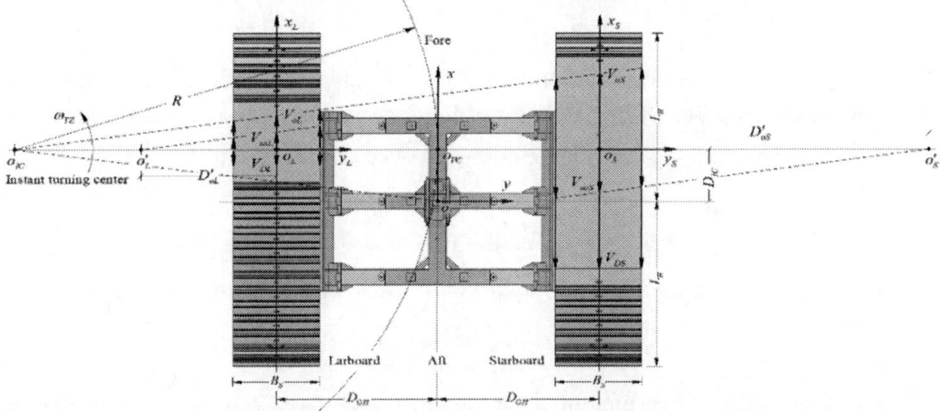

Figure 3 Instantaneous velocity center of SMT during a steady-state steering

According to the definition of the slipping rate i_*, the slipping rate of the tracks, can be expressed as follows:

$$i_* = \left(V_{D*} - V_{o*}\right)/V_{D*} \tag{4}$$

The traction velocity V_{oL} of the tracks can be expressed as,

$$\begin{cases} V_{oL} = \left(R - D_{GH}\right)\omega_{TZ} \\ V_{oS} = \left(R + D_{GH}\right)\omega_{TZ} \end{cases} \tag{5}$$

The relationship between lateral shift of the instantaneous velocity center and the slipping rate can be expressed as,

$$\begin{cases} D'_{oL} = \left(R - D_{GH}\right)\dfrac{i_L}{1-i_L} \\ D'_{oS} = \left(R + D_{GH}\right)\dfrac{i_S}{1-i_S} \end{cases} \tag{6}$$

By analyzing Equations, the following results were obtained:

- when $i<0$, the track is subject to a slip resistance, at which time the instantaneous center of the shear rate is near the turning center of the tracks;
- when $i>0$, the track is subject to a slip drive, at which time the instantaneous center of the shear rate is away from the turning center.

3.2 Shear Displacement of Steady-state Steering

Due to the poor bearing capacity of bottom sediments and the existence of slip drive and slip resistance, the center of gravity CG often fails to coincide with o_{VC}, the projection of the turning center o_{IC} on the x axis, resulting in the eccentric distance D_{IC}. The included angle β between the movement direction of the vehicle body and the ox axis of the body-fitted coordinates is defined as the sideslip angle, and the shear rate at points o_L and o_S on the larboard and starboard tracks.

$$\begin{cases} V_{soL} = R_{SP}\omega_{DL} - \omega'_{TZ}\left(R' - D_{TG}\right) \\ V_{soS} = R_{SP}\omega_{DS} - \omega'_{TZ}\left(R' + D_{TG}\right) \end{cases} \tag{7}$$

As shown in Figure 3, $dA_L(x_{AL}, y_{AL})$ is a random infinitesimal element on the larboard track. According to the kinematic geometry shown in the figure, the turning radius of this point $R_{AL} = \sqrt{(R' - D_{TG})^2 + x_{AL}^2}$, the turning linear velocity $V_{AL} = R_{AL}\omega'_{TZ}$, then $V_{AL/xL}$ and $V_{AL/yL}$, the components of linear velocity of dA_L in the coordinate system $o_L x_L y_L$ can be expressed as follow,

$$\begin{cases} V_{AL/xL} = \omega'_{TZ} R_{AL}\cos\psi_{AL} = \omega'_{TZ}\left(R' - D_{TG} + y_{AL}\right) \\ V_{AL/yL} = \omega'_{TZ} R_{AL}\sin\psi_{AL} = \omega'_{TZ} x_{AL} \end{cases} \tag{8}$$

Without regard to the overall deformation and assembly clearance of the tracks, the component of linear velocity of a random point on the rib band of the larboard track in the coordinate system $o_L x_L y_L$ $V_{TL/xL} = R_{SP}\omega_{DL}$, $V_{TL/yL} = 0$; and the coordinate component of shear rate of dA_L is,

$$\begin{cases} V_{sxL} = V_{TL/xL} - V_{AL/xL} = R_{SP}\omega_{DL} - \omega'_{TZ}\left(R' - D_{TG} + y_{AL}\right) \\ V_{syL} = V_{TL/yL} - V_{AL/yL} = -\omega'_{TZ} x_{AL} \end{cases} \tag{9}$$

At this time, $\Delta\psi_{AL}$, the turning angle of the larboard track, can be expressed as,

$$\Delta\psi_{AL} = \omega_{TZ} t_{AL} = \frac{L_H - D_{IC} - y_{AL}}{R_{SP}\omega_{DL}} \cdot \omega_{TZ} \tag{10}$$

In Equation (9), the shear displacement component of each axis in the coordinate system $o_L x_L y_L$ can be obtained by integrating time t_{AL},

$$\begin{cases} s_{xL} = \left[\left(L_H - D_{IC}\right) - y_{AL}\right] \cdot \left[1 - \dfrac{\omega'_{TZ}\left(R' - D_{TG} + y_{AL}\right)}{R_{SP}\omega_{DL}}\right] \\ s_{yL} = \left[\left(L_H - D_{IC}\right) - y_{AL}\right] \cdot \dfrac{\left(-\omega'_{TZ} x_{AL}\right)}{R_{SP}\omega_{DL}} \end{cases} \tag{11}$$

As shown in Figure 3, $dA_S(x_{AS}, y_{AS})$ is a random infinitesimal element on the starboard track. The shear displacement component can be obtained based on the following equation,

$$\begin{cases} s_{xS} = \left[\left(L_H - D_{IC}\right) - y_{AS}\right] \cdot \left[1 - \dfrac{\omega'_{TZ}\left(R' + D_{TG} + y_{AS}\right)}{R_{SP}\omega_{DS}}\right] \\ s_{yS} = \left[\left(L_H - D_{IC}\right) - y_{AS}\right] \cdot \dfrac{\left(-\omega'_{TZ} x_{AS}\right)}{R_{SP}\omega_{DS}} \end{cases} \tag{12}$$

3.3 Shear Rate and Shear Displacement at Random Points on the Starboard Track

ω_{DL} and ω_{DS} are the angular velocity of the driving wheel on the larboard and starboard tracks, respectively. The geometrical relationship between the central point velocity (o_L, o_S), turning angular velocity and turning radius of the larboard and starboard tracks of the SMT-CS during steady-state steering was analyzed to obtain the turning radius and turning angular velocity of the vehicle body of the SMT-CS.

1) In an ideal state without considering the slipping rate and lateral shift of instantaneous center of the tracks, the turning radius R and turning angular velocity ω_{TZ} satisfy the following equations,

$$\begin{cases}(\omega_{DS}-\omega_{DL})\cdot R=(\omega_{DS}+\omega_{DL})\cdot D_{GH}\\ \omega_{TZ}=(\omega_{DS}+\omega_{DL})\cdot\dfrac{R_{SP}}{2R}=(\omega_{DS}-\omega_{DL})\cdot\dfrac{R_{SP}}{2D_{GH}}\end{cases} \tag{13}$$

According to the above equation:

- When $\omega_{DS}=\omega_{DL}$, the turning radius $R\to\infty$, and the tracks go straight;
- When $\omega_{DS}\cdot\omega_{DL}=0$, the track is locked at one side, and $R=\pm D_{GH}$;
- When $\omega_{DS}=-\omega_{DL}$, the larboard and starboard tracks are driven forward and backward, the vehicle body rotates in situ, and $R=0$.

The above three working conditions can be regarded as special working conditions of steady-state steering, which are relatively simple to solve and analyze; here the steady-state steering of non-special working conditions was defined as conventional steering. For ideal steady-state conventional steering:

- When the driving velocity of the larboard and starboard tracks (ω_{DS} and ω_{DL}) are fixed, the turning radius R is proportional to the half gauge of the chassis D_{GH}, and the turning angular velocity ω_{TZ} is inversely proportional to D_{GH};
- When the gauge of the chassis $2D_{GH}$ remains constant, the turning radius is proportional to $(\omega_{DS}+\omega_{DL})/(\omega_{DS}-\omega_{DL})$, the ratio of the sum of the driving forces of the starboard and larboard tracks to the difference between the two driving forces, and the turning angular velocity is proportional to $(\omega_{DS}-\omega_{DL})$, the difference between the driving forces of the starboard and larboard tracks.

$$\begin{cases}(\omega_{DS}-\omega_{DL})\cdot R=(\omega_{DS}+\omega_{DL})\cdot D_{GH}\\ \omega_{TZ}=(\omega_{DS}+\omega_{DL})\cdot\dfrac{R_{SP}}{2R}=(\omega_{DS}-\omega_{DL})\cdot\dfrac{R_{SP}}{2D_{GH}}\end{cases} \tag{14}$$

For a left turn in conventional steady-state steering, $\omega_{DL}\neq 0$, the ratio of the driving force of starboard track to that of larboard track $K_{Dr}=\omega_{DS}/\omega_{DL}$, then Equation (13) can be expressed as follows using K_{Dr},

$$\begin{cases}(K_{Dr}-1)\cdot R=(K_{Dr}+1)\cdot D_{GH}\\ \omega_{TZ}=\dfrac{R_{SP}}{2D_{GH}}\cdot(K_{Dr}-1)\cdot\omega_{DL}\end{cases} \tag{15}$$

2) In conventional steering, considering the slip drive and slip resistance of the tracks, the relationship between the lateral shift of instantaneous center of larboard and starboard tracks (D'_{oL} and D'_{oS}) and the turning radius R' as well as the turning angular velocity ω'_{TZ} was solved; based on the geometrical relationship between relevant parameters, the driving forces of larboard and starboard tracks ($V_{DL}=R_{SP}\omega_{DL}$ and $V_{DS}=R_{SP}\omega_{DS}$), the following equation can be worked out,

$$\begin{cases}(\omega_{DS}-\omega_{DL})R'=(\omega_{DS}+\omega_{DL})\cdot D_{GH}+(\omega_{DL}D'_{oS}-\omega_{DS}D'_{oL})\\ \omega'_{TZ}=(\omega_{DS}-\omega_{DL})\cdot\dfrac{R_{SP}}{2D_{GH}+(D'_{oS}-D'_{oL})}\end{cases} \tag{16}$$

For a left turn in conventional steady-state steering, $\omega_{DL} \neq 0$, by substituting K_{Dr} into the above equation, the following equation can be obtained,

$$\begin{cases} (K_{Dr}-1)\cdot R' = (K_{Dr}+1)\cdot D_{GH} + (D'_{oS} - K_{Dr}D'_{oL}) \\ \omega'_{TZ} = \dfrac{R_{SP}}{2D_{GH} + (D'_{oS} - D'_{oL})}\cdot (K_{Dr}-1)\cdot \omega_{DL} \end{cases} \quad (17)$$

3) In conventional steering, considering the slipping rate i, according to the geometrical relationship between relevant parameters and Equation (14), the turning radius R' and the turning angular velocity ω'_{TZ} can be calculated as follows,

$$\begin{cases} [\omega_{DS}(1-i_S) - \omega_{DL}(1-i_L)]\cdot R' = [\omega_{DS}(1-i_S) + \omega_{DL}(1-i_L)]\cdot D_{GH} \\ \omega'_{TZ} = [\omega_{DS}(1-i_S) + \omega_{DL}(1-i_L)]\cdot \dfrac{R_{SP}}{2R'} = [\omega_{DS}(1-i_S) - \omega_{DL}(1-i_L)]\cdot \dfrac{R_{SP}}{2D_{GH}} \end{cases} \quad (18)$$

For a left turn in conventional steady-state steering, $\omega_{DL} \neq 0$, by substituting K_{Dr} into the above equation, the following equation can be obtained,

$$\begin{cases} [K_{Dr}(1-i_S) - (1-i_L)]\cdot R' = [K_{Dr}(1-i_S) + (1-i_L)]\cdot D_{GH} \\ \omega'_{TZ} = \dfrac{R_{SP}}{2D_{GH}}\cdot [K_{Dr}(1-i_S) - (1-i_L)]\cdot \omega_{DL} \end{cases} \quad (19)$$

4. Conclusion

(1)In this study, a kinetic model was established based on a force analysis of the SMT-CS during steady-state steering. The computational analysis shows that the slipping rate i of the tracks has an influence on steering performance of the chassis during steady-state steering.

(2)According to the geometrical relationship between parameters of the chassis during steady-state steering, the relational expression between the slipping rate i, drive ratio K_{Dr} and the actual turning radius R' was derived.

Acknowledgements

This work was supported by Research on Operation Technology of Cross-sea High Voltage DC Submarine Cable and Key Technologies of Detection Robot, which was funded by the Technology Projects of State Grid Corporation of China. This support is very gratefully acknowledged.

References

[1] Kim H, Hong S, Choi J, et al. Dynamic analysis of underwater tracked vehicle on extremely soft soil by using Euler parameters, Proceedings of The Sixth ISOPE Ocean Mining Symposium, Vol. 8 (2015) No.23, p. 141-148.

[2] Dai Yu, Liu Shaojun. Modeling and acquisition path simulation analysis of a new submarine crawler mining vehicle. Journal of applied foundation and engineering science, Vol. 2 (2014) No.28, p. 400-413.

[3] Zhang Tao, Dai Yu, Liu Shaojun, Chen Jun, Huang Zonghua. Multi-body dynamic modeling and walking performance simulation analysis of deep-sea crawler collector. Chinese journal of mechanical engineering, Vol. 6 (2015) No.51, p. 173-180.

[4] Wong J Y, Preston Thomas j. On the characterization of the shear stress-displacement relationship of terrain. Journal of Terramechanics. Vol. 4 (1983) No.19, p. 225-234.

[5] Choi J, Hong S, Kim h. w. An Experimental Study on Steering Performance of Cohesive Vehicle on Soft Soil by DOE (Design of Experiment) Using Orthogonal Arrays. (2004) p. 103-105.

[6] Edwin P, Shankar K, Kannan k. Soft soil track interaction modeling in single rigid body tracked vehicle models. Journal of Terramechanics, Vol. 3 (2018) No.77, p. 1-14.

Force Balanced Axial Piston Motor Power Loss and Leakage Analysis

WEN Desheng, SUI Guangdong*, LV Shijun, GAO Junfeng and LI Dexiong

College of Mechanical Engineering, Yanshan University, Qinhuangdao

*75343856@qq.com

Abstract. Aiming at the axial imbalance force when the traditional axial piston motor works, the force balanced axial piston motor utilized the concept of double stator to improve the structure of the traditional axial piston motor. Because of the special structure, the motor can not only achieve the axial force balance, but also output a variety of torque and speed. The working efficiency of the motor was an important indicator for evaluating the performance of the motor. Therefore, the power loss and leakage of the motor were analyzed in detail, and the prototype was tested. The experimental results showed that: (1) When the corresponding two single motors in the motor worked at the same time, theoretically, the leakage between the distribution cylinder and the cylinder accounts for the largest percentage of total leakage, reaching 56%. When the whole motor worked normally, the total leakage reaches 9.55L/min theoretically, and the leakage between the plunger and the plunger hole accounts for the largest percentage of the total leakage, which is 44% of the total leakage, and the distribution cylinder and cylinder the percentage of leakage between bodies has decreased, but its specific leakage has not changed. (2) The volumetric efficiency of the motor decreased as the pressure at the inlet of the motor increased, and the mechanical efficiency and overall efficiency increased accordingly. (3) When the motor inlet pressure was the same, the volumetric efficiency of two single motors at the same time was lower than the volumetric efficiency of the entire motor.

1. Introduction

The conventional axial piston motors can be divided into swash plate type and inclined axis type. Both of these motors have large axial unbalance forces, which greatly affect the service life of the motor [1]. In order to improve the performance of the motor and extend its life, a plurality of thrust bearings is generally assembled on the shaft to balance the axial imbalance force. The force-balanced axial piston motor proposed in this paper uses the "double stator" idea to achieve the balance of axial force by changing the structure of the motor [2]. In this paper, the main leakage forms and ways of the motor are analyzed, and the basic mathematical models of various leaks are calculated. The power loss is briefly explained. Through the design of the experimental system diagram, the prototype of the motor is tested and the data is carried out. The acquisition laid the foundation for the next step to improve the motor structure and performance.

2. Structure and working principle of force balanced axial piston motor

2.1 the structure of force balanced axial piston motor

The force balanced axial piston motor has made several corresponding designs based on the structure of the traditional axial piston motor [3]:(1) The motor has two corresponding swash plates, which are called double stators, and the swash plates were respectively located on both sides of the motor through-going shaft, and were connected by the sliding bearing and the through-going shaft of the motor. (2) Since the radial dimension is relatively small, the through-going shaft and the cylinder are designed as one body, collectively referred to as a through-going shaft. (3) The inner surface of the distribution shell is provided with a waist-shaped distribution groove, and its main function is to distribute the flow into the oil passage of the cylinder. (4) The inlet and outlet ports of the motor are all disposed on the surface of the casing of the motor, and have two oil inlets and two oil outlets respectively, and the diameters of the two are substantially the same, and the motor will be distributed from the casing. Figure 1 is a structural schematic diagram of a force balanced axial piston motor.

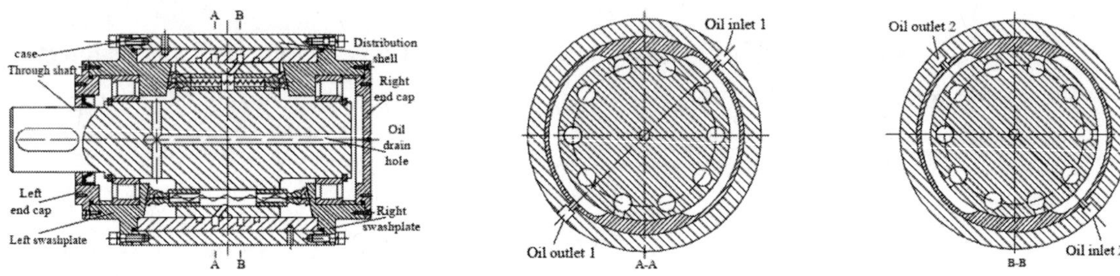

Figure 1. Schematic diagram of the force balanced axial piston motor

2.2 Working principle of force balanced axial piston motor

When the high pressure oil enters the waist distribution groove of the distribution case through the two inlet ports, there are 10 pairs of plungers working at the same time, that is, two single motors work at the same time. At this time, the high-pressure oil in the waist-shaped oil distribution tank enters the plunger cavity through the oil-passing hole, so that the plunger moves axially with respect to the through-going shaft. The plunger acts on the surface of the swash plate through the sliding shoe, and the surface of the swash plate generates a corresponding reaction force to the sliding shoe, which generates a radial component force to the plunger to generate a torque with respect to the through-going shaft. When the high-pressure oil entered the waist-shaped distribution groove of the distribution shell through only one inlet port, only five pairs of plungers that were spaced apart generate torque to the through-going shaft, that is, only one single motor operated separately [4].

3. Power loss analysis of force balanced axial piston motor

3.1 Power loss caused by viscous friction between the swash plate and the shoe of the plunger head

The shoe mounted on the head of the plunger performs an elliptical motion on the surface of the swash plate. Figure 2 shows the loss of friction power of the swash plate of the shoe.

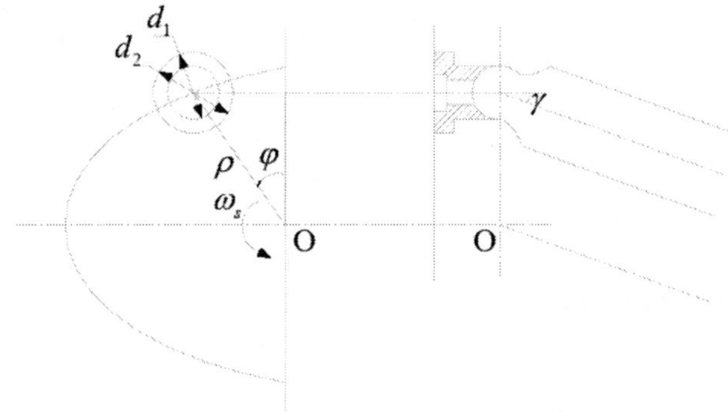

Figure 2. Analysis of friction power loss of sliding shoes

According to Figure 2, the viscous friction power $\Delta W_{1\mu}^{'}$ when a single shoe slides along the surface of the swash plate is:

$$\Delta W_{1\mu}^{'} = \frac{\mu_0 v_s^2}{h} \frac{\pi}{4}(d_2^2 - d_1^2) \quad (1)$$

From the formula $v_s = \omega_s \cdot \rho$ can be obtained:

$$\Delta W_{1\mu}^{'} = \frac{\pi\mu_0(d_2^2 - d_1^2)}{4h}\omega_s^2\rho^2 \quad (2)$$

Where: μ_0 is the dynamic viscosity coefficient of the hydraulic oil; v_s is the sliding line speed of the shoe along the surface of the swash plate; h is the gap between the bottom surface of the shoe and the surface of the swash plate; ω_s is the sliding angular velocity of the shoe around the point O in the figure; ρ is the polar diameter of the running track of the shoe; d_1, d_2 is the inner and outer diameter of the sealing strip at the bottom of the shoe.

The above formula (2) is the viscous friction power loss between the single shoe and the swash plate. Since there are a plurality of plunger shoes in the motor, it is assumed that the number of plungers of the motor is z. When the half of the number of plungers on one side of the motor is an odd number, when there are $z_0 = \frac{\frac{z}{2}+1}{2}$ plungers in the high pressure zone of the motor 1, the power loss $\Delta W_{1\mu 1}$ caused by the viscous friction between the shoe and the swash plate is:

$$\Delta W_{1\mu 1} = \frac{(\frac{z}{2}+1)\pi\mu_0(d_2^2 - d_1^2)}{8h}\omega_s^2\rho^2 \quad (3)$$

The high pressure zone of the motor 2 on the same side has $z_0 = \frac{\frac{z}{2}-1}{2}$ plungers, and the power loss $\Delta W_{1\mu 2}$ caused by the viscous friction between the shoe and the swash plate is:

$$\Delta W_{1\mu 2} = \frac{(\frac{z}{2}-1)\pi\mu_0(d_2^2 - d_1^2)}{8h}\omega_s^2\rho^2 \quad (4)$$

Therefore, when the half of the number of plungers on the motor side is an odd number, the power loss $\Delta W_{1\mu}$ caused by the viscous friction between the shoe and the swash plate is:

$$\begin{aligned}\Delta W_{1\mu} &= \Delta W_{1\mu 1} + \Delta W_{1\mu 2} \\ &= \frac{z\pi\mu_0(d_2^2 - d_1^2)}{8h}\omega_s^2\rho^2\end{aligned} \quad (5)$$

According to the formula:

$$\rho = R\sqrt{\sin^2\varphi + \frac{\cos^2\varphi}{\cos^2\gamma}} \quad (6)$$

$$\omega_s = \frac{\cos\gamma}{\cos^2\varphi + \cos^2\gamma\sin^2\varphi}\omega \quad (7)$$

Where: γ is the swashplate inclination angle; φ is the rotation angle of the plunger; ω is the angular velocity when the cylinder rotates; R is the plunger distribution circle radius.

Substituting the above formula (6) and formula (7) into the equation (5), the power loss caused by the viscous friction between the shoe and the swash plate on one side of the motor is:

IOP Conf. Series: Journal of Physics: Conf. Series **1267** (2019) 012062 doi:10.1088/1742-6596/1267/1/012062

$$\Delta W_{1\mu} = \frac{z\pi\mu_0(d_2{}^2 - d_1{}^2)}{8h}\omega_s{}^2\rho^2 = \frac{z\pi\mu_0\omega^2R^2(d_2{}^2 - d_1{}^2)}{8h(\cos^2\varphi + \sin^2\varphi\cos^2\gamma)} \tag{8}$$

3.2 Power loss between the plunger and the cylinder due to viscous friction

Assuming that the axial direction of the plunger is parallel to the axial direction of the cylinder, the following briefly analyzes the gap leakage between the plunger and the cylinder. Figure 3 shows the analysis of the friction power loss between the plunger and the cylinder.

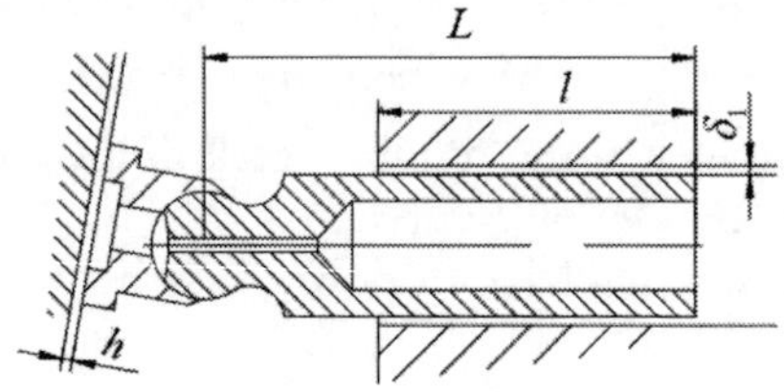

Figure 3. Analysis of friction power loss between plunger and cylinder

According to the knowledge of fluid mechanics, when the bottom of the plunger is filled with high-pressure oil, the viscous friction between the single plunger and the cylinder is:

$$\Delta F_2 = \mu_0\frac{v}{\delta_1}l\pi d \tag{9}$$

$$l = L - R(1+\cos\varphi)\tan\gamma \tag{10}$$

Where: v is the axial movement speed of the plunger relative to the cylinder; δ_1 is the radial clearance between the plunger and the plunger hole in the cylinder; l is the length of the plunger and the length of the plunger hole in the cylinder; d is the diameter of the plunger; L is the geometric length of the plunger.

Since there were a plurality of plungers in the motor, the average frictional power loss $\Delta W_{2\mu}$ of one side of the motor caused by the viscous friction of the plunger and the plunger hole in the cylinder is:

$$\Delta W_{2\mu} = \frac{1}{\pi}\int_0^{\pi} z\Delta F_2 v \mathrm{d}\varphi \tag{11}$$

Where $v = R\omega\tan\gamma\sin\varphi$, the axial movement speed v of the plunger relative to the cylinder is substituted into the above formula (10) and integrated:

$$\Delta W_{2\mu} = \frac{\pi\mu_0 zld\omega^2R^2\tan^2\gamma}{2\delta_1} \tag{12}$$

Since the motor displacement is $V = \pi d^2 zR\tan\gamma$, after substituting the above formula (11):

$$\Delta W_{2\mu} = \frac{V^2\mu_0 l\omega^2}{4\pi d^2 z\delta_1} \tag{13}$$

4. Leakage analysis and prototype leakage calculation of force balanced axial piston motor

4.1 Motor leakage analysis

(1) Leakage loss between the shoe and the swash plate

The leakage between the surface of the shoe and the swash plate is analyzed. It is assumed that the pressure of the high-pressure hydraulic oil that passes into the cavity at the bottom of the plunger is p_1, and the pressure of the hydraulic oil after the leakage between the shoe and the swash plate is p_2, such as Figure 4 shows.

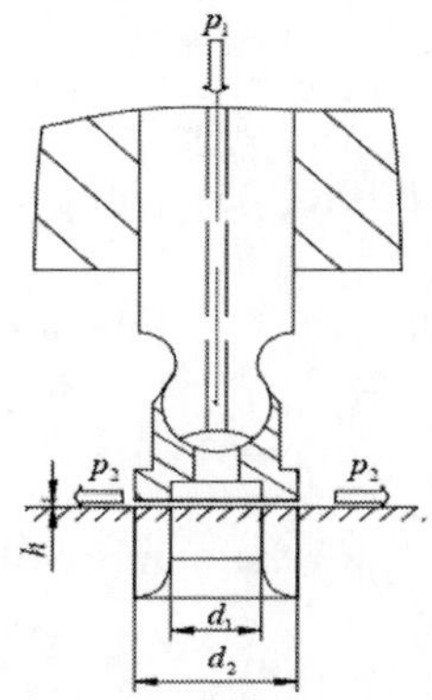

Figure 4. Leakage analysis diagram between the shoe and the swash plate surface

As can be seen from Figure 4, the gap between the surface of the shoe and the swash plate belongs to the parallel plate gap, and the leakage flow between the two parallel plates is:

$$\Delta q_1^{'} = \frac{\pi h^3 \Delta p}{6\mu_0 \ln(d_2 / d_1)} \tag{14}$$

Where: Pressure difference $\Delta p = p_1 - p_2$.

Due to the particularity of the motor structure, the two sides of the motor cylinder can be divided into two single motors, namely a motor 1 and a motor 2. Since the torque ripple of the odd number of motors is smaller than the number of plungers, it is determined that the number of the single motor plungers is an odd number, and the number of the plunger shoes of the motor 1 in the pressure oil zone is $\frac{\frac{z}{2}+1}{2}$. The total amount of leakage of the swash plate surface of the motor 1 in contact with the bottom of the shoe is:

$$\Delta q_1 = \frac{\frac{z}{2}+1}{2} \cdot \Delta q_1^{'} = \frac{(\frac{z}{2}+1)\pi h^3 \Delta p}{12\mu_0 \ln(d_2 / d_1)} \tag{15}$$

The number of plunger shoes in the pressure zone of the motor 2 is $\frac{\frac{z}{2}-1}{2}$, and the total leakage of the surface of the swash plate in contact with the bottom of the shoe is:

$$\Delta q_2 = \frac{\frac{z}{2}-1}{2} \cdot \Delta q_1^{'} = \frac{(\frac{z}{2}-1)\pi h^3 \Delta p}{12\mu_0 \ln(d_2 / d_1)} \tag{16}$$

Therefore, the total leakage at the contact between the sliding shoe on the motor side and the swash plate is:

$$\begin{aligned} \Delta Q_1 &= \Delta q_1 + \Delta q_2 \\ &= \frac{z\pi h^3 \Delta p}{12\mu_0 \ln(d_2 / d_1)} = \frac{z\pi h^3 (p_1 - p_2)}{12\mu_0 \ln(d_2 / d_1)} \end{aligned} \tag{17}$$

(2) Leakage loss between the plunger and the plunger hole in the motor

As shown in Figure 3, assuming that the plungers are unbiased in the cylinder plunger bores, their gap leakage amount is equal to the concentric cylindrical annular gap flow. Assuming that the hydraulic oil pressure after leakage through the gap between the plunger and the cylinder is p_3, the leakage amount $\Delta q_3^{'}$ is:

$$\Delta q_{3i}' = \frac{\pi d \delta_1^3 \Delta p}{12 \mu_0 l_i} \tag{18}$$

Where: $\Delta p = p_1 - p_3$; l_i is the length of the i-th plunger in the high pressure zone and the plunger hole in the cylinder.

The adjacent plunger angle of the motor is β, so the amount of leakage between the plunger and the plunger bore should be a function of period β. The number of plungers in which the motor 1 is in the high pressure oil zone is $m_1 = \frac{\frac{z}{2}+1}{2}$, and the number of plungers in which the motor 2 is in the high pressure oil zone is $m_2 = \frac{\frac{z}{2}-1}{2}$. It can be seen that the number of plungers in the high pressure zone on one side of the motor is: $m = m_1 + m_2$, that is $m = \frac{z}{2}$. The total gap leakage between the plunger and the plunger hole on the motor side is:

$$\Delta Q_2' = \sum \frac{\pi d^2 \delta_1^3 (p_1 - p_3)}{12 \mu_0 l_i} \tag{19}$$

Since the length l between the plunger and the plunger bore is a function of the cylinder angle φ, the total leakage is:

$$\Delta Q_2 = \frac{\pi d^2 \delta_1^3 (p_1 - p_3)}{12 \mu_0} \sum_{i=1}^{m} \frac{1}{L - R(1+\cos\varphi_i)\tan\gamma} \tag{20}$$

Where: φ_i is the corner of the i-th plunger.

As can be seen from equation (19), the amount of leakage between the plunger and the plunger bore is a function of the cylinder angle φ, which varies with the cylinder angle φ.

(3) Leakage loss between the distribution tube and the outer surface of the cylinder

When the high-pressure oil is introduced into the distribution casing, as shown in Figure 5, the high-pressure oil pressures of the two oil-passing zones of the distribution casing are all p_1, and no pressure difference is generated between them. The end pressure p_4 at the end of the distribution shell and the cylinder must be much smaller than the p_1 in the high pressure zone, thus creating a pressure differential Δp. Due to the influence of the pressure difference, certain leakage will inevitably occur between them. In order to facilitate the calculation, the leakage between the two is regarded as the gap leakage between the parallel plates [5, 6].

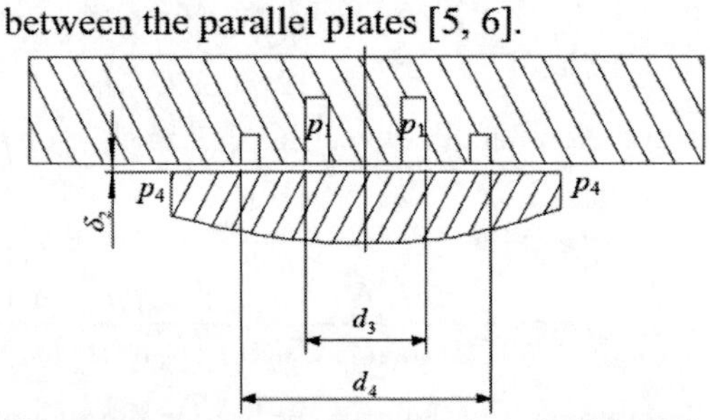

Figure 5. Leakage analysis diagram between the distribution shell and the outer surface of the cylinder

According to the gap leakage theory between parallel plates, the leakage amount ΔQ_3 of the inner surface of the distribution shell and the outer surface of the cylinder can be obtained as:

$$\Delta Q_3 = \frac{\pi \delta_2^3 (p_1 - p_4)}{6 \mu_0 \ln(d_4 / d_3)} \tag{21}$$

Where: δ_2 is the small gap between the outer diameter of the cylinder and the inner diameter of the distribution shell; d_3 and d_4 are the inner and outer diameters of the distribution port of the distribution housing.

Since there are plungers on both sides of the cylinder of the motor, when calculating the sum of the leakage between the shoe and the swash plate surface and the leakage of the plunger and the cylinder hole, the other side of the motor needs to be considered. Inside, the total leakage ΔQ of the motor is:

$$\begin{aligned}\Delta Q &= 2\Delta Q_1 + 2\Delta Q_2 + \Delta Q_3 \\ &= \frac{z\pi h^3(p_1 - p_2)}{6\mu_0 \ln(d_2/d_1)} + \frac{\pi d^2 \delta_1^3 (p_1 - p_3)}{12\mu_0} \cdot \\ &\sum_{i=1}^{m}\left[\frac{1}{L - R(1+\cos\varphi_i)\tan\gamma}\right] + \frac{\pi \delta_2^3 (p_1 - p_4)}{6\mu_0 \ln(d_4/d_3)}\end{aligned} \tag{22}$$

4.2 Leakage calculation of motor test prototype

The main design parameters of the motor are: Dynamic viscosity coefficient of hydraulic oil μ_0 is $3\times10^{-2}\,\text{Pa}\cdot\text{s}$; Inner diameter of the sealing strip at the bottom of the shoe d_1 is 20mm; Outer diameter of the sealing strip at the bottom of the shoe d_2 is 28mm; Number of motor plungers z is 10; Pressure difference ΔP is 8MPa; Geometric length of the plunger L is 85mm; Plunger diameter d is 20mm; Inner diameter of the distribution port of the distribution housing d_3 is 40mm; Inner and outer diameters of the distribution port of the distribution housing d_4 is 40mm; Inner and outer diameters of the distribution port of the distribution housing d_4 is 70mm; Swashplate inclination γ is 16°; Small gap between the outer diameter of the cylinder and the inner diameter of the distribution tube δ_2 is 0.052mm; Radial clearance between the plunger and the plunger bore in the cylinder δ_1 is 0.02mm; Plunger distribution circle radius R is 48mm.

Since the motor is constantly rotating, the leakage created by the gap between the plunger and the plunger bore is a function of the cylinder angle φ. This leak will vary depending on the corner of the cylinder. According to the formula of the leak, it can be inferred that when the cylinder angle $\varphi=0$, the corresponding plunger leakage amount is the largest. Here, we only count the case of the largest amount of leakage. Due to the structural specificity of the motor, there is only leakage when the left and right motor sides are operated and leakage when the left and right motors are simultaneously operated. Table 1 shows the theoretical leakage amount when the left and right motors are operated by only one side of the motor, and Table 2 shows the theoretical leakage amount when the left and right motors are simultaneously operated.

Table 1. Theoretical leakage of the left and right motors when only one side of the motor is working

Leak form	Leakage (L/min)	Percentage
Leak between the shoe and the swash plate	0.80	12%
Leak between the plunger and the plunger hole	2.12	32%
Leakage between the distribution shell and the cylinder	3.72	56%
Total leakage	6.64	100%

Table 2. Theoretical leakage when the left and right motors are working at the same time

Leak form	Leakage (L/min)	percentage
Leak between the shoe and the swash plate	1.63	17%
Leak between the plunger and the plunger hole	4.20	44%
Leakage between the distribution shell and the cylinder	3.72	39%
Total leakage	9.55	100%

The theoretical flow rate of the motor can be calculated from the theoretical formula $Q_{tm} = Vn$ of the theoretical flow of the motor. n is the rated speed of the motor. The rated speed of the motor is n=800r / min . The total displacement V=172mL / r of the motor is the theoretical flow Q_{tm} of the motor: Q_{tm}=Vn=$1.376\times10^5$$m$L / min . According to the definition of motor volumetric efficiency to calculating the volumetric efficiency η_{VW} of the motor in combination with the amount of motor leakage sought:

$$\begin{aligned}\eta_{VW} &= \frac{Q_{tm}}{Q_{tm}+\Delta Q}\times 100\% \\ &= \frac{1.376\times 10^5}{1.376\times 10^5+9.55\times 10^3}\times 100\% \\ &=93.5\%\end{aligned} \tag{23}$$

5. Experiment

In order to verify the feasibility of the working principle of the force balance axial piston motor and the rationality of the structure, the experimental test platform of the plunger motor was built, and the performance parameter data was collected and analyzed. Figure 6 is a physical view of a part of the motor, and Figure 7 is a system diagram of the test bench of the motor.

Figure 6. Physical diagram of some parts of the motor

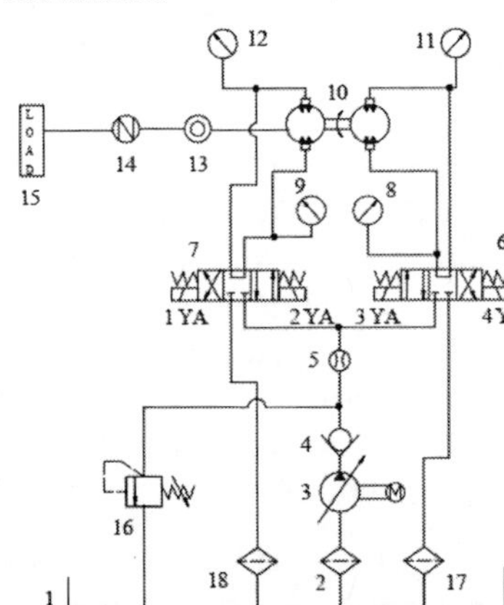

Figure 7. Piston motor test system diagram 1. Fuel tank 2. 17,18 Oil filter 3. Fuel supply pump 4. Check valve 5. Flow meter 6,7. Three-position four-way electromagnetic reversing valve 8,9,11,12. Pressure gauge 10. Plunger motor 13,14. Torque speed tester 15. Load 16. Overflow valve

In the experimental schematic diagram 7, the motor under test is directly connected to the load, which refers to the load pump, which can load the motor under test. The variable pump 3 of the motor supplies high pressure oil. The relief valve 16 regulates the inlet flow of the motor. By adjusting the working pressure of the load pump in the system, the oil inlet of the motor can get different pressure inputs. The inlet and outlet pressures of the motor are measured by pressure gauges 8, 9, 11, and 12, respectively. The rotational speed and torque of the motor at different pressures can be collected by the torque speed testers 13, 14. Table 3 shows the main design parameters of the test prototype.

Table 3. Main design parameters of the test prototype

Experimental maximum speed	800r/min
Experimental maximum pressure	8MPa
Displacement when two single motors are working	86mL/r
Displacement of the entire motor during operation	172mL/r

Due to the small size of the test bed fuel tank, the motor running speed is lower during the actual experiment to verify the theoretical accuracy. The performance parameters of the motor in the above two connection modes are recorded through experiments, and the volumetric efficiency, mechanical efficiency and total efficiency of the motor during actual operation are calculated according to the performance parameters. The specific data are shown in Tables 4 and 5 below.

Table 4. Experimental data of the motor on the left and right sides working on only one side(working mode 1)

Import pressure(MPa)	Input traffic(L/min)	Rotating speed(r/min)	Torque (N·m)	Volumetric efficiency η_v(%)	Mechanical efficiency η_m(%)	Total efficiency η(%)
1	10.0	117	7.1	93.6	52.9	49.5
2	9.7	112	9.8	90.2	64.7	58.4
3	9.4	109	16.3	88.7	77.0	68.3
4	9.1	106	24.7	87.4	80.6	70.4
5	8.8	102	39.5	85.2	83.2	70.9
6	8.5	98	59.3	83.8	84.1	70.7

Table 5. Experimental data of the simultaneous operation of the left and right motors (working mode 2)

Import pressure(MPa)	Input traffic(L/min)	Rotating speed(r/min)	Torque (N·m)	Volumetric efficiency ηv(%)	Mechanical efficiency ηm(%)	Total efficiency η(%)
1	10.0	61	13.6	92.6	53.6	49.6
2	9.7	58	21.1	89.1	64.6	57.8
3	9.4	56	33.5	86.9	73.0	63.4
4	9.1	54	50.7	85.4	74.9	64.0
5	8.8	52	76.5	83.7	80.6	67.5
6	8.5	98	59.3	83.8	84.1	70.7

It can be obtained from Tables 4 and 5. Under the operation of two different connection modes, the volumetric efficiency of the motor decreases as the inlet pressure of the motor increases, and the mechanical efficiency and total efficiency gradually increase. The reason is mainly because when the oil pressure is low, the relative sliding surface between the friction pairs fails to form a pressure oil film. At this time, the friction pair is in a dry friction or a semi-dry friction state, and the friction loss is more serious. When the oil pressure rises, a pressure oil film is formed between the friction pairs, and the pressure oil film greatly reduces the friction between the respective moving parts, thereby improving the mechanical efficiency of the motor. Figure 8 below shows the efficiency curve for the two connection modes of the motor.

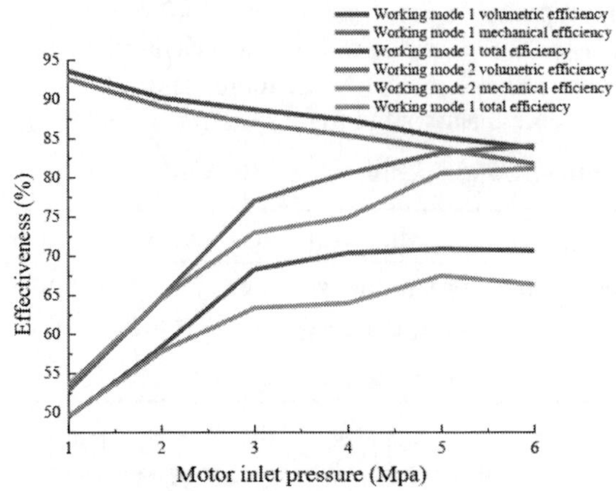

Figure 8. Motor efficiency graph

The working mode 1 indicates the efficiency curve when the left and right motors operate only on one side of the motor. Working mode 2 indicates the efficiency curve when the left and right motors are working at the same time. As can be seen from the above figure, when the motor inlet pressure is the same, the volumetric efficiency of only one side of the motor is lower than that of the left and right side motors. The main reason is that the displacement of the left and right motors when working together is twice that of the former, that is, the degree of displacement is increased to a greater extent than the amount of leakage. Therefore, in comparison, the volume efficiency of the motor on one side is low. As the inlet pressure of the motor increases, the volumetric efficiency of the two working modes will decrease correspondingly, the mechanical efficiency will increase accordingly, and the total efficiency will show a highest point in the process of gradual change.

6. Conclusion

(1) When only one side of the motor is operating, the leakage between the distribution tube and the cylinder is the main leakage in the motor. When the left and right motors are working at the same time, the leakage between the plunger and the plunger hole accounts for the largest percentage of the total leakage.

(2) The volumetric efficiency of the motor decreases as the inlet pressure of the motor increases, and the mechanical efficiency and overall efficiency gradually increase.

(3) When the motor inlet pressure is the same, the volumetric efficiency when only one side of the motor is operated is lower than that when the left and right side motors are operated together. The research results in this paper have a strong guiding significance and reference value for the design and use of the force balance axial piston motor.

References

[1] Shi Hu and Mei Xuesong *Some New Features and Development Trend of Modern Hydraulic Drive Technology*. Machine Tools & Hydraulics, 2017, 45(23): 158-166.

[2] Wen Desheng, Shang Xudong, Lv Shijun, Shi Zizhou, Pan Weiyuan and Wang Lei. *Torque Ripple Analysis of Differential Connection of Double Stator Hydraulic Motor*. Engineering Science and Technology, 2017, 49(5): 178-184.

[3] Liu Qiaoyan, Wen Desheng and Gao Junfeng. *Analysis of Double Stator Force Balanced Axial Piston Pump and Its Flow Volatility*. China Mechanical Engineering, 2017, 28(2): 228-232.

[4] Chen Gong. Research on principle of radial double stator piston motor. *Yanshan University*, 2015.

[5] Wang Xingjian, Lin Siru, Wang Shaoping, He Zhaomin and Zhang Chao. *Remaining useful life prediction based on the Wiener process for an aviation axial piston pump*. Chinese Journal of Aeronautics, 2016, 29(3): 779-788.

[6] Xu Rui and Gu Lichen. *Global Coupling Dynamic Modeling of Axial Piston Pumps*. Journal of Agricultural Machinery, 2016, 47(1): 369-376.

IOP Conf. Series: Journal of Physics: Conf. Series **1267** (2019) 012063 doi:10.1088/1742-6596/1267/1/012063

Event-triggered Group Consensus for First-order Multi-agent Systems Based on Directed Topology

Shaolei Zhou, Shuailei Wang*, Shi Yan and Xueyuan Zhao

Naval Aviation University, Yantai 264001, Shandong, PR China

*201011231807@mail.bnu.edu.cn

Abstract. Event-triggered group consensus problem is addressed in this paper. Based on the assumption named the in-degree balance, a directed first-order multi-agent system is constructed, which consists of two subgroups. To analyze the problem, the Laplacian matrix of the initial system is decomposed to be two matrices with less row and column dimension. Then the initial state vector can be substituted by a new one, each element in which is the error of two adjacent elements of the initial state vector. Then a new reduced-order system is constructed and the consensus problem is converted to a stability problem. With the designed Lyapunov function, the validity of the proposed algorithm is proved. The lower-bound of the inter-event time interval is also presented. Simulation results are presented to demonstrate the effectiveness of the protocol.

1. Introduction

Multi-agent systems have been applied broadly to areas such as robots[1, 2], unmanned vehicles[3] and computer science[4], which in return attracts more interests from researchers. In recent years, scholars have put much attention to consensus problem in multi-agent systems, and have done research in more particular aspects like group consensus[5-8] for example.

Consensus means that all the agents in the system converge to a certain state. And talking about group consensus, there exist at least two groups in the system, and agents in the same group converge to a same state, while those belonging to different groups may converge to a different stable one. In early research, Yu and Wang[5] proposed first-order group consensus problem based on undirected topology, and converted the fixed topology into switching topology by double-tree-form transformation[6], which can construct a reduced-order system, and communication delays are also taken into consideration. Feng[7] studied a system consisting of agents with double-integrator dynamics. With a special matrix whose zero eigenvalue's algebraic multiplicity is four, the system can reach time-varying consensus. Sufficiency and necessity analysis is presented. Miao and Ma[8] proposed an algorithm with nonlinear input constraints for continuous-time/discrete-time systems, the validity of which was verified by Lyapunov function and LaSalle's invariance principle.

Since a large number of studies are based on the assumption that the agents can detect and communicate with neighbours all the time, while in reality making the electronic equipment work keep communicating costs much, some scholars put forward an event-triggered strategy[9-11]. With this algorithm, agents can update their control inputs at discrete time instants, which can dramatically decrease the cost. Studies utilizing this strategy are still few. Tu et al[12] did research on leader-following multi-agent systems and put their attention on fixed and switching topologies with event-triggering mechanism. Ma et al[13] designed even-triggered functions for centralized and

IOP Conf. Series: Journal of Physics: Conf. Series **1267** (2019) 012063 doi:10.1088/1742-6596/1267/1/012063

decentralized conditions, and analysed the distributed event-triggered problem with a triggering function depending on the number of each agent's neighbours[14]. Yan et al[15] studied the switching topology with disturbance. These above studies are all based on undirected topology.

Inspired by the existing studies, this paper further investigate the event-triggered group consensus for first-order multi-agent systems based on directed topology. The system consists two subgroups, and a centralized event-triggered function is designed. The directed Laplacian matrix of the system is decomposed for the convenience of analysis. A Lyapunov function is constructed and a sufficient condition is presented. Simulation results verify the validity of the proposed protocol.

The rest of the paper is as follows. Section 2 presents some preliminaries and introduced some lemmas. Section 3 gives the theoretical result and section 4 gives the simulation counterpart.

2. Problem Formulation

A multi-agent system can be described by a directed or undirected graph $G=(\mathcal{V}, \mathcal{E}, A)$. Considering n agents in the system, there is a set of nodes $\mathcal{V}=\{v_1, v_2, \cdots, v_n\}$ consisting of n elements, and a corresponding set of edges $\mathcal{E} \subseteq \mathcal{V} \times \mathcal{V}$. An edge of the graph G is described as $e_{ij}=(v_i, v_j)$. An undirected graph is connected if there is a path between any two distinct nodes, and if any two nodes can be linked by a directed path, the directed graph can also be called connected. In a directed graph, if there is only one node that has no parent and all the other nodes has exactly one parent, and the special node can connect any other nodes by the paths, then it is called the root. If the topology of the system is undirected, a symmetrical adjacency matrix $\boldsymbol{A}=[a_{ij}]$ is an n order square matrix. And if the topology is directed, normally the adjacency matrix $\boldsymbol{A}$ is asymmetrical. The Laplacian matrix $\boldsymbol{L}=[l_{ij}]$ can be derived from $\boldsymbol{A}$ by the following way: $l_{ij}=-a_{ij}$ if $i \neq j$, and $l_{ii}=\sum_{j=1, j\neq i}^{n} a_{ij}$. It is obvious that $\boldsymbol{L}$ has a right eigenvector $[1,1,\cdots,1]_{1\times n}^{\mathrm{T}}$ and 0 is the corresponding eigenvalue. The set of neighbours of agent i is written as $N_i=\{v_j \in \mathcal{V} | (v_i, v_j) \in \mathcal{E}\}$.

To analyse the group consensus problem, suppose that there are $n+m$ agents in the system, firstly the agents may be listed in a rank by their index $\mathrm{I}=\{1,2,\cdots,n+m\}$. The former n agents belong to the first group called $\mathcal{I}_1$, and the latter m agents belong to the second group called $\mathcal{I}_2$. Then the adjacency matrix $\boldsymbol{A}$ is a block matrix $\boldsymbol{A}=\begin{bmatrix} \boldsymbol{A}_{11} & \boldsymbol{A}_{21} \\ \boldsymbol{A}_{12} & \boldsymbol{A}_{22} \end{bmatrix}_{(n+m)\times(n+m)}$ where $\boldsymbol{A}_{11}$ ($\boldsymbol{A}_{22}$) is the inner adjacency matrix in $\mathcal{I}_1$ ($\mathcal{I}_2$), and $\boldsymbol{A}_{12}$ ($\boldsymbol{A}_{21}$) stands for the links from $\mathcal{I}_2$ ($\mathcal{I}_1$) to $\mathcal{I}_1$ ($\mathcal{I}_2$).

Suppose that all the agents in the system follow the following dynamics

$$\dot{\boldsymbol{x}}_i=\boldsymbol{u}_i, \forall i \in \mathrm{I} \tag{1}$$

in which $\boldsymbol{x}_i, \boldsymbol{u}_i \in \mathbb{R}$.

Lemma 1. The multi-agent system is said to reach group consensus[6] if the states of agents satisfy

$$\begin{cases} \lim\limits_{t\to\infty} \left\| \boldsymbol{x}_i - \boldsymbol{x}_j \right\| = 0, \forall i, j \in \mathcal{I}_1 \\ \lim\limits_{t\to\infty} \left\| \boldsymbol{x}_i - \boldsymbol{x}_j \right\| = 0, \forall i, j \in \mathcal{I}_2 \end{cases} \tag{2}$$

Lemma 2. For a graph Laplacian matrix $\boldsymbol{L} \in \mathbb{R}^{N\times N}$ and a full row rank matrix $\boldsymbol{E} \in \mathbb{R}^{(N-1)\times N}$ defined as follows

$$\boldsymbol{E}=\begin{bmatrix}1 & -1 & 0 & \cdots & 0\\ 0 & 1 & -1 & \ddots & \vdots\\ \vdots & \ddots & \ddots & \ddots & 0\\ 0 & \cdots & 0 & 1 & -1\end{bmatrix} \tag{3}$$

There exists a matrix $\boldsymbol{M}\in\mathbb{R}^{N\times(N-1)}$ such that $\boldsymbol{L}=\boldsymbol{ME}$ [16].

Lemma 3(Cauchy-Schwarz Inequation).

$$\sum_{k=1}^{n}a_k b_k \le \left(\sum_{k=1}^{n}a_k^2\right)^{1/2}\left(\sum_{k=1}^{n}b_k^2\right)^{1/2} \tag{4}$$

Lemma 4. The Frobenius-norm of a matrix is compatible to the 2-norm of a vector, that is

$$\|\boldsymbol{Ax}\|_2 \le \|\boldsymbol{A}\|_F\,\|\boldsymbol{x}\|_2 \tag{5}$$

Where $\|\cdot\|_2$ is the 2-norm and $\|\cdot\|_F$ is the Frobenius-norm[17].

Assumption 1. The system is in-degree-balanced[5] if

$$\begin{cases}\sum_{j=n+1}^{n+m}a_{ij}=0, i\in\mathcal{I}_1\\ \sum_{j=1}^{n}a_{ij}=0, i\in\mathcal{I}_2\end{cases} \tag{6}$$

To solve the event-triggered group consensus problem, a new protocol is proposed

$$\begin{cases}\boldsymbol{u}_i=\sum_{j=1}^{n}a_{ij}\left[\boldsymbol{x}_j(t_k)-\boldsymbol{x}_i(t_k)\right]+\sum_{j=n+1}^{n+m}a_{ij}\boldsymbol{x}_j(t_k), i\in\mathcal{I}_1\\ \boldsymbol{u}_i=\sum_{j=n+1}^{n+m}a_{ij}\left[\boldsymbol{x}_j(t_k)-\boldsymbol{x}_i(t_k)\right]+\sum_{j=1}^{n}a_{ij}\boldsymbol{x}_j(t_k), i\in\mathcal{I}_2\end{cases} \tag{7}$$

Where $a_{ij}\ge 0$ is the element in $\boldsymbol{A}$, and t_k is the last event time instant when the control input $\boldsymbol{u}_i$ updates, and $\boldsymbol{x}_i(t_k)$ is the reference state of agent i.

For each agent, the state measure error vector[11] is defined as

$$\boldsymbol{e}_i=\boldsymbol{x}_i(t_k)-\boldsymbol{x}_i, t\in[t_k,t_{k+1}) \tag{8}$$

List all the variables in vector $\boldsymbol{x}=[\boldsymbol{x}_1,\boldsymbol{x}_2,\cdots,\boldsymbol{x}_{n+m}]^{\mathrm{T}}$, $\boldsymbol{x}(t_k)=[\boldsymbol{x}_1(t_k),\boldsymbol{x}_2(t_k),\cdots,\boldsymbol{x}_{n+m}(t_k)]^{\mathrm{T}}$, and $\boldsymbol{e}=[\boldsymbol{e}_1,\boldsymbol{e}_2,\cdots,\boldsymbol{e}_{n+m}]^{\mathrm{T}}$, then $\boldsymbol{e}=\boldsymbol{x}(t_k)-\boldsymbol{x}$.

Based on Lemma 2, the Laplacian matrix $\boldsymbol{L}$ can be decomposed

$$\boldsymbol{L}=\hat{\boldsymbol{M}}\hat{\boldsymbol{E}} \tag{9}$$

in which $\hat{\boldsymbol{E}}=\begin{bmatrix}\boldsymbol{E}^{\mathcal{I}_1}_{(n-1)\times n} & \boldsymbol{0}\\ \boldsymbol{0} & \boldsymbol{E}^{\mathcal{I}_2}_{(m-1)\times m}\end{bmatrix}$. And use another variable $\tilde{\boldsymbol{x}}_i$

$$\begin{cases}\tilde{\boldsymbol{x}}_i=\boldsymbol{x}_i-\boldsymbol{x}_{i+1}, i=1,2,\cdots,n-1\\ \tilde{\boldsymbol{x}}_i=\boldsymbol{x}_i-\boldsymbol{x}_{i+1}, i=n+1,n+2,\cdots,n+m-1\end{cases} \tag{10}$$

to substitute $\boldsymbol{x}_i$, then an $n+m-2$ dimensional vector $\tilde{\boldsymbol{x}}=[\tilde{\boldsymbol{x}}_1,\tilde{\boldsymbol{x}}_2,\cdots,\tilde{\boldsymbol{x}}_{n-1},\tilde{\boldsymbol{x}}_{n+1},\tilde{\boldsymbol{x}}_{n+2},\cdots,\tilde{\boldsymbol{x}}_{n+m-1}]^{\mathrm{T}}$ is obtained. In fact, by Lemma 2 and equation(6), it is obvious that

$$\tilde{\boldsymbol{x}}=\hat{\boldsymbol{E}}\boldsymbol{x} \tag{11}$$

Similarly, we get $\tilde{\boldsymbol{e}}=\hat{\boldsymbol{E}}\boldsymbol{x}$ and $\tilde{\boldsymbol{x}}(t_k)=\hat{\boldsymbol{E}}\boldsymbol{x}(t_k)$.

Based on measurement error, a triggering function $f(\boldsymbol{x},\boldsymbol{e})$ would be defined in the next section.

3. Theoretical results

In this section, on the foundation of graph theories and matrix theories, and referring to the analysis in [11], group consensus of multi-agent system based on directed topology will be considered.

Given the system dynamics (1) and control protocol (7), the system will reach group consensus if it is triggered when

$$\|\tilde{\boldsymbol{e}}\|_2=\alpha\frac{\|\hat{\boldsymbol{E}}\hat{\boldsymbol{M}}\tilde{\boldsymbol{x}}\|_2}{\|\hat{\boldsymbol{E}}\hat{\boldsymbol{M}}\|_F} \tag{12}$$

where $0<\alpha<1$.

Proof: Under assumption 1, it is obvious that

$$\sum_{j=n+1}^{n+m} a_{ij}\boldsymbol{x}_i=0 \tag{13}$$

Then equation (4) can be rewritten as

$$\begin{cases} \boldsymbol{u}_i=\sum_{j=1}^{n+m} a_{ij}\left[\boldsymbol{x}_j(t_k)-\boldsymbol{x}_i(t_k)\right], i\in\mathcal{I}_1 \\ \boldsymbol{u}_i=\sum_{j=1}^{n+m} a_{ij}\left[\boldsymbol{x}_j(t_k)-\boldsymbol{x}_i(t_k)\right], i\in\mathcal{I}_2 \end{cases} \tag{14}$$

which means

$$\dot{\boldsymbol{x}}_i=\boldsymbol{u}_i=-\boldsymbol{L}\boldsymbol{x}(t_k) \tag{15}$$

And by equation(8), it can obtain that

$$\dot{\tilde{\boldsymbol{x}}}=-\hat{\boldsymbol{E}}\hat{\boldsymbol{M}}\tilde{\boldsymbol{x}}(t_k) \tag{16}$$

A candidate Lyapunov function can be written as

$$V=\frac{1}{2}\tilde{\boldsymbol{x}}^{\mathrm{T}}\boldsymbol{M}^{\mathrm{T}}\boldsymbol{E}^{\mathrm{T}}\tilde{\boldsymbol{x}} \tag{17}$$

Then the derivative of V is

$$\begin{aligned} \dot{V}&=\tilde{\boldsymbol{x}}^{\mathrm{T}}\hat{\boldsymbol{M}}^{\mathrm{T}}\hat{\boldsymbol{E}}^{\mathrm{T}}\dot{\tilde{\boldsymbol{x}}} \\ &=-\tilde{\boldsymbol{x}}^{\mathrm{T}}\hat{\boldsymbol{M}}^{\mathrm{T}}\hat{\boldsymbol{E}}^{\mathrm{T}}\hat{\boldsymbol{E}}\hat{\boldsymbol{M}}\tilde{\boldsymbol{x}}(t_k) \\ &=-\tilde{\boldsymbol{x}}^{\mathrm{T}}\hat{\boldsymbol{M}}^{\mathrm{T}}\hat{\boldsymbol{E}}^{\mathrm{T}}\hat{\boldsymbol{E}}\hat{\boldsymbol{M}}(\tilde{\boldsymbol{x}}+\tilde{\boldsymbol{e}}) \\ &=-\tilde{\boldsymbol{x}}^{\mathrm{T}}\hat{\boldsymbol{M}}^{\mathrm{T}}\hat{\boldsymbol{E}}^{\mathrm{T}}\hat{\boldsymbol{E}}\hat{\boldsymbol{M}}\tilde{\boldsymbol{x}}-\tilde{\boldsymbol{x}}^{\mathrm{T}}\hat{\boldsymbol{M}}^{\mathrm{T}}\hat{\boldsymbol{E}}^{\mathrm{T}}\hat{\boldsymbol{E}}\hat{\boldsymbol{M}}\tilde{\boldsymbol{e}} \end{aligned} \tag{18}$$

Since $\tilde{\boldsymbol{x}}^{\mathrm{T}}\hat{\boldsymbol{M}}^{\mathrm{T}}\hat{\boldsymbol{E}}^{\mathrm{T}}\hat{\boldsymbol{E}}\hat{\boldsymbol{M}}\tilde{\boldsymbol{x}}=\|\hat{\boldsymbol{E}}\hat{\boldsymbol{M}}\tilde{\boldsymbol{x}}\|_2^2$, and based on Lemma 3, $\tilde{\boldsymbol{x}}^{\mathrm{T}}\hat{\boldsymbol{M}}^{\mathrm{T}}\hat{\boldsymbol{E}}^{\mathrm{T}}\hat{\boldsymbol{E}}\hat{\boldsymbol{M}}\tilde{\boldsymbol{e}}\le\|\hat{\boldsymbol{E}}\hat{\boldsymbol{M}}\tilde{\boldsymbol{x}}\|_2\|\hat{\boldsymbol{E}}\hat{\boldsymbol{M}}\tilde{\boldsymbol{e}}\|_2$, so that

$$\dot{V} \le -\left\|\hat{E}\hat{M}\tilde{x}\right\|_2^2 + \left\|\hat{E}\hat{M}\tilde{x}\right\|_2 \left\|\hat{E}\hat{M}\right\|_F \left\|\tilde{e}\right\|_2 \tag{19}$$

So that if condition (12) is satisfied and $\left\|\hat{E}\hat{M}\tilde{x}\right\|_2 \ne 0$, it can obtain that

$$\dot{V} \le (\alpha - 1)\left\|\hat{E}\hat{M}\tilde{x}\right\|_2^2 \tag{20}$$

Because $0 < \alpha < 1$, it is obvious that $\dot{V}$ is negative.

Given the system dynamics (1), the control protocol (7) and the error function (12), the inter-event time interval $t_{k+1} - t_k$ has a lower bound τ which is strictly positive

$$\tau = \frac{\alpha}{(1+\alpha)\left\|\hat{E}\hat{M}\right\|_F} \tag{21}$$

Proof: Based on equation (12), it can obtain that the time derivate of $\frac{\left\|\tilde{e}\right\|_2}{\left\|\hat{E}\hat{M}\tilde{x}\right\|_2}$ is

$$\begin{aligned}
&\frac{\mathrm{d}}{\mathrm{d}t}\frac{\left\|\tilde{e}\right\|_2}{\left\|\hat{E}\hat{M}\tilde{x}\right\|_2} \\
&= \frac{\tilde{e}^{\mathrm{T}}\dot{\tilde{e}}}{\left\|\tilde{e}\right\|_2\left\|\hat{E}\hat{M}\tilde{x}\right\|_2} - \frac{\left\|\tilde{e}\right\|_2\left(\hat{E}\hat{M}\tilde{x}\right)^{\mathrm{T}}\left(\hat{E}\hat{M}\dot{\tilde{x}}\right)}{\left\|\hat{E}\hat{M}\tilde{x}\right\|_2^3} \\
&\le \frac{\left\|\tilde{e}\right\|_2\left\|\dot{\tilde{e}}\right\|_2}{\left\|\tilde{e}\right\|_2\left\|\hat{E}\hat{M}\tilde{x}\right\|_2} + \frac{\left\|\tilde{e}\right\|_2\left\|\hat{E}\hat{M}\tilde{x}\right\|_2\left\|\hat{E}\hat{M}\dot{\tilde{x}}\right\|_2}{\left\|\hat{E}\hat{M}\tilde{x}\right\|_2^3} \\
&= \frac{\left\|\dot{\tilde{e}}\right\|_2}{\left\|\hat{E}\hat{M}\tilde{x}\right\|_2} + \frac{\left\|\tilde{e}\right\|_2\left\|\hat{E}\hat{M}\right\|_F\left\|\dot{\tilde{x}}\right\|_2}{\left\|\hat{E}\hat{M}\tilde{x}\right\|_2^2} \\
&\le \left(1 + \frac{\left\|\tilde{e}\right\|_2\left\|\hat{E}\hat{M}\right\|_F}{\left\|\hat{E}\hat{M}\tilde{x}\right\|_2}\right)^2
\end{aligned} \tag{22}$$

Let

$$y(t, y_0) = \frac{\left\|\tilde{e}\right\|_2}{\left\|\hat{E}\hat{M}\tilde{x}\right\|_2} \tag{23}$$

It is obvious that $\dot{y} > 0$, $\dot{y} \le \left(1 + \left\|\hat{E}\hat{M}\right\|_F y\right)^2$, and $y(t, y_0) \le \bar{y}(t, y_0)$ where $\dot{\bar{y}} = \left(1 + \left\|\hat{E}\hat{M}\right\|_F \bar{y}\right)^2$. Since

IOP Conf. Series: Journal of Physics: Conf. Series **1267** (2019) 012063 doi:10.1088/1742-6596/1267/1/012063

$$\bar{y} = \frac{\left\|\hat{E}\hat{M}\right\|_F t + 1 - c}{\left\|\hat{E}\hat{M}\right\|_F c - \left\|\hat{E}\hat{M}\right\|_F^2 t} \tag{24}$$

in which c is a constant. Because $y_0 = 0$, it can obtain that $c = 1$, and $\bar{y} = t\left(1 - \left\|\hat{E}\hat{M}\right\|_F t\right)^{-1}$.

Based on

$$\bar{y}(\tau, 0) = \frac{\tau}{1 - \left\|\hat{E}\hat{M}\right\|_F \tau} = \frac{\alpha}{\left\|\hat{E}\hat{M}\right\|_F} \tag{25}$$

we know that

$$\tau = \frac{\alpha}{(1+\alpha)\left\|\hat{E}\hat{M}\right\|_F} \tag{26}$$

and the proof is complete.

4. Simulation results

The effectiveness of the previous theoretical results will be justified by computer simulation results.

Consider a multi-agent system of $3+4$ agents, meaning that the first group contains 3 agents and the second one contains 4. And the Laplacian matrix of the system is as follows.

$$\boldsymbol{L} = \begin{bmatrix} 0 & 0 & 0 & -1 & 0 & 1 & 0 \\ -1 & 1 & 0 & 0 & 0 & -1 & 1 \\ 0 & -1 & 1 & -1 & 1 & 0 & 0 \\ 1 & 0 & -1 & 0 & 0 & 0 & 0 \\ 0 & 0 & 0 & -1 & 1 & 0 & 0 \\ -1 & 1 & 0 & 0 & -1 & 1 & 0 \\ 0 & 0 & 0 & 0 & 0 & -1 & 1 \end{bmatrix} \tag{27}$$

And the corresponding $\hat{\boldsymbol{M}}$ and $\hat{\boldsymbol{E}}$ are

$$\hat{\boldsymbol{M}} = \begin{bmatrix} 0 & 0 & 0 & -1 & -1 & 0 \\ -1 & 0 & 0 & 0 & 0 & -1 \\ 0 & -1 & 0 & -1 & 0 & 0 \\ 1 & 1 & 0 & 0 & 0 & 0 \\ 0 & 0 & 0 & -1 & 0 & 0 \\ -1 & 0 & 0 & 0 & -1 & 0 \\ -1 & -1 & 0 & 0 & 0 & -1 \end{bmatrix}, \hat{\boldsymbol{E}} = \begin{bmatrix} 1 & -1 & 0 & 0 & 0 & 0 & 0 \\ 0 & 1 & -1 & 0 & 0 & 0 & 0 \\ 0 & 0 & 1 & -1 & 0 & 0 & 0 \\ 0 & 0 & 0 & 1 & -1 & 0 & 0 \\ 0 & 0 & 0 & 0 & 1 & -1 & 0 \\ 0 & 0 & 0 & 0 & 0 & 1 & -1 \end{bmatrix} \tag{28}$$

The initial state vector of the agents is $x(0) = [17.6, 4.6, 82.4, 4.35, 60.56, 32.415, 49.6]^{\mathrm{T}}$, and the parameter $\alpha = 0.1$. Figure 1 shows the changing process of the agents' state. It is obvious that the former 3agents converge to one same state and the latter 4 agents converge to another different one.

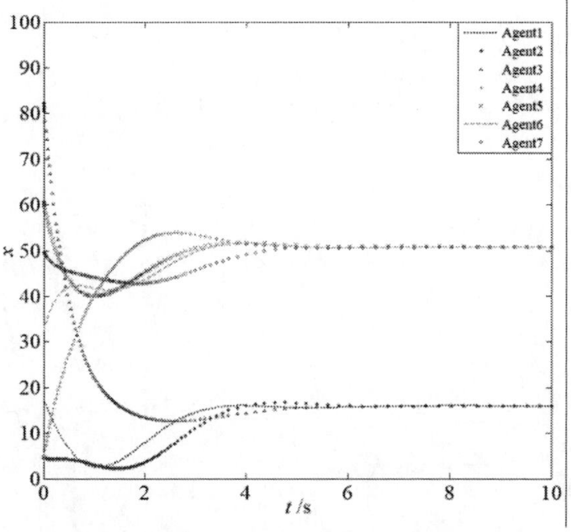

Figure 1. The state converging process.

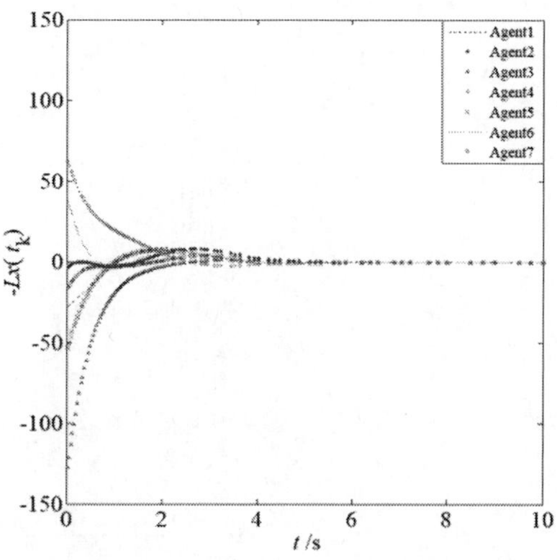

Figure 2. The control inputs of agents.

Figure 2 shows the control inputs $-\boldsymbol{Lx}(t_k)$ of the agents. Since it is assumed that the agents have great ability so that they can change their state at any rate, the control inputs at the beginning can be very large. But in practical applications the ability of agents is limited, so the future studies should concentrate on event-triggered consensus of limited-ability multi-agent systems.

The event time series is presented in figure 3. At the beginning, since there exists a great difference among the states of agents, the system triggers quite frequently, which results into the fact that in figure 3, the former time intervals between triggering time instants is very short. After a converging process the states of agents don't vary that much, so the intervals gradually increase to a fairly large value.

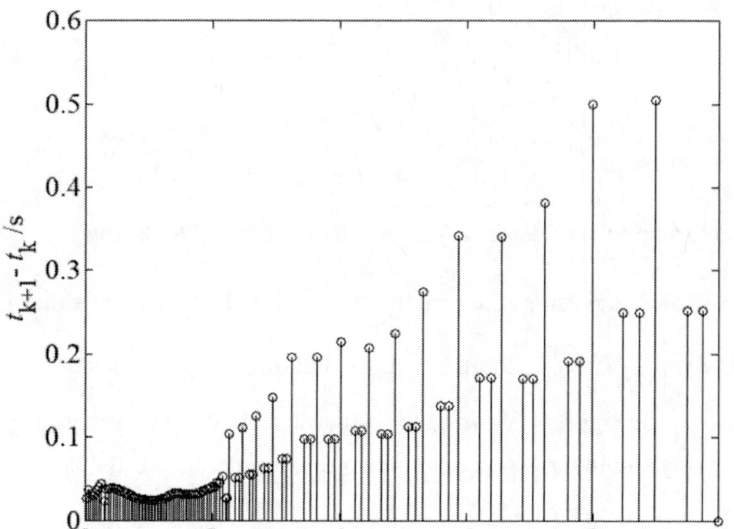

Figure 3. Figure with short caption (caption centred).

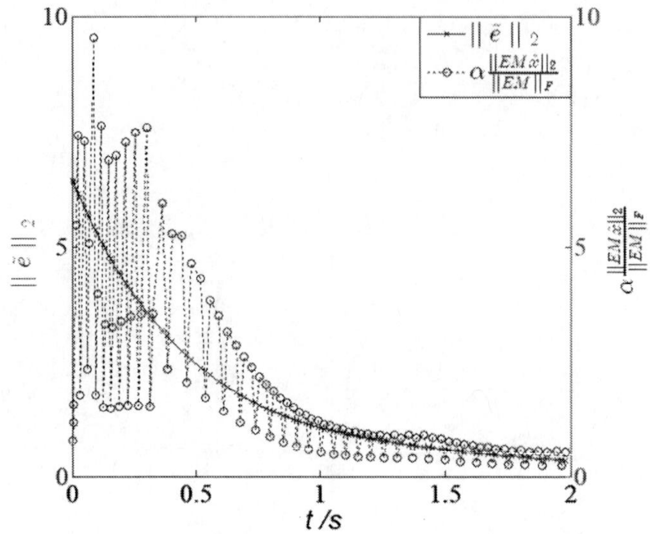

Figure 4. Norm of error vector and the reference value during the former 2 seconds.

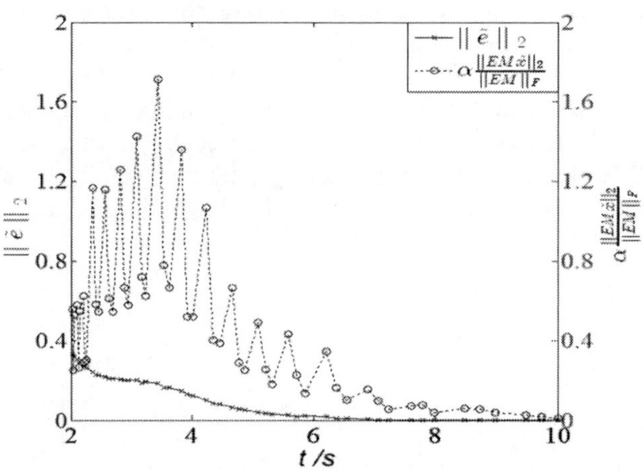

Figure 5. Norm of error vector and the reference state value during the latter 8 seconds.

Figure 4 and 5 show the norm of the error vector $\tilde{e}$ and the reference value $\alpha\left\|\hat{E}\hat{M}\hat{x}\right\|_2/\left\|\hat{E}\hat{M}\right\|_F$. When $\|\tilde{e}\|_2 > \alpha\left\|\hat{E}\hat{M}\hat{x}\right\|_2/\left\|\hat{E}\hat{M}\right\|_F$, the triggering condition is satisfied, then the agents update their reference state $x_i(t_k)$. In figure 4, during the former 2 seconds, the system triggers more frequently than in figure 5 during the latter 8 seconds, which is consistent with the conclusion we obtain from figure 3.

5. Conclusions

The event-triggered group consensus of multi-agent system is considered in this article, which is based on a first-order directed topology. The agents are separated into 2 groups, and all the agents have a same triggering algorithm. Theoretical analysis is supported by simulation results, and the validity of the proposed protocol is verified. Studies in the future will concentrate on the limited-ability multi-agent system and the application of the proposed protocol in it.

IOP Conf. Series: Journal of Physics: Conf. Series **1267** (2019) 012063 doi:10.1088/1742-6596/1267/1/012063

References

[1] S. J. Trujillo, V. L. Ruiz, N. Esparza, J. A. Riley, K. C. Chu, and W. Bokhari, "Solar tracking system employing multiple mobile robots," ed: Google Patents, 2016.

[2] D. Panagou, D. Stipanovic, and P. Voulgaris, "Distributed Coordination Control for Multi-Robot Networks Using Lyapunov-Like Barrier Functions," IEEE Transactions on Automatic Control, vol. 61, no. 3, pp. 617-632, 2016.

[3] Z. Obdrzalek, "Mobile agents in multi-agent UAV/UGV system," in International Conference on Military Technologies, 2017.

[4] A. Shamsoshoara, M. Khaledi, F. Afghah, A. Razi, and J. Ashdown, "Distributed Cooperative Spectrum Sharing in UAV Networks Using Multi-Agent Reinforcement Learning," arXiv preprint arXiv:1811.05053, 2018.

[5] J. Yu and L. Wang, "Group consensus of multi-agent systems with undirected communication graphs," in Asian Control Conference, 2009. ASCC 2009. 7th. IEEE, 2009, pp. 105-110.

[6] J. Yu and L. Wang, "Group consensus in multi-agent systems with switching topologies and communication delays," Systems & Control Letters, vol. 59, no. 6, pp. 340-348, 2010.

[7] Y. Feng, S. Xu, and B. Zhang, "Group consensus control for double - integrator dynamic multiagent systems with fixed communication topology," International Journal of Robust and Nonlinear Control, vol. 24, no. 3, pp. 532-547, 2014.

[8] G. Miao and Q. Ma, "Group consensus of the first-order multi-agent systems with nonlinear input constraints," Neurocomputing, vol. 161, pp. 113-119, 2015.

[9] Åarzén and Karl-Erik, "A simple event-based PID controller," IFAC Proceedings Volumes, vol. 32, no. 2, pp. 8687-8692, 1999.

[10] D. V. Dimarogonas and K. H. Johansson, "Event-triggered cooperative control," in Control Conference (ECC), 2009 European. IEEE, 2009, pp. 3015-3020.

[11] D. V. Dimarogonas and K. H. Johansson, "Event-triggered control for multi-agent systems," 2009.

[12] Z. Tu, D. Zhang, X. Xia, and H. Yu, "Event-triggered group consensus of leader-following multi-agent systems with nonlinear dynamics," in Control Conference, 2016, pp. 7885-7890.

[13] H. Ma, D. Liu, W. Ding, F. Tan, ánd L. Chao, "Centralized and decentralized event-triggered control for group consensus with fixed topology in continuous time," Neurocomputing, vol. 161, no. C, pp. 267-276, 2015.

[14] H. Ma, W. Ding, D. Liu, and L. Chao, "Continuous-time group consensus using distributed event-triggered control," in Seventh International Conference on Advanced Computational Intelligence, 2015.

[15] C. Yan, Y. Mei, and L. Cheng, "Event-triggered tracking control for couple-group multi-agent systems," Journal of the Franklin Institute, vol. 354, no. 14, p. S0016003217303307, 2017.

[16] S. Zhou, L. Wei, Q. Wu, and G. Yin, "Leaderless Consensus of Linear Multi-agent Systems: Matrix Decomposition Approach," in International Conference on Intelligent Human-machine Systems & Cybernetics, 2015.

[17] B. Shi and M. Gai, Introduction to Matrix Analysis with Applications. Beijing: National Defence Industry Press, 2010.

IOP Conf. Series: Journal of Physics: Conf. Series **1267** (2019) 012064 doi:10.1088/1742-6596/1267/1/012064

ABB Robotic Arm Offline Programming System

Zhe Xu[1, a] and Yuanjiang Liu[1, b]

[1]Beijing University of Technology, Beijing, China

[a]xuzhe@bjut.edu.cn; [b]411914912@qq.com

Abstract: The offline programming technology of industrial robots is more and more widely used because it can effectively improve work efficiency. A DXF drawing file is a file generated by CAD that can serve as a bridge for data exchange with other software. Through experiments, this paper combines CAD-generated DXF drawing files with ABB robotic arm to realize data extraction of CAD drawings, and automatic generation and storage of robotic arm running programs, which are imported into the robot arm, so that the robot arm can be made according to CAD. The trajectory on the drawing performs the welding function. The simulation of path planning was carried out under the Robot studio software developed by ABB Robotics to meet the experimental requirements. The realization of this off-line programming system achieves the following two purposes. One is to assist non-control professional teaching, to help understand the simulation and actual production and processing of industrial robot arm; the second is to improve the efficiency of industrial robot arm programming and realize full automation programming.

1. Introduction

With the development of economic globalization and the intensification of market competition, the demand for industrial robot technology and automation equipment in China's manufacturing industry has grown rapidly. Among them, the application of industrial robots in welding is the most common[1]. The laboratory has a multi-disciplinary robotic arm project that combines materials, electromichanical, environmental energy and control. The industrial robot (ABB robot arm) is used as a tool to weld or laser weld some metal materials.

With the increase of robot control speed and precision, the application range of robotic arc welding is becoming more and more extensive[2]. At present, most of the domestic welding robots are working in the teaching and reproduction mode. The main technical problems in the actual production and application of the teaching and reproducing robots are [3]: (1) The online teaching programming process of the robot is cumbersome and inefficient. . (2) The accuracy of teaching is determined entirely by the experience of the teacher, and it is difficult to achieve a satisfactory teaching effect for complex paths. (3) For some applications that need to make real-time decisions based on external information, they cannot be made at any time. The teaching reproduction requires the production to stop the online programming of the welding robot. The contradiction between this production method and programming will be more and more, and the effective way to solve the contradiction is to use offline programming. Therefore, the research and application of arc welding robot offline programming technology [4] is an inevitable trend to improve the level of welding automation.

2. DXF file

The system takes CAD drawings as input and processes it to obtain the executable file of the robot arm, which is then imported into the robot arm, so that the robot arm works according to the trajectory on the CAD drawing. In CAD, there are three formats for file saving: DWG, DWS, DWT, DXF.

DXF (Drawing Exchange Format) is a CAD data file format developed by Auto-desk to exchange CAD data between Auto-CAD and other software [5]. DXF is an open vector data format that can be divided into two categories: ASCII format and binary format; ASCII has good readability, but it occupies a large space; binary format occupies less space and reads faster. Since Auto-CAD is now the most popular CAD system, DXF is also widely used as a de facto standard. Most CAD systems can read or output DXF files [6]. Therefore, research and analysis of DXF file structure has important significance.

2.1. Introduction of the DXF file

The complete DXF file consists of six segments and an end marker, each segment starting with a group code of 0 and a corresponding group value SECTION, followed by a group code and a group value representing the segment name, each piece of content being a group of elements The code and the group value are composed. The corresponding group code 0 of the group value ENDSEC indicates the end of the segment, and the entire DXF file end flag is marked with the group code 0 and the group value EOF[7]. The names, uses and contents of the six segments are shown in Figure 1.

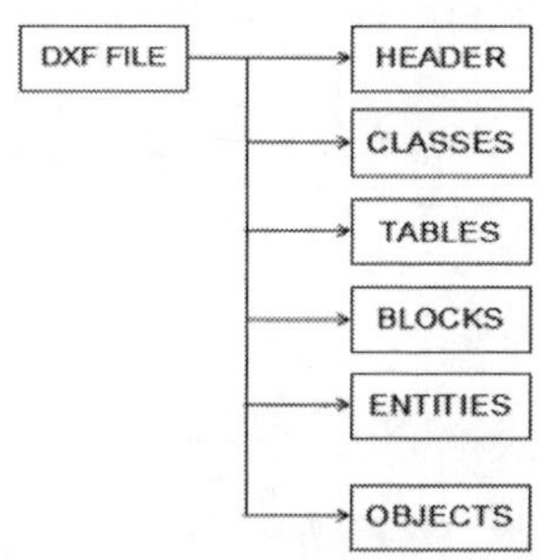

Figure 1. DXF file structure

1) HEADER SECTION:with more than six authors:This section contains basic information about the graph and records the settings of the various variables associated with the graph. It consists of a database version number and a number of system variables, each of which includes a variable name and its group value.

2) CLASSES SECTION:This section contains information about the application-defined classes, and instances of these classes are contained in the database for segments, segments, and segments. Class definitions are fixed in the hierarchy of the class

3) TABLES SECTION:This section contains definitions for the following symbol tables: application identifier, block reference table, dimension style sheet, layer table, line style table, text style sheet, user coordinate system table, view table, and viewport configuration table.

4) BLOCKS SECTION:This section contains the block definition and the graphic entity information that makes up each block reference in the drawing.

5) ENTITIES SECTION:This section contains entity (primitive) information that makes up the graph, including block references.

6) OBJEECTS SECTION:This section contains non-graphic objects in the graph, and all those non-symbolic, symbolic, and non-primitive objects are stored in this area.

2.2. Processing of DXF files

Set the position of the target data according to the key field in the DXF file, ENTITY is the entity field, all the track data is stored in the entity, and the C language technology can be used to extract the data value of straight lines, polylines, arcs, splines, etc..

```
CFileDialog opendlg (TRUE,_T("*"),_T("*.dxf"),OFN_OVERWRITEPROMPT, _T("All
      files(*.*;)|*.*||"),NULL);
            if (opendlg.DoModal()==IDOK)
            { CString filename=opendlg.GetPathName();
               SetDlgItemText(IDC_EDIT_path,filename);}
            CString path;
            GetDlgItemText(IDC_EDIT_path,path);
            stream = fopen(path, "r");
            if(!stream)
              {Temp.Format(_T(" Failed to open file"));}
            else
              {Temp.Format(_T("Open successfully"));}
```

2.3. Extract the line coordinates in the DXF file

```
fscanf(stream,"%d",&code);
fscanf(stream,"%s",codevalue);
if(code == 0 && strcmp(codevalue,"LINE")==0)
{ entity1 = (EntityData *)malloc(DATASIZE);
  strcpy(entity1->id,codevalue);
  fscanf(stream,"%d",&code);
  while(code)
   {//Save coordinate values}
}
```

2.4. Extract the polyline coordinates in the DXF file

```
fscanf(stream,"%d",&code);
fscanf(stream,"%s",codevalue);
if(code == 0 && strcmp(codevalue,"POLYLINE")==0)
{entity1  = (EntityData *) malloc(DATASIZE);
 strcpy(entity1->id,codevalue);
 fscanf(stream,"%d",&code);
 while(code)
   {//Save coordinate values}
}
```

2.5. Extract the arc coordinates in the DXF file

```
fscanf(stream,"%d",&code);
fscanf(stream,"%s",codevalue);
if(code ==0 && strcmp(codevalue,"ARC")==0)
{entity1 = (EntityData *)malloc(DATASIZE);
 strcpy(entity1->id,codevalue);
 fscanf(stream,"%d",&code);
 while(code)
   {//Save coordinate values}
}
```

2.6. Extract the B-spline curve in the DXF file

```
fscanf( stream,"%d", &code );
fscanf( stream,"%s", codevalue );
if( code ==0&&strcmp(codevalue,"SPLINE") == 0 )
{entity1 = (EntityData *)malloc(DATASIZE);
  strcpy(entity1->id,codevalue);
  fscanf(stream,"%d",&code);
  while(code)
    { //Save coordinate values }
}
```

3. Implementation of offline programming system

The robot offline programming system is based on the MFC framework in Visual Studio. MFC is the abbreviation of Microsoft Foundation Class Library. It is a C++ class library implemented by Microsoft Corporation. It mainly encapsulates most of the Windows API functions and contains an application. Framework to reduce the workload of application developers.

This off-line programming system acts as an intermediate bridge connecting the work path to the ABB arm of the work tool, as shown in Figure 2.

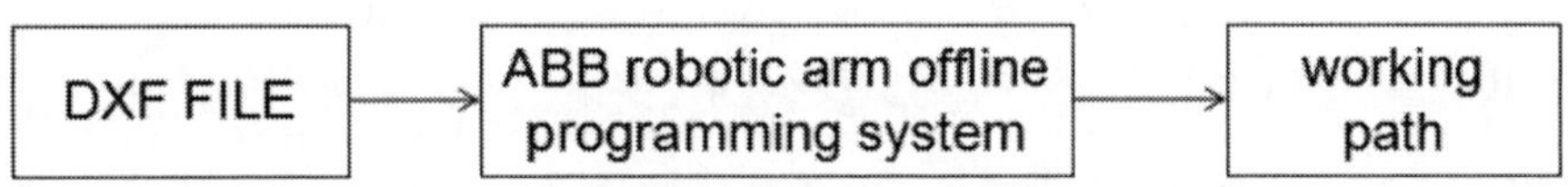

Figure 2. System structure

Figure 3 shows the flow chart of the entire system.

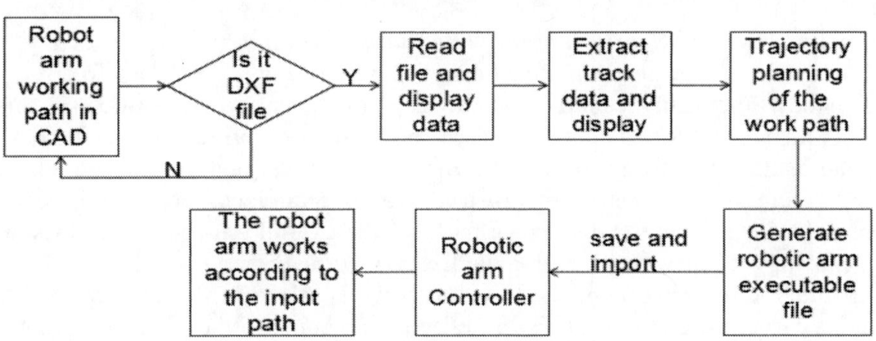

Figure 3. System flow chart

3.1. Conversion algorithm of arc coordinate data

The arc information is stored in ENTITIES, and after the program scans to "ARC", all useful information related to the arc is saved. In the DXF file, the position information of the arc is saved in the file in the format specified by the CAD software. The DXF file records the coordinates of the center point of the arc, the starting arc of the arc, and the arc of the end point of the arc. The parameters that need to be passed in the MOVC command that the robot arm can recognize are the coordinates of the three points on the arc. Value, these three points include the starting point coordinates of the arc, the midpoint coordinates of the arc, and the end point coordinates of the arc. So in this process, the geometry of the arc is involved. The geometry corresponding to the starting point, end point, center and radius of the arc is shown in Figure 4.

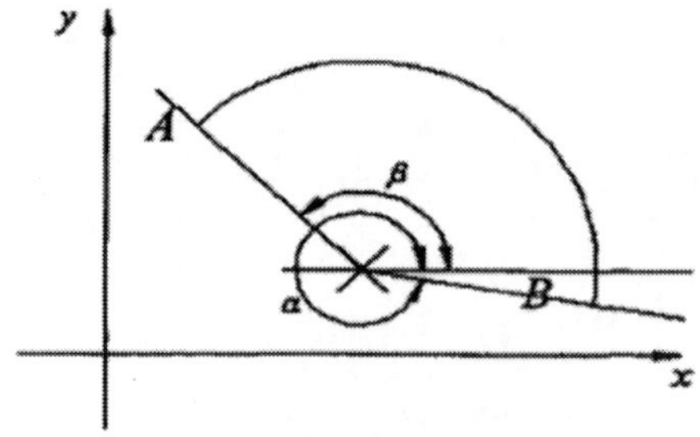

Figure 4. Schematic diagram of Arc

The arc is stored in the DXF file in the form of a center (x, y, z), a starting angle α, and an ending angle β. According to the geometric relationship of the arc, the following transformation can be performed: the arc is stored in the DXF file in the form of a center (x, y, z), a starting angle α, and an ending angle β. According to the geometric relationship of the arc, you can make the following conversion:

$$x1 = x + R\cos\alpha,\ \ y1 = y + R\sin\alpha,\ \ z1 = z$$

$$x2 = x + R\sin\beta,\ \ y2 = y + R\sin\beta,\ z2 = z$$

$$x3 = x + R\cos((\alpha + \beta)/2),\ y3 = y + R\sin((\alpha + \beta)/2),\ z3 = z$$

This algorithm calculates the circular arc information extracted from the DXF file (ie, the x, y, and z coordinates of the center point of the arc and the starting arc of the arc, and the ending arc) according to the corresponding relationship in the geometry. The three-point coordinate value[8] on the arc (that is, the x, y, and z coordinate values of the starting point, midpoint, and end point of the arc).

3.2. Conversion and Discrete Algorithm of B-Spline Curve Coordinate Data

The B-spline method has the powerful function of expressing and designing free-form curves and surfaces, and is one of the mainstream methods of shape mathematical description. In addition, the B-spline method is the current international standard for geometric definition of industrial products-the rational B-spline method (NURBS). basis. The B-spline method combines all the advantages of the Bezier method, including geometric invariance, affine invariance, etc., while overcoming the shortcomings of the Bezier method due to the fact that the overall representation does not have local properties[9] (moving a control vertex will Will affect the entire curve).

In the DXF file, the information of the B-spline curve is stored in the physical segment SPLINE. To get a B-spline curve, the curve must be discretized. According to the x, y, and z coordinates of the control points in the DXF file, the new x, y, and z coordinate values of the discrete points can be calculated. Taking four control points P1, P2, P3, and P4 and setting an equal-point u, the new discrete point coordinate values can be calculated according to the following formula.

```
n1=(-u*u*u+3*u*u-3*u+1)/6;
n2=(3*u*u*u-6*u*u+4)/6;
n3=(-3*u*u*u+3*u*u+3*u+1)/6;
n4=(u*u*u)/6;
cx=n1*p1(x)+n2*p2(x)+n3*p3(x)+n4*p4(x);
cy=n1*p1(y)+n2*p2(y)+n3*p3(y)+n4*p4(y);
cz=n1*p1(z)+n2*p2(yz+n3*p3(z)+n4*p4(z);
```

IOP Conf. Series: Journal of Physics: Conf. Series **1267** (2019) 012064 doi:10.1088/1742-6596/1267/1/012064

3.3. System implementation

The Figure 5 shows a functional interface implemented for this system.

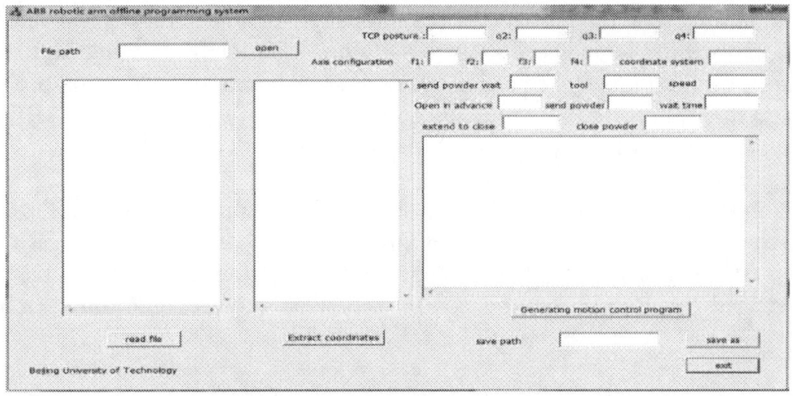

Figure 5. System function interface

1) Click on the "Open" control and select the file path to open via the dialog.

2) Click on the "Read" control to display the text information in the DXF file in the dialog box.

3) Click the "Extract Coordinates" control to extract and display the coordinates of the trajectory data in the DXF file in the dialog box.

4) q1-q4 is the attitude information, f1-f4 is the axis configuration, the workpiece coordinate system is used to determine the coordinate system, and the powder feeding waiting time to the powder input part is used to determine the value.

5) Click on the "Generate Executable" control to generate a command program that automatically moves into the arm and welds.

6) Enter the file save path and click the "Save As" control to save the generated program to the specified path.

4. Software simulation and experiment

The Robot Studio software is a set of software developed by ABB Robotics specifically for simulation with the ABB robotic arm. It is unique in that it has no translation phase in the process of downloading to the actual controller.

Using DXF files as a bridge for data exchange, the target data in the DXF file can be extracted and saved to an array for use by reading and writing files in the C language and the DXF file format. Finally, the data is written into the program in a certain format and saved. An executable file of the robot arm can be generated by the offline programming system, and the generated executable file is imported into the robot studio for simulation.

Figure 6 is a planar path drawn in CAD, saved in DXF format.Figure 7 is a path simulation performed in Robot Studio. The DXF file generated in Figure 1 can be used to generate an executable program of the robot arm through the robot offline programming system. Then the generated program is simulated in the virtual environment, and the same path as the DXF file can be obtained.Figure 8 is a verification of the motion control program generated by the system with the actual robotic arm.

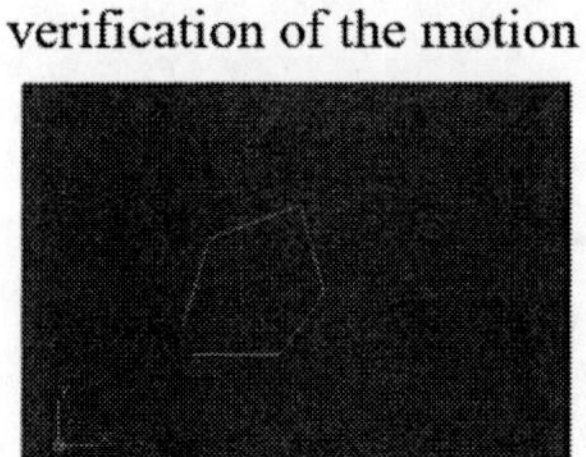

Figure 6. Plane path in CAD

Figure 7. Virtual simulation in Robot Studio

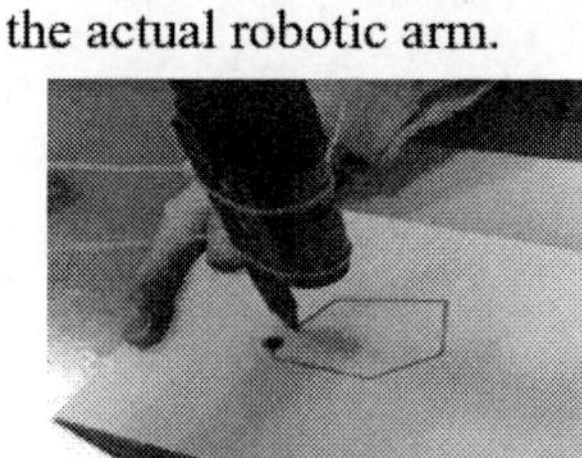

Figure 8. Experimental verification

5. Conclusion

The Through the simulation and experimental verification of Robot Studio software, the feasibility of this off-line programming system is proved. The established trajectory drawn by the CAD drawing can be simulated and applied to the actual robot arm welding. Input DXF drawing files, through the ABB robotic arm offline programming system, you can directly generate the execution program of the robot arm operation, and then import the executable program file into the laboratory ABB IRB 1600 robot arm, the arm can be welded according to the expected path. The actual experiment proved to be feasible.

In the realization process of the robot arm offline programming system, the key point is the extraction of the trajectory coordinate data in the DXF file. The partial trajectory coordinate data needs to be converted and discrete according to the geometric relationship, such as the trajectory of the circular trajectory and the B-spline curve. Coordinate data requires coordinate transformation and dispersion.

Acknowledgment

This work was supported by the "National Key R&D Program of the Ministry of Science and Technology (2018YFC1900801)".

References

[1] Ilana Nisky,Ferdinando A,Amir Karniel. Analytical study of perceptual and motor transparency in bilateral teleoperation[J].IEEE Transactions on Human-Ma-chine Systems.2013,43(6) : 570 – 582.
[2] Song Yixu, Yang Hongjun, Lv Hongbo.Intelligent con-trol for a robot belt grinding system[J].IEEE Tranca-tions on Control Systems Technology.2013,21(3) : 716 –724.
[3] H.V.Brussel.About CIM,Concurrent Engineering and Meehatronies[J].Advaneed Manufacturing Teehnology.1994(9):23–30
[4] Matin Kortus. An Alternative Approach to off-line Programming. Industrial Robot, 1993.20(4):17–20
[5] Bi jan Shirinzadeh. Task Planning and off-line Programming. Industrial Robot, 1996.23(5):4–5
[6] Xie Yuanyuan,Wang Chongchang.Realization of Digital Topographic Map Coordinate System Transformation in the CAD[J].Geomatics &Spatial Information Technology,2018,41(6):192-197.
[7] Lin Junjian,Zhang Xiaojin,Lin yueke. Design of Offline-programming and Simulation System for Industrial Robot Based on XNA[J]. Machine & Hydraulics, 2017, 5(3): 24-27.
[8] Wang Xiaoyu. Coordinate conversion of DXF graphic files[J].Geospatial Information, 2018,16(6):108-111.
[9] Wang Guitang,Xiong Feng,Chen Yongbin. Automatic path generation in 2-D detection based on DXF file[C]. 2017 IEEE 13th International Conference on Electronic Measurement & Instruments, 2017.

IOP Conf. Series: Journal of Physics: Conf. Series **1267** (2019) 012065 doi:10.1088/1742-6596/1267/1/012065

Study on Testing Environment Simulation Method for Thermal Flux Density of Aerothermal

Ju shichao

Shanghai Aircraft Design and Research Institute, COMAC, No.5188 Jinke Road, Shanghai, China

symseer@163.com

Abstract. This paper presents the use of forced convection heat transfer approach to near-space thermal environment of hypersonic aerodynamic ground simulation test method for surface heat flux with a great gradient and temperature gradient of the tip wedge (rn=1.5mm) shape component level thermal test . In the test device test section, two different high subsonic speed airflow were nested within each of the two vents from the exhaust, the impact of the specimen heated to test sharp wedge surface heat flow and temperature distribution in line with a hypersonic flight state aerodynamic heat distribution patterns. Numerical results show that under this scenario, the specimen rear stagnation point heat flux and heat flux can be adjusted two ways to adjust the air flow; test methods within a certain accuracy-related parts for the superb aerodynamic vehicle thermal environment simulation.

1. Introduction

The hypersonic vehiclewill suffer severe aerodynamic heating due to its interaction with gases during its flight or return to the atmosphere[1]. Usually when the flight Mach number Ma > 3, the air temperature of the aircraft's head can reach over 400 °C ; when the flight speed reaches 8 Mach, the temperature of the aircraft's nose cone can reach 1800 °C, and the temperature of other parts will be above 600 °C [2]; when the speed of Ma = 8 is flying at 27 km, the temperature of the nose cone is about 1800 °C, and the temperature of the wing or the leading edge of the tail is as high as 1450 °C [3]. In order to avoid the damage of aircraft caused by surface high temperature caused by aerodynamic heat, it is necessary to analyze and test the heat-carrying capacity of the structure.

Thermal protection research of hypersonic vehicle thermal protection system can be divided into two types: thermal environment research and thermal protection research. Thermal protection research simulates the thermal environment based on the conclusions of thermal environment research, and then carries out the overall layout of thermal protection system and the selection of thermal protection materials and thermal protection structure[4]. The aerothermal simulation includes a series of radiation heating equipment and airflow heating test equipment. The main equipment are high temperature and high speed wind tunnel, arc heating wind tunnel, gas heating device, high temperature structure wind tunnel and so on. High-temperature and high-speed wind tunnel is used to verify the calculation method of aerothermal environment of aircraft, to provide thermal load control parameters for radiation heating simulation test, to study the heat transfer relationship between surface protrusions of aircraft structure and high-speed airflow, and arc wind tunnel is used to study the high-temperature performance test of small-size typical structural materials. The high temperature structure wind tunnel, which adopts the thermal test method of airflow heating structure, has a wide range of applications. It mainly studies the aerodynamic shape of aircraft, high temperature structural materials, ramjet performance under simulated high speed flight environment, thermal performance of thermal shield, etc. At present, it is set up in NASA Langley Center.

At present, there are some shortcomings in the domestic ground simulation test system of structural thermal environment, such as the limited size of the specimen and the short working time of the system, and the limited maximum simulation temperature. In order to meet the progress of the thermal protection system and material of hypersonic vehicle, it is necessary to develop the thermal environment simulation test technology and strengthen the test capability of ground equipment in order to ensure the research of hypersonic vehicle, especially its key technology. At the same time, the thermal environment problem challenges the thermal environment simulation test. The development of high temperature, high heat flux gradient and high precision, reliable and stable thermal environment ground simulation test system is of great significance to the progress of thermal protection technology for hypersonic vehicle.

Aiming at the characteristics of aerothermal local high temperature of hypersonic vehicle and the insufficiency of ground test equipment for thermal environment of traditional structure, this paper presents a thermal environment simulation method for shock heating of specimens by using two different subsonic high-temperature and high-speed airflow ejected from two nested nozzles, and preliminarily verify by means of numerical simulation.

2. Thermal Test Principle of Shape Components

When a hypersonic vehicle is flying in full orbit, the surface heat flow and transient temperature distribution of the whole aircraft can be obtained by using the experimental method, which consumes a lot of money and is not suitable for the initial design stage of an aircraft. Generally, the simple engineering test method is adopted in the design of components stage, which decomposes the whole aircraft into some simple geometric elements and tests each element separately. As shown in Figure 1, a complex aircraft is decomposed into spheres, cones, swept cylinders and wedges. The interference between simple shapes can be corrected by using wind tunnel and flight test data according to the actual situation by using the surface heat flow test method with mature simple shapes.

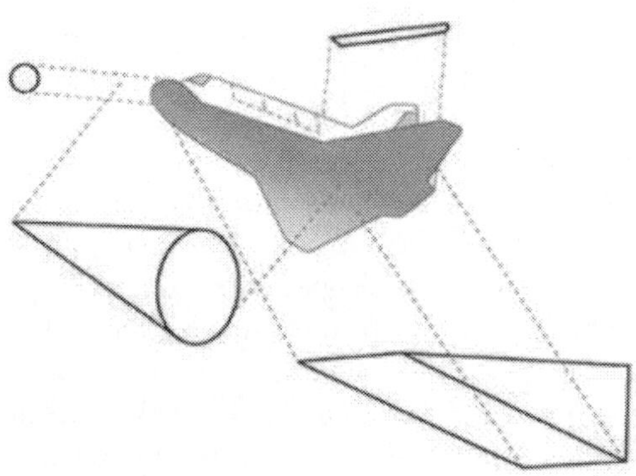

Figure 1. Typical decomposition of complex aircraft configuration

2.1. Aerodynamic heat of wedge-shaped parts

The flight thermal environment of hypersonic vehicle is mainly due to the high temperature environment generated on the outer surface of the aircraft by the action of aerodynamic heat, and the heating of the aircraft by the airflow passing through the outer surface of the aircraft has the characteristics of local large heat flux density and large area medium heat flux density. For the typical hypersonic component tip wedge, it can be considered as composed of the head cylinder and the rear plate. For the cylindrical region, the heat flux distribution varies greatly. The front heat flux is the stagnation point heat flux, which is the high heat flux region. For the flat plate region, the heat flux changes little, and the heat flux is smaller than the front stagnation point heat flux, which is the medium heat flux region [5], as shown in Figure 2.

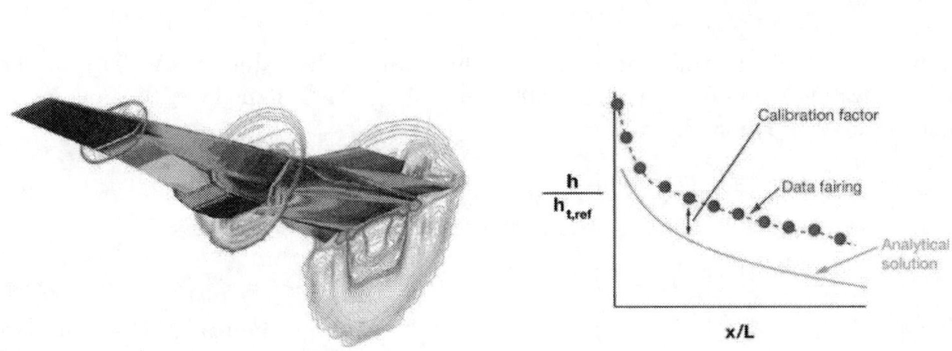

Figure 2. Wedge heat-transfer model.

2.2. *Metal Heat-proof structure*

Metal thermal protection system has many advantages [6], such as high toughness, good impact resistance, high reuse rate and low cost. It has great development potential in large area thermal protection of hypersonic vehicles. Because of the need of high-speed cruise, the front of X-43 is designed very sharp to reduce resistance. For the need of aircraft shape, a large part of the front part of the fuselage is made of solid tungsten material, which simplifies the design of thermal protection in this area. Carbon-carbon leading edge and side strip are used to protect the edge with the highest heating capacity of tungsten wedge head, while relatively short flight time can prevent most of the surface of tungsten wedge head from being heated. Aerodynamic heat of wedge-shaped components has the characteristics of large heat flux gradient, which will cause temperature gradient and uneven thermal expansion in the structure, and thus generate great thermal stress in the front of the structure.

Therefore, in order to avoid the damage due to the surface high temperature caused by aerodynamic heat to the wedge components of aircraft, the heat-carrying capacity of the structure must be analyzed and tested, and the key point is to simulate the corresponding heat flux gradient and temperature gradient.

3. Aerothermal Simulation means of wedge-shaped parts

3.1. *High enthalpy gas equipment*

High enthalpy gas thermal simulation test equipment is a kind of economical and convenient thermal environment simulation equipment. Experiments have proved that it has enough thermal simulation ability and can simulate the thermal environment of hypersonic vehicle components in a certain precision range. The thermal simulation experiment of hypersonic vehicle is expensive and the experimental ability is insufficient, which meets the urgent need of the development of hypersonic vehicle. The high temperature gas thermal simulation scheme uses the high temperature gas generated by the combustion chamber of the gas to emit and eject the normal temperature and pressure air from the gas nozzle through the high temperature pipeline. The cold and hot air flows through the momentum exchange mixing to blow and heat the sample. The test tail gas after cooling treatment enters the noise reduction tower and discharges into the outdoor atmosphere.

High temperature gas simulates hypersonic aerodynamic heat, which has the following characteristics. The gas outlet velocity is usually subsonic, that is, using subsonic high temperature gas flow to simulate the heating of high speed gas flow. Because of the different nature of the two flow patterns, there are also some differences in the thermal effect of the specimen. As mentioned earlier, hypersonic heating has local large heat flow and large area medium heat flow. The cumulative effect of heating is reflected in the large temperature gradient of the specimen. For gas heating, when the stagnation point heat flux reaches the required value, the temperature speed of gas is often required to be higher, but the heat flux density at the back of the specimen is much higher than that at the high aerodynamic heating value. As a result, the temperature rise at the back of the specimen is very fast,

IOP Conf. Series: Journal of Physics: Conf. Series **1267** (2019) 012065 doi:10.1088/1742-6596/1267/1/012065

and the temperature gradient within the specimen can not meet the requirements. The distribution of heat flux of gas heating with 40 mm nozzle 1800K and Ma=6, H=30Km, Tw=300K on the surface of the specimen is shown in Figure 3.

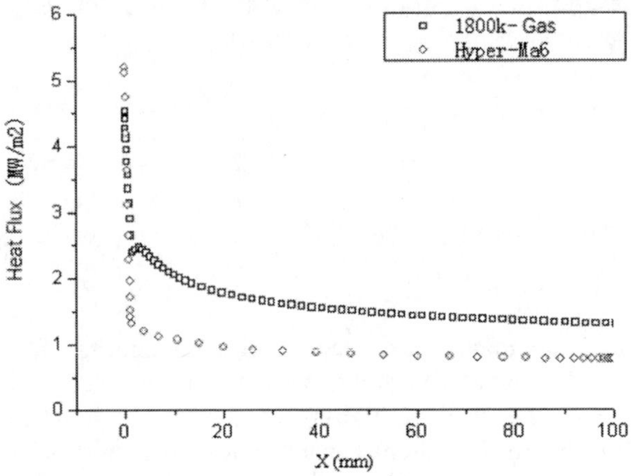

Figure 3. Distribution of Heat Flux Density of Gas Heating and Super Aerodynamic Heating on the Surface of the Specimen

3.2. The engineering calculation of convection heating

For approximate calculation of surface heat transfer of wedge-shaped parts with gas impingement tip, the calculation formula of cross-flow cylinder can be selected in the cylinder part, and the calculation Because the front cylinder is in laminar flow and the angle is smaller than, the fitting formula of laminar heat transfer around the cylinder is adopted.

$$\frac{h_\phi d_f}{\lambda} = K_3[1.14 + \mathrm{Re}_d^{1/2}\,\mathrm{Pr}^{0.4}(1 - \left|\frac{\phi}{90}\right|^3)],\ -80° < \phi < 80° \qquad (1)$$

$$\mathrm{Re}_d = \frac{\rho_\infty u_{g\infty} d_f}{\mu} \qquad (2)$$

The airfoil specimens are 9 degree flat plates formed horizontally behind the circular arc, and the fitting formula of plate heat flow is used.

laminar flow: $$Nu_x = 0.332\,\mathrm{Re}_x^{-1/2}\,\mathrm{Pr}^{1/3} \qquad (3)$$

turbulence flow: $$Nu_x = 0.0296\,\mathrm{Re}_x^{0.8}\,\mathrm{Pr}^{1/3} \qquad (4)$$

$$Nu_x = \frac{h_x x}{\lambda};\ \mathrm{Re}_x = \frac{\rho u_{gx} \cdot x}{\mu} \qquad (5)$$

The qualitative temperature in the above formula is taken as the reference Eckert temperature T*, and the turbulence correction factor is related to the turbulence. For a smooth plate without pressure gradient, when Rex=300000-500000, the boundary layer changes from laminar to turbulent. From the above formula, the required air temperature and velocity can be calculated according to the required simulated heat flux of aerodynamic heat.

3.3. Two-way airflow heating scheme

From the above analysis, it can be seen that it is difficult to simulate a large temperature gradient with a single gas nozzle. For hypersonic aerodynamic heating, including heat conduction and friction heating, both occur in the boundary layer. From the static temperature distribution of hypersonic flow field, it can be seen that the temperature in the boundary layer at stagnation point is much higher than that in the flat area. For gas flow heating, the temperature difference in the boundary layer at stagnation point is not very large compared with that in the flat area.

Therefore, the heat flux field on the whole specimen can be divided into two zones to simulate separately, i.e. high temperature and high speed air flow is used to simulate the high heat flux area of the head, and the medium heat flux in the flat plate area is simulated by the gas in a lower state. A small nozzle with the same size as the head of the specimen is added to the center of the original large nozzle, and the air temperature in the small nozzle is also simulated. At the same time, the original nozzle emits gas with lower temperature.

4. Quasi-static heat flow simulation

4.1. Testing Environment condition

The maximum aerodynamic heating heat flux of hypersonic vehicle usually occurs in the accelerated climbing stage. Because of the fast climbing speed of the vehicle, it can climb to the altitude of 60,000 meters in less than 2 minutes.

4.2. Simulation results and analysis

When the width of the small nozzle is 4 mm and the distance between the small nozzle and the head of the specimen is 8 mm, the gas flow temperature is 1800K, the velocity is Ma=0.7, the temperature of the cold cover flow is 1000K, the velocity is Ma=0.4, and the temperature of the specimen is 300K, the cold wall heat flow on the surface of the specimen can be obtained. The calculated temperature field is shown in Figure 4.

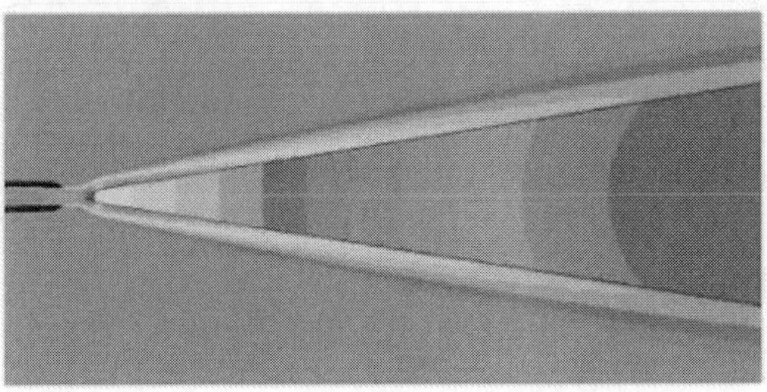

Figure 4. temperature field of the Specimen

The heat flux distribution on the surface of the specimen is shown in Figure 5.

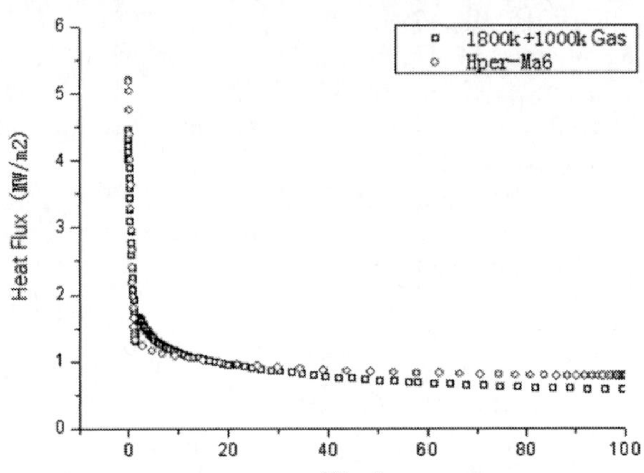

Figure 5. Surface heat flux distribution of double-channel air impingement heating

It can be seen from Figure 5 that the surface heat flux of the specimen can decrease rapidly in the direction of the airflow and tend to be higher than that of the aerodynamic heat by using the method of

IOP Conf. Series: Journal of Physics: Conf. Series **1267** (2019) 012065 doi:10.1088/1742-6596/1267/1/012065

small nozzle and cooling hood flow. In the transition zone from the head of the specimen to the plate area, the gas heating rate is higher than that of the high superheat flux. However, the high and super heat flux in the head region is relatively high. Because of the existence of thermal diffusion, the heating effect of the specimen is basically equivalent, and the influence on the temperature distribution behind the specimen will be very small.

5. Transient aerothermal simulation

In order to simulate the transient aerodynamic heat with high temperature gas impact test pieces, it is necessary to know the real aerodynamic heating environment in flight, specifically the aerodynamic heating heat flux density and temperature at each position of the aircraft under flight envelope. Then, according to the obtained heating heat flux density and temperature, the temperature flow of gas is usually selected by numerical calculation method to simulate the thermal environment of the test pieces. Whether the errors between the surface temperature distribution and the true value of a given specimen satisfy the requirements of thermal test or not, the experiment can be carried out, otherwise the gas state needs to be adjusted and the method of numerical calculation revised.

6. Conclusions

In the two-way air impingement heating scheme, two different air streams are ejected from each other's nested nozzles to impinge on the wedge specimen. By setting reasonable air flow parameters, the surface heat flow and temperature distribution of the wedge in the test can conform to the law of aerodynamic heat distribution in a hypersonic flight state. The numerical results show that the heat flux density at the standstill point and at the rear of the specimen can be adjusted by adjusting the two airflow paths. The experimental method can simulate the aerothermal environment of the relevant components of the hypersonic vehicle in a certain precision. The research shows that the method is feasible in engineering and has potential in the thermal simulation of larger-scale components. In transient simulation, the basic process of coordinated control of experimental parameters and the theory of integrated control need to be further explored.

- Hypersonic aerothermal heat flux density at the standstill point and at the rear of the specimen can be adjusted by adjusting the two airflow paths.
- The basic process of coordinated control of experimental parameters and the theory of integrated control need to be further study.

References

[1] Sachin Kumar, Shripad P. Mahulikar Aero-thermal analysis of lifting body configurations in hypersonic flow, Acta Astronautica. 126 (2016) 382–394.

[2] R. Savino, G. Russo, V. D'Oriano, M. Visone, M. Battipede, P. Gili, Performances of a small hypersonic airplane (HyPlane), Acta Astronaut. 115 (2015) 338–348.

[3] John J. Bertin, Russel M. Cummings. Critical Hypersonic Aerothermodynamic Phenomena[J].Annu.Rev.Fluid.Mech,2006.38:129-157.

[4] Rivers H K, Glass D E. Advances in hot-structure development[R]. NASA-2006.

[5] J. Liu, F. Ding, W. Huang, L. Jin, Novel approach for designing a hypersonic gliding–cruising dual waverider vehicle, Acta Astronaut. 102 (2015) 81–88.

[6] Glass D E. Ceramic matrix composite (CMC) thermal protection system (TPS) and hot structure for hypersonic vehicles[R]. AIAA-2008-2682.

IOP Conf. Series: Journal of Physics: Conf. Series **1267** (2019) 012066 doi:10.1088/1742-6596/1267/1/012066

The Design and Application of *FEU: Automation Monitoring Unit for 5G base stations*

Hong Li[1,a], Huazhao Liu[2,b] and Baiyi Yan[1,c]

[1]China Tower Co, Ltd., Beijing, China
[2]China Communication Service Application and Solution Technology Co, Ltd., Guangzhou, China

E-mail: [a]lihong@chinatowercom.cn, [b]liuhuazhao@ccssoft.com.cn, [c]yanby@chinatowercom.cn

Abstract: In order to realize the co-construction, sharing and intensive management of millions of stations across the network, supporting 5G go into operation efficiently, China Tower Group has developed a new generation of intensive and distributed operation maintenance monitoring platform over the inherited mobile base stations from the three major telecom operators. Based on the development and the application of this platform, FEU (an intelligent Power and Environment Monitoring Unit) is proposed, which uses the dynamic shared library method to embed the edge gateway to realize the edge computing capability of China Tower's intelligent operation and maintenance platform. It effectively solves the normalized monitoring of heterogeneous FSU and massive sites, constructs the intensive operation maintenance capability for China Tower and lays a solid foundation for the tower-based IoT [1,2] expansion business.

1. Introduction
As a vital components of communication system, mobile base stations can provide system interface and wireless function to mobile terminal. In recent years, by the quick development of communication technology, especially the 4G network, there is a great increasing of the mobile base stations. By 2020, 5G network is planning to go into commercial usage, which bring greater demand for constructing more mobile base stations, however, the features of our base stations are decentralization, rugged environment and unstable power supply. For the development of mobile communication, it is even more important to have real time monitoring the power and environment of base stations. It depends on the cooperation of locale FSU, centralized O&M monitoring platform and operation team. Thereinto, centralized O&M monitoring platform is the key to enhance the stability, quality and efficiency of all the base stations.

2. Distributed Architecture Design of Mobile Station Monitoring Platform

2.1. The Current Situation of Power Environment Monitoring Platform
The operators usually use 3-tier-architecture power supply environment system, which is deployed according to the classification of group company, province company and city company. Over 95% of the base stations access with wire (a few use wireless GPSK), and most of them use the method of central analysis [3]. After CHINA TOWER is established, it took over all the base stations from three Chinese main operator. As the number of base stations reaches 2 million and locates all over 31 provinces of China, CHINA TOWER have built a core R&D team and cooperate with equipment

vendors to quickly achieve production support ability. The power supply monitoring system use centralize architecture and the network management system is deployed in the headquarter with distributed nodes for calculation. The ways of transmission are wireless (mainly 3G and 4G). The signal data performance and remote control of base stations directly access to FSU. It adopts End-Analysis method [4,5] and no longer access to OMC directly.

As the equipment of FSU have a large amount of types, those function interfaces vary considerably. The building of 5G network and the development of IoT accelerate extending the types and number of those equipment, which generate serious requirement for the compatibility of FSU system. Frequent upgrading is necessary for operation and maintenance of base stations. For the vendors vary in their technique, it is complicated to reform, upgrade and combined adjusting their interfaces. The loose relevance between FSU and SC platform cause the complexity to locate problem, which make the platform difficult to control the key functions, like registering FSU, reporting alarm, collecting performance, remote controlling and so on [3].

2.2. The Standardizing Design and Application of Intensive Distributed Platform According to FSU Interface

At the beginning, China Tower integrate the standard of power supply environment monitoring from China Telecom, China Mobile and China Unicom and make a unified monitoring standard and a cross-platform-interface technical proposal [2]. After investigation and communication with 16 domestic main FSU vendor, China Tower reform and upgrade their B Interface of power supply environment monitoring. Then, they formed the architecture of Automatic Intensive and self-adapted monitoring platform [2,6] , as shown in Figure 1, after a lot of test on the platform in the experimental network and the typical base stations.

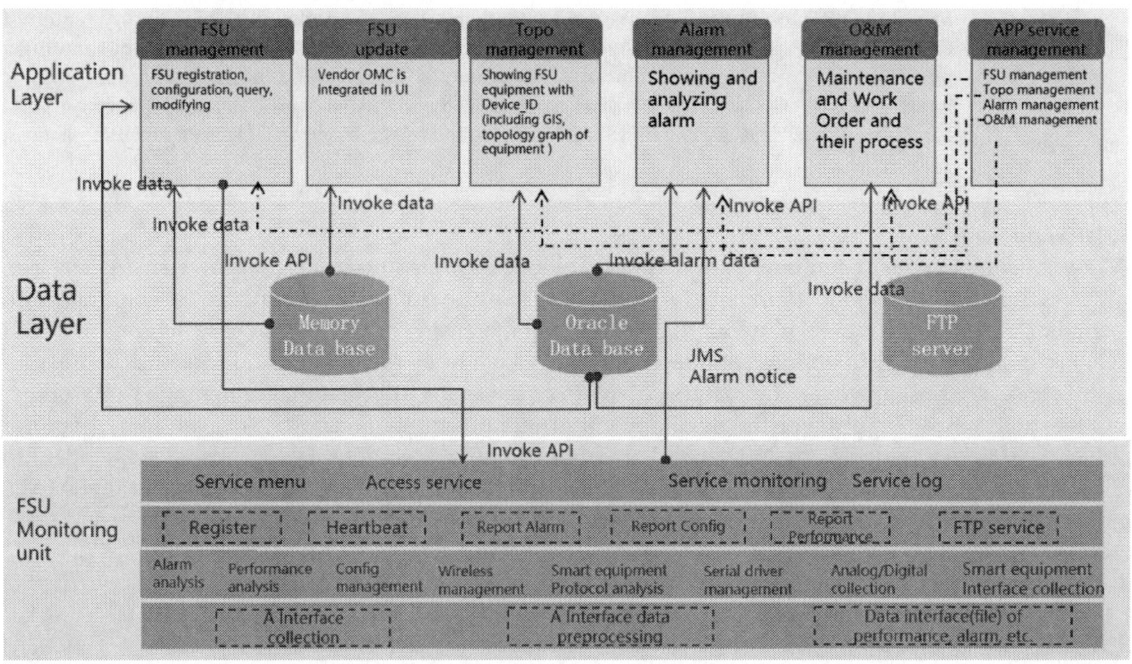

Figure 1. Architecture of the Tower's power supply environment intensive monitoring platform base on standard FSU interface

This architecture has worked more than 1 year, helping China Tower to take over all the existing base stations from 3 main operators in China. Now, China Tower can collect resource data, like performance, alarm and analysis, and remotely control the base stations, like door opening, camera control and battery measurement, which gives China Tower the Intensive Management Ability.

In this architecture, FSU works as a basic element of the whole monitoring network. By install embedded system and application into terminal-automatic-unit equipment, it can manage and deal with the monitoring data. The main business functions of FSU are controlling the collection of data, interacting with superior monitoring platform and, according to specific hardware, making integrated design with hardware and software. As the types and system software vary in different FSU equipment, although there is unified standard, the monitoring centre is difficult to maintain, analysis and locating the collecting data, especially alarm data due to the process to deal with data vary [7]. If it is necessary to upgrade the FSU software, each type of equipment need to develop its own application, which cause the low flexibility and high cost.

3. The Automatization of FEU Edge Node Calculating

As 5G network incoming, base stations expanding and IoT technique developing [4,8], there are more and more sensors put into monitoring, which cause an explosive growth in the capacity and difficulty of the platform operating. More than ten types of the FSU equipment and millions of them need to upgrade their interfaces. It is quiet a difficult job to update the version depend on different area and vendor, which make system updating a huge amount of work and the software lacks of robustness. Therefore, it is necessary to reform and unified this system. The iTower Core R&D Team of China Tower spent more than three months to analyse the structure of platform software [2], the business process, the function of Edge Node calculation, the core chip of FSU and the algorithm of embedded system [9,10]. Combining the Functional Specification of intensive monitoring platform, this team designed a kind of automatic embedded access unit, FEU, which move the terminal calculating function into FSU operation system and make it possible for each base station to calculate as a distributed unit. Based on this, China Tower plan to reform the access terminal and start the FSU Automatizing Project. The structure after reformed is shown in Figure 2.

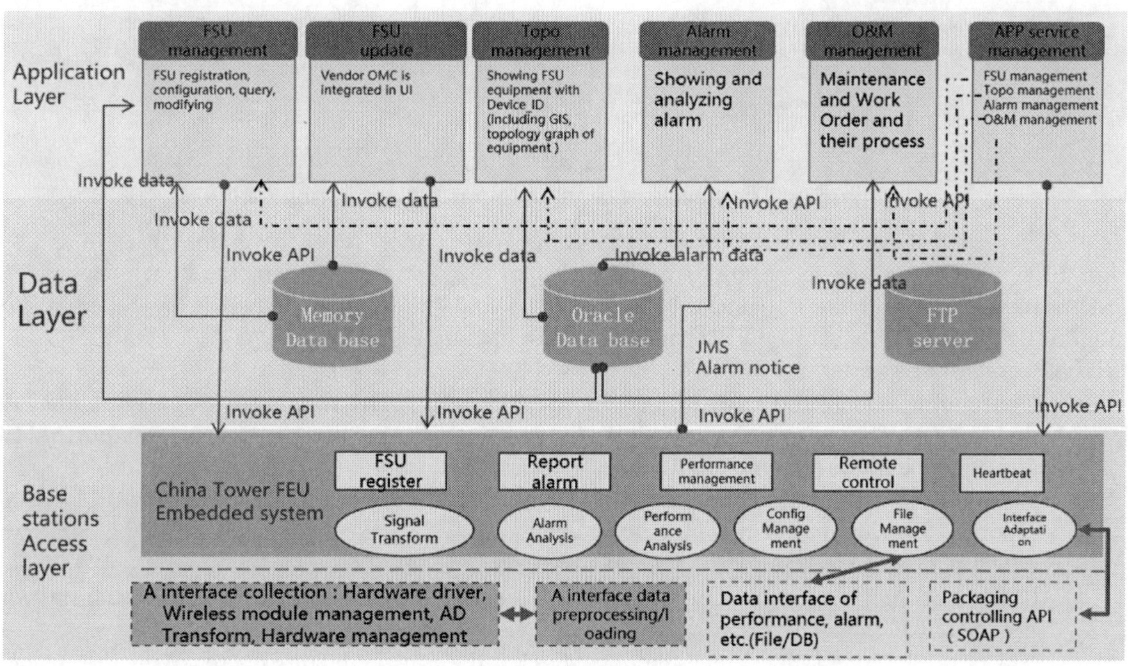

Figure 2. Architecture of the Tower's power supply environment intensive monitoring platform base on automatic FEU

3.1. Design of FEU Software Architecture

FEU includes all the functions of B interface. After FSU collects and delivers data from the equipment by A interface to FEU, FEU deal with the data depending on the kind and then update performance,

IOP Conf. Series: Journal of Physics: Conf. Series **1267** (2019) 012066 doi:10.1088/1742-6596/1267/1/012066

report alarm, store the data as history and so on. Wireless Network is connected by FSU. When the connecting status (online/offline) of VPN has changed, FSU need to notify FEU with all the registering information, like the VPN address, IMSI, network standard and so on. As soon as FEU judge it is Heartbeat Timeout, it will inform FSU and FSU will deal with it following the Triple-Reboot standard. The process is shown in Figure 3.

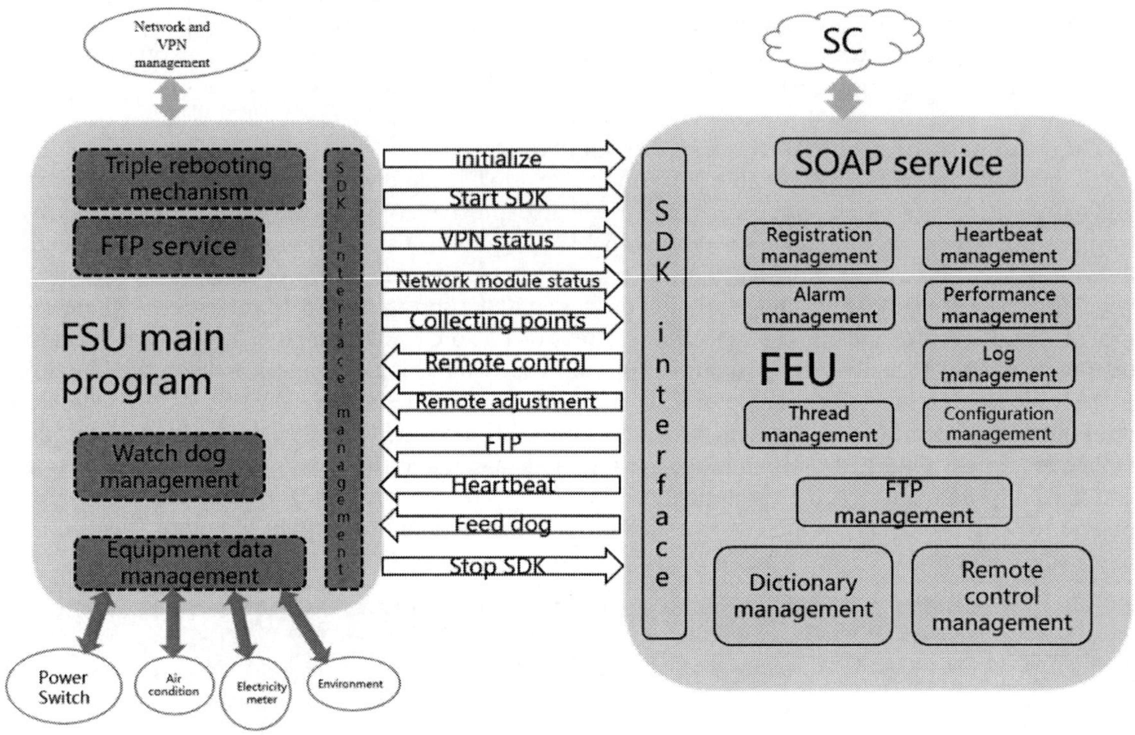

Figure 3. The scheme design of automatic monitoring unit

FEU is provided as Dynamically Shared Lib to the FSU vendor. As each vendor varies in its CPU chip, FEU need to compile according to the Cross Link Compilation Environment of each vendor. The FSU software can load the dynamic lib by the feu.h and libfeu.so, and then it can interact with monitoring platform after start-up FEU function by the operation process.

FEU API consist of 4 parts:

- Configuring management, which set the parameter for SDK running by Configuration File, including the storing space offer to SDK, the space for IPC picture, the base information of FSU and equipment, the information of collecting point and so on.
- Call-back function, which includes modifying FTP service parameter and control orders, reporting for Heartbeat Timeout, getting version of FSU, getting the status of VPN, correcting clock, obtaining the resource of system and so on.
- Control of SDK running, which can start or stop the SDK function, distribute initial memory for FSU, load the configure parameter, create instant and so on.
- Business Interaction Managing Function, which is the core model of FEU receiving the business monitoring function from FSU, including the real-time-remote-measurement data, the real-time data of remote signal related to remote measurement, the real-time data of remote signal, the real-time data of remote control and submit function, the informing about the status of VPN, getting the register status of FEU and so on.

After the platform issues a remote controlling order, FEU distributes it to FSU by Call-back function and FSU operates the controller and returns the result to FEU, and then FEU feed it back to the platform.

3.2. The Applying Effects of Automatic FEU

The applying of automatic FEU has 6 main benefits:

- Unified management of B interface, which makes the monitoring data more accurate and reliable.
- High expansibility of function, which can fast upgrade and deploy in any environment.
- Accuracy of the monitoring data, which help to locate the problem quickly.
- Compatibility to most FSU equipment, suitable for old and new equipment.
- Self-development, which reduce the cost to purchase FSU.
- Automazation of monitoring.

China Tower first apply automatic unit FEU into the power supply environment monitoring system of far-away base stations. By now, there are over 300,000 base stations which have installed FEU. As upper interface is uniform, the version of software is the same and format of reported data is totally unified, it is simple to install, upgrade to support the new function of base stations and the operation, maintenance and management is standard, which make the system stable and reliable. This project archive the goal of base stations unified management, automatic deployment and simple expansion. FEU design the model depending on the demand, dividing the software highly related to hardware and interface of equipment into independent part. Designing the applications which interacts with monitoring centre and manage the data as individual components, FEU can unify the process of delivering monitoring data and ensure the data is complete and reliable.

4. Conclusion

This paper presents a new system architecture for automatic monitoring unit. It packages the interaction with platform and business deal as a function model FEU for FSU to interact with platform. It is independent of the application dependent on hardware and highly compatible. It is benefit to use model design for the development of power supply environment monitoring standard. As the incoming of NB-IoT and 5G network, in the future, the standard scene of FEU will be more detailed and, focusing on different access signal, FEU will put more calculating function into the edge layer, which can make the business service of platform more and more automatic.

References

[1] Bo Chen, Zhihui Gan 2016 The research of the commercial value and networking scheme of NB-IoT *Mobile Communication.07-15*

[2] Huazhao Liu 2016 The design and application of big data platform for IDC internet behaving analysis *Telecom Technique.Jan*

[3] Xin Gong, Zhiju Wang 2017 The technical proposal to reform power-supply environment monitoring unit of stock base stations *Telecom Technique.01-25*

[4] Huoming Cheng, Linsu Shi, Shan Gao, Meng Jin, Luda Miao 2016 The Summary of D2D Technology for 5G communication network *Communication Journal.07-25*

[5] SCUT.Wenjian Xu 01/16/2015 The design and realization of monitoring digital base stations base on SNMP

[6] Yin Ling, Yun Qiu 2017 The method to monitor the network behaviour of IoT Terminal base on internet data *Telecom Science.12-20*

[7] Huoming Cheng, Linsu Shi, Shan Gao, Meng Jin, Luda Miao 2017 The Expectation of China Tower's Big Data Application *Communication Journal.07-20*

[8] 3GPP TR 23.720. Study on architecture enhancements for Cellular Internet of Things (Release 13)[S]. 2016

[9] Wei Liu, Jun Liu 2017 The research of distributed and dynamic resource management in the network of time-delay-sensitive sensors *Communication Journal.07-25*

[10] Hong Zhang, Yufeng Wang 2017 The frame, key technique and application of NB-IoT base on Fog Calculation *ZTE Communication Technique.01-05*

IOP Conf. Series: Journal of Physics: Conf. Series **1267** (2019) 012067 doi:10.1088/1742-6596/1267/1/012067

A Novel Network Traffic Prediction Method Based on RMPM

Jianjun Wu[1, a], Weijun Gong[1, b] and Zhen Shang[1, c]

[1]College of Information Technology and Communication, Hexi University, Zhangye, 734000, China.

[a]wujj@hxu.edu.cn; [b]gongwj@hxu.edu.cn; [c]shangzh@hxu.edu.cn

Abstract. The network traffic prediction plays the key role in congestion control and bandwidth allocation. A variety of traditional learning models such as artificial neural networks (ANN) have been applied in prediction. To avoid the drawbacks of traditional models for prediction, a novel robust minimax probability machine (RMPM)-based traffic prediction method is proposed in this paper. The prediction performance is tested on two different types of traffic data, Ethernet data flow and MPEG4 video flow, at the timescale 1. The experiments demonstrate that the proposed method attains satisfactory performance in prediction accuracy. Therefore, the proposed method can be used for congestion control or bandwidth allocation, to meet the user QOS requirements.

1. Introduction

The network data streams exhibit uncertainty and time-varying, and it is difficult to capture the rules of change. Through network traffic prediction and other means, it can predict the size of data traffic in the future, which is of great practical significance for network planning and design, performance analysis and network management [1].

In order to improve the prediction accuracy, a nonlinear model such as artificial neural network (ANN) is commonly used for network traffic prediction [2]. However, the artificial neural network has problems such as overfitting, too large sample size and too small local values, which limits its application. In addition to neural networks, other machine learning models are also used for network traffic prediction. For example, Hong Fei proposed a multi-scale network traffic prediction model based on wavelet, which can achieve multi-scale accurate prediction [3]. Zhang Yinglu proposed a network traffic prediction model based on support vector machine, and support vector machine relied on genetic algorithm to optimize the parameters. The experimental results show that the prediction accuracy is higher than that of traditional neural network [4]. In addition, Dai Yue et. al. researched the chaotic characteristics of network traffic sequence, and then realized network traffic prediction by means of phase space reconstruction theory [5].

The minimax probability machine (hereinafter referred to as MPM) is a new learning model that has emerged in recent years. It achieves classification [6] and regression prediction by minimizing the upper bound of the generation error [7] and other functions. Since the minimum and maximum probability machine has good characteristics, the model has been successfully used in medical diagnosis and face recognition [8]. However, some parameters in the MPM, such as the mean vector and the covariance matrix, are estimated by learning samples. When the number of samples is too small, the estimation of the mean vector and the covariance matrix will bring a large error, which will affect the accuracy of the classification. To this end, C. H. Hoi and Lanckriet proposed a robust

IOP Conf. Series: Journal of Physics: Conf. Series **1267** (2019) 012067 doi:10.1088/1742-6596/1267/1/012067

minimax probability machine, named RMPM to solve this problem [9-10].

This paper is based on the good characteristics of RMPM and using it for network traffic prediction. Firstly, we give an accurate description of the network traffic prediction problem. Then the basic principles of MPM and RMPM are briefly reviewed. Finally, the prediction methods proposed in this paper are used to predict the two types of network traffic such as Ethernet data stream Ethernet and MPEG4 video stream. The prediction results verify the effectiveness of the proposed method.

2. Network traffic prediction problem

Suppose there is a piece of network traffic data $s(n)$, it can be seen as a time series. Based on past several observations $s(n-1)$, $s(n-2),\ldots s(n-d)$ of the time series, and it can predict the current value $s(n)$ with a predictive model. The corresponding mathematical formula is

$$s(n)=f\left[s(n-1),s(n-2),\cdots,s(n-d)\right] \tag{1}$$

The parameter d is called the embedding dimension, and a function f is a mapping from a d-dimensional space to a one dimensional space.

Assume a learning sample $\mathbf{x}_n \equiv \left(s(n-1),s(n-2),\cdots,s(n-d)\right)^T$ ($n=1,2,\cdots,N$) and corresponding target value $t_n \equiv s(n)$, Eq. (1) can be abbreviated as

$$t_n = f(\mathbf{x}_n),\quad n=1,2,\cdots,N \tag{2}$$

It can be seen from Eq. (2) that the main task of the prediction model is to obtain N learning samples based on the d dimension and the corresponding target value estimation function $f(\mathbf{x})$.

3. Robust MiniMax Probability Machine

Given N learning samples $\mathbf{x}_1,\mathbf{x}_2,\cdots\mathbf{x}_N$ and their corresponding target values $t_1,t_2,\cdots t_N$, Now write the function $f(\mathbf{x})$ to be solved

$$f(\mathbf{x})=\mathbf{w}^T\mathbf{x}+b_r=w^{(1)}x^{(1)}+w^{(2)}x^{(2)}+\cdots+w^{(d)}x^{(d)}+b_r \tag{3}$$

Where $\mathbf{w}=\left(w^{(1)},w^{(2)},\cdots,w^{(d)}\right)^T$ is the weight vector. The vector $\mathbf{x}=\left(x^{(1)},x^{(2)},\cdots,x^{(d)}\right)^T$ is the total of n learning samples. The parameter b_r is a constant. In addition, the total number of $t_1,t_2,\cdots t_N$ is t, It's a one-dimensional variable.

In order to get $\mathbf{w}$ and b_r in Eq. (3), we let

$$\min\left[\max P\left(\left|f(\mathbf{x})-t\right|\geq\varepsilon\right)\right] \text{ or } \min\left[SupP\left(\left|f(\mathbf{x})-t\right|\geq\varepsilon\right)\right] \tag{4}$$

The minimax decision rules is used in Above Eq.(4) . Now rewrite it into the following equivalent form, it can clarify its physical meaning better.

$$\max\left[\min P\left(\left|f(\mathbf{x})-t\right|\leq\varepsilon\right)\right] \text{ or } \max\left[InfP\left(\left|f(\mathbf{x})-t\right|\leq\varepsilon\right)\right] \tag{5}$$

Suppose there is a tube with a radius of ε (referred to as ε tubes), it is obtained by shifting the learning samples by ε in the positive and negative directions of the t-axis. According to $\left|f(\mathbf{x})-t\right|\leq\varepsilon$, $t-\varepsilon\leq f(\mathbf{x})\leq t+\varepsilon$ is obtained (for the nth sample, there is $t_n-\varepsilon\leq f(\mathbf{x}_n)\leq t_n+\varepsilon$). If $P_\varepsilon = P\left(\left|f(\mathbf{x})-t\right|\leq\varepsilon\right)$, then the physical meaning of P_ε is the probability that the function $f(\mathbf{x})$ is located in the ε-tube. The physical meaning of $InfP\left(\left|f(\mathbf{x})-t\right|\leq\varepsilon\right)$ in Eq.(5) is the lower bound of the

IOP Conf. Series: Journal of Physics: Conf. Series **1267** (2019) 012067 doi:10.1088/1742-6596/1267/1/012067

probability that $f(\mathbf{x})$ is located in the ε-tube, and the whole Eq. (5) is to maximize the lower bound, which indicates that the function $f(\mathbf{x})$ is as far as possible in the ε-tube.

If the variable α is introduced, the Eq. (5) can also be rewritten into the following constraint optimization form

$$\begin{cases} \max \alpha \\ s.t.\ \ InfP\left(\left|f(\mathbf{x})-t\right|\le\varepsilon\right)\ge\alpha \end{cases} \tag{6}$$

The physical meaning of α in the above equation is the same as the physical meaning of $InfP\left(\left|f(\mathbf{x})-t\right|\le\varepsilon\right)$, so the parameter α is the lower bound of P_{ε}, and satisfying $\alpha \le P_{\varepsilon}$.

It is very difficult to solve Eq. (5) or Eq. (6) directly. Now we will convert $f(\mathbf{x})$ to a classification problem and then solve it [4]. Specifically, it is utilized

$$\begin{cases} \min\limits_{\mathbf{a}}\left(\left\|\mathbf{R}_{\tilde{\mathbf{x}}\tilde{\mathbf{x}}}^{1/2}\mathbf{a}\right\|_2+\left\|\mathbf{R}_{\tilde{\mathbf{y}}\tilde{\mathbf{y}}}^{1/2}\mathbf{a}\right\|_2\right) \\ s.t.\ \ \mathbf{a}^T\left(\tilde{\boldsymbol{\mu}}_{\mathbf{x}}-\tilde{\boldsymbol{\mu}}_{\mathbf{y}}\right)=1 \end{cases} \tag{7}$$

and

$$b=\mathbf{a}^T\boldsymbol{\mu}_{\tilde{\mathbf{x}}}-\kappa_*\sqrt{\mathbf{a}^T\mathbf{R}_{\tilde{\mathbf{x}}\tilde{\mathbf{x}}}\mathbf{a}}=\mathbf{a}^T\boldsymbol{\mu}_{\tilde{\mathbf{y}}}+\kappa_*\sqrt{\mathbf{a}^T\mathbf{R}_{\tilde{\mathbf{y}}\tilde{\mathbf{y}}}\mathbf{a}} \tag{8}$$

Firstly we calculate the parameters $\mathbf{a}=\left(a^{(1)},a^{(2)},\cdots,a^{(d)},a^{(d+1)}\right)^T$ and b, and then substituting the parameters $\mathbf{a}$ and b into Eq. (9)

$$\begin{cases} w^{(i)}=-a^{(i+1)}/a^{(i)} \\ b_r=b/a^{(1)} \end{cases} \tag{9}$$

To get $\mathbf{w}$ and b_r. Where $\hat{\boldsymbol{\mu}}_{\mathbf{x}}$, $\mathbf{R}_{\mathbf{yy}}$, $\hat{\boldsymbol{\mu}}_{\mathbf{y}}$, $\mathbf{R}_{\mathbf{xx}}$ in Eq. (7) and Eq. (8) is the mean and covariance matrix of the two types of samples respectively.

In order to improve the robustness of estimates $\boldsymbol{\mu}_{\mathbf{x}}$, $\boldsymbol{\mu}_{\mathbf{y}}$, $\mathbf{R}_{\mathbf{xx}}$ and $\mathbf{R}_{\mathbf{yy}}$ based on the learning samples, it is assumed that the range of values of these parameters is known, and the value is allowed to fluctuate within the range of values. So define the set

$$S_{\mathbf{x}}=\left\{\left(\boldsymbol{\mu}_{\mathbf{x}},\mathbf{R}_{\mathbf{xx}}\right):\left(\boldsymbol{\mu}_{\mathbf{x}}-\hat{\boldsymbol{\mu}}_{\mathbf{x}}\right)^T\mathbf{R}_{\mathbf{xx}}^{-1}\left(\boldsymbol{\mu}_{\mathbf{x}}-\hat{\boldsymbol{\mu}}_{\mathbf{x}}\right)\le\nu^2,\left\|\mathbf{R}_{\mathbf{xx}}-\boldsymbol{\Sigma}_{\mathbf{xx}}\right\|_{Fro}\le\rho\right\} \tag{10}$$

$$S_{\mathbf{y}}=\left\{\left(\boldsymbol{\mu}_{\mathbf{y}},\mathbf{R}_{\mathbf{yy}}\right):\left(\boldsymbol{\mu}_{\mathbf{y}}-\hat{\boldsymbol{\mu}}_{\mathbf{y}}\right)^T\mathbf{R}_{\mathbf{yy}}^{-1}\left(\boldsymbol{\mu}_{\mathbf{y}}-\hat{\boldsymbol{\mu}}_{\mathbf{y}}\right)\le\nu^2,\left\|\mathbf{R}_{\mathbf{yy}}-\boldsymbol{\Sigma}_{\mathbf{yy}}\right\|_{Fro}\le\rho\right\} \tag{11}$$

In the Eq. (10) and Eq. (11) above, $\|\cdot\|_{Fro}$ represents the Frobenius norm, It satisfies the $\|\mathbf{A}\|_{Fro}^2=Tr\left(\mathbf{A}^T\mathbf{A}\right)$, $\nu\ge0$ and $\rho\ge0$ are two constants.

4. Experiments

The two network data streams used for prediction in the experiment are Ethernet and MPEG4. Among them, Ethernet is the test data collected by the research institute named Bellcore Morristontown, and MPEG4 is the video data stream of the medium-quality version of "Star Wars 4". The data stream above is first processed as the number of bytes passed per unit time, and then aggregated data having a time scale of 1 second and 5 seconds, respectively, is formed, and only aggregated data of a scale of 1 second is used herein.

4.1. Ethernet network traffic data prediction

Before predicting Ethernet network traffic, the shilling embedded dimension d is equal to 3, and then 200 learning samples and 150 test samples are constructed based on the traffic data. When using RMPM prediction, the parameter ε contained in RMPM is taken as 0.2, and the mean square error (MSE) is used to measure the prediction performance.

The predicted results are shown in Figure 1. Figure 1 is divided into upper and lower subgraphs. The solid line in the above figure represents the true value of Ethernet network traffic, and the dotted line represents the predicted value of network traffic. The solid line in the figure below represents the prediction error. As can be seen from Figure 1, the predicted curve of the Ethernet network traffic is well matched to the actual curve. The calculation results show that the mean square error (MSE) of this prediction is 5.307×10^{-3}.

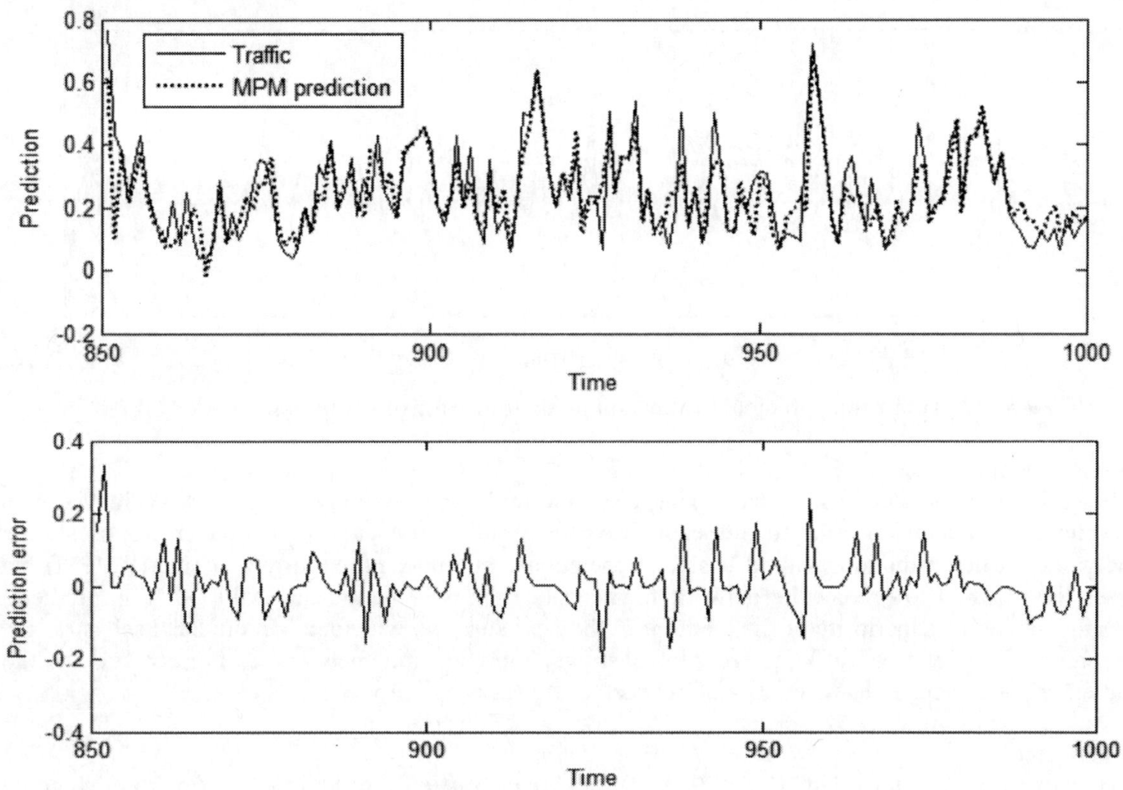

Figure 1. True value Ethernet network traffic data, the predicted value and the prediction error

4.2. MPEG4 network traffic data prediction

When using RMPM to predict MPEG4 network traffic, the parameter value is the same as the previous experiment, that is $d=3$. The number of learning samples and the number of test samples are still 200 and 150 respectively, but the parameter ε is changed to 0.4. The predicted results are shown in Figure 2. Similar to Figure 1, the Figure 2 is also divided into upper and lower subgraphs. The solid line in the upper graph represents the true value of MPEG4 network traffic, and the dashed line represents the predicted value of network traffic. The solid line in the following figure represents the prediction error. It can also be seen from Figure 2 that the prediction curve of MPEG4 network traffic matches well with the actual curve. The calculation results show that the mean square error of this prediction is

3.182×10^{-3}, which means that the prediction accuracy of MPEG4 network traffic is higher than that of Ethernet.

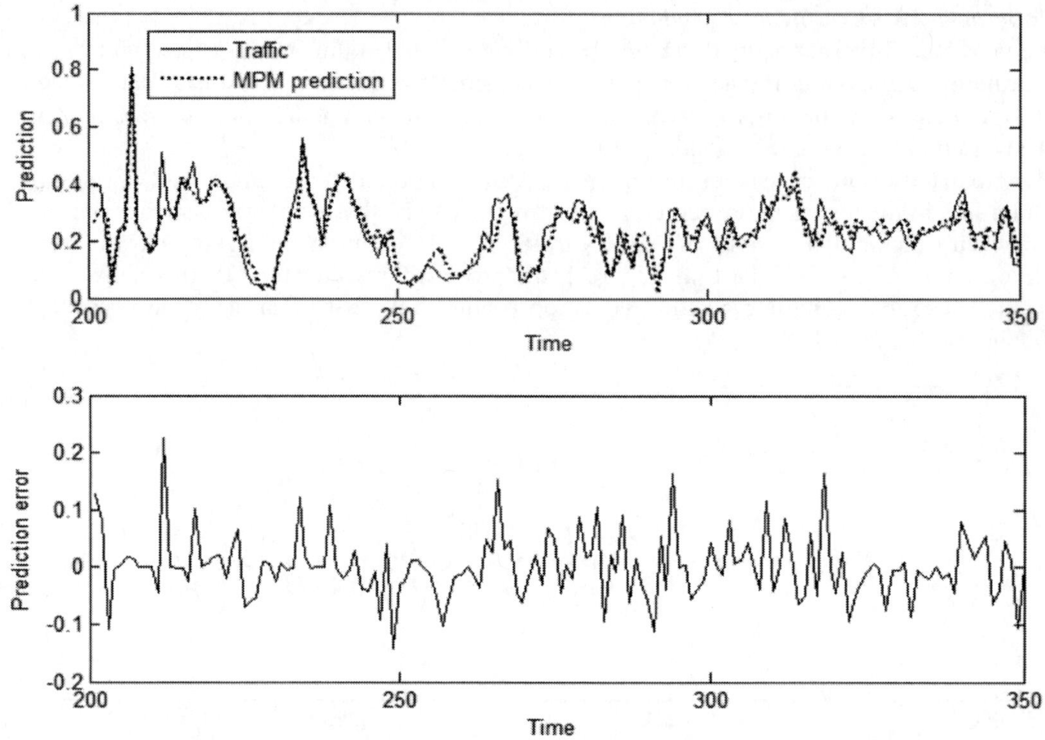

Figure 2. True value, predicted value and prediction error of MPEG4 network traffic data

5. Conclusions

Network traffic prediction is a challenging task, and traditional learning models have achieved some results in prediction. In order to further improve the prediction accuracy, this paper proposes a new network traffic prediction method based on the robust minimax probability machine (RMPM), and uses this method to predict the network traffic data such as Ethernet and MPEG4. The prediction results verify the superiority of the method. In the next step, we will focus on the local self-similarity research of the network traffic, in order to construct a more accurate network traffic prediction method based on the essential characteristics of network traffic data.

References

[1] Paxson, V. and Floyd, S, 1995, Wide-Area Traffic: the Failure of Poisson Modeling, IEEE/ACM Transactions on Networking, vol. 3, pp.226-244.

[2] Xin Yao, Manfred Fischer and Gavin Brown, 2001 Neural network ensembles and their application to traffic flow prediction in telecommunication networks, IJCNN' 2001, vol.1 pp.693-698.

[3] Hong Fei, and Wu Zhimei, 2006, Multiscale Network Traffic Prediction Model Based on Wavelet, Journal of Computer Science, pp. 88-91.

[4] Zhang Yinglu, 2008, Network traffic forecasting based on genetic algorithm optimization support vector machine, Computer Science, Vol. 35, pp.177-179.

[5] Dai Yue, 2008, Research on chaotic characteristics of network traffic and prediction algorithm of network traffic. Jiangnan University, Master's Degree Thesis.

[6] G. R. G. Lanckriet, L. E. Ghaoui, C. Bhattacharyya, and M. I. Jordan , 2002, Minimax probability machines. in Advances in Neural Information Processing Systems (NIPS) 14, T. G. Dietterich, S. Becker, and Z. Ghahramani editors, Cambridge, MA: MIT Press.
[7] T. R. Strohmann and G. Z, 2002, Grudic, A Formulation for minimax probability machine regression, In Advances in Neural Information Processing Systems (NIPS) 14, T. G. Dietterich, S. Becker, and Z. Ghahramani editors, Cambridge, MA: MIT Press.
[8] K. Z. Huang, H. Q. Yang, and I. King et al, 2004, Biased minimax probability machine for medical diagnosis, Tech. Rep., Dep. Computer Science and Engineering, The Chinese University of Hong Kong.
[9] C. H. Hoi and M. R. Lyu, 2004, Robust face recognition using minimax probability machine, Tech. Rep., Dep. Computer Science and Engineering, The Chinese University of Hong Kong.
[10] G. R. G. Lanckriet, L. E. Ghaoui, C. Bhattacharyya, and M. I. Jordan, 2002, A robust minimax approach to classification, Journal of Machine Learning Research, Vol.3, pp.555-582.

Design of By-Pass Excitation Cable Force Sensor

Zhou Yishu[1], An Ping[2], Cao Jinjin[1], Zhang Jinning[1*] and Liu Chang[1]

[1]Research Institute of Highway Ministry of Transport No. 8 Xitucheng Road, Haidian District, Beijing

[2]China National Accreditation Service for Conformity Assessment

[*]Corresponding Author E-mail: jinning.zhang@foxmail.com

Abstract. With the continuous progress of bridge technology and the requirements for bridge aesthetics, cable technology is increasingly widely used in long-span bridges. Due to the long-term environment of alternating stress, corrosion and wind-induced vibration, it is easy to cause local fatigue and damage of the cables, which endangers the safety of the entire bridge structure. The sleeve cable force sensor based on magnetic flux method is widely used in practical engineering among many measurement methods. Existing sleeve cable force sensors need to be installed at the beginning of construction, so it is difficult to reinstall them in the follow-up, and its long-term performance is also difficult to evaluate and measure online. To solve this problem, this paper introduces the design of a kind of magnetic flux cable force sensor with by-pass excitation, and presents its measuring principle and design structure. The designed by-pass excitation cable force sensor is expected to solve the problem of the existing sleeve sensor.

1. Introduction

As the bearing member of various large buildings and structures, steel cables are the most important part of the whole structure, and the stress status plays an extremely important role in the overall structure safety [1, 2]. Real-time measurement of cable force is the main means to judge whether the bridge structure is safe. At present, many measurement methods such as vibrating string, fiber optic strain, pressure gauge and magnetic flux have been developed [3-7] to realize long-term monitoring and measurement of cable force. Among them, the cable force sensor based on magnetic flux method is widely used in practical engineering due to its excellent sensing performance and long-term working stability [8-10].

The existing magnetic flux cable force sensors are all designed in a sleeve type, that is, both the excitation coil and the induction coil are wound and nested on the same insulating hollow sleeve, the tested cable passes through the sleeve, and the excitation magnetic field directly magnetizes the tested cable, and its magnetization characteristics are obtained from the measuring coil for measurement and analysis [11]. The sensor has simple structure, uniform magnetic field, good linearity and accurate measurement result. However, due to the sleeve design, it needs to be installed at the beginning of bridge construction. Once completed, it is difficult to carry out follow-up installation. In addition, for long-term service bridges, it is difficult to disassemble and calibrate the measurement performance of sensors after long-term operation. Therefore, there is an urgent need for a kind of high-precision sensor that can be applied to the field installation at a later stage, to install the cable force monitoring system for the built bridge, and to carry out field measurement and calibration for the existing sensor.

This paper introduces the measurement principle and design structure of a kind of bypass cable force sensor with double bypass excitation.

2. Principle of magnetic flux cable force measurement

The ferromagnetic materials will be magnetized in the magnetic field, and when it is subjected to mechanical force, strain is generated internally, resulting in the change of permeability μ. Permeability μ is the ratio of magnetic flux density B to magnetic field intensity H. In the presence of stress, the permeability will change, and the relationship between the change value and the stress can be expressed as:

$$\Delta\mu=-2\lambda_m\left(\frac{\mu_1}{B_m}\right)^2\sigma \quad (1)$$

$\Delta\mu$ is the permeability variation, λ_m is the magnetostriction coefficient in the state of magnetization saturation, μ_1 is the permeability without external force, B_m is the magnetic induction intensity in the state of magnetization saturation, σ is the stress generated inside a component. From formula (1):

$$\sigma=-\frac{B_m^2}{2\lambda_m\mu_1^2}\Delta\mu \quad (2)$$

As can be seen from formula (2), $\Delta\mu$ is proportional to σ, and the stress value can be calculated by measuring $\Delta\mu$. Since it is difficult to measure the permeability directly, the magnetization characteristics of ferromagnetic materials can be studied by using the principle of electromagnetic induction. That is, the alternating magnetic field is generated around the material by the excitation coils, and induced electromotive force generated by inductive measurement of the measuring coil is used to calculate the variation of permeability [12]:

$$\Delta\mu=\frac{\Delta B}{\Delta H}=\mu_0\left[1+\frac{S_0}{S_f}\left(\frac{V_{out}}{V_0}-1\right)\right] \quad (3)$$

Among them, S_0 is the total cross-sectional area of the coil, S_f is the cross-sectional area of the cable under test, V_{out} is the average output voltage after integrating the induced voltage with time, and V_0 is the output voltage when the cable is unloaded. The integral voltage is measured by experiment, and the permeability variation can be calculated from formula (3), and then the stress value can be obtained from formula (2).

3. Design of bypass excitation sensor

Figure 1. 15.2 mm steel cable for test

In this paper, the system designed for the Φ15.2 mm prestressed steel cable, which is the most widely used prestressed steel cable specification. The cable consists of seven steel wires, including six side wires and one middle wire. The diameter of side wires is 5.025 mm and the diameter of middle wire is 5.15 mm (Figure 1).

In order to facilitate subsequent disassembly and assembly, it is necessary to change the excitation mode of the sleeve type. Compared with sleeve sensors, bypass excitation has lower relative efficiency due to the problems of air reluctance and magnetic leakage. Four excitation coils are used to increase the magnetization intensity, and more turns are designed to facilitate the magnetization. Because the number of measuring coil is less, it can be manually wound on site.

The sensor is mainly composed of bypass core, cover plate and excitation coil arranged on the bypass core. The 3D structure of the designed sensor is shown in Figure 2.

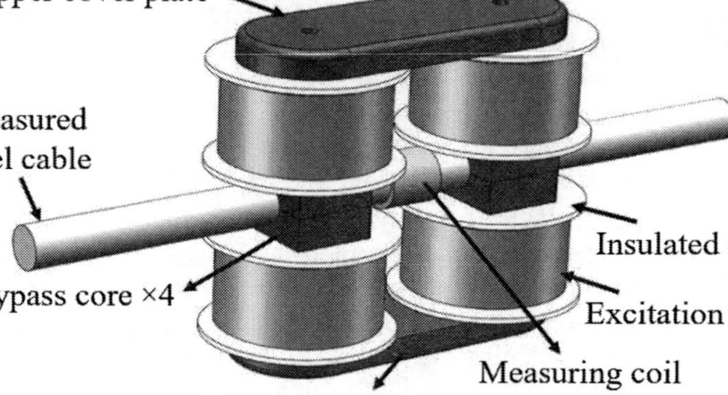

Figure 2. The 3D structure of the designed sensor

3.1 Bypass core

Bypass core is used to connect and fix the measured steel cable, and as the core of the excitation coil, it is the key part of the magnetic circuit. The bypass core is processed with pure electrical iron to ensure good magnetic conductivity. Φ15.6 mm installation space is designed for Φ15.2 mm measured steel cables to ensure a certain system error margin. At the same time, excessive gap leading to excessive magnetoresistance should be avoided. The structure of bypass core is shown in Figure 3(a).

3.2 Cover plate

The cover plates are used to fix the iron core and conduct magnetic conduction as part of the magnetic circuit in the system. The system adopts double bypass excitation mode, requiring two cover plates with the same structure design. To ensure good magnetic conductivity, the cover plates are processed with pure electrical iron. The structure of cover plate is shown in Figure 3(b).

3.3 Insulated coil tube

The insulated coil tube is nested on the bypass core, and is made of insulating epoxy resin material. It can be winded separately in advance, and then installed and wiring during field measurement. The structure of insulated coil tube is shown in Figure 3(c).

3.4 Sensor assembly

The sensor is arranged on both sides of the cable in a symmetrical structure. About 1600 turns of high strength enameled wire of Φ0.6 mm is wound around the coil tube. When the sensor is assembled on the cable, magnetorheological grease is used to reduce the magnetoresistance introduced by air between the bypass core and the cable. The structure of sensor assembly is shown in Figure 4.

IOP Conf. Series: Journal of Physics: Conf. Series **1267** (2019) 012068 doi:10.1088/1742-6596/1267/1/012068

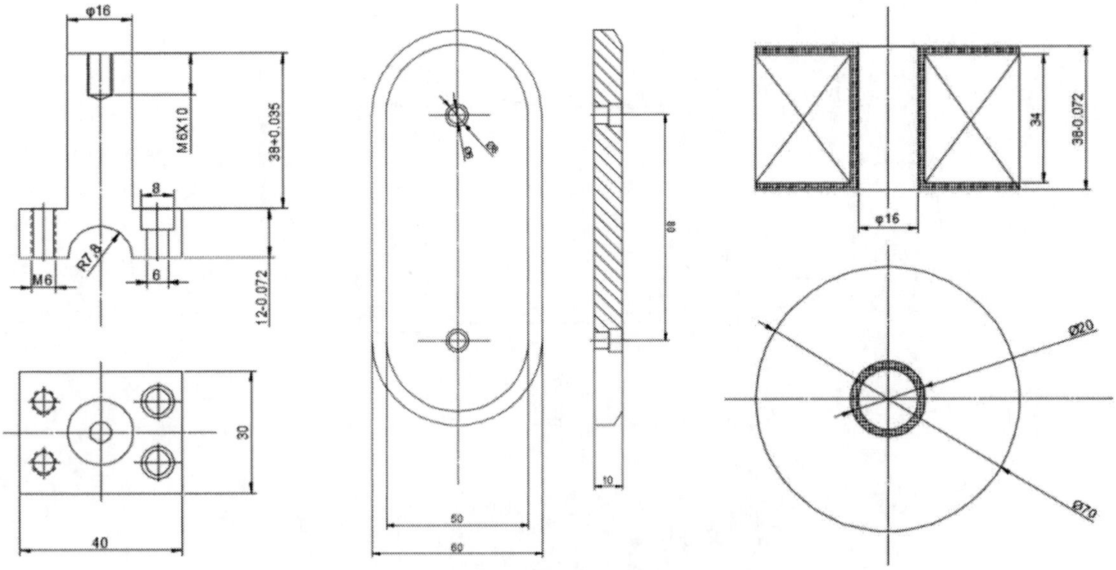

(a) bypass core (b) cover plate (c) insulated coil tube

Figure 3. The design of key components

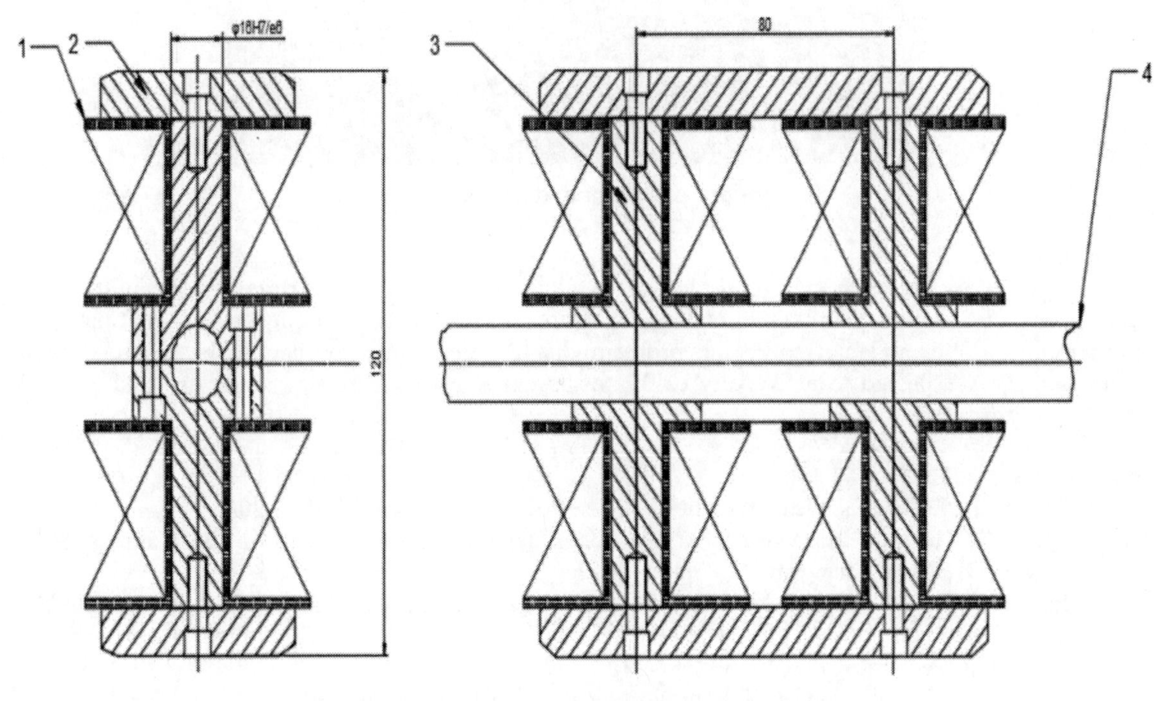

Figure 4. The structure of sensor assembly

It is necessary to ensure that the magnetic field generated by the four coils of the sensor is in the same direction when passing through the cable. Figure 5 shows the wiring method of sensor coil. Figure 6 shows the sensor and its installation.

IOP Conf. Series: Journal of Physics: Conf. Series **1267** (2019) 012068 doi:10.1088/1742-6596/1267/1/012068

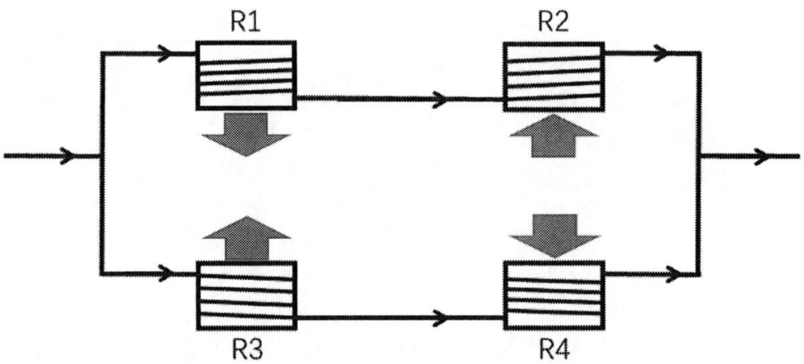

Figure 5. Wiring method of sensor coil

Figure 6. Sensor and its installation

4. Conclusion

This paper introduces the structural design of a magnetic flux cable force sensor with bypass excitation. The sensor adopts the bypass structure, which is convenient for installation at the measurement field, and can solve the problem that the sleeve sensor cannot be installed in the follow-up. It is expected to effectively calibrate the existing sleeve sensor in service and solve the problem of on-site calibration.

References

[1] Wang M L, Wang G. Measurement of stresses of large steel cables [M]. 2003.
[2] Wang G. The Development of a Cable Force Detecting System based on the Magneto Elastic Effect [D]. Xi'an: Chang'an University. 2017.
[3] Park S H, Kim J W, Nam M J, et al. Magnetic Flux Leakage Sensing-Based Steel Cable NDE Technique Incorporated on a Cable Climbing Robot for Bridge Structures [J]. Advances in Science & Technology, 2013, 83:217-222.
[4] Zhou X F, Xu J F, Liang L. Design and Analysis of Fiber Bragg Grating Cable Tension Sensor [J]. Chinese Journal of Sensors and Actuators, 2007, 20(5): 1007-1011.
[5] Jiang X, Jing Y Z, Zhang H C. Vibrating Wire Sensor based on Sweeping Frequency to Excite Single Coil [J]. Transducer and Microsystem Technologies, 2001, 20(5): 22-4.
[6] Park S, Kim J W, Moon D J. NNon-contact Main Cable NDE Technique for Suspension Bridge using Magnetic Flux-based BH Loop Measurements[C]. International Society for Optics and Photonics, 2015, 9437: 94370M.

[7] Wu K X, Liu K M, Yang J X. Measuring System of Cable Tension Based on Frequency Method [J]. China Journal of Highway and Transport, 2006, 19(2): 62-6.
[8] Cappello C, Zonta D, Laasri H A, et al. Calibration of Elasto-Magnetic Sensors on In-Service Cable-Stayed Bridges for Stress Monitoring [J]. Sensors, 2018, 18(2): 466.
[9] Deng N C, Long Y, Sun L M, et al. Magnetic Flux Sensor and its Application in Bridge Engineering [J]. Prestress Technology, 2008, 02: 17-20.
[10] Lenz J, Edelstein A S. Magnetic Sensors and Their Applications [J]. IEEE Sensors journal, 2006, 6(3): 631-649.
[11] Liu X L. Design Method and Experiment of Magneto-elastic Cable Tension Sensor [D]; Chong Qing: Chong University, 2011.
[12] Li D F, Wang G, Yang K B. Design and Simulation of Magnetic-elastic Cable Tension Detecting System [J]. Transducer and Microsystem Technologies, 2017, 36(09): 67-9+76.

Kinematic Analysis of Treatment Table Robot in Proton Radiotherapy Project

Yu Xiong[1,2], Qingxi Yang[1*], Sheng Yan[1,2], Yuntao Song[1,2] and Jianghua Wei[1]

[1]Institute of Plasma Physics, Chinese Academy of Sciences , Hefei, 200031, China
[2]University of Science and Technology of China, Hefei, 200031, China. No.350, Shushan Lake Road, Luyang District, Hefei City, Anhui Province

[*]qxyang@ip.ac.cn

Abstract. This paper introduces a multi-link robot, named treatment table robot (TTR), which has been developed as an important part in protan therapy system SC200. TTR is aimed to carry and move patient's body during treatment, so as to change the space location and poseture of patient'tumour to ensure the proton beem can always focus on it. To realize the requirement, the TTR has been designed as a 6-DOF robot. Based on the builded D-H model, the forward and inverse kinematics are studied. Then the joint trajectory planning has been researched. Finally,ADAMS software was enployed to simulate the kinematic of the TTR, which verified the feasibility of the kinematic algorithm.

1. Introduction

Protan therapy is thinked as a better way to cure cancer compared with X-Ray, the advantage of this method is its accuracy and lesser side effect on normal tissue[1].SC200 is a integrated protan therapy system developed by Hefei ion medical center. Proton therapy contains the following steps: (1)Superconducting Cyclotron,create and accelerate proton beams,(2)Beam Transport System,which focuses and shapes the beam and guides it to the treatment room,(3)Treatment system,the rotating gantry and treatment table robot adjusted position and poseture to cooperatively complete established stratified scanning irradiation plan of the tumour,(4)Control system, handle various data and manage the whole system.

As shown in figure 1, The robot, named treatment table robot (TTR),as an important component of treatment system, is installed on the ground of Patient Treatment Room,. In order to associatively adjust the position and poseture of patient's tomour with the rotation of gantry, the TTR should have the ability of moving along and rotating around all the xyz-axis, based on this request, the TTR has been designed as a 6 DOF serial-link robot.so, the accompanying issue, such as the structure and mathematic model of TTR, the motion planning method of robot to response the instruction of control system, is what we need investigate in this paper[2].

IOP Conf. Series: Journal of Physics: Conf. Series **1267** (2019) 012069 doi:10.1088/1742-6596/1267/1/012069

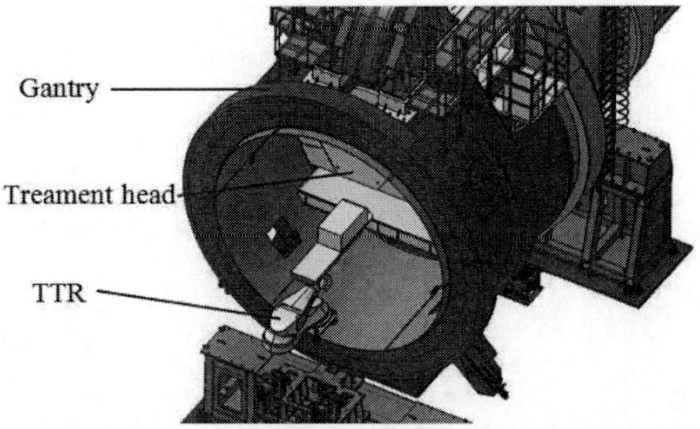

Figure 1. The location of TTR in tratement room

2. Structure and D-H model

As shown in figure 2, The TTR contains seven arms and six joints, the base arm is fixed on ground, the terminal arm is designed as a platform for patient to lie flat, all of the arms connected one by one to form the special configuration of joints. Joints 1and 2 make the TTR can move on the horizontal plane widely, the joints 4 and 6 modify the heading and roll,The joints 3 and 5 change both the position of TTR in vertical direction and the pitch. The unique structure makes the TTR can fulfill the adjustment of position and posrture of patient's tomour during proton medical therapy.

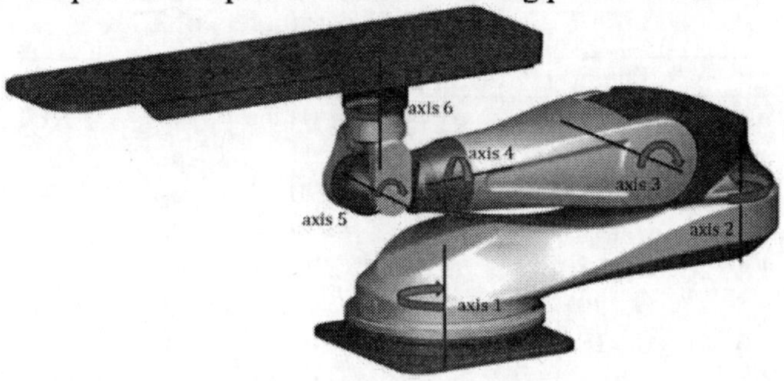

Figure 2. Joint structure of TTR

Specialy, the z-axis of joint 4,5 and 6 has been designed as perpendicularly intersected with eachother, so that accord with the Pieper norm.as we know, when the 6 D.O.F manipulator meet the Pieper principle, the rotate degree of joint 1,2,3 mainly influence the final coordinate point's location, and the rotate degree of joint 4,5 and 6 have more impact on the ultimate arm's posture. Based on this feature, we can do some reasonable adjustment of the joint's coordinate system when we set up the D-H model of TTR to make inverse kenimatic problem more easy .

As shown in figure 3, the D-H model of TTR is established, each joint's Cartesian coordinate system has been formulated by referring to D-H convention [3]. If we want to adjust the location and poseture of tumour's coordinate system, the first step is identified the location and poseture of tumour's coordinate frame in global reference coordinate sysytem, then the homogeneous coordinate transition model of joints just like a bridge to realize it.

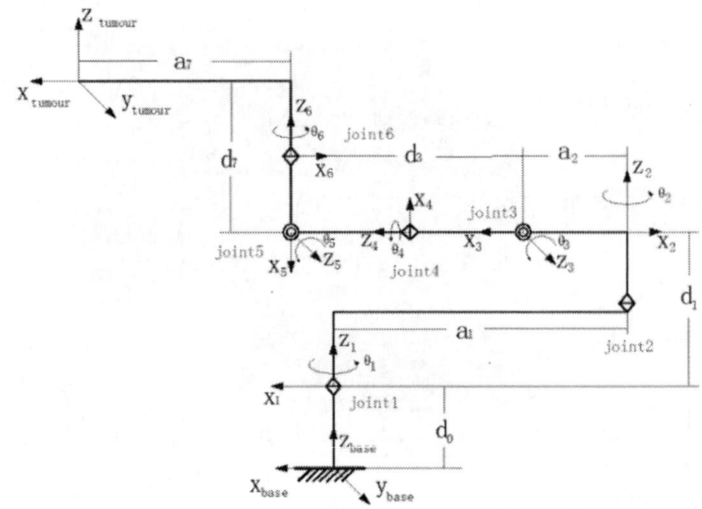

Figure 3. The D-H model established for TTR

The homogeneous coordinate transition model of joints have been accomplished in the order of theta--alpha--d--a.What's more, the frame 2 has been moved some distance along its z axes from joint 2,and moved the frame 4,6 to joint 5 along its z axes, that ensured the homogeneous coordinate transfered in certain order, and made the D-H paramaters more simplified, so as to derive the forward and inverse kinematics more easy. The standard D-H parameters[4] of TTR as shown in Table 1.

Table 1. Standard D-H paramaters of TTR.

Link	Theta(°)	d(mm)	Alpha(°)	a(mm)	Range(°)
base	0	d_0	0	0	NULL
1	θ_1 (180)	d_1	0	a_1	+/- 220
2	θ_2 (180)	0	-90	a_2	+/- 220
3	θ_3 (-90)	d_3	-90	0	-45~18
4	θ_4 (180)	0	-90	0	+/- 10
5	θ_5 (90)	0	-90	0	+/- 28
6	θ_6 (180)	0	0	0	+/- 165
tumour	0	a_7	0	d_7	NULL

(a_1= 1200,a_2=360,d_1=563,d_3=1165.5,a_7=1581 and d_7=604.5)

3. Kinematic analysis

3.1. Forward kinematic

The objective of forward kinimatic is identified the location and poseture of tumour's coordinate frame in global reference coordinate sysytem(GCS) when we already knew the rotation angle of each joint. According to D-H convention and transformation theory between neighboring D-H coordinate system, the tumour's position and poseture in GCS can be ascertained as *equation* 1.

$$ {}^{tumour}_{base}T = A_{base}(A_i)^6 A_{tumour} = A_{base} T_{table} A_{tumour} \quad (i=1.2.3...6) \tag{1} $$

The transformation matrix of the base and tumour are varied with different installation site and paitient,so we fucous on the constant part--the robot. as mentioned, the transformation order between neighboring D-H coordinate system is theta--alpha--d--a,then the transformation matrix of two joints could be expressed as *equation* 2.in which θ_i,a_i,d_i and α_i are the D-H parameters of link i, c ≡ cos and s ≡ sin, the same abbreviations are used in the following formulae.

$$ A_i = Rot(z,\theta_i)Rot(x,\alpha_i)Trans(0,0,d_i)Trans(a_i,0,0) $$

$$A_i = \begin{bmatrix} C(\theta_i) & -C(\alpha_i)S(\theta_i) & S(\alpha_i)S(\theta_i) & a_iC(\theta_i)+d_iS(\alpha_i)S(\theta_i) \\ S(\theta_i) & C(\alpha_i)C(\theta_i) & -S(\alpha_i)C(\theta_i) & a_iS(\theta_i)-d_iS(\alpha_i)C(\theta_i) \\ 0 & S(\alpha_i) & C(\alpha_i) & d_iC(\alpha_i) \\ 0 & 0 & 0 & 1 \end{bmatrix} \quad (2)$$

With the parameters in Table.1,forward kinematics can be solved by multiplying each homogeneous transformation matrix between consecutive links.The formula shown as *equation* 3

$$T_{table} = (A_i)^6 = \begin{bmatrix} n(\theta) & o(\theta) & a(\theta) & p(\theta) \\ 0 & 0 & 0 & 1 \end{bmatrix} = \begin{bmatrix} n_x & o_x & a_x & p_x \\ n_y & o_y & a_y & p_y \\ n_z & o_z & a_z & p_z \\ 0 & 0 & 0 & 1 \end{bmatrix} \quad (3)$$

where $\boldsymbol{\theta}$ is the (6x1) vector of joint variables, and $\boldsymbol{n}$, $\boldsymbol{o}$, $\boldsymbol{a}$ are the unit vectors of the tumour frame and $\boldsymbol{p}$ is the position vector respect to the origin of the base frame,details as equation 4.

$$\begin{cases} n_x = C6(C12S3S5 + S12C5S4 + C12C3C4C5) - S6(S12C4 - C12C3S4) \\ o_x = -C6(S12C4 - C12C3S4) - S6(C12S3S5 + S12C5S4 + C12C3C4C5) \\ a_x = C12C5S3 - S5(S12S4 + C12C3C4) \\ n_y = S6(C12C4 + S12C3S4) + C6(S12S3S5 - C12C5S4 + S12C3C4C5) \\ o_y = C6(C12C4 + S12C3S4) - S6(S12S3S5 - C12C5S4 + S12C3C4C5) \\ a_y = S5(C12S4 - S12C3C4) + S12C5S3 \\ n_z = C6(C3S5 - C4C5S3) - S3S4S6 \\ o_z = -S6(C3S5 - C4C5S3) - C6S3S4 \\ a_z = C3C5 + C4S3S5 \end{cases} \quad (4)$$

$$\begin{cases} p_x = a_2C12 + a_1C1 - d_3C12S3 \\ p_y = a_2S12 + a_1S1 - d_3S12S3 \\ p_z = d_1 - d_3C3 \end{cases}$$

3.2. Inverse kinematic

The inverse kinematics is required to find the angles of the six revolute joints while the the tumour's position and poseture in GCS is given, and the inverse kinematics is the fundation of trajectory planning.

Based on the Eq.4,the analytic solution can be derived,the inverse kinematics of the TTR is solved in two stages[5],in the first stage,we get the theta 1,2,3 using the position vector $[p_x\ p_y\ p_z]$, The steps are as follow.

1.From the equation p_z, we can infer that

$$\theta_3 = -a\cos[(d_1 - p_z)/d_3] \quad (5)$$

2.Extracted cos(θ_1) and sin(θ_2) from the equation p_x p_y, and obtained sum of the squares,then we get

$$\begin{cases} k_1S12 + k_2C12 = k_3 \\ k_1 = 2p_y(d_3S3 - a_3) \\ k_2 = 2p_x(d_3S3 - a_2) \\ k_3 = a_1^2 - p_x^2 - p_y^2 - (d_3S3 - a_2)^2 \\ \theta_1 + \theta_2 = \text{atan2}(k_3, \pm(k_1^2 + k_2^2 - k_3^2)^{1/2}) - \text{atan2}(k_2, k_1) \end{cases} \quad (6)$$

3.Plug the $\theta_1 + \theta_2$ into p_x p_y ,we get

$$\begin{cases} k_4 = \sin(\theta_1) = (p_y - a_2S12 + d_3S12S3)/a_1 \\ k_5 = \cos(\theta_1) = (p_x - a_2C12 + d_3C12S3)/a_1 \\ \theta_1 = a\tan2(k_4, k_5) \ ; \ \theta_2 = (\theta_1 + \theta_2) - \theta_1 \end{cases} \quad (7)$$

In the second stage,defined the matrix $R=A_4A_5A_6$ and matrix $L=inv(A_3)inv(A_2)inv(A_1)T_{table}$, then each item of L and R are corresponding equal.

4.based on $L_{33}=R_{33}$,we get

$$\begin{cases} k_6 = -C5 = a_x S1S2S3 - a_x C1C2S3 - a_y C1S2S3 - a_y C2S1S3 - a_z C3 \\ \theta_5 = a\cos(-k_6) \end{cases} \tag{8}$$

5.based on $L_{31}+L_{32}=R_{31}+R_{32}$, we get

$$\begin{cases} k_7 S6 - k_7 C6 = k_8 \\ k_7 = S5 \\ k_8 = n_x S1S2S3 - n_x C1C2S3 - n_y C1S2S3 - n_y C2S1S3 + o_x S1S2S3 - o_x C1C2S3 - o_y C1S2S3 - o_y C2S1S3 - o_z C3 - n_z C3 \\ \theta_6 = a\sin\left[k_8 / \sqrt{2} k_7\right] + \pi / 4 \end{cases} \tag{9}$$

6.based on $L_{13}+L_{23}=R_{13}+R_{23}$, we get

$$\begin{cases} k_7 S4 + k_7 C4 = k_9 \\ k_9 = a_x C1C2C3 - a_z S3 + a_y C1C3S2 + a_y C2C3S1 - a_x C3S1S2 + a_x S12 - a_y C12 \\ \theta_4 = a\sin(k_{10} / \sqrt{2} k_9) - \pi / 4 \end{cases} \tag{10}$$

Normally, the inverse kinematic could be solved from the formula 5-10 when ***a o n p*** are given,each angle of rotation can directly obtain a closed-form solution.

3.3. Joint trajectory planning

Regarding the motion of robots,we often need to discuss the trajectory planning of joint space and the motion planning of Cartesian space[6].For TTR,there are no other special requirements except speed limit about the movement in space,the rotation angle range limit has already prevented the collision of TTR with gantry, what we need to do is connect the two discrete joint angle values which obtained by the inverse solution algorithm with smooth curves,to drive TTR movement from one point to anather. In this paper,the joint angle space has been planned by 5th polynomial as equation 11.

$$\theta(t) = a_0 + a_1 t + a_2 t^2 + a_3 t^3 + a_4 t^4 + a_5 t^5 \tag{11}$$

For given beginning and end points,when we difined the speed and acceleration limit of rotation, the 5th polynomial's parameters can be derived by equation 12.

$$\begin{cases} a_0 = \theta_{begin} \quad ; \quad a_1 = \dot{\theta}_{begin} \quad ; \quad a_2 = \ddot{\theta}_{begin} / 2 \\ a_3 = \left[20\theta_{end} - 20\theta_{begin} - (8\dot{\theta}_{end} + 12\dot{\theta}_{begin})\, t_{total} - (3\ddot{\theta}_{begin} - \ddot{\theta}_{end})\, t_{total}^{2} \right] / 2t_{total}^{3} \\ a_4 = \left[30\theta_{begin} - 30\theta_{end} + (14\dot{\theta}_{end} + 16\dot{\theta}_{begin}) t_{total} + (3\ddot{\theta}_{begin} - 2\ddot{\theta}_{end}) t_{total}^{2} \right] / 2t_{total}^{4} \\ a_5 = \left[12\theta_{end} - 12\theta_{begin} - (6\dot{\theta}_{end} + 6\dot{\theta}_{begin}) t_{total} - (\ddot{\theta}_{begin} - \ddot{\theta}_{end}) t_{total}^{2} \right] / 2t_{total}^{5} \end{cases} \tag{12}$$

Further,we need determined the time(t_{total}) of motion between two points by caculating the interval of the joints with the rotation speed limit of the joints.

4. Kinematic simulation

4.1. Joint position curve

In order to test the kinematics algorithm,selected the motion between two position for simulation analysis. As shown in figure 4,the position 1 is assigned for patient to get on the table,the position and poseture in GCS is *[R(z,-60°),-1245,-2020,930]*, and the position 2 is assigned for patient to complete CBCT imaging,the position and poseture in GCS is *[E,2700,0,1400]*.

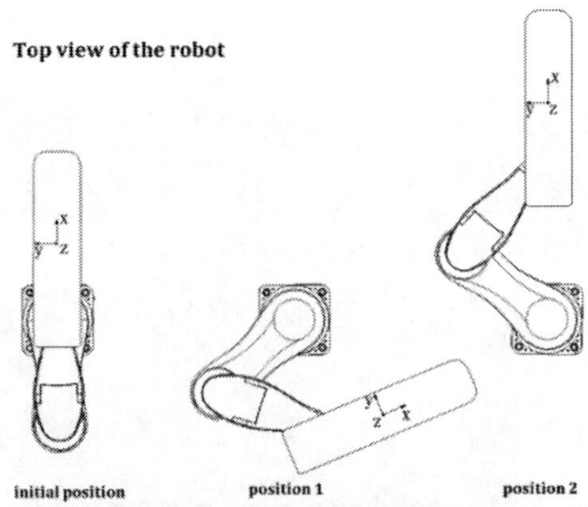

Figure 4. Position indication

Based on the given [***a o n p***] and the inverse kinematic algorithm, the corresponding joint angle can be calculated by the use of Matlab. as shown in Table.2.

Table 2. Joint angle of position.

position	θ_1(°)	θ_2(°)	θ_3(°)	θ_4(°)	θ_5(°)	θ_6(°)
1	156.26	74.16	-72.39	180	72.39	249.57
2	81.8	226.79	-95.79	180	95.79	231.41

According to Table 2,we can find that rotational stroke of joint 2 is the biggest,to ensure smooth movement and take care of patient's feeling,limited the speed of joint rotation with no more than 10 deg/s,simply assumed that the joint rotate at average speed 5deg/s,the whole motion time is nearly 31 second.what's more,the speed and acceleration of each joint's rotation must be zero to reduce the impact of startting and stopping on TTR.so,the 5th polynomial of each joint can be derived as equation 13.

$$\begin{cases} \theta_1(t)=156.2628-0.02499t^3+0.001209t^4-0.0000156t^5 \\ \theta_2(t)=74.1637+0.05123t^3-0.002479t^4+0.00003199t^5 \\ \theta_3(t)=-72.3932-0.00785t^3+0.00038t^4-0.0000049t^5 \\ \theta_5(t)=72.3932+0.00785t^3-0.00038t^4+0.0000049t^5 \\ \theta_6(t)=249.569-0.00609t^3+0.0002949t^4-0.0000038t^5 \end{cases} \tag{13}$$

Plot the joint position curve(a) ,the velocity(b) and acceleration(c) in figure 5.as shown,the curve of velocity and acceleration are continuous without mutation, the velocity of joint's rotation is less than 10deg/s,and the speed and acceleration of joint rotation start and stop are zero, it meet the planning expectation.

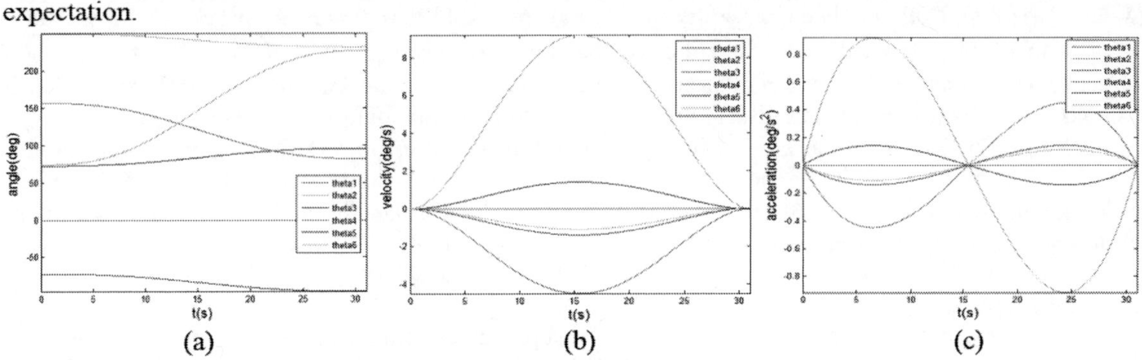

Figure 5. Rotation related curve of joints

4.2. Joint position curve

In order to test whether the joint position curve can drive TTR from position 1 to position 2 as we want,a kinematic model is established in ADAMS software as shown in figure 6 [7].

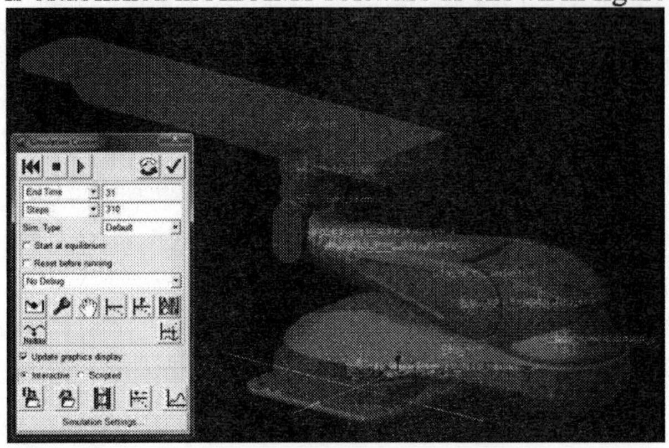

Figure 6. Kinematic model in ADAMS

Applied curves in equation 13 as the drive function for each joint to the kinematic model,the entrie exercise process taked 31s ,Then,tracked the position change in each axis direction of the assumed tumour coordinate system in the GCS.As figure 7 shown,the track curve is continuous and smooth,and the position of the beginning and the end exactly reached our planned location.so the PTP movement can be considered successfull.

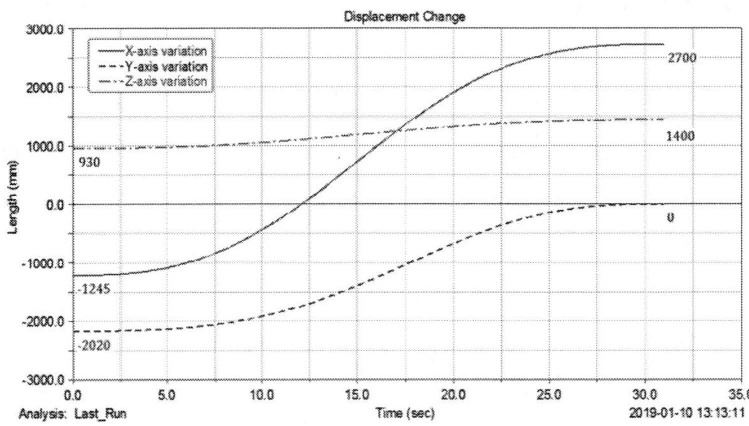

Figure 7. Coordinate track during the simulation from position 1 to 2

5. Conclution

A 6-DOF robot TTR has been presented as an indispensable component of protan therapy system SC200, to carrying patient' s body to adjust location and poseture in GCS as protan therapy plan requsted.The structure and D-H model are established,and the whole kinematic moduls are coded in Matlab. Two position points are selected to test the motion planning algorithm,Then,the drive function of each joint are added to the kinematic model which is setted up in ADAMS to implement motion simulation.The TTR achieves point-to-point motion smoothly and accurately,so the kinematic planning algorithm this paper offered is practical and reliable.Further research as dynamic studies and structure strength analysis are in the works.

Acknowledgment

The work presented in this paper received financial support from fund of *Joint Design of Sino-Russian Superconducting Proton Accelerator (1604b0602005)* and *Design of 200Mev Advanced Superconducting Proton Accelerator (1503062029)* in china. The authors are grateful to the anonymous workers for their helpful comments and suggestions to improve the presentation of the paper.

References

[1] Bussiere MR,JA Adams 2003 Treatment planning for conformal proton radiation therapy, *Techanol Cancer Res Treat* **2**(5) 389-399.
[2] Bruno Siciliano, Lorenzo Sciavicco, Luigi Villani, Giuseppe Oriolo, 2000 *Robotic: Modelling, Planning and Control*, Springer, Roman.
[3] https://en.wikipedia.org/wiki/Denavit_Hartenberg_parameters
[4] Richard S Hartenberg,Jacques Denavit 1995 A Kinematic Notation for Lower Pair Mwchanisms Based on Matrices, *Journal of Applied Mechanics* **22** 215-221.
[5] Zhou Ke 2013 Research on the Design and Trajectory Planning of 6-DOF Industrial Manipulators. Harbin Insitute of Technology.
[6] Cai Zixing 2009 *Robotics*.Tsinghua University Press,Beijin.
[7] Fu Mengyu 2017 The Kinematics and Dynamics Simulation of Spider Robots Based on ADAMS, *FMSMT*.

IOP Conf. Series: Journal of Physics: Conf. Series **1267** (2019) 012070 doi:10.1088/1742-6596/1267/1/012070

Design of Management and Control System of Aero-engine Nozzle Processing Line

Kai Chang[1,2,*], DeChao Zheng[1,2], Jian Yin[1,2] and Guangchao Tan[1,2]

[1] Intelligent Detection and Equipment Department, Shenyang Institute of Automation Chinese Academy of Sciences, Shenyang 110016, China.
[2]Institutes for Robotics and Intelligent Manufacturing, Chinese Academy of Sciences, Shenyang 110016, China.

[*]E-mail: changkai@sia.cn

Abstract. This paper introduces a kind of management and control system of machining production line for aero-engine nozzle. The main logistics is based on the logic control of PLC, which can dispatch a robot to realize the automatic transfer of the parts between workstations, automatic feeding and unloading of processing and testing equipment. Workstations are equipped with RFID to detect the identification of each nozzle. The system can reconstruct the process route according to the nozzle type, associate processing and test data of the nozzle with their identification number to establish the processing file. It realizes data exchanges between management and control systems by OPC interface, achieves centralized monitoring and control of the equipment of the whole line and improves the informatization level of workshop.

1. Introduction

At present, China's aviation manufacturing equipment is developing at a high speed. In most aeronautical parts processing factories, although multi-axis high-precision CNC equipment is introduced, the flow of semi-processed products between devices are still carried manually. It is inefficient and inconvenient to adjust the processing parts between processes, resulting in the accumulation of semi-finished products which occupies a large amount of space. Therefore, the demand for establishing automatic mechanical processing logistics transmission line is increasing day by day.

Because of the particularity of aviation products, the production process based on small and medium batches need to deal with a variety of processing objects. So, the production information changes frequently. It is very important to grasp the information of parts for the production management and control system.

Based on the nozzle processing project of an aero-engine company, this paper introduces a management and control system for a flexible machining production line, which can realize the automation of feeding and unloading, centralized control and monitoring of processing process and the establishment of parts processing files.

2. The layout of machining production line and design goal of management and control system

2.1. The layout of machining production line

The layout of machining production line for an kind of aero-engine nozzle is shown in figure 1. The processing flow is: putting the blank into the upper rotating silo manually → automatic marking → processing in turning-milling compound center → milling center → cleaning and drying → measuring in CMM → putting finished product to the lower rotating silo automatically. In order to maintain production rhythm of turning-milling and milling process cycle, two milling centers are arranged in machining production Line. The robot which can clamp the nozzle automatically with ground track realizes the automation of main logistics.

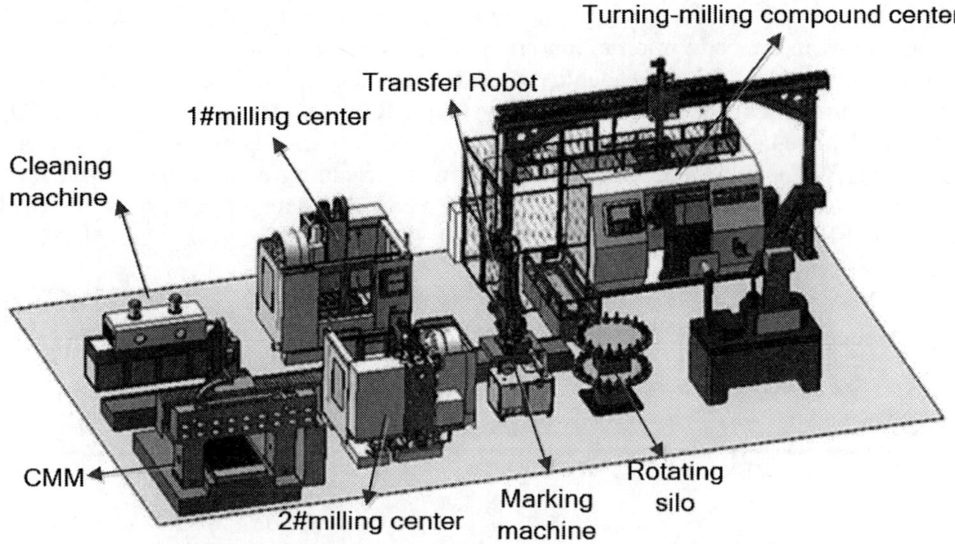

Figure 1. The layout of machining production line.

2.2. The design goal of management and control system

Based on the layout above, we need to establish a reliable system of management and control which can support the operation of a flexible machining production line. According to the technical agreement, it needs following 4 functions:

- The control system can dispatch the robot to realize the automatic transfer between all workstations of machining production line.
- The control system can identify the code of parts at each station which can be uploaded to the management system.
- The management system can realize real-time acquisition of production status data, reconstruct the process route according to the identification of parts and centralized display the status of all equipment.
- The management system can establish the process files of nozzle parts. The processing and testing key data of each production can be traced through unique identification code.

3. Design and implementation of management and control System

3.1. Hardware design of control System

The core of the main logistics control system is PLC 315-2PN/DP. Through field bus PROFIBUS and RS232 serial communication interface, it can interact with 6-DOF robot, three CNC machines (840D SL CNC system), cleaning machine, marking machine and CMM. It can dispatch robot to realize material placement for a station equipment, and then notify the equipment to start. When the equipment completes a work cycle, the completion signal of processing or testing will be fed back to main logistics control system, and then it will dispatch the robot to grab the parts to the next workstation to complete the full automation operation of the whole logistics.

IOP Conf. Series: Journal of Physics: Conf. Series **1267** (2019) 012070 doi:10.1088/1742-6596/1267/1/012070

In order to automatically collect the key data of each workstation which will be correlated with the code of nozzle number, RFID read-write head are configured in marking machine, CNC machine, cleaning machine and CMM workstations. The RFID carrier tag is embedded in the accompanying tooling of each nozzle component. Each RFID read-write head has PROFIBUS bus interface which can be connected to the main control PLC for read-write programming. When the robot grabs the blank parts to the marking machine station (1 # workstation), the management system transmits the information of flow number to the main control PLC and marking machine. After the marking process, the control system will notify the robot to grab the blank to the RFID read-write head of this workstation and write the code information into the RFID carrier embedded in accompanying tooling to achieve the unification of nozzle identification number and storage object of RFID carrier. When the nozzle parts arrive at a workstation, control system will read nozzle code stored in RFID carrier tag through the read-write head of this workstation and upload it to management system [1]. According to nozzle type, management system can switch processing, cleaning or measuring program to achieve flexible production. It can also bind the processing or measuring data information of the workstation to the nozzle number and complete the establishment of product processing files. Thus, the control system network established is shown in figure 2.

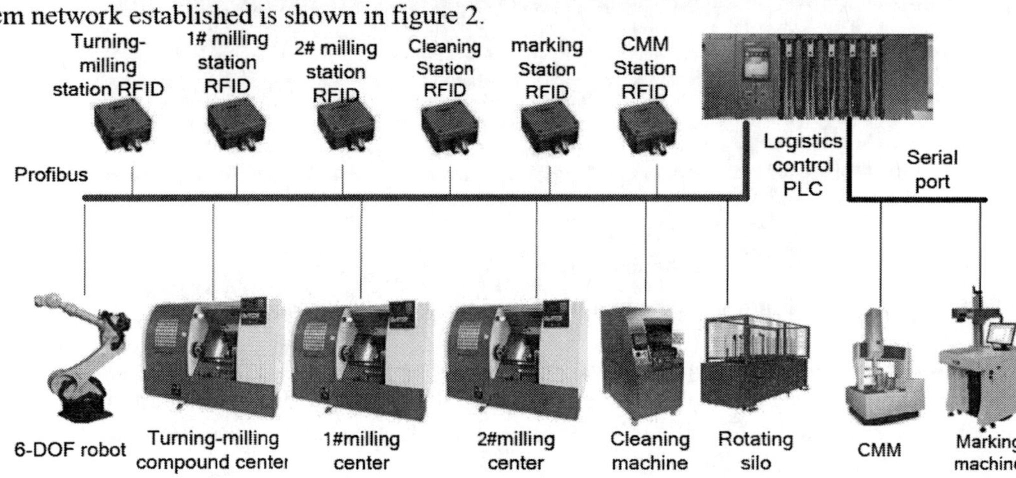

Figure 2. Hardware network diagram of control system.

3.2. Hardware design of management system

The management system is based on the IPC in the central control room. It uses industrial Ethernet to connect production units, electronic Kanban and other terminals together. Through OPC technology, the data interface between the production units and information Layer is opened, that realizes the interaction of production information [2]. The overall network structure of the management system is shown in figure 3.

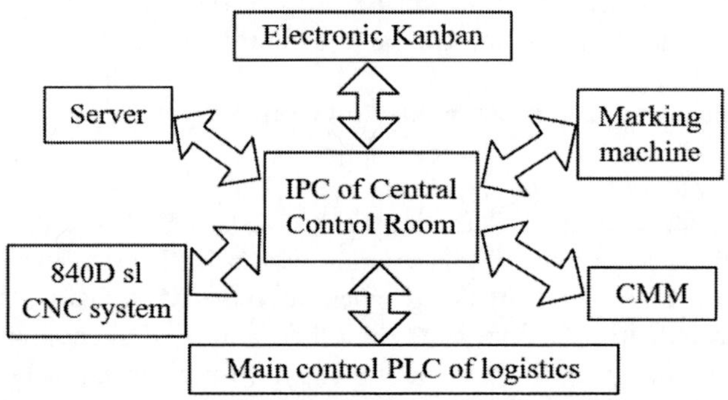

Figure 3. The network structure of the management system.

Managers can set production plan information in the IPC software interface of the centralized control room. After the production plan is formed, the system transmits the coding information of the nozzle to the marking machine, which engraves the unique coding identification for each nozzle component in the production process. The IPC platform of centralized control room can interact with each CNC machine and logistics main control PLC through OPC interface. So the management system can grasp the real-time status information of each equipment and realize Andon management by electronic Kanban [3]. The system can store the number of nozzles, the processing and testing data in the database of the server. By inputting the processing time or the number of the relevant batches of nozzles into the IPC software interface, the production information record of a certain nozzle can called out to be viewed.

3.3. Software design of control system

The realization of the control system mainly includes two parts: the logic control program of the main control PLC, the feeding and unloading motion control program of 6-DOF robot with seventh axis. Robot motion control program is mainly composed of the feeding and unloading motion instruction set and robot state self-checking module [4].

The main control PLC program is mainly composed of robot operation scheduling module, equipment status module, equipment communication module and RFID read-write module. Robot operation scheduling module can dispatch the robot to grab nozzle parts, finish automatic feeding and unloading cooperated with all devices and complete the main logistics operation according to the process flow. The equipment status module is responsible for collecting the production status signals of running, waiting, abnormal alarm and overtime alarm which will be uploaded to the management system used in production process monitoring and Andon management. The equipment communication module can receive the ready and running status word signals of the equipment and transmit the start or stop control word signals to the equipment through PROFIBUS bus and serial port to realize automatic feeding and uploading with the cooperation of each equipment and 6-DOF robot, automatic start of processing or testing program. The RFID read-write module can read the number of nozzle parts which are being processed or tested at a certain workstation and notify the management system to switch the processing, testing or cleaning procedures according to the nozzle type. When the processing or testing of a workstation is completed, the relevant process data are saved in the database with the nozzle number. Figure 4 is a basic flow chart for a workstation.

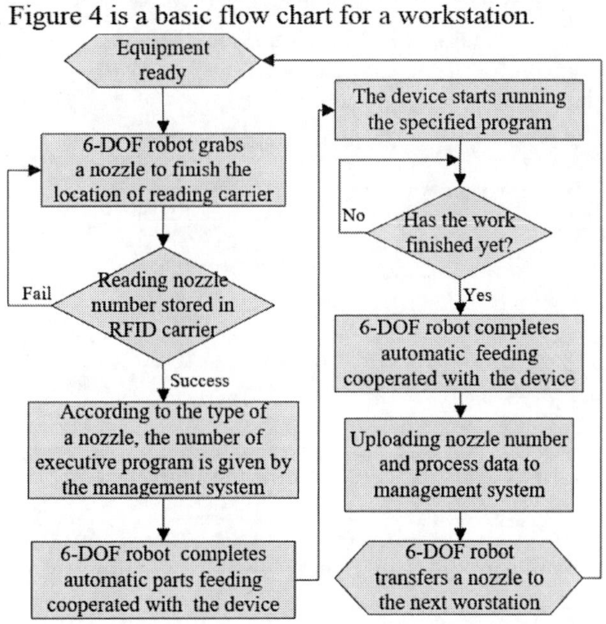

Figure 4. Basic flow chart of a workstation program.

3.4. Software design of management system

The overall implementation of management system is based on C/S architecture. It uses C# language to write the client program of management system and SQL Server 2014 database to store nozzle number information with related process data sets. The software is designed with modular structure, mainly including four functional modules: data acquisition, equipment status monitoring, Andon management and quality data traceability. The data acquisition module provides data support for the implementation of other modules by means of collecting status information of processing equipment and testing equipment, processing quality data and main logistics status etc. through the OPC interface. Equipment status monitoring module can display the running status information of each processing or testing equipment on the IPC client and electronic Kanban terminals. The operation of the whole line is unattended except the feeding and unloading for the rotating silo. Therefore, the management system uses Andon management module to realize the horn call and the display electronic Kanban of emergency, which will prompt managers to remove failures and resume production as soon as possible. The quality data traceability module can uniquely correlate the production quality data collected with the nozzle number to form the production process files. In figure5, the picture shows the main interface of IPC client in centralized control room programmed. In figure6, the picture shows the milling center monitoring interface of IPC in centralized control room programmed.

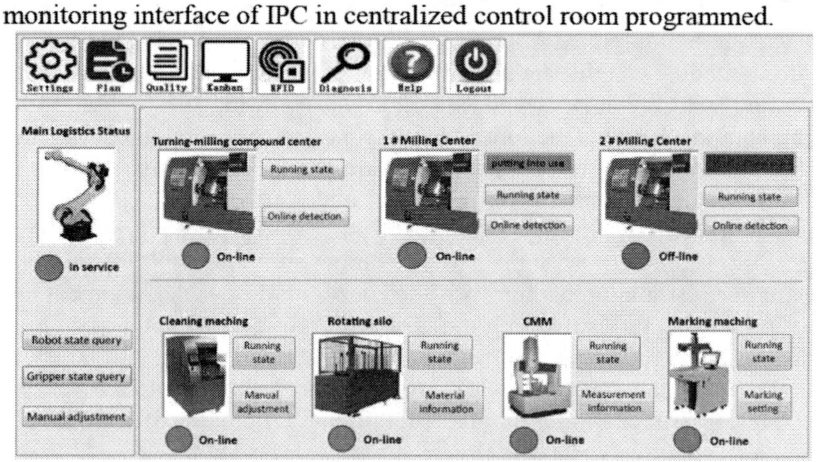

Figure 5. Main interface of IPC.

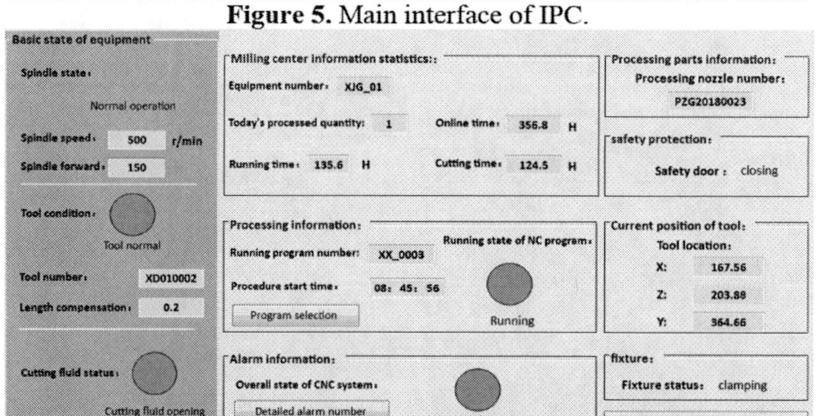

Figure 6. Milling center monitoring interface of IPC.

4. Debugging

For the robot, we start to debug in the virtual environment by offline simulation software to test the transfer time and optimize the track of the robot. It will reduce the downtime caused by the adjustment in the field [5].

For the debugging of the management system, we set up OPC server on IPC, test database function of storing and reading data and separately call each functional module to test operation result by the IPC client and adjust accordingly.

5. Conclusion

This paper describes a design method of the management and control system of machining production line for aero-engine nozzle. The system has realized the integration of control and management in the production process of aero-engine nozzle and improved the level of automation and informatization for the process of machining. Now, the system has been put into use in the processing of some aero-engine nozzles. Compared with the initial situation, the whole production line reduce staffing by 6 people; production efficiency has increased by 30%; the area of material caching between devices has decreased by 12 ㎡. It creates good economic benefits for enterprises. At present, the second phase project, the upgrade and reform for management system, has been started. It will increase some functions such as equipment running status report, SPC analysis, cutting tool management and so on. It will take a firm step to improve production measures, increase equipment utilization, promote product quality and realize digital management of the whole production line.

Acknowledgments

This work was supported by National Key R&D Program of China(No. 2017YFB1303701), the Project of Intelligent Manufacturing Integrated Standardization and New Model Application, Chongqing Artificial Intelligence Technology Innovation Major Thematic Projects (No. zdyf0027), Made-in-China 2025 Sichuan Action Fund Project(2017ZZ003) and K.C.Wong Education Foundation.

References

[1] Hongtao Zhao, Hui Ma and Jingchun Yi 2009 *Techniques of Automation and Application* (Harbin: China Automation Society) vol 28 pp 52-54
[2] Jiang Han, Zhipeng Li and Daoyang Yu 2008 *Manufacturing Automation* (Beijing: Beijing Press and Publishing Board) vol.8 pp19-22, 53
[3] Chenggang Lu, Shuhong Chen and Ping Chen 2009 *Chinese Journal of Scientific Instrument* (Beijing: China Instrument and Instrument Society) vol 10 pp 56-59
[4] Kai Chang, Dechao Zheng and Hongfei Bai 2015 *Manufacturing Automation* (Beijing: Beijing Press and Publishing Board) vol.11 pp 58-60
[5] Wei Li, Naiguang lu and Mingli Dong 2018 *Robot* (Shenyang: China National Book Import and Export Corporation) vol 3 pp 301-308

Optimization of Potato Soil Transportation Separation Mechanism Based on Discrete Element Method and TRIZ Theory

Haifeng Xie, Guohua Gao, Boxing Tian, Bowen Li, Sai Zhang and Juan Huang

College of Mechanical Engineering and Applied Electronics Technology, Beijing University of Technology, Beijing 100124, China

E-mail: 1058038658@qq.com

Abstract. In this paper, based on the problem of poor operation of sweet potato harvester, the discrete element method and TRIZ theory are used to optimize the key mechanism. Based on the discrete element method, EDEM software was used to simulate the operation process of the potato soil conveying and separating mechanism, and the optimal combination of working parameters was obtained. The linear velocity of the rod lift was 2.3 m/s, the inclination angle was 18 degrees, and the amplitude was 8 mm. Field experiments were carried out to verify the results. Through the problems in simulation and field experiment, based on TRIZ theory, the improvement scheme is proposed by using physical contradiction, and the two-stage potato soil transport separation mechanism is obtained. The simulation results show that the operation effect is improved obviously. Finally, the optimization of potato soil transport and separation mechanism is completed by determining the optimal working parameters and improving the structure innovation.

1. Introduction

Although China is a big country for sweet potato production and cultivation, its mechanized production technology is very backward, and the harvesting process is prone to problems such as low quality of work [1-3]. As the main operating mechanism of sweet potato harvester, the potato soil conveying and separating mechanism is the key to affect the harvesting effect. At present, the rod-type potato soil conveying and separating mechanism is the most widely used, and the harvesting effect is good. The structure is as shown in Figure 1, which comprises the separating grid rod composed of rods, the driving wheel, the shaking wheel and many more [4]. The working principle is as follows: during the traveling process, the sweet potato harvester excavates the potato soil mixture and transports it back to the rod elevator, and the rod rotates under the driving of the driving wheel and generates vibration effect through the shaking wheel. The soil was broken and screened to the ground, so that the potato pieces were separated from the soil, and the potato pieces were collected by hand to collect the chips.

In this paper, the existing bar-type sweet potato harvester as a reference object, as shown in Figure 2, using discrete element method to simulate the operation process of potato soil conveying and separating mechanism, optimize the parameters of the mechanism, and verify by field experiments[5].

IOP Conf. Series: Journal of Physics: Conf. Series **1267** (2019) 012071 doi:10.1088/1742-6596/1267/1/012071

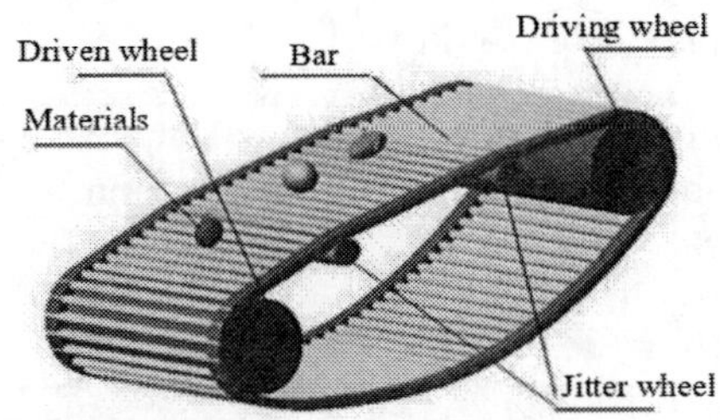

Figure 1. The stem lift conveyer[6]

Figure 2. Sweet potato harvesting prototype

2. EDEM simulation of operation process of potato soil conveying and separating mechanism

2.1. *Selection of Particle Contact Model*

Particle contact model is an important foundation of discrete element technology. For different simulation objects, different contact models need to be selected. In the process of sweet potato harvesting, sweet potato and soil are separated by collision and contact between lifter and potato-soil mixture. In addition, because of the role of liquid bridge between soil particles and cohesion in the contact area, Hertz-Mindin and JKR Cohesion contact model is adopted. As shown in Figure 3, the contact model is also called "elastic-damping-sliding-friction contact". Mechanics Model.

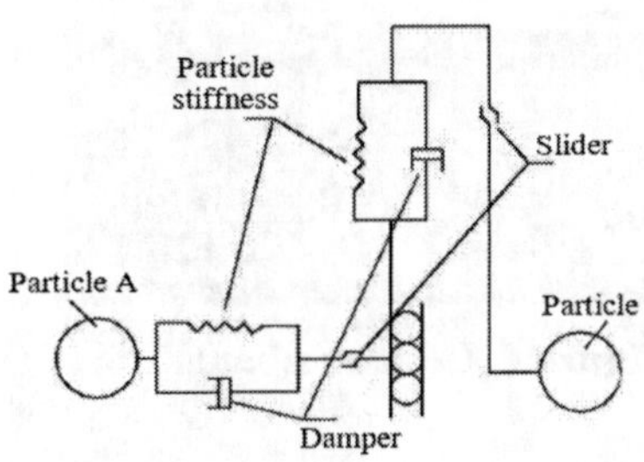

Figure 3. The contact model of particles

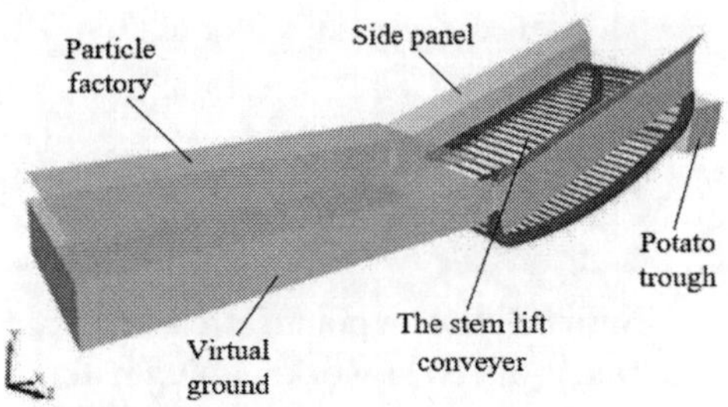

Figure 4. Simplified mechanism model

2.2. *Establishment of simulation model*

The mechanism model is established by referring to the existing sweet potato harvester as shown in Figure 2, and the simplified model as shown in Figure 4 is obtained by using Solidworks. Because of the complex shape of soil particles, the soil is simplified into spheres, which are composed of spheres of different sizes and shapes. The reference object of sweet potato simulation model is Potato 32.

2.3. *Setting Discrete Element Simulation Parameters*

Through density measurement test and static compression test of sweet potato, ring knife test, direct shear test and triaxial test of soil, the physical and mechanical properties parameters of sweet potato and soil were measured, and the simulation parameters as shown in Table 1. The working parameters of the sweet potato harvester are shown in Table 2. Through the design of single factor simulation test, the speed, inclination and amplitude of the rod line are changed under certain conditions. The optimal working parameters are obtained through the analysis of the simulation results.

Table 1. Discrete element simulation parameters

Material	Parameter		
	Density ($g\cdot cm^{-3}$)	Poisson's ratio	Shear modulus (Pa)
Sweet potato	1.02	0.32	1.03×10^{6}
Soil	1.21	0.38	1.07×10^{6}
Steel	7.80	0.3	7.0×10^{10}

Table 2. Working parameter setting

Factor	Velocity (m/s)	Dip angle (°)	Amplitude (mm)	Bar diameter (mm)	Bar spacing (mm)
	1.9	18	8		
Level	2.1	21	10	16	55
	2.3	24	12		

2.4. Analysis of simulation results

The simulation results show that the interaction force between sweet potato and soil is negligible. It can be concluded that the damage of sweet potato is mainly caused by the collision force between sweet potato and machine. Therefore, the damage degree of sweet potato is defined by comparing the relative magnitude of collision normal force under different working parameters, as shown in Figure 5. Combining with the simulation conveying efficiency, the optimum working parameters are finally selected, that is, the linear velocity of the rod lift is 2.3m/s, the inclination angle is 18 degrees, and the amplitude is 8mm.

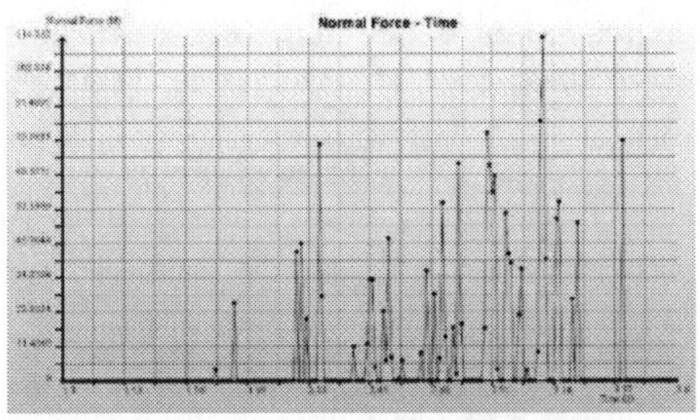

Figure 5. Sweet potato stress diagram under different working parameters

Figure 6. Commissioning site

3. Field test verification

Field test refers to "NY/T 648-2002 Technical Specification for Quality Evaluation of Potato Harvester" to provide quality evaluation index, test method and evaluation rules for field test of sweet potato harvester. In order to observe the change of potato-soil mixture material and the movement of sweet potato more clearly in the working process of potato-soil conveying and separating mechanism, a high-speed camera was used to record the working process of potato-soil conveying and separating mechanism, as shown in Figure 6.

In this field experiment, the linear velocity of the elevator was taken as the variable, and the casualty rate and the bare rate of the experimental results were calculated and compared with the simulation results. The field test results are shown in Table 3.

Table 3. Field test results

Factor (v)	Potato injury rate (%)	Bare potato rate (%)
1.9m/s	5.9	98.3
2.1m/s	5.1	97.7
2.3m/s	6.4	96.4

The results showed that there was no linear relationship between wounded rate, bare rate and linear velocity. When the linear velocity is 2.1m/s, the casualty rate and the bare rate of potato are better than the other two working conditions. The experimental results are consistent with the simulation results, which proves the feasibility of using discrete element method to simulate the operation process of potato soil conveying and separating mechanism.

4. Innovative design of potato soil conveying and separating mechanism

The optimal working parameter combination of the potato soil transport separation mechanism was determined in the previous paper. However, during the simulation test and field experiment research analysis, it was found that the damage of sweet potato mainly occurred in the collision between the cone and the lift, and the sweet potato rolled. The state is random and uncontrollable, so there will still be some probability of damage during the harvesting process. In order to further reduce the damage rate of sweet potato and enhance the stability of the transport separation process, it is necessary to carry out innovative design of its structure to improve the research.

Therefore, this section will reduce the casualty rate of potato soil transport and separation mechanism, enhance the stability of operation as the optimization objective, use TRIZ theory to carry out system function analysis, and find the key issues, combined with the separation principle of physical contradictions in TRIZ theory, put forward a new optimization scheme[7-9].

4.1. System component analysis

Component analysis is to summarize and classify the components contained in the potato soil conveying and separating mechanism system and the supersystem, list and analyze the components. The main elements of the technical system include rods, side plates, driving wheels, driven wheels and dithering wheels, elements of soil and sweet potatoes interacting with the mechanism, auxiliary tool elements such as rack, shovel, reducer and belt, and component analysis is shown in Table 4.

Table 4. System Component Analysis

Engineering system	major function	Component	Supersystem Component
Potato soil conveying and separating mechanism	Transportation and Separation of Potato Soil	Bar Side panel Jitter wheel Driving wheel Driven wheel Sweet potato Soil	Frame Digging shovel Reducer Belt Tractor

4.2. Establishment of Functional Model

Based on the interaction between the above components, the total functions of the system and the sub-functions of each component of the system, the implementation of the functions and whether harmful effects are produced are defined, as shown in Figure 7.

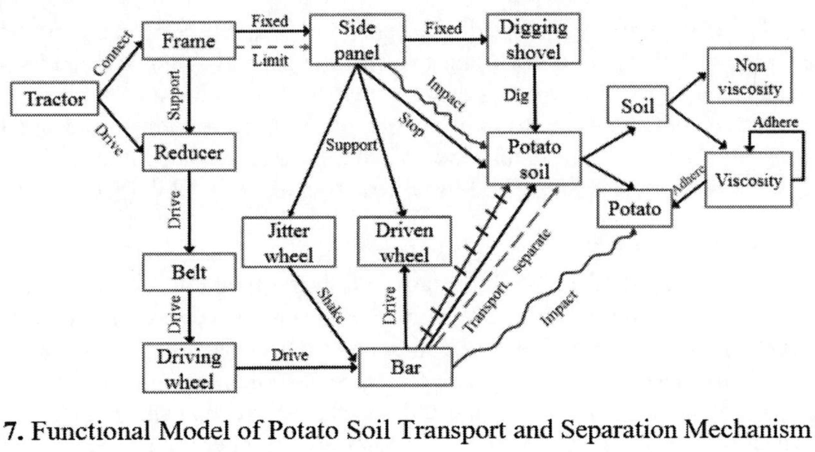

Figure 7. Functional Model of Potato Soil Transport and Separation Mechanism System

In the function diagram, the arrow points to the active object, the straight line indicates the normal function, the broken line indicates the insufficient function, the wavy line indicates the harmful function, and the straight line plus the vertical short line indicates the excessive function.

4.3. Problem description and analysis

4.3.1. Defining physical contradictions Based on the description of the above problems, the physical contradiction is put forward: the ability of crushing and screening of potato soil conveying and separating mechanism is both large and small. Higher screening capacity of crushed soil is to ensure a higher rate of bare potatoes, while smaller screening capacity is to ensure that sweet potatoes can be protected by soil and reduce the rate of wounded potatoes. It is difficult to achieve this at the same time for the current potato soil transport and separation mechanism.

4.3.2. Solving Physical Contradictions According to the problems that need to be solved in the actual design of potato soil conveying and separating mechanism, the above inventive principles are screened and analyzed. Finally, the separation principle is selected as the guiding idea for the innovative design of potato soil conveying and separating mechanism.

According to the principle of partition, the original potato soil transport and separation mechanism is innovatively designed. As shown in figure 8, the two-stage potato soil transport and separation mechanism is designed, which separates the functions of soil crushing, potato soil separation and transportation. The first and second stages of potato soil transport and separation mechanism respectively accomplish the above two functions.

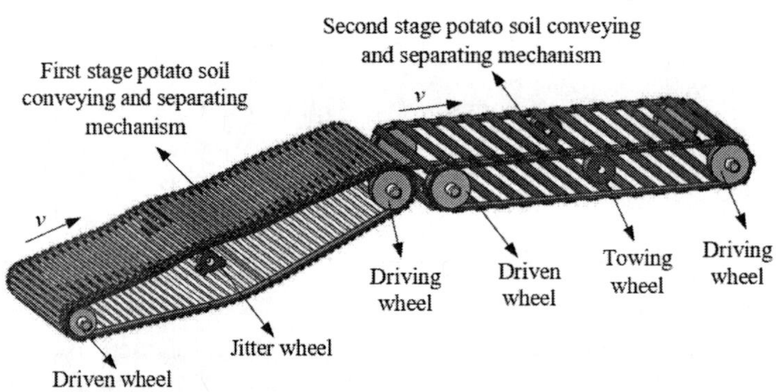

Figure 8. Two-stage potato soil conveying and separating mechanism

Among them, the first stage of potato soil transport and separation mechanism mainly completes soil crushing and transfers the mixture of potato soil to the second stage of potato soil transport and separation mechanism. Its operation process ensures that the soil is crushed while not being sifted in large quantities, so that the sweet potato can be effectively buffered and protected in the transport process. According to the previous research, the working parameters are set as follows: the rod linear velocity is 2.3m/s, the amplitude is 8mm, and the inclination angle is 18 degrees. However, in order to ensure that the soil is not sifted out in large quantities, the rod spacing of the first potato soil conveying and separating mechanism is reduced from 55mm to 35mm, and the rod diameter is 14mm. In addition, the effective length of the working face is reduced to about 3/4 of the original length, i.e. 1.2m, according to the photography and analysis of the working process of the potato soil conveying and separating mechanism by the high-speed camera. At the same time, the problem of the overall longitudinal size growth caused by the two-stage structure is avoided to some extent.

The second stage of potato soil conveying and separating mechanism mainly completes the soil sieving and the backward transportation of sweet potatoes. Its operation process does not need soil crushing. Therefore, the shaking wheel is replaced by the pulley to keep the working face running smoothly, thus avoiding the damage of sweet potato caused by collision. In order to further reduce the damage of sweet potato, rubber sleeve was wrapped on the outside of the rod. Its outer diameter was 16 mm, the clearance of the rod was 55 mm, and the length of the working face was 0.8 m.

4.4. Feasibility verification of improved scheme based on discrete element method

Based on the simplified model and working parameters, the discrete element simulation of the two-stage potato soil conveying and separating mechanism was carried out to verify the actual effect of the new scheme. Through the analysis and comparison of the simulation results before and after the improvement, it can be seen that the operation effect of the improved two-stage potato soil conveying and separating mechanism has been improved obviously, the relative damage probability of sweet potato has been greatly reduced, the average loss rate of the first stage operation mechanism is as high as 73.9%, and the average loss rate of the second stage operation mechanism is 11.0%, which verifies the feasibility of the improved scheme.

5. Conclusions

Based on the discrete element method, the optimal working parameters of the potato soil transport and separation mechanism are obtained, that is, the linear velocity of the rod elevator is 2.3 m/s, the inclination angle is 18 degrees, and the amplitude is 8 mm. At the same time, the simulation results are verified by field experiments, which verifies the accuracy and feasibility of the simulation. Based on TRIZ theory, the mechanism was innovatively designed, and a two-stage potato soil conveying and separating mechanism was obtained. The simulation results show that the relative damage probability of sweet potato is greatly reduced. The average loss rate of the first stage operation mechanism is 73.9%, and that of the second stage operation mechanism is 11.0%.

Acknowledgments

This research work is supported by China's national key R&D project and the Ministry of Science and Technology's innovative methods, projects number 2016YFD0701604-03 and SQ2016IM03640007, and the titles of the projects are the Innovative Design of Key Mechanisms for High-efficiency and Low-loss Transport Separation of Sweet Potato and Systematic Application and Demonstration of Intelligent Manufacturing for Agricultural Machinery Equipment..

References

[1] Ma B, Hu L L, Xu L Y, Tian L J, Ji F L and Wang B, "Status about sweet potato planting and its production machinery in China", *Journal of CAM*, Vol. 34, issue 1, pp. 42—46, 2013.

[2] Hu L L, Tian L J, Ji F L, Zhang H J, Wang B and Liu M J, "Analysis and discussion on the restriction causations for harvesting mechanization of sweet potato in China", *CAM*, issue 3, pp. 16—18, 2011.

[3] Wang B, Hu L L, Hu Z C, Tian L J, Ji F L and Ma B, "Study on the damage mechanism of potato soil in chain rod elevator", *Journal of CAU*, Vol. 19, issue 2, pp. 174-180, 2014.

[4] Zhang J and Wei H A, "Movement simulation of soil and potato separation of 4M-2 type potato combine harvester", *Journal of GAU*, Vol. 44, issue 3, pp. 153-157, 2009.

[5] Yang J W, Sun H N and Zhang Z Q, "A review on fundamentals of distinct element method and its applications in agricultural engineering realm", *XDSP*, issue 15, pp. 28-33, 2015.

[6] Zhang J and Wei H An, "Moving simulation on potato and soil separating by 4M-2 potato-harvester", *Journal of GAU*, Vol. 44, issue 3, pp. 153-157, 2009.

[7] Mei C A, Ng K W and Ahmad S A, "An engineering design support tool based on triz", *Adv. VISIP*, pp. 115-127, 2013.

[8] Li T, "Retraction note: applying triz and ahp to develop innovative design for automated assembly systems", *International JAMT*, Vol. 77, issue 9-12, pp. 2289-2289, 2015.

[9] Jiang J and Li Y, "Application of triz to develop new antistatic materials", *Asian JC*, issue 24, pp. 4429-4432, 2014.

IOP Conf. Series: Journal of Physics: Conf. Series **1267** (2019) 012072 doi:10.1088/1742-6596/1267/1/012072

Calibration and Measurement Study on Density of Selective Laser Sintered 3D Printing Powder Bed

Jianqiang Li, Guohua Gao, Shuo Zhang and Yunsong Du

College of Mechanical Engineering and Applied Electronics Technology, Beijing University of Technology, Beijing 100124,China

E-mail: lijianqiang0906@163.com

Abstract. The commonly used method for measuring the density of a selective laser sintered powder bed is to extract a certain volume of powder bed powder, measure the mass of the powder, and then calculate the density of the powder bed using the mass formula. Aiming at the shortcomings of the existing methods, such as complex operation and low measurement accuracy, this paper proposes a method based on plumb bob method for selective laser sintering powder bed density measurement. The method can measure the density of the powder bed without causing large damage to the powder bed, and the measurement efficiency and precision are higher than the existing measurement methods. In this test, the most commonly used ABS powder was used, and its density was calibrated using the plumb method. The function of the depth between the depth of the plumb bobbin and the density of the powder bed was fitted to the density of the powder bed. The measurements provide data support.

1. Introduction

The existing method for measuring the density of the powder bed of the selective laser sintering powder layer is mainly to extract a certain volume of the powder layer powder, weigh the powder quality of the extracted powder layer, and use the mass formula to calculate the overall density of the powder layer[1]. The main disadvantage of this method is that it is extremely difficult to extract a certain volume of powder layer powder. During the extraction process, the extraction tool and powder effect change the volume of the powder layer. It is difficult to ensure the volume and unextracted powder of the powder layer consistent before.

Therefore, there is a need for a more reliable and accurate method for measuring the powder density of the powder layer. This paper proposes a powder density measurement method based on the plumb method, which can carry out the powder bed density measurement without causing major damage to the powder bed. Firstly, a series of density powder beds are prefabricated in the powder box, and the device is used for density calibration. The steel balls are freely dropped into the powder bed, and the density of the powder bed is measured by the depth of the steel balls falling into the powder bed. After the calibration is completed, the density is determined. The mathematical relationship between the density of the powder bed and the depth of the steel ball falling into the powder bed can be fitted[2-3].

When the measurement of the density of the powder bed is required, the steel ball is dropped into the powder bed from the same height as the calibration, and then the depth of the steel ball falling into the powder bed is measured, and the depth is brought into the above mathematical relationship to calculate the powder bed density. The method is convenient and quick, and can improve the efficiency and precision of the powder bed density measurement.

2. Powder bulk density and tap density measurement

According to the previous literature research, it can be known that the powder bed of the selective laser sintering equipment has a powder bed density that is not less than the bulk density of the powder and is not greater than the tap density of the powder. Since it is necessary to artificially produce a series of powder beds of different densities in the calibration of the density of the powder bed, it is necessary to measure the bulk density and the tap density of the test powder before the calibration, and the artificially produced powder bed density interval is loose density to tap density[4].

The measuring equipment used in this paper is BT-1000 powder comprehensive characteristic tester, electronic scale and constant volume container,as shown in figure 1.

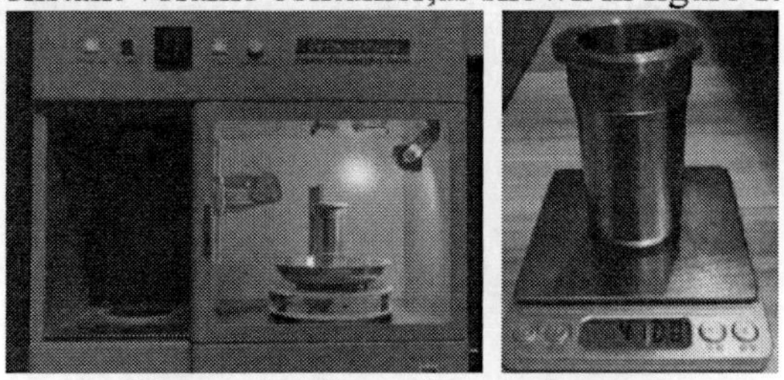

Figure 1. Measuring equipment

2.1. Measurement of bulk density

The bulk density refers to the powder under the specified conditions and naturally filled with a specific container state, that is, the mass per unit volume when the powder is loosely packed, often expressed in g/cm^3, which is a process property of the powder. The bulk density is measured by the vibrating funnel method, that is, the powder is placed in a funnel with a vibrating device, and vibrated under certain conditions, and the powder is freely dropped from the funnel hole by a certain height by means of vibration, and filled with a container filled with a constant volume of 100 ml..

Continuous measurement 5 times and weigh the 100ml container and the mass of the loaded powder, expressed in G_x, as shown in figure 2. And record these five measurements in Table 1. The mass of the 100ml container G_0 is 410g, as shown in figure 1.

Table 1. Bulk density measurement data

Times/x	1	2	3	4	5
Mass G_x/g	456.8	456.9	457.7	456.6	456.4

After 5 measurements, the average mass of the filled powder container was taken:

$$G = \frac{G_1 + G_2 + G_3 + G_4 + G_5}{5} = 456.88\text{g} \quad (1)$$

Figure 2. Measurement of bulk density

The calculated bulk density is:

$$\rho_a = \frac{G - G_0}{100} = \frac{456.88 - 410}{100} = 0.468\text{g/cm}^3 \tag{2}$$

2.2. Measurement of tap density

The tap density refers to the mass per unit volume measured after the powder is vibrated in a specific container under specified conditions,the specified conditions is that the stroke is 3mm and vibration frequency is 100 times/min - 300 times/min[5]. That is, the mass per unit volume of the powder compaction filling, often expressed in g/cm^3, is a process property of the powder. There are two methods for measuring the tap density of a powder: a fixed mass method and a fixed volume method. This measurement uses a fixed volume method,that is, the powder after tapping is filled with a fixed volume container. as shown in figure 3

Figure 3. Measurement of tap density

Continuous measurement 5 times and weigh the 100ml container and the mass of the loaded powder, expressed in Gx', as shown in figure 3. And record these five measurements in Table 2.

Table 2 Tap density measurement data

Times/x	1	2	3	4	5
Mass $G_x^{'}$/g	474.9	474.4	474.4	474.4	474.4

After 5 measurements, take the average of the mass after full shaking:

$$G^{'} = \frac{G_1^{'} + G_2^{'} + G_3^{'} + G_4^{'} + G_5^{'}}{5} = 474.5g \tag{3}$$

The calculated tap density is:

$$\rho_b = \frac{G^{'} - G_0^{'}}{100} = \frac{474.5 - 410}{100} = 0.645\text{g/cm}^3 \tag{4}$$

After the above measurement, the selective laser sintering of the ABS powder printed in this test has a bulk density of 0.468 g/cm^3 and a tap density of 0.645 g/cm^3, which is a prefabrication for subsequent selective laser sintering of different density powder beds. Provide a density interval reference range.

3. Powder bed density calibration and measurement principle based on plumb method

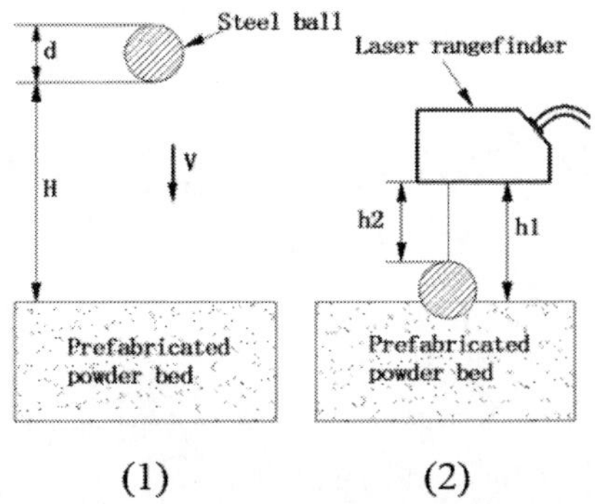

Figure 4. Description of measurement principle

As shown in figure 4, the prefabricated powder bed in the figure 4 is a powder bed which has been prefabricated by a pre-preparation. The steel ball shown in Figure 4(a) is released from the height H, and falls into the pre-powder bed as shown in Figure 4(b), the diameter of the steel ball is counted as "d". The distance to the prefabricated powder bed was measured using a laser range finder, calculated as "h_1", the distance to the highest point of the steel ball, calculated as "h_2". Then the steel ball falls into the depth of the powder bed, which is counted as h:

$$h = d - (\mathrm{h}_1 - \mathrm{h}_2) \tag{5}$$

By prefabricating a series of powder beds of different densities, the principle is used to measure the depth h of the steel balls falling into different density powder beds, and the density of the powder bed is measured by the depth "h" of the steel balls falling into the powder bed to achieve the purpose of calibrating the powder beds of different densities. After the calibration is completed, the relationship between the density of the powder bed and the depth of the steel ball falling into the powder bed and the mathematical relationship can be fitted.

When the measurement of the density of the powder bed is required, the steel ball is dropped into the powder bed from the same height as the calibration, and then the depth of the steel ball falling into the powder bed is measured, and the depth is brought into the mathematical relationship described above to calculate the density of the powder bed.

4. Test bench design and data measurement

4.1 The structure of the test bench and its key components

As shown in Figure 5, the overall structure of the test bench is mainly composed of a frame, a steel ball release device, a powder box, a powder box pedestal, a slide rail, a rotating shaft , a laser range finder and a laser range finder mounting bracket. The powder box is placed on the powder box pedestal, and the powder box pedestal is connected to the slider on the slide rail by bolts. The laser range finder is mounted on the Laser range finder mounting bracket that can be rotated about the rotating shaft.

The steel ball release device is composed of a bracket and a steering gear, and the steering gear has two working positions as shown in figure 6. In the A position, the steering gear prevents the steel ball from falling, and this state is also the normal state of the steering gear. When the release switch is pressed, the steering gear quickly moves to the B position, and the steel ball falls at the same time and falls into the powder bed.

In order to achieve high-precision distance measurement[6-8], the high-precision distance measuring sensor selected for this test bench is the LDS-S-20 standard laser displacement sensor. The measuring range is 20mm; the measuring accuracy is 1μm.

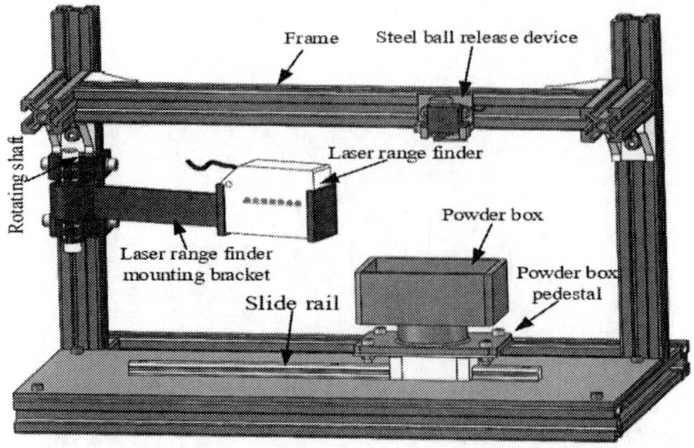

Figure 5. The overall structure of the test bench

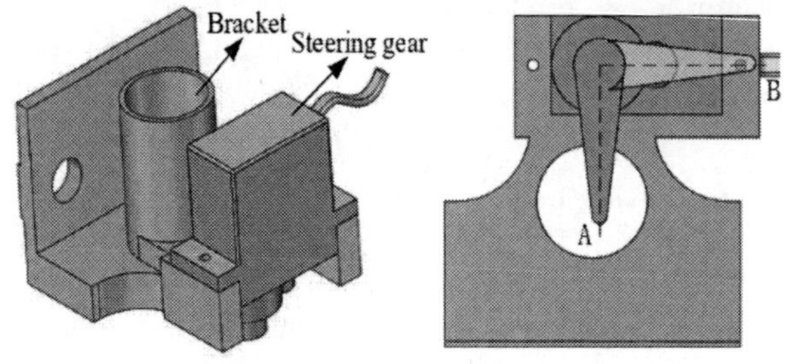

Figure 6. Steel ball release device

4.2 Data measurement

First, the powder bed of the bulk density and the tap density was prefabricated using the powder box in the Figure5, and the container had a net weight of 50 g and a volume of 198 cm^3. The density of the pre-formed powder bed can be obtained by subtracting the net weight of the powder box divided by the volume of the powder box. Calculated as follows:

$$\rho = \frac{M - m}{V}, \quad g/cm^3 \tag{6}$$

In the above formula:The M is the weight of the powder box after tapping;the m is the net weight of the powder box;the V is volume of the powder box.

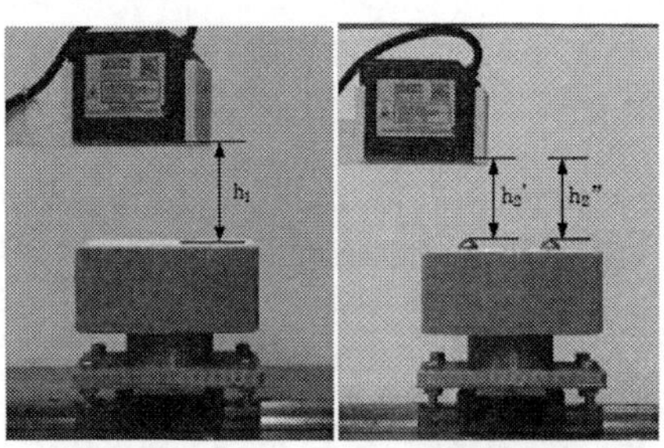

Figure 7. Data measurement

The powder box in which the powder bed was previously prepared was placed on a test stand, and the distance h_1 of the surface of the powder bed was measured using a laser distance measuring sensor[8]. Then adjust the position of the powder box and make the two steel balls fall into the powder bed respectively, and measure the distances h_2' and h_2'' of the highest points of the two steel balls respectively, and take the average value as h_2 [9],shown in figure 7.

Table 3. Test data

NO.	Test Data		
	Powder bed density ρ/(g/cm³)	(h_1-h_2)/mm	Depth value h/mm
1	0.498	-1.405	13.405
2	0.508	-0.866	12.866
3	0.572	6.862	5.138
4	0.577	7.658	4.342
5	0.589	8.305	3.695
6	0.600	8.775	3.225
7	0.610	9.280	2.720
8	0.623	9.385	2.615
9	0.637	9.682	2.318

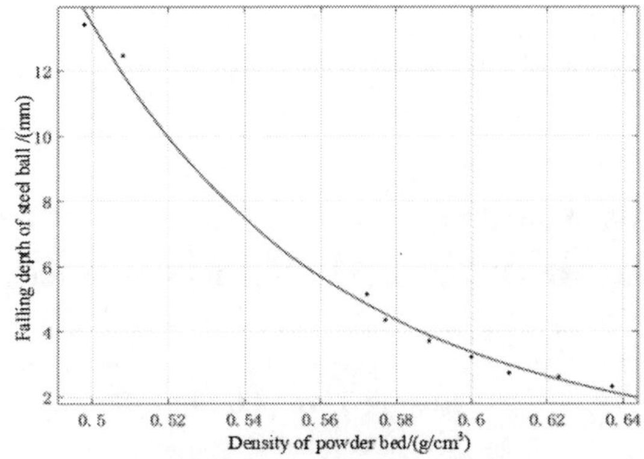

Figure 8. Relationship lines

5. Conclusions

Through the above test methods and data measurement methods and then our completing the prefabrication of nine species of powder beds of different densities and their related data measurements, the measured or calculated data are recorded in Table 3. The pre-fabricated bed density ρ and the steel ball falling into the powder bed depth h are introduced into the Matlab software to produce a curve of ρ and h, as shown in Figure 8. The curve fitting method is Power2, and the mathematical expression is a power function with a constant term. The mathematical expression is $\rho=0.1047h^{-7.05}-0.4752$. These test data and methods provide data support for the subsequent measurement of the density of a series of powder beds.

Acknowledgments

This research work is supported by Hunan Enterprise Association of Science and Technology Association Small and Medium Enterprises TRIZ Promotion Demonstration Application, project number 2017IM050100.

References

[1] Wang Chong-lin.On the measurement of density for p/m materials[J]. Powder Metallurgy Industry,2010, 20(6):1-4.
[2] Sheng Anlian,Gu Bingqi. Instantaneous impulsion method for rapid measurement of subgrade compactness[J]. China Civil Engineering Journal, 1998(1):67-71.
[3] Xue Liang.Determination and error analysis of power density[J]. China Measurement Technology, 2004, 30(6).
[4] Yan Cuiping,Zhang Mingxing and Lin Longyuan.Effect of graphite particle size on tap bulk density[J]. Non-Metallic Mines, 2015(3).
[5] Cui Xiang-long,Xu Bing,Zhang Yi, et al .Application of quality by design in granulation process for ginkgo leaf tablet (1): comprehensive characterization of granule properties[J].China Journal of Chinese Materia Medica, 2017(6).
[6] Shi Zhi-yong,Pan Xiao-sheng,Zhang Qian. High-precision pulsed laser measuring distance by time delay method[J]. Optics and Precision Engineering, 2014, 22(2).
[7] YAN J X. Laser Principles and Techniques [M].Beijing:Higher Education Press,2011.(in Chinese)
[8] GAO L K,SONG W.Laser Measuring Distance[M].Beijing:People's Railway Press,1977.(in Chinese)
[9] JI Wei-xing ,ZHANG Lu-lu, CHEN Juan , et al .Polynomial Fitting Acceleration Method Based on Time-Series Segmentation[J]. Transactions of Beijing Institute of Technology,2018(5).

IOP Conf. Series: Journal of Physics: Conf. Series **1267** (2019) 012073 doi:10.1088/1742-6596/1267/1/012073

The Innovative Design of Seedling Transplanting Combination Mechanism with TRIZ

Boxing Tian[1], Guohua Gao[1], Yongzhi Liang[1], Kai Wang[1], Bowen Li[1] and Guoxiong Lin[2]

[1] College of Mechanical Engineering and Applied Electronics Technology, Beijing University of Technology, Beijing 100124, China
[2] State Grid Beijing maintenance company, State Grid Corporation of China, Beijing 100031, China

E-mails: tianboxing@126.com, ggh6768@126.com, 420204090@qq.com, 164783613@qq.com, leowang1206@163.com, @qq.com and 834831436@qq.com

Abstract. Aiming at the problem of poor gathering effect of seedling transplanting combination mechanism that is designed independently, the intensive study has been done by using the main tools of TRIZ: the functional analysis, the physical contradiction and technical contradiction, the Su-Field model. And four feasible solutions have been found during the intensive study. Comparing the advantages and disadvantages of solutions and Taking the actual engineering situation into consideration, the final solution is obtained by collecting the advantages of the former solutions. The final solution can solve the poor gathering problem very well.

1. Introduction

The modern industry of flowers and plants, which started to develop since 1980s in china, is developing rapidly during the 30 years [1]. In 2012, the total area of flowers and plants production in China was 1120.3 thousand hectares, the total sales amount was 12.0772 billion yuan and the total export value was 533million USD. The sales and export growth rates are both more than 10% at a stable growth trend [2]. Currently, the most domestic planting of potted flower in greenhouse are in factory form, which requires multiple transplanting operations in the planting process. However, the task of transplanting and combinating seedling is usually done by artificial, which leads to the low efficiency. It's very important to invent a highly efficient and reliable transplanting combination device.

In order to study the scientific principles of human inventions and solve the technical problems, the research staff, led by S. Alt shuller G, summed up the basic principles of TRIZ (TRIZ stands for the theory of inventive problem solving) after analyzing and studying the world's nearly 2500000 high level inventions patents since 1946 [3]. The TRIZ process of solving problems: at first, transform the problem of professional field to the standard problem of TRIZ, then get the general solution by the standard problem , finally find the special solution by the general solution (as shown in Figure 1). The theory of TRIZ can't directly provide the solution to the special problem, but it can open up our minds to find the solutions [4].

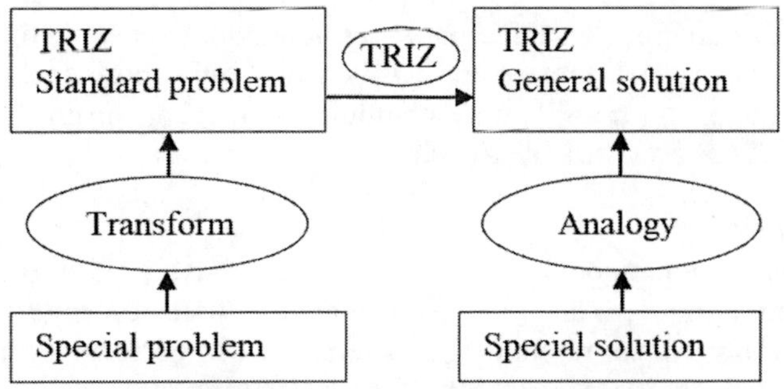

Figure 1. The TRIZ process of solving

2. The Background of Project

2.1. The transplanting

Figure 2. The transplanting method

Figure 3. The transplanting seedling

The transplanting seedling is planted in the tray with specification 15 × 7 in this project (as shown in Figure 2). The seedling is transplanted to the 110mm pots in a method of one group combined by three strains when it's cultivated for a period of time(as shown in Figure 3).

2.2. The process of transplanting combination

There are four steps to transplant seedling from trays to pots (as shown in Figure 4).

- The grippers grab three strains of seedling.
- The combination of three strains of seedling by using the combination mechanism.
- The three strains of seedling are transplanted in the 110mm pots.
- The compacting mechanism compacts the soil media in the 110mm pots.

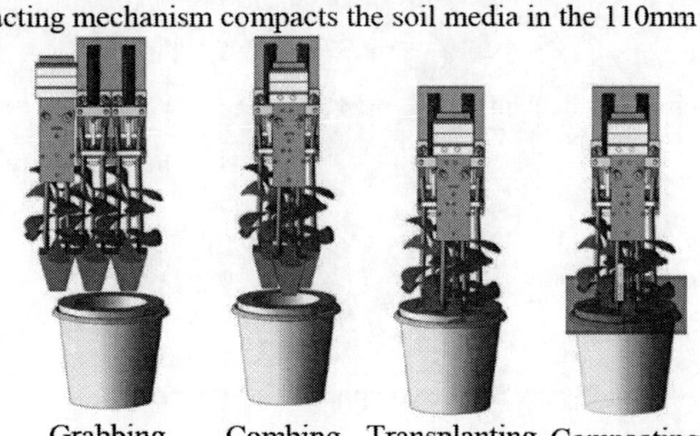

Figure 4. The four steps of seedling transplanting

3. The Analysis Of TRIZ

The classic TRIZ, mainly combined by three parts: the functional analysis, the physical and technical contradictions, the Su-Field model, is applied to solve the failure problems of existing mechanism. The failure problems of seedling transplanting combination mechanism in this project belong to the field, where the tools of TRIZ are available to help.

3.1. The functional analysis

The functional analysis is combined by three parts: the analysis of component, the analysis of interaction and the function model. The component refers to a part of a engineering system or a super system. And the engineering system refers to the system that can perform a certain function, the super system refers to the system that is consisted by the engineering system and other systems [5]. The engineer system in this project is the system of seedling transplanting combination device. The engineer system is consisted by the components: the combination mechanism, the transplanting gripper, the compacting mechanism, the stem of seedling, the matrix of seedling. And the components of the super system include the pots, the soil cavity of pots(as shown in Table1). The system of seedling transplanting combination mechanism is as shown in figure 5.

Table1. The components

The system	Components
The engineering system	the combination mechanism the transplanting gripper the compacting mechanism the stem of seedling the matrix of seedling
The super system	the pots the soil cavity of pots

The interaction analysis (as shown in TABLE II) refers to the relationship between components. If it is a "+", it means the components have a kind of relationship. If it is a "-", it means there is no relationship between presentation components.

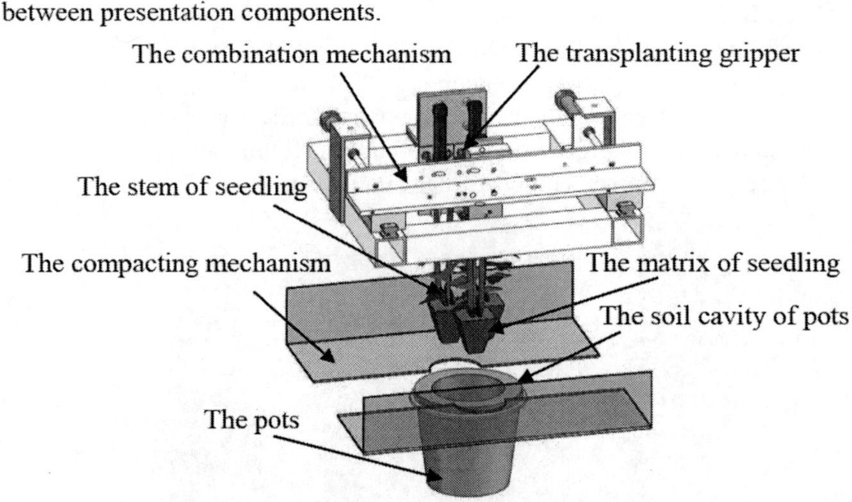

Figure 5. The components of system

In Table2, “a” stands for the combination device, “b” stands for the transplanting gripper, “c” stands for the stem of seedling, “d” stands for the matrix of seedling, “e” stands for the soil cavity of pots, “f” stands for the compacting device, “g” stands for the pots.

Table2. The analysis of interaction

	a	b	c	d	e	f	g
a	\	+	-	-	-	-	-
b	+	\	+	+	-	-	-
c	-	+	\	-	-	-	-
d	-	+	-	\	+	+	-
e	-	-	-	+	\	+	+
f	-	-	-	+	+	\	-
g	-	-	-	-	+	-	\

The function model (as shown in figure 6) describes the function of the components, as well as performance and costs. In the function model, the function is divided into two categories: useful and harmful functions (" ∿∿→ "). Useful function is divided into normal function (" → "), insufficient function ("- - -→ "), excessive function (" ⇉ ") .

According to the functional model, the problem of TRIZ is proposed: the engineering system wants the space of three strains of seedling small to improve the effect of ornamental and the area of the matrix of three strains of seedling small to avoid scattering the soil cavity of pots.

The problem of TRIZ should be highlighted to form a sharp contradiction:

- If the space of three strains of seedling is smaller, then the effect of gather and ornamental will be better, but the area of the matrix of three strains of seedling will get bigger to scatter the soil cavity of pots.

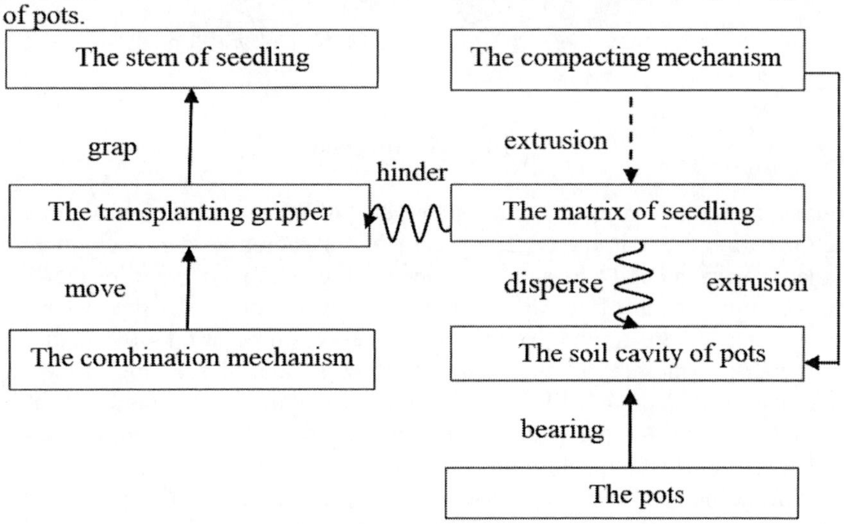

Figure 6. The function model

If the space of three strains of seedling is bigger, then the area of the matrix of three strains of seedling won't get bigger and the soil cavity of pots won't scattered, but the effect of gather and ornamental will get worse.

3.2. The physical and technical contradiction

The physical conflict means the two sides of a single physical parameter of an object are reasonable and required in the problem. The bigger and smaller space of three strains of seedling are wanted in the engineering system. The bigger space of three strains of seedling and the smaller area of the matrix of three strains of seedling are required to facilitate the seedling fall into pots. The smaller space of three strains of seedling are required to improve the ornamental after the seedling falling into the pots. The separation principles are available to give the general solutions for The physical contradiction. The separation principles include the spatial separation principle, the time separation principle, the separation principle based on the relationship, the separation principle based on system. The spatial separation principle means the contradiction exists in different places of a engineering system. The time separation principle means the contradiction exists in different time. The separation principle based on relationship means the components of contradiction come from the super system. the separation principle based on system means the contradiction exist in the engineering system and the super system [6].

According to the engineering system, the first solution (as shown in figure 7) is proposed by using the time separation principle: the combination mechanism only complete the task of combination and transplanting, the compacting mechanism is designed to complete the task of gathering the stems of three strains of seedling and compacting the matrix [7].

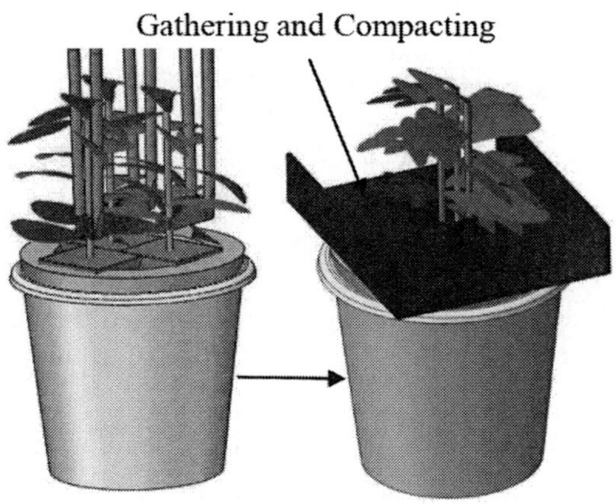

Figure 7. The first solution

The technical contradiction means that a parameter of the system is improved, which will lead to the deterioration of another parameter. The technical contradiction in this engineering system is that the improvement of ornamental (the smaller space of three strains of seedling) lead to the bigger area of the matrix of three strains of seedling, which will scatter the soil cavity of pots. The improving parameter is the area of the moving object (the space of three strains of seedling), the worsening parameter is the shape (the area of the matrix of three strains of seedling). Querying the matrix of technical contradiction (as shown in Table3), the four principles are available for finding the solutions. The four principles are No.4 the asymmetry-adding principle, No.5 the combination principle, No.29 the principle of pneumatic and hydraulic, No.34 the abandon and regenerate principle. No.4 the asymmetry-adding principle means the symmetrical objects can be turned into asymmetrical objects or the degree of asymmetry can be increased. No.5 the combination principle means the same or similar objects can be combined. No.29 the principle of pneumatic and hydraulic means the fixed part of the object can be replaced by gas and fluid. No.34 the abandon and regenerate principle means the parts, who complete its function, can be abandoned by using the methods of dissolution and evaporation or can be changed into the useful one in the working process.

Table3. The matrix of technical contradiction

Worsening feature→ Improving feature↓	No.12 the shape
No.5 the area of the moving object	No.5 No.34 No.29 No.4

Figure 8. The second solution

The second solution (shown in figure 8) is obtained by using the No.4 principle. The shape of soil cavity, which is the symmetrical circular, is changed into asymmetric shape. It won't expand the area of the soil matrix of the three strains of seedling in the process of transplanting, and it will facilitate the seedling fall into the pots.

The third solution (as shown in figure 9) is obtained by using the No.34 principle. The matrix of seedling is the growing environment of seedling before the transplanting operation. The soil matrix of pots can replace the mission of the matrix of seedling after the transplanting operation. So the matrix of seedling can be broke up to shrink the space of the three strains of seedling and improve the ornamental.

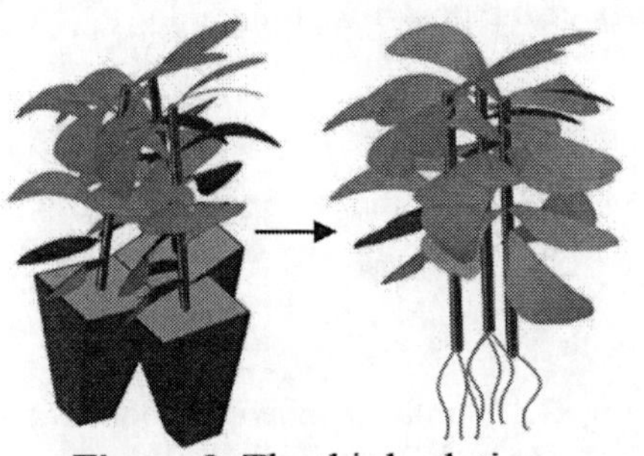

Figure 9. The third solution

3.3. The Su-Field model

The Su-Field model, which is short for (the Substance–Field model), means the system is consisted of substances and fields. The substance means the parts owning static quality in the engineering system, for example: transplanting gripper, combination mechanism, etc. The field refers the general concept of energy, for example: the mechanical field, the electric field, the magnetic field, the gravity field, the biological field, etc. An effective Su-Field model must have two substances beneficial function and a field at least. The field builds the interaction between substances [8]. The Su-Field model of engineering system is shown in figure10. The seedling is transplanted into the pots by the grippers under the mechanical force F1. The soil matrix of pots (S1) is too loose to fix the transplanting seedling (S3). The fourth solution (as shown in Figure 11) is obtained by using the Su-Field model. In order to improve the strength of soil matrix, The substance water (S3) is added in the process of transplanting.

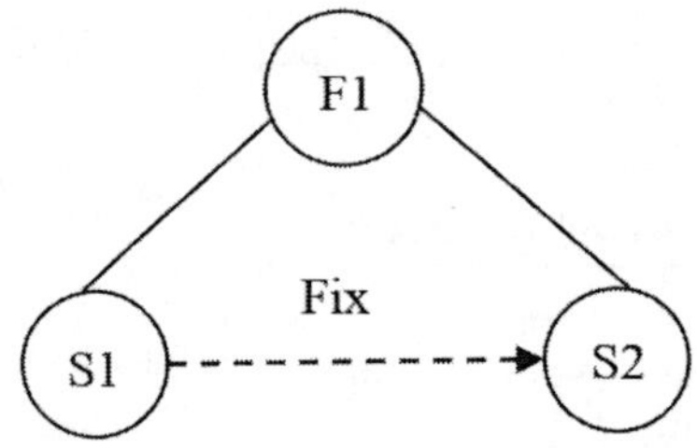

Figure 10. The Su-Field model

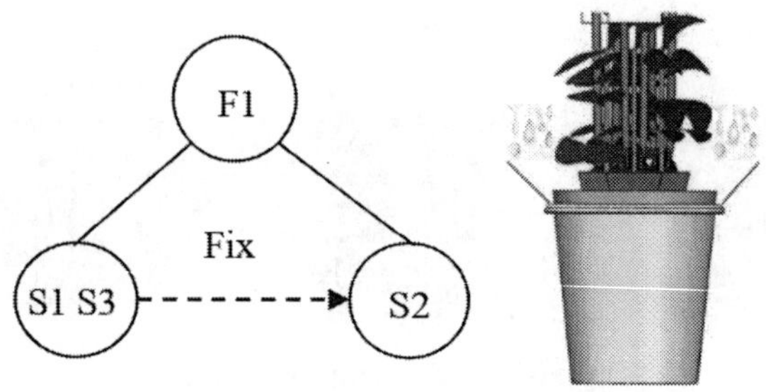

Figure 11. The fourth solution

4. Conclusions

The four solutions are obtained by using the TRIZ tools to solve the failure problems of the seedling transplanting combination mechanism. Comparing the advantages and disadvantages of four solutions and according to the key problem "the poor gathering effect of seedling transplanting combination mechanism", the final solution is decided. The matrix of seedling is removed in gripping process (according to the analysis of the technical contradiction), which will facilitate the combination of seedling (according to the analysis of the physical contradiction). The substance water is added to improve the strength of the matrix of the pots after the seedling transplanting operation.

Acknowledgments

This research work is supported by the Ministry of Science and Technology's innovative methods, project number SQ2016IM03640007, and the title of the project is Systematic Application and Demonstration of Intelligent Manufacturing Innovation Method for Agricultural Machinery Equipment

References

[1] Cheng T R, Wang J, Zhang Q, "Strategic thinking about the development of innovative flower industry in China," *China Garden*, vol. 2, pp. 73-78, 2013.

[2] Qiao D H, "Flower production to achieve steady growth-the national flower of statistical data analysis in 2012," *China Flowers & Horticulture*, vol. 15, pp. 26-30, 2013.

[3] Xiong K F, Zhang H, Cui P, "The research of TRIZ theory in china," *Packing Engineering*, vol. 11, pp. 221-223, 2009.

[4] Gao C Q, Huang K Z, Wang G F, Zhang Y, "Research the special solution by the general solution of TRIZ theory," *The Chinese mechanical engineering*, vol. 17, pp. 84-88, January,2006.

[5] Lu X M, Zhang F Y, Zhang Q Q, "Product innovation design based on the theory of TRIZ and functional analysis," *Machinery Design & Manufacture*, vol. 12, pp. 255-257, December,2010.

[6] Zhou Y Q, "The Contradiction theory in TRIZ and its application," *Science and Management*, vol. 3, pp. 15-17, 2010.

[7] Liu G T, Yuan C J, Dai Y G, "The application of technology contradiction solve matrix based on TRIZ," *Machinery Design & Manufacture*, vol. 8, pp. 63-65, August,2007.

[8] Zhou C Q, Peng W, "Evolution and application TRIZ theory of Su-field model," *Light Industry machinery*, vol. 28, pp. 63-65, April,20

IOP Conf. Series: Journal of Physics: Conf. Series **1267** (2019) 012074 doi:10.1088/1742-6596/1267/1/012074

Tire-Pavement System Modelling and Triaxial Contact Stress Analysis

Guo Minrui[1, 2, *]

[1]College of Automotive and Transportation Engineering, Wuhan University of Science and Technology, Wuhan 430000, China
[2]Huanghuai University, Zhumadian 463000, China

[*]mr669988@163.com

Abstract. In the pavement design, the tire load is usually assumed to be a circular or rectangular evenly distributed load, which is actually not like this. In order to provide data reference for the development of tire-pavement contact stress intelligent test instrument, it is necessary to model the tire-pavement system, and construct tire-pavement 3D finite element model using large finite element software. Firstly, establish three-dimensional model of tire and road surface, and impose certain loads and constraints. Secondly, the triaxial contact stress of the tire-pavement under static conditions is analyzed. Then the triaxial stress under traction conditions is analyzed, and the triaxial stress under different conditions of use of the tire is compared and analyzed. The results show that as the load of the tire gradually increases, the maximum vertical contact stress and the maximum longitudinal stress gradually decrease, but the maximum lateral pressure gradually increases. When the inflation pressure of the tire is gradually increased, the contact stresses in the three directions are correspondingly increased. However, the maximum vertical contact stress and maximum longitudinal stress increase more obviously than the maximum transverse contact stress. In addition, the transverse contact stress could be ignored under traction conditions.

1. Introduction
As an important part of the vehicle, the tire carries the load transfer between the vehicle and the pavement. The problem of contact with the road surface seriously affects the driving performance of the car [1]. Under static conditions, there is a triaxial stress distribution between the tire pavements, and under the traction condition, the characteristics of the stress distribution will change to a large extent [2,3]. Therefore, ABAQUS analysis software was used to establish the finite element model of truck tire and pavement respectively, and the three-dimensional model of tire-road contact was established [4]. This paper analyzes the triaxial stress distribution characteristics of the tire-pavement contact, compares the variation rules of the triaxial stress distribution of the tire under different service conditions, in order to provide a data reference for the development of the tire pavement contact stress intelligent test instrument.

2. Establishment of finite element model
With the significant increase in the number of vehicles transported by overload, the vehicle load is also increasing, resulting in shortened tire life and road damage, which requires a targeted study of the

load of truck tires. According to the national standard GB/T2977-2016, the 315/70R22.5 general-purpose radial tires for trucks are selected [5].

2.1. Model of the rubber part of the tire

The rubber material is a typical superelastic material that exhibits a distinct nonlinear elastic response, and its stress-strain relationship is described by the strain potential energy [6]. Common strain potential energy functions include polynomial model, Ogden model, Arruda-Boyce model and Van der Waals model, among which polynomial models are the most common, which can be expressed as:

$$U=\sum_{i+j=1}^{N} C_{ij}(I_1-3)^i(I_2-3)^j+\sum_{i=1}^{N}\frac{1}{D_i}(J-1)^{2i} \tag{1}$$

Where U is the strain potential energy, N is the polynomial order, C_{ij} is the material parameter, describes the shear characteristics, I_1 and I_2 are the distortion metrics of the material, J is the elastic volume ratio, D_i is the material parameter, and introduces compressibility. If N is equal to 1, the initial shear modulus μ_0 and the bulk modulus K_0 of the material are:

$$\mu_0=2(C_{01}+C_{10}) \tag{2}$$

$$K_0=\frac{2}{D_1} \tag{3}$$

Formula (1) could be simplified to:

$$U=C_{10}(I_1-3)+C_{01}(I_2-3)+\frac{1}{D_1}(J-1)^2 \tag{4}$$

Formula (4) is the compressible model applied to the rubber part of the tire of this paper. It can better simulate the material properties of small strain and medium strain, and is closer to the experimental data.

2.2. Model generation

Firstly, a two-dimensional section model of the tire is established. After symmetry and rotation, the generated three-dimensional model of the tire is shown in figure 1. Since the deformation of the tire is much larger than the deformation of the road surface, it can be assumed that the road surface is a rigid road surface [7,8], and a three-dimensional model of the tire contacting the road surface is shown in figure 2.

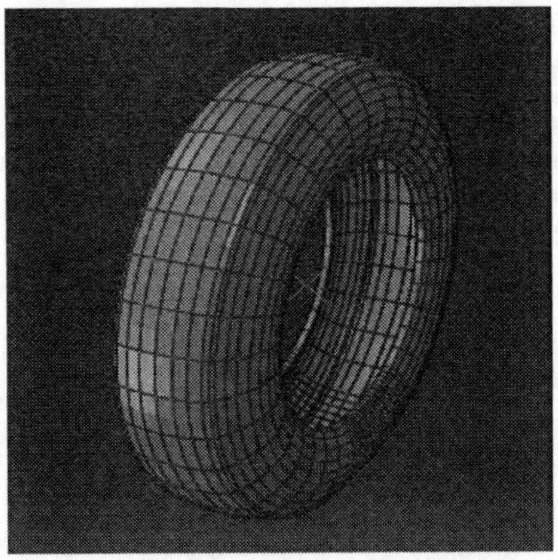

Figure 1. 3D model of the tire.

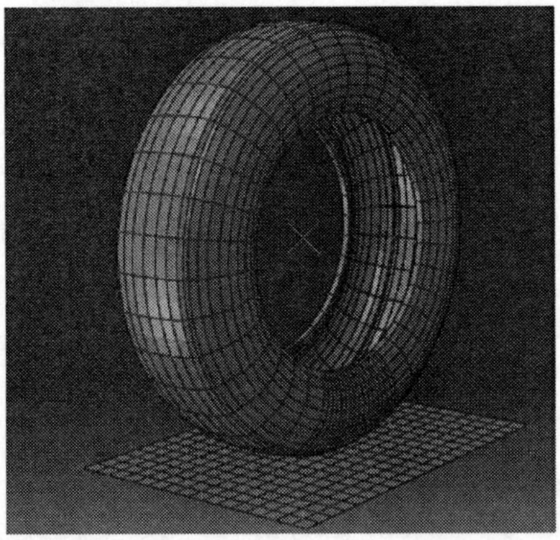

Figure 2. The tire-pavement three-dimensional model of contact.

3. Stress analysis of tire - pavement under static conditions

When the tire comes into contact with the road surface, the tread in the contact patch is compressed and subjected to compressive stress, and the side part of the tire is stretched and subjected to tensile stress, resulting in uneven distribution of normal pressure in the contact patch between the tire and the road surface. In addition, different tire material poisson ratio and special tire pattern structure promote the generation of shear stress [9,10]. This is the main reason for the normal contact stress, transverse stress and longitudinal stress between tire and pavement.

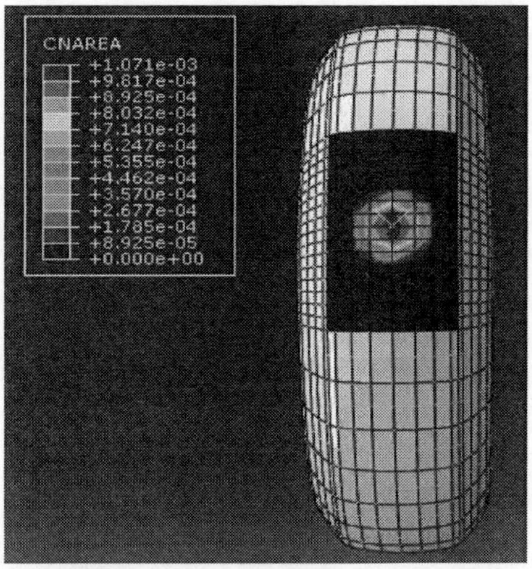

Figure 3. Tire-pavement contact area nephogram

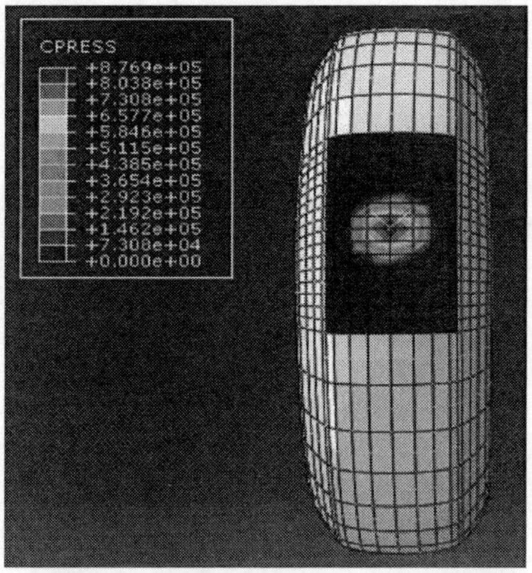

Figure 4. Normal stress nephogram

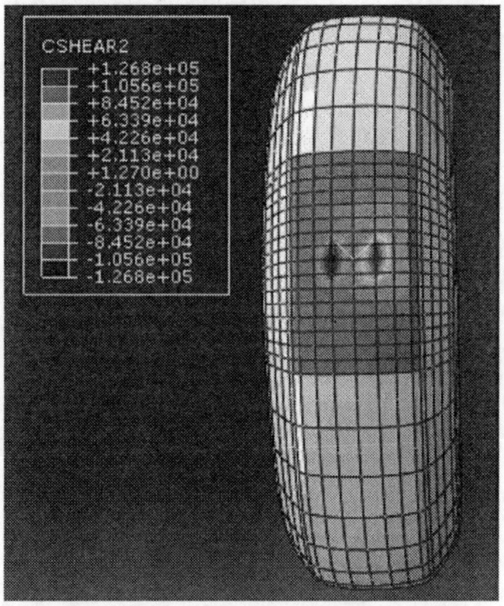

Figure 5. Transverse stress nephogram

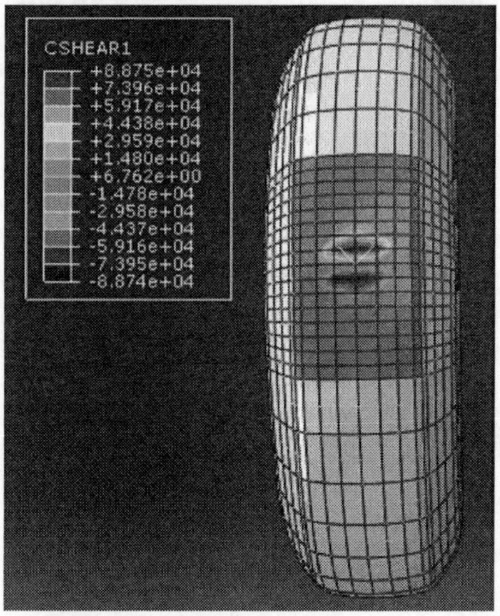

Figure 6. Longitudinal stress nephogram

In the analysis in this section, besides the pneumatic pressure of P= 600kPa, the vertical load of F=8kN is applied to the tire, and the friction coefficient between the tire and the road surface is 0.6. After calculation and analysis, figure 3 shows the cloud diagram of contact area between tires and road surface. CNAREA is used in ABAQUS to represent the cloud diagram of contact area. It can be seen that the contact area in the middle red area is the largest, and the farther away it is, the smaller it is. As shown in Figure 4, the normal contact pressure between tires and pavement is not uniformly distributed in the structural design specification of asphalt pavement. In ABAQUS analysis, CPRESS

IOP Conf. Series: Journal of Physics: Conf. Series **1267** (2019) 012074 doi:10.1088/1742-6596/1267/1/012074

is used to represent the normal contact stress. The maximum vertical stress is 876.9 kPa, and the positive value indicates that the vertical stress is compressive stress, and the vertical downward value indicates a positive sign. Figure 5 shows the distribution cloud diagram of transverse stress (CSHEAR2) in the tyre-pavement contact. The maximum transverse stress is +126.8kpa and -126.8kpa, respectively. Figure 6 shows the distribution cloud diagram of the longitudinal stress (CSHEAR1) on the tyre-to-pavement contact. The maximum longitudinal stress is +88.75kPa and -88.75 kPa, respectively. The longitudinal stress values have positive or negative, and the lateral center line of the tire grounding mark is symmetrically distributed.

4. Stress analysis of tire - pavement under traction condition

In the process of full traction, the tire-road contact stress is similar to the normal contact pressure under the static condition, but shear stress changes greatly. Take the linear velocity V of the tire to be 10km/h (i.e., V=2.778m/s), and simulate the contact stress of traction condition through different rotational angular velocities of the tire. Figure 7 shows the longitudinal force corresponding to different rotational angular velocities of the tire. When the angular velocity w of the tire rotation is 6 rad/s, the tire is in a fully traction state. Figure 8 shows a cloud diagram of the normal contact stress between tire and pavement. The maximum normal contact pressure is 860.2 kPa, and the size and distribution characteristics are similar to the static state. Figure 9 shows a cloud diagram of the transverse contact stress between tire and pavement. The maximum transverse contact stress is +12.97 kPa. The lateral contact stress is not completely balanced. This is because the tire is in a state of complete traction and the tire is sliding on the ground. However, the lateral contact stress is small and could be neglected.Figure 10 shows tire-pavement cloud diagram of the longitudinal contact stress, the largest longitudinal contact stress of 516.1 kPa, the direction of the longitudinal contact stress are negative, mainly because under the completely traction, the power output of the engine is transmitted to the tires, the ground exerts a forward driving force on the tire, and the ground is subjected to the rearward force of the tire, so the longitudinal contact stress is negative [11].

4.1. Influence of different tire loads on contact triaxial stress

Tire-road contact stress is affected by various factors such as tire load, tire inflation pressure, friction coefficient, road material and road surface unevenness excitation. In the traction condition, the linear speed of the center of the tire is 10km/h, the angular velocity of the tire rotation is w=6rad/s, and the inflation pressure is 600 kPa. Table 1 shows the influence of different loads on the triaxial stress of the tire. When the load of the tire is gradually increased, the maximum vertical contact stress and the maximum longitudinal stress gradually increase, however, the maximum lateral pressure gradually increases. Compare with vertical and longitudinal stresses, regardless of the maximum transverse stress, are always negligible. This is clearly seen from the ratio of the maximum contact stress.

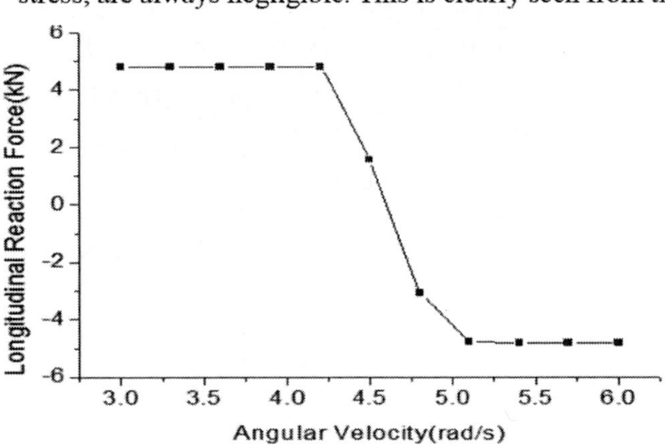

Figure 7. Longitudinal force at different angular velocities

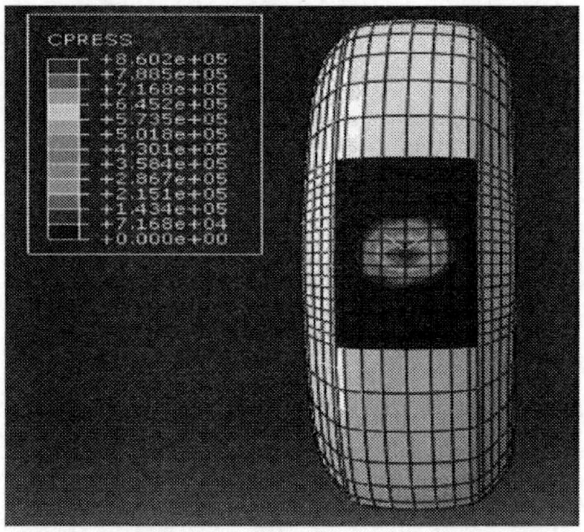

Figure 8. Normal stress nephogram

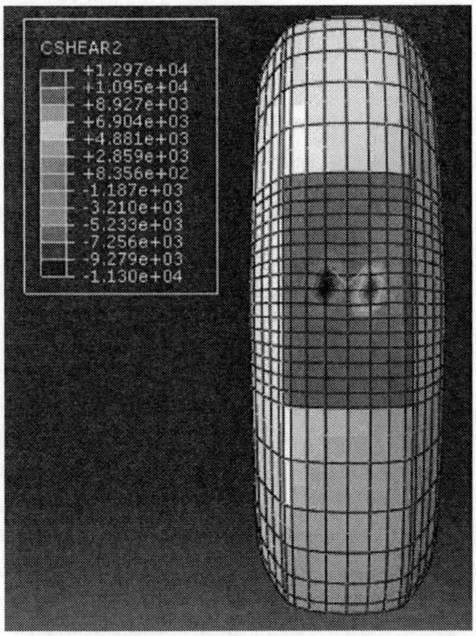

Figure 9. Transverse stress nephogram

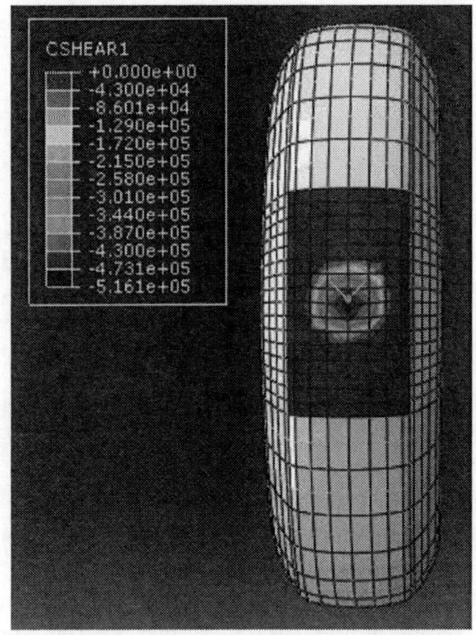

Figure 10. Longitudinal stress nephogram

Table 1. Analysis of the influence of different tire loads on triaxial stress.

Full traction	Load/kN	Maximum contact stress(kPa)			Maximum stress ratio
		Vertical	Transverse	Longitudinal	
V=10km/h w=6rad/s P=600kPa	8(12.5% of the standard load)	860	12.97	516.1	1:0.015: 0.600
	12.5(50% of the standard load)	835.4	16.43	501.3	1:0.020:0.600
	25(the standard load)	833.4	32.45	500	1:0.039:0.599

4.2. Influence of different inflation pressures of tires on contact triaxial stress

In the traction condition, the linear velocity of the center of the tire is V=10km/h, the angular velocity of the tire rotation is w=6rad/s, and when the load is F=25kN, as shown in table 2, the influence of different inflation pressures on the triaxial stress of the tire is analyzed. When the inflation pressure of the tire gradually increases, the contact stress in three directions increases correspondingly, but the maximum vertical contact stress and the maximum longitudinal stress are more obvious than the maximum lateral contact stress. As can be seen from tables 1 and 2, the lateral pressure value is always negligible regardless of changes in tire inflation pressure and tire load. Moreover, the vertical stress is approximately equal to the longitudinal stress multiplied by the friction coefficient, indicating that the contact area between the tire and the pavement is a sliding working area under the full traction condition.

IOP Conf. Series: Journal of Physics: Conf. Series **1267** (2019) 012074 doi:10.1088/1742-6596/1267/1/012074

Table 2. Analysis of the influence of different inflation pressures on triaxial stress.

Full traction	Tire inflation pressure /kPa	Maximum contact stress(kPa) Vertical	Transverse	Longitudinal	Maximum stress ratio
V=10km/h	500	753.3	32.41	451.8	1:0.043:0.599
w=6rad/s	600	833.4	32.45	500	1:0.039:0.600
F=25kN	700	934.9	32.72	560.9	1:0.035:0.600

5. Conclusion

Under static conditions and full traction conditions, the normal contact stress between the tire and the road surface is similar in magnitude and distribution, but the lateral contact stress and the longitudinal contact stress are very different. Under the full traction condition, the lateral contact stress is basically negligible, and the longitudinal contact stress is much larger than the longitudinal stress in the static state. The effects of different tire inflation pressures and different tire loads on the triaxial stress magnitude and distribution under traction conditions were analyzed and compared. When the load of the tire is gradually increased, the maximum vertical contact stress and the maximum longitudinal stress are gradually decreased, but the maximum lateral pressure is gradually increased. When the inflation pressure of the tire gradually increases, the contact stress in three directions increases correspondingly. However, the maximum vertical contact stress and maximum longitudinal stress increase more obviously than the maximum transverse contact stress.

Acknowledgments

This work is supported by curriculum reform and resource construction project of application-oriented courses in Huanghuai University. (No.2018XJGLX0419)

References

[1] Miao Yu;Guoxiong Wu; Lingyun Kong and Yu Tang 2017 *Appl. Sci.* **7** p 1123
[2] Lu Y J and Yang S P 2010 *Applied Mathematical Modelling*, **34** p 2698
[3] Zhao L and Ziran R L 2012 *Tire Science and Technology*,**40** p 83
[4] Wang H and Al-Qadi I L 2012 *International Journal of Pavement Engineering*,**13** p 310
[5] Shang Yongning, Shen Yude and Huang Keke2016 *China National Standardization Administration Committee* p 1
[6] Song Yong and Liang Yanlong 2014 *Journal of Highway and Transportation* **31** p112
[7] Wang Y S and Wu J 2009 *2nd Int. Conf. on Intelligent Computation Technology and Automation, IEEE Computer Society* p 566
[8] Anghelache, G and Moisescu, R 2012 Veh. Syst.Dyn **50** p 1747
[9] Hambleton, J. P and Drescher, A 2009 J. Terramechs **46** p 35
[10] Hao Wang; Imad L. Al-Qadi; Silvia Portas; Mauro Coni 2013 Transport. Res.Rec.**8** P 76
[11] Ghoreishy, M.H.R.;Malekzadeh, M.; Rahimi, H 2007 Iran.Polym.J. **16** p 540

IOP Conf. Series: Journal of Physics: Conf. Series **1267** (2019) 012075 doi:10.1088/1742-6596/1267/1/012075

Maloperation Prevention Filter for Touchscreen Used in Robot Arm

L. Yang[1,*], T. Ichikawa[1], Y. Miyawaki[1], R. Sakamoto[2], N. Kato[1], K. Yano[1] and S. Shimada[3]

[1]Graduate School of Engineering, Mie University, 1577 Kurimamachiya-cho, Tsu-city, Mie, 514-8507, Japan
[2]Mie University Hospital, 2-174 Edobashi, Tsu-city, Mie, 514-8507, Japan
[3]Technotools Corporation, Higashi-Naganuma, Inagi-city, Tokyo, 206-0802, Japan

[*]ljyang@robot.mach.mie-u.ac.jp

Abstract. In recent years, robot arms are being developed for self-reliance supporting for those who are unable to move without using an electric wheelchair or live a bedridden life. For example, mounting a robot arm on an electric wheelchair or bedside could help the limb disabled people to perform daily activities without nursing. However, the existing robot arms are mainly using traditional operation interfaces like keyboards or joysticks. Under different scenarios, some patients with upper limb disabilities such as contracture cannot operate such interfaces smoothly. In this study, in order to reduce the difficulty in operating robot arms, we developed a new kind of operation system using smartphone touchscreen as screen joystick operating interface. Meanwhile, the system includes a filter which detects and adjusts maloperation due to the capacitive touchscreen panel with high maloperation rate. Finally, an operating test was conducted by a patient with upper limb disabilities, and the effectiveness of this study was shown from the evaluation of the trajectory of robot hands from the initial position to the aim position. In the future, we are planning to develop an innovate operation system for a robot arm which is mounted on the wheelchair, with the effectiveness confirm by several operation tests.

1. Introduction

Among the disabled people in Japan, around 50% of them are with limb disability. Among the people with limb disability, robot arms are developed for self-reliance supporting for those who are unable to move without using a wheelchair or live a bedridden life [1][2][3]. To counter with the issue, practice like mounting a robot arm to wheelchair or bedside could help the limb disabled people to do some basic things without nursing such as holding a cup or picking up items from the floor.

Currently, the extant robot arms are mainly using the traditional operation interface like keyboards or joysticks. Under different scenarios, some patients with upper limb disabilities cannot operate such interfaces smoothly. For this problem, some new type of operating interface to operate robot arms was developed recently. However, even the number of keys was reduced, for patients with upper limb disability, it is still hard to handle. In addition, an interface using the 3D mouse as the input device which does not require the mode switching operation has been developed [4]. However, in the case of cervical spinal cord injured persons, it is difficult for them to grasp the object due to the contracture of the wrist joint which leads to limited range of motion of the finger joints. On the other hand, interfaces which

IOP Conf. Series: Journal of Physics: Conf. Series **1267** (2019) 012075 doi:10.1088/1742-6596/1267/1/012075

operate robot arms by recognizing voice commands are also being developed [5]. But this pattern can only recognize limited simple commands which are not sufficient for practical use.

As we mentioned above, since the range of motion of the finger joints is restricted by the contracture of the wrist joint, the person with cervical spinal injury can't touch the screen with the fingertip, but can use the second joint of the little finger shown in Figure 1. Therefore, it is difficult for him to hold the interface, and it is necessary to design an interface that enables simple operation for the person with cervical cord injury. In this study, in order to reduce the difficulty in operating robot arms, we developed a new kinds of operation system which use smartphone touchscreen as the operating interface.

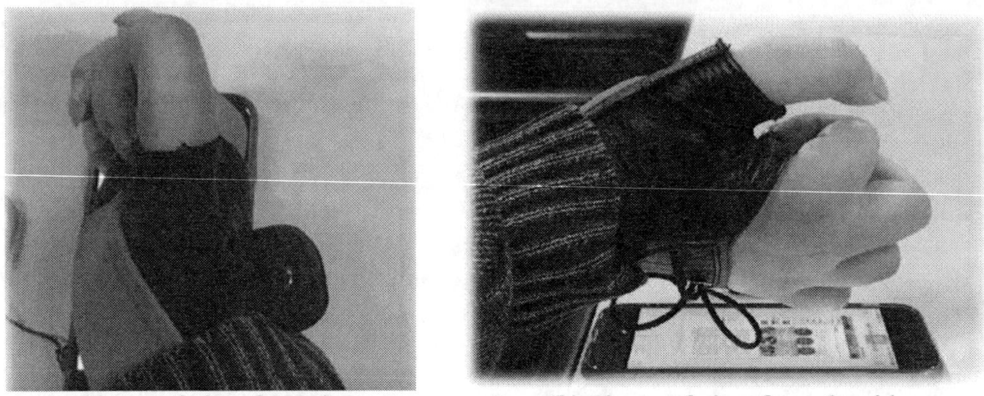

(a) Figure of view from above (b) Figure of view from the side

Figure 1. Operation of a smartphone by cervical spinal cord injured person

2. Specifications of operating system

The configuration of the robot arm operation system in this study is shown in Figure 2. The operating device is Apple's iPhone 7. The operation screen is displayed in the web browser of the device which is connecting with WebIOPi, a software supported by Raspberry_Pi. When the operation was detected, the input information will be sent to the Raspberry_Pi. After this, by using I^{2}C interface of WebIOPi, the input information will be sent to Digispark, which could be used as a USB gamepad emulator. Therefore, the system allows us to control the robot arm that was designed to be operated by a gamepad.

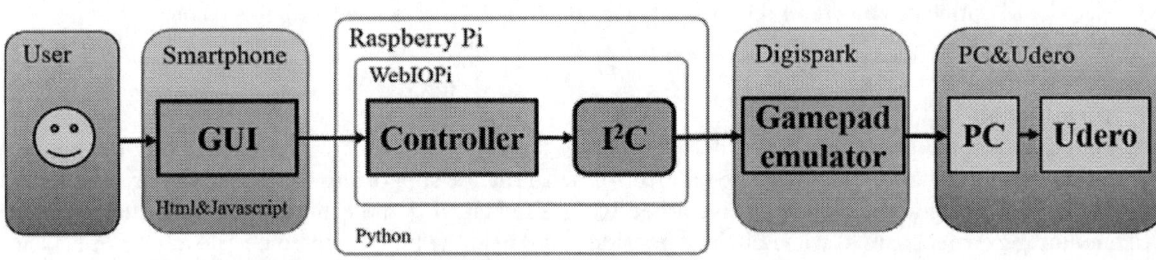

Figure 2. Configuration of the robot arm operation system

Figure 3 shows the GUI (graphical user interface) used in this study which was developed by HTML and JavaScript. Meanwhile, the Table 1 shows the specification of developed operation screen.

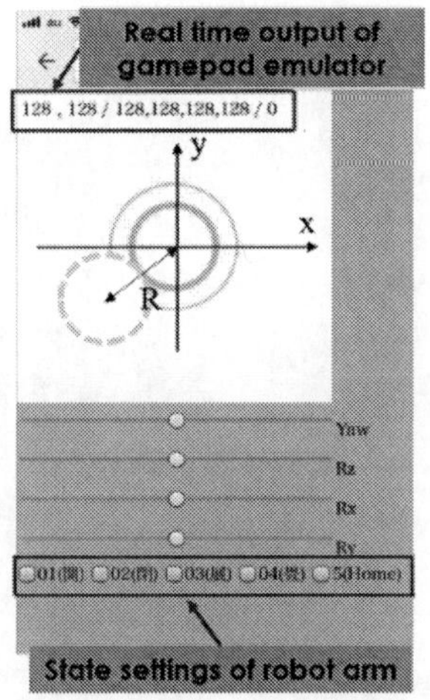

Figure 3. GUI used in this study

Table 1. Specification of operation screen	
Range of movement R	14.6[mm]
Operable range	54.4[mm]×54.4[mm]
Range of coordinates x	-50[px]$\leq x \leq$ 50[px]
Range of coordinates y	-50[px]$\leq y \leq$ 50[px]

As the figure shows, the origin of dragging range is fixed. The horizontal move of robot hand is operated by the dragging operation using the button in the centre of the screen, the vertical movement of robot arm is performed by the dragging operation of slider. And the state settings of robot arm are displayed on the bottom of the screen. On the other hand, the real time input of gamepad emulator is displayed in the upper part of GUI. When the joystick is in the neutral state, the real time input value of gamepad emulator will stay at the median as (128, 128).

3. Operation verification

3.1 Operation check
In order to verify the operation of proposed interface, we did a verification experiment shown in Figure 4. During the operation check, the robot arm was mounted on origin. Meanwhile, as the task of experiment, the operator should operate the robot arm by using the proposed interface to move the robot hand from initial position to target position. The robot arm used in this experiment was Udero [6], which is developed by Technotools, shown in Figure 5. The specifications of Udero is shown in Table 2.

3.2 Introduce of maloperation filter
During the operation check, we noted that because of the operating system is using capacitive touchscreen panel as the interface, an unconscious touch from operator will lead the movement of robot arm immediately. It is dangerous in practical use. In order to solve this problem, we decided to introduce a maloperation filter into the proposed system.

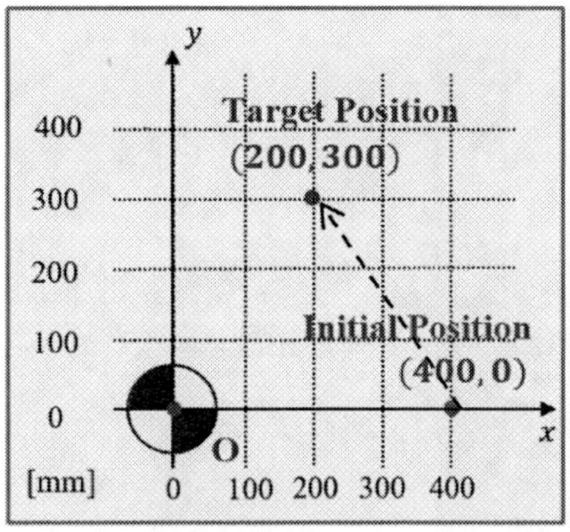

Figure 4. Experimental task

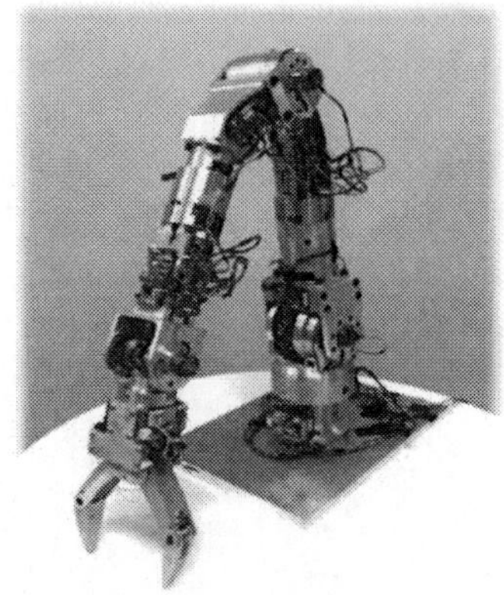

Figure 5. Robot arm "Udero"

Table 2. Specification of Udero

DOF	4(Arm) + 3(Wrist) + 1(Hand)
Max Reach	885 [mm]
Weight	6.0 [kg]
Max Payload	1.0 [kg]
Max Rotation Speed	60[deg/s]

3.2.1 The algorithm of maloperation filter. In this study, we avoid the maloperation by judging the displacement between the first touching point and the origin of operable range. If the displacement between the first touching point and the origin exceed the threshold, the operation will be judged as maloperation and the joystick will maintain the neutral state. Here, we set the threshold as the 20% of the radius of operable range. Figure 6 shows the flowchart of maloperation prevention filter.

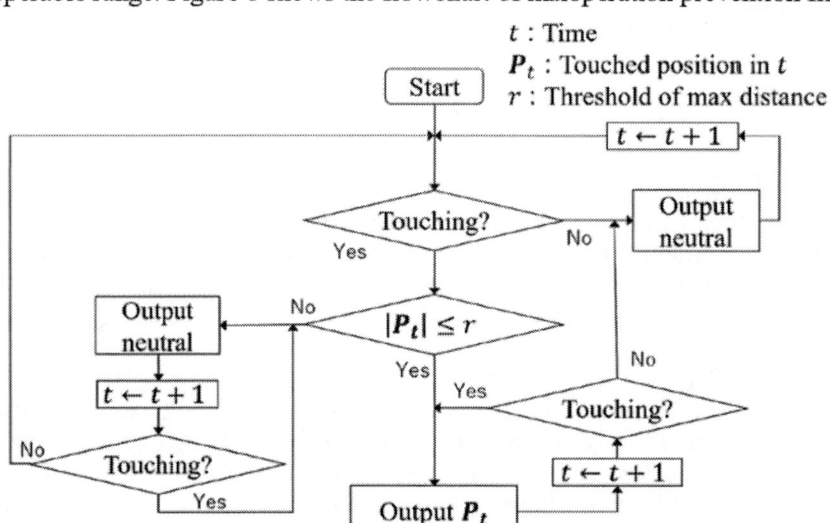

Figure 6. Flowchart of maloperation prevention filter

3.2.2 Performance of maloperation prevention filter. Figure 7 shows operation screen of the operation system with and without the maloperation filter. The real time input of gamepad emulator is enclosed by the red box. In the case of the operation system without filter, we can note that the coordinate will be updated immediately even the first touch point is far away from the origin. By contrast, in the case of the operation system with maloperation filter, if the first touch point is far away from the origin, the coordinate of touch point will not be updated and the joystick will keep the neutral state.

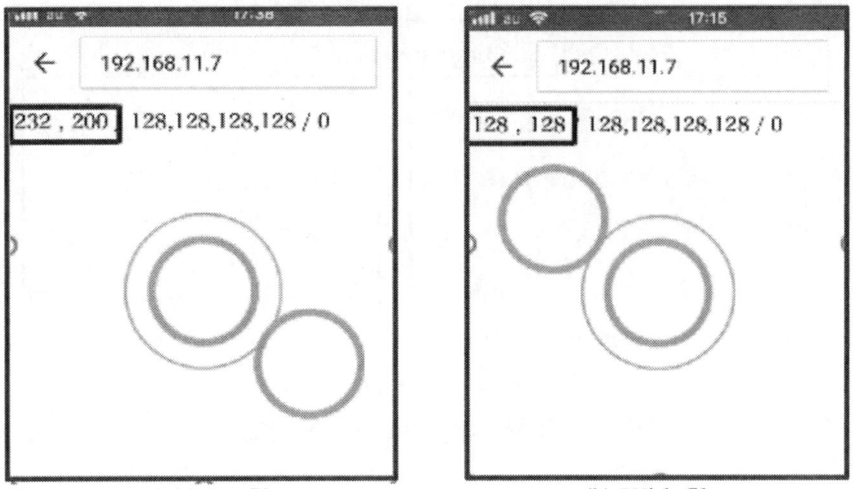

(a) Without filter (b) With filter

Figure 7. Comparison of operation screenshots

4. Comparative operation experiment

4.1 Experimental outline

The experimental subject was a patient with cervical spinal injury, and the experiments were conducted three times by operating the robot arm hand from start point to the target point (same as Figure 4). We measured the trajectory of robot hand in real time. Also, as an operation condition, in order to simulate the maloperation, the operator should place the smartphone on his knee without looking at the touch screen during the operation. As a procedure of the experiment, experiments by the operation system without the filter were performed after the experiment of the operation system with filter. In addition, experiments were conducted in a condition that the subject were sufficiently accustomed to the operation.

4.2 Experimental result

Figure 8 shows the results of comparative operation experiment in 3 times respectively. In Figure 8, the brown line represents the moving trajectory of robot hand in case of operation system without the maloperation filter. The blue line represents the move track of robot hand in case of the operation system with the maloperation filter. And the red dash line represents the ideal moving trajectory.

According to the results shown in Figure 8, we note that the proposed operation system can suit the patient with a cervical spinal injury. Particularly, by using the proposed maloperation filter, the real trajectory is close to the ideal trajectory.

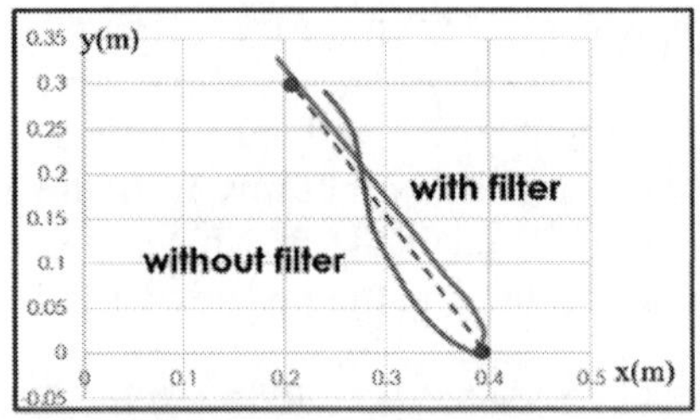

(a) The first operation

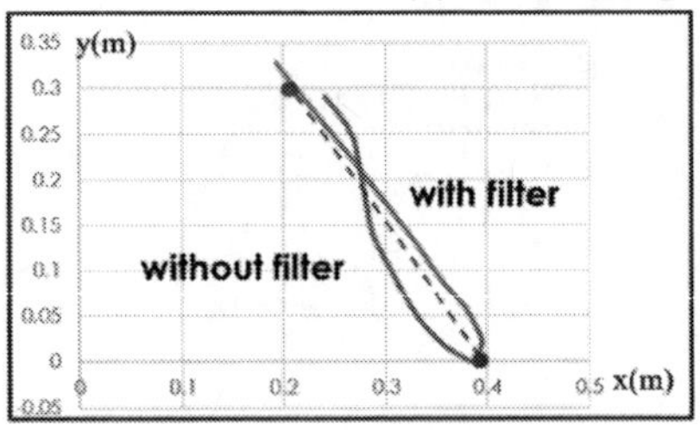

(b) The second operation

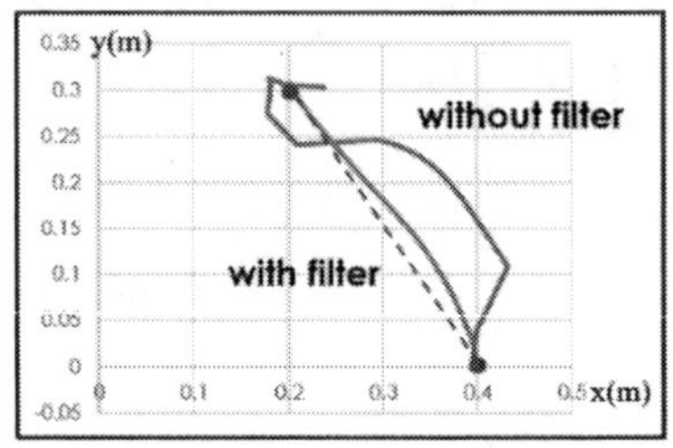

(c) The third operation

Figure 8. Experiment results

5. Conclusion

By using robot arm, patients with upper limb dysfunction such as cervical spinal cord injury are able to conduct daily activities independently. However, the existing robot arms are mainly using the traditional operation interface like keyboards or joysticks which may be difficult for them to handle. In order to solve this problem, in this study, we proposed a new kind of operation system which use smartphone as the operating interface. On the other hand, we also proposed a maloperation filter which could avoid the maloperation caused by unconscious touch. Finally, we conducted a comparative operation experiment to verify the effectiveness of this research. As the result of experiment, the proposed operation system is suit for patients with upper limb dysfunction and the maloperation filter is also functional. As the future work, we are planning to develop an innovate operation system for robot arm which is mounted on a wheelchair, with the effectiveness confirming by several operation tests.

Reference

[1] Kenji Chihara, Katsutoshi Fujii: Development of a High-Performance Electric Wheelchair, *Research report of Gifu Prefectural research institute of information technology*, pp.51-56 (2007)

[2] Yuki Suga, Wei Wang, Shigeki Sugano: Development of Light-Weight Robot Arm Mounted on Wheel Chair with Storage Mechanism, *The Robotics and Mechatronics Conference*,1A1-14D (2009)

[3] Natsuki Yamanobe, Yujin Wakita: Daily Life Support for Persons with Upper-limb Disabilities by Robotic Arm, *The 24th Annual Conference of the Japanese Society for Artificial Intelligence*, 1H2-NFC3b-10 (2010)

[4] Kousuke Ichikawa, Tetsusi Oka, Keisuke Matsusima: The evaluation of manipulator operation interface using 3D mouse, *The 78th National Convention of IPSJ*, 2Y-01 (2016)

[5] Hairong Jiang, Ting Zhang, Juan Wachs, Bradley Duerstock: Enhanced control of a wheelchair-mounted robotic manipulator using 3-D vision and multimodal interaction, *Computer Vision and Image Understanding*, Vol.149, pp.21-31 (2016)

[6] Eito Tanaka, Tomoya Ichikawa, Ryota Sakamoto, Ken`ichi Yano, Shintaro Shimada: 7-DOF robot arm "Udero" for self-reliance support of disabled persons, *The 17th SICE System Integration Division Annual Conference*, (2016)

An Automatic Electricity Meter Reading System Based on Neural Vision Location

Jin Yin[1], Pengfei Xia[1], Jingsong He[1,*], Li Gu[1] and Kejun Yang[2]
[1]University of Science and Technology of China, Hefei, China
[2]Anhui Nari Jiyuan Electric Power System Tech CO., LTD, Hefei, China

[*]Email: hjss@ustc.edu.cn

Abstract. Remote meter automatic reading and smart meter reading is a vital part of smart grid and has been widely concerned. Traditional manual method is manpower consuming. Developing an automatic electricity ether reading system is necessary in practice. Two particular aspects in this task should be concerned. One is different-perspective problem, which causes by the variations of camera position and leads to pool display localization. Another is information diversity problem. Namely, the electrical meter used in this paper needs to recognize effective information from diversity information. This paper describes a novel automatic electricity meter reading system, which can automatically locate and recognise effective information. For the camera perspective variability problem, we utilize a new localization method based on neural network. Via actual meter monitoring experiment, the result shows our system can efficient recognition and achieve 97% accuracy.

1. Introduction

With the rapid development of industrial automation and intelligent instruments, the power industry is facing following challenges, including lack of automated analysis, huge resources consuming, etc. Therefore, smart grid gradually become a research hotspot [1, 2]. Remote meter automatic reading and smart meter reading, as a vital part of smart grid, are widely concerned. Traditional method of manual reading is manpower consuming, especially in practical application, where meter-data need to be all-round monitored. So, it is necessary to develop an automatic electricity meter reading system which can apply in real-world.

There are two particular aspects in the process of electricity meter reading. One is different-angel problem, which means pool display localization effect owing to the variations of camera position(level), shot lengths and angles, as shown in Figure 1.a. It is apparent that every dial in images has its own unique angle and dimension. Especially, the localization task is important in the process of recognition because small deviation will affect final recognition performance. Another challenge in electricity meter reading tasks is information-diversity problem. That is to say, the electrical meter used in this paper is not like house's electrical meter mentioned in [3,4,5]. House's electrical meter only focus on energy consumed, however current electrical meter contains variable information and not all diversity information is useful, so the major aspect of digit recognition system lies in recognizing effective information from diversity information, as shown in Figure 1.b.

For above aspects, there are several works for the location and recognition. Previously, information extraction from images and video usually make full use of image features, such as edges, strokes and corners feature [6,7,8]. In the past few years, due to the effective application of neural networks, some work extract information via neural networks [9]. Similarly, we use semantic segmentation to solve the camera visual variability problem in this paper.

The semantic segmentation is a core computer vision task, which predicted the category of per-pixel that it belongs to. On the basis of encoder-decoder structure, U-net appends a useful supplementary structure to increase segmentation accuracy. It has been demonstrated an excellent capability in various image analysis tasks, such as remote sensing image analysis [10] and medical image analysis [11].

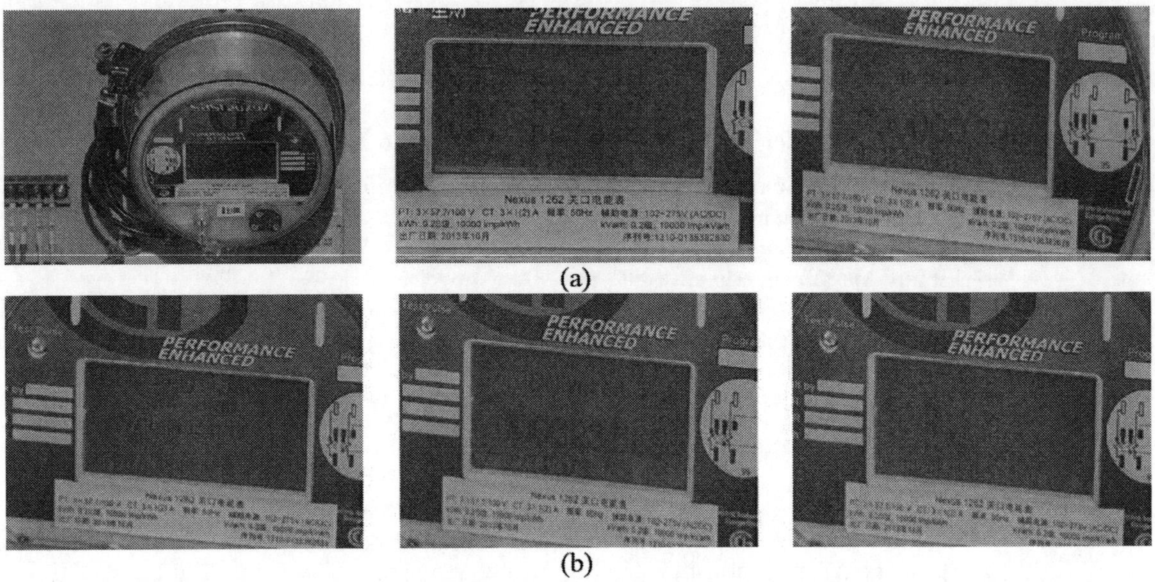

Figure 1. Different dial in real-world.

The first row is different picture captured from different video. This is different-angel problem. The second row is different picture captured from same video but different time. This is information-diversity problem.

In this paper, we proposed a novel automatic electricity meter reading system based on neural vision location, which consists of two major stages: automatic location system and digit recognition system. The proposed automatic location system utilizes semantic segmentation to accurately determine display coordinates. Comparing with other methods, semantic segmentation has high precision and efficiency, this is mainly because semantic segmentation is less affected by other interference information, such as dial information. The digit recognition method is to read the corresponding digit displayed in reading area. The key features of the developed system are:

- It is effective to solve the problem having risen from the camera orientation.
- In the process of digit recognition, our system can automatically select effective information from diversity information collected.

2. Automatic localization system

The important aspect in the process of automatic electricity meter reading is accurate localization of reading area. Considering the outstanding performance of U-net in various semantic tasks, including biomedical image [12] and remote sensing [13] image we choose U-net structure as our segmentation network. Semantic segmentation and determined coordinates are mainly units in automatic localization system.

2.1. Semantic segmentation

2.1.1. Network. U-net structure is a symmetric structure with a contracting path(left path) and an expansive path(right path). Following basic U-net, contracting path is composed of continuous two 3×3 convolutions module, each convolution followed by activation function and a 2×2 max-pooling operation with stride 2. After two 3×3 convolution module, the number of feature channel is doubled. Similarly, every 3×3 convolutions modules in expensive path, each convolution followed by activation function and a 2×2 up-sampling. After 3×3 convolution module, the number of feature channel is halved which concatenate with corresponding feature map from contracting path, which can propagate context information from lo-level features to high-level features. This is important to edge-preserving. In our system, we shrink channel numbers and use parametric rectified linear unit (PRelu) instead of rectified linear unit (Relu) owing to "dying Relu" problem.

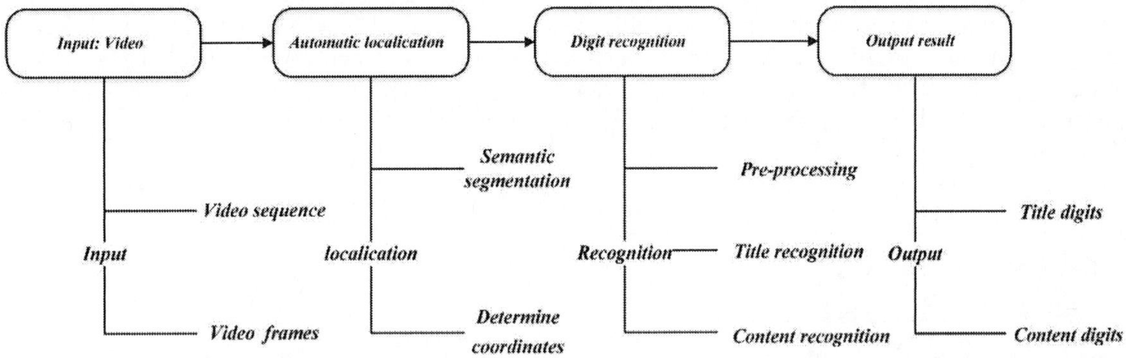

Figure 2. The framework of entire system.

2.1.2. Training process. For our training process, we set the "batchsize" to 16 and use batch normalization [14] to accelerate training. Since there are only few training samples for our system, it is essential to use appropriate data pre-processing and augmentation method. Considering very little training data available, we primarily partition entire image (1280×720) into multiple parts, whose resolution are 128×128. New images are generated by synthesizing three new rotations (90, 180, 270), random flipping, random adding noise and random blurring. The augmented images are used to train the network with the stochastic gradient descent.

Due to segmentation image produced by semantic segmentation presents lack of precision near boundaries of the reading area, the separation border is enhanced using weighted cross entropy. The basic cross-entropy for binary classification can be calculated by [15]:

$$C = -\frac{1}{n}\sum_x [ylnp + (1-y)ln(1-p)] \tag{1}$$

Where $y \in \{0,1\}$, $y = 1$ means the ground-truth class is consistent with predict class. $p \in [0,1]$ specifies estimated probability for the ground-truth class.The notable property of cross-entropy loss is same weight to all pixels. We consider address boundary problems by means of up-weights the loss assigned to boundary pixels. The proposed enhanced cross-entropy loss as:

$$EC = -\frac{1}{n}\sum_x \alpha[ylnp + (1-y)ln(1-p)] \tag{2}$$

In practice α may be treated as a parameter. In our system, we set $\alpha = 5$.

2.2. Determine coordinates

As in the previous section, The reading area is indicated using U-net structure and edge enhancement method. In order to segment display area from entire dial image, we consider determining coordinates by minimum bounding rectangle (MBR).

IOP Conf. Series: Journal of Physics: Conf. Series **1267** (2019) 012076 doi:10.1088/1742-6596/1267/1/012076

3. Digital recognition system

Digital recognition system is to identify different information displayed on different occasions. The system mainly consists of three different units: pre-processing, title digit recognition and content digit recognition. An overview of digital recognition system is depicted in Figure 3.

3.1. Pre-processing

In order to make further processing more conveniently, we used the following strategy to optimize the segmentation results: reading area belonging to diverse camera perspective have different shapes, therefore we performed region-correction step to rectify reading area to a fixed size. Then execute with image segmentation to separate content and title, because only the area where the content and title are the regions of interesting. The title digit represents the information categories, which is up to 19 classes but only 11 classes are effective. A curial step, during digital recognition system, is to distinguish title information. This detail is important because from the result of title information, effective content information can be judged.

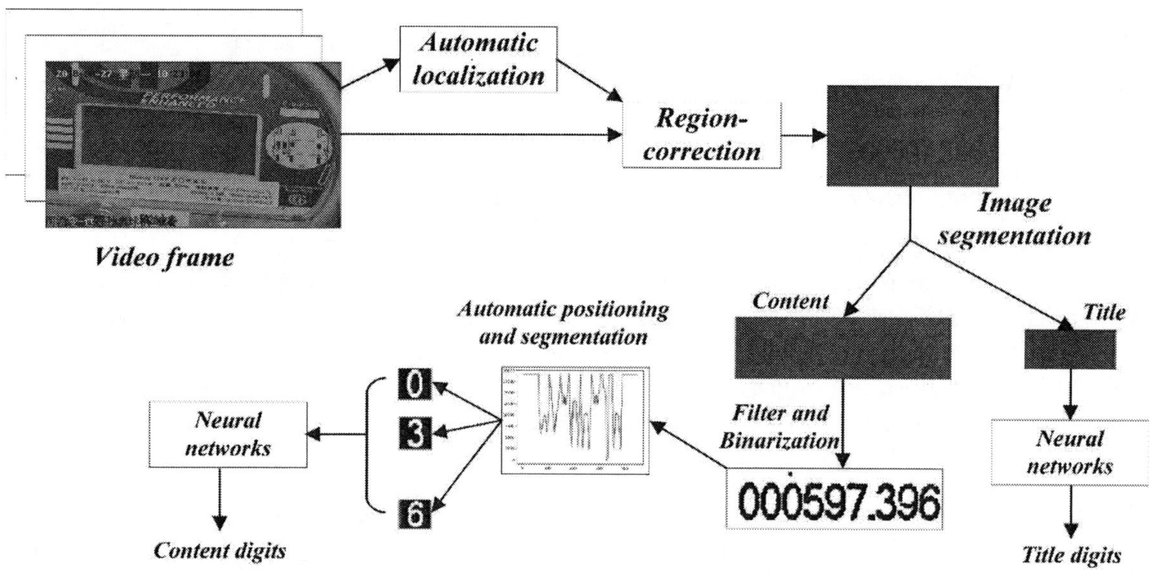

Figure 3. The process of digit recognition.

3.2. Title digit recognition

Through the image pre-processing, the content region and title region have been extracted. Then title digit in the regions of interesting is identified firstly. The numeric are in the range of 001 and 019. Obviously in this situation, classification using neural works but recognition after digit segmentation can improve speed. Especially, The scale of the rectangle should be converted to fixed scale before classification. Figure 3 shows the process of title digit recognition.

3.3. Content digit recognition

The content digit, stands in contrast to title digit recognition to contain a small number, is more challenging. Focus on this problem, these steps are applied in order to eliminate noise and improve accuracy.

Firstly, the image process of filtering and binarization are performed on content image to get the binary image of the content, then shrink area where the digits are to reduce interference.

Secondly, evaluate if the information displayed on content area is valid. In order to ensure the validity of the information, we consider the consistency of title digit and corresponding content digit format. For example, the valid digit format (see Figure 3) contains nine digits and a decimal point, of which six numbers exits before the decimal point. To judge the validity of digit format, we construct a column histogram which counter the total value of pixels for each column in binary image. By this way, can we obtain the number of black pixel blocks and validity of digit format.

Thirdly, segment each digit and train a neural network to classify. Knowing the title digit and validity of digit format, the automatic positioning and segmentation can be utilized to obtain each digits. Similarly, the position is determined by a row histogram and a column histogram. By analyzing the histogram, the exact location of the digits is obtained. Based on above result, each digit can be segmented and converted to uniform scale. Afterwards, a neural network model is trained to classify each digit.

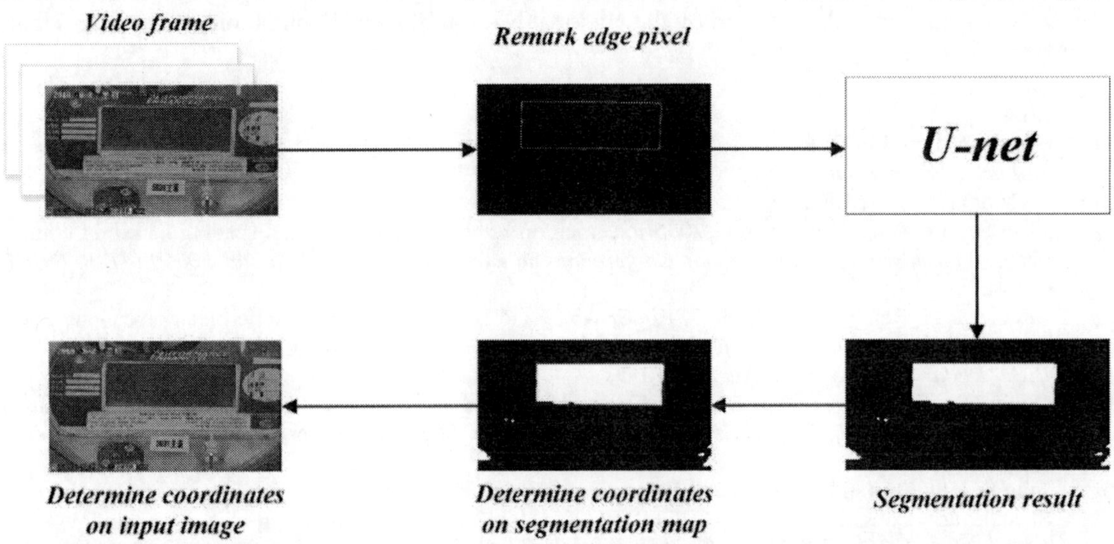

Figure 4. The process of digit recognition.

4. Experiments

To demonstrate the performance of proposed system, we test our system on five video sequences with a 1280×720 resolution camera collected from real-world, the first image frame from video is used to automatically locate. The details of our experiments are specified in this section, including each part of the test and overall result.

The U-net is trained by stochastic gradient descent (SGD) on Keras. The first step is to remark edge pixels only the area where the digits are where the digits are. After training 50 epochs, we determine coordinates after obtaining segmentation map from U-net model. To exhibit our contribution, we show the process of semantic segmentation as follows.

For following digit recognition problem, we tested the digit detection strategy presented in Section 3. In order to provide a practical system, we create a human-computer interaction. Figure 5 shows the finally experiment result, yellow rectangles means information class, green rectangles represents valid information. The obtained overall accuracy is 97% and the speed is approximately 10frame/s. The devices used for our experiments as follows, GPU:GTX 1060 (3GB); CPU:Intel Core i7-7700.

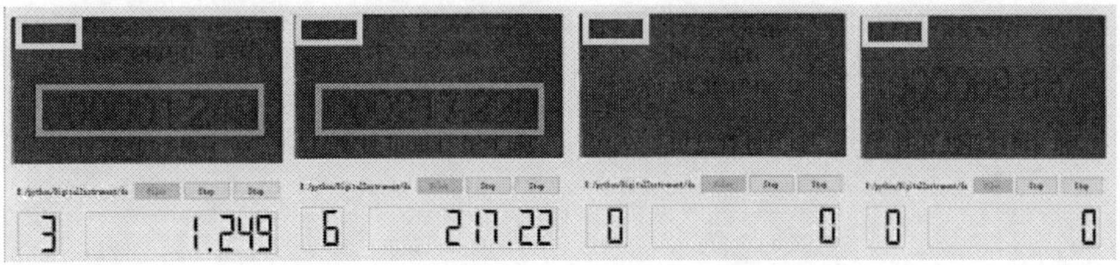

Figure 5. The final experiment result.

5. Conclusion

The current development trend in electric power industry is towards intelligent direction. Aim to provide a practical tool, we proposed a novel electricity meter reading system. The proposed system, including automatic localization system and digital recognition system, is able to automatically detect the area of interest and recognize effective information. Experimental results show that our system has ability to automatically recognise and its accuracy can achieve 97%.

Acknowledgments

This study was funded and supported by the National Natural Science Foundation of China through Grant No.61273315.

Reference

[1] Gungor V C, Sahin D, Kocak T, Ergut S, Buccella C, Cecati C and Hancke G P 2011 *IEEE transactions on Industrial informatics* 7 529-539

[2] Farhangi H 2010 IEEE *power and energy magazine* **8**.

[3] Zhao S, Li B, Yuan J and Cui G 2005 Research on remote meter automatic reading based on computer vision *Transmission and Distribution Conference and Exhibition: Asia and Pacific, 2005 IEEE/PES (IEEE)* pp 1-4.

[4] Elrefaei L A, Bajaber A, Natheir S, AbuSanab N and Bazi M 2015 Automatic electricity meter reading based on image processing *Applied Electrical Engineering and Computing Technologies (AEECT), 2015 IEEE Jordan Conference on* (IEEE) pp 1-5

[5] Gupta N and Shukla D 2016 Design of embedded based automated meter reading system for real time processing *Electrical, Electronics and Computer Science (SCEECS), 2016 IEEE Students' Conference on* (IEEE) pp 1-6.

[6] Jung K, Kim K I and Jain A K 2004 *Pattern recognition* **37** 977-997.

[7] Shen H and Coughlan J 2006 Finding text in natural scenes by figure-ground segmentation *Pattern Recognition, 2006. ICPR 2006. 18th International Conference on* vol 4 (IEEE) pp 113-118

[8] Kim K, Byun H, Song Y, Choi Y W, Chi S, Kim K K and Chung Y 2004 Scene text extraction in natural scene images using hierarchical feature combining and verification *Pattern Recognition, 2004. ICPR 2004. Proceedings of the 17th International Conference on* vol 2 (IEEE) pp 679-682.

[9] Vanetti M, Gallo I and Nodari A 2013 *Pattern Recognition Letters* **34** 519-526.

[10] Lu K, Sun Y and Ong S H 2018 Dual-resolution u-net: Building extraction from aerial images *2018 24th International Conference on Pattern Recognition (ICPR)* (IEEE) pp 489-494

[11] Oktay O, Schlemper J, Folgoc L L, Lee M, Heinrich M, Misawa K, Mori K, McDonagh S, Hammerla N Y, Kainz B *et al.* 2018 *arXiv preprint arXiv:1804.03999*

[12] Alom M Z, Hasan M, Yakopcic C, Taha T M and Asari V K 2018 *arXiv preprint arXiv: 1802.06955.*

[13] Zhang Z, Liu Q and Wang Y 2018 *IEEE Geoscience and Remote Sensing Letters*

[14] Ioffe S and Szegedy C 2015 *arXiv preprint arXiv:1502.03167*

[15] Lin T Y, Goyal P, Girshick R, He K and Doll´ar P 2018 *IEEE transactions on pattern analysis and machine intelligence*

IOP Conf. Series: Journal of Physics: Conf. Series **1267** (2019) 012077 doi:10.1088/1742-6596/1267/1/012077

PD Control at the Hopf Bifurcation Point of a Neuron System with Inertia and Delay

Shuo Shi[1,*], **Huaifei Wang**[1] **and Min Xiao**[1]

[1]College of Automation, Nanjing University of Posts and Telecommunications, Nanjing, China

[*]2312081035@qq.com

Abstract. In this paper, a two-neuron system with inertia and delay is proposed firstly. a PD controller is then applied to the system for the purpose of improving its dynamical performance. Through the mathematical transformation, we extend the system to a four-dimensional one with only time delays. With the help of the associated characteristic equation of the mathematical model, suffcient conditions for ensuring the system stability are proposed. Furthermore, with the time delay as the bifurcation parameter, relevant requirements for the generation of Hopf bifurcation are derived. Then a series of numerical simulatiosns are carried out to justify the theoretical analysis and it is found that the application of PD control scheme helps to advance the bifurcation point dramatically through a slight adjustment of the controller parameters.

1. Introduction

A growing number of attention has been paid to the neuron system ever since the simplified neural network was initially proposed by Hopfield in [1] and it has been applied successfully in different engineering fields like pattern recognition, signal and image processing and optimization cryptography [2]. Biological experiments on the human brain have demonstrated that neurons interact with each other, displaying various dynamical phenomena, such as oscillations, bifurcation and chaos [3, 4, 5, 6]. Inertial neuron networks (INNs) were first proposed in [7] by Babcock and Westervelt. It is found in practical use that the intervention of the inertia parts helps with the unordered search of memory. INNs also bear a profound bioengineering background. For example, the squid axon can be described as a mathematical model containing a phenomenological inductance. Besides, in the semicircular canals of some organisms, the membrane of a hair cell can be modeled by the equivalent circuits consisting of inductance. Such networks with inductance are usually referred as inertial networks.

It is known that Hopf bifurcations serve as the strategy for coping with various dynamic properties of nonlinear systems More detailed information about the performance of periodic solutions around the equilibrium can be obtained after the Hopf bifurcation analysis. Recently, bifurcation control has aroused tremendous concern for researchers from different disciplines. Ever since the initial work done by Ott et al. [8], diverse bifurcation control schemes have been proposed [9, 10, 11], and these control strategies have been successfully applied to different nonlinear systems. By contrast, although the dynamics of inertial neural networks have been discussed, the control of such systems is rarely covered. Therefore, in this paper, we attempt to apply PD control scheme to adjust the dynamical performance of neuron systems with inertia. For one thing, PD control has been extensively applied in the robotic manipulation, indicating its excellence in industrial engineering. For another thing, in terms of Hopf bifurcation and bifurcation control, complex networks under the PD controller have been investigated in recent years. Tang et al. [12] applied the fractional-order PD control to a congestion control system. Xiao et al. [13] addressed the problem of fractional-order PD control in a small-world networks and demonstrated that the onset of Hopf bifurcations can be altered via the proposed PD controller by setting proper control parameters. Motivated by the previous works, we adopt a PD controller to a two-neuron system with inertia and delay for the purpose of altering the system dynamics.

2. Model Description

Inspired by the work done in [14], we simplify the model and propose a two-neuron system with inertia and delays, which is described as follows:

$$\begin{cases} \ddot{u}_1(t) = -\dot{u}_1(t) - a_1 u_1(t) + v_1 f(u_2(t-\tau)), \\ \ddot{u}_2(t) = -\dot{u}_2(t) - a_2 u_2(t) + v_2 f(u_1(t-\tau)). \end{cases} \tag{1}$$

where a1 > 0; a2 > 0 indicate respectively the stabilities of internal neuron processes. v1 < 0; v2 < 0 stand for the interaction parameters between two neurons. ≥ 0 is the coupled time delay and f(·) is the activation function.

It is obvious that system (1) has a unique equilibrium (u1,u2) = (0,0). Therefore, when adopting a PD controller to the system, it turns into

$$\begin{cases} \ddot{u}_1(t) = -\dot{u}_1(t) - a_1 u_1(t) + v_1 f(u_2(t-\tau)) + Kpu_1(t) + Kd\dot{u}_1(t), \\ \ddot{u}_2(t) = -\dot{u}_2(t) - a_2 u_2(t) + v_2 f(u_1(t-\tau)) + Kpu_2(t) + Kd\dot{u}_2(t). \end{cases} \tag{2}$$

Then let $u_1(t) = x_1(t), \dot{u}_1(t) = x_2(t), u_2(t) = x_3(t), \dot{u}_2(t) = x_4(t)$, system (2) turns to be four-dimensional described as follows:

$$\begin{cases} \dot{x}_1(t) = x_2(t), \\ \dot{x}_2(t) = -x_2(t) - a_1 x_1(t) + v_1 f(x_3(t-\tau)) + Kpx_1(t) + Kdx_2(t). \\ \dot{x}_3(t) = x_4(t), \\ \dot{x}_4(t) = -x_4(t) - a_2 x_3(t) + v_2 f(x_1(t-\tau)) + Kpx_3(t) + Kdx_2(t). \end{cases} \tag{3}$$

3. Stability and Bifurcation Analysis

In this section, the stability will be studied for controlled system (3) and relevant conditions for the existence of Hopf bifurcation will be obtained.

Linearizing system (3) at the equilibrium leads to:

$$\begin{cases} \dot{x}_1(t) = x_2(t), \\ \dot{x}_2(t) = (Kp - a_1)x_1(t) + (Kd - 1)x_2(t) + v_1 f'(0)x_3(t-\tau), \\ \dot{x}_3(t) = x_4(t), \\ \dot{x}_4(t) = (Kp - a_2)x_3(t) + (Kd - 1)x_4(t) + v_2 f'(0)x_1(t-\tau). \end{cases} \tag{4}$$

Therefore, the associated characteristic equation of system (17)

$$\begin{vmatrix} \lambda & -1 & 0 & 0 \\ a_1 - Kp & \lambda - Kd + 1 & -v_1 f'(0)e^{-\lambda\tau} & 0 \\ 0 & 0 & \lambda & -1 \\ -v_2 f'(0)e^{-\lambda\tau} & 0 & a_2 - Kp & \lambda - Kd + 1 \end{vmatrix} = 0, \tag{5}$$

Namely

$$\lambda^4 + \beta_1\lambda^3 + \beta_2\lambda^2 + \beta_3\lambda + \beta_4 e^{-2\lambda\tau} + \beta_5 = 0, \tag{6}$$

where $\beta_1 = 2(1-Kd), \beta_2 = (1-Kd)^2 + a_1 + a_2 - 2Kp, \beta_3 = (1-Kd)(a_1 + a_2 - 2Kp), \beta_4 = -v_1 v_2 f'^2(0), \beta_5 = (a_1 - Kp)(a_2 - Kp)$.

Let $\lambda = i\omega = \omega(\cos\omega\tau + i\sin\omega\tau)(\omega > 0)$ be a root of Eq. (6), then it can be obtained that

$$\begin{cases} \omega^4 - \beta_2\omega^2 + \beta_5 = -\beta_4 \cos 2\omega\tau, \\ -\beta_1\omega^3 + \beta_3\omega = \beta_4 \sin 2\omega\tau. \end{cases} \tag{7}$$

Taking the square on both sides of the two equations in Eq. (7) and adding them up lead to

$$\omega^8 + A_1\omega^6 + A_2\omega^4 + A_3\omega^2 + A_4 = 0, \tag{8}$$

in which $A_1 = \beta_1^2 - 2\beta_2, A_2 = \beta_2^2 + 2\beta^5 - 2\beta_1\beta_3, A_3 = \beta_3^2 - 2\beta_2\beta_5, A_4 = \beta_5^2 - \beta_4^2$. Here we label $M(\omega) = \omega^8 + A_1\omega^6 + A_2\omega^4 + A_3\omega^2 + A_4$ for further use.

It is obvious that if A4 < 0, Eq. (8) owns no less than one positive root. From the first equation of Eq. (7), we can obtain that

$$\tau_n^{(j)} = \frac{1}{2\omega_n}\{\arccos\frac{-\omega_n^4 + \beta_2\omega_n^2 - \beta_5}{\beta_4} + 2j\pi\}, \tag{9}$$

where n = 1; 2; 3; 4; j = 0; 1; 2; · · · .

Given that $\tau = \tau_n^{(j)}$, (7) owns a pair of purely imaginary roots $\pm i\omega_n$. It is also clear from (9) that the sequence $\{\tau_n^{(j)}\}_{j=0}^{\infty}$ is monotone increasing and $\lim_{j\to\infty}\tau_n^{(j)} = +\infty$, $n = 1,2,3,4$.

We then make the following assumption that

$$\tau_0 = \tau_{n0}^{(0)} = \min_{n\in\{1,2,3,4\}}\{\tau_n^{(0)}\}, \quad \omega_0 = \omega_{n0}. \tag{10}$$

In order to establish of system (1) when τ=0 we put forward the following conditions.

$$D_1=4>0 \tag{11}$$

$$D_2 = \begin{vmatrix} 2 & 1 \\ a_1 + a_2 & a_1 + a_2 + 1 \end{vmatrix} > 0, \tag{12}$$

$$D_3 = \begin{vmatrix} 2 & 1 & 0 \\ a_1 + a_2 & a_1 + a_2 + 1 & 2 \\ 0 & a_1a_2 - v_1v_2 f'^2(0) & a_1 + a_2 \end{vmatrix} > 0. \tag{13}$$

Lemma 1. If conditions of (11)-(13) are satisfied, then the positive equilibrium (u1; u2) of the system (1) is asymptotically stable when = 0.

Proof 1. If time delay and the PD controller are removed from the system, Eq. (17) turns to

$$\lambda^4 + 2\lambda^3 + (a_1 + a_2 + 1)\lambda^2 + (a_1 + a_2)\lambda + a_1a_2 - v_1v_2 f'^2(0) = 0. \tag{14}$$

According to (11)-(13), it is obvious that the well-known Routh-Hurwits criterion is met in Eq. (14), indicating that all roots of Eq. (14) have negative real parts. Therefore, the positive equilibrium (u1; u2) of the system (1) is asymptotically stable when = 0. This ends the proof.

It is notable that the results in Lemma 1 are only sufficient conditions for guaranteeing the stability of system (1).In order to further probe into the the conditions for the generation of Hopf bifurcation, we give the following lemma.Lemma 2 Let $\lambda(\tau) = \gamma(\tau) + i\omega(\tau)$ be the solution of Eq. (6) near $\tau = \tau_j$, which satisfies that $\gamma(\tau_j) = 0, \omega(\tau_j) = \omega_0$, then the transversality condition puts that $Re\left[\frac{d\lambda}{d\tau}\right]_{\tau=\tau_0,\omega=\omega_0} \neq 0$, in which ω_0 represents the critical frequency and τ_0 the bifurcation point.

Proof 2 Differentiating Eq. (6) in terms of τ on both sides leads to

$$\left[\frac{d\lambda}{d\tau}\right]^{-1} = \frac{4\lambda^3 + 3\beta_1\lambda^2 + 2\beta_2\lambda + \beta_3}{2\beta_4\lambda e^{-2\lambda\tau}} - \frac{\tau}{\lambda} \tag{15}$$

Then we assume that λ=iω_0, and it can be obtained that

$$Re\left[\frac{d\lambda}{d\tau}\right]^{-1}_{\tau=\tau_0}=\frac{1}{2\beta_4}Re\left[\frac{e^{2\lambda\tau}(4\lambda^3+3\beta_1\lambda^2+2\beta_2\lambda+\beta_3)}{\lambda}\right]=\frac{M'(\omega)}{4\beta_4\omega^2}>0$$

Therefore, the transversality condition holds. This completes the proof. Then we summarize the following theorem based on the discussion above.

Theorem 1 For system (2), the following results can be determined:

(i) The controlled system (2) is asymptotically stable at the equilibrium (u_1, u_2) when $\tau\in[0,\tau_0)$

(ii) System (2) undergoes a Hopf bifurcation at the equilibrium when $\tau=\tau_0$, i.e., and a class of periodic solutions bifurcate around the equilibrium when passes through τ_0.

4. Simulation Examples.

In this section, several numerical examples are illustrated to verify the efficiency and feasibility of the previous theoretical analysis.

Example 1 In this example, we propose a specific two-neuron inertial model without controller as that of system (1) in which $a_1 = 1$, $a_2 = 1.2$, $v_1 = v_2 = -1$ and the activation function is selected as $f(\cdot) = \tanh(\cdot)$. The time delay is chosen as the bifurcation parameter and the mathematical model can be described as follows:

$$\begin{cases}\ddot{u}_1(t)=-\dot{u}_1(t)-u_1(t)-\tanh(u_2(t-\tau)),\\ \ddot{u}_2(t)=-\dot{u}_2(t)-1.2u_2(t)-\tanh(u_1(t-\tau)).\end{cases}\tag{16}$$

It is known that the equilibrium point of system (16) is (0,0). After careful calculation, we can obtain that the bifurcation point is τ_0 = 1.464. Figure 1 shows that when the selected $\tau=1.223<\tau_0$, the uncontrolled system (1) remains asymptotically stable around the equilibrium.

As in the figure, x(t) stands for $\dot{u}_1(t)$, y(t) for $u_1(t)$, z(t) for $\dot{u}_2(t)$, and s(t) for $u_2(t)$. Whereas, when $\tau=1.664>\tau_0$, system (1) turns unstable and the Hopf bifurcation is generated at the equilibrium subsequently, as displayed in Figure 2.

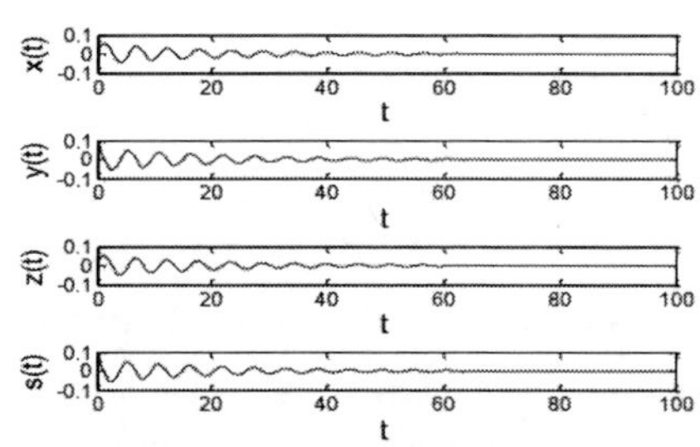

Figure 1. Waveform plot of system (1) with a1 = 1, a2 = 1:2, v1 = v2 = −1 and f(·) = tanh(·). The equilibrium is asymptotically stable when = 1.223 < = 1.464.

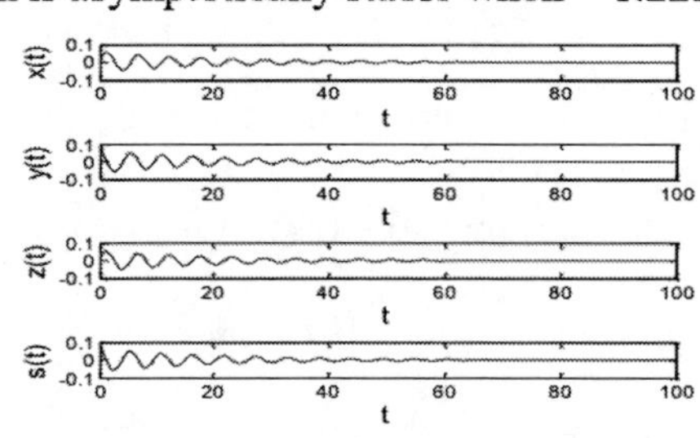

Figure 2. Waveform plot of system (1) with $a_1 = 1$, $a_2 = 1.2$, $v_1 = v_2 = -1$ and $f(\cdot) = \tanh(\cdot)$. A periodic oscillation bifurcate around the equilibrium when τ = 1.664 >τ_0 = 1.464.

Example 2 In this example, we apply a PD controller to the delayed neuron system with inertia and the specific mathematical model is described as follows in which a1 = 1, a2 = 1.2, v1 = v2 = −1, Kp = −1.5, and Kd = −0.5.

$$\begin{cases} \dot{x}_1(t) = x_2(t), \\ \dot{x}_2(t) = -2.5x_1(t) - 0.5x_2(t) - \tanh(x_3(t-\tau)), \\ \dot{x}_3(t) = x_4(t), \\ \dot{x}_4(t) = -2.7x_3(t) - 0.5x_4(t) - \tanh(x_1(t-\tau)). \end{cases} \tag{17}$$

After calculation, it can be obtained that the bifurcation point of (2) is $\tau_0 = 0.615$. Viewed from the three dimensional phase plot, it can be seen from Figure 3,4 that when $\tau = 0.455 < \tau_0$, the curve is bound to converge to the equilibrium in a long given time while when $\tau = 0.723 > \tau_0$, a limit circle will appear, implying the existence of Hopf bifurcation.

From the simulation results displayed above, it can be concluded that the application of PD controller can greatly advance the bifurcation point of the original system, which corroborates the efficiency and feasibility of PD controller.

Furthermore, we investigate into the relationship between controller parameters and bifurcation point and the results shown in Figure 5 and 6 indicate that both Kp and Kd remain monotonically decreasing with relevance to τ_0, respectively.

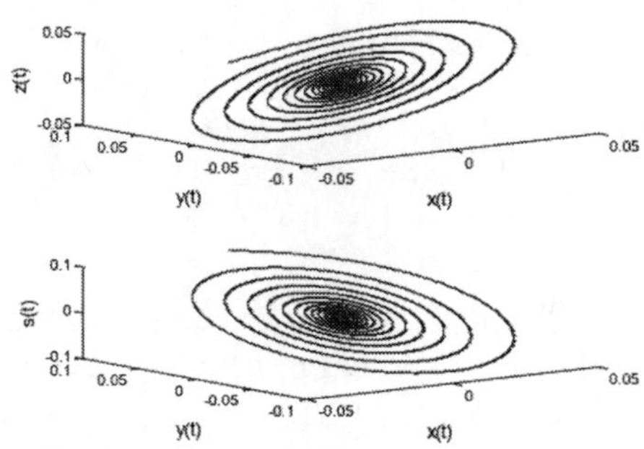

Figure 3. Phase portraits of system (2): projection on x − y − z and x−y −s, with a1 = 1, a2 = 1.2, v1 = v2 = −1, Kp = −1.5, Kd = 0.5 when $\tau = 0.423 < \tau_0 = 0.615$.

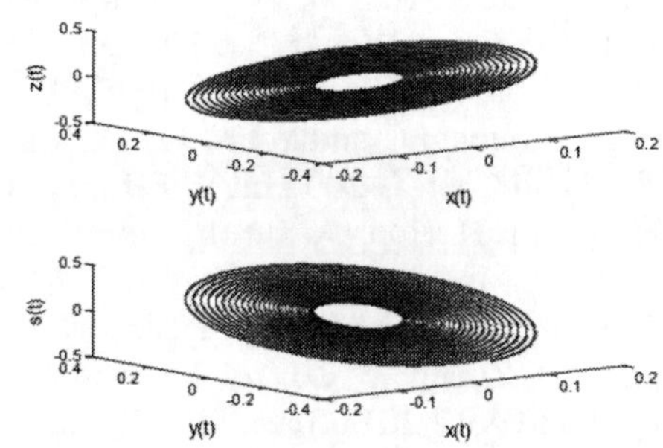

Figure 4. Phase portraits of system (2): projection on x − y − z and x−y −s, with a1 = 1, a2 = 1.2, v1 = v2 = −1,Kp = −1.5, Kd = 0.5 when $\tau = 0.723 > \tau_0 = 0.615$

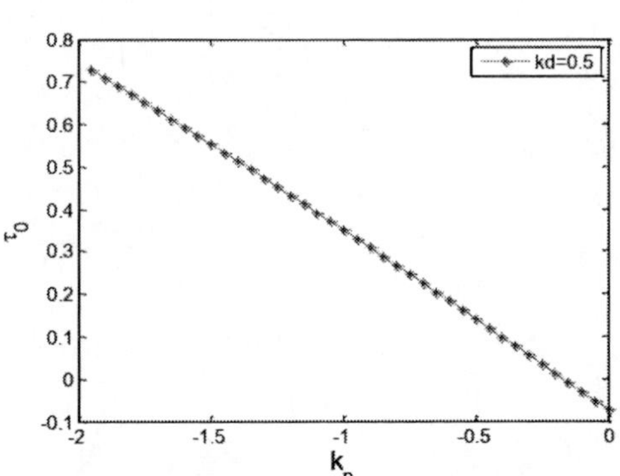

Figure 5. Relationship between Kp and τ_0.

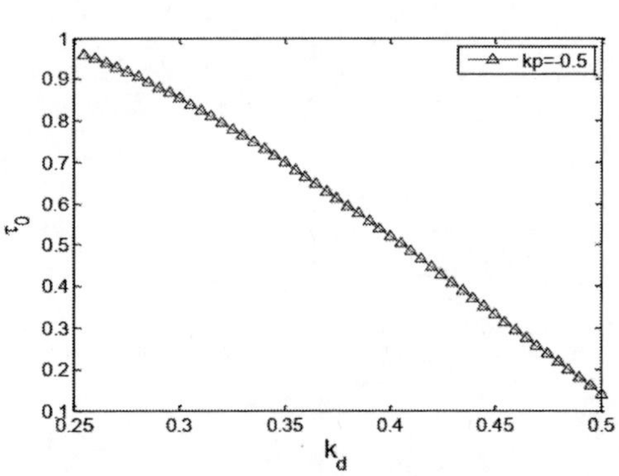

Figure 6. Relationship between Kd and τ_0.

5. Conclusion

In this paper, we have introduced a two-neuron system with inertia and delay. Then we have managed to employ a PD controller to interfere with the dynamical behavior of the neural system. A stability criterion has been obtained and relevant conditions for generating Hopf bifurcation has also been proposed. Moreover, it has been justified through numerical simulations that the bifurcation point can be advanced drastically by altering the controller parameters. Since the original is system two-dimensional, future work lies in that PD controller may be applied to high-dimensional neural networks with inertia and delays.

Acknowledgements

This work is supported in part by the National Natural Science Foundation of China (No. 61573194), the Natural Science Foundation of Jiangsu Province of China (No. BK20181389), the Key Project of Philosophy and Social Science Research in Colleges and Universities in Jiangsu Province (Grant No. 2018SJZDI142), Postgraduate Research & Practice Innovation Program of Jiangsu Province (No.KYCX18 0924

References

[1] Hopfield J J 1984 Proc. Natl. Acad. Sci. 81 3088
[2] Xie WD, Fu Z J and Xie W F 2016 Intell. Autom. Soft. Co. 22 111
[3] Xiao M, Zheng W and Cao J 2013 Neurocomputing 99 206
[4] Huang C, Meng Y, Cao J, Alsaedi A and Alsaadi F 2017 Chaos Solitons Fractals 100 31
[5] Xu W, Cao J, Xiao M, Ho DWC and Wen G 2015 IEEE Trans Cybern. 45 2224
[6] Xiao M, Zheng W and Cao J 2013 IEEE Trans Neural Netw. Learn. Syst. 24 118
[7] Babcock K and Westervelt R 1987 Physica D 28 305
[8] Ott E, Grebogi C and Yorke J A 1990 Phys. Rev. Lett. 64 1196
[9] Yu P and Chen G 2004 Int. J. Bifurcat. Chaos 14 1683
[10] Nguyen L H, Hong K S 2012 Phys. Lett. A 376 442
[11] Chen G, Moiola J L and Wang H O 2000 Int. J. Bifurcat. Chaos 10 511
[12] Tang Y H, Xiao M,Jiang G P,Lin J X,Cao J D and Zheng W X 2017Nonlinear Dynam. 90 2185
[13] Xiao M, Zheng W X, Lin J X, Jiang G P, Zhao L D, Cao J D 2017 J. Franklin Inst. 354 7643
[14] Ge J and Xu J 2013 Sci. China Technol. Sci. 56 2299

On the Equivalent Position Workspace for a Coal Gangue Picking Robot

P Liu[1,2,*], H W Ma[1], X H Zhang[1] and X Z Qiao[1]

[1] School of Mechanical Engineering, Xi'an University of Science and Technology, Xi'an, 710054, China
[2] Key Laboratory of Electronic Equipment Structure Design, Ministry of Education, Xi'an, 710000, China

[*]Corresponding author E-mail address: 200304405liupeng@163.com

Abstract. Coal gangue separation is of key important for green mining. Robotic separation of coal and gangue based on the machine vision system, which is called a coal gangue picking robot with a gangue grab driven by four cables, is developed in this study. Here, this paper focuses on the equivalent position workspace, within which a coal gangue picking robot possesses the identical stability, to find out any more information about the structural stability for a coal gangue picking robot. First, the kinematic and kinetostatic models of the coal gangue picking robot are presented for analyzing the effects of it on the structural stability for the coal gangue picking robot. And moreover, a non-iterative polynomial-based optimization algorithm with the proper optimal objective function is presented based on the convex optimization theory, in which the cable with the minimum cable tension at any pose is determined. Then, three position performance indices are proposed to show the important effects on the structural stability for the coal gangue picking robot in a specified region of the workspace. Besides, a new workspace, the Equivalent Position Workspace (EPW), is introduced. Finally, the theoretical relationship between the two performance indices and the stability is corrected based on simulation results. The research has important guiding significance and practical value for coal gangue robotic separation.

Keywords: Cable-driven Parallel Robot; Coal Gangue Picking Robot; Stability

1. Introduction

Coal is the major fossil fuel in China and continues to play a pivotal role in the energy sector. With the widely use of mechanized coal mine production and the increasingly extraction of thin seam, large number of gangue mixed raw coal has greatly increased, which affects the efficiency of coal preparation and increases the cost of preparation. Meanwhile gangues are stacked on the ground after preparation, which become the hazard source for environment. Various separation methods, in order to reduce the damage to environment of gangue, have been proposed based on the physical property differences between coal and gangue. The separation of gangue from coal can not only improve the quality of raw coal, decrease the cost of preparation, but also provide materials to the gangue filling underground and realize the real green mining [1]. But the separation of coal and gangue depends mainly on the manual gangue picking. The main motivation for moving toward the robotic separation of coal and gangue as the future coal utilizing technique is the environmental and other advantages over the conventional manual gangue picking. And furthermore, it is a promising technology as it is a combination of the image processing, information sensing and electromechanical integration. However, this technology is still in a developing stage in the worldwide.

Parallel mechanisms are used in industrial applications requiring high load carrying capacity, high velocity, or precise positioning. In the family of parallel mechanisms, cable-driven parallel robots allow for larger workspaces and can produce large accelerations [2]. Therefore, these advantages make the cable-driven parallel robots a promising alternative to the rigid-link mechanisms in many industrial applications, such as medical rehabilitation [3], wind tunnel experiment [4], astronomical observation [5], material transportation [6] and other fields [7]. With advantages of high speed, high payload and good flexibility, a cable-driven robot is a necessary choice in applications for driving forces and institutional quality limited. An important application of the cable-driven robots is in the manipulation task with high payload, and indeed, a coal gangue picking robot developed here is a redundantly cable-driven parallel manipulator with high load capacity to realize roboticized separation of coal and gangue.

The coal gangue picking robot is based on the machine vision system, which is an integrated technology using computer image processing technology and analysis technology to identify the target of various patterns of coal and gangue. It mainly contains machine vision and robotic separation process. It is mainly based on coal and gangue physical feature differences and image recognition to realize separation, which is suitable for preliminary separation of coal and gangue. Scholars mainly probed into a general machine vision approach for on-line estimation of coal and gangue, the robotic separation of coal and gangue has not been studied adequately. Song et al. [8] proposed an online intelligent refuse picking system based on dual—energy X—ray recognition and 3D mechanical gripper grabbing target refuse, and moreover, the working principle, design principle and specific structural design of the picking system are discussed. In order to separate coal gangue from lump coal, the paper investigated the fundamental characteristics of coal images and gangue images on the basis of digital image processing technology. The image processing results are obtained, which showed that the coal and gangue can been distinguished based on their grayscale histogram[9]. Reddy et al. [10] proposed a system identifies the gangue by analyzing the characteristics of the image, and moreover, the research results indicated that the comparison of coal images and gangue images could successfully achieve the real-time automatic separation of gangue. Zhang et al. [1] proposed an underground pneumatic separation of coal and gangue based on the machine vision system, and moreover, the key factors affecting coal gangue pneumatic separation are analyzed. But the equipment is just applicable to the gangue with large diameter ($\geqslant$ 50 mm). From above, to the best of our knowledge, the researchers have done lots of research on the recognition of coal and gangue, the robotic separation of coal and gangue has not been studied adequately. Liu et al. [11] investigated the minimum cable tension distributions in the workspace for cable-based parallel robots, and furthermore, the relationship between the stability and the minimum cable tensions is explained clearly. But the effect of the position of the end-effector on the stability of cable-based parallel robots is not considered. In this work, the equipment of robotic separation of coal and gangue is presented and the equivalent position workspace for the coal gangue picking robot is discussed. The workspace with equivalent positions for the coal gangue picking robot is addressed. This study will help to motivate both applied and theoretical research work on the roboticized separation of coal and gangue.

2. Modeling of the coal gangue picking robot

2.1 Kinematic model of the coal gangue picking robot

The coal gangue picking robot proposed in this paper is made of a gangue grab attached to a cable-driven parallel robot (Fig. 1). The each pulley can be attached to the ceiling of a mast. As a result, the gangue grab moves freely in every direction because the cables can be shortened and lengthened controlled by four servo motors mounted to the fixed base. The servo motors receive control commands from the central controller. It is worth noting that the platform is assimilated to a unique point mass. As shown in Fig.2, a base reference frame, noted *OXYZ*, is attached to the fixed base, where O is the origin point. While a mobile reference frame, noted $O'X'Y'Z'$, is attached to the mobile gangue grab, where O' is the reference point. The position $\boldsymbol{P}=(x, y, z)^{\mathrm{T}}$ of the mobile platform is given by the vector connecting O to O'. Point B_i, at which the ith cable (i=1, 2, 3, 4) enters the pulley, is assumed to be fixed to the base frame. Obviously, vectors $\boldsymbol{B}_i=[B_{i,x}, B_{i,y}, B_{i,z}]^{\mathrm{T}}$ are constant position vectors connecting O to B_i in $OXYZ$.

Referring to Fig.2, the ith cable length ρ_i, compatible with a given position $\boldsymbol{P}=(x, y, z)^{\mathrm{T}}$, is given by

$$\rho_i=\|\boldsymbol{B}_i-\boldsymbol{P}\|_2=\sqrt{(B_{i,x}-x)^2+(B_{i,y}-y)^2+(B_{i,z}-z)^2},\ i=1,2,3,4 \tag{1}$$

When expressed in the base frame, the unit vector along cable $i$is noted $\boldsymbol{u}_i$ and can be written as:

$$\boldsymbol{u}_i=(\boldsymbol{B}_i-\boldsymbol{P})/\rho_i,\ i=1,2,3,4 \tag{2}$$

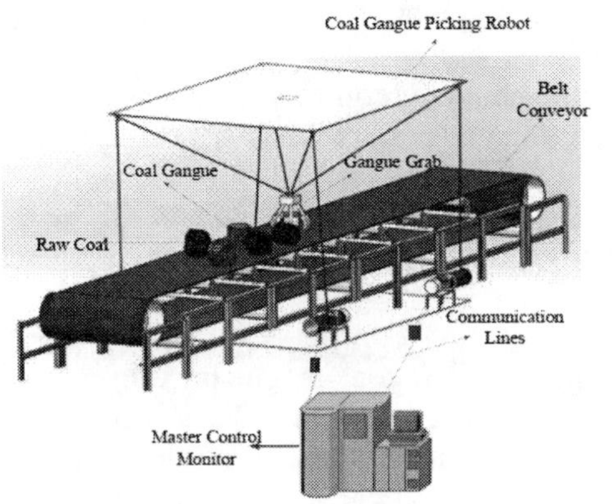

Figure 1. Schematic of a coal gangue picking robot

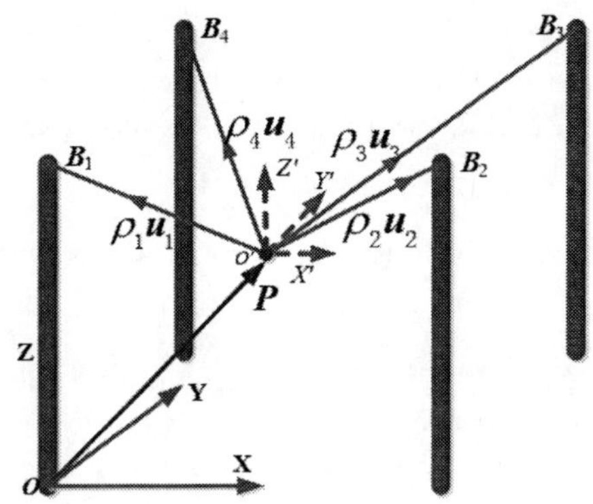

Figure 2. Kinematic notation of a coal gangue picking robot

2.2 Kinetostatic model of the coal gangue picking robot

In order to maintain positive cable tension, the coal gangue picking robot must be actuated redundantly due to the unilateral nature of cables. Thus, for the coal gangue picking robot with 4 cables, there will be a total of 3 unknowns which are cable tension for each cable. And therefore, the kinetostatic model of the coal gangue picking robot, using Newton method, can be developed based on [12]:

$$\boldsymbol{JT}+\boldsymbol{W}=\boldsymbol{0} \tag{3}$$

where $\boldsymbol{W}\in\mathbb{R}^3$ is the external wrench on the gangue grab; $\boldsymbol{J}=[\boldsymbol{J}_1,\ \boldsymbol{J}_2,\boldsymbol{J}_3,\ \boldsymbol{J}_4]\in\mathbb{R}^{3\times4}$ is the structure matrix of the coal gangue picking robot. And the vector $\boldsymbol{J}_i$ is defined as

$$\boldsymbol{J}_i=\boldsymbol{u}_i,\ i=1,2,3,4 \tag{4}$$

$\boldsymbol{T}=[T_1, T_2, T_3, T_4]^{\mathrm{T}}$ is the vector consisting of all cable tensions.

$$\boldsymbol{T}_{s,min}\le\boldsymbol{T}\le\boldsymbol{T}_{s,max} \tag{5}$$

where the lower bound of the cable tension $\boldsymbol{T}_{s,\min}=[T_{1,\min}, T_{2,\min}, T_{3,\min}, T_{4,\min}]^{\mathrm{T}}$ is required to keep cables taut; while upper bound of the cable tension $\boldsymbol{T}_{s,\max}=[T_{1,\max}, T_{2,\max}, T_{3,\max},, T_{4,\max}]^{\mathrm{T}}$ is limited by the output torques of the servo motors and the maximum tension the cable can withstand without breaking. Form Eq. (5), the cable tensions may exist infinite solutions because the number of cables is larger than the degree of freedom of the gangue grab. Therefore, in order to obtain a unique solution, the minimum variance is used to optimize while using (5) as a constraint. Mathematically, the determination of the cable tensions can be formulated as follows [13]:

$$min\left(\frac{1}{4}\left[\sum_{i=1}^{4}\left(t_i - E(\boldsymbol{T})\right)^2\right]\right)$$
$$subject \;\; to \;\; \boldsymbol{JT} + \boldsymbol{W} = \boldsymbol{0}$$
$$and$$
$$\boldsymbol{T}_{s,min} \leq \boldsymbol{T} \leq \boldsymbol{T}_{s,max} \tag{6}$$

where $E(\boldsymbol{T}) = \frac{t_1 + t_2 + t_3 + t_4}{4}$ is the arithmetic mean value of the vector ***T***. According to convex optimal theory, cable tensions are determined uniquely.

The minimum cable tension $T_{\min}$ at any position can be obtained when cable tensions are determined uniquely. And moreover, the minimum cable tension $T_{\min}$ can be expressed as:

$$T_{min} = min(\boldsymbol{T}) \tag{7}$$

where $min(\bullet)$ is the smallest element in a vector ***T***.

Here, the cable having minimum cable tension when the gangue grab is at a specified position is determined, that is to say, the weakest constraint direction, the gangue grab may be unstable, can be obtained.

3. The three position performance indices and EPW generation algorithm

3.1 The position performance indices

Based on the solutions of the kinematic and kinetostatic models of the coal gangue picking robot, three position performance indices are discussed in more detail at a later stage. In order to make clear the physical meaning of the three position performance indices, the schematic diagram relating to them is shown in Fig.3, in which the outmost box is the task space of the coal gangue picking robot. Without loss of generality, the optional position in the workspace is denoted by *P*; the vertical midline of the workspace is denoted by *a*; the horizontal plane on which the specified position is located is highlighted with green; the intersection of the vertical midline and the horizontal plane is denoted by *Q*; the intersection of the vertical midline and the top of the workspace is denoted by *M*; the intersection of the vertical midline and the top of the task space is denoted by M'; the angles at the position *P*, *Q* and *M* between the cable having the minimum cable tension and the horizontal plane are denoted by θ_P, θ_Q and θ_M respectively.

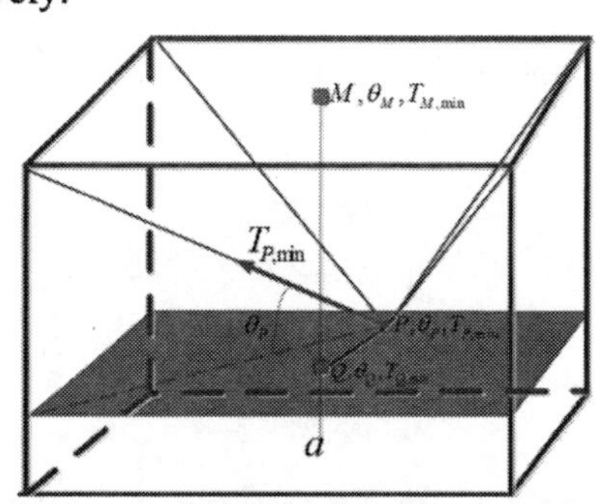

Figure 3. Schematic diagram relating to the three position performance indices

In detail, three position performance indices having an important effect on the stability of the coal gangue picking robot are proposed to show how far the specified position is away from the center and the top of the workspace. The three position performance indices, $\Re_{s_-}$, $\Re_{s_\perp}$, and $\Re_s$ can be respectively expressed as:

$$\Re_{s_\perp} = \frac{\tan\theta_M}{\tan\theta_Q} \tag{8}$$

$$\Re_{s_-} = \frac{\tan\theta_P}{\tan\theta_Q} \tag{9}$$

$$\Re_s = \Re_{s_\perp} \times \frac{1}{\Re_{s\_}} \tag{10}$$

Comprehending the definitions of the three position performance indices leads to some conclusions that the performance index $\Re_{s_\perp}$ shows the distributions of the equivalent positions along Z-direction; the performance index $\Re_{s\_}$ is a function of the distance between the present position and the center position of the horizontal plane with the same elevation along Z-direction, therefore showing distributions of the equivalent positions along horizontal direction; And furthermore, the performance index $\Re$ is proposed to describe the comprehensive distributions of the equivalent positions over the whole workspace. It is perfectly clear that the distributions of the equivalent positions over the whole workspace can be described based on the three performance indices.

3.2 EPW generation algorithm

Furthermore, EPW is composed of the positions of the gangue grab having the equivalent effect on the stability of the coal gangue picking robot. And thus, the workflow of EPW generation algorithm can be summarized as follows:

(1) Input the real-time gangue grab position $\boldsymbol{X}_i$ in the workspace (i=1, 2,..., N. N is the total number of positions); H is the height of the towers; the external wrench $\boldsymbol{F}_e(\boldsymbol{X}_i)$; the gravity $\boldsymbol{W}_g(\boldsymbol{X}_i)$ of the gangue grab.
(2) Calculate the structure matrix $\boldsymbol{J}(X_i)$ and the null space matrix Null($\boldsymbol{J}(X_i)$) of it.
(3) Determine the cable tension $\boldsymbol{T}^*(\boldsymbol{X}_i)$, using convex programming theory and Eq. (6).
(4) Obtain the minimum cable tension $T_{\min}(\boldsymbol{X}_i)$ using Eq. (7), and find the cable having minimum cable tension when the gangue grab is at a the position $\boldsymbol{X}_i$.
(5) Calculate the the length of the cable $\rho_{min,i}$ at the position P, Q and M and the angles at the position P, Q and M between the cable having the minimum cable tension and the horizontal plane.
(6) Calculate three position performance indices using the Eqs. (8)-(10) respectively and go to the next position.
(7) Judge $\boldsymbol{X}_i$ is whether the last position. If not, go to (1) and solve three position performance indices for the next position $\boldsymbol{X}_{i+1}$; if it is, record the three position performance indices and the position $\boldsymbol{X}_i$ and moreover, depict the equipotential curves and the equivalent position workspace.

4. Simulation Examples

4.1 Description of the coal gangue picking robot

An example is now presented in order to illustrate the equivalent position workspace and the relationship between the three position performance indices and the stability of the coal gangue picking robot. The mass of the gangue grabm_c=10kg, the bound of the cable tension $\boldsymbol{T}_{s,\max}=[300\ 300\ 300\ 300]^T$ N, $\boldsymbol{T}_{s,\min}=[10\ 10\ 10\ 10]^T$ N. And meanwhile, the origin of the fixed base reference frame $OXYZ$ is located at the bottom of mast1. The position vectors that the cables enter the pulleys are: $\boldsymbol{A}_1=[0\ 0\ 23]^T$dm, $\boldsymbol{A}_2=[37\ 0\ 23]^T$ dm, $\boldsymbol{A}_3=[37\ 40\ 23]^T$ dm, $\boldsymbol{A}_4=[0\ 40\ 23]^T$ dm. Therefore, the midpoint of the horizontal surface is at x=18.5dm and y=20dm.

4.2 Results and discussion

The three position performance indices above are computed using the Eq. (8), (9) and (10) in this section for the coal gangue picking robot. And the results are displayed in Fig.4. As expected, the position performance index $\Re_\perp$ increases as z increases along the vertical midline shown in Fig.4(a). It is worth noting that the shape of the curve is a nonlinear function of the elevations. It can be seen from Fig.4(b) that the positions where gangue grab stays hold the bigger $\Re_{s\_}$ are the ones being close to the center of every horizontal plane. That is because the upper and central positions possess bigger T_{min}

than others, thereby improving the stiffness and stability of the robot. In addition, the position performance index $\Re_s$ is depicted from Fig.4(c) to Fig.4 (f). With (c) and (d) together, it is evident that the curves that have the same position performance index $\Re_s$ are at the same elevation along *Z*-direction because they are front and lateral views of $\Re_s$ in the workspace respectively. A further view of the figure curves indicates that the interior positions where gangue grab stays retain bigger $\Re_s$ than others because of the uniformity of the cable tensions. And moreover, a specified value of $\Re$, which is a equipotential surface over the whole workspace, is displayed in Fig.4(e), and while the planform of the equipotential surface is described. Of course, it is worth noting that the positions within the surface retain position performance index $\Re_s$ being bigger than 0.4. The colours are worthy of note that represent the elevation along *Z*-direction. Moreover, one of its features is the symmetry about *X*-direction and *Y*-direction. It should be pointed that the equipotential surface with the bigger position performance index $\Re_s$ is the ones being higher than the surface in Fig.4(e). As expected, further insight into the integral shape of EPW can be obtained from them.

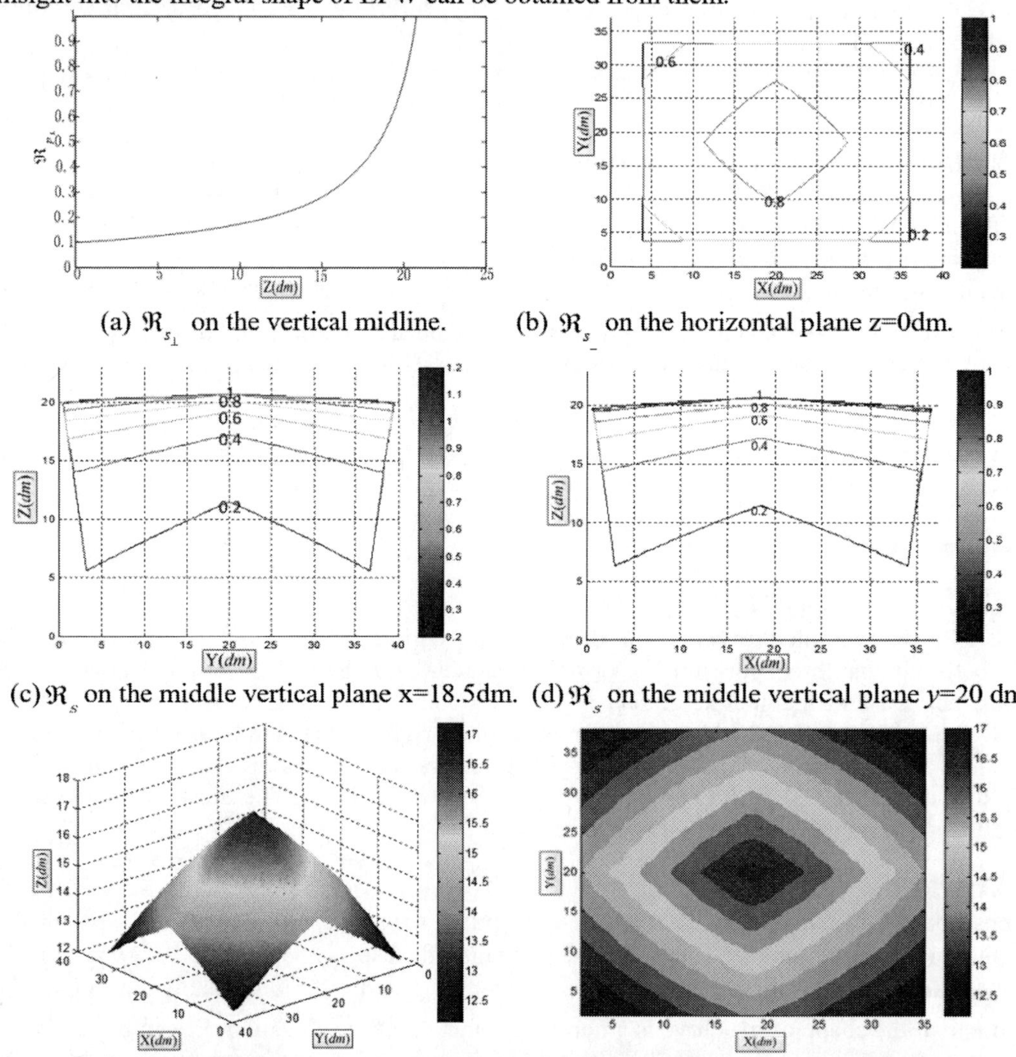

(a) $\Re_{s_\perp}$ on the vertical midline. (b) $\Re_s$ on the horizontal plane z=0dm.

(c) $\Re_s$ on the middle vertical plane x=18.5dm. (d) $\Re_s$ on the middle vertical plane *y*=20 dm.

(e) Equipotential surface $\Re_s = 0.4$ (f) Planform of the equipotential surface $\Re_s = 0.4$

Figure 4. The three position performance indices in the workspace.

As expected, further insight into the integral shape of EPW can be obtained from them. From above it can be concluded that the combination of all the curves in Fig.4 (a)-(d) is in conformity with Fig.4(e) and (f) that displays $\Re_s = 0.4$ in the workspace.

5. Conclusions

Robotic separation of coal and gangue, which is called a coal gangue picking robot with a gangue grab driven by four cables, is developed in this study. The equivalent position workspace of the coal gangue picking robot is discussed using the three position performance indices based on the kinematic and kinetostatic models of the robot. Solutions to the problems of the effect of the position of the gangue grab on the stability of the robot presented to a typical coal gangue picking robot. Simulation results of the three position performance indices show that as the three performance indices increase, so do the effect of the position of the gangue grab on the stability of the robot, therefore improving the stability of the robot. In the whole workspace, the positions when the gangue grab stays in the center and on top of the workspace possess bigger position performance indices than others, leading to have better stability. It is important to notice that the positions of the gangue grab have great important effect on the stability of the coal gangue picking robot.

Acknowledgments

The research is supported by Open Fund of Key Laboratory of Electronic Equipment Structure Design (Ministry of Education) in Xidian University. The authors gratefully acknowledge the financial support of Special Scientific Research Program Project of Shaanxi Provincial Education Department under Grant No. 18JK0506.

References

[1] Zheng K., Du C., and Li J., et al. Underground pneumatic separation of coal and gangue with large size (≥ 50 mm) in green mining based on the machine vision system[J]. Powder Technology, 2015, 278: 223-233.

[2] Barbazza L., Zanotto D., and Rosati G., et al. Design and optimal control of an underactuated cable-driven micro–macro robot[J]. IEEE Robotics and Automation Letters, 2017, 2(2): 896-903.

[3] Mao, Y., and Agrawal, S. K., Design of a Cable-Driven Arm Exoskeleton (CAREX) for Neural Rehabilitation[J]. IEEE Trans. on Robotics, 2012, 28(4): 922-931.

[4] Zheng Y. Q., and Zhao S. H., Research Survey of Technique about Wire-Driven Parallel Suspension Systems Used in Forced Oscillation Experiments in Low-Speed Wind Tunnels for Obtaining Dynamic Derivatives[J]. Informatics in Control, Automation and Robotics, 2012, 133: 131-138.

[5] Yao R., Zhu W. B., and Sun C. H. et al., Pose Planning for the Feed Support System of FAST[J]. Advances in Mechanical Engineering, Volume 2014, Article ID 209167: 1-9.

[6] Oh S. R., Ryu J. C., and Agrawal S. K. Dynamics and control of a helicopter carrying a payload using a cable-suspended robot[J]. Journal of Mechanical Design, 2006, 128(5): 1113-1121.

[7] Tang X. Q., An Overview of the Development for Cable-Driven Parallel Manipulator [J].Advances in Mechanical Engineering, Volume 2014, Article ID 823028: 1-9.

[8] Song W. G., Tao Y. D., and Qi C. J. et al., Design and research of intelligent robot refuse picking system [J]. Coal processing and comprehensive utilization, 2018, (9): 5-10.

[9] Wang R., and Liang Z. Automatic separation system of coal gangue based on DSP and digital image processing[C]//Photonics and Optoelectronics (SOPO), 2011 Symposium on. IEEE, 2011: 1-3.

[10] Reddy K. G. R., Tripathy D. P. Separation of Gangue From Coal Based on Histogram Thresholding [J]. International Journal of Technology Enhancements and Emerging Engineering Research, 2013, 1(4): 31-34.

[11] Liu P., Qiu Y. Y., Su Y., et al. On the Minimum Cable Tensions for the Cable-based Parallel

Robots [J]. Journal of Applied Mathematics, vol. 2014, Article ID 350492, 2014.
[12] Gosselin C., Grenier M. On the determination of the force distribution in overconstrained cable-driven parallel mechanisms[J]. Meccanica, 2011, 46(1): 3-15.
[13] Liu P., Qiu Y. Y., Su Y. A new hybrid force-position measure approach on the stability for a camera robot [J]. Proceedings of the Institution of Mechanical Engineers, Part C: Journal of Mechanical Engineering Science:2016, 230(14): 2508 - 2516.

Fault Tolerant Consensus with Multiagent Systems for Distributed Coordinated Control Algorithm in the Energy Internet Network

Chung-Hung Hsieh and Yee Ming Chen*

Department of Industrial Engineering and Management, Yuan Ze University Tao-Yuan, Taiwan, R.O.C.

* corresponding email:chenyeeming@saturn.yzu.edu.tw

Abstract. The future power system can be considered as an aggregation of controllable distributed systems devices that coordinate with each other through the Internet of Things (IoT), energy internet (EI) new paradigms arise. In EI environments, agreement among multiple computers is essential for power generation, transmission, distribution, and consumption etc. applications. Coordinated control and security protection play important roles in the application of a distributed computing environment. A consensus approach can provide fault tolerance and prevent system errors and attacks, thereby providing the system with strong security in distributed computing environment. In this paper, a distributed Kalman filter with consensus coordinated control algorithm for multiagent systems (MAS) was proposed to investigate the energy internet with links failure by cyber attack. The proposed algorithm is evaluated in the energy internet through a MAS model using MATLAB software.

Keywords—multiagent systems, consensus, energy internet

1. Introduction

The growing synergy between Internet of Things (IoT) and energy internet (EI) brings the potential of integrating more distributed generation and renewables in a cost effective way, enabling peer to peer trading of energy, e.g. electricity and heat. As part of the IoT, EI keeps the grid stable by balancing the power generated from renewables with the electricity that is consumed — by electric vehicles, for example. EI is, therefore, defined as an integrated dynamic network infrastructure based on standard and interoperable communication protocols that interconnect the energy network with the Internet allowing units of energy (locally generated, stored, and forwarded) to be dispatched when and where it is needed. With the significant increasing amount of distributed energy resources in the EI, researchers are interested in applying consensus based distributed control in EI related problems[1], such as economic dispatch [2], demand response [3], and plug-in electric vehicles charging coordination [4] etc. In those distributed applications, controllers (usually low-cost single-board microcontrollers) coordinate with neighboring peers and make decisions for local controllable devices [5]. In many realistic EI networks, due to the complexity of systems and undesirable cyber attacks or disturbance, the occurrence of sensors controllers (generically referred to as 'nodes') or communication link failure is inevitable[6]. There exists a cyber attack on sensors or a network exchanging data between sensors and controllers, which may be subject to maliciously destroy in common communication setting or wireless communication one[7]. This kind of phenomena is usually implemented by cyber attackers

IOP Conf. Series: Journal of Physics: Conf. Series **1267** (2019) 012079 doi:10.1088/1742-6596/1267/1/012079

with the aim of the enormous economy benefits. For real world energy internet network, representative examples of cyber attacks include an attack on Slammer worm on Davis Besse power plant in Ohio, U.S. [8], and recent Stuxnet worm targeted many industrial control systems[9].

As is well known, the main purpose of consensus based distributed control is to design a suitable control protocol such that the states of a team of agents can reach some common features. Obviously, the consensus issues with cyber attacks deserve adequate research attention simply because of the high requirements of security and reliability in EI environments, and this gives rise to the main motivation for our current research. For multiagent systems (MAS) under cyber attacks, a typical phenomenon is that the communication link among agents could be destroyed or compromised by adversaries in a random manner from the defenders' perspective. When an attack occurs, the connectivity of network topology cannot be guaranteed and the resulting varying topology issue poses essential difficulties for the consensus control. Motivated by the above discussion, this paper is concerned with aim to design a distributed Kalman filter with consensus concept update approach to compensate for the undesirable effects of the failure links or nodes and the consensus property is retained after the recovery process in energy internet consensus control of networked multiagent systems for distributed coordination. The organization of the remaining part is presented as follows. In Section II, distributed Kalman filter and consensus coordinated control approach are introduced. In Section III, a distributed Kalman filter with consensus coordinated control algorithm for multi-agent networks are presented and numerical simulations are given to show the effectiveness of the approach results in Section IV. In the final section, concluding remarks are drawn.

2. Distributed Kalman Filter

In this section, we discuss the distributed Kalman filter. The distributed Kalman filtering that relies on communicating state estimates between neighboring nodes and refer to it as local node using Kalman filter.

2.1. Local Kalman Filtering

Assume that node i only receives information from its neighbors. In each local node Kalman filtering, let be the set of neighbours of node on graph G. Each node of the EI network communicates its measurement, covariance information, and observation matrix with its neighbours. For node, we assume that the information flow from neighbouring nodes to node is prohibited if there is no nodes except for its neighbours exist. Therefore, node can use a Kalman filter that only utilizes the observation vectors and observation matrices of the nodes in. Then, we have the iterations of node i in local node Kalman filtering as

$$y^i = \sum_{j\in J_i} H_j^T R_j^{-1} z_j = \sum_{j\in J_i} y_j \tag{1}$$

$$S^i = \sum_{j\in J_i} H_j^T R_j^{-1} H_j = \sum_{j\in J_i} S_i \tag{2}$$

$$\hat{x}_i = \bar{x}_i + M_i\left(y^i - S^i\bar{x}\right) \tag{3}$$

$$M_i = \left(P_i^{-1} + S^i\right)^{-1} \tag{4}$$

$$P_i^+ = AM_iA^T + BQB^T \tag{5}$$

$$\bar{x}_i^+ = A\hat{x}_i \tag{6}$$

Where y^i and S^i are local aggregate information vector and matrix, respectively and node i locally computes both y^i and S^i.{. +} is the update operation.

2.2. Consensus Coordinated Control Approach

Consider a team of agents to agree on specific consensus states, and at any discrete-time instant, the communication EI topology between agents can be described by the graph, the graph G is undirected, is the vertex set, and is the edge set. In the consensus coordinated control approach, each agent in the network maintains a local node copy of the consensus state and updates this value using its neighbours' consensus states according to the consensus coordinated control rule:

$$\zeta_i[\tau+1]=\zeta_i[\tau]+\sum_{j=1}^{n}\beta_{ij}[\tau](\zeta_i[\tau]-\zeta_i[\tau]) \tag{7}$$

where τ indicates the consensus filter iteration step. To choose the weights $\beta_{ij}[\tau]$. We can use the Metropolis weights which preserves the averaging in consensus filters and can be computed by

$$\beta_{ij}[\tau]=\begin{cases}(1+\max\{d_i[\tau],d_j[\tau]\})^{-1}\\ 1-\sum_{(i,l)\in E(\tau)}\beta_{il}[\tau]\\ 0\end{cases} \tag{8}$$

where $d_i[\tau]$ is the degree of agent i in the graph $G[\tau]$. Arrange the local consensus states into the vector $\zeta[\tau]=[\zeta_1^T[\tau],\ldots,\zeta_n^T[\tau]]^T$,and define the matrix $(B[\tau])_{ij}=\beta_{ij}[\tau]$ for $i\neq j$; otherwise $(B[\tau])_{ii}=1-\sum_{(i,l)\in E(\tau)}\beta_{il}[\tau]$, and we can rewrite the update in (7) as

$$\zeta[\tau+1]=(B[\tau]\otimes I)\zeta[\tau] \tag{9}$$

where I is the appropriate size identity matrix and $\otimes$ denotes the matrix Kronecker product. The ij th element of $B[\tau]$ in (9) satisfies the following four conditions:

(1) $(B[\tau])_{ij}\geq 0$

(2) $\sum_i (B[\tau])_{ij}=1$

(3) $\sum_j (B[\tau])_{ij}=1$

(4) each nonzero entry is both uniformly upper and lower bounded.

3. Distributed Kalman Filter with Consensus Coordinated Control Algorithm

We now present the distributed Kalman filter with consensus coordinated control algorithm in this section. The Kalman filter uses consensus coordinated control (7) on the state estimate in a distributed Kalman filter, where each node maintains a local node Kalman filter. Corresponding to (7), let $\zeta_i[\tau]=\bar{x}_i$ be the prior estimate at time τ and $\zeta_i[\tau+1]=\bar{x}_i^c$ be fused prior estimate; each node fuses the prior estimates from its neighbours according to the rule:

$$\bar{x}_i^c=\bar{x}_i+\sum_{j\in N_i}\beta_{ij}[\tau](\bar{x}_j-\bar{x}_i) \tag{10}$$

Using the fused prior estimate $\bar{x}_i^c$, the filter estimate at node i could be implemented by

$$\hat{x}_i = \bar{x}_i^c + M_i\left(y^i - S^i\bar{x}_i^c\right) = \bar{x}_i + M_i\left(y^i - S^i\bar{x}_i\right) + \left(I - M_i S^i\right)\sum_{j\in N_i}\beta_{ij}[\tau]\left(\bar{x}_j - \bar{x}_i\right) \tag{11}$$

The distributed Kalman filter with consensus coordinated control algorithm is summarized in the following as

Initialization (for agent (i.e. node i)):

$$\bar{x}_i = x(0) \quad P_i = P_0$$
$$\tau = 1 \qquad \tau_p = \tau + T_p$$

Loop {Local iteration on node i }

(1) Consensus coordinated control approach

$$\bar{x}_i^c = \bar{x}_i + \sum_{j\in N_i}\beta_{ij}[\tau]\left(\bar{x}_j - \bar{x}_i\right)$$
$$y^i = \sum_{j\in J_i} H_j^T R_j^{-1} z_j$$
$$S^i = \sum_{j\in J_i} H_j^T R_j^{-1} H_j$$
$$\tau \leftarrow \tau + 1$$

(2) If new observations are taken then the distributed Kalman consensus state estimate are updated

$$\hat{x}_i = \bar{x}_i^c + M_i\left(y^i - S^i\bar{x}_i^c\right) = \bar{x}_i + M_i\left(y^i - S^i\bar{x}_i\right) + \left(I - M_i S^i\right)\sum_{j\in N_i}\beta_{ij}[\tau]\left(\bar{x}_j - \bar{x}_i\right)$$

$$M_i = \left(P_i^{-1} + S^i\right)^{-1}$$

(3) If time for a predication step (i.e., $\tau = \tau_p$) then prediction step

$$P_i^+ = AM_iA^T + BQB^T$$
$$\bar{x}_i^+ = A\hat{x}_i$$
$$\tau_p = \tau + T_p$$

End loop

where τ is the time index for the consensus coordinated control approach and $T_p \in Z^+$ is the time interval between prediction updates. One-time step $k\text{-}1 \rightarrow k$ is equivalent to T_p time steps of the consensus time index $\tau \rightarrow \tau + 1$; that is, for each node, the information exchanges between neighbouring nodes occurred faster than the prediction update step. The three steps in distributed Kalman filter prediction, local filter estimate, and consensus coordinated control update sequential.

4. Numerical Simulations

By utilizing these generalized agents in combination, an effective MAS model can be created. A simple example is shown in Figure 1. In this energy internet, there are six generation sources (e.g. wind turbine, photovoltaics, or diesel generator etc.). The Energy Internet bus is connected to both the local generations and prosumers. The MAS is comprised of generation/prosumer agents assigned to each generator; in the interconnected state, each agent manages its assigned asset according to assigned objectives.

IOP Conf. Series: Journal of Physics: Conf. Series **1267** (2019) 012079 doi:10.1088/1742-6596/1267/1/012079

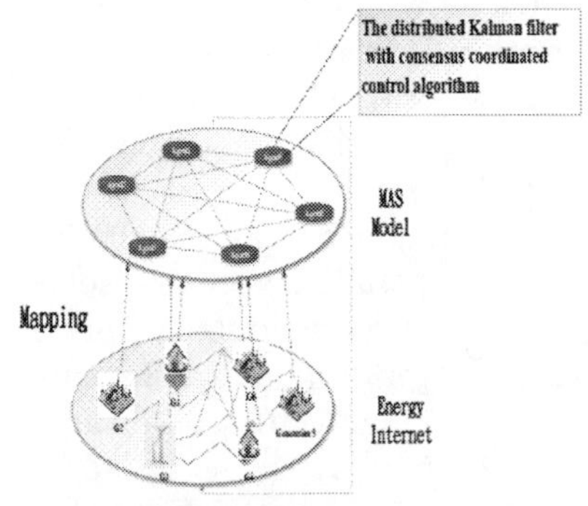

Agent(Genration no.)	P_{ini}[Mw]	P_{min} [Mw]	P_{max}[Mw]
1	450	400	500
2	300	250	400
3	100	50	200
4	400	300	500
5	250	200	300
6	150	100	200

Figure 1. Example energy internet with assigned MAS agents. Dashed lines show communication; solid lines power flow

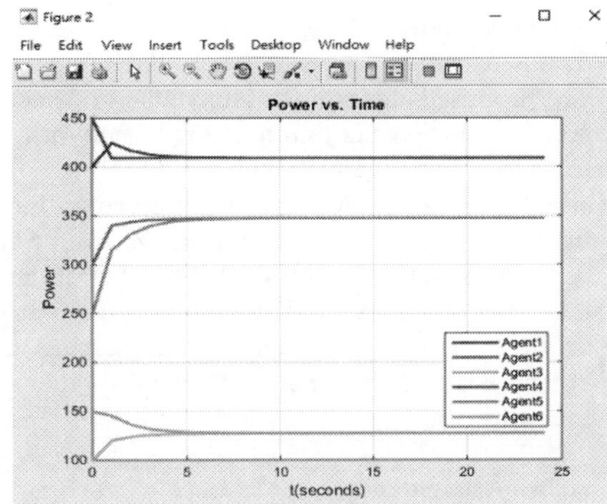

Figure 2. The state estimation iterations of time instant for EI network

In this EI network example, under a communication protocol and link connected network nodes, the distributed Kalman filter with consensus coordinated control algorithm can be achieved for multi-agent networks with quantization. The characteristics of the generators are modelled using the simplified synchronous generator from Matlab in the following two case studies which carried out to evaluate the performance of the proposed algorithm. It's noted that it takes a longer time for the generators to reach a consensus due to the inertia of the generators. After about 2 sec, the generators reach the same consensus as shown in Figure 2. In the second case study of the simulations, we focus on the performance of fault tolerance for investigation the EI with links failure by cyber attack. For cyber attacks conduction, proposed algorithm is used to find EI generated power. Suppose some links fail at time t = 28 and the other agents had to compensate the shortage in the generated power after 32 sec (see Figure 3). The recovery process can be estimated to be able to converge after 40 sec (see Figure 4). Thus, the algorithm can be applied to large scale EI network.

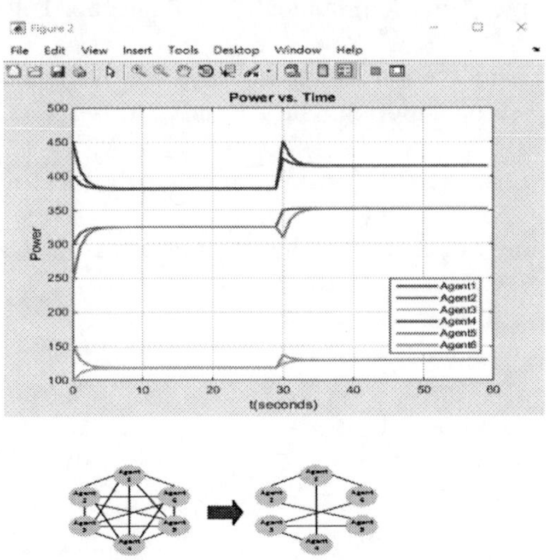

Figure 3. Generation power compensation subject to link failure caused cyber attack

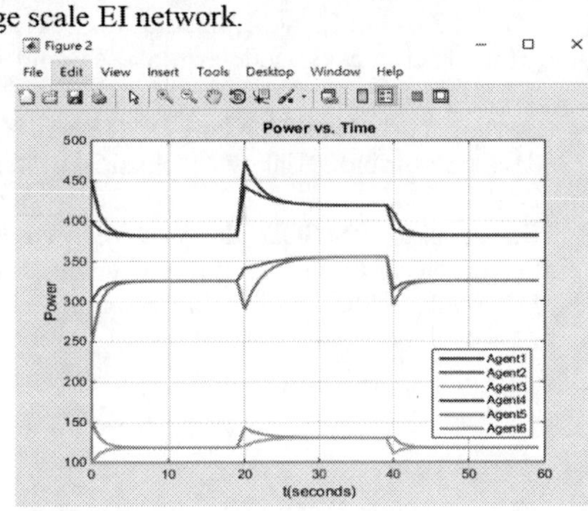

Figure 4. The scenario of the EI network from links failure to recovery

IOP Conf. Series: Journal of Physics: Conf. Series **1267** (2019) 012079 doi:10.1088/1742-6596/1267/1/012079

5. Conclusions

In this paper, detection of cyber attack or fault has been considered on a network of Energy Internet. The proposed algorithm not only able to detect a fault /cyber attack but also successfully tolerance and recovery nodes after failure in the EI network. Furthermore, the proposed algorithm has been proposed to safely and automatically recovery the faulty node under cyber attack. Finally, a numerical simulation cases study have been given with a typical example of six nodes in a hexagon EI network with a possible node and communication cyber attacks. Finally, a numerical case study has demonstrated that the residual generated at the monitoring node able to successfully detect and isolate the cyber attack. Also, the faulty node recovery has been shown eff ectively to maintain the quality of service accordingly. Future work includes extension of the proposed algorithm to handle more complex attack patterns and applying the approach for other types of EI network coordination missions.

Acknowledgments

This research work was sponsored by the Ministry of Science and Technology, R.O.C., under project number MOST107-2221-E-155-038.

References

[1] R. Olfati-Saber, J. A. Fax, and R. M. Murray, "Consensus and cooperation in networked multi-agent systems," Proceedings of the IEEE, vol. 95, no. 1, pp. 215–233, 2007.

[2] S. Yang, S. Tan, and J.-X. Xu, "Consensus based approach for economic dispatch problem in asmart grid," IEEE Transactions on Power Systems, vol. 28, no. 4, pp. 4416–4426, 2013.

[3] C. Zhao, J. Chen, and J. He, "Consensus-Based Energy Management in Smart Grid With Transmission Losses and Directed Communication," IEEE Transactions on Smart Grid, vol. 8, no. 5, pp. 2049–2061, 2017.

[4] N. Rahbari-Asr and M. Y. Chow, "Cooperative Distributed Demand Management for Community Charging of PHEV/PEVs Based on KKT Conditions and Consensus Networks," IEEE Transactions on Industrial Informatics, vol. 10, no. 3. pp. 1907–1916, 2014.

[5] M. H. Cintuglu, T. Youssef, and O. A. Mohammed, "Development and Application of a Real-Time Testbed for Multiagent System Interoperability : A Case Study on Hierarchical Microgrid Control," IEEE Transactions on Smart Grid, vol. 9, no. 3, pp. 1759–1768, 2018.

[6] Xu Li; Xiaohui Liang; Rongxing Lu; Xuemin Shen; Xiaodong Lin; Haojin Zhu, "Securing smart grid: cyber attacks, countermeasures, and challenges," in Communications Magazine, IEEE , vol.50, no.8, pp.38-45, August 2012.

[7] D. Ding, Q.-L. Han, Y. Xiang, X. Ge, and X.-M. Zhang, "A survey on security control and attack detection for industrial cyber-physical systems," Neurocomputing, vol. 275, pp. 1674–1683, Jan. 2018.

[8] S. Kuvshinkova, SQL Slammer Worm Lessons Learned for Consideration by the Electricity Sector, North Amer. Elect. Rel. Council, Atlanta, GA, USA, 2003.

[9] J. Farwell and R. Rohozinski, "Stuxnet and the future of cyber war," Survival, vol. 53, no. 1, pp. 23–40, 2011

Integrated Modelling and Static Experiment of a Novel Tandem Ducted Fan Flying Robot

Lei Liu[1, a], **Changle Xiang**[1, b], **Bin Xu**[1, *] **and Wei Fan**[1, c]

[1]School of Mechanical Engineering, Beijing Institute of Technology, Beijing, 100081, China

[a]liulei159159@126.com, [b]xiangcl@bit.edu.cn, [*]bitxubin@bit.edu.cn, [c]fanweixx@gmail.com

Abstract. This paper puts forward a novel tandem ducted fan flying robot which has nice passivity and compact structure aiming at operating in complicated and dangerous environment. The structure of its ducted coaxial twin-rotor with reverse rotation is unconventional and make its aerodynamic characteristics much more complex. To solve the problem, the modelling method of the traditional open rotors with the blade element theory and the momentum theory is extended to the ducted coaxial twin-rotor structure, and the integrated dynamic model of the ducted fan flying robot is established. After the modelling, a static lift experiment has been carried out, and the aerodynamic model of the ducted coaxial twin-rotor structure is modified based on the test results.

1. Introduction

Nowadays, autonomous unmanned flying robots have become hotspots of production and research [1]. On the one hand, the highly developed autonomous flight control technology reduces the difficulty of operating the flying robot and reduces its operating costs for production; on the other hand, more and more researchers choose unmanned flying robots as the research object of advanced technology because of their superior performance.

With the rapid development of society and technology, people's expectations for unmanned flying robots are getting higher and higher. For example, one might expect an unmanned aerial vehicle to enter an indoor environment, or to replace or assist related personnel in other more complex work environments [2-3]. This makes the operating environment of unmanned flying robots more and more challenging. At present, relatively mature unmanned flying robots use lift structures of fixed-wings or open rotors. The fixed-wing unmanned aerial robot is suitable for relatively long-term high-speed conditions, but it cannot hover, which makes it less functional in complex environments. Open-rotor unmanned aerial robots, such as helicopters, rotorcraft and multi-axis aircraft, are the most common and widely used unmanned flying robots on the market. However, the innate structure of the open rotor greatly reduces the adaptability of this type of flying robot to complex operating environments, and the relatively poor safety has become a major limitation of this type of flying robot in actual working conditions.

In order to improve the applicability of unmanned flying robots in complex operating environments, ducted fan unmanned aerial robots have become a better choice. Compared with the traditional open-rotor flying robot, the ducted fan flying robot has many unique advantages, including good safety, higher load capacity, and smaller structure size under the same conditions [4]. These advantages make the ducted fan flying robot capable of operating in an unknown, complex, and dangerous environment.

The multi-domain vehicle research group of Beijing Institute of Technology has been exploring for a long time in the research of ducted fan flying robots. The development and iteration of prototypes have also been carried out on the structure of the ducted fan flying robot. Whether the versatile robot can obtain good work quality, in addition to the advanced controller structure and algorithm, has an important relationship with the robot's own configuration and layout.

Here we introduce a novel tandem ducted fan flying robot shown in figure 1 which has nice passivity and compact structure aiming at operating in complicated environment and narrow space. It is mainly composed of a lift system, rudders, landing gears, and a control system. The lift system consists of two ducted fans with coaxial rotors which rotate in opposite directions in the same duct. The pitch of the ducted fan flying robot is achieved by the speed difference of the front and rear ducted rotors. The roll is achieved by swinging the rudders. The yaw is achieved by the speed difference of the two coaxial rotors in the same duct.

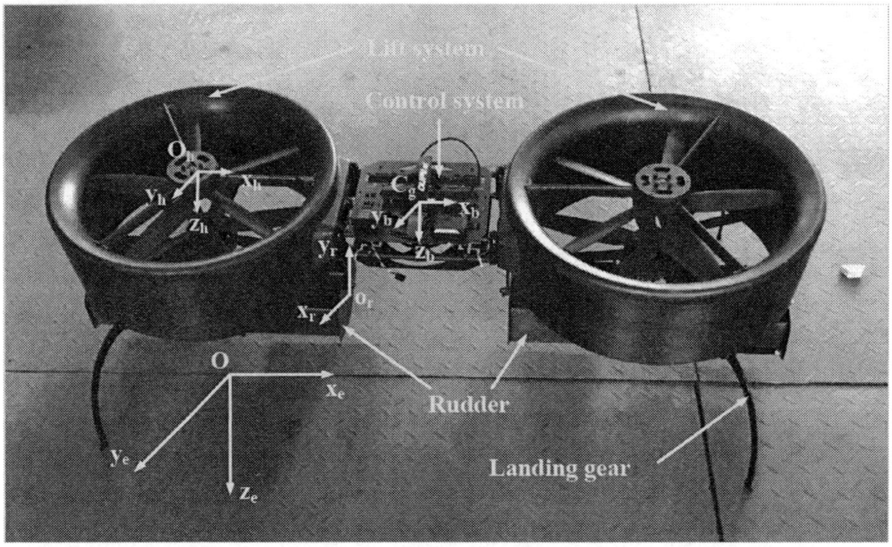

Figure 1. The novel tandem ducted fan flying robot

The research object of this paper is the ducted coaxial twin-rotor structure. At present, the aerodynamic model research of the ducted coaxial twin-rotor structure is not rich enough. In the paper [5], the aerodynamic model of the open coaxial counter-rotating rotors is analyzed by using the blade element momentum theory (BEMT). In the paper [6], the Froude conservation theory is applied to calculate the aerodynamic characteristics of the ducted single-rotor structure. In this paper, the methods in the above articles are combined and improved, and the BEMT modelling method is extended to the ducted coaxial twin-rotor structure.

2. Integrated modelling

2.1. Rigid-body Kinematic and dynamic model

The coordinate systems of the flying robot are shown in figure 1. $O-x_e y_e z_e$ is the earth coordinate system C_e. $C_g-x_b y_b z_b$ is the body coordinate system C_b and its origin is located at the center of gravity of the flying robot. $O_h-x_h y_h z_h$ is the hub coordinate system and it is based on the rotation center of the rotor. $O_r-x_r y_r$ is the rudder coordinate system and its origin is located at the swing center of the main rudder. The hub coordinate system and the rudder coordinate system are fixedly connected to C_b, and move along with C_b with respect to C_e. C_e can be converted to C_b by the standard Euler rotation and translation. The transformation matrix from C_e to C_b is

$$\mathbf{R}_b^e=\begin{bmatrix}\cos\theta\cos\psi & \cos\theta\sin\psi & -\sin\theta\\ \sin\theta\cos\psi\sin\varphi-\sin\psi\cos\varphi & \sin\theta\sin\psi\sin\varphi+\cos\psi\cos\varphi & \cos\theta\sin\varphi\\ \sin\theta\cos\psi\cos\varphi+\sin\psi\sin\varphi & \sin\theta\sin\psi\cos\varphi-\cos\psi\sin\varphi & \cos\theta\cos\varphi\end{bmatrix} \tag{1}$$

We define the position and attitude vector of the flying robot in C_e as

$$\begin{cases}\mathbf{P}=[x\ y\ z]^T\\ \mathbf{\Phi}=[\varphi\ \theta\ \psi]^T\end{cases} \tag{2}$$

and the velocity and angular velocity vector of the flying robot in C_b as

$$\begin{cases}\mathbf{V}=[U\ V\ W]^T\\ \mathbf{\Omega}=[p\ q\ r]^T\end{cases} \tag{3}$$

The kinematic model of the flying robot can be described as

$$\begin{cases}\mathbf{V}=\mathbf{R}_b^e\dot{\mathbf{P}}\\ \mathbf{\Omega}=\mathbf{W}\dot{\mathbf{\Phi}}\end{cases} \tag{4}$$

where $\mathbf{W}$ is given as

$$\mathbf{W}=\begin{bmatrix}1 & 0 & -\sin\theta\\ 0 & \cos\varphi & \cos\theta\sin\varphi\\ 0 & -\sin\varphi & \cos\theta\cos\varphi\end{bmatrix} \tag{5}$$

The dynamic model of the flying robot can be described by Newton-Euler equations as

$$\begin{cases}\dot{\mathbf{V}}=-\mathbf{\Omega}\times\mathbf{V}+\dfrac{1}{\mathrm{m}}\mathbf{F}\\ \dot{\mathbf{\Omega}}=\mathbf{I}^{-1}(\mathbf{M}-\mathbf{\Omega}\times\mathbf{I}\times\mathbf{\Omega})\end{cases} \tag{6}$$

where m and $\mathbf{I}$ are the total mass and inertia, $\mathbf{F}$ and $\mathbf{M}$ are the resultant force and moment of the flying robot. $\mathbf{F}$ includes the forces generated by the rotors and rudders, additional lift of the duct, and the gravity. The moments generated by rotors, rudders and ducts constitute $\mathbf{M}$.

2.2. Duct thrust model

The structures of two ducts of the flying robot are identical, and the front duct is selected as the research object. The modelling process of the rear duct is the same as that of the front duct, so its derivation process is not described. The aerodynamic characteristics of the ducted rotors are very complicated. In order to simplify the modelling and derivation process, we first give the assumption that the whole structure works in the axial flow condition, and the airflow is the ideal fluid, and the airflow state does not change with time [7-8].

2.2.1. Duct inlet lift model. Apply the momentum theorem to the gas at the inlet region of the duct:

$$F_{in}+P_{in}S_{in}-P_{du1}S_1=\rho V_{i1}S_1V_{i1}-\rho V_{in}S_{in}V_{in} \tag{7}$$

where S_{in} and S_1 are the area of the duct entrance and the upper rotor disc respectively, P_{in} and V_{in} are the air pressure and airflow velocity at the duct entrance, P_{du1} is the air pressure at the upper surface

IOP Conf. Series: Journal of Physics: Conf. Series **1267** (2019) 012080 doi:10.1088/1742-6596/1267/1/012080

of the upper rotor disc, ρ is the air density, and V_{i1} is the induced velocity of the air at the upper rotor. Now we define the equivalent inlet ratio h_i and outlet ratio h_e of the ducted rotors as

$$\begin{cases} h_i = \dfrac{S_{in}}{S_1} \\ h_e = \dfrac{S_{out}}{S_2} \end{cases} \tag{8}$$

where S_{out} and S_2 are the area of the duct outlet and the lower rotor disc respectively. Selecting the duct inlet and upper surface of the upper rotor disc as the control surface, the Bernoulli equation is

$$P_{in} + \frac{1}{2}\rho V_{in}^2 = P_{du1} + \frac{1}{2}\rho V_{i1}^2 \tag{9}$$

Including the pressure outside the duct, the net lift of the duct inlet is

$$F_{in1} = F_{in} + P_{in}(S_{in} - S_1) = \frac{1}{2}\rho S_1 \left(\frac{h_i - 1}{h_i}\right)^2 V_{i1}^2 \tag{10}$$

2.2.2. Duct outlet resistance model. The derivation of the duct outlet resistance is similar to that of the duct inlet lift, so the net resistance of the duct outlet is obtained as

$$F_{out1} = F_{out} - P_{out}(S_{out} - S_2) = -\frac{1}{2}\rho S_2 \left(\frac{h_e - 1}{h_e}\right)^2 (V_{i1} + V_{i2})^2 \tag{11}$$

where V_{i2} is the induced velocity of the air at the lower rotor.

2.2.3. Thrust of the front duct. Combining the inlet lift and outlet resistance of the duct, we get the thrust of the front duct

$$F_{d1}=F_{in1}+F_{out1} \tag{12}$$

2.3. Coaxial twin-rotor model

2.3.1. Upper rotor model. The momentum theorem model of the upper rotor characterizes the rotor's lift as the pressure difference generated by the pressure jump between the upper and lower surfaces of the rotor. Therefore, it is only need to obtain the upper and lower surface pressure of the rotor to obtain the rotor lift. The pressure difference between the upper and lower surfaces of the upper rotor is obtained by using Bernoulli's equation and mass conservation theory:

$$\Delta P_{up1} = \frac{1}{2}\rho V_{i1}^2 \tag{13}$$

In order to facilitate subsequent derivation, we define the differential element dx_h of the blade distributed along the blade extension in the plane of the rotor disc, as shown in Figure 2.

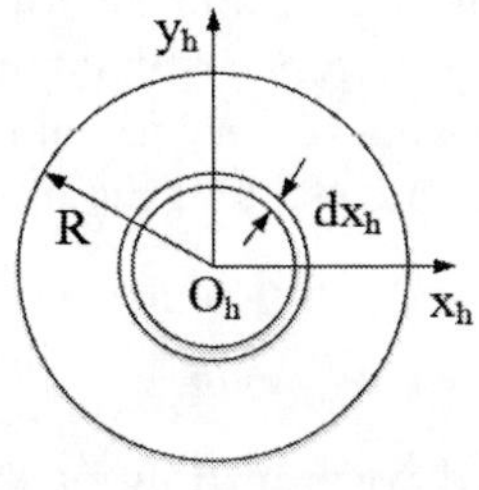

Figure 2. Element dx_h

Taking dx_h as the calculation object, the lift of the blade element ring is

$$dF_1 = \Delta P_{up1} dS_1 = \pi \rho V_{i1}^2 x_h dx_h \tag{14}$$

Here we define the equivalent radius r_1 and r_2 of the rotor and the inflow ratio λ_1, λ_2 and λ :

$$\begin{cases} r_1 = \dfrac{x_h}{R_1}, r_2 = \dfrac{x_h}{R_2} \\ \lambda_1 = \dfrac{V_{in} + V_{i1}}{\omega_1 R_1}, \lambda_2 = \dfrac{V_{i1} + V_{i2}}{\omega_2 R_2}, \lambda = \dfrac{V_{i1}}{\omega_2 R_2} \end{cases} \tag{15}$$

where R_1 and R_2 are the radius of the upper and lower rotor, ω_1 and ω_2 are the rotating speed of the upper and lower rotor. The dimensionless lift of the blade element ring can be re-expressed as

$$dC_{F1} = \frac{dF_1}{\pi \rho R_1^2 (\omega_1 R_1)^2} = \lambda_1^2 r_1 dr_1 \tag{16}$$

The dimensionless lift and torque of the upper rotor can also be calculated by using the blade element theory:

$$\begin{cases} dC_{F1} = \dfrac{1}{2}\sigma\left[C_{l\alpha 1}(\theta_1 r_1^2 - \lambda_1 r_1) - C_{d1}\lambda_1 r_1\right] dr_1 \\ dC_{M1} = \dfrac{1}{2}\sigma\left[C_{d1} r_1^2 + C_{l\alpha 1}\lambda_1(\theta_1 r_1 - \lambda_1)\right] dr_1 \end{cases} \tag{17}$$

where C_{d1} is the drag coefficient, $C_{l\alpha 1}$ is the slope of the lift characteristic curve, θ_1 is the mounting angle of the upper rotor, and σ is the rotor solidity. Now we derive the aerodynamic model of the upper rotor based on the momentum theorem, and obtain a set of relationship between rotor lift and inflow ratio as expressed in equation (16). Another set of relationship between rotor lift and inflow ratio based on the blade element theory is shown in equation (17). Combining these two sets of relationships, we get a general solution for the equivalent inflow ratio of the upper rotor disc:

$$\lambda_1 = \frac{1}{4}\sigma C_{l\alpha 1}\left[\left(1 + \frac{8\theta_1 r_1}{\sigma C_{l\alpha 1}}\right)^{1/2} - 1\right] \tag{18}$$

Then the dimensioned lift and torque of the upper rotor can be calculated by

$$\begin{cases} F_1 = \pi \rho R_1^2 (\omega_1 R_1)^2 \int_0^1 \dfrac{1}{2}\sigma\left[C_{l\alpha 1}(\theta_1 r_1^2 - \lambda_1 r_1) - C_{d1}\lambda_1 r_1\right] dr_1 \\ M_1 = \pi \rho R_1^3 (\omega_1 R_1)^2 \int_0^1 \dfrac{1}{2}\sigma\left[C_{d1} r_1^2 + C_{l\alpha 1}\lambda_1(\theta_1 r_1 - \lambda_1)\right] r_1 dr_1 \end{cases} \tag{19}$$

2.3.2. lower rotor model. The derivation of the lift of the lower rotor based on the momentum theorem is similar to that of the upper rotor. The difference is that the lower rotor works in the downward wash airflow of the upper rotor. Therefore, it is necessary to count the air inflow velocity when calculating the lift of the lower rotor. So the dimensionless lift of the blade element ring of the lower rotor is

$$dC_{F2} = \frac{dF_2}{\pi\rho R_2^2(\omega_2 R_2)^2} = \left(\frac{1}{h_e^2\lambda_2^2} - \lambda^2\right) r_2 dr_2 \tag{20}$$

Similarly, considering the downward wash airflow of the upper rotor, the dimensionless lift and torque of the lower rotor calculated by the blade element theory are

$$\begin{cases} dC_{F2} = \frac{1}{2}\sigma\left[C_{la2}(\theta_2 r_{d2}^2 - \lambda_{r2} r_{d2}) - C_{d2}\lambda_{r2} r_{d2}\right] dr_{d2} \\ dC_{M2} = \frac{1}{2}\sigma\left[C_{d2} r_{d2}^2 + C_{la2}\lambda_{r2}(\theta_2 r_{d2} - \lambda_{r2})\right] dr_{d2} \end{cases} \tag{21}$$

Combining equation (20) and (21), the solution for the inflow ratio of the lower rotor is given as

$$\lambda_2 = \frac{1}{4} h_e^2 \sigma C_{la2} \left\{ \left[\left(\frac{4\lambda}{h_e \sigma C_{la2}} \right)^2 + \frac{8\theta_2 r_2}{h_e^2 \sigma C_{la2}} + 1 \right]^{1/2} - 1 \right\} \tag{22}$$

Then the dimensioned lift and torque of the lower rotor can be calculated by

$$\begin{cases} F_2 = \pi\rho R_2^2(\omega_2 R_2)^2 \int_0^1 \frac{1}{2}\sigma\left[C_{la2}(\theta_2 r_2^2 - \lambda_2 r_2) - C_{d2}\lambda_2 r_2\right] dr_2 \\ M_2 = \pi\rho R_2^3(\omega_2 R_2)^2 \int_0^1 \frac{1}{2}\sigma\left[C_{d2} r_2^2 + C_{la2}\lambda_2(\theta_2 r_2 - \lambda_2)\right] r_2 dr_2 \end{cases} \tag{23}$$

where C_{d2} is the drag coefficient, C_{la2} is the slope of the lift characteristic curve, θ_2 is the mounting angle of the lower rotor.

2.4. Rudder model

There are two sets of rudders below the front duct, each of which consists of a main rudder and two auxiliary rudders. The two sets of rudders are linked and their deflection angles are the same. For the rudders under the front duct, the lift and resistance can be obtained as follows:

$$\begin{cases} F_{r1} = \frac{1}{h_e^2}\rho\theta_r\left[C_{l\theta r1} S_{mr}(\omega_2 R_2 \lambda_2)^2 + 2C_{l\theta r2} S_{ar}(\omega_2 R_2 \lambda_2)^2\right] \\ T_{r1} = \frac{1}{h_e^2}\rho\left[C_{dr1} S_{mr}(\omega_2 R_2 \lambda_2)^2 + 2C_{dr2} S_{ar}(\omega_2 R_2 \lambda_2)^2\right] \end{cases} \tag{24}$$

where θ_r is the rudder deflection angle, $C_{l\theta r1}$ and C_{dr1} are lift coefficient and drag coefficient for the main rudder, $C_{l\theta r2}$ and C_{dr2} are lift coefficient and drag coefficient for the auxiliary rudder, S_{mr} and S_{ar} are the windward area of the main rudder and auxiliary rudder.

3. Static experiment and model correction

It is worth noting that we have simplified the aerodynamics of the structure during the modelling. In addition, the ducted rotors are pneumatically special, and there are very complicated aerodynamic interferences between the two rotors in the same duct. These aerodynamic disturbances are often difficult to accurately describe by mathematical methods. These factors have brought some errors to aerodynamic modelling of the ducted rotors [9]. Therefore, we designed and carried out the static experiment of the ducted twin-rotor, introduced the aerodynamic interference factor, and corrected the aerodynamic weight of the duct and the upper and lower rotor.

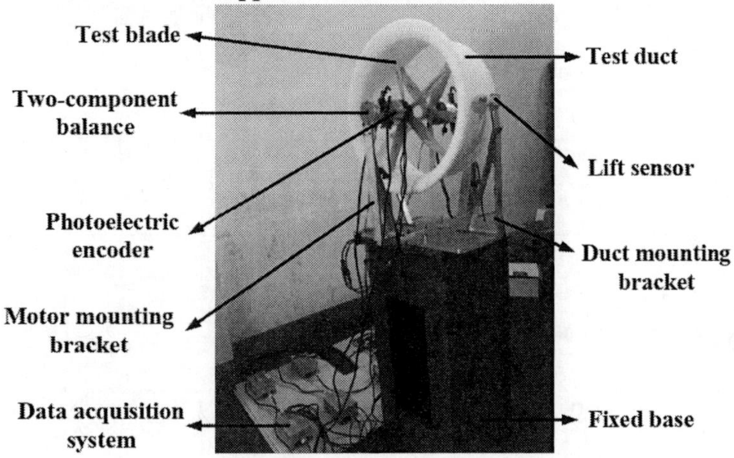

Figure 3. Static test system of the ducted twin-rotor

The static test system is mainly composed of test ducts, test rotors, fixed base, sensors and data acquisition systems, as shown in figure 3. Each of the ducted rotor rotation centers is equipped with a photoelectric encoder and a two-component balance to measure the real-time rotating speed, lift and torque of each rotor. On the outer wall symmetrical with the center of the duct, two lift sensors are installed to measure the additional lift generated on the duct. The data acquisition system includes amplifiers, capture cards and electronic governors. The amplifiers output a voltage change signal of the two-component balance and the lift sensor, and perform the corresponding amplitude gain for subsequent processing. The electronic governors are used to collect the real-time voltage and current of each motor. The capture cards are used to collect the voltage and current signals mentioned above and calculate the final results which we want to collect.

By analyzing the data collected by the static test, we correct the lift of the upper rotor and lower rotor and the duct thrust calculated from our model:

$$\begin{cases} F_{1*} = c_{F1} F_1 \\ F_{2*} = c_{F2} F_2 \\ F_{d1*} = c_{Fd1} F_{d1} \end{cases} \tag{25}$$

where c_{F1}, c_{F2} and c_{Fd1} are the correction coefficients. Figure 4 shows the comparison of the theoretical thrusts of the upper and lower rotor and the duct of the modified ducted twin-rotor model and static test results. It can be obtained in figure 4 that we get the ideal correction effect of the model when the speed of the ducted rotor is less than 5000 rpm. When the rotor speed is above 5000 rpm, even if it is corrected, there is still some error between the theoretical calculation results and the test results. This is because the performance of each pneumatic component in the test is not only affected by its own structure but also by the motor. When the rotor speed is too high, the output power and current of the motor increases, and the temperature rises obviously, which would have an impact on the test results. Considering that the motor of the ducted fan flying robot mainly works at 4000~5000rpm, the model correction is considered to be effective.

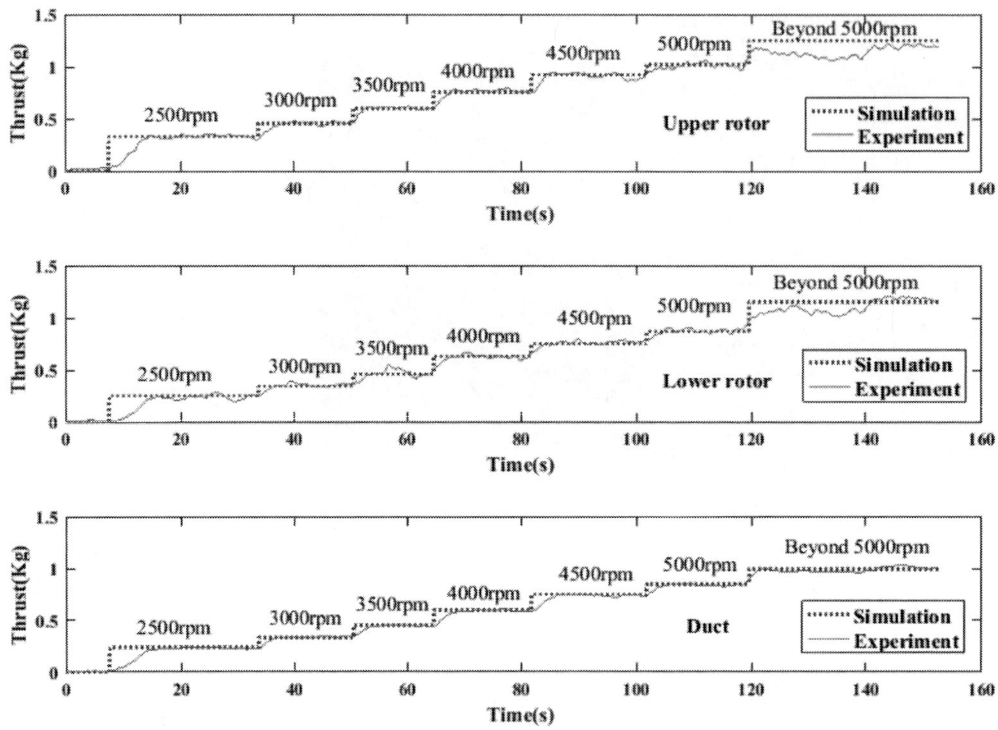

Figure 4. Comparison of the modified theoretical thrusts and static test results

4. Conclusion

This paper proposes a novel ducted fan flying robot whose structure is compact aiming at operating in complicated environment and narrow space. Considering that the structure of its ducted coaxial twin-rotor with reverse rotation is unconventional, the modelling method of the traditional open rotors with the blade element theory and the momentum theory is extended to the ducted coaxial twin-rotor structure, and the integrated dynamic model of the ducted fan flying robot has been founded. After the modelling, we designed and carried out the static experiment of the aerodynamic structure of the ducted twin-rotor, introduced the aerodynamic interference factor, and corrected the aerodynamic weight of the duct and the upper and lower rotor. At last we obtain the ideal correction effect and get a more accurate model of the ducted fan flying robot. Our future work will focus on control system designing and corresponding flight experiments of the ducted fan flying robot.

Acknowledgments

This work is supported by National Natural Science Foundation of China under Grant 51505031.

References

[1] Beard R W and McLain T W 2012 *Small unmanned aircraft: Theory and practice* Princeton university press

[2] Khosiawan Y and Nielsen I 2016 A system of UAV application in indoor environment *Production & Manufacturing Research*, *4*(1), 2-22

[3] Grzonka S, Grisetti G and Burgard W 2012 A fully autonomous indoor quadrotor *IEEE Transactions on Robotics*, *28*(1), 90-100

[4] Wang Z, Liu Z, Fan N and Guo M 2013 Flight dynamics modeling of a small ducted fan aerial vehicle based on parameter identification *Chinese Journal of Aeronautics*, *26*(6), 1439-1448

[5] Bohorquez F 2007 *Rotor hover performance and system design of an efficient coaxial rotary wing micro air vehicle* (Doctoral dissertation)

[6] Pflimlin J M, Binetti P, Trouchet D, Soueres P and Hamel T 2007 Aerodynamic modeling and practical attitude stabilization of a ducted fan uav. In *2007 European Control Conference* pp 4023-4029

[7] Ohanian O, Gelhausen P and Inman D 2010 A compact method for modeling the aerodynamics of ducted fan vehicles. In *48th AIAA Aerospace Sciences Meeting Including the New Horizons Forum and Aerospace Exposition* p 1052

[8] Ko A, Ohanian O and Gelhausen P 2007 Ducted fan UAV modeling and simulation in preliminary design. In *AIAA modeling and simulation technologies conference and exhibit* p 6375

[9] Naldi R, Gentili L, Marconi L and Sala A 2010 Design and experimental validation of a nonlinear control law for a ducted-fan miniature aerial vehicle *Control Engineering Practice, 18*(7), 747-760

IOP Conf. Series: Journal of Physics: Conf. Series **1267** (2019) 012081 doi:10.1088/1742-6596/1267/1/012081

Adaptive Integral Sliding Mode Guidance Law with Impact Angle Constraint Considering Autopilot Lag

X B Li[1,*], G R Zhao[1], S Liu[1] and X Han[1]

[1]Naval Aviation University, 188 Er-ma Road,Yantai, China

[*]Corresponding author's e-mail: lixiaobaohjhy@163.com

Abstract. For the interception problem of maneuvering targets, considering impact angle constraint and first-order autopilot lag, an adaptive integral sliding mode guidance law is designed. A new nonsingular finite-time integral sliding mode surface is constructed, which consists of LOS angle, LOS angular rate and LOS angular acceleration, and ensures that the states of the guidance system converge to zero on the sliding mode surface in a finite time. Moreover, an adaptive law is designed to estimate the upper bound of the target's unknown acceleration. The Lyapunov stability theory proves that the guidance system can converge to zero strictly in a finite time with the proposed guidance law. Finally, the simulation results verify the effectiveness of the designed guidance law.

1. Introduction

The autopilot dynamics of the missile during the terminal guidance process is not ideal and has a delay effect. When designing the guidance law without considering autopilot lag, the actual miss distance, impact angle and other guidance indicators will not have a good performance. For the interception of a maneuvering target, the guidance law will be significantly weakened, and may even fail to intercept the target [1]. Therefore, it is of great practical significance to consider the dynamic characteristics of the autopilot when designing the terminal guidance law.

The dynamic characteristics of the missile's autopilot can be regarded as a first-order inertia link. The related guidance law research methods include optimal control [2], sliding mode control [3] and Lyapunov stability theory [4]. However, the research on the combination of missile autopilot lag and finite time convergence is still rare. In [5], a nonsingular terminal sliding mode surface is designed by using LOS angular rate and LOS angular acceleration, and a finite time convergent guidance law with a first-order autopilot lag is proposed. But it does not consider the constraint of impact angle because the terminal guidance process will become more complicated when considering the dynamic characteristics of the autopilot, and the guidance system becomes high-order equation with three state variables when considering impact angle constraint. In [6], a finite time convergent guidance law based on two terminal sliding mode surfaces solve the problem of impact angle constraint, but the solution of two terminal sliding mode surfaces is complex and the amount of calculation is large.

In order to solve the terminal guidance problem of intercepting maneuvering targets, this paper studies a guidance law with impact angle constraint considering the dynamics of autopilot. When considering a first-order autopilot lag, a new type of terminal sliding mode surface is constructed. By introducing the integral term to avoid the singularity problem and using the characteristics of a class of nonlinear finite-time stable systems, the proposed guidance law ensures the guidance system converge to zero in a finite time, the design of a single sliding mode is not only simple and effective, but also meets the requirements of impact angle constraint.

2. Problem description and related lemma

The terminal guidance process can be simplified to be performed in a two-dimensional plane as shown in Figure 1. r and q represent the distance between the missile M and target T and the LOS angle. γ_M and γ_T represent the flight path angles of the missile and target, respectively. a_M and a_T represent the normal accelerations of the missile and target. V_M and V_T represent the speeds and are constant.

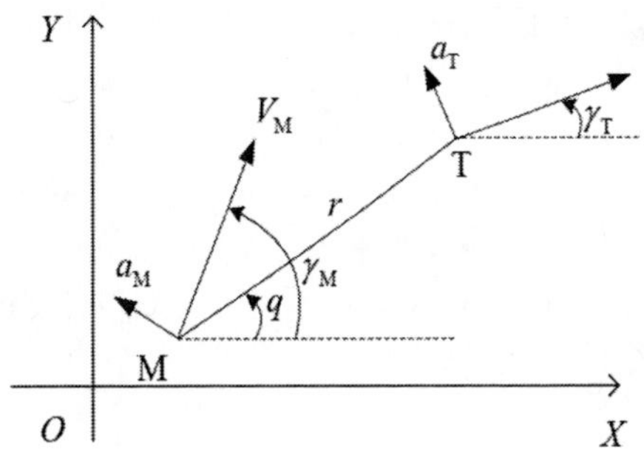

Figure 1. Missile and target engagement geometry of terminal guidance

The engagement kinematics of the guidance system can be written as

$$\dot{r} = V_T\cos(\gamma_T - q) - V_M\cos(\gamma_M - q) \tag{1a}$$

$$r\dot{q} = V_T\sin(\gamma_T - q) - V_M\sin(\gamma_M - q) \tag{1b}$$

Differentiating eqs. (1) with respect to time yields

$$\ddot{r} = r\dot{q}^2 - a_{Mr} + a_{Tr} \tag{2a}$$

$$\ddot{q} = \left(-2\dot{r}\dot{q} - a_{Mq} + a_{Tq}\right)/r \tag{2b}$$

where $a_{Mr} = a_M\sin(q-\gamma_M)$ and $a_{Tr} = a_T\sin(q-\gamma_T)$ denote the accelerations of the missile and target along the LOS, $a_{Mq} = a_M\cos(q-\gamma_M)$ and $a_{Tq} = a_T\cos(q-\gamma_T)$ denote the accelerations of the missile and target normal to the LOS.

The missile's autopilot dynamics are usually described by a first-order term

$$\dot{a}_{Mq} = \left(u - a_{Mq}\right)/\tau \tag{3}$$

where τ is the time constant of the autopilot and u is the guidance command to the autopilot.

The missile's impact angle θ_{imp} constraint can be transformed into the terminal LOS angle q_f constraint. Assume q_d is the desired terminal LOS angle, $x_1 = q - q_d$ is the LOS angle tracking error, $x_2 = \dot{x}_1$ is the LOS angular rate, $x_3 = \dot{x}_2$ is the LOS angular acceleration. Eqs. (2) can be written as

$$\dot{x}_1 = x_2 \tag{4a}$$

$$\dot{x}_2 = x_3 \tag{4b}$$

$$\dot{x}_3 = f + bu + d/r \tag{4c}$$

where $f = \left(-2\ddot{r}x_2 - 3\dot{r}x_3 + a_{Mq}/\tau\right)/r$, $b = -1/\tau r$, $d = \dot{a}_{Tq}$.

Assumption 1. The target's acceleration a_T and its derivative $\dot{a}_T$ satisfy $|a_T| \le \Delta_1$, $|\dot{a}_T| \le \Delta_2$, where Δ_1 and Δ_2 are two positive constants.

In Eq. (4), d can be regarded as the unknown disturbance caused by the target acceleration. It is known from Assumption 1 that there is an upper bound Δ that satisfies $|d| \le \Delta$. By designing the guidance command u to ensure x_1 and x_2 converge to the origin in a finite time, so that the missile can hit the target with the desired terminal LOS angle q_d considering the first-order autopilot lag.

Lemma 1 [7]. For a continuous positive definite function $V(t)$, if $\dot{V}(t) \le -\alpha V(t) - \beta V(t)^{\gamma}$, $\forall t > t_0$, where $\alpha > 0, \beta > 0$ and $0 < \gamma < 1$, $V(t)$ will converge to the equilibrium point in finite time t_f given by

$$t_f \le t_0 + \frac{1}{\alpha(1-\gamma)} \ln \frac{\alpha V(t_0)^{1-\gamma} + \beta}{\beta} \tag{5}$$

Lemma 2 [8]. For a chain system

$$\dot{y}_1 = y_2,\ \dot{y}_2 = y_3,\ \ldots,\ \dot{y}_{m-1} = y_m,\ \dot{y}_m = u \tag{6}$$

Suppose $D(x) = x^m + k_m x^{m-1} + \cdots + k_2 x + k_1$ is a Hurwitz polynomial and $k_1, k_2, \cdots, k_m > 0$, then the system (6) is finite time stable with the feedback control:

$$u = -k_1 sig^{\alpha_1} y_1 - k_2 sig^{\alpha_2} y_2 - \cdots - k_m sig^{\alpha_m} y_m \tag{7}$$

where $\alpha_{i-1} = \alpha_i \alpha_{i+1} / (2\alpha_{i+1} - \alpha_i)$, $\alpha_m = \alpha$, $\alpha_{m+1} = 1$, $\alpha \in (0,1)$, $i = 2,3,\cdots,m$.

3. Adaptive integral sliding mode guidance law design

3.1. Nonsingular integral sliding mode surface

The guidance system (4) with impact angle constraint and first-order autopilot lag contains three state variables. However, the traditional terminal sliding mode surface is usually composed of two state variables to achive the finite time convergence on the sliding mode surface. In order to avoid designing two sliding mode surfaces as in [6], the LOS angle tracking error, LOS angular rate and LOS angular acceleration are considered together. A new type of terminal sliding mode surface containing three state variables is proposed in [9] and threse state variables eventually converge to zero in a finite time. The sliding mode surface is as follows:

$$s = x_3 + k_2 sig^{\alpha_2} x_2 + k_1 sig^{\alpha_1} x_1 \tag{8}$$

where $k_1, k_2 > 0$, $0 < \alpha_1 < 1$, $\alpha_2 = 2\alpha_1/(1+\alpha_1)$. Because $0 < \alpha_1, \alpha_2 < 1$, there is a problem of singularity in the sliding mode surface (8), and no further research is implemented on this problem in [9]. To slove the singularity problem, an integrated sliding mode without autopilot lag is proposed in [10]. Inspired by these, a new type of nonsingular integral sliding surface is constructed:

$$s = x_3 + \int_0^t \left(k_1 sig^{\alpha_1} x_1 + k_2 sig^{\alpha_2} x_2 + k_3 sig^{\alpha_3} x_3 \right) dt \tag{9}$$

where $\alpha_1 = \alpha_2 \alpha_3 / (2\alpha_3 - \alpha_2)$, $\alpha_2 = \alpha_3/(2-\alpha_3)$, $0 < \alpha_3 < 1$ and $k_1, k_2, k_3 > 0$ satisfy Hurwitz conditions.

3.2. Guidance law design and stability analysis

Theorem 1. For the guidance system (4) with the sliding mode surface (9), if the guidance law u is designed as

$$u = -\frac{1}{b}\left(f + \frac{\sigma \hat{\Delta} \operatorname{sign}(s)}{r} + k_1 sig^{\alpha_1} x_1 + k_2 sig^{\alpha_2} x_2 + k_3 sig^{\alpha_3} x_3 + \alpha s + \beta sig^{\gamma} s \right) \tag{10}$$

where $\alpha, \beta > 0$, $\sigma > 1$, $0 < \gamma < 1$, $\hat{\Delta}$ is an adaptive estimate of Δ, and the adaptive law is

$$\dot{\hat{\Delta}} = \sigma |s| / r,\ \hat{\Delta}(0) > 0 \tag{11}$$

Then the following conclusions are established:

1) The sliding variable s and estimation error $\tilde{\Delta}=\Delta-\hat{\Delta}$ are bounded;

2) The sliding variable s converge to zero in a finite time;

3) The guidance states x_1, x_2 converge to zero in a finite time.

Proof. The proving process is divided into the following three steps.

Step 1. Define a Lyapunov function $V=\frac{1}{2}s^2+\frac{1}{2}\tilde{\Delta}^2$, and according to Eqs. (9)-(11), the derivative of V can be written as follows:

$$\begin{aligned}\dot{V}&=s\dot{s}+\tilde{\Delta}\dot{\tilde{\Delta}}\\&=s\left(f+bu+\frac{d}{r}+k_1 sig^{\alpha_1}x_1+k_2 sig^{\alpha_2}x_2+k_3 sig^{\alpha_3}x_3\right)-\tilde{\Delta}\dot{\hat{\Delta}}\\&=\frac{ds}{r}-\frac{\sigma\hat{\Delta}|s|}{r}-\alpha s^2-\beta|s|^{\gamma+1}-\frac{\sigma|s|\left(\Delta-\hat{\Delta}\right)}{r}\\&\le\frac{\Delta|s|(1-\sigma)}{r}-\alpha s^2-\beta|s|^{\gamma+1}\\&\le-\alpha s^2-\beta|s|^{\gamma+1}\end{aligned}\tag{12}$$

Since $\dot{V}\le 0$, it's clear that the sliding variable s and the estimation error $\tilde{\Delta}$ are bounded.

Step 2. Define a Lyapunov function $V_1=\frac{1}{2}s^2$. Differentiating V_1 and substituting Eqs. (9)(10) gives

$$\begin{aligned}\dot{V}_1&=s\left(f+bu+\frac{d}{r}+k_1 sig^{\alpha_1}x_1+k_2 sig^{\alpha_2}x_2+k_3 sig^{\alpha_3}x_3\right)\\&=\frac{ds}{r}-\frac{\sigma\hat{\Delta}|s|}{r}-\alpha s^2-\beta|s|^{\gamma+1}\\&\le\frac{|s|\left(\Delta-\sigma\hat{\Delta}\right)}{r}-\alpha s^2-\beta|s|^{\gamma+1}\end{aligned}\tag{13}$$

Since $\hat{\Delta}(0)\ge 0$ and $\dot{\hat{\Delta}}\ge 0$, there is $\hat{\Delta}(t)\ge\hat{\Delta}(0)\ge 0$. Choose $\hat{\Delta}(0)$ large enough and σ satisfying $\sigma\ge 1+\sqrt{\hat{\Delta}^2(0)+\eta}/\hat{\Delta}(0)$, where $\eta\ge 0$. Then

$$\begin{aligned}\Delta-\sigma\hat{\Delta}&\le\Delta-\hat{\Delta}(0)-\sqrt{\hat{\Delta}^2(0)+\eta}\\&\le\left|\tilde{\Delta}(0)\right|-\sqrt{\hat{\Delta}^2(0)+\eta}\\&\le\hat{\Delta}(0)-\sqrt{\hat{\Delta}^2(0)+\eta}\\&\le 0\end{aligned}\tag{14}$$

Therefore, it can be obtained that $\dot{V}_1\le-\alpha s^2-\beta|s|^{\gamma+1}\le-2\alpha V-2^{\frac{\gamma+1}{2}}\beta V^{\frac{\gamma+1}{2}}$. According to Lemma1, the integral sliding variable s will converge to zero in finite time.

Step 3. It is known that $s=\dot{s}=0$ on the sliding mode surface. Differentiating s gives

$$\dot{x}_3=-k_1 sig^{\alpha_1}x_1-k_2 sig^{\alpha_2}x_2-k_3 sig^{\alpha_3}x_3\tag{15}$$

According to Lemma 2, the states x_1, x_2, x_3 of the guidance system on the sliding surface (9) eventually converge to zero in finite time. This completes the proof.

The guidance law (10) is discontinuous which may cause the chattering problem because of the signum function sign(s). So sign(s) can be replaced with a continuous sigmoid function approximately:

$$\text{sgmf}(s)=\begin{cases}\text{sign}(s), & |s|>\varepsilon \\ \left(1-e^{-s/\varepsilon}\right)/\left(1+e^{-s/\varepsilon}\right), & |s|\le\varepsilon\end{cases} \tag{16}$$

Hence, the guidance law in (10) is modified as:

$$u=-\frac{1}{b}\left(f+\frac{\sigma\hat{\Delta}\text{sgmf}(s)}{r}+k_1 sig^{\alpha_1}x_1+k_2 sig^{\alpha_2}x_2+k_3 sig^{\alpha_3}x_3+\alpha s+\beta sig^{\gamma}s\right) \tag{17}$$

The guidance law (17) is called a finite time convergent adaptive integral sliding mode guidance law with autopilot lag (ALAISMG).

4. Numerical simulations

In this section, the effectiveness of the proposed guidance law is demonstrated through numerical simulations. The initial positions for the missile and the target are x_{M0}=−3km, y_{M0}= 10km and x_{T0}= 0m, y_{T0}= 0m. The initial flight path angles are γ_{M0}=−30deg and γ_{T0}= 135deg. The initial velocities are V_M= 600m/s, V_T=200m/s. The desired terminal LOS angle of −80deg (q_d=−80deg) are presented and the maximum guidance command in autopilot loop of the missile is 40g ($u_{\max}$=−40g), g is the acceleration of gravity (g= 9.8m/s^2). The parameters of ALAISMG are chosen as $\alpha=20$, $\beta=0.5$, $\alpha_1=0.57$, $\alpha_2=0.67$, $\alpha_3=0.8$, $k_1=0.4$, $k_2=0.8$, $k_3=1.2$, $\gamma=0.8$, $\tau=0.5$, $\sigma=2$, $\hat{\Delta}(0)=50$.

4.1 Performance under different maneuvering targets

In order to fully analyze the performance of ALAISMG, considering three different maneuvering targets:1) cosine maneuvering a_T=5g cos ($4\pi/t$); 2) constant maneuvering a_T=5g; 3) step maneuvering a_T=5g, t <5s; a_T=−5g, t ≥5s. Simulation results are shown in Figure 2.

Figure 2(a) shows that the missile can intercept different maneuvering targets with ALAISMG successfully. Figure 2(b)(c) shows that LOS angle q and LOS angular rate $\dot{q}$ can converge to q_d and zero in a finite time respectively and short peaks occur in the early stage which cause the normal accelerations of the missile to peak. It can be seen from Figure 2(d) that in the case of the limiting of the guidance command u, all of a_{Mq} rapidly increase to near 30g in the early stage and then decrease rapidly as $\dot{q}$ converge to zero. Besides, a_{Mq} curves are smooth and there are no chattering problems.

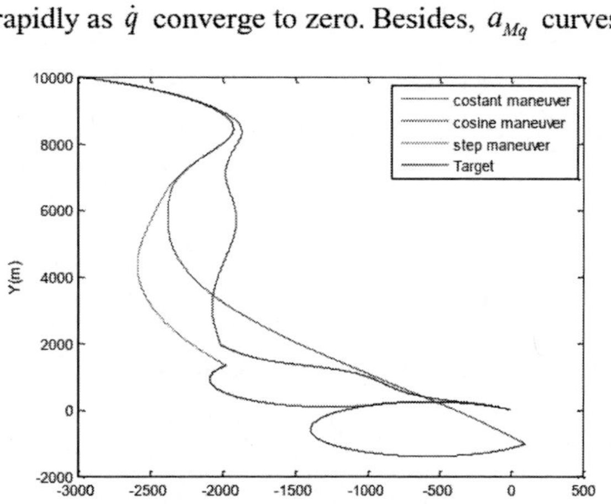

(a) Trajectories of missile and target

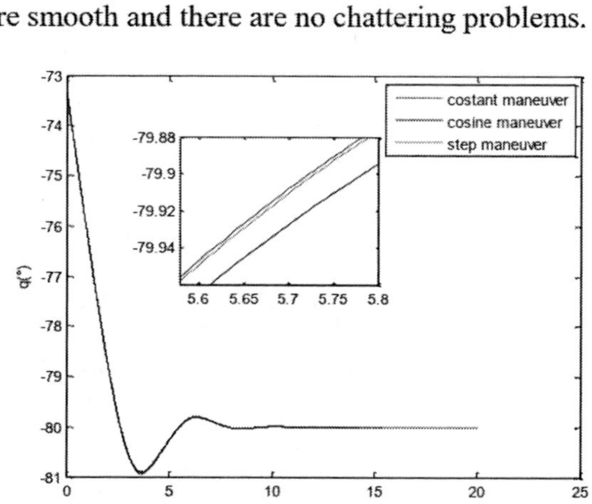

(b) LOS angle

IOP Conf. Series: Journal of Physics: Conf. Series **1267** (2019) 012081 doi:10.1088/1742-6596/1267/1/012081

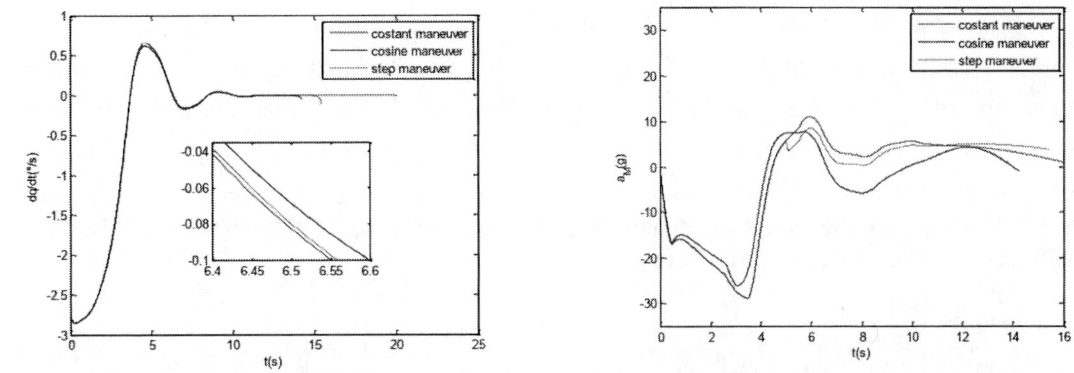

(c) LOS angular rate (d) Missile normal acceleration

Figure 2. Terminal guidance simulation for intercepting different maneuvering targets

4.2 Comparison with other guidance laws

In order to demonstrate the superiority of ALAISMG, comparisons with a adaptive integral sliding mode guidance law (AISMG) in [10] and a sliding mode guidance with first-order autopilot lag (ALSMG) in [11] are made in this subsection. The acceleration of target is selected as a_T=5g cos(4π/t).

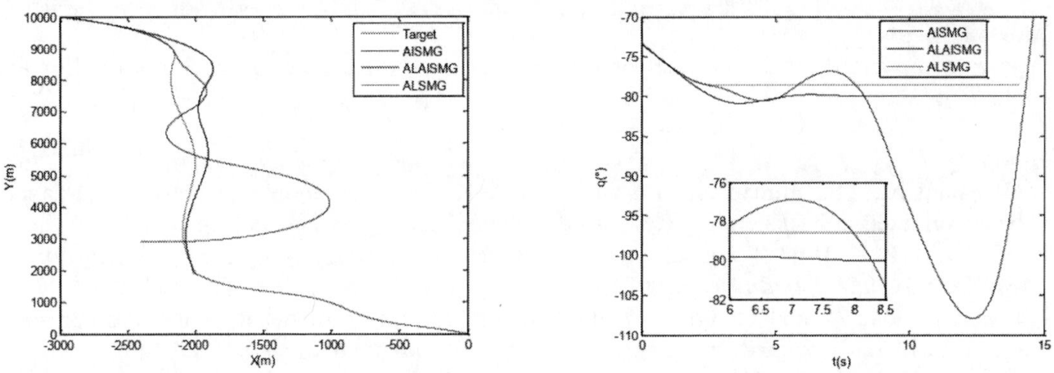

(a) Trajectories of missile and target (b) LOS angle

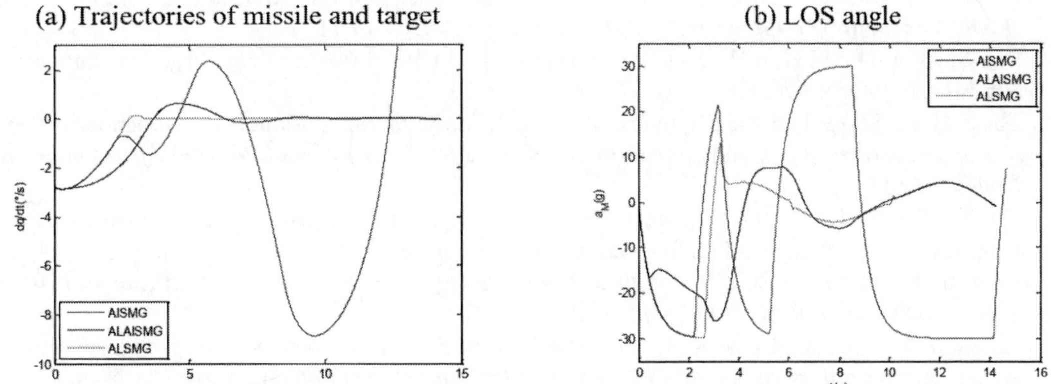

(c) LOS angular rate (d) Missile normal acceleration

Figure 3. Terminal guidance simulation comparison of different guidance laws

Figure 3(a) shows that when considering autopilot lag, both ALAISMG and ALSMG can enable the missile to intercept the target successfully, while AISMG fails to intercept the target which demonstrates that autopilot lag is a key factor affecting the terminal guidance performance. Figure

3(b)(c) show that AISMG fails to satisfy the finite time convergence characteristics of q and $\dot{q}$. Although q and $\dot{q}$ in ALSMG has a faster convergence rate than ALAISMG, q in ALSMG can't converge to the desired LOS angle q_d in finite time because ALSMG does not consider the impact angle constraint. Figure 3(d) shows the normal acceleration a_{Mq} curves of the missile. It can be seen that the curve of AISMG changes sharply and is always saturated. The curves of ALAISMG and ALSMG fluctuate in the early stage and decrease to near zero in the later stage. Besides, the ALAISMG curve is gentler and smoother.

5. Conclusions

A terminal guidance law with first-order autopilot lag is proposed, which enables the interception of maneuvering targets with a specified impact angle. A new type of integral sliding mode composed of LOS angle tracking error, LOS angular rate and LOS angular acceleration is constructed. The guidance proposed can ensure that the LOS angle tracking error and the LOS angular rate converge to zero in a finite time, which solves the problem of impact angle constraint and autopilot lag simultaneously on the basis of a single sliding surface. The simulation results show that the guidance law designed in this paper can intercept different maneuvering targets with the desired impact angle. Compared with the existing guidance laws, the feasibility of the guidance law proposed is further verified.

Acknowledgments

This paper is supported by the Project for the National Natural Science Foundation of China under Grant No. 61473306.

Reference

[1] Golestani M, Mohammadzaman I and Vali A R 2014 Finite-time convergent guidance law based on integral backstepping control *Aerospace Science and Technology* **39** 370

[2] Aggrawal R K 1998 Terminal guidance algorithm for ramjet-powered missiles *Journal of Guidance, Control, and Dynamics* **21** 862

[3] Koren A, Idan M and Golan O M 2008 Integrated sliding mode guidance and control for a missile with on-off actuators *Journal of Guidance, Control, and Dynamics* **31** 204

[4] Chaw D Y, Chol J Y 2003 Adaptive nonlinear guidance law considering control loop dynamics *IEEE Transactions on Aerospace and Electronic Systems* **39** 1134

[5] Sun S, Zhou D and Hou W T 2013 A guidance law with finite time convergence accounting for autopilot lag *Aerospace Science and Technology* **25** 132

[6] Zhou H B, Song J H and Song S M 2018 Sliding mode guidance law considering missile dynamic characteristics and impact angle constraints *International Journal of Automation and Computing* **15** 218

[7] Yu S H, Yu X H and Shirinzadeh B 2005 Continuous finite time control for robotic manipulators with terminal sliding mode *Automatica* **41** 1957

[8] Bhat S, Bernstein D 2005 Geometric homogeneity with applications to finite time stability *Mathematics of Control Signals and Systems* **17** 101

[9] Zhang Y X, Liu J and Zou K S 2016 Sliding-mode guidance law with impact angle constraint accounting for autopilot lag *28th Chinese Control and Decision Conference Yinchuan* 637

[10] Zhao B, Zhou J and Lu X D 2017 Adaptive integral sliding mode guidance law considering impact angelconstraint *Control and Decision* **32** 1695

[11] Song J H, Song S M 2015 Adaptive sliding mode guidance law with inputconstraints and autopilot lag *Journal of Chinese Inertial Technology* **23** 339

IOP Conf. Series: Journal of Physics: Conf. Series **1267** (2019) 012082 doi:10.1088/1742-6596/1267/1/012082

Control Strategy Comprehensively Considering Fuel Economy and Drive Comfort of Duel-Axis-Parallel PHEV

K S Xi[1], Y Zou[1,*], Y F Zhou[2], Z G Zhang[3] and H H Lei[3]

[1]National Engineering Laboratory for Electric Vehicles, School of Mechanical Engineering, Beijing Institute of Technology, NO.5 Zhongguancun south Street, Haidian , Beijing, China

[2]The Experimental High School Attached to Beijing Normal University, Jia NO.14 Erlong Road, Xicheng, Beijing, China

[3]Zhengzhou Foguang Electric Power Equipment CO., LTD, NO.50 Dongqing Street, High-tech development zone, Zhengzhou, China

[*]Corresponding author e-mail: zouyuanbit@vip.163.com

Abstract. In this paper, in order to comprehensively consider the economy and comfort of the special duel-axis-parallel PHEV, a vehicle control strategy composed of a shift strategy based on dynamic programming(DP), a rule-based energy management strategy and a dynamic coordinated control strategy based on motor torque compensation is proposed. A vehicle model which can show the dynamic characteristics of shifting and engine start – stop process is constructed for simulations. Shift strategy is extracted from the optimization results obtained by dynamic programming(DP). The energy management strategy takes the fuel economy of engine as a major consideration and the rule is determined by characteristics of engine, motor and battery. In the dynamic coordinated control strategy, an open-loop motor torque control based on clutch torque estimation is presented. Simulation is carried out on typical driving cycles. And the results indicate that the proposed vehicle control strategy can improve the fuel economy and reduce longitudinal impact caused by shifting and operated mode changing.

1. Introduction

Nowadays problems about energy exhaustion and air pollution are increasingly serious. As a good solution to these problems, new energy vehicles has received unprecedented attention [1]. PHEV which has the advantages of electric vehicles and traditional hybrid electric vehicles has achieved great development in recent years. Energy management for PHEV which is aimed at getting long driving distances through reasonable power distribution of power sources is a hot research topic. In addition, consumers have more requirements on the comfort of PHEV with the development of new energy vehicles.

There are mainly two categories of power management methods: rule-based power management and optimization based power management. Rule-based strategies are widely adopted in practical application because of reliability [2]. Instantaneous optimization control strategy has high real-time performance, like equivalent consumption minimization strategy (ECMS) [3]. Global optimal power management methods normally rely on the knowing of future driving cycle, such as dynamic programming (DP) [4, 5]. At the same time, aiming at improving the dynamic performance of PHEV during mode switching and gear shifting, some control methods which make full use of the advantages of multiple power sources are put forward [6, 7].

The remainder of this paper is organized as follows. In Section 2, the model of the powertrain system applied for simulation is described. In Section 3, vehicle control strategy composed of shift strategy, energy management and dynamic coordinated control is proposed. In Section 4, in order to prove the improvements in fuel economy, the Vehicle control strategy is compared with DP. And by comparing with the strategy without dynamic coordination control, it is shown that the drive comfort can be significantly improved by using dynamic coordination control. In Section 5, conclusions are given.

2. Model of Powertrain system

The structure of the duel-axis-parallel PHEV in this paper is shown in Figure 1.

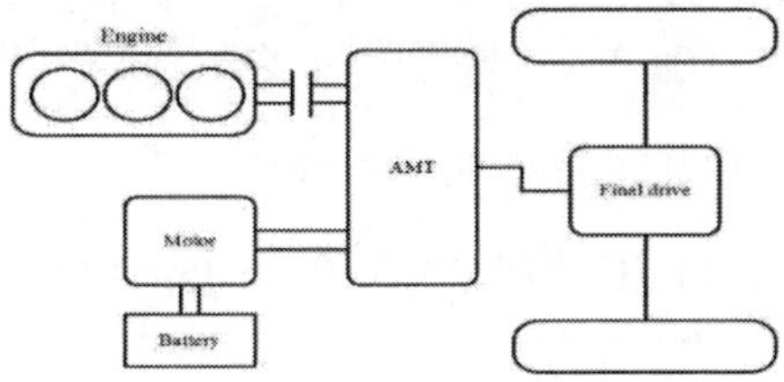

Figure 1. The structure of the duel-axis-parallel PHEV.

It is mainly composed of engine, power battery pack, motor, clutch, AMT with double shaft input and final drive. The engine and motor are connected to the AMT through two shafts. The engine torque and motor torque are coupled internally in the AMT by a variety of different gear combinations. After analyzing the characteristics of the structure of duel-axis-parallel PHEV, the following relationship can be obtained:

$$T_{req} = T_e I_{g\_e} I_0 + T_m I_{g\_m} I_0 \tag{1}$$

T_{req} (Nm) is vehicle required torque. T_e (Nm) is engine torque. T_m (Nm) is motor torque. I_{g_e} is transmission ratio of the engine road . I_{g_m} is transmission ratio of the motor road. I_0 is transmission ratio of final drive.

Engine, motor and battery are all modeled using normal methods. What needs special introduction is the method of clutch modelling. The three working states of clutch are separation, slip and engagement. The calculation formula of the transmission torque T_c of the clutch in three working states is as follows:

$$T_c = \begin{cases} 0 & disengaged \\ T_f & slipping\ and\ T_f \le T_{c\_\max} \\ T_e & engaged\ and\ T_e \le T_{c\_\max} \\ T_{c\_\max} & T_e > T_{c\_\max}\ or\ T_f > T_{c\_\max} \end{cases} \tag{2}$$

The calculation formula of the dynamic friction torque of the clutch T_f is as follows:

$$T_f = \mathrm{sign}(\omega_e - \omega_c) Z \mu_d F_N R_c + \lambda_e T_e, \lambda_e = \mathrm{e}^{-1.07|\omega_e - \omega_c| R_c} \tag{3}$$

ω_e (rad/s) is speed of engine. ω_c is speed of clutch driven shaft(rad/s). Z is number of friction surface. μ_a is dynamic friction coefficient. F_N (N) is clutch pressure. R_c (m) is equivalent friction radius. λ_e is impact factor of input torque.

3. Vehicle control strategy

3.1. Shift strategy

PHEV system can be discretized into a discrete dynamic system represented by the following formula:

$$x(k)=\begin{bmatrix} SOC(k) \\ g_e(k) \\ g_m(k) \end{bmatrix}, u(k)=\begin{bmatrix} T_{e\_req}(k) \\ shift_e(k) \\ shift_m(k) \end{bmatrix} \tag{4}$$

After analyzing the powertrain system, three state variables are selected. First one is state of charge of the battery $SOC(k)$ .Second one is engine gear $g_e(k)$ .Third one is motor gear $g_m(k)$ .Control vector is consists of engine required torque $T_{e\_req}(k)$, engine shift command $shift_e(k)$ and motor shift command $shift_m(k)$.

$$g_e(k+1)=\begin{cases} 5 & g_e(k)+shift_e(k)>5 \\ 1 & g_e(k)+shift_e(k)<1 \\ g_e(k)+shift_e(k) & otherwise \end{cases} \tag{5}$$

$$shift_e(k)=\begin{cases} -1 & downshift \\ 1 & upshift \\ 0 & otherwise \end{cases} \tag{6}$$

$g_m(k)$ and $shift_m(k)$ are similar to $g_e(k)$ and $shift_e(k)$.

In this paper, regardless of engine emissions, cost-to-go function and instantaneous cost function are defined as follows:

$$J_\pi(x_0)=\lim_{N\to\infty} E\left\{\sum_{k=0}^{N-1} L(x(k),u(k))\right\} \tag{7}$$

$$L(x(k),u(k))=L_{fuel}(k)+\alpha L_{gs\_e}(k)+\beta L_{gs\_m}(k) \tag{8}$$

N is the duration of the driving cycle. Drive comfort will be reduced if shifting frequently, therefore an additional cost function is adopted to avoid frequent shift. This additional cost function is as follows, α and β are the weighting factors.

$$\alpha L_{gs\_e}(k)=\alpha\left|shift_e(k)\right|, \beta L_{gs\_m}(k)=\beta\left|shift_m(k)\right| \tag{9}$$

After the above analysis, we can get the objective function of dynamic programming optimization as follows:

$$J_\pi(x_0)=\lim_{N\to\infty} E\left\{\sum_{k=0}^{N-1}\left[L_{fuel}(k)+\alpha\left|shift_e(k)\right|+\beta\left|shift_m(k)\right|\right]\right\} \tag{10}$$

The system constraint is as follows:

$$\begin{cases} \omega_{e\_\min} \le \omega_e(k) \le \omega_{e\_\max} \\ \omega_{m\_\min} \le \omega_m(k) \le \omega_{m\_\max} \\ SOC_{\min} \le SOC(k) \le SOC_{\max} \\ T_{e\_\min}(\omega_e(k)) \le T_e(k) \le T_{e\_\max}(\omega_e(k)) \\ T_{m\_\min}(\omega_m(k),SOC(k)) \le T_m(k) \le T_{m\_\max}(\omega_m(k),SOC(k)) \end{cases} \tag{11}$$

According to the optimization results of DP, the working point of the optimal gear position is drawn in Figure 2. The dividing line between the different points is the shift rule curve.

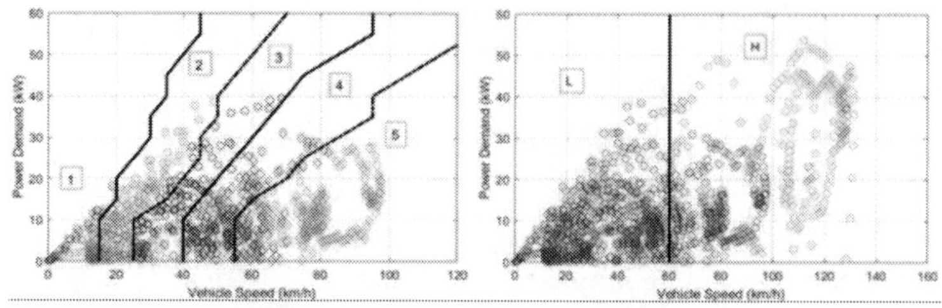

Figure 2. Optimization results of DP.

In order to avoid frequent shifting, the curve above is shifted to the left to get the downshift curve. The final upshift and downshift curves are shown in Figure 3:

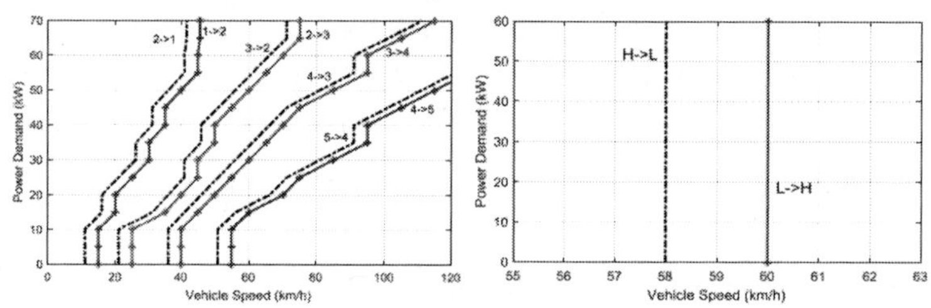

Figure 3. Upshift and downshift curves.

3.2. Energy management strategy

Energy management in this paper is the CD-CS mode which is widely used in PHEV. CD mode and CS mode are divided into several submodes, including motor drive mode, engine drive mode, motor assisted drive mode, engine assisted drive mode and charge mode. The conditions for switching between modes are predefined. Changes in external input and vehicle status variables will lead to switching between modes. There is a specific torque allocation strategy according to the characteristics of the engine and motor based on experiments in each submode, which makes the engine and motor work in a high efficiency range.

3.3. Dynamic coordinated control strategy

Actual output torque of the engine and clutch pressure signal can be obtained by sensors. It is difficult to control engine torque. In order to prevent engine torque from changing too much in a short time, control strategy of engine torque based on slope restrictions is adopted. Then the collected engine torque signal and clutch pressure signal are used to calculate the clutch transfer torque which the torque of motor needed is determined by.

During the mode switching process with engine start, the motor drives the vehicle and starts the engine at the same time. The control strategy is as follows:

$$\begin{cases} T_{e\_req} = 0 & \omega_e < 1000 \\ T_{e\_req} = \min(k_e t, T_{e\_tar}(t_f)) & \omega_e \geq 1000 \end{cases}$$
$$T_{m\_req} = \frac{T_{req} - T_c I_{g\_e} I_0}{I_{g\_m} I_0} \tag{12}$$

During the mode switching process with engine stop, the control strategy is as follows:

$$T_{e\_req} = \max((T_{e\_tar}(t_0) - k_e t), 0)$$
$$T_{m\_req} = \frac{T_{req} - T_c I_{g\_e} I_0}{I_{g\_m} I_0} \tag{13}$$

During the process of shifting, engine torque drops to zero at slope k_e at the beginning. then shift operation takes place. When shifting is completed, engine torque rises to target torque at slope k_e .The motor compensates the torque during the whole process.

$$\begin{cases} T_{e\_req} = \max((T_{e\_tar}(t_0) - k_e t), 0) & 0 < t \le t_1 \\ 0 & t_1 < t \le t_2 \\ T_{e\_req} = \min(k_e(t - t_2), T_{e\_tar}(t_f)) & t_2 < t \le t_f - t_0 \end{cases}$$
$$T_{m\_req} = \frac{T_{req} - T_c I_{g\_e} I_0}{I_{g\_m} I_0} \tag{14}$$

$T_{e\_tar}(t_0)$ is the engine target torque at the beginning of the dynamic coordinated control. $T_{e\_tar}(t_f)$ is the engine target torque at the end of the dynamic coordinated control. t is the time since the dynamic coordinated control occurs. T_{e_req} is engine required torque. T_{m_req} is motor required torque.

4. Simulation results

Because DP is a kind of global optimization method, which means the policy given by DP can be considered as the optimal policy. So DP based PHEV energy management can serve as a bench mark of fuel economy for comparison with simulation results of proposed method.4 × WLTP are used as the experimental driving cycle in simulation. The initial SOC is set to 0.95 and the final SOC is set to 0.26.

The simulation results of DP method and proposed method are shown in Table 1. The cost of proposed method one driving cycle takes is 6.43% higher than the optimal results calculated by DP. Because the cost is close between proposed method and DP method, it can be considered that the control strategy presented in this paper has satisfactory fuel economy.

Table 1. Results comparison of two methods.

	Driving cycle	Mileage	Cost	Comparison
DP method	4WLTP	93.07km	2.80L/100km	6.43% higher than
Proposed method	4WLTP	93.07km	2.98L/100km	DP

In order to objectively evaluate the driving performance of vehicles, the longitudinal jerk of vehicles is used as the evaluation index of driving performance in this paper. Jerk is defined as the inverse of the longitudinal acceleration of a vehicle. The formula is as follows:

$$J = \frac{da}{dt} = \frac{d^2v}{dt^2} \tag{15}$$

Take the mode switching process with engine start as an example. Figure 4 shows the simulation results without dynamic coordinated control when the engine starts. Output torque can't meet the demand torque without the compensating torque provided by the motor which causes the speed fluctuation. And the maximum jerk is up to 15 m/s^3 .Figure 5 shows the simulation results with dynamic coordinated control. The output torque can be maintained stable and the maximum jerk is 4 m/s^3 . By comparing the results in two simulations, it is clear that the dynamic performance has been improved.

IOP Conf. Series: Journal of Physics: Conf. Series **1267** (2019) 012082 doi:10.1088/1742-6596/1267/1/012082

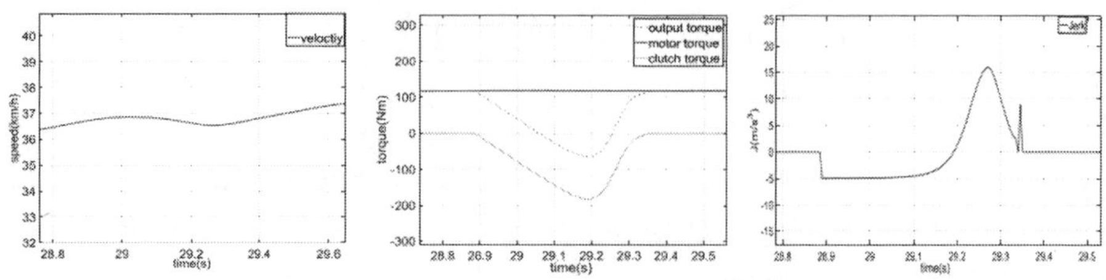

Figure 4. Results without dynamic coordinated control.

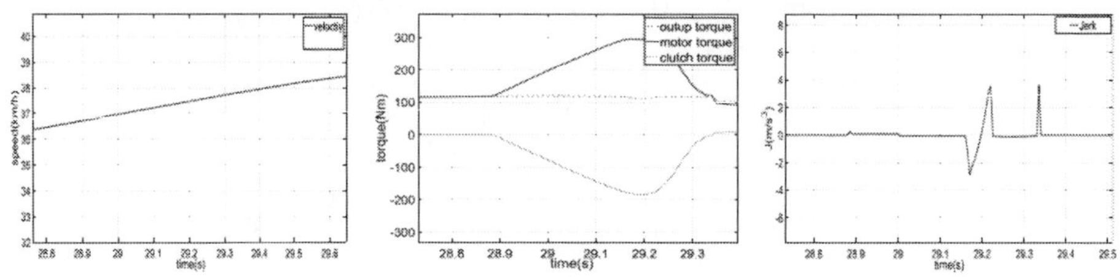

Figure 5. Results with dynamic coordinated control.

5. Conclusions

This paper presents a control strategy for PHEV which consists of the shift strategy extracted from optimization results obtained by DP, the energy management strategy based on rules determined by experiments and the dynamic coordinated control strategy based on motor torque compensation. A specific model which can reflect the transient and steady-state performance of the PHEV is constructed for simulations. The simulation results illustrate that the proposed control strategy has satisfactory fuel economy and can notably improve the power performance and drive comfort during the process of mode switching and shifting.

Acknowledgments

This work was financially supported by SAIC Motor Commercial Vehicle Co., Ltd. Technical Center (Contract No.7200025538).

References

[1] Ferrero E, Alessandrini S and Balanzino A 2016 Impact of the electric vehicles on the air pollution from a highway *Appl. Energy* **169** 450–9

[2] Peng J K, He H W and Xiong R 2016 Rule based energy management strategy for a series–parallel plug-in hybrid electric bus optimized by dynamic programming *Appl. Energy* **185** 1633–43

[3] Geng B, Mills J K and Sun D 2011 Energy management control of micro turbine-powered plug-in hybrid electric vehicles using the telemetry equivalent consumption minimization strategy *IEEE Trans. Veh. Technol.* **60** 4238–48

[4] Wang X M, He H W, Sun F C and Zhang J L 2015 Application study on the dynamic programming algorithm for energy management of plug-in hybrid electric vehicles *Energies* **8** 3225–44

[5] Liu J M and Peng H 2008 Modeling and Control of a Power-Split Hybrid Vehicle *IEEE Trans. Control Syst. Technol.* **16** 1242–51

[6] Davis R I and Lovenz R D 2003 Engine torque ripple cancellation with an integrated starter alternator in a hybrid electric vehicle: implementation and control *IEEE Trans. Ind. Appl.* **39** 1765–74

[7] Hwang H S, Yang D H, Choi H K, Kim H S and Hwang S H 2011 Torque control of engine clutch to improve the driving quality of hybrid electric vehicles *Int. J. Automot. Technol.* **12** 763–8

IOP Conf. Series: Journal of Physics: Conf. Series **1267** (2019) 012083 doi:10.1088/1742-6596/1267/1/012083

Short-Term Photovoltaic Generation Forecasting Based on LVQ-PSO-BP Neural Network and Markov Chain Method

Xiran Wang[1,*], Xue Ma[2], Suhua Lou[1], Fei Peng[2] and Song Wu[1]

[1]State Key Laboratory of Advanced Electromagnetic Engineering and Technology (College of Electrical and Electronic Engineering, Huazhong University of Science and Technology), Wuhan 430074, Hubei Province, China

[2]State Grid Qinghai Electric Power Company Economic Research Institute, Xining 810000, Qinghai Province, China

[*]E-mail: 1131250221@qq.com

Abstract. With the rapid development of solar photovoltaic generation, the effective prediction of photovoltaic is of great significance to mitigate its impact on power system. According to the analysis of main factors which affect power output of photovoltaic system, a short-term power forecasting model based on back propagation(BP) neutral network and LVQ-PSO-BP neural network and Markov chain method was established. The weather is clustered and distinguished by using learning vector quantization(LVQ) and the particle swarm optimization(PSO) is used to optimize BP neural network weights and thresholds, improving forecasting network training speed. Finally, daily predictive value is corrected by Markov chain method to improve short-term photovoltaic generation forecasting precision. The simulation results indicate that the proposed method can accelerate the speed of searching optimums, improving the classification accuracy of weather types and the precision of the photovoltaic generation output effectively.

1. Introduction

Photovoltaic output is affected by many meteorological factors, such as solar irradiance and atmospheric temperature. The uncertainty of photovoltaic output brings a series of problems to power balance and safe operation of power system. Therefore, the effective prediction of photovoltaic output is of great significance for realizing the combined operation of photovoltaic and traditional power sources.

Reference[1]-[2]establishes the short-term photovoltaic generation forecasting model without irradiation by adopting the back propagation(BP) neural network with humidity, temperature and other meteorological factors. However, the BP neural network is sensitive to the initial weight and is prone to converge on local optimum. Reference[3]-[4] establish the photovoltaic output model based on support vector machine(SVM)considering the factors affecting the output of photovoltaic. However, the selection of model parameters had a great influence on the prediction effect and the method only using a single model to predict different weather types will also reduced the prediction accuracy.

In order to solve these problems, a short-term photovoltaic generation forecasting method is proposed based on LVQ-PSO-BP neural network and error correction of Markov theory. First, the weather is clustered and distinguished by using learning vector quantization(LVQ) and then particle swarm optimization(PSO) is used to optimize back propagation(BP) neural network weights and thresholds, improving forecasting network training speed. Finally, the daily predictive value is corrected by Markov chain method to improve short-term photovoltaic generation forecasting precision.

2. Correlational analyses between photovoltaic output and meteorological factors

The photovoltaic output is related to meteorological factors such as solar radiation intensity, and temperature. If all the meteorological factors are taken as input variables of the prediction model, the complexity of the model will increase and the convergence speed will slow down. Therefore, in this paper we use the grey correlation coefficient method to figure out the meteorological factors which have great correlation with the output of photovoltaic output.

Table 1. Result of correlation of meteorological factors and PV generation output.

VARIABLE	RA	TEMP	HR	SPD	STP	KT	PRCP	P_{solar}
RA	1	0.765	0.650	0.684	0.705	0.715	0.514	0.720
TEMP	0.764	1	0.682	0.692	0.689	0.612	0.553	0.709
HR	0.645	0.678	1	0.752	0.654	0.544	0.604	0.590
SPD	0.684	0.693	0.756	1	0.657	0.556	0.618	0.598
STP	0.705	0.690	0.660	0.657	1	0.710	0.470	0.666
KT	0.715	0.613	0.550	0.556	0.710	1	0.407	0.783
PRCP	0.514	0.554	0.610	0.618	0.470	0.407	1	0.447
P_{solar}	0.720	0.709	0.596	0.598	0.666	0.783	0.447	1

From the Table 1 we can see that the maximum correlation between output power and clear sky index of photovoltaic system is 0.783; the correlation with solar radiation intensity is 0.720, followed by temperature correlation, which is 0.709. Therefore, we select the key factors such as clear sky index, solar radiation intensity and temperature as input data for short-term prediction of photovoltaic output.

3. Forecasting model of photovoltaic generation

3.1. Weather clustering based on LVQ

The structure of LVQ neural network is shown in Figure 1. LVQ neural network consists of input layer, competitive layer and output layer. When a sample enters the LVQ network, the neurons in the competitive layer generate the winning neurons through the competition rule and the winning neurons have a state of 1 and the others are 0. The output neurons connected with the winning neurons are 1 and the others are 0. Each output neuron represents different classes, and the neuron with state 1 gives the classes of samples so as to achieve the purpose of sample classification.

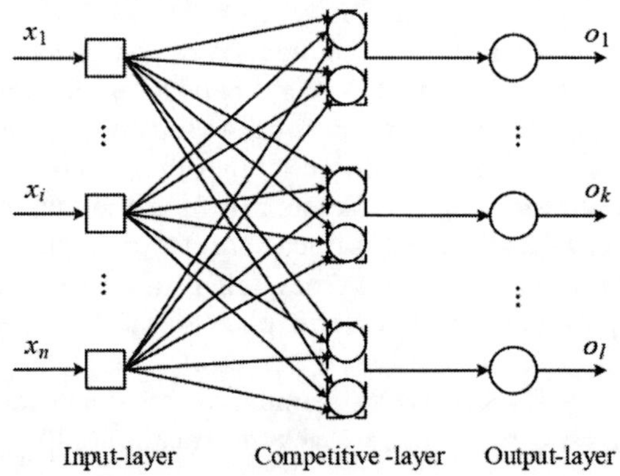

Figure 1. The structure of LVQ neural network.

According to the classification criteria of weather types formulated by the China Meteorological Administration (CMA), the weather conditions can be divided into 33 different types [5]. In view of the fact that too many classifications will lead to few samples and complicated calculation process, this paper divides 33 weather types into four types: sunny, cloudy, overcast and rainy.

3.2. PSO-BP neural network

Although BP neural network has good ability to approximate non-linear mapping, it's essentially a gradient descent method. BP neural network has a slow convergence speed and it's prone to converge on local optimum. Therefore, this paper uses PSO algorithm to optimize BP neural network by training the network to get the initial weights and thresholds. On this basis, BP algorithm is used to further refine and optimize the network parameters until the optimal network parameters are obtained.

The process of PSO algorithm optimizing the BP neural network is as follows:

1) Design the structure of BP neural network;

2) Determine the number of particles *N*, threshold c_1 、 c_2, inertia weight ω.

3) Optimize the network with PSO. The position vector of PSO includes the connection weight and threshold of BP network.

4) Take the optimized network weights and thresholds as the initial weights and thresholds of BP algorithm.

5) Implement BP neural network optimization until the network achieves the performance index. Save the connection weights and thresholds and carry out prediction simulation.

3.3. Error correction based on Markov chain

Markov theory combines the initial state of the system and predicts the state of the system at a certain time in the future through the transition probability matrix. It is suitable for tracking and predicting the random discrete time series. Considering that abrupt weather may affect the prediction results of photovoltaic power station output, this paper corrects the prediction values of BP neural network based on multi-order weighted Markov chain model, so as to ensure the prediction accuracy of abrupt weather. The detailed process of multi-order weighted Markov chain can refer to the reference [6].

4. Training and evaluation of model

4.1. LVQ clustering network training

Step 1: Set the input data and Initialization. Determine the input vector, learning efficiency η and iteration times. In this paper, the input data include clear sky index, solar radiation intensity, temperature and precipitation. Initialize the weights with smaller random values and normalize the weights and input vectors.

Step 2: Find the winning neuron j^*. Calculate the scalar product of the input samples and the connecting weight, the largest of which can be served as the winning neuron j^*. Step 3: update the weights. Adjust the weights according to whether the classification is correct or not.

If the classification is correct, the weight matrix moves in the direction of the input vector, the formula of which can be expressed as:

$$\omega_{j^*}^1(t+1)=\omega_{j^*}^1(t)+\eta(t)\left[P-\omega_{j^*}^1(t)\right] \tag{1}$$

If the classification error occurs, the weight matrix moves away from the input vector, the formula of which can be expressed as:

$$\omega_{j^*}^1(t+1)=\omega_{j^*}^1(t)-\eta(t)\left[P-\omega_{j^*}^1(t)\right] \tag{2}$$

Step 4: determine whether it convergences. Determine whether the number of iterations reaches the preset value. If the maximum number of iterations is not reached, then go back to step 3, otherwise the calculation will be completed.

After the training, using the trained network to identify the weather types of the day with undefined labels to determine the appropriate prediction model.

4.2. PSO-BP neural network training

BP neural network uses a three-layer network structure with a single hidden layer. The input layer is meteorological data with 18 neuron nodes, including daily clear sky index, minimum/maximum temperature and solar radiation intensity at 7:00-21:00. The output layer is the photovoltaic output power with 15 neuron nodes. And the optimal number of nodes in the hidden layer is determined to be 12 by the equation (3).

$$p=\sqrt{\mathrm{m}+\mathrm{n}}+a \tag{3}$$

where m is the number of neuron nodes in the output layer; n is the number of neuron nodes in the output layer; a is the constant in the range of $[1,10]$.

According to the structure of BP neural network, there are 12 neurons in the hidden layer and 15 neurons in the output layer, so the total number of connection weights W_{ij} are 216(18×12) and the number of weights W_{jk} are 180(12×15).In addition, there are 12 thresholds θ_j and 15 thresholds α_k.Therefore, the space dimensions of initialization swarm are 423, which includes all the connection weights and thresholds of BP neural network.

In this paper, the absolute value of prediction error is used as the individual fitness function to train the error sum of samples, which can be expressed as in equation (4). Then search for the optimal particle by iteration calculation as the initial connection weight and threshold of BP neural network.

$$F=\mathrm{g}\left[\sum_{k=1}^{\mathrm{n}}\mathrm{abs}(Z_k-Y_k)\right] \tag{4}$$

where g is constant coefficient; n is the number of nodes in the output layer; Z_k is the expected output of the output node K; Y_k is the forecasting output power of the output node K.

PSO has the better local search ability with the smaller inertia weight and the better global search ability with the larger inertia weight.

4.3. Evaluation indicators for predicted performance evaluation indicators

In this paper, we use Mean Absolute Percent Error (MAPE) and Root Mean Square Error (RMSE) to evaluate the accuracy of photovoltaic power generation forecasting.

$$\mathrm{MAPE}=\frac{1}{\mathrm{N}}\sum_{i=1}^{\mathrm{N}}\left|\frac{P_{pi}-P_{mi}}{P_{mi}}\right|\times100\% \tag{5}$$

$$\mathrm{RMSE}=\frac{1}{P_{\mathrm{N}}}\left(\frac{1}{\mathrm{N}}\sum_{i=1}^{\mathrm{N}}(P_{pi}-P_{mi})^2\right)^{\frac{1}{2}} \tag{6}$$

where P_{pi} is the forecasting output power at i-th time point; P_{mi} is the actual power at i-th time point of the photovoltaic power station; N is the total number of predicted samples.

5. Results and discussion

5.1. Weather clustering based on LVQ

In the case study, the historical meteorological data are provided from January 2014 to December 2014 in a certain area of Qinghai Province .Taking summer as an example, the actual effective data consist 92 days, of which there are 31 sunny days, 26 cloudy days, 15 overcast days and 20 rainy days respectively. We adopt the cross-validation method to carry out four clustering calculations with dividing training set and test set in proportion of 3:1.

The comparison results of weather clustering based on K-means algorithm and LVQ network are

IOP Conf. Series: Journal of Physics: Conf. Series **1267** (2019) 012083 doi:10.1088/1742-6596/1267/1/012083

shown in Table 2.We can see that the accuracy of weather clustering by K-means algorithm is lower than that of LVQ network and the sample data can easily deviate from other clustering centers when using K-means.

Table 2. Comparison of clustering results of LVQ and K-means.

Actual weather	K-means cluster				LVQ cluster			
	Sunny	Cloudy	Overcast	Rainy	Sunny	Cloudy	Overcast	Rainy
Sunny	19	5	7	0	31	0	0	0
Cloudy	9	11	6	0	2	24	0	0
Overcast	0	4	8	3	0	0	14	1
Rainy	0	2	1	17	0	1	2	17

5.2. Discussions of simulation results

Taking the photovoltaic power forecasting model in summer as an example, select the sub-models of corresponding weather types and input the meteorological data of the predicted day into the prediction system shown in Figure 1 to obtain the forecasting output power. In order to verify the effectiveness of the proposed method, we adopt the traditional BP neural network and PSO-BP neural network as comparison models. Table 3 lists the errors of three forecasting models.

Table 3. Prediction errors of different models.

Model	BP		PSO-BP		PSO-BP-Markov	
Weather	MAPE	RMSE	MAPE	RMSE	MAPE	RMSE
Sunny	0.0815	0.1452	0.0463	0.0948	0.0373	0.0551
Cloudy	0.109	0.1746	0.0955	0.1499	0.0873	0.1045
Overcast	0.3685	0.3975	0.1247	0.2459	0.1077	0.1489
Rainy	0.4476	0.6771	0.1903	0.2962	0.1279	0.1781

Figure 2 shows the forecast results of typical weather types in summer. Comparing the prediction results of the three models, we can see that the prediction effect of model 3 is better than that of model 2 and model 1. The forecasting errors of the three models are relatively small but the forecasting accuracy in cloudy day decreases. In overcast and rainy days, the fluctuation of photovoltaic output makes it more difficult to predict. The forecasting output of model 1 deviates from the actual output seriously with the error increasing sharply. Compared with model 1, the forecasting result of model 2 is improved but still not ideal. However, the model 3 based on error correction of Markov chain is sensitive to the variation of error, which can accurately track the drop of photovoltaic output and further improve the forecasting accuracy of photovoltaic output.

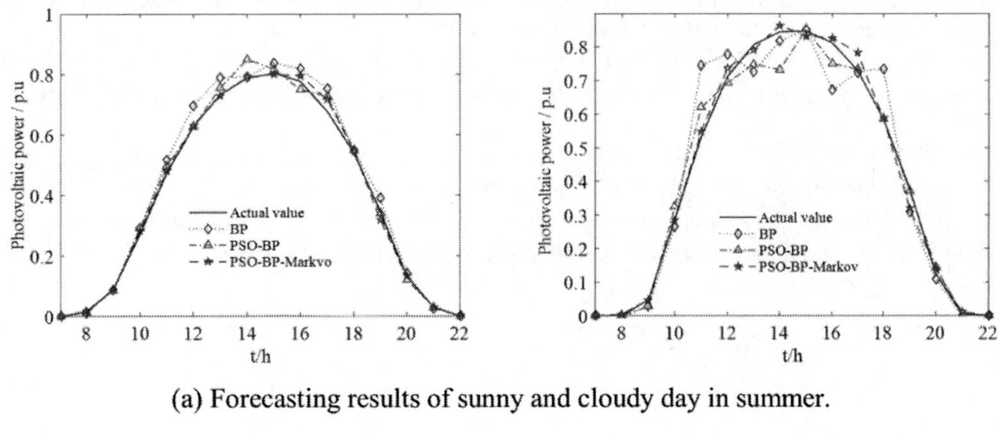

(a) Forecasting results of sunny and cloudy day in summer.

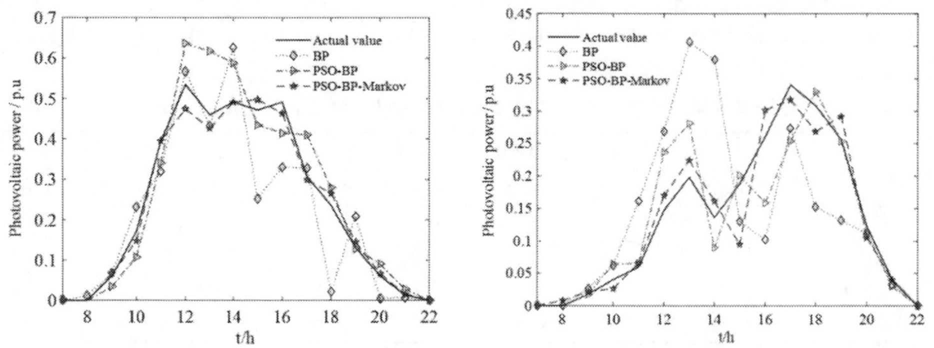

(b) Forecasting result of overcast and rainy day in summer.

Figure 2. Forecast results of typical weather types in summer.

6. Conclusion

This paper analyses the correlation between photovoltaic generation system and many meteorological factors and establishes a short-term photovoltaic output forecasting model based on LVQ-PSO-BP neural network and Markov chain error correction. We adopt the LVQ network to cluster all weather types according to season, which effectively improves the clustering accuracy of each weather type. And the PSO algorithm is used to optimize the initial weights and thresholds of BP neural network, which effectively overcomes the deficiencies of the traditional BP neural network of slow convergence speed and easily falling into local optimum. Moreover, we adopt the multi-order Markov chain correct the forecasting errors for the further improvement of the forecasting accuracy, making up for the shortcomings of the low accuracy prediction of traditional methods. Therefore, the short-term forecasting model proposed in this paper, which is based on LVQ-PSO-BP network and Markov chain error correction, has a better accuracy and applicability.

Acknowledgement

This paper is supported by State Grid Corporation Science and Technology Project (522830170009).

References

[1] Lorenz E, Hurka J, Heinemann D, et al. Irradiance forecasting for the power prediction of grid connected photovoltaic systems[J]. IEEE Journal of Selected Topics in Applied Earth Observations and Remote Sensing, 2009, 2(1):2-10

[2] Amrouche B, Pivert X L. Artificial neural network based daily local forecasting for global solar radiation[J]. Applied Energy, 2014, 130(5): 333-341

[3] Shi J, Lee W J, Liu Yongqian, et al. Forecasting power output of photovoltaic systems based on weather classification and support vector machines[J]. IEEE Transactions on Industry Applications, 2012, 48(3):1064-1069

[4] Gao K L, Liu J M and Xu R Z 2011 A hybrid security situation prediction model for information network based on support vector machine and particle swarm optimization[J]. Power System Technology 35(4):176-182

[5] WANG S X, ZHANG N. Short-term output forecast of photovoltaic based on a grey and neural network hybrid model[J]. Automation of Electric Power Systems, 2012, 36(19):37-41

[6] T Jin, D Changhong, Y Wei,et al. Ultra-short-term Photovoltaic Power Forecasting in Microgrid Based on Adaboost Clustering[J]. Automation of Electric Power Systems,2017, 41(21): 33-39

IOP Conf. Series: Journal of Physics: Conf. Series **1267** (2019) 012084 doi:10.1088/1742-6596/1267/1/012084

The Optimal Control Problems of Exponentially Damped System

Guozhong Xiu[1,*], Shi Bao[1], Gang Zhou[1] and Liying Wang[1]

[1]Institute of System Science and Mathematics, Naval Aeronautical University, Yantai 264001, P.R.China

*E-mail: xiuguozhong2013@163.com

Abstract. The optimal control problems of exponentially damped system are considered in this paper. Exponentially damped system involves convolution integrals over exponentially decaying functions, which is used as damping models in viscoelastically damped structures. The dynamic constraint is in the form of a differential equation that includes integer derivatives and the "integral" term. The performance index considered is a function of both the state and the control variables. The traditional state-space approach is extended to the optimal control problems of exponentially damped system, A direct numerical technique is used to solve the resulting equations. Numerical simulations are provided to illustrate the above control design.

1. Introduction

Viscoelastic materials are widely used to control or reduce vibrations and sound radiation in aerospace structures, industrial machines, civil engineering structures, etc. Of many nonviscously damping models, the convolution integral model is possibly the most general model within the scope of linear analysis. The integral constitutive models are derived based on the materials properties of stress relaxation and creep applying Boltzmann's superposition principle. The stress relaxation functions and creep functions are memory and hereditary kernels in the integral constitutive equations. They can be expressed by a series of exponential functions [1,2], power-law functions [3], Mittage-Leffler functions [3], or other types of functions. The integral constitutive relations can be readily used as damping models in viscoelastically damped structures. Equations of motion of such systems are a set of coupled second-order integro-differential equations. The presence of the "integral" term makes the vibration analysis and control design more complicated than the classical ones. The integral type damping models may be also called nonviscously damping models and the corresponding oscillators are called nonviscously damped oscillators.

Researches on nonviscously damped oscillators are mainly concentrated on two types: one is the exponentially damped oscillators where the damping force are expressed by exponentially fading memory kernel; and the other is the fractional-order oscillators where the viscoelastic relaxation functions are characterized by power-law functions or Mittage-Leffler functions. S. Adhikari and his colleagues have systematically investigated the structural dynamics with exponentially damped models, including dynamics of exponentially damped single degree of freedom and multi-degree-of-freedom systems, identification and quantification of damping in [4-6].

In recent years, Various fractional control techniques have been proposed, such as CRONE control, fractional PID control, fractional sliding control [7], fractional adaptive control [8], fractional optimal control [9], etc. However, control designs for exponentially damped systems are very limited. This paper is going to propose the optimal control problems of exponentially damped system, Furthermore, This control method can also be applied to N-order exponentially damped system containing multiple integral teams.

2. State-space representation

To describe constitutive relation for viscoelastic material, the following nonviscous damping model that depend on the past history of motion via convolution integral over kernel function is adopted.

$$\sigma(t)=\int_0^t G(t-\tau)\mathrm{d}\varepsilon(t) \tag{1}$$

Where, $\sigma(t)$ is the stress, $\varepsilon(t)$ is the stain, $G(t)$ is the stress relaxation function. Here, we will use a damping model for which the kernel function is exponentially damped model.

$$G(t)=\sum_{k=1}^{n} C_k \mu_k e^{-\mu_k t} \tag{2}$$

Where, $\mu_k \in \mathrm{R}^+$ are known as the relaxation parameters, and n denotes the number of relaxation parameters used to described the damping behavior. $C_k \in \mathrm{R}^{N\times N}$ are the damping coefficient matrices. We consider such n-order exponentially damped system.

$$\begin{aligned}&\frac{\mathrm{d}^n x(t)}{\mathrm{d}t^n}+a_{n-1}\frac{\mathrm{d}^{n-1}x(t)}{\mathrm{d}t^{n-1}}+\cdots+a_1\frac{\mathrm{d}x(t)}{\mathrm{d}t}+c_m\int_0^t \mu_m e^{-\mu_m(t-\tau)}\dot{x}(\tau)\mathrm{d}\tau\\&+\cdots+c_1\int_0^t \mu_1 e^{-\mu_1(t-\tau)}\dot{x}(\tau)\mathrm{d}\tau+a_0x(t)=u(t)\end{aligned} \tag{3}$$

Where, n is an integer. Assuming $x=x_1$, $\frac{\mathrm{d}x}{\mathrm{d}t}=x_2$, $\frac{\mathrm{d}^2x}{\mathrm{d}t^2}=x_3,\cdots,$ $\frac{\mathrm{d}^{n-1}x}{\mathrm{d}t^{n-1}}=x_n$. the above equation can be expressed in the following form

$$\begin{aligned}&\frac{\mathrm{d}}{\mathrm{d}t}\begin{bmatrix}x_1\\x_2\\\vdots\\x_n\end{bmatrix}+\begin{bmatrix}0&0&\cdots&0\\0&0&\cdots&0\\\vdots&\vdots&\vdots&\vdots\\c_1&0&0&0\end{bmatrix}\int_0^t \mu_1 e^{-\mu_1(t-\tau)}\begin{bmatrix}\dot{x}_1\\\dot{x}_2\\\vdots\\\dot{x}_n\end{bmatrix}\mathrm{d}\tau+\cdots+\begin{bmatrix}0&0&\cdots&0\\0&0&\cdots&0\\\vdots&\vdots&\vdots&\vdots\\c_{m-1}&0&0&0\end{bmatrix}\int_0^t \mu_{m-1} e^{-\mu_{m-1}(t-\tau)}\begin{bmatrix}\dot{x}_1\\\dot{x}_2\\\vdots\\\dot{x}_n\end{bmatrix}\mathrm{d}\tau\\&+\begin{bmatrix}0&0&\cdots&0\\0&0&\cdots&0\\\vdots&\vdots&\vdots&\vdots\\c_m&0&0&0\end{bmatrix}\int_0^t \mu_m e^{-\mu_m(t-\tau)}\begin{bmatrix}\dot{x}_1\\\dot{x}_2\\\vdots\\\dot{x}_n\end{bmatrix}\mathrm{d}\tau=\begin{bmatrix}0&1&0&\cdots&0\\0&0&1&\cdots&0\\\vdots&\vdots&\vdots&\vdots&\vdots\\-a_0&-a_1&-a_2&\cdots&-a_n\end{bmatrix}\begin{bmatrix}x_1\\x_2\\\vdots\\x_3\end{bmatrix}+\begin{bmatrix}0\\0\\\vdots\\1\end{bmatrix}u\end{aligned} \tag{4}$$

So we can abbreviate it into the following form

$$\begin{aligned}&\dot{X}+P_1\int_0^t \mu_1 e^{-\mu_1(t-\tau)}\dot{X}(\tau)\mathrm{d}\tau+\cdots+P_{m-1}\int_0^t \mu_{m-1} e^{-\mu_{m-1}(t-\tau)}\dot{X}(\tau)\mathrm{d}\tau\\&+P_m\int_0^t \mu_m e^{-\mu_m(t-\tau)}\dot{X}(\tau)\mathrm{d}\tau=AX+bu\end{aligned} \tag{5}$$

It is easy to see that the above system is similar to integer order system, but the existence of integral terms makes the calculation more difficult.

3. The optimal control problem of exponentially damped system

In this section, we consider a single-input multivariable system. The dynamic constraints are expressed in the above state-space formulation, but only one "integral" term is restricted. The performance index function is expressed as follows

$$J(u)=\frac{1}{2}\int_0^{t_f}\left(X^{\mathrm{T}}QX+ru^2\right)\mathrm{d}t \tag{6}$$

Subject to the dynamic constraint

$$\dot{X}+P\int_0^t \mu e^{-\mu(t-\tau)}\dot{X}(\tau)\mathrm{d}\tau = AX+bu \tag{7}$$

And the boundary conditions as $X(0)=X_0$, $X(t_f)$ is free and t_f is fixed. Where, Q is a weighted matrix related to the state vector, r is the weighted coefficient associated with the control signal. Here, $X(t)$ and $u(t)$ are n-dimensional state vector and a control input respectively. $A\in \mathrm{R}^{n\times n}$, $b\in \mathrm{R}^{n\times 1}$. We assume that the following formula holds

- $Q_{n\times n}$ is a nonnegative matrix.
- $r>0$
- $P_{n\times n}=\begin{bmatrix}0 & 0 & \cdots & 0\\ 0 & 0 & \cdots & 0\\ \vdots & \vdots & \vdots & \vdots\\ k & 0 & \cdots & 0\end{bmatrix}$, where, k is a nonzero real number.

Assuming that $\lambda(t)$ is the Lagrange multiplier. We construct the augmented performance index as

$$J_a(u)=\int_0^{t_f}[\frac{1}{2}(X^{\mathrm{T}}QX+ru^2)+\lambda^{\mathrm{T}}(AX+bu-\dot{X}-P\int_0^t \mu e^{-\mu(t-\tau)}\dot{X}(\tau)\mathrm{d}\tau)]\mathrm{d}t \tag{8}$$

Setting the first variation of (8), we get

$$\begin{aligned}\delta J_a(u)=\int_0^{t_f}&[(QX+A^{\mathrm{T}}\lambda)^{\mathrm{T}}\delta X-\lambda^{\mathrm{T}}\delta\dot{X}-\lambda^{\mathrm{T}}P\delta\int_0^t \mu e^{-\mu(t-\tau)}\dot{X}(\tau)\mathrm{d}\tau\\ &+(ru+b^{\mathrm{T}}\lambda)\delta u+(AX+bu-\dot{X}-P\int_0^t \mu e^{-\mu(t-\tau)}\dot{X}(\tau)d\tau)^{\mathrm{T}}\delta\lambda]\mathrm{d}t\end{aligned} \tag{9}$$

The second and third terms on the right side of the upper formula are calculated separately and the following results are obtained.

$$\begin{aligned}\int_0^{t_f}(-\lambda^{\mathrm{T}}\delta\dot{X})\mathrm{d}t&=-\int_0^{t_f}\lambda^{\mathrm{T}}\frac{\mathrm{d}\delta X}{\mathrm{d}t}\mathrm{d}t=-\int_0^{t_f}\lambda^{\mathrm{T}}\mathrm{d}\delta X=-\lambda^{\mathrm{T}}\delta X|_0^{t_f}+\int_0^{t_f}\delta X\mathrm{d}\lambda^{\mathrm{T}}\\ &=-\lambda^{\mathrm{T}}(t_f)\delta X(t_f)+\lambda^{\mathrm{T}}(0)\delta X(0)+\int_0^{t_f}\dot{\lambda}^{\mathrm{T}}\delta X\mathrm{d}t\end{aligned} \tag{10}$$

Because $X(0)=X_0$ is specified, so $\delta X(0)=0$, but $X(t_f)$ is not specified, so we require $\lambda(t_f)=0$. We can get

$$\int_0^{t_f}-\lambda^{\mathrm{T}}\delta\dot{X}\mathrm{d}t=\int_0^{t_f}\dot{\lambda}^{\mathrm{T}}\delta X\mathrm{d}t \tag{11}$$

$$\begin{aligned}\int_0^{t_f}\left[-\lambda^{\mathrm{T}}P\delta\int_0^t \mu e^{-\mu(t-\tau)}\dot{X}(\tau)\mathrm{d}\tau\right]\mathrm{d}t&=\int_0^{t_f}\left[-\mu e^{-\mu t}\lambda^{\mathrm{T}}P\int_0^t e^{\mu\tau}\delta\dot{X}(\tau)\mathrm{d}\tau\right]\mathrm{d}t\\ &=\int_0^{t_f}\int_0^t e^{\mu\tau}\delta\dot{X}(\tau)\mathrm{d}\tau\mathrm{d}\int_t^{t_f}\mu\lambda^{\mathrm{T}}Pe^{-\mu\tau}\mathrm{d}\tau\\ &=-\int_0^{t_f}\int_t^{t_f}\mu\lambda^{\mathrm{T}}Pe^{-\mu\tau}\mathrm{d}\tau\left[e^{\mu t}\delta\dot{X}(t)-\delta\dot{X}(0)\right]\mathrm{d}t\\ &=-\int_0^{t_f}e^{\mu t}\int_t^{t_f}\mu\lambda^{\mathrm{T}}Pe^{-\mu\tau}\mathrm{d}\tau\mathrm{d}\delta X(t)\\ &=\int_0^{t_f}\delta X(t)\mathrm{d}e^{\mu t}\int_t^{t_f}\mu\lambda^{\mathrm{T}}Pe^{-\mu\tau}\mathrm{d}\tau\\ &=\int_0^{t_f}[\mu^2 e^{\mu t}\int_t^{t_f}\lambda^{\mathrm{T}}Pe^{-\mu\tau}\mathrm{d}\tau-\mu\lambda^{\mathrm{T}}P]\delta X(t)\mathrm{d}t\end{aligned} \tag{12}$$

Substitute the above calculation results into (9)

$$\begin{aligned}\delta J_a(u)=\int_0^{t_f}&[(QX+A^{\mathrm{T}}\lambda+\dot{\lambda}+\mu^2 e^{\mu t}P^{\mathrm{T}}\int_t^{t_f}\lambda e^{-\mu\tau}\mathrm{d}\tau-\mu P^{\mathrm{T}}\lambda)^{\mathrm{T}}\delta X\\ &+(ru+b^{\mathrm{T}}\lambda)\delta u+(AX+bu-\dot{X}-P\int_0^t \mu e^{-\mu(t-\tau)}\dot{X}(\tau)\mathrm{d}\tau)^{\mathrm{T}}\delta\lambda]\mathrm{d}t\end{aligned} \tag{13}$$

IOP Conf. Series: Journal of Physics: Conf. Series **1267** (2019) 012084 doi:10.1088/1742-6596/1267/1/012084

Minimization of $J_a(u)$ (and hence minimization of $J(u)$ requires the coefficients of $\delta\lambda$, δX and δu in equation (9) be zero. So we can get

$$\dot{X}+P\int_0^t \mu e^{-\mu(t-\tau)}\dot{X}(\tau)\mathrm{d}\tau = AX+bu \tag{14}$$

$$\dot{\lambda}+\mu^2 e^{\mu t}P^{\mathrm{T}}\int_t^{t_f}\lambda e^{-\mu\tau}\mathrm{d}\tau-\mu P^{\mathrm{T}}\lambda+QX+A^{\mathrm{T}}\lambda=0 \tag{15}$$

$$ru+b^{\mathrm{T}}\lambda=0 \tag{16}$$

With the boundary conditions as $X(0)=X_0$, $\lambda(t_f)=0$. (17)

So, equation (14), (15), (16) with the boundary conditions (17) constitute a boundary value problem of optimal control problems for exponentially damped system. We can use a direct numerical technique to solve. Here we only discuss the optimal control problem with one "integral" term in the dynamical constraint. The same method can be used to solve the optimal control problem with multiple "integrals" in the dynamical constraint. If the single input system is replaced by the multi-input system, this method can also solve the problems.

4. Numerical example

When the integral constitutive equation (1) is applied to model the dynamics of structures incorporated with viscoelastic dampers, the equation of motion is

$$m\ddot{x}(t)+c\int_0^t G(t-\tau)\dot{x}(\tau)\mathrm{d}\tau+kx(t)=u(t) \tag{18}$$

$$m\ddot{x}(t)+c\int_0^t \mu e^{-\mu(t-\tau)}\dot{x}(\tau)\mathrm{d}\tau+kx(t)=u(t) \tag{19}$$

Where, m is the mass, k is the stiffness, c is the damping coefficient, $u(t)$ is the input force acting on the system, $x(t)$ is the displacement of m. Defining the following state vectors: $x=x_1$, $\dot{x}=x_2$. Then the equation of (18) is transformed into the following equation in the state space

$$\begin{bmatrix}\dot{x}_1\\ \dot{x}_2\end{bmatrix}+\begin{bmatrix}0 & 0\\ 0 & \frac{c}{m}\end{bmatrix}\int_0^t \mu e^{-\mu(t-\tau)}\begin{bmatrix}\dot{x}_1\\ \dot{x}_2\end{bmatrix}\mathrm{d}\tau=\begin{bmatrix}0 & 1\\ -\frac{k}{m} & 0\end{bmatrix}\begin{bmatrix}x_1\\ x_2\end{bmatrix}+\begin{bmatrix}0\\ \frac{1}{m}\end{bmatrix}u \tag{20}$$

For the convenience of the following calculation, we can take $m=1, c=1$, $k=1$, and then the equation can be abbreviated to the following form

$$\dot{X}+P\int_0^t \mu e^{-\mu(t-\tau)}\dot{X}(\tau)\mathrm{d}\tau = AX+bu \tag{21}$$

Where, $X=\begin{bmatrix}x_1\\ x_2\end{bmatrix}$, $P=\begin{bmatrix}0 & 0\\ 1 & 0\end{bmatrix}$, $A=\begin{bmatrix}0 & 1\\ -1 & 0\end{bmatrix}$, $b=\begin{bmatrix}0\\ 1\end{bmatrix}$.

Next, we discuss the optimal control problem with exponentially damped terms in dynamic constraint. The performance index function is expressed as follows

$$J(u)=\frac{1}{2}\int_0^{t_f}\left(X^{\mathrm{T}}QX+ru^2\right)\mathrm{d}t \tag{22}$$

Subject to the dynamic constraint

$$\dot{X}+P\int_0^t \mu e^{-\mu(t-\tau)}\dot{X}(\tau)\mathrm{d}\tau = AX+bu \tag{23}$$

The given initial condition as $X_0=\begin{bmatrix}1 & 0\end{bmatrix}^{\mathrm{T}}$, $X(t_f)$ is free and $t_f=1$, where $Q=\begin{bmatrix}1 & 0\\ 0 & 1\end{bmatrix}$, $r=1$.

Optimal conditions for this problem are given by

$$\dot{X}+\begin{bmatrix}0&0\\1&0\end{bmatrix}\int_0^t \mu e^{-\mu(t-\tau)}\dot{X}(\tau)\mathrm{d}\tau=\begin{bmatrix}0&1\\-1&0\end{bmatrix}X+\begin{bmatrix}0\\1\end{bmatrix}u \tag{24}$$

$$\dot{\lambda}+\mu^2 e^{\mu t}\begin{bmatrix}0&1\\0&0\end{bmatrix}\int_t^{t_f}\lambda e^{-\mu\tau}\mathrm{d}\tau=\mu\begin{bmatrix}0&1\\0&0\end{bmatrix}\lambda-\begin{bmatrix}1&0\\0&1\end{bmatrix}X-\begin{bmatrix}0&-1\\1&0\end{bmatrix}\lambda \tag{25}$$

$$u=-\begin{bmatrix}0&1\end{bmatrix}^{\mathrm{T}}\lambda \tag{26}$$

$$X_0=\begin{bmatrix}1&0\end{bmatrix}^{\mathrm{T}},\lambda(1)=\begin{bmatrix}0&0\end{bmatrix}^{\mathrm{T}} \tag{27}$$

Equation (24), (25), (26), (27) constitute a two point boundary value problem. To solve equations (24) to (26) we divide the entire time domain into N equal domains. The size of each domain is given as $h=\frac{1}{N}$. The equation can be expressed as

$$\frac{X^m-X^{m-1}}{h}+\begin{bmatrix}0&0\\1&0\end{bmatrix}\int_0^t \mu e^{-\mu(t-\tau)}\frac{X^m-X^{m-1}}{h}\mathrm{d}\tau=\begin{bmatrix}0&1\\-1&0\end{bmatrix}X^m-\begin{bmatrix}0&0\\0&1\end{bmatrix}\lambda^m \quad m=1,2,\cdots,N \tag{28}$$

$$\frac{\lambda^m-\lambda^{m+1}}{h}+\mu^2 e^{\mu t}\begin{bmatrix}0&1\\0&0\end{bmatrix}\int_t^1 \lambda^m e^{-\mu\tau}\mathrm{d}\tau=\mu\begin{bmatrix}0&1\\0&0\end{bmatrix}\lambda^m-\begin{bmatrix}1&0\\0&1\end{bmatrix}X^m-\begin{bmatrix}0&-1\\1&0\end{bmatrix}\lambda^m \quad m=N-1,\cdots,0 \tag{29}$$

Where, X^i and λ^i are the numerical approximations of state vector $X(t)$ and $\lambda(t)$ at node i.

Now let's turn the vector equation above into a scalar form, and assume that $\mu=1$

$$\frac{x_1^m-x_1^{m-1}}{h}=x_2^m,\ m=1,2,\cdots,N \tag{30}$$

$$\frac{x_2^m-x_2^{m-1}}{h}+e^{-t}\int_0^t e^{\tau}\frac{x_1^m-x_1^{m-1}}{h}\mathrm{d}\tau=-x_1^m-\lambda_2^m,\ m=1,2,\cdots,N \tag{31}$$

$$\frac{\lambda_1^m-\lambda_1^{m+1}}{h}+e^t\int_t^1 \lambda_2^m e^{-\tau}\mathrm{d}\tau=2\lambda_2^m-x_1^m,\ m=N-1,N-2,\cdots,0 \tag{32}$$

$$\frac{\lambda_2^m-\lambda_2^{m+1}}{h}=-x_2^m-\lambda_1^m,\ m=N-1,N-2,\cdots,0 \tag{33}$$

Equation (30) to (33) form 4n number of equations in terms of 4n unknowns. These could be solved directly using any linear equation solver. Once $\lambda=\begin{bmatrix}\lambda_1&\lambda_2\end{bmatrix}^{\mathrm{T}}$ is known, control u can be computed using equation (26).

Figures 1 and 2 show the convergence of the state variables and Figure 3 shows the convergence of control variable as a function of time for $\mu=0.2$. From these figures it is clear to see that solutions convergence as N is increased. However, accuracy of the numerical method can be improved.

5. Conclusion

In this paper, we consider the optimal control problem with exponential damping dynamic. We generalize the traditional state space method to obtain the state space representation of n-order differential equations with m integral terms, and use it as dynamic constraints for optimal control design. Through complex calculation, we find the necessary condition to satisfy the minimum performance index, which is the solution of a Two Point Boundary Value problem and can be obtained by numerical simulation directly. Then we take the second-order exponential damper system as an example to design the optimal control, and get the Two Point Boundary Value problem which satisfies the optimal conditions. Then we simulate it by the method of precise integration. The image shows that the analytical solution is consistent with the numerical solution.

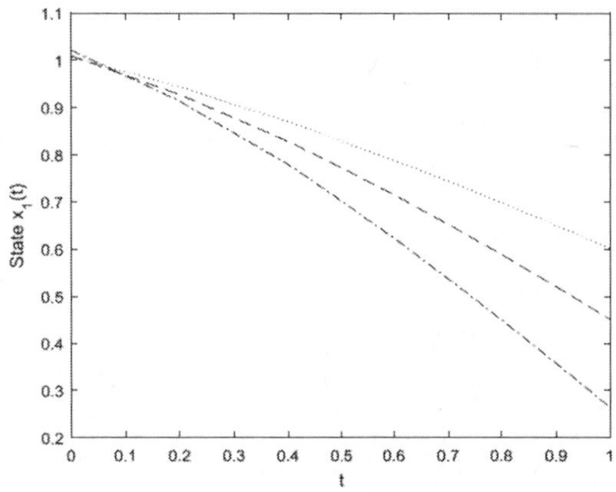

Figure 1. Convergence of function $x_1(t)$ ($\cdots N=60$, $— N=80$, $\cdot - N=100$, $\mu=0.2$)

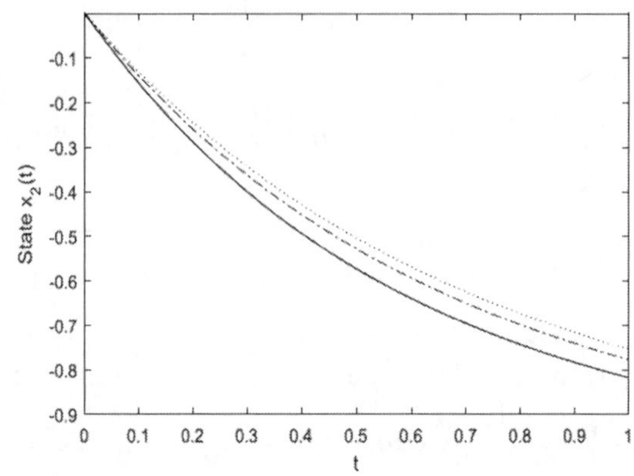

Figure 2. Convergence of function $x_2(t)$ ($\cdots N=60$, $\cdot - N=80$, $— N=100$, $\mu=0.2$)

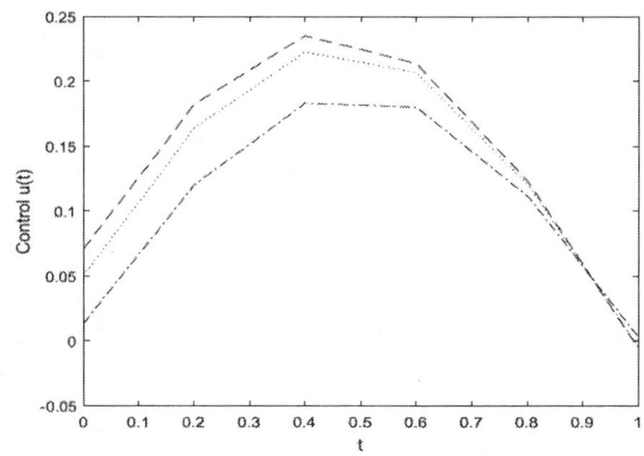

Figure 3. Convergence of function $u(t)$ ($— N=60$, $\cdots N=80$, $\cdot - N=100$, $\mu=0.2$)

References

[1] Li Li and Yujin Hu 2014 State-space method for viscoelastic systems involving general damping model *AIAA Journal*

[2] S Adhikari, N Wagner 2003 Analysis of Asymmetric Nonviscously damped linear dynamic systems *Journal of Applied mechanics* chapter 6 pp 885–893

[3] Mainardi F 2010 Fractional Calculus and Waves in Linear Viscoelasticity *Imperial College Press* 368

[4] Adhikari S 2013 Structural Dynamic Analysis with Generalized Damping Models

[5] Adhikari S 2008 Dynamic Response Characteristics of a Nonviscously Damped Oscillator pp 148–155

[6] N Wagner, S Adhikari 2003 Symmetric state-space method for a class of nonviscously damping systems. *AIAA Journal* Vol.41, No.5.

[7] Jian Yuan, Bao Shi 2015 Sliding mode control of fractional Lorenz-Stenflo Hyperchaotic system *Journal of Discontinuity Nonlinearity and Complexity*

[8] Jian Yuan, Bao Shi 2013 Adaptive sliding mode control of a novel class of fractional chaotic systems *Advances in Mathematical Physics*

[9] Agrawal OP 2003 A general formulation and solution scheme for fractional optimal control problems *Nonlinear Dynamics* 38: 323-337

Research on the Guidance Law for UAV Rendezvous Based on Virtual Leader

J X Wen[1*], G R Zhao[1], X B Li[1] and X Han[1]

[1]Naval Aviation University, 188 Er-ma Road, Yantai, China

[*]Corresponding author's e-mail: wenjiaxinhjhy@163.com

Abstract. The rendezvous of UAVs as a first step in the formation of a UAV is crucial. In this paper, a maneuvering system based on virtual leader is proposed. Firstly, the relative kinematics model of the virtual leader and UAV is established, and then a proportional guidance law with falling angle constraint is designed to make the UAV. Finally, the desired attitude angle can be achieved, and the speed of the UAV can be controlled, so that the UAV can track the virtual leader at a set speed and maintain a relative distance, thereby achieving the purpose of formation flying. The simulation results show that this method can effectively guide the formation of UAVs and has certain engineering application value.

1. Introduction

A single UAV will inevitably be subject to performance limitations during mission execution, such as limited attack range and insufficient reconnaissance capabilities. And the formation of the formation by rendezvous the assembly will make up for these shortcomings of the single UAV, improve the success rate of execution tasks and the ability to resist emergencies [1]. In this context, the research on the issue of UAV formation has become a hot topic, including formation, formation maintenance, formation transformation, formation disbanding and so on [2]. At present, there are many researches on the related issues of UAV formation flight and formation control at home and abroad [3-5], and there are few related studies on formation assembly.

Among these problems in the rendezvous of formations, the focus is on the strategic aspects of assembly [6, 7], and the guidance laws for specific assembly are rarely involved [8, 9]. In the data reviewed by the author, the literature [10] proposes the assembly method of the virtual leader, and discusses its advantages. The rendezvous guidance law designed in [9], although considering only the two-dimensional plane, is based on the visual line guidance law. It is strong and simple to implement. Therefore, based on this, this paper extends it to 3D and proposes the guidance law for the virtual machine to track down the virtual leader. Multiple UAVs can be assembled under the guidance of the same virtual leader, and the virtual leader is adopted. It can avoid the impact of physical long-term problems on the overall formation and improve the success rate of successful formation. The design of the guidance law of this paper also draws on the airborne refueling of drones and the proportional guidance law with falling angle constraints used in missiles [11-13]. The guidance law has the advantages of strong practicability, high accuracy and simple structure, and has certain reference value for the assembly of UAVs.

2. Problem description

Aggregation formation refers to the process in which multiple UAVs arrive at designated areas from

IOP Conf. Series: Journal of Physics: Conf. Series **1267** (2019) 012085 doi:10.1088/1742-6596/1267/1/012085

different locations and form a certain formation according to requirements. The virtual leader based assembly proposed in this paper is to design a virtual leader for the UAV. The virtual leader flight is scheduled to fly, and each UAV is guided by the virtual leader. At the same time, each UAV is required to maintain a certain relative distance between virtual leader. Therefore, from the perspective of guidance, it is necessary to control the relative position of the UAV and the virtual leader. Firstly, the relative kinematics model of the virtual leader and the UAV is given, and then the design of the guidance law is carried out. Figure 1 and Figure 2 show kinematics analysis of the UAV and the virtual leader.

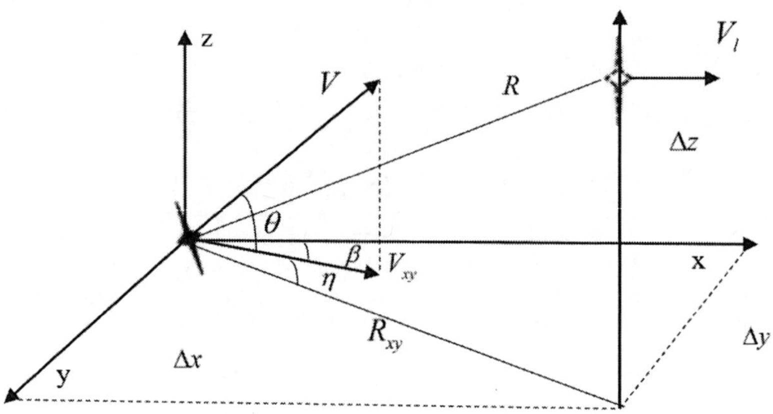

Figure 1. UAV and virtual leader three-dimensional movement

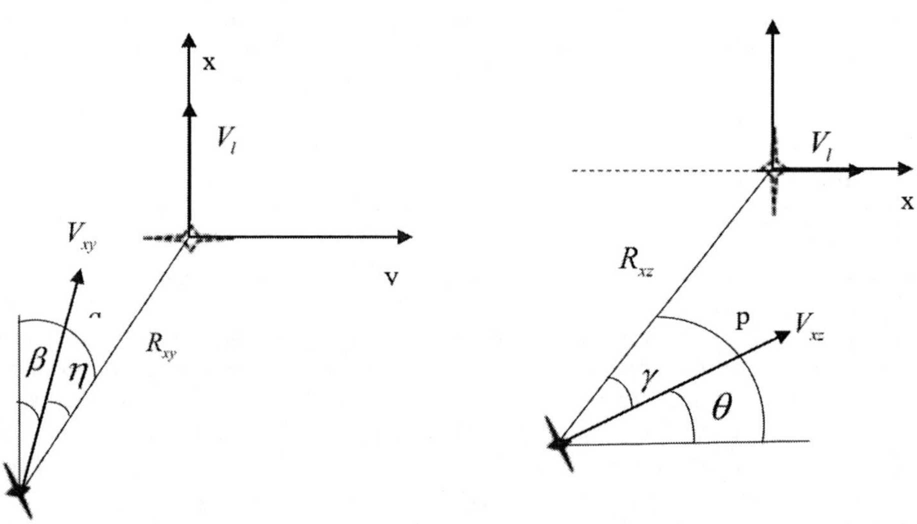

Figure 2. UAV and virtual leader horizontal and vertical movement

It can be seen from Figure 1 that the virtual leader is flying in the $+X$ direction at the speed V_l, the UAV is located behind the virtual leader, the speed is in the direction of the V and coincides with the direction of the heading and the speed in the horizontal plane is V_{xy}; The pitch angle is positive, and the general range is -90°~+90°. At the same time, as shown in Figure 2, γ is the angle between V_{xz} and R_{xz} in the vertical plane, which is called the vertical deviation angle; R is the relative distance between the UAV and the virtual leader. It is projected on the horizontal plane as R_{xy}, and the vertical plane is projected as R_z; β is the heading angle, which is positive in the $+X$ direction and negative on the left side; η is V_{xy} and The angle between R_{xy}, called the line of sight deviation angle, is defined as positive on the right side in the V_{xy} direction and negative on the left side. In Figure 2, p and q are target line angles in the horizontal and vertical directions, respectively.

Therefore, establish the kinematics equation as follows

$$\Delta\dot{x} = V_l - V\cos\theta\cos\beta \quad (1)$$

$$\Delta\dot{y} = -V\cos\theta\sin\beta \quad (2)$$

$$\Delta\dot{z} = -V\sin\theta \quad (3)$$

where $\Delta\dot{x}$, $\Delta\dot{y}$ and $\Delta\dot{z}$ denote the Speed difference of the UAV and the virtual leader along the three-dimensional.

$$q = \arctan\frac{\Delta y}{\Delta x} \quad (4)$$

$$p = \arctan\frac{\Delta z}{R_{xz}} \quad (5)$$

$$\dot{q} = \frac{\Delta\dot{y}\Delta x - \Delta y\Delta\dot{x}}{\Delta x^2 + \Delta y^2} \quad (6)$$

$$\dot{p} = \frac{\Delta\dot{z}R_{xy} + \Delta z\dot{R}_{xy}}{R^2} \quad (7)$$

$$\dot{R}_{xy} = \frac{(\Delta\dot{x}\Delta x + \Delta y\Delta\dot{y})}{\sqrt{\Delta x^2 + \Delta y^2}} \quad (8)$$

where $\dot{q}$ and $\dot{p}$ denote the rate of the target line angles in the horizontal and vertical directions along respectively.

3. Guidance law design

The purpose of the guidance law design is to make the horizontal and vertical target line deviation angles between the UAV and the virtual leader reach p_c and q_c respectively, and the relative distance between the UAV and the virtual leader is R_c and the speed is consistent. Therefore, the following guidance laws are designed:

$$\dot{\beta} = k_1\dot{q} + k_2(q - q_c) \quad (9)$$

The above formula is the horizontal guidance law, where k_1, k_2 are guidance parameters, q_c is the expected horizontal target line deviation angle.

$$\dot{\theta} = k_3\dot{p} + k_4(p - p_c) \quad (10)$$

The above formula is the vertical guidance law, where k_3 , k_4 are guidance parameters, p_c is the expected vertical target line deviation angle.

$$\dot{V} = k_5\left(V_l - V\right) + k_6\left(R - R_c\right) \quad (11)$$

The above formula is speed control law, the controller controls the speed based on the relative distance. , k_5 , k_6 are guidance parameters, R_c is the expected distance.

4. Numerical simulations

Based on the space particle model of UAV and virtual leader, this paper simulates the design of the rendezvous guidance law. According to the previously designed guidance law and the established relative kinematics model, the relevant parameters are set and simulated by MATLAB programming.

4.1 Simulation model

The sensor characteristics are neglected, and the UAV flight control system is characterized by a first-order equivalent link. The time constant is 0.1s. The schematic block diagram of the simulation model is shown in Figure 3.

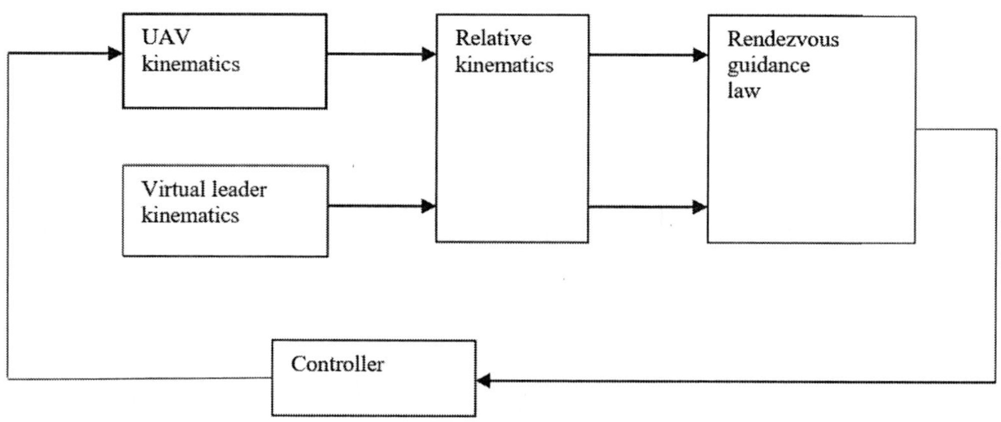

Figure 3. Simulation model structure

4.2 Simulation condition

Suppose there are two UAVs tracking the same virtual leader to form a formation. The initial position of the virtual leader is $(0,0,0)$, the initial speed is $80m/s$, and the initial attitude angle is $(\beta_0,\theta_0)=(0°,0°)$. The initial position of the UAV 1 is $(-3000,-3000,-3000)m$, and the initial speed is $60m/s$. The initial attitude angle is $(\beta_1,\theta_1)=(40°,40°)$; the initial position of the UAV 2 is $(-3000,3000,-3000)m$, the initial speed is $60m/s$, the initial attitude angle is $(\beta_2,\theta_2)=(-40°,40°)$; the maximum speed and maximum acceleration of the UAV is $150m/s$, $10m/s^2$, and the desired attitude angle of the UAV is $(\beta_c,\theta_c)=(0°,0°)$, expectation The distance is $R_c=50m$; the relevant guidance parameter is $k_1=0.02$, $k_2=0.2$, $k_3=0.02$, $k_4=0.2$, $k_5=0.05$, $k_6=0.15$.

4.3 Simulation results and analysis

The rendezvous process of the drone takes two UAVs to track the same virtual leader as an example.

The attitude angle change, the speed change, and the relative distance change take the UAV 2 as an example. The relevant simulation curves are shown in Figures 4-8.

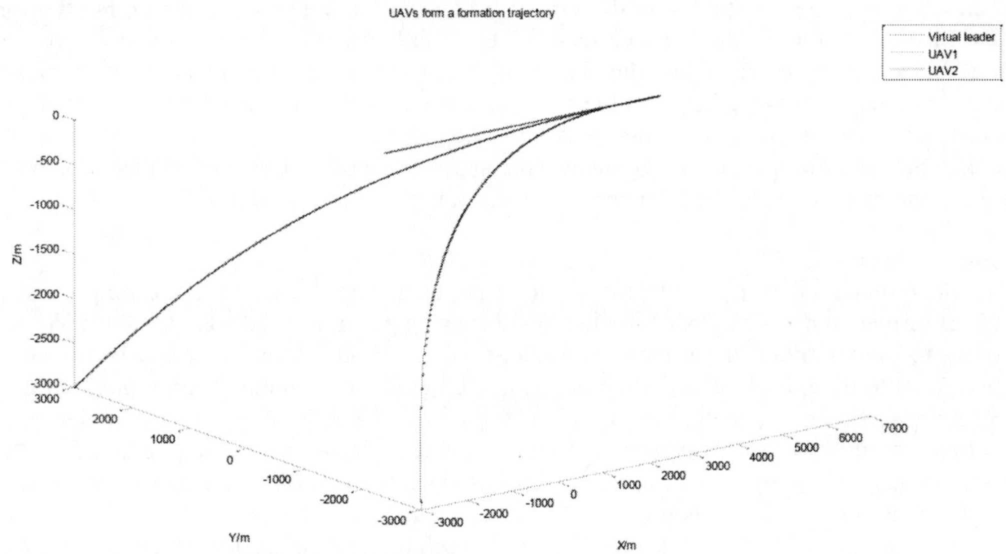

Figure 4. UAVs form a formation trajectory

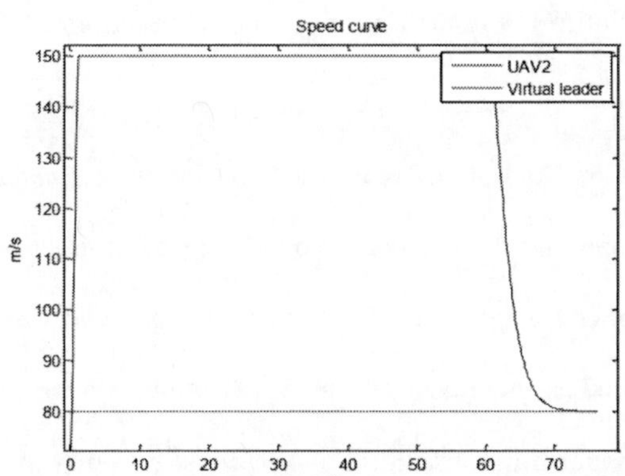

Figure 5. Speed curve

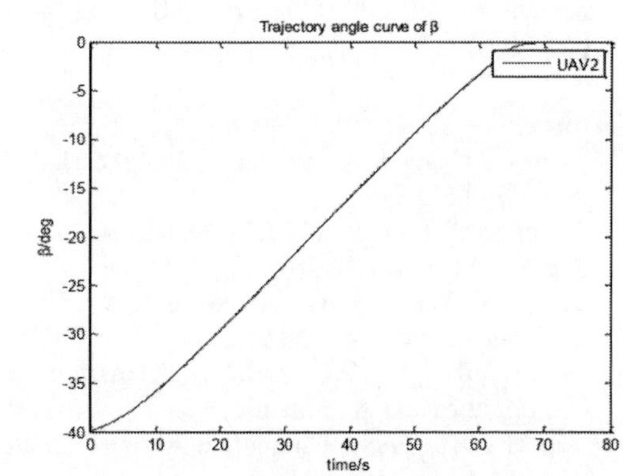

Figure 6. Trajectory angle curve of β

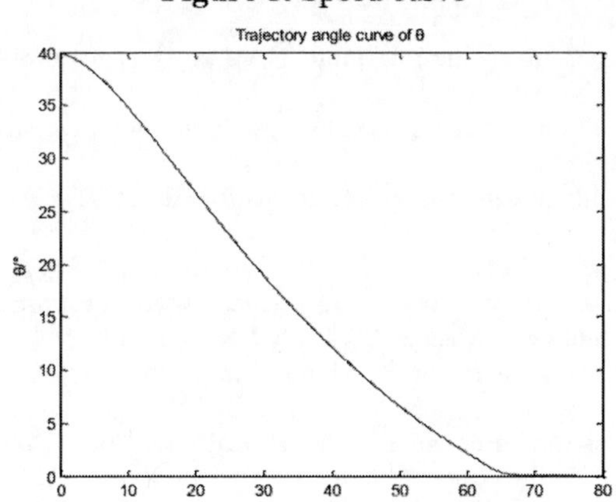

Figure 7. Trajectory angle curve of θ

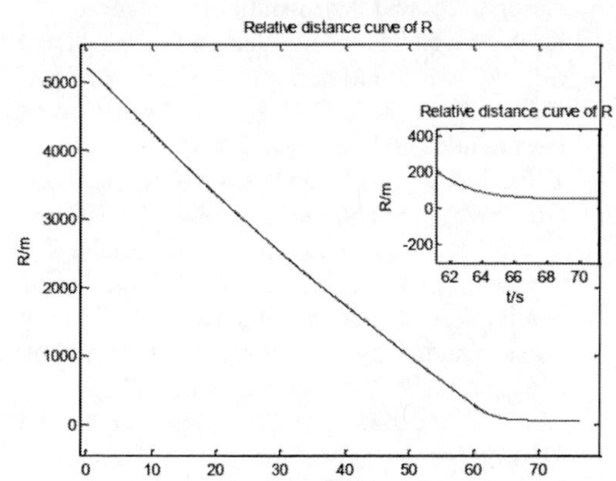

Figure 8. Relative distance curve of R

It can be seen from Figure 4 that the two UAVs form a stable formation under the guidance of the virtual leader, and the relative distance between the UAV and the virtual leader remains stable, and the speed direction is also consistent with the virtual leader, and in the process of forming the formation. The flight path is gentle. As can be seen from Figure 6 and Figure 7, the attitude angle of the UAV 2 finally reaches the desired value and the change is stable without abrupt change. As can be seen from Figure 8, the relative distance between the UAV and the virtual leader reaches the desired value. The above simulation results verify the validity of this guidance law.

The simulation of this paper is based on the spatial particle model. The general guidance law research is based on the particle model, so the research results have certain credibility.

5. Conclusions

Through the research of this paper, the following main conclusions are obtained: Aiming at the guidance problem of formation of multi-UAV, a method of clustering guidance based on virtual leader is proposed. According to the relative motion relationship between UAV and virtual leader, The guidance law is designed. It is verified by simulation that the generated guidance command can better guide the UAV to form a formation and maintain the formation flight, and the form is simple and easy to understand and realize. At the same time, the introduction of virtual leader as a reference point of assembly can replace the real-world long-term assembly method, reduce the risk of accidents caused by real leader, and improve the success rate of assembly. Of course, the above conclusions are based on the simplified model. To be further applied to the project, more in-depth research is needed on a more detailed model.

Acknowledgments

This paper is supported by the Project for the National Natural Science Foundation of China under Grant No. 61473306.

Reference

[1] Zhu X , Zhang X X and Yan M D 2016 UAV Formation Control Strategy Based on Consensus *Computer Simulation* **33** 30

[2] Wang G Q, Luo H and Hu X X 2013 A Survey on Coordinated UAV Formation Management *Electronics Optics & Control* **20** 48

[3] Deng W, Wang X M and Wang X Y 2011 Controller Design of UAVs Formation Keep and Change *Computer Simulation* **28** 73

[4] Ni L Q 2016 UAVs Cooperative Formation Flight based on Intelligent Technology (Nanjing University of Aeronautics and Astronautics)

[5] Liu H 2015 UAVs Unmanned Aerial Vehicle Formation Robust Control and Reconstruction (National University of Defense Technology)

[6] Li Q 2015 UAVs Cooperative Formation Flight based on Intelligent Technology (Nanjing University of Aeronautics and Astronautics)

[7] Fu X W, Cui H J and Gao X G 2015 Distributed solving method of multi-UAV rendezvous problem *Control and Engineering and Electronics* **8** 1797

[8] Ji L L 2008 UAVs Cooperative Formation Flight based on Intelligent Technology (Nanjing University of Aeronautics and Astronautics)

[9] Tahk MJ, Park CS and Ryoo CK 2005 Line-of-sight guidance laws for formation flight *Journal of Guidance, Control, and Dynamics* **28** 708

[10] Charles R P 2007 The virtual UAV leader *Tactical Missile Technology AIAA Conference and Exhibit* 2785

[11] Gu W J, Lei J W and Pan C P 2006 Design of the Climbing Trajectory Using Virtual Target's Proportional Navigation Method with the Control of Terminal Azimuth of a Missile *Flight Dynamics* **24** 43

[12] Yuan B and Yang J 2015 Design of UAV Autonomous Guidance Law for Rrendezvous in Aerial Refueling *Navigation Positioning and Timing* **2** 25

[13] Ma D B and Zhang D 2015 Research on mathematics model of biased proportional navigation *Tactical Missile Technology* **3** 90

Multi-Identification and Motion Control of a Two-Axis Pan-Tilt Platform

Yihui Fan[1], Xingyan Zheng[1], Bin Ye[1*], Yuzhe Yang[1] and Da Zhao[2]
[1]School of Information and Control Engineering, China University of Mining and Technology
[2]School of Mechatronic Engineering, China University of Mining and Technology

[*]Corresponding e-mail: yebin@cumt.edu.cn

Abstract. The two-axis pan-tilt platform (PTP) has been widely used in industrial control systems. The control performance of the PTP is sharply affected by the load variations, which are more significant on the pitch axis. Moreover, the PTP control system is sensitive to sensor noise and the friction moment. To achieve rapid and precise control, we build a pan-tilt platform model using the method of multiple system identification and design a speed controller and position controller on both the yaw and the pitch axes. The proposed method avoids the need to measure a large number of mechanical and electrical parameters. The experimental result shows that the proposed control system has small overshoot, fast response and suitable robustness. Relative to a controller designed and adjusted by experience, the designed system increases the bandwidth by approximately 50% on the yaw axis, and the settling time is decreased.

1. Introduction

The pan-tilt platform (PTP) is well suited for a wide range of applications in real life, including surveillance, border patrol, and entertainment [1]. In addition, it is widely used in unmanned aerial systems for aerial photography [2] and target tracking [3]. A PTP system is typically driven by motors and is thus influenced by load variations and gravity torques, among other perturbations [4]. The need for a rapid and precise PTP control system is widespread. Thus, developing an accurate PTP model is vital in the design of control systems.

Existing PTP models can be divided into two types: linear models and nonlinear models. In the linear models, the centrifugal, Coulomb friction and Coriolis forces are ignored; however, they are incorporated in the nonlinear model [5]. Mechanism modeling and system identification are the two basic methods to build up a model [6]. Currently, PTP design is based primarily on mechanism modeling, and system identification methods are barely used. In the mechanism modeling of the PTP system, it is typically necessary to obtain the back electromotive force constant, torque constant of the motor, and selected physical and electrical parameters such as the moment of inertia of the PTP, which increases the challenge of developing the model. Load changes also affect the model [7]. In addition, the model accuracy may rely on the accuracy of the parameter measurement. Some parameters are hard to obtain; thus, a large deviation potentially exists between the obtained model and real model [8,9]. In [10], a biomimetic control strategy of an on-board pan-tilt-zoom camera is presented. This system can compensate the deflection caused by the flight platform and thus enhance the system performance. In [11], the active disturbance rejection control technique is applied to the tracking control of a pan-tilt camera to improve the performance of the pan-tilt camera platform. An implementation of two fuzzy logic controllers working in parallel for a pan-tilt camera platform on an unmanned aerial vehicle is presented in [12], and the controllers have shown favorable performances in real flights for static objects. In [13], the harmonic drive model of the pan-tilt platform is put forward, which consists of a linear spring and a load inertia to represent the compliance between load side angles and the motor, this approach can better incorporate the impact of friction on the system, thus improving the accuracy of the system. A control method based on self-adaptive PID, used in a uniaxial turntable control system, is presented in [14]. The controller can solve the nonlinear and time-varying band problems of turntable system.

With the number of controllers increasing, the inconsistency between the obtained PTP model and the real model is expected to accumulate [14]. In addition, after each new controller is added, this also means control links increase, the disturbance factor is introduced due to the sensor of the parameter measurement, which is not conducive to continuing to design the controller.

The PTP considered here can realize 360° rotation around the yaw axis. Consequently, the motion interval is relatively large, and the stability requirement for the model is higher. In this paper, we treat the PTP on the pitch and the yaw axis as two independent systems, modeling them separately and designing different controllers for them. In this paper, we propose to use a multi-identification approach to develop the model instead of using mechanism modeling. By considering the PTP on each axis, we first identify its model and then establish a speed controller based on the model. The system with the speed controller should be re-identified, and the position controller is designed according to the new system to complete the PTP control system. Since the yaw axis is not disturbed by the gravity torque and can be stabilized by itself, the yaw axis in the model is obtained by using the open-loop identification method. On the pitch axis, the deflection away from the equilibrium point(the angle when tilt parallels to the horizontal plane) arising from the gravity torque requires external forces to maintain stability. Therefore, the pitch axis in the model is obtained by closed-loop identification. The experimental results show that the control system has the advantages of both flexibility and strong adaptability.

2. Motivation and Contribution

(i) The multiple identifications we use can identify the interference factors introduced by each new controller. In the next controller design, the model can be made closer to the real situation, and the influence of the interference amplified on the system will not be increased due to the increase of the controllers' number

(ii) The system identification method we use can reduce the index measurement work, and integrate the difficult-to-measure parameters links into the transfer function of the system by identification, which solves the problem that the model modeling is inaccurate.

(iii) Pitch and Yaw are treated as two independent controlled objects, and they are decoupled. The Pitch uses closed-loop identification to solve the interference of gravity on Pitch. While the Yaw uses open-loop identification to reduce the difficulty of system identification.

3. Overall structure of the control system

3.1. Pan-tilt platform

A two-axis PTP is a two-degree-of-freedom rotary platform that consists of a horizontally rotatable yaw axis and a vertical pitch axis. Two motors are used to drive the yaw axis and the pitch axis independently. Figure 1 shows the PTP system considered in the paper. The pitch motor is a RM6623 brushless DC motor, and the yaw motor is a DM9015 brushless DC motor. The basic parameters of the motor used in the design control system are listed in Table 1. By comparing the parameters of the PTT motor and the actual needs, we believe that this motor can meet the basic requirements of this PTP design. The DM9015 disc motor has a through hole in the center, where the slip ring could get through. In this way, we can make it as a 360° pan/tilt while providing electric power and signal to the pitch motor, which makes the PTP more flexible.

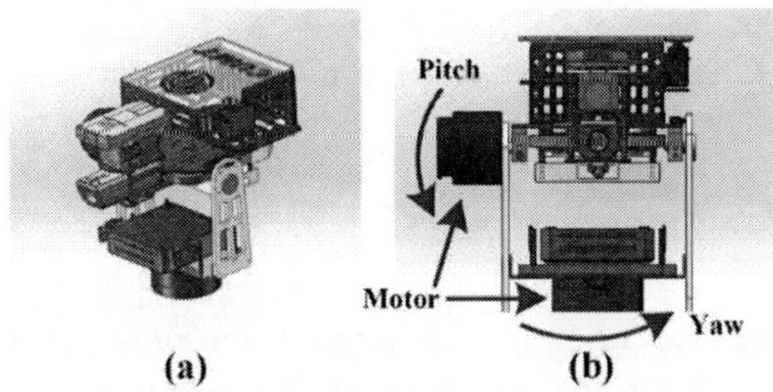

Figure 1. (a) PTP 3D figure and (b) illustration of the PTP.

Table 1. Motor parameters.

Item	RM6623	DM9015
Weight	605g	560g
Torque Constant	0.38N•m/A	0.53N•m/A
Speed Constant	25RPM/V	19.5RPM/V
Torque	2N•m	5.93N•m
Stall Current	5.3A	11.2A
Rated Rotational Speed	600RPM	390RPM

The through hole designed in the middle of the motor can accommodate a conductive sliding ring to realize 360° omnidirectional rotation of the yaw axis, which increases the flexibility of PTP significantly. We use the STM32F405 microcontroller unit as the computing unit.

3.2. Control System Structure

The control diagram for the PTP is shown in Figure 2, where $G(s)$ is the transfer function of a controller and θ denotes the angular displacement. Here, θ_{in} is the target value of the input angle, and θ_{out} is the output of the current angle value. The motor angular velocity ω_{gyro} is obtained as an angular velocity feedback by the gyroscope. The pan/tilt position $\theta_{encoder}$ is obtained by measuring the motor with a magnetic encoder, and the signal is normalized to the dimension of θ_{in} by unit conversion as an output. The two axes of the PTP are independently controlled, and each axis uses a double loop controller that contains a speed loop and a position loop. The output of the speed controller is sent to the current controller, which contains an integrated current loop. The angular velocity is measured by a gyroscope mounted on the PTP and fed back to the speed controller. The positional information is fed back by the magnetic encoder mounted on the motor.

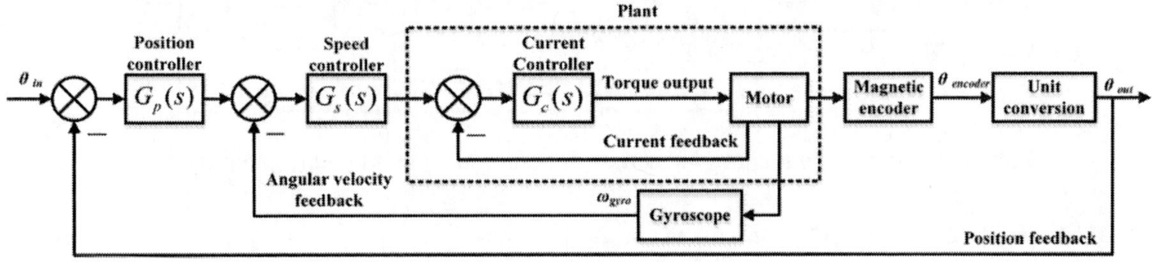

Figure 2. Continuous system control scheme. The control scheme includes a position controller $G_p(s)$ and a speed controller $G_s(s)$. The DC motor contains an integrated current controller $G_c(s)$.

In the actual applications, the angular velocity feedback from the gyroscope cannot be directly compared with the output of the position controller. Since the angular velocity of gyro feedback is not measured in degrees, unit conversion is required:

$$\omega = \omega_{gyro} \div 32.768 \tag{1}$$

where ω is the angular velocity in the angle system, ω_{gyro} is the angular velocity measured by the gyroscope, and 32.768 corresponds to 32.768 LSB/(°/s). The PTP controller is typically implemented with digital devices; thus, the continuous system shown in Figure 2 can be discretized with a sampling period of 0.001 seconds.

Since the current loop is integrated in the current controller, the current controller can be integrated with the motor. After the PTP is physically constructed, each axis is artificially set at 0 degrees, but the value of the motor encoder is not 0; thus, we have to subtract the initial value. Since the encoder feeds back an absolute position and the feedback value is within [0, 8191], we need to consider the situation of zero crossing. To couple the dimensions of θ_{in} and θ_{out}, we convert the absolute position feedback of the encoder into an incremental form $\theta_{encoder}$. Finally, the incremental form value of the encoder feedback is converted into the incremental form value of the output angle θ_{out} by unit conversion:

$$\theta_{out} = \theta_{encoder} \div 22.756 \tag{2}$$

where 22.756 corresponds to 22.756 LSB/degree. We also convert the input to incremental form θ_{in} to achieve unit negative feedback in the position closed loop. In the continuous system, the angular velocity can be used to calculate the angle by integration. Since the angle incremental form value can be fed back by the gyroscope, the unit of the gyroscope feedback is converted to degrees per second. The integral element should be transferred into $\frac{0.0305}{s}(\frac{1}{s} \div 32.768 = \frac{0.0305}{s})$.

The sample time equals 0.001 seconds, we can discretize the system with a zero-order holder. The integral element now changes to $\frac{3.052\times10^{-5}}{z-1}$. The actual PTP control system in the discrete form is shown in Figure 3. In Figure 3, $G_0(z)$ is the plant that we need to identify. The discrete controllers $G_s(z)$ and $G_p(z)$ are the ones that we are going to design. Here, the Φ_{speed} is a closed loop with a speed controller, and the Φ_{speed} linked with a integral section consists the Φ_{core}.

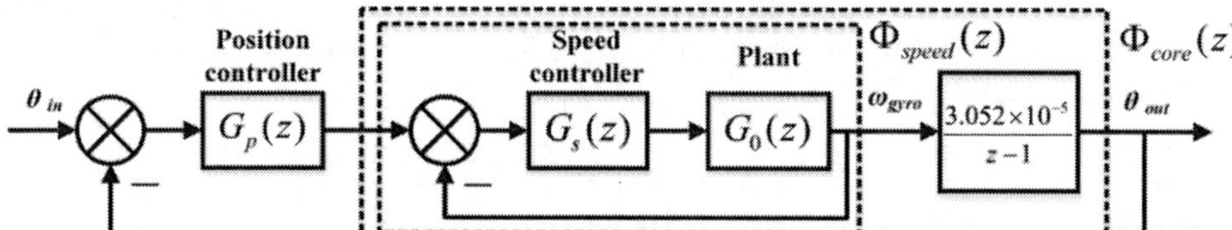

Figure 3. Discrete PTP control scheme containing the position and speed controller.

4. Modeling and design of the controllers on the yaw axis

4.1. Modeling

The horizontal rotation around the yaw axis is not affected by the gravity torque factor and can be stabilized without external force. Therefore, the controller can be designed by using the open-loop system identification method. Input the yaw axis motor with a sinusoidal signal:

$$Signal = A\sin(2\pi ft) \tag{3}$$

where the amplitude A is set to 4000, which is determined according to the range of data defined in the current controller protocol. The frequency f is changed from 1 Hz to 500 Hz, and each frequency lasts for 20 cycles. Since the frequency bandwidth of the input signal is large, the amplitude variation of the PTP in the high-frequency portion is small; thus, we adjust the frequency f to increase almost exponentially. At the same time, it is necessary to guarantee that the input signal corresponding to each frequency starts from the zero phase.

Because the amplitude of the PTP changes little in the high-frequency part, the vibration frequency is faster, and the collected data may not reflect the true condition of the system due to the signal acquisition frequency and the lower resolution of the sensor. In this case, the frequency f can be appropriately reduced to improve the accuracy of the identification.

The gyroscope angular velocity feedback is acquired as a system output. By referencing other papers describing the mechanism model [1,5,14] and using the root locus analysis [15,16], we find that the number of the zeros of the PTP model is 0 and the number of poles is 2. The input and output are systematically identified by system identification. The discrete transfer function of the yaw axis is obtained, and the sample time is 0.001 seconds:

$$G_0^{Yaw}(z) = \frac{0.0006369}{1-1.898z^{-1}+0.8986z^{-2}} \tag{4}$$

4.2. Speed Controller Design

The PID controller is simple, reliable, and easy to implement on MCU; thus, it is widely used in industrial fields. The controller of this paper needs to be implemented on MCU, and the system is required to be stable and reliable; therefore, the PID controller is adopted. Because the PTP fast response speed is required and the deviation is small on this basis, besides, gyroscope's feedback frequency is high, so that only the proportional element in the speed controller is required. Its transfer function reads

$$G_s^{Yaw}(z) = 14.3211 \tag{5}$$

4.3. Position Controller Design

Here, we introduce the speed loop controller $G_s^{Yaw}(z)$ into the control system. When a new section is added to the control model, the error increases between the idealized model and the real model. Inconsistency inevitably arises between the closed-loop transfer function $\Phi_{speed}^{Yaw}(z)$ and the real model due to various uncertain factors. In addition, model augmentations and an interference factor may be introduced to the entire system. Therefore, re-identification should be performed to identify a new model and design the position controller based on it.

To design the position controller, the measurements by the magnetic encoder are read and converted as the feedback of the system. In Figure 3, an integral element is introduced between ω_{gyro} and θ_{out}, namely, $\frac{3.052\times10^{-5}}{z-1}$, to increase the number of poles. The result indicates that the new model has 0 zeros and 3 poles. After the re-identification of the speed closed-loop together with the integral element, we obtain the discrete transfer function:

$$\Phi_{core}^{Yaw}(z) = \frac{8.48\times10^{-7}}{1-2.836z^{-1}+2.701z^{-2}-0.8646z^{-3}} \tag{6}$$

The position controller $G_p^{Yaw}(z)$ can be designed based on the transfer function $\Phi_{core}^{Yaw}(z)$, where we use the integral separation PI controller

$$G_p^{Yaw}(z) = \frac{1971-1935z^{-1}}{1-z^{-1}} \tag{7}$$

The integral element works when the bias $|\theta_{in} - \theta_{out}| \le 2^{\circ}$, and the output of the integral element is limited to [-3000,3000].

4.4. System Performance Analysis

To analyze the performance of the designed control system, we input a sinusoidal signal similar to Equation (3). For comparison, we build a set of controllers by experience. With varying amplitude *A* and frequency f, we obtain the step response, bode diagram and tracking curves shown in Figure 4 and Figure 5. The reference signal shown in Figure 5 is the same as that in Equation (3).

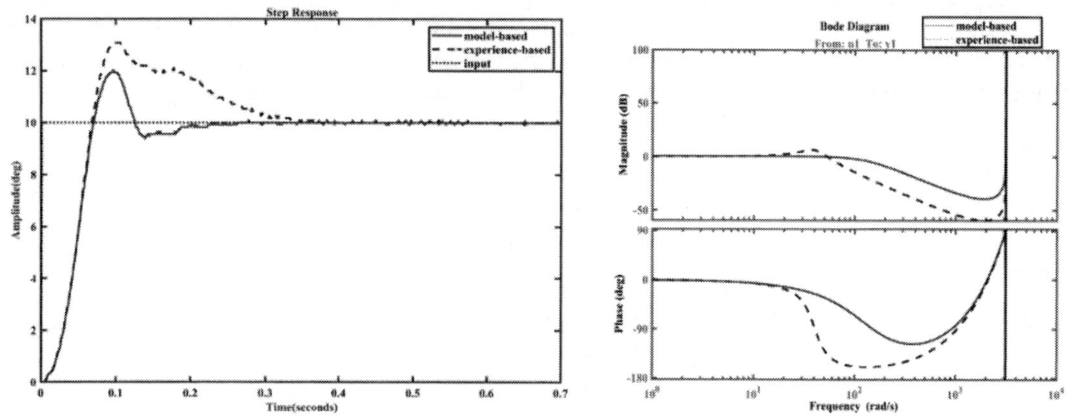

Figure 4. Step response and Bode diagram of the control system on the yaw axis.

Table 2. Performance comparison between the model-based and experience-based control system on the yaw axis.

Item	Model-based	Experience-based
Bandwidth	120rad/s	63 rad/s
Rise time	0.071s	0.068s
Settling time	0.181s	0.282s
Overshoot	20.2%	32.5%

As shown in Figure 4, the model-based controller can increase the bandwidth of the system substantially and reduce the settling time. The results indicate that the rise time is essentially unchanged, while the overshoot is decreased. Figure 5 shows that when the system is operating at low frequency, the performance of the model-based and the experience-based controllers is almost identical. In this stage, the input signal can be well matched. In Table 2, with increasing frequency, the lag time of model-based controller is slightly longer, the precision is slightly decreased, but the performance is clearly better than that of the experience-based controller. On the yaw axis, the method of this paper greatly reduces the settling time and overshoot, and improves the tracking performance. The yaw control system established by system identification is robust and fast in response and exhibits strong tracking performance.

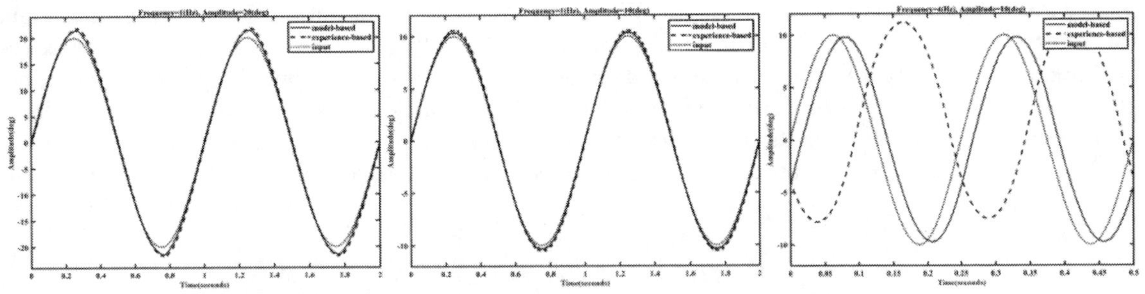

Figure 5. Tracking curve of the control system on the yaw axis.

5. Pitch axis Modeling and design of motion systems

5.1. Modeling

Due to the influence of gravity torque, the controller can maintain stability only by applying an external force. Therefore, the controller is designed using a closed-loop system identification. The unsimplified pitch control system in the discrete form is shown in Figure 6. Here, $H(z)$ is a nonlinear gravity compensation component, which could be calculated with the cross-angle between the PTP and the ground surface(θ_{out}) and the motor torque curve function. Through the motor torque curve function, we can know the relationship between the motor torque increment and the input increment.

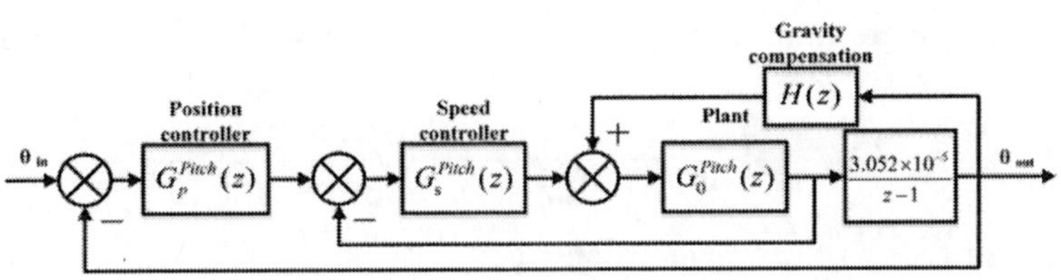

Figure 6. Discrete control structure on the pitch axis.

Since the pitch axis is affected by gravity torque, the speed loop should be compensated according to the current angle. The pitch axis rotating angle ranges from -30° to 30°, when it is 0°, the force of the gravity torque acting on the rotating shaft is $T_0 = mgr$, where mg is the gravity on the center of gravity of the pitch and r is the distance from the center of gravity to the axis of rotation. When the pitch rotation angle is 30°, the force of gravity torque acting on the rotating shaft $T_0 = mgr * \cos(30^\circ)$. Since mg and r are relatively small and the torque is barely changed under the rotation angle of the PTP, we do not adopt the gravity torque compensation element $H(z)$. Therefore, the pitch control structure we adopt is similar to the yaw control structure shown in Figure 3. Note that although the control system structure on the pitch axis is the same as that on the yaw axis, the identification procedures are completely different due to the gravity torque on the pitch axis.

We develop the speed closed-loop control system based on experience, that is, we design a suitable speed controller $G_s^{*Pitch}(z)$ and drive the PTP to a stable state. Here, we obtain $G_s^{*Pitch}(z) = 2$. Then, we develop the model of the speed closed-loop system by system identification. The input to the system is a sinusoidal signal with amplitude A=3000,

$$Signal = A\sin(2\pi ft) + B \tag{8}$$

where the frequency f also increases exponentially and B is the compensation value. Since the center of gravity of the PTP is not on the axis of rotation, the PTP slides in one direction under the influence of gravity torque. Because there are mechanical limits on the pitch axis, the feedback data are invalidated when the PTP is out of range. When the tilt angle is too large, the compensation value forces the pitch axis deflection in the horizontal direction. When the tilt is close to the horizontal position, the compensation value B is reduced or canceled. In this paper, when the pitch angle exceeds 20°, B equals 500; when the pitch angle is less than -20°, B equals -300. The angular velocity feedback of the gyroscope is read as the output of the system. The input and output are systematically identified to obtain the discrete transfer function of the control system. The sample time is 0.001 seconds as well, and we obtain

$$\Phi_{speed}^{Pitch}(z)=\frac{0.009593}{1-1.885z^{-1}+0.8954z^{-2}} \tag{9}$$

Together with $\Phi_{speed}^{Pitch}(z)$ and $G_s^{*Pitch}(z)$, $G_0^{Pitch}(z)$ is calculated as

$$G_0^{Pitch}(z)=\frac{0.009593}{1.981-3.771z^{-1}+1.791z^{-2}} \tag{10}$$

5.2. Controller Design

Using the transfer function $G_0^{Pitch}(z)$, we design the speed controller $G_s^{Pitch}(z)$, that is,

$$G_s^{Pitch}(z)=1.7572 \tag{11}$$

Similarly, we first design a suitable position controller $G_p^{*Pitch}(z)=1000$ by experience, and we define the entire Figure 3 as $\Phi_{all}^{Pitch}(z)$. By using the input signal in Equation (3) with the amplitude A=20, we identify the discrete transfer function of the entire control system with 0 zeros and 3 poles and obtain

$$\Phi_{all}^{Pitch}(z)=\frac{0.01068}{1\text{-}0.7992z^{-1}+0.9946z^{-2}+0.8045z^{-3}} \tag{12}$$

From $\Phi_{all}^{Pitch}(z)$ and $G_p^{*Pitch}(z)$, we can calculate

$$\Phi_{core}^{Pitch}(z)=\frac{1.068\times10^{-5}}{1\text{-}0.7992z^{-1}+0.9946z^{-2}+0.8045z^{-3}} \tag{13}$$

Based on the re-identified transfer function $\Phi_{core}^{Pitch}(z)$, the position controller on the pitch axis is designed as a integral separation PI controller

$$G_p^{Pitch}(z)=\frac{1618\text{-}1523z^{-1}}{1\text{-}z^{-1}} \tag{14}$$

The integral element works when $|\theta_{in}-\theta_{out}|\le 3^{\circ}$, and the output of integral element is constrained to [-3000,3000].

5.3. System Performance Analysis

The step response and Bode diagram are shown in Figure 7. The model-based controller can greatly increase the bandwidth of the system, significantly reduce the settling time, and reduce overshoot. Since the pitch axis is subjected to considerable external interference, the model-based controller is very important for system performance improvement. When the system operates at low frequency, the performance of the model-based controller is similar to that of the experience-based controller, which is shown in Figure 8. As the frequency increases, the lag time of the model-based controller becomes slightly longer, the accuracy decreases slightly, but the performance of the experience-based controller decreases significantly, particularly in overshoot. The performance comparison is shown in Table 3. The pitch axis control system established by system identification is robust and fast in response and exhibits strong tracking performance.

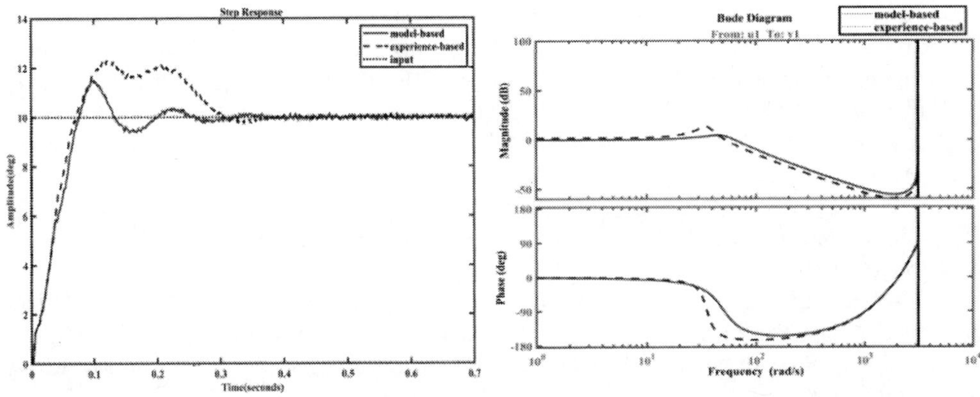

Figure 7. Step response and Bode diagram of the control system on the pitch axis.

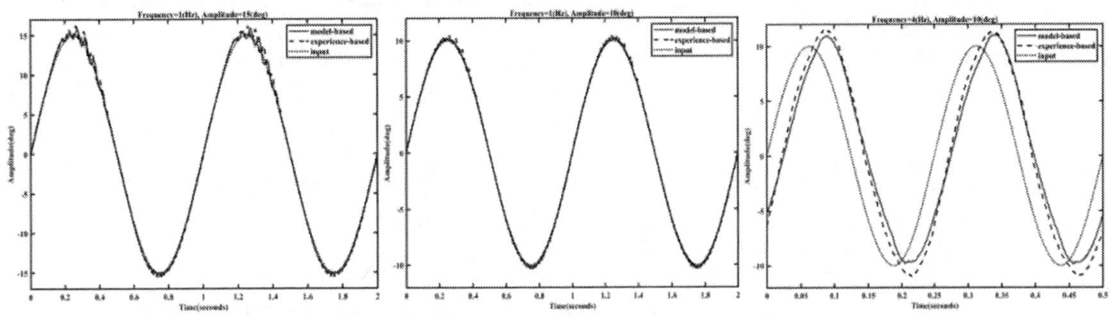

Figure 8. Tracking performance on the pitch control system.

Table 3. Performance comparison between the model-based and experience-based control system on the Pitch axis.

Item	Model-based	Experience-based
Bandwidth	70rad/s	57 rad/s
Rise time	0.075s	0.071s
Settling time	0.175s	0.283s
Overshoot	14.9%	23.7%

6. Conclusions

This paper proposes a simple and efficient method of system identification and controller design for a two-axis pan-tilt platform. The model established by system identification does not need to identify the parameters such as the resistance and response lag of the PTP, which is beneficial to the subsequent controller design. The entire control system is designed with an inside-out approach, ensuring that each element is optimized to meet the requirements so that the entire controller can provide suitable performance.

Acknowledgments

This work is supported in part by the Science and Technology Program of Xuzhou (Grant No. KC18069) and the Laboratory Open Funds of China University of Mining and Technology (Grant No. 20180217).

IOP Conf. Series: Journal of Physics: Conf. Series **1267** (2019) 012086 doi:10.1088/1742-6596/1267/1/012086

References

[1] I. S. Sarwar and A. M. Malik, Modeling, analysis and simulation of a Pan Tilt Platform based on linear and nonlinear systems, 2008 IEEE/ASME International Conference on Mechtronic and Embedded Systems and Applications, Beijing, pp. 147-152 (2008)

[2] G. Binkun, Research on Intelligent Control Algorithm for UAV Three-axis Self-stable Pan-tilt, Diss, JiMei University (2013)

[3] Z. Xin, Y. Fang, G. Zhang and H. Shen, Experiment platform for pan-tilt control of a small scale autonomous helicopter, Proceedings of the 29th Chinese Control Conference, Beijing, pp. 3544-3547 (2010)

[4] Naung, A. Schagin, H. L. Oo, K. Z. Ye and Z. M. Khaing, Implementation of data driven control system of DC motor by using system identification process, 2018 IEEE Conference of Russian Young Researchers in Electrical and Electronic Engineering (EIConRus), Moscow, pp. 1801-1804 (2018)

[5] I. S. Sarwar, J. Iqbal, and A. M. Malik, Modeling, analysis and motion control of a pan tilt platform based on linear and nonlinear systems, WSEAS Transactions on Systems and Control 4.8, pp. 389-398 (2009)

[6] H. Zhong-xu, Q. Xiao-hong, L. Min, Z. Zhi and Z. Chuan-xin, Mixed Boiler-turbine Coordinated Control System Mathematical Model Based on Mechanism Modelling and Parameter Identification, 2006 1ST IEEE Conference on Industrial Electronics and Applications, Singapore, pp. 1-5 (2006)

[7] Z. Song, Y. Lin, X. Mei and G. Jiang, A Novel Inertia Identification Method for Servo System Using Genetic Algorithm, 2016 International Conference on Smart Grid and Electrical Automation (ICSGEA), Zhangjiajie, pp. 22-25 (2016)

[8] R. Fung and W. Yang, System identification of an induction motor, 2017 International Conference on Applied System Innovation (ICASI), Sapporo, pp. 1068-1071 (2017)

[9] W. Tang, Z. Liu and Q. Wang, DC motor speed control based on system identification and PID auto tuning, 2017 36th Chinese Control Conference (CCC), Dalian, pp. 6420-6423 (2017)

[10] S. Xie, J. Luo, P. Xie, Z. Gong and H. Zou, Biomimetic Control of Pan-tilt-zoom Camera Mounted on an Autonomous Helicopter, 2007 International Conference on Mechatronics and Automation, Harbin, pp. 2003-2008 (2007)

[11] H. Chen, X. Zhao and M. Tan, A novel pan-tilt camera control approach for visual tracking, Proceeding of the 11th World Congress on Intelligent Control and Automation, Shenyang, pp. 2860-2865 (2014)

[12] M. A. Olivares-Méndez, P. Campoy, C. Martínez and I. Mondragón, A pan-tilt camera Fuzzy vision controller on an unmanned aerial vehicle, 2009 IEEE/RSJ International Conference on Intelligent Robots and Systems, St. Louis, MO, pp. 2879-2884 (2009)

[13] S. Sharp, A. Wicks, A. Ordys and G. Collier, Modelling of a pan and tilt servo system, Proceedings of 2012 UKACC International Conference on Control, Cardiff, pp. 546-550 (2012)

[14] Aihua Zhang, and Jianfei NI, Uniaxial turntable control system based on fuzzy self-adaptive PID, Journal of Heilongjiang University of Science & Technology (2015)

[15] W. R. Evans, Control System Synthesis by Root Locus Method, Transactions of the American Institute of Electrical Engineers, vol. 69, no. 1, pp. 66-69 (1950)

[16] N. S. Nise, Control systems engineering, Wiley student edition, Fourth edition, pp. 515-524 (2004)

IOP Conf. Series: Journal of Physics: Conf. Series **1267** (2019) 012087 doi:10.1088/1742-6596/1267/1/012087

Two-Stage Backward Elimination Method for Neural Networks Model Reduction

Xiaoquan Tang[1] **and Long Zhang**[2]

[1] School of Artificial Intelligence and Automation, Huazhong University of Science and Technology, Wuhan 430074, China

[2] School of Electrical and Electronic Engineering, The University of Manchester, Manchester, M13 9PL, United Kingdom

E-mail: [1]xqtang2016@hust.edu.cn; [2]long.zhang@manchester.ac.uk

Abstract. The single-hidden-layer neural networks (NN) has been widely used for complex system identification. However, the hidden neurons are often determined by trial-and-error method and the amount of neurons is usually large. This commonly leads to over-fitting problem and the training process is time consuming. In this paper, we propose a two-stage backward elimination (TSBE) method to obtain a parsimonious network with fewer hidden neurons but remains a good performance and saves training time. In the first stage, neural networks with a predetermined number of hidden neurons is trained based on stochastic gradient decent (SGD) algorithm with part of training data and Least absolute shrinkage and selection operator (Lasso) is applied for dropping redundant neurons leading to a simplified neural model. In the second stage, the remaining training data is used to update the parameters of the simplified neural model. A simulation example is used to validate and show that the novel approach gives a more compressed model and higher level of accuracy comparing with the recently proposed pruning-based method.

1. Introduction

System identification is widely used for modelling dynamic systems. Most real-world systems are highly nonlinear. Therefore, neural networks is often used to describe the unknown nonlinearities in the system. This is achieved by directly processing raw measurement data of system and automatically learning the description of the system [1]. When building neural networks, determining the amount of hidden neurons and hidden layers is needed. Currently, the widely used method for setting the hidden-neuron numbers is the trial-and-error which generally sets the amount of hidden neurons without any prior, then tests the performance of obtained model. The process mentioned above is usually time consuming. In addition, it often produces an overly complex network with a huge amount of hidden neurons. According to the parsimony principle, a simpler model is preferable to a more complex model when two models produce similar accuracy. The main reason is that the over complex model may require large computations and cause over-fitting problem with a poor generalization performance.

In order to obtain a simplified network, the network prunning method removes the neurons whose weights are smaller than a pre-determined threshold. Recently, an improved pruning method is proposed which is a two-step learning method and has attracted lots of attentions [2]. Pruning or those related methods either calculate sensitivity of error to the removal connections or add penalty to cost function leading to a parsimonious networks [3]. However, most of sensitivity-based methods can not remove those correlated connections. Further, it may be difficult to set a proper threshold to drop redundant connections, especially when all the connection weight are similar and the number of small weights is only few or very limited.

Another class of related works randomly reduce the complexity of neural networks during the training process. Dropout is the most popular one of these model reduction methods [4]. Although it has attracted significant attentions, dropout may not finally provide a parsimonious model. In addition, a class of methods focus on reducing the storage requirements of the neural networks where the low-rank approximation is commonly used, although its approximation performance may not be guaranteed [5]. A recent work called Dropneuron which can drop neurons and connections simultaneously by regularising all incoming connections' weights and outgoing connections' weights to be zeros with combining the pruning method[6]. Although Dropneuron can obtain a more parsimonious networks, it has the same problems with pruning-based methods mentioned above, namely, it is difficult to set a suitable threshold and remove correlated connections. More recently, a novel method is proposed to perform pruning and feature selection simultaneously using sparse group Lasso penalty [7]. It can be viewed as a method based on forward selection method which generally begins with an empty neural model, then gradually adds input neurons using feature selection methods.

In this paper, we propose a novel two-stage backward elimination method to obtain a simplified networks by directly dropping neurons with the aid of Lasso. In the first stage, the predetermined neural networks are trained using SGD with part of training data and then Lasso method is applied for dropping those redundant hidden neurons to obtain a simplified networks. The dropping method can be treated as backward elimination method as it can simplify neural networks by removing redundant neurons from output layer to input layer. In the second stage, updating the parameters of selected neurons is carried out with the remaining train data.

2. Nonlinear system identification via neural networks

With statistical learning algorithms, neural networks can approximate any nonlinear system, even just with a single hidden layer. A large number of neurons can effectively reduce training error. However, a complex neural networks that have too many unnecessary neurons could have very poor testing performance.

A nonlinear system could be described as follows:

$$y(t) = \sum_{i=1}^{M} \beta_i \phi_i(w_i x(t) + b_i) + e(t) \tag{1}$$

here $y(t) \in \mathbf{R}^N, x(t) \in \mathbf{R}^{r \times N}$ are neural networks output and input, and w_i is the weights between i^{th} hidden neuron and neurons of input layer with $w_i \in \mathbf{R}^r$. M represents the amount of hidden neurons and b_i are bias. Suppose $\phi_i(x) = 1/(1 + \exp(-x))$, where x is the input variable.

We rewrite equation (1) as matrix form:

$$\mathbf{y} = \mathbf{P}\boldsymbol{\Theta} + \boldsymbol{\Xi} \tag{2}$$

where

$$\mathbf{y} = [\ y(1), y(2), \cdots, y(N)\ ]^T, \boldsymbol{\Theta} = [\ \beta_1, \beta_2, \cdots, \beta_M\ ]^T, \boldsymbol{\Xi} = [\ e(1), e(2), \cdots, e(N)\ ]^T \tag{3}$$

and

$$\mathbf{P} = \begin{bmatrix} \phi_1(w_1 x(1) + b_1) & \phi_2(w_2 x(1) + b_2) & \cdots & \phi_M(w_M x(1) + b_M) \\ \phi_1(w_1 x(2) + b_1) & \phi_2(w_2 x(2) + b_2) & \cdots & \phi_M(w_M x(2) + b_M) \\ \vdots & \vdots & \vdots & \vdots \\ \phi_1(w_1 x(N) + b_1) & \phi_2(w_2 x(N) + b_2) & \cdots & \phi_M(w_M x(N) + b_M) \end{bmatrix} \tag{4}$$

For convenience, $\mathbf{P}$ can be re-writen as

$$\mathbf{P} = [P_1, P_2, \cdots, P_M] \tag{5}$$

The paper deals with neuron dropping, the resultant model only have a proportion of neurons. Therefore, here, define $\mathbf{P}_\varepsilon = [P_{i_1}, P_{i_2}, \cdots, P_{i_k}]$ for the set $\varepsilon = \{i_1, i_2, \cdots, i_k\} \subseteq \{1, 2, \cdots, M\}$ and $\mathbf{P}_{-\varepsilon}$ contains those components not in ε.

3. A two-stage backward elimination method

3.1. Drop redundant neurons with lasso

For a single hidden layer network, the model output could be described as a linear combination of hidden neurons. Lasso can be used for selecting important neurons and dropping those unimportant neurons to make the networks to be simplified as Figure 1. More specifically, Lasso algorithm minimizes the residual sum of squares by

$$\widehat{\mathbf{\Theta}} = \arg\min \frac{1}{2} \sum_{t=1}^{N} (y(t) - \sum_{i=1}^{M} \mathbf{\Theta}_i \phi_i(w_i x(t) + b_i))^2 + \lambda \sum_{i=1}^{M} |\mathbf{\Theta}_i| \tag{6}$$

which can be re-writen as matrix format:

$$\widehat{\mathbf{\Theta}} = \arg\min \frac{1}{2} \|\mathbf{y} - \mathbf{P}\mathbf{\Theta}\|_2^2 + \lambda \|\mathbf{\Theta}\|_1 \tag{7}$$

To write lasso solution with a more explicit form, we turn to the Karush-Kuhn-Tucker (KKT) optimal conditions for problem (7), which can be expressed as[8]:

$$\mathbf{P}^T(\mathbf{y} - \mathbf{P}\widehat{\mathbf{\Theta}}) = \lambda\gamma$$
$$\gamma_i \in \begin{cases} \{sign(\widehat{\mathbf{\Theta}}_i)\} & if\ \widehat{\mathbf{\Theta}}_i \neq 0 \\ \text{[-1,1]} & if\ \widehat{\mathbf{\Theta}}_i = 0 \end{cases}, for\ i = 1, 2, \cdots, M. \tag{8}$$

$\widehat{\mathbf{\Theta}}$ is the solution of lasso problem (7) if and only if $\widehat{\mathbf{\Theta}}$ satisfies equation (8). Now, we know the necessary conditions of obtaining the solution, while how to calculate $\widehat{\mathbf{\Theta}}$ in a more explicit form. First we need to define the equicorrelation set

$$\varepsilon = \{i \in \{1, 2, \cdots, M\} : |\mathbf{P}_i^T(\mathbf{y} - \mathbf{P}\widehat{\mathbf{\Theta}})| = \lambda\} \tag{9}$$

Second, define the equicorrelation signs as

$$s = sign(\mathbf{P}_\varepsilon^T(\mathbf{y} - \mathbf{P}\widehat{\mathbf{\Theta}})) \tag{10}$$

We use $\widehat{\mathbf{\Theta}}_{-\varepsilon} = 0$ for the solution $\widehat{\mathbf{\Theta}}$ of problem (7) with the definition of the subgradient γ. Therefore, we have

$$\mathbf{P}_\varepsilon^T(\mathbf{y} - \mathbf{P}_\varepsilon\widehat{\mathbf{\Theta}}_\varepsilon) = \lambda s \tag{11}$$

which can be expressed as

$$\mathbf{P}_\varepsilon^T\mathbf{P}_\varepsilon\widehat{\mathbf{\Theta}} = \mathbf{P}_\varepsilon^T(\mathbf{y} - (\mathbf{P}_\varepsilon^T)^+\lambda s) \tag{12}$$

Here $(\mathbf{P}_\varepsilon^T)^+$ means the Moore-Penrose pseudoinverse of $\mathbf{P}_\varepsilon^T$. Due to $\mathbf{P}_\varepsilon\widehat{\mathbf{\Theta}}_\varepsilon = \mathbf{P}\widehat{\mathbf{\Theta}}$, then

$$\mathbf{P}\widehat{\mathbf{\Theta}} = \mathbf{P}_\varepsilon(\mathbf{P}_\varepsilon)^+(\mathbf{y} - (\mathbf{P}_\varepsilon^T)^+\lambda s) \tag{13}$$

The solution $\widehat{\Theta}$ can be calculated, and the important nodes are selected and redundant neurons are dropped since $\widehat{\Theta}_\varepsilon \neq 0$ and $\widehat{\Theta}_{-\varepsilon} = 0$.

The Lasso solutions provide some unimportant hidden nodes dropped to obtain a simplified networks. The solution of lasso problem (7) may not be unique for the obtained fully connected networks. In other words, one may get different solutions $\widehat{\Theta}$ using different Lasso solvers. Consider a simple example, suppose there are 5 neurons, namely $\widehat{\Theta}^1 = (-0.8, 0.6, 0.3, 0.5, 0)^T$, $\widehat{\Theta}^2 = (-0.8, 0.7, 0.4, 0, 0)^T$. The index set of first solution is $\{1, 2, 3, 4\}$ and that for the second solution is $\{1, 2, 3\}$. It is not easy to know which solution is optimal or unique. Here, we provide the conditions where Lasso can have unique solution.

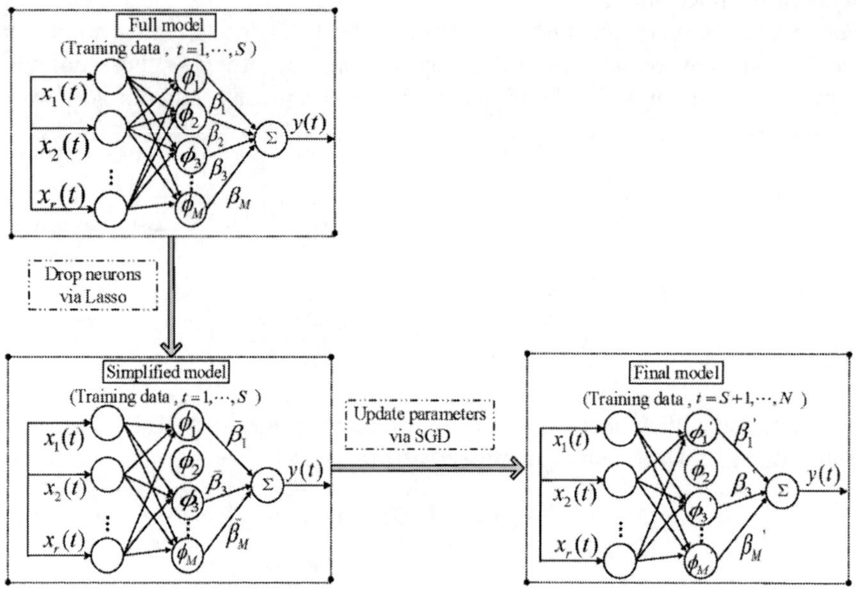

Figure 1. The main idea of TSBE.

In [8], it is pointed that if the null space of $\mathbf{P}_\varepsilon$ is 0 or equivalently if the $rank(\mathbf{P}_\varepsilon) = |\varepsilon|$, then the solution of problem (7) is unique which is given as

$$\widehat{\Theta}_{-\varepsilon} = 0 \ \ and \ \ \widehat{\Theta}_\varepsilon = (\mathbf{P}_\varepsilon^T \mathbf{P}_\varepsilon)^{-1} (\mathbf{P}_\varepsilon^T \mathbf{y} - \lambda s) \tag{14}$$

This unique solution is proved in the literature [8]. We pay attention to if $rank(P_\varepsilon) = |\varepsilon|$ in our proposed algorithm so that the solution is unique.

Proof: w_i are generated randomly in the process of initialization before training the neural networks. In the minimization process, we apply SGD algorithms to iteratively update parameters by [9]

$$W_t = W_{t-1} - \eta \nabla_W L(f(x(i); W), y(i)) \tag{15}$$

where $L(f(x(i); W), y(i)) = \frac{1}{2}(y(i) - \sum_{j=1}^{M} \beta_j \phi_j(w_j x(i) + b_j))^2$ is the loss function and $\nabla_W L(f(x(i); W), y(i))$ represents the partial information of the loss function. $f(x(i); W)$ is a nonlinear function with

$$f(x(i); W) = \sum_{j=1}^{M} \beta_j \phi_j(w_j x(i) + b_j) \tag{16}$$

IOP Conf. Series: Journal of Physics: Conf. Series **1267** (2019) 012087 doi:10.1088/1742-6596/1267/1/012087

where $f(x(i);W)$ and $y(i)$ represent the predicted output and real output of the system, respectively. W is the set of weights and biases with $W = \{w, b\}$. It is worth pointing out that W_{t-1} represents the set of previous parameters at the step $t-1$ and W_t is set of the updated parameters. Then we can assume that $\phi_j(w_j x(t) + b_j) \neq \phi_{j'}(w_{j'} x(t) + b_{j'})$ with any $j \neq j'$ $(1 \leq j \leq M$ and $1 \leq j' \leq M)$, namely, $\mathbf{P}_j \neq \mathbf{P}_{j'}$ with any $j \neq j'$. Therefore, the condition for unique solution is that the column vectors of $\mathbf{P}$ can be full-rank, namely, $rank(\mathbf{P}_\varepsilon) = \varepsilon$.

3.2. Update parameters of significant neurons

In the second stage of the proposed method, we update parameters including weights and biases with SGD method, which is a popular algorithm for training models in machine learning. When using SGD, the learning rate η should be tuned carefully since a worse value may make the obtained model perform worse in new data. The main procedure of TSBE is given as follows:

Algorithm 1 : TSBE

Stage 1: Drop redundant neurons with lasso
1:Initiallization
2:**for** $i = 1, \cdots, S$ **do**
3: $W_t = W_{t-1} - \eta \nabla_W L(f(x(i);W), y(i))$
4:**end for**
5:Drop neurons: $\widehat{\Theta}_\varepsilon = (\mathbf{P}_\varepsilon^T \mathbf{P}_\varepsilon)^{-1}(\mathbf{P}_\varepsilon^T \mathbf{y} - \lambda s)$ and $\widehat{\Theta}_{-\varepsilon} = 0$
Stage 2: Update parameters (W') of significant neurons
6:**for** $i = S+1, \cdots, N$ **do**
7: $W'_t = W'_{t-1} - \eta \nabla_{W'} L(f(x(i);W'), y(i))$
8: **end for**

4. Simulation example

Consider the following nonlinear example:

$$y(t) = -0.8y(t-2) + 0.65u(t-1) - 0.05u(t-3)y(t-1) - 0.2u^3(t-1) + e(t) \tag{17}$$

here $y(t)$, $u(t)$, $e(t)$ represent system output, input and noise. A uniformly distributed white noise is used to excite the nonlinear system with $u(t) \in [-1, 1]$. And Gaussian noise is used to disturb the system with signal-to-noise ratio $15dB$.

Here we select $\{u(t-1), u(t-2), u(t-3), y(t-1), y(t-2), y(t-3), y(t-4)\}$ as the model input for identification. Generating 1800 samples for identification with 80% of total data being training data and the rest being test data. Suppose that the amount of initial neurons (size1) is 50. Then, compare the proposed method with Dropneuron on the efficiency of model simplification, performance and training time.

Table 1. Results for simulation example

	size1	size2	sparsity	train	test	time(s)
$TSBE(0.001)$	50	6.24	0.1448	0.0061	0.0077	148
$TSBE(0.005)$	50	4.86	0.1172	0.0066	0.0082	118
$Dropneuron(0.1)$	50	5.76	0.0561	0.0402	0.0471	414
$Dropneuron(0.01)$	50	17.41	0.1905	0.0117	0.0134	413
NN	50	50	1	0.0159	0.0182	181

In order to validate the efficiency of TSBE method, Monte Carlo simulation is carried out with 100 repetitions. In addition, in order to show the effect of parameter λ on model reduction when using TSBE, two different values, namely, $\lambda = 0.001$ and $\lambda = 0.005$ are chosen. To make a fair comparison, the predetermined threshold for model reduction using Dropneuron are also given two values, namely 0.1 and 0.01. The training time, the average number of selected hidden neurons (size2) and sparsity that is defined as the amount of remained connections divided by the amount of full connections are used to compare the performance of TSBE, Dropneuron and NN. From Table 1, one can see that a large threshold can get a simple network when using Dropneuron method, but the accuracy is poor. In addition, although NN method has a similar performance with TSBE, the scale of networks is larger. Therefore, one can say that TSBE obtains a more parsimonious model with high accuracy. At the same time, the training time of the novel method is shorter than two other algorithms.

5. Conclusion

The neural networks with single hidden layer has been used to identify nonlinear dynamical systems and performs well in many practical applications. However, the amount of hidden neurons is often defined randomly, therefore, the size of network is often large and many hidden neurons of the complex network can be dropped. In this paper, the proposed two-stage backward elimination method can simplify neural networks using Lasso and can give a satisfied accuracy with fewer hidden neurons. We validate the proposed method with a nonlinear example and the higher level of prediction error and shorter training time shows that the novel TSBE method is effective and efficient.

References

[1] Lecun Y., Bengio Y. and Hinton G., Deeping learning. *Nature*, 2015, 521: 436-444.
[2] Han S., Pool J., Tran J. and Dally W., Learning both weights and connections for efficient neural network. In *Advances in Neural Information Processing Systems*, 2015, 1135-1143.
[3] Reed R., Pruning algorithms-a survey. *IEEE Transactions on Neural Networks*, 1993, 4(5): 740-747.
[4] Srivastava N., Hinton G., Krizhevsky A., Sutskever I. and Salakhutdinov R., Dropout: A Simple Way to Prevent Neural Networks from Overfitting. *Journal of Machine Learning Research*, 2014, 15: 1929-1958.
[5] Sainath T. N., Kingsbury B., Sindhwani V., Arisoy E. and Ramabhadran B., Low-rank matrix factorization for deep neural network training with high-dimensional output targets. In *IEEE International Conference on Acoustics, Speech and Signal Processing*, 2013, 6655-6659.
[6] Pan W., Dong H. and Guo Y., DropNeuron: Simplifying the Structure of Deep Neural Networks. *arXiv preprint arXiv: 1606.07326, 2016.*
[7] Scardapane S., Comminiello D., Hussain A. and Uncini A., Group Sparse Regularization for Deep Neural Networks. *Neurocomputing*, 2017, 241: 81-89.
[8] Tibshirani R. J., The lasso problem and uniqueness. *Electronic Journal of Statistics*, 2013, 7(1): 1456-1490.
[9] Goodfellow I., Bengio Y. and Courvile A., Deep learning. *MIT press, 2016.*

Disturbance Observer Based Sliding Mode Control for Marine Diesel Engine

Xuemin Li, Yufei Liu*, Yunlong Yang, Jian Zhang, Runzhi Wang and Chunyue Feng

College of Power and Energy Engineering, Harbin Engineering University, Harbin, China

*lyf232911@163.com

Abstract. In this work, a disturbance observer is designed to obtain the total disturbance including the unknown part, and the error system of the disturbance observer is proved to be finite-time convergence. Based on the total disturbance, a sliding mode controller with exponential reaching rule is proposed for the speed control system of a marine diesel engine. The performance of the proposed controller and the disturbance observer is tested by the simulations on a marine diesel engine model. It shows that the proposed disturbance observer can achieve satisfactory results, and both the control performance and the robustness of the DOSMC are superior to the PID controller.

1. Introduction

Owing to the advantages of high-efficient and economical power installations, the diesel engines have been widely used in daily life and industrial production. In the marine industry, diesel engine is mainly used as the main engine and electric generator. Speed control is the most typical control mode in marine diesel engine as it provides a nearly linear and expected relationship between the speed setting of engine and the ship speed, even though many researchers consider it causes unnecessary load disturbances [1]. The speed control is required to be accurate and smooth, because the speed fluctuation and overshoot will reduce the service life of propulsion system, or even paralyze the entire system [2]. However, the diesel engine is such a complex system that comprises of multiple subsystems, and all subsystems are nonlinear and difficult to model accurately. Besides, the working environment of marine main engine is very harsh. Therefore, the speed control problem for the diesel engine is troublesome, and an accurate and robust speed controller is necessary.

A few decades ago, electronic governors were first designed which make the speed control of diesel engines flexible. From then on, Proportion-Integration-Differentiation (PID) has been widely used in the speed control of diesel engine because of simplicity and reliability [3]. However, the PID controller with the same parameters cannot achieve the best control effect under all conditions for the nonlinear system like the diesel engine [4]. Thus, some researchers modify the PID parameters by combining with different algorithms, such as fuzzy control logic [4], H-infinity [5], genetic algorithm [6], etc. Although these methods can expand the scope of the applicable conditions of PID controller, they cannot fundamentally solve the limitations of PID controller and the PID would lose its advantages of simple and easy implementation. With the advent of high pressure common rail (HPCR) system, fuel injection control becomes more precise, flexible and real-time. Therefore, the application of advanced control algorithms can significantly improve the performance of the marine diesel engine.

IOP Conf. Series: Journal of Physics: Conf. Series **1267** (2019) 012088 doi:10.1088/1742-6596/1267/1/012088

And, some researchers have experimented with different algorithms, such as neural networks control [7], model predictive control [8], active disturbance rejection control [9], etc. Sliding mode control (SMC) is also used in diesel engine speed control by some scholars due to its good robustness. Yuan and his collaborators design discrete sliding mode variable structure controller and multi-sliding surface controller, and both controllers have good control effect [10, 11]. However, such methods rely on the accurate model. Khan and Li use the second order sliding mode controller to control engine speed [12, 13], which no longer requires accurate diesel model, but the robustness is decreased. A simple SMC is designed by Zhang for speed control of generators, but the control performance is not satisfactory [14]. Zhang uses the RBF neural network to approximate the diesel engine model parameters, based on which a sliding mode controller is developed [15]. Nevertheless, this method greatly increases the computation and the convergence time is uncertain.

Considering the above situation, this work develops a disturbance observer based sliding mode controller (DOSMC). A finite time stable disturbance observer (DO) is proposed to obtain the current diesel engine disturbances quickly and accurately. Then, a sliding mode controller is designed based on the observer to realize accurate speed control. The designed DOSMC is easy to implement and has satisfactory control performance. Meanwhile, it maintains the robustness of sliding mode control.

2. Marine Diesel Engine Model and Problem Formulation

In this study, a mean value engine model (MVEM) is introduced for studying the speed control system of marine diesel engine. The MVEM consists of engine body and subsidiary systems include turbocharger system, fuel supply system, lubrication system, etc. Figure 1 shows the subsystems of a diesel engine.

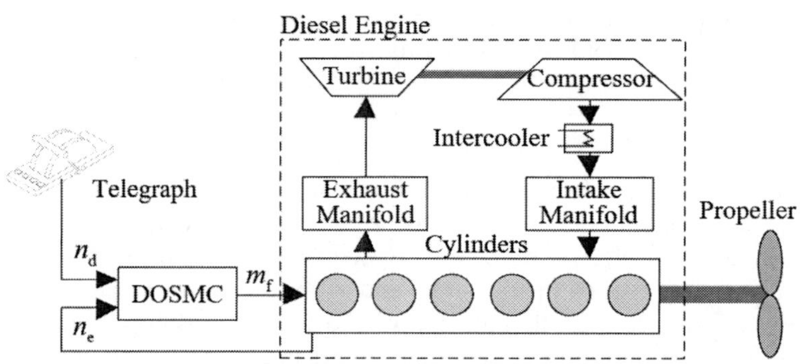

Figure 1. Structure of the marine diesel engine in a ship.

According to the desired speed n_d given by the telegraph and the actual speed n_e, a suitable fuel delivery per cycle per cylinder m_f is calculated by the controller to ensure the diesel engine speed track the target speed quickly and accurately. The indicated torque M_i produced by fuel combustion can be calculated by the following equation

$$M_i = \frac{m_f q_{LHV} \eta_i N_{cyl}}{2\pi N_{st}}, \quad (1)$$

where q_{LHV} represents the low heating value of fuel, η_i denotes the gross indicated efficiency, N_{cyl} denotes the number of cylinders, N_{st} denotes the number of strokes with $N_{st} = 1$ for two-stroke and $N_{st} = 2$ for four-stroke.

In addition to generating indicated torque, diesel engine itself also consumes some torque internally, for pumping loss and friction are not evitable. The pumping torque M_a and friction torque M_f can be calculated as follows

$$M_a = \frac{V_d}{4\pi}\left(p_{em} - p_{im}\right), \tag{2}$$

$$M_f = \frac{V_d}{4\pi}10^5\left(k_1 n_e^2 + k_2 n_e + k_3\right), \tag{3}$$

where V_d is the total displacement, p_{em} and p_{im} respectively denote the pressures in the exhaust manifold and the intake manifold, and k_1, k_2, k_3 are experimental constants.

As shown in the figure 1, the load torque of marine diesel engine mainly comes from propeller, which consists of two parts, i.e., the propeller torque M_p and the distraction torque M_d. Usually, M_d is irregular and much smaller than M_p. M_p can be simulated as

$$M_p = kn_e^2, \tag{4}$$

where k is a coefficient depends on the propeller torque coefficient, the density of water, and the propeller diameter.

Above all, the dynamic equation of the whole propulsion system including diesel engine, main shafting, and propeller can be described as

$$\frac{dn_e}{dt} = \frac{30}{\pi J}\left(M_i - M_a - M_f - M_p - M_d\right), \tag{5}$$

where J is the total rotational inertia of propulsion system.

Combining equations (1)-(5), the dynamic equation model of the propulsion system can be described as

$$\dot{x} = f(x) + g(t)u + d(t), \tag{6}$$

where system state variable $x = n_e$, control input variable $u = m_f$, $f(x)$, $g(t)$ and disturbance $d(t)$ are given as follows

$$f(x) = -\frac{30}{\pi J}(M_f + M_p),\ g(t) = \frac{15 q_{LHV} \eta_i N_{cyl}}{\pi^2 J N_{st}},\ d(t) = -\frac{30}{\pi J}(M_p + M_d) \tag{7}$$

In general, M_f and M_p can be measured in an experimental environment. However, in the harsh working environment of marine diesel engines, there is always a deviation between the measured value and the actual value. Moreover, the gross indicated efficiency η_i is affected by many factors and difficult to measure. But it usually changes in a small range when the engine is running normally. Considering all these unknown parts as disturbance, if the total interference is denoted as $D(t)$, system (6) can be rewritten as

$$\dot{x} = f_0(x) + g_0(t)u + D(t), \tag{8}$$

where $f_0(x)$ and $g_0(t)$ denote the known part of $f(x)$ and $g(t)$.

3. Controller Design

3.1. Disturbance Observer Design

For the design of the controller, a disturbance observer is designed in this subsection.

Assuming that $\dot{D}(t) = h(t)$ and $h(t) < \bar{h}$. Treating D as an additional system state variable and letting $z_1 = x$, $z_2 = D$, the original system in (8) can be described as

$$\begin{cases}\dot{z}_1 = f_0(x) + g_0(t)u + z_2 \\ \dot{z}_2 = h(t)\end{cases} \tag{9}$$

which z_1 and z_2 are observable. And a disturbance observer is designed to be

$$\begin{cases}\dot{\hat{z}}_1 = f_0(x) + g_0(x)u + \hat{z}_2 - \theta\rho_1\operatorname{sign}(\hat{z}_1 - z_1) \\ \dot{\hat{z}}_2 = -\theta^2\rho_2\operatorname{sig}^{1/2}(\hat{z}_1 - z_1)\end{cases} \tag{10}$$

where $\operatorname{sig}^\gamma(x) = |x|^\gamma \operatorname{sign}(x)$, $\hat{z}_1$ and $\hat{z}_2$ are observations of z_1 and z_2, respectively.

Define observation error $e_1 = \hat{z}_1 - z_1$ and $e_2 = \hat{z}_2 - z_2$. In view of equations (9) and (10), one can obtain an observation error system as

$$\begin{cases}\dot{e}_1 = e_2 - \theta\rho_1\operatorname{sign}(e_1) \\ \dot{e}_2 = -\theta^2\rho_2\operatorname{sig}^{1/2}(e_1) - h(t)\end{cases} \tag{11}$$

It can be proved that the error system converges in finite time. And, the proof process is given below.

Define auxiliary variable $\varepsilon_i = e_i\theta^{1-i-\bar{\theta}}$, $j=1,2$, $0<\bar{\theta}<1$, the observation error system (11) can be rewritten as

$$\begin{cases}\dot{\varepsilon}_1 = -\theta^{1-\bar{\theta}}\rho_1\operatorname{sign}(\varepsilon_1) + \theta\varepsilon_2 \\ \dot{\varepsilon}_2 = -\theta^{1-\bar{\theta}/2}\rho_2\operatorname{sig}^{1/2}(\varepsilon_1) - \theta^{-(1+\bar{\theta})}h(t)\end{cases} \tag{12}$$

Consider a Lyapunov function candidate as

$$V_e = \frac{1}{2}\boldsymbol{\vartheta}^{\mathrm{T}}\boldsymbol{P}\boldsymbol{\vartheta}$$

where $\boldsymbol{P}$ is a positive definite symmetric matrix, and auxiliary variable $\boldsymbol{\vartheta} = \left[\operatorname{sig}^{1/2}(\varepsilon_1) \quad \varepsilon_2\right]^{\mathrm{T}}$.

Differentiating the above Lyapunov function, one can get

$$\dot{V}_e = \boldsymbol{\vartheta}^{\mathrm{T}}\boldsymbol{P}\begin{bmatrix}\frac{1}{2}|\varepsilon_1|^{-1/2}\left(-\theta^{1-\bar{\theta}}\rho_1\operatorname{sign}(\varepsilon_1) + \theta\varepsilon_2\right) \\ -\theta^{1-\bar{\theta}/2}\rho_2\operatorname{sig}^{1/2}(\varepsilon_1) - \theta^{-(1+\bar{\theta})}h(t)\end{bmatrix} = 2\boldsymbol{\vartheta}^{\mathrm{T}}\boldsymbol{P}\begin{bmatrix}0 \\ -h(t)\theta^{-(1+\bar{\theta})}\end{bmatrix} - |\varepsilon_1|^{-1/2}\boldsymbol{\vartheta}^{\mathrm{T}}\boldsymbol{Q}\boldsymbol{\vartheta}$$

where $\boldsymbol{Q} = \boldsymbol{A}_\vartheta^{\mathrm{T}}\boldsymbol{P} + \boldsymbol{P}\boldsymbol{A}_\vartheta$ and $\boldsymbol{A}_\vartheta = \begin{bmatrix}-\theta^{1-\bar{\theta}}\rho_1/2 & \theta \\ -\theta^{1-\bar{\theta}/2}\rho_2/2 & 0\end{bmatrix}$ which is obviously a Hurwitz matrix.

In view of the definition of V_e, one can get $|\varepsilon_1|^{-1/2} \le \|\boldsymbol{\vartheta}\| \le \lambda_{\min}^{-1/2}(P)V_e^{1/2}$. Therefore,

$$\dot{V}_e \le 2\boldsymbol{\vartheta}^{\mathrm{T}}\boldsymbol{P}\begin{bmatrix}0 \\ -h(t)\theta^{-(1+\bar{\theta})}\end{bmatrix} - \lambda_{\min}^{-1/2}(\boldsymbol{P})V_e^{1/2}\frac{\lambda_{\min}(\boldsymbol{Q})}{\lambda_{\max}(\boldsymbol{P})} \le -(c_1 - c_2)V_e^{1/2},$$

where $c_1 = \lambda_{\min}^{-1/2}(\boldsymbol{P})\lambda_{\max}^{-1}(\boldsymbol{P})\lambda_{\min}(\boldsymbol{Q})$ and $c_2 = 2\bar{h}\lambda_{\max}(\boldsymbol{P})\lambda_{\min}^{-1}(\boldsymbol{P})\theta^{-(1+\bar{\theta})}$. There is always a θ that makes $\theta^{1+\bar{\theta}} > 2\bar{h}\lambda_{\max}^2(\boldsymbol{P})\lambda_{\min}^{-1/2}(\boldsymbol{P})\lambda_{\min\min}^{-1}(\boldsymbol{Q})$, and let $c_1 - c_2 > 0$. According to literature [16], it can be known that the observation error converges to zero in a finite time, and the convergence time $T_o \le 2V_e(\vartheta_0)^{1/2}(c_1 - c_2)^{-1}$. So, the finite-time stability of the system (12) is proved.

3.2. Sliding Mode Controller Design

Based on the proposed disturbance observer, a sliding mode controller is designed.

Firstly, a control law is designed as

$$u = g_0^{-1}(t)\left[v - f_0(x) - \hat{D}(t)\right], \tag{13}$$

where $\hat{D}(t) = \hat{z}_2$ is the observed value of the total disturbance $D(t)$. One can get from the previous proof that in a finite time there will be $D(t)\text{-}\hat{D}(t) \approx 0$. Then, substituting equation (13) for system (8), a simple first order linear system can get as

$$\dot{x} = v \tag{14}$$

And, the sliding mode controller is proposed for this system. The structure of the entire controller is shown in figure 2.

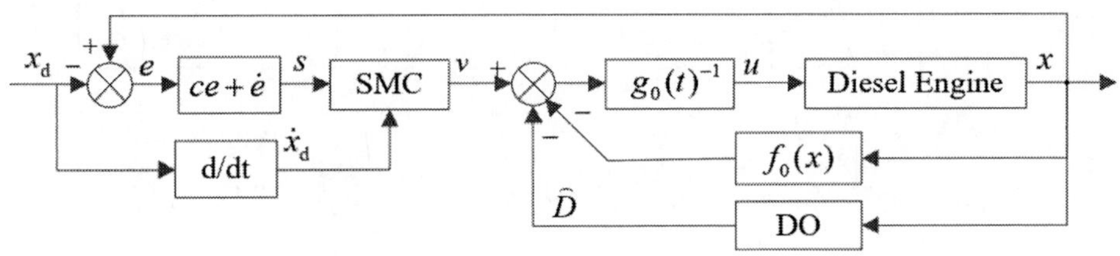

Figure 2. Structure of the proposed DOSMC.

As show, the error between the actual engine speed x and the reference value x_d is defined as $e = x - x_d$. The sliding surface is chosen as $s = ce + \dot{e}$, $c > 0$. The speed control of diesel engine is required to follow the target value quickly and avoid large overshoot. Thus, an exponential reaching rule is selected, which is

$$\dot{s} = -\varepsilon \text{sign}(s) - ks, \tag{15}$$

where $\varepsilon, k > 0$.

For the subsystem (14), the first time derivative of s is obtained

$$\dot{s} = c\dot{e} + \ddot{e} = cv + \dot{v} - c\dot{x}_d - \ddot{x}_d, \tag{16}$$

For marine diesel engine, a slope limiter for the desired speed x_d is employed to avert the rapid change of engine speed which could cause damage to the propulsion system [17]. Thus $\dot{x}_d$ is zero or a constant, and $\ddot{x}_d = 0$. Combining equation (15) and (16), one can get

$$\dot{v} = c\dot{x}_d - cv - \varepsilon \text{sign}(s) - ks. \tag{17}$$

Substitute controller law (17) into system (12), it definitely satisfies

$$s\dot{s} = -\varepsilon|s| - ks^2 \le 0. \tag{18}$$

In conclusion, the designed control law for the speed control system (8) can be described as

$$\begin{cases} u = g_0^{-1}(t)\left[v - f_0(x) - \hat{D}(t)\right] \\ \dot{v} = c\dot{x}_d - cv - \varepsilon \text{sign}(s) - ks \end{cases} \tag{19}$$

4. Simulation and Analysis

In order to get an objective comparison, a classical PID controller is designed to compare with the proposed DOSMC. The simulation processes include starting, accelerating, decelerating, suddenly loading and unloading. The slope limit of the desired speed is 400 rpm/s.

4.1. Disturbance Observer Performance

In order to avoid the early stage of the starting process, the DO begins to calculate at about 0.7s when the speed reaches the switching value of closed-loop control. Figure 3 shows the overall tracking performance of the DO. It can be observed that the DO converges rapidly although the disturbance drastically changes because of the large deviation between the fixed $g_0(t)$ and the actual $g(t)$. The DO keeps a good tracking performance in the following conditions as well. Moreover, the rate of convergence is satisfactory during loading and unloading. In addition, faster tracking performance or smoother effects can be implemented by adjusting the parameters of observer.

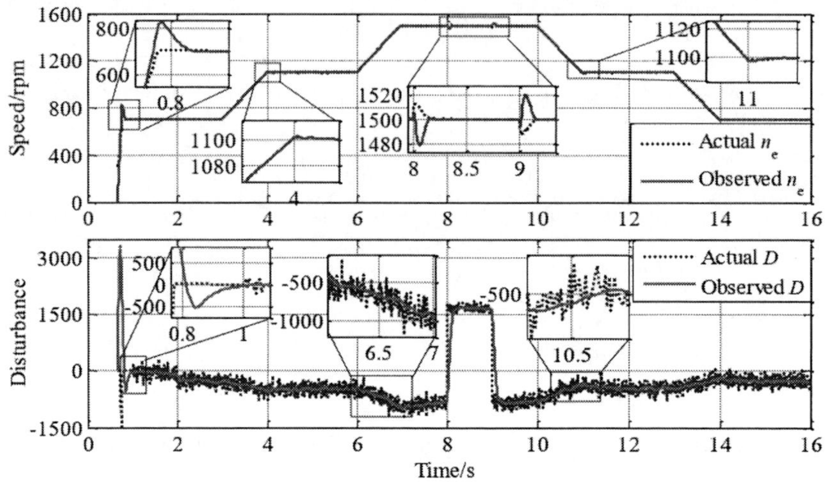

Figure 3. The overall tracking performance of the disturbance observer.

4.2. Sliding Mode Controller Performance

The comparative results between the PID and DOSMC are illustrated in figure 4.

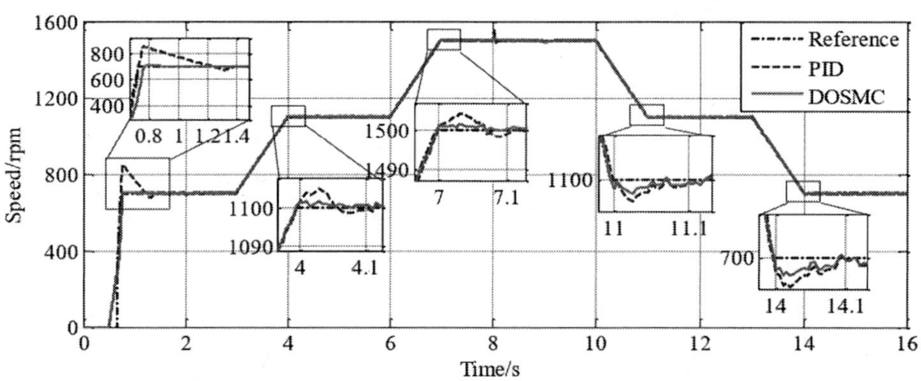

Figure 4. The control performance comparison between PID and DOSMC.

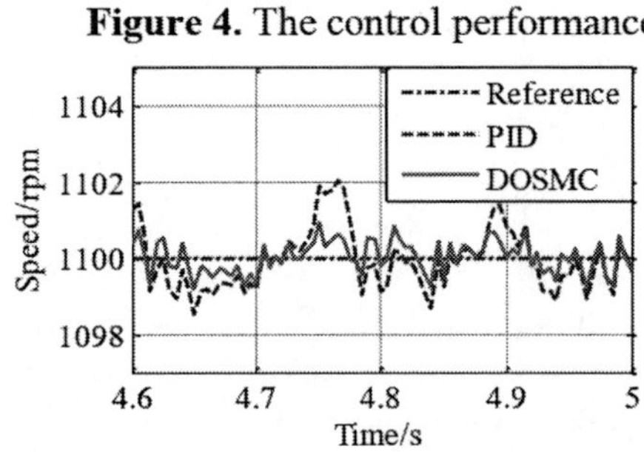

Figure 5. Steady state speed comparison

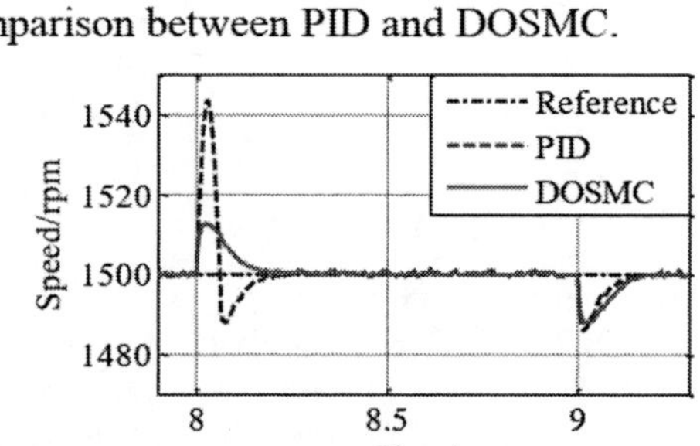

Figure 6. Speed comparison during loading and unloading.

IOP Conf. Series: Journal of Physics: Conf. Series **1267** (2019) 012088 doi:10.1088/1742-6596/1267/1/012088

In the initial stage, the diesel engine speed rises rapidly under the action of starting torque, and the controllers start work since the engine speed reaches to 300 rpm. The parameters of the two controllers are adjusted for acceleration and deceleration conditions. Hence, the control performances of the both controllers are satisfying during acceleration and decelerate conditions. However, the control effect of PID controller is not as good as that of the DOSMC. The DOSMC controller also achieves good result under the starting condition, but the PID controller performs poorly.

In this study, a Gaussian noise is employed to simulate the unknown disturbance M_d, so there are speed fluctuations in the steady state. Figure 5 gives the steady state speed comparison of the two controllers. It shows that the speed fluctuation in the DOSMC is smaller than that in the PID. The control performance of two controllers in the face of 50% sudden unloading and loading is shown in figure 6. Both controllers achieve a similar effect when loading, but the performance of the PID controller is unsatisfactory during unloading.

As can be seen above, the DOSMC controller can keep good control performance during all working conditions, and its control performance is better than the PID.

It is well known that the same diesel engine may be used for different propellers, and thc moment of inertia of the propulsion system may change with time or working conditions. This is a challenge for the robustness of the controllers. Here, the simulations with different moment of inertia are carried out, and the results are given in figure 7 and figure 8. PID1 and DOSMC1 are the simulation results with the original moment of inertia and PID2 and DOSMC2 are the results under a changed moment of inertia. Obviously, the control effect of the PID controller becomes worse when the rotary inertia is changed. Nevertheless, the DOSMC can always maintain good control performance. Therefore, it can be indicated that the DOSMC is more robust than the PID controller.

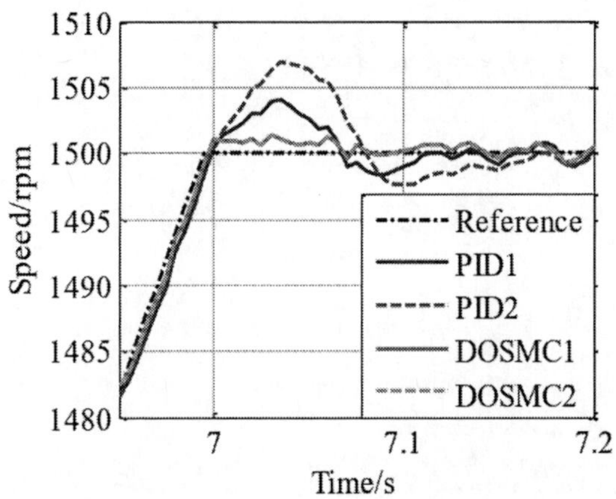

Figure 7. Speed response with different moment of inertia in acceleration process.

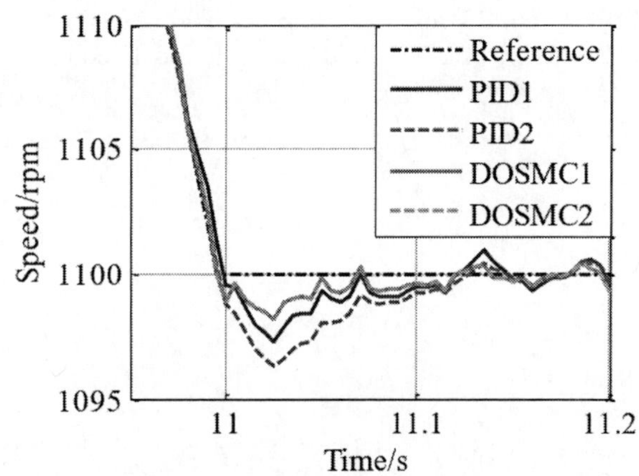

Figure 8. Speed response with different moment of inertia in deceleration process.

5. Conclusion

In this paper, a disturbance observer with finite time convergence is designed, and a DOSMC based on the disturbance observer is proposed for the speed control of a marine diesel engine. In order to assess the control performance of the proposed method, extensive comparatively simulations are carried out on a MVEM by comparing with a conventional PID controller. The tested operation conditions include starting, acceleration, deceleration, loading and unloading. Besides, the robustness of the two controllers is compared by changing the model parameter. The simulation results demonstrate that the designed disturbance observer has good performance, and the proposed DOSMC can effectively improve the control effect and has better robustness than the PID.

IOP Conf. Series: Journal of Physics: Conf. Series **1267** (2019) 012088 doi:10.1088/1742-6596/1267/1/012088

References

[1] Geertsma R D, Negenborn R R, Visser K and Hopman J J 2017 *Appl. Energy* **194** 30

[2] Guo Y, Li W, Yu S, Han X, Yuan Y, Wang Z and Ma X 2017 *Mechanical Systems and Signal Processing* **94** 1

[3] Lee S H, Yim J S, Lee J H and Sul S K 2008 *IEEE Industry Applications Society Meeting*, October 5-9,2008, Alberta,Canada p. 159-163

[4] McGowan D J, Morrow D J and Fox B 2006 *IEEE Trans. Energy Convers.* **21** 476

[5] Xiros N I 2004 *Proc. Inst. Mar. Eng. [H]* 3

[6] Mohamed F A and Koivo H N *International Conference on Future Power Systems*, November 18,2005, Amsterdam,Netherlands, p. 4

[7] Ouladsine M, Bloch G and Dovifaaz X 2004 *J. Intell. Robot. Syst.* **41** 157

[8] Wang R, Li X, Liu Y, Fu W, Liu S and Ma X 2018 *Math. Probl. Eng.* 3252653

[9] Pan W, Zhou Y and Han Y *2010 IEEE International Conference on Automation and Logistics, ICAL 2010*, p. 528-532

[10] Yuan Y, He Y, Cai L and Mao X 2017 *Proc. Inst. Mech. Eng. Part I-J Syst Control Eng.* **231** 367

[11] Yuan Y, Zhang M, Chen Y and Mao X 2018 *Trans. Inst. Meas. Control* **40** 22

[12] Li X, Ahmed Q and Rizzoni G 2017 *J. Mar. Eng. Technol.* **16** 1

[13] Khan M K, Goh K B and Spurgeon S K 2003 *Asian Journal of Control* **5** 614

[14] Zhang G-C *Proceedings - 2010 2nd WRI Global Congress on Intelligent Systems, GCIS 2010*, p. 31-34

[15] Zhang J, Lan H and Sun Y *International Conference on Material and Manufacturing, ICMM 2011, September 7-9, 2011*, Jinzhou, Liaoning, China, p. 1190-1193

[16] Bhat S P and Bernstein D S 2000 *SIAM J. Contr. Opti.* **38** 751

[17] Mizythras P, Boulougouris E and Theotokatos G 2018 *Ocean Eng.* **149** 383

Polytopic Model Weighted H_∞ Output Feedback Control of Switched Lpv System Based on Mdadt

Tao Wang[1] and Xuehai Wang[1,*]

[1]School of Electrical Engineering, Southwest Jiaotong University, Chengdu 610031, China

[*]Email: xnjtdx216510@163.com

Abstract. This paper provides and validates a class of weighted H_∞ output feedback controllers of switched LPV (Linear Parameter Varying) systems based on the MDADT (Modal Dependent Average Dwell Time) method. First of all, the article uses a multi-lyapunov function to convert the linear parameter varying system into a polytopic model, and designs a switching law that can guarantee the system to be globally uniform and exponentially stable on the MDADT method. It can be shown that when the state is stable and can be measured in real time, the switching law and weighted H_∞ controllers for each subsystem under the MDADT limit can be calculated. Finally, a numerical example is given to verify the feasibility of this method and the stability of the switched LPV system.

1. Introduction

Switched system is a sort of hybrid system [1] and generally composed of several subsystems. Each subsystem has different parameters and can be switched one by one. The system can be operated stably by choosing a right switching law and useful controllers. However, variable parameters of switched system are nonlinear in some real working places [2]. Thus methods like Jacobian-linearization are usually used to convert the system into LPV (Linear Parameter Varying) model [3]. Bei Lu proposed a new conception that combined the method of LPV control and the idea of switched system. This new proposal has given a new approach of switching LPV control based on Lyapunov function [4].

The state feedback controllers of switched LPV system based on ADT (Average Dwell Time) [5] is not easy to implement in real world. Moreover, ADT demands each subsystem sharing the same time lower limitation. Thus MDADT (Modal Dependent Average Dwell Time) as an advanced conception evolved from ADT is provided by Xudong Zhao in [6]. MDADT is better than ADT because MDADT is used for each subsystem but not for the whole switched system. Each subsystem has each time limitation by each mode of MDADT but not the same limitation at all.

This paper is interested in switched LPV system.There are three main contributions. Firstly, the polytopic model is used to replace the modeling of a linear variable-parameter system. The subsystem corresponding to each vertex is used to represent the entire system, and the problem of system parameter changes is solved. Secondly, the sufficient conditions for the existence of MDADT-based output feedback controller are given, and the correctness of the condition is proved. Finally, the design method of output feedback controller for disturbance suppression is given and a numerical simulation example is given to prove our results.

2. Multicellular Decomposition of Switched LPV System

Given a complete switched LPV systems

$$\begin{cases} \dot{x}(t)=A_{\sigma(t)}(\rho)x(t)+B_{1\cdot\sigma(t)}(\rho)w(t)+B_{2\cdot\sigma(t)}(\rho)u(t) \\ y(t)=C_{1\cdot\sigma(t)}(\rho)x(t)+D_{11\cdot\sigma(t)}(\rho)w(t)+D_{12\cdot\sigma(t)}(\rho)u(t), \\ z(t)=C_{2\cdot\sigma(t)}(\rho)x(t)+D_{21\cdot\sigma(t)}(\rho)w(t)+D_{22\cdot\sigma(t)}(\rho)u(t) \end{cases} \tag{1}$$

where $x(t)\in R^n$ and $u(t)\in R^r$ denote the state vector and the input vector respectively, and $y(t)\in R^p$ denotes the controlled output vector. $w(t)\in R^q$ is the disturbance input vector, $\rho(t)=[\rho_1(t),\rho_2(t),\cdots,\rho_s(t)]^T$ is the time varying parameter vector that can be measured in real time and has the solution in the compact set. $A_{\sigma(t)}(\rho), B_{1\cdot\sigma(t)}(\rho), B_{2\cdot\sigma(t)}(\rho), C_{1\cdot\sigma(t)}(\rho), C_{2\cdot\sigma(t)}(\rho), D_{11\cdot\sigma(t)}(\rho), D_{11\cdot\sigma(t)}(\rho), D_{12\cdot\sigma(t)}(\rho), D_{21\cdot\sigma(t)}(\rho)$, $D_{22\cdot\sigma(t)}(\rho)$ are the functions of time-varying parameter matrix $\rho(t)$. $\sigma(t)$ denotes the control signal of the switched system and $\sigma(t):[t_0,\ \infty)\to N=\{1,2,\ldots,N\}$. N is a set of numbers, each number stands for a subscript of subsystem. For the switched LPV system under the control of switching signal $\sigma(t)$, when $\sigma(t)=i$, subsystem i is activated.

Definition 1: For the vector of variable parameters $\rho\in P\subset R^r$, $\rho(t)=[\rho_1(t),\rho_2(t),\ldots,\rho_s(t)]^T$, there is a polytopic description as $Co\{\omega_m, m=1,2,\ldots r\}:=\{\sum_{m=1}^{r}\alpha_m\omega_m:\alpha_m\geq 0,\sum_{m=1}^{r}\alpha_m=1\}$ about this vector. Then for each LPV subsystem i of the switched LPV system, when $\sigma(t)=i$, the subsystem is activated, the polytopic model is as followings

$$\begin{bmatrix} A_{\sigma(t)}(\rho) & B_{1\cdot\sigma(t)}(\rho) & B_{2\cdot\sigma(t)}(\rho) \\ C_{1\cdot\sigma(t)}(\rho) & D_{11\cdot\sigma(t)}(\rho) & D_{12\cdot\sigma(t)}(\rho) \\ C_{2\cdot\sigma(t)}(\rho) & D_{21\cdot\sigma(t)}(\rho) & D_{22\cdot\sigma(t)}(\rho) \end{bmatrix}:=\sum_{m=1}^{r}\alpha_{\sigma(t)\cdot m}(t)\begin{bmatrix} A_{\sigma(t)\cdot m} & B_{1\cdot\sigma(t)\cdot m} & B_{2\cdot\sigma(t)\cdot m} \\ C_{1\cdot\sigma(t)\cdot m} & D_{11\cdot\sigma(t)\cdot m} & D_{12\cdot\sigma(t)\cdot m} \\ C_{2\cdot\sigma(t)\cdot m} & D_{21\cdot\sigma(t)\cdot m} & D_{22\cdot\sigma(t)\cdot m} \end{bmatrix} \tag{2}$$

Then, process of designing controllers of each switched LPV subsystem represented by their vertices with the help of MDADT is given. For each vertex m of (2), according to the properties of the variable parameters, $\rho(t)\in P\subset R^r$, P is the varying region of the system parameters. The control signal is $\sigma(t):[t_0,\ \infty)\to N=\{1,2,\ldots,N\}$. When $\sigma(t)=i$, subsystem i gets started, weighted H_∞ output feedback controllers $C\,u(t)=K_{i\cdot m}(\rho)y(t)$ for each subsystem i can be designed as follows

$$\begin{cases} \dot{\hat{x}}_{i\cdot m}(t)=A_{K\cdot i\cdot m}\hat{x}_{i\cdot m}(t)+B_{K\cdot i\cdot m}y(t) \\ u(t)=C_{K\cdot i\cdot m}\hat{x}_{i\cdot m}(t)+D_{K\cdot i\cdot m}y(t) \end{cases} \quad \begin{bmatrix} A_{K\cdot i}(\rho) & B_{K\cdot i}(\rho) \\ C_{K\cdot i}(\rho) & D_{K\cdot i}(\rho) \end{bmatrix}:=\sum_{m=1}^{r}\alpha_{\sigma(t)\cdot m}(t)\begin{bmatrix} A_{K\cdot i\cdot m} & B_{K\cdot i\cdot m} \\ C_{K\cdot i\cdot m} & D_{K\cdot i\cdot m} \end{bmatrix} \tag{3}$$

$\hat{x}_i(t)$ denotes the state vector of controllers, $A_{K\cdot i}(\rho)$, $B_{K\cdot i}(\rho)$, $C_{K\cdot i}(\rho)$, $D_{K\cdot i}(\rho)$ are parameter-dependent matrices of the controllers of each subsystem, (3) denote matrices of controllers in each vertex m of each subsystem i. After the decomposition of polytopic model, $A_{K\cdot i\cdot m}$, $B_{K\cdot i\cdot m}$, $C_{K\cdot i\cdot m}$, $D_{K\cdot i\cdot m}$ all have no variable parameters and can be calculated offline. After adding these controllers, $x_{i\cdot m}(t)$ denotes state of vertex m of of subsystem i, $\hat{x}_{i\cdot m}(t)$ denotes controller state of vertex m of subsystem i, closed-loop switched LPV system model converted from the open-loop original is as follow

$$\begin{cases} \dot{\xi}_{i\cdot m}(t)=A_{Cl\cdot i\cdot m}\xi_{i\cdot m}(t)+B_{Cl\cdot i\cdot m}\omega(t) \\ \dot{z}_{i\cdot m}(t)=C_{Cl\cdot i\cdot m}\xi_{i\cdot m}(t)+D_{Cl\cdot i\cdot m}\omega(t) \end{cases}, \tag{4}$$

where $A_{cl\cdot i\cdot m}$, $B_{cl\cdot i\cdot m}$, $B_{cl\cdot i\cdot m}$, $B_{cl\cdot i\cdot m}$ donate closed-loop switched LPV system parameter matrix and

$$\xi_{i\cdot m}(t)=\begin{bmatrix} x_{i\cdot m}(t) \\ \hat{x}_{i\cdot m}(t) \end{bmatrix} \quad A_{Cl\cdot i\cdot m}=\begin{bmatrix} A_{i\cdot m}+B_{2\cdot i\cdot m}D_{K\cdot i\cdot m}C_{2\cdot i\cdot m} & B_{2\cdot i\cdot m}C_{K\cdot i\cdot m} \\ B_{K\cdot i\cdot m}C_{2\cdot i\cdot m} & A_{K\cdot i\cdot m} \end{bmatrix} \quad B_{Cl\cdot i\cdot m}=\begin{bmatrix} B_{1\cdot i\cdot m}+B_{2\cdot i\cdot m}D_{K\cdot i\cdot m}D_{21\cdot i\cdot m} \\ B_{K\cdot i\cdot m}D_{21\cdot i\cdot m} \end{bmatrix} \tag{5}$$

$$C_{Cl\cdot i\cdot m}=\begin{bmatrix} C_{1\cdot i\cdot m}+D_{12\cdot i\cdot m}D_{K\cdot i\cdot m}C_{2\cdot i\cdot m} & D_{12\cdot i\cdot m}C_{K\cdot i\cdot m} \end{bmatrix} \tag{6}$$

$$D_{Cl\cdot i\cdot m}=D_{11\cdot i\cdot m}+D_{12\cdot i\cdot m}D_{K\cdot i\cdot m}D_{21\cdot i\cdot m} \tag{7}$$

3. Polytopic model weighted H_∞ Output Feedback Controllers Based on MDADT

Lemma 1: [7] For the switched system (1), $0<\alpha_0<1, \mu>1, \gamma>0$ are given constants. Assuming there exist a positive definite continuous differentiable function $V_{\sigma(t)}$: $R^n \to R$, $\sigma(t)\in N$ that satisfies the zero initial condition $V_{\sigma(t)}=0, \forall(i,j)\in N\times N, i\neq j, V_i(x(t))\le \mu_i V_j(x(t)), \forall i\in N$, we easily get in (9) by defining disturbances to be $\Gamma(t)=y^{\mathrm T}y-\gamma^2 w^{\mathrm T}w$

$$\dot V_i(x(t))+\Gamma(t)\le -\alpha_i V_i(x(t)) \tag{8}$$

Then from the definition of MDADT the system is globally uniformly asymptotically stable, and for any switching signal satisfying $N_{i\cdot m}(t,T)\le N_{0\cdot i}+(T-t)/\tau_{a\cdot i}$ in [8], the signal can be obtained as

$$\tau^*_{a\cdot i}\ge \tau_{a\cdot i}=\frac{\ln\mu_i}{\alpha_i} \tag{9}$$

Subsystem i corresponding to each vertex m of the system (1) has a $\tau_{a\cdot i\cdot m}$ and satisfies MDADT. The $\tau_{a\cdot i\cdot m}$ represents the average of the dwell time between any two consecutive handovers within each subsystem. MDADT cannot be less than this constant. This section, we design controllers of each vertex of subsystems ruled by MDADT and make the switched LPV system stable. Next, we make a concise deduction and proof.

Theorem 1. For each vertex m of subsystem i, and given constants $0<\alpha_i<1$, $\mu_i>1, \gamma>0$, the switching system (1) is globally uniformly exponentially stable and has output feedback controllers with performance γ, if there exist $X_{Cl\cdot i}$ such that

$$\begin{bmatrix} A_{Cl\cdot i\cdot m}{}^T X_{Cl\cdot i}+X_{Cl\cdot i}A_{Cl\cdot i\cdot m}+\alpha_i X_{Cl\cdot i} & X_{Cl\cdot i}B_{Cl\cdot i\cdot m} & C_{Cl\cdot i\cdot m}{}^T \\ B_{Cl\cdot i\cdot m}{}^T X_{Cl\cdot i} & -\gamma I & D_{Cl\cdot i\cdot m}{}^T \\ C_{Cl\cdot i\cdot m} & D_{Cl\cdot i\cdot m} & -\gamma I \end{bmatrix}\le 0 \tag{10}$$

$$X_{Cl\cdot i}\le \mu_i X_{Cl\cdot j} \tag{11}$$

Proof: For each vertex m of subsystem i, assuming its quadratic form Lyapunov function

$$V_i=x^T X_i x,\ \ \forall\sigma(t)=i\in N \tag{12}$$

When there is no disturbance $\omega(t)=0$, we can easily get $\dot V_i(x)+\alpha_i V_i(x)=x^T\Lambda x$, and define $\Lambda=A_{Cl\cdot i}(\rho)^T X_{Cl\cdot i}+X_{Cl\cdot i}A_{Cl\cdot i}(\rho)+\alpha_i X_{Cl\cdot i}$, where $A_{Cl\cdot i}(\rho)$ is a closed-loop system matrix, therefore, we can get the system is possible to be stable if $\Lambda<0$. When there is disturbances in the system, we have

$$\begin{aligned} &\dot V_i(x)+\alpha_i V_i(x)+y^T y-\gamma^2 w^T w \\ &=\sum_{m=1}^{r}\alpha_{\sigma(t)\cdot m}(t)\begin{bmatrix} x^T & w^T\end{bmatrix}\begin{bmatrix} A^T_{Cl\cdot i\cdot m}X_{Cl\cdot i}+X_{Cl\cdot i}A_{Cl\cdot i\cdot m}+\alpha_i X_{Cl\cdot i} & X_{Cl\cdot i}B_{Cl\cdot i\cdot m} & C^T_{Cl\cdot i\cdot m} \\ * & -\gamma I & D^T_{Cl\cdot i\cdot m} \\ * & * & -\gamma I\end{bmatrix}\begin{bmatrix} x \\ w\end{bmatrix} \\ &=\sum_{m=1}^{r}\alpha_{\sigma(t)\cdot m}(t)\begin{bmatrix} x^T & w^T\end{bmatrix}\Theta\begin{bmatrix} x \\ w\end{bmatrix}\end{aligned} \tag{13}$$

From theorem (1), if $\Theta<0$, the system will be stable, and $\tau^*_{a\cdot i\cdot m}\ge\tau_{a\cdot i\cdot m}=\ln\mu_i/\alpha_i$

So (10) and (11) must be verified with a feasible solution. Next solve a set of LMIs by using elimination. Introduce following matrices to simplify (10)

$$K_{i\cdot m}=\begin{bmatrix} A_{K\cdot i\cdot m} & B_{K\cdot i\cdot m} \\ C_{K\cdot i\cdot m} & D_{K\cdot i\cdot m}\end{bmatrix}\ A_{0\cdot i\cdot m}=\begin{bmatrix} A_{i\cdot m} & 0 \\ 0 & 0\end{bmatrix}\ B_{0\cdot i\cdot m}=\begin{bmatrix} B_{i\cdot m} \\ 0\end{bmatrix}\ C_{0\cdot i\cdot m}=\begin{bmatrix} C_{1\cdot i\cdot m} & 0\end{bmatrix}\ \overline{B}_{i\cdot m}=\begin{bmatrix} 0 & B_{2\cdot i\cdot m} \\ I & 0\end{bmatrix} \tag{14}$$

$$\overline{C}_{i\cdot m}=\begin{bmatrix} 0 & I \\ C_{2\cdot i\cdot m} & 0\end{bmatrix}\ \overline{D}_{12\cdot i\cdot m}=\begin{bmatrix} 0 & D_{12\cdot i\cdot m}\end{bmatrix}\ \overline{D}_{21\cdot i\cdot m}=\begin{bmatrix} 0 \\ D_{21\cdot i\cdot m}\end{bmatrix} \tag{15}$$

IOP Conf. Series: Journal of Physics: Conf. Series **1267** (2019) 012089 doi:10.1088/1742-6596/1267/1/012089

Then we can get

$$\begin{bmatrix} \Xi & X_{Cl\cdot i}(B_{0\cdot i\cdot m}+\bar{B}_{i\cdot m}K_{i\cdot m}\bar{D}_{21\cdot i\cdot m}) & (C_{0\cdot i\cdot m}+\bar{D}_{12\cdot i\cdot m}K_{i\cdot m}\bar{C}_{i\cdot m})^T \\ * & -\gamma I & (D_{11\cdot i\cdot m}+\bar{D}_{12\cdot i\cdot m}K_{i\cdot m}\bar{D}_{21\cdot i\cdot m})^T \\ * & * & -\gamma I \end{bmatrix} \le 0, \tag{16}$$

where $\Xi=(A_{0\cdot i\cdot m}+\bar{B}_{i\cdot m}K_{i\cdot m}\bar{C}_{i\cdot m})^T X_{Cl\cdot i}+X_{Cl\cdot i}(A_{0\cdot i\cdot m}+\bar{B}_{i\cdot m}K_{i\cdot m}\bar{C}_{i\cdot m})+\alpha_i X_{Cl\cdot i}$. In (9), $X_{Cl\cdot i}$ and $K_{i\cdot m}$ can be solved.

Define assistant matrices $H_{Cl\cdot i\cdot m}$, $M_{X_{Cl\cdot i\cdot m}}$, $N_{i\cdot m}$, $T_{Cl\cdot i\cdot m}$

$$H_{Cl\cdot i\cdot m}=\begin{bmatrix} A_{0\cdot i\cdot m}{}^T X_{Cl\cdot i}+X_{Cl\cdot i}A_{0\cdot i\cdot m}+\alpha_i X_{Cl\cdot i} & X_{Cl\cdot i}B_{0\cdot i\cdot m} & C_{0\cdot i\cdot m}{}^T \\ * & -\gamma I & D_{11\cdot i\cdot m}{}^T \\ * & * & -\gamma I \end{bmatrix}, M_{X_{Cl\cdot i\cdot m}}=\begin{bmatrix}\bar{B}_{i\cdot m}X_{Cl\cdot i} & 0 & \bar{D}^T_{12\cdot i\cdot m}\end{bmatrix}, N_{i\cdot m}=\begin{bmatrix}\bar{C}_{i\cdot m} & \bar{D}_{21\cdot i\cdot m} & 0\end{bmatrix} \tag{17}$$

$$T_{Cl\cdot i\cdot m}=\begin{bmatrix} X^{-1}_{Cl\cdot i} & 0 & 0 \\ 0 & I & 0 \\ 0 & 0 & I \end{bmatrix} H_{Cl\cdot i\cdot m}\begin{bmatrix} X^{-1}_{Cl\cdot i} & 0 & 0 \\ 0 & I & 0 \\ 0 & 0 & I \end{bmatrix}=\begin{bmatrix} A_{0\cdot i\cdot m}{}^T X^{-1}_{Cl\cdot i}+X^{-1}_{Cl\cdot i}A_{0\cdot i\cdot m}+\alpha_i X^{-1}_{Cl\cdot i} & B_{0\cdot i\cdot m} & X^{-1}_{Cl\cdot i}C_{0\cdot i\cdot m}{}^T \\ * & -\gamma I & D_{11\cdot i\cdot m}{}^T \\ * & * & -\gamma I \end{bmatrix} \tag{18}$$

Then (16) can be rewritten as

$$H_{Cl\cdot i\cdot m}+M_{X_{Cl\cdot i}}{}^T K_{i\cdot m}N_{i\cdot m}+N_{i\cdot m}{}^T K_{i\cdot m}{}^T M_{X_{Cl\cdot i\cdot m}}<0 \tag{19}$$

Using the properties of nuclear space, $M^{\perp}_{X_{Cl\cdot i\cdot m}}$ and $N^{\perp}_{i\cdot m}$ are matrices constructed by the kernel space basis vectors of the corresponding matrices. to solve the problem more easily, using $T_{Cl\cdot i\cdot m}$, we can get

$$\begin{cases} \bar{M}^{\perp}_{X_{Cl\cdot i\cdot m}}{}^T T_{Cl\cdot i\cdot m}\bar{M}^{\perp}_{X_{Cl\cdot i\cdot m}}<0 \\ N^{\perp}_{i\cdot m}{}^T H_{Cl\cdot i\cdot m}N^{\perp}_{i\cdot m}<0 \end{cases} \tag{20}$$

Since the matrix $X_{Cl\cdot i\cdot m}$ is a real symmetric matrix with dimensions $(n+n_K)\times(n+n_K)$, n and n_K denote orders of system model and controller, the decompose $X_{Cl\cdot i}$ and $X^{-1}_{Cl\cdot i}$:

$$X_{Cl\cdot i}=\begin{bmatrix} X_i & X_{2\cdot i} \\ X^T_{2\cdot i} & X_{3\cdot i}\end{bmatrix},\ X^{-1}_{Cl\cdot i\cdot m}=\begin{bmatrix} Y_i & Y_{2\cdot i} \\ Y^T_{2\cdot i} & Y_{3\cdot i}\end{bmatrix} \tag{21}$$

Taking back (21) into $H_{Cl\cdot i\cdot m}$ and $T_{Cl\cdot i\cdot m}$ and defining $L_{0\cdot i\cdot m}$ and $L_{c\cdot i\cdot m}$ are matrices constructed by column vectors which are arbitrary sets of basis vectors in space $\ker(\begin{bmatrix}C_{2\cdot i\cdot m} & D_{21\cdot i\cdot m}\end{bmatrix})$ and $\ker(\begin{bmatrix}B^T_{2\cdot i\cdot m} & D^T_{12\cdot i\cdot m}\end{bmatrix})$ respectively. We can get

$$\begin{bmatrix} L_{c\cdot i\cdot m} & 0 \\ 0 & I\end{bmatrix}^T \begin{bmatrix} A_{i\cdot m}Y_i+Y_iA^T_{i\cdot m}+\alpha_i Y_i & Y_iC^T_{1\cdot i\cdot m} & B_{1\cdot i\cdot m} \\ * & -\gamma I & D_{11\cdot i\cdot m} \\ * & * & -\mu I \end{bmatrix}\begin{bmatrix} L_{c\cdot i\cdot m} & 0 \\ 0 & I\end{bmatrix}\le 0 \tag{22}$$

$$\begin{bmatrix} L_{0\cdot i\cdot m} & 0 \\ 0 & I\end{bmatrix}^T \begin{bmatrix} A_{i\cdot m}X_i+X_iA^T_{i\cdot m}+\alpha_i X_i & X_iB_{1\cdot i\cdot m} & C^T_{1\cdot i\cdot m} \\ * & -\gamma I & D^T_{11\cdot i\cdot m} \\ * & * & -\mu I \end{bmatrix}\begin{bmatrix} L_{0\cdot i\cdot m} & 0 \\ 0 & I\end{bmatrix}\le 0 \tag{23}$$

$$X_{Cl\cdot i}\le \mu_i X_{Cl\cdot j} \tag{24}$$

$$\tau^*_{a\cdot i\cdot m}\ge \tau_{a\cdot i\cdot m}=\frac{\ln \mu_i}{\alpha_i} \tag{25}$$

After verifying the feasibility of $X_{Cl\cdot i}$, solutions of $A_{K\cdot i\cdot m}$, $B_{K\cdot i\cdot m}$, $C_{K\cdot i\cdot m}$, $D_{K\cdot i\cdot m}$ can be solved. The controllers of the each subsystem are obtained as (26).

$$K_{i\cdot m}:=\sum_{m=1}^{r}\alpha_{\sigma(t)\cdot m}(t)\begin{bmatrix} A_{K\cdot i\cdot m} & B_{K\cdot i\cdot m} \\ C_{K\cdot i\cdot m} & D_{K\cdot i\cdot m}\end{bmatrix} \tag{26}$$

There exist various MDADTs for each vertex m of each switched LPV subsystem, and the system can be exponentially stable as long as the average ADT of each subsystem is larger than MDADT. If the MDADTs are chosen the same by its property, their different vertices of the subsystem can be switched synchronously and the shape of the system can be maintained. For polytopic model of switched LPV system with multiple vertices that has the same parameter range, by using the same MDADT the whole switched LPV system can realize exponential stability.

4. Simulation

In this section MDADT will be calculated and weighted H_∞ output feedback controllers will be obtained. A numerical example with two is adopted and the initial stateInitial state $x_0 = [4\ \ -7]^T$.

Subsystem 1

$$A_1 = \begin{bmatrix} -3-0.1\rho(t) & -1 \\ -2 & -4 \end{bmatrix}, B_{1\cdot1} = \begin{bmatrix} 1 \\ 1.5 \end{bmatrix}, B_{2\cdot1} = \begin{bmatrix} 1 \\ 1.5 \end{bmatrix}, C_{1\cdot1} = \begin{bmatrix} 0 & 1 \\ -1 & -1 \end{bmatrix}, C_{2\cdot1} = \begin{bmatrix} 1 & 0 \\ 0 & 1 \end{bmatrix}, D_{11\cdot1} = D_{12\cdot1} = D_{22\cdot1} = D_{22\cdot1} = \begin{bmatrix} 0 \\ 0 \end{bmatrix} \quad (27)$$

Subsystem 2

$$A_2 = \begin{bmatrix} -2-0.1\rho(t) & -1.5 \\ -2 & -6.5 \end{bmatrix}, B_{1\cdot2} = \begin{bmatrix} 1.2 \\ 1.3 \end{bmatrix}, B_{2\cdot2} = \begin{bmatrix} 0.62 \\ 0.88 \end{bmatrix}, C_{1\cdot2} = \begin{bmatrix} 0 & 1 \\ -1 & 1 \end{bmatrix}, C_{2\cdot2} = \begin{bmatrix} 1 & 0 \\ 0 & 1 \end{bmatrix}, D_{11\cdot2} = D_{12\cdot2} = D_{21\cdot2} = D_{22\cdot2} = \begin{bmatrix} 0 \\ 0 \end{bmatrix} \quad (28)$$

Variable parameter vector $\rho(t)=|\sin(t)|$, $\rho(t)\in(0,1)$. We decompose the system according to the section 2 to get the Polytopic model, and then get its vertex controller as follows

Vertex 1 controller for subsystem 1

$$A_{K\cdot1\cdot1} = \begin{bmatrix} -2.0011 & 0 \\ 0 & -2.0011 \end{bmatrix}, B_{K\cdot1\cdot1} = \begin{bmatrix} 0 & 0 \\ 0 & 0 \end{bmatrix}, C_{K\cdot1\cdot1} = \begin{bmatrix} 0 & 0 \end{bmatrix}, D_{K\cdot1\cdot1} = \begin{bmatrix} 3.6218 & -0.0544 \end{bmatrix} \quad (29)$$

Vertex 2 controller for subsystem 1

$$A_{K\cdot1\cdot2} = \begin{bmatrix} -2.0042 & 0 \\ 0 & -2.0042 \end{bmatrix}, B_{K\cdot1\cdot2} = \begin{bmatrix} 0 & 0 \\ 0 & 0 \end{bmatrix}, C_{K\cdot1\cdot2} = \begin{bmatrix} 0 & 0 \end{bmatrix}, D_{K\cdot1\cdot2} = \begin{bmatrix} 4.1262 & 0.0066 \end{bmatrix} \quad (30)$$

Vertex 1 controller for subsystem 2

$$A_{K\cdot2\cdot1} = \begin{bmatrix} -5.2573 & 0 \\ 0 & -5.2573 \end{bmatrix}, B_{K\cdot2\cdot1} = \begin{bmatrix} 0 & 0 \\ 0 & 0 \end{bmatrix}, C_{K\cdot2\cdot1} = \begin{bmatrix} 0 & 0 \end{bmatrix}, D_{K\cdot2\cdot1} = \begin{bmatrix} -5.1347 & 3.7732 \end{bmatrix} \quad (31)$$

Vertex 2 controller for subsystem 2

$$A_{K\cdot2\cdot2} = \begin{bmatrix} -5.0205 & 0 \\ 0 & -5.0205 \end{bmatrix}, B_{K\cdot2\cdot2} = \begin{bmatrix} 0 & 0 \\ 0 & 0 \end{bmatrix}, C_{K\cdot2\cdot2} = \begin{bmatrix} 0 & 0 \end{bmatrix}, D_{K\cdot2\cdot2} = \begin{bmatrix} -5.231 & 4.2468 \end{bmatrix} \quad (32)$$

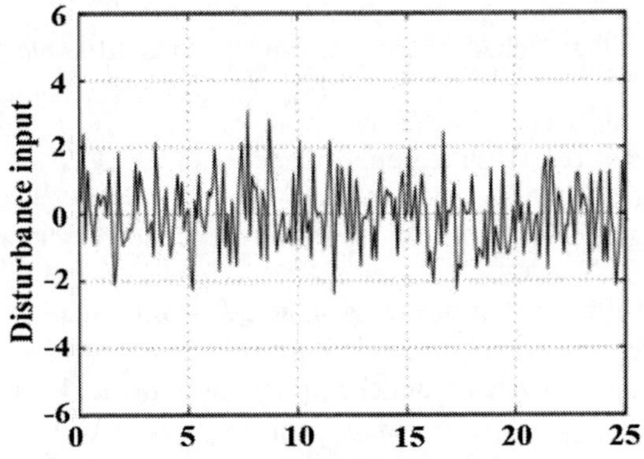

Figure 1. Disturbance Input

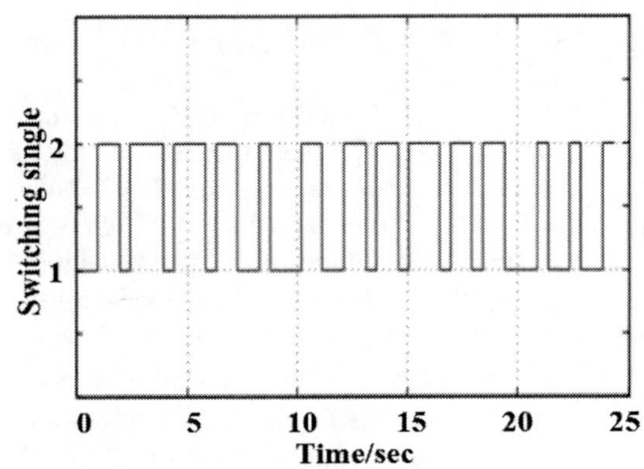

Figure 2. Switching Single

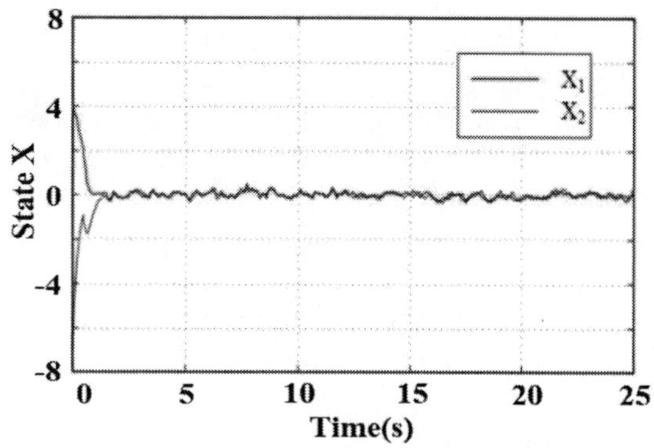

Figure 3. State response of the system

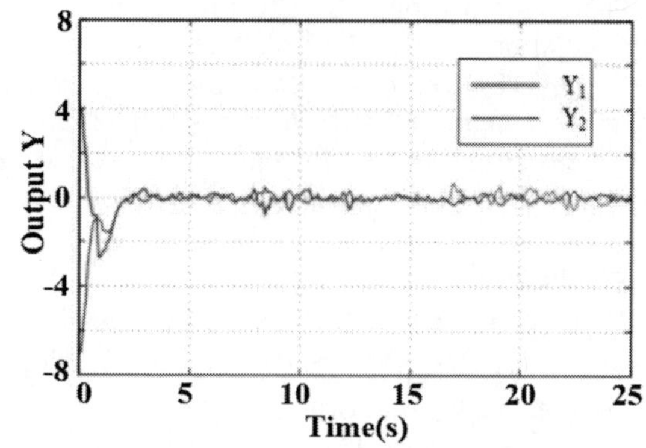

Figure 4. Output of the system

The external disturbance signal of the system is shown in Figure 1. Let $a_1 = a_2 = a = 0.5$, $\mu_1 = 1.22$ $\mu_2 = 1.98, \gamma = 1$ and MDADT of each vertex subsystem $\tau_{a\text{-}1}=\tau_{a\text{-}1\text{-}1}=\tau_{a\text{-}1\text{-}2}=0.3977$, $\tau_{a\text{-}2}=\tau_{a\text{-}2\text{-}1}=\tau_{a\text{-}2\text{-}2}=1.3662$ In this time, the switching single of the system is shown in Figure 2, the state response and the output of the system is shown in Figure 3 and Figure 4. Therefore, the polytopic model weighted H_∞ output feedback control based on MDADT proposed in this paper can make the switched LPV system stable.

5. Conclusions

The purpose of this paper is to design weighted H_∞ output feedback controllers of switched LPV system based on MDADT. For the switched LPV system, LPV part of the system can be divided into different vertices of each switched subsystem by using polytopic model. Then design output feedback controllers for each vertex based on MDADT of each subsystem so that the system can switch synchronously and maintain its shape and the region of variable parameters. Finally all vertices are integrated into a complete system and the exponential stability of the system can be achieved because of the switching signal restricted by MDADT. The feasibility of this method is verified by simulation. Comparing with the state feedback controller, this approach has a better practical significance in the working places and is easier to implement.

Acknowledgement

This work is supported by the National Natural Science Foundation of China (51477146, 61603312)

References

[1] Sun Z, Ge S S 2005 Analysis and synthesis of switched linear control systems. *Automatica*, 41(2):181-195.

[2] Liberzon D 2003 Switching in Systems and Control. Berlin: Birkhauser.

[3] Baranyi P 2004 TP model transformation as a way to LMI-based controller design. IEEE *Transaction on Industrial Electronics*, 51(2):387-400.

[4] Lu B, Wu F 2004 Switching LPV control designs using multiple parameter-dependent Lyapunov functions. *Automatica*, 40(11): 1973-1980.

[5] Haspanha J P, Morse A S 1995 Stability of switched systems with average dwell time. P*aper presented at the Proceeding of 38th IEEE Conference on Decision and Control*, Phoenix.

[6] Zhao X D, Liu H 2013 Weighted H_∞ performance analysis of switched linear systems with mode-dependent average dwell time. *International Journal of Systems Sciences*, 44(11):2130-2139.

[7] Lu Q, Zhang L, Karimi H R, Shi Y 2013 H_∞ control for asynchronously switched LPV systems with mode-dependent average dwell time. *IET Control Theory and Applications.* 7(5):673-683.

[8] Zhao X D, Liu H 2013 Weighted H_∞ performance analysis of switched linear systems with mode-dependent average dwell time. *International Journal of Systems Sciences*, 44(11):2130-21

IOP Conf. Series: Journal of Physics: Conf. Series **1267** (2019) 012090 doi:10.1088/1742-6596/1267/1/012090

An Improved Strategy for Machine Comprehension Based upon Dynamic Memory Networks

Yuchen Sun and Min Zhu*

School of Computer Science and Software Engineering, East China Normal University, Shanghai, 200062, China

*yc_sun1002@foxmail.com

Abstract. Recently many complicated reading comprehension models have been proposed. Most of these models are constantly improved from a basic model. However, most of the innovations are at the interaction layer. In this work, we analyse the shortcomings of Dynamic Memory Networks and its extended model. Then we decide to put forward some different opinions that making some improvements in the embedding layer based upon these models. First of all, this method uses pre-training to get character embedding. Our model uses CNN to train the character vector, and then we splicing the character vector and the word vector. Meanwhile, we validate the model with the bAbI dataset. Our results show that this method can improve the interaction between context and question words.

1. Introduction

Machine reading comprehension is actually similar to the problems encountered by human reading comprehension. However, in order to reduce the difficulty of the task, many of the current machine reading comprehension studies exclude world knowledge, use artificially constructed relatively simple data sets and answer some relatively simple questions. Give an article that requires machine understanding and the corresponding questions and answers. More common forms of tasks include synthetic question answering (QA), Cloze-style queries, and multiple-choice questions.

Synthetic question answering is an artificially constructed article formed by several simple facts, and gives corresponding questions. It requires the machine to read and understand the content of the article and make some reasoning to get the correct answer. The correct answer is a keyword or entity in the article. The question answering system is designed to accurately find the information the user is looking for, not the document or fragment that contains the answer. A special form of question answering system, the extractive QA, is the answer to the question directly extracted from the document.

Neural network based approaches have made great strides in text classification and image (Krizhevsky et al., 2012; Socher et al., 2013)[1][2]. However, it was not until these years that progress was made on more complex tasks that required logical reasoning. This success is mostly based on the addition of memory and attention components to complex neural networks. For example, a memory network can reason about facts written in natural language or (subjects, relation, objects) triples. The attention mechanism is a successful component in machine translation (Bahdanau et al., 2015; Luong et al., 2015)[3][4].

Dynamic Memory Network+ (DMN+) is a neural network model with memory components and attention mechanisms (Xiong et al., 2016)[5]. The DMN+ has the not bad results in question answering that require reasoning based on facts during training, sentiment analysis and part-of-speech tagging.

We improve the accuracy of question and answer by analysing DMN components, especially input modules. We propose a new input module that uses a convolutional neural network to train character sequences. The goal is to allow the model to model the word-by-word interaction between the question and the context.

2. Model

Although the original model baseline can achieve a good result, it still has several shortcomings. First, it does not capture the combination of language, making it more difficult to detect reasonable answers. Second, the semantics of the problem is greatly simplified by the representation of its expected answer type and scalar word problem characteristics.

We propose and compare the embedding- layer: the input representation. The final extended model achieved higher precision on the bAbI-10k dataset (Weston et al., 2015)[6] without supporting the facts. The purpose of this design choice is to improve the accuracy of the model in processing the data set. Embedding are described and computed in detail in the following.

2.1. Input Model

In the following, we denote the hidden dimensionality of the model by n, the question tokens by $Q = \left(q_1, \ldots, q_{L_Q}\right)$, and the context tokens by $X = \left(x_1, \ldots, x_{L_X}\right)$. The task of embedding layer is to represent tokens x as their corresponding n-dimensional representations $\boldsymbol{x}$. Typically, this is typically done by mapping each word x to its corresponding word embedding $\boldsymbol{x}^w$ (*lookup-embedding*) using the embedding matrix E, s.t. $\boldsymbol{x}^w = Ex$. Then each word is embedded by encoding their corresponding character sequence $\boldsymbol{x}^c = (c_1, \ldots, c_{L_X})$ with C, s.t. $\boldsymbol{x}^c = C(x_c)$ (*char-embedding*). In this work, we use a convolutional neural network for C of filter width 5 with max-pooling over time (Seo et al., 2017)[7]. This method splicing the word vector and the word vector, and finally embedding becomes $\boldsymbol{x} = [\boldsymbol{x}^w; \boldsymbol{x}^c] \in \mathbb{R}^d$.

The process of embedding character vectors is as follows. The char representation is obtained by convolving the neural network by treating the characters of the word as a sequence. Then concat with the corresponding word embedding. For example, the word "one", [o, n, e] constitutes a sequence. This sequence maps characters to n-dimensional embedding after the char embedding layer. Enter embedding into the convolutional neural network to get an m-dimensional char representation. Then it is contacted together with the word embedding corresponding to the word 'one'. Word embedding is pre-trained, and the char representation is randomly initialized and adjusted to follow the network training. We build simple convolutional character based embedding for words with a fixed filter and size. After the convolution max-pooling over characters is employed for each filter. If word sequences are given, these will be embedded with the newly created embedding.

The structure of the input module is shown in Figure 1. The sentence reader uses a position encoder and the input fusion layer uses a bi-directional GRU. The output of an encoding scheme is entered into the sentence encoding f_i taking the final embedding $\left[w_1^i, \ldots, w_{M_i}^i\right]$, where M_i is the length of the sentence.

The sentence reader can adapt to any form of coding scheme. So we chose the position code (Sukhbaatar et al., 2015)[8] and compared with their work. GRU and LSTM are also considered, but require more computing resources. And if no pre-processing is done (such as reconstructing the original sentence) then it is easy to over-fitting.

The scheme of position coding is as follows, sentence representation is procedure by $f_i = \sum_M^{j=1} l_j \circ w_j^i$, where $\circ$ is an element-by-element multiplication. l_j is a column vector having a structure $l_{jd} = (1 - j/M) - (d/D)(1 - 2j/M)$, where d represents an embedded index and D represents a dimensional embedding.

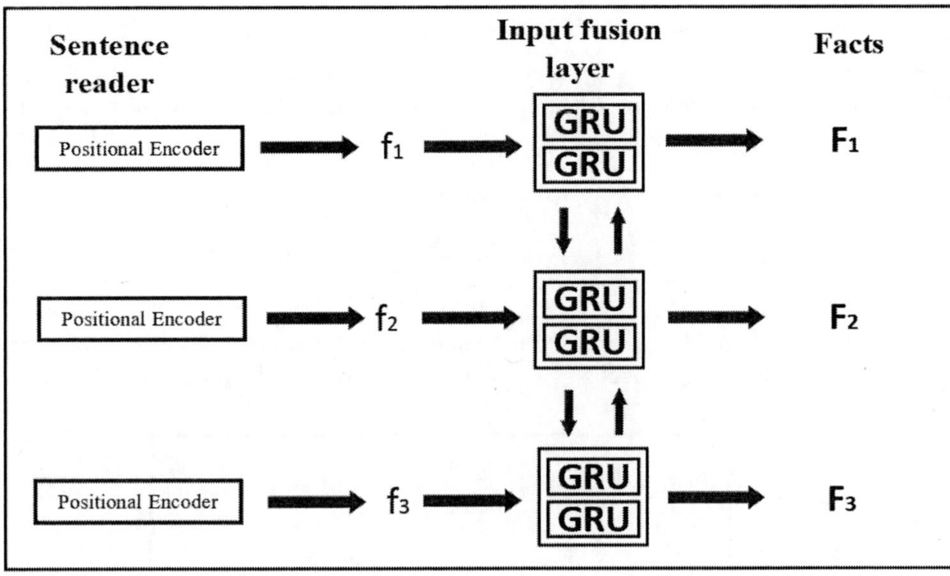

Figure 1. The input module with a "fusion layer". The sentence reader encodes the sentence, and the information flows between the sentences through the bi-directional GRU.

The input fusion layer takes these input facts. And use bi-directional GRU to ensure the exchange of information between them.

$$\overrightarrow{f_i} = GRU_{fwd}(f_i, \overrightarrow{f_{i-1}}) \quad (1)$$

$$\overleftarrow{f_i} = GRU_{bwd}(f_i, \overleftarrow{f_{i+1}}) \quad (2)$$

$$\overleftrightarrow{f_i} = \overleftarrow{f_i} + \overrightarrow{f_i} \quad (3)$$

Where f_i represents the input fact of the time step i, the hidden state of the forward GRU of the time step i is $\overrightarrow{f_i}$, and the hidden state of the backward GRU of the time step i is $\overleftarrow{f_i}$. This allows $\overleftrightarrow{f_i}$ to be influenced by contextual information from future and past facts.

We have studied a number of coding schemes for sentence readers, including but not limited to GRU, LSTM and the position coding schemes (Sukhbaatar et al., 2015)[8]. In order to make the experiment simple and effective, we finally chose the position coding scheme. GRU and LSTM are also taken into account. But more computational resources are needed, and if no pre-processing is done (such as reconstructing the original sentence) then it is easy to over-fitting.

2.2. Semantic Memory Module

Semantic memory consists of two parts, the stored word concepts and the facts about them. We initialize the embedding to a vector as described in the previous section. This module includes a gazetteer or other form of explicit knowledge base, but we do not intend to use them in the work described in this article.

2.3. Question Module

This module maps the problem to the representation, which can then be entered into the input module to query for a specific fact. We have questions that consist of sequences of T_Q words $W_t{}^Q$. We compute a hidden state for each via $q_t = GRU(L[w_t{}^Q], q_{t-1})$, where the GRU and the embedding weight use the same input module. The question vector is finally defined as q = q_{T_Q} .

2.4. Episodic Memory Module

As shown in Figure 1, the story memory module retrieves information from the input facts $\overleftrightarrow{F} =$

$[\overleftrightarrow{f_1}, \dots, \overleftrightarrow{f_N}]$ provided to it because of the focus on a subset of these facts. We associate a single scalar value (attention gate g_i^t) with each fact $\overleftrightarrow{f_i}$ one by one to achieve this concern, during the pass t. This is calculated by allowing the interaction between facts and problem representations and episode memory states.

$$z_i^t = \left[\overleftrightarrow{f_i}\circ q; \overleftrightarrow{f_i}\circ m^{t-1}; \left|\overleftrightarrow{f_i} - q\right|; \left|\overleftrightarrow{f_i} - m^{t-1}\right|\right] \quad (4)$$

$$Z_i^t = W^{(2)}\tanh\left(W^{(1)}z_i^t + b^{(1)}\right) + b^{(2)} \quad (5)$$

$$g_i^t = \frac{\exp(Z_i^t)}{\sum_{k=1}^{M_i}\exp(Z_k^t)} \quad (6)$$

Formula above, where m^{t-1} is the previous episode memory, q is the original question, $\overleftrightarrow{f_i}$ is the ith fact, $\circ$ is the element-wise product, $|\cdot|$ is the element-wise absolute value, and ; represents concatenation of the vectors.

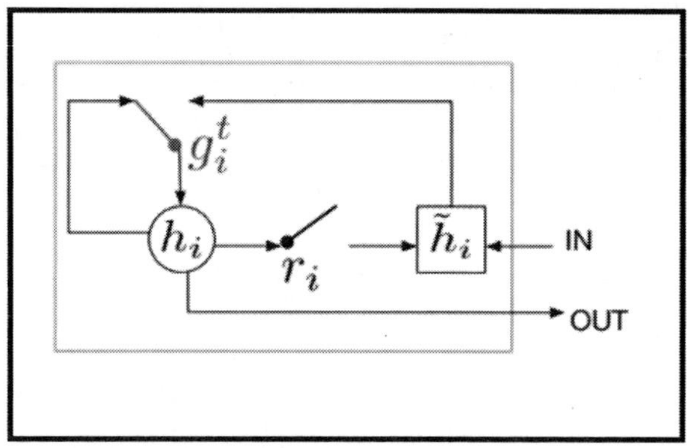

Figure 2. The attention-based GRU model.

2.4.1. Attention based GRU. We want to note that the mechanism is sensitive to the location and ordering of input facts $\overleftrightarrow{F}$ under more complex query conditions. In this case, RNN has a big advantage, except that they cannot be used in Equation (6).

$$h_i = g_i^t\circ\tilde{h}_i + \left(1 - g_i^t\right)\circ h_{i-1} \quad (7)$$

As shown in Figure 2, an important consideration is that g_i^t is a scalar and its activation function is softmax. In contrast, $u_i \in \mathbb{R}^{nH}$ is a vector whose activation function is sigmod. Although no further research was done, the softmax activation in Equation (6) was replaced with sigmod. This will result in $g_i^t \in \mathbb{R}^{nH}$ generating contextual vector c^t which can be used to update the context memory state m^t, we use the final hidden state of GRU based-attention.

2.4.2. Episode Memory Updates. Whenever we pass the attention mechanism, we hope to update the episode memory m^{t-1} with the newly constructed context vector c^t, which in turn produces m^t. The ReLU layer is adopted for the memory update, and compute the new episode memory state by

$$m^t = ReLU(W^t[m^{t-1}; c^t; q] + b) \quad (8)$$

Formula above, where ; is the concatenation operator, $W^t \in \mathbb{R}^{nH\times nH}$, $b \in \mathbb{R}^{nH}$, and nH is the hidden size.

3. Experiment

3.1. Dataset

In order to analyse our proposed model improvement and compare them to the original model architecture, we use the bAbI dataset.

IOP Conf. Series: Journal of Physics: Conf. Series **1267** (2019) 012090 doi:10.1088/1742-6596/1267/1/012090

We use the bAbI-10k English dataset (Weston et al., 2015)[6] to assess the accuracy of the improved model. This comprehensive dataset contains 20 different tasks. Each example contains a set of facts, a question, an answer, and a supporting fact that can reason the answer. The dataset has two different sizes, which refers to the number of training samples per task: bAbI-1k and bAbI-10k. The early experiment found that if a smaller bAbI-1k dataset was used (Sukhbaatar et al., 2015)[8], the lowest error rate was on average three times higher than bAbI-10k.

3.2. Analysis of improvement

Most reading comprehension systems are now built in the form of top-down. The structure of a typical neural network model contains an embedding-, encoding-, interaction- and answer layer (Wang and Jiang, 2017; Yu et al., 2017; Xiong et al., 2017; Seo et al., 2017; Yang et al., 2017; Wang et al., 2017)[9][10][11][7][12][13]. That is to say, a very complicated structure was proposed from the beginning. Then use the method called "ablation study" to continuously reduce some module configurations to verify the idea.

So we reproduce the experiment in DMN+ (Xiong et al., 2016)[5]. And some improvements have been made in the embedding- layer. The details of the improvements are shown in the previous section. We found that additional character embedding has a significant improvement on the results.

3.3. Comparison to the baseline using bAbI-10k

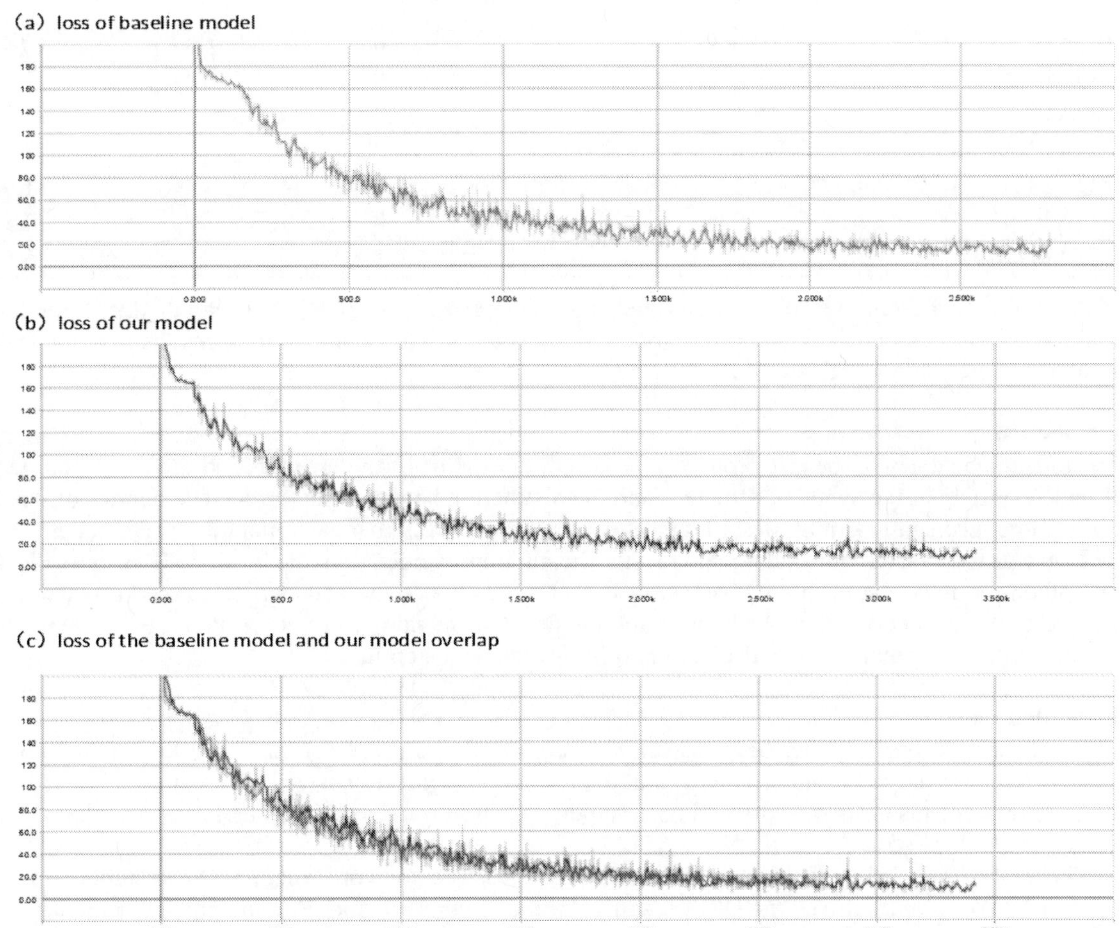

Figure 3. (a) The loss of baseline model, (b) the loss of our improved model, and (c) overlapping (a) and (b).

IOP Conf. Series: Journal of Physics: Conf. Series **1267** (2019) 012090 doi:10.1088/1742-6596/1267/1/012090

The original DMN architecture is proposed (Kumar et al. 2016)[14]. The original model did not make any improvements. Based on the DMN model, someone made some improvements (Xiong et al. 2016)[5]. DMN+ uses a unique set of weights for each pass and a linear layer with a ReLU activation to compute the memory update. Finally, DMN+ (Extended model), based upon DMN+, make some improvements in the embedding- layer.

As shown in the figure 3, the loss of our model is smaller than the loss of the baseline. This result also verifies from the side that the improved model is better than baseline.

Table 1. Test error rates of various models on the bAbI-10k dataset. The other unmentioned bAbI task achieved 0 error across all models.

Task	DMN	DMN+	DMN+ (Extended model)
	bAbI English 10k dataset		
QA2	36.0	6.1	5.5
QA3	42.2	22.8	15.8
QA5	0.1	0.5	0.4
QA6	35.7	0.0	0.0
QA7	8.0	2.4	2.2
QA8	1.6	0.0	0.0
QA9	3.3	0.0	0.0
QA10	0.6	0.0	0.0
QA14	3.6	8.8	1.1
QA16	55.1	45.3	42.5
QA17	39.6	4.2	3.8
QA18	9.3	2.1	1.9
QA20	1.9	0.0	0.1
Mean error	11.8	7.1	5.6

As can be seen from the table 1 on mean error, our improved model is better than the DMN model in all tasks, and is superior to the DMN+ model in most tasks. The degradation of results in a few tasks is subject to further research and improvement.

4. Conclusion

We do some analysis of the erroneous results of DMN+. Most of the errors found are due to the lack of understanding of the syntactic structure and the distinction between fine-grained semantics of words with different semantic similarities. Many other errors come from manual annotation preferences.

The extra character embedding has a significant improvement on the results. It is possible to strengthen the interrelationship between words and words. Once the encoder has some knowledge of the question, the encoder can selectively track the problem-related information. It can let the sentence get the context information. It is also very helpful for the next step fusion.

5. Related work

The early model dynamic memory network was proposed, a unified neural network framework that processes input sequences and problems, and then generates relevant answers (Kumar et al., 2015)[14]. Some improvements to its memory and input modules, based upon dynamic memory network (DMN+) (Xiong et al., 2016)[5].

The embedding layer of FastQA connects word vectors to word vectors (Weissenborn et al., 2017)[15]. We have improved the DMN+ with reference to this method. For comparison with baseline, we still use bAbI-10k English (Weston et al., 2015)[6], a synthetic dataset which features 20 different tasks.

References

[1] Krizhevsky A, Sutskever I and Hinton G E 2012 Imagenet classification with deep convolutional neural networks *Advances in Neural Information Processing Systems* pp 1097-1105
[2] Socher R, Perelygin A, Wu J, Chuang J, Manning C D, Ng A and Potts C 2013 Recursive deep models for semantic compositionality over a sentiment treebank *Conference on Empirical Methods in Natural Language Processing* pp 1631-1642
[3] Bahdanau D, Cho K and Bengio Y 2015 Neural machine translation by jointly learning to align and translate *International Conference of Learning Representation*
[4] Luong M T, Pham H and Manning C D 2015 Effective approaches to attention-based neural machine translation *Conference on Empirical Methods in Natural Language Processing*
[5] Xiong C, Merity S and Socher R 2016 Dynamic memory networks for visual and textual question answering *International conference on machine learning* pp 2397-2406
[6] Weston J, Bordes A, Chopra S and Mikolov T 2015 Towards ai-complete question answering: a set of prerequisite toy tasks *arXiv preprint arXiv:1502.05698*
[7] Seo M, Kembhavi A, Farhadi A and Hajishirzi H 2017 Bi-directional attention flow for machine comprehension *International Conference of Learning Representation*
[8] Sukhbaatar S, Weston J and Fergus R 2015 End-to-end memory networks *Advances in neural information processing systems* pp 2440-2448
[9] Wang S and Jiang J 2017 Machine comprehension using match-lstm and answer pointer *International Conference of Learning Representation*
[10] Yu Y, Zhang W, Hasan K, Yu M, Xiang B and Zhou B 2017 End-to-end reading comprehension with dynamic answer chunk ranking *arXiv preprint arXiv:1610.09996*
[11] Xiong C, Zhong V and Socher R 2017 Dynamic coattention networks for question answering *International Conference of Learning Representation*
[12] Yang Z, Dhingra B, Yuan Y, Hu J, Cohen W W and Salakhutdinov R 2017 Words or characters? fine-grained gating for reading comprehension *International Conference of Learning Representation*
[13] Wang Z, Mi H, Hamza W and Florian R 2017 Multi-perspective context matching for machine comprehension *arXiv preprint arXiv:1612.04211*
[14] Kumar A, Irsoy O, Ondruska P, Iyyer M, Bradbury J, Gulrajani I and Socher R 2016 Ask me anything: dynamic memory networks for natural language processing *International Conference on Machine Learning* pp 1378-1387
[15] Weissenborn D, Wiese G and Seiffe L 2017 Making neural qa as simple as possible but not simpler *Conference on Comutational Natural Language Learning* pp 271-280

IOP Conf. Series: Journal of Physics: Conf. Series **1267** (2019) 012091 doi:10.1088/1742-6596/1267/1/012091

Complex Dynamic Behaviors of A Congestion Control System under A Fractional-order PD Control

Qiu Lu[1], Huaifei Wang[1], Min Xiao[1,*] and Lingzhi Si[1]
[1]College of Automation, Nanjing University of Posts and Telecommunications, Nanjing, China
E-mail: candymanxm2003@aliyun.com

Abstract. This paper proposed an original fractional-order proportional-derivative (PD) feedback controller which is designed to control the Hopf bifurcation caused by the congestion control system. The proposed $PD^{\frac{1}{n}}$ controller has the different order with the original congestion system. The proposed fractional-order PD controller has the different order with the controlled system and the communication delay is selected as the bifurcation parameter. Then the conditions of the stability and Hopf bifurcation are obtained by analyzing its characteristic equation and the stability domain can be extended under the adjustment of appropriate control gain parameters and the order. Therefore, the congestion system becomes controllable and the desirable behaviors can be realized. Finally, numerical simulations are carried out to testify the validity of the theoretical analysis in the designed fractional-order PD controller.

1. Introduction

With the rapid development of network technology, the number of Internet users are keeping rising, and the concomitant possibility that the resource capacity of computer networks fails to meet the needs of Internet users. The phenomenon of computer network congestion has become increasingly serious. Therefore, there have been a great of achievements in the research of congestion control system [1, 2, 3]. In this paper, we choose the integer-order congestion control system with time delay as our controlled model which was introduced in [4] and it can be expressed as the following differential equation:

$$\dot{x}(t) = kx(t)\left[\frac{1}{x(t-\tau)} - c\right], \tag{1}$$

where the variable x is the price at the link and τ denotes the communication delay respectively. Meanwhile, $k > 0$ is the gain parameter and the scalar c is the capacity of this model. Obviously, it can be obtained that system (1) has a sole equilibrium $x^* = \frac{1}{c}$. Some control schemes were designed to control the undesirable Hopf bifurcation so that we can adjust the initiation of bifurcation and achieve desirable dynamical characteristic of the controlled system [5, 6, 7].

Fractional-order calculus has a wide range of applications in physics, biology and engineering and it extends the traditional concept of differential and integral calculus. It has been found that fractional-order derivatives has superiority of memory and hereditary effects in mathematical modeling, which guarantees a better global performance of the controlled system [8]. Therefore, abundant significative scientific achievements have been achieved in stability analysis [9], bifurcations [10], and chaos [11].

In recent years, researches have an increasing attention on applying PID controller to control the Hopf bifurcation and a lot of important research results have been achieved [12, 13]. However, few studies have considered that the order of the controller is different from that of the controlled model. Therefore, we propose a fractional-order PD controller to control the integer-order congestion control model in this paper. The proposed controller can be represented in the following form:

$$u(t) = k_p(x(t) - x^*) + k_d \left[{}_0^C D_t^{\frac{1}{n}} (x(t) - x^*) \right], \tag{2}$$

where k_p and k_d respectively represent the proportional control parameter and the derivative control parameter and n is a natural number.

The rest of this paper is organized as follows. Section 2 presents the mathematical model based on fractional-order PD controller and linearizes the controlled model at equilibrium and discuss the impacts of order and gain parameters on stability and Hopf bifurcation and the transversality condition is proved. In Section 3, some numerical simulation results are carried out to verify our results. Finally, Section 4 concludes this paper.

2. Model Description and Linearize Controlled Model at Equilibrium

Applying the fractional-order controller (2) to system (1), we can achieve a controlled fractional system as follows:

$$\begin{aligned} \dot{x}(t) = {} & kx(t)\left[\frac{1}{x(t-\tau)} - c\right] + u(t) \\ = {} & kx(t)\left[\frac{1}{x(t-\tau)} - c\right] + k_p(x(t) - x^*) + k_d \left[{}_0^C D_t^{\frac{1}{n}} (x(t) - x^*) \right]. \end{aligned} \tag{3}$$

Obviously, the equilibrium of system (3) still locate in x^* of system (1). Let

$$\begin{aligned} x_1(t) &= x(t) - x^*, \\ x_2(t) &= {}_0^C D_t^{\frac{1}{n}} x_1(t), \\ x_3(t) &= {}_0^C D_t^{\frac{1}{n}} x_2(t), \\ &\cdots \\ x_n(t) &= {}_0^C D_t^{\frac{1}{n}} x_{n-1}(t). \end{aligned} \tag{4}$$

We can derive

$$\begin{aligned} {}_0^C D_t^{\frac{1}{n}} x_n(t) &= {}_0^C D_t^{\frac{1}{n}} \left[{}_0^C D_t^{\frac{1}{n}} x_{n-1}(t) \right] \\ &= {}_0^C D_t^{\frac{2}{n}} x_{n-1}(t) = \cdots = {}_0^C D_t^{\frac{n}{n}} x_1(t) = \dot{x}_1(t) = \dot{x}(t) \\ &= k\left[x_1(t) + x^*\right]\left[\frac{1}{x_1(t-\tau) + x^*} - c\right] + k_p x_1(t) + k_d x_2(t), \end{aligned} \tag{5}$$

Therefore, system (3) can translate into a n-dimensional commensurate-order system. We linearize the controlled system at the equilibrium point and can achieve the associated characteristic equation of system as follows:

$$s - k_d s^{\frac{1}{n}} - (k_p - cke^{-s\tau}) = 0. \tag{6}$$

IOP Conf. Series: Journal of Physics: Conf. Series **1267** (2019) 012091 doi:10.1088/1742-6596/1267/1/012091

2.1. The case of a fractional-order PD controller ($n \geq 2$) without delays ($\tau = 0$)
When $n \geq 2$ and $\tau = 0$, (6) can be switched to

$$s - k_d s^{\frac{1}{n}} - k_p + ck = 0. \tag{7}$$

Lemma 1 *If $k_d < 0$ and $k_p < ck$, the roots of (7) are all in the left half of the complex plane.*

Proof 1 *Let $s = Ae^{\theta i} = A(\cos\theta + i\sin\theta)$, A is the modulus and θ argument of s on the complex plane. Therefore, (7) can be separated the real and imaginary parts. Obviously, if $k_d < 0$, $k_p < ck$ and $\theta \in [-\frac{\pi}{2}, \frac{\pi}{2}]$, (7) has no roots. It means that all roots of (7) are located on on the left part of the complex plane. The proof is complete.*

Theorem 1 *Assume that $n \geq 2$, $k_p < ck$ and $k_d < 0$ hold, system (3) without delay is asymptotically stable at the equilibrium point $x^* = \frac{1}{c}$.*

2.2. The case of a fractional-order PD controller ($n \geq 2$) with delays ($\tau > 0$)
When $n > 2$ and $\tau > 0$, letting $s = i\omega$ be a purely imaginary root of (6). Separating the real and imaginary parts and adding the squares of two parts, we can obtain

$$(ck)^2 = k_p^2 + 2k_p k_d \omega^{\frac{1}{n}} \cos\frac{\pi}{2n} + \omega^2 - 2k_d \omega^{\frac{1}{n}+1} \sin\frac{\pi}{2n} + (k_d \omega^{\frac{1}{n}})^2. \tag{8}$$

Let $\omega^{\frac{1}{n}} = z$, then $\omega = z^n$. Therefore, (8) can be formulated as

$$z^{2n} - 2k_d \sin\frac{\pi}{2n} z^{n+1} + k_d^2 z^2 + 2k_p k_d \cos\frac{\pi}{2n} z + k_p^2 - (ck)^2 = 0. \tag{9}$$

Lemma 2 *When $-ck < k_p < ck$, $k_d < 0$ hold, (9) has real positive roots. When $k_p < -ck$, $k_d < 0$, (9) has not a purely imaginary root for any τ.*

Proof 2 *We define a function $H(z)$ as follows*

$$H(z) = z^{2n} - 2k_d \sin\frac{\pi}{2n} z^{n+1} + k_d^2 z^2 + 2k_p k_d \cos\frac{\pi}{2n} z + k_p^2 - (ck)^2. \tag{10}$$

When $k_d < 0$ and $k_p < 0$, then $H'(z) > 0$ is permanent for $z \in [0, +\infty)$, which means that $H(z)$ is a monotone increasing function.

(1) When $k_p < -ck$, $k_d < 0$, then $k_p^2 - (ck)^2 > 0$. Therefore, $H(0) > 0$. Moreover, (9) has no real positive roots.

(2) When $-ck < k_p < 0$, $k_d < 0$, then $k_p^2 - (ck)^2 < 0$. Therefore, $H(0) < 0$ and the $H(z)$ has a unique zero z_0. Moreover, (9) has real positive roots.

When $k_d < 0$ and $0 \leq k_p < ck$, then $H(0) < 0$ and the $H(z)$ has also a unique zero z_0. This ends the proof.

Therefore, we can derive

$$\omega_{02} = z_0^n, \tau_{02} = \frac{1}{\omega_{02}} \arccos \frac{k_p + k_d \omega_{02}^{\frac{1}{n}} \cos\frac{\pi}{2n}}{ck}. \tag{11}$$

Lemma 3 *Let $s(\tau) = \kappa(\tau) + i\omega(\tau)$ be the root of (6) near $\tau = \tau_{02}$ satisfying $\kappa(\tau_{02}) = 0$, $\omega(\tau_{02}) = \omega_{02}$. Then the following transversality condition of (6) holds:*

$$Re\left[\frac{ds}{d\tau}\right]_{\substack{s=i\omega_{02}\\ \tau=\tau_{02}}} > 0. \tag{12}$$

Proof 3 *Differentiating (6) on both sides with respect of τ, we can deduce*

$$Re\left[\frac{d\tau}{ds}\right] = Re\left[\frac{1-\frac{k_d}{n}s^{\frac{1}{n}-1}}{cke^{-s\tau}s}\right]_{\substack{s=i\omega_{02}\\ \tau=\tau_{02}}} = \frac{g(\omega_{02})}{(ck)^2\omega_{02}}, \tag{13}$$

where

$$g(\omega) = \omega - k_d(\frac{1}{n}+1)\sin\frac{\pi}{2n}\omega^{\frac{1}{n}} + k_p\frac{k_d}{n}\cos\frac{\pi}{2n}\omega^{\frac{1}{n}-1} + \frac{k_d^2}{n}\omega^{\frac{2}{n}-1}. \tag{14}$$

According to (8), we can suppose

$$G(\omega) = k_p^2 + 2k_pk_d\omega^{\frac{1}{n}}\cos\frac{\pi}{2n} + \omega^2 - 2k_d\omega^{\frac{1}{n}+1}\sin\frac{\pi}{2n} + (k_d\omega^{\frac{1}{n}})^2 - (ck)^2. \tag{15}$$

Differentiating (15) on both sides with respect of ω, we can achieve

$$G'(\omega) = 2g(\omega). \tag{16}$$

$G'(\omega) > 0$ is permanent based on the condition of $-ck < k_p < ck$ and $k_d < 0$. Therefore, $Re\left[\frac{d\tau}{ds}\right]_{\substack{s=i\omega_{02}\\ \tau=\tau_{02}}} > 0$. Hence, $Re\left[\frac{ds}{d\tau}\right]_{\substack{s=i\omega_{02}\\ \tau=\tau_{02}}} > 0$. This ends the proof.

Theorem 2 *Assume that $n \geq 2$, $-ck < k_p < ck$ and $k_d < 0$ hold, system (3) is asymptotically stable at the equilibrium point $x^* = \frac{1}{c}$ for all $\tau \in [0, \tau_{02})$ and undergoes a Hopf bifurcation at the equilibrium $x^* = \frac{1}{c}$ when $\tau = \tau_{02}$. Suppose that $n \geq 2$, $k_p < -ck$ and $k_d < 0$ are met, system (3) is asymptotically stable at the equilibrium point $x^* = \frac{1}{c}$ for all $\tau \geq 0$.*

3. Numerical Simulation and Analysis

In this section, we illustrate the validity and practicability of theoretical analyses which are obtained in the previous section by numerical simulations and discuss the influence of the control gain parameters of PD controller on system stability in this paper. For the sake of comparison, we consider system (1) with the same parameters $k = 0.01$ and $c = 50$ which are used in [4]. Therefore, system (1) has the unique equilibrium $x^* = \frac{1}{c} = 0.02$.

3.1. The case of without control

We first consider the original congestion control system (1) and the time delay τ is chosen as the bifurcation parameter. By (6), it can be obtained $\tau_{00} = \frac{\pi}{2ck} = \pi$ without control ($k_p = k_d = 0$). System (1) is asymptotically stable at the equilibrium point x^* when $\tau \in [0, \tau_{00})$ and undergoes a Hopf bifurcation when $\tau > \tau_{00}$. These properties are shown in Figures. 1 and 2.

3.2. The case of a fractional-order PD controller ($n \geq 2$)

For advancing the bifurcation of system (3), we take the $k_p = 0.25$, $k_d = -0.01$, $n = 3$ and obtain $\tau_{02} = 3.08$ by Lemma 2.2. It can be seen from Theorem 2.2 that system (3) is asymptotically stable at the equilibrium point x^* for $\tau \in [0, \tau_{02})$ and undergoes a Hopf bifurcation at the equilibrium x^* when $\tau > \tau_{02}$. Figure 3 displays the stable x^* of system (3) when $\tau = 2.45 < \tau_{02} = 3.08$ and a periodic oscillation occurs through a Hopf bifurcation when $\tau = 3.10 > \tau_{02} = 3.08$ as shown in Figure 4. Note that $\tau_{02} = 3.08 < \tau_{00} = \pi$.

We make $k_p = -0.15$, $k_d = -0.05$ and $n = 3$ in order to delay the Hopf bifurcation. By Lemma 2.2, we have $\tau_{02} = 4.0$. According to Theorem 2.2, system (3) is asymptotically stable at the equilibrium point x^* for all $\tau \in [0, \tau_{02})$ and undergoes a Hopf bifurcation at the equilibrium x^* when $\tau \geq \tau_{02}$. The numerical simulations for different values of the delay are graphically displayed in Figures 5 and 6.

IOP Conf. Series: Journal of Physics: Conf. Series **1267** (2019) 012091 doi:10.1088/1742-6596/1267/1/012091

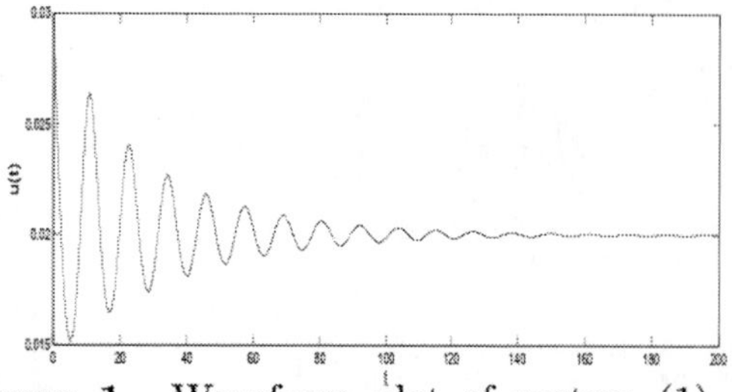

Figure 1. Waveform plot of system (1) with $k = 0.01$, $c = 50$, $k_p = 0$ and $k_d = 0$. System (1) is asymptotically stable when $\tau = 2.80 < \tau_{00} = \pi$.

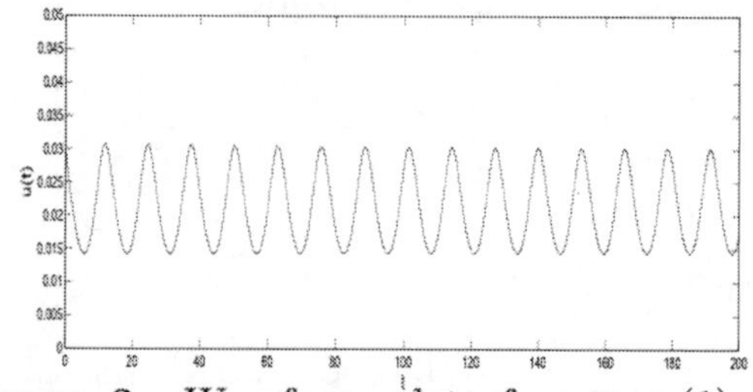

Figure 2. Waveform plot of system (1) with $k = 0.01$, $c = 50$, $k_p = 0$ and $k_d = 0$. A periodic oscillation bifurcates when $\tau = 3.20 > \tau_{00} = \pi$.

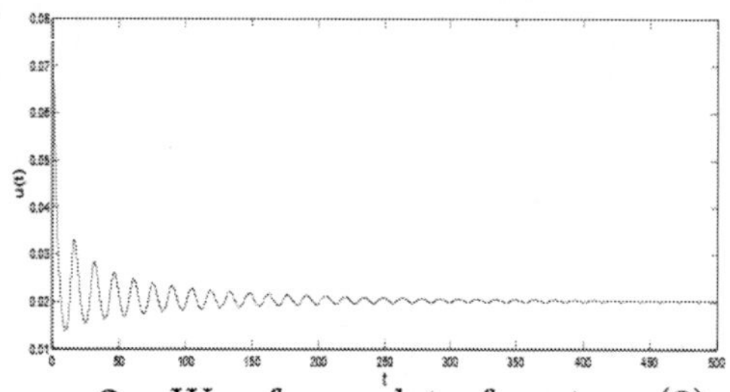

Figure 3. Waveform plot of system (3) with $n = 3$, $k = 0.01$, $c = 50$, $k_p = 0.25$ and $k_d = -0.01$. System is asymptotically stable for when $\tau = 2.45 < \tau_{02} = 3.08$.

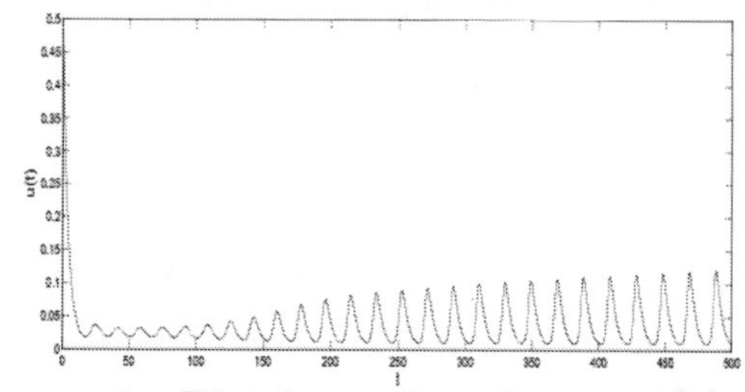

Figure 4. Waveform plot of system (3) with $n = 3$, $k = 0.01$, $c = 50$, $k_p = 0.25$ and $k_d = -0.01$. A periodic oscillation bifurcates when $\tau = 3.10 > \tau_{02} = 3.02$.

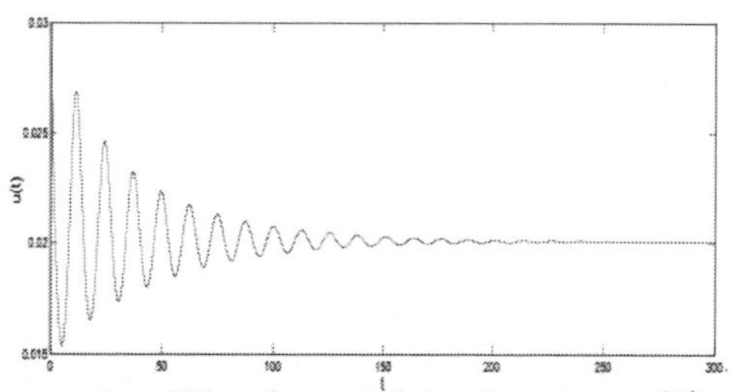

Figure 5. Waveform plot of system (3) with $n = 3$, $k = 0.01$, $c = 50$, $k_p = -0.15$ and $k_d = -0.05$. System is asymptotically stable for when $\tau = 3.80 < \tau_{02} = 4.0$.

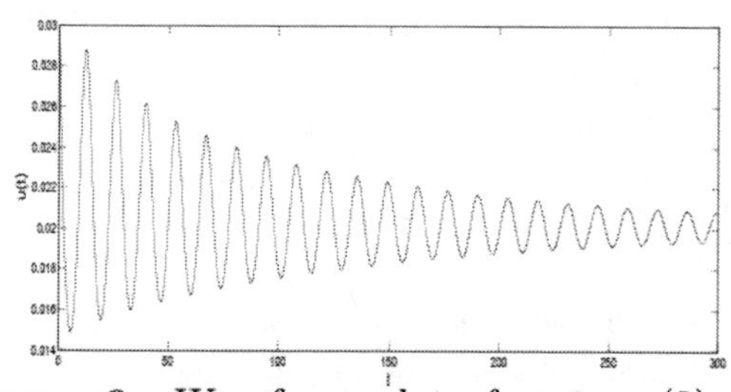

Figure 6. Waveform plot of system (3) with $n = 3$, $k = 0.01$, $c = 50$, $k_p = -0.15$ and $k_d = -0.05$. A periodic oscillation bifurcates when $\tau = 4.20 > \tau_{02} = 4.0$.

3.3. The influence of k_p, k_d and n on the critical value τ_0

Figure 7 describes the relationship of k_p and τ_0 with $k_d = -0.1$. We conclude that the values of τ_0 is declining with the rise of k_p. k_d and τ_0 have a similar relationship shown in Figure 8. According to the above analysis, we discover that n also have a great influence on the change of the bifurcation point τ_0. Table 1 displays the influence of n on the bifurcation point τ_0 when $k_p = 0.25$, $k_d = -0.1$, $k = 0.01$, and $c = 50$. Regardless of the case of integer-order ($n = 1$), we draw a conclusion that the bifurcation point τ_0 monotonically declines with the rise of n ($n \geq 2$). We can draw a conclusion from Table 1 that the controlled system is difficult to maintain stable in a larger total delay, and are more likely to have a Hopf bifurcation when the total delay τ varies.

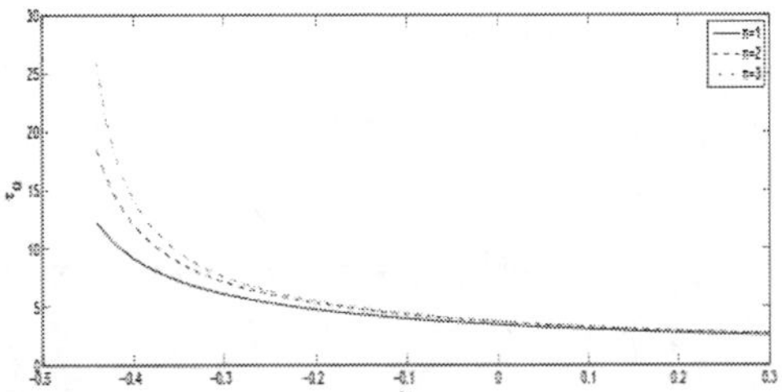

Figure 7. The relationship between k_p and τ_0 with $k_d = -0.1$.

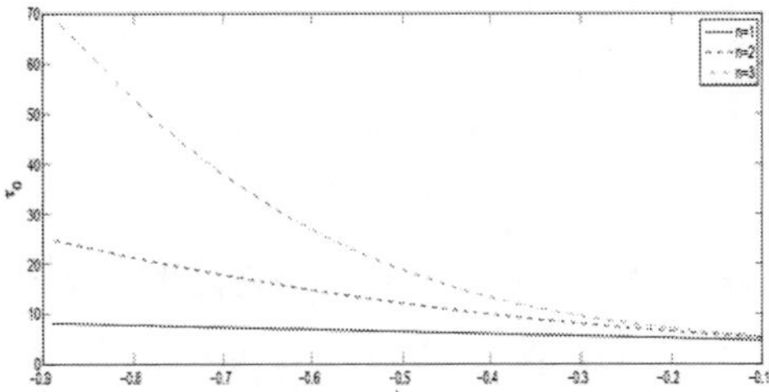

Figure 8. The relationship between k_d and τ_0 with $k_p = -0.2$.

Table 1. Influence of n on the value of the bifurcation point τ_0 when $k_p = 0.25$, $k_d = -0.1$.

Fractional-order parameter n	Bifurcation point τ_0
1	3.3484
2	3.4337
3	3.3983
4	3.3635
5	3.3369
6	3.3167
7	3.3011

4. Conclusion

In the beginning, in order to better control the Hopf bifurcation, we consider the advantages of the fractional-order PD controller. Furthermore, through a series of mathematical substitutions, we have expanded the one-dimensional controlled network with different orders to n-dimensional one with the same order. Meanwhile, the dynamic behavior of the controlled network is obtained by analyzing the stability problem of the n-dimensional network with same order. Finally, it is found that the Hopf bifurcation point of the controlled system can be effectively delayed and advanced. Despite the limitations of Caputo derivatives, recent studies have shown that it still plays a vital role in many physical systems.

Acknowledgements

This work is supported in part by the National Natural Science Foundation of China (No. 61573194), the Natural Science Foundation of Jiangsu Province of China (No. BK20181389), and the Key Project of Philosophy and Social Science Research in Colleges and Universities in Jiangsu Province (Grant No. 2018SJZDI142).

References

[1] Zhang X and Papachristodoulou A 2015 *IEEE Trans. Autom. Control* **60** 522
[2] Azadegan M, Beheshti MTH and Tavassoli B 2015 *Int. J. Control Autom. Syst.* **13** 764
[3] Kelly F P, Maulloo A K and Tan D K H 1998 *J. Oper. Res. Soc.* **49** 237
[4] Raina G 2005 *IEEE Trans. Autom. Control* **50** 1135
[5] Xiao M and Cao J D 2007 *J. Math. Anal. Appl.* **332** 1010
[6] Liu F, Wang H O and Guan Z H 2012 *Nonlinear Anal. Real World Appl.* **12** 1466
[7] Liu C L and Tian Y P 2008 *Chaos Solit. Fract.* **35** 878
[8] Elwakil A S 2010 *IEEE Circuits Syst. Mag.* **10** 40
[9] Lim Y, Oh K and Ahn H 2103 *IEEE Trans. Autom. Control* **58** 1062
[10] Xiao M, Zheng W X, Jiang G P and Cao J D 2015 *IEEE Trans. Neural Netw. Learn. Syst.* **26** 3201
[11] Hartley T, Lorenzo C and Qammer H 1995 *IEEE Trans. Circuits Syst. I, Fundam. Theory Appl.* **42** 485
[12] Xiao M, Zheng W X, Lin J X, Jiang G P, Zhao L D and Cao J D 2017 *J. Franklin I* **354** 7643
[13] Tang Y H, Xiao M, Jiang G P, Lin J X, Cao J D and Zheng W X 2017 *Nonlinear Dyn.* **90** 2185

IOP Conf. Series: Journal of Physics: Conf. Series **1267** (2019) 012092 doi:10.1088/1742-6596/1267/1/012092

Design and Simulation of PI Controller with Two Degrees of Freedom

Wang Yuyu

Xi'an Aeronautical Polytechnic Institute

E-mail: yuyuer826@126.com

Abstract. Aiming at the fully coupled problem of steady state and dynamic performance of a PI controller with one degree of freedom, two kinds of two DOF PI controllers were designed based on the setpoint filter structure and the setpoint feed-forward structure, and the design method of compensation link was introduced.On this basis, a variable structure two-degree-of-freedom PI controller was designed with reference to the idea of variable gain PI controller.Finally, the simulation results show that the controller can not only simplify the process of setting control parameters, but also take into account the steady-state performance, dynamic performance and anti-jamming performance of the system, and realize the function of output fast, no overshoot, no error tracking input signal changes.

1. Introduction

The parameter adjustment of traditional one-degree-of-freedom PI controller can affect the dynamic and steady-state performance of the system at the same time. If the parameters are adjusted according to the optimal dynamic performance, the steady-state performance of the system will become worse. If the parameters are tuned according to the optimal steady-state performance, the dynamic performance of the system will deteriorate, so we often use trial-and- error or compromise method to solve [1] [2] when tuning the parameters of conventional PI controllers. Based on this, if we can design another compensation link, and use this link to ensure that the system can obtain better dynamic and dynamic performance at the same time, we can realize the two-degree-of-freedom control of the system [3]. The variable structure two-degree- of-freedom PI controller designed in this paper optimizes many problems existing in traditional PI controller. It is simple, practical and reliable. It is very suitable for high performance speed servo system and provides a practical and effective new method for the field of speed servo control.

2.2-DOF PI Controller

The typical two-degree-of-freedom set-value filter controller structure and set-value feed-forward structure are combined with PI control. Taking the first-order system as an example, the two-degree-of-freedom PI control structure diagrams are obtained as shown in figure 1 and 2 below.

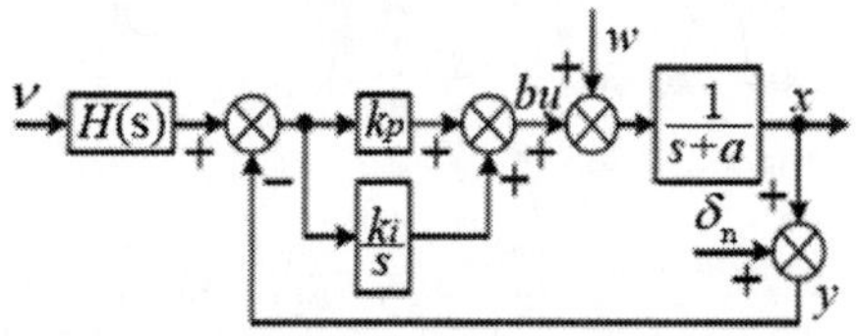

Figure 1. Two-degree-of-freedom set value filter PI controller

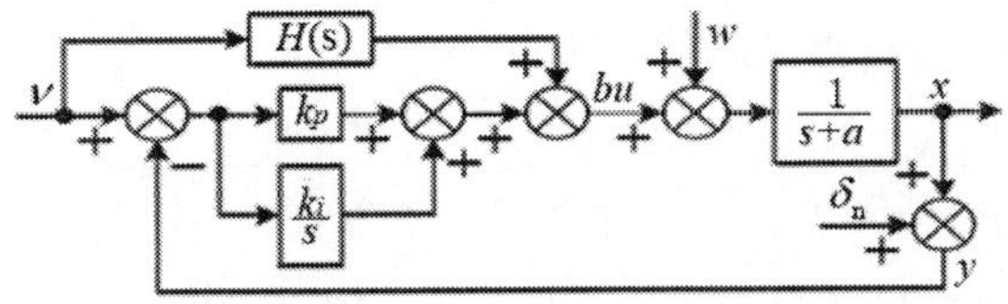

Figure 2. Two-degree-of-freedom set-point feedforward PI controller

As can be seen from Figure 1, the output of the two-degree-of-freedom set-value filter PI control system is as follows:

$$x(s)=\frac{H(s)C(s)P(s)}{1+C(s)P(s)}v(s)+\frac{P(s)}{1+C(s)P(s)}w(s)-\frac{C(s)P(s)}{1+C(s)P(s)}\delta_n(s)$$
$$=\frac{(k_p s+k_i)H(s)}{s^2+(k_p+a)s+k_i}v(s)+\frac{s}{s^2+(k_p+a)s+k_i}w(s)-\frac{k_p s+k_i}{s^2+(k_p+a)s+k_i}\delta_n(s) \tag{1}$$

In particular, when a=0, the output of the control system can be simplified to:

$$x(s)=\frac{(k_p s+k_i)H(s)}{s^2+k_p s+k_i}v(s)+\frac{s}{s^2+k_p s+k_i}w(s)-\frac{k_p s+k_i}{s^2+k_p s+k_i}\delta_n(s) \tag{2}$$

For the same reason, Figure 2 shows that the output of the set-point feedforward system is as follows:

$$x(s)=\frac{H(s)C(s)P(s)}{1+C(s)P(s)}v(s)+\frac{P(s)}{1+C(s)P(s)}w(s)-\frac{C(s)P(s)}{1+C(s)P(s)}\delta_n(s)$$
$$=\frac{k_p s+k_i+sH(s)}{s^2+(k_p+a)s+k_i}v(s)+\frac{s}{s^2+(k_p+a)s+k_i}w(s)-\frac{k_p s+k_i}{s^2+(k_p+a)s+k_i}\delta_n(s) \tag{3}$$

In particular, when a=0, the output of the control system can be simplified to:

$$x(s)=\frac{k_p s+k_i+sH(s)}{s^2+k_p s+k_i}v(s)+\frac{s}{s^2+k_p s+k_i}w(s)-\frac{k_p s+k_i}{s^2+k_p s+k_i}\delta_n(s) \tag{4}$$

From the above analysis, it can be seen that the steady-state performance will not be affected when the dynamic performance is adjusted by the compensation link H(s).Therefore,the two-degree-of-freedom PI controller first uses the requirements of steady-state performance to design parameters, and then uses the requirements of dynamic performance to design function H (s) [4][5].

3. Variable Structure Two Degree of Freedom PI Controller

For a two-degree-of-freedom PI controller with set-value filter, when the input is a time-varying given signal, in order to obtain good steady-state performance, the transfer function of the system can be designed according to equation (2).

$$\Phi_r(s)=\frac{x(s)}{v(s)}=\frac{(k_p s+k_i)H_1(s)}{s^2+k_p s+k_i}=1 \tag{5}$$

Thus, the set value filter function H1 (s) can be obtained as:

$$H_1(s)=\frac{s^2+k_p s+k_i}{k_p s+k_i} \tag{6}$$

In order to make the input and output of the system without overshoot under step action, the transfer function is designed as a first or second order low pass filter link.

$$\Phi_r(s)=\frac{x(s)}{v(s)}=\frac{(k_p s+k_i)H_2(s)}{s^2+k_p s+k_i}=\frac{mk_p}{s+mk_p}\text{ or }\frac{mk_p s+k_i}{s^2+k_p s+k_i} \tag{7}$$

Thus, the set-value filter function H2(s) can be obtained as:

$$H_2(s)=\frac{(s^2+k_p s+k_i)m}{s^2+(mk_p+k_i/k_p)s+mk_i} \quad \text{or} \quad \frac{mk_p s+k_i}{k_p s+k_i} \tag{8}$$

In the formula, m is a positive constant and can be set artificially.

Similarly, for a set-value feedback controller, when the input of the system is a time-varying given signal, in order to obtain good steady-state performance [6], according to equation (4) the transfer function is designed as:

$$\Phi_r(s)=\frac{x(s)}{v(s)}=\frac{k_p s+k_i+sH_1(s)}{s^2+k_p s+k_i}=1 \tag{9}$$

Thus, the corresponding set-value filter function H1 (s) can be obtained as follows:

$$H_1(s)=s \tag{10}$$

In order to make the output of the system without overshoot under step action, the transfer function is designed as a first-order or second-order low-pass filter link.

$$\Phi_r(s)=\frac{x(s)}{v(s)}=\frac{k_p s+k_i+sH_2(s)}{s^2+k_p s+k_i}=\frac{mk_p}{s+mk_p} \quad \text{or} \quad \frac{mk_p s+k_i}{s^2+k_p s+k_i} \tag{11}$$

Thus, the set-value filter function H2(s) can be obtained as:

$$H_2(s)=\frac{(m-1)k_p s-k_i}{s+mk_p} \quad \text{or} \quad (m-1)k_p \tag{12}$$

From the previous calculation, the compensating links H (s) corresponding to the two types of controllers under the action of two input signals can be obtained as shown in Table 1 below [7].

Table 1 H(s) corresponding to two types of controllers under the action of two input signals

Given Type	Time-varying Given	Step giving	
Design Transfer Function	$\Phi_r(s)=\frac{x(s)}{v(s)}=1$	$\Phi_r(s)=\frac{x(s)}{v(s)}=\frac{mk_p}{s+mk_p}$	$\Phi_r(s)=\frac{x(s)}{v(s)}=\frac{mk_p s+k_i}{s^2+k_p s+k_i}$
H(s) in Set Value Filter Controller	$\frac{s^2+k_p s+k_i}{k_p s+k_i}$	$\frac{m(s^2+k_p s+k_i)}{s^2+(mk_p+k_i/k_p)s+mk_i}$	$\frac{mk_p s+k_i}{k_p s+k_i}$
H(s) in Set Value Feedforward Controller	s	$\frac{(m-1)k_p s-k_i}{s+mk_p}$	$(m-1)k_p$

In order to make the system have good dynamic performance under the action of two different types of input signals, a variable structure two-degree-of-freedom PI controller is designed based on the two-degree-of-freedom PI controller, which can distinguish the input type according to the given change rate, and then according to the results of table 1, the appropriate compensation link H(s) is selected. Its control structure is shown in figure 3 and 4 [8][9].

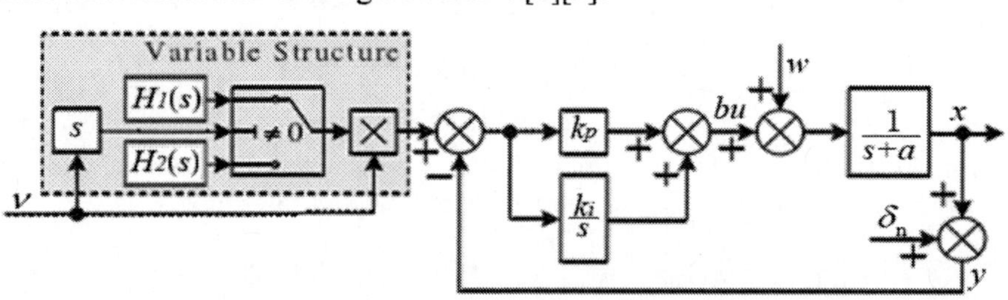

Figure 3. Set-value filtered variable structure two-degree-of-freedom PI controller

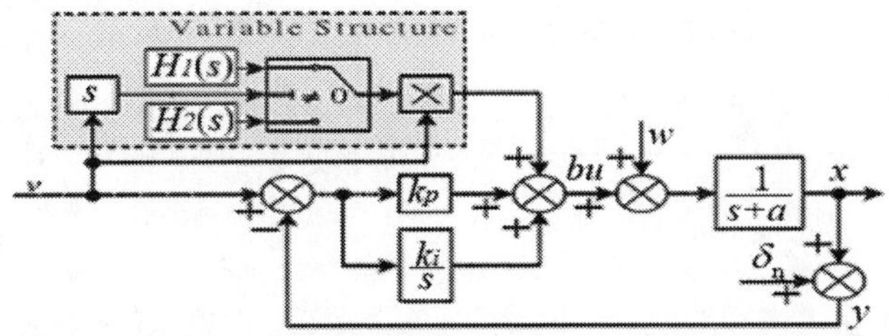

Figure 4. Set-point feedforward variable structure two-degree-of-freedom PI controller

After the first control cycle, the given change rate of time-varying is not constant at 0. At this time, H1 (s) is chosen as the compensation link, which can realize the error-free tracking of the system for the given continuous change. The change rate given by the step is 0 after the first control cycle. When the compensation link is H2 (s), the preset trajectory of the step response of the system can be achieved [10] [11].

The structure block diagram of the variable structure two-degree-of-freedom PI controller applied to the speed loop of permanent magnet synchronous motor is shown in figure 5 below.

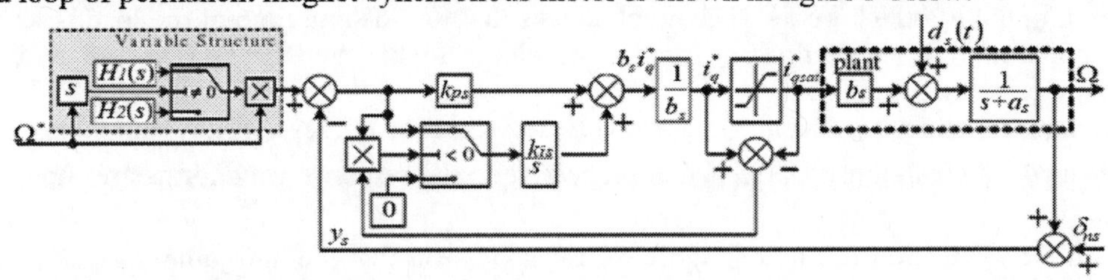

(a) Setting value filter variable structure two-DOF PI controller for speed Ring

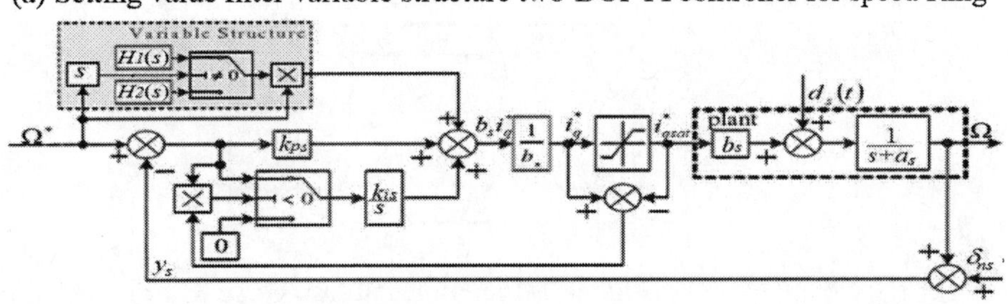

(b) Setting value feedforward variable structure two degree of freedom PI controller for speed ring

Figure 5. Variable structure two-degree-of-freedom PI controller for speed loop

4. Simulation analysis

Because the step response compensation of the set-value feedforward controller is simpler than that of the filter one, the following simulation analysis is carried out with the feed-forward controller as an example. Next, in order to verify the effectiveness of the proposed controller, the response characteristics and stability of the controller under step signal are simulated and analyzed, and the step response transfer function is designed as a second-order low-pass filter.

The sampling period and calculation step of speed loop and current loop are both 0.1ms. Both current loops are controlled by PI with a bandwidth of 2000 rad/s and a current limit of 9A. The PI parameters in the controller are designed in the same way as those in the previous paper, taking $k_{ps} = 2\zeta\omega_n$ $k_{is} = \omega_n^2$.

The values of m = 0 and ω_n are 60, 80 and 100respectively, and the step input is 80r/min start-up without load. The response waveform is shown in Figure 6 (a).

ω_n takes 60, m takes 1, 2 and 3, respectively, and the step input is still 80r/min. The response waveform is shown in Figure 6 (b) below.

From the simulation results of Figure 6, we can see that the dynamic performance of the system is related to the values of ω_n and m.

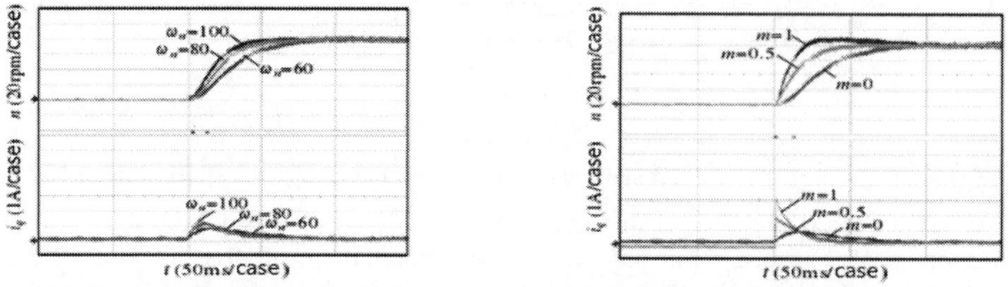

(a) Step response under different n (b) Step response under different m

Figure 6. Stepr esponse waveform of 80r/min no-load starting

In order to verify the anti-jamming performance of the system, under the action of step signal, the motor is first allowed to run without load at 80r/min speed, and then the load is loaded and unloaded as shown in the **Figure 7** below. The graph shows that the loading time of the load is longer than the unloading time, and the unloading process is closer to the step change, so we only choose the unloading process to simulate.

When m is 0 and a is 1, 2 and 3, respectively, the response curves are shown in Figure 8 (a).

When ω_n is 60 and m is 0, 0.5 and 1, respectively, the response waveforms are shown in Figure 8 (b) below.

From the simulation results of Figure 8, it can be seen that the anti-jamming performance of the system is only related to ω_n, but not to m.

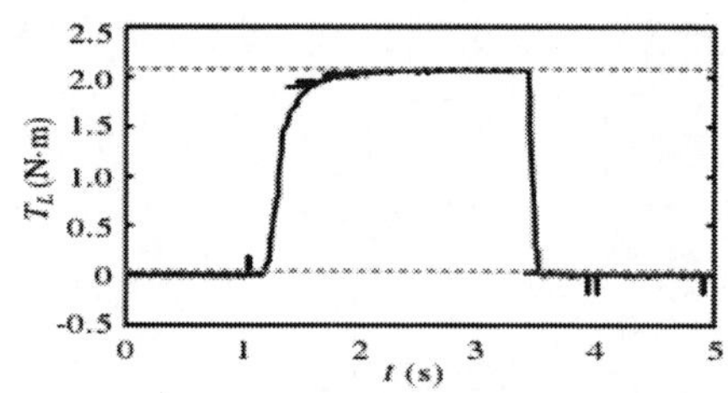

Figure 7. Experimental load curve

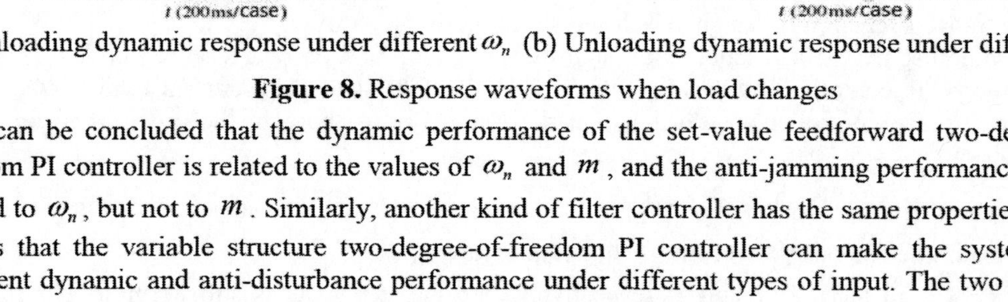

(a) Unloading dynamic response under different ω_n (b) Unloading dynamic response under different m

Figure 8. Response waveforms when load changes

It can be concluded that the dynamic performance of the set-value feedforward two-degree-of-freedom PI controller is related to the values of ω_n and m, and the anti-jamming performance is only related to ω_n, but not to m. Similarly, another kind of filter controller has the same properties, which proves that the variable structure two-degree-of-freedom PI controller can make the system have excellent dynamic and anti-disturbance performance under different types of input. The two kinds of performance can be taken into account, and the tuning process of control parameters becomes simple, and the engineering practicability is strong [12].

5. Concluding Remarks

The variable structure two-degree-of-freedom PI controller designed in this paper can not only simplify the tuning process of control parameters, but also take into account the tracking and anti-disturbance performance. It can also achieve fast step response without overshoot, and can track continuously changing input without error.

The variable structure two-degree-of-freedom PI controller designed in this paper is simple, practical and reliable. It is very suitable for high performance speed servo system, and provides a practical and effective new method for speed servo control field.

Fund projects

Ministry of Education Vocational College Teaching Research Project 2018: 2018LXA0056; 2018 College-level Teaching Reform Project: 18XHJG-030

Reference

[1] D.Knittel, E. Laroche, D. Gigan, et al. Tension Control for Winding Systems with Two-degrees-of-freedom H∞Controllers[J]. IEEE Transactions on Industry Applications, 2003,39(1):113-120.

[2] Guo Yuanbo, Zhou Xin, Zhang Xiaohua et al. Auto-disturbance Rejection Control of Three-phase Voltage-source Pulse Width Modulation Rectifier[J]. Power System Automation, 2011, 35 (16): 87-93.

[3] Li Ming, Cheng Qiming, Chen Gen et al. Two-degree-of-freedom Control of Permanent Magnet Synchronous Servo Motor [J]. Motor and Control Applications, 2014, 41 (10): 1-5.

[4] Zhang Jinggang. Two-degree-of-freedom Control [M].Beijing: Electronic Industry Press, 2010.

[5] Yang Ming, Zhang Yang, Cao He jinsheng et al. AC Servo System Controller Parameter Self-tuning and Optimization [J].Journal of Motor and Control,2010, 14 (12): 29-34.

[6] Wang Hongjia, Yang Ming, Niuli et al. Optimum Design Method of Speed Controller for Permanent Magnet AC servo System [J].Motor and Control Science Newspaper, 2012, 16 (2): 25-31.

[7] Luda, Zhao Guangzhou, Quyilong et al. Permanent Magnet Synchronous Motor Parameter-free Tuning ADRC [J]. Journal of Electrical Technology, 2013, 28 (3): 27-34.

[8] Zhang Jinggang, Liu Zhiyun, Peirun. Two-degree-of-freedom Internal Model Control of AC Servo System [J]. Journal of Electrical Technology, 2002, 17 (04): 45-48.

[9] Gao Zhiqiang. Exploration of ADRC Thought [J]. Control Theory and Application, 2013, 30 (12): 1498-1510.

[10] L.Shihua, Z.Kai, L.Huixian. A Composite Speed Controller Based on A Second-order Model of Permanent Magnet Synchronous Motor System[J].Transactions of the Institute of Measurement and Control, 2011,33(5):522-541.

[11] Harnefors L, Saarakkala S E, Hinkkanen M.Speed Control of Electrical Drives UsingClassical Control Methods[J].IEEE Transactions on Industry Applications, 2013,49(2):889-898.

[12] Chen Zhen,Geng Jie.Time-varying Sliding Mode Control based on Integral Speed Regulation System of Permanent Magnet Synchronous Motor [J]. Electrical Technology Journal, 2011, 26 (6): 56-61.

IOP Conf. Series: Journal of Physics: Conf. Series **1267** (2019) 012093 doi:10.1088/1742-6596/1267/1/012093

Design and Simulation of Variable Structure PI Controller

Liu Shaojun[1] and Wang Yuyu[1,*]

[1]Xi'an Aeronautical Polytechnic Institute

[*]E-mail: 80709047@qq.com

Abstract. Aiming at the poor performance of IP control system in tracking continuously varying input signals, a variable structure PI controller was proposed, which can eliminate overshoot of step response and track continuously varying input signals without error. The controller performs as an IP controller for step input signals and a PI controller for continuously changing input signals. Firstly, the control performance defects of traditional IP controller were analyzed, and then the structure of variable structure PI controller was built. Finally, the simulation results show that the VSPI and IP control systems have the same control effect when the step signal is input. Compared with PI control system, the VSPI and IP control systems not only have no overshoot but also have better response stability. When the input is a continuous sinusoidal signal, VSPI and PI can track the input signal almost without error, and the control effect is obviously better than IP control. The anti-jamming performance of the system is inversely proportional to the bandwidth value, but has nothing to do with the control mode.

1. Introduction

When the speed loop of permanent magnet synchronous motor is controlled by PI, the output response will overshoot for step input. As a result, document [1] uses the method of arranging transition process to eliminate overshoot, but the form of process function is complex; document [2] uses output differential negative feedback to reduce overshoot, but this method increases measurement noise due to the use of differentiators. For this reason, the pseudo-differential feedback control and IP control methods proposed in [3] and [4,5] are essentially identical, both of which can suppress overshoot of step response well, but at the same time, the steady-state error of input response with continuous variation will be increased. The traditional controller only considers the condition of step input signal. When the input signal is continuously changing, it needs to be redesigned. Although IP controller can eliminate overshoot of step response, it also makes tracking performance worse. In order to improve the adaptability of the controller to various inputs, a variable structure PI controller is proposed in this paper. It can automatically select different control modes according to different types of input signals to achieve optimal control.

2. IP Controller

The speed loop IP controller using the pre-stop integral anti-saturation method [6] is shown in Figure 1 below.

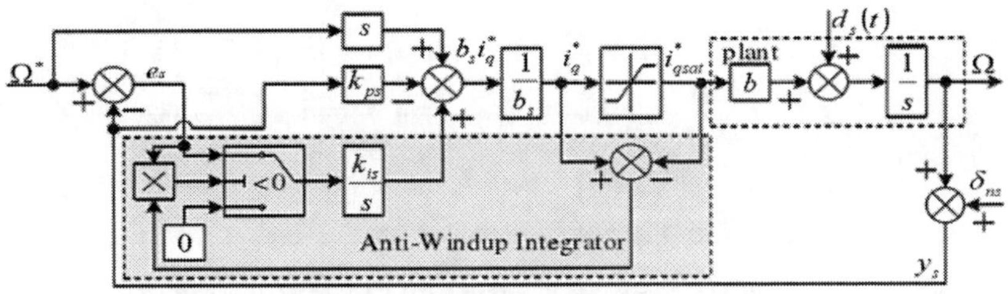

Figure 1. IP controller block diagram

As can be seen from Figure 1, the expression of the control quantity of the system is as follows:

$$i_q^* = \frac{\dot{\Omega}^* - k_{ps} y_s + k_{is} \int (\Omega^* - y_s) dt}{b_s} \tag{1}$$

In the formula, k_{ps} is the proportional coefficient of the controller; k_{is} is the integral coefficient of the controller, b_s is the control gain, i_q^* is the given value of the cross-axis current, Ω is the mechanical angular velocity, and y_s is the measured value of the mechanical angular velocity.

Regardless of the effects of feedback speed filtering, current loop bandwidth and control saturation, formula (1) is substituted into the state equation of angular velocity of the mechanism, and the output of the system is as follows:

$$\Omega(s) = \frac{s^2 + k_{is}}{s^2 + k_{ps} + k_{is}} \Omega^*(s) + \frac{s}{s^2 + k_{ps} + k_{is}} d_s(s) - \frac{k_{ps} s + k_{is}}{s^2 + k_{ps} + k_{is}} \delta_{ns}(s) \tag{2}$$

The current amplitude and bandwidth of the current loop are limited, so when the step signal is input, its input differential is a pulse signal, and the system does not respond to the signal, resulting in the input differential feed-forward is basically ineffective. The output of the system will be changed to:

$$\Omega(s) = \frac{k_{is}}{s^2 + k_{ps} + k_{is}} \Omega^*(s) + \frac{s}{s^2 + k_{ps} + k_{is}} d_s(s) - \frac{k_{ps} s + k_{is}}{s^2 + k_{ps} + k_{is}} \delta_{ns}(s) \tag{3}$$

It can be seen from equation (3) that the system is a typical second-order system. $s^2 + k_{ps} s + k_{is}$ is the characteristic polynomial $s^2 + 2\zeta\omega_n s + \omega_n^2$ of the typical second-order system, then $k_{ps} = 2\zeta\omega_n$, $k_{is} = \omega_n^2$, ω_n is the natural frequency without damp and ζ is the damping ratio.

As long as the value of the damping ratio is greater than 1, the step response of the system can be achieved without overshoot. Compared with the step response of the system, there will be overshoot PI controller under any damping condition, which is obviously more advantageous in suppressing overshoot.

But when the input signal changes continuously, the input differential feed forward can take effect because the system can respond to the differential of the signal. Formula (2) shows that there are steady-state errors in the IP control system at this time. Compared with the PI control system which continuously tracks the change of input signal without error, its control performance is worse than that of the PI control system.

3. Variable Structure PI Controller

Based on the defect of the above IP control system in tracking continuously varying input signals[7] [8], a new variable structure controller is proposed. Its control structure is shown in Figure 2 below. The system is an IP controller when the step signal is input, and a PI controller when the continuous change signal is input. Therefore, the system can not only eliminate the overshoot of the step response, but also track the continuous change of the input signal accurately.

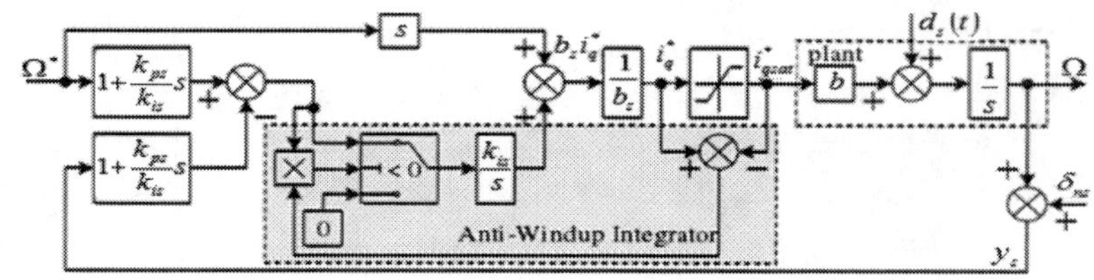

Figure 2 Variable Structure PI Control System

If the input is a step signal, the control quantity generated by the given differential term is large, which will lead to saturation in the first cycle. At this time, the integral input is zero, and the anti-integral saturation algorithm takes effect[9] [10]. Since the given differential term after the second cycle is zero, the control quantity will not be saturated. The output of the corresponding adder is the input of the integral link [11], and the controller can be equivalent to the control structure shown in Figure 3 below.

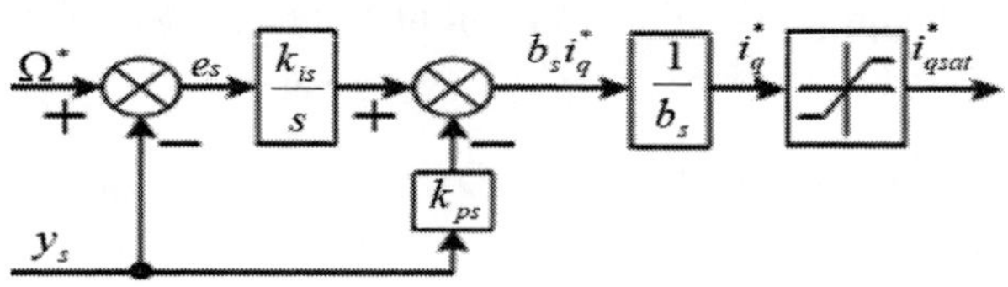

Figure 3 Equivalent structure of VSPI with step input

4. Simulation Analysis

Next, in order to verify the effectiveness of the proposed controller, the response characteristics and stability of the controller under the action of step signal and sinusoidal signal are simulated and analyzed respectively.

4.1 Step Signal Action

No-load startup experiments of VSPI, PI and IP control systems with input step signals of 80rpm and 800rpm are carried out respectively. The output dynamic response waveforms are shown in Figure 4. From the graph, the dynamic response curves of VSPI and IP control systems basically coincide. Compared with PI control system, the response of VSPI and IP control systems is not only non-overshoot, but also more stable.

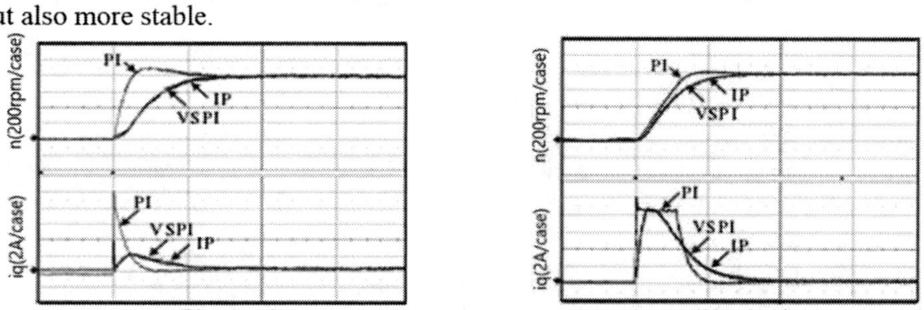

(a) Dynamic response at 80 rpm (b) Dynamic response at 800 rpm

Figure 4 Dynamic response of no-load starting at 80rpm and 800rpm

In order to further verify the effect of system bandwidth ω_n on speed response, it is assumed that the step input signal is 800rpm and the system bandwidth ω_n is 80, 160 and 320, respectively. The final speed response curve is shown in Figure 5.

Obviously, the larger the value of ω_n, the better the response speed and tracking performance of the system. In this case, the rapidity and overshoot are mutually unified [12].

IOP Conf. Series: Journal of Physics: Conf. Series **1267** (2019) 012093 doi:10.1088/1742-6596/1267/1/012093

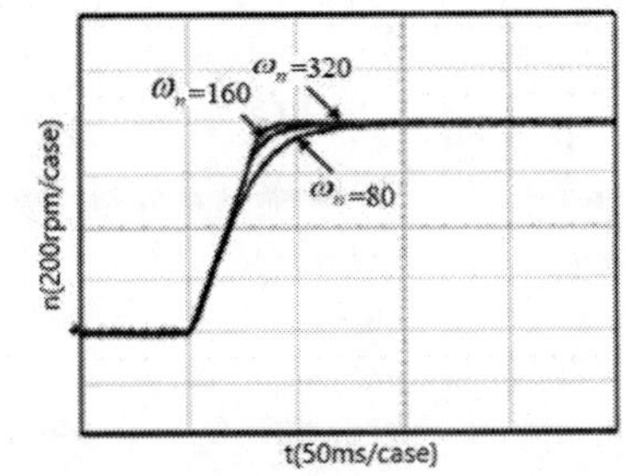

Figure 5 Speed response of idle start under different ω_n

4.2 Sinusoidal Signal Action

The no-load start experiments of VSPI, PI and IP control systems under sinusoidal input signal are carried out. The amplitude of the signal is 500 rpm, the frequency is 5 Hz, and the ω_n is 80. The response waveform is shown in Figure 6.

From the graph, the error ranges of VSPI and PI control are 5 rpm and 6 rpm respectively. The speed response curves of VSPI and PI control coincide almost completely, while the response error range of IP control is 340 rpm, so VSPI and PI control are obviously better than IP control.

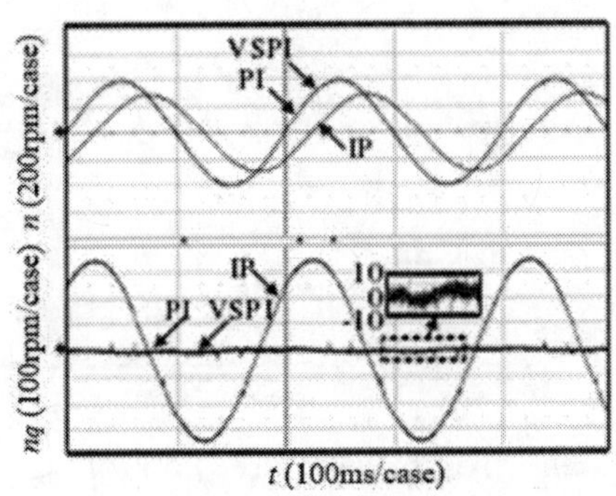

Figure 6 Speed response under sinusoidal setting

4.3 Anti-jamming Analysis

In order to test the anti-jamming performance of VSPI control system, when the system runs stably at 800rpm, the bandwidth ω_n is 80 and 160, respectively. The unloading experiment of the system is carried out. The loading curve is shown in Figure 7 below, and the response waveform is shown in Figure 8 below.

It can be seen from the graph that the larger the value of ω_n, the smaller the fluctuation of speed caused by load, the smaller the anti-interference performance, but the greater the system noise [13].

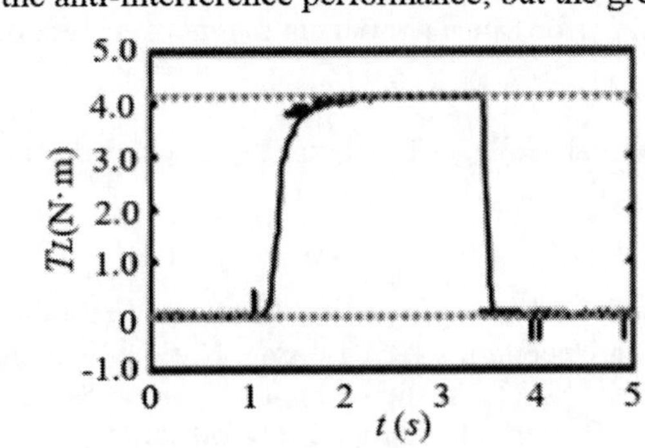

Figure 7 Experimental Load Curve

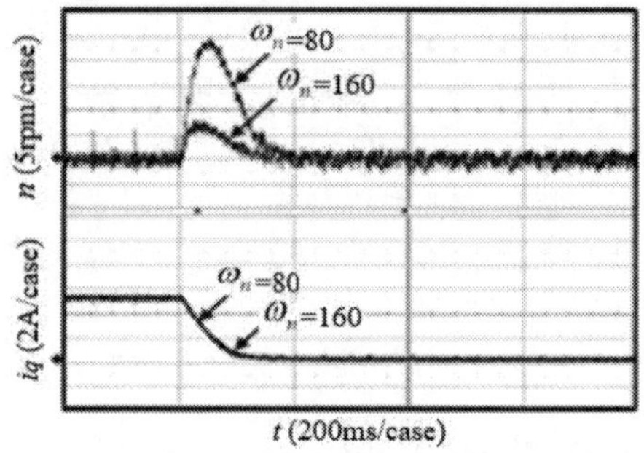

Figure 8 Unloading dynamic response under different ω_n

In order to verify the anti-jamming performance of three different control systems, taking ω_n as 80, three different control systems are loaded according to the curve shown in Figure 7, and their respective response waveforms are shown in Figure 9. From the graph, when the bandwidth ω_n is the same, the response curves of the three control systems almost coincide, which shows that the anti-jamming performance of the three control systems is the same when the bandwidth is the same.

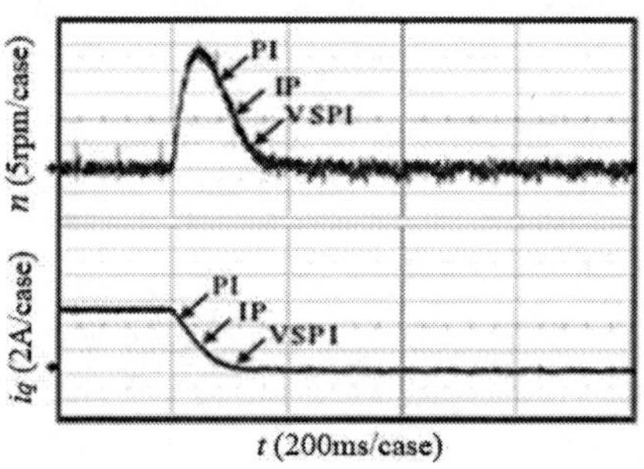

Figure 9 Unloading dynamic response of three control systems under the same ω_n

5. Concluding Remarks

In order to obtain the optimal control effect under the action of multiple input signals, this paper designs a variable structure PI control based on the principle of traditional IP control, which effectively solves the contradiction between overshoot and rapidity of step response, and greatly improves the tracking performance of continuous change input response. The controller has simple structure design, convenient parameter setting and can adjust the control law adaptively according to different types of input speed signals. Finally, the comparative simulation results verify the superiority of the variable structure PI control proposed in this paper, which opens up new methods and ideas for the speed servo system of high performance permanent magnet synchronous motor.

Fund projects

Ministry of Education Vocational College Teaching Research Project 2018: 2018LXA0056; 2018 College-level Teaching Reform Project: 18XHJG-030

Reference

[1] Han Jingqing. Active Disturbance Rejection Control Technology-Control Technology for Estimating Compensation Uncertainty [M]. Beijing: National Defense Industry Press, 2008.
[2] Yang Ming, Zhang Yang, Cao He Jinsheng, et al.AC Servo System Controller Parameter Self-tuning and Optimization [J]. Journal of Motor and Control, 2010, 14 (12): 29-34.
[3] Li Guangquan, Ge Hongjuan, Liu Tianxiang, etc.Pseudo-differential Feedback Control of Permanent Magnet Synchronous Motor Speed Control System[J].Journalof Electrical

Technology, 2010, 25 (08): 18-23.
[4] Guo Xinhua, Wen Xuhui, Zhao Feng and others. New IP Speed Control of Permanent Magnet Synchronous Motor based on Electromagnetic Torque Feedback Compensation [J]. China Journal of Electrical Engineering, 2010, 30 (27): 7-13.
[5] Wang Hongjia, Yang Ming, Niuli et al. Speed of Permanent Magnet AC Servo System Optimum Design Method of Degree Controller[J].Motor and Control Science Newspaper, 2012, 16 (2): 25-31. [6]Wang Lina, Zhu Hongyue, Yang Zongjun. Permanent Magnet Synchronous Motor Tuning Parameter Tuning Method of PI Controller for Speed System [J].Electrical Technology Journal of Academy, 2014, 29 (05): 104-117.
[7] Luda, Zhao Guangzhou, Quyilong et al. Permanent Magnet Synchronous Motor Parameter-free Tuning ADRC[J].Journal of Electrical Technology, 2013, 28 (3): 27-34.
[8] Zhang Jinggang, Liu Zhiyun, Peirun. Two-degree-of-freedom Internal Model Control of AC servo system [J].Journal of Electrical Technology, 2002, 17 (04): 45-48.
[9] Gao Zhiqiang. Exploration of ADRC thought [J]. Control Theory and Application, 2013, 30 (12): 1498-1510.
[10] Luda, Zhao Guangzhou, Quyilong et al. Permanent Magnet Synchronous Motor Active Disturbance Rejection Controller with Parameter Tuning [J].Journal of Electrical Technology,2013, 28 (3): 27-34.
[11] Harnefors L, Saarakkala S E, Hinkkanen M. Speed Control of Electrical Drives UsingClassical Control Methods[J].IEEE Transactions on Industry Applications, 2013,49(2):889-898.
[12] Chen Zhen, Geng Jie, Liu Xiangdong. Time-varying Sliding Mode Control based on Integral Speed Regulation System of Permanent Magnet Synchronous Motor [J]. Electrical Technology Journal, 2011, 26 (06): 56-61.
[13] Li Zheng, Hu Guang, Cui Jiarui and others. Permanent Magnet Synchronous Motor Tuning Integral Sliding Mode Variable Structure Control for Speed Systems [J].China Journal of Electrical Engineering, 2014, 34 (03): 431-437.

LM Algorithm Neural Network Predictive Control of FlexRay Bus System

LIU Zhi-chao[2] and WANG Yi[1,2,*]

[1]Guizhou City Vocational College, 550025 Guiyang, China
[2]School of Physics and Electronic Science, Guizhou Normal University, 550025 Guiyang, China

[*]Corresponding Email: wyigz@126.com

Abstract. In view of the uncertainty of data transmission due to the irregular data flow in the network and the limitation of network bandwidth resources, which makes the control performance and stability of FlexRay network decrease when data is transmitted at high speed and the reliability and safety of the control system can not be guaranteed, this paper proposes a method based on the Levenberg-Marquardt (LM) algorithm for neural network predictive control of the FlexRay bus. The method can predict the next moment operating state of the car network according to the current moment working state of the FlexRay car network to adaptively adjust task workload to adapt to changes in vehicle network system load and improve the reliability and stability of FlexRay network control system. The simulation results show that the neural network predictive control has good adaptability and robustness, which improves the control performance of the FlexRay automotive networked control system effectively.

1. Preface

FlexRay vehicle network system is a very complicated distributed control system. The complex control tasks in the system transmit data in a time-sharing manner. However, due to the irregularly changing data traffic and limited resource bandwidth in the network, the message transmission is uncertain. Thus, the system control performance of FlexRay network will be weakened in high-speed transmission of information, which can not guarantee the reliability and security control system.

Although many researchers at home and abroad have made a series of researches on FlexRay bus network control system from different aspects [1-4] and they have achieved some research results, there are still some problems in the control performance of the system when the bus is transmitting data at high speed. Among the existing control methods, the model predictive control is a predictive control method based on the mathematical model. When the model mismatches, the result of the control output can not meet the control requirements. The traditional BP neural network is prone to fall into a partial minimum, and the convergence rate is slow. Fuzzy adaptive control performance is poor, so the control accuracy can not meet the system requirements.

The LM algorithm combines the advantages of both gradient descent and Gauss-Newton algorithms, both has local convergence and global characteristic[5]. Therefore, the modified BP neural network based on LM algorithm can be used as the prediction model of the vehicle-mounted FlexRay network to overcome the shortcomings of the traditional BP neural network in terms of convergence speed and falling into partial minima. Meanwhile, the neural network predictive control is applied to

FlexRay Network performance tracking control, making full use of neural network strong nonlinear mapping ability and predictive control of rolling optimization and feedback correction mechanism, which will be more suitable to the uncertainty of complex control system and the actual situation of time-varying, using Network training algorithm to conduct rolling optimization towards the weight of network weight and the network threshold in the control process, in order to improve the prediction accuracy and system stability.

2. FlexRay network predictive control theory and mathematical model

By sampling the information of network state at the current time, the theory uses the acquired information to calculate the utilization rate of network resources J, forecasting the utilization rate of network resources at the next moment, and adapting the workload of the task online adaptively to adapt to the load with dynamic changes in the FlexRay vehicle network system. The prediction controller compares the network utilization pre-evaluation value $\overline{J}$ with the maximum utilization setting J_{sp}. If $J < J_{SP}$, it indicates that the network load is light,and the data transmission task can work stably in the previous state; if $J \geq J_{SP}$, it indicates that the network load is heavy or it is overloaded. At this time, the load of the system should be reduced by increasing the task sampling period adaptively so as to maintain the network stable operation of the entire control system.

In the FlexRay network communication cycle, the static segment is responsible for the sending and receiving of periodic messages, which is the time period that must be included in the communication cycle and can account for more than 70% of the total communication cycle length. Therefore, in this paper, the bandwidth utilization of the static segment is approximated as the entire network bandwidth utilization . The unused bandwidth is divided into two parts by the FlexRay frame format. One part is the load that must be added by the frame coding specified in the FlexRay protocol and the other part is the one that can not be used effectively due to the long static fixed time slot. Among them, the additional load of the agreement is:

$$U_1 = P \times \sum_{i=1}^{n} \left(\frac{L_i}{L} + 1\right) = (P + BSS \times L) \times N \tag{1}$$

Where P represents the frame header of data frame and the total number of bits after encoding at the end part of the frame, the provisions of $P = 113b$;

$$N = \sum_{i=1}^{n} \left(\frac{L_i}{L} + 1\right) \tag{2}$$

N represents the number of static time slots. L_i is the length of the i signal. L is the length of valid data in the static frame. BSS indicates the number of bits in the start sequence of bytes, and specifies $BSS = 2b$.

The number of bits that are not utilized due to fixed static time slots during one communication cycle is:

$$U_2 = (8 + BSS) \times (L \times N - \sum_{i=1}^{n} L_i) \tag{3}$$

The load that has not been effectively used is:

$$\Delta U = U1 + U2 \tag{4}$$

The number of bits of the entire static segment cycle length :

$$U = \sum_{i=1}^{n} \left(\frac{L_i}{L_{opt}} + 1\right) \times \left[P + \left(8 + BSS\right) \times L\right] \tag{5}$$

Then, the effective bandwidth utilization is:

IOP Conf. Series: Journal of Physics: Conf. Series **1267** (2019) 012094 doi:10.1088/1742-6596/1267/1/012094

$$J = 1 - \frac{\Delta U}{U} \tag{6}$$

Maximizing the effective bandwidth utilization is equivalent to minimizing the ineffective bandwidth utilization, which is:

$$\min\{\frac{\Delta U}{U}\} \tag{7}$$

The following constraints apply: The number of slots in the static segment must be greater than or equal to the number of signals transmitted; the optimized payload length must be less than the maximum payload size and the bytes of the optimal payload size must be even, which is:

$N \geq n$

$2 \leq L_{opt} \leq 254$

$$L_{opt} = 2 \times \text{nearest}(\frac{L}{2}) \tag{8}$$

$$L_{opt} = \sqrt{K \cdot \frac{D}{n}} \tag{9}$$

Where $K = \frac{P}{8+2BSS}$, and D is the sum of all signal lengths.

3. Neural network predictive control

The predict model based on neural network is the model of using neural networks to approximate the controlled object , to establish a predictive signal neural network model after training . Model prediction is to establish the neural model of the controlled object first, then using the controller to predict the performance of the neural network at the next moment. The structure of the neural network predictive control system is shown in Figure 1.

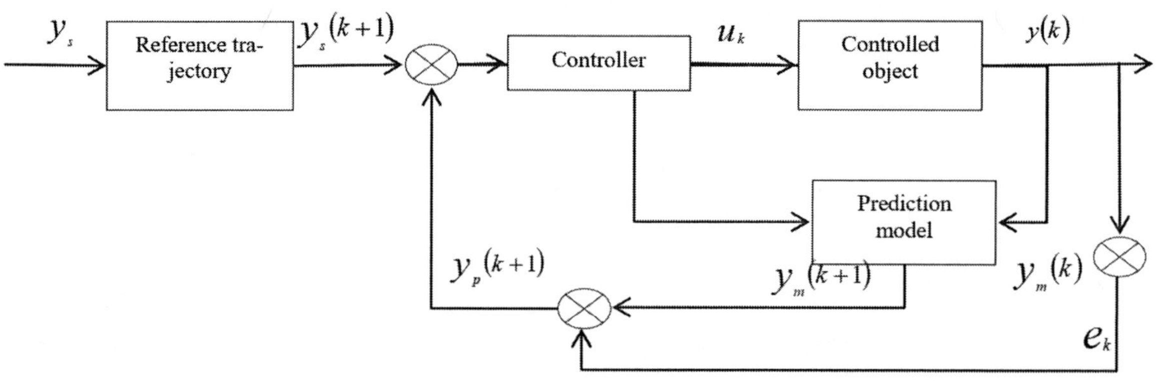

Figure 1. Neural predictive control system structure

In the figure, $u(k)$ is the control signal at the current time,and $y(k)$ is the actual output of the controlled object at the current time. $y_m(k)$ is the predicted output at the current time. $y_m(k+1)$ is the predicted output of the prediction model at time $k+1$,and $y_p(k+1)$ is the predicted output after feedback correction at time $k+1$. y_s is the reference input of the controlled object,and $y_s(k+1)$ is the reference input at $k+1$ moment,that is ,the desired output of the controlled,which is J_{sp} in this control system.

3.1. System Identification

The model prediction is firstly to train the neural network model, then the neural network model uses the current input and output to predict the output value at the next moment, and takes the prediction error as the training signal of the neural network.The error between the actual output and the prediction output of the model is:

$$e_m(k) = y(k) - y_m(k) \tag{10}$$

The model prediction output after feedback correction is:

$$y_p(k+1) = y_m(k+1) + he_m(k) \tag{11}$$

h is the error correction factor.

3.2. Model prediction

Before performing model prediction, the reference trajectory is usually input. Based on the prediction output and the reference trajectory, the control signal is determined by establishing rolling optimization performance index . The index function of the most optimized performance is:

$$J = \sum_{j=1}^{P}[y_r(k+j) - y_p(k+j)]^2 + \lambda\sum_{j=1}^{M}[u(k+j-1) - u(k+j-2)]^2 \tag{12}$$

Where, P is predicting the length of the time domain; M is controlling the length of the time domain; ρ is tracking error; λ is the weighted coefficient of controlled quantity.

3.3. The optimization of neural network *system based on LM algorithm*

Generally, the initial weights and thresholds of BP neural network are two arbitrary values between [0, 1], but the weights and thresholds obtained in this way are very apt to fall into many unfavorable conditions, such as the partial minimum, the convergence rate being slow, many iteration steps and so on. Therefore, this paper uses LM algorithm to optimize the weights and thresholds of BP neural network to improve the identification accuracy of the model. The main idea of LM algorithm is as follows: To reduce the non-optimal singularity problem in practice, making the characteristic near the extremum point being approximately quadratic property to accelerate the speed of convergence when the objective function tends to be the optimal point, a small positive number u is generally added to each element in the main diagonal of the coefficient matrix $J^T J$,which is:

$$\left(J^T J + uI\right) p = -J^T e \tag{13}$$

$$w(k+1) = w^k + ap \tag{14}$$

Where e is the error; p is the forward direction; w is the connection weight; J is the first derivative of the objective function with respect to the connection weight; a is the adjustment step; u is the damping factor; and I is the unit matrix.

The choice of learning algorithm depends on the size of u .The following formula is an LM algorithm for updating parameters:

$$\Delta w = \left(J^T J + uI\right)^{-1} \bullet J^T e \tag{15}$$

When the value of u is larger, the LM algorithm approximates the gradient descent method: when the value of u is smaller, it is the Gauss-Newton algorithm.

The procedures of optimizing the BP neural network with the LM algorithm and building the model based on which is shown in Figure 2.

IOP Conf. Series: Journal of Physics: Conf. Series **1267** (2019) 012094 doi:10.1088/1742-6596/1267/1/012094

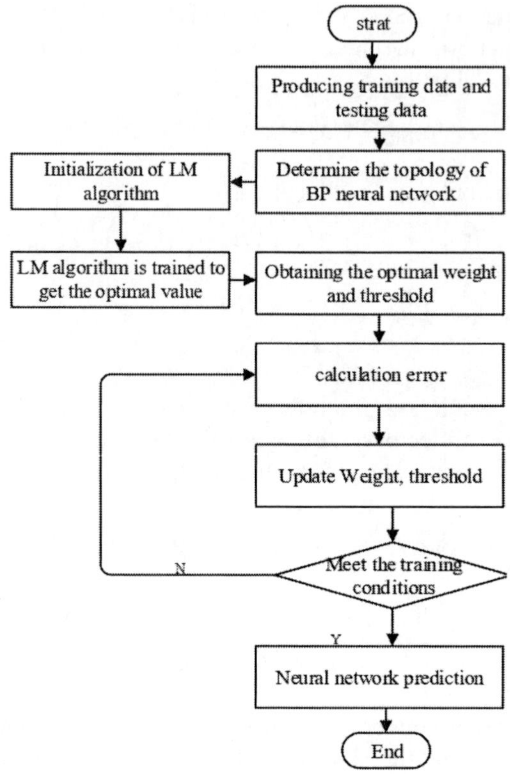

Figure 2. LM-BP neural network flow chart

4. Experimental Simulation and Analysis

In this paper, neural network prediction controller module provided by the neural network toolbox in MATLAB is simulated by Simulink. The first step is system identification. First, the data needed for neural network training is generated, setting the random input signal and the structure of the neural network In the model identification dialog, using LM algorithm neural network and established network resource utilization model to generate 5000 sets of training data. After setting the number of iterations and the training algorithm, the training of the neural network is conducted. Figure 3 is the result of the training, and Figure 4 is the legal data. In these two figures, Input represents the input signal; Plant Outptu represents the controlled object output; Error is the error signal, and NN Output represents the output of the neural network model.

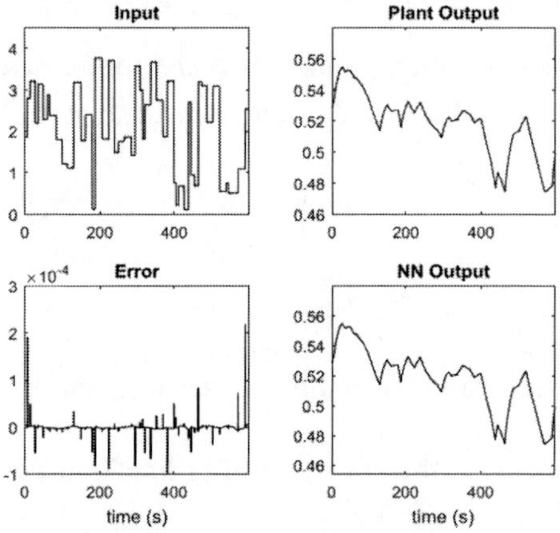

Figure 3. Training data

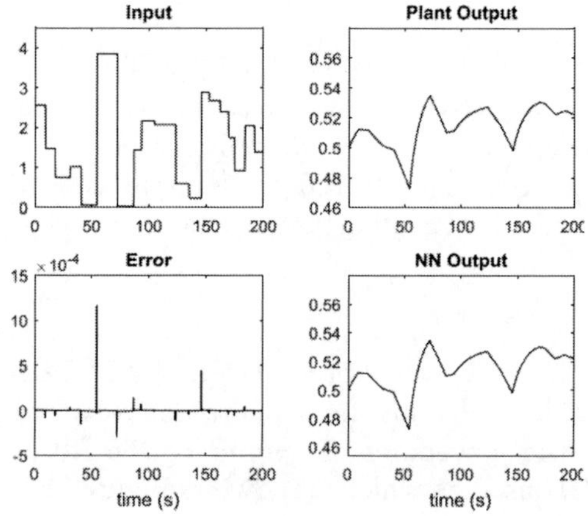

Figure 4. Legal Data

After the neural network model is established, predicting the future output of the network bandwidth utilization J by the neural network prediction controller, the optimization algorithm controls the system input through the predicted value. The simulation results are shown in Figure 5, and the relevant parameters of neural network predictive controller are: $P = 7$, $M = 4$, $\rho = 0.03$, $\lambda = 0.5$. Data 1 represents the actual output of the system and data2 represents the reference input. Predictive control results show that the predictive control system can well realize the predictive control of FlexRay vehicle network and improve the control performance of FlexRay network control system effectively.

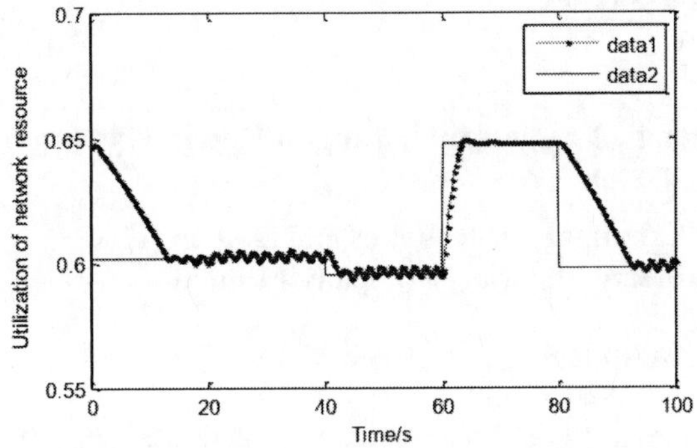

Figure 5. Neural network predictive control system simulation

5. Concluding remarks

Based on the mathematical model of the network utilization of FlexRay bus network, the neural network predictive controller is used to track and control the task workload. Combining the LM algorithm to predict the network resource utilization of the vehicle network control system, which makes up for the defects of traditional neural network falling into Local minimization easily and slow convergence, the speed of network convergence has been improved. The simulation results show that the neural network predictive control can effectively predict the network utilization of the FlexRay vehicle network system and ensure the safe and reliable transmission of information in the vehicle network system by adjusting the task workload in real time and improve the reliability and stability of FlexRay network control system.

Acknowledgements

This research was financially supported by National Natural Science Foundation of China (Grant No.61462015).

References

[1] Rui ZHAO, Guihe QIN, Tiehu FAN. Bus cycle optimization for FlexRay communication protocol[J] .2010,27 (10): 3847-3850.In Chinese
[2] Xiang Xinan .FlexRay bus static section parameters optimization and online control system [D]. Hunan University .2016.In Chinese
[3] Zeng H, Natale Di, Ghosal M. Schedule optimization of time-triggered systems communicating over the FlexRay static segment [J]. IEEE Transactions on Industrial Informatics, 2011, 7 (1): 1-17.
[4] Kang M, Park K, Jeong M. Frame packing for minimizing the bandwidth consumption of the FlexRay static segment [J]. IEEE Transactions on Industrial Electronics, 2013, 60 (9): 4001-4008.
[5] Qiao ZHANG,Gui-shi DENG.Levenberg-Maequardt Application of Neural Networks in Reliability Evaluation of Coal Mine Workers[J].Journal of Dalian University of Technology.2015,55(4):424-430.In Chinese

IOP Conf. Series: Journal of Physics: Conf. Series **1267** (2019) 012095 doi:10.1088/1742-6596/1267/1/012095

Semantic Segmentation based Dense RGB-D SLAM in Dynamic Environments

Jianbo Zhang[1], Yanjie Liu*[1], Junguo Chen[2], Liulong Ma[1], Dong Jin[1] and Jiao Chen[1]

[1]State Key Laboratory of Robotics and System (HIT)
[2]China Petroleum Materials Company Limited

E-mail: 17S008048@stu.hit.edu.cn

Abstract. Visual Simultaneous Location and Mapping (SLAM) based on RGB-D has developed as a fundamental capability for intelligent mobile robot. However, most of existing SLAM algorithms assume that the environment is static and not suitable for dynamic environments. This is because moving objects in dynamic environments can interfere with camera pose tracking, cause undesired objects to be integrated into the map. In this paper, we modify the existing framework for RGB-D SLAM in dynamic environments, which reduces the influence of moving objects and reconstructs the background. The method starts by semantic segmentation and motion points detection, then removing feature points on moving objects. Meanwhile, a clean and accurate semantic map is produced, which contains semantic information maintenance part. Quantitative experiments using TUM RGB-D dataset are conducted. The results show that the absolute trajectory accuracy and real-time performance in dynamic scenes can be improved.

1. Introduction

For navigation purposes, visual SLAM is widely used in many robotic applications. In the past years, RGB-D based visual SLAM is extensively researched due to the emergence of economical depth cameras such as Kinect2. In order to estimate the camera's ego-motion and perceive environmental information, the SLAM system based on RGB-D can directly use the color and depth information provided by the camera. Lots of advanced SLAM algorithms have been developed and achieve good results.

In order to simplify the problem, most of current visual SLAM systems assume that the environment is static. It means that the change of image information depends only on the motion of the camera itself. However, in the actual environment, there are inevitably dynamic objects, such as walking people, moving tables and chairs, etc. Some algorithms (such as random sampling consistency, RANSAC) can be used to handle the outliers introduced by small moving objects. However, when the dynamic object is relatively large in the image, it will introduce errors into the SLAM system. Then, the system gets the wrong pose information, which affects the accurate construction of the map. Besides, most of the semantic maps built by the system are less accurate and lack of effective semantic information maintenance [1].

Improving the performance of the SLAM system in dynamic environments is very important for mobile robots. SLAM uses the continuous image sequence output by the camera to estimate the pose. Dynamic objects in the scene can lead to mismatches in feature matching, affecting pose estimation accuracy. Therefore, Dynamic objects in the scene should be removed. Some methods are adopted to reduce the impact of moving objects, for instance, moving objects can be inspected and removed using scene flow in order to avoid mismatches in the alignment step [2]. But this problem has not been solved very well.

Semantic map can provide rich environmental information for robot navigation. The traditional semantic map construction mainly uses the support vector, CRF and other machine learning methods to detect and segment the scene, but the performance and accuracy are limited. At present, the deep learning method provides a new way of constructing a semantic map, which can achieve better performance.

In this paper, we are concerned about real-time detection of dynamic objects to reduce the impact of moving objects. Meanwhile, the system should provide a semantic octo-tree map, which contains accurate semantic information.

The contribution of this paper are as follows:

- We propose to use ENet to obtain pixel-level semantic segmentation of the image. Then, moving objects can be detected by combining semantic segmentation with moving consistency check method, so that the SLAM algorithm will not extract features on them. Compared with ORB, the accuracy and robustness of the system are improved. We achieve a balance of system accuracy and real-time performance.
- We build a semantic octo-tree map in the map creation thread. In order to avoid fusing moving objects into map, the dense semantic octo-tree map uses semantic information and log-odds score method to filter out them [1]. Meanwhile, this thread combines the semantic information of different viewpoints to achieve multi-view consistent semantic segmentation.

The paper is organized as follows. We review state-of-art visual SLAM in dynamic environments as well as several approaches about semantic SLAM in Section 2. Then Section 3 presents the overall structure of the whole SLAM system, that how to detect moving objects on images and update semantic information. Section 4 provide our evaluation and results of performance of the system against dynamic environments, where we compare the method against related works. Section 5 concludes our work and future works for research.

2. Related work

2.1. Moving objects detection

Most SLAM systems are based on the assumption of static environment. However, there are inevitably dynamic objects in the actual environment, which will affect the accuracy and robustness of SALM system.

To address this problem, special measure should be taken to enhance its robustness and accuracy in dynamic environments. In most SLAM systems, dynamic objects are classified as noise, separated out before doing SLAM, therefore neither included in the map nor used for pose estimation. There are several SLAM systems that can detect dynamic objects more specifically. P.Ochs et al. [3] detect dynamic object by long term point trajectory analysis without priori model. The method is robust, but the time cost is very expensive. Jens et al. [4] combine optical flow and depth information to detect moving objects. However, optimization methods need to be used to reduce processing complexity and minimize the high computational time. Besides, Wangsiripitak and Murray [5] deal with dynamic objects by tracking known 3D objects in the scene. The method can detect those a priori dynamic objects, but it can't detect changes made by static objects, such as a chair pushed by a person, or a ball thrown by someone.

2.2. Semantic SLAM

In most SLAM algorithms, the position of the robot and the map point information are only dense or sparse geometric points in the space. By estimating the position of these spatial points, we can obtain relatively accurate position information, but can't get higher-level semantic information. Thus, the rich semantic information in the environment can't be used in complex task. Semantic information can help SLAM better understand the scene and get more information about the scene, thus improving the accuracy of positioning and map construction.

So semantic map has become a popular research domain in recent years. Galindo et al. [6] divide the map into spatial map and conceptual map, organized in a hierarchical manner, and merge the spatial map nodes with the conceptual map nodes. With the development of visual target detection algorithms based on deep neural networks, we can use more accurate semantic information to construct semantic maps. J.McCormack et al. [7] propose a dense 3D semantic mapping method using convolutional neural networks by combining CNN with ElasticFusion. Their work focus on building maps using semantic information, while semantic information is not used in other parts. Besides, many methods have been proposed to fuse semantic information from different perspectives to perform multi-view consistent semantic information. Li et al. [8] propose to fuse the semantics information by using bayesian fusion on key frames from their SLAM backend.

In this paper, semantic information is not only used to generate semantic octree maps, but also utilized to filter out moving object in the process of tracking in dynamic environments [1]. And we will fuse semantic information from different perspectives.

3. System description

In this section, first, the framework of the system is presented. Second, a robust moving object segmentation method will be introduced in details, which enhance the robustness and accuracy of dense SLAM in dynamic environments. We use a real-time semantic segmentation network named ENet and optical flow to detect moving objects. Finally, we present the method to create semantic octo-tree map, which includes semantic information maintenance.

3.1. Framework of the system

ORB-SLAM2 performs well in most practical scenarios, so the system is improved based on ORB-SLAM2 to provide global feature-based SLAM in dynamic environments. we modify the existing framework [1] for RGB-D SLAM in dynamic environments, and the overview of the system is shown in Figure 1.

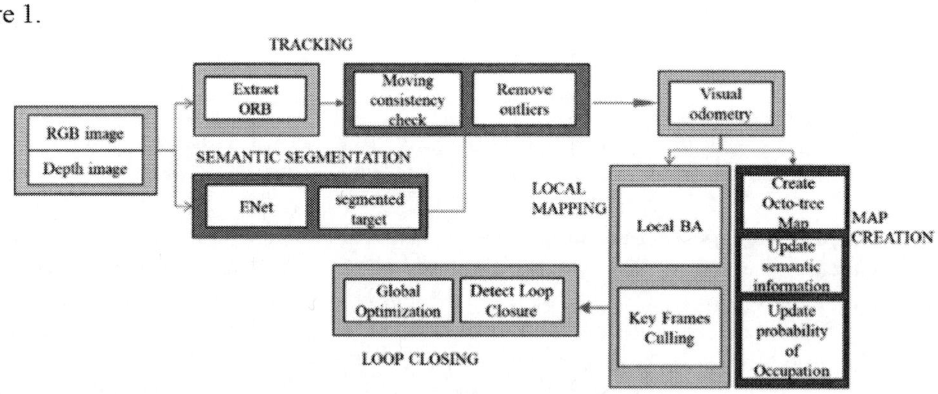

Figure 1. The framework of system

Five threads run in parallel in the system: tracking, semantic segmentation, local mapping, loop closing, and map creation. Local mapping and loop closing are the same as ORB-SLAM2. The RGB image captured by Kinect2 is utilized to semantic segmentation thread and moving consistency check simultaneously. Moving objects are detected by combining the potential outliers from moving consistency check and pixel-wise semantic label from semantic segmentation thread, then the ORB feature points located in moving objects will be filtered out. Finally, the system calculates the transformation matrix by matching the stable ORB feature points [1].

The system is based on ROS (Robot Operating System). ROS is an information delivery system in which many parts can work together by communicating output information.

IOP Conf. Series: Journal of Physics: Conf. Series **1267** (2019) 012095 doi:10.1088/1742-6596/1267/1/012095

3.2. Moving objects detection

This system combines semantic segmentation with moving consistency check to detect dynamic objects, using the contours extracted by semantic segmentation thread and motion information provided by moving consistency check thread.

To extract the contours of dynamic objects, we use a CNN to obtain a pixel-level semantic segmentation of the image. Considering the application of the system in the actual scene, the system should achieve a balance between accuracy and real-time. So we use ENet to process images in our experiments, which is one of the most advanced semantic segmentation networks. ENet consists of a main branch and an additional branch with a convolution kernel, and finally performs pixel-level addition and fusion. ENet is 18 times faster, has 79 less parameters, requires 75 less FLOPs, provides similar or better accuracy to other existing models [9]. The input to ENet is the RGB raw image and the idea is to segment those classes that may be dynamic or moveable, such as people, bicycles, cars, motorcycles etc.

Dataset plays an important role in the training process of semantic segmentation network, which will affect the accuracy of segmentation. This article combines the original pascal-voc2012 dataset with the voc_aug dataset provided by B. Hariharan et al. for our semantic segmentation task. The Adam optimization algorithm is used to train the network, which allows ENet to quickly converge using GTX 1080Ti GPU. The training is performed in two phases: First, we only train the encoder to classify the downsampled regions of the input image, then add the decoder and train the network for upsampling and pixel classification. Learning rate of 5e-3 and L2 weight decay of 2e-4, along with batch size of 8 consistently provided the best results. Training result is shown in Figure 2, Figure 3.

Figure 2. RGB image

Figure 3. The semantic segmentation

The test result demonstrates that ENet has similar accuracy, but better real-time performance compared with Segnet. The inference time of ENet per frame is 23.86ms, while Segnet is 40ms.

We can detect moving points by moving consistency check based on sparse optical flow method. Firstly, optical flow pyramid is calculated to get the matched points in the current frames. Secondly, the fundamental matrix is calculated by using RANSAC with the most inliers. Then, the epipolar line in the current frame can be found by using the fundamental matrix. Finally, the key points will be determined to be moving if the distance from a matched point to its epipolar line is more than a threshold [2]. Therefore, we can detect the dynamic points in the current frame.

This system combines semantic segmentation with moving consistency check to detect dynamic objects. If the number of dynamic points in the contours of a segmented object is more than a certain threshold, this object is regard as a dynamic object. Then remove the feature points located in dynamic regions. The experimental results are shown in the Figure 4, Figure 5.

Figure 4. ORB feature points

Figure 5. Removing outliers

3.3. Dense semantic map building

The octo-tree map saves a lot of storage space compared to the point cloud map. In addition, the octo-tree map can be used to query occupancy information and used for navigation. Therefore, the system creates a separate thread to build a dense 3D octo-tree map using new key frames from tracking and segmentation result. In this process, semantic information is not only used to filter out dynamic objects but also to create semantic map [1]. When we look at a scene from a different point of view, we may get different semantic results. We combine the semantic information of different viewpoints to achieve multi-view consistent semantic segmentation. The system uses max fusion to fuse semantics information of different measurements.

When max fusion is performed, the voxel only includes the semantic color with the highest confidence generated by the ENet and its highest confidence. When inserting the point cloud into an octo-tree map, if the voxel has new semantic result, we fuse two semantics in this way [10]. The pseudo code is shown in Algorithm 1.

Algorithm 1 max fusion

1. function max fusion(sem_1, sem_2)
2. if $sem_1.color == sem_2.color$ then $sem_{fusion}.color \leftarrow sem_1.color$
3. $sem_{fusion}.confidence \leftarrow (sem_1.confidence + sem_2.confidence)/2$
4. else
5. if $sem_1.confidence > sem_2.confidence$ then $sem_{fusion} > sem_1$
6. else $sem_{fusion} > sem_2$
7. $sem_{fusion}.confidence \leftarrow sem_{fusion}.confidene \times 0.9$
8. return sem_{fusion}

4. Result and evaluation

In this section, to evaluate the effectiveness of the system, we tested the real-time performance and the accuracy of the system in dynamic environments using public TUM RGB-D dataset. The dataset has a ground truth of the camera trajectory, which is used to measure both relative and absolute drift. All experiments are performed on a computer with Intel I7 CPU, GTX 1080Ti GPU, and 16GB memory.

The system is based on ORB-SLAM2, so we compared Metrics Absolute Trajectory Error(ATE) that stands for the global quality of the trajectory. We evaluate the values of RMSE, Mean Error, and Standard Deviation (S.D.) of the ATE in the result. And the improvement values are calculated as shown: improvement=$(m-n)/m \times 100\%$, where m is the value of ORB-SLAM2, n is the value of the system. The comparison results are shown in Table 1. The improvement of the system can reach 78% on average. Besides, the system is compared with previous methods to demonstrate the advantages of system. The results in comparison to previous SLAM are shown in Table 2. As we can see from the results, the system performs well in dynamic environments.

Table 1. Result of Absolute trajectory error (ATE)

sequence	ORB-SLAM2(cm)			The system(cm)			improvement		
	RMSE	Mean	S.D.	RMSE	Mean	S.D.	RMSE	Mean	S.D.
fr3_walking_xyz	73.8	64.5	38.1	32.6	28.9	15.1	55%	55%	54%
fr3_walking_half	76.8	69.9	31.7	10.9	5.6	9.3	85%	92%	71%
fr3_walking_static	39.3	36.1	15.7	2.1	1.4	1.6	94%	96%	90%

IOP Conf. Series: Journal of Physics: Conf. Series **1267** (2019) 012095 doi:10.1088/1742-6596/1267/1/012095

Table 2. RMSE of absolute trajectory error for the system in comparison to previous SLAM in dynamic environments

sequence	ATE RMSE (cm)				
	VO-SF [11]	ElasticFusion [12]	Co-fusion [13]	Motion removal [14]	The system
fr3_walking_static	32.7	29.3	55.1	6.5	**2.1**
fr3_walking_xyz	87.4	90.6	69.6	9.3	32.6
fr3_walking_half	48.2	48.6	75.6	12.5	**10.9**

Real-time performance is an important indicator to evaluate SLAM system. The time required of each module of the system is compared with a work which uses Segnet to segment the image. The result is shown in Table 3. Because the real-time semantic segmentation network ENet is used, the real-time performance of the system is relatively improved. The average computation time per frame of the system is 60ms when the resolution of the image is 360×360, which is less than SLAM system adopting other semantic segmentation network. Compared to the system, the work [2] requires 85ms per frame. Compared with no real-time dynamic SLAM system, the system has considerable advantages in real-time performance.

Table 3. Time analysis (image resolution: 360*360)

	ORB extraction	Moving consistency check	Semantic segmentation
ENet	9.4ms	14.9ms	37ms
Segnet	9.1ms	16ms	60ms

5. Conclusion

In this paper, we modify a real-time SLAM system that can handle dynamic environments. The system combines a real-time semantic segmentation network ENet with moving consistency check to handle dynamic objects in scene, then the feature points located in dynamic regions will be filtered out to improve the accuracy of the system. Besides, we build a semantic octo-tree map in the map creation thread. Meanwhile, this thread combines the semantic information of different viewpoints to achieve multi-view consistent semantic segmentation. The real-time performance and accuracy of the system are improved considerably, especially the real-time performance. Besides, we get a map with more accurate semantic information.

Acknowledgment

This work was supported by National Key R&D Program of China, Provincial Funding for National Key R&D Program (Task), Support by Self-Planned Task (NO. SKLRS201813B) of State Key Laboratory of Robotics and System (HIT).

References

[1] Yu C, Liu Z, Liu X, Xie F, Yang Y and Wei Q 2018 Ds-slam: a semantic visual slam towards dynamic environments *IEEE/RSJ International Conference on Intelligent Robots and Systems*

[2] Bakkay M, Majdi A and Zagrouba E 2015 Dense 3D SLAM in dynamic scenes using kinect *7th Iberian Conference IbPRIA, Pattern Recognition and Image Analysis, Lecture Notes in Computer Science*

[3] Ochs P, Malik J and Brox T 2014 Segmentation of moving objects by long term video analysis *IEEE Transactions on Pattern Analysis & Machine Intelligence* **36(6)** 1187-1200

[4] Klappstein J, Vaudrey T, Rabe C, Wedel A and Klette R 2008 Moving object segmentation using optical flow and depth information *Lecture Notes in Computer Science* **5414** 611-623

[5] Wangsiripitak S and Murray D 2009 Avoiding moving outliers in visual SLAM by tracking moving objects *IEEE International Conference on Robotics & Automation*

[6] Galindo C, Saffiotti A, Coradeschi S, Buschka P, Fernandez-Madrigal J and Gonzalez J 2005 Multi-hierarchical semantic maps for mobile robotics *IEEE/RSJ International Conference on Intelligent Robots & Systems*

[7] Mccormac J, Handa A, Davison A and Leutenegger S 2016 Semanticfusion: dense 3d semantic mapping with convolutional neural networks.

[8] Li X and Belaroussi R 2016 Semi-dense 3d semantic mapping from monocular slam

[9] Paszke A, Chaurasia A, Kim S and Culurciello E 2016 Enet: a deep neural network architecture for real-time semantic segmentation

[10] Zhang X 2018 Real-time voxel based 3D semantic mapping with a hand held RGB-D camera

[11] Jaimez M, Kerl C, Gonzalez-Jimenez J and Cremers D 2017 Fast odometry and scene flow from RGB-D cameras based on geometric clustering *IEEE International Conference on Robotics & Automation.*

[12] Whelan T, Salas-Moreno R, Glocker B, Davison A and Leutenegger S 2016 Elasticfusion. *International Journal of Robotics Research*

[13] Rünz M and Agapito L 2017 Co-fusion: real-time segmentation, tracking and fusion of multiple objects

[14] Sun Y, Ming L and Meng Q 2017 Improving rgb-d slam in dynamic environments: a motion removal approach *Robotics & Autonomous Systems* **89(Complete)** 110–122

Sliding Mode Attitude Control of Quadrotor UAV Based on Proportional Integral Observer

Zitong Wang[1] and Youhong Wan[1]

[1]College of Automation, Nanjing University of Posts and Telecommunications, Nanjing, China

E-mail: 496669242@qq.com

Abstract. This paper designs a sliding mode self-pose attitude controller for a quadrotor unmanned aerial vehicle (UAV) system based on proportional integral observer, which also considers the problem of external disturbances. In this paper, firstly, considering the influence of external disturbance, a proportional integral observer is developed to contains the ratio of the estimated error and the integral loopobtain, and obtain both the estimation of the system state and unknown input. Then, the stability of the state estimation error of the proportional integral observer is proved. Thirdly, based on the results of the proportional integral observer, a sliding mode attitude controller is designed. This method is simple and easy to implement, which can improve the system robustness and eliminate the chattering phenomenon existing in the traditional sliding mode control. Finally, the simulation results show that the UAV system achieves better control effect and dynamic performance.

1. Introduction

The quadrotor UAV has a simple structure, strong relative load capacity, vertical take-off and landing, and good maneuverability. It has been widely used in military and civilian fields, such as aerial reconnaissance, communication, electronic interference, resource detection, and forest fire prevention, border patrols, etc. With the rapid development of science and technology, the automation and intelligence of the quadrotor UAV have significantly improved, becoming a hot spot for global development. Therefore, research on UAV flight control technology has also received great attention and development.

Quadrotor UAV attitude and speed control is the basis of UAV flight control, and its control performance greatly affects the safe flight efficiency of UAVs[1, 2]. In recent years, many methods have been applied to the quadrotor UAV flight control[3, 4]. For example, PID control, gain adjustment control, feedback linearization method, neural network control, backstepping control etc[5]. Among them the PID control method can satisfy the basic flight operation control of the quadrotor UAV[6, 7], however, when deviating from the balance point by disturbance, the control performance will not be guaranteed. The feedback linearization method utilizes full state feedback to make the linear relationship between the input and output of the new system[8, 9], this method has high precision requirements for the mathematical model of the research object. The dynamic inverse control based on the neural network method can use the neural network to identify the system when the UAV is difficult to obtain an accurate mathematical model[10, 11].

The above control methods computational complexity of the quadrotor UAV model is high. It is required that all state variables of the system are measurable which are often affected by

IOP Conf. Series: Journal of Physics: Conf. Series **1267** (2019) 012096 doi:10.1088/1742-6596/1267/1/012096

factors such as gusts and load changes in actual flight, and some system variables are difficult to be accurately measured. So the above methods are limited in practical applications.

In conclusion, the control method based on proportional integral state observer is proposed to solve such problems. The integral term is introduced in the proportional integral state observer, not only can all states of the UAV be estimated by output feedback, but also the unknown input interference of the system can be simultaneously estimated. Moreover, the proportional integral observer observes the value of the disturbance and compensates the sliding mode controller, which solves the problem of chattering in the sliding mode, in which calculation is simple and thus is easy to implement in the engineering. Combined with the sliding mode attitude controller[12] the system can obtain satisfactory dynamic quality.

The rest of this paper is organized as follows: Section 2 establish the quadrotor UAV dynamics model with Euler angle as the attitude description. Section 3 presents the design of the proportional integral observer to estimate the system states and disturbance. Section 4 design the sliding mode attitude control with the system state estimation value. The controller suppresses the disturbance and realizes the high-precision control of the attitude of the quadrotor UAV, which makes the attitude error tend to be asymptotically stable. The simulation results are shown in Section 5. Finally, Section 6 includes the conclusion of this paper.

2. Establishment of Dynamic Model for Quadrotor UAV

Quadrotor UAV is typical under-drive model. This paper mainly studies the attitude control of the quadrotor UAV, and does not consider the attitude angle greater than 90°. Therefore, the Euler angle should be used to describe the space attitude of the quadrotor UAV and model it.

Suppose the Euler angles of x, y and z around the X, Y and Z axes are $\Theta = \begin{bmatrix}\phi & \theta & \psi\end{bmatrix}^T$, and the space rotation changes in the order of X, Y and Z axes respectively. Where ϕ is the roll angle, the angle around which the body rotates around its own X axis. θ is the pitch angle, the angle at which the body rotates around its own Y axis. ψ is the yaw angle, the angle at which the body rotates around its own Z axis.

The quadrotor UAV is a rigid body motion, and the center of gravity of the aircraft is at the origin of the body coordinate. The dynamic model of the quadrotor UAV is as follows:

$$\begin{cases} \ddot{\phi} = -\frac{lK_1}{I_x}\dot{\phi} + u_1 + d_1, \\ \ddot{\theta} = -\frac{lK_2}{I_y}\dot{\theta} + u_2 + d_2, \\ \ddot{\psi} = -\frac{lK_3}{I_z}\dot{\psi} + u_3 + d_3. \end{cases} \tag{1}$$

where l represents the distance from each rotor to the center of mass of the drone, $d_i(i = 1, 2, 3)$ is the unknown external disturbance, meeting the conditions simultaneously: $d_i \leq ||d||$; $K_i(i = 1, 2, 3)$ is the drag coefficient; I_x, I_y, I_z are the three-axis moment of inertia of the aircraft body respectively; u_1, u_2, u_3 are the control inputs of roll, pitch and yaw direction respectively.

Now, the roll angle is analyzed firstly, and the other two angles can be analyzed by a similar method. The system model is deformed as follows:

The roll angle ϕ can be organized into the following system model form:

$$\begin{cases} \dot{x} = Ax + Bu + Dd \\ y = Cx + Ed \end{cases} \tag{2}$$

where $x = \begin{bmatrix}\phi & \dot{\phi}\end{bmatrix}^T$, $A = \begin{bmatrix}0 & 1 \\ 0 & -\frac{lK_1}{I_x}\end{bmatrix}$, $B = \begin{bmatrix}0 & 1\end{bmatrix}^T$, $d = \begin{bmatrix}0 & d_1\end{bmatrix}^T$, $d_1 \leq ||d||$, $C = \begin{bmatrix}c & 0\end{bmatrix}$, $E = \begin{bmatrix}0 & 1\end{bmatrix}$. D is the unit matrix, $u = u_1$. Here, y is the output of the system which denotes the measurable variable, and d is the existing disturbance.

IOP Conf. Series: Journal of Physics: Conf. Series **1267** (2019) 012096 doi:10.1088/1742-6596/1267/1/012096

3. Design of Proportional Integral State Observer

For the above system formula (2), a proportional integral observer of the following form can be designed:

$$\begin{cases} \dot{\hat{x}} = A\hat{x} + Bu + K_P(y - \hat{y}) + Df \\ \dot{f} = k_I(y - \hat{y}) \end{cases} \tag{3}$$

where $\hat{y} = C\hat{x}$. K_P and k_I represent the proportional gain matrix and integral coefficient of the proportional integral observer respectively. f is introduced integral action.

Defining $e = \hat{x} - x$ as the state estimation error that can be converted into the following form:

$$\begin{bmatrix} \dot{e} \\ \dot{f} \end{bmatrix} = A_e \begin{bmatrix} e \\ f \end{bmatrix} + \begin{bmatrix} K_P E - D \\ k_I E \end{bmatrix} d \tag{4}$$

where $A_e = \begin{bmatrix} A - K_p C & D \\ -k_I C & 0 \end{bmatrix}$.

Convert formula (4) to:

$$\dot{e}_f = A_e e_f + B_e d \tag{5}$$

where $e_f = \begin{bmatrix} e \\ f \end{bmatrix}$, $B_e = \begin{bmatrix} K_P E - D \\ k_I E \end{bmatrix}$.

Theorem 1 For the system (5), if a Lyapunov function $V(e_f)$ can be found and $\dot{V}(e_f)$ is negative, then the system is asymptotically stable. That is:

$$\dot{V}(e_f) \leq 0 \tag{6}$$

Proof 1 Set positive quadratic function $V(e_f) = e_f^T P e_f$ as Lyapunov function, make:

$$A_e^T P + P A_e = -I \tag{7}$$

The result of first-order derivation of $V(e_f)$ is as follows:

$$\begin{aligned} \dot{V}(e_f) &= \dot{e}_f^T P e_f + e_f^T P \dot{e}_f \\ &= e_f^T (A_e^T P + P A_e) e_f + e_f^T P B_e d + d^T B_e^T P e_f \\ &= -e_f^T e_f + 2 e_f^T P B_e f \\ &\leq -e_f^T e_f + \tfrac{1}{4} e_f^T e_f + 4 \|PB_e\|_2^2 d^2 \\ &\leq -\frac{3}{4\lambda_{\max(P)}} V + 4 \|PB_e\|_2^2 \lambda_{\max(P)}^2 d^2 \end{aligned} \tag{8}$$

In formula (8), we defined:

$$a = -\frac{3}{4\lambda_{\max(P)}}, b = 4 \|PB_e\|_2^2 \lambda_{\max(P)}^2 d^2 \tag{9}$$

where d is bounded, a and b are constants, $\lambda_{\max(P)}$ is the maximum eigenvalue of P. Therefore, the formula (8) can be written as:

$$\begin{aligned} \dot{V}(e_f) &\leq -aV + b \\ &\leq \left[V(0) - \tfrac{b}{a}\right] e_{xp}^{-at} + \tfrac{b}{a} \end{aligned} \tag{10}$$

Note that $\lim_{t\to\infty}(V(0)-\frac{b}{a})e_{xp}^{-at}\to 0$, so formula (10) can be expressed as:

$$\dot{V}(e_f)\leq\frac{b}{a} \tag{11}$$

Considering the formula (9), that we can get that a is a normal number and b is a negative number. In the end we can get:

$$\dot{V}(e_f)\leq 0 \tag{12}$$

Therefore, the system is asymptotically stable.

It can be seen from the above that since P is a positive definite symmetric matrix, in formula (7), A_e satisfies the Huewitz matrix condition, so its eigenvalue is in the left-half plane. Thus the range of K_p , k_I can be derived from the result of the eigenvalue in the left negative half plane.

4. Design of Sliding Mode Attitude Controller

Define the state variable of the quadrotor UAV state space as:$X=[\phi\quad\dot{\phi}\quad\theta\quad\dot{\theta}\quad\psi\quad\dot{\psi}]$, where $x_1=\phi, x_2=\dot{\phi}, x_3=\theta, x_4=\dot{\theta}, x_5=\psi, x_6=\dot{\psi}$.

Define the control variable $U=[u_1\quad u_2\quad u_3]^T$, according to formula (1) and the definition of state variables available quadrotor UAV state space as follows:

$$\begin{cases}\dot{x}_1=\dot{\phi}=x_2\\ \dot{x}_2=\ddot{\phi}=a_1\dot{\phi}+b_1u_1+d_1=a_1x_2+b_1u_1+d_1\\ \dot{x}_3=\dot{\theta}=x_4\\ \dot{x}_4=\ddot{\theta}=a_2\dot{\theta}+b_2u_2+d_2=a_2x_4+b_2u_2+d_2\\ \dot{x}_5=\dot{\psi}=x_6\\ \dot{x}_6=\ddot{\psi}=a_3\dot{\psi}+bu_3+d_3=a_3x_6+b_4u_3+d_3\end{cases} \tag{13}$$

where $a_1=-\frac{lK_1}{I_x}, b_1=1, a_2=-\frac{lK_2}{I_y}, b_2=1, a_3=-\frac{lK_3}{I_z}, b_3=1$.

This chapter will use the rolling angle ϕ as an example to combine the above-mentioned proportional integral observer to design the sliding mode attitude controller, which the control rate can be derived. The other two angles can be analyzed by a similar method. Combined with formula (3) and formula (13), the state space of the roll angle is described as follows:

$$\begin{cases}\dot{\hat{x}}_1=\hat{x}_2\\ \dot{\hat{x}}_2=a_1\hat{x}_2+b_1u_1+k_{P\phi}(y_1-\hat{y}_1)+f_1\end{cases} \tag{14}$$

where $y_1=cx_1+d$, $\hat{y}_1=c\hat{x}_1$, $\dot{f}_1=k_{I\phi}(y_1-\hat{y}_1)$,$k_{P\phi}$ and $k_{I\varphi}$ represent the proportional and integral coefficients of the proportional integral observer, respectively.

The sliding mode function is designed for formula (14) as follows:

$$s_1=c_1x_{1e}+\dot{x}_{1e} \tag{15}$$

where $c_1>0$, x_{1e} is the tracking error of the roll angle and expressed as follows:

$$x_{1e}=\hat{x}_1-x_{1d} \tag{16}$$

where x_{1d} is roll angle expectation.

The sliding mode function has the following formula for the roll angle ϕ subsystem:

$$\begin{cases} \dot{s}_1 &= c_1\dot{x}_{1e} + \ddot{x}_{1e} = c_1\dot{x}_{1e} + \dot{\hat{x}}_2 - \ddot{x}_{1d} \\ &= c_1\dot{x}_{1e} + a_1\hat{x}_2 + b_1u_1 + k_{P\phi}(y_1 - \hat{y}_1) + f_1 - \ddot{x}_{1d} \end{cases} \tag{17}$$

Let b_1 be 1, then the sliding mode control rate is designed as follows:

$$u_1 = -c_1\dot{x}_{1e} - a_1\hat{x}_2 + \ddot{x}_{1d} - k_1s_1 - \eta_1\text{sgn}(s_1) \tag{18}$$

where $\eta_1 \geq k_{P\phi}(y_1 - \hat{y}_1) + f_1, k_1 > 0$.

So we can get the following formula (19):

$$\dot{s}_1 = -k_1s_1 - \eta_1\text{sgn}(s_1) + k_{P\phi}(y_1 - \hat{y}_1) + f_1 \tag{19}$$

Assume that the Lyapunov function is $V_1 = \frac{1}{2}s_1^2$, thus the following formula (20):

$$\begin{cases} \dot{V}_1 &= s_1\dot{s}_1 = -k_1{s_1}^2 - \eta_1|s_1| + \{k_{P\phi}(y_1 - \hat{y}_1) + f_1\}s_1 \\ &\leq -k_1{s_1}^2 \leq 0 \end{cases} \tag{20}$$

It can be seen from the above equation that the s_1 index converges, so that the roll angle ϕ index converges.

Similarly, the sliding mode control rates of pitch angle θ and yaw angle ψ are designed as follows:

$$\begin{cases} u_2 = -c_2\dot{x}_{2e} - a_2\hat{x}_4 + \ddot{x}_{2d} - k_2s_2 - \eta_2\text{sgn}(s_2) \\ u_3 = -c_3\dot{x}_{3e} - a_3\hat{x}_6 + \ddot{x}_{3d} - k_3s_3 - \eta_3\text{sgn}(s_3) \end{cases} \tag{21}$$

Among them:

$$\begin{cases} \eta_2 \geq k_{p\theta}(y_2 - \hat{y}_2) + f_2, k_2 > 0 \\ \eta_3 \geq k_{p\psi}(y_3 - \hat{y}_3) + f_3, k_3 > 0 \\ s_2 = c_2x_{2e} + \dot{x}_{2e}, x_{2e} = \hat{x}_3 - x_{2d}, c_2 > 0 \\ s_3 = c_3x_{3e} + \dot{x}_{3e}, x_{3e} = \hat{x}_5 - x_{3d}, c_3 > 0 \end{cases} \tag{22}$$

5. Simulation

The simulation module is built in MATLAB to carry out the whole system simulation experiment. The model related parameters used by the quadrotor UAV are valued as: $I_x = 0.02kg \cdot m^2, I_y = 0.02kg \cdot m^2, I_z = 0.04kg \cdot m^2$.

The initial state of the system state variable is: the initial Euler angle is (0.4,0.2,0)rad, the expectation Euler angular is (1.2,1,0.5)rad, and the initial Euler angular velocity is (0,0,0)rad/s.

The sliding mode controller parameters based on the proportional integral observer design have a direct impact on the performance of the flight control system. The larger the parameter value is, the stronger the control ability is, but at the same time it will cause a large control error. If the parameter is too small, then the system is prone to shock. Combined with the control ability and control performance requirements of the micro quadrotor UAV actuator, the parameters of the attitude controller are repeatedly adjusted through simulation experiments. The final selected attitude angle control parameters are as valued as: $c_1 = 6, c_2 = 8, c_3 = 6, K_{P\phi} = \begin{bmatrix} 0 \\ 1 \end{bmatrix}, K_{P\theta} = \begin{bmatrix} 0 \\ 4 \end{bmatrix}, K_{P\psi} = \begin{bmatrix} 0 \\ 1 \end{bmatrix}, k_{I\phi} = 1, k_{I\theta} = 5, k_{I\psi} = 9$.

We can get the output of each attitude angle of the system as shown in Figure 1. According to Figure 1., the attitude angle of the quadrotor UAV is stable near the desired attitude angle,and the error is very small, and it can be seen from the figure that the system response time has reached a steady state within 0.5 seconds to 1 seconds. Figure 2 and Figure 3, Figure 4 and Figure 5, Figure 6 and Figure 7 are the comparison of the true value and the observed value of each angular velocity, at the same time, the true value and observed value of the interference are also compared when $d = \begin{bmatrix} 0 & d_1 \end{bmatrix}$, $d = \begin{bmatrix} 0 & d_2 \end{bmatrix}$, $d = \begin{bmatrix} 0 & d_3 \end{bmatrix}$. They basically stabilized in 1 second, and the observation effect is very good and the experimental results are satisfactory.

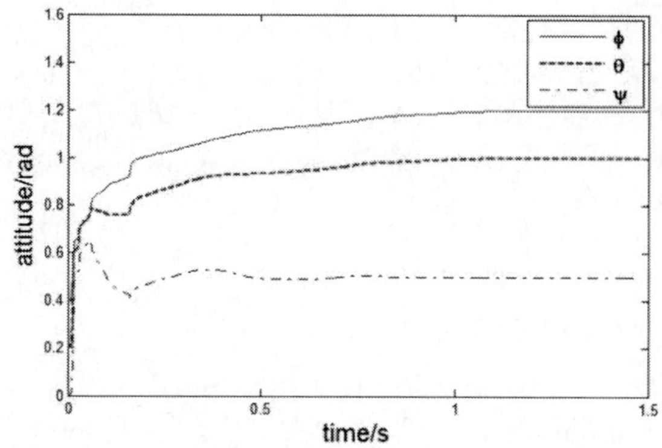

Figure 1. Rolling angle, pitch angle, yaw angle

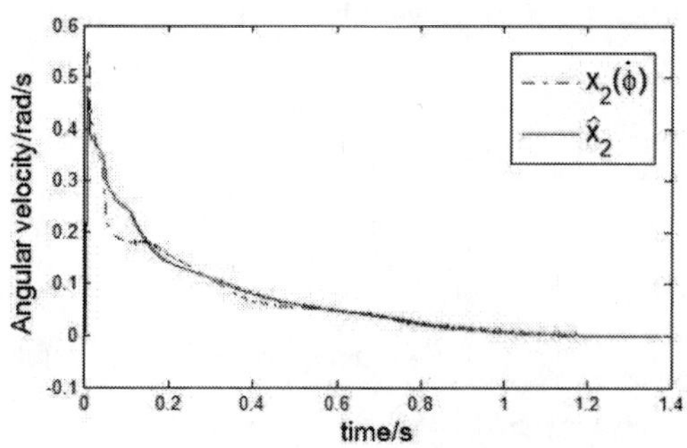

Figure 2. Simulation results of x_2, $\hat{x}_2$, when $d_1 = 1$

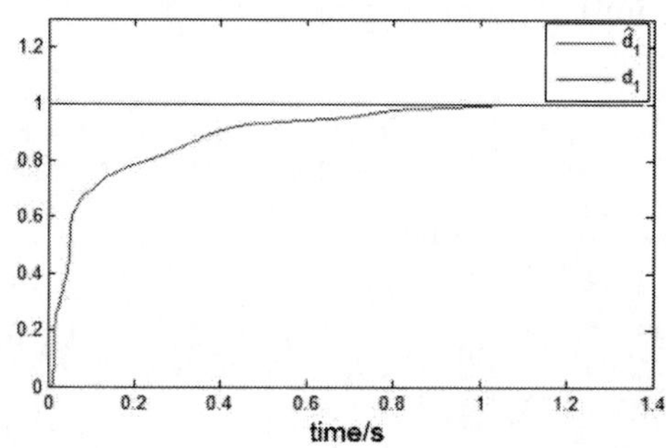

Figure 3. Simulation results of d_1, $\hat{d}_1$, when $d_1 = 1$

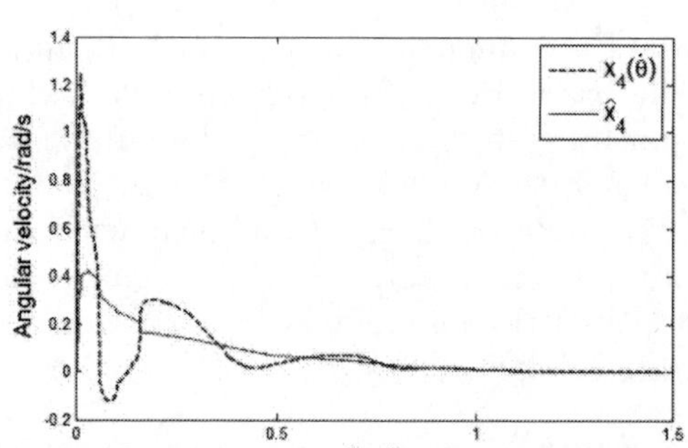

Figure 4. Simulation results of x_4, $\hat{x}_4$, when $d_2 = 1$

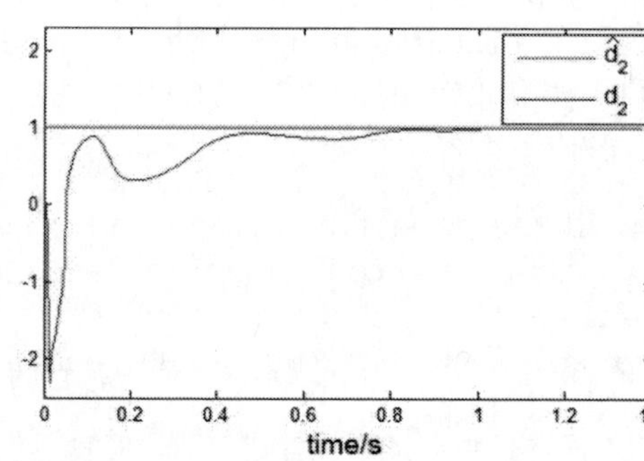

Figure 5. Simulation results of d_2, $\hat{d}_2$, when $d_2 = 1$

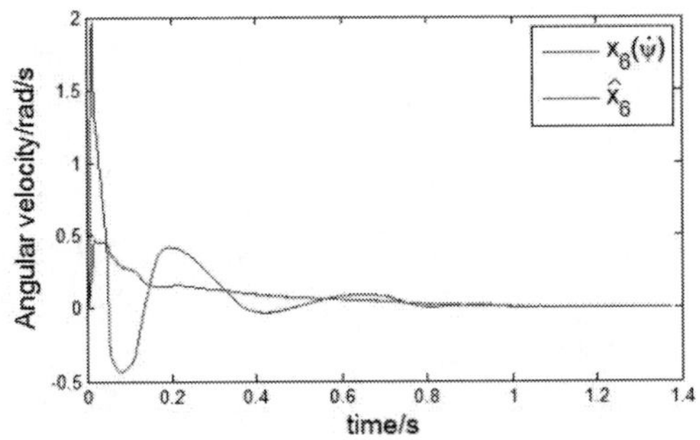

Figure 6. Simulation results of x_6, $\hat{x}_6$, when $d_3 = 1$

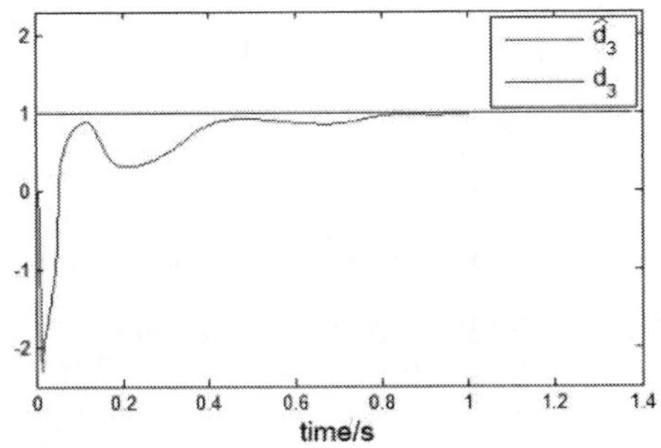

Figure 7. Simulation results of d_3, $\hat{d}_3$, when $d_3 = 1$

6. Conclusion

In this paper, an attitude control method for a UAV quadrotor with external disturbance has been proposed. The main contribution here is to propose a proportional integral observer combined with sliding mode control method to design the corresponding attitude controller based on the attitude angle model in the dynamic model of the quadrotor UAV. Through the proof in this paper, the proportional integral observer estimates the state of the system through partial feedback and estimates the unknown input interference of the system. In addition, the sliding mode attitude controller improves the system robustness, and the proportional integral observer can compensate for unknown interference and eliminate the chattering phenomenon existing in the traditional sliding mode. Finally, through the simulation experiment, the effectiveness of the method is verified by observing the simulation experiment graph.

Acknowledgements

This work was supported by the National Natural Science Foundation of China [grant number 61374180, 61603196]; the Natural Science Foundation of Jiangsu Province of China [grant number BK20150851]; the Project funded by China Postdoctoral Science Foundation [grant number 2015M581842]; the Project funded by Jiangsu Province Postdoctoral Science Foundation [grant number 1601259C]; the Sponsored by NUPTSF [grant number NY215011],the Postgraduate Research & Practice Innovation Program of Jiangsu Province [grant number 2015M581842].

References

[1] Ivn Gonzlez, Salazar S, Torres J, et al. 2013 *Journal of Intelligent & Robotic Systems* **70** 93
[2] Zou A M, Kumar K D, Hou Z G, et al. 2011 *IEEE Transactions on Systems, Man and Cybernetics* **41** 0
[3] Tayebi 2008 *IEEE Transactions on Automatic Control* **53** 1516
[4] Nicol C, Macnab C J B, Ramirez-Serrano A. 2011 *Mechatronics* **21** 927
[5] Ha C S, Zuo Z, Choi F B, et al. 2014 *Robotics and Autonomous Systems* **62** 1305
[6] Haoqin S, Zhan H, Xiaoxiang B, et al. 2015 *Procedia Engineering* **99** 873
[7] Xing-Hua L U, Shi-Wei Z, Wen-Quan Y U, et al. 2017 *Transducer & Microsystem Technologies* **36** 116
[8] Ducard G, Geering H P. 2008 *IEEE/RSJ International Conference on Intelligent Robots & Systems.* **64** 360
[9] Lotufo M A, Colangelo L, Perez-Montenegro C, et al. 2016 *IFAC-PapersOnLine* **49** 266
[10] Boudjedir H, Bouhali O, Rizoug N. 2014 *PAdvanced Robotics* **28** 14
[11] Xiangsong K, Xurui C, Jiansheng G. 2016 *Cybernetics and Information Technologies* **16** 15
[12] Mofid O, Mobayen S. 2018 *Isa Trans* **72** 1

IOP Conf. Series: Journal of Physics: Conf. Series **1267** (2019) 012097 doi:10.1088/1742-6596/1267/1/012097

Lunar Rover Path Planning Based on Comprehensive Genetic Algorithm Based on Slip Prediction

Zhou Lanfeng[*,a], Yang Lina and Fang Hua[*,b]
School of Computer Science and Information Engineering, Shanghai Institute of Technology, Shanghai, China

[*a]lfzhou@sit.edu.cn; [b]fanghuame@163.com

Abstract. Aiming at the problem that the lunar rover path planning algorithm generally has slow convergence rate, falls into local optimal solution, neglects the mutual applicability of environment modeling technology and path planning algorithm, a comprehensive genetic algorithm based on virtual three-dimensional model is proposed. By setting the genetic factor unchanged, the terrain comprehensive cost function is added to set the fitness function. Using genetic algorithm and improved comprehensive genetic algorithm, after 100 simulation experiments, the improved comprehensive genetic algorithm has good search performance, fast convergence and high stability. The key words of this paper are as follows: path planning, terrain comprehensive cost function, genetic algorithm.

1. Introduction

The lunar rover is an effective tool for moonwatch detection. During its inspection, an important task is to carry out path planning. Path planning is divided into map-based global path planning and sensor-based partial path planning. The so-called global path planning is to find the global optimal path according to one or several optimization criteria (such as the shortest path, the least time, etc.) when the environment is completely known. Path planning research began in the 1970s[1], and from that time back to the present, path planning algorithms of various forms have been proposed and are recognized. Contains both traditional A* algorithms[2], Genetic algorithm[3], Also comes with some new intelligent algorithms, such as Ant Colony Algorithm[4], Neural network algorithm[5], Particle swarm optimization[6], As far as the genetic algorithm is concerned, the search efficiency is low, the obstacle avoidance effect is not good, and the algorithm complexity is too large, resulting in an inability to obtain an excellent solution.

In this paper, the existing genetic algorithm is easy to fall into the local optimal solution, comprehensively consider the lunar terrain environment information, and integrate the pass-through cost function by adding the slip factor prediction of the terrain factor[7]. Using MATLAB simulation experiment, keeping the terrain parameters unchanged, adjusting the slip passability function parameters, comparing 100 experimental simulation results, the integrated genetic algorithm evolutionary algebra is less than the traditional genetic algorithm. The improved genetic algorithm improves the smoothness of the lunar rover path planning, shortens the length of the path, and saves the time and evolution of the path planning.

2. Integrated genetic algorithm

2.1. Establishment of genetic algorithm path planning model

In this paper, the genetic coding uses real number coding. For each chromosome in the genetic algorithm, there is one solution. In the path planning process, the gene consists of several intermediate points through the starting point, the end point and the path planning[8]. The start point(X_1,Y_1,Z_1), the end point(X_n,Y_n,Z_n), and the $n-2$ intermediate point shown in figure 1 constitute one chromosome carrying path planning information from the start point to the target point.

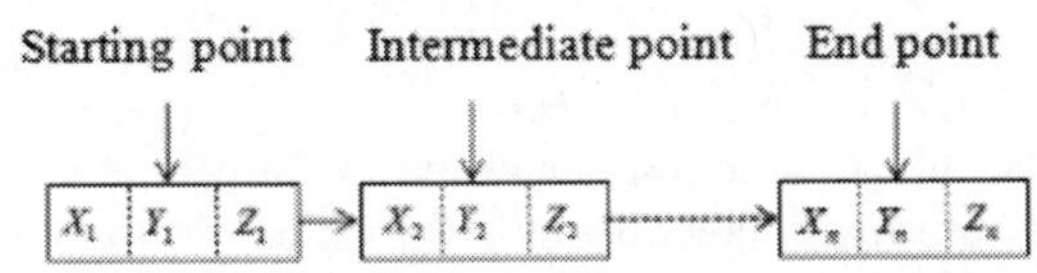

Figure 1. One chromosome with n genes in path planning.

Therefore, a complete robot path is an ordered combination of several line segments, using the (P)express, as in formula (1):

$$P=\sum P_i \tag{1}$$

P_i Representing the i vector representation of a segment of a straight line.

2.2. Initial population generation

In the process of generating chromosomes, starting from the starting point, randomly select the transitive nodes of the neighbor nodes of the current node, add their coordinates to the chromosome, and so on, until a target point is found, and one chromosome is generated[9]. Repeat the above operation to reach the initial target population and terminate the loop.

2.3. Genetic manipulation (selection, crossover, mutation operator design)

The genetic algorithm was proposed by Professor J. Holland of the University of Michigan in the United States. It is an adaptive algorithm for bionic global optimization probability search[10]. The preset parameters required by the algorithm include initial population number, maximum evolution algebra, crossover probability and mutation probability.generally include three processes of selection, intersection and mutation[11]. In the process of algorithm evolution, the crossover and mutation probability affect the local part of the algorithm. Global search capabilities, fixed crossover probabilities, and mutation probabilities do not guarantee that the algorithm converges faster and searches for global optimal solutions[12]. Through continuous iteration, the optimal solution of the problem is finally obtained.

- This algorithm adopts elitist selection strategy, that is, replacing some poor individuals with some excellent individuals[13].
- Cross-operation is a process simulation of genetic recombination in nature, and is the main method for genetic algorithms to generate new individuals[14].In this paper, some of the paper uses partial mapping to perform chromosome crossover operations to generate two random numbers m, n, the exchange of gene fragments between m and n on the X, Y chromosome.Cross probability formulas such as (2):

$$P_c = P_{cmax} - \frac{(P_{cmax} - P_{cmin})(F_c - F_{min})}{F_{max} - F_{min}} \tag{2}$$

In the form: P_c Indicates the crossover probability in dynamic path planning; P_{cmax} Indicates the maximum value of the crossover probability; P_{cmin} Indicates the minimum value of the crossover probability; F_{max} Indicates the maximum fitness value in the population; F_{min} Indicates the minimum fitness value; F_c Indicates the larger fitness value of the two individuals to cross.

- The mutation operation simulates the process of mutation in the natural world[15]. During the iterative process, the genes on the chromosome are randomly mutated to produce new individuals. Variation probability formula such as (3):

IOP Conf. Series: Journal of Physics: Conf. Series **1267** (2019) 012097 doi:10.1088/1742-6596/1267/1/012097

$$P_m = P_{m\max} - \frac{(P_{m\max} - P_{m\min})(F_m - F_{\min})}{F_{\max} - F_{\min}} \tag{3}$$

In the form: P_m Indicates the probability of mutation in dynamic path planning; $P_{m\max}$ Indicates the maximum value of the mutation rate; $P_{m\min}$ Minimum value indicating the probability of mutation; $F_{\max}$ Indicates the maximum fitness value in the population; $F_{\min}$ Minimum fitness value; F_m Indicates the fitness value of the individual to be mutated.

3. Design of fitness function for slip prediction

In the genetic algorithm, the fitness function is a criterion for judging the individual's ability to survive in the population, which is determined according to the objective function. In solving the optimal path, the larger the value of the fitness function, the better. In the path planning of this paper, the shortest path and the smallest evolutionary algebra are the primary fitness functions. In addition, this article adds a terrain passivity cost function. The topographical passivity function is a function to evaluate the ability of topographical passivity. This function takes into account factors such as terrain slope, terrain roughness, and obstacles. In this paper, the slip factor is added to the consideration factor to form a terrain passivity cost function based on slip prediction. Using this function, you can evaluate the passability of the path between two points. Here, the above terrain comprehensive cost function $f(p,n)$ is incorporated into the fitness function of the genetic algorithm, and the integrated genetic algorithm is used for path planning.

The terrain comprehensive cost function for setting the node p -to-node n gene fragment is as follows:

$$f(p,n) = f_1 \times f_{trav}(p,n) + f_2 \times f_{risky}(p,n) + f_3 \times f_{guide}(p,n) + f_4 \times f_{smooth}(p,n) \tag{4}$$

In the form: $f_{trav}(p,n)$ is the passability cost function from node p to node n (generally 0 or 1); $f_{risky}(p,n)$ is the potential risk cost function; $f_{guide}(p,n)$ is the cost function of path directivity, which is related to the slope angle of the terrain; $f_{smooth}(p,n)$ is the path smoothness cost function takes a fixed constant value; f_1, f_2, f_3, f_4 is the weight of $f_{trav}(p,n)$, $f_{risky}(p,n)$, $f_{guide}(p,n)$, and $f_{smooth}(p,n)$ respectively,they are generally treated as the same fixed constant value. Then, the extendable slip-through passability cost function $f(p,n)$ is added to the fitness function of the genetic algorithm to form a new fitness function $f'(p,n)$. The formula called the comprehensive fitness function $f'(p,n)$.The formula is as follows:

$$f'(p,n) = f(p,n) + f_0(p,n) \tag{5}$$

$f'(p,n)$ is the fitness function of the function of the integrated genetic algorithm; $f(p,n)$ is the terrain comprehensive cost function; $f_0(p,n)$ is the fitness function based on the genetic algorithm.

4. Simulation experiment analysis

This paper uses a PC, based on the Matlab 2014 environment to create a virtual three-dimensional terrain environment $(21 \times 21 \times 21)$, virtual three-dimensional space abstraction into the grid, set the starting point and end point and height value of the search path. Set the initial population to 30, and the maximum evolutionary algebra to 200 generations. As can be seen from Figure 2, the three-dimensional genetic algorithm path planning is more intuitive and effective than the two-dimensional grid model.

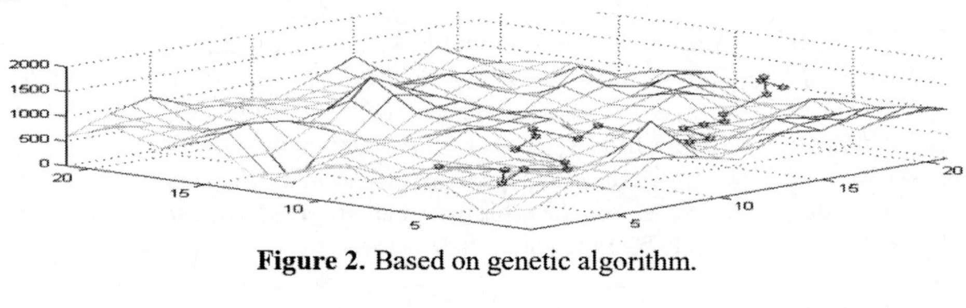

Figure 2. Based on genetic algorithm.

In this paper, genetic algorithm and integrated genetic algorithm (IGA), one evolutionary algebra and comprehensive fitness change diagram, as shown in Figure 3 and Figure 4. It can be seen from Figure 3 that the basic genetic algorithm has evolved into the local optimal solution after about 10 generations, the optimal solution is updated after the 50th generation, and the optimal solution is obtained once in the 140th generation. The comprehensive fitness value is 130. It can be seen from the path plan of the integrated genetic algorithm in Figure 4 that the optimal solution is obtained once in 20 evolutionary algebras, and the comprehensive fitness value is 125. The path planning time is shortened to some extent, and the work efficiency is improved.

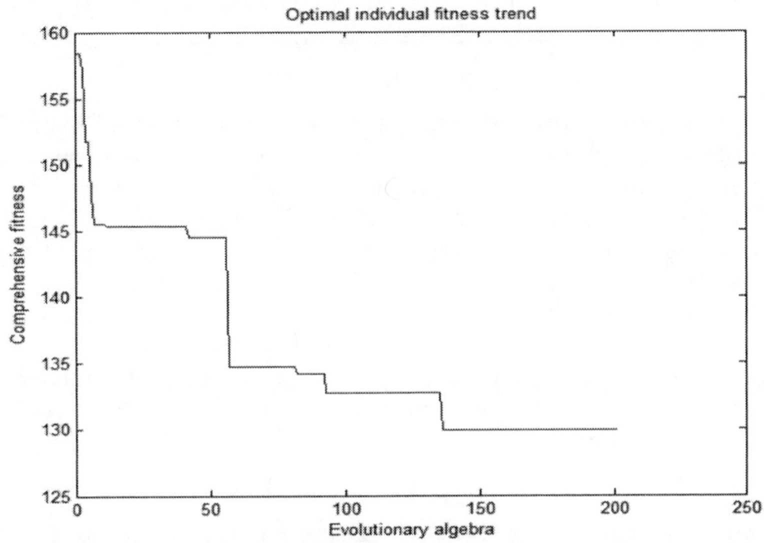

Figure 3. Genetic algorithm.

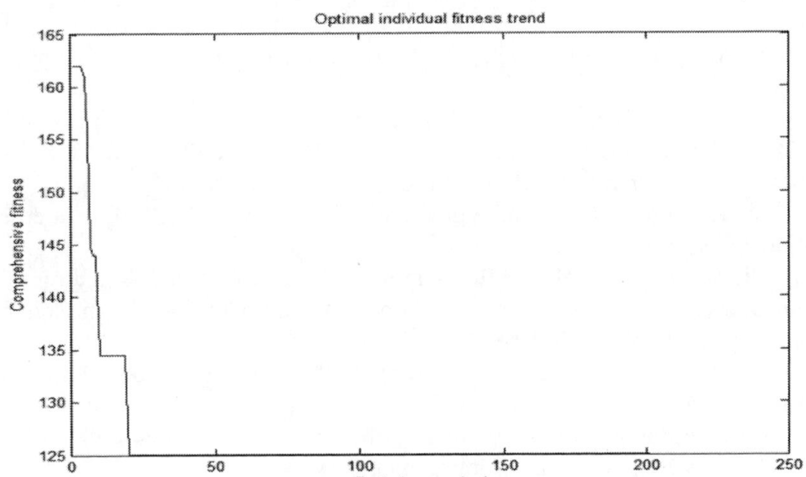

Figure 4. Integrated genetic algorithm.

In order to further prove the rationality of the algorithm and avoid the influence of accidental factors on the algorithm, the two path planning algorithms are tested 100 times respectively. The statistical experimental results are shown in Table 1. It can be seen from Table 1 that the average maximum path of the integrated genetic algorithm is 34.7% shorter than the maximum path of the genetic algorithm, the average fitness value is increased by 12.4%, and the average evolutionary algebra is reduced by 49.2%.

Table 1.performance comparison between genetic algorithm and improved genetic algorithm.

Algorithm name	Average maximum path(/km)	Average comprehensive fitness	Average evolutionary algebra(/times)
Genetic algorithm	49.6	120.40	128
Integrated genetic algorithm	32.4	135.38	65

5. Conclusion

In order to solve the problem of slow convergence and easy to fall into the local optimal solution in the path planning problem of genetic algorithm, the terrain synthesis cost function is introduced to design the fitness function. The crossover operator and mutation operator are obtained dynamically according to the comprehensive fitness, and a path planning algorithm based on comprehensive genetic algorithm is proposed. Through the experimental simulation in the three-dimensional simulation environment, the analysis of the results, the improved synthetic genetic algorithm in search of the path, the time efficiency compared with the basic genetic algorithm has been improved. The results show that the improved algorithm can achieve a good path planning effect under certain conditions. In reality, the terrain is very complex, Path planning based on one algorithm alone can not achieve ideal results. 3D environment reconstruction is the focus of path planning, so constructing complex dynamic path planning is the next research focus.

Acknowledgments
This work was supported by the National Natural Science Foundation of China(41671402).

References

[1] Anish Pandey, Dayal R. Parhi. Optimum path planning of mobile robot in unknown static and dynamic environments using Fuzzy-Wind Driven Optimization algorithm[J]. Defence Technology,2017.

[2] Ren Yujie, Fu Lixia, Zhang Yong, Mao Jianlin. Dynamic Path Planning of Robot Based on Smooth A* Artificial Potential Field Method[J].Software Guide, 2018,17(01):8-10.

[3] N. Weerayuth , N. Chaiyaratana.Closed-loop time-optimal path planning using a multi-objective diversity control oriented genetic algorithm[P]. Systems, Man and Cybernetics, 2002 IEEE International Conference on,2002.

[4] Cai Wen,Yansong Deng. Global path planning of Multi-robot Fish based on adaptive ant colony algorithm in dynamic environment[P]. 4th International Conference on Renewable Energy and Environmental Technology (ICREET 2016),2017.

[5] Gong Mengmeng. Research on path planning method of mobile robot based on neural network [D]. Harbin Institute of Technology, 2017.

[6] Lu Dan. Application of particle swarm optimization in path planning of mobile robots [D]. Wuhan University of Science and Technology, 2009.

[7] Ha, C., Yoon, J., Kim, C., Lee, Y., Kwon, S., & Lee, D. (2018). Teleoperation of a platoon of distributed wheeled mobile robots with predictive display. Autonomous Robots, 42(8), 1819-1836.

[8] Ha C, Yoon J, Kim C, et al. Teleoperation of a platoon of distributed wheeled mobile robots with predictive display[J]. Autonomous Robots, 2018, 42(8):1819-1836.

[9] Ha, Changsu, et al. "Teleoperation of a platoon of distributed wheeled mobile robots with predictive display." Autonomous Robots42.8(2018):1819-1836.

[10] WANG X P, CAO L M.Genetic Algorithm Theory, Application and Software Implementation[M].Xi'an: Xi'an Jiaotong University Press, 2002:2-5.

[11] Wu, Meihua, et al. "Path planning of mobile robot based on improved genetic algorithm." Chinese Automation Congress 2018.

[12] Huang Chao. Research on path planning of AGV system based on wireless network [D]. Shanghai: East China University of Science and Technology, 2015.

[13] Hong Qu,Ke Xing,Takacs Alexander.An improved genetic algorithm with co-evolutionary strategy for global path planning of multiple mobile robots[J].Neurocomputing,2013(120):509-517.

[14] Yuan Guangming, Yan Yunfei, Ding Chengjun, Zhang Peng. AGV Path Planning Based on Improved Genetic Algorithm[J].Journal of Beijing Union University,2018,32(01):65-69.

[15] Wu M, Chen E, Shi Q, et al. Path planning of mobile robot based on improved genetic algorithm[C]// Chinese Automation Congress. 2018.

A SLAM Algorithm Based on Multi-Constraint Learning Model

Haibin Shi*, Menghao Guo, Yuanbin Zou and Zhi Xu

College of Information Science and Engineering, Northeastern University

*E-mail: shihaibin@ise.neu.edu.cn

Abstract. Unsupervised Learning based SLAM algorithm has lately drawn significant attention for its potential in label-free leaning ability and robustness to camera parameters and environmental variations. In order to achieve better robustness and accuracy, a multi-constraint learning model is proposed. In contrast to traditional geometry-based methods, multi-constraint unsupervised learning models optimize the photometric consistency over image sequences by warping one view into another; make the Network learning more geometrically information. A lot of experiments on the KITTI data set show that our model is superior to previous unsupervised methods and has comparable results with the supervised method.

1. Introduction

Simultaneous Localization and Mapping (SLAM) [1] is a robot equipped with vision, laser, visual odometry and other sensors, to achieve self-localization process of constructing a map of an unknown environment at the same time, play a key role in autonomous robot navigation tasks. Although the sensors used by SLAM have many types such as LIDAR and vision, the processing generally consists of two parts: front-end inter-frame estimation and back-end optimization. The front-end inter-frame estimation solves the motion estimation of the robot in the time interval of acquiring the two frames of sensor information before and after [2, 3], while the back-end optimization solves the optimization problem of the historical trajectory after the robot detects the path closed loop [4].

Due to the different sensors, the SLAM algorithm is divided into LIDAR SLAM and visual SLAM. Among them, with the rapid development of visual algorithms and hardware, visual SLAM has received wide attention because of the large amount of information received by visual sensors and the wide range of applications.

The visual SLAM algorithm can be reclassified into two groups based on the technology and framework [5]: geometry-based and learning-based methods. Geometry-based methods are usually solved by shallow features or luminosity re-projection and online error minimization. Learning-based method has lately drawn significant attention for its potential in label-free leaning ability and robustness to camera parameters and environmental variations.

The traditional visual SLAM algorithm, for example, ORB-SLAM [6] uses ORB features, and feature matching and relocation based on ORB descriptions have better viewing angle invariance. In addition, the feature matching of the new 3D points is more efficient, so the scene can be expanded more timely. But the constructed map is a sparse point cloud map. Only a part of the feature points in the image is retained as a key point, and it is fixed in the space for positioning, and it is difficult to depict the existence of obstacles in the map.

Kendall et al. [7] first use the deep learning method on the SLAM algorithm directly regress the camera's world pose from RGB images with the convolutional neural network. However, the cost of collecting ground truth poses limits the application of such methods.

The first architecture that achieved unsupervised learning of ego-motion from the video is SfMLeaner proposed by Zhou et al [8]. SfMLeaner takes consecutive temporal images to predict both depth and ego-motion with view synthesis as supervision. However, similar to geometric approaches, SfMLeaner can only observe ego-motion in a relative scale from monocular image.

The main idea behind them is to optimize the photometric consistency over image sequences by warping one view into another. It is a well-known fact that simply minimizing such an error might not effectively capture the relation between pixels. Further, minimizing the photometric loss does not ensure proper pixel correspondences, which is a key factor for accurate depth and pose estimations. In contrast, we present a novel multi-constraint learning SLAM algorithm with depth warp constraints loss based on unsupervised deep convolutional neural networks. We trained two convolutional network end-to-ends to calculate depth and ego-motion from a continuous, unlabeled pair of images. The ego-motion is estimated using image projection constraints and depth warp constraints as supervisory information.

2. Multi-constraint learning SLAM algorithm

The basis of our approach is based on the nature of the 3D scene geometry. The intuitive explanation is that most natural scenes consist of rigid static surfaces, i.e. roads, houses, trees, etc. The 2D image motion project between video frames can be determined entirely by depth structure and camera motion. Traditional methods have been able to calculate depth from binocular images and then use 2D image projection to construct VO. Therefore, we use a deep convolution network to grasp the above intuition, build our deep learning network. Our deep learning network consists of three parts:

1) By constructing a convolutional neural network called Depth net, left and right photometric consistency is used to determine the left and right photometric twist error loss function, called Spatial Photometric Loss. At the same time, a corresponding Depth Map is obtained from the left and right image.

2) By constructing a convolutional neural network called Pose net, 2D image projection constraints of two consecutive frames is used to determine Temporal Photometric Loss, and at the same time, ego-motion is learned from the Pose net.

3) Based on the convolutional neural network of Depth net and Pose net, the depth images of two consecutive frames predicted by part 1 is used to determine a Depth Warp Loss by depth warp Constraints to assist the constraints Spatial Photometric Loss and Temporal Photometric Loss.

In summary, we make the following contributions:

(1) An unsupervised two-part cascade network structure to learn the depth map and ego-motion separately without being affected by scale ambiguities.

(2) Multi-constraint learning SLAM algorithm using all the constraints available in spatial and temporal image pairs to improve the accuracy of predicts ego-motion.

(3) A novel depth warp constraints is proposed, to adaptively solve the rigid flow and object motion of the scene, which significantly improves depth and visual odometry accuracy.

3. Three partial constraints

This section introduce the Network Architecture of SLAM algorithm based on multi-constraint (shown in Figure. 1), which learning $T_{21} \in SE3$ from two consecutive frames and learning the depth image D_1 and D_2 from the left and right images of the binocular camera at the same time. Our example uses a divide-and-conquer strategy. A new cascade structure consisting of three partial constraints is designed to adaptively solve the rigid flow and object motion of the scene. Therefore, the global constraints can be gradually improved, making our complete learning process a decomposed and easy to learn way.

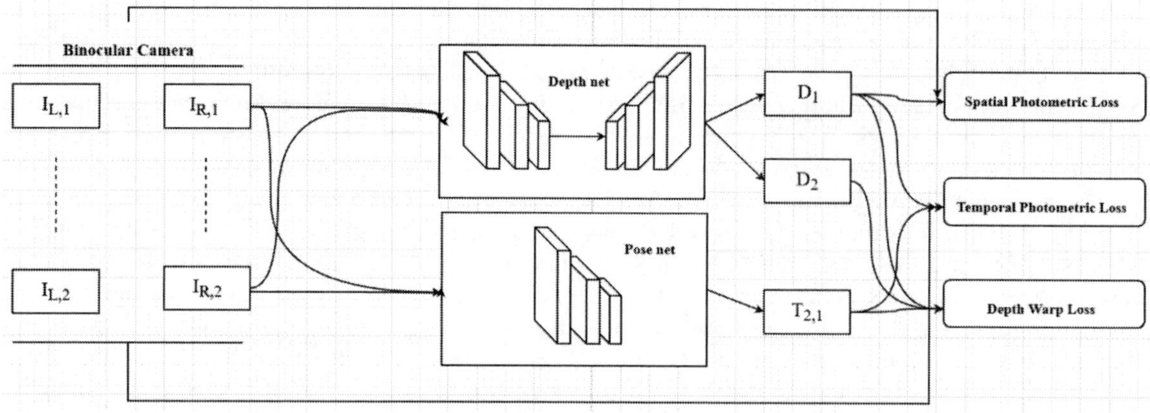

Figure 1. Algorithm Framework.

3.1. Spatial Photometric Loss

The image reconstruction loss between the warp view $I'_{R,1}$ and the real view $I_{R,1}$ is calculated as a supervised signal for training Depth net. The image construction loss called Spatial Photometric Loss is represented by the following formula:

$$L_{SP}=\sum(|I_{R,1}-I'_{R,1}|+|I_{R,2}-I'_{R,2}|) \quad (1)$$

We can use polar line geometry to obtain the projected coordinates from the reference view $I_{L,1}$ warp to the view $I'_{R,1}$. $p_{L,1}$ is the homogeneous coordinate of the pixel in the reference view $I_{L,1}$. Warp view $I'_{R,1}$ can be synthesized from the reference view $I_{L,1}$ using the differentiable bilinear interpolation mechanism (warping) proposed in [9].

$$p'_{R,1}=(K*(T_{LR}*(D_1*K^{-1})))*p_{L,1} \quad (2)$$

$$p'_{R,2}=(K*(T_{LR}*(D_2*K^{-1})))*p_{L,2} \quad (3)$$

3.2. Temporal Photometric Loss

The image reconstruction loss between the warp view $I'_{R,2}$ and the real view $I_{R,2}$ is calculated as a supervised signal for training Depth net and Pose net. The image construction loss called Temporal Photometric Loss is represented by:

$$L_{TP}=\sum(|I_{R,2}-I'_{R,2}|) \quad (4)$$

We can use the direct method to obtain the projected coordinates $p'_{R,2}$ from the reference view $I_{R,1}$ to the warp view $I'_{R,2}$. $p'_{R,2}$ is the homogeneous coordinate of the pixel in the reference view $I'_{R,2}$. Warp view $I'_{R,2}$ can be synthesized from the reference view $I_{R,1}$ using the differentiable bilinear interpolation mechanism (warping) proposed in [9].

$$p'_{R,2}=(K*(T_{21}*(D_2*K^{-1})))*p_{R,1} \quad (5)$$

3.3. Depth Warp Loss

The image reconstruction loss between the warp view $D_2^{'}$ and the real view D_2 is calculated as a supervised signal for training Depth net and Pose net. The image construction loss called Depth Warp Loss is represented by:

$$L_{DW}=\sum(|D_2 - D_2^{'}|) \tag{6}$$

We can use the constraint formula [10] between the depth maps corresponding with two consecutive frames to obtain the projected coordinates from the reference view D_1 warp to the view $D_2^{'}$.

3.4. Finally Loss

In summary, finally loss function contains two distinguishable operations that allow for gradient propagation for Depth net and Pose net training. These two operations are polar line geometry transformations and warp. The former defines the correspondence between pixels in two views, while the latter synthesizes the image by distorting the real-time view. The final loss function becomes:

$$L=\lambda_{SP}L_{SP} + \lambda_{TP}L_{TP} + \lambda_{DW}L_{DW} + \lambda_{ds}L_{ds} \tag{7}$$

Where, λ is the loss weight of each loss item and is obtained through training and fine tuning. In order to obtain smooth depth prediction, following the method adopted by Zhan et al. [11], we encourage depth to be localized by introducing edge-aware smoothness. Edge-aware smoothing loss is the formula:

$$L_{ds}=\sum_{m,n}^{W,H}\left\| \partial_x D_{m,n} \right| e^{-|\partial_x I_{m,n}|} + \left| \partial_y D_{m,n} \right| e^{-|\partial_y I_{m,n}|}\right| \tag{8}$$

4. Experiments and Results

In this section, we present a wide range of experiments for evaluating the performance of our proposed framework. We compare our approach in the KITTI dataset [12, 13] with the state of the art SLAM algorithm.

4.1. Experiments

The network models were implemented with the Caffe [14] framework and trained with NVIDIA TITAN Xp GPUs and Intel Core i7 2.7GHz CPU. Adam optimizer [15] was employed to train the networks for up to 20000 epochs with parameter $\beta_1 = 0.9$, $\beta_2 = 0.999$ and $\varepsilon = 10^{-8}$. The initial learning rate for all trained networks is 0.001, and we will manually reduce when training losses converge. Our work does not involve any data expansion. In the loss weights in our final loss function, we have empirically found that the main parameters in stability training are:

$$\begin{bmatrix} \lambda_{SP} \\ \lambda_{TP} \\ \lambda_{DW} \\ \lambda_{ds} \end{bmatrix} = \begin{bmatrix} 1 \\ 1 \\ 0.01 \\ 10 \end{bmatrix} \tag{9}$$

Since our network only needs paired binocular camera data for training, and the test uses the image of the monocular camera for testing, we use the raw data part of the KITTI dataset, the original image size is 1242x375 pixels. However, because the picture quality is too high, increasing the training cost, we adjust the Caffe input data parameter during training. The input image size is 608x160 pixels.

As mentioned in the previous paragraph, our framework uses monocular data for testing. To evaluate our SLAM algorithm performance and compare it with previous method [8, 11, 16], we follow [8] by test the Pose net on the official KITTI Odometry training set. The details of the test set are: The KITTI Odometry Split [12, 13] contains 11 driving sequences with publicly available ground truth camera poses. We trained our system on the Odometry Split in accordance with [8] (no fine-tuning of Eigen Split). The split of sequence 00-08 (sequence 03 is not available in KITTI raw data) is used for training, while 09-10 is used for evaluation. The training set consists of 8 sequences with 19,600 time stereo pairs.

4.2. Result and Analysis

Our unsupervised method is superior to previous unsupervised methods and has comparable results with the supervised method, which reflects the effectiveness and advantages of our approach. The large error in our frame-to-frame rotation estimation results in a larger gradual drift, which should be fixed by the beam. A visual comparison of the estimated trajectories of our method can be seen in Figure 2, Figure 3.

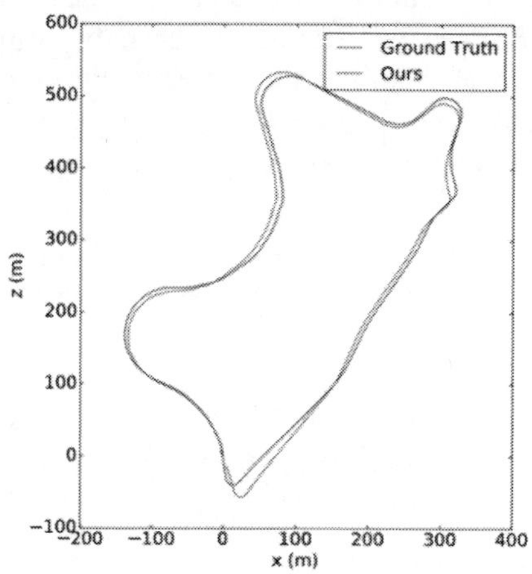

Figure 2. Figure with Seq. 09

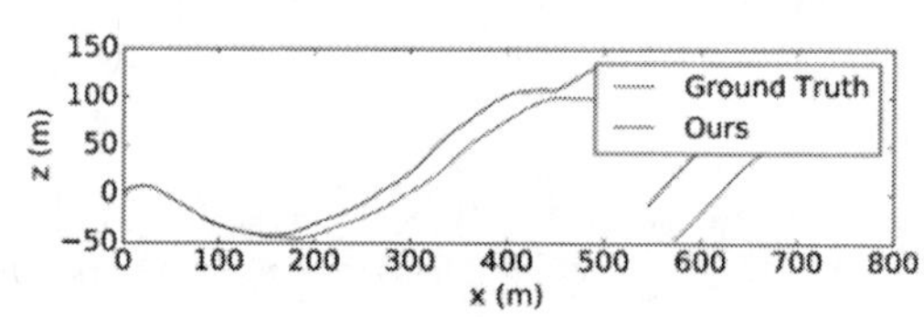

Figure 3. Figure with Seq. 10

5. Conclusion

We propose an unsupervised learning SLAM algorithm based on multi-constraint learning model, the depth map is predicted by the Depth net based on Spatial Photometric constraints in the first part, and the ego-motion is predicted by the Pose net based on Temporal Photometric constraints in the second part. We innovatively add depth warp constraints on the two-part cascade network, which further enables the cascade network to capture the relation between pixels, thereby improving the accuracy of predicting ego-motion. Our unsupervised method is superior to previous unsupervised methods and has comparable results with the supervised method, which reflects the effectiveness and advantages of our approach.

But there are still many challenges for our framework that we need to solve. The experimental results also illustrate the dependence of the proposed algorithm on the training data. Especially when the inter-frame speed of the image sequence is too fast, the algorithm error is large. The reason is that the lack of high-speed training samples in the training set causes the estimated rotation error to be large.

Acknowledgments
This work is supported by National Natural Science Foundation (NNSF) of China under Grant 61873306.

References

[1] J. J. Leonard and H. F. Durrant-Whyte, "Mobile robot localization by tracking geometric beacons," IEEE Transactions on robotics and Automation, vol. 7, no. 3, pp. 376–382, 1991.
[2] Scaramuzza D, Faundorfer F. Visual odometry: Part I: the first 30 years and fundamentals. IEEE Robotics and Automation Magazine, 2011, 18(4): 80−92
[3] Fraundorfer F, Scaramuzza D. Visual odometry: Part II: matching, robustness, optimization, and applications. IEEE Robotics and Automation Magazine, 2012, 19(2):78−90
[4] Kummerle R, Grisetti G, Strasdat H, et al. g2o: A general frame work for graph optimization [C]//IEEE International Conference on Robotics and Automation. Piscataway, USA: IEEE, 2011: 3607-3613.
[5] Tomasz M. (2016, January 13). The Future of Real-Time SLAM and Deep Learning vs SLAM. Retrieved September 4, 2018, from http://www.computervisionblog.com/2016/01/why-slam-mattersfuture-of-real-time.html
[6] Mur-Artal, R.,Montiel,J.M.M.,&Tardos,J.D.(2015).ORB-SLAM: a versatile and accurate monocular SLAM system. IEEE Transactions on Robotics, 31(5), 1147-1163.
[7] Kendall, A., Grimes, M., & Cipolla, R. (2015). Posenet: A convolutional network for real-time 6-dof camera relocalization. In Proceedings of the IEEE international conference on computer vision (pp. 2938-2946).
[8] Zhou, T., Brown, M., Snavely, N., & Lowe, D. G. (2017, July). Unsupervised learning of depth and ego-motion from video. In CVPR (Vol. 2, No. 6, p. 7).
[9] M. Jaderberg, K. Simonyan, A. Zisserman, et al. Spatial transformer networks. In Advances in Neural Information Processing Systems, pages 2017–2025, 2015.
[10] Fang Z, Scherer S. Real-time onboard 6DoF localization of an indoor MAV in degraded visual environments using a RGB-D camera[C]// 2015 IEEE International Conference on Robotics and Automation (ICRA). IEEE, 2015.
[11] Zhan H, Garg R, Weerasekera C S, et al. Unsupervised Learning of Monocular Depth Estimation and Visual Odometry with Deep Feature Reconstruction [J]. 2018.
[12] A. Geiger, P. Lenz, and R. Urtasun. Are we ready for autonomous driving? The kitti vision benchmark suite. In Conference on Computer Vision and Pattern Recognition (CVPR), 2012.
[13] A. Geiger, P. Lenz, C. Stiller, and R.Urtasun. Vision meets robotics: The kitti dataset. International Journal of Robotics Research (IJRR), 2013.
[14] Y. Jia, E. Shelhamer, J. Donahue, S. Karayev, J. Long, R. Gir-shick, S. Guadarrama, and T. Darrell. Caffe: Convolutional architecture for fast feature embedding. In Proceedings of the 22nd ACM international conference on Multimedia, pages 675–678. ACM, 2014.
[15] D. Kingma and J. Ba. Adam: A method for stochastic optimization. arXiv preprint arXiv: 1412.6980, 2014.
[16] R. Mur-Artal, J. M. M. Montiel, and J. D. Tardos. Orb-slam: a versatile and accurate monocular slam system. IEEE Transactions on Robotics, 31(5):1147–1163, 2015.

IOP Conf. Series: Journal of Physics: Conf. Series **1267** (2019) 012099 doi:10.1088/1742-6596/1267/1/012099

The Architecture of a Driverless Robot Car Based on EyeBot System

Shuangquan Sun[1], Jingwen Zheng[1], Zihan Qiao[2], Shanqi Liu[3], Zihan Lin[1]and Thomas Bräunl[4,*]

[1]Dept. of Auto., University of Sci. & Tech. of China, China
[2]Dept. of EE, Zhejiang University, China
[3]Dept. of CSE, Zhejiang University, China
[4]Dept. of EECE, The University of Western Australia

[*]Thomas Bräunl: thomas.braunl@uwa.edu.au, +61 8 6488 1763

Abstract. This paper presents the system architecture of a driverless robot car, designed to participate in the Carolo-Cup, a competition regarding automated model vehicles. We describe an implementation that completed the different tasks in this competition on our EyeBot platform in detail. EyeBot has one RGB camera as well as three infrared distance sensors, and it's powered by Raspberry Pi 3B. We developed the lane detection algorithm using the OpenCV library and completed the traffic sign recognition task based on SVM which can be used offline. Experimental and simulation results recorded in real-time are also reported. The test result showed that our programs can run at high speed to achieve stable motion control in real time and complete all the tasks in the competition. The highlight of our work is that the whole system can run on a platform with limited computing resources.

1. Introduction

Autonomous vehicles have been the subject of some research words in fields including robotics, computer vision, and machine learning. The well-known DARPA Grand Challenge [1] has triggered a great deal of research by confronting academics with real practical situations. Other competitions have also been set up, such as the Intelligent Vehicle Future Challenge (IVFC) [2], the Audi Autonomous Driving Cup [3] and the Carolo-Cup [4].

In this article, we present the complete system architecture of the driverless robot car that completed the different scenarios in Carolo-Cup. In section 2, the model car platform EyeBot [5], the software EyeSim [6], the operating system RoBIOS [7], the test track and some preparation work are presented. Section 3 and 4 gives a detailed view of our system architecture, covering all the implemented modules: lane detection using OpenCV [8] and traffic sign recognition. These functions are discussed separately in the following sections, including algorithms and experimental results. Finally, Section 5 concludes the paper and provides an overview of future work.

2. Platform & preparation work

2.1. EyeBot & EyeSim & RoBIOS

EyeBot is the name of a robot family developed by UWA. We used SoccerBot, a member of the EyeBot family which is shown in figure 1. It is powered by a Raspberry Pi 3B in combination with a self-developed IO-Board EyeBot7, which handles all motor and sensor control and communicates with the Raspberry Pi via USB. EyeBot7 and RoBIOS provide an API to assist in robot program design.

EyeSim is a multi-robot, multi-tasking simulation software that allows realistic simulation of mobile robots on MacOS, Windows, and Linux. Its snapshot is shown in Figure 1. EyeSim uses the Unity [9] physics engine and also provides VR applications for Oculus Go [10] and HTC Vive [11].

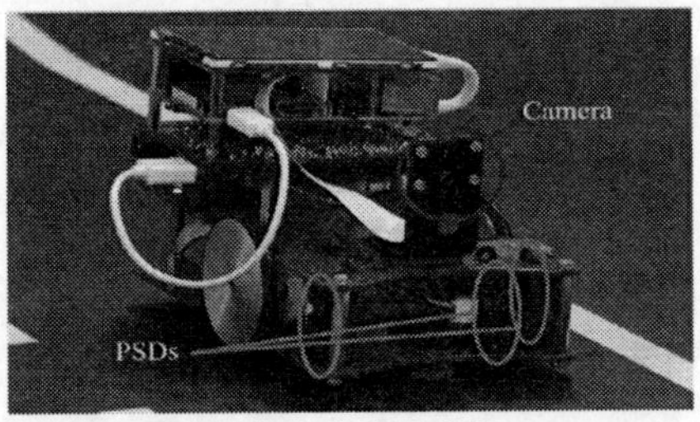

Figure 1. SoccerBot S4 and its sensors.

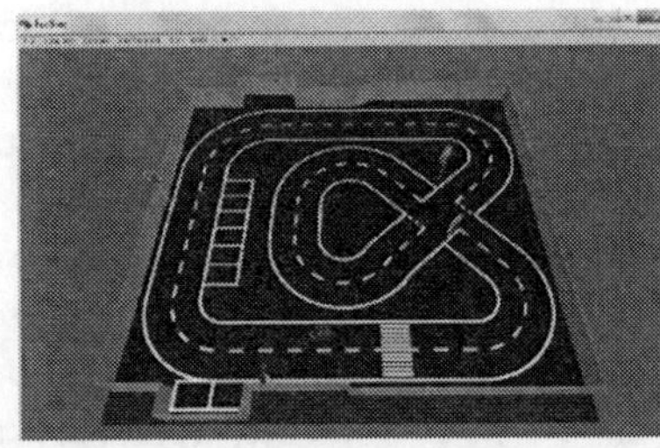

Figure 2. Snapshot of EyeSim: Simulation Windows.

EyeBot's operating system is called RoBIOS (Robot Basic Input-Output System). It comprises a user interface for robot applications on the Raspberry Pi, a set of libraries for various I/O and sensor data processing functions, plus numerous demonstration applications.

2.2. Indoor testbed & simulation

We built an indoor testbed for research, and due to the smaller size of SoccerBot robots and limited space in the laboratory, the sizes of the track and traffic signs were halved. This track consists of a double-lane road, six curves, an intersection and two parking lots. We also built a similar traffic environment in the EyeSim simulator for testing. Figure 3 shows the test track in reality and simulation.

Figure 3. Test track in reality and simulation.

3. System architecture

In order to perform all the tasks required in Carolo-Cup, we need to bring several sensors and processing modules into play. Figure 4 shows the global architecture of the system implemented on the experimental platform. A single computer running a Linux OS was used to handle the different sub-modules in multiple threads.

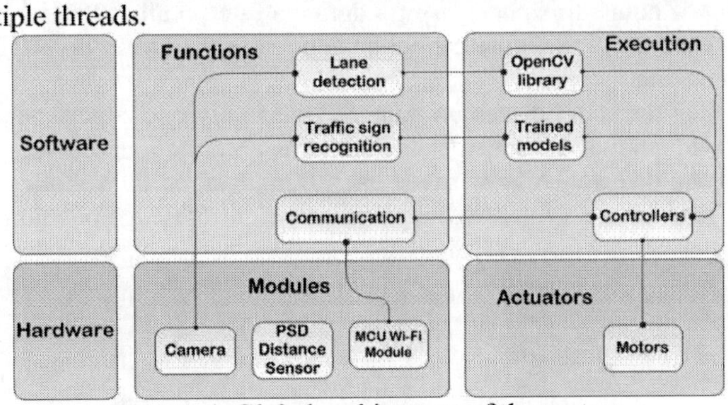

Figure 4. Global architecture of the system.

IOP Conf. Series: Journal of Physics: Conf. Series **1267** (2019) 012099 doi:10.1088/1742-6596/1267/1/012099

3.1. Lane detection based on OpenCV

Lane detection is a fundamental task in autonomous driving. In this section, we discuss the method we used to detect lanes running on the EyeBot platform using OpenCV. The method is modified due to the limitation of computing power, while it maintains good flexibility for different scenarios like a straight road, curves, crosswalk, and intersection. Our algorithm can run at around 30fps, which is fast enough to achieve a satisfying lane following ability. Our lane detection algorithm contains the following steps:

- Steps of lane detection. Images captured from the camera [Figure 5] need to be preprocessed for features extraction. Several techniques are used:
- Set Region of Interest (ROI): remove the top part of the original image that is useless for lane detection to accelerate the image processing speed. [Figure 6]
- Greyscale Transform: Convert color image to grayscale image.
- Gaussian Blur: This step reduces noise in the image.

Figure 5. Image captured

Figure 6. The region of interest.

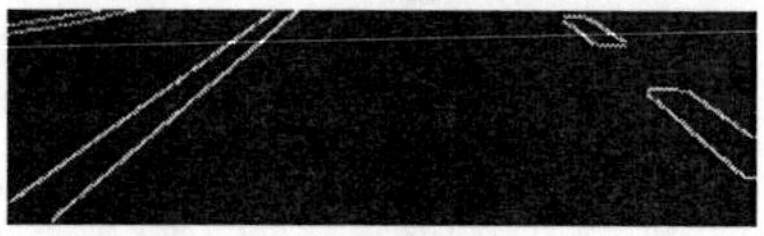

Figure 7. Canny edge detection.

Then we search for specific edges in the image to extract the road feature. We use:

- Canny Edge Detection. It is a computational approach to detect edges in an image [12]. We used the result of canny edge detection to find edges of the lane. [Figure 7]
- Hough Transform. It is a method can be used for line detection and general curve fitting [13]. We use it to find all possible straight lines to determine the two inner edges of the lane.

The final step is feature validation. Considering the limited field-of-view of the camera, we implemented a new way to determine the left and right edges of the lane. First, select the possible lines of the left edge and right edge by the slope of the lines. Then, for each part of the image, calculate the distance from the line to the top left or top right spot of the region of interest which is shown in Figure 8. The lines with the longest distance to the left or right corner spot are considered as the inner edge of the lane markings. Figure 9 demonstrated the result of lane detection.

Figure 8. Edges (blue), distance to spot (red).

Figure 9. Determined inner edge of the lane.

3.1.1. Detection of different scenes. As for a curved road, one of the most common methods is curve fitting. Curve fitting is a mathematical technique that is widely used in engineering applications. It consists of fitting a set of points to a curve using Lagrange Interpolation Polynomial [14]. However, it is computationally intensive and therefore cannot be applied to the EyeBot platform while maintaining a reasonably high frame rate.

In our scenario, when the robot drives on a curve, often only one edge is visible in the camera as Figure 10 and Figure 11 show. We consider the curve as a continuous polyline. The information of angle and position of the detected edge is saved each frame. When the robot drives into a curve and one edge of the lane becomes invisible, we assume that the current edge position is the same as in the last detected position.

Figure 10. Visible solid line for the right curve.

Figure 11. Visible dash line for the left curve.

To detect the crosswalk, we use the number of lines detected in a frame. When the crosswalk is in the sight of the camera as Figure 12 shows, the number of detected lines will increase to around fifteen, and after passing the crosswalk, it will reduce to normal level.

To detect the intersection, as Figure 13 shows, when the robot arrives at the intersection (① and ②), the number of detected lines reduced to zero due to the temporarily missing edges on the two sides. Based on this information, we can decide if there is an intersection ahead.

In an experiment, we let the car drive from the blue spot to the red spot on the left in figure 14 and printed the number of detected left or right edges in a graph. From the graph, we can see a sharp peak when the robot reaches the crosswalk.

Figure 12. Detected lines before the crosswalk.

Figure 13. Image before the intersection.

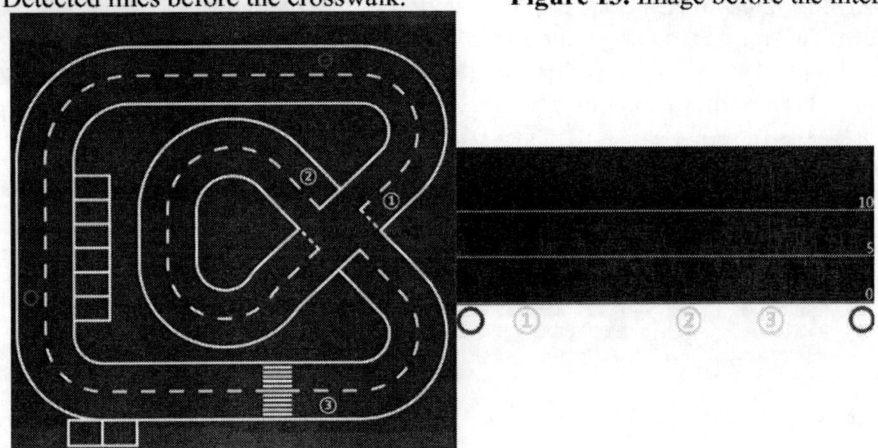

Figure 14. Test track and number of lines detected from blue spot to red spot.

3.2. Traffic sign recognition task

3.2.1. The HOG feature. There are many classical methods which can complete the traffic sign recognition task. The most common of hand-crafted features are SIFT [15], HOG [16], LBP [17], etc. After describing traffic signs, classifiers are learned based on the feature representing for traffic sign detection and classification, such as support vector machines (SVM) [18] and random forest classifier [19], [20].

For our application, we found the HOG feature can gain the acceptable accuracy for our driverless simulation task after particular optimization of the current scene. Moreover, its reaction time is short enough for real-time needs. The libraries that the algorithm relies on are limited, so it is easy to be installed on the Raspberry Pi. Based on these reasons, we chose the HOG feature to perform the traffic sign detection and recognizing task.

3.2.2. The influence of different parameters in SVM. To optimize the algorithm for our needs, we must change some parameters and need to add some methods for block normalization. This can help with generalization and the speed of the function. We mainly modified these parameters to adapt the model to our situation — the C for the SVM (penalty term), and thresholding the singular values of the filters and the size of the scan windows.

Firstly, for the C of the SVM, the below pictures tell how much it should bc to avoid misclassifying each training example. In our situation, the parameter C can be large, because our training datasets do not have too many noises. The accuracy sensitivity to C shows in Figure 15.

IOP Conf. Series: Journal of Physics: Conf. Series **1267** (2019) 012099 doi:10.1088/1742-6596/1267/1/012099

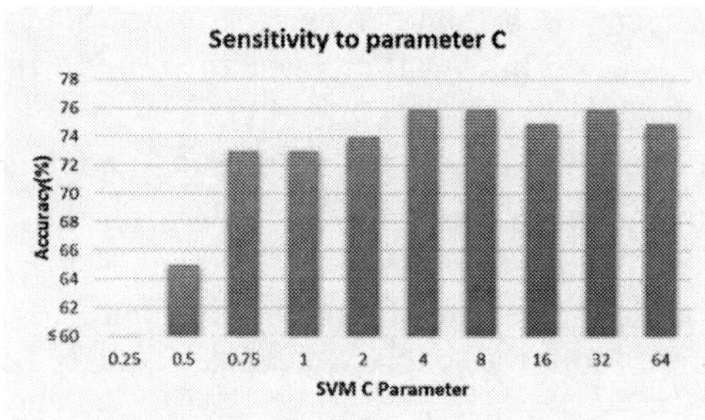

Figure 15. Sensitivity to C.

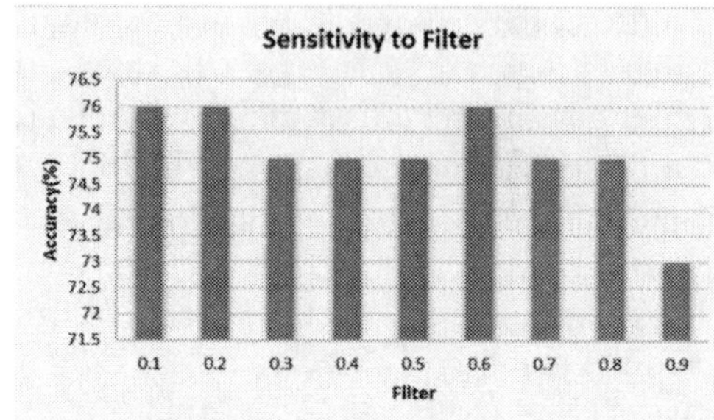

Figure 16. Sensitivity to Filter.

As for the threshold of the filter, we keep it small, because we want to keep more details. The accuracy sensitivity to the filter shows in Figure 16.

The windows size is more complicated as it depends on the input image size. In our situation, the input image size is 320x240 (QVGA). So, the size of the scanning window should be smaller than 80. However, when the size of scanning windows is too small, the performance of the model deteriorates, especially in the presence of noise in the training dataset. Figure 17 shows window size sensitivity. Figure 18 shows the sensitivity to norm.

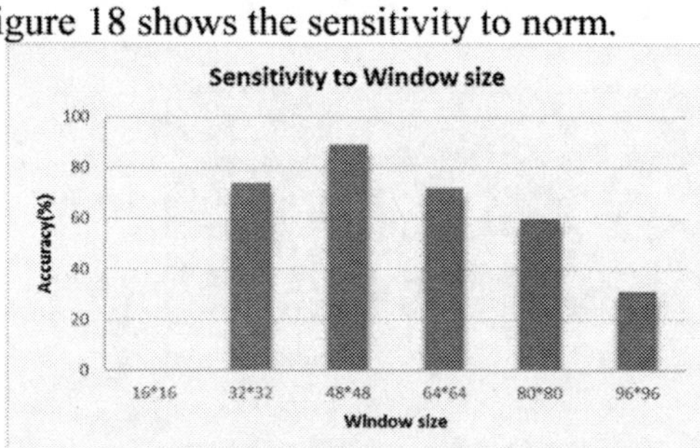

Figure 17. Window sensitivity.

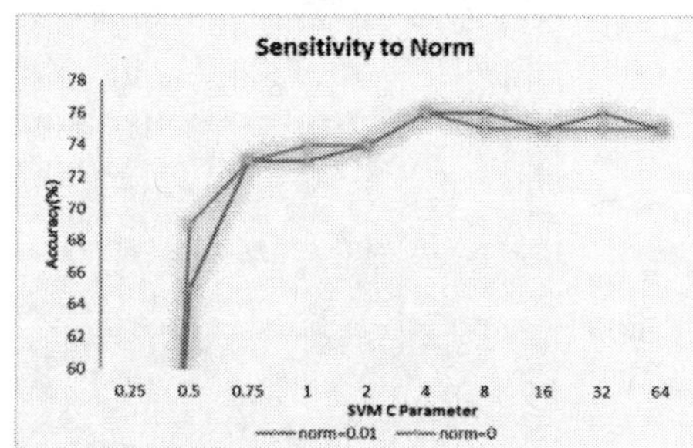

Figure 18. Sensitivity to Norm.

4. Experimental results and discussions

4.1. Traffic sign recognition

4.1.1. HOG feature models. The result images of our HOG feature models are shown in Figure 19. They primarily reflect the shapes and appearances of the given traffic sign images, so we presume the models we trained are correct.

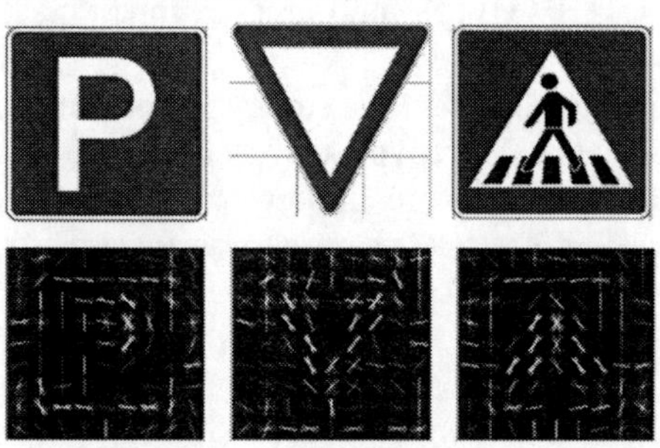

Figure 19. The HOG of some traffic signs.

Traffic signs	Precision	Recall	F
Stop	82.61%	61.29%	70.37%
Pedestrian	100%	73.33%	84.62%
Park	100%	50.00%	66.67%
Speed limit	100%	63.64%	77.78%
Speed limit cancel	100%	42.11%	59.26%
Give way	100%	60.00%	75.00%

Figure 20. The performance of the model.

4.1.2. The recognition results. Moreover, the accuracy of our model is tested in a testing dataset we build based on our map; the main performance is measured by the accuracy, recall, and F-measure, the result of our model is shown in Figure 20 and Figure 21.

Figure 21. The recognition results.

4.1.3. Execution time. The most important part of this model is the executing time, as real-time performance is required for the robots. The average time of executing the main control loop is 340ms. However, the time varies from 240ms to 408ms, which is quite stable for most of the time. Combining the accuracy and executing time results, we can assume that this model meets our soft real-time and accuracy requirements for this application.

5. Conclusion and Future Work

The system architecture described in this paper was successfully implemented and demonstrated within the context of Carolo-Cup 2018, both on real EyeBot/SoccerBot robots, as well as in the EyeSim simulation environment. The implemented modules worked reliably and were capable of handling complex scenarios, despite the limited computational and memory resources of the chosen hardware.

In the future, we plan to improve the robustness and accelerate the speed of the recognition process which is crucial for racing competition. Besides, considering the basic functions we have, we can implement the communication function into our system. The possibilitics offered by wireless communications for an automated distributed understanding of driving scenes are numerous and still largely unexplored.

Acknowledgments
This work was supported by The University of Western Australia. The authors would like to thank Alex Arnold, Kai Li Lim, Nicholas Burleigh, and Michael Finn.

References

[1] Buehler, M. , Iagnemma, K. , & Singh, S. . (1973). The 2005 DARPA grand challenge: the great robot race. Industrial Robot An International Journal, 35.
[2] NovAtel Inc. *Driverless in China | Velocity Magazine | NovAtel. [Online],* Available: https://www.novatel.com/tech-talk/velocity/velocity-2015/driverless-in-china/
[3] AUDI AG 2018, *June Audi Autonomous Driving Cup 2018 Rulebook. [Online],* Available: https://www.audi-autonomous-driving-cup.com/wp-content/uploads/2018/06/2018-06-13_en_Rulebook_2018_V2.0EU.pdf
[4] Technische Universität Braunschweig, *Carolo-Cup Regulations 2018. [Online],* Available: https://wiki.ifr.ing.tu-bs.de/carolocup/system/files/Carolo-Cup%20Regulations_1.pdf
[5] Braunl T 1999 EyeBot: a family of autonomous mobile robots *ICONIP'99. ANZIIS'99 & ANNES'99 & ACNN'99. 6th Intl. Conf. on Neural Information Proc., pp.* 645-9.
[6] Braunl Thomas, *EyeSim VR - Unity Based EyeBot Simulator. [Online].* Available: http://robotics.ee.uwa.edu.au/eyesim/
[7] Thomas Bräunl, Remi Keat, Marcus Pham, *RoBIOS - Mobile Robot Library. [Online].* Available: http://robotics.ee.uwa.edu.au/eyebot7/Robios7.html
[8] OpenCV team. (2018). *OpenCV library. [Online].* Available: https://opencv.org/
[9] Unity Technologies. (2018). *Unity. [Online].* Available: https://unity3d.com/
[10] Oculus. *Oculus Go. [Online].* Available: https://www.oculus.com/go/
[11] HTC Corporation. (2011-2018) *VIVE™ Australia | Discover Virtual Reality Beyond Imagination. [Online].* Available: https://www.vive.com/au/

[12] Canny J 1986 A computational approach to edge detection *IEEE Transactions on Pattern Analysis and Machine Intelligence* **PAMI-8(6)** 679–98
[13] *Duda, R. O. , & Hart, P. E. . (1972). Use of the hough transformation to detect lines and curves in pictures. Communications of the ACM, 15(1), 11-15.*
[14] Yang Y 2013 Polynomial curve fitting and lagrange interpolation *Mathematics and Computer Education* **47(3)** 224-30
[15] Meuter M, Nunn C, Gormer S M, Muller-Schneiders S and Kummert A 2011 A decision fusion and reasoning module for a traffic sign recognition system *IEEE Transactions on Intelligent Transportation Systems* **12(4)** 1126-34
[16] Dalal N and Triggs B 2005 Histograms of oriented gradients for human detection *2005 IEEE Computer Society Conference on Computer Vision and Pattern Recognition (CVPR'05)*
[17] Ojala T, Pietikainen M and Maenpaa T 2002 Multiresolution gray-scale and rotation invariant texture classification with local binary patterns *IEEE Transactions on Pattern Analysis and Machine Intelligence* **24(7)** 971-87
[18] Greenhalgh J and Mirmehdi M 2012 Real-time detection and recognition of road traffic signs *IEEE Transactions on Intelligent Transportation Systems* **13(4)** 1498-506.
[19] Zaklouta F and Stanciulescu B 2012 Real-time traffic-sign recognition using tree classifiers *IEEE Transactions on Intelligent Transportation Systems* **13(4)** 1507-14
[20] Guo J, Lu J, Qu Y and Li C 2018 Traffic-sign spotting in the wild via deep features 2018 IEEE Intelligent Vehicles Symposium (IV) 120-5

Research and Development of High Precision Robot Automatic Docking System Based on mxAutomation

Dechao Zheng[1, 2], Rui Guo[1, 2], Kai Chang[1, 2] and Jian Yin[1, 2]

[1] Intelligent Detection and Equipment Laboratory, Shenyang Institute Of Automation ,Chinese Academy Of Sciences, No.114Nanta Street, Shenhe District, Shenyang, Liaoning Province, P.R.China
[2] Institutes for Robotics and Intelligent Manufacturing, Chinese Academy of Sciences, Shenyang 110016, China

zhengdechao@sia.cn

Abstract. In this paper we designs an automatic docking system to simulate the precision docking of the key components in aerospace assembly line. By using the robot to make the actual measurement of the experimental workpiece, we figure out the spatial coordinate system of the workpiece which correspond to the robot world coordinate system. This coordinate system is used to automatically control the robot to calculate the motion trajectory and complete the precision docking work. This system is characterized by the use of Siemens PLC to control the KUKA robot dynamically through the KUKA-TIA-Library interface, which can meet the requirements of flexibility and customization in aerospace industry.

1. Introduction

In aerospace manufacturing, objects such as rockets and aircraft are very large in size and require a lot of assembly space. Usually, the main body is not the whole processing, but needs to be manufactured in sections, and then in the final assembly link to assemble the large parts together [1]. The traditional docking methods mainly relies on tooling, standard templates, samples, gauges and process compensation to ensure the accurate relative position of large size components. This kind of docking method is difficult to locate the benchmark, difficult to adjust by manual, low accuracy, and greatly affected by human factors, resulting in long manufacturing cycle, poor reliability, so the conventional docking method is difficult to meet its requirements.In this paper, a set of high-precision dynamic and automatic robot auto-docking system based on mxAutomation control is designed to improve and optimize the assembly methods.

2. Composition of the control system

The system is mainly consists of robot control system, PLC control system, superior system, and docking experimental parts [2]. The robot control system adopts KR 6 R700 sixx small industrial robot of Kuka company, supporting KR C4 Compact controller, with a load of 6kg and a repetitive positioning accuracy of + / - 0.03mm. The PLC control system adopts Siemens s7-1515-2pn PLC, with 500KB code storage and 3MB data storage area, bit instruction execution time up to 30ns. The superior system adopts Advantech IPC610 drawer type industrial controller, which is equipped a 24 "industrial display. The specific configuration is shown in 'Figure 1':

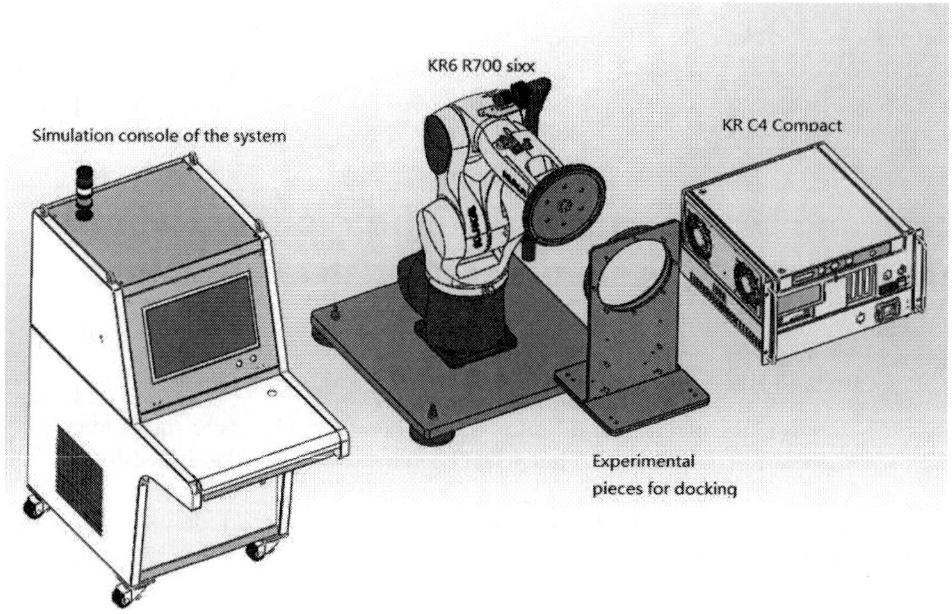

Figure 1. Composition of simulation docking system.

The system adopts Siemens Profinet communication bus and use the Siemens PLC as the main station controller. The KRC4 Compact integrate a Profinet slave station module inside and worked as a slave station on the system bus.The programming software adopts the latest version of TIA Portal V15, and KUKA. PLC mxAutomation library is installed. The specific hardware model and software version are shown in the 'Table 1':

Table 1. The configuration list of control system.

Name of the element	Specific model	Remarks
Simatic Control	S7-1515-2PN	Firmware version V2.5
PLC memory card	6ES7954-8LE03-0AA0	12M
Kuka robot	KR 6 R700 SIXX	
Robot controller	KRC4 Compact	Firmware version V2.5
Industrial computer	IPC 610	
Configuration software	TIA Portal	Software version V15
Library files	KUKA-TIA Library	Library version V2.13
Kuka external control interface	Kuka.PLC.mxAutomation	Software version V2.13
Kuka slave station module	Kuka.Profinet Device	Software version V3.2
Kuka configuration file	Kuka GSD file for KRC4	Software version V2.3.1
Kuka programming software	Work Vistual	Software version V4.0.29

3. Configuration and application of Kuka.PLC. mxAutomation

As an optional software package of Kuka robot, Kuka.PLC.mxAutomation [3] is used to convert the programming instructions of PLC into KRL language of KUKA robot and submit it to the robot for execution. Through this software package, the programmer can control the robot to perform absolute and relative positioning movement, definition and switching of tool coordinate system and base coordinate system dynamically and flexibly. The establishment of point positions and coordinate systems of these motions is carried out dynamically, and it is not necessary to teach or define them in advance in the robot.Through this interface, up to 5 robots can be controlled at the same time.

When using the software package, it is necessary to configure the robot controller side and PLC controller side separately. On the robot side, the software package is installed on the robot controller in the form of KOP installation package through WorkVistual software [4]. After installation, we configuredrobot controller as Profinet device and selected the corresponding communication I/O Numbers are 2032 (254 bytes), see 'Figure 2' for details.

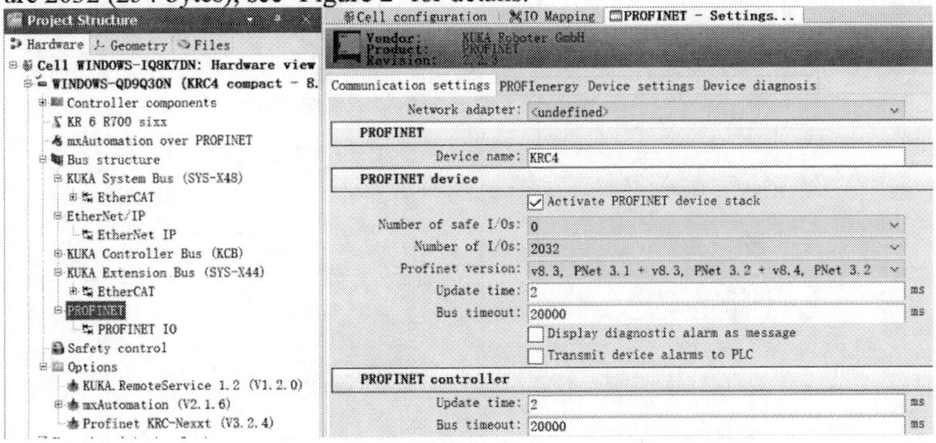

Figure 2. Configuration of KUKA robot in WorkVistual.

In the configuration software of PLC, through the installation of Kuka-Tia-Library file, and in accordance with the provisions of the use of Library file function programming operation, PLC can achieve the external automatic control function of the robot. In the HMI software simulation interface of PLC, you can start, stop, reset, set the running speed of the robot, manual JOG mode, manual teach the point and save to PLC's data blocks, also can read real-time robot position information, status information and other important parameters of the robot. The HMI interface look like 'Figure 3'

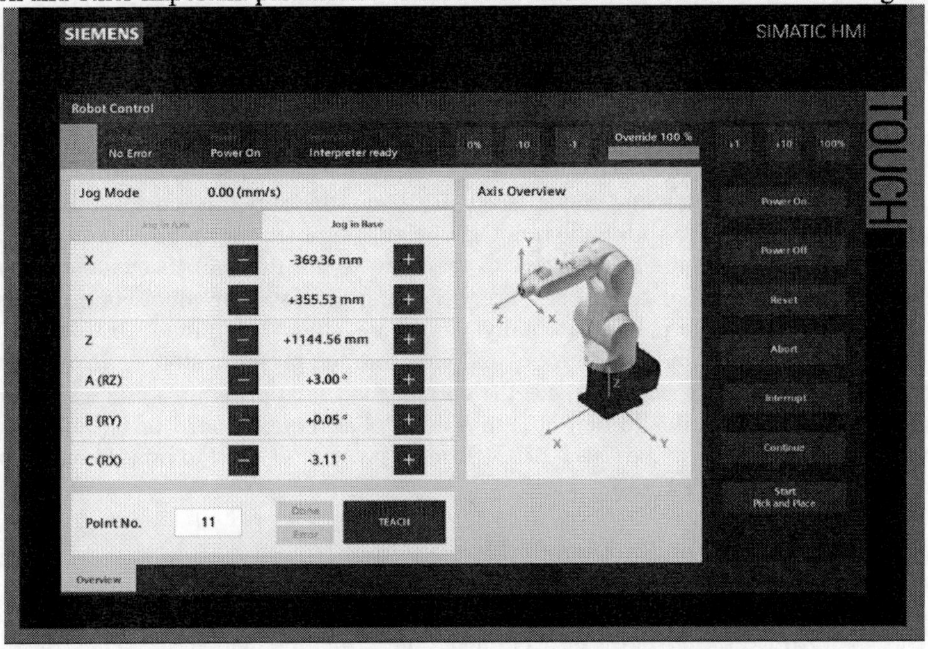

Figure 3. The HMI software simulation interface.

4. Control of the KUKA robot in TIA Portal

When using Siemens PLC to program control KUKA robot, it must follow a certain process. The flow chart in 'Figure 4' briefly illustrates the control logic:

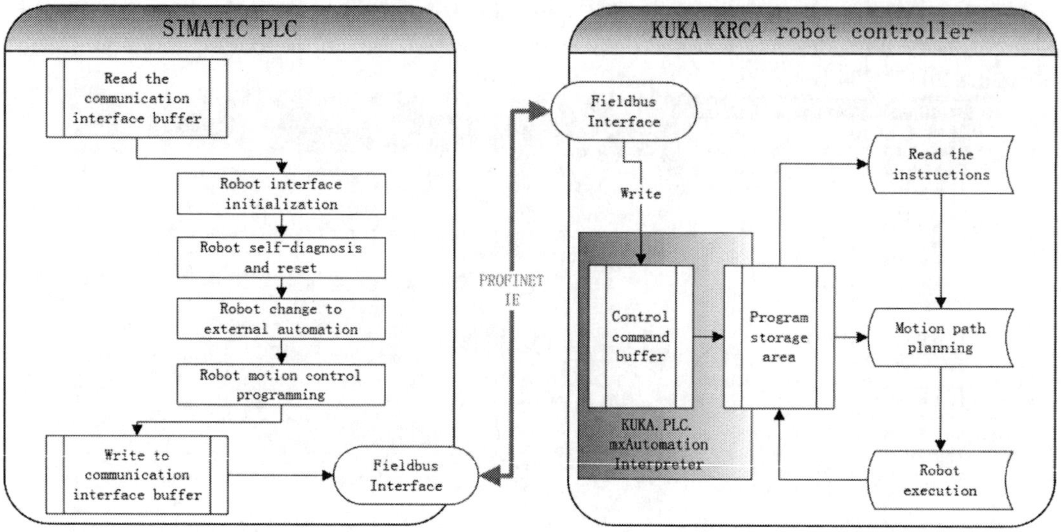

Figure 4. The control flow chart of the PLC and robot controller.

The starting program of the PLC must use KRC_ReadAxisGroup() function to read the information of the robot from thecommunication interface's receiver cache to the local data block. Similarly, the ending program must use the KRC_WriteAxisGroup() function to write all control programs to the communication interface's sender cache. The effective programming interval is between these two functions, and all control and motion instructions need to be written between these two functions.

At the beginning of the robot control program, we use KRC_Initialize () function to initialize mxA interface, then using KRC_ReadMxaStatus () to check the real-time Status of the interface, when the parameters of the Status is 3 or more which indicated interfaces on both sides are showed normal. Otherwise, KRC_ReadKrcError (), KRC_ReadMxa_Error () and KRC_Diag () need to be run to detect robot KRC program error, robot mxA interface data error and robot self-diagnostic error, respectively. According to the feedback of detection results, KRC_Message_Reset () can be run to successfully reset the fault information, and then KRC_AutoExt () function can be run (at this time, the robot needs to switch to the external automatic mode AUT EXT on the robot teacher). When the external automatic mode is started successfully, the robot can be programmed for motion control.

In a single scanning cycle, the PLC writes all the work to the Profinet IE bus and passes it to the Profinet bus module of the robot controller. After receiving the data, the robot controller first run the Kuka.PLC.mxAutomation interpreter internallyto interpret the transmitted PLC data into KRL language that can be operated by KUKA robot and pass into the program storage area of the robot. At this point, the robot interpreter starts to reads the commands in the program storage area and controls the operation of the robot. It should be noted that the global motion speed of the robot needs to be limited in the control process, so the KRC_SetOverride() function is used to limit the axial motion and linear motion to an acceptable range.

5. The experimental process of precision docking system

In the construction of this experimental system, the robot is mainly used to docking the two parts with very small clearanceautomatically and accurately, which is difficult to achieve by ordinary teaching function.The stationary docking piece is a circular hole structure, which simulates the flange hole commonly used in aviation assembly field, and is fixed on the ground by a right-angle steel plate base. The mobile docking piece is a disc structure, simulating the lining ring in the supporting flange hole, which can be directly installed in the center of the robot flange plate and ensure that the connection between the two center points is perpendicular to the disk. The specific size is shown in the 'Figure 5'.

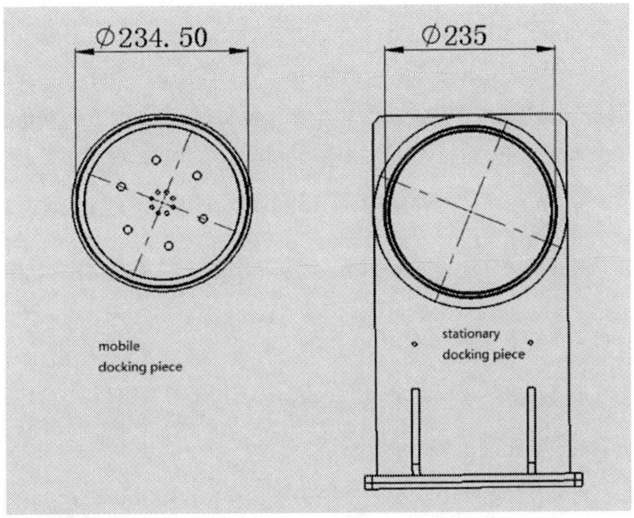

Figure 5. The dimension of the docking pieces.

As you can see the fit clearance between is only 0.5 mm, and it requires the mobile docking piece to be vertically inserted along the center of stationary docking piece to complete the precision assembly. So we set the center of the stationary docking piece as the origin of the robot base coordinate, the horizontal plane where the whole circular hole is located is set as the x-y plane of the coordinates, the line perpendicular to the x-y plane and passing through the origin serves as the z-axis of the coordinate. This coordinate system serves as the workpiece coordinate system (Base coordinate system) for robot docking [5].

In order to obtain the spatial expression of the circular hole of the stationary docking piece, firstly, a known probe tool is installed at the end of the robot flange, and the physical parameters of the tool are set in the tool coordinate system of the robot. Then, the robot is controlled to touch the probe to some points on the edge of the circular hole, and the recorded this points to the robot controller. This measurement process is the result of manual instruction, and some human errors will be introduced. After the subsequent introduction of the measurement system, the errors can be gradually eliminated and automated [6].

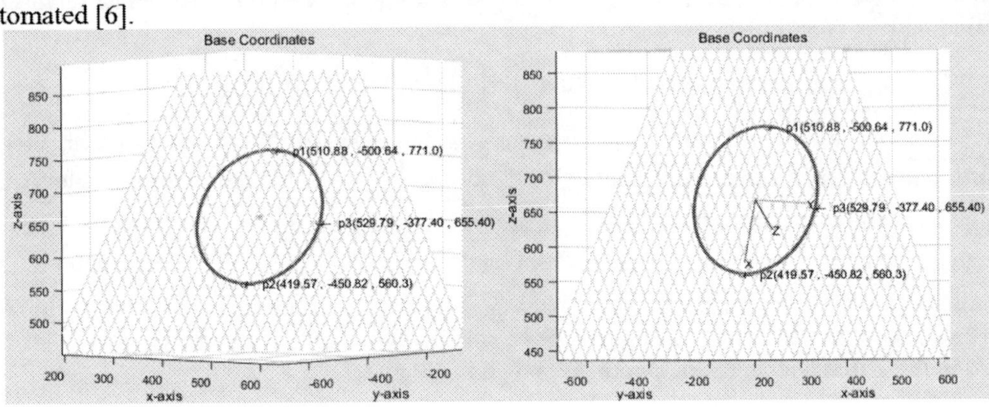

(a) (b)

Figure 6. The robot base coordinate system.

The 'Figure 6' above shows three spatial coordinate points p1, p2 and p3 found on the circular hole plane by manual instruction. The coordinate values of these points are all based on the values of the world coordinate system of the robot. The plane where the circular hole is located and the position of the center point o of the circular hole are fitted by these three points, which serves as the origin of the base coordinate system.

With the origin as the center, p1 points to p2 as the vector $\vec{x}$, p1 points to p3 as the middle vector $\vec{m}$. After the vector $\vec{x}$ cross multiplied by vector $\vec{m}$, we can get the normal vector $\vec{z}$ of the plane, then we use vector $\vec{x}$ cross multiplied by vector $\vec{z}$ to get the vector $\vec{y}$. Finally, by standardizing the three axes, the specific operation results can be established as follows:

$$\vec{x} = (\vec{p_2} - \vec{p_1}) / norm(\vec{p_2} - \vec{p_1}) = [-0.3886, 0.212, -0.8967] \tag{1}$$

$$\vec{m} = (\vec{p_3} - \vec{p_1}) / norm(\vec{p_3} - \vec{p_1}) = [0.1112, 0.7248, -0.6799] \tag{2}$$

$$\vec{z} = (\vec{x} \times \vec{m}) / norm(\vec{x} \times \vec{m}) = [-0.7289, -0.5245, -0.44] \tag{3}$$

$$\vec{y} = (\vec{z} \times \vec{x}) / norm(\vec{z} \times \vec{x}) = [0.5636, 0.8246, -0.0493] \tag{4}$$

Rotation matrix of the base coordinate system:

$$\text{Rot} = \begin{bmatrix} \vec{x}' & \vec{y}' & \vec{z}' \end{bmatrix} = \begin{bmatrix} -0.3886 & 0.5636 & 0.7289 \\ 0.212 & 0.8246 & -0.5245 \\ -0.8967 & -0.0493 & -0.44 \end{bmatrix} \tag{5}$$

According to the position of the center 'O' of the circular hole, the homogeneous transformation matrix of the base coordinate system can be established:

$$\vec{O} = [465.5653, -475.2322, 665.6352] \tag{6}$$

$$\text{T} = \begin{bmatrix} Rot & \vec{O}' \\ 0 & 1 \end{bmatrix} = \begin{bmatrix} -0.3886 & 0.5636 & 0.7289 & 465.5653 \\ 0.212 & 0.8246 & -0.5245 & -475.2322 \\ -0.8967 & -0.0493 & -0.44 & 665.6352 \\ 0 & 0 & 0 & 1 \end{bmatrix} \tag{7}$$

In Matlab, the function tr2eul in Robotics Toolbox is used to obtain the Euler angle required by the corresponding robot controller:

$$eul = tr2eul(T, \text{'deg'}) = [-35.7398, 116.1018, -3.1467] \tag{8}$$

In KRL language, when defining a workpiece coordinate system (Base coordinate system), it is necessary to input 6 parameters: the position of the origin $[x, y, z]$, and the rotation angle $[R_x, R_y, R_z]$ which are relative to the robot world coordinate system. According to the above calculation, the coordinate value of origin position 'O' can be assigned to $[x, y, z]$, and the eul assigned to $[R_x, R_y, R_z]$, so that every time the robot makes precise docking, it can make accurate docking along the Z axis in the workpiece coordinate system. Using the above calculation results, a complete docking process was achieved in the actual docking experiment, as shown in 'Figure 7'.

Figure 7. Actual docking verification process.

6. Conclusion

In the simulated docking experiment, the control system dynamically controls the robot to successfully complete the precision docking. By means of manual measurement, some characteristic points of the docking workpiece in the robot coordinate system are found, and the workpiece coordinate system of the robot is established according to these characteristic points, so as to complete a precision docking work. In the following work, the method of manually measuring feature points will be upgraded to the method of robot vision measurement, which will greatly improve the accuracy and stability of the system and can be applied to the field of aerospace assembly.

Acknowledgments

This work was supported by National Key R&D Program of China(No. 2017YFB1303701), the Project of Intelligent Manufacturing Integrated Standardization and New Model Application, Chongqing Artificial Intelligence Technology Innovation Major Thematic Projects (No. zdyf0027), Made-in-China 2025 Sichuan Action Fund Project(2017ZZ003) and the Project of Intelligent Manufacturing Integrated Standardization and New Model Application 2017.

References

[1] Liu wei, Liu feng, NI yang yong, Bai xue and Li fei 2018 *Astronautical Systems Engineering Technology* **2** 33-36
[2] Yao li zhou, Fang wei and Zhu xiao jin, 2018 *Instrumentation Technology* **9** 9-12
[3] Li shi cheng, Xu shi bao, Zhang yun jian, Cheng xu and Wang shi gang 2017 **36** 75-78

IOP Conf. Series: Journal of Physics: Conf. Series **1267** (2019) 012100 doi:10.1088/1742-6596/1267/1/012100

[4] Wu hang 2017 research of active compliant technology and application software development for handling and assembly robot *Southeast University*
[5] John J. Craig 2005 *Introduction to Robotics: Mechanics and Control chapter* 2 pp 14-29
[6] A. Glamnik and R. Safaric 2012 *9th International Conference on Remote Engineering and Virtual Instrumentation* control of kuka kr 5 robot with a haptic device

IOP Conf. Series: Journal of Physics: Conf. Series **1267** (2019) 012101 doi:10.1088/1742-6596/1267/1/012101

Position Measurements of the Space Manipulator based on Monocular Microscope Vision

Wu Yanmiao[1,*], Wang Ke[2], Chen Han[2] and Lao Guanqing[2]

[1] University of Chinese Academy of Sciences, Beijing 100049, P. R. China
[2] Key Laboratory of Space Utilization, Technology and Engineering Center for Space Utilization, Chinese Academy of Sciences, Beijing 100094, P. R. China

[*]Email: wuyanmiao16@csu.ac.cn

Abstract. In this paper, an inspection manipulator with a micro-camera at the end was designed for in-situ microscopic observation of samples on the material exposure platform. In order to achieve the positioning requirements, a novel position measurement method for the manipulator was proposed based on monocular microscope vision. First, a planar target was designed for feature detection. Next, the measurement task was performed in two steps respectively. On the image plane, a motion-based active calibration method was used to measure the translation. For the rotation around Z axis, the relationship between the image blur b and the rotation θ was fitted to estimate the rotation from images. In final, the proposed approach was tested in term of accuracy and robustness in ground experimental conditions. The focusing error of the piezoelectric motor was about 1μm and the obtained average alignment accuracies were less than 10μm in X,Y, and 0.002° around Z, respectively, which satisfied the positioning requirements of the in-situ observation.

1. Introduction

The China Space Station is an important experiment base for China's space science and new technology research. Considering the threats of space environment to the reliability and safety of all robotic space exploration missions [1], advance in the space environmental effects on materials and devices has been considered as a crucial program for China Space Station Construction. The material exposure device, which is mounted outside the China Space Station, is an observation platform developed for the study of the space environment. And an inspection manipulator is designed to drive the microscope camera to observe the samples on the exposed surface in-situ. In this case, success in meeting the observation requirements will rely on the positioning accuracy of the manipulator.

Generally, the position measurement methods in the space environment can be simply classified into three categories [2-5]: (1) Multiple sensor fusion approach integrates information from multiple sources, especially active sensing technology, such as 2D-PSD (Position Sensitive Detector), Laser Rangefinder, and Structured lighting, etc. It is widely used in the occasion with a large detection range, even though the hardware cost is generally high. ESA (European Space Agency) designed a geostationary orbit restorer to capture abandoned satellites in orbit, where the laser ranging and active vision measures were used [6]; (2) Stereo vision belongs to a passive measurement method. The principle is recovering the 3D information of the object from multiple images. In space environment, it commonly occurs in spacecraft surface inspection and close manipulator on-orbit, but the time consuming correspondence problems still exist. *Jasiobedski* et al. [7] proposed a stereo–

IOP Conf. Series: Journal of Physics: Conf. Series **1267** (2019) 012101 doi:10.1088/1742-6596/1267/1/012101

vision-camera-based vision system to measure pose and track satellites; (3) Monocular vision is the most simple measurement method, which is frequently integrated into the small robot system. It has the advantages of simple structure and flexible operation. The Space Station Remote Manipulator System (SSRMS or Canadarm2) [8], European Robotic Arm (ERA) [9], Japanese Experiment Module Remote Manipulator System (JEMRMS) [10] and most other space manipulators have monocular camera mounted on the manipulator end-effector. While, the accuracy of monocular vision is mainly limited by the measurement error along the optical axis [11].

Micro-vision is a new technology for precision measurement that combines computer vision and microscopic imaging. And its geometric scale of the object can range from millimeter to micrometer, so that it has been widely used in robot micro-operation [12]. In order to realize the position measurement with the least hardware resources, a novel position precision measurement method based on the monocular microscope vision system is proposed, which provides maximum utility and flexibility for the observation experiments.

2. Inspection manipulator system

On the material exposure device, the modular experiment containers for mounting and exposing samples is plugged on the four sides of a base plate. To have a global observation of the base, the inspection manipulator is designed as a 4-DOF cylindrical manipulator. As figure 1 shows, the J_1, J_2, J_3 of manipulator correspond to the r, θ, z in cylindrical coordinate and the J_3 is the degree of freedom of rotation around Z axis. More details about the main components and the specific parameters of the manipulator can be seen in table 1.

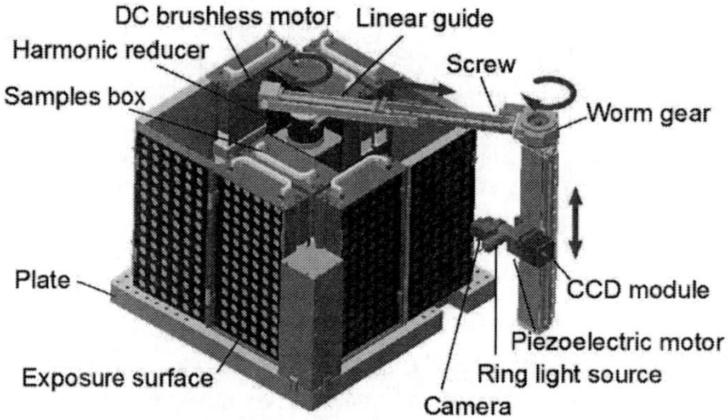

Figure 1. Main components of Space Inspection Manipulator system

Table 1. Main structural and performance parameters of four joints

	J_1	J_2	J_3	J_4
Type	revolute	prismatic	revolute	prismatic
Mass	1.93kg	3.06kg	5.58kg	
Range	0~360°	<223mm	0~360°	<385mm
Reduction ratio	1/100	2mm/rev	1/100	2mm/rev

3. Position measurement method

Comparing to conventional vision, microscope vision systems usually have some special imaging characteristics, including small view depth, small view field, and neglectable lens distortion, which are mainly determined by the magnification factor of lens. As discussed in [13], the view depth is neglectable comparing to the object distance, thus, the object distance can be considered as a constant. Obviously, in the condition that the image is clear, the sensitive freedoms of the microscope vision

IOP Conf. Series: Journal of Physics: Conf. Series **1267** (2019) 012101 doi:10.1088/1742-6596/1267/1/012101

system are two translation motions on the image plane and one rotation motion around Z axis, respectively. Therefore, the measurement work is performed in the corresponding two steps.

3.1 Design of target

Selecting an appropriate target can not only simplify the difficulty of feature extraction, but also improve the accuracy of visual calibration[14]. In this paper, a planar target is designed as shown in the figure 2, which includes a “cross” in the center and a series of triangles pointing to the center. The size of target is 1mm×1mm and the distribution of targets A,B,C,D on an exposure surface is shown in figure 3.

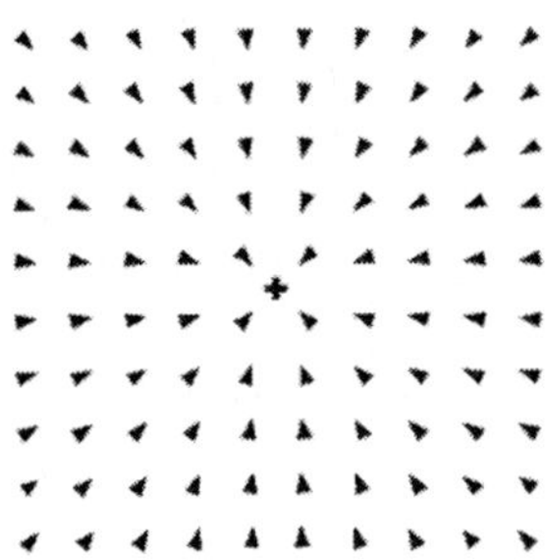

Figure 2. Planar target design

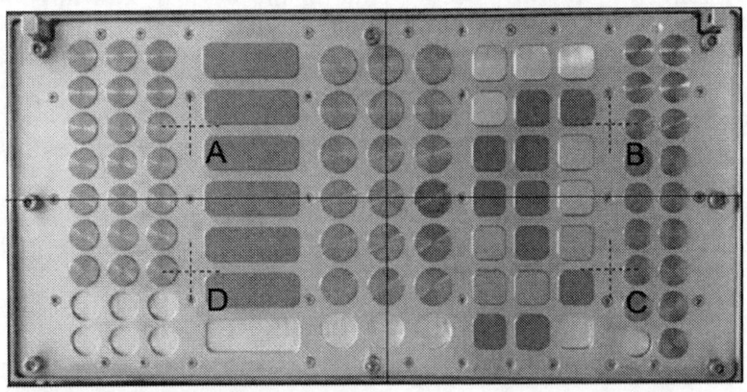

Figure 3. Distribution of targets on an exposure surface

Limited by the small view field of microscope vision system, it is likely that only a few target's features are captured, which will affect the precision of position measurement. Thus, the manipulator is driven to move approach “cross” based on the features in images. The main steps are as follows: (1) Detect all closed triangle in the feature detection area (450pixels×450pixels) and get their centroid coordinates; (2) Select the triangle nearest to the center of image; (3) Move in a fixed step of 0.1mm in the direction of the altitude vector of the triangle; and Repeat (1)~(3) until the “cross” appears in the view.

3.2 Translation on the Image Plane

The measurement of the translation motion on the image plane need to be done continuously and quickly, so the traditional pattern-based calibration method is not suitable. Instead, a motion-based active calibration method is used. Since the image plane is parallel to the exposure surface, the measurements satisfy the plane constraint. Assume that $[\Delta x \quad \Delta y]^T$ is the relative position vector in Cartesian space, $[\Delta u \quad \Delta v]^T$ is relative pixel vector in image space, the relationship can be expressed as follows.

$$\begin{bmatrix} \Delta x \\ \Delta y \end{bmatrix} = \begin{bmatrix} J_{11} & J_{12} \\ J_{21} & J_{22} \end{bmatrix} \begin{bmatrix} \Delta u \\ \Delta v \end{bmatrix} \tag{1}$$

where, $J_{ii}\ (i=1,2)$ is the element of the image Jacobean matrix.

The image Jacobean matrix can be solved by form an equation group:

$$\begin{bmatrix} \Delta x_1 & \Delta x_2 & \cdots & \Delta x_n \\ \Delta y_1 & \Delta y_2 & \cdots & \Delta y_n \end{bmatrix} = \begin{bmatrix} J_{11} & J_{12} \\ J_{21} & J_{22} \end{bmatrix} \begin{bmatrix} \Delta u_1 & \Delta u_2 & \cdots & \Delta u_n \\ \Delta v_1 & \Delta v_2 & \cdots & \Delta v_n \end{bmatrix} \tag{2}$$

The least square solution of (2) is:

$$J = AB^T \left(BB^T\right)^{-1} \tag{3}$$

where, $A = \begin{bmatrix} \Delta x_1 & \Delta x_2 & \cdots & \Delta x_n \\ \Delta y_1 & \Delta y_2 & \cdots & \Delta y_n \end{bmatrix}, B = \begin{bmatrix} \Delta u_1 & \Delta u_2 & \cdots & \Delta u_n \\ \Delta v_1 & \Delta v_2 & \cdots & \Delta v_n \end{bmatrix}$

3.3 Rotation around the Z Axis

Based on the optical imaging model shown in Figure 4, when f (focal length), u (distance of object from the principal) and v (distance of the focused image from lens) satisfy (4), the image is focused. And defocus will occur when object is outside the focal plane. Specially for the microscope vision system, a slight displacement along the optical axis direction may cause image blur, which is an important cue of rotation estimation from microscopic image.

$$\frac{1}{u} + \frac{1}{v} = \frac{1}{f} \tag{4}$$

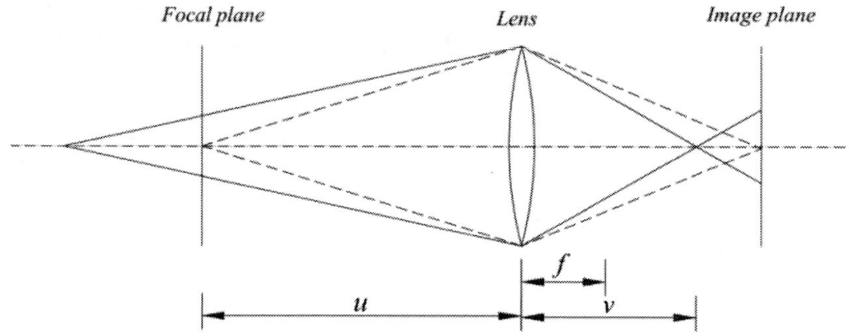

Figure 4. Optical imaging model of camera

In order to find the relationship between the image blur b and the rotation θ, the CPBD (Cumulative Probability of Blur Detection) proposed by *Marzilianno* et al, is adopted for blur detection in this paper. The principle is utilizing the edge width together with the concept of just noticeable blur (JNB) to obtain the fuzzy probability [15], and it has advantages including: near real-time, low computational complexity and good performance independent of the image content [16].

4 Experiment and Result

4.1 Experiment System

Experiments are conducted with the inspection manipulator system, which consists of an air-bearing table, an operation platform, a manipulator and a microscope vision system, as shown in figure 5. The microscope lens is OLYMPUS LMPLFLN20x (WD=12mm, Mob=20, FOV=0.6mm×0.8mm) and the camera is selected to RZIMAGE RZUH-500C CCD module(PPI=2592X1944,PX=2.2μm×2.2μm).

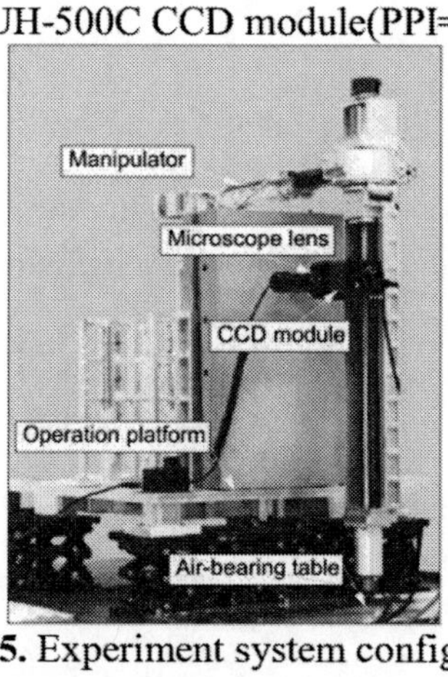

Figure 5. Experiment system configuration

IOP Conf. Series: Journal of Physics: Conf. Series **1267** (2019) 012101 doi:10.1088/1742-6596/1267/1/012101

4.2 Results and analysis

The measurement process includes two stages, which are both automatically controlled. The first one can be taken as the initialization of the work, where the manipulator is controlled with motion commands. When the target is captured, a piezoelectric motor is used to drive the micro-camera to achieve auto-focus based on the climbing hill algorithm, whose positioning accuracy is about 1μm. During the later experiment, the object distance is not changed.

And to capture more features of the target, the manipulator is moved approach the center of image with the method of image recognition process in section 3.1, as shown in figure 6.

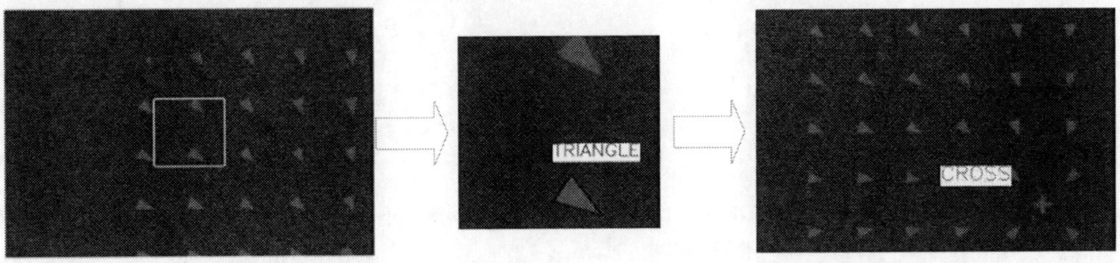

Figure 6. Image recognition process in the initialization of the measurement

The second stage is the main work of the measurement, which is completed in two steps: translation measurement on the image plane and rotation measurement around Z axis.

On the image plane, the microscope vision system is first calibrated with the method in section 3.2. When the camera is adjusted to the surface of the clear view, 4 groups of data are recorded in different positions on a plane, as listed in A and B in (5) and (6). Thus, the Jacobean matrix J is computed as given in (7). After the calibration, the manipulator is moved to a lot of points, where the relative positions measured in experiment are compared with the calculated value. And the statistical results can be seen in table 2. The maximum measuring error is less than 10μm in any direction. Figure 7 shows 10 sets of them.

$$A=\begin{bmatrix} 0.3257 & 0.2098 & -0.3810 & 0.0463 \\ -0.2322 & 0.2145 & -0.1219 & -0.2964 \end{bmatrix} mm \tag{5}$$

$$B=\begin{bmatrix} 843 & 587 & -1030 & 88 \\ -705 & 653 & -372 & -894 \end{bmatrix} pixel \tag{6}$$

$$J=\begin{bmatrix} 0.3746 & -0.0147 \\ -0.0005 & 0.3300 \end{bmatrix} \tag{7}$$

The second step is the rotation θ measurement around Z axis, and the method is to fit the relationship between the image blur b and the rotation θ. First, a target A is selected, where only joint 3 is driven both clockwise and counterclockwise to capture the corresponding blur images and other joints stay still. The range is from -0.04° to 0.04° with a scanning step of 0.005°. Then, the image blur b is calculated based on the CPBD algorithm. And the same operation is carried out at other two targets B,C. From the curve shown in figure.8, the relationship between the image blur b and the rotation θ is approximate to a quadratic curve, therefore, the fitting function is obtained as in (8). Furthermore, the error analysis is made and it is shown in table 2.

$$b=701.1\theta^2+3.765\theta+0.0176 \tag{8}$$

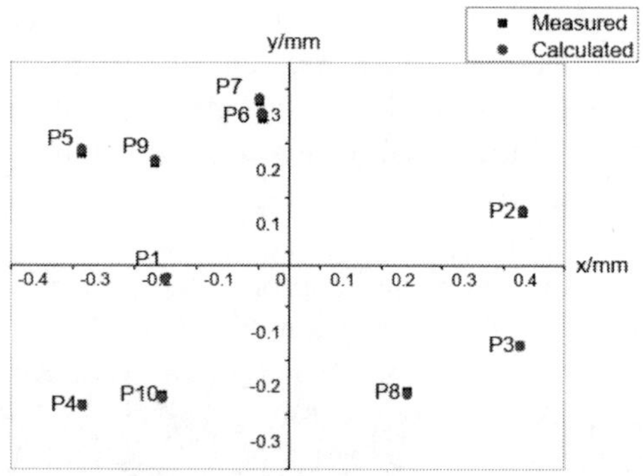

Figure 7. Comparison between the measured position and the calculated position

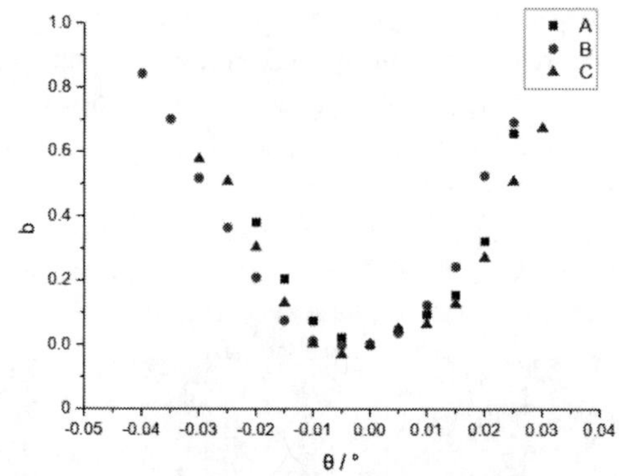

Figure 8. The curve of the image blur b with the rotation θ

Table 2. Error analysis results.

	AVE	MSE
Translation in X,Y axis /mm	0.0089	0.0154
Rotation around the Z axis /°	0.002	0.001

5 Conclusion

In this paper, a novel position measurement method of space manipulator based on the monocular microscopic vision and artificial planar target was studied. And a measurement strategy including two stages was proposed. In the first stage, the micro-camera was driven to focus and the manipulator was controlled to approach the target. In the second stage, the measurements were performed in two steps: 1) the translation on the image plane was measured with the motion-based active calibration method; 2)the rotation around Z axis was obtained by fitting the relationship between the image blur b and the rotation θ. Within the allowable range of error, the ground experiment results showed that the focusing error of the piezoelectric motor along the optical axis was about 1μm, and the average position measurement error in X,Y axis and around the Z axis were less than 10μm, 0.002°, respectively, which satisfied the positioning requirements of the in-situ observation for the exposure platform.

References

[1] Sheila A. Thibeault, Stuart A. Cooke, Melissa P. Ashe, et al. MISSE-X: An ISS External Platform for Space Environmental Studies in the Post-Shuttle Era [EB/OL]. https://ntrs.nasa.gov/archive/nasa/casi.ntrs.nasa.gov/20110008059.pdf.

[2] Hao GangTao, Du Xiaoping. Advances in Optical Measurement of Position and Pose for Space Non-Cooperative Target [J]. *Laser and optoelectronics progress*, 2013,8(28):1-9.

[3] Fuyuto t, heihachiro k, shin'ichiron. Motion Estimation to a Failed Satellite on Orbit using Stereo Vision and 3D Model Matching[C]. *International conference on control automation robotics & vision (ICARCV)*,2006.1-8.

[4] Yu Han Zhang, Xiujie Liu, Lingyu Wang, Shuo Song Shenmin. Relative dynamics estimation of non-cooperative spacecraft with unknown orbit elements and inertial tensor[J]. *CHINESE JOURNAL OF AERONAUTICS*,2016,29(2):479-491.

[5] Yang Guanglie, Zhao Guowei, Luo Kun. A position calculating method of binocular vision based on datum optimization[J].*Chinese space science and technology*.2018,38(2):1-8.

IOP Conf. Series: Journal of Physics: Conf. Series **1267** (2019) 012101 doi:10.1088/1742-6596/1267/1/012101

[6] T. Yasaka and E. W. Ashford, ''GSV: An approach toward space system servicing,'' Earth Space Rev., vol. 5, no. 2, pp. 9–17, 1996.

[7] P. Jasiobedski, M. Greenspan, and G. Roth, ''Pose determination and tracking for autonomous satellite capture,'' in Proc. Int. Symp. Artif. Intell., Robotics Autom. Space, 2001, pp. 1–9.

[8] MD Robotics, "Mobile Servicing System - Data Sheet," Brampton, Ontario, Canada: MD Robotics, 2002.

[9] Knape B. and Berger A., "Development of a Remote Tank Inspection (RTI) Robotic System," *Proc. of the 4th ANS Topical Meeting, Robotics and Remote Systems*, 1991:470-481.

[10] Mori H., Iwata T. and Oda M., "An Overview of Japan's Space Automation and Robotics Activities," *Proceedings of the 1992 Artificial Intelligence, Robotics and Automation in Space (i-SAIRAS92),* 1992:19-24.

[11] Chao Zhichao, Fu Sihua, Jiang Guangwen, etc. Mono Camera and Laser Rangefinding Sensor Position-Pose Measurement System[J]. *ACTA OPTICA SINICA*, 2011,31(3):1-7.

[12] Zhai Binbin. Research on Technology of 3D Reconstruction of Rotational Image Sequences Based on Microscopic Vision[D]. Zhejiang University of Technology for the Degree of Master.

[13] De Xu, Fudong Li, Zhengtao Zhang, etc. Characteristic of Monocular Microscope Vision and Its Application on Assembly of Micro-Pipe and Micro-Sphere[J]. *Proceedings of the 32nd Chinese Control Conference,*2013:5758-5763.

[14] Wang Tianyu, Dong Wenbo, Wang Zhenyu. Position and orientation measurement system based on monocular vision and fixed target[J]. *Infrared and Laser Engineering*,2017,46(4):1-8.

[15] HUANG Xiaomei, TIAN Xuhong. No-reference Sharpness Assessment for Blurred Image[J]. *Journal of Chinese Computer Systems*,2015,36(5):1117-1121.

[16] Narvekar N D, Karam L J. A no-reference image blur metric based on the cumulative probability of blur detection (CPBD)[J]. *IEEE Transactions on Image Processing*, 2011, 20(9):2678-2683.

Kinematic and Static Analysis and Experimental Study for Dual-arm Continuum Robot

Guohua Gao, Yuhang Zheng and Hao Wang

College of Mechanical Engineering and Applied Electronics Technology, Beijing University of Technology, Beijing 100124, China

E-mail: 18801493998@163.com

Abstract. The continuum robot, which is different from the traditional rigid robot, consists of a flexible structure, and becomes a research hotspot in the field of robots, because of its good flexibility and adaptability in the complex environment. According to the advantages of the continuum robot, it could be applied in medical, exploration and agricultural operations. This paper proposed a cable-driven dual-arm continuum robot with three-degree-of-freedom. The kinematics of posture parameters and driving parameters, as well as the statics, was established. And the dual-arm coordinated reachable work space was described according to the principle of constant curvature deformation, on the basis of which, the theoretical motion model of the continuum robot was simulated using MATLAB, and experiments of the posture of the dual-arm continuum robot was conducted on the prototype to verify the accuracy of the theoretical kinematics and statics model.

1. Introduction

Robots can be devided into three categories which are discrete robots, squat robots, and continuum robots according to the structure format[1]. The comlexity of the work environment robots faced with is increasing with the development of science and technology. The continuum robot is becoming the hotpot of the robot field in recent years with its advantages of flexibility and self-adaptivity. The continuum robots are inspired by bionics[2-5]. And the flexible characteristic of continuum robots stems from their elastic parts, such as elastic rods[6-7], TiNi wires[8] and concentric tubes[9-10], etc. This paper presented a dual-arm continuum robot which possesses 3 DOFs(degree of freedom), of which each arm is driven by 3 wires installed symetrically, and the flexible shell is adopted to support the posture. On the basis of a single arm, the theory of wire driving kinematics and statics are established according to principle of constant curvature deformation. The correctness and feasibility of the theoretical model shall be verified by theoretical and experimental analysis.

As the Figure 1 a) shows, driving wires, base, segments and flexible shells consist of the main body of the continuum robot, among which the flexible shell is made up of flexible materials. And the flexible, which supports the posture of the continuum robot and keeps it bending with constant curvature, is designed as a hollow shape. As is show in Figure 1 b), a single arm of the continuum manipulator, of which the length is 300mm and diameter is 100mm, is composed of 3 flexible shells with the same length, each flexible shell is connected by a segment. There are three holes, which is designed for installing the driving wires, in the flexible shell of which the radius of graduated circle is 40mm. And the elastic deformation of the driving wire as tiny as can be neglected.

IOP Conf. Series: Journal of Physics: Conf. Series **1267** (2019) 012102 doi:10.1088/1742-6596/1267/1/012102

The streching, bending and rotating movement of the continuum can be achieved by driving the 3 driving wires with different motion amount.

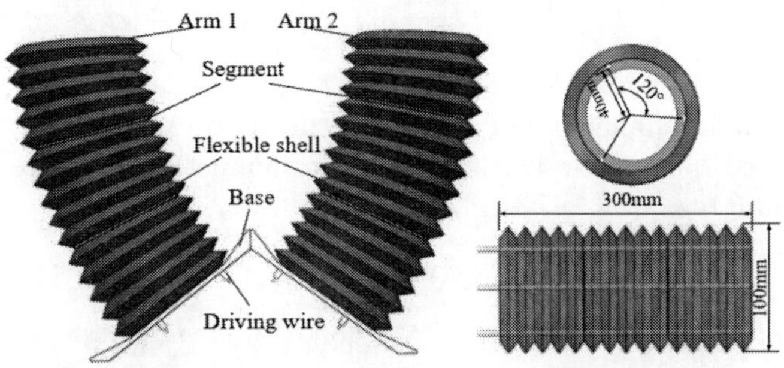

a) Single Arm b) Dual Arm

Figure 1. The schematic of continuum robot

2. Kinematics

According to the design, the motion form and principle of the two continuum manipulators are the same. The kinematics research can be based on one of them, and then the corresponding coordinate conversion can be obtained. Figure 2 shows the bending of the continuum manipulator in *Oxz* plane

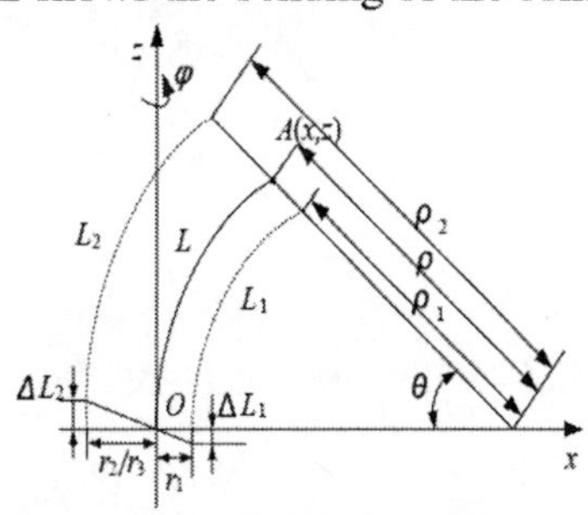

Figure 2. Bending of the flexible continuum manipulator

The variables and parameters in Figure 2 are: L_i is the lenghth of each driving wire, θ is the bending angle, ρ_i is the radius of curvature of each driving wire, r_i is the distance of each to the O point, ΔL_i is the change amount of each driving wire, φ is the rotating angle, A is the free-end.

According to the arc formula, the coordinate of A canbe abtained:

$$\begin{bmatrix} x \\ y \\ z \end{bmatrix} = \frac{L}{\theta} \cdot \begin{bmatrix} (1-\cos\theta)\cdot\cos\varphi \\ (1-\cos\theta)\cdot\sin\varphi \\ \sin\theta \end{bmatrix} \tag{2.1}$$

Equation (2.1) is abtained under the condition that the initial state of the continuum manipulator is coincide with z axi. Therefor, coordinate transformantion is needed according to the structure, and the homogeneous transformation is:

$$\text{T=Rot}(y, \frac{\pi}{4})[x, y, z, 1]^{\text{T}} \tag{2.2}$$

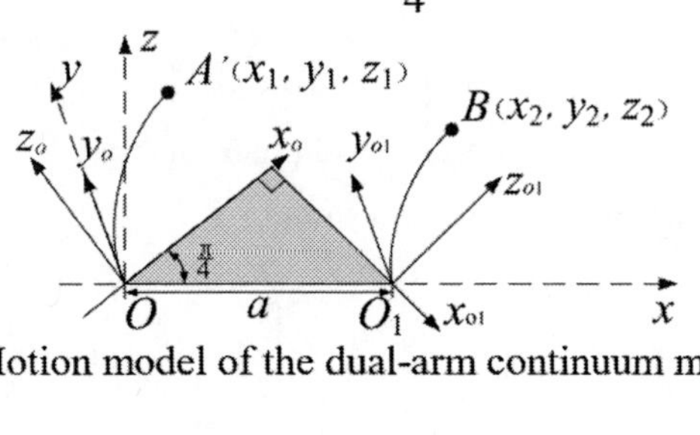

Figure 3. Motion model of the dual-arm continuum manipulator

IOP Conf. Series: Journal of Physics: Conf. Series **1267** (2019) 012102 doi:10.1088/1742-6596/1267/1/012102

The coordinate of A' is abtaibed:

$$\begin{bmatrix} x_1 \\ y_1 \\ z_1 \end{bmatrix} = \frac{\sqrt{2}L}{2\theta} \cdot \begin{bmatrix} (1\text{-}\cos\theta)\cdot\sin\varphi + \sin\theta \\ \sqrt{2}(1\text{-}\sin\theta)\cdot\sin\varphi \\ \sin\theta\text{-}(1\text{-}\cos\theta)\cdot\sin\varphi \end{bmatrix} \tag{2.3}$$

Simplify the continuum manipulator in Figure 1 into the form in Figure 3.

A' and B are the free-ends of the two continuum manipulators respectively. The homogeneous transformation from A' to B is:

$$T_1 = \text{Trans}(0,0,a)\text{Rot}\left(y,\text{-}\frac{\pi}{4}\right)\cdot[x_1,y_1,z_1,1]^T \tag{2.4}$$

The coordinate of B is abtained:

$$\begin{bmatrix} x_2 \\ y_2 \\ z_2 \end{bmatrix} = \frac{\sqrt{2}L}{2\theta} \cdot \begin{bmatrix} (1\text{-}\cos\theta)\cdot\sin\varphi\text{-}\sin\theta + \sqrt{2a} \\ \sqrt{2}(1\text{-}c\theta)\cdot\sin\varphi \\ (1\text{-}\cos\theta)\cdot\sin\varphi + \sin\theta \end{bmatrix} \tag{2.5}$$

And the posture of the continuum manipulator is:

$$\begin{cases} x = \frac{L}{\theta}\cos\alpha + \frac{L}{\theta} \\ z = \frac{L}{\theta}\sin\alpha \\ \alpha = [\pi\text{-}\theta, \pi] \end{cases} \tag{2.6}$$

According to equations (2.4) and (2.5), the work space of the robot can be abtained which is shown in Figure 4.

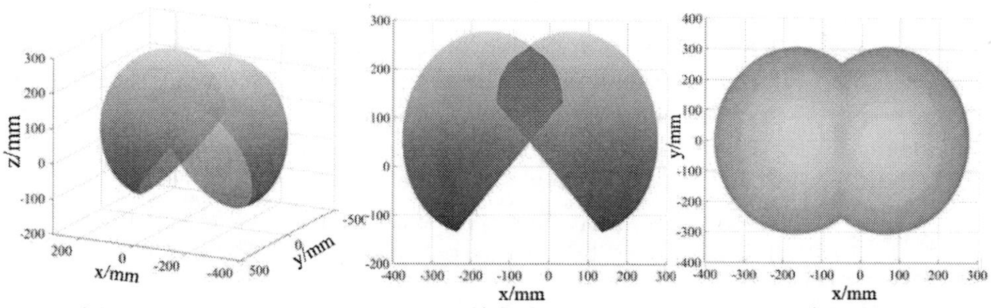

Figure 4. Work space of the continuum robot

3. Statics

Figure 5 shows the posture and mechanical analysis of the continuum manipulator, the method of dividing the continuum manipulator into multi-segments is adopted, therefore the coordinate of the free-end of each segment is:

$$\begin{bmatrix} x_i \\ y_i \end{bmatrix} = r_i \cdot \begin{bmatrix} 1\text{-}\cos\theta_i \\ \sin\theta_i \end{bmatrix} \tag{3.1}$$

The homogeneous transformation from $\{O_{i\text{-}1}\}$ to $\{O_i\}$ is:

IOP Conf. Series: Journal of Physics: Conf. Series **1267** (2019) 012102 doi:10.1088/1742-6596/1267/1/012102

$$ {}^{O_{i-1}}_{O_i}\mathrm{T} = \begin{bmatrix} \cos\theta_i & -\sin\theta_i & x_i \\ \sin\theta_i & \cos\theta_i & y_i \\ 0 & 0 & 1 \end{bmatrix} \tag{3.2} $$

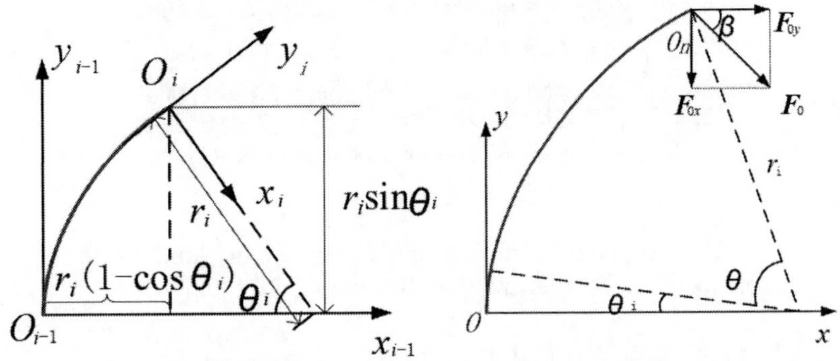

a) The bending posture b) The mechanical analysis

Figure 5. The overall mechanical analysis schematic diagram of the flexible continuum manipulator

$$ \boldsymbol{F}_{0i}^{o} = \mathrm{Trans}F \cdot \boldsymbol{F}_{0}^{o} \tag{3.3} $$

where

$$ \mathrm{Trans}F = \begin{bmatrix} 1 & 0 & r_i(1-\cos\theta) \\ 0 & 1 & r_i\sin\theta \\ 0 & 0 & 1 \end{bmatrix} \tag{3.4} $$

Expressed the force in $\{O_i\}$:

$$ \boldsymbol{F}_{0i}^{O_i} = {}^{O}_{O_i}\mathrm{T}_i^{-1} \times \boldsymbol{F}_{0i}^{o} = (\prod_{i=1}^{n} {}^{O_{i-1}}_{O_i}\mathrm{T})^{-1} \cdot \boldsymbol{F}_{0i}^{o} \tag{3.5} $$

The moment of the force:

$$ \boldsymbol{M}_{F_0} = \overrightarrow{O_iO_n} \times \boldsymbol{F}_{0i}^{O_i} \tag{3.6} $$

$$ \overrightarrow{O_iO_n} = \left[r_i(1-\cos\theta), r_i\sin\theta, 0 \right]^{\mathrm{T}} \tag{3.7} $$

$$ \boldsymbol{M}_{F_0} = \frac{(EI)_i}{r_i} \tag{3.8} $$

Thus the coordinate of the free-end of each segment can be calculated by equations from (3.1) to (3.8), by which the posture of the continuum manipulator can be obtained.

4. Experiment

4.1 Experiment platform introduction

The experiment platform includes a cherckerboard as calibration scale, several standard weight for providing external load, and a high speed camera for taking pictures of postures of the continuum manipulator, see Figure 6.

IOP Conf. Series: Journal of Physics: Conf. Series **1267** (2019) 012102 doi:10.1088/1742-6596/1267/1/012102

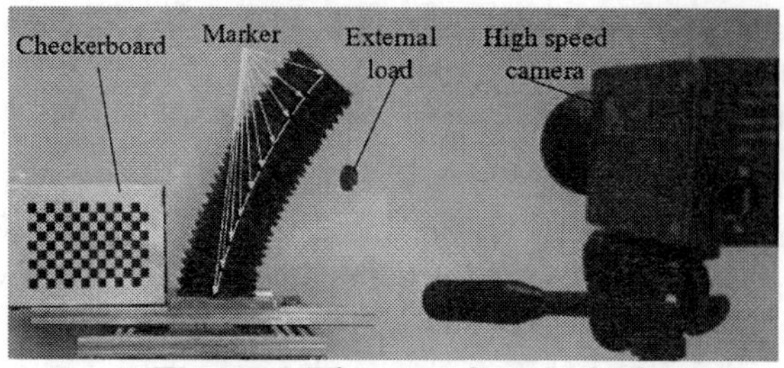

Figure 6. The experiment platform

4.2 Experiment for kinematics

In the experiment, let the bending angle be π/6, π/3, π/2 respectively, and the postures are abtained as shown in Figure 7, the errors of each experiment are listed in Table 1 as well.

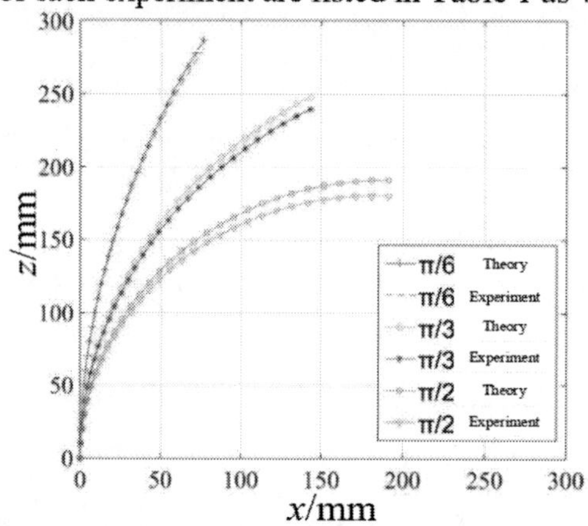

Figure 7. Contrast between theoretical and experimental posture of kinematics

Table 1. Maximum error of each experiment

Bending angle	π/6	π/3	π/2
Error	1.7%	3.4%	5.8%

4.3 Experiment for statics

In the experiment, let the bending angle be π/6, π/3, π/2 respectively, then the external load from 1N to 5N is applied to each expriemnt. And the postures are abtained as shown in Figure 8, the errors of each experiment are listed in Table 2 as well.

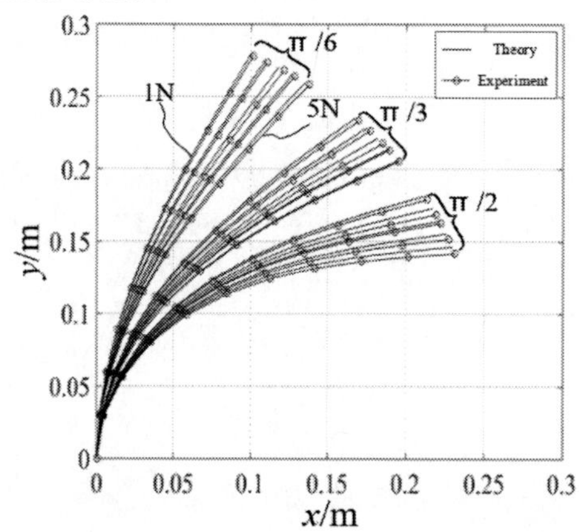

Figure 8. Contrast between theoretical and experimental posture of statics

Table 2. Maximum error of each experiment

Bending angle	External load				
	1N	2N	3N	4N	5N
π/6	0.888%	0.746%	1.533%	0.667%	1.080%
π/3	0.631%	2.055%	1.546%	2.407%	2.119%
π/2	1.081%	3.060%	3.686%	1.099%	1.354%

From the verification experiments above, as for the kinematics, the errors, which increase with the icrement of bengding angle, are no more than 5.8%. As for statics, the theory fits well with the results of experiments, and all the errors, which are no more than 3.686%, change with no obvious trend.

5.Conclusions

This paper presented a continuum robot that can achieve dual-arm motion. A mathematical model based on the principle of constant curvature is proposed to establish the basic deformation motion of the robot, which can achieve precise control, as well as statics of applying external load at the free-end. And experimental platform of three-degree-of-freedom flexible continuum robot was built for verifacation experiments. According to the verification experiments aimed at the kinematics and statics, maximum errors of them are less than 5.8% and 3.686% respectively.

Acknowledgments

We would like to acknowlege the help of Yaozong Zhang, Cong Liu and Pengyu Wang in theoretical analysis and experiment accomplishment, as well as to the help of Haifeng Xie in typesetting of this paper. And this work is supported by the National Natural Science Foundation of China with the Grant No. 51675011.

References

[1] Robinson G, Davis J B C. Continuum robots - a state of the art[C]// IEEE International Conference on Robotics and Automation, 1999. Proceedings. IEEE, 2002:2849-2854 vol.4.
[2] Seok S, Onal C D, Cho K J, et al. Meshworm: A peristaltic soft robot with Antagonistic nickel titanium coil actuators[J]. IEEE/ASME Transactions on Mechatronics, 2013, 18(5):1485-1497.
[3] Shepherd R F, Ilievski F, Choi W, et al. A multi-gait soft robot[J]. 2011..
[4] Lin H T, Leisk G G, Trimmer B. GoQBot: a caterpillar-inspired soft-bodied rolling robot. bioinspiration & biomimetics, 2011, 6(2):026007
[5] Nakamaru S, Maeda S, Hara Y, et al. Development of novel self-oscillating gel actuator for achievement of chemical robot// Intelligent Robots and Systems, 2009. IROS 2009. IEEE/RSJ Interna-tional Conference on. IEEE, 2009: 4319-4324
[6] Jones B A , Walker I D. Kinematics for multisection continuum robots[J]. IEEE Transactions on Robotics, 2006, 22(1):43-55.
[7] Simaan N, Taytor R, and Flint P. A dexterous system for laryngeal surgery, in Proc. IEEE Int. Conf. Robot. Autom., New Orleans, LA, Apr. 2004, pp. 351–357.
[8] Camarillo D B, Milne C F, Carlson C R, et al. Mechanics modeling of tendon-driven continuum manipulators[J]. IEEE Transactions on Robotics, 2008, 24(6):1262-1273.
[9] Webster R, Romano J, and Cowan N. Mechanics of Precurved-Tube Continuum Robots[J]. IEEE Transactions on Robotics, 2009, 25(1):67-78.
[10] Dpont P, Lock J, Itkowitz B, et al. Design and control of concentric-tube Robots[J]. IEEE Trans Robot, 2010, 26(2):209-225.